MITSUBISHI PICK-UP 1983-9... MANUAL

CHILTON'S

CEO	Rick Van Dalen
President	Dean F. Morgantini, S.A.E.
Vice President–Finance	Barry L. Beck
Vice President–Sales	Glenn D. Potere
Executive Editor	Kevin M. G. Maher, A.S.E.
Manager–Consumer Automotive	Richard Schwartz, A.S.E.
Manager–Professional Automotive	Richard J. Rivele
Manager–Marine/Recreation	James R. Marotta, A.S.E.
Manager–Electronic Fulfillment	Will Kessler, A.S.E., S.A.E.
Production Specialists	Brian Hollingsworth, Melinda Possinger
Project Managers	Thomas A. Mellon, A.S.E., S.A.E., Christine L. Sheeky, S.A.E., Todd W. Stidham, A.S.E., Ron Webb
Schematics Editors	Christopher G. Ritchie, A.S.E., S.A.E., S.T.S., Stephanie A. Spunt
Editor	George B. Heinrich III, A.S.E., S.A.E.

CHILTON *Automotive Books*

PUBLISHED BY **W. G. NICHOLS, INC.**

Manufactured in USA
© 1995 W. G. Nichols, Inc.
1020 Andrew Drive
West Chester, PA 19380
ISBN 0-8019-8666-4
Library of Congress Catalog Card No. 94-069448
6789012345 9876543210

Chilton is a registered trademark of Cahners Business Information, a division of Reed Elsevier, Inc., and has been licensed to W. G. Nichols, Inc.

www.chiltononline.com

Contents

1 GENERAL INFORMATION AND MAINTENANCE

- 1-2 HOW TO USE THIS BOOK
- 1-3 TOOLS AND EQUIPMENT
- 1-5 SERVICING YOUR VEHICLE SAFELY
- 1-13 ROUTINE MAINTENANCE
- 1-32 FLUIDS AND LUBRICANTS
- 1-50 JACKING
- 1-54 SPECIFICATIONS CHARTS

2 ENGINE PERFORMANCE AND TUNE-UP

- 2-2 TUNE-UP PROCEDURES
- 2-7 FIRING ORDERS
- 2-11 ELECTRONIC IGNITION
- 2-17 IGNITION TIMING
- 2-20 VALVE LASH
- 2-21 IDLE SPEED AND MIXTURE ADJUSTMENT
- 2-23 SPECIFICATIONS CHARTS

3 ENGINE AND ENGINE REBUILDING

- 3-2 BASIC ELECTRICITY
- 3-4 ENGINE ELECTRICAL
- 3-25 ENGINE MECHANICAL
- 3-29 SPECIFICATIONS CHARTS
- 3-173 EXHAUST SYSTEM

4 EMISSION CONTROLS

- 4-2 GASOLINE EMISSION CONTROLS
- 4-20 DIESEL EMISSION CONTROLS
- 4-21 CARBURETED CONTROLS
- 4-22 INJECTION CONTROLS
- 4-68 TROUBLE CODES
- 4-80 VACUUM DIAGRAMS

5 FUEL SYSTEM

- 5-2 CARBURETED FUEL SYSTEM
- 5-18 SPECIFICATIONS CHARTS
- 5-19 MULTI-PORT FUEL INJECTION SYSTEM (MFI)
- 5-39 DIESEL FUEL SYSTEM
- 5-46 FUEL TANK

6 CHASSIS ELECTRICAL

- 6-2 UNDERSTANDING ELECTRICAL SYSTEMS
- 6-11 SRS (AIR BAG)
- 6-13 HEATER AND A/C
- 6-54 CRUISE CONTROL
- 6-78 INSTRUMENTS
- 6-87 LIGHTING
- 6-103 WIRING DIAGRAMS

Contents

7 DRIVE TRAIN

- 7-2 MANUAL TRANSMISSION
- 7-21 CLUTCH
- 7-35 AUTOMATIC TRANSMISSION
- 7-68 TRANSFER CASE
- 7-70 DRIVELINE
- 7-75 FRONT DRIVE AXLE
- 7-91 REAR AXLE

8 SUSPENSION AND STEERING

- 8-2 WHEELS
- 8-2 FRONT SUSPENSION
- 8-22 SPECIFICATIONS CHARTS
- 8-23 REAR SUSPENSION
- 8-27 STEERING

9 BRAKE SYSTEMS

- 9-2 BRAKE SYSTEM
- 9-18 FRONT DISC BRAKES
- 9-26 REAR DRUM BRAKES
- 9-33 REAR DISC BRAKES
- 9-38 PARKING BRAKE
- 9-47 4-WHEEL ABS
- 9-52 REAR WHEEL ABS
- 9-57 SPECIFICATIONS CHARTS

10 BODY AND TRIM

- 10-2 EXTERIOR
- 10-29 INTERIOR
- 10-71 SPECIFICATIONS CHARTS

GLOSSARY

- 10-76 GLOSSARY

MASTER INDEX

- 10-81 MASTER INDEX

SAFETY NOTICE

Proper service and repair procedures are vital to the safe, reliable operation of all motor vehicles, as well as the personal safety of those performing repairs. This manual outlines procedures for servicing and repairing vehicles using safe, effective methods. The procedures contain many NOTES, CAUTIONS and WARNINGS which should be followed, along with standard procedures to eliminate the possibility of personal injury or improper service which could damage the vehicle or compromise its safety.

It is important to note that repair procedures and techniques, tools and parts for servicing motor vehicles, as well as the skill and experience of the individual performing the work vary widely. It is not possible to anticipate all of the conceivable ways or conditions under which vehicles may be serviced, or to provide cautions as to all possible hazards that may result. Standard and accepted safety precautions and equipment should be used when handling toxic or flammable fluids, and safety goggles or other protection should be used during cutting, grinding, chiseling, prying, or any other process that can cause material removal or projectiles.

Some procedures require the use of tools specially designed for a specific purpose. Before substituting another tool or procedure, you must be completely satisfied that neither your personal safety, nor the performance of the vehicle will be endangered.

Although information in this manual is based on industry sources and is complete as possible at the time of publication, the possibility exists that some car manufacturers made later changes which could not be included here. While striving for total accuracy, Nichols Publishing cannot assume responsibility for any errors, changes or omissions that may occur in the compilation of this data.

PART NUMBERS

Part numbers listed in this reference are not recommendations by Nichols Publishing for any product brand name. They are references that can be used with interchange manuals and aftermarket supplier catalogs to locate each brand supplier's discrete part number.

SPECIAL TOOLS

Special tools are recommended by the vehicle manufacturer to perform their specific job. Use has been kept to a minimum, but where absolutely necessary, they are referred to in the text by the part number of the tool manufacturer. These tools can be purchased, under the appropriate part number, from your local dealer or regional distributor, or an equivalent tool can be purchased locally from a tool supplier or parts outlet. Before substituting any tool for the one recommended, read the SAFETY NOTICE at the top of this page.

ACKNOWLEDGMENTS

Nichols Publishing expresses appreciation to Mitsubishi Motor Sales of America for their generous assistance.

Nichols Publishing would like to express thanks to all of the fine companies who participate in the production of our books. Hand tools supplied by Craftsman are used during all phases of our vehicle teardown and photography. Many of the fine specialty tools used in our procedures were provided courtesy of Lisle Corporation. Lincoln Automotive Products (1 Lincoln Way, St. Louis, MO 63120) has provided their industrial shop equipment, including jacks (engine, transmission and floor), engine stands, fluid and lubrication tools, as well as shop presses. Rotary Lifts, the largest automobile lift manufacturer in the world, offering the biggest variety of surface and in-ground lifts available (1-800-640-5438 or www.Rotary-Lift.com), has fulfilled our shop's lift needs. Much of our shop's electronic testing equipment was supplied by Universal Enterprises Inc. (UEI).

No part of this publication may be reproduced, transmitted or stored in any form or by any means, electronic or mechanical, including photocopy, recording, or by information storage or retrieval system, without prior written permission from the publisher.

FLUIDS AND LUBRICANTS
  AUTOMATIC TRANSMISSION 1-39
  BODY DRAIN HOLES 1-46
  BRAKE AND CLUTCH MASTER
    CYLINDERS 1-44
  CHASSIS GREASING 1-45
  COOLING SYSTEM 1-42
  DOOR HINGES 1-46
  DRIVE AXLES 1-40
  ENGINE 1-34
  ENGINE OIL
    RECOMMENDATIONS 1-33
  FLUID DISPOSAL 1-32
  FUEL RECOMMENDATIONS 1-32
  HOOD LATCH AND HINGES 1-46
  LOCK CYLINDERS 1-46
  MANUAL STEERING GEAR 1-45
  MANUAL TRANSMISSION 1-38
  POWER STEERING PUMP 1-44
  TRANSFER CASE 1-39
  WHEEL BEARINGS 1-47
HOW TO USE THIS BOOK 1-2
JACKING AND HOISTING
  EMERGENCY JACKING 1-50
  SERVICE JACKING 1-51
JUMP STARTING
  JUMP STARTING
    PRECAUTIONS 1-52
  JUMP STARTING PROCEDURE 1-52
ROUTINE MAINTENANCE
  AIR CLEANER 1-13
  AIR CONDITIONING SYSTEM 1-26
  BATTERY 1-22
  BELTS 1-24
  CV-BOOTS 1-26
  EVAPORATIVE CANISTER 1-19
  FUEL FILTER 1-15
  HOSES 1-26
  PCV VALVE 1-19
  TIMING BELT 1-26
  TIRES AND WHEELS 1-30
  WINDSHIELD WIPERS 1-29
SERIAL NUMBER IDENTIFICATION
  CHASSIS NUMBER 1-8
  ENGINE 1-9
  TRANSMISSION 1-13
  UNDERHOOD LABELS 1-13
  VEHICLE IDENTIFICATION
    NUMBER 1-6
  VEHICLE INFORMATION CODE
    PLATE 1-7
  VEHICLE SAFETY CERTIFICATION
    LABEL 1-8
SERVICING YOUR VEHICLE SAFELY
  DO'S 1-6
  DON'TS 1-6
SPECIFICATIONS CHARTS
  CAPACITIES 1-53
  ENGINE IDENTIFICATION 1-12
  MAINTENANCE INTERVALS 1-53
  RECOMMENDED LUBRICANTS 1-53

  VEHICLE IDENTIFICATION 1-7
TOOLS AND EQUIPMENT
  SPECIAL TOOLS 1-5
TOWING THE VEHICLE 1-49
TRAILER TOWING
  COOLING 1-49
  HITCH WEIGHT 1-48
  TRAILER WEIGHT 1-48
  WIRING 1-49

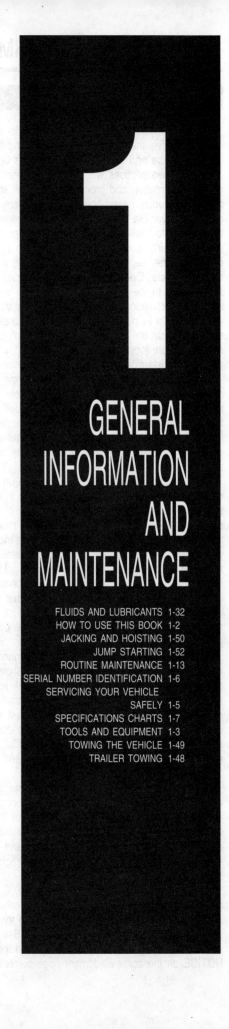

1

GENERAL INFORMATION AND MAINTENANCE

FLUIDS AND LUBRICANTS 1-32
HOW TO USE THIS BOOK 1-2
JACKING AND HOISTING 1-50
JUMP STARTING 1-52
ROUTINE MAINTENANCE 1-13
SERIAL NUMBER IDENTIFICATION 1-6
SERVICING YOUR VEHICLE
SAFELY 1-5
SPECIFICATIONS CHARTS 1-7
TOOLS AND EQUIPMENT 1-3
TOWING THE VEHICLE 1-49
TRAILER TOWING 1-48

1-2 GENERAL INFORMATION AND MAINTENANCE

HOW TO USE THIS BOOK

Chilton's Repair Manual for Mitsubishi Pick-up trucks and Montero is designed to help you understand the inner workings of your vehicle and save you money on its upkeep.

The first two sections will be the most used, since they contain information and procedures on both maintenance and tune-ups. Studies have shown that a properly tuned and maintained vehicle can get better gas mileage (which translates into lower operating costs) and periodic maintenance can catch minor problems before they turn into major repair bills. The other sections deal with the more complex systems of your vehicle. Operating systems from engine through brakes are covered to the extent that the average do-it-yourselfer becomes mechanically involved. It will give you the detailed instructions to help you change your own brake pads and shoes, tune-up the engine, replace spark plugs and filters, and do many more jobs that will save you money, give you personal satisfaction and help you avoid expensive problems.

A secondary purpose of this book is a reference guide for owners who want to understand their vehicle and/or their mechanics better. In this case, no tools at all are required. Knowing just what a particular repair job requires in parts and labor time will allow you to evaluate whether or not you're getting a fair price quote and help decipher itemized bills from a repair shop.

Before attempting any repairs or service on your vehicle, read through the entire procedure outlined in the appropriate section. This will give you the overall view of what tools and supplies will be required. There is nothing more frustrating than having to walk to the bus stop on Monday morning because you were short one gasket on Sunday afternoon. So read ahead and plan ahead. Each operation should be approached logically and all procedures thoroughly understood before attempting any work. Some special tools that may be required can often be rented from local automotive jobbers or places specializing in renting tools and equipment. Check the yellow pages of your phone book.

All sections contain adjustments, maintenance, removal and installation procedures, and some overhaul procedures. When overhaul is not considered practical, we tell you how to remove the failed part and then how to install the new or rebuilt replacement. In this way, you at least save the labor costs. Backyard overhaul of some components (such as the alternator or water pump) is just not practical, but the removal and installation procedure is often simple and well within the capabilities of the average vehicle owner.

Two basic mechanic's rules should be mentioned here. First, whenever the LEFT side of the vehicle or engine is referred to, it is meant to specify the DRIVER'S side of the vehicle. Conversely, the RIGHT side of the vehicle means the PASSENGER'S side. Second, all screws and bolts are removed by turning counterclockwise, and tightened by turning clockwise, unless otherwise noted.

Safety is always the most important rule. Constantly be aware of the dangers involved in working on or around an automobile and take proper precautions to avoid the risk of personal injury or damage to the vehicle. See the entry in this section, Servicing Your Vehicle Safely, and the SAFETY NOTICE on the acknowledgment page before attempting any service procedures and pay attention to the instructions provided. There are 3 common mistakes in mechanical work:

1. Incorrect order of assembly, disassembly or adjustment. When taking something apart or putting it together, performing steps in the wrong order usually just costs you extra time; however it CAN break something. Read the entire procedure before beginning disassembly. Do everything in the order in which the instructions say you should do it, even if you can't immediately see a reason for it. When you're taking apart something that is very intricate (for example a carburetor), you might want to draw a picture of how it looks when assembled at one point in order to make sure you get everything back in its proper position. We will supply exploded views whenever possible, but sometimes the job requires more attention to detail than an illustration provides. When making adjustments (especially tune-up adjustments), perform them in order. One adjustment often affects another and you cannot expect satisfactory results unless each adjustment is made only when it cannot be changed by any other.

2. Overtorquing (or undertorquing) nuts and bolts. While it is more common for overtorquing to cause damage, undertorquing can cause a fastener to vibrate loose and cause serious damage, especially when dealing with aluminum parts. Pay attention to torque specifications and utilize a torque wrench in assembly. If a torque figure is not available remember that if you are using the right tool for the job, you will probably not have to strain yourself to get a fastener tight enough. The pitch of most threads is so slight that the tension you put on the wrench will be multiplied many times in actual force on what you are tightening. A good example of how critical torque is can be seen in the case of spark plug installation, especially where you are putting the plug into an aluminum cylinder head. Too little torque can fail to crush the gasket, causing leakage of combustion gases and consequent overheating of the plug and engine parts. Too much torque can damage the threads or distort the plug, which changes the spark gap at the electrode. Since more and more manufacturers are using aluminum in their engine and chassis parts to save weight, a torque wrench should be in any serious do-it-yourselfer's tool box.

There are many commercial chemical products available for ensuring that fasteners won't come loose, even if they are not torqued just right (one common brand is Loctite®). If you're worried about getting something together tight enough to hold, but loose enough to avoid mechanical damage during assembly, one of these products might offer substantial insurance. Read the label on the package and make sure the product is compatible with the materials, fluids, etc. involved before choosing one.

3. Crossthreading. This occurs when a part such as a bolt is screwed into a nut or casting at the wrong angle and forced, causing the threads to become damaged. Crossthreading is more likely to occur if access is difficult. It helps to clean and lubricate fasteners, to start threading with the part to be installed going straight in, and to use your fingers. If you encounter resistance, unscrew the part and start over again at a different angle until it can be inserted and turned several times without much effort. Keep in mind that many parts, especially spark plugs, use tapered threads so

GENERAL INFORMATION AND MAINTENANCE 1-3

that gentle turning will automatically bring the part you're threading to the proper angle if you don't force it or resist a change in angle. Don't put a wrench on the part until it's been turned in a couple of times by hand. If you suddenly encounter resistance and the part has not seated fully, don't force it. Pull it back out and make sure it's clean and threading properly.

Always take your time and be patient; once you have some experience, working on your vehicle may become an enjoyable hobby.

TOOLS AND EQUIPMENT

▶ See Figures 1, 2, 3, 4, 5, 6, 7, 8, 9, 10 and 11

Naturally, without the proper tools and equipment it is impossible to properly service your vehicle. It would be impossible to catalog each tool that you would need to perform each and every operation in this book. It would also be unwise for the amateur to rush out and buy an expensive set of tools on the theory that he may need one or more of them at sometime.

The best approach is to proceed slowly, gathering together a good quality set of those tools that are used most frequently. Don't be misled by the low cost of bargain tools. It is far better to spend a little more for better quality. Forged wrenches, 6 or 12-point sockets and fine tooth ratchets are by far preferable to their less expensive counterparts. As any good mechanic can tell you, there are few worse experiences than trying to work on a car or truck with bad tools. Your monetary savings will be far outweighed by frustration and mangled knuckles.

Begin accumulating those tools that are used most frequently; those associated with routine maintenance and tune-up. In addition to the normal assortment of screwdrivers and pliers, you should have the following tools for routine maintenance jobs (your Mitsubishi uses metric fasteners):

• Metric wrenches and sockets, and combination open end/box end wrenches in sizes from 3mm to 19mm; and a spark plug socket.

If possible, buy various length socket drive extensions. One break in this department is that the metric sockets available in the U.S. will all fit the ratchet handles and extensions you may already have (¼ in., ⅜ in., and ½ in. drive).

• Jackstands for support.
• Oil filter wrench.
• Oil filter spout for pouring oil.
• Grease gun for chassis lubrication.
• Hydrometer for checking the battery.
• A container for draining oil.
• Many rags for wiping up the inevitable mess.

In addition to the above items there are several others that are not absolutely necessary, but handy to have around. These include absorbent gravel or oil dry, a transmission fluid funnel and the usual supply of lubricants, antifreeze and fluids, although these can be purchased as needed. This is a basic list for routine maintenance, but only if your personal needs and desires can accurately determine your list of tools.

The second list of tools is for tune-ups. While the tools involved here are slightly more sophisticated, they need not be outrageously expensive. There are several inexpensive tachometer/dwell meters on the market that are every bit as good for the average mechanic as a professional model. Just be sure that the meter scale goes to at least 1,200-1,500 rpm on the tach scale and that it works on both 4-cylinder and 6-cylinder engines. A basic list of tune-up equipment could include:

• Tach/dwell meter
• Spark plug wrench
• Timing light (a timing light that works from the vehicle's battery is best).
• Wire spark plug gauge/adjusting tools
• Set of feeler blades

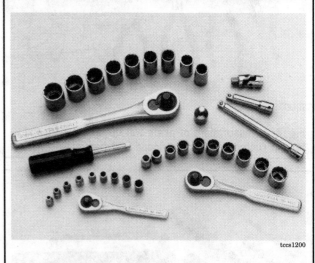

Fig. 1 Various metric wrenches and sockets

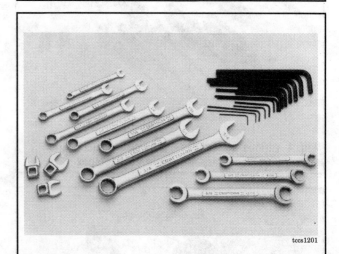

Fig. 2 An assortment of combination wrenches will be necessary

1-4 GENERAL INFORMATION AND MAINTENANCE

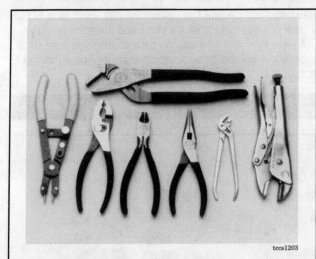

Fig. 3 An assortment of pliers will be handy, especially for old rusted parts and stripped bolt heads

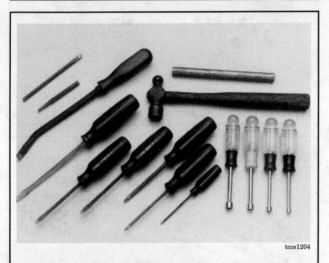

Fig. 4 Various screwdrivers, a hammer, chisels and prybars are necessary to have in your toolbox

Fig. 5 A hydraulic floor jack and a set of jackstands are essential for lifting and supporting the vehicle

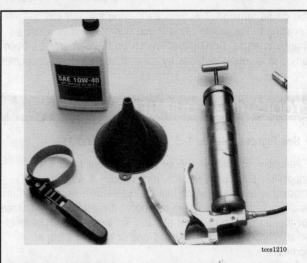

Fig. 6 A few inexpensive lubrication tools will make regular service easier

In addition to these basic tools, there are several other tools and gauges you may find useful. These include:
- A compression gauge. The screw-in type is slower to use, but eliminates the possibility of a faulty reading due to escaping pressure.
- A manifold vacuum gauge
- A test light
- A Digital Volt-Ohmmeter (DVOM). This meter allows direct testing of electrical components and grounds.

As a final note, you will probably find a torque wrench necessary for all but the most basic work. There are three types of torque wrenches available: deflecting beam type, dial indicator and click type. The beam and dial indicator models are perfectly adequate, although the click type models are more precise, and allow the user to reach the required torque without having to assume a sometimes awkward position to read the scale. No matter what type of torque wrench you purchase, have it calibrated periodically to ensure accuracy.

A torque specification for each fastener will be given in the procedure in any case that a specific torque value is required.

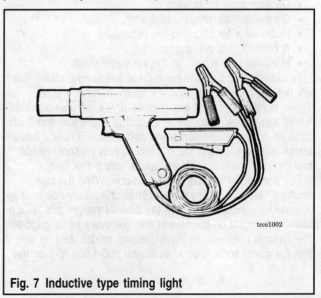

Fig. 7 Inductive type timing light

GENERAL INFORMATION AND MAINTENANCE

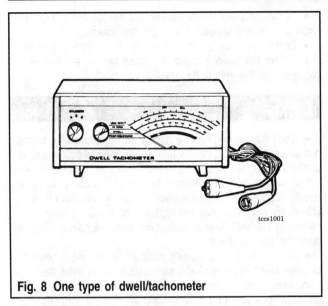

Fig. 8 One type of dwell/tachometer

Fig. 10 Useful tools for servicing spark plugs: a set of feeler blades, a spark plug gap gauge/adjusting tool, spark plug wrench and socket

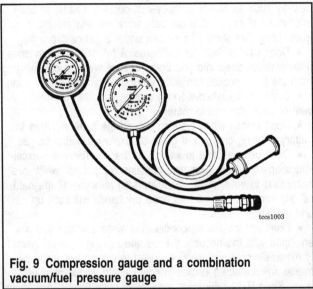

Fig. 9 Compression gauge and a combination vacuum/fuel pressure gauge

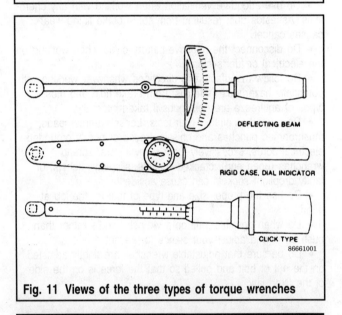

Fig. 11 Views of the three types of torque wrenches

If no torque specifications are given, use the following values as a guide, based upon fastener size:

Bolts marked 6T
6mm bolt/nut: 5-7 ft. lbs. (7-10 Nm)
8mm bolt/nut: 12-17 ft. lbs. (16-23 Nm)
10mm bolt/nut: 23-34 ft. lbs. (31-46 Nm)
12mm bolt/nut: 41-59 ft. lbs. (56-80 Nm)
14mm bolt/nut: 56-76 ft. lbs. (76-103 Nm)

Bolts marked 8T
6mm bolt/nut: 6-9 ft. lbs. (8-12 Nm)
8mm bolt/nut: 13-20 ft. lbs. (18-27 Nm)
10mm bolt/nut: 27-40 ft. lbs. (37-54 Nm)
12mm bolt/nut: 46-69 ft. lbs. (62-94 Nm)
14mm bolt/nut: 75-101 ft. lbs. (102-137 Nm)

Special Tools

Normally, the use of special factory tools is avoided for repair procedures, since these are not readily available for the do-it-yourselfer mechanic. When it is possible to perform the job with more commonly available tools, it will be pointed out, but occasionally, a special tool was designed to perform a specific function and should be used. Before substituting another tool, you should be convinced that neither your safety nor the performance of the vehicle will be compromised. Special tools are available at your local Mitsubishi dealer or at your local parts store or jobber market.

SERVICING YOUR VEHICLE SAFELY

It is virtually impossible to anticipate all of the hazards involved with automotive maintenance and service, but care and common sense will prevent most accidents.

The rules of safety for mechanics range from "don't smoke around gasoline", to "use the proper tool for the job." The trick

1-6 GENERAL INFORMATION AND MAINTENANCE

to avoiding injuries is to develop safe work habits and take every possible precaution.

Do's

- Do keep a fire extinguisher and first aid kit within easy reach.
- Do wear safety glasses or safety goggles when cutting, drilling, grinding or prying. If you wear glasses for the sake of vision, wear safety goggles over your regular glasses.
- Do shield your eyes whenever you work around the battery. Batteries contain sulfuric acid. In case of contact with the eyes or skin, flush the area with water or a mixture of water and baking soda, then immediately get medical attention.
- Do use safety stands for any under-truck service. Jacks are for raising vehicles; Jackstands are for making sure the vehicle stays raised until you want it to come down. Whenever the vehicle is raised, block the wheels remaining on the ground and set the parking brake.
- Do use adequate ventilation when working with any chemicals. Asbestos dust resulting from some brake lining wear causes cancer.
- Do disconnect the negative battery cable when working on the electrical or fuel system.
- Do follow manufacturer's directions whenever working with potentially hazardous materials. Both brake fluid and most types of antifreeze are poisonous if taken internally.
- Do properly maintain your tools. Loose hammerheads, mushroomed punches and chisels, frayed or poorly grounded electrical cords, excessively worn screwdrivers, spread wrenches (open end), cracked sockets, slipping ratchets, or faulty droplight sockets can cause accidents.
- Do use the proper size and type of tool for the job at hand.
- Do when possible, pull on a wrench handle rather than push on it, and adjust your stance to prevent a fall.
- Do be sure that adjustable wrenches are tightly adjusted on the nut or bolt and pulled so that the force is on the side of the fixed jaw.
- Do select a wrench or socket that fits the nut or bolt. The wrench or socket should sit straight, not cocked.
- Do strike squarely with a hammer. Avoid glancing blows.
- Do set the parking brake and block the wheels if work requires that the engine be running.

Don'ts

- Don't run an engine in a garage or anywhere else without proper ventilation — EVER! Carbon monoxide is poisonous. It is absorbed by the body 400 times faster than oxygen. It takes a long time to leave the human body and you can build up a deadly supply of it in your system by simply breathing in a little every day. You may not realize that you are slowly poisoning yourself. Always use power vents, windows, fans or open the garage doors.
- Don't work around moving parts while wearing a necktie or other loose clothing. Short sleeves are much safer than long, loose sleeves. Hard-toed shoes with neoprene soles protect your toes and give a better grip on slippery surfaces. Jewelry such as watches, fancy belt buckles, beads, or body adornment of any kind is not safe while working around a truck. Long hair should be hidden under a hat or cap.
- Don't use pockets for toolboxes. A fall or bump can drive a screwdriver deep into you body. Even a wiping cloth hanging from the back pocket can wrap around a spinning shaft or fan.
- Don't smoke when working around gasoline, cleaning solvent or other flammable material.
- Don't smoke when working around the battery. When the battery is being charged, it gives off explosive hydrogen gas.
- Don't use gasoline to wash your hands. There are excellent soaps available. Gasoline contains compounds which are hazardous to your health. Gasoline also removes all the natural oils from the skin so that bone dry hands will suck up oil and grease.
- Don't service the air conditioning system unless you are equipped with the necessary tools and training. The refrigerant, is extremely cold and when exposed to the air, will instantly freeze any surface it comes in contact with, including your eyes. Keep R-12 refrigerant away from open flames, poisonous gas will be produced if refrigerant burns.

SERIAL NUMBER IDENTIFICATION

Vehicle Identification Number

▶ See Figures 12 and 13

The Vehicle Identification Number (VIN) is located on a plate attached to the left front of the dash panel so it can be seen through the windshield when you stand beside the vehicle, in front of the driver's door. The letters and numbers in the VIN digits can be interpreted according to their positions in the sequence as follows:
1. Manufacturing country.
2. Make.
3. Vehicle type.
4. Type of seat belt system, weight class of truck/multi-purpose vehicle.
5. Vehicle line.
6. Trim code/price class.
7. Body type.
8. Engine displacement.
9. Check digit — a special letter or number code used to verify the serial number. This contains no useful information for the vehicle owner.
10. Model year
    - D = 1983
    - E = 1984
    - F = 1985
    - G = 1986
    - H = 1987
    - J = 1988
    - K = 1989
    - L = 1990
    - M = 1991

GENERAL INFORMATION AND MAINTENANCE 1-7

- N = 1992
- P = 1993
- R = 1994
- S = 1995

11. Plant where the vehicle was built.
12. On 1985 and earlier vehicles, the 12th digit represents the transmission code, followed by a 5-digit sequential (serial) number. For 1986 and all later vehicles, all 6 digits are the sequential number.

Vehicle Information Code Plate

► See Figure 14

The vehicle information code plate is riveted onto the cowl top outer panel in the engine compartment. The plate shows model code, engine model, transmission model and body color code.

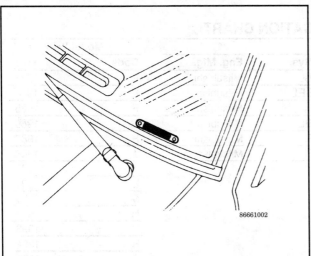

Fig. 12 The VIN is located on a plate on the front of the dash panel

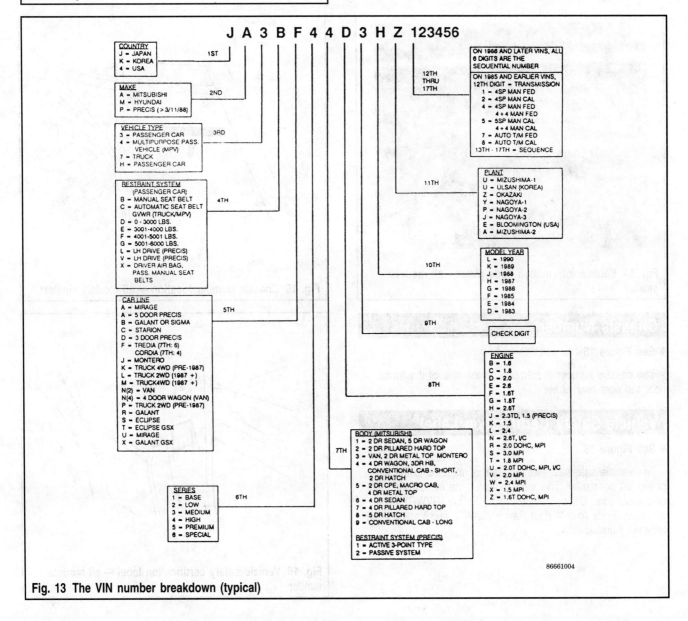

Fig. 13 The VIN number breakdown (typical)

1-8 GENERAL INFORMATION AND MAINTENANCE

VEHICLE IDENTIFICATION CHART

Engine Code						Model Year	
Code	Liters	Cu. In. (cc)	Cyl.	Fuel Sys.	Eng. Mfg.	Code	Year
D	2.0	122 (1997)	4	2 BBL	Mitsubishi	D	1983
J	2.3	143 (2346)	4	DIESEL	Mitsubishi	E	1984
L, W, G	2.4	147 (2350)	4	MFI	Mitsubishi	F	1985
E	2.6	156 (2555)	4	2 BBL	Mitsubishi	G	1986
S, H	3.0	181 (2972)	6	MFI	Mitsubishi	H	1987
M	3.5	213 (3497)	6	MFI	Mitsubishi	J	1988
						K	1989
						L	1990
						M	1991
						N	1992
						P	1993
						R	1994
						S	1995

2 BBL - Two barrel carburetor
MFI - Multi-port Fuel Injection

Fig. 14 Vehicle information code plate — all models similar

Chassis Number

▶ See Figure 15

The chassis number is stamped on the side of the frame near the right rear wheel.

Vehicle Safety Certification Label

▶ See Figure 16

The vehicle safety certification label is attached to the face of the left door pillar. This label indicates the month and year of manufacture, Gross Vehicle Weight Rating (GVWR), front and rear Gross Axle Weight Rating (GAWR) and Vehicle Identification Number (VIN).

Fig. 15 Chassis number location — all models similar

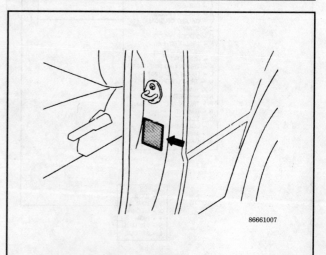

Fig. 16 Vehicle safety certification label — all models similar

GENERAL INFORMATION AND MAINTENANCE 1-9

Engine

▶ See Figures 17, 18, 19, 20, 21, 22, 23, 24 and 25

The engine model is stamped at the right rear of the top of the cylinder block. The engine serial number is stamped near the engine model number.

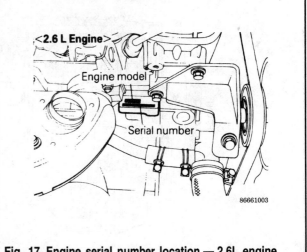

Fig. 17 Engine serial number location — 2.6L engine (other 4-cylinder engines similar)

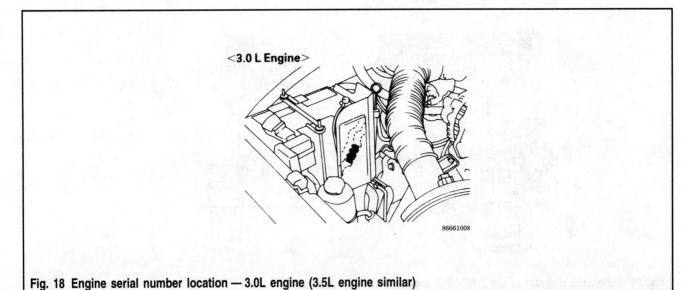

Fig. 18 Engine serial number location — 3.0L engine (3.5L engine similar)

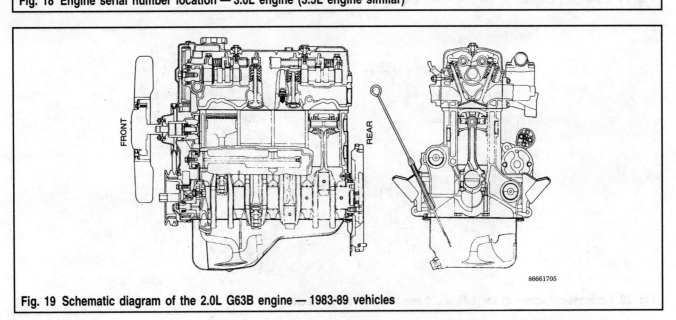

Fig. 19 Schematic diagram of the 2.0L G63B engine — 1983-89 vehicles

1-10 GENERAL INFORMATION AND MAINTENANCE

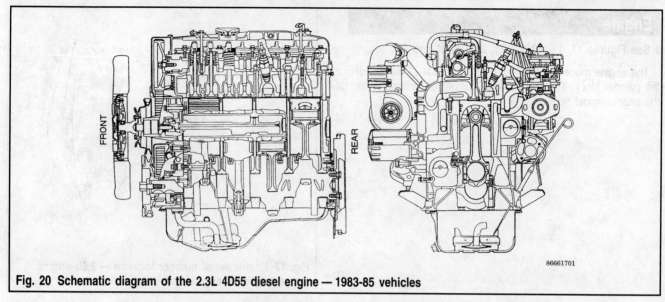

Fig. 20 Schematic diagram of the 2.3L 4D55 diesel engine — 1983-85 vehicles

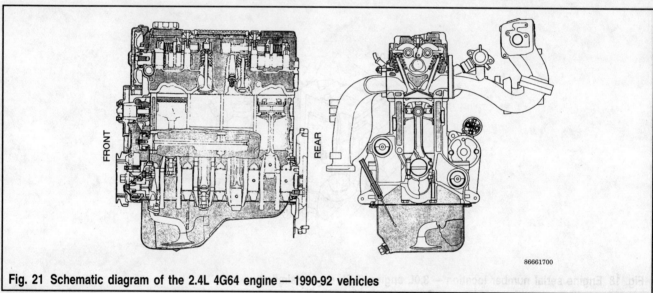

Fig. 21 Schematic diagram of the 2.4L 4G64 engine — 1990-92 vehicles

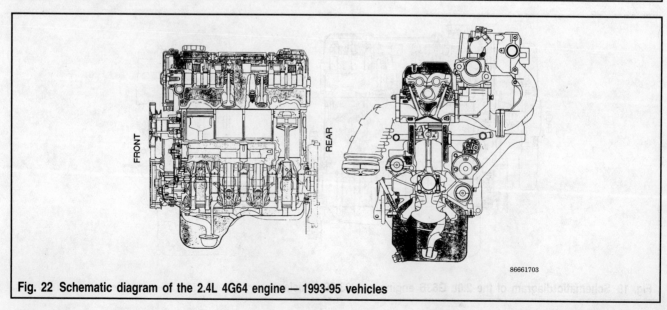

Fig. 22 Schematic diagram of the 2.4L 4G64 engine — 1993-95 vehicles

GENERAL INFORMATION AND MAINTENANCE 1-11

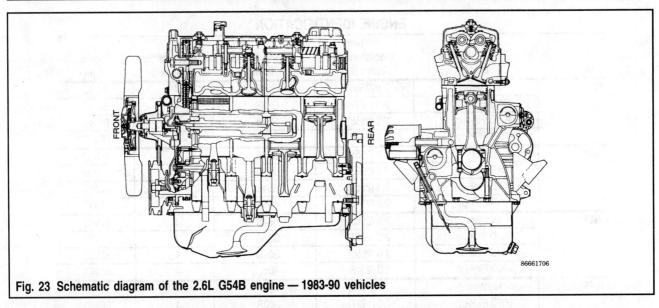

Fig. 23 Schematic diagram of the 2.6L G54B engine — 1983-90 vehicles

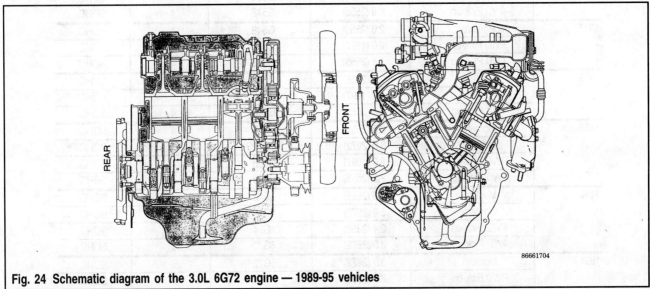

Fig. 24 Schematic diagram of the 3.0L 6G72 engine — 1989-95 vehicles

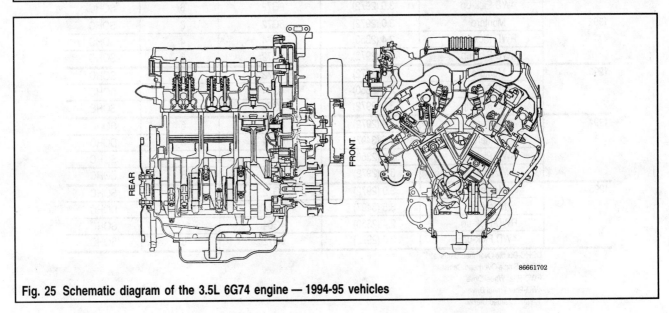

Fig. 25 Schematic diagram of the 3.5L 6G74 engine — 1994-95 vehicles

1-12 GENERAL INFORMATION AND MAINTENANCE

ENGINE IDENTIFICATION

Year	Model	Engine Displacement Liters (cc)	Engine Series (ID/VIN)	No. of Cylinders	Engine Type
1983	Montero	2.6 (2555)	G54B	4	SOHC
	RWD Pick-up	2.0 (1997)	G63B	4	SOHC
	RWD Pick-up	2.6 (2555)	G54B	4	SOHC
	4WD Pick-up	2.3 (2346)	4D55	4	SOHC [1]
1984	Montero	2.6 (2555)	G54B	4	SOHC
	RWD Pick-up	2.0 (1997)	G63B	4	SOHC
	RWD Pick-up	2.6 (2555)	G54B	4	SOHC
	4WD Pick-up	2.3 (2346)	4D55	4	SOHC [1]
1985	Montero	2.6 (2555)	G54B	4	SOHC
	RWD Pick-up	2.0 (1997)	G63B	4	SOHC
	RWD Pick-up	2.6 (2555)	G54B	4	SOHC
	4WD Pick-up	2.3 (2346)	4D55	4	SOHC [1]
1986	Montero	2.6 (2555)	G54B	4	SOHC
	RWD Pick-up	2.0 (1997)	G63B	4	SOHC
	4WD Pick-up	2.6 (2555)	G54B	4	SOHC
1987	Montero	2.6 (2555)	G54B	4	SOHC
	RWD Pick-up	2.0 (1997)	G63B	4	SOHC
	4WD Pick-up	2.6 (2555)	G54B	4	SOHC
1988	Montero	2.6 (2555)	G54B	4	SOHC
	RWD Pick-up	2.0 (1997)	G63B	4	SOHC
	4WD Pick-up	2.6 (2555)	G54B	4	SOHC
1989	Montero	2.6 (2555)	G54B	4	SOHC
	Montero	3.0 (2972)	6G72	6	SOHC
	RWD Pick-up	2.0 (1997)	G63B	4	SOHC
	4WD Pick-up	2.6 (2555)	G54B	4	SOHC
1990	Montero	2.6 (2555)	G54B	4	SOHC
	Montero	3.0 (2972)	6G72	6	SOHC
	RWD Pick-up	2.4 (2350)	4G64	4	SOHC
	4WD Pick-up	3.0 (2972)	6G72	6	SOHC
1991	Montero	3.0 (2972)	6G72	6	SOHC
	RWD Pick-up	2.4 (2350)	4G64	4	SOHC
	4WD Pick-up	3.0 (2972)	6G72	6	SOHC
1992	Montero	3.0 (2972)	6G72	6	SOHC
	RWD Pick-up	2.4 (2350)	4G64	4	SOHC
	4WD Pick-up	3.0 (2972)	6G72	6	SOHC
1993	Montero	3.0 (2972)	6G72	6	SOHC
	RWD Pick-up	2.4 (2350)	4G64	4	SOHC
	4WD Pick-up	3.0 (2972)	6G72	6	SOHC
1994	Montero	3.0 (2972)	6G72	6	SOHC
	Montero	3.5 (3497)	6G74	6	DOHC
	RWD Pick-up	2.4 (2350)	4G64	4	SOHC
	4WD Pick-up	3.0 (2972)	6G72	6	SOHC
1995	Montero	3.0 (2972)	6G72	6	SOHC
	Montero	3.5 (3497)	6G74	6	DOHC
	RWD Pick-up	2.4 (2350)	4G64	4	SOHC
	4WD Pick-up	3.0 (2972)	6G72	6	SOHC

DOHC-Double Overhead Camshaft
SOHC-Single Overhead Camshaft
RWD-Rear Wheel Drive
4WD-Four Wheel Drive
[1] Turbocharged Diesel

GENERAL INFORMATION AND MAINTENANCE

Transmission

On all models, the basic transmission model number is stamped on the vehicle information code plate, located on the top, outer cowl panel in the engine compartment.

Underhood Labels

▶ See Figures 26, 27 and 28

Along with the previous labels, the labels located on the underside of the hood are very helpful with vehicle identification and vehicle maintenance information. Three underhood labels are: vacuum hose installation/adjusting screw location label, important vehicle information label (which contains information for vehicle driveability and maintenance), and engine coolant label.

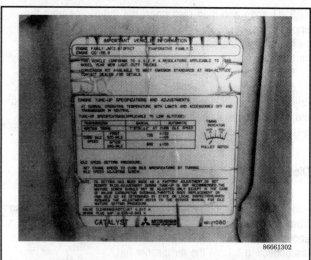

Fig. 27 The important vehicle information label contains information used for timing and carburetor adjustments

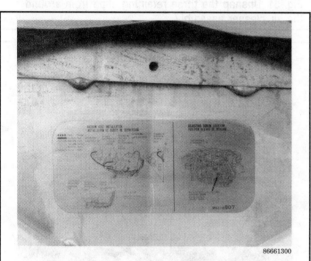

Fig. 26 The vacuum hose installation/adjusting screw location label, located on the underside of the hood

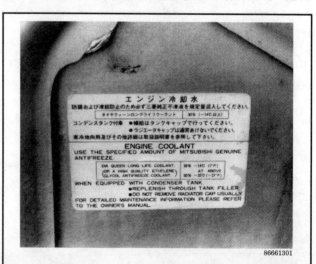

Fig. 28 The engine coolant label contains information useful when working on the cooling system

ROUTINE MAINTENANCE

Air Cleaner

An air cleaner is used to keep airborne dirt and dust out of the air flowing through the engine. This material, if allowed to enter the engine, would form an abrasive compound in conjunction with the engine oil and drastically shorten engine life. For this reason, you should never run the engine without the air cleaner in place except for a very brief period if required for trouble diagnosis. You should also be sure to use the proper replacement part to avoid poor fit and consequent air leakage.

Proper maintenance is important since a clogged air filter will allow an ever decreasing amount of air to enter the engine and, therefore, will increasingly enrich the fuel/air mixture, causing poor fuel economy, a drastic increase in emissions and even serious damage to the catalytic converter system.

The maximum maintenance interval on all models is 30,000 miles (48,000 km). Maximum efficiency is maintained by changing the filter more often. Diesel engines are particularly sensitive to partly clogged air filters and will lose performance if not well maintained. In all cases the air cleaner element must be replaced; it cannot be cleaned.

FILTER REPLACEMENT

▶ See Figures 29, 30, 31, 32, 33, 34, 35, 36 and 37

4-Cylinder Gasoline Engines

1. Remove the wing nut with your fingers. Pliers should be used only if it has been tightened excessively.
2. Disconnect the vacuum hose connected to the top of the air cleaner assembly.

1-14 GENERAL INFORMATION AND MAINTENANCE

3. Unsnap and then disconnect the three clips from the upper body.
4. Remove the top of the air cleaner and remove the cartridge.
5. Wipe out the filter housing with a clean rag.
6. Install the new cartridge squarely so that it seats properly. Install the air cleaner lid over the wing nut stud, aligning the arrow on the top with the arrow on the air intake tube. Install the wing nut finger-tight.
7. Hook, then lock all clips and hoses.

Diesel Engine

1. Unsnap the three finger clips and unhook the lower ends. Pull the top of the air cleaner upward or toward you and away from the air cleaner body. The flex hose will give you enough freedom of motion to permit you to do this without disconnecting it from the air cleaner top.
2. Pull the filter cartridge out of the air cleaner body.
3. Install the new filter so that it seats properly in the air cleaner body. Then, install the top squarely on the lower body,

Fig. 29 Disconnect the top vacuum hose from the air cleaner assembly — 2.6L engine shown

Fig. 30 Remove the wing nut from the top of the air cleaner lid

Fig. 31 Unsnap the three retaining clips from around the air cleaner assembly

Fig. 32 Lift the air cleaner lid off of the assembly to gain access to the filter

Fig. 33 Remove the old air filter; discard and replace with a new one if necessary

GENERAL INFORMATION AND MAINTENANCE

hook the clips and lock them. If any of the clips are hard to lock, recheck the position of the top on the lower body. If it is not positioned properly, it will not seal effectively and the clips may be damaged.

6-Cylinder Gasoline Engines

1. Loosen the hose clamp that connects the intake flexible hose to the air cleaner cover. Separate the hose from the cover.
2. Unfasten the air flow sensor multi-connectors.
3. Release the air cleaner cover clips. Move the flexible air intake out of the way to allow you to remove the air cleaner cover. Remove the air cleaner cover carefully because it contains the air flow sensor.
4. Remove the air cleaner element from the housing and wipe the housing clean with a rag.
5. Place the new filter element into the housing making sure that it seats properly.
6. Replace the air cleaner cover and plug the air flow sensor multi-connectors in. Connect the air intake hose to the air cleaner cover and tighten the hose clamp.

Fuel Filter

There are three types of fuel filters:
- Carbureted engines use a conventional, inline fuel filter with ordinary clamps and lines.
- Fuel injected engines use a special, high pressure filter with banjo fittings. Note that the pressure in the system must be relieved before attempting to remove the filter in this type of system.
- The diesel engine uses a filter assembly that not only filters sediment and water from the fuel but also, under certain conditions, warms the fuel to prevent fuel line blockage. The filter housing is equipped with a water level sensor which will illuminate a dash light, warning the operator that the level of trapped water within the filter has risen to capacity. The filter then needs to be drained as soon as possible, after the dash light has illuminated.

The fuel filter must be replaced at least every five years or 50,000 miles (80,000 km). This is a maximum replacement

Fig. 35 Air intake hose mounting location on V-6 Pickups and Monteros

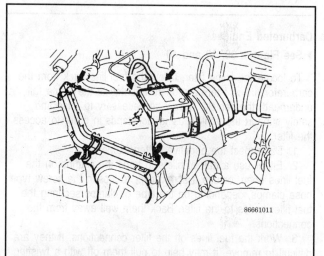
Fig. 36 The retaining clips and air sensor connector locations

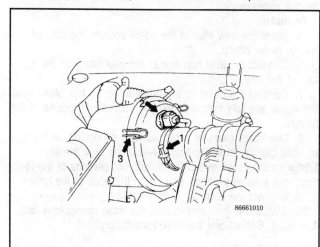
Fig. 34 Air filter location on most V-6 Pick-ups and Monteros

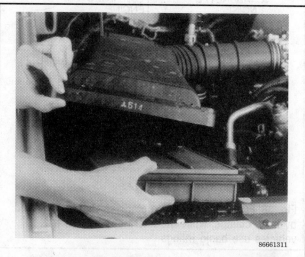
Fig. 37 Remove the old air filter; discard and replace with a new one if necessary

1-16 GENERAL INFORMATION AND MAINTENANCE

period. However, since the amount of dirt and, in the case of a diesel engine, water in the fuel varies greatly, the filter should be replaced whenever you suspect it to be clogged. Typically, the symptoms of a clogged filter are a lack of engine performance under full throttle conditions, especially at high rpm, even though the engine operates normally under moderate driving conditions. With severe clogging, the filter may cause poor running under virtually every operating condition but idle. One way to check out the filter is to follow the steps below for removal and then drain the contents out of the filter through the inlet. Drain the filter into an aluminum can. While it is normal for a light concentration of particles or a few drops of water to be trapped in the fuel on the inlet side of the filter, large amounts of water and heavy concentrations of dirt indicate dirty fuel and, probably, a clogged filter. If you find a great deal of dirt trapped in the filter, it is a good idea not only to replace the filter, but to have the fuel tank drained or pumped out as well.

REMOVAL & INSTALLATION

Carbureted Engines

▶ See Figures 38, 39 and 40

To locate the inline filter, follow the fuel line back from the carburetor. Inline filters are often mounted to the frame rail underneath the vehicle. It may be necessary to raise and safely support the vehicle using jackstands in order to access the filter.

1. Disconnect the negative battery cable.
2. Either use a pair of pliers to force the clamps on the fuel lines open, or use a screwdriver to loosen the screw type hose clamps, depending on the type of clamps securing the fuel filter hoses to the filter. Back them well away from the connections.
3. Work the fuel lines off the filter connections. If they are difficult to remove, it may help to pull them off with a twisting motion. Remove the filter from its mounting clip.

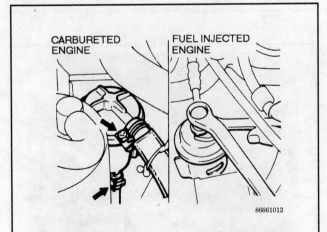

Fig. 38 Fuel filters on carbureted vehicles are secured by hose clamps, while the filters on fuel injected vehicles use banjo fittings

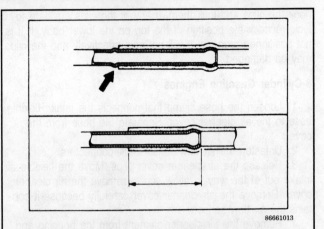

Fig. 39 When connecting the fuel lines to the filter, make sure the fuel lines are installed far enough onto the connections. The line should overlap the connection about 1 inch (25mm)

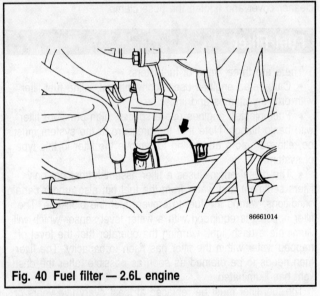

Fig. 40 Fuel filter — 2.6L engine

4. Inspect the fuel lines for cracks or breaks and replace them if necessary.

To install:

5. Install the new filter in the same position the old one was in in the clamp.
6. Connect the inlet fuel line to the inlet fitting on the bottom of the filter.
7. Connect the outlet to the outlet fitting on top. Make sure the hoses are fully installed over the bulged-out portions of the fittings.
8. Use either pliers or a screwdriver, depending on the type of hose clamps, to fasten the hose clamps over the filter fittings so they are beyond the bulged-out sections of the fittings, but a small distance away from the ends of the hoses.
9. Connect the negative battery cable.
10. Start the engine and inspect the hose connections for fuel leaks. Correct any fuel leak immediately.

GENERAL INFORMATION AND MAINTENANCE   1-17

Fuel Injected Engines

♦ See Figures 41, 42 and 43

➡Some Montero vehicles are equipped with a fuel pump access cover. Take out the carpet in the cargo area and lift up the floor cover to remove the fuel pump access cover. Unplug the fuel pump harness connector, then start and run the engine until it stalls to reduce the fuel pressure.

To locate the inline filter, follow the fuel line back from the throttle body. Inline filters are often mounted to the frame rail underneath the vehicle. It may be necessary to raise and safely support the vehicle using jackstands in order to access the filter.

1. First, you MUST reduce the pressure in the fuel system; refer to Section 5 for this procedure.
2. Unfasten the negative battery cable from the battery.
3. Remove the air cleaner assembly; refer to the air cleaner filter removal procedure described earlier in this section.
4. Using an open-end wrench to hold the fuel filter stationary, loosen the bolt for the banjo type connector on top of the fuel filter with a box wrench. Perform the same procedure to the inlet connector on the bottom of the filter.
5. Remove the bolt or nuts attaching the filter to the bracket and remove it.

To install:

6. To install the new filter, reverse the above procedure. A torque wrench is recommended to tighten the bolts for the fuel line banjo fittings. If the banjo fitting washers are damaged, replace them. Tighten the outlet fitting to 18-25 ft. lbs. (24-34 Nm) and the inlet fitting to 25 ft. lbs. (34 Nm).
7. Reconnect the negative battery cable, then start the engine and check for leaks.

Diesel Engines

♦ See Figures 44, 45, 46 and 47

The fuel filter element (cartridge) is contained within the filter canister which must be removed to gain access. The filter

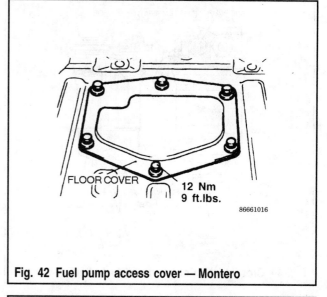

Fig. 42 Fuel pump access cover — Montero

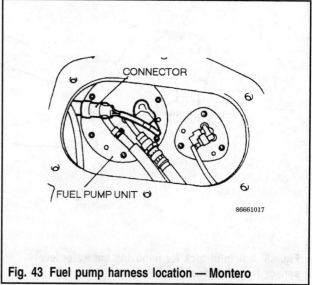

Fig. 43 Fuel pump harness location — Montero

canister is located in the engine compartment and is positioned between the fuel tank and the feed pump in the fuel system.

1. Label and unplug all electrical connectors running to the filter canister. These connectors are for the water level sensor, the fuel heater and the fuel temperature sensor.
2. Carefully unfasten the fuel hoses at the fuel filter. Have a supply of rags handy to catch overflow from the hoses.

➡While not as flammable as gasoline, diesel fuel is slippery, smelly and very capable of staining anything it touches. Prevent spillage whenever possible and mop up spilled fluid immediately. Cat litter or similar products are ideal for dealing with puddles on the floor.

3. Remove the two bolts holding the filter to the body and remove the filter canister.
4. Remove the protector and bracket from the canister.
5. Screw the filter out of the canister body by hand. Carefully remove the water level sensor and the drain plug from the cartridge. It may be handy to lightly clamp the unit in a vise to aid removal, but do not damage the sensor or the housing.

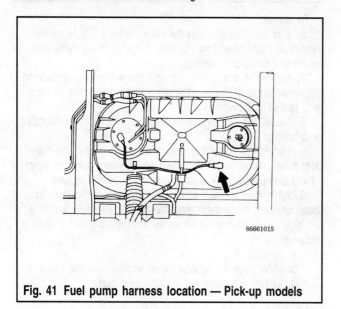

Fig. 41 Fuel pump harness location — Pick-up models

1-18 GENERAL INFORMATION AND MAINTENANCE

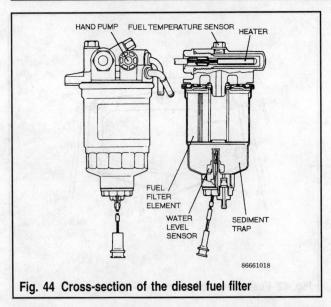

Fig. 44 Cross-section of the diesel fuel filter

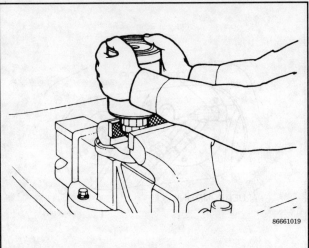

Fig. 45 A helpful trick for removing the water level sensor from the diesel fuel filter

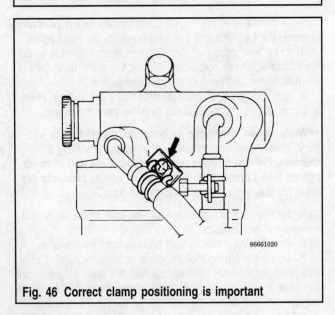

Fig. 46 Correct clamp positioning is important

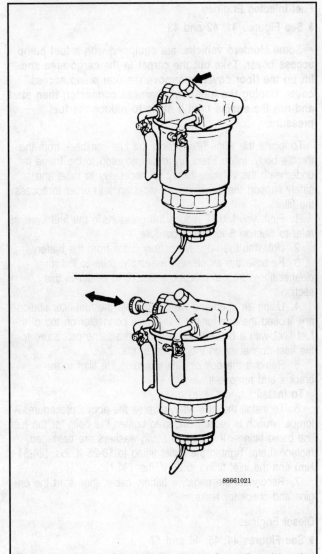

Fig. 47 Location of the air bleed port (upper) and hand pump (lower) for bleeding the diesel fuel system

To install:

6. It is possible to clean the filter element with kerosene. However, replacement with a new cartridge is highly recommended instead of cleaning.

7. Install the drain plug and and the water level sensor on the new cartridge. Tighten the drain plug to 3 ft. lbs. (4 Nm) and the water sensor to 9 ft. lbs. (12 Nm).

8. Screw the cartridge onto the body. Install the protector and bracket.

9. Install filter assembly onto the vehicle. Install the main fuel hoses. When tightening the clamps, make sure the heads of the clamp bolts face away from the body of the filter.

10. Whenever the fuel supply has run out or the lines have been opened, the system must be bled to eliminate air. The air bleed plug projects at an angle from the top of the filter housing.

   a. Loosen it.
   b. Have rags handy and place some under the bleed port; as fuel will come out of the bleed port.

GENERAL INFORMATION AND MAINTENANCE

    c. The knob for the hand pump is located on the side of the filter body. Unscrew it and pull the pump lever out of the housing.

    d. Work the hand pump and watch the fuel coming out of the air bleed plug; when there are no air bubbles mixed with the fuel, tighten the air bleed plug.

    e. Continue pumping until the operation of the pump lever feels stiff or heavy during each stroke.

    f. Push the pump lever all the way in and turn it to the right to lock it in place.

11. Refasten the wiring connectors to the pump housing terminals. Make certain each is firmly seated.

DRAINING WATER FROM THE DIESEL FUEL FILTER

♦ See Figure 48

When water accumulates in the fuel filter, the fuel-water separator light will come on. This indicates that the filter has reached its safe capacity and must be drained. Even if the light has not come on, the wise owner will drain the filter with every oil change. This simple procedure can be done with the filter on the vehicle and can prevent severe engine damage or failure.

1. Use the proper sized wrench to loosen the drain plug on the bottom of the fuel filter.
2. The knob for the hand pump is located on the side of the filter body. Unscrew it and pull the pump lever out of the housing.
3. Pump the hand pump until the water is expelled and fuel is being pumped out.
4. Push the pump lever all the way in and turn it to the right to lock it in place.

PCV Valve

➡ The Positive Crankcase Ventilation (PCV) system should be checked and cleaned at least every five years or 50,000 miles (80,000 km) on average. Inspection and service should be performed any time a malfunction is observed or suspected. Severe usage conditions may require shorter intervals — refer to the owner's manual for your truck.

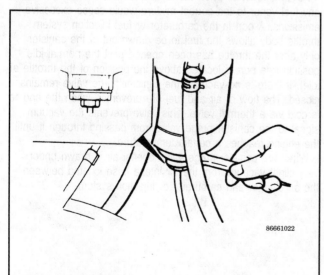

Fig. 48 Draining water from the diesel fuel filter

To prevent combustion blow-by gasses from entering the atmosphere, all Mitsubishi gasoline engines use a closed type crankcase ventilation system. The system uses a PCV valve that is threaded into the valve cover. The PCV system supplies fresh air from the air cleaner through a metered orifice in the PCV valve to the crankcase. Inside the crankcase, fresh air is mixed with the blow-by gasses. This mixture passes through the PCV valve and into the air induction system.

The diesel engine does not use a PCV system; the crankcase vapors are instead routed through a breather hose and pipe where they are introduced into the intake airstream. After passing through the turbocharger, they enter the engine to be reburned.

REMOVAL & INSTALLATION

♦ See Figures 49, 50, 51, 52 and 53

1. Use a wrench to unscrew the PCV valve at the valve cover. It may be necessary to disconnect the hose at the outer end of the PCV to avoid twisting it too much (see the first part of the next step). If necessary, reconnect the PCV hose. Then, start the engine and run it at idle. You should hear a hissing sound as air is drawn into the valve. Place your finger over the inlet. You should be able to feel a strong vacuum. If either test is failed, the valve will have to be replaced. Stop the engine.
2. Using a pair of pliers, pinch open the clamp that fastens the hose to the outer end of the PCV and then slide the clamp back a few inches. Pull the valve out of the hose. Then, blow through the threaded end of the valve. If air will pass freely through the valve, the valve is okay. If not, try to flush it out with spray solvent and if that doesn't work, replace it. PCV valves are not serviceable.
3. Clean both PCV hoses by spraying a safe solvent, such as a fuel injection safe throttle body cleaner and gum cutter solvent, through them. Inspect the hoses for cracks or excessive stiffness, and replace, as necessary.
4. Install the old PCV valve or replace it with a new one in reverse order. Be careful not to crossthread the valve into the aluminum valve cover, and do not overtighten it.

Evaporative Canister

♦ See Figures 54 and 55

➡ The evaporative emission control system should be serviced at least every five years or 50,000 miles (80,000 km) on average. This service interval may vary with some models and years — refer to the owner's manual for your truck.

This particular system is only applicable to gasoline engines. The heart of this system is a charcoal canister located in the engine compartment. Fuel vapor that collects in the carburetor float bowl or gas tank and which would ordinarily be released to the atmosphere is stored in the canister because of the attraction between the charcoal and it.

In order to restore the ability of the charcoal to hold fuel, fresh air is drawn through the charcoal under certain operating conditions, thus drawing the fuel back out and burning it in the combustion chambers.

1-20 GENERAL INFORMATION AND MAINTENANCE

Fig. 49 Disconnect the vacuum hose from the PCV valve located in the right-hand rocker arm cover

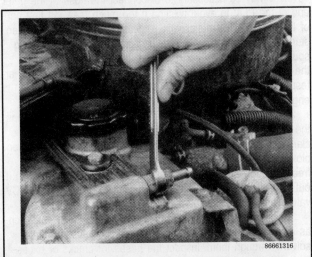

Fig. 50 Use a wrench or socket to loosen the PCV valve and remove it

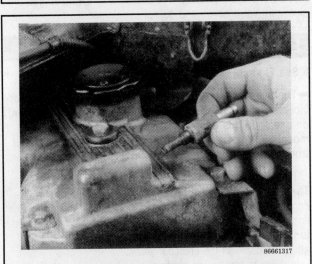

Fig. 51 Once the PCV valve is removed, vacuum can be tested and the valve can be cleaned

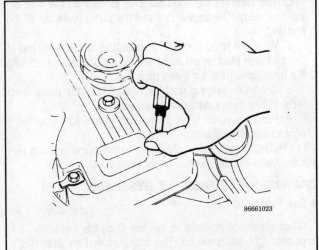

Fig. 52 Test for vacuum at the PCV valve with the engine idling

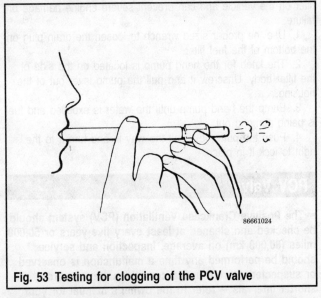

Fig. 53 Testing for clogging of the PCV valve

At idle speed, or when the engine is cold, the addition of any fuel vapor to the correct mixture would cause excessive emissions. A port in the carburetor or fuel injection system throttle body allows the fuel to be drawn out of the canister only after the throttle has been opened past the normal idle position (the port is located above the position of the throttle at idle). If there is no vacuum, the canister purge valve remains closed. The flow of air and fuel are prevented when the engine is cold via a thermal valve. This valve prevents the vacuum signal to the canister purge valve from passing through it until the engine reaches a certain temperature.

When the canister purge valve opens, air is drawn under very slight vacuum from the air intake hose located between the air cleaner and carburetor or injection system.

GENERAL INFORMATION AND MAINTENANCE 1-21

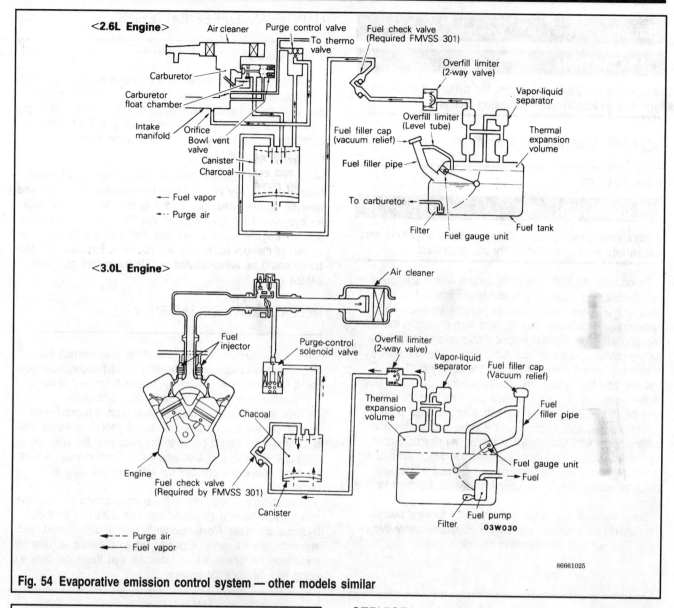

Fig. 54 Evaporative emission control system — other models similar

Fig. 55 The charcoal canister with the air cleaner, carburetor, and fuel tank hose mounting locations — 1989 Montero shown; other models similar

SERVICE

Every five years or 50,000 miles (80,000 km), the Evaporative Emissions System (EES) should be inspected as follows:

1. Disconnect the fuel vapor vent line at the vapor/liquid separator on the fuel tank and at the canister, blow it clean with compressed air, or have this done at a repair shop. Remove the filler cap and check the seal for cracks or breaks. If necessary, replace the cap.
2. Replace the canister (if necessary — refer to the owner manual for service mileage interval). To do this, mark and then disconnect all hoses. Unclamp and remove the canister. **To Install:**
3. Position the new or cleaned old canister back into the original position and refasten the clamps, which hold it in place.
4. Reconnect the hoses to their corresponding fittings.
5. Attach the filler cap, either the new one or the old one if still in adequate condition, and connect the fuel vapor vent line to the vapor/liquid separator and canister.

1-22 GENERAL INFORMATION AND MAINTENANCE

Battery

> ✴✴**CAUTION**
>
> Keep flame or sparks away from the battery! It gives off explosive hydrogen gas while being charged.

GENERAL MAINTENANCE

▸ See Figure 56

> ✴✴**CAUTION**
>
> Always wear goggles when cleaning the battery. Acid may splash into your eyes if they are not protected!

Periodically, clean the top of the battery with a solution of baking soda and water using a stiff bristle brush. This will neutralize acid and clean corrosion from the terminals and battery case, which will keep the acid from damaging the battery and will allow a better electrical connection between the battery cables and the battery. Since this solution neutralizes acid, make sure it does not get into the battery itself.

If any acid has spilled onto the battery tray, clean this area in the same way. If paint has been removed from the tray, wire brush the area and paint it with a rust-resistant paint.

Remove the cable ends, clean the cable end clamps and battery posts, and reconnect and tighten the clamps. Apply a thin coat of petroleum jelly to the terminals, which will help to retard corrosion. The terminals can be cleaned with a wire brush or with an inexpensive terminal cleaner designed for this purpose.

Some batteries were equipped with a felt terminal washer. This should be saturated with engine oil approximately every year. This will also help to retard corrosion.

MAINTENANCE-FREE BATTERIES

▸ See Figure 57

The factory-installed battery is a maintenance-free type on all the Mitsubishi models covered by this manual. That means that you'll never have to remove caps (there aren't any) to add water. But a yearly inspection and cleaning of the battery, connections, and battery mountings is recommended to guarantee maximum reliability.

➡On a sealed maintenance-free sealed battery a built-in hydrometer or eye is used for checking the fluid level and specific gravity readings. If your battery is equipped with an eye, use it for checking the condition of the battery by determining the color of the eye. Refer to the battery case or owner manual for further instructions. Replacement batteries could be either sealed maintenance-free or a non-sealed type.

REPLACEMENT BATTERIES

Many replacement batteries are of the maintenance free type. For these batteries, follow the procedures described earlier. If the replacement battery you have purchased is not a maintenance free type, follow these easy procedures.

Check the battery fluid level at least once a month, more often in hot weather or during extended periods of travel. The electrolyte level should be up to the bottom of the split ring in each cell. If the level is low, add water. Distilled water is best for this purpose, but ordinary tap water can be used in a pinch.

At least once a year, check the specific gravity of the battery with a hydrometer. It should be between 1.20-1.26 on the hydrometer's scale. Most importantly, all the cells should read approximately the same. If one or more cells read significantly lower than the others, it's an indication that these low cells are going bad. Replace the battery.

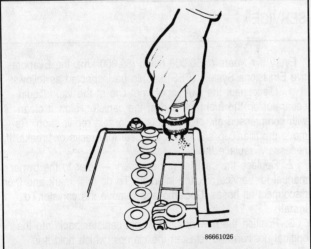

Fig. 56 Clean the battery posts with a wire brush or terminal cleaner made for that purpose

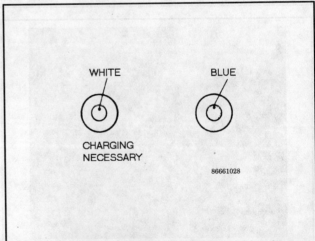

Fig. 57 The battery charge indicator, located on the top surface, changes from blue (when adequately charged) to white when the battery is undercharged

GENERAL INFORMATION AND MAINTENANCE

If water is added during freezing weather, the vehicle should be driven several miles to allow the electrolyte and water to mix. Otherwise the battery could freeze.

Filling the Battery

Batteries should be checked for proper electrolyte level at least once a month or more frequently. Keep a close eye on any cell or cells that are unusually low or seem to constantly need water — this may indicate a battery on its last legs, a leak, or a problem with the charging system.

Top up each cell to the bottom of the split ring, or, if the battery has no split ring, about ⅜ in. (9.5mm) above the tops of the plates. Use distilled water where available, or ordinary tap water, if the water in your area isn't too hard. Hard water contains minerals that may slowly damage the plates of your battery.

CABLES AND CLAMPS

▶ See Figures 58 and 59

Once year, the battery terminal posts and the cable clamps should be cleaned. Loosen the clamp bolts (you may have to brush off any corrosion with a baking soda and water solution if they are really messy) and remove the cables, negative cable first. On batteries with posts on top, the use of a battery clamp puller is recommended. It is easy to break off a battery terminal if a clamp gets stuck without the puller. These pullers are inexpensive and available in most auto parts stores. Side terminal battery cables are secured with a bolt.

The best tool for battery clamp and terminal maintenance is a battery terminal brush. This inexpensive tool has a female ended wire brush for cleaning terminals, and a male ended wire brush inside for cleaning the insides of battery clamps. When using this tool, make sure you get both the terminal posts and the insides of the clamps nice and shiny. Any oxidation, corrosion or foreign material will prevent a sound electrical connection and inhibit starting or charging. If your battery has side terminals, there is also a cleaning tool available for these.

Before installing the cables, remove the battery hold-down clamp or strap and remove the battery. Inspect the battery casing for leaks or cracks (which unfortunately can only be fixed by buying a new battery). Check the battery tray, wash it off with warm soapy water, rinse and dry. Any rust on the tray should be sanded away, and the tray given at least two coats of a quality anti-rust paint. Replace the battery, and install the hold-down clamp or strap, but do not overtighten.

Reinstall your clean battery cables, negative cable last. Tighten the cables on the terminal posts snugly; do not overtighten. Wipe a thin coat of petroleum jelly or grease all over the outside of the clamps. This will help to inhibit corrosion.

Finally, check the battery cables themselves. If the insulation of the cables is cracked or broken, or if the ends are frayed,

Fig. 58 Clean the inside of the clamps with a wire brush or a special tool

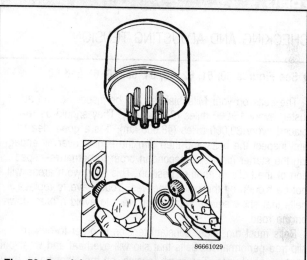

Fig. 59 Special tools are also available for cleaning the posts and clamps on side terminal batteries

replace the cable with a new cable of the same length or gauge.

✱✱CAUTION

Batteries give off hydrogen gas, which is explosive. DO NOT SMOKE around the battery! The battery electrolyte contains sulfuric acid. If you should splash any into your eyes or skin, flush with plenty of clear water and get immediate medical help.

CHARGING AND REPLACEMENT

Charging a battery is best done by the slow charging method (often called trickle charging), with a low amperage charger. Quick charging a battery can actually "cook" the battery, damaging the plates inside and decreasing the life of the battery drastically. Any charging should be done in a well ventilated area away from the possibility of sparks or flame. The

1-24 GENERAL INFORMATION AND MAINTENANCE

cell caps (not found on maintenance-free batteries) should be unscrewed from their cells, but not removed.

If the battery must be quick-charged, check the cell voltages and the color of the electrolyte a few minutes after the charge is started. If cell voltages are not uniform or if the electrolyte is discolored with brown sediment, stop the quick charging in favor of a trickle charge. A common indicator of an overcharged battery is the frequent need to add water to the battery.

When it becomes necessary to replace the battery, be sure to select a new battery with a cold cranking power rating equal to or greater than the battery originally installed. Deterioration, embrittlement and just plain aging of the battery cables, starter motor and associated wires makes the battery's job all the more difficult in successive years. The slow increase in electrical resistance over time makes it prudent to install a new battery with a greater capacity than the old. Details on battery removal and installation are covered in Section 3.

Belts

CHECKING AND ADJUSTING TENSION

▶ See Figures 60, 61, 62, 63, 64, 65, 66, 67 and 68

The belts on your Mitsubishi should be inspected and adjusted every 15,000 miles (24,000 km). They should be replaced every 30,000 miles (48,000 km). It is a good idea to first inspect the belts and then turn the engine over by engaging the starter for a split second in order to permit re-inspection of the belts in another position. This way, worn areas will not be hidden by the pulley sheaves. You'll have to replace belts that show severe wear if you want to avoid a breakdown on the road.

Belts must run under constant tension in order to ensure slip-free performance. Belts that slip will overheat and wear out at a very high rate. Too much tension, on the other hand, may cause premature failure of alternator, water pump, or power steering pump bearings.

Estimate tension by pressing a belt halfway between pulleys with thumb pressure (about 20 lbs. or 9 kg). The belt should deflect downward 1/4-3/8 in. (6-10 mm). Note that "deflection" does not refer to play or droop, but stretch. If the belt requires adjustment, use wrenches at both ends of the alternator or power steering pump support bolt and loosen it. Then loosen the adjusting bolt that passes through a slotted bracket. Pull (pry with an appropriate prybar, only if necessary) the alternator or power steering pump away from the engine block to create tension and hold it in position while you tighten the adjusting bolt.

On some later model vehicles, belt tension is adjusted by loosening the tension pulley locknut, turning the adjusting bolt on the tension pulley. Once you've reached the proper belt tension, tighten the locknut to retain the adjustment.

➡If possible, do not pry on the component. If you cannot get enough tension on the belt without prying, make sure to pry with an appropriate prybar squarely on the front of the housing right where the adjusting bolt passes through it. After tightening the adjusting bolt, tighten the support bolt and nut.

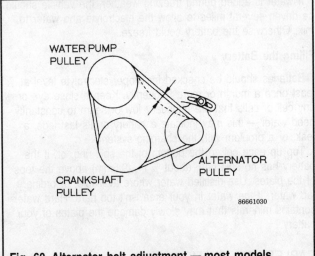

Fig. 60 Alternator belt adjustment — most models similar

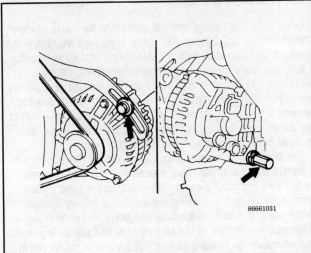

Fig. 61 Alternator belt adjustment support and adjusting bolts

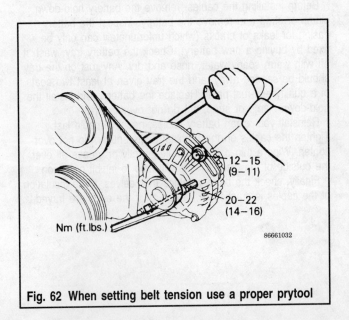

Fig. 62 When setting belt tension use a proper prytool

GENERAL INFORMATION AND MAINTENANCE   1-25

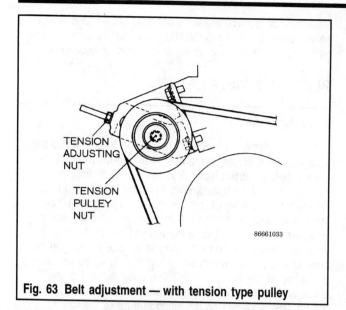

Fig. 63 Belt adjustment — with tension type pulley

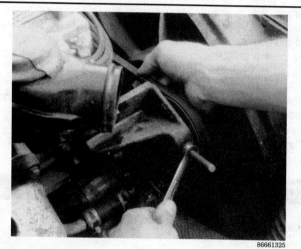

Fig. 64 Adjusting the drive belt using the tension adjusting nut

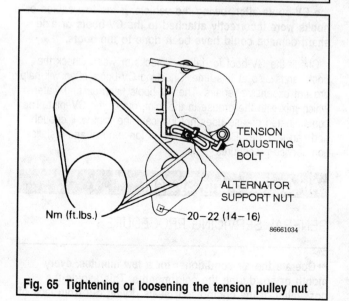

Fig. 65 Tightening or loosening the tension pulley nut

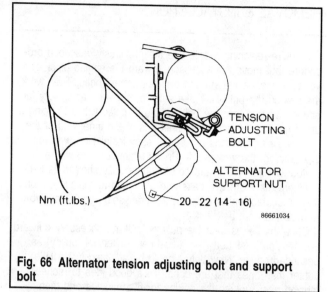

Fig. 66 Alternator tension adjusting bolt and support bolt

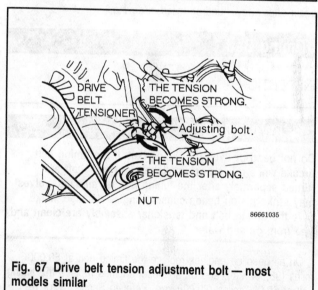

Fig. 67 Drive belt tension adjustment bolt — most models similar

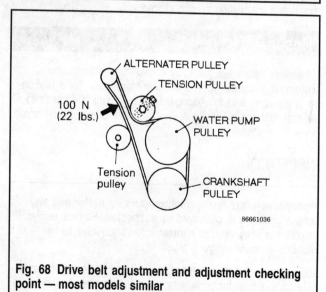

Fig. 68 Drive belt adjustment and adjustment checking point — most models similar

GENERAL INFORMATION AND MAINTENANCE

REMOVAL & INSTALLATION

If you're replacing a belt, use the previously described procedure, but move the accessory toward the engine block after you've loosened the support and adjustment bolts. Then work the belt off the pulleys. Make sure to move the accessory far enough so that the new belt can be installed without forcing it on. If moving the accessory until it is at the inner end of the adjusting slot still does not permit easy installation, you've probably got the wrong belt. Prying a new belt on with a prytool will damage it and could substantially shorten its life.

Where there are two belts and you have to replace the one that is located farther back on the crankshaft, you'll have to remove the front belt first.

Once the belt is over the pulleys, pull the accessory outward and tension it (slotted type adjustment or tension pulley) as described earlier. New belts should be adjusted just a little tighter than ones that are used. It's a good idea to recheck tension after running the engine for five minutes and then again after a few hundred miles of driving.

Timing Belt

INSPECTION

※※WARNING

Do not bend or twist the timing belt. If the timing belt breaks while driving or crankshaft and camshaft are turned separately after the timing belt is removed, valves may strike piston heads causing engine damage. Make sure the timing belt and tensioner assembly are clean and free from oil and water.

On all gasoline engines replace the timing belt at 60,000 miles (96,000 km), and on all diesel engines replace the timing belt at 50,000 miles (80,000 km). Refer to Section 3 for removal and installation service procedures.

Hoses

Radiator hoses are generally of two constructions, the preformed (molded) type, which is custom made for a particular application, and the spring-loaded type, which is made to fit several different applications. Heater hoses are all of the same general construction.

INSPECTION

→Inspection and replacement should be performed any time a problem is observed or suspected. Severe usage conditions may require shorter intervals. Refer to the owner's manual for your truck.

Cooling system and heater hoses should be inspected carefully once a year for brittleness, cracks, softening, bulging, and heat damage from nearby exhaust system parts. Mitsubishi recommends that you replace all cooling system and heater hoses, every five years or 50,000 miles (80,000 km) on some early models. Refer to your owner's manual, if necessary.

REMOVAL & INSTALLATION

▶ See Figure 69

There is usually no problem in replacing hoses. Just make sure you allow the engine to cool and then drain the cooling system before starting work. Also make sure to fully loosen clamps and pull them back from the connections before attempting to disconnect hoses. If hoses are tough to break loose from their connections, a twisting motion often works best. If they have been on a hot-running engine for a long time (and you know you are going to replace them), you might want to make a cut with a sharp knife going in the direction of the length of the hose. Cut until you pass the end of the connector on the block or engine accessory. This will loosen the grip the hose has on the fitting and make it easier to remove. Be careful not to cut deeply into relatively soft aluminum fittings. When the hoses are in place on the connections, position the hose clamps about ¼ in. (6mm) from the end of the hose and tighten them down. Hose clamps do not require that much force to form a good seal.

As a final note, don't throw those old hoses away. If they have not been cut or damaged, shake them out to remove the coolant, seal them in a plastic bag and store them in your truck. Old hoses will serve as good short term repair if by chance one of your hoses breaks. Of course, if you keep up on your maintenance, you may never need them.

CV-Boots

INSPECTION

→When vehicle is raised and safely supported during other general maintenance check the CV-boots. Also check the CV-boots after towing the vehicle. If the tow cables or hooks were incorrectly attached to the CV-boots or axle shaft, damage could have been done to the boots.

Check the CV-boot for fatigue, cracks or wear. Check the boot bands also at this time. Checking CV-boots often will help prevent expensive repairs. The CV-boots keep air and water, which mix with the grease in the joint, out of the CV-joint. The contaminated grease does not lubricate the joint well enough and can leak out. Refer to Section 7 for removal and installation service procedures.

Air Conditioning System

GENERAL SERVICING PROCEDURES

→Operate the air conditioner for a few minutes, every month or so, during the cold months. This avoids the possibility of the compressor seals drying out from lack of lubrication.

GENERAL INFORMATION AND MAINTENANCE

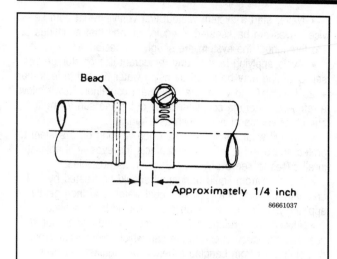

Fig. 69 Position the hose clamp so that it is about 1/4 inch (6mm) from the end of the hose

➡ To properly discharge and charge the A/C system a special charging system, quick connectors, complete training in refrigerant recycling and service procedures and a certification license is absolutely necessary. DO NOT VENT ANY AMOUNT (EVEN SMALL AMOUNTS — UNDER PENALTY OF LAW) OF REFRIGERANT INTO THE ATMOSPHERE.

The most important aspect of air conditioning service is the maintenance of a pure and adequate charge of refrigerant in the system. A refrigeration system cannot function properly if a significant percentage of the charge is lost. Leaks are common because the severe vibration encountered in a truck can easily cause a sufficient cracking or loosening of the air conditioning fittings. As a result, the extreme operating pressures of the system forces refrigerant out.

The problem can be understood by considering what happens to the system as it is operated with a continuous leak. Because the expansion valve regulates the flow of refrigerant to the evaporator, the level of refrigerant there is fairly constant. The receiver/drier stores any excess refrigerant, and so a loss will first appear there as a reduction in the level of liquid. As this level nears the bottom of the vessel, some refrigerant vapor bubbles will begin to appear in the stream of liquid supplied to the expansion valve. This vapor decreases the capacity of the expansion valve very little as the valve opens to compensate for its presence. As the quantity of liquid in the condenser decreases, the operating pressure will drop there and throughout the high side of the system. As the refrigerant continues to be expelled, the pressure available to force the liquid through the expansion valve will continue to decrease, and, eventually, the valve's orifice will prove to be too much of a restriction for adequate flow even with the needle fully withdrawn.

At this point, low side pressure will start to drop, and a severe reduction in cooling capacity, marked by freeze-up of the evaporator coil, will result. Eventually, the operating pressure of the evaporator will be lower than the pressure of the atmosphere surrounding it, and air will be drawn into the system wherever there are leaks in the low side.

Because all atmospheric air contains at least some moisture, water will enter the system and mix with the refrigerant and the oil. Trace amounts of moisture will cause sludging of the oil, and corrosion of the system. Saturation and clogging of the filter/drier, and freezing of the expansion valve orifice will eventually result. As air fills the system to a greater and greater extend, it will interfere more and more with the normal flows of refrigerant and heat.

SERVICE PRECAUTIONS

A list of general precautions that should be observed while servicing the system include:
• You must be certified to work on air conditioning systems in order to buy any refrigerant.
• It is illegal to service an air conditioning system unless certified and trained to do so.
• Keep all tools as clean and dry as possible.
• Thoroughly purge the service gauges and hoses of air and moisture before connecting them to the system. Keep them capped when not in use.
• Thoroughly clean any refrigerant fitting before disconnecting it, in order to minimize the entrance of dirt into the system.
• Plan any operation that requires opening the system beforehand in order to minimize the length of time it will be exposed to open air. Cap or seal the open ends to minimize the entrance of foreign material.
• When adding oil, pour it through an extremely clean and dry tube or funnel. Keep the oil capped whenever possible. Do not use oil that has not been kept tightly sealed.

✼✼WARNING

Use the correct refrigerant only. R-12 and R-134a must never be mixed, even in the smallest amounts. These two refrigerant systems are not compatible, mixing these will result in damage to your air conditioning system.

• Completely evacuate any system that has been opened to replace a component, other than when isolating the compressor, or that has leaked sufficiently to draw in moisture and air. This requires evacuating air and moisture with a good vacuum pump for at least one hour.

➡ If a system has been open for a considerable length of time it may be advisable to evacuate the system for up to 3 hours.

• Use a wrench on both halves of a fitting that is to be disconnected, so as to avoid placing torque on any of the refrigerant lines.

PREVENTIVE MAINTENANCE CHECKS

Antifreeze

In order to prevent heater core freeze-up during A/C operation, it is necessary to maintain permanent type antifreeze protection of +15°F (-9°C) or lower. A reading of -15°F (-26°C) is

GENERAL INFORMATION AND MAINTENANCE

ideal since this protection also supplies sufficient corrosion inhibitors for the protection of the engine cooling system.

➡ **The same antifreeze should not be used longer than the manufacturer specified.**

Radiator Cap

For efficient operation of an air conditioned truck's cooling system, the radiator cap should have a holding pressure which meets manufacturer's specifications. A cap which fails to hold this pressure should be replaced.

Condenser

Any obstruction of or damage to the condenser configuration will restrict the air flow which is essential to its efficient operation. It is therefore, a good rule to keep this unit clean and in proper physical shape.

➡ **Bug screens are regarded as obstructions.**

Condensation Drain Tube

This is normally a single molded drain tube which expels the condensation, that accumulates on the bottom of the evaporator housing, into the engine compartment.

If this tube is obstructed, the air conditioning performance can be restricted and condensation buildup can spill over onto the vehicle's floor.

SAFETY PRECAUTIONS

➡ **Check with your local authorities before attempting to service you vehicle's A/C system. In most areas it is illegal to purchase R-12 or R-134a and service the system unless you are a certified technician.**

❄❄CAUTION

The compressed refrigerant used in the air conditioning system expands into the atmosphere at a temperature of -21.7°F (-29.8°C) or lower. This will freeze any surface, including your eyes, that it contacts. In addition, R-12 refrigerant decomposes into a poisonous gas in the presence of a flame. Do not open or disconnect any part of the air conditioning system, near open flame.

- Avoid contact with a charged refrigeration system, even when working on another part of the air conditioning system or vehicle. If a heavy tool comes into contact with a section of copper tubing or a heat exchanger, it can easily cause the relatively soft material to rupture.
- When it is necessary to apply force to a fitting which contains refrigerant, as when checking that all system couplings are securely tightened, use a wrench on both parts of the fitting involved, if possible. This will avoid putting torque on refrigerant tubing. (It is advisable, when possible, to use tube or line wrenches when tightening these flare nut fittings.)
- Never start a system without first verifying that both service valves are backseated, if equipped, and that all fittings are throughout the system are snugly connected.
- Avoid applying heat to any refrigerant line or storage vessel. Charging may be aided by using water heated to less than +125°F (+51°C) to warm the refrigerant container. Never allow a refrigerant storage container to sit out in the sun, or near any other source of heat, such as a radiator.
- Always wear safety goggles when working on a system to protect the eyes. If refrigerant contacts the eyes, it is advisable in all cases to see a physician as soon as possible.
- Frostbite from liquid refrigerant should be treated by first gradually warming the area with cool water, and then gently applying petroleum jelly. A physician should be consulted.
- Always keep refrigerant can fittings capped when not in use. Avoid sudden shock to the can which might occur from dropping it, or from banging a heavy tool against it. Never carry a can in the passenger compartment of a truck
- Always completely discharge and recover the system before painting the vehicle (if the paint is to be baked on), or before welding anywhere near the refrigerant lines.

TEST GAUGES

Most of the service work performed in air conditioning requires the use of a set of two gauges, one for the high (head) pressure side of the system, the other for the low (suction) side.

The low side gauge records both pressure and vacuum. Vacuum readings are calibrated from 0-30 in. Hg (0-101 kPa) and the pressure graduations read from 0 to no less than 60 psi (413 kPa).

The high side gauge measures pressure from 0 to at least 600 psi (4134 kPa).

Both gauges are threaded into a manifold that contains two hand shut-off valves. Proper manipulation of these valves and the use of the attached test hoses allow the user to test high and low side pressures.

The manifold valves are designed so that they have no direct effect on gauge readings, but serve only to provide for, or cut off, flow of refrigerant through the manifold. During all testing and hook-up operations, the valves are kept in a closed position to avoid disturbing the refrigeration system.

INSPECTION

❄❄CAUTION

The compressed refrigerant used in the air conditioning system expands into the atmosphere at a temperature of -21.7°F (-29.8°C) or lower. This will freeze any surface, including your eyes, that it contacts. In addition, the refrigerant decomposes into a poisonous gas in the presence of a flame. Do not open or disconnect any part of the air conditioning system.

GENERAL INFORMATION AND MAINTENANCE 1-29

Sight Glass Check

♦ See Figures 70 and 71

You can safely make a few simple checks to determine if your air conditioning system needs service. The tests work best if the temperature is warm (about +70°F/21°C).

➡If your vehicle is equipped with an aftermarket air conditioner, the following system check may not apply. You should contact the manufacturer of the unit for instructions on systems checks.

1. Clean the sight glass, which is located on the top of the refrigerant receiver canister in the engine compartment.
2. Start the vehicle's engine.
3. Push the air conditioning button to operate the compressor, place the blower switch to HIGH, and move the temperature lever to the extreme left.
4. After operating for a few minutes in this manner, check the sight glass
    a. If the sight glass is clear, the magnetic clutch is engaged, the compressor discharge line is warm, and the compressor inlet line is cool, the system has a full charge.
    b. If the sight glass is clear, the magnetic clutch is engaged, and there is no significant difference between the compressor inlet and the discharge lines, the system has most of its refrigerant charge.
    c. If the sight glass is clear and the magnetic clutch is disengaged, the clutch is faulty or the system is out of refrigerant. Take the vehicle to a certified mechanic for servicing.
    d. If the sight glass shows bubbles or foam, the system could be low on refrigerant or the receiver is restricted. Occasional foam or bubbles are normal when the ambient temperature is above 110°F (43°C) or below 70°F (21°C). Take the vehicle to a certified mechanic for servicing.

✱✱WARNING

If it is determined that the system has a leak, it should be corrected as soon as possible. Leaks may allow moisture or contaminants to enter and cause a very expensive problem.

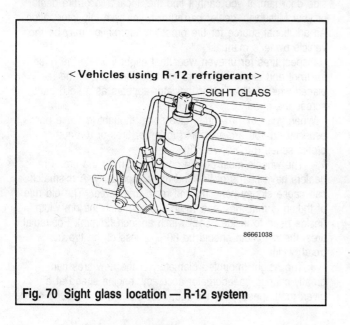

Fig. 70 Sight glass location — R-12 system

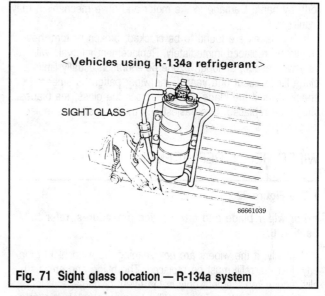

Fig. 71 Sight glass location — R-134a system

DISCHARGING, EVACUATING AND CHARGING THE SYSTEM

To properly discharge and charge the A/C system a special charging system, quick connectors, complete training in refrigerant recycling/special service procedures, and a certification license is absolutely necessary. DO NOT VENT ANY AMOUNT (EVEN SMALL AMOUNTS-UNDER PENALTY OF LAW) OF REFRIGERANT INTO THE ATMOSPHERE.

If you are not properly trained and certified, take the vehicle to a certified mechanic in order to have the system evacuated and discharged. Immediately after finishing work on the vehicle, take the truck back to the mechanic to have the system recharged as soon as possible.

LEAK TESTING

Some leak tests can be performed with a soapy water solution. There must be at least a ½ lb. charge in the system for a leak to be detected. The most extensive leak tests are performed with either a Halide flame type leak tester or the more preferable electronic leak tester.

In either case, the equipment is expensive, and, the use of a Halide detector can be **extremely** hazardous!

Windshield Wipers

Intense heat from the sun, snow and ice, road oils and the chemicals used in windshield washer solvents combine to deteriorate the rubber wiper refills. The refills should be replaced about once a year or whenever the blades begin to streak or chatter.

For maximum effectiveness and longest element life, the windshield and wiper blades should be kept clean. Dirt, tree sap, road tar and so on will cause streaking, smearing and blade deterioration if left on the windshield. It is advisable to wash the windshield carefully with a commercial glass cleaner at least once a month. Wipe off the rubber blades with a wet rag afterwards. Do not attempt to move the wipers back and

1-30 GENERAL INFORMATION AND MAINTENANCE

forth by hand! Damage to the motor and drive mechanism may result.

If the blades are found to be cracked, broken or torn they should be replaced immediately. Replacement intervals will vary with usage, although ozone deterioration usually limits blade lift to about one year. If the wiper pattern is smeared or streaked, or if the blade chatters across the glass, the blades should be replaced. It is easiest and most sensible to replace them in pairs.

WIPER REFILL REPLACEMENT

◆ See Figure 72

➡ For wiper blade and arm service procedures, refer to Section 6.

Normally, if the wipers are not cleaning the windshield properly, just the refill must be replaced. The blade and arm usually require replacement only in the event of damage. It is necessary (except on Tridon refills) to remove solely the arm or the blade to replace the refill (rubber part), though you may have to position the arm higher on the glass. You can do this by turning the ignition switch **ON** and operating the wipers. When they are positioned where they are accessible, turn the ignition switch **OFF**.

There are several types of refills and your vehicle could have any kind, since aftermarket blades and arms may not use exactly the same type refill as the original equipment. The original equipment wiper elements can be replaced as follows:

1. Lift the wiper arm off the glass.
2. Depress the release lever on the center bridge and remove the blade from the arm.
3. Lift the tab and pinch the end bridge to release it from the center bridge.
4. Slide the end bridge from the wiper blade and the wiper blade from the opposite end bridge.
5. Install a new element and be sure the tab on the end bridge is down to lock the element in place. Check each release point for positive engagement.

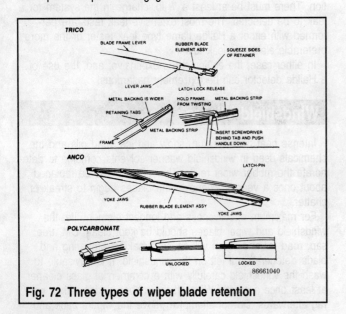

Fig. 72 Three types of wiper blade retention

Most Trico styles use a release button that is pushed down to allow the refill to slide out of the release jaws. The new refill slides in and locks in place. Some Trico refills are removed by locating where the metal backing strip or the refill is wider. Insert a small prytool blade between the frame and the metal backing strip. Press down to release the refill from the retaining tab.

The Anco style is unlocked at one end by squeezing 2 metal tabs, and the refill is slid out of the frame jaws. When the new refill is installed, the tabs will click into place, locking the refill.

The polycarbonate type is held in place by a locking lever that is pushed downward out of the groove in the arm to free the refill. When the new refill is installed, it will lock in place automatically.

The Tridon refill has a plastic backing strip with a notch about an inch from the end. Hold the blade (frame) on a hard surface so that the frame is tightly bowed. Grip the tip of the backing strip and pull up while twisting counterclockwise. The backing strip will snap out of the retaining tab. Do this for the remaining tabs until the refill is free of the arm. The length of these refills is molded into the end and they should be replaced with identical types.

No matter which type of refill you use, be sure that all of the frame claws engage the refill. Before operating the wipers, be sure that no part of the metal frame is contacting the windshield.

Tires and Wheels

◆ See Figures 73, 74 and 75

Inspect the tires regularly for wear and damage. Remove stones or other foreign particles which may be lodged in the tread. If tread wear is excessive or irregular it could be a sign of front end problems, or simply improper inflation.

The inflation should be checked at least once per month and adjusted if necessary. The tires must be cold (driven less than one mile) or an inaccurate reading will result. Do not forget to check the spare tire.

The correct inflation pressure for your vehicle can be found on a decal mounted in either the glove box or on the driver's side door jam. If you cannot find the decal a local tire dealer or your Mitsubishi dealer can furnish you with this information. An additional source for tire pressure information may be the vehicle owner's manual.

Inspect tires for uneven wear that might indicate the need for front end alignment or tire rotation. Tires should be replaced when a tread wear indicator appears as a solid band across the tread.

When you buy new tires, give some thought to these points, especially if you are switching to larger tires or to another profile series (50, 60, 70, 78):

• The wheels must be the correct width for the tire. Tire dealers have charts of tire and rim compatibility. A mismatch can cause sloppy handling and rapid tread wear. The old rule of thumb is that the tread width should match the rim width (inside bead to inside bead) within an inch (25mm). For radial tires, the rim width should be 80% or less of the tire (not tread) width.

• The height (mounted diameter) of the new tires can greatly change speedometer accuracy, engine speed at a given road speed, fuel mileage, acceleration, and ground clear-

GENERAL INFORMATION AND MAINTENANCE

ance. Tire makers furnish full measurement specifications. Speedometer drive gears may be available from Mitsubishi dealers for correction.

➡ **Dimensions of tires marked the same size may vary significantly, even among tires from the same maker.**

• The spare tire should be usable, at least for low speed operation, with the new tires.
• There shouldn't be any body interference when loaded, on bumps, or in turning.

The only sure way to avoid problems with these points is to stick to tire and wheel sizes available as factory options.

TIRE ROTATION

◆ See Figure 76

Rotation is recommended every 12,000 miles (7,500 km) or so, to obtain maximum tire wear. Snow tires sometimes have directional arrows molded into the sidewall of the tire: the arrow shows the direction of rotation. They will wear very rapidly and studded tires will lose their studs if their rotational direction is reversed.

➡ **Mark the wheel position or direction of studded snow tires before removing them.**

If your truck is equipped with tires having different load ratings on the front and the rear, the tires should not be rotated front to rear. Rotating these tires could affect tire life (the tires with the lower rating will wear faster, and they could become overloaded), upsetting truck handling.

TIRE DESIGN

For maximum satisfaction, tires should be used in sets of five. Mixing different types (radial, bias-belted, fiberglass belted) should be avoided. Conventional bias tires are constructed so that the cords run bead-to-bead at an angle. Alternate plies run at an opposite angle. This type of construction gives rigidity to both tread and sidewall. Bias-belted tires are

Fig. 75 A penny works well for checking tread depth; when the top of Lincoln's head is visible, it's time for new tires

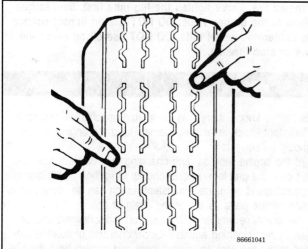

Fig. 73 Tread wear indicators will appear when the tire is worn out

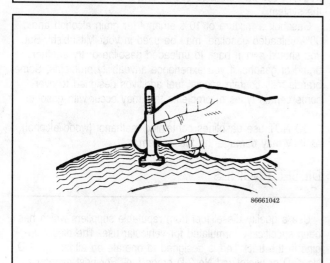

Fig. 74 Tread depth can also be checked with an inexpensive gauge

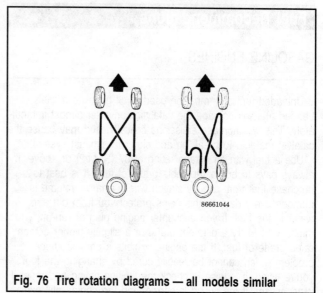

Fig. 76 Tire rotation diagrams — all models similar

similar in construction to conventional bias ply tires. Belts run at an angle and also at a 90 degree angle to the bead, as in the radial tire. Tread life is improved considerably over the conventional bias tire. The radial tire differs in construction, but instead of the sidewall plies running at an angle of 90 degree to each other, they run at an angle of 90 degree to the bead. This gives the tread a great deal of rigidity and the sidewall a great deal of flexibility and accounts for the characteristic bulge associated with radial tires.

Radial tires are recommended for use on all Mitsubishi trucks. If they are used, tire sizes and wheel diameters should be selected to maintain ground clearance and tire load capacity equivalent to the minimum specified tire. Radial tires should always be used in sets of five, but in an emergency radial tires can be used with caution on the rear axle only. If this is done, both tires on the rear should be of radial design.

➡ **Radial tires should never be used on ONLY the front axle.**

STORAGE

Store the tires at the proper inflation pressure if they are mounted on wheels. Keep them in a cool dry place, on their sides. If the tires are stored in the garage or basement, do not let them stand on a concrete floor; set them on strips of wood.

CARE OF SPECIAL WHEELS

An aluminum or special type wheel may be porous and leak air. Locate the leak by inflating the tire to 40 psi and submerging the tire/wheel assembly in water. Mark the leak areas. Remove the tire from the wheel and scuff the inside rim surface with 80 grit sandpaper. Apply a thick layer of adhesive sealant or equivalent to the leak area and allow to dry for approximately 6 hours.

Aluminum is very susceptible to the action of alkalies often found in various detergents and on road/sea salts. If the wheels have been exposed to these types of compounds, wash the the wheels with a special wheel cleaner or mild soap and water. Do not use harsh detergents or solvents or else the protective coating may be damaged. If steam is used to clean the vehicle, do not direct the steam at the wheels.

➡ **When changing an aluminum or special type wheel rimmed tire, finger tighten the lug nuts first, then torque them to the proper value. DO NOT use an impact wrench to tighten the wheel nuts. DO NOT use oil on either the nut or stud threads.**

FLUIDS AND LUBRICANTS

Fluid Disposal

Used fluids such as engine oil, transmission fluid, antifreeze and brake fluid are hazardous wastes and must be properly disposed. Before draining any fluids, consult with local authorities; in many areas, waste oil, etc. is accepted as a part of a recycling program. A number of service stations and auto parts stores are also accepting waste fluids for recycling.

Familiarize yourself with the recycling centers policies before draining any fluids, as many will not accept different fluids that have been mixed together, such as oil and antifreeze.

Fuel Recommendations

GASOLINE ENGINES

Unleaded fuel only must be used in all gasoline models. Leaded fuel will damage the catalytic converter almost immediately. This will increase emissions critically, and may cause the catalyst material to break up and clog the exhaust system.

Use a fuel with an octane rating of 87 (R+M/2) or above. It always pays to buy a reputable brand of fuel. It is best to purchase fuel from a busy station where a large volume is pumped every day, as this helps protect you from dirt and water in the fuel. If you encounter engine ping or run-on, you might want to try a different brand or a slightly higher octane rated grade of fuel. If the engine exhibits a chronic knock problem which cannot be readily cured by changing the fuel you're using, it is wise to check the ignition timing and reset it, if necessary. If the timing is correct and the engine still exhibits severe knock, there may be internal mechanical problems. Persistent knock is severely damaging to the engine and should be corrected.

If the engine runs on, and changing to a different fuel does not cure the problem, routine checks of ignition timing and idle speed should be made. Persistent run-on can be damaging to such engine parts as the timing chain or belt.

Mitsubishi recommends against the "indiscriminate use of fuel system cleaning agents." Occasional use of solvents added to the fuel tank to remove gum and varnish from the fuel system is permissible; however, continuous use or extremely frequent use can damage gasket and diaphragm parts used in the system.

Gasohol, a mixture of 10% ethanol (or grain alcohol) and 90% unleaded gasoline, may be used in your Mitsubishi. But, you should switch back to unleaded gasoline or try another brand of gasohol if you experience driveability problems. Some brands may contain special fuel additives designed to overcome certain types of problems that may occur with gasohol use.

DO NOT use gasolines containing methanol (wood alcohol), as they may damage the fuel system.

DIESEL ENGINES

Use a quality diesel fuel from reputable suppliers which has been specifically formulated for vehicular use. The diesel engine in the truck line is designed to operate on either No. 1-D, No. 2-D or winterized No 2-D grade fuel. For best economy and performance, a No. 2-D fuel should be used at all temperatures above 20°F. At lower temperatures, either 1-D or win-

GENERAL INFORMATION AND MAINTENANCE 1-33

terized 2-D should be used to avoid wax (paraffin) plugging the fuel filter. This can result in the engine not starting on a cold morning.

To avoid severe damage to the engine, do not mix gasoline with the diesel fuel. Do not use household heating oil or any diesel oil intended for use in marine or industrial engines.

The fuel filler cap and lid have labels stating **DIESEL**. In the event of gasoline being pumped into the tank by mistake, DO NOT attempt to start the engine, even to leave the pump area. If the vehicle can be towed and the tank drained before the engine is started, great damage and expense will be avoided.

No solvents or additives should ever be mixed with the diesel fuel. Damage to seals, gaskets and the injection pump can result.

Engine Oil Recommendations

▶ See Figures 77 and 78

Use only quality oils. Never use straight mineral or non-detergent oils, that is, oils not equipped with special cleaning agents. You must not only choose the grade of oil, but the viscosity number. Viscosity refers to the thickness of the oil. It's actually measured by how rapidly it flows though a hole of calibrated size. Thicker oil flows more slowly and has higher viscosity numbers, such as SAE 40 or 50. Thinner oil flows more easily and has lower numbers, such as SAE 10 or 20.

Mitsubishi recommends the use of what are called "multigrade" oils. These are specially formulated to change their viscosity with a change in temperature, unlike straight grade oils. The oils are designated by the use of two numbers, the first referring to the thickness of the oil, relative to straight mineral oils, at a low temperature such as 0°F (-18°C). The second number refers to the thickness, also relative to straight mineral oils, at high temperatures typical of highway driving (200°F (93°C)). These numbers are preceded by the designation "SAE", representing the Society of Automotive Engineers which sets the viscosity standards. For example, use of an SAE 10W-40 oil would give nearly ideal engine operation under almost all operating conditions. The oil would be as thin as a straight 10 weight oil at cold cranking temperatures, and as thick as a straight 40 weight oil at hot running conditions.

Note that diesel engines require different oils. This is because the oil gets much hotter, due to running with the high compression rates used in diesel engines. Be careful to adhere to all recommendations strictly for the longest possible engine life and best service. Oil recommendations are as follows.

NORMALLY ASPIRATED ENGINES

- Temperatures ranging from -20° to 60°F (-29° to 16°C): SAE 5W-20, 5W-30, 5W-40
- Temperatures ranging from -10° to 120°F (-23° to 49°C): SAE 10W-30, 10W-40, 10W-50
- Temperatures ranging from 32° to 120°F (0° to 49°C): SAE 20W-20, 20W-40, 20W-50

➡ **5W-20 is not recommended for sustained high speed operation regardless of the weather.**

FLUIDS AND LUBRICANTS
Recommended Lubricants

Engine Oil	SE or SF ①
Manual Transmission/Transaxle	GL-4
Automatic Transmission/Transaxle	DEXRON® II
Rear Axle	GL-5 ②
Wheel Bearings	NLGI Grade #2 EP Multipurpose Grease
Chassis Grease	NLGI Grade #2 Multipurpose Grease
Brake Fluid	DOT3
Clutch Fluid	DOT3
Manual Steering	GL-4
Power Steering	DEXRON® II
Antifreeze	ethylene glycol
Hinges	engine oil

① Applies to 1983–84 models and 1985 Cordia/Tredia. All 1985 models except Cordia/Tredia and all 1986–89 models use SF or SF/CC.
② Applies to standard rear axles only. For limited slip differentials, use Mitsubishi Gear Oil Part #8149630EX, STA-LUBE® API GL-5 High Performance Gear Oil, or equivalent.

Fig. 77 Factory recommended fluids and lubricants for Mitsubishi trucks and Monteros

DIESEL ENGINES

- Temperatures ranging from -20° to 50°F (-29° to 10°C): SAE 5W-30
- Temperatures ranging from -4° to 104°F (-20° to 40°C): SAE 10W-30
- Temperatures ranging from 5° to 120°F (-15° to 49°C): SAE 15W-40
- Temperatures ranging from 14° to 120°F (-10° to 49°C): SAE 20W-40
- Temperatures ranging from 32° to 104°F (0° to 40°C): SAE 30, 20W-40, 15W-40

➡ **5W-30 oils should be used only in extremely cold areas, where the temperature is consistently below freezing when starting. 5W oils are not recommended for sustained high speed or high rpm driving.**

1-34 GENERAL INFORMATION AND MAINTENANCE

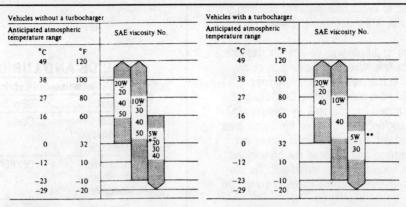

Fig. 78 Oil viscosity standards for both normally aspirated and turbocharged (diesel) engines. Adherence to these standards helps insure long engine life, easier starting and economical operation in cold weather

SYNTHETIC OIL

There are excellent synthetic and fuel-efficient oils available that, under the right circumstances, can help provide better fuel mileage and better engine protection. However, these advantages come at a price, which can be more than the cost per quart of conventional motor oils.

Before pouring any synthetic oils into your vehicle's engine, you should consider the condition of the engine and the type of driving you do. Also, check the manufacturer's warranty conditions regarding the use of synthetics.

Generally, it is best to avoid the use of synthetic oil in both brand new and older, high mileage engines. New engines require a proper break-in, and the synthetics are so slippery that they can hinder this. Most manufacturers recommend that you wait at least 5,000 miles before switching to a synthetic oil. Conversely, older engines are looser and tend to use more oil. Synthetics will slip past worn parts more readily than regular oil. If your truck already leaks oil (due to worn parts and bad seals or gaskets), it will leak more with a slippery synthetic inside.

Consider your type of driving. If most of your accumulated mileage is on the highway at higher, steadier speeds, a synthetic oil will reduce friction and probably help deliver fuel mileage. Under such ideal highway conditions, the oil change interval can be extended, as long as the oil filter will operate effectively for the extended life of the oil. If the filter can't do its job for this extended period, dirt and sludge will build up in your engine's crankcase, sump, oil pump and lines, no matter what type of oil is used. If using synthetic oil in this manner, you should continue to change the oil filter at the recommended intervals.

Trucks used under harder, stop-and-go, short hop circumstances should always be serviced more frequently, and for these trucks, synthetic oil may not be a wise investment. Because of the necessary shorter change interval needed for this type of driving, you cannot take advantage of the long recommended change interval of most synthetic oils.

Engine

✲✲CAUTION

The EPA warns that prolonged contact with used engine oil may cause a number of skin disorders, including cancer! You should make every effort to minimize your exposure to used engine oil. Protective gloves should be worn when changing the oil. Wash your hands and any other exposed skin areas as soon as possible after exposure to used engine oil. Soap and water, or waterless hand cleaner should be used.

OIL LEVEL CHECK

♦ See Figures 79 and 80

As often as you stop for fuel, check the engine oil as follows:

1. Park the vehicle on a level surface (if the vehicle is not level, the reading will not be completely accurate).
2. If the vehicle has been running, stop the engine and allow it to sit for a full three minutes. If the engine is cold, check the oil before starting it. It does not matter whether the oil is hot or cold, as long as it has had time to drain out of the engine itself and into the oil pan.
3. Open the hood and locate the dipstick (refer to the owner's manual, if necessary). It consists of a ring-like handle running into a tube which is connected to the engine. Pull the dipstick out and wipe all the oil off the bottom with a clean rag. If this is not done, you will not get an accurate reading of the oil level.
4. Re-insert the dipstick and make sure it goes all the way into the tube. Then, pull it out and read it on the side with the oil level scale (two lines, two dots, etc.).

   a. If the oil level is above the lower line, although the oil level is high enough, you should still add enough oil to bring the level up to the upper mark. Usually the amount of oil needed to bring the level from the lower mark to the upper

GENERAL INFORMATION AND MAINTENANCE

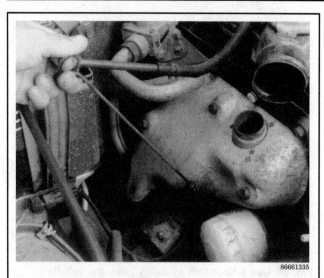

Fig. 79 Use the oil dipstick to check the oil level

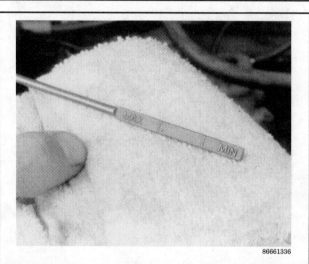

Fig. 80 Fill the engine slowly with oil until the oil level registers between the MAX and MIN marks

mark is one quart, however you should fill the oil slowly and check often. It is important not to overfill the engine.

b. If it is right near or at the lower line, add oil slowly (⅓ or ½ quart) and check the level often. Fill the oil up to the upper mark, but do not overfill.

c. If the oil is below the lower line, add oil, ⅓ or ½ quart at a time, until the level is at the upper mark. A beginning level below the lower line indicates that either you are not checking the oil level frequently enough or that the engine is using too much oil.

➡ Running the engine with the oil below the lower line may contribute to excessive heat and dirt in the oil, and will leave you with insufficient reserve to allow for normal oil consumption — you could run out on the road. However, you should not add oil to the point where the level is significantly above the upper line. Under these conditions, the rotating crankshaft will cause the oil to foam, which can be damaging to the engine and will sometimes cause valve train noise.

5. To add oil, unscrew the cap on the valve cover on top of the engine and pour the oil into the engine. Avoid letting any dirt get into the engine, and make sure to reinstall the cap before starting the engine.

OIL AND FILTER CHANGE

▶ See Figures 79, 80, 81, 82, 83, 84, 85, 86, 87, 88, 89, 90 and 91

➡ It may be a good idea to look under the vehicle, before starting any service procedure, to orientate yourself with the necessary components and locations.

The mileage figures given in the Maintenance Intervals chart are the factory recommended intervals for oil and filter changes assuming you are driving your vehicle under average conditions. Make sure to read the footnote concerning decreasing the change interval for certain types of driving and adhere to the recommendations.

While Mitsubishi recommends changing the oil filter only every other oil change (after the initial change), we recommend changing the filter every time because of the extra insurance this provides. Not only will this help guarantee that the oil filter will work effectively (the filter bypasses dirty oil directly to engine parts when it gets saturated with dirt); but changing the filter removes at least an additional quart of dirty oil from the engine oil passages. Purchase oil which conforms to all the specifications listed earlier. Make sure that the filter you purchase is specified for your particular vehicle model, year, and engine type. The filter must be able to withstand 256 psi (1764 kPa) to conform to factory specifications.

Always drain the oil after the engine has been running long enough to bring it to operating temperature. It's best to actually drive the vehicle until the temperature gauge reaches normal operating temperature to help ensure the oil will be as warm as possible. Hot oil will flow out of the oil pan more easily and will keep contaminants in suspension so that they will be removed with the oil instead of staying in the pan. You will need a large capacity oil pan — usually about 6 qts. (5.68L). Just make sure the capacity of the pan is greater than the oil pan and filter as shown in the Capacities Chart. You will also need

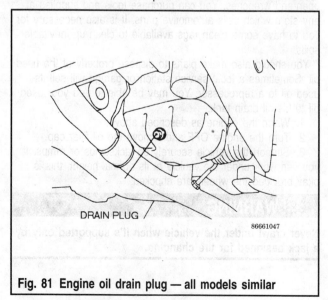

Fig. 81 Engine oil drain plug — all models similar

1-36 GENERAL INFORMATION AND MAINTENANCE

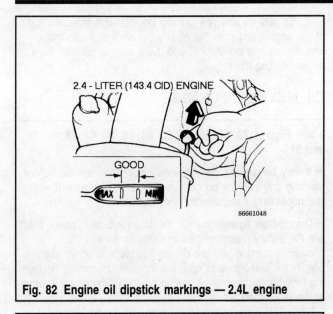

Fig. 82 Engine oil dipstick markings — 2.4L engine

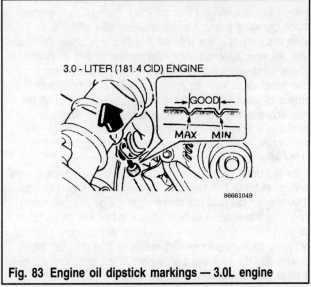

Fig. 83 Engine oil dipstick markings — 3.0L engine

Fig. 84 Use a floor jack to safely raise the vehicle — not a bumper or scissors jack

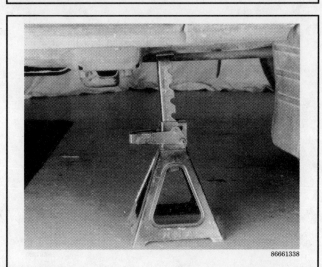

Fig. 85 Jackstands are a must to securely support a raised vehicle and create a safer work environment

a strap wrench to loosen the filter, and an ordinary set of open-end wrenches. You can purchase tools and supplies at any store which sells automotive parts. It is also necessary for you to have some clean rags available to clean up inevitable spills.

You should also make plans to dispose properly of the used oil. Sometimes a local service station or garage will sell its used oil to a reprocessor. You may be able to add your used oil to his oil drain tank.

1. Warm the engine as described above.
2. Turn the engine **OFF** and remove the oil filler cap.
3. Support the vehicle securely on jackstands or ramps. If you can work under the vehicle at its normal height, this is okay provided the wheels are chocked.

✶✶WARNING

Never crawl under the vehicle when it's supported only by a jack designed for tire changing.

4. Place the drain pan under the oil pan. It should be located where the stream of oil running out of the drain hole will run into the pan — not just directly below the drain hole.
5. Loosen the drain plug using a box wrench or a ratchet, short extension and socket. Turn the plug out slowly by hand, using a rag to shield your fingers from the hot oil. By keeping inward pressure on the plug with your fingers as you unscrew it, oil won't escape past the threads and you can remove it without being burned by hot oil.
6. Quickly withdraw the drain plug and move your hands out of the way, but make sure you keep hold of the plug so that it does not drop into the pan. Wipe the plug with a clean rag. Put it in a safe place — one where it won't get kicked or bumped out of sight. As the oil drains, the stream may shift as the level in the pan changes. Keep your eye on the stream and shift the pan as needed.
7. Allow the oil to drain completely in the pan, then install and carefully tighten the drain plug. Use a new drain plug washer if necessary. Be careful not to overtighten the drain

GENERAL INFORMATION AND MAINTENANCE 1-37

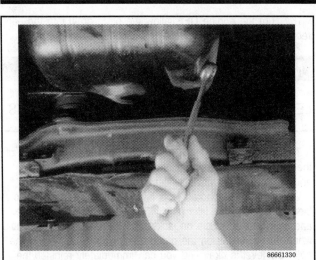

Fig. 86 Loosen the oil pan drain plug — make sure a drain container is in position

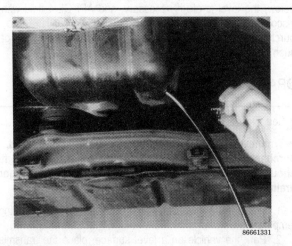

Fig. 87 Try to keep the oil pan plug from dropping into the drain pan — be careful of hot oil, which can cause painful burns

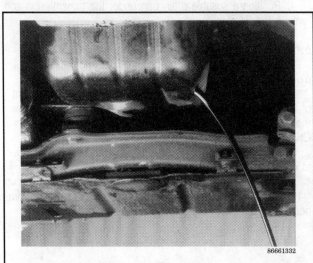

Fig. 88 Allow the engine oil to drain completely and move the drain pan to keep it under the flow of dirty oil

plug, otherwise you will be buying a new pan or a trick replacement plug for stripped threads.

8. Move the pan under the oil filter. Use a strap-type or cap-type wrench to loosen the oil filter. Cover your hand with a rag and spin the filter off by hand; turn it slowly. Keep in mind that it's holding about one quart of dirty, hot oil. Empty the filter into the drain pan.

➡ **If the oil filter cannot be loosened by conventional methods, punch a hole through both sides at the mounting base of the filter, insert a punch and use it to break the oil filter loose. After the oil filter is loosened, remove the oil filter from the engine with a oil filter type wrench or by hand.**

9. With a clean rag, wipe off the filter adapter on the engine block. Make sure that no lint from the rag remains on the adapter as it could clog an oil passage. Also make sure the rubber gasket from the old filter did not remain on the adapter.

10. Using your finger, apply a film of new oil to the rubber gasket on the top of the new oil filter. Read the directions on the side of the filter, or on the box it came in, to ascertain how tightly it should be installed. Carefully screw the new filter onto the oil filter mounting pad. If the filter becomes immediately difficult to turn, it is probably crossthreaded. Remove the filter and continue to install it until the filter goes on and turns easily. Once the threads do start, turn the filter gently until it just touches the engine block; it will suddenly get harder to turn at this point.

11. You may want to mark the filter at this point so you'll know just how far you turn it. By hand, turn it an additional $1/2$-$3/4$ turn, or as specified by the filter manufacturer. If the filter is turned past this point, the rubber gasket may leak.

12. Wipe the drain plug area on the oil pan and carefully reinstall the drain plug. Just as with the filter, be careful not to crossthread the plug. It will easily turn well past the point where the threads have started if it's not crossthreaded.

13. Lower the vehicle back down to the ground.

14. Pour in oil to the full capacity of the oil pan and filter, as specified in the Capacities Chart. Reinstall the filler cap. Just as a precaution, remove the dipstick and check the oil level. If the oil level is slightly over the upper mark on the dipstick, this

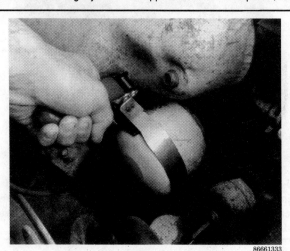

Fig. 89 Use an oil filter wrench to loosen the old filter, but not to tighten the new one

1-38 GENERAL INFORMATION AND MAINTENANCE

Fig. 90 Before installing a new oil filter, coat the rubber gasket with a film of clean oil

is alright because the new filter has not yet been filled with oil. After the engine has been run, however, the oil level should have dropped to below the upper mark.

15. Once the engine has enough oil, start the engine, preferably without touching the throttle, as there will be no oil pressure for 10 seconds or more while the oil pump fills the filter and engine oil passages. Allow the engine to idle at the lowest possible speed until the oil light goes out or the gauge shows that oil pressure has been established. If you do not get oil pressure within 15 to 30 seconds, stop the engine and investigate. Once oil pressure is established, leave the engine running and inspect the filter and oil plug for leaks. If there is slight leakage around the filter, you might want to try to tighten it just a bit more to stop the leaks. Usually, if you've tightened it properly, the only cause of leakage is a defective filter or gasket, which would have to be replaced before you drive the truck.

Fig. 91 Fill the engine with the amount of oil specified in the Capacities Chart at the end of this section. Be careful not to overfill the engine, which could result in damaged engine parts

16. Turn the engine **OFF**, allow the oil to drain into the pan, recheck the level, and add oil as needed.

Manual Transmission

LEVEL CHECK AND FLUID RECOMMENDATIONS

The manual transmission filler plug is located on the side of the unit. Be careful to identify the drain plug on the bottom and differentiate it from the filler plug on the side.

With the vehicle parked on a level surface and the wheels blocked, remove the filler plug. It may be necessary to raise and support the vehicle on jackstands if the plug is inaccessible when the vehicle is on the ground. If the fluid runs out of the hole, the level is okay and you can immediately reinstall the plug. If fluid does not run out, you can stick your finger into the hole to check the level. As long as its level is at or near the bottom of the hole (no more than about ¼ inch (6mm) below) the level is ok. If necessary, add API GL-4 fluid with a viscosity of SAE 80W or SAE 75W-85W. You may want to purchase a syringe-like device for adding small quantities of such fluid from your local automotive parts store.

DRAIN AND REFILL

▶ See Figure 92

➡**Whenever you plan to change the transmission fluid, purchase a new drain plug gasket ahead of time. This item should be replaced at each fluid change to ensure a good drain plug seal.**

1. Warm the engine until it has reached normal operating temperature.
2. Park the vehicle on a level surface, place the transmission in Neutral, apply the parking brake and block the rear wheels.
3. It may be necessary to raise and support the vehicle on jackstands to access the drain plug on the transmission.
4. Position a suitable drain pan under the transmission and remove the fill plug.
5. Remove the drain plug and allow the oil to drain completely.
6. Wipe off the threads of the drain plug and install the plug with a new gasket.
7. Through the filler plug opening, add the proper amount of fluid (see the Capacities Chart) until the level is just below the bottom of the opening. If a small amount runs out, that's OK.
8. Lower the vehicle if it was necessary to raise it in order to drain the transmission.
9. After wiping the threads, install and tighten the filler plug.
10. Drive the vehicle to allow the oil to reach operating temperature and then check the drain and fill plugs for leaks.

GENERAL INFORMATION AND MAINTENANCE 1-39

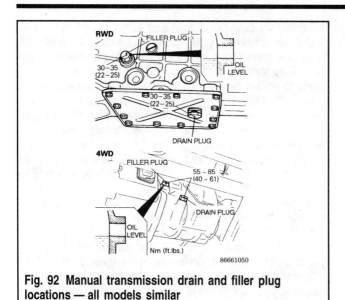

Fig. 92 Manual transmission drain and filler plug locations — all models similar

Automatic Transmission

LEVEL CHECK AND FLUID RECOMMENDATIONS

➡ Do not overfill the transmission. An electronic fluid level sensor, available on some vehicles, provides a convenient means of checking the transmission fluid level. allow the engine to idle and observe the transmission warning lamp for 30 seconds. If lamp remains OFF, it is not necessary to proceed further. If the lamp glows, the fluid level should be checked manually.

Check the automatic transmission fluid every 30,000 miles (48,000 km). The fluid level in these units is very sensitive to heat and must only be checked when the unit is at normal operating temperature. Drive the vehicle until a few miles after the engine has reached operating temperature. Then, with the engine idling, engage every gear selector position for a few seconds. Finally, put the gear selector in the Neutral position and securely engage the handbrake.

Remove the dipstick, wipe it, reinstall it all the way, and remove it again. Fluid level should be between the two marks on the stick. If the fluid level is at the lower mark, gradually add one pint to bring the level even with the upper mark on the dipstick. Do not overfill the transmission as this will cause foaming of the fluid and operating problems.

You should add Dexron®II automatic transmission fluid only. Turn the engine **OFF** and add the fluid with a small funnel through the transmission dipstick tube. Mitsubishi recommends against the use of any transmission additives whatsoever.

DRAIN, REFILL AND FILTER REPLACEMENT

▶ See Figures 93 and 94

➡ Most trucks do not have a torque converter drain plug. No attempt should be made to drain the torque converter. If the vehicle is equipped with an oil pan drain plug, remove the drain plug (replace the gasket, if so equipped, upon installation) and drain the fluid. Reinstall the drain plug and tighten to 13-17 ft. lbs. (18-23 Nm). Refer to the service procedure for reference only.

It's best to change the fluid with the transmission hot. Drive the vehicle until the transmission is warmed up; a few miles after the engine temperature gauge reaches normal levels. Then follow these steps:

1. Support the vehicle securely on jackstands or ramps. It should be reasonably level.
2. There is no drain plug on most Mitsubishi automatic transmission oil pans. To remove the pan, loosen all the pan bolts a few turns, but do not completely remove and then place a large container under the oil pan. Tap one corner of the pan with a soft hammer to break the seal.
3. Once the pan seal is broken loose, support it, remove all the bolts, and then tilt it to one side to drain the fluid.

➡ Check the fluid in the drain pan, it should always be a bright red color. It if is discolored (brown or black), or smells burnt, serious transmission troubles, probably due to overheating, should be suspected. The transmission should be inspected by a qualified service technician to locate the cause of the burnt fluid.

4. Remove the attaching bolts and remove the filter assembly. This is located on the underside of the transmission in the area covered by the transmission pan. Most filters are held in place by four bolts, however the actual number of bolts may be different. Strainers may be cleaned in a safe solvent and air dried. Foam filters must be replaced.
5. Install the filter or strainer and tighten the bolts alternately (diagonally) in several stages.
6. Clean all the gasket surfaces thoroughly. Then, put the pan and new gasket in position with bolt holes aligned. Reinstall the bolts, tightening them only very gently with your fingers.
7. Tighten the bolts, using a diagonal pattern, in several stages to 12 ft. lbs. (17 Nm). Pour fluid in cautiously until it reaches the lower mark on the dipstick. Refer to the Capacities Chart for the amount you will need. Start the engine, put the gear selector in each of the positions for several seconds, and then go back to Neutral. Check the fluid level again and make sure it's above the lower mark. Drive the vehicle until the transmission is hot; then add fluid gradually until it has reached the full mark.

PAN AND FILTER SERVICE

Refer to Section 7 for complete service procedures.

Transfer Case

▶ See Figure 95

The 4-wheel drive family of vehicles use a transfer case between the transmission and the driveshaft. This case contains a series of gears and chains which must be properly lubricated. Although the transfer case is connected to the transmission, the two units have separate oil supplies. Do not make the mistake of thinking that draining or filling one unit takes care of both.

1-40 GENERAL INFORMATION AND MAINTENANCE

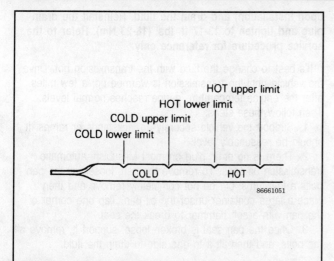

Fig. 93 Automatic transmission dipstick — all models similar

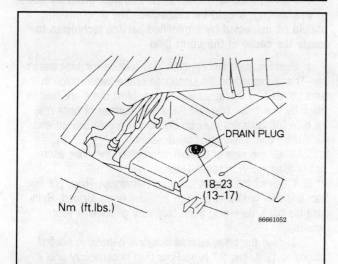

Fig. 94 Automatic transmission drain plug, if so equipped

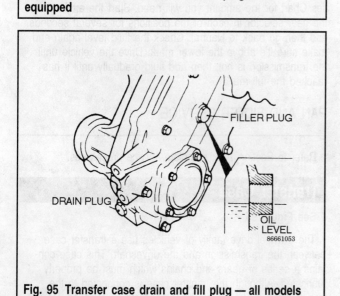

Fig. 95 Transfer case drain and fill plug — all models similar

The transfer case has a drain plug at the lower edge and a fill plug about halfway up the case. Fluid level is checked with the vehicle parked on a level surface and the wheels blocked. It may be necessary to raise and support the vehicle on jackstands to access the drain and fill plugs on the transfer case. Remove the filler plug. If the fluid runs out of the hole, the level is okay and you can immediately reinstall it. If fluid does not run out, you can stick your finger into the hole to check the level. As long as it is at or near the level of the bottom of the hole (no more than about ¼ inch below) the level is ok. If necessary, add API GL-4 fluid of viscosity SAE 80W or SAE 75W-85W. This fluid specification does not change if the vehicle is equipped with an automatic transmission. To drain the fluid follow the service procedure below.

1. Park the vehicle on a level surface.
2. It may be necessary to raise and support the vehicle on jackstands to access the drain plug on the transmission.
3. Position a suitable drain pan under the transfer case drain plug and remove the fill plug.
4. Remove the drain plug and allow the oil to drain completely.
5. Wipe off the threads of the drain plug and install the plug with a new gasket.
6. Through the filler plug opening, add fluid (API GL-4, SAE 80W or SAE 75W-85W) until the level is just below the bottom of the opening. If a small amount runs out, that's OK.
7. After wiping the threads, install and tighten the filler plug.
8. Drive the vehicle to allow the oil to reach operating temperature and then check the drain, fill plugs for leaks.

Drive Axles

LEVEL CHECK AND FLUID RECOMMENDATIONS

➡Inspection and replacement should be performed any time a problem is observed or suspected. Severe usage conditions may require shorter intervals — refer to the owner's manual for your truck.

The fluid level must be checked every 30,000 miles (48,000 km) on conventional differentials. On limited slip units, change the fluid every 30,000 miles (48,000 km). To check the fluid level, first make sure the wheels are chocked and then simply remove the plug from the side of the unit. Place a drain pan under the drain and fill plugs to catch any spillage. It may be necessary to raise and support the vehicle on jackstands to access the drain and fill plugs on the differential unit. If the fluid runs out or you can feel with your finger that it's not more than ¼ inch (6mm) below the level of the hole, the level is satisfactory. Otherwise, add the following specified fluid with a syringe-like device you can buy at a local auto parts store.

For the limited slip differential use Hypoid gear Mitsubishi genuine gear oil part number 8149630EX or equivalent. For the conventional axle, the API classification is GL-5 or higher, and viscosity recommendations are as follows:

• Above -10°F (-23°C): SAE 90, SAE 85W-90, or SAE 80W-90
• -10°F (-23°C) to -30°F (-34°C): SAE 80W, SAE 80W-90
• Below -30°F (-34°C): SAE 75W

GENERAL INFORMATION AND MAINTENANCE

DRAIN AND REFILL

♦ See Figures 96, 97, 98, 99, 100, 101 and 102

1. Make sure the wheels are chocked. Install a drain pan under the drain plug, located in the bottom of the differential housing. Place a drain pan under the differential unit to catch the fluid from the drain plug.
2. If there is dirt or build up around the plugs, use a wire brush to remove the dirt. This will help keep contaminents out of the differential fluid.
3. Remove the fill plug from the side of the unit, making sure to retain or replace the fill plug washer.
4. Remove the drain plug, keeping it in your fingers.
5. When the fluid has drained completely, replace the drain plug, making sure the gasket is retained on the plug.
6. Add fluid until it runs out of the filler plug hole, and install the filler plug.

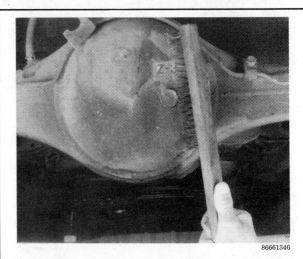

Fig. 98 Use a wire brush to clean the area around the fill plug to keep dirt out of the differential housing

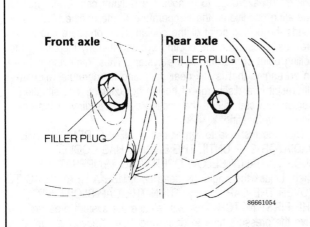

Fig. 96 Drain and fill plug locations — conventional differential

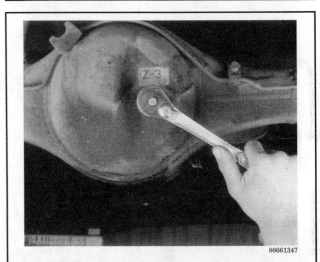

Fig. 99 Loosen and remove the fill plug, located on the back cover of the differential unit

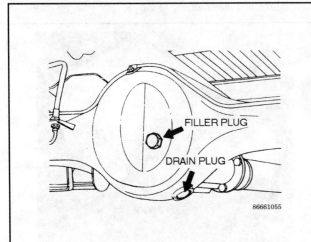

Fig. 97 Drain and fill plug locations — limited slip differential

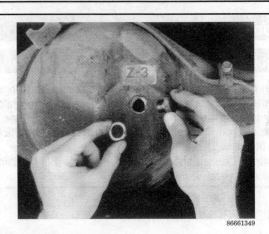

Fig. 100 When removing and installing the fill and drain plugs on the differential units, make sure to either save the old washers (if sufficient for reuse) or install new ones

1-42 GENERAL INFORMATION AND MAINTENANCE

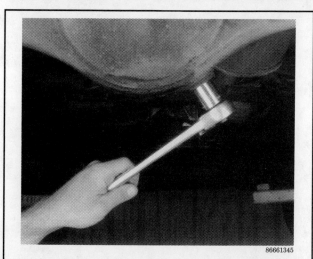

Fig. 101 Unscrew the drain plug to drain the differential unit oil

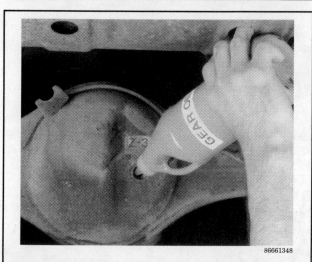

Fig. 102 After installing the drain plug, fill the unit with clean gear oil until a little runs out of the fill plug hole

Cooling System

✱✱CAUTION

When draining coolant, keep in mind that cats and dogs are attracted by ethylene glycol antifreeze, and are quite likely to drink any that is left in an uncovered container or in puddles on the ground. This will prove fatal in sufficient quantity. Always drain the coolant into a sealable container. Coolant should be reused unless it is contaminated or several years old.

FLUID RECOMMENDATIONS

➡ Propylene glycol and ethylene glycol antifreeze are not compatible and must not be mixed together. Refer to the label on antifreeze/coolant container if necessary.

Mitsubishi recommends the use of a quality ethylene glycol coolant containing corrosion inhibitors, and approved for use with all the metals in the engine, including aluminum. Mitsubishi recommends against the use of additional cooling system additives as they may be incompatible with the coolant itself.

LEVEL CHECK

▶ See Figures 103, 104, 105 and 106

Coolant level should always be checked by simply looking at the level in the overflow tank. If there is a coolant leak, the level in the tank will drop as the coolant is drawn back into the radiator. If there is no coolant visible in the tank, the coolant level needs to be checked by removing the radiator cap. Also, if there are indications that the engine is overheating, it is wise to check the level in the radiator, as a defective cap may permit the level to drop there even though there is an ample supply in the overflow system.

One of the reasons for the use of an overflow system is that it is often hazardous to remove the radiator cap. Under normal operating conditions, the temperature of the engine block exceeds the boiling point of the coolant. The pressure cap increases pressure in the cooling system, thereby increasing the boiling point of the fluid in the system. Thus, removing the cap on an engine that is still near operating temperature will usually result in a discharge of boiling hot coolant. The situation worsens substantially if the cooling system is dirty or the engine is overheating.

Thus the first rule to remember is: NEVER REMOVE THE RADIATOR CAP UNTIL THE ENGINE HAS COOLED SUBSTANTIALLY BELOW OPERATING TEMPERATURE. Then, unless the engine is dead cold, USE A HEAVY RAG TO COVER THE CAP. Finally, TURN THE CAP SLOWLY TO THE FIRST NOTCH. This will release the system pressure. Give the pressure time to drop, and then remove the cap. If the radiator cap is badly worn, the pressure may not be released. This is why it is necessary to use the rag, cool the engine first, and still proceed cautiously and slowly.

Once the cap has been removed, you can add a 50-50 mix of antifreeze and water slowly to the radiator until the level

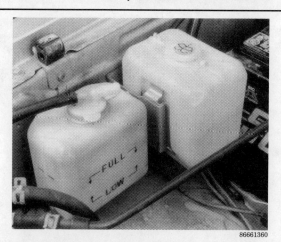

Fig. 103 Simply check the overflow system level by comparing the fluid level with the FULL and LOW level indicators on the overflow container

GENERAL INFORMATION AND MAINTENANCE

increases. Once the level is near the top of the radiator tank, start the engine. Let the engine idle until the thermostat opens (you'll see coolant flow through the top of the radiator tank and the upper radiator hose will become hot). Air in the system will be expelled at this point, causing the level of coolant to drop again. Keep adding coolant until the level remains near the top of the radiator. Then, install the cap, add coolant to the overflow tank until it is well past the lower mark and install the overflow tank cap as well.

DRAINING, FLUSHING AND REFILLING

▶ See Figure 107

It's best to drain the cooling system when the engine is warm (but has cooled to well below operating temperature for safety) to assist in complete removal of old coolant and any

Fig. 104 Once the engine has cooled down and the pressure in the system is relieved, remove the radiator cap

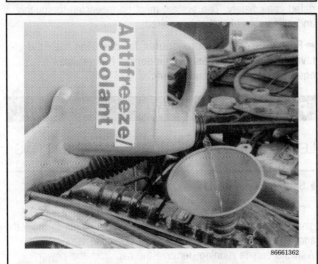

Fig. 105 With the radiator cap removed, fill the system with a 50-50 mixture of water and antifreeze

Fig. 106 Once the system is completely filled, fill the overflow reservoir to the FULL mark

suspended material. Follow the procedure below to help ensure you will not be burned by hot coolant.

1. Loosen the drain cock or remove the drain plug located on the bottom of the radiator tank. This will begin the draining process and relieve pressure. DON'T START OUT BY REMOVING THE CAP! Just make sure you are well away from the direction the coolant flow will take when you remove the plug, and then remove it.

✲✲CAUTION

Draining hot water can cause painful burns. Be careful when doing this.

2. Once coolant flow out of the bottom of the radiator has slowed, remove the radiator cap to vent the system. Then, remove the coolant drain plug from the side of the engine block.

3. When all the coolant has drained, replace the plugs. If the hoses need attention, this is an ideal time to replace them. Slowly fill the system with water until the water level reaches

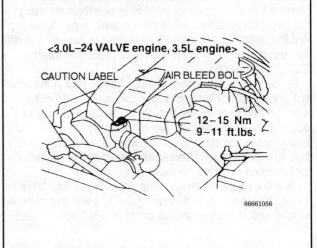

Fig. 107 Cooling system air bleed bolt location — 6-cylinder engines

1-44 GENERAL INFORMATION AND MAINTENANCE

the top of the radiator tank. Start the engine and run it at idle. When the thermostat opens and water begins to flow through the top of the radiator, add more water as necessary until the engine is full. Shut the engine OFF and again carefully remove both drain plugs. Repeat the process of filling the system with water and draining it until drained water is clear.

➡If the system cannot be cleaned out effectively this way, you may want to buy a reverse flushing kit at your local parts store and use it. An alternative is the use of a chemical cleaner; if you need to use one of these, just make sure it is compatible with the use of aluminum engine parts and that you follow the directions on the can carefully to ensure that you do not damage your engine.

4. Coat both drain plugs with sealer and install them snugly. Look up the coolant capacity in the Capacities Chart, later in this section. See the chart on the side of the antifreeze container in order to calculate just how much antifreeze is required to protect the system down to the lowest expected temperature in your area. Pour the antifreeze in first. Then, follow up with clean water until the level reaches the top of the radiator tank. You can also pre-mix the antifreeze and water before filling the system. Finally, follow the steps at the end of the preceding procedure for filling the system with coolant after checking the level.

➡If equipped, always open the air relief plug or bolt before filling the cooling system. This service procedure is for bleeding the trapped air in the cooling system. Only when the cooling system is bled properly can the correct amount of coolant be added to the system. The 3.0L (24 valve) and 3.5L engines have a caution label near the air bleed bolt or plug.

Brake and Clutch Master Cylinders

LEVEL CHECK AND FLUID RECOMMENDATIONS

▶ See Figures 108 and 109

The brake fluid reservoir on top of the master cylinder is clear so that you can check the fluid level without removing the top. If the level falls below the low mark, wipe off the cap and the outside of the reservoir, then remove the cap and refill the reservoir up to the MAX mark. Use brake fluid meeting DOT 3 standards only.

As the vehicle ages, the presence of brake fluid in the reservoir will sometimes leave a residue that makes it difficult to read the fluid level reliably through the plastic. It is wise to occasionally check the fluid level by removing the cap to make sure you can accurately determine whether or not additional fluid is required. Always be careful to keep dirt out of the system when removing the cap.

It is normal for the level to drop gradually, as fluid takes up lost volume in the system occurring due to brake pad wear. A sudden loss of a significant amount of fluid should be investigated immediately. The system must be inspected thoroughly and all causes of leakage repaired.

➡Follow these same procedures for checking the fluid level in the clutch master cylinder on the vehicles equipped with a hydraulic clutch.

Fig. 108 The master cylinder brake fluid can be checked either through the transparent master cylinder reservoir or by removing the cap

Fig. 109 Use only approved DOT 3 brake fluid and be careful not to let dirt get into the master cylinder reservoir

Power Steering Pump

LEVEL CHECK AND FLUID RECOMMENDATIONS

▶ See Figures 110, 111, 112 and 113

Park the vehicle on a flat, level surface and leave the engine running. Turn the steering wheel several times to the left and right to raise the fluid to normal operating temperature. With the engine running, remove the cap from the power steering fluid reservoir. This unit is separate from the pump but connected to it by hoses. There are MAX and MIN marks on

GENERAL INFORMATION AND MAINTENANCE 1-45

the dipstick connected to the cap. Add DEXRON®II automatic transmission fluid to the reservoir to bring the level up to the maximum.

At this time, you should inspect the hoses for signs of leakage or cracking. If either condition is noted, the defective hose or hoses should be replaced.

Manual Steering Gear

Mitsubishi vehicles with manual steering have a permanently lubricated steering box. Lubricant level checks or changes are not required.

Chassis Greasing

♦ See Figure 114

➡ If the vehicle is driven in extremely sandy conditions, lubricating the driveshaft assembly is a must!

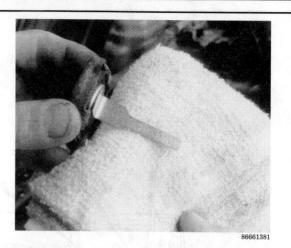

Fig. 112 The dipstick attached to the underside of the cap provides markings, with which the fluid level can be checked

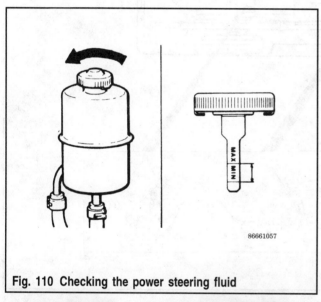

Fig. 110 Checking the power steering fluid

Fig. 111 After turning the wheel back and forth several times, remove the power steering pump cap to check the fluid level

Fig. 113 Use only new, clean power steering fluid to fill the pump

Trucks and Monteros do have lubrication grease fittings at various locations within the steering and front suspension. These fittings should be lubricated with every oil change and after any off-pavement driving in water or sand. You'll need a grease gun filled with Multipurpose grease SAE J310 (NLGI No. 2) and a large rag. Inexpensive grease guns are available almost everywhere and the newer styles use grease cartridges which eliminate the mess of loading the grease into the gun.

1. Raise and safely support the vehicle on jackstands. Ramps can be used, but access is easier when the suspension hangs down.
2. Locate each grease nipple and wipe it clean of dust and grime. Fit the grease gun end onto each nipple. Apply just enough grease into the fitting to fill the dust seal — but not until it comes out of the dust seal covering the joint.
3. Remove the gun from the fitting and use the rag to wipe away excess grease at the fitting.
4. Continue with each fitting. Depending on the year and model, grease fittings may be found on the lower ball joint and steering joint (left and right), the upper ball joint (left and right),

1-46 GENERAL INFORMATION AND MAINTENANCE

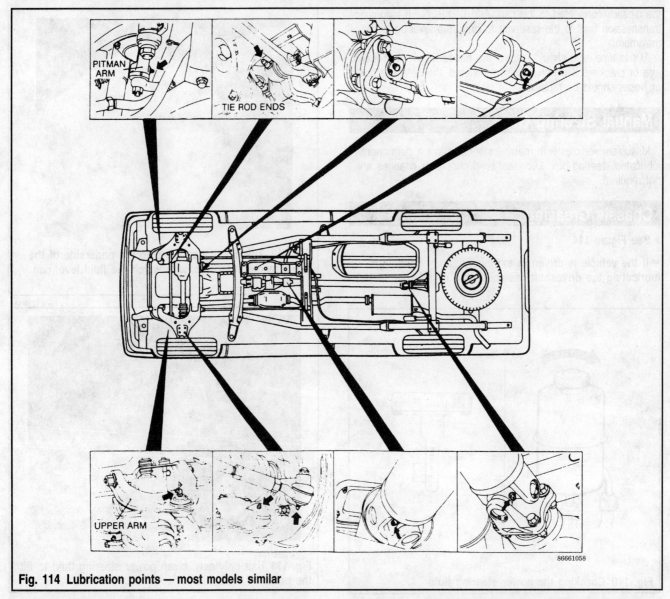

Fig. 114 Lubrication points — most models similar

the tie rod ends (left and right) and the pivot for the pitman (steering) arm. Additionally, some trucks have fittings at both ends of the driveshaft running to the rear wheels.

5. Lower the vehicle to the ground.

Lock Cylinders

Apply graphite lubricant sparingly through the key slot. Insert the key and operate the lock several times to be sure that the lubricant is worked into the lock cylinder.

Hood Latch and Hinges

Clean the latch surfaces, then apply clean engine oil to the latch pilot bolts and the spring anchor. Also lubricate the hood hinges with engine oil. Use a chassis grease to lubricate all the pivot points in the latch release mechanism.

Door Hinges

The gas tank filler door and truck doors should be wiped clean and lubricated with clean engine oil once a year. The door lock cylinders and latch mechanisms should be lubricated periodically with a few drops of graphite lock lubricant or a few shots of silicone spray.

Body Drain Holes

Be sure that the drain holes in the doors and rocker panels are cleared of obstruction. A small screwdriver can be used to clear them of any debris.

GENERAL INFORMATION AND MAINTENANCE 1-47

Wheel Bearings

REMOVAL, PACKING & INSTALLATION

➡ The following procedures are for the non-driven wheels on 2wd vehicles only. Since all the wheels on 4wd vehicles are driven, periodic greasing of the bearings is not necessary. For wheel bearing procedures on 4wd vehicles, please refer to Section 8.

The wheel bearings of the non-driven wheels on trucks, the front wheels on 2-wheel drive Pick-ups, should be repacked with Multipurpose Grease NLGI Grade #2 E.P. or equivalent grease every 2 years or 30,000 miles (48,000 km). The best way to accomplish this is to combine the repacking operation with brake repairs. In other words, if brake linings require attention, always repack the wheel bearings associated with the repair at the same time to avoid repeating the operation at the specified interval. Of course, if brake linings last longer than this interval, wheel bearings should be repacked as a discrete operation.

Before handling the bearings, there are a few things that you should remember to do and not to do.

Remember to DO the following:
- Remove all outside dirt from the housing before exposing the bearing.
- Treat a used bearing as gently as you would a new one.
- Work with clean tools in clean surroundings.
- Use clean, dry canvas gloves, or at least clean, dry hands.
- Clean solvents and flushing fluids are a must.
- Use clean paper when laying out the bearings to dry.
- Protect disassembled bearings from rust and dirt. Cover them with a clean cloth or plastic bag.
- Use clean rags to wipe bearings.
- Keep the bearings in oil-proof paper or plastic when they are to be stored or are not in use.
- Clean the inside of the housing before replacing the bearing.

Do NOT do the following:
- Do not work in dirty surroundings.
- Do not use dirty, chipped or damaged tools.
- Try not to work on wooden work benches or use wooden mallets.
- Don't handle bearings with dirty or moist hands.
- Do not use gasoline for cleaning; use a safe solvent.
- Do not spin-dry bearings with compressed air. They will be damaged.
- Do not spin dirty bearings.
- Avoid using cotton waste or dirty cloths to wipe bearings.
- Try not to scratch or nick bearing surfaces.
- Do not allow the bearing to come in contact with dirt or rust at any time.

➡ Wheel bearing service in this section covers only removal and installation for the non-driven wheels (front wheels) on 2wd trucks and Monteros. For more complete service procedures and 4-wheel drive, refer to Section 8. Refer to that section before any work or adjustments are performed.

2-Wheel Drive

♦ See Figure 115

➡ To perform this procedure, you'll need three special tools and access to a large press to insert the bearings. You will also need bearing grease and a new cotter pin for each wheel assembly.

1. Remove the hub and brake disc; separate the disc from the hub as described in Section 9.
2. Remove the oil seal and inner bearing. If the bearings are in good condition, free of flats, gouges, scores etc., they may be cleaned, repacked and reused. If the bearing must be replaced, the bearing races must also be replaced. With a brass drift and hammer, knock out the outer races for both the inner and outer bearings. In doing this, work from above and knock the bearing out the bottom of the hub; then turn the hub over and repeat the process for the other bearing.
3. Use MB990938-01, or an equivalent driving handle, and, for the outer bearing, MB990927-01 and, for the inner bearing, MB990931-01 or equivalent race driving tools. Press each bearing race in from the top with the appropriate tool. The wider part of the race goes upward and the contour of the race fits that of the special tool. Races must be pressed in until the lower surface contacts the ridge in the hub designed to retain them.
4. Pack the bearing with grease meeting SAEJ310A NLGI grade #2 EP standards. Use a liberal amount of grease and occasionally press the bearing into the palm of your hand to make sure that the grease passes all the way through. Also pack the inner contours of the hub and the hub cap with the grease.
5. Press fit a new oil seal into the inner diameter of the hub with the MB990938-01 or equivalent and MB990931-01 or equivalent, but using the FLAT surface against the outer surface of the seal, rather than the contoured surface used to press in the inner bearing. The seal must end up flush with the inner surface of the hub. Apply grease to the lip of the oil seal.
6. Install the rotor to the hub and reinstall the hub to the steering knuckle. Adjust the wheel bearings — refer to Knuckle and Spindle removal and installation procedures in Section 8.

1-48 GENERAL INFORMATION AND MAINTENANCE

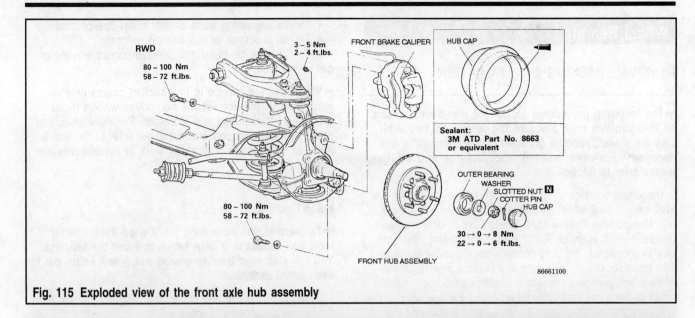

Fig. 115 Exploded view of the front axle hub assembly

TRAILER TOWING

➡ Due to the amount of vehicle coverage (years, model, engines, axles, etc.) in this manual, always refer to your owner's manual for trailer towing weight limit specifications and or additional information.

Factory trailer towing packages are available on most Mitsubishi trucks. However, if you are installing a trailer hitch and wiring on your truck, there are a few things which you ought to know.

Trailer Weight

▶ See Figure 116

Trailer weight is the first, and most important, factor in determining whether or not your vehicle is suitable for towing the trailer you have in mind. The horsepower-to-weight ratio should be calculated. The basic standard is a ratio of 35:1. That is, 35 pounds of Gross Vehicle Weight (GVW) for each horsepower.

To calculate this ratio, multiply your engine's rated horsepower by 35, then subtract the weight of the vehicle, including passengers and luggage. The resulting figure is the ideal maximum trailer weight that you can tow.

Hitch Weight

There are three kinds of hitches: bumper mounted, frame mounted, load equalizing.

Bumper mounted hitches are those which attach solely to the vehicle's bumper. Many states prohibit towing with this type of hitch, when it attaches to the vehicle's stock bumper, since it subjects the bumper to stresses for which it was not designed. Aftermarket rear step bumpers, designed for trailer towing, are acceptable for use with bumper mounted hitches.

Frame mounted hitches can be of the type which bolts to two or more points on the frame, plus the bumper, or just to several points on the frame. Frame mounted hitches can also

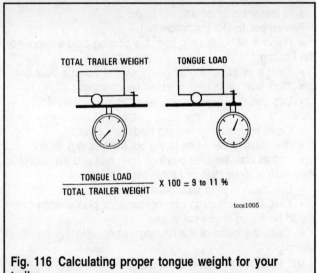

Fig. 116 Calculating proper tongue weight for your trailer

be of the tongue type, for Class I towing, or, of the receiver type, for Classes II and III.

Load equalizing hitches are usually used for large trailers. Most equalizing hitches are welded in place and use equalizing bars and chains to level the vehicle after the trailer is hooked up.

Of course, bolt-on hitches are the most common, since they are relatively easy to install.

Check the gross weight rating of your trailer. Tongue weight is usually figured as 10% of gross trailer weight. Therefore, a trailer with a maximum gross weight of 2000 lbs. (900 kg) will have a maximum tongue weight of 200 lbs. (90 kg). Class I trailers fall into this category. Class II trailers are those with a gross weight rating of 2000-3500 lbs. (900-1600 kg), while Class III trailers fall into the 3500-6000 lb. (1600-2730 kg) category. Class IV trailers are those over 6000 lbs. (2730 kg) and are for use with fifth wheel trucks only.

GENERAL INFORMATION AND MAINTENANCE

When you've determined the hitch that you'll need, follow the manufacturer's installation instructions, exactly, especially when it comes to fastener torques. The hitch will be subjected to a lot of stress and good hitches come with hardened bolts. Never substitute an inferior bolt for a hardened bolt.

Wiring

Wiring your Mitsubishi truck for towing is fairly easy. There are a number of good wiring kits available and these should be used, rather than trying to design your own. All trailers will need brake lights and turn signals as well as tail lights and side marker lights. Most states require extra marker lights for overly wide trailers. Also, most states have recently required back-up lights for trailers, and most trailer manufacturers have been building trailers with back-up lights for several years.

Additionally, some Class I, most Class II and just about all Class III trailers will have electric brakes. Add to this number an accessories wire, to operate trailer internal equipment or to charge the trailer's battery, and you can have as many as seven wires in the harness.

Determine the equipment on your trailer and buy the wiring kit accordingly. The kit will contain all the wires needed, plus a plug adapter set which includes the female plug, mounted on the bumper or hitch, and the male plug, wired into, or plugged into the trailer harness.

When installing the kit, follow the manufacturer's instructions. The color coding of the wires should be standard throughout the industry.

One final point, the best kits are those with a spring loaded cover on the vehicle mounted socket. This cover helps prevent dirt and moisture from corroding the terminals. Never let the vehicle socket hang loosely. Always mount it securely to the bumper or hitch.

Cooling

ENGINE

One of the most common, if not THE most common, problem associated with trailer towing is engine overheating.

With factory installed trailer towing packages, a heavy duty cooling system is usually included. Heavy duty cooling systems are available as optional equipment on most Mitsubishi trucks, with or without a trailer package. If you have one of these extra-capacity systems, you shouldn't have any overheating problems.

If you have a standard cooling system, without an expansion tank, you'll definitely need to get an aftermarket expansion tank kit, preferably one with at least a 2 quart capacity. These kits are easily installed on the radiator's overflow hose, and come with a pressure cap designed for expansion tanks.

Another helpful accessory is a flex fan. These fans are large diameter units which are designed to provide more airflow at low speeds, with blades that have deeply cupped surfaces. The blades flex, or flatten out, at high speed, when less cooling air is needed. These fans are far lighter in weight than stock fans, requiring less horsepower to drive them. Also, they are far quieter than stock fans.

If you do decide to replace your stock fan with a flex fan, note that if your truck has a fan clutch, a spacer between the flex fan and water pump hub will usually be needed.

Aftermarket engine oil coolers are helpful for prolonging engine oil life and reducing overall engine temperatures. Both of these factors increase engine life. While not absolutely necessary in towing Class I and some Class II trailers, they are recommended for heavier Class II and all Class III towing. Engine oil cooler systems consist of an adapter, screwed on in place of the oil filter, a remote filter mounting pad and a multi-tube, finned heat exchanger, which is mounted in front of the radiator or air conditioning condenser.

TRANSMISSION

An automatic transmission is usually recommended for trailer towing. Modern automatics have proven reliable and, of course, easy to operate, in trailer towing.

The increased load of a trailer, however, causes an increase in the temperature of the automatic transmission fluid. Heat is the worst enemy of an automatic transmission. As the temperature of the fluid increases, the life of the fluid decreases. It is essential, therefore, that you install an automatic transmission cooler.

The cooler, which consists of a multi-tube, finned heat exchanger, is usually installed in front of the radiator or air conditioning compressor, and hooked inline with the transmission cooler tank inlet line. Follow the cooler manufacturer's installation instructions.

Select a cooler of at least adequate capacity, based upon the combined gross weights of the truck and trailer.

Cooler manufacturers recommend that you use an aftermarket cooler in addition to, and not instead of, the present cooling tank in your radiator. If you do want to use it in place of the radiator cooling tank, get a cooler at least two sizes larger than normally necessary.

➡ **A transmission cooler can, sometimes, cause slow or harsh shifting in the transmission during cold weather, until the fluid has a chance to come up to normal operating temperature. Some coolers can be purchased or retrofitted with a temperature bypass valve which will allow fluid flow through the cooler only after the fluid has reached operating temperature.**

TOWING THE VEHICLE

➡ **When towing, make sure the the transmission, axles, steering system and powertrain are in good order. If any unit is damaged, a dolly should be used.**

When the steering wheel can be unlocked with the ignition key, towing should be accomplished by hooking cables to the towing hooks under the front of the chassis. As long as vehicles with manual transmissions can be put in Neutral, (4-wheel

1-50 GENERAL INFORMATION AND MAINTENANCE

drive: Select Neutral and 2H; unlock freewheeling hubs) there is no mileage limit to the distance they may be towed or limit as to maximum speed. Automatic transmission equipped vehicles must not be towed faster than 19 mph (30 kmph) or for a greater distance than 15 miles (24 km), as certain transmission parts will not be lubricated. To tow these vehicles a greater distance, raise the drive wheels off the ground or transport the vehicle on a flatbed.

If the ignition key is not available, the vehicle should be towed with the front wheels off the ground because of the locked steering wheel.

JACKING AND HOISTING

➡ Refer to the illustrations for correct placement of a floor jack or safety stands. A good hydraulic floor jack is worth the investment when working on your vehicle.

Emergency Jacking

▶ See Figure 117

Mitsubishi Pick-ups and Monteros use a cartridge or bottle jack. This is a jack with a heavier lifting capacity. Do not attempt the lift the vehicle with a scissors jack borrowed from a car.

1. Block the wheel opposite to the one being removed in front of and to the rear of the tire. Apply the handbrake and make sure the engine is **OFF**. Make sure the jack will be supported on a hard surface.
2. Jacking points are as follows:
   - Montero front: either of two reinforced positions (inboard or outboard) on the transmission support crossmember.
   - Montero rear: on the rear axle, just inboard of the shock absorber mount.
   - Pick-up front: on the frame rail, just behind the transmission support crossmember.
   - Pick-up rear: on the frame rail just ahead of the spring mount (preferred) or on the rear axle, inboard of the spring.
3. Pry off the center wheel cap by using the wheel nut wrench.
4. Loosen the lug nuts with the wrench. Don't remove them yet.
5. Tighten the relief valve on the jack by using the slotted end of the jack handle.
6. Fit the jack handle into the holder pump up and down to partially raise the jack ram. Push the jack into position at the jacking point nearest the flat tire. Make certain the groove in the top of the ram aligns with the contact point.
7. Continue elevating the jack until the flat tire is clear of the ground.

✱✱WARNING

The jack is hydraulic and the ram is a two-stage type. When both rams are raised and the stop mark of the upper ram becomes visible, stop jacking immediately. Further extension of the ram may damage the jack.

8. Install the new wheel, and then install and slightly tighten all the lug nuts.
9. Remove the lug wrench from the jack holder and place it on the relief valve.
10. SLOWLY loosen the valve, allowing the vehicle to settle to the ground. Do not release the jack suddenly and allow the vehicle to crash to the ground; damage to wheels, tires and suspension may result.
11. Tighten the lug nuts and install the center or hub cap.
12. Remove the jack. Occasionally, the upper ram will not retract enough allow the jack to come out. If this happens, get in the vehicle and rock the body from side to side with some vigor; the motion will compress the ram the needed fractions of an inch.
13. Securely stow the jack and the flat tire in the vehicle.
14. Remove the wheel chock.

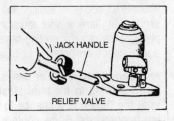
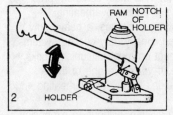
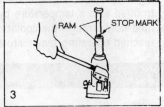
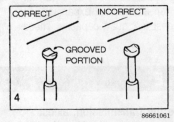

Fig. 117 When using a bottle jack, tighten the relief valve (1), raise the ram by pumping the holder (2), do not exceed the stop marks (3) and align the groove with the jack point (4)

GENERAL INFORMATION AND MAINTENANCE 1-51

Service Jacking

► See Figures 118, 119, 120, 121 and 122

➡When jacking the vehicle, make sure the transmission is in Park (automatic) or Neutral (manual), the parking brake is engaged, and the wheels left on the ground are blocked.

To raise the vehicle:

1. Place the transmission in Park, if automatic, or in Neutral, if manual. Apply the parking brake and block the wheels not been lifted off the ground.
2. Locate the correct jacking points on the frame of the vehicle, slide the floor jack under this position and raise the jacking pad on the jack slowly. When the jacking pad makes contact with the frame, check to make sure the pad is squarely under the jacking point.
3. Raise the vehicle slowly off the ground, keeping watch for any instability of the vehicle. If the vehicle starts to move or wobble, lower it to the ground and make sure the brakes are applied, the wheels are blocked and the jack is securely under the frame jacking point.
4. Continue to raise the vehicle until there is enough room under the vehicle, so that the jackstands will fit under the vehicle's frame. Set the jackstands in place, making sure that the jackstands are squarely under the correct frame jacking points, and secure the jackstands' height.

✱✱CAUTION

Securing the jackstands' height is very important, otherwise the jackstands could break or shift under the full weight of the vehicle and personal injury or even death could result.

5. Slowly lower the vehicle onto the jackstands and make sure that the vehicle is stable. If the vehicle is not stable, raise it again and reposition the jackstands to offer a more secure base.

To lower the vehicle:

6. Raise the vehicle with the floor jack until the weight of the vehicle is off the jackstands and there is enough to remove the jackstands from underneath the vehicle.
7. Cautiously remove the jackstands from their positions under the vehicle.
8. Slowly lower the vehicle to the ground.

Fig. 118 Unless a hoist is available, a floor jack is recommended to raise the vehicle

Fig. 119 When using a floor jack, the frame, crossmembers and axles provide stable and safe positions for lifting a vehicle

Fig. 120 Use a floor jack only to jack a vehicle; use jackstands to support the vehicle once it has been raised to the desired height

1-52 GENERAL INFORMATION AND MAINTENANCE

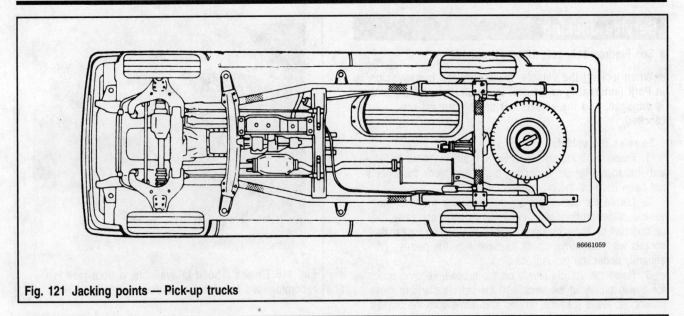

Fig. 121 Jacking points — Pick-up trucks

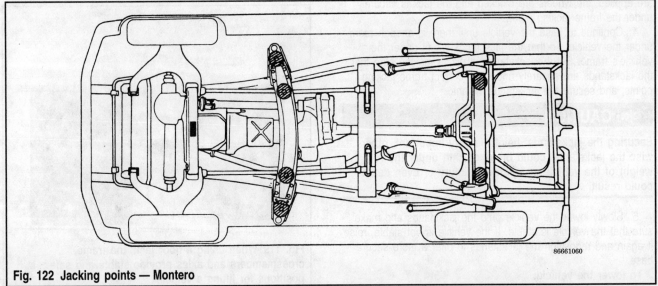

Fig. 122 Jacking points — Montero

JUMP STARTING

Jump starting is the only way to start an automatic transmission model with a weak battery, and is the preferred on a manual transmission models.

Jump Starting Precautions

1. Be sure that both batteries are of the same voltage. All vehicles covered by this manual and most vehicles on the road today utilize a 12 volt charging system.
2. Be sure that both batteries are of the same polarity (have the same ground terminal); in most cases, it is NEGATIVE.
3. Be sure that the vehicles are not touching or a short could occur.
4. On serviceable batteries, be sure the vent cap holes are not obstructed.
5. Do not smoke or allow sparks anywhere near the batteries.
6. In cold weather, make sure the battery electrolyte is not frozen. This can occur more readily in a battery that has been in a state of discharge.
7. Do not allow electrolyte to contact your skin or clothing.

Jump Starting Procedure

▶ See Figure 123

✶✶CAUTION

Do not attempt this procedure on a frozen battery, it will probably explode. The battery in the other vehicle must be a 12 volt, negatively grounded one. Do not attempt to jump start your vehicle with a 24 volt power source; serious electrical damage will result.

GENERAL INFORMATION AND MAINTENANCE

1. Make sure that the voltages of the 2 batteries are the same. Most batteries and charging systems are of the 12 volt variety.
2. Pull the jumping vehicle (with the good battery) into a position so the jumper cables can reach the dead battery and that vehicle's engine. Make sure that the vehicles do NOT touch.
3. Place the transmissions of both vehicles in Neutral or Park, as applicable, then firmly set their parking brakes.

➡ If necessary for safety reasons, both vehicle's hazard lights may be operated throughout the entire procedure without significantly increasing the difficulty of jumping the dead battery.

4. Turn all lights and accessories off on both vehicles. Make sure the ignition switches on both vehicles are turned to the **OFF** position.
5. Cover the battery cell caps with a rag, but do not cover the terminals.
6. Make sure the terminals on both batteries are clean and free of corrosion or proper electrical connection will be impeded. If necessary, clean the battery terminals before proceeding.
7. Identify the positive (+) and negative (-) terminals on both battery posts.
8. Connect the first jumper cable to the positive (+) terminal of the dead battery, then connect the other end of that cable to the positive (+) terminal of the booster (good) battery.
9. Connect one end of the other jumper cable to the negative (-) terminal of the booster battery and the other cable clamp to an engine bolt head, alternator bracket or other solid, metallic point on the dead battery's engine. Try to pick a ground on the engine that is positioned away from the battery in order to minimize the possibility of the 2 clamps touching should one loosen during the procedure. DO NOT connect this clamp to the negative (-) terminal of the bad battery.

✱✱CAUTION

Be very careful to keep the jumper cables away from moving parts (cooling fan, belts, etc.) on both engines.

10. Check to make sure that the cables are routed away from any moving parts, then start the donor vehicle's engine. Run the engine at moderate speed for several minutes to allow the dead battery a chance to receive some initial charge.
11. With the donor vehicle's engine still running slightly above idle, try to start the vehicle with the dead battery. Crank the engine for no more than 10 seconds at a time and let the starter cool for at least 20 seconds between tries. If the vehicle does not start in 3 tries, it is likely that something else is also wrong or that the battery needs additional time to charge.
12. Once the vehicle is started, allow it to run at idle for a few seconds to make sure that it is properly operating.
13. Turn on the headlights, heater blower and, if equipped, the rear defroster of both vehicles in order to reduce the severity of voltage spikes and subsequent risk of damage to the vehicles' electrical systems when the cables are disconnected.
14. Carefully disconnect the cables in the reverse order of connection. Start with the negative cable that is attached to the engine ground, then the negative cable on the donor battery. Disconnect the positive cable from the donor battery and finally, disconnect the positive cable from the formerly dead battery. Be careful when disconnecting the cables from the positive terminals not to allow the alligator clips to touch any metal on either vehicle or a short and sparks will occur.

➡ It is recognized that some or all of the precautions outlined in this procedure are often ignored with no harmful results. However, the procedure outlined is recommended to best assure safety to the people and vehicles involved.

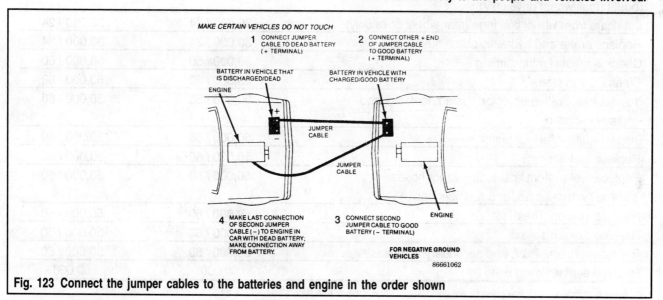

Fig. 123 Connect the jumper cables to the batteries and engine in the order shown

1-54 GENERAL INFORMATION AND MAINTENANCE

Maintenance Intervals

Intervals are miles or miles/months [1]

Maintenance Item	Montero	Pick-up
Change engine oil and oil filter	7,500 / 12 [2]	7,500 / 12 [2]
Check & adjust idle speed	15,000	15,000
Check & adjust valve clearance	15,000 [3]	15,000 [3]
Check drive belts	15,000	15,000
Check A/C system and drive belt	15,000 / 12	15,000 / 12
Check brake and hydraulic clutch fluid level and system for leaks	15,000 / 12	15,000 / 12
Check coolant condition and check cooling system for leaks	15,000 / 12	15,000 / 12
Check exhaust system connections and inspect for excessive corrosion	15,000 / 12	15,000 / 12
Inspect front and rear brake linings and brake hoses	15,000 / 12	15,000 / 12
Inspect power steering fluid level, hoses, and belt	15,000 / 12	15,000 / 12
Change nonslip rear drive axle fluid	30,000	15,000 / 12
Change automatic transmission fluid (drain & fill transfer case if four-wheel drive)	30,000	30,000
Check conventional rear drive axle fluid level	30,000	30,000
Replace air cleaner	30,000	30,000
Replace drive belts	30,000	30,000
Replace spark plugs	30,000	30,000
Check manual transmission (and transfer case fluid if four-wheel drive)	30,000	30,000 / 24
Change engine coolant	30,000 / 24	30,000 / 24
Inspect ball joint, steering linkage seals, and driveshaft boots (four-wheel drive only)	30,000 / 24	30,000 / 24
Lubricate front wheel bearings (rear-wheel drive only)	30,000 / 24	30,000 / 24
Replace brake and hydraulic clutch fluid	30,000 / 24	30,000 / 24
Check & adjust ignition timing	50,000 / 60	50,000 / 60
Clean PCV system	50,000 / 60	50,000 / 60
Inspect evaporative emission system for leaks and replace canister	50,000 / 60	50,000 / 60
Inspect fuel system for leaks	50,000 / 60	50,000 / 60
Replace fuel filter	50,000 / 60	50,000 / 60
Replace fuel system hoses, fuel vapor hoses, cooling system hoses, and fuel filler cap	50,000 / 60	50,000 / 60
Replace ignition wires	50,000 / 60	50,000 / 60
Replace oxygen sensor	50,000 / 60	50,000 / 60
Replace vacuum, PCV, and secondary air hoses	50,000 / 60	50,000 / 60
Replace engine timing belt	60,000 [4]	60,000 [4]

1 When both mileage and time intervals are given, service at whichever one occurs first.
2 Under severe conditions (short trips at cold temperatures, driving in heavy dust, or towing a trailer), change the engine oil and filter every 3 months or 3,000 miles, whichever occurs first. Service the air cleaner filter, PCV system, and spark plugs at more frequent intervals.
3 Jet valve only - adjust as required
4 All engines except for the 2.6L and 2.3L engines (these engines are equipped with timing chains)

GENERAL INFORMATION AND MAINTENANCE 1-55

CAPACITIES

Year	Model	Engine ID/VIN	Engine Displacement Liters (cc)	Engine Oil with Filter (qts.)	Transmission (pts.) 4-Spd	Transmission (pts.) 5-Spd	Transmission (pts.) Auto. 14	Transfer Case (pts.)	Drive Axle Front (pts.)	Drive Axle Rear (pts.)	Fuel Tank (gal.)	Cooling System (qts.)
1983	Montero	G54B	2.6 (2555)	6.1	4.6	4.6	14.4	4.6	2.3	3.8	15.9	8.5
	Pick-up	G63B	2.0 (1997)	4.2	4.4	1	14.4	4.6	2.3	3.2 10	15.1	9.5
	Pick-up	4D55	2.3 (2346)	5.0	4.4	1	14.4	4.6	2.3	3.2 10	18.0	9.5
	Pick-up	G54B	2.6 (2555)	5.2	4.4	1	14.4	4.6	2.3	3.2 10	18.0	8.5
1984	Montero	G54B	2.6 (2555)	6.1	4.6	4.6	14.4	4.6	2.3	3.8	15.9	8.5
	Pick-up	G63B	2.0 (1997)	4.2	4.4	1	14.4	4.6	2.3	3.2 10	15.1	9.5
	Pick-up	4D55	2.3 (2346)	5.9	4.4	1	14.4	4.6	2.3	3.2 10	18.0	9.5
	Pick-up	G54B	2.6 (2555)	2	4.4	1	14.4	4.6	2.3	3.2 10	18.0	8.5
1985	Montero	G54B	2.6 (2555)	6.1	4.6	4.6	14.4	4.6	2.3	3.8	15.9	8.5
	Pick-up	G63B	2.0 (1997)	4.2	4.4	1	14.4	4.6	2.3	3.2 10	15.1	9.5
	Pick-up	4D55	2.3 (2346)	5.9	4.4	1	14.4	4.6	2.3	3.2 10	18.0	9.5
	Pick-up	G54B	2.6 (2555)	2	4.4	1	14.4	4.6	2.3	3.2 10	18.0	8.5
1986	Montero	G54B	2.6 (2555)	6.1	4.6	4.6	14.4	4.6	2.3	3.8	15.9	8.5
	Pick-up	G63B	2.0 (1997)	4.2	4.4	1	14.4	4.6	2.3	3.2 10	15.1	9.5
	Pick-up	G54B	2.6 (2555)	2	4.4	1	14.4	4.6	2.3	3.2 10	18.0	9.5
1987	Montero	G54B	2.6 (2555)	5.3	4.7	4.7	15.2	4.7	2.3	3.8	15.9	8.5
	Pick-up	G63B	2.0 (1997)	4.2	-	1	14.4	4.7	2.3	2.7 5	11	9.5
	Pick-up	G54B	2.6 (2555)	4	-	1	14.4	4.7	2.3	2.7 5	15	9.5
1988	Montero	G54B	2.6 (2555)	5.0	4.7	4.7	15.2	4.7	2.3	3.8	15.9	8.5
	Pick-up	G63B	2.0 (1997)	3.5	-	1	14.4	4.7	2.3	2.7 5	3	9.5
	Pick-up	G54B	2.6 (2555)	4.0	-	1	14.4	4.7	2.3	2.7 5	3	9.5
1989	Montero	G54B	2.6 (2555)	5.5	4.7	4.7	-	4.7	2.3	3.8	15.9	8.5
	Montero	G72B	3.0 (2972)	5.5	5.3	5.3	15.2	4.7	2.3	5.5	6	9.5
	Pick-up	G63B	2.0 (1997)	3.5	-	1	14.4	4.7	2.3	3.2 10	3	9.5
	Pick-up	G54B	2.6 (2555)	4.0 12	-	1	14.4	4.7	3.2	3.2 10	3	9.5
1990	Montero	G54B	2.6 (2555)	6.0	4.6	4.6	-	4.7	2.3	3.8	15.9	8.5
	Montero	6G72	3.0 (2972)	5.5	13	13	15.2	4.7	2.3	5.5	6	9.6
	Pick-up	4G64	2.4 (2350)	5.0	-	13	14.8	-	-	3.2	7	9
	Pick-up	6G72	3.0 (2972)	5.5	13	13	-	4.7	5.5	5.5	8	8.9
1991	Pick-up	4G64	2.4 (2350)	4.2	-	13	14.8	-	-	3.2	7	9
	Pick-up	6G72	3.0 (2972)	4.9	13	13	-	4.6	5.5	5.2	8	8.9
	Montero	6G72	3.0 (2972)	5.5	13	13	15.2	4.6	2.3	5.2	6	9.6
1992	Pick-up	4G64	2.4 (2350)	4.0	-	13	14.8	4.6	-	3.2	7	9
	Pick-up	6G72	3.0 (2972)	4.9	13	13	-	4.6	2.4	5.2	15.9	8.9
	Montero	6G72	3.0 (2972)	5.5	13	13	15.2	4.8	2.6	5.2	24.3	10.0
1993	Pick-up	4G64	2.4 (2350)	4.0	-	13	14.8	-	-	3.2	7	9
	Pick-up	6G72	3.0 (2972)	4.9	-	13	-	4.6	2.4	5.2	15.9	8.9
	Montero	6G72	3.0 (2972)	5.5	-	13	15.2	4.8	2.6	5.2	24.3	10.0
1994	Pick-up	4G64	2.4 (2350)	4.0	-	13	14.8	-	-	3.2	7	9
	Pick-up	6G72	3.0 (2972)	4.9	-	13	-	4.6	2.4	5.2	15.9	8.9
	Montero	6G72	3.0 (2972)	5.5	-	13	15.2	4.8	2.6	5.2	24.3	10.0
	Montero	6G74	3.5 (3497)	5.5	-	13	15.2	5.2	2.6	6.6	24.3	10.0
1995	Pick-up	4G64	2.4 (2350)	4.0	-	13	14.8	NA	NA	3.2	13.7	9
	Montero	6G72	3.0 (2972)	5.5	-	13	15.2	4.8	2.6	5.2	24.3	10.0
	Montero	6G74	3.5 (3497)	5.5	-	13	15.2	5.2	2.6	6.6	24.3	10.0

All measurements for liquids are approximate; fill all fluids slowly and check their level often.
NA - Not Available

1 2WD: 4.9 pts.
  4WD: 4.6 pts.
2 2WD: 5.2 qts.
  4WD: 6.1 qts.
3 2WD std. body: 13.7 gals.
  2WD long body: 18.2 gals.
  4WD std. body: 15.7 gals.
  4WD long body: 19.8 gals.
4 2WD: 4.5 pts.
  4WD: 6.1 pts.
5 Limited slip: 3.2 pts
6 2 door: 19.8 gals.
  4 door: 24.3 gals.
7 Std. body: 13.7 gals.
  Long body: 18.2 gals.
8 Std. body: 15.7 gals.
  Long body: 19.8 gals.
9 Manual transmission: 6.3 qts.
  Automatic transmission: 6.4 qts.
10 Limited Slip: 3.8
11 2WD std. body: 11.4
   2WD long body: 18.2
12 4WD: 5.5
13 2WD: 4.9, 4WD: 5.2
14 Caution - Usually less fluid is needed since some fluid remains in the torque converter when the transmission is emptied; fill with 10 pints, warm up the vehicle, then check the fluid level and fill as necessary
15 4WD std. body: 15.7
   4WD long body: 19.8

RECOMMENDED LUBRICANTS

Item	Recommended lubricants
Engine oil	API classification SG ECII or SG/CD ECII (For further details, refer to SAE viscosity number)
Manual transmission	API classification GL-4, SAE 75W-90 or 75W-85W
Automatic transmission	DIAMOND ATF SP or equivalent
Transfer case <4WD>	API classification GL-4, SAE 75W-90 or 75W-85W
Front axle <4WD>	API classification GL-5 or higher(For further details, refer to SAE viscosity number)
Rear axle — Coventional differential	API classification GL-5 or higher(For further details, refer to SAE viscosity number)
Rear axle — Limited-slip differential	Mitsubishi Genuine Gear Oil Part No. 8149630 EX or equivalent
Manual steering	API classification GL-4 or highter, SAE 90
Power steering	Automatic Transmission Fluid "DEXRON II"
Brakes	DOT3 or DOT4
Clutch <4WD>	DOT3 or DOT4
Hood lock catch, door lock strikers, seat adjusters, rear gate lock, parking brake cable mechanism	Multipurpose grease NLGI No. 2
Engine coolant	DIA QUEEN LONG-LIFE COOLANT (Part No. 0103044) or HIGH QUALITY ETHYLENE GLYCOL ANTIFREEZE COOLANT
Door hinges, rear gate hinges	Engine oil

ELECTRONIC IGNITION
  ADJUSTMENTS 2-11
  COMPONENT REPLACEMENT 2-15
  DESCRIPTION AND
    OPERATION 2-11
  DIAGNOSIS AND TESTING 2-12
FIRING ORDERS 2-7
IDLE SPEED AND MIXTURE
  ADJUSTMENT
  IDLE SPEED ADJUSTMENT 2-21
  MIXTURE ADJUSTMENT 2-22
IGNITION TIMING
  GENERAL INFORMATION 2-17
SPECIFICATIONS CHARTS
  DIESEL ENGINE TUNE-UP
    CHART 2-23
  GASOLINE ENGINE TUNE-UP
    CHART 2-24
TUNE-UP PROCEDURES
  SPARK PLUG WIRES 2-6
  SPARK PLUGS 2-2
VALVE LASH
  GENERAL INFORMATION 2-20

2

ENGINE PERFORMANCE AND TUNE-UP

ELECTRONIC IGNITION 2-11
FIRING ORDERS 2-7
IDLE SPEED AND MIXTURE
  ADJUSTMENT 2-21
IGNITION TIMING 2-17
SPECIFICATIONS CHARTS 2-23
TUNE-UP PROCEDURES 2-2
VALVE LASH 2-20

2-2 ENGINE PERFORMANCE AND TUNE-UP

TUNE-UP PROCEDURES

An engine tune-up is a service designed to restore the maximum capability of power, performance, economy and reliability in an engine, and, at the same time, assure the owner of a complete check and more lasting results in efficiency and trouble-free performance. Engine tune-up becomes increasingly important each year, to ensure that pollutant levels are in compliance with federal emissions standards.

The extent of an engine tune-up is usually determined by the length of time since the previous service, although the type of driving and general mechanical conditioning of the engine must be considered. Specific maintenance should also be performed at regular intervals, depending on operating conditions.

Troubleshooting is a logical sequence of procedures designed to lead the owner to the particular cause of trouble. Service usually comprises two areas; diagnosis and repair. While the apparent cause of trouble, in many cases, is worn or damaged parts, performance problems are less obvious. The first job is to locate the problem and cause. Once the problem has been isolated, refer to the appropriate section for repair, removal or adjustment procedures.

It is advisable to read the entire section before beginning a tune-up, although those who are more familiar with tune-up procedures may wish to go directly to the instructions.

It should be noted that the diesel engine does not require tune-ups in the conventional sense. The diesel does not use spark ignition to fire the fuel (it uses compression ignition) so it has no spark plugs, distributor, or ignition wires. On the other hand diesels are much more sensitive to clogged air and fuel filters and should have their engine oil and filter changed more frequently as well. Diesel owners simply need to redefine "tune-up" to be the inspection and replacement of fluids and filters on a regular basis.

A complete tune-up inspection should be performed every 30,000 miles (48,000 km) or 24 months, whichever comes first. This interval should be halved (as a rule of thumb on certain items) if the truck is operated under severe conditions, such as trailer towing, prolonged idling, continual stop and start driving, or if starting or running problems are noticed. It is assumed that the routine maintenance described in Section 1 has been kept up, as this will have a decided effect on the results of a tune-up inspection.

A tune-up inspection for models covered in this manual may consist of the following:

- Inspecting the drive belts
- If necessary, checking and adjusting valve clearance
- Cleaning the air filter housing and replacing the air filter element.
- If equipped, replacing the PCV filter.
- Inspecting or replacing the fuel filter assembly.
- Inspecting all fuel and vapor lines.
- Checking or replacing the distributor cap, rotor and ignition wires.
- Replacing the spark plugs and making all necessary engine adjustments.
- Always refer to the Maintenance Interval Chart for additional service information.

Spark Plugs

▶ See Figures 1, 2, 3 and 4

Spark plugs ignite the air and fuel mixture in the cylinder as the piston reaches the top of the compression stroke. The controlled explosion that results forces the piston down, turning the crankshaft and the rest of the drivetrain.

The average life of a spark plug is about 30,000 miles (48,000 km). This is, however, dependent on a number of factors: the mechanical condition of the engine, the type of fuel used, the driving conditions and the driver.

When you remove the spark plugs, check their condition. They are a good indicator of the condition of the engine. A small deposit of light tan or gray material on a spark plug that has been used for any period of time is to be considered normal. Any other color, or abnormal amounts of deposit, indicates that there may be something wrong in the engine.

When a spark plug is functioning normally or, more accurately, when the plug is installed in an engine that is functioning properly, the plugs can be taken out, cleaned, regapped, and reinstalled in the engine without doing the engine any harm.

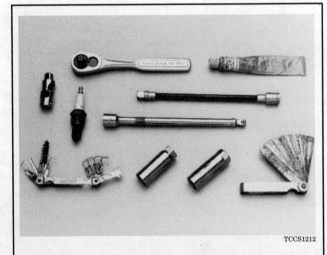

Fig. 1 A variety of tools and gauges are needed for spark plug service

➡Some later model vehicles, may use a platinum type spark plug. If the vehicle has platinum type plug installed, these plugs are usually marked with circular blue lines and are not to be cleaned and regapped.

There are several reasons why a spark plug will foul and you can learn which is at fault by just looking at the plug.

ENGINE PERFORMANCE AND TUNE-UP

GAP BRIDGED

IDENTIFIED BY DEPOSIT BUILD-UP CLOSING GAP BETWEEN ELECTRODES.

CAUSED BY OIL OR CARBON FOULING. REPLACE PLUG, OR, IF DEPOSITS ARE NOT EXCESSIVE THE PLUG CAN BE CLEANED.

OIL FOULED

IDENTIFIED BY WET BLACK DEPOSITS ON THE INSULATOR SHELL BORE ELECTRODES.

CAUSED BY EXCESSIVE OIL ENTERING COMBUSTION CHAMBER THROUGH WORN RINGS AND PISTONS, EXCESSIVE CLEARANCE BETWEEN VALVE GUIDES AND STEMS, OR WORN OR LOOSE BEARINGS. CORRECT OIL PROBLEM. REPLACE THE PLUG.

CARBON FOULED

IDENTIFIED BY BLACK, DRY FLUFFY CARBON DEPOSITS ON INSULATOR TIPS, EXPOSED SHELL SURFACES AND ELECTRODES.

CAUSED BY TOO COLD A PLUG, WEAK IGNITION, DIRTY AIR CLEANER, DEFECTIVE FUEL PUMP, TOO RICH A FUEL MIXTURE, IMPROPERLY OPERATING HEAT RISER OR EXCESSIVE IDLING. CAN BE CLEANED.

NORMAL

IDENTIFIED BY LIGHT TAN OR GRAY DEPOSITS ON THE FIRING TIP.

PRE-IGNITION

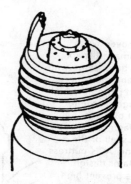

IDENTIFIED BY MELTED ELECTRODES AND POSSIBLY BLISTERED INSULATOR. METALIC DEPOSITS ON INSULATOR INDICATE ENGINE DAMAGE.

CAUSED BY WRONG TYPE OF FUEL, INCORRECT IGNITION TIMING OR ADVANCE, TOO HOT A PLUG, BURNT VALVES OR ENGINE OVERHEATING. REPLACE THE PLUG.

OVERHEATING

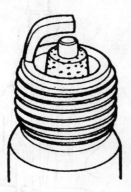

IDENTIFIED BY A WHITE OR LIGHT GRAY INSULATOR WITH SMALL BLACK OR GRAY BROWN SPOTS AND WITH BLUISH-BURNT APPEARANCE OF ELECTRODES.

CAUSED BY ENGINE OVERHEATING, WRONG TYPE OF FUEL, LOOSE SPARK PLUGS, TOO HOT A PLUG, LOW FUEL PUMP PRESSURE OR INCORRECT IGNITION TIMING. REPLACE THE PLUG.

FUSED SPOT DEPOSIT

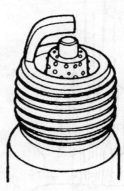

IDENTIFIED BY MELTED OR SPOTTY DEPOSITS RESEMBLING BUBBLES OR BLISTERS.

CAUSED BY SUDDEN ACCELERATION. CAN BE CLEANED IF NOT EXCESSIVE, OTHERWISE REPLACE PLUG.

Fig. 2 Inspect the spark plug to determine engine running conditions

2-4 ENGINE PERFORMANCE AND TUNE-UP

Tracking Arc
High voltage arcs between a fouling deposit on the insulator tip and spark plug shell. This ignites the fuel/air mixture at some point along the insulator tip, retarding the ignition timing which causes a power and fuel loss.

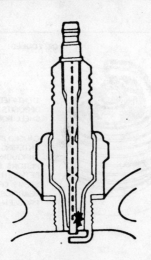

Wide Gap
Spark plug electrodes are worn so that the high voltage charge cannot arc across the electrodes. Improper gapping of electrodes on new or "cleaned" spark plugs could cause a similar condition. Fuel remains unburned and a power loss results.

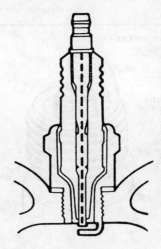

Flashover
A damaged spark plug boot, along with dirt and moisture, could permit the high voltage charge to short over the insulator to the spark plug shell or the engine. AC's buttress insulator design helps prevent high voltage flashover.

Fouled Spark Plug
Deposits that have formed on the insulator tip may become conductive and provide a "shunt" path to the shell. This prevents the high voltage from arcing between the electrodes. A power and fuel loss is the result.

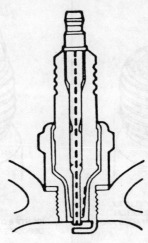

Bridged Electrodes
Fouling deposits between the electrodes "ground out" the high voltage needed to fire the spark plug. The arc between the electrodes does not occur and the fuel air mixture is not ignited. This causes a power loss and exhausting of raw fuel.

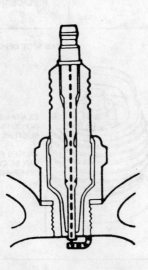

Cracked Insulator
A crack in the spark plug insulator could cause the high voltage charge to "ground out." Here, the spark does not jump the electrode gap and the fuel air mixture is not ignited. This causes a power loss and raw fuel is exhausted.

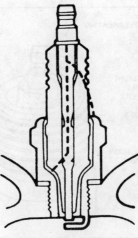

Fig. 3 Used spark plugs which show damage may indicate engine problems

ENGINE PERFORMANCE AND TUNE-UP

There are many spark plugs suitable for use in your engine and are offered in a number of different heat ranges. The amount of heat which the plug absorbs is determined by the length of the lower insulator. The longer the insulator the hotter the plug will operate; the shorter the insulator, the cooler it will operate. A spark plug that absorbs (or retains) little heat and remains too cool will accumulate deposits of lead, oil, and carbon, because it is not hot enough to burn them off. This leads to fouling and consequent misfiring. A spark plug that absorbs too much heat will have no deposits, but the electrodes will burn away quickly and, in some cases, pre-ignition may result. Pre-ignition occurs when the spark plug tips get so hot that they ignite the air/fuel mixture before the actual spark fires. This premature ignition will usually cause a pinging sound under conditions of low speed and heavy load. In severe cases, the heat may become high enough to start the air/fuel mixture burning throughout the combustion chamber rather than just to the front of the plug. In this case, the resultant explosion will be strong enough to damage pistons, rings, and valves.

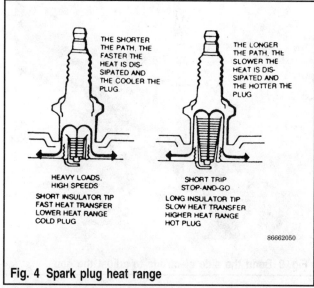

Fig. 4 Spark plug heat range

In most cases the factory recommended heat range is correct; it is chosen to perform well under a wide range of operating conditions. However, if most of your driving is long distance, high speed travel, you may want to install a spark plug one step colder than standard. If most of your driving is of the short trip variety, when the engine may not always reach operating temperature, a hotter plug may help burn off the deposits normally accumulated under those conditions.

REMOVAL

▶ See Figures 5, 6 and 7

➡ On some later models engines, removal of engine air intake plenum and other components will be necessary.

1. Always keep track of the spark plug cable routing and plug wire bracket locations. Number the spark wires so that you won't cross them when you replace them.

2. Remove the wire from the end of the spark plug by grasping the wire by the rubber boot. If the boot sticks to the plug, remove it by twisting and pulling at the same time. DO NOT pull wire itself or you will damage the core.

3. Use a spark plug socket or equivalent to loosen all of the plugs about two turns.

➡ Remove the spark plugs when the engine is cold, if possible, to prevent damage to the threads. If removal of the plugs is difficult, apply a few drops of penetrating oil or silicone spray to the area around the base of the plug, and allow it a few minutes to work.

4. If compressed air is available, apply it to the area around the spark plug holes. Otherwise, use a rag or a brush to clean the area. Be careful not to allow any foreign material to drop into the spark plug holes.

5. Remove the plugs by unscrewing them the rest of the way from the engine.

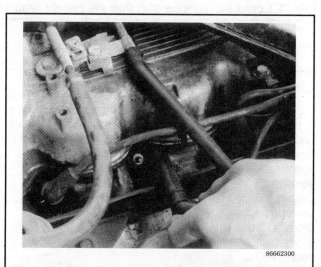

Fig. 5 Twist and pull on the rubber boot to remove the spark plug wires. Never pull on the wire itself

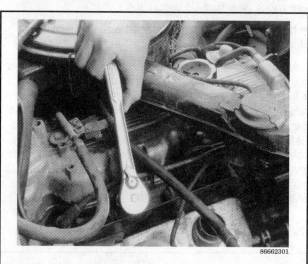

Fig. 6 An extension can be useful in reaching the spark plugs

2-6 ENGINE PERFORMANCE AND TUNE-UP

Fig. 7 Extract the old spark plug from the cylinder head

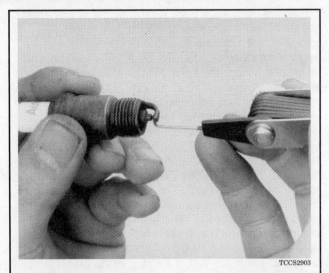

Fig. 8 Check the spark plugs with a wire feeler gauge

INSPECTION

▶ See Figures 8 and 9

Check the plugs for deposits and wear. If they are not going to be replaced, clean the plugs thoroughly. Remember that any kind of deposit will decrease the efficiency of the plug. Plugs can be cleaned on a spark plug cleaning machine, which can sometimes be found in service stations, or you can do an acceptable job of cleaning with a stiff brush. If the plugs are cleaned, the electrodes must be filed flat. Use an ignition points file, not an emery board or the like, which will leave deposits. The electrodes must be filed perfectly flat with sharp edges; rounded edges reduce the spark plug voltage by as much as 50%.

Check spark plug gap before installation. The ground electrode (the L-shaped one connected to the body of the plug) must be parallel to the center electrode and the specified size wire gauge (see Tune-Up Specifications chart).

➡ NEVER adjust the gap on a used platinum type spark plug.

Always check the gap on new plugs as they are not always set correctly at the factory. Do not use a flat feeler gauge when measuring the gap, because the reading will be inaccurate. A wire type gapping tool is the correct way to check the gap. Wire gapping tools usually have a bending tool attached. Use it to adjust the side electrode until the proper distance is obtained. Absolutely never bend the center electrode. Also, be careful not to bend the side electrode too far or too often as it may weaken and break off within the engine, requiring removal of the cylinder head to retrieve it.

INSTALLATION

1. Lubricate the threads of the spark plugs with a drop of oil. Install the plugs and tighten them by hand first. Take care not to cross-thread them.

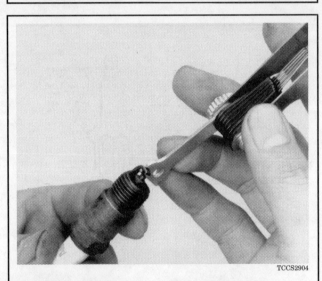

Fig. 9 Bend the side electrode to adjust the gap

2. Tighten the spark plugs with spark plug socket. Do not apply the same amount of force you would use for a bolt; just snug them in. If a torque wrench is available, tighten to 15 ft. lbs. (25 Nm).

3. Install the wires on their respective plugs. Make sure the wires are firmly connected. You will be able to feel them click into place. Check the spark plug cable routing and always make sure the plug wires are in the correct plug wire bracket.

Spark Plug Wires

CHECKING AND REPLACEMENT

▶ See Figure 10

At every tune-up inspection, visually inspect the spark plug cables for burns cuts, or breaks in the insulation. Check the boots and the nipples on the distributor cap and coil. Replace any damaged wiring.

ENGINE PERFORMANCE AND TUNE-UP 2-7

Every 50,000 miles (80,000 km) or 60 months, replace the spark plug wires or at the least check the resistance of the wires with an ohmmeter. Wires with excessive resistance will cause misfiring, and may make the engine difficult to start in damp weather.

To check resistance, remove the distributor cap, leaving the wires attached. Connect one lead of an ohmmeter to an electrode within the cap; connect the other lead to the corresponding spark plug terminal (remove it from the plug for this test). Replace any wire which shows a resistance over 30,000 ohms. Test the high tension lead from the coil by connecting the ohmmeter between the center contact in the distributor cap and either of the primary terminals of the coil. If resistance is more than 25,000 ohms, remove the cable from the coil and check the resistance of the cable alone. Anything over 15,000 ohms is cause for replacement. It should be remembered that resistance is also a function of length; the longer the cable, the greater the resistance. Thus, if the cables on your truck are longer than the factory originals, resistance will be higher, quite possibly outside these limits.

When installing new cables, replace them one at a time to avoid mix-ups. Start by replacing the longest one first. Install the boot firmly over the spark plug. Route the wire over the same path as the original. Insert the nipple firmly into the

Fig. 10 Some factory spark plug wires have numbers already printed on them (arrow): in this case labeling is not necessary

tower on the cap or the coil. Check the spark plug cable routing and always make sure the plug wires are in the correct plug wire bracket.

FIRING ORDERS

▸ See Figures 11, 12, 13, 14 and 15

➡ To avoid confusion, always replace spark plug wires one at a time.

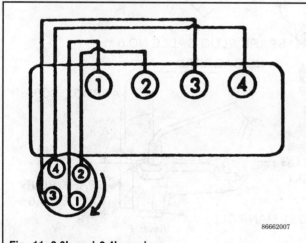

Fig. 11 2.0L and 2.4L engines
Firing order: 1-3-4-2
Distributor rotation: Clockwise

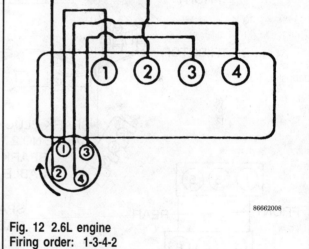

Fig. 12 2.6L engine
Firing order: 1-3-4-2
Distributor rotation: Clockwise

2-8 ENGINE PERFORMANCE AND TUNE-UP

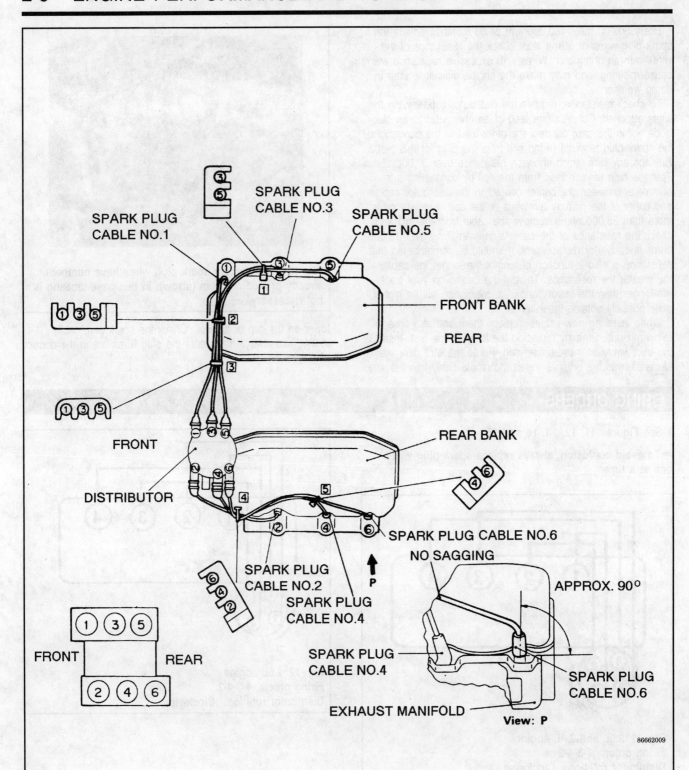

Fig. 13 3.0L (12 valve) engine
Firing order: 1-2-3-4-5-6
Distributor rotation: Counterclockwise

ENGINE PERFORMANCE AND TUNE-UP 2-9

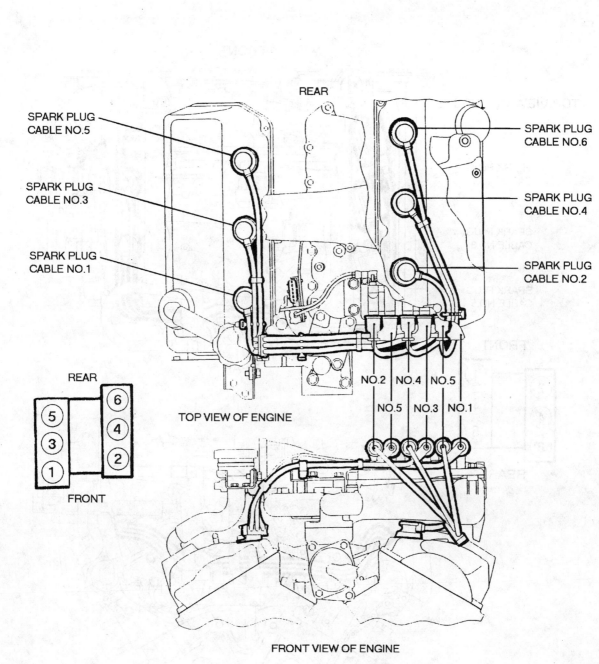

Fig. 14 3.0L (24 valve) engine
Firing order: 1-2-3-4-5-6
Distributor rotation: Distributorless

2-10 ENGINE PERFORMANCE AND TUNE-UP

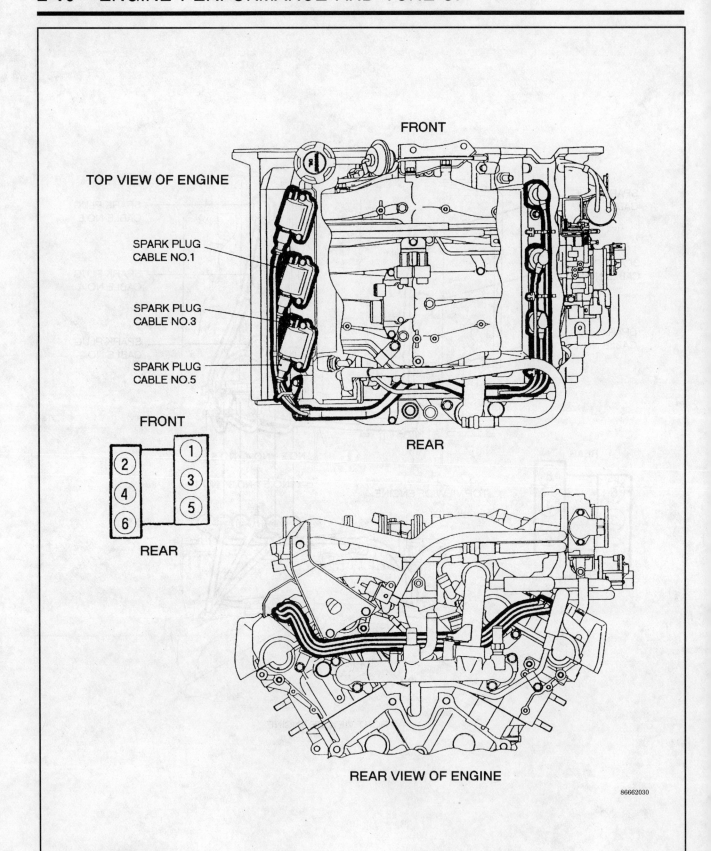

Fig. 15 3.5L DOHC engine
Firing order: 1-2-3-4-5-6
Distributor rotation: Distributorless

ENGINE PERFORMANCE AND TUNE-UP

ELECTRONIC IGNITION

Description And Operation

On vehicles equipped with a conventional distributor, when the ignition switch is turned **ON** the battery voltage is applied to ignition coil primary winding. As the distributor shaft rotates, the igniter opens and closes the circuit repeatedly causing the ignition coil primary winding current to flow through the ignition coil negative terminal and igniter to ground or be interrupted.

The action induces high voltage in the ignition coil secondary winding. from the ignition coil, the secondary winding current produced flows through the distributor and spark plug to ground, thus causing ignition in each cylinder.

The 3.0L (24 valve) and 3.5L engines utilize a distributorless ignition system which is controlled by the Electronic Control Module (ECM). The ECM uses signals (electrical voltage pulses) from various components (i.e. crankshaft position sensor, throttle position sensor, etc.) to control the timing of the engine to match any conditions the engine is encountering. The crankshaft position sensor on the 3.0L (24 valve) engine is located down by the crankshaft under the timing belt cover. The crankshaft position sensor on the 3.5L engine is located under the upper left-hand timing belt cover secured to the cylinder head. Refer to Section 4 for more indepth procedures of the individual components of this system.

Adjustments

RELUCTOR GAP

2.6L Engine
♦ See Figure 16

On this engine, the reluctor gap can only be checked. Doing so is not a matter of routine maintenance, but usually is necessary only if the distributor is overhauled. However, too tight a fit between the signal rotor and the pickup stator could cause rotation of the distributor to damage the parts involved or produce an incorrect signal.

In some cases, severe wear or incorrect original manufacturing tolerances in the distributor bearings could cause wear of these two pieces that resembles damage due to too tight a fit.

1. Remove the distributor cap.
2. Remove the two bolts that secure the rotor in place.
3. Rotate the engine using a wrench on the front pulley until the three vertical stator pieces on the distributor line up with three of the rotor lobes.
4. Using a non-magnetic feeler gauge of 0.008 in. (0.2 mm) thickness, insert it straight down between the stator and rotor., As long as the gauge can be inserted and moved easily, the parts are okay; the gap may be slightly wider than the 0.008 in. (0.2 mm) specification.
5. Install the cap and rotor.

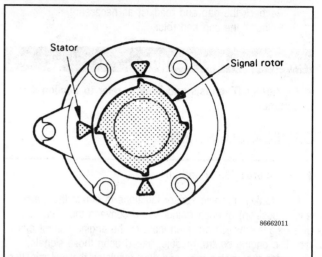

Fig. 16 The reluctor gap position in the 2.6L engine distributor

2.0L Engine
♦ See Figure 17

While service is not required as a part of normal maintenance, if you have worked on the distributor or if you suspect the reluctor gap might be incorrect because of ignition problems, you can check and adjust it.

1. Remove the distributor cap and rotor.
2. Rotate the engine (you can use a large socket wrench on the bolt that attaches the front pulley — but only turn the engine clockwise) so that one of the prongs of the rotor is directly across from the igniter pickup.
3. Insert a non-magnetic feeler gauge of 0.031 in. (0.8 mm) between the prong and pickup. If the gap is correct, there will be a very slight drag. If the gauge fits loosely or cannot be inserted, loosen both mounting screws and insert the gauge. Slowly adjust the gap by pivoting the igniter assembly on the left screw and rotating it at the right side, where it's slotted.

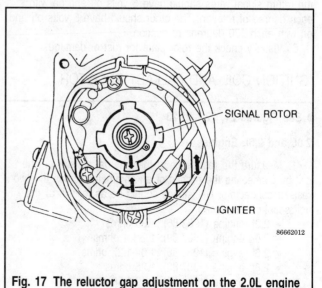

Fig. 17 The reluctor gap adjustment on the 2.0L engine

2-12 ENGINE PERFORMANCE AND TUNE-UP

When the gauge is just being touched by the prong and the pickup, tighten the right side screw first and then the screw on the left.

4. Recheck the gap and readjust as necessary.
5. Reinstall the cap and rotor.

Diagnosis And Testing

➡ For further Diagnosis and Testing refer to Section 4 in this manual.

CRANK ANGLE SENSOR

▶ See Figures 18, 19 and 20

The crankshaft angle sensor functions to detect the crank angle (position) of each cylinder, and converts this data into pulse signals, which are then input to the engine control module. The engine control module, based upon those signals, calculates the engine rpm, and also regulates the fuel injection timing and the ignition timing.

The power for the crankshaft angle sensor is supplied from the MFI relay and is grounded to the vehicle body. The crankshaft angle sensor, by intermitting the flow (to ground) of the 5 volt signal applied from the engine control module, produces pulse signals.

This part is easy to test, but must be removed from the vehicle to do so.

1. First test the circuit to the sensor. Unplug the sensor connector, turn the ignition **ON** and check for 12 volts between one of the terminals and ground. The other terminals are return signals to the ECU, do not use them for a voltmeter ground.
2. Turn the ignition **OFF**.
3. Remove the distributor assembly and reconnect the plug to the crank angle sensor.
4. Peel back the insulation-wrap just far enough to connect a voltmeter to the wires and ground the coil high tension wire.
5. With the voltmeter connected to a good chassis ground, turn the ignition **ON** and slowly rotate the distributor. One of the return signal wires should have 5 volts off and on with each degree of rotation. The other should have 5 volts off and on with each 120 degrees of rotation.
6. Visually check the rotor plate for rust or damage.

IGNITION COIL AND BALLAST RESISTOR

▶ See Figures 21, 22 and 23

2.0L and 2.6L Engines

1. Measure the resistance of the external resistor on the coil by connecting the probes of an ohmmeter across the two resistor connectors. Resistance should be as follows or the unit should be replaced.
   - 2.0L engine (1983-86): 1.35 ohms
   - 2.0L engine (1987-89): 1.2-1.4 ohms
   - 2.6L engine (1983-86): 1.04-1.27 ohms
   - 2.6L engine (1987-88): 1.25 ohms
   - 2.6L engine (1989-90): 1.12-1.38 ohms

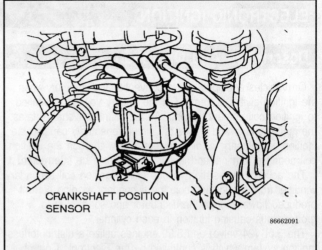

Fig. 18 The crankshaft position sensor location on the 3.0L (12 valve) engine

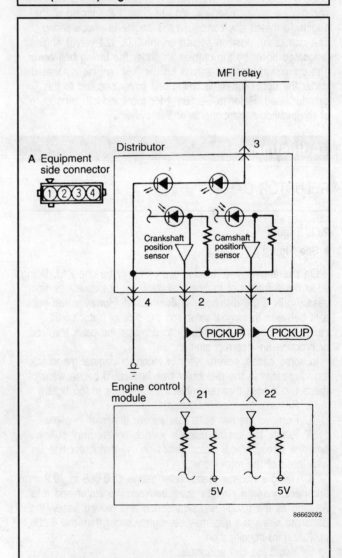

Fig. 19 The crankshaft and camshaft position sensor wiring schematic for the 3.0L (12 valve) engine

ENGINE PERFORMANCE AND TUNE-UP

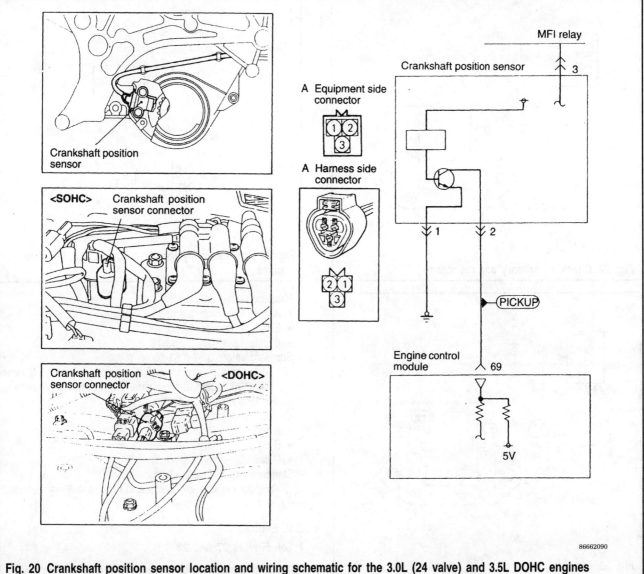

Fig. 20 Crankshaft position sensor location and wiring schematic for the 3.0L (24 valve) and 3.5L DOHC engines

2. Measure the resistance across the coil primary circuit by connecting an ohmmeter between the (+) and (−) coil connectors. Resistance must be approximately as follows or the coil should be replaced.
   - 2.0L engine (1983-86): 1.2 ohms
   - 2.0L engine (1987-89): 1.08-1.32 ohms
   - 2.6L engine (1983-86): 1.04-1.27 ohms
   - 2.6L engine (1987-88): 1.25 ohms
   - 2.6L engine (1989-90): 1.12-1.38 ohms

3. Set your ohmmeter to the x1000 scale and measure the resistance between the connector inside the coil tower and the coil (+) terminal. This is secondary resistance and must be as follows, or the unit should be replaced.
   - 2.0L engine (1983-86): 1.2 kilohms
   - 2.0L engine (1987-89): 1.08-1.32 kilohms
   - 2.6L engine (1983-86): 7.1-9.6 kilohms
   - 2.6L engine (1987-88): 11.0 kilohms
   - 2.6L engine (1989-90): 9.4-12.7 kilohms

4. Inspect the unit for oil leaks and cracks in the coil tower, and replace it if any defects are noted.

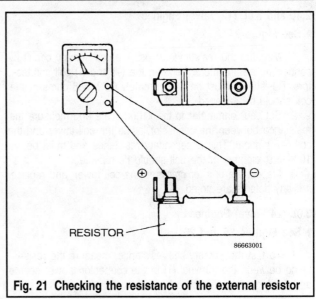

Fig. 21 Checking the resistance of the external resistor

2-14 ENGINE PERFORMANCE AND TUNE-UP

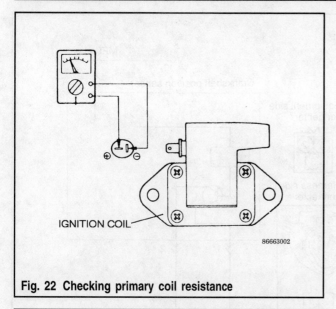

Fig. 22 Checking primary coil resistance

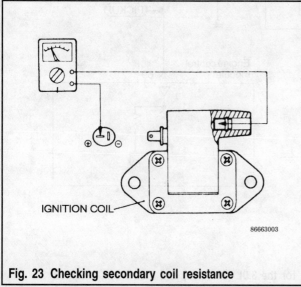

Fig. 23 Checking secondary coil resistance

2.4L and 3.0L (12 Valve) Engines
♦ See Figure 24

1. Measure the resistance across the coil primary circuit by connecting an ohmmeter between the (+) and (-) coil connectors. Resistance must be approximately 0.72-0.88 ohms or the coil should be replaced.
2. Set your ohmmeter to the x1000 scale and measure the resistance between the connector inside the coil tower and the coil (+) terminal. This is secondary resistance and must be 10.3-13.9 kilohms, or the unit should be replaced.
3. Inspect the unit for cracks in the coil tower, and replace it if any defects are noted.

3.0L (24 Valve) Engines
♦ See Figures 25 and 26

1. To test the primary coil resistance, measure the resistance between the terminal (1) of the connector (power source) and the terminals of each coil.

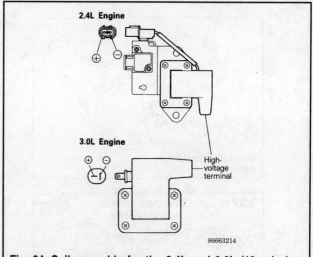

Fig. 24 Coil assembly for the 2.4L and 3.0L (12 valve) engines

Measurement points:
- A coil: Terminals 1 and 11
- B coil: Terminals 1 and 13
- C coil: Terminals 1 and 12

Standard value: 0.67-0.81 ohms

2. To test the secondary coil resistance, switch the ohmmeter to the x1000 scale and measure the resistance between each terminal for a high pressure.

Measurement points:
- A coil
- B coil
- C coil

Standard value: 11.3-15.3 kilohms

3. Inspect the unit for cracks and replace it if any defects are noted.

3.5L Engines
♦ See Figures 27 and 28

1. To test the primary coil resistance, measure the resistance between the terminals for each cylinder (No. 1 & No. 4, No. 2 & No. 5, No. 3 & No. 6) of the ignition coil.

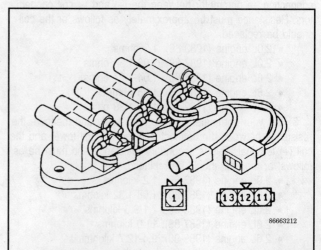

Fig. 25 The primary coil configuration and resistance check — 3.0L (24 valve) engine

ENGINE PERFORMANCE AND TUNE-UP 2-15

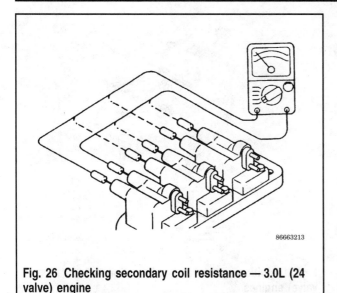

Fig. 26 Checking secondary coil resistance — 3.0L (24 valve) engine

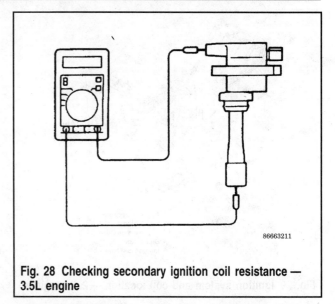

Fig. 28 Checking secondary ignition coil resistance — 3.5L engine

Standard value: 0.69–0.85 ohms

2. To test the secondary coil resistance, switch the ohmmeter to the x1000 scale and measure the resistance between the high voltage terminals for each cylinder (No. 1 & No. 4, No. 2 & No. 5, No. 3 & No. 6) of the ignition coil as shown in the illustration.

Standard value: 15.3–20.7 kilohms

3. Inspect the unit for cracks and replace it if any defects are noted.

Component Replacement

REMOVAL & INSTALLATION

Distributor Cap and Rotor

1. The distributor cap is held on by two screws or clips. Release them with a screwdriver and lift the cap straight up and off, with the wires attached.

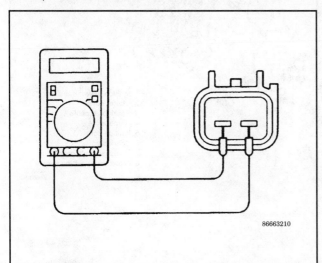

Fig. 27 Checking the primary ignition coil resistance — 3.5L engine

2. Inspect the cap for cracks, carbon tracks, or a worn center contact.
3. Pull the ignition rotor straight up to remove (remove screws if so equipped). Replace it if its contacts are worn, burned, or pitted. Do not file the contacts. To replace, press it firmly onto the shaft. It only goes on one way, so be sure it is fully seated (install screws if so equipped).
4. Inspect the plug wires for cracks or brittleness, replace them if necessary. Transfer the wires from the old distributor cap to the new one. Be sure to place them on the correct terminals.
5. Install the distributor cap and tighten the two screws or engage clips properly.

Ignition Coil and Ballast Resistor

▶ See Figures 29, 30, 31 and 32

1. Disconnect the negative battery cable.
2. Remove any components necessary to access the coil.
3. Tag and disconnect all electrical leads.
4. Remove any mounting bolts then the ignition coil.

To install:

5. Install the coil assembly in position and tighten the mounting bolts.
6. Connect all wires to the coil assembly.
7. Install any components removed during disassembly.
8. Connect the negative battery cable.

Crank Angle Sensor

Please refer to Section 4 for removal and installation of the crank angle sensor.

2-16 ENGINE PERFORMANCE AND TUNE-UP

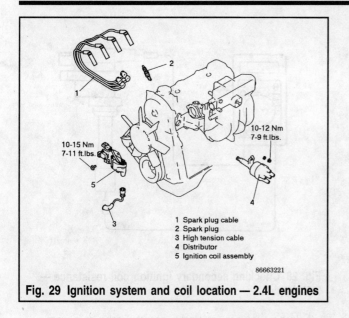

Fig. 29 Ignition system and coil location — 2.4L engines

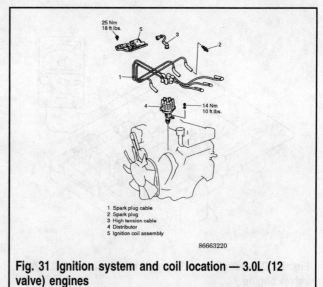

Fig. 31 Ignition system and coil location — 3.0L (12 valve) engines

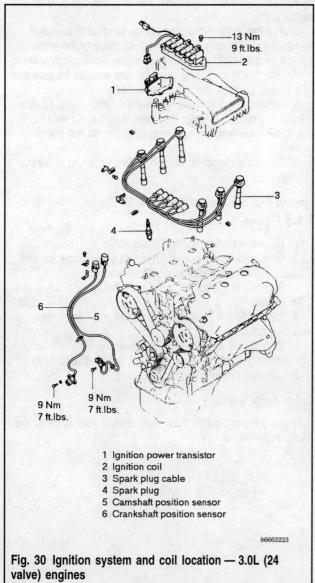

Fig. 30 Ignition system and coil location — 3.0L (24 valve) engines

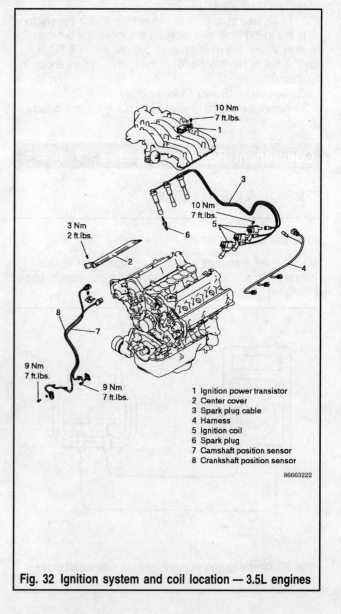

Fig. 32 Ignition system and coil location — 3.5L engines

ENGINE PERFORMANCE AND TUNE-UP 2-17

IGNITION TIMING

It is wise to check the ignition timing at every tune-up, although timing varies little with electronic ignition systems. Mitsubishi permits a tolerance of 2° on either side of the timing setting. Most engines run at their best and with maximum resistance to detonation if the timing is as close as possible to the setting.

General Information

On all vehicles equipped with adjustable ignition timing and an adjustable idle speed the two processes of adjusting the idle speed and the ignition timing are very closely related and affect each other closely. When doing either of these adjustments, it is usually necessary to adjust each in a back and forth manner since they affect each other. For example, if the ignition timing is adjusted the idle speed could increase past what is recommended. This in turn could affect actual ignition timing during normal use of the vehicle, which would affect the vehicle's driveability. Therefore, the idle speed would have to be adjusted, which could in turn affect the ignition timing. Eventually, after adjusting each back and forth a few times to reach the desired specifications, they will meet the specifications of both simultaneously. Most often they need only to be adjusted a few times before both are correct.

INSPECTION AND ADJUSTMENT

2.0L and 2.6L Engines
▶ See Figures 33 and 34

1. Drive the vehicle until the engine is hot; the temperature gauge indicates normal operating temperature.
2. Leave the engine idling, apply the handbrake and put the transmission in Neutral (manual) or Park (automatic). Turn off all accessories.

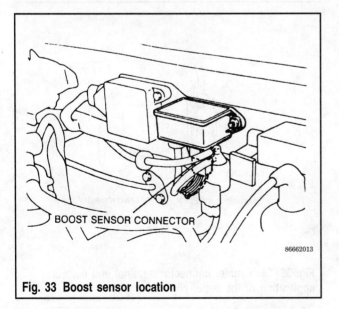

Fig. 33 Boost sensor location

3. Install a tachometer, connecting the red lead to the (-) terminal of the coil and the black lead to a clean ground (the battery negative terminal works well) or as otherwise detailed by the manufacturer of the tachometer. Verify that the engine idle speed is correct. If not adjust it. See the idle adjustment procedure later in this section.
4. Stop the engine and connect a timing light according to the manufacturer's instructions.
5. If the front pulley and timing marks are dirty, wipe them clean with a rag. If they are hard to see, you might want to put a small drop of white paint on both the timing scale and groove in the pulley. You may have to turn the engine over using a wrench on the bolt in the front of the crankshaft pulley to do this. Check the timing specifications on the Tune-up Charts or on the engine compartment sticker. Make sure all wires are clear of any rotating parts and the light is in a secure place. Make sure you haven't left any rags or tools near the engine.

➡ If the tune-up specifications on the Vehicle Emission Control Information sticker in the engine compartment of your Mitsubishi disagree with the Tune-Up Specifications chart in this section, the figures on the sticker must be used. The sticker often reflects changes made during the production run.

6. If you are performing the timing procedure at high altitudes (more than 3,900 ft. above sea level), disconnect the pressure sensor electrical connector (if so equipped), located just across from the distributor wires at the top of the cap, before stopping the engine. The sensor is a box bolted to the fender with a vacuum hose connected to the bottom of the unit. Also disconnect the vacuum hose with the white stripes that is connected to the lowermost portion of the distributor vacuum advance, if so equipped. Plug the end of the hose.
7. Start the engine and allow it to idle. Point the timing light at the mark on the front cover and read the timing by noting the position of the groove in the front pulley in relation to the timing mark or scale on the front cover. If the timing is incorrect, loosen the distributor mounting bolt or nut. Turn the distributor slightly clockwise to retard the timing or counterclockwise to advance it (advance means turning to a setting representing more degrees before top dead center). When the timing is correct, tighten the distributor mounting bolt and verify that the setting has not changed. If necessary, readjust the position of the distributor until the setting is correct after the bolt or nut is tight.
8. If so equipped, reconnect the white striped vacuum hose or the pressure sensor connector. Recheck the timing. The timing should advance at high altitudes.
9. Turn the engine off, disconnect the timing light and tachometer.

2.4L and 3.0L (12 Valve) Engines
▶ See Figures 35, 36 and 37

For the timing adjustment procedure on these vehicles, you will need a jumper wire approximately two feet long with alligator clips on both ends. You will also need a paper clip. Make

2-18 ENGINE PERFORMANCE AND TUNE-UP

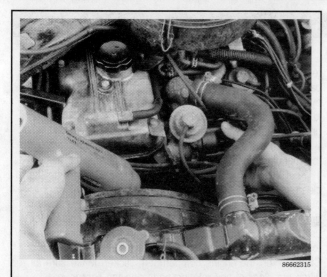

Fig. 34 Rotate the distributor to adjust the timing

sure that you have these items on hand before attempting to check the timing.

1. Drive the vehicle until the engine is hot; the temperature gauge indicates normal operating temperature. Apply the handbrake and put the transmission in Neutral (manual) or Park (automatic). Turn off all accessories.
2. Stop the engine and connect a timing light in accordance with the manufacturer's instructions.
3. Trace the wire that runs from the primary side of the ignition coil to the noise filter. You will find a single terminal harness connector between these two points or locate the engine speed detection connector — refer to the illustrations for location on your model. Insert a paper clip through the connector on either the female or the male side where the wire enters the connector, do not separate the connector. The paper clip must make full contact with the surface of the connector terminal and must be inserted at the proper angle or you will not be able to get it out.
4. Once the paper clip is in place, connect a tachometer to it.
5. Start the engine and check the idle speed. Adjust as necessary.
6. Stop the engine and turn the ignition switch **OFF**.
7. Connect a lead wire with alligator clips to the ignition adjustment terminal (refer to the illustration), and ground it.
8. Start the engine and allow it to idle.
9. Illuminate the timing marks on the crankshaft pulley with the timing light. Basic ignition timing should be between 3-7° BTDC.
10. If the timing is not as specified, loosen the distributor mounting nut and turn the distributor to the right to retard the timing or to the left to advance it.
11. Once the timing is set to specifications, tighten the mounting nut, being careful not to disturb the distributor.
12. Stop the engine and disconnect the jumper wire from the ignition timing connector and ground.
13. Start the engine and run it at idle speed.

➡If the tune-up specifications on the Vehicle Emission Control Information sticker in the engine compartment of your Mitsubishi disagree with the specifications in this

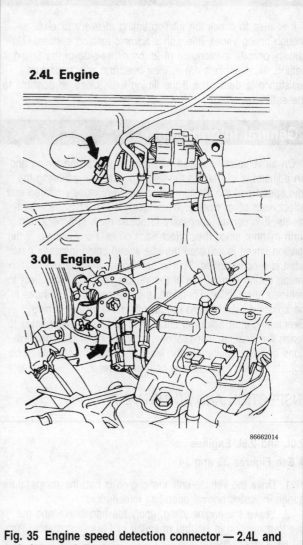

Fig. 35 Engine speed detection connector — 2.4L and 3.0L (12 valve) Engines

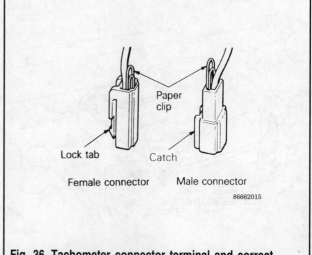

Fig. 36 Tachometer connector terminal and correct application of the paper clip

ENGINE PERFORMANCE AND TUNE-UP 2-19

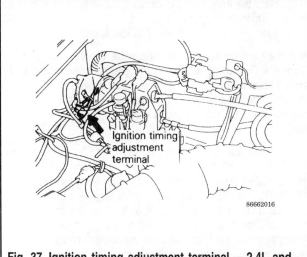

Fig. 37 Ignition timing adjustment terminal — 2.4L and 3.0L (12 valve) Engines

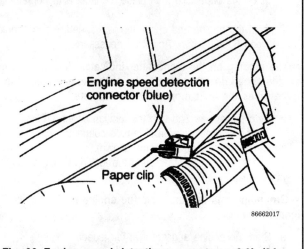

Fig. 38 Engine speed detection connector — 3.0L (24 valve) and 3.5L Engines

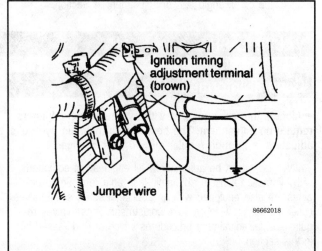

Fig. 39 Ignition timing adjustment terminal — 3.0L (24 valve) and 3.5L Engines

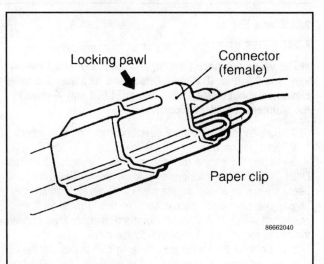

Fig. 40 The correct location for inserting the paper clip — 2.4L, 3.0L and 3.5L models

section, the figures on the sticker must be used. The sticker often reflects changes made during the production run.

14. Check the timing again with the timing light. The timing should be 8° BTDC for the 2.4L and 15° BTDC for the 3.0L (12 valve) engines. This is called actual ignition timing. If the timing is not exactly 8° or 15°, do not be alarmed. This value may vary depending on what mode the computer is in at the time the base adjustment was made. If you do not see a change, check the base timing again. If the base timing is correct, the engine is in time and no further adjustment is necessary.

➡The ignition timing may vary from the specification even under normal operating conditions and is automatically further advanced at higher altitudes.

3.0L (24 Valve) and 3.5L Engines
♦ See Figures 38, 39 and 40

For the timing adjustment procedure on these vehicles, you will need a jumper wire approximately two feet long with alligator clips on both ends. You will also need a paper clip. Make sure that you have these items on hand before attempting to check the timing.

1. Drive the vehicle until the engine is hot; the temperature gauge indicates normal operating temperature. Apply the handbrake and put the transmission in Neutral (manual) or Park (automatic). Turn off all accessories.
2. Stop the engine and connect a timing light in accordance with the manufacturer's instructions.
3. Trace the wire that runs from the primary side of the ignition coil to the noise filter. You will find a single terminal harness connector between these two points or locate the engine speed detection connector — refer to the illustrations for location on your model. Insert a paper clip through the connector on either the female or the male side where the wire enters the connector. Do not disconnect the two wires. The paper clip must make full contact with the surface of the connector terminal and must be inserted at the proper angle or you will not be able to get it out.

2-20 ENGINE PERFORMANCE AND TUNE-UP

4. Once the paper clip is in place, connect a tachometer to it.
5. Start the engine and check the idle speed, which should be between 600-800 rpm. Adjust as necessary.

➡ **The reading on the tachometer indicates 1/3 of the actual engine speed. In other words, the actual engine speed is three times the indication on the tachometer.**

6. Stop the engine and turn the ignition switch **OFF**.
7. Disconnect the waterproof female connector from the ignition timing adjustment connector (brown).
8. Use a jumper wire to ground the ignition timing adjustment terminal.

➡ **Grounding this terminal sets the engine to the basic ignition timing.**

9. Start the engine and run it at idle speed.
10. Illuminate the timing marks on the crankshaft pulley with the timing light. Basic ignition timing should be 2-8° BTDC.
11. If the ignition is not within the standard value range, refer to the crank angle sensor adjustment in Section 4.
12. Stop the engine and disconnect the jumper wire from the ignition timing connector and ground.
13. Start the engine and run it at idle speed.

➡ **If the tune-up specifications on the Vehicle Emission Control Information sticker in the engine compartment of your Mitsubishi disagree with the specifications in this section, the figures on the sticker must be used. The sticker often reflects changes made during the production run.**

14. Check the timing again with the timing light. The timing should be 15° BTDC for the 3.0L (24 valve) and 3.5L engines. This is called actual ignition timing. If the timing is not exactly 15° BTDC, do not be alarmed. This value may vary depending on what mode the computer is in at the time the base adjustment was made. If you do not see a change, check the base timing again. If the base timing is correct, the engine is in time and no further adjustment is necessary.

➡ **The ignition timing may vary from the specification even under normal operating conditions and is automatically further advanced at higher altitudes.**

VALVE LASH

General Information

➡ **Models that are equipped with hydraulic lash adjusters require no adjustment. The best way to maintain hydraulic adjusters is through regular oil and filter changes.**

Valve lash must be adjusted on all engines not equipped with hydraulic (oil-filled) automatic lash adjusters. Some engines are also equipped with an additional set of small valves called jet valves which require adjustment. These valves require special adjustment procedures which are addressed later in this section.

ADJUSTMENT

2.3L Diesel Engine
▸ See Figure 41

➡ **The valve clearance (hot engine) for intake and exhaust valves is 0.010 in. (0.25mm). Check and or adjust the valve clearance about every 16,000 miles (24,000 km) — refer to the Maintenance Interval chart.**

1. Run engine until normal operating condition is reached.
2. Stop engine and turn off the ignition.
3. Rotate the engine until piston in No. 1 cylinder is at Top Dead Center (TDC) on the compression stroke. To do this, use a wrench or socket on the crankshaft pulley to rotate the engine in a clockwise direction until the timing mark on the crankshaft pulley or crankshaft vibrational damper lines up with the timing tab on the front of the engine block.
   a. Remove the rocker arm cover to gain access to the intake and exhaust valves.
   b. Observe the No. 1 cylinder valve springs. If either of the two valve springs is compressed, or either of the two valve springs is at a different height from the other one, then the No. 1 cylinder is not on its compression stroke. If this is the case, rotate the engine one more crankshaft revolution until the timing marks once again line up.
4. Loosen the locknuts on the intake and exhaust valves of cylinder No. 1, the intake valve of cylinder No. 2 and the exhaust valve of cylinder No. 3 (labelled A in diagram).
5. Adjust the clearances to 0.010 in. (0.25 mm) with a feeler gauge. Retighten the locknuts.
6. Use the same method previously described to rotate the engine until the No. 4 piston is at TDC on its compression stroke.
7. Loosen the locknuts on the intake and exhaust valves of cylinder No. 4, the intake valve of cylinder No. 3 and the exhaust valve of cylinder No. 2.
8. Adjust the clearances to 0.010 in. (0.25 mm) with a feeler gauge. Retighten the locknuts.

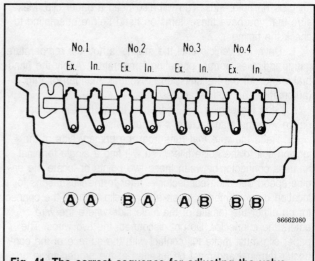

Fig. 41 The correct sequence for adjusting the valve clearances — 2.3L diesel engine

ENGINE PERFORMANCE AND TUNE-UP 2-21

9. Reinstall the rocker arm cover.
10. Check idle speed and readjust if necessary. Check for any leaks from around the rocker arm gasket.

2.0L and 2.6L Engines

These engines are equipped with jet valves and hydraulic lash adjusters. The lash adjusters eliminate the need for periodic intake and exhaust valve adjustments.

1. The first step is to warm the engine to operating temperature. Since the engine will cool as you work, it's best to actually drive the vehicle at least several miles in order to heat the internal parts to maximum temperatures. Drive some distance after the temperature gauge has stabilized. Then, stop the vehicle, turn off the engine, and block the wheels.
2. Remove the rocker arm cover as described in Section 3.
3. Recheck the torque for the cylinder head assembly bolts. Observe the sequence in the appropriate illustration (see Section 3).
4. Rotate the engine until piston No. 1 cylinder is on the compression stroke at Top Dead Center (TDC). To do this, use a wrench or socket on the crankshaft pulley to rotate the engine in a clockwise direction until the timing mark on the crankshaft pulley or crankshaft vibrational damper lines up with the timing tab on the front of the engine block.
    a. Observe the No. 1 cylinder valve springs. If either of the two valve springs is compressed, or either of the two valve springs is at a different height from the other one, then the No. 1 cylinder is not on its compression stroke. If this is the case, rotate the engine one more crankshaft revolution until the timing marks once again line up.
5. Loosen the jet valve locknut and back off on the adjusting screw a small amount. Insert a 0.10 in. (2.5 mm) feeler gauge between the rocker and the end of the jet valve stem. Check the clearance by moving the feeler gauge in and out. A slight drag should be felt. If the clearance is not as specified, turn the adjusting screw either in or out until the clearance is correct. Hold the screw to prevent it from moving and tighten the locknut. Check the clearance again to make sure it didn't change when the locknut was tightened.
6. Depending on what cylinder you started with, bring the next cylinder in the firing order to TDC and repeat the adjustment procedure as described in Steps 5 and 6. Repeat this procedure until all the valves are adjusted.
7. Install the rocker arm cover using a new gasket as required.
8. Check the timing and idle speed and adjust as necessary. Check for any leaks.

2.4L, 3.0L and 3.5L Engines

The newer 2.4L, 3.0L and 3.5L engines are not equipped with jet valves, thus, because of hydraulic lash adjusters, do not have any valves that require periodic adjustment. The only time the valves on these engines would have to be adjusted is in the case of reassembly after dismantling the engine head. For this procedure refer to section 3.

IDLE SPEED AND MIXTURE ADJUSTMENT

On carbureted engines, idle speed is adjusted periodically to compensate for engine wear and to ensure smooth operation. Idle mixture adjustments are not required as a matter of routine, but only when major carburetor work is required.

Fuel injected engines employ an idle speed control which compensates for all normal variations in idle speed. This requires adjustment only if a major fuel system part is replaced. For further information, see Section 5 in this manual.

Idle Speed Adjustment

On all vehicles equipped with adjustable ignition timing and an adjustable idle speed the two processes of adjusting the idle speed and the ignition timing are very closely related and affect each other closely. When performing either of these adjustments, it is usually necessary to adjust each in a back and forth manner since they affect each other. For example, if the ignition timing is adjusted the idle speed could increase past what is recommended. This in turn could affect actual ignition timing during normal use of the vehicle, which would affect the vehicle's driveability. Therefore, the idle speed would have to be adjusted, which could in turn affect the ignition timing. Eventually, after adjusting each back and forth a few times to reach the desired specifications, they will meet both specifications simultaneously. Most often they only need to be adjusted a few times before both are correct.

CARBURETED ENGINES

▶ See Figures 42 and 43

➡**The throttle valve adjusting screw should not be tampered with unless the carburetor has been rebuilt. This screw is preset and determines the relationship between the throttle valve and the free lever, and has been accurately set at the factory. If this setting is disturbed, other service adjustments cannot be done accurately. Also the improper setting (throttle valve opening) will increase the exhaust gas temperature during deceleration, which in turn will reduce the life of the catalyst greatly and deteriorate the exhaust gas cleaning performance. It will also effect fuel consumption and engine braking.**

1. Turn all lights and accessories to the OFF position.
2. Connect a tachometer and timing light to the engine.
3. Place the transmission in Neutral and disconnect the electric cooling fan.
4. Start and warm up the engine at idle speed for a few minutes. On 1987-89 vehicles, press the gas pedal once to disengage the fast idle.

✱✱WARNING

Since the cooling fasn is disconnected, be extremely cautious about the length of time the engine is allowed to idle. Check the temperature gauge often and turn the ignition OFF if the engine starts to overheat.

ENGINE PERFORMANCE AND TUNE-UP

5. Check the ignition timing and adjust as necessary.
6. Check the idle speed. If it does not meet specifications, adjust the idle speed using the idle speed adjusting screw, which is located closest to the primary throttle valve shaft and labeled SAS 1 in the illustration.
7. After the idle speed is set, stop the engine and reconnect the cooling fan connector. Disconnect the tachometer and the timing light.

FUEL INJECTED ENGINES

These vehicles are equipped with an Idle Speed Control (ISC) system which controls the idle speed automatically, making adjustment unnecessary. For more information on the ISC system, see Section 5. To check the curb idle speed, proceed as follows:

1. Turn all lights and accessories to the off position and place the transmission in Neutral (manual) or Park (automatic). Place the steering wheel in the straight ahead position.
2. Start the engine and warm it up for a few minutes.
3. Locate the engine speed detection connector between the noise filter and the primary side of the ignition coil. Insert a paper clip through the harness side to short it and connect a tachometer to the paper clip. Refer to the ignition timing procedure for details.
4. Run the engine at idle speed and check the ignition timing. If not within specification, adjust it.
5. Race the engine at 2,000-3,000 rpm for five seconds then allow the engine to run at idle for two minutes.
6. Read the tachometer and check the idle speed.

➡ **If idle speed is not within the specified limits, check the ISC system. Adjustment of this system is usually unnecessary — refer to Section 4.**

7. Stop the engine and disconnect the tachometer and the timing light.

DIESEL ENGINES

▶ See Figure 44

➡ **Use of a diesel (magnetic) tachometer is required. A conventional tachometer as used on gasoline engines will not work. When connecting the diesel tachometer, follow the manufacturer's instructions carefully.**

1. Make certain that the lights and all electrical accessories are turned off.
2. Set the parking brake and place the shift selector in Neutral.
3. Connect the tachometer.
4. Run the engine until the coolant warms to normal operating temperature.
5. Once the engine is completely warmed up, run the engine at 2000-3000 rpm for at least 5 seconds.
6. Allow the engine to run at idle for 2 minutes.
7. Read the idle speed on the tachometer. If the idle is not within specifications, adjust the idle speed screw. Loosen the locknut, then turn the adjusting screw with a small screwdriver.

➡ Do not disturb other screws or linkages.

8. When the proper idle speed is achieved, hold the screw in position and tighten the locknut.
9. Stop the engine and remove the tachometer.

Mixture Adjustment

➡ **All carburetors have tamper-resistant idle mixture adjusting screws. This setting has been performed at the factory. Neither removal of the plug or adjustment of the mixture screw is required during service unless a major carburetor overhaul is performed.**

CARBURETED ENGINES

1. Remove the carburetor from the engine as described in Section 5. The idle mixture screw is located in the base of the carburetor, just to the left of the PCV hose. Mount the carburetor, carefully, in a soft-jawed vise, protecting the gasket surface, and with the mixture adjusting screw facing upward.

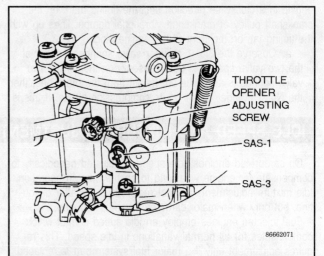

Fig. 42 When adjusting the idle speed on carbureted engines adjust only SAS-1 screw

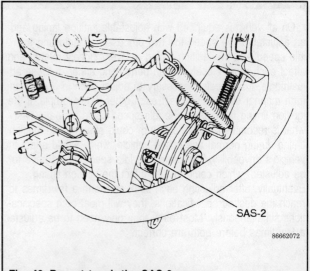

Fig. 43 Do not touch the SAS-2 screw

ENGINE PERFORMANCE AND TUNE-UP 2-23

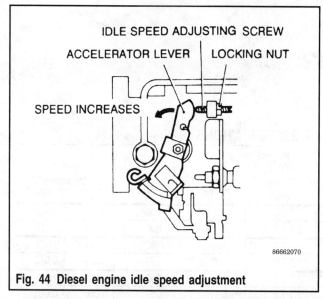

Fig. 44 Diesel engine idle speed adjustment

2. Drill a ⁵⁄₆₄ in. (2 mm) hole through the casting from the underside of the carburetor. Make sure that this hole intersects the passage leading to the mixture adjustment screw just behind the plug. Now, widen that hole with a ⅛ in. (3.18 mm) drill bit.

3. Insert a blunt punch into the hole and tap out the plug. Install the carburetor on the engine and connect all hoses, lines, etc.

4. Start the engine and run it at fast idle until it reaches normal operating temperature. Make sure that all accessories are OFF and the transmission is in Neutral (manual) or Park (automatic). Turn the ignition switch **OFF** and disconnect the battery ground cable for about 5 seconds, then reconnect it. Disconnect the oxygen sensor.

5. Start the engine and run it for at least 5 seconds at 2,000-3,000 rpm. Then allow the engine to idle for about 2 minutes.

6. Connect a tachometer and allow the engine to operate at the curb idle speed. Adjust it, if necessary. Insert a CO meter probe into the exhaust pipe. A reading of 0.1-0.3% is necessary. Adjust the mixture screw to obtain the reading. If, during this adjustment, the idle speed is varied more than 100 rpm in either direction, reset the idle speed and readjust the CO until both specifications are met simultaneously. Shut off the engine, reconnect the oxygen sensor and install a new concealment plug.

FUEL INJECTED ENGINES

Fuel injected trucks use a electronic fuel injection system which is controlled by a series of temperature, altitude and air flow sensors which feed information into a Electronic Control Unit (ECU). The control unit then relays an electronic signal to the injector nozzle(s), which allows a predetermined amount of fuel into the combustion chamber.

DIESEL ENGINE TUNE-UP SPECIFICATIONS

Year	Engine ID/VIN	Engine Displacement Liters (cc)	Initial Injection Timing 1	Delivery Valve Opening Pressure psi (kPa)	Injection Nozzle Pressure psi (kPa)	Idle Speed (rpm) 2	Valve Clearance in. (mm) Intake 2	Exhaust 2
1983	4D55	2.3 (2346)	5° ATDC	306 (2,108)	1,707 (11,768)	700-800	0.010 (0.25)	0.010 (0.25)
1984	4D55	2.3 (2346)	5° ATDC	306 (2,108)	1,707 (11,768)	700-800	0.010 (0.25)	0.010 (0.25)
1985	4D55	2.3 (2346)	5° ATDC	306 (2,108)	1,707 (11,768)	700-800	0.010 (0.25)	0.010 (0.25)

NOTE: The Vehicle Emission Control Information label often reflects specification changes made during production. The label figures must be used if they differ from those in this chart.

ATDC - After Top Dead Center

1 Degree of timing measured at .0394 in. (1 mm) of plunger stroke.

2 Hot engine

ENGINE PERFORMANCE AND TUNE-UP

GASOLINE ENGINE TUNE-UP SPECIFICATIONS

Year	Engine ID/VIN	Engine Displacement Liters (cc)	Spark Plugs Gap (in.)	Ignition Timing (deg.) MT	Ignition Timing (deg.) AT	Fuel Pump (psi)	Idle Speed (rpm) MT	Idle Speed (rpm) AT	Valve Clearance In.	Valve Clearance Ex.
1983	G63B	2.0 (1997)	0.039-0.043	5B	5B	2.7-3.7	700	700	HYD 1	HYD
	G54B	2.6 (2555)	0.039-0.043	10B	10B	3.5-3.8	850	850	0.006 1	0.010
1984	G63B	2.0 (1997)	0.039-0.043	5B	5B	2.7-3.7	700	700	HYD 1	HYD
	G54B	2.6 (2555)	0.039-0.043	10B	10B	3.5-3.8	850	850	0.006 1	0.010
1985	G63B	2.0 (1997)	0.039-0.043	5B	5B	2.7-3.7	700	700	HYD 1	HYD
	G54B	2.6 (2555)	0.039-0.043	10B	10B	3.5-3.8	850	850	0.006 1	0.010
1986	G63B	2.0 (1997)	0.039-0.043	5B	5B	2.7-3.7	600-800	600-800	HYD 1	HYD
	G54B	2.6 (2555)	0.039-0.043	10B	10B	3.5-3.8	750-950	750-950	HYD 1	HYD
1987	G63B	2.0 (1997)	0.039-0.043	5B	5B	2.7-3.7	600-800	600-800	HYD 1	HYD
	G54B	2.6 (2555)	0.039-0.043	10B	10B	3.5-3.8	750-950	750-950	HYD 1	HYD
1988	G63B	2.0 (1997)	0.035-0.039	8B	8B	2.7-3.7	600-800	600-800	HYD 1	HYD
	G54B	2.6 (2555)	0.035-0.039	10B	10B	3.5-3.8	750-950	750-950	HYD 1	HYD
1989	G63B	2.0 (1997)	0.039-0.043	5B	5B	2.7-3.7	600-800	600-800	HYD	HYD
	G54B	2.6 (2555)	0.039-0.043	10B	10B	3.5-3.8	750-950	750-950	HYD 1	HYD
	6G72	3.0 (2972)	0.039-0.043	5B	5B	38	600-800	600-800	HYD	HYD
1990	4G64	2.4 (2350)	0.039-0.043	5B	5B	38	650-850	650-850	HYD	HYD
	G54B	2.6 (2555)	0.039-0.043	7B	7B	2.8-4.2	650-850	650-850	HYD	HYD
	6G72	3.0 (2972)	0.039-0.043	5B	5B	38	600-800	600-800	HYD	HYD
1991	4G64	2.4 (2350)	0.039-0.043	5B	5B	38	650-850	650-850	HYD	HYD
	6G72	3.0 (2972)	0.039-0.043	5B	5B	38	600-800	600-800	HYD	HYD
1992	4G64	2.4 (2350)	0.039-0.043	3-7B	3-7B	38	650-850	650-850	HYD	HYD
	6G72	3.0 (2972)	0.039-0.043	3-7B	3-7B	38	600-800	600-800	HYD	HYD
1993	4G64	2.4 (2350)	0.039-0.043	3-7B	3-7B	38	650-850	650-850	HYD	HYD
	6G72	3.0 (2972)	0.039-0.043	3-7B	3-7B	38	600-800	600-800	HYD	HYD
1994	4G64	2.4 (2350)	0.039-0.043	3-7B	3-7B	38	650-850	650-850	HYD	HYD
	6G72	3.0 (2972)	0.039-0.043	3-7B	3-7B	34	600-800	600-800	HYD	HYD
	6G74	3.5 (3496)	0.039-0.043	2-8B	2-8B	38	600-800	600-800	HYD	HYD
1995	4G64	2.4 (2350)	0.039-0.043	3-7B	3-7B	38	600-800	600-800	HYD	HYD
	6G72	3.0 (2972)	0.039-0.043	3-7B	3-7B	38	600-800	600-800	HYD	HYD
	6G74	3.5 (3497)	0.039-0.043	2-8B	2-8B	38	600-800	600-800	HYD	HYD

NOTE: The Vehicle Emission Control Information label often reflects specification changes made during production. The label figures must be used if they differ from those in this chart.

B - Before top dead center

HYD - Hydraulic

1 Jet valve clearance: 0.010

86662C01

BASIC ELECTRICITY
  BATTERY, STARTING AND
    CHARGING SYSTEMS 3-3
  UNDERSTANDING BASIC
    ELECTRICITY 3-2
ENGINE ELECTRICAL
  ALTERNATOR 3-7
  BATTERY 3-12
  DISTRIBUTOR 3-4
  SENDING UNITS AND
    SENSORS 3-23
  STARTER 3-14
  VOLTAGE REGULATOR 3-12
ENGINE MECHANICAL
  CAMSHAFTS AND BEARINGS 3-146
  CHECKING ENGINE
    COMPRESSION 3-26
  COMBINATION MANIFOLD 3-81
  CORE (FREEZE) PLUGS 3-166
  CRANKSHAFT AND MAIN
    BEARINGS 3-168
  CYLINDER HEAD AND GASKET 3-98
  ENGINE 3-39
  ENGINE FAN 3-89
  ENGINE OIL COOLER 3-88
  ENGINE OVERHAUL TIPS 3-25
  EXHAUST MANIFOLD 3-76
  FLYWHEEL AND RING GEAR 3-171
  HYDRAULIC LASH
    ADJUSTERS 3-114
  INTAKE MANIFOLD 3-68
  OIL PAN 3-137
  OIL PUMP AND SILENT
    SHAFTS 3-139
  PISTONS AND CONNECTING
    RODS 3-157
  RADIATOR 3-84
  REAR MAIN SEAL 3-166
  ROCKER ARMS AND SHAFTS 3-62
  THERMOSTAT 3-66
  TIMING BELT AND COVERS 3-117
  TIMING CHAIN AND COVER 3-134
  TURBOCHARGER 3-82
  VALVE COVER (ROCKER ARM
    COVER) 3-58
  VALVE GUIDES 3-114
  VALVE SEATS 3-113
  VALVES AND VALVE SPRINGS 3-109
  WATER PUMP 3-90
EXHAUST SYSTEM
  CATALYTIC CONVERTER 3-174
  CENTER PIPE 3-174
  FRONT PIPE 3-174
  INSPECTION 3-173
  SAFETY PRECAUTIONS 3-173
  SPECIAL TOOLS 3-173
  TAILPIPE AND MUFFLER 3-174
SPECIFICATIONS CHARTS
  ALTERNATOR SPECIFICATIONS 3-13
  CAMSHAFT SPECIFICATIONS 3-34
  CRANKSHAFT AND CONNECTING
    ROD SPECIFICATIONS 3-36
  GENERAL ENGINE
    SPECIFICATIONS 3-29
  PISTON AND RING
    SPECIFICATIONS 3-32
  STARTER SPECIFICATIONS 3-22
  TORQUE SPECIFICATIONS 3-38
  VALVE SPECIFICATIONS 3-30

3

ENGINE AND ENGINE REBUILDING

BASIC ELECTRICITY 3-2
ENGINE ELECTRICAL 3-4
ENGINE MECHANICAL 3-25
EXHAUST SYSTEM 3-173
SPECIFICATIONS CHARTS 3-13

ENGINE AND ENGINE REBUILDING

BASIC ELECTRICITY

Understanding Basic Electricity

For any electrical system to operate, it must make a complete circuit. This simply means that the power flow from the battery must make a full circle. When an electrical component is operating, power flows from the battery to the components, passes through the component (load) causing it to function, and returns to the battery through the ground path of the circuit. This ground may be either another wire or the actual metal part of the car depending upon how the component is designed.

Perhaps the easiest way to visualize this is to think of connecting a light bulb with two wires attached to it to the battery. If one of the two wires was attached to the negative post of the battery and the other wire to the positive (+) post, the light bulb would light and the circuit would be complete. Electricity could follow a path from the battery to the bulb and back to the battery. Its not hard to see that with longer wires on our light bulb, it could be mounted anywhere on the car. Further, one wire could be fitted with a switch so that the light could be turned on and off at will. Various other items could be added to our primitive circuit to make the light flash, become brighter or dimmer under certain conditions or advise the user that it's burned out.

Some automotive components are grounded through their mounting points. The electrical current runs through the chassis of the vehicle and returns to the battery through the ground (-) cable; if you look, you'll see that the battery ground cable connects between the battery and the body of the car.

Every complete circuit must include a "load" (something to use the electricity coming from the source). If you were to connect a wire between the two terminals of the battery (DON'T do this) without the light bulb, the battery would attempt to deliver its entire power supply from one pole to another almost instantly. This is a short circuit. The electricity is taking a short cut to get to ground and is not being used by any load in the circuit. This sudden and uncontrolled electrical flow can cause great damage to other components in the circuit and can develop a tremendous amount of heat. A short in an automotive wiring harness can develop sufficient heat to melt the insulation on all the surrounding wires and reduce a multiple wire cable to one sad lump of plastic and copper. Two common causes of shorts are broken insulation (thereby exposing the wire to contact with surrounding metal surfaces or other wires) or a failed switch (the pins inside the switch come out of place and touch each other).

Some electrical components which require a large amount of current to operate also have a relay in their circuit. Since these circuits carry a large amount of current (amperage or amps), the thickness of the wire in the circuit (wire gauge) is also greater. If this large wire were connected from the load to the control switch on the dash, the switch would have to carry the high amperage load and the dash would be twice as large to accommodate wiring harnesses as thick as your wrist. To prevent these problems, a relay is used. The large wires in the circuit are connected from the battery to one side of the relay and from the opposite side of the relay to the load. The relay is normally open, preventing current from passing through the circuit. An additional, smaller wire is connected from the relay to the control switch for the circuit. When the control switch is turned on, it grounds the smaller wire to the relay and completes its circuit. The main switch inside the relay closes, sending power to the component without routing the main power through the inside of the car. Some common circuits which may use relays are the horn, headlights, starter and rear window defogger systems.

It is possible for larger surges of current to pass through the electrical system of your car. If this surge of current were to reach the load in the circuit, it could burn it out or severely damage it. To prevent this, fuses, circuit breakers and/or fusible links are connected into the supply wires of the electrical system. These items are nothing more than a built-in weak spot in the system. It's much easier to go to a known location (the fusebox) to see why a circuit is inoperative than to dissect 15 feet of wiring under the dashboard, looking for what happened.

When an electrical current of excessive power passes through the fuse, the fuse blows and breaks the circuit, preventing the passage of current and protecting the components.

A circuit breaker is basically a self repairing fuse. It will open the circuit in the same fashion as a fuse, but when either the short is removed or the surge subsides, the circuit breaker resets itself and does not need replacement.

A fuse link (fusible link or main link) is a wire that acts as a fuse. One of these is normally connected between the starter relay and the main wiring harness under the hood. Since the starter is the highest electrical draw on the car, an internal short during starting could direct about 130 amps into the wrong places. Consider the damage potential of introducing this current into a system whose wiring is rated at 15 amps and you'll understand the need for protection. Since this link is very early in the electrical path, it's the first place to look if nothing on the car works but the battery seems to be charged and is properly connected.

Electrical problems generally fall into one of three areas:
- The component that is not functioning is not receiving current.
- The component is receiving power but not using it or using it incorrectly (component failure).
- The component is improperly grounded.

The circuit can be can be checked with a test light and a jumper wire. The test light is a device that looks like a pointed screwdriver with a wire on one end and a bulb in its handle. A jumper wire is simply a piece of wire with alligator clips on each end. If a component is not working, you must follow a systematic plan to determine which of the three causes is the villain.

1. Turn on the switch that controls the inoperative item.

➡ Some items only work when the ignition switch is turned ON.

2. Disconnect the the power supply wire from the component.

3. Attach the ground wire on the test light to a good metal ground.

ENGINE AND ENGINE REBUILDING 3-3

4. Touch the end probe of the test light to the power wire; if there is current in the wire, the light in the test light will come on. You have now established that current is getting to the component.

5. Turn the ignition or dash switch **OFF** and reconnect the wire to the component.

If the test light did not go on, then the problem is between the battery and the component. This includes all the switches, fuses, relays and the battery itself. The next place to look is the fusebox; check carefully either by eye or by using the test light across the fuse clips. The easiest way to check is to simply replace the fuse. If the fuse is blown, and upon replacement, immediately blows again, there is a short between the fuse and the component. This is generally (not always) a sign of an internal short in the component. Disconnect the power wire at the component again and replace the fuse; if the fuse holds, the component is the problem.

If all the fuses are good and the component is not receiving power, find the switch for the circuit. Bypass the switch with the jumper wire. This is done by connecting one end of the jumper to the power wire coming into the switch and the other end to the wire leaving the switch. If the component comes to life, the switch has failed.

✼✼WARNING

Never substitute the jumper for the component. The circuit needs the electrical load of the component. If you bypass it, you will cause a short circuit.

Checking the ground for any circuit can mean tracing wires to the body, cleaning connections or tightening mounting bolts for the component itself. If the jumper wire can be connected to the case of the component or the ground connector, you can ground the other end to a piece of clean, solid metal on the car. Again, if the component starts working, you've found the problem.

A systematic search through the fuse, connectors, switches and the component itself will almost always yield an answer. Loose and/or corroded connectors, particularly in ground circuits, are becoming a larger problem in modern cars. The computers and on-board electronic (solid state) systems are highly sensitive to improper grounds and will change their function drastically if one occurs.

Remember that for any electrical circuit to work, ALL the connections must be clean and tight.

Battery, Starting and Charging Systems

BASIC OPERATING PRINCIPLES

Battery

The battery is the first link in the chain of mechanisms which work together to provide cranking of the automobile engine. In most modern cars, the battery is a lead/acid electrochemical device consisting of six 2v subsections (cells) connected in series so the unit is capable of producing approximately 12v of electrical pressure. Each subsection consists of a series of positive and negative plates held a short distance apart in a solution of sulfuric acid and water.

The two types of plates are of dissimilar metals. This causes a chemical reaction to be set up, and it is this reaction which produces current flow from the battery when its positive and negative terminals are connected to an electrical appliance such as a lamp or motor. The continued transfer of electrons would eventually convert the sulfuric acid to water, and make the two plates identical in chemical composition. As electrical energy is removed from the battery, its voltage output tends to drop. Thus, measuring battery voltage and battery electrolyte composition are two ways of checking the ability of the unit to supply power. During the starting of the engine, electrical energy is removed from the battery. However, if the charging circuit is in good condition and the operating conditions are normal, the power removed from the battery will be replaced by the alternator which will force electrons back through the battery, reversing the normal flow, and restoring the battery to its original chemical state.

Starting System

The battery and starting motor are linked by very heavy electrical cables designed to minimize resistance to the flow of current. Generally, the major power supply cable that leaves the battery goes directly to the starter, while other electrical system needs are supplied by a smaller cable. During starter operation, power flows from the battery to the starter and is grounded through the car's frame and the battery's negative ground strap.

The starting motor is a specially designed, direct current electric motor capable of producing a great amount of power for its size. One thing that allows the motor to produce a great deal of power is its tremendous rotating speed. It drives the engine through a tiny pinion gear (attached to the starter's armature), which drives the very large flywheel ring gear at a greatly reduced speed. Another factor allowing it to produce so much power is that only intermittent operation is required of it. Thus, little allowance for air circulation is required, and the windings can be built into a very small space.

The starter solenoid is a magnetic device which employs the small current supplied by the start circuit of the ignition switch. This magnetic action moves a plunger which mechanically engages the starter and closes the heavy switch connecting it to the battery. The starting switch circuit consists of the starting switch contained within the ignition switch, a transmission neutral safety switch or clutch pedal switch, and the wiring necessary to connect these in series with the starter solenoid or relay.

The pinion, a small gear, is mounted to a one way drive clutch. This clutch is splined to the starter armature shaft. When the ignition switch is moved to the **START** position, the solenoid plunger slides the pinion toward the flywheel ring gear via a collar and spring. If the teeth on the pinion and flywheel match properly, the pinion will engage the flywheel immediately. If the gear teeth butt one another, the spring will be compressed and will force the gears to mesh as soon as the starter turns far enough to allow them to do so. As the solenoid plunger reaches the end of its travel, it closes the contacts that connect the battery and starter and then the engine is cranked.

ENGINE AND ENGINE REBUILDING

As soon as the engine starts, the flywheel ring gear begins turning fast enough to drive the pinion at an extremely high rate of speed. At this point, the one-way clutch begins allowing the pinion to spin faster than the starter shaft so that the starter will not operate at excessive speed. When the ignition switch is released from the starter position, the solenoid is de-energized, and a spring pulls the gear out of mesh interrupting the current flow to the starter.

Some starters employ a separate relay, mounted away from the starter, to switch the motor and solenoid current on and off. The relay replaces the solenoid electrical switch, but does not eliminate the need for a solenoid mounted on the starter used to mechanically engage the starter drive gears. The relay is used to reduce the amount of current the starting switch must carry.

Charging System

The automobile charging system provides electrical power for operation of the vehicle's ignition system, starting system and all the electrical accessories. The battery serves as an electrical surge or storage tank, storing (in chemical form) the energy originally produced by the engine-driven alternator. The system also provides a means of regulating output to protect the battery from being overcharged and to avoid excessive voltage to the accessories.

The storage battery is a chemical device incorporating parallel lead plates in a tank containing a sulfuric acid/water solution. Adjacent plates are slightly dissimilar, and the chemical reaction of the two dissimilar plates produces electrical energy when the battery is connected to a load such as the starter motor. The chemical reaction is reversible, so that when the generator is producing a voltage (electrical pressure) greater than that produced by the battery, electricity is forced into the battery, and the battery is returned to its fully charged state.

Newer automobiles use alternating current generators or alternators, because they are more efficient, can be rotated at higher speeds, and have fewer brush problems. In an alternator, the field rotates while all the current produced passes only through the stator winding. The brushes bear against continuous slip rings. This causes the current produced to periodically reverse the direction of its flow. Diodes (electrical one way valves) block the flow of current from traveling in the wrong direction. A series of diodes is wired together to permit the alternating flow of the stator to be rectified back to 12 volts DC for use by the vehicles's electrical system.

The voltage regulating function is performed by a regulator. The regulator is often built in to the alternator; this system is termed an integrated or internal regulator.

ENGINE ELECTRICAL

Distributor

REMOVAL

◆ See Figures 1, 2, 3, 4, 5, 6, 7 and 8

➡ You will need a new O-ring for installation of the distributor.

1. Disconnect the battery ground cable.
2. Disconnect the retaining clips or screws and pull the distributor cap and seal off the distributor.
3. Position the cap and wires away from the distributor. You may find this procedure easier if you label and remove all spark plug cables and the ignition coil cable from the distributor cap.
4. Matchmark the location of the distributor on the cylinder block. A good place to do this is near where the mounting bolt passes through the slot.
5. Disconnect and label the vacuum advance line, if so equipped.
6. Disconnect and label the distributor wiring connector.
7. Matchmark the relationship between the tip of the distributor rotor and the distributor body.

➡ Later model distributors have alignment marks on the flange, housing and gear. Aligning these marks is more accurate than aligning the rotor with a homemade mark.

8. Remove the mounting nut or bolt and pull the distributor carefully out of the engine. Note the direction and degree to which the rotor turns as you pull it out. Mark the location of the rotor after it has turned, also.

Fig. 1 Use paint or a grease pencil to mark the relationship between the distributor and the engine block

9. Remove the old O-ring from the distributor shaft and replace it with a new one before reinstalling the distributor.

INSTALLATION

Engine Not Disturbed

This condition exists if the engine has not been rotated while the distributor was removed.

1. Position the distributor so that the distributor and block matchmarks are lined up.

ENGINE AND ENGINE REBUILDING 3-5

Fig. 2 Mark the position of the rotor in relation to the distributor housing, both before and after pulling the distributor out of the engine

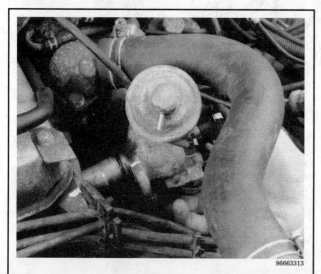

Fig. 3 Pull the distributor slowly out of the engine

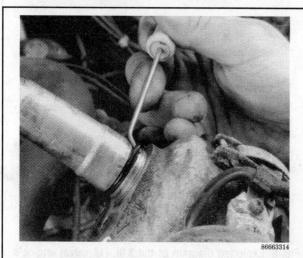

Fig. 4 Remember to replace the O-ring on the distributor shaft before reinstallation

2. Position the rotor so that it lines up with the matchmark on the distributor body after the distributor was pulled part way out.

3. Insert the distributor into the block until the gears at the bottom engage and begin turning the rotor. If there is resistance, turn the rotor back and forth slightly so the gears mesh. Once the gears engage and inserting the distributor causes the rotor to turn, push the distributor in until it seats and the rotor is lined up with the first mark made on the body.

4. Install the distributor mounting bolt and tighten it finger-tight.

5. Reconnect the vacuum advance line, if so equipped and distributor wiring connector, and reinstall the gasket and cap.

6. Reconnect the negative battery cable.

7. Adjust the ignition timing as described in Section 2.

8. Tighten the distributor mounting bolt securely.

Engine Disturbed

This condition exists when the engine has been rotated with the distributor removed.

1. If the engine has been rotated while the distributor was out, you will have to first put the engine on No. 1 cylinder at Top Dead Center (TDC) of the compression stroke. You can either remove the valve cover or No. 1 spark plug to determine engine position. Rotate the engine with a socket wrench on the nut at the center of the front pulley in the normal direction of rotation. Either feel for air being expelled forcefully through the spark plug hole or watch for the engine to rotate up to the top center mark without the valves moving (both valves will be closed or all the way up). If the valves are moving as you approach TDC or there is no air being expelled through the plug hole, turn the engine another full turn until you get the appropriate indication as the engine approaches TDC position.

2. Start the distributor into the block with the matchmarks between the distributor body and the block lined up.

3. Turn the rotor slightly until the matchmarks on the bottom of the distributor body and the bottom of the distributor shaft near the gear are aligned.

4. Insert the distributor all the way into the block. If you have trouble getting the distributor and camshaft gears to

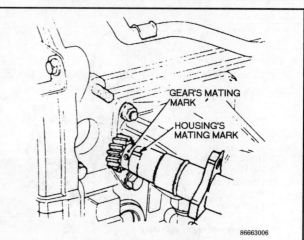

Fig. 5 If the distributor has matchmarks already on it, set the engine to TDC No. 1 cylinder and align the housing mark with the gear mating mark

3-6 ENGINE AND ENGINE REBUILDING

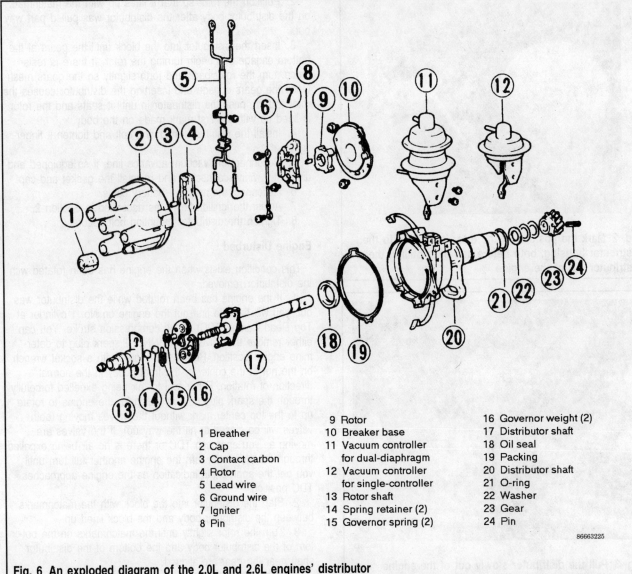

1 Breather	9 Rotor	16 Governor weight (2)
2 Cap	10 Breaker base	17 Distributor shaft
3 Contact carbon	11 Vacuum controller for dual-diaphragm	18 Oil seal
4 Rotor		19 Packing
5 Lead wire	12 Vacuum controller for single-controller	20 Distributor shaft
6 Ground wire		21 O-ring
7 Igniter	13 Rotor shaft	22 Washer
8 Pin	14 Spring retainer (2)	23 Gear
	15 Governor spring (2)	24 Pin

Fig. 6 An exploded diagram of the 2.0L and 2.6L engines' distributor

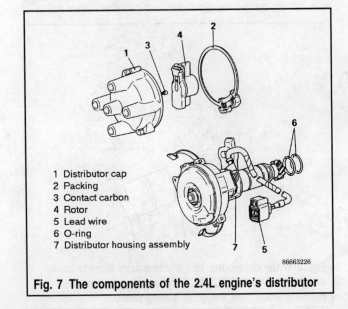

1 Distributor cap
2 Packing
3 Contact carbon
4 Rotor
5 Lead wire
6 O-ring
7 Distributor housing assembly

Fig. 7 The components of the 2.4L engine's distributor

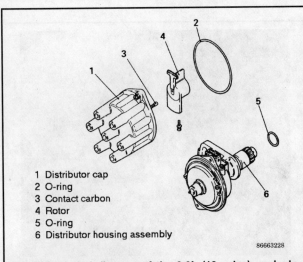

1 Distributor cap
2 O-ring
3 Contact carbon
4 Rotor
5 O-ring
6 Distributor housing assembly

Fig. 8 Exploded diagram of the 3.0L (12 valve) engine's distributor

ENGINE AND ENGINE REBUILDING 3-7

mesh, turn the rotor back and forth very slightly until the distributor can be inserted easily. If the rotor is not lined up with the position of No. 1 plug terminal, you'll have to pull the distributor back out slightly, shift the position of the rotor appropriately, and then reinstall it.

5. Align the matchmarks between the distributor and block.
6. Install the distributor mounting bolt and tighten it finger-tight.
7. Reconnect the vacuum advance line, if so equipped and distributor wiring connector, and reinstall the gasket and cap.
8. Reconnect the negative battery cable.
9. Adjust the ignition timing as described in Section 2.
10. Tighten the distributor mounting bolt securely.

Alternator

The alternator charging system is a negative (-) ground system which consists of an alternator, a regulator, a charge indicator, a storage battery, wiring connecting the components, and a fuse link wire.

The alternator is belt-driven from the engine. Energy is supplied from the alternator/regulator system to the rotating field through two brushes to two slip rings. The slip rings are mounted on the rotor shaft and are connected to the field coil. This energy supplied to the rotating field from the battery is called excitation current and is used to initially energize the field to begin the generation of electricity, the excitation current comes from its own output rather than the battery.

The alternator produces power in the form of alternating current. The alternating current is rectified by 6 diodes into direct current. The direct current is used to charge the battery and power the rest of the electrical system.

When the ignition key is turned on, current flows from the battery, through the charging system indicator light on the instrument panel, to the voltage regulator, and to the alternator. Since the alternator is not producing any current, the alternator warning light comes on. When the engine is started, the alternator begins to produce current and turns the alternator light off. As the alternator turns and produces current, the current is divided in two ways; part to the battery to charge the battery and power the electrical components of the vehicle, and part is returned to the alternator to enable it to increase its output. In this situation, the alternator is receiving current from the battery and from itself. A voltage regulator is wired into the current supply to the alternator to prevent it from receiving too much current which would cause it to put out too much current. Conversely, if the voltage regulator does not allow the alternator to receive enough current, the battery will not be fully charged and will eventually go dead.

The battery is connected to the alternator at all times, whether the ignition key is turned on or not. If the battery were shorted to ground, the alternator would also be shorted. This would damage the alternator. To prevent this, a fuse link is installed in the wiring between the battery and the alternator. If the battery is shorted, the fuse link is melted, protecting the alternator.

ALTERNATOR PRECAUTIONS

To prevent damage to the alternator and regulator, the following precautionary measures must be taken when working with the electrical system.

1. Never reverse the battery connections. Always check the battery polarity visually. This is to be done before any connections are made to ensure that all of the connections correspond to the battery ground polarity of the truck.
2. Booster batteries must be connected properly. Make sure the positive cable of the booster battery is connected to the positive terminal of the battery which is getting the boost.
3. Disconnect the battery cables before using a fast charger; the charger has a tendency to force current through the diodes in the opposite direction for which they were designed.
4. Never use a fast charger as a booster for starting the truck.
5. Never disconnect the voltage regulator while the engine is running, unless as noted for testing purposes.
6. Do not ground the alternator output terminal.
7. Do not operate the alternator on an open circuit with the field energized. Make sure that all connections within the circuit are clean and tight.
8. Do not attempt to polarize the alternator.
9. Disconnect the battery cables and remove the alternator before using an electric arc welder on the truck.
10. Protect the alternator from excessive moisture. If the engine is to be steam cleaned, cover or remove the alternator.
11. Disconnect the battery terminals when performing any service on the electrical system. This will eliminate the possibility of accidental reversal of polarity.
12. Do not use test lamps of more than 12 volts for checking diode continuity.

REMOVAL & INSTALLATION

➡ Some vehicles may have an alternator cover that must be removed before removing this assembly. On some of the 4-cylinder engines, the A/C compressor assembly may have to be removed for working access.

2.0L and 2.6L Engines

WITHOUT AIR CONDITIONING

▸ See Figure 9

1. Turn ignition switch to **OFF** and disconnect both battery cables.
2. Loosen the support bolt and adjusting bolt, and then shift the alternator toward the engine so belt tension is relaxed. Remove the belt. On some models it will be necessary to loosen the power steering pump and belt and then remove the pump from its brackets.
3. Note locations of all connectors. Make a drawing, if necessary. Unplug electrical connectors and unscrew fastening nuts for terminal type connectors. Clean any dirty connections.
4. Remove the adjusting bolt. Remove the nut from the rear of the mounting bolt. Note that in the next step, shims located between the alternator and the rear of the engine front cover may fall out. Retain these shims for re-use. Supporting

3-8 ENGINE AND ENGINE REBUILDING

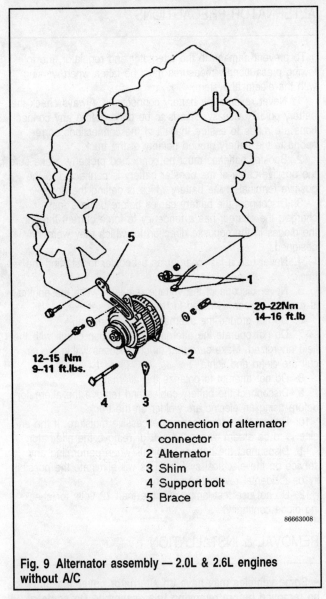

Fig. 9 Alternator assembly — 2.0L & 2.6L engines without A/C

1. Connection of alternator connector
2. Alternator
3. Shim
4. Support bolt
5. Brace

the alternator, slide the mounting bolt out and then remove the unit from the vehicle.

➡ Some models may not be equipped with mounting shims.

To install:

5. Position the assembly so that the mounting bolt can be inserted. If the same alternator is being re-used, insert the shims between the rear of the front cover and the rear hinge of the alternator body. Install the mounting bolt. If the alternator is a different one, you'll have to measure the clearance between the rear of the front cover and the section of alternator body that fits against it. Pull the alternator forward and use a flat feeler gauge. If the clearance exceeds 0.008 in. (0.2mm), insert shim(s) thick enough so there is a slight friction fit and they will not fall out when you let go of them. Put the shims between the front of the front cover and the rear of the front hinge on the alternator body. Make sure you have installed shims of adequate thickness (or that the clearance is within specification). Otherwise, the alternator body may be-

come damaged. Install the mounting bolt if it is not already in place, and then install the mounting bolt nut loosely.

6. Install the adjusting bolt, and then loosely install the nut at the rear of it. Install the belt and tighten it as described in Section 1. Tighten the adjusting bolt to 9-10 ft. lbs. (12-13 Nm) and the mounting bolt and nut to 15-18 ft. lbs. (20-24 Nm).

7. Reinstall the power steering pump and belt if it was removed. Adjust the belt to the proper tension. Tighten the power steering pump bolts to 18 ft. lbs. (24 Nm).

WITH AIR CONDITIONING

▶ See Figure 10

1. Have the air conditioning system discharged by a trained and certified mechanic using a recovery/recycling machine.
2. Turn the ignition **OFF** and disconnect both battery cables.

> **✳✳CAUTION**
>
> Only perform the following procedure AFTER the system has been discharged by a trained and certified mechanic. Doing otherwise may result in personal harm, environmental damage and damage to the vehicle's components.

3. Disconnect and label the discharge hose from the compressor. Cap the compressor port and plug the line immediately.
4. Disconnect and label the suction hose from the compressor. Cap the compressor port and plug the line immediately.
5. Disconnect and label the wiring connector to the compressor.
6. Loosen the belt tension adjuster, the lockbolt and the pivot bolt. Remove the belt, the lockbolt and the pivot bolt from the compressor. Support the compressor and remove it from the engine.
7. Loosen the upper and lower alternator bolts, release the belt tension for the alternator belt, pivot the alternator toward the engine and remove the belt from the alternator.
8. Label and disconnect the wiring connections at the back of the alternator.
9. Remove the upper bolt. Remove the lower bolt while supporting the alternator. Be prepared to catch the bolt shims, which will fall out when the bolt is removed. Remove the alternator from its mount.

➡ Some models may not be equipped with mounting shims.

To install:

10. Install the alternator in its original position and insert the lower support bolt, but do not install the nut. Install the upper bolt just enough to hold the bolt in place.
11. Push the alternator forward (towards the front of the truck) and insert shims between the front leg of the alternator and the front case (mount). There should be enough shims to hold themselves in place when you let go. The alternator should still be movable with light pressure.
12. Tighten the upper and lower mounting bolts until they are finger-tight.
13. Install the alternator belt and adjust it to the correct tension. For this procedure, see Section 1. Tighten the upper mounting bolt to 10 ft. lbs. (13 Nm) and the lower mounting bolt to 16 ft. lbs. (28 Nm). Connect the wiring to the alternator.

ENGINE AND ENGINE REBUILDING 3-9

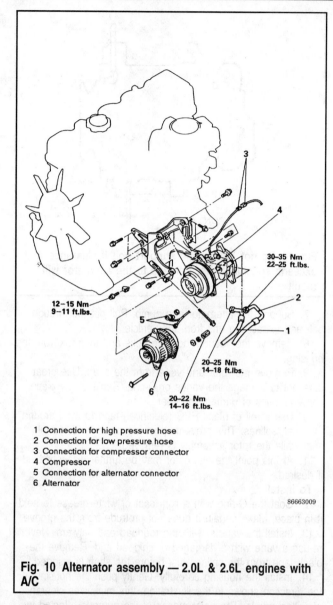

1. Connection for high pressure hose
2. Connection for low pressure hose
3. Connection for compressor connector
4. Compressor
5. Connection for alternator connector
6. Alternator

Fig. 10 Alternator assembly — 2.0L & 2.6L engines with A/C

14. Reinstall the compressor. Install the belt over the pulley (making sure the belt is still correctly mounted on the other pulleys) and gently snug the belt by moving the compressor on its mounts. As soon as there is a little tension on the belt, tighten the compressor mounting bolts.
15. Set the belt to its final tension by tightening the adjuster screw on the compressor. Do not pry on the compressor to adjust the belt.
16. Connect the compressor wiring harness.
17. Remove the plugs and caps from the A/C fittings. Connect the suction hose to its port on the compressor and then connect the discharge hose. Tighten the fitting until finger-tight, making very certain they are clean and correctly threaded. Tighten the suction hose to 24 ft. lbs. (32 Nm) and tighten the discharge hose to 16 ft. lbs. (28 Nm).
18. Reconnect the cables to the battery.
19. Double check the drive belts for proper tension and adjust as necessary. If a belt is replaced during this procedure, it will need to be rechecked after 50-100 miles (80-160 km) of driving.

20. Take the vehicle immediately back to the mechanic to have the A/C system evacuated and recharged. The sooner this is done, the less chance of having the system become contaminated.

Diesel Engine

▶ See Figures 11, 12, 13, 14, 15 and 16

1. Make certain the ignition switch is **OFF** and disconnect the battery cables.
2. Disconnect the cable from terminal B of the alternator.
3. Remove the electrical connector from the back of the alternator.
4. Label and disconnect the oil hose, vacuum hose and return hose from the vacuum pump. Due to difficult access, it is usually easier to disconnect the oil return hose at the oil pan.
5. Remove the brace bolt and the support bolt nut; loosen the alternator.
6. Remove the drive belt from the alternator pulley.

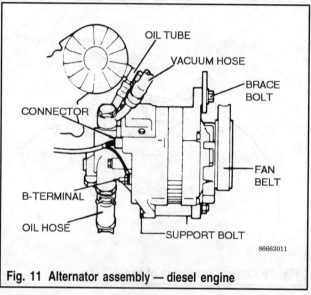

Fig. 11 Alternator assembly — diesel engine

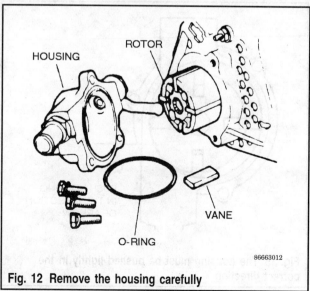

Fig. 12 Remove the housing carefully

3-10 ENGINE AND ENGINE REBUILDING

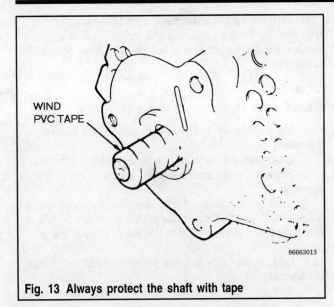

Fig. 13 Always protect the shaft with tape

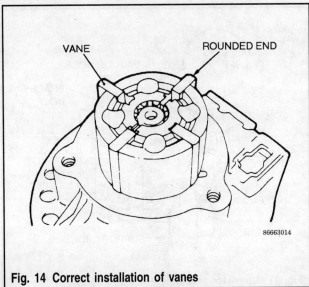

Fig. 14 Correct installation of vanes

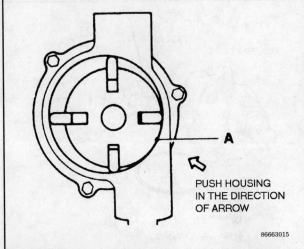

Fig. 15 The housing must be pushed lightly in the correct direction

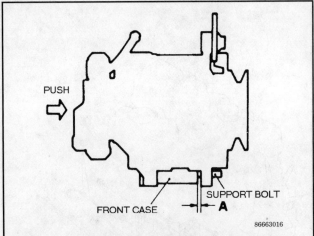

Fig. 16 Do not tighten the support bolt if clearance A exceeds specifications. Damage to the alternator will result

7. Support the alternator and pump. Pull out the through-bolt and extract the unit from the vehicle.
8. Remove the vacuum housing bolts and then take out the housing.
9. Remove the rotor and vane from the shaft. Use great care not to damage the vanes or O-ring. Replace any piece showing signs of damage or wear.
10. Use a roll of black PVC (electrical) tape to wind around the shaft splines. This prevents possible damage to the oil seal while the rotor assembly is removed.
11. At this point the alternator may be further disassembled, if desired.

To install:
12. Coat the O-ring with a light coat of white grease to hold it in place. Make certain it does not protrude from the groove.
13. Install the vane(s) with the rounded end outward. Never re-use a vane with a damaged or chipped end. Remove the black tape from the splines just before the shaft is inserted.
14. Install the housing carefully. Gently push the housing in the direction shown in the illustration, this minimizes clearance at critical points. The performance of the pump is affected by the installation of the housing.
15. Tighten the bolts evenly while keeping light pressure.
16. If it is necessary to bench test the alternator after reassembly, make certain the pump is properly lubricated before turning the alternator. Take great care to insure the vacuum pump is never operated without lubrication.
17. Install the oil return hose to the vacuum pump and install the alternator in position. Once the unit is in place, it is impossible to attach the hose properly. Insert the through-bolt from the rear.
18. Install the drive belt onto the pulley; install the brace (upper) bolt.
19. Push the alternator towards the front of the engine and check for proper clearance between the alternator leg and the

ENGINE AND ENGINE REBUILDING 3-11

front case. If clearance is greater than 0.008 in. (0.2mm) insert spacers to reduce clearance.

> **✱✱WARNING**
>
> If the support bolt is tightened without reducing the clearance, the alternator leg may be broken off.

20. Connect the oil delivery tube, the vacuum hose and the electrical connector.
21. Connect the wiring to terminal **B**.
22. Adjust the belt tension correctly.
23. Connect the battery cables to their perspective terminals.

2.4L and 3.0L (12 Valve) Engines

▶ See Figures 17 and 18

1. Turn ignition switch **OFF** and disconnect both battery cables.
2. Loosen the support bolt and the brace bolt, then loosen the adjusting bolt to shift the alternator toward the engine so that belt tension is relaxed. Remove the belt from the alternator pulley. If the alternator belt needs replacing it will be necessary to loosen the power steering pump and belt and then remove the pump from its brackets.
3. Note locations of all connectors. Make a drawing, if necessary. Unplug electrical connectors and unscrew fastening nuts for terminal type connectors. Clean any dirty connections.
4. Remove the adjusting bolt. Remove the nut from the rear of the support bolt. Supporting the alternator, slide the support bolt out and then remove the unit from the vehicle.

To install:

5. Position the assembly so that the support bolt can be inserted. Install the support bolt and attach the support bolt nut loosely.
6. Install the adjusting bolt assembly and brace bolt, and then loosely install the nut at the rear of the brace bolt. Install the belt and tighten it as described in Section 1. Tighten the adjusting bolt to 9-10 ft. lbs. (12-13 Nm) and the support bolt and nut to 15-18 ft. lbs. (20-24 Nm).
7. Adjust the belt to the proper tension.

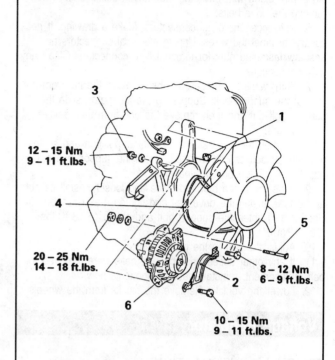

1. Generator adjustment bolt
2. Generator cover
3. Generator brace bolt
4. Generator drive belt
5. Generator support bolt
6. Generator

Fig. 18 Alternator assembly components and location — 3.0L (12 valve) engines

3.0L (24 Valve) and 3.5L engines

▶ See Figure 19

1. Turn ignition switch to **OFF** and disconnect both battery cables.

➡ Since the alternator on these engines is located on the lower right hand (passenger) side of the engine, removal and installation of the alternator will require access from both the top and bottom of the engine compartment.

2. Apply the parking brake and block the rear wheels.
3. Raise the front of the vehicle with a floor jack, using the proper frame positions.
4. Support the vehicle on the frame or transmission crossmember, with jackstands. Before working under the vehicle, make sure the vehicle is stable on the jackstands. If it is not, reposition the jackstands so that it is.
5. Loosen the support and brace bolts, then shift the alternator toward the engine so that belt tension is relaxed. Remove the belt from the alternator pulley. If the alternator belt needs replacing it will be necessary to loosen the power steer-

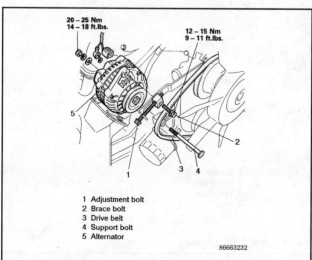

1. Adjustment bolt
2. Brace bolt
3. Drive belt
4. Support bolt
5. Alternator

Fig. 17 Alternator assembly components and location — 2.4L engines

3-12 ENGINE AND ENGINE REBUILDING

ing pump, along with any other belt-driven accessories, and remove the drive belts.

6. Note locations of all connectors. Make a drawing, if necessary, or label the wires. Unplug electrical connectors and unscrew fastening nuts for terminal type connectors. Clean any dirty connections.

7. Withdrawal the brace bolt, then remove the nut from the rear of the support bolt. Supporting the alternator, slide the support bolt out and then remove the unit from the vehicle.

To install:

8. Position the alternator in its original position and so that the support bolt can be inserted. Install the support bolt and attach the support bolt nut loosely.

9. Slide the brace bolt into place and loosely install its nut at the rear. Install the drive belt and tighten it as described in Section 1. Tighten the support bolt and nut to 15-18 ft. lbs. (20-24 Nm).

10. Adjust the belt to the proper tension.

11. If any other accessories or their belts were removed for belt replacement, reinstall these components now.

12. Lower the vehicle. Remove the blocks from the wheels.

Voltage Regulator

REMOVAL & INSTALLATION

In all cases, the solid-state voltage regulator is contained within the alternator. This allows compact, efficient design and function. However, because of individual parts costs and the complexity of disassembling and reassembling the alternator, it would be both economically advantageous and time efficient to replace the entire alternator when the regulator malfunctions.

➡ All voltage regulators are internal electronic type. The ampere rating is from 45 — 75 amps depending on year, model and engine. The regulated voltage is 14.1-14.7 volts on all models.

Battery

REMOVAL & INSTALLATION

1. Turn the ignition switch **OFF**.
2. Disconnect the negative battery terminal, then disconnect the positive battery terminal. Be careful to keep the wrench from connecting the two battery terminals when loosening the nuts that tighten the battery connectors. If the connectors are difficult to pull off, either use a special puller designed for this purpose or carefully pry the halves of the connectors apart with a prytool. Be careful; the connectors are very soft metal and are easily damaged.
3. Loosen and remove the retaining nuts, remove the battery mounting bracket, and remove the battery.

✶✶CAUTION

Battery fluid contains sulfuric acid, capable of causing skin and eye burns. Keep the battery upright at all times. Wear eye protection when working with batteries. Wear heavy rubber gloves if any leakage is evident.

4. Clean the surface of the battery of grease and dust and inspect it for cracks while it is out of the vehicle. If the case is cracked, the battery must be replaced. Clean the battery box or tray with a solution of baking soda and water to remove corrosion. Clean the top of the battery with the same solution, rinse it with clear water and dry it with clean rags. Make sure that none of the baking soda/water solution leaks into the battery.

➡ If the battery is to be charged while out of the vehicle, place it on wooden blocks or a workbench. Do not charge the battery while it is sitting on a concrete floor. The concrete will absorb heat and prevent the battery from charging.

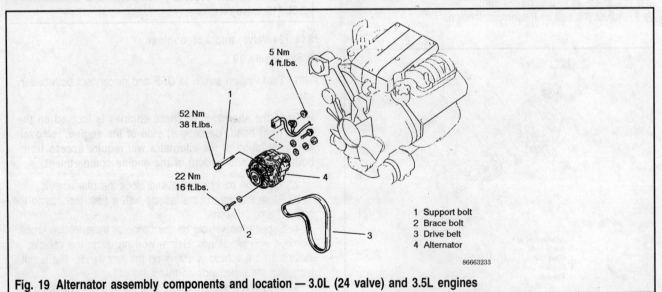

Fig. 19 Alternator assembly components and location — 3.0L (24 valve) and 3.5L engines

1 Support bolt
2 Brace bolt
3 Drive belt
4 Alternator

ENGINE AND ENGINE REBUILDING 3-13

ALTERNATOR SPECIFICATIONS

Year	Model	Engine	Ampere Rating	Regulator	Regulated Voltage @ 68°F
1983	Montero	G54B	45	Internal Electronic	13.9-14.9
	Pick-up	G63B	45	Internal Electronic	13.9-14.9
	Pick-up	G54B	45	Internal Electronic	13.9-14.9
	Pick-up	4D55	50	Internal Electronic	13.9-14.9
1984	Montero	G54B	50 [1]	Internal Electronic	13.9-14.9
	Pick-up	G63B	45	Internal Electronic	13.9-14.9
	Pick-up	G54B	45	Internal Electronic	13.9-14.9
	Pick-up	4D55	50	Internal Electronic	13.9-14.9
1985	Montero	G54B	50 [1]	Internal Electronic	13.9-14.9
	Pick-up	G63B	45	Internal Electronic	14.1-14.7
	Pick-up	G54B	45	Internal Electronic	14.1-14.7
	Pick-up	4D55	50	Internal Electronic	14.1-14.7
1986	Montero	G54B	50 [1]	Internal Electronic	13.9-14.9
	Pick-up	G63B	45	Internal Electronic	14.1-14.7
	Pick-up	G54B	45	Internal Electronic	14.1-14.7
1987	Montero	G54B	50	Internal Electronic	13.9-14.9
	Pick-up	G63B	45	Internal Electronic	14.1-14.7
	Pick-up	G54B	45	Internal Electronic	14.1-14.7
1988	Montero	G54B	50	Internal Electronic	13.9-14.9
	Pick-up	G63B	45	Internal Electronic	14.1-14.7
	Pick-up	G54B	45	Internal Electronic	14.1-14.7
1989	Montero	G54B	50	Internal Electronic	13.9-14.9
	Montero	6G72	75	Internal Electronic	13.9-14.9
	Pick-up	G63B	45	Internal Electronic	14.1-14.7
	Pick-up	G54B	45	Internal Electronic	14.1-14.7
1990	Montero	G54B	50	Internal Electronic	13.9-14.9
	Montero	6G72	75	Internal Electronic	13.9-14.9
	Pick-up	4G64	45 [2]	Internal Electronic	13.9-14.9
	Pick-up	6G72	65	Internal Electronic	13.9-14.9
1991	Montero	6G72	75	Internal Electronic	13.9-14.9
	Pick-up	4G64	40	Internal Electronic	13.9-14.9
	Pick-up	6G72	65	Internal Electronic	13.9-14.9
1992	Montero	6G72	75	Internal Electronic	13.9-14.9
	Pick-up	4G64	40	Internal Electronic	13.9-14.9
	Pick-up	6G72	65	Internal Electronic	13.9-14.9
1993	Montero	6G72	75	Internal Electronic	13.9-14.9
	Pick-up	4G64	40	Internal Electronic	13.9-14.9
	Pick-up	6G72	65	Internal Electronic	13.9-14.9
1994	Montero	6G72	90	Internal Electronic	13.9-14.9
	Montero	6G74	90	Internal Electronic	13.9-14.9
	Pick-up	4G64	40	Internal Electronic	13.9-14.9
	Pick-up	6G72	65	Internal Electronic	13.9-14.9
1995	Montero	6G72	90	Internal Electronic	13.9-14.9
	Montero	6G74	90	Internal Electronic	13.9-14.9
	Pick-up	4G64	40	Internal Electronic	13.9-14.9

[1] 55 amps w/ Automatic Transmission
[2] 60 amps w/ Power-Up option

3-14 ENGINE AND ENGINE REBUILDING

To install:

5. Reposition the battery in the battery tray and install the mounting brackets and nuts. Make sure the battery is mounted tightly to prevent it from sliding around or vibration damage. Clean the battery terminal posts and cable clamps with a brush designed for this purpose.

6. Connect the cables to the battery terminals. When connecting the cables, make sure that the positive cable is attached to the positive battery terminal (+) and the negative cable is fastened to the negative terminal (-). Tighten the connectors securely. Then coat both connectors with petroleum grease.

Starter

DIAGNOSIS

Starter Won't Crank The Engine

1. Dead battery.
2. Open starter circuit, such as:
   a. Broken or loose battery cables.
   b. Inoperative starter motor solenoid.
   c. Broken or loose wire from ignition switch to solenoid.
   d. Poor solenoid or starter ground.
   e. Bad ignition switch.
3. Defective starter internal circuit, such as:
   a. Dirty or burnt commutator.
   b. Stuck, worn or broken brushes.
   c. Open or shorted armature.
   d. Open or grounded fields.
4. Starter motor mechanical faults, such as:
   a. Jammed armature end bearings.
   b. Bad bearings, allowing armature to rub fields.
   c. Bent shaft.
   d. Broken starter housing.
   e. Bad starter drive mechanism.
   f. Bad starter drive or flywheel-driven gear.
5. Engine hard or impossible to crank, such as:
   a. Hydrostatic lock, water in combustion chamber.
   b. Crankshaft seizing in bearings.
   c. Piston or ring seizing.
   d. Bent or broken connecting rod.
   e. Seizing of connecting rod bearings.
   f. Flywheel jammed or broken.

Starter Spins Freely, Won't Engage

1. Sticking or broken drive mechanism.
2. Damaged ring gear.

REMOVAL & INSTALLATION

▶ See Figure 20

The starter motor is located on either the lower right hand (passenger) or left hand (driver) side of the engine, bolted to the transmission housing with two mounting bolts.

1. Disconnect the negative battery cable.

➡**Removal and installation of the starter motor will require access from both the top and bottom of the engine compartment.**

2. Apply the parking brake and block the rear wheels.
3. Raise the front of the vehicle with a floor jack, using the proper frame positions.
4. Support the vehicle on the frame or transmission crossmember, with jackstands. Before working under the vehicle, make sure the vehicle is stable on the jackstands. If it is not, reposition the jackstands so that it is.
5. Label and disconnect all wiring connectors at the starter.
6. Slowly loosen the two starter mounting bolts until they are loose enough to take out by hand. Since the starter motor is uncommonly heavy for its size, it is important to support it well with one hand, while removing the two mounting bolts with the other hand. If the starter is not adequately supported, it could fall causing injury. Remove the starter from the vehicle.

To install:

7. Clean the surfaces of the starter motor flange and the flywheel housing where the starter attaches.
8. Reinstall the starter and install the retaining bolts. Tighten the bolts to 16-23 ft. lbs. (22-31 Nm).
9. Connect the wiring to the starter, making sure the terminals and connectors are clean and tight.
10. Lower the truck to the ground
11. Connect the negative battery cable.

SOLENOID REPLACEMENT

▶ See Figures 21, 22 and 23

Under some circumstances, the solenoid or magnetic switch may fail. This can be indicated by a failure of the starter to engage or produce adequate power. To test solenoid failure, disconnect the heavy starter motor wire at the **M** terminal on the solenoid. Run a jumper wire from the battery (+) terminal to the **S** terminal of the solenoid. If possible, use a remote starter control with a switch built into it. If this is not available, touch the wire to the **S** terminal briefly but do not leave the solenoid engaged for more than 10 seconds.

If the solenoid now engages in a positive manner, inspect the wiring and ignition switch for defects. (You have proven that the solenoid works if it gets the correct message.) If not, the solenoid should be replaced. You can test the solenoid switch itself by pulling the coil wire out of the distributor cap, having someone engage the starter, and then measuring the voltage at both the **B** and **M** terminals of the solenoid (**M** terminal wire connected). If voltage is close to battery voltage at the **B** terminal, but drops significantly at the **M** terminal with

ENGINE AND ENGINE REBUILDING 3-15

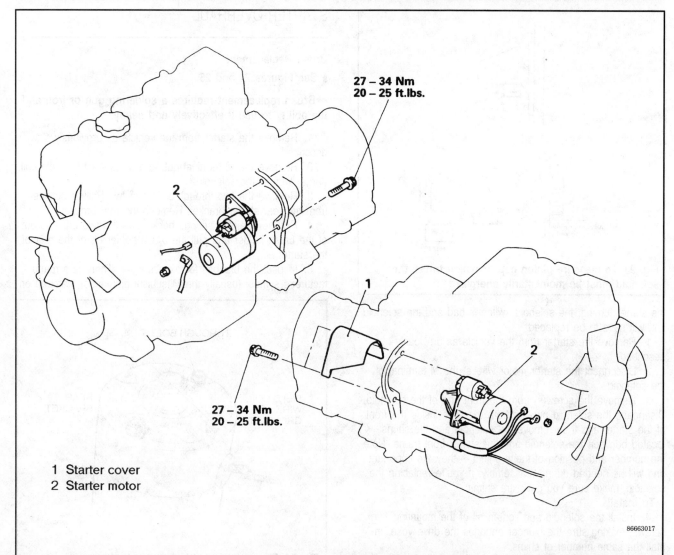

1 Starter cover
2 Starter motor

Fig. 20 Common starter location and installation — the starter is a fairly heavy component, be careful when lifting it out of the vehicle

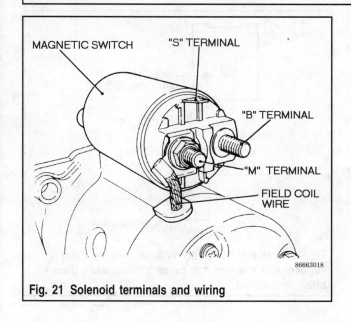

Fig. 21 Solenoid terminals and wiring

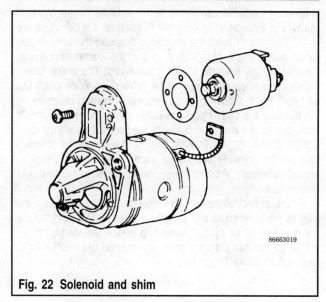

Fig. 22 Solenoid and shim

3-16 ENGINE AND ENGINE REBUILDING

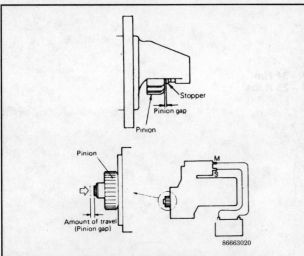

Fig. 23 To measure pinion gap or pinion travel, the solenoid must be momentarily energized

the starter turning, the solenoid switch is bad and the solenoid unit will have to be replaced.

1. Remove the starter from the vehicle as previously described.
2. Disconnect the starter motor wire at the **M** terminal of the solenoid.
3. Remove the screw(s) from the front end of the solenoid. Disengage the solenoid plunger from the yoke inside the front of the starter and then remove the solenoid and the shims located between the solenoid and the starter front frame. Note the number and position of these shims — they are important and will be needed during reassembly. If you're replacing the solenoid, make sure you get extra shims.

To install:

4. Install the solenoid and tighten all of the mounting screws, making sure the plunger engages the drive yoke. Install the same number of shims.
5. Energize the solenoid by running jumper wires — including a switch if possible — from the (+) terminal of a 12 volt battery to the **S** terminal of the solenoid and from the (-) terminal of the battery to the **M** terminal. Make certain the field coil wire is disconnected from the **M** terminal. Quickly measure (in 10 seconds or less) the clearance between the front of the pinion gear and the stop in front of it in the starter front frame. De-energize the solenoid before it overheats. The pinion gear should be pushed back against the drive mechanism when you do this. On reduction gear starters you'll have to measure the distance the pinion gear assembly travels when you shift it back and forth. Use a flat feeler gauge; correct clearance is 0.020-0.079 in. (0.50-2.00mm). Change the number of shims between the solenoid and starter frame to correct the clearance if necessary. Adding shims decreases the clearance, and vice-versa.
6. Disconnect all test wiring hook-ups. Connect the field coil wire to the **M** terminal and reinstall the starter. Make certain the matching faces of the starter and engine are clean; any grit or grease can act as a shim and change the position of the starter relative to the engine.

STARTER OVERHAUL

Brush Replacement

▶ See Figures 24 and 25

➡ Brush replacement requires a soldering gun or iron and the ability to use it effectively and neatly.

1. Remove the starter from the vehicle as previously described.
2. Remove the **M** terminal nut, and disconnect the field coil (large) wire at the solenoid.
3. Remove the two through-bolts and two Phillips screws from the rear starter bracket. Remove the rear bracket.
4. Pry the retaining springs back, slide the two brushes out of the brush holder and pull the brush holder off of the rear of the starter.
5. Inspect the brushes for excess wear. There is a manufacturer's symbol (usually the Mitsubishi diamonds) stamped on

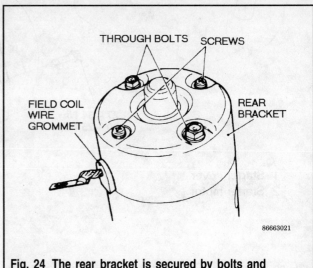

Fig. 24 The rear bracket is secured by bolts and screws

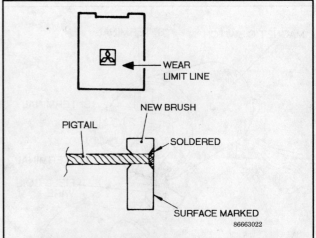

Fig. 25 Note that the solder is flush with the outer surface. Any overflow can cause binding when the brush is installed

ENGINE AND ENGINE REBUILDING   3-17

the side of each. If the brush is worn to the bottom of the emblem it should be replaced.

6. To replace the brush, it must be crushed with a pair of pliers to crack it where the wiring pigtail passes through the brush. Be careful not to damage the wiring pigtail in doing this. Use sandpaper to sand the end of the pigtail smooth. Also sand the outer surface of the last 0.25 in. (6mm) or so of the pigtail wire until it is bright and free of corrosion.

To install:

7. Insert the pigtail into the hole in the new brush until the flat end of the pigtail just reaches the opposite end of the hole in the brush. Insert the wire from the unmarked side of the brush. The brush and pigtail must be brought to just the right temperature for the solder to run in between the brush and pigtail. Make sure solder does not get onto the outer surface of the brush, as this could cause it to bind in the brush holder later.

8. Install the brushes into the holders.

9. Assemble the rear case of the starter and install bolts and screws. Connect the field coil wire to the M terminal of the solenoid.

10. Reinstall the starter.

Drive Replacement

The starter drive may need to be replaced if the starter motor turns but does not engage properly. If, in removing the drive, damage to the starter pinion gear is noted, the flywheel ring gear should also be inspected. If there is significant damage to the ring gear, the flywheel will have to be replaced, too.

When diagnosing apparent starter problems, test the battery first. A low battery or weak connections can prevent the required amount of current from getting to the starter motor. Also check the solenoid before condemning the starter drive itself; a solenoid that does not engage properly may cause the same symptoms.

DIRECT DRIVE STARTER

◆ See Figure 26

1. Remove the starter and remove the solenoid.
2. Remove the 2 through-bolts and 2 screws from the rear bracket. Remove the rear bracket.
3. Pry back the retaining rings and slide the 2 brushes out of the brush holder. Remove the brush holder and the yoke assembly.
4. Remove the washer from the rear of the armature. Remove the field coil assembly from the front frame. Remove the spring retainer, spring, and spring seat from the starter front frame.
5. Separate the armature from the front bracket by first pulling the armature back out of the front bearing and then shifting the armature so the starter drive is pulled out of the yoke. Make sure you don't lose the washer located in the front frame.
6. Invert the armature so the starter drive is on top and rest the rear of the armature on a solid surface. Use a deep well socket wrench that is just slightly larger than the diameter of the armature shaft to press the snap-ring collar back. Install the socket over the top of the shaft and then press it downward or tap it very lightly to force the ring downward. Once the snap-ring is exposed, use snap-ring pliers to open it until it will slide upward, out of the groove and off the shaft. Pull the starter drive and snap-ring collar upward and off the armature shaft.

7. With the starter disassembled, do not immerse parts in cleaning solvent. The yoke and field coil assembly will be damaged. Wipe the parts with a clean cloth. The overrun clutch is packed with lubricant which will be washed out by any solvent or fluid.

8. Use a micrometer or caliper to measure the diameter of the commutator. If below the limit, the commutator must be replaced.

To install:

9. Lightly coat the front of the armature shaft with high temperature grease. Install the starter drive, snap-ring collar, and snap-ring. Make sure the snap-ring seats in its groove. Then use a puller to pull the snap-ring collar up and over the snap-ring until the bottom of the collar touches the snap-ring.

10. Place the washer in position in the front frame. Insert the armature through the lever and yoke. Make certain the armature is correctly seated in the bearing.

11. Install the spring seat, spring and spring retainer.

12. Install the field coil assembly to the front frame. Place the washer on the rear of the armature.

13. Install the brush holder and yoke assemblies and install the brushes.

14. Position the rear bracket and install the 2 screws and 2 through-bolts.

15. Install the solenoid and connect the field coil wire.

REDUCTION GEAR STARTER

◆ See Figures 27, 28, 29, 30, 31 and 32

1. Remove the starter from the vehicle as described above. Remove the solenoid.
2. Remove the two through-bolts and two Phillips screws from the rear starter bracket. Remove the rear bracket.
3. Pry the retaining springs back and slide the two brushes out of the brush holder. Pull the brush holder off the rear of the starter. Remove the field coil (yoke) assembly from the front frame. Remove the armature.
4. Remove the pinion shaft end cover from the center frame. Measure the clearance between the spacer and center cover and record it. If the pinion shaft is replaced, you'll have to insert or subtract spacer washers until the clearance is the same as that recorded. Use a screwdriver to remove the retaining clip and then remove the washers. Remove the retaining bolt and then separate the center frame from the front frame.
5. Remove the spring retainer and spring for the yoke from the front frame. Then remove the washer, reduction gear, shift yoke lever, and two lever supports.
6. Turn the front frame so the pinion gear is at the top and support it securely. Use a socket that fits tightly over the pinion shaft to force the snap-ring collar (stop ring) downward. Tap the socket lightly at the top or use a press to do this. Use a small prytool to work the snap-ring out of its groove and remove it from the shaft (a pair of small snapring pliers will make this procedure much easier). Remove the collar. Remove the pinion and the spring behind it from the shaft.

3-18 ENGINE AND ENGINE REBUILDING

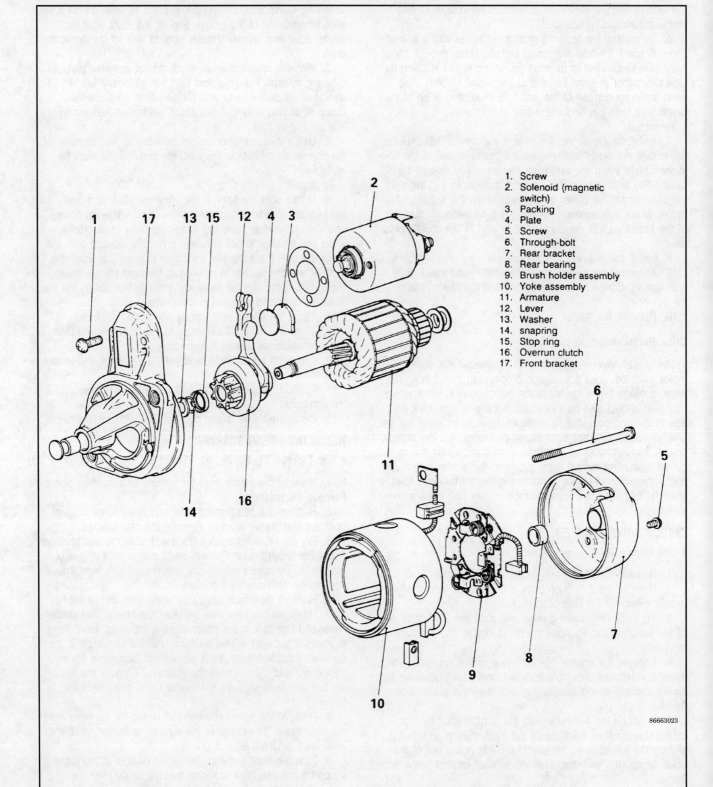

1. Screw
2. Solenoid (magnetic switch)
3. Packing
4. Plate
5. Screw
6. Through-bolt
7. Rear bracket
8. Rear bearing
9. Brush holder assembly
10. Yoke assembly
11. Armature
12. Lever
13. Washer
14. snapring
15. Stop ring
16. Overrun clutch
17. Front bracket

Fig. 26 An exploded view of a direct drive starter

ENGINE AND ENGINE REBUILDING 3-19

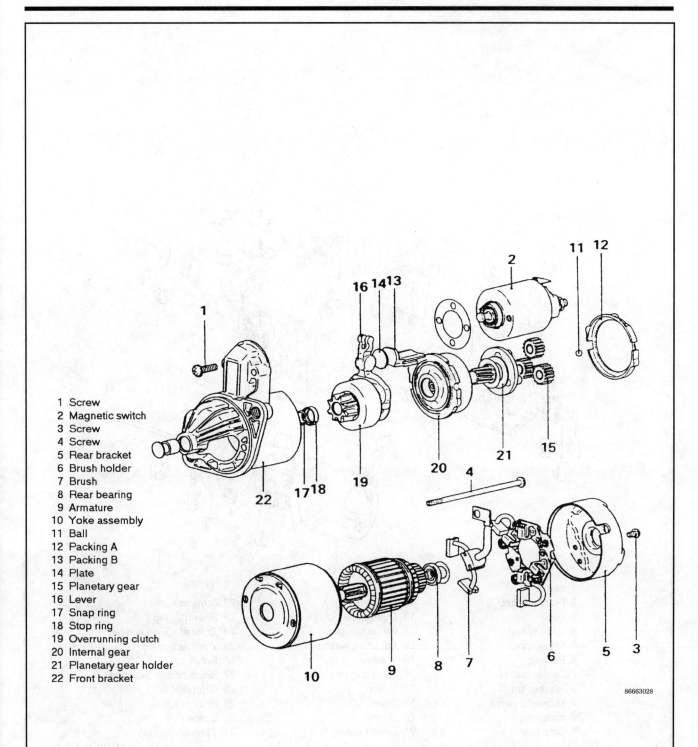

1 Screw
2 Magnetic switch
3 Screw
4 Screw
5 Rear bracket
6 Brush holder
7 Brush
8 Rear bearing
9 Armature
10 Yoke assembly
11 Ball
12 Packing A
13 Packing B
14 Plate
15 Planetary gear
16 Lever
17 Snap ring
18 Stop ring
19 Overrunning clutch
20 Internal gear
21 Planetary gear holder
22 Front bracket

Fig. 27 An exploded view of a reduction drive starter

3-20 ENGINE AND ENGINE REBUILDING

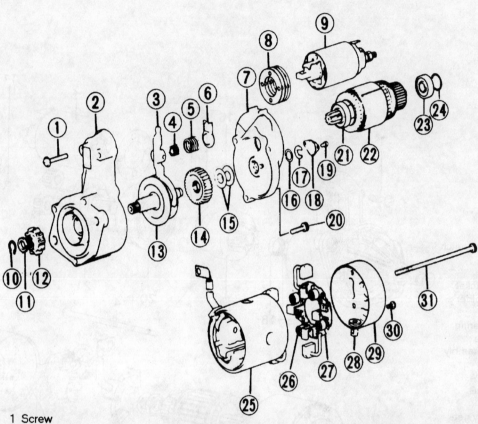

1 Screw
2 Front bracket
3 Lever
4 Inner spring
5 Outer spring
6 Packing
7 Center braket
8 Washer set
9 Magnetic switch
10 Snap ring
11 Stop ring
12 Pinion
13 Overrunning clutch
14 Reduction gear
15 Adjusting washer
16 Washer
17 Retaining ring
18 Cover
19 Screw (2)
20 Screw
21 Front bearing
22 Armature
23 Rear bearing
24 Conical spring washer
25 Yoke assembly
26 Brush spring
27 Brush holder assembly
28 Grommet
29 Rear bracket
30 Screw
31 Through bolt (2)

86663029

Fig. 28 Diesel engine starter is similar, but has more power

ENGINE AND ENGINE REBUILDING   3-21

7. Pull the lever and pinion shaft assembly out of the rear of the front frame. Replace the pinion if its teeth are damaged (check the flywheel ring gear as well). Replace the overrunning clutch if the pinion gear is damaged or if the one-way action of the clutch is not precise.

8. Do not immerse parts in cleaning solvent. The yoke and field coil assembly will be damaged. Wipe the parts with a clean cloth. The overrun clutch is packed with lubricant which will be washed out by any solvent or fluid.

9. Use a micrometer or caliper to measure the diameter of the commutator and compare to the specifications chart. If below the limit, the commutator must be replaced.

To install:

10. Lightly coat the front of the armature shaft with high temperature grease. Install the pinion shaft, spring, gear and stop ring. Install the snapring and use a puller to seat the stop ring over the snapring.

11. Place the washer in position in the front frame. Insert the armature through the lever and yoke. Make certain the armature is correctly seated in the bearing.

12. Install the lever and pinion shaft assembly with the reduction gear into the rear of the front frame. Make certain all the springs, spacers and washers are present and in the correct order. When the clearance is correct, install the cover and its small screw.

13. Fit the center frame onto the shaft and install the washer and retaining clip. Note that the clearance must be corrected by changing the thickness or number of washers if the overrunning clutch and pinion shaft assembly have been replaced. Double check the clearance and install or remove shims as necessary.

14. Install the armature into the yoke (field coil). Install the brush holder and install the brushes, making sure they are properly seated and do not bind in the holders.

15. Install the rear bracket and the small screws. Assemble the motor to the drive and install the through-bolts.

16. Install the solenoid and connect the field coil wire to the M terminal.

17. Reinstall the starter.

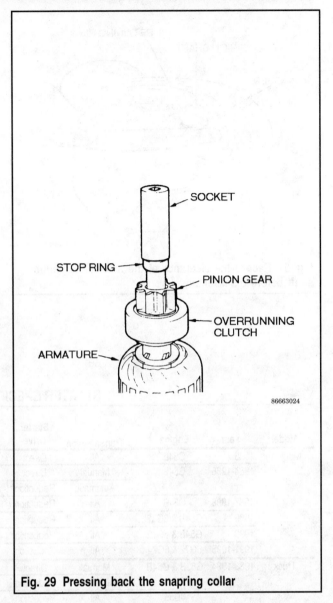

Fig. 29 Pressing back the snapring collar

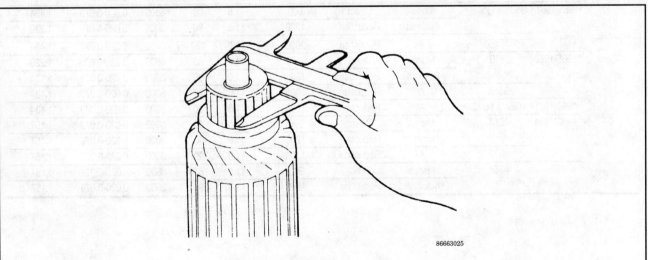

Fig. 30 Measuring commutator diameter — take measurements at 2 or 3 points and use smallest diameter as reference

3-22 ENGINE AND ENGINE REBUILDING

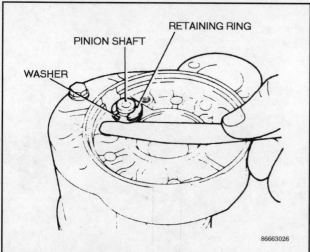

Fig. 31 Record the clearance at the rear of the pinion shaft before disassembly

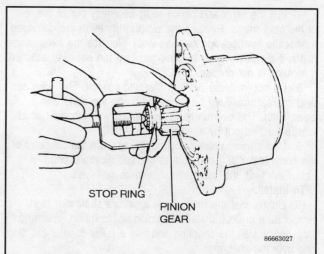

Fig. 32 Use a small puller to bring the stop ring over the snapring

STARTER SPECIFICATIONS CHART

Model	Year	Engine	Transmission	Starter Drive	Nominal Watts @ 12.0V	Free Running Test Volts	Free Running Test Max. Amps	Free Running Test Minimum RPM	Pinion Gap or Travel (in.)	Commutator Minimum Diameter (in.)
Montero	1983	G54B	All	Direct	900	11.5	60	6500	0.020-0.079	1.484
	1984-1986	G54B	Manual	Direct	900	11.5	60	6500	0.020-0.079	1.484
			Automatic	Reduction	1200	11.5	100	3000	0.020-0.079	1.220
	1987-1988	G54B	All	Reduction	1200	11.0	90	3000	0.020-0.079	1.134
	1989	G54B & 6G72	All	Reduction	1200	11.0	90	3000	0.020-0.079	1.134
	1990	G54B & 6G72	All	Reduction	1200	11.0	90	3000	0.020-0.079	1.134
	1991-1995	6G72 & 6G74	All	Reduction	1200	11.0	90	3000	0.020-0.079	1.157
Truck	1983-1984	G63B & G54B	Manual	Direct	900	11.5	60	6500	0.020-0.079	1.484
			Automatic	Reduction	1200	11.5	100	3000	0.020-0.079	1.220
		4D55	All	Reduction	2000	11.0	130	4000	0.020-0.079	1.484
	1985-1987	G63B & G54B	Manual	Direct	900	11.5	60	6500	0.020-0.079	1.484
			Automatic	Reduction	1200	11.5	90	3000	0.020-0.079	1.220
		4D55	All	Reduction	2000	11.0	130	4000	0.020-0.079	1.484
	1988-1989	G63B	Manual	Direct	900	11.5	60	6600	0.020-0.079	1.220
			Automatic	Reduction	1200	11.0	90	3000	0.020-0.079	1.134
		G54B	All	Reduction	1200	11.0	90	3000	0.020-0.079	1.134
	1990-1991	4G64 & 6G72	All	Reduction	1200	11.0	90	3000	0.020-0.079	1.134
	1992-1994	4G64	Manual	Direct	900	11.5	60	6600	0.020-0.079	1.157
			Automatic	Reduction	1200	11.0	90	3000	0.020-0.079	1.157
		6G72	All	Reduction	1200	11.0	90	3000	0.020-0.079	1.157
	1995	4G64	Manual	Direct	900	11.5	60	6600	0.020-0.079	1.157
			Automatic	Reduction	1200	11.0	90	3000	0.020-0.079	1.157

1 1985 only

ENGINE AND ENGINE REBUILDING

Sending Units and Sensors

REMOVAL & INSTALLATION

Engine Coolant Temperature Gauge Sensor

▶ See Figures 33, 34, 35, 36 and 37

The engine coolant temperature gauge sensors on Mitsubishi trucks and Monteros are all located on the intake manifold, except for the 4D55 diesel engine (1983-85 models), which is located on the cylinder head. See the diagrams for the specific model location.

1. Disconnect the negative battery cable.
2. Drain the cooling system.
3. Disconnect the electrical connector from the sensor.
4. Unscrew the sensor from the intake manifold or cylinder head.

To install:

5. Clean the sensor's threads to remove any corrosion, dirt or other contaminants.
6. Apply a liquid sealant to the threads of the sensor and gently screw into its hole. Make sure not to crossthread the sensor.
7. Tighten the sensor to the correct torque (see figure) and refasten the sensor's wiring.
8. Refill the cooling system as described in Section 1.
9. Reconnect the cable to the negative battery terminal.

Oil Pressure Sender

▶ See Figures 38 and 39

The oil pressure sender unit is located in the same area on all Mitsubishi engines. It is located on the lower front of the engine, either mounted onto the engine vertically on the oil filter galley or horizontally next to the oil filter itself.

1. Disconnect the negative battery cable.
2. Unfasten the electrical connector from the oil pressure sender.
3. Remove the oil pressure sender from the engine.

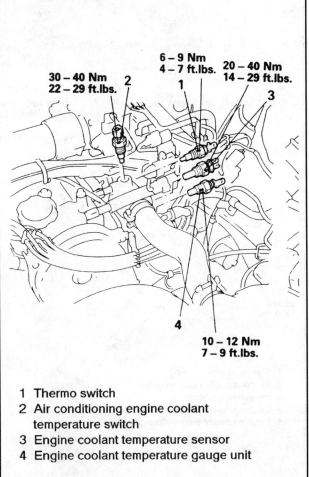

1. Thermo switch
2. Air conditioning engine coolant temperature switch
3. Engine coolant temperature sensor
4. Engine coolant temperature gauge unit

Fig. 34 Coolant system sensor locations — 2.4L engines

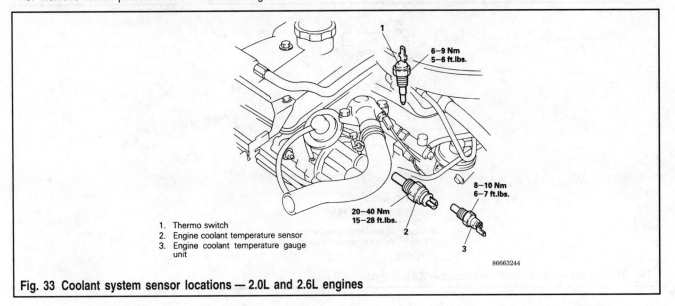

1. Thermo switch
2. Engine coolant temperature sensor
3. Engine coolant temperature gauge unit

Fig. 33 Coolant system sensor locations — 2.0L and 2.6L engines

3-24 ENGINE AND ENGINE REBUILDING

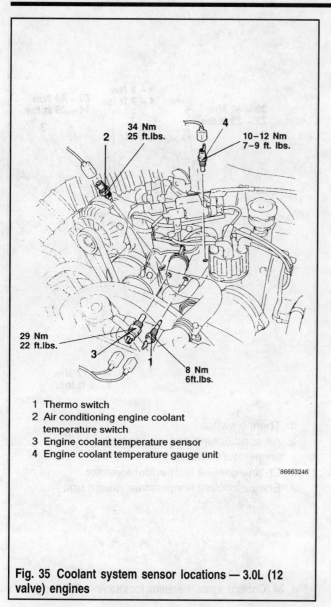

1. Thermo switch
2. Air conditioning engine coolant temperature switch
3. Engine coolant temperature sensor
4. Engine coolant temperature gauge unit

Fig. 35 Coolant system sensor locations — 3.0L (12 valve) engines

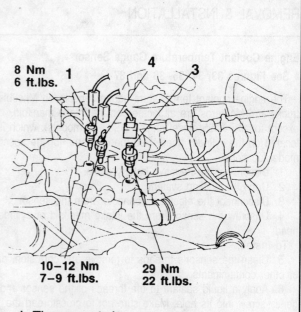

1. Thermo switch
2. Air conditioning engine coolant temperature switch
3. Engine coolant temperature sensor
4. Engine coolant temperature gauge unit

Fig. 36 Coolant system sensor locations — 3.0L (24 valve) engines

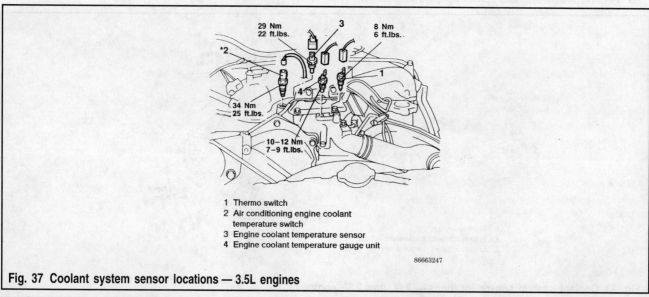

1. Thermo switch
2. Air conditioning engine coolant temperature switch
3. Engine coolant temperature sensor
4. Engine coolant temperature gauge unit

Fig. 37 Coolant system sensor locations — 3.5L engines

ENGINE AND ENGINE REBUILDING 3-25

To install:

4. Clean the sensor's threads of all dirt, oil and other contaminants.

5. Apply a thin film of a liquid sealant to the threads, making sure not to use so much that some could drip into the engine and perhaps block an oil galley.

6. Tighten the oil pressure sensor snugly into its mounting hole. Do not crossthread the sensor.

7. Reconnect the sensor's wiring, and hook the negative battery cable back up to the battery.

8. Check the engine oil level.

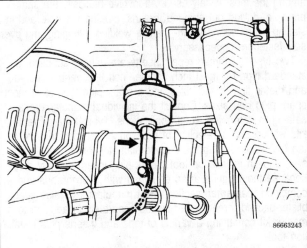

Fig. 38 Disconnecting the oil pressure gauge wiring — 2.0L and 2.6L engines

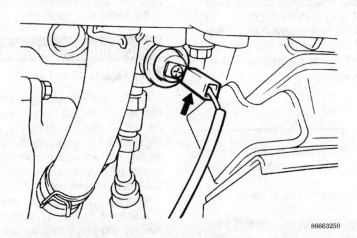

Fig. 39 Disconnecting the oil pressure gauge wiring — 2.3L diesel engine

ENGINE MECHANICAL

Engine Overhaul Tips

Most engine overhaul procedures are fairly standard. In addition to specific parts replacement procedures and specifications for your individual engine, this section is also a guide to accepted rebuilding procedures. Examples of standard rebuilding practices are shown and should be used along with specific details concerning your particular engine.

Competent and accurate machine shop services will ensure maximum performance, reliability and engine life. In most instances it is more profitable for the do-it-yourself mechanic to remove, clean and inspect the component, buy the necessary parts and deliver these to a shop for actual machine work.

On the other hand, much of the rebuilding work (crankshaft, block, bearings, piston rods, and other components) is well within the scope of the do-it-yourself mechanic. Patience, proper tools, and common sense coupled with a basic understanding of the engine can yield satisfying and economical results.

TOOLS

The tools required for an engine overhaul or parts replacement will depend on the depth of your involvement. With a few exceptions, they will be the tools found in a mechanic's tool kit (see Section 1). More in-depth work will require any or all of the following:.

- A dial indicator (reading in thousandths) mounted on a universal base
- Micrometers and telescope gauges
- Jaw and screw-type pullers
- Gasket scrapers; the best are wood or plastic
- Valve spring compressor

3-26 ENGINE AND ENGINE REBUILDING

- Ring groove cleaner
- Piston ring expander and compressor
- Ridge reamer
- Cylinder hone or glaze breaker
- Plastigage®
- Engine stand

The use of most of these tools is illustrated in this section. Many can be rented for a one-time use from a local parts jobber or tool supply house specializing in automotive work.

Occasionally, the use of special tools is called for. See the information on Special Tools and Safety Notice in the front of this manual before substituting another tool.

INSPECTION TECHNIQUES

Procedures and specifications are given in this section for inspecting, cleaning and assessing the wear limits of most major components. Other procedures such as Magnaflux® and Zyglo® can be used to locate material flaws and stress cracks. Magnaflux® is a magnetic process applicable only to ferrous (iron and steel) materials. The Zyglo® process coats the material with a fluorescent dye penetrant and can be used on any material. Checks for suspected surface cracks can be more readily made using spot check dye. The dye is sprayed onto the suspected area, wiped off and the area sprayed with a developer. Cracks will show up brightly.

OVERHAUL TIPS

Aluminum has become extremely popular for use in engines, due to its low weight. Observe the following precautions when handling aluminum parts:

- Never hot tank aluminum parts (the caustic hot tank solution will eat the aluminum.).
- Remove all aluminum parts (identification tag, etc.) from engine parts prior to the tanking.
- Always coat threads lightly with engine oil or anti-seize compounds before installation to prevent seizure.
- Never overtighten bolts or spark plugs, especially in aluminum threads.

Stripped threads in any component can usually be repaired using any of several commercial repair kits (Heli-Coil®, Microdot®, Keenserts®, etc.).

When assembling the engine, any parts that will be in frictional contact must be pre-lubed to provide lubrication at initial start-up. Any product specifically formulated for this purpose can be used, but engine oil is not recommended as a pre-lube.

When semi-permanent (locked, but removable) installation of bolts or nuts is desired, threads should be cleaned and coated with Loctite® or other similar, commercial non-hardening sealant.

REPAIRING DAMAGED THREADS

◆ See Figures 40, 41, 42, 43 and 44

Several methods of repairing damaged threads are available. Heli-Coil® (shown here), Keenserts® and Microdot® are among the most widely used. All involve basically the same principle--drilling out stripped threads, tapping the hole and installing a pre-wound insert--making welding, plugging and oversize fasteners unnecessary.

Two types of thread repair inserts are usually supplied: a standard type for most inch coarse, inch fine, metric course and metric fine thread sizes and a spark plug type to fit most spark plug port sizes. Consult the individual manufacturer's catalog to determine exact applications. Typical thread repair kits will contain a selection of pre-wound threaded inserts, a tap (corresponding to the outside diameter threads of the insert) and an installation tool. Spark plug inserts usually differ because they require a tap equipped with pilot threads and a combined reamer/tap section. Most manufacturers also supply blister-packed thread repair inserts separately in addition to a master kit containing a variety of taps and inserts plus installation tools.

Before effecting a repair to a threaded hole, remove any snapped, broken or damaged bolts or studs. Penetrating oil can be used to free frozen threads. The offending item can be removed with locking pliers or with a screw or stud extractor. After the hole is clear, the thread can be repaired, as shown in the series of accompanying illustrations.

Checking Engine Compression

◆ See Figures 45 and 46

A noticeable lack of engine power, excessive oil consumption and/or poor fuel mileage measured over an extended period are all indicators of internal engine wear. Worn piston rings, scored or worn cylinder bores, leaking head gaskets, sticking or burnt valves and worn valve seats are all possible culprits here. A check of each cylinder's compression will help you locate the problems.

As mentioned in the Tools and Equipment section of Section 1, a screw-in type compression gauge is more accurate that the type you simply hold against the spark plug hole, although it takes slightly longer to use. It's worth it to obtain a more accurate reading.

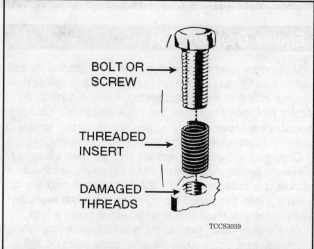

Fig. 40 Damaged bolt holes can be repaired with thread repair inserts

ENGINE AND ENGINE REBUILDING

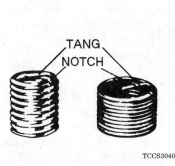

Fig. 41 Standard thread repair insert (left) and spark plug thread insert (right)

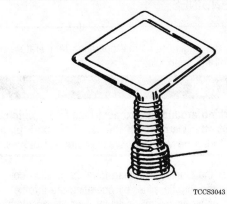

Fig. 44 Screw the threaded insert onto the installation tool until the tang engages the slot — Screw the insert into the tapped hole until it is 1/4-1/2 turn below the top surface. After installation break off the tang with a hammer and punch

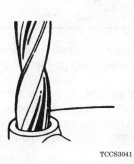

Fig. 42 Drill out the damaged threads with specified drill bit — Be sure to drill completely through the hole or to the bottom of a blind hole

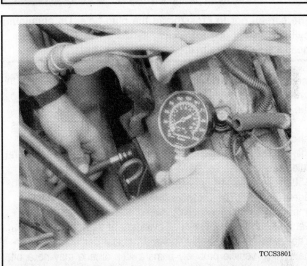

Fig. 45 The screw-in type compression gauge is more accurate

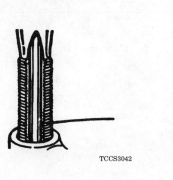

Fig. 43 With the tap supplied, tap the hole to receive the thread insert — Keep the tap well oiled and back it out frequently to avoid clogging the threads

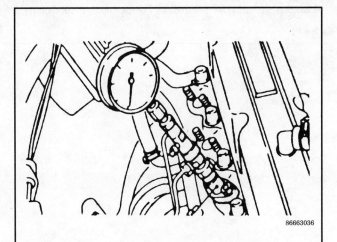

Fig. 46 Diesel engines require a special compression gauge adaptor

ENGINE AND ENGINE REBUILDING

GASOLINE ENGINES

1. Warm up the engine to normal operating temperature.
2. Remove all the spark plugs.

> **✳✳CAUTION**
>
> Be careful working around the exhaust manifolds, which will be very hot after warming up the car. The spark plugs themselves will also be very hot. Wear protective gloves.

3. Disconnect the high tension lead from the ignition coil.
4. Fully open the throttle either by operating the throttle linkage by hand or by having an assistant floor the accelerator pedal.
5. Screw the compression gauge into the No.1 spark plug hole until the fitting is snug.

> **✳✳WARNING**
>
> Be careful not to crossthread the plug hole. On aluminum cylinder heads use extra care, as the threads in these heads are easily ruined.

6. While you read the compression gauge, ask an assistant to crank the engine two or three times in short bursts using the ignition switch.
7. Read the compression gauge at the end of each series of cranks, and record the highest of these readings. Repeat this procedure for each of the engine's cylinders. As a general rule, new engines will have compression on the order of 150-170 psi (1034-1171 kPa). This number will decrease with age and wear. The pressure that your test shows is not as important as the evenness between all the cylinders. Many vehicles run very well with all cylinders at 105 psi (723.5 kPa). The lower number simply shows a general deterioration internally. This vehicle probably burns a little oil and may be a bit harder to start, but, based on these numbers, doesn't warrant an engine tear-down yet.   Compare the highest reading of all the cylinders. Any variation of more than 10% should be considered a sign of potential trouble. For example, if your compression readings for cylinders 1 through 4 were: 135 psi (930 kPa), 125 psi (861 kPa), 90 psi (620 kPa) and 125 psi (861 kPa), it would be fair to say that cylinder number three is not working efficiently and is almost certainly the cause of your oil burning, rough idle or poor fuel mileage.
8. If a cylinder is unusually low, pour a tablespoon of clean engine oil into the cylinder through the spark plug hole and repeat the compression test. If the compression comes up after adding the oil, it appears that the cylinder's piston rings or bore are damaged or worn. If the pressure remains low, the valves may not be seating properly (a valve job is needed), or the head gasket may be blown near that cylinder. If compression in any two adjacent cylinders is low, and if the addition of oil doesn't help the compression, there might be leakage past the head gasket. Oil and coolant in the combustion chamber can result from this problem. There may be evidence of water droplets on the engine dipstick when a head gasket has blown.
9. After the compression check is complete, install the spark plugs and hook their perspective spark plug cables back up.

DIESEL ENGINES

Checking cylinder compression on diesel engines is basically the same procedure as on gasoline engines except for the following:

1. A special compression gauge adaptor suitable for diesel engines (because these engines have much greater compression pressures) must be used.
2. Remove the injector tubes and remove the injectors from each cylinder.

➡ **Don't forget to remove the washer underneath each injector; otherwise, it may get lost when the engine is cranked.**

3. When fitting the compression gauge adaptor to the cylinder head, make sure the bleeder of the gauge (if equipped) is closed.
4. When reinstalling the injector assemblies, install new washers underneath each injector.

ENGINE AND ENGINE REBUILDING

GENERAL ENGINE SPECIFICATIONS

Year	Engine ID/VIN		Engine Displacement Liters (cc)	Fuel System Type	Net Horsepower @ rpm	Net Torque @ rpm (ft. lbs.)	Bore x Stroke (in.)	Compression Ratio	Oil Pressure @ rpm	
1983	G63B		2.0 (1997)	2 BBL	93 @ 5200	108 @ 3000	3.35 x 3.46	8.5:1	11 @ idle	1
	4D55		2.3 (2346)	DSL	84 @ 4200	136 @ 2500	3.59 x 3.54	21.0:1	56 @ 2000	
	G54B	2	2.6 (2555)	2 BBL	105 @ 5000	136 @ 2500	3.59 x 3.54	8.2:1	11 @ idle	1
1984	G63B		2.0 (1997)	2 BBL	93 @ 5200	108 @ 3000	3.35 x 3.46	8.5:1	11 @ idle	1
	4D55		2.3 (2346)	DSL	84 @ 4200	136 @ 2500	3.59 x 3.54	21.0:1	56 @ 2000	
	G54B	2	2.6 (2555)	2 BBL	105 @ 5000	136 @ 2500	3.59 x 3.54	8.2:1	11 @ idle	1
1985	G63B		2.0 (1997)	2 BBL	93 @ 5200	108 @ 3000	3.35 x 3.46	8.5:1	11 @ idle	1
	4D55		2.3 (2346)	DSL	84 @ 4200	136 @ 2500	3.59 x 3.54	21.0:1	56 @ 2000	
	G54B	3	2.6 (2555)	2 BBL	105 @ 5000	136 @ 2500	3.59 x 3.86	8.2:1	11 @ idle	1
	G54B	4	2.6 (2555)	2 BBL	108 @ 5000	142 @ 2500	3.59 x 3.86	8.7:1	11 @ idle	1
1986	G63B		2.0 (1997)	2 BBL	90 @ 5000	108 @ 3000	3.35 x 3.46	8.5:1	11 @ idle	1
	G54B	3	2.6 (2555)	2 BBL	105 @ 5000	139 @ 2500	3.59 x 3.86	8.2:1	11 @ idle	1
	G54B	4	2.6 (2555)	2 BBL	108 @ 5000	142 @ 2500	3.59 x 3.86	8.7:1	11 @ idle	1
1987	G63B		2.0 (1997)	2 BBL	90 @ 5500	109 @ 3500	3.35 x 3.46	8.5:1	49 @ 2000	
	G54B	2	2.6 (2555)	2 BBL	109 @ 5000	142 @ 2500	3.59 x 3.86	8.7:1	56 @ 2000	
1988	G63B		2.0 (1997)	2 BBL	90 @ 5500	109 @ 3500	3.35 x 3.46	8.5:1	49 @ 2000	
	G54B	3	2.6 (2555)	2 BBL	109 @ 5000	142 @ 2500	3.59 x 3.86	8.7:1	56 @ 2000	
	G54B	4	2.6 (2555)	2 BBL	109 @ 5000	193 @ 3000	3.59 x 3.86	8.7:1	56 @ 2000	
1989	G63B		2.0 (1997)	2 BBL	90 @ 5500	109 @ 3500	3.35 x 3.46	8.5:1	49 @ 2000	
	G54B	2	2.6 (2555)	2 BBL	109 @ 5000	142 @ 2500	3.59 x 3.86	8.7:1	56 @ 2000	
	6G72		3.0 (2972)	MFI	143 @ 5000	168 @ 2500	3.59 x 2.99	8.9:1	40 @ 2000	
1990	4G64		2.4 (2350)	MFI	116 @ 5000	132 @ 3500	3.41 x 3.94	8.5:1	41 @ 2000	
	G54B	2	2.6 (2555)	2 BBL	109 @ 5000	142 @ 3000	3.59 x 3.86	8.7:1	50-64 @ 2000	
	6G72	2	3.0 (2972)	MFI	143 @ 5000	168 @ 2500	3.59 x 2.99	8.9:1	30-80 @ 2000	
1991	4G64		2.4 (2350)	MFI	116 @ 5000	136 @ 3500	3.41 x 3.94	8.5:1	41 @ 2000	
	6G72	4	3.0 (2972)	MFI	143 @ 5000	168 @ 2500	3.59 x 2.99	8.9:1	30-80 @ 2000	
	6G72		3.0 (2972)	MFI	151 @ 5000	174 @ 4000	3.59 x 2.99	8.9:1	30-80 @ 2000	
1992	4G64		2.4 (2350)	MFI	116 @ 5000	136 @ 3500	3.41 x 3.94	8.5:1	41 @ 2000	
	6G72	2	3.0 (2972)	MFI	151 @ 5000	174 @ 4000	3.59 x 2.99	8.9:1	30-80 @ 2000	
1993	4G64		2.4 (2350)	MFI	116 @ 5000	136 @ 3500	3.41 x 3.94	8.5:1	41 @ 2000	
	6G72	2	3.0 (2972)	MFI	151 @ 5000	174 @ 4000	3.59 x 2.99	8.9:1	30-80 @ 2000	
1994	4G64		2.4 (2350)	MFI	116 @ 5000	136 @ 3500	3.41 x 3.94	8.5:1	41 @ 2000	
	6G72	2	3.0 (2972)	MFI	151 @ 5000	174 @ 4000	3.59 x 2.99	8.9:1	30-80 @ 2000	
	6G74	4	3.5 (3496)	MFI	215 @ 5500	228 @ 3000	3.66 x 3.38	9.5:1	30-80 @ 2000	
1995	4G64		2.4 (2350)	MFI	116 @ 5000	136 @ 3500	3.41 x 3.94	8.5:1	41 @ 2000	
	6G72	2	3.0 (2972)	MFI	177 @ 5500 [5]	188 @ 4500 [6]	3.59 x 2.99	9.0:1 [7]	30-80 @ 2000	
	6G74	4	3.5 (3497)	MFI	215 @ 5000	228 @ 3000	3.66 x 3.38	9.5:1	30-80 @ 2000	

2 BBL - Two Barrel Carburetor
DSL - Diesel
MFI - Multi-port Fuel Injection
1 Minimum value
2 Both Montero and Pick-up
3 Pick-up only
4 Montero
5 California: 168@5500
6 California: 183@4500
7 9.5:1 for Montero SR

VALVE SPECIFICATIONS

Year	Engine ID/VIN	Engine Displacement Liters (cc)	Seat Angle (deg.)	Face Angle (deg.)	Spring Test Pressure (lbs. @ in.)	Spring Installed Height (in.)	Stem-to-Guide Clearance (in.) Intake	Stem-to-Guide Clearance (in.) Exhaust	Stem Diameter (in.) Intake	Stem Diameter (in.) Exhaust
1983	G63B	2.0 (1997)	44.0-44.5	45.0-45.5	62.0 @ 1.591	1.591	0.0010-0.0022	0.0020-0.0035	0.3150	0.3150
	4D55	2.3 (2346)	44.0-44.5	45.0-45.5	62.0 @ 1.591	1.591	0.0012-0.0024	0.0020-0.0035	0.3150	0.3150
	G54B	2.6 (2555)	44.0-44.5	45.0-45.5	62.0 @ 1.591	1.591	0.0010-0.0024	0.0020-0.0035	0.3150	0.3150
1984	G63B	2.0 (1997)	44.0-44.5	45.0-45.5	62.0 @ 1.591	1.591	0.0010-0.0022	0.0020-0.0035	0.3150	0.3150
	4D55	2.3 (2346)	44.0-44.5	45.0-45.5	62.0 @ 1.591	1.591	0.0012-0.0024	0.0020-0.0035	0.3150	0.3150
	G54B	2.6 (2555)	44.0-44.5	45.0-45.5	62.0 @ 1.591	1.591	0.0012-0.0024	0.0020-0.0035	0.3150	0.3150
1985	G63B	2.0 (1997)	44.0-44.5	45.0-45.5	62.0 @ 1.591	1.591	0.0010-0.0022	0.0020-0.0035	0.3150	0.3150
	4D55	2.3 (2346)	44.0-44.5	45.0-45.5	61.0 @ 1.591	1.591	0.0012-0.0024	0.0020-0.0035	0.3150	0.3150
	G54B	2.6 (2555)	44.0-44.5	45.0-45.5	62.0 @ 1.591	1.591	0.0012-0.0024	0.0020-0.0035	0.3150	0.3150
1986	G63B	2.0 (1997)	44.0-44.5	45.0-45.5	72.0 @ 1.591	1.591	0.0010-0.0022	0.0020-0.0035	0.3150	0.3150
	G54B	2.6 (2555)	44.0-44.5	45.0-45.5	72.0 @ 1.591	1.591	0.0012-0.0024	0.0020-0.0035	0.3150	0.3150
1987	G63B	2.0 (1997)	44.0-44.5	45.0-45.5	72.0 @ 1.591	1.591	0.0010-0.0022	0.0020-0.0035	0.3150	0.3150
	G54B	2.6 (2555)	44.0-44.5	45.0-45.5	72.0 @ 1.591	1.591	0.0012-0.0024	0.0020-0.0035	0.3150	0.3150
1988	G63B	2.0 (1997)	44.0-44.5	45.0-45.5	72.0 @ 1.591	1.591	0.0010-0.0022	0.0020-0.0035	0.3150	0.3150
	G54B	2.6 (2555)	44.0-44.5	45.0-45.5	72.0 @ 1.591	1.591	0.0012-0.0024	0.0020-0.0035	0.3150	0.3150
1989	4G63	2.0 (1997)	44.0-44.5	45.0-45.5	72.0 @ 1.591	1.591	0.0010-0.0022	0.0020-0.0035	0.3150	0.3150
	G54B	2.6 (2555)	44.0-44.5	45.0-45.5	72.0 @ 1.591	1.591	0.0012-0.0024	0.0020-0.0035	0.3150	0.3150
	6G72	3.0 (2972)	44.0-44.5	45.0-45.5	74.0 @ 1.591	1.591	0.0012-0.0024	0.0020-0.0035	0.3140	0.3130
1990	4G64	2.4 (2350)	44.0-44.5	45.0-45.5	73.0 @ 1.590	1.591	0.0012-0.0024	0.0020-0.0035	0.3100	0.3100
	G54B	2.6 (2555)	44.0-44.5	45.0-45.5	73.0 @ 1.590	1.591	0.0012-0.0024	0.0020-0.0035	0.3140	0.3130
	6G72	3.0 (2972)	44.0-44.5	45.0-45.5	74.0 @ 1.591	1.591	0.0012-0.0024	0.0020-0.0035	0.3134-0.3140	0.3122-0.3130
1991	4G64	2.4 (2350)	44.0-44.5	45.0-45.5	73.0 @ 1.590	1.591	0.0012-0.0024	0.0020-0.0035	0.3100	0.3100
	6G72	3.0 (2972)	44.0-44.5	45.0-45.5	74.0 @ 1.591	1.591	0.0012-0.0024	0.0020-0.0035	0.3134-0.3140	0.3122-0.3130

ENGINE AND ENGINE REBUILDING 3-31

VALVE SPECIFICATIONS

Year	Engine ID/VIN	Engine Displacement Liters (cc)	Seat Angle (deg.)	Face Angle (deg.)	Spring Test Pressure (lbs. @ in.)	Spring Installed Height (in.)	Stem-to-Guide Clearance (in.) Intake	Stem-to-Guide Clearance (in.) Exhaust	Stem Diameter (in.) Intake	Stem Diameter (in.) Exhaust
1992	4G64	2.4 (2350)	44.0-44.5	45.0-45.5	73.0 @ 1.590	1.591	0.0008-0.0024	0.0020-0.0035	0.3150	0.3110
	6G72	3.0 (2972)	44.0-44.5	45.0-45.5	72.5 @ 1.590	1.591	0.0012-0.0024	0.0020-0.0035	0.3150	0.3110
1993	4G64	2.4 (2350)	44.0-44.5	45.0-45.5	73.0 @ 1.590	1.591	0.0008-0.0024	0.0020-0.0035	0.3150	0.3110
	6G72	3.0 (2972)	44.0-44.5	45.0-45.5	72.5 @ 1.590	1.591	0.0012-0.0024	0.0020-0.0035	0.3150	0.3110
1994	4G64	2.4 (2350)	44.0-44.5	45.0-45.5	73.0 @ 1.590	1.591	0.0008-0.0024	0.0020-0.0035	0.3150	0.3110
	6G72	3.0 (2972)	44.0-44.5	45.0-45.5	72.5 @ 1.590	1.591	0.0012-0.0024	0.0020-0.0035	0.3150	0.3110
	6G74	3.5 (3496)	44.0-44.5	45.0-45.5	52.9 @ 1.490	1.490	0.0008-0.0020	0.0020-0.0035	0.2600	0.2560
1995	4G64	2.4 (2350)	44.0-44.5	45.0-45.5	73.0 @ 1.590	1.591	0.0008-0.0024	0.0020-0.0035	0.3150	0.3110
	6G72	3.0 (2972)	44.0-44.5	45.0-45.5	72.5 @ 1.590	1.591	0.0012-0.0024	0.0020-0.0035	0.3150	0.3110
	6G74	3.5 (3497)	44.0-44.5	45.0-45.5	52.9 @ 1.490	1.490	0.0008-0.0020	0.0020-0.0035	0.2600	0.2560

ENGINE AND ENGINE REBUILDING

PISTON AND RING SPECIFICATIONS
All measurements are given in inches.

Year	Engine ID/VIN	Engine Displacement Liters (cc)	Piston Clearance	Ring Gap			Ring Side Clearance		
				Top Compression	Bottom Compression	Oil Control	Top Compression	Bottom Compression	Oil Control
1983	G63B	2.0 (1997)	0.0008-0.0016	0.0100-0.0180	0.0080-0.0160	0.0080-0.0200	0.0020-0.0040	0.0010-0.0020	SNUG
	4D55	2.3 (2346)	0.0016-0.0024	0.0100-0.0160	0.0100-0.0160	0.0100-0.0180	0.0010-0.0020	0.0010-0.0030	0.001-0.003
	G54B	2.6 (2555)	0.0008-0.0016	0.0100-0.0150	0.0100-0.0180	0.0120-0.0240	0.0020-0.0040	0.0010-0.0020	SNUG
1984	G63B	2.0 (1997)	0.0008-0.0016	0.0100-0.0180	0.0080-0.0160	0.0080-0.0200	0.0020-0.0040	0.0010-0.0020	SNUG
	4D55	2.3 (2346)	0.0016-0.0024	0.0100-0.0160	0.0100-0.0160	0.0100-0.0180	0.0010-0.0020	0.0010-0.0030	0.001-0.003
	G54B	2.6 (2555)	0.0008-0.0016	0.0120-0.0180	0.0100-0.0150	0.0120-0.0240	0.0020-0.0040	0.0010-0.0020	SNUG
1985	G63B	2.0 (1997)	0.0008-0.0016	0.0100-0.0180	0.0080-0.0160	0.0080-0.0280	0.0020-0.0040	0.0010-0.0020	SNUG
	4D55	2.3 (2346)	0.0016-0.0024	0.0100-0.0160	0.0100-0.0160	0.0100-0.0180	0.0010-0.0020	0.0010-0.0030	0.001-0.003
	G54B	2.6 (2555)	0.0008-0.0016	0.0120-0.0180	0.0100-0.0150	0.0120-0.0240	0.0020-0.0040	0.0010-0.0020	SNUG
1986	G63B	2.0 (1997)	0.0008-0.0016	0.0100-0.0180	0.0080-0.0160	0.0080-0.0280	0.0020-0.0040	0.0010-0.0020	SNUG
	G54B	2.6 (2555)	0.0008-0.0016	0.0120-0.0180	0.0100-0.0150	0.0120-0.0240	0.0020-0.0040	0.0010-0.0020	SNUG
1987	G63B	2.0 (1997)	0.0004-0.0012	0.0100-0.0160	0.0080-0.0140	0.0080-0.0280	0.0010-0.0030	0.0008-0.0024	SNUG
	G54B	2.6 (2555)	0.0008-0.0016	0.0120-0.0180	0.0100-0.0160	0.0120-0.0320	0.0020-0.0035	0.0008-0.0024	SNUG
1988	G63B	2.0 (1997)	0.0004-0.0012	0.0100-0.0160	0.0080-0.0140	0.0080-0.0280	0.0010-0.0030	0.0008-0.0024	SNUG
	G54B	2.6 (2555)	0.0008-0.0016	0.0120-0.0180	0.0100-0.0160	0.0120-0.0320	0.0020-0.0035	0.0008-0.0024	SNUG
1989	4G63	2.0 (1997)	0.0008-0.0016	0.0100-0.0150	0.0080-0.0140	0.0080-0.0280	0.0010-0.0030	0.0008-0.0024	SNUG
	G54B	2.6 (2555)	0.0008-0.0016	0.0120-0.0180	0.0100-0.0160	0.0120-0.0320	0.0020-0.0035	0.0008-0.0024	SNUG
	6G72	3.0 (2972)	0.0008-0.0016	0.0120-0.0180	0.0100-0.0160	0.0080-0.0270	0.0012-0.0035	0.0008-0.0024	SNUG
1990	4G64	2.4 (2350)	0.0010-0.0020	0.0098-0.0157	0.0079-0.0157	0.0079-0.0276	0.0012-0.0028	0.0008-0.0024	SNUG
	G54B	2.6 (2555)	0.0008-0.0016	0.0120-0.0180	0.0100-0.0160	0.0118-0.0315	0.0020-0.0035	0.0008-0.0024	SNUG
	6G72	3.0 (2972)	0.0010-0.0020	0.0118-0.0177	0.0098-0.0157	0.0080-0.0270	0.0012-0.0035	0.0008-0.0024	SNUG
1991	4G64	2.4 (2350)	0.0010-0.0020	0.0098-0.0157	0.0079-0.0157	0.0079-0.0276	0.0012-0.0028	0.0008-0.0024	SNUG
	6G72	3.0 (2972)	0.0010-0.0020	0.0118-0.0177	0.0098-0.0157	0.0080-0.0280	0.0012-0.0035	0.0008-0.0024	SNUG

ENGINE AND ENGINE REBUILDING 3-33

PISTON AND RING SPECIFICATIONS

All measurements are given in inches.

Year	Engine ID/VIN	Engine Displacement Liters (cc)	Piston Clearance	Ring Gap			Ring Side Clearance		
				Top Compression	Bottom Compression	Oil Control	Top Compression	Bottom Compression	Oil Control
1992	4G64	2.4 (2350)	0.0010-0.0020	0.0079-0.0138	0.0079-0.0157	0.0079-0.0276	0.0012-0.0028	0.0008-0.0024	SNUG
	6G72	3.0 (2972)	0.0010-0.0020	0.0118-0.0177	0.0177-0.0236	0.0079-0.0236	0.0012-0.0028	0.0008-0.0024	SNUG
1993	4G64	2.4 (2350)	0.0010-0.0020	0.0098-0.0157	0.0177-0.0236	0.0079-0.0236	0.0012-0.0024	0.0012-0.0024	SNUG
	6G72	3.0 (2972)	0.0010-0.0020	0.0118-0.0177	0.0177-0.0236	0.0079-0.0236	0.0012-0.0028	0.0008-0.0024	SNUG
1994	4G64	2.4 (2350)	0.0010-0.0020	0.0098-0.0157	0.0177-0.0236	0.0079-0.0236	0.0012-0.0024	0.0012-0.0024	SNUG
	6G72	3.0 (2972)	0.0010-0.0020	0.0118-0.0177	0.0177-0.0236	0.0079-0.0236	0.0012-0.0028	0.0008-0.0024	SNUG
	6G74	3.5 (3496)	0.0010-0.0020	0.0118-0.0177	0.0177-0.0236	0.0039-0.0138	0.0012-0.0028	0.0008-0.0024	SNUG
1995	4G64	2.4 (2350)	0.0004-0.0012	0.0098-0.0157	0.0177-0.0236	0.0079-0.0236	0.0012-0.0024	0.0012-0.0024	SNUG
	6G72	3.0 (2972)	0.0010-0.0020	0.0118-0.0177	0.0177-0.0236	0.0079-0.0236	0.0012-0.0028	0.0008-0.0024	SNUG
	6G74	3.5 (3497)	0.0010-0.0020	0.0118-0.0177	0.0177-0.0236	0.0039-0.0138	0.0012-0.0028	0.0008-0.0024	SNUG

CAMSHAFT SPECIFICATIONS
All measurements given in inches.

Year	Engine ID/VIN	Engine Displacement Liters (cc)	Journal Diameter					Elevation		Bearing Clearance	Camshaft End-Play
			1	2	3	4	5	In.	Ex.		
1983	G63B	2.0 (1997)	1.339	1.339	1.339	1.339	1.339	1.661	1.661	0.0020-0.0040	0.0040-0.0080
	G54B	2.6 (2555)	1.339	1.339	1.339	1.339	1.339	1.673	1.673	0.0020-0.0040	0.0040-0.0080
	4D55	2.3 (2346)	1.181	1.181	1.181	1.181	1.181	1.461	1.461	0.0020-0.0040	0.0040-0.0080
1984	G63B	2.0 (1997)	1.339	1.339	1.339	1.339	1.339	1.661	1.661	0.0020-0.0040	0.0040-0.0080
	G54B	2.6 (2555)	1.339	1.339	1.339	1.339	1.339	1.673	1.673	0.0020-0.0040	0.0040-0.0080
	4D55	2.3 (2346)	1.181	1.181	1.181	1.181	1.181	1.461	1.461	0.0020-0.0040	0.0040-0.0080
1985	G63B	2.0 (1997)	1.339	1.339	1.339	1.339	1.339	1.657	1.657	0.0020-0.0040	0.0040-0.0080
	G54B	2.6 (2555)	1.339	1.339	1.339	1.339	1.339	1	1	0.0020-0.0040	0.0040-0.0080
	4D55	2.3 (2346)	1.181	1.181	1.181	1.181	1.181	1.461	1.461	0.0020-0.0040	0.0040-0.0080
1986	G63B	2.0 (1997)	1.339	1.339	1.339	1.339	1.339	1.657	1.657	0.0020-0.0040	0.0040-0.0080
	G54B	2.6 (2555)	1.339	1.339	1.339	1.339	1.339	1	1	0.0020-0.0040	0.0040-0.0080
1987	G63B	2.0 (1997)	1.339	1.339	1.339	1.339	1.339	1.656	1.656	0.0020-0.0035	0.0040-0.0080
	G54B	2.6 (2555)	1.339	1.339	1.339	1.339	1.339	1.669	1.669	0.0010-0.0020	0.0040-0.0080
1988	G63B	2.0 (1997)	1.340	1.340	1.340	1.340	1.340	1.656	1.656	0.0020-0.0035	0.0040-0.0080
	G54B	2.6 (2555)	1.340	1.340	1.340	1.340	1.340	1.670	1.670	0.0020-0.0035	0.0040-0.0080
1989	G54B	2.6 (2555)	1.340	1.340	1.340	1.340	1.340	1.670	1.670	0.0020-0.0035	0.0040-0.0080
	6G72	3.0 (2972)	1.340	1.340	1.340	1.340	—	1.624	1.624	0.0020-0.0035	0.0020-0.0080
	G63B	2.0 (1997)	1.340	1.340	1.340	1.340	1.340	1.656	1.656	0.0020-0.0035	0.0040-0.0080
1990	4G64	2.4 (2350)	1.340	1.340	1.340	1.340	1.340	1.669	1.669	0.0020-0.0035	0.0040-0.0080
	G54B	2.6 (2555)	1.340	1.340	1.340	1.340	1.340	1.670	1.670	0.0020-0.0035	0.0040-0.0080
	6G72	3.0 (2972)	1.340	1.340	1.340	1.340	—	1.624	1.624	0.0020-0.0035	0.0020-0.0080
1991	4G64	2.4 (2350)	1.340	1.340	1.340	1.340	1.340	1.669	1.669	0.0020-0.0035	0.0040-0.0080
	6G72	3.0 (2972)	1.340	1.340	1.340	1.340	—	1.624	1.624	0.0020-0.0035	0.0020-0.0080
1992	4G64	2.4 (2350)	1.340	1.340	1.340	1.340	1.340	2	2	0.0020-0.0040	0.0020-0.0080
	6G72	3.0 (2972)	1.340	1.340	1.340	1.340	—	1.620	1.620	0.0020-0.0040	0.0020-0.0080

ENGINE AND ENGINE REBUILDING

CAMSHAFT SPECIFICATIONS
All measurements given in inches.

Year	Engine ID/VIN	Engine Displacement Liters (cc)	Journal Diameter					Elevation		Bearing Clearance	Camshaft End-Play
			1	2	3	4	5	In.	Ex.		
1993	4G64	2.4 (2350)	1.340	1.340	1.340	1.340	1.340	1.669	1.669	0.0020-0.0040	0.0020-0.0080
	6G72	3.0 (2972)	1.340	1.340	1.340	1.340	—	1.620	1.620	0.0020-0.0040	0.0020-0.0080
1994	4G64	2.4 (2350)	1.340	1.340	1.340	1.340	1.340	1.669	1.669	0.0020-0.0040	0.0020-0.0080
	6G72	3.0 (2972)	1.340	1.340	1.340	1.340	—	1.374	1.374	0.0020-0.0040	0.0020-0.0080
	6G74 [3]	3.5 (3496)	1.022	1.022	1.022	1.022	—	1.390	1.370	0.0020-0.0040	0.0020-0.0080
1995	4G64	2.4 (2350)	1.340	1.340	1.340	1.340	1.340	1.669	1.669	0.0020-0.0040	0.0020-0.0080
	6G72	3.0 (2972)	1.340	1.340	1.340	1.340	—	1.620	1.620	0.0020-0.0040	0.0020-0.0080
	6G74 [3]	3.5 (2972)	1.022	1.022	1.022	1.022	1.022	1.390	1.370	0.0020-0.0040	0.0020-0.0080

1. Montero: 1.669; Truck: 1.673
2. ID mark D: 1.67
   ID mark AR: 1.753
3. Double overhead camshaft

86663c07

CRANKSHAFT AND CONNECTING ROD SPECIFICATIONS

All measurements are given in inches.

Year	Engine ID/VIN	Engine Displacement Liters (cc)	Crankshaft				Connecting Rod		
			Main Brg. Journal Dia.	Main Brg. Oil Clearance	Shaft End-Play	Thrust on No.	Journal Diameter	Oil Clearance	Side Clearance
1983	G63B	2.0 (1997)	2.244	0.0008-0.0020	0.0020-0.0071	3	1.7720	0.0008-0.0020	0.0040-0.0100
	4D55	2.3 (2346)	2.598	0.0008-0.0020	0.0020-0.0071	3	2.0866	0.0008-0.0020	0.0040-0.0100
	G54B	2.6 (2555)	2.362	0.0008-0.0020	0.0020-0.0071	3	2.0866	0.0008-0.0020	0.0040-0.0100
1984	G63B	2.0 (1997)	2.244	0.0008-0.0020	0.0020-0.0071	3	1.7720	0.0008-0.0020	0.0040-0.0100
	4D55	2.3 (2346)	2.598	0.0008-0.0020	0.0020-0.0071	3	2.0866	0.0008-0.0020	0.0040-0.0100
	G54B	2.6 (2555)	2.362	0.0008-0.0020	0.0020-0.0071	3	2.0866	0.0008-0.0020	0.0040-0.0100
1985	G63B	2.0 (1997)	2.244	0.0008-0.0020	0.0020-0.0071	3	1.7720	0.0008-0.0020	0.0040-0.0100
	4D55	2.3 (2346)	2.598	0.0008-0.0020	0.0020-0.0071	3	2.0866	0.0008-0.0020	0.0040-0.0100
	G54B	2.6 (2555)	2.362	0.0008-0.0020	0.0020-0.0071	3	2.0866	0.0008-0.0020	0.0040-0.0100
1986	G63B	2.0 (1997)	2.244	0.0008-0.0020	0.0020-0.0071	3	1.7720	0.0008-0.0020	0.0040-0.0100
	G54B	2.6 (2555)	2.362	0.0008-0.0020	0.0020-0.0071	3	2.0866	0.0008-0.0020	0.0040-0.0100
1987	G63B	2.0 (1997)	2.244	0.0008-0.0020	0.0020-0.0071	3	1.7720	0.0008-0.0020	0.0040-0.0100
	G54B	2.6 (2555)	2.362	0.0008-0.0020	0.0020-0.0071	3	2.0866	0.0008-0.0020	0.0040-0.0100
1988	G63B	2.0 (1997)	2.244	0.0008-0.0020	0.0020-0.0071	3	1.7720	0.0008-0.0020	0.0040-0.0100
	G54B	2.6 (2555)	2.362	0.0008-0.0020	0.0020-0.0071	3	2.0866	0.0008-0.0020	0.0040-0.0100
1989	4G63	2.0 (1997)	2.244	0.0008-0.0020	0.0020-0.0071	3	1.7720	0.0008-0.0020	0.0039-0.0098
	G54B	2.6 (2555)	2.362	0.0008-0.0020	0.0020-0.0071	3	2.0900	0.0007-0.0022	0.0039-0.0098
	6G72	3.0 (2972)	2.362	0.0008-0.0019	0.0020-0.0098	3	1.9690	0.0008-0.0019	0.0039-0.0098
1990	4G64	2.4 (2350)	2.244	0.0008-0.0020	0.0020-0.0071	3	1.7720	0.0008-0.0020	0.0039-0.0098
	G54B	2.6 (2555)	2.362	0.0008-0.0020	0.0020-0.0071	3	2.0870	0.0008-0.0020	0.0039-0.0098
	6G72	3.0 (2972)	2.362	0.0008-0.0020	0.0020-0.0098	3	1.9690	0.0008-0.0019	0.0039-0.0098
1991	4G64	2.4 (2350)	2.244	0.0008-0.0020	0.0020-0.0071	3	1.7720	0.0008-0.0020	0.0039-0.0098
	6G72	3.0 (2972)	2.362	0.0008-0.0020	0.0020-0.0098	3	1.9690	0.0008-0.0019	0.0039-0.0098

CRANKSHAFT AND CONNECTING ROD SPECIFICATIONS
All measurements are given in inches.

Year	Engine ID/VIN	Engine Displacement Liters (cc)	Crankshaft Main Brg. Journal Dia.	Crankshaft Main Brg. Oil Clearance	Crankshaft Shaft End-Play	Crankshaft Thrust on No.	Connecting Rod Journal Diameter	Connecting Rod Oil Clearance	Connecting Rod Side Clearance
1992	4G64	2.4 (2350)	2.244	0.0008-0.0020	0.0020-0.0071	3	1.7720	0.0008-0.0020	0.0039-0.0098
	6G72	3.0 (2972)	2.362	0.0008-0.0020	0.0020-0.0098	3	1.9690	0.0008-0.0019	0.0039-0.0098
1993	4G64	2.4 (2350)	2.244	0.0008-0.0020	0.0020-0.0071	3	1.7720	0.0008-0.0020	0.0039-0.0098
	6G72	3.0 (2972)	2.362	0.0008-0.0020	0.0020-0.0098	3	1.9690	0.0008-0.0019	0.0039-0.0098
1994	4G64	2.4 (2350)	2.244	0.0008-0.0020	0.0020-0.0071	3	1.7720	0.0008-0.0020	0.0039-0.0098
	6G72	3.0 (2972)	2.362	0.0008-0.0020	0.0020-0.0098	3	1.9690	0.0008-0.0019	0.0039-0.0098
	6G74	3.5 (3496)	2.520	0.0008-0.0020	0.0020-0.0098	3	2.1650	0.0012-0.0020	0.0039-0.0098
1995	4G64	2.4 (2350)	2.244	0.0008-0.0020	0.0020-0.0071	3	1.7720	0.0008-0.0020	0.0039-0.0098
	6G72	3.0 (2972)	2.362	0.0008-0.0020	0.0020-0.0098	3	1.9690	0.0008-0.0019	0.0039-0.0098
	6G74	3.5 (3497)	2.520	0.0008-0.0020	0.0020-0.0098	3	2.1650	0.0012-0.0020	0.0039-0.0098

ENGINE AND ENGINE REBUILDING

TORQUE SPECIFICATIONS
All readings in ft. lbs.

Year	Engine ID/VIN	Engine Displacement Liters (cc)	Cylinder Head Bolts	Main Bearing Bolts	Rod Bearing Bolts	Crankshaft Damper Bolts	Flywheel Bolts	Manifold Intake	Manifold Exhaust	Spark Plugs	Lug Nut
1983	G63B	2.0 (1997)	73-79	38	37	80-94	94-101	11-14	11-14	15-21	1
	4D55	2.3 (2346)	76-83 8	55-61	33	—	94-101	11-14	11-14	—	1
	G54B	2.6 (2555)	73-79 2	55-61	34	80-94	94-101	11-14	11-14	15-21	1
1984	G63B	2.0 (1997)	73-79	38	37	80-94	94-101	11-14	11-14	15-21	1
	4D55	2.3 (2346)	76-83 8	55-61	33	—	94-101	11-14	11-14	—	1
	G54B	2.6 (2555)	73-79 2	55-61	34	80-94	94-101	11-14	11-14	15-21	1
1985	G63B	2.0 (1997)	73-79	38	37	80-94	94-101	11-14	11-14	15-21	1
	4D55	2.3 (2346)	76-83 8	55-61	34	—	94-101	11-14	11-14	—	1
	G54B	2.6 (2555)	73-79 2	55-61	33	80-94	94-101	11-14	11-14	15-21	1
1986	G63B	2.0 (1997)	73-79	38	37	80-94	94-101	11-14	11-14	15-21	1
	G54B	2.6 (2555)	73-79 2	55-61	33	80-94	94-101	11-14	11-14	15-21	1
1987	G63B	2.0 (1997)	73-79	38	37	80-94	94-101	11-14	11-14	18	3
	G54B	2.6 (2555)	73-79 4	55-61	33	80-94	94-101	11-14	11-14	18	3
1988	G63B	2.0 (1997)	73-79	38	37	80-94	94-101	11-14	11-14	18	3
	G54B	2.6 (2555)	73-79 4	55-61	33	80-94	94-101	11-14	11-14	18	3
1989	4G63	2.0 (1997)	72-80	36-40	36-38	80-94	94-101	13-18	18-22	18	3
	G54B	2.6 (2555)	73-79 4	55-61	33	80-94	94-101	11-14	11-14	18	3
	6G72	3.0 (2972)	73-79	55-61	38	109-115	53-55	11-14	11-16	18	65-80
1990	4G64	2.4 (2350)	73-79	37-39	37-38	80-94	94-101	11-14	11-14	18	87-101
	G54B	2.6 (2555)	73-79 4	55-61	33	80-94	94-101	11-14	11-14	18	72-87
	6G72	3.0 (2972)	73-79	55-61	38	108-116	53-55	11-14	11-16	18	3
1991	4G64	2.4 (2350)	73-79	37-39	37-83	80-94	94-101	11-14	11-14	18	87-101
	6G72	3.0 (2972)	73-79	65-72	38	130-137	53-55	11-14	11-16	18	3
1992	4G64	2.4 (2350)	80	38	14	87	98	11-14	11-14	18	3
	6G72	3.0 (2972)	80	57	38	136	54	11-14	11-16	18	3
1993	4G64	2.4 (2350)	5	18 6	14.5 6	87	98	11-14	11-14	18	65-80
	6G72	3.0 (2972)	80	57	38	136	54	11-14	11-16	18	3
1994	4G64	2.4 (2350)	5	18 6	14.5 6	87	98	11-14	11-14	18	65-80
	6G72	3.0 (2972)	80	57	38	136	54	11-14	11-16	18	3
	6G74	3.5 (3496)	80	54	38	136	54	13	33	18	72-87
1995	4G64	2.4 (2350)	5	18 6	14.5 6	87	98	11-14	11-14	18	3
	6G72 7	3.0 (2972)	80	67	38	134	54	16	22	18	3
	6G74	3.5 (3497)	80	54	38	134	54	13	33	18	3

1 Steel wheels: 50-57 ft. lbs.
Aluminum wheels: 66-81 ft. lbs.
2 No. 11 bolt: 11-15 ft. lbs.
3 Montero: 72-87 ft. lbs.
Pick-up: 87-101 ft. lbs.
4 M8 bolts: 11-16 ft. lbs.
5 Step 1: 54 ft. lbs.
Step 2: Loosen completely, then tighten to 14.5 ft. lbs. plus an additional 1/4 turn
Step 3: Tighten an additional 1/4 turn
6 Tighten to value, loosen and retighten to value
7 24 valve, SOHC
8 Cold specifications

ENGINE AND ENGINE REBUILDING

Engine

REMOVAL & INSTALLATION

➡ Elevate and safely support the vehicle as needed to gain access to any components. Depending on tools, arm length and agility, each operation may be easier from above or below. All wires and hoses should be labeled at the time of removal. The amount of time saved during reassembly makes the extra effort well worthwhile.

2-Wheel Drive Pick-up (2.0L and 2.6L Engines) and 4-Wheel Drive Pick-up (2.6L Engines)

▶ See Figures 47, 48, 49, 50 and 51

1. If the vehicle is equipped with air conditioning, take the vehicle to a trained and certified mechanic to discharge the air conditioning system using a recycling/recovery machine.
2. Matchmark and remove the hood. Disconnect the negative battery cable at the battery.
3. Elevate and safely support the truck on jackstands. Remove the undercovers and shields below the engine.
4. Drain the engine coolant.

✳✳CAUTION

When draining the coolant, keep in mind that cats and dogs are attracted by ethylene glycol antifreeze, and are quite likely to drink any that is left in an uncovered container or in puddles on the ground. This will prove fatal in sufficient quantity. Always drain the coolant into a sealable container. Coolant should be reused unless it is contaminated or several years old.

5. Drain the engine and transmission oil.
6. Remove the air cleaner assembly.
7. If the vehicle is equipped with air conditioning, remove the upper and lower fan shrouds from the radiator.
8. Disconnect the upper and lower hose from the radiator and remove the radiator.
9. Disconnect the battery ground cable from the left-hand side of the engine block.
10. Disconnect the coil wire from the distributor cap.
11. Label and disconnect the water temperature sensor (with automatic transmission), the water temperature gauge sender and the starter motor connector.
12. Disconnect the positive battery cable from the starter.
13. Disconnect the transmission harness, the oxygen sensor connection, the wiring to the alternator and the oil pressure sender wire.
14. Disconnect the accelerator cable.
15. Carefully disconnect both hoses from the power steering pump. Do not damage the hose and mop up any spilled fluid.
16. Disconnect both air conditioning hoses from the compressor. Carefully move the hoses out of the way but do not crimp or crease the hoses. Cap the compressor nozzles and the lines.
17. Disconnect the brake booster vacuum hose.
18. Disconnect the exhaust pipe from the bottom of the exhaust manifold. Use wire or string to support the pipe to the side. Do not let the pipe hang of its own weight.
19. Carefully label and disconnect the fuel lines, breather and vacuum lines and the canister lines.

✳✳CAUTION

On vehicles equipped with fuel injection systems, relieve the fuel system pressure before loosening or removing any fuel lines.

20. Inside the vehicle, remove the shifter assembly (manual transmission) or disconnect the automatic shift selector linkage.

➡ When removing the manual shifter, remove the mounting bolts from the stopper plate and remove the lever assembly with the stopper plate.

21. Remove the driveshaft, or both driveshafts if the vehicle is equipped with 4-wheel drive. Before disconnecting the rear flange from the axle, scribe matchmarks on both the shaft and axle flange so that the shaft may be reinstalled in its original position. If the driveshaft is the 3-joint type, disconnect the center carrier from the floor pan of the vehicle. After the flange is free at the rear, slide the shaft out of the transmission housing. Immediately plug or cover the transmission; do not allow dirt or foreign matter to enter the transmission case. Do the same for the front driveshaft.
22. Install the hoisting equipment and make sure it is securely fastened to the engine. Draw tension on the hoist just enough to support the weight of the motor without elevating it.

✳✳CAUTION

The truck is supported on stands. Tension the hoist slowly and do not disturb the position of the truck on the supports.

23. Position a floor jack under the transmission (No.2) crossmember and adjust it to just touch to crossmember for support.
24. Remove the outer bolts holding the crossmember to the vehicle. Carefully remove the two center bolts holding the crossmember to the transmission case. Lower the jack slowly to remove the crossmember from the vehicle.
25. Double check for any remaining cables, hoses, wiring or lines still running to the engine and/or transmission and remove them. Check that any item which was moved out of the way is truly in a safe location.
26. Place the floor jack with a large block of wood under the transmission case for support. The jack will need to be adjusted and eventually removed during the engine removal.
27. Double check the security of the lifting apparatus. Make certain the cables are tensioned properly. Remove the nut, lockwasher and spacer from each motor mount.
28. Elevate the hoist slowly and raise the engine and transmission assembly out of the truck. The transmission will need to be lowered and the assembly lifted out on an angle. Take your time; clearances may be close and the unit is heavy. As soon as the engine is clear, mount it on a stand or support it on wooden blocks. Do not allow it to rest on the oil pan or lie on its side. Never leave an engine hanging from a hoist. The

3-40 ENGINE AND ENGINE REBUILDING

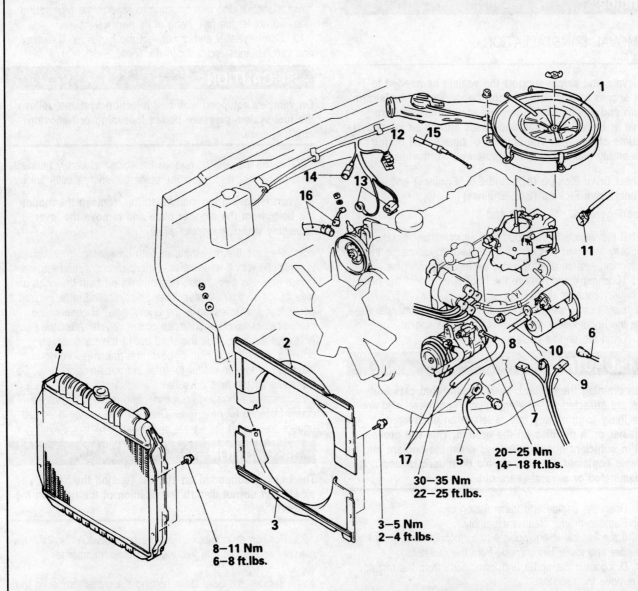

1. Air filter
2. Radiator upper shroud (Vehicles with an air conditioner)
3. Radiator lower shroud (Vehicles with an air conditioner)
4. Radiator
5. Battery ground cable connection
6. High tension cable
7. Water temperature sensor connector (Vehicles with an automatic transmission)
8. Water temperature gauge unit connector
9. Startor motor connector
10. Battery positive cable connection
11. Transmission harness connector
12. Oxygen sensor connector
13. Alternator connector
14. Oil pressure switch connector
15. Accelerator cable connection
16. Power steering oil pipe connection
17. Air conditioner compressor pipe connection

Fig. 47 Preliminary engine removal components — 2.0L Engines

ENGINE AND ENGINE REBUILDING   3-41

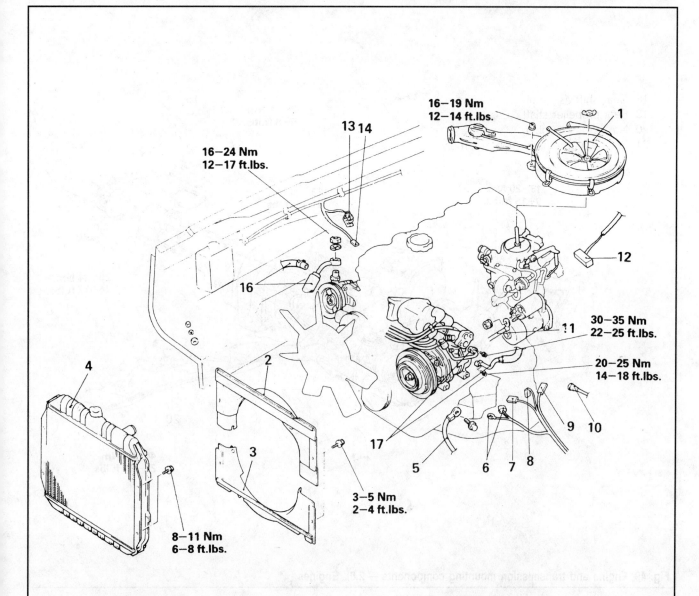

1. Air cleaner
2. Radiator upper shroud
   (Vehicles with an air conditioner)
3. Radiator lower shroud
   (Vehicles with an air conditioner)
4. Radiator
5. Battery ground cable connection
6. Alternator connector
7. Water temperature unit connector
   (Vehicles with an automatic transmission)
8. Water temperature gauge unit connector
9. Starter motor connector
10. High tension cable
11. Battery positive cable connection
12. Transmission harness connector
13. Oxygen sensor connector
14. Oil pressure gauge unit connector
15. Accelerator cable connection
16. Power steering oil pipe connection
17. Air conditioner compressor pipe connection

Fig. 48 Preliminary engine removal components — 2.6L Engines

3-42 ENGINE AND ENGINE REBUILDING

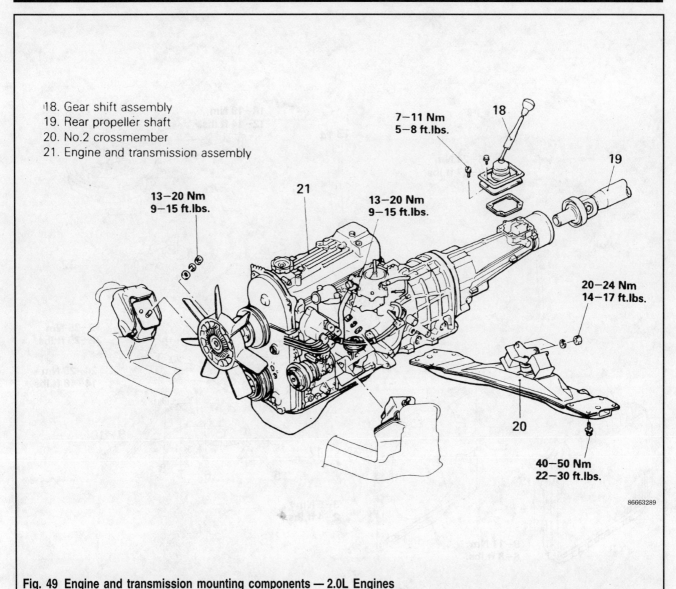

Fig. 49 Engine and transmission mounting components — 2.0L Engines

transmission should be disconnected from the engine as soon as convenient.

To install:

29. After repairs, make certain the engine and transmission are fully reassembled before installation. All components removed with the engine out of the truck should be in place before reinstallation.

30. Install the engine into the truck and lower it until the bolt holes for the mounts align with the brackets. Use the floor jack and wood block to support and guide the transmission into place as necessary. Install the mount nuts, tightening them to 14 ft. lbs (19 Nm).

31. Using the floor jack as necessary for support, install the No. 2 crossmember. Tighten the body bolts to 41 ft. lbs. (55 Nm) and the nuts holding the crossmember to the transmission to 17 ft. lbs. (24 Nm). When the mounts and crossmember are securely fastened, the floor jack and the engine lifting apparatus may be removed from the vehicle.

32. Install one or both driveshafts, depending on whether the vehicle is 2-wheel drive or 4-wheel drive. Remove the plug or cover from the transmission and insert the driveshaft. If the driveshaft is of the 3-joint type, connect the center support and tighten the nuts to 26 ft. lbs. (35 Nm). Turn the differential flange until the mark made during removal aligns with the mark on the shaft flange. Install the retaining bolts and tighten to 41 ft. lbs. (55 Nm).

33. Reinstall the manual shifter assembly with new gaskets or reconnect the automatic shift linkage.

34. Connect the fuel, vacuum, breather and canister lines to their correct locations. Make certain lines are not crimped or split. Route the lines clear of any hot surfaces or moving parts.

35. Use a new gasket with new nuts and connect the exhaust pipe to the manifold.

36. Connect the brake booster vacuum hose.

37. Using a new O-ring inside each line, connect the air conditioning lines to the compressor. Tighten the fittings carefully to 18 ft. lbs. (25 Nm).

38. Connect the power steering hoses to the pump. Tighten the retaining nut.

39. Connect the accelerator cable.

40. Connect the wiring to the transmission, the oxygen sensor, the alternator and oil pressure sender.

ENGINE AND ENGINE REBUILDING 3-43

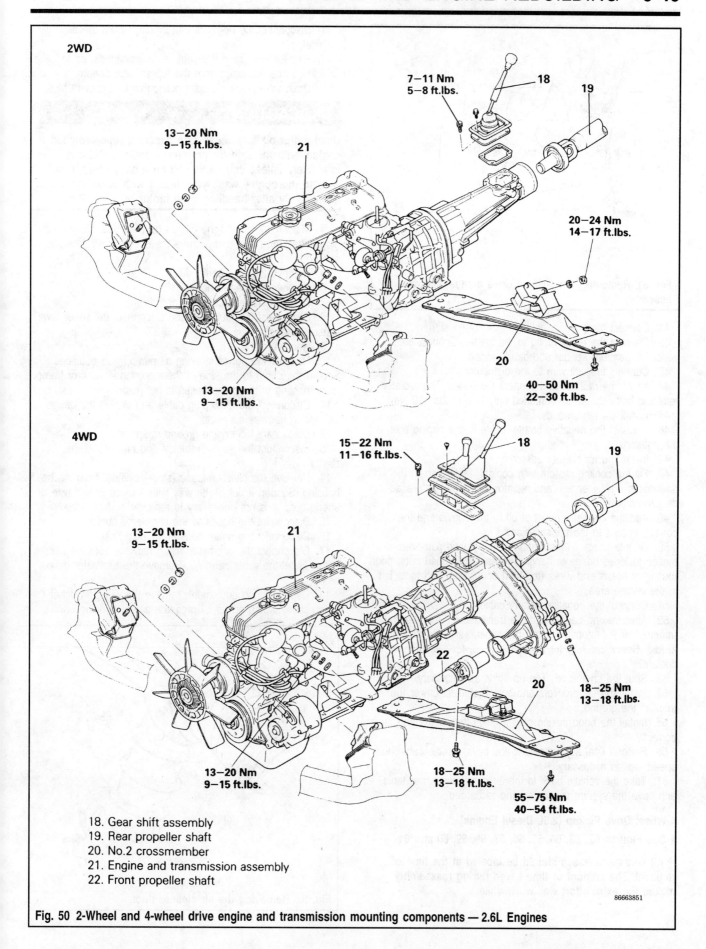

Fig. 50 2-Wheel and 4-wheel drive engine and transmission mounting components — 2.6L Engines

3-44 ENGINE AND ENGINE REBUILDING

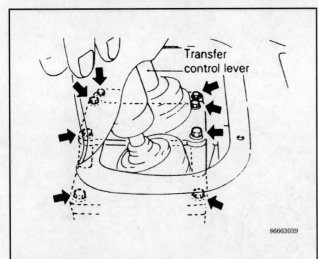

Fig. 51 Removing the 4-wheel drive manual shifter assembly

41. Connect the positive battery cable to the starter.
42. Connect the other wiring to the starter. Connect the water temperature sender and gauge wiring.
43. Connect the coil wire to the distributor.
44. Install the radiator and connect the hoses. The radiator retaining bolts should be tightened only to 8 ft. lbs. (10 Nm).
45. Install the fan shrouds.
46. Connect the negative battery cable to the engine (not to the battery).
47. Install the air cleaner assembly.
48. Fill the cooling system with coolant. Make certain the draincocks on the engine and radiator are closed before adding the fluid.
49. Add the correct amount of oil to the engine and the correct amount to the transmission.
50. Double check all installation items, paying particular attention to loose hoses or hanging wires, untightened nuts, poor routing of hoses and wires (too tight or rubbing) and tools left in the engine area.
51. Connect the negative battery cable.
52. After making certain that the transmission is in Neutral (manual) or **P** (automatic), start the engine, allowing it to run at idle. Check carefully for any leaks of oil, fuel, vacuum or coolant.
53. Shut the engine off. Top up fluids as necessary.
54. Install the undercovers and splash shields. Lower the truck to the ground.
55. Install the hood, taking care to align the seams and latch correctly.
56. Perform final adjustments to the belts, throttle cable, idle speed etc. as necessary.
57. Take the vehicle back to the air conditioning mechanic and have the system evacuated and recharged.

4-Wheel Drive Pickup (2.3L Diesel Engine)

▶ See Figures 52, 53, 54, 55, 56, 57, 58, 59, 60 and 61

➡ All wires and hoses should be labeled at the time of removal. The amount of time saved during reassembly makes the extra effort well worthwhile.

1. Disconnect the negative battery cable from the battery terminal.
2. Raise and support the vehicle on jackstands, so that the vehicle can be accessed from the top and the bottom.
3. Drain the engine oil, transmission oil and coolant fluid.

✽✽CAUTION

Used motor oil may cause skin cancer if repeatedly left in contact with the skin for prolonged periods. Although this is unlikely unless you handle oil on a daily basis, it is wise to thoroughly wash your hands with soap and water immediately after handling used motor oil.

4. Matchmark around the hinges on the hood. Remove the hinge attaching bolts, and remove the hood from the hinges and body.
5. Remove the air cleaner duct.
6. Remove the heater hoses.
7. Disconnect the accelerator and throttle cable.
8. Disconnect the fuel lines and disconnect the water level sensor connector.
9. Remove the fuel filter.
10. Remove the power steering oil pump, if so equipped. Secure the pump to the side of the inner fender well or frame, disconnecting of the pump hoses is not necessary.
11. Disconnect the glow plug cable and unplug the gauge unit wiring harness connector.
12. Disconnect the engine ground strap.
13. Disconnect the starter motor wiring harness at the starter.
14. Remove the clutch release (slave) cylinder from the bell housing. Support it out of the way with a piece of stiff wire or strong cord. It is not necessary to remove the fluid hose.
15. Disconnect the hoses to the engine oil cooler.
16. Disconnect the brake booster vacuum hose.
17. Disconnect the alternator wiring harness from the alternator. Label the wires carefully. Remove the alternator drive belt.
18. Depending on equipment, remove the wiring to either the oil pressure switch or the oil pressure gauge sending unit.

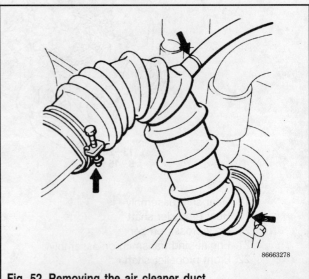

Fig. 52 Removing the air cleaner duct

ENGINE AND ENGINE REBUILDING   3-45

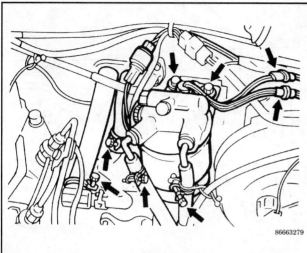

Fig. 53 Heater hose, fuel hose and fuel filter location and connections

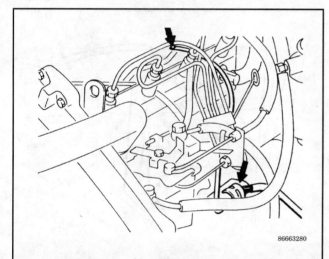

Fig. 54 Glow plug cable and gauge unit harness connectors locations

19. Disconnect the radiator hoses from the engine and radiator.
20. Remove the radiator assembly.
21. Disconnect the exhaust pipe at the first joint under the truck. Support the remainder of the system with wire. Remove the bolts holding the front pipe to the turbocharger and remove the front pipe.
22. Remove the undercovers. On 4-wheel drive vehicles, remove the protective skid plates under the engine and transfer case.
23. Disconnect the speedometer cable from the transmission.
24. Disconnect the reverse light switch at the transmission. If 4-wheel drive, disconnect the 4WD indicator switch harness also.

25. Matchmark and remove the driveshafts:
   a. Make mating marks on the flange yoke and differential companion flange.

➡For 4-wheel drive models, before proceeding with the removal procedure, set the free-wheeling hubs to the unlocked position, and also set the transfer case gear shift lever to the 2H position.

   b. Remove the bolts connecting the flange yoke to the differential companion flange, and for rear-wheel drive models, remove the nuts attaching the center bearing assembly.
   c. Remove the propeller (driveshaft) shaft by pulling it out.

➡When the propeller shaft is pulled out of the transmission extension housing or the transfer case, note that transmission oil or transfer case oil may leak depending upon the vehicle position and angle. When removing the propeller shaft, be careful not to damage the oil seal lip and see that no foreign substance is present in the lip area. Also use care to keep the oil seal clean and free of dust.

26. Remove the gearshift lever assembly.
27. Use a floor jack to support the transmission. Remove the rear insulator (mount) from the transmission.
28. Remove the crossmember.
29. On 4-wheel drive models, the transfer case must be supported by a second floor jack or adjustable stand. Remove the transfer case mounting bracket and the insulator.
30. Remove the plate from the side frame. Remove the mounting bracket from the transfer case.
31. Position and secure a chain or overhead engine hoist. Attach it to the lift points and draw tension on the chain or cable to support the engine without lifting it.
32. Remove the engine mount nuts from the front engine mounts. Double check completely around the motor (and below it) for any wire, cable, hose, or linkage still running to the body. At this point, the engine should be completely isolated from the vehicle.
33. Elevate the hoist, raising the engine and transmission as a unit. Push downward on the rear of the transmission; remove the engine/transmission diagonally.
34. Once clear of the vehicle, the unit should be carefully lowered onto wooden blocks. Once the transmission is removed, the engine should be mounted to a secure stand, allowing it to be worked on.

To install:
35. Position the engine/transmission unit with the hoist and lower it into the vehicle. The front mounts should engage and the transmission and/or transfer case should rest on the jacks.
36. Install the front mounts.
37. On 4-wheel drive vehicles, install the mounting bracket to the transfer case and tighten the bolts and nuts to 27 ft. lbs. (36 Nm).
38. If 4-wheel drive, install the transfer case mounting bracket and insulator.
39. Install the rear crossmember and the transmission mount (insulator). Tighten the bolts holding the crossmember to the body to 36 ft. lbs. (48 Nm) The bolts holding the transmission insulator to the crossmember should be tightened to 13.5 ft. lbs. (18 Nm).
40. When the transmission (and transfer case) is properly attached to the supports, the jacks may be removed. The

3-46 ENGINE AND ENGINE REBUILDING

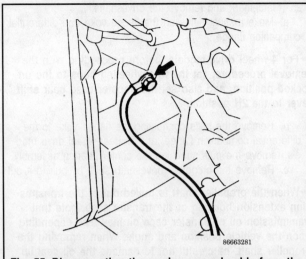
Fig. 55 Disconnecting the engine ground cable from the engine block

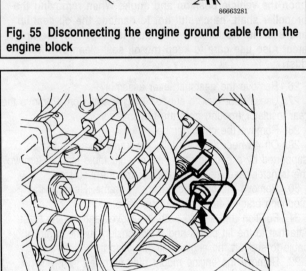
Fig. 56 The starter motor wiring harness connections and location

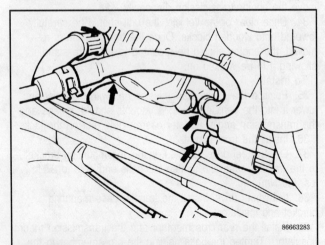
Fig. 57 Disconnecting the engine oil cooler hoses, the brake booster vacuum hose, and the alternator wiring harness

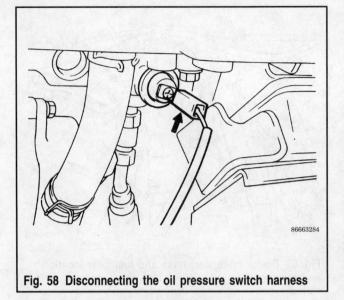
Fig. 58 Disconnecting the oil pressure switch harness

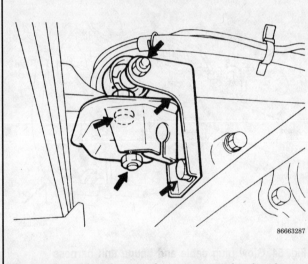
Fig. 59 Detaching the transfer case mounting bracket

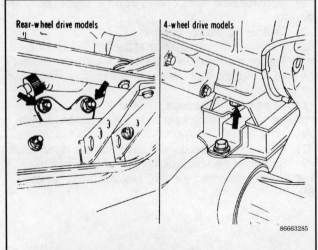
Fig. 60 Rear insulation transmission mounts on both 2-wheel drive and 4-wheel drive models

ENGINE AND ENGINE REBUILDING

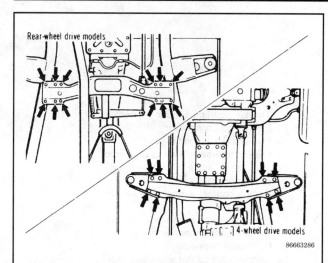

Fig. 61 Transmission crossmember mounts for both 2-wheel drive and 4-wheel drive vehicles

engine hoist chain may also be removed once the engine and transmission is firmly mounted to the vehicle.

41. Install the gearshift lever assembly.
42. Install the driveshaft(s).
43. Connect the wiring to the transmission. Make certain the connectors are tight and the wiring is correctly routed out of the way of moving parts.
44. Install the speedometer cable.
45. Install the skid plates and/or undercovers.
46. Use new gaskets and install the front exhaust pipe. It will be easier to attach the pipe loosely to the system and then to the turbocharger. Tighten the pipe-to-turbocharger nuts. Tighten the pipe-to-pipe joint under the vehicle.
47. Install the radiator assembly. Connect the hoses to the engine. Use new clamps as necessary.
48. Connect the wiring harnesses to the alternator, the oil pressure switch and or the oil pressure gauge sender.
49. Connect the brake booster vacuum hose.
50. Connect the engine oil cooler lines.
51. Install the clutch release cylinder. Adjust the clutch.
52. Install the starter motor wiring harness; connect the engine ground harness to the engine block.
53. Install the power steering pump. Tighten the bolts to 14 ft. lbs. (18 Nm).
54. Connect the fuel lines. Install the fuel filter and connect the water level sensor connector.
55. Connect the glow system harness and the gauge unit harness.
56. Connect the throttle cable and adjust it.
57. Connect the heater hoses. Install the air cleaner ductwork.
58. Double check all installation items, paying particular attention to loose hoses or hanging wires, untightened nuts, poor routing of hoses and wires (too tight or rubbing) and tools left in the engine area.
59. Fill the cooling system with the correct amount of coolant.
60. Install the proper amount of engine oil. If not already done, replace the oil filter.
61. Bleed the fuel system.
62. Connect the negative battery cable.
63. Start the engine, following the correct starting procedures. The engine may crank longer than usual; this is normal.
64. Allow the engine to run at idle. While it is warming up, check carefully for any leaks of oil, fuel, vacuum or coolant. Shut the engine off and attend to any leaks immediately. Allow the engine to cool completely before working on the cooling system.
65. Reinstall the hood and align it with the marks made during removal. Close the hood and check the alignment of the seams.

2-Wheel Drive Pick-up (2.4L Engines)

▶ See Figures 62 and 63

➡ When performing this procedure you will need a new fuel line O-ring, a new exhaust manifold to front exhaust pipe gaskets, new oil, transmission fluid, and antifreeze.

1. If the vehicle is equipped with air conditioning, first have the A/C system discharged by a trained and certified mechanic using a recovery/recycling machine.
2. Matchmark and remove the hood.
3. Elevate and safely support the truck on jackstands. Remove the undercovers and shields below the engine and transmission.

➡ The transmission assembly must be removed before the engine can be removed. Please refer to Section 7 — Drive Train for detailed instructions on the removal of these components.

4. Depressurize the fuel system by adhering to the following steps:
   a. Disconnect and label the fuel pump harness wiring plug at the fuel tank rear side.
   b. Start the engine and after it stops by itself, turn the ignition switch OFF.
5. Disconnect the negative battery cable from the battery.
6. Drain the cooling system.

➡ While removing or disconnecting any wires or hoses, make sure to label them. This will make reassembly much easier.

7. Drain the engine and transmission fluid.
8. Disconnect the cooling system overflow, upper radiator, and lower radiator hoses.
9. Remove the upper and lower radiator shrouds.
10. Disconnect the automatic transmission oil cooler hose from the radiator, if the vehicle is so equipped.
11. Remove the radiator from the engine compartment.
12. Disconnect the exhaust pipe from the exhaust manifold.
13. Remove the transmission assembly — refer to Section 7.
14. Remove the air cleaner duct hose.
15. Disengage the accelerator cable connection.
16. Unfasten the throttle control cable connection, if the vehicle is equipped with an automatic transmission.
17. Disconnect and label the wiring connector to the compressor.
18. Loosen the belt tension adjuster, the lockbolt and the pivot bolt. Remove the belt, the lockbolt and the pivot bolt from the compressor. Support the compressor and remove it from the bracket and hook it with wire or rope to the body side.

3-48 ENGINE AND ENGINE REBUILDING

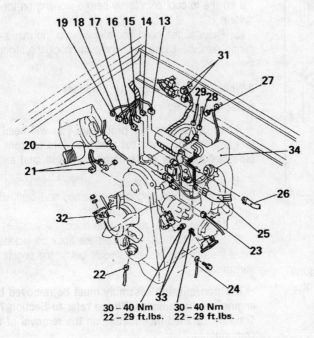

13 Throttle position sensor connector
14 Ignition coil connector
15 Power transistor connector
16 EGR temperature sensor connector
17 Engine coolant temperature gauge unit connector
18 Engine coolant temperaure sensor connector
19 Thermal switch connector <A/T>
20 Oxygen sensor connector
21 Generator connector
22 Oil pressure gauge unit connector
23 A/C engine coolant temperature switch connector <A/C>
24 Ground cable connection
25 Emission control vacuum hose connection
26 Brake booster vacuum hose connection
27 Ground cable connection
28 Idle speed control motor connector
29 Idle speed control motor position sensor connector
30 Idle air control motor connector
31 Control wiring harness connector
32 Heat protector
33 Engine mounting bolt
34 Engine assembly

Fig. 62 Component identification — 2.4L Engines non-California models

Move the compressor with the high-pressure hose and low-pressure hose still connected.

19. Loosen the power steering pump and remove the drive belt. Remove the power steering oil pump from the bracket and fasten it to the body side with strong wire or cord. Move the power steering oil pump with the pressure hose and return hose still connected.

20. Unfasten and label the cruise control vacuum hose connection, if so equipped.

21. Disconnect and label the high pressure fuel hose and remove the O-ring. Cover the fuel pipe line with rags after relieving the pressure, since fuel pressure may still be present.

22. Disconnect and label the fuel return hose connection.

23. Remove the water hose connection.

24. Remove and label the throttle position sensor, ignition coil, power transistor, EGR temperature sensor, engine coolant temperature gauge unit, engine coolant temperature sensor, thermal switch (if equipped with an automatic transmission), oxygen sensor, alternator, oil pressure gauge unit, and air conditioning engine coolant temperature switch connectors.

25. Unscrew and label the ground cable connection from the lower left hand side of the engine block.

26. Disengage and label the emission control vacuum hose connection.

27. Remove and label the brake booster vacuum hose connection.

28. Disconnect and label the ground cable from the firewall above the back of the engine.

29. Disconnect and label the idle speed control motor and the idle speed control motor position sensor wires.

30. Remove and label the idle air control motor connector.

31. Disconnect and label the control wiring harness.

32. Install the hoisting equipment and make sure it is securely fastened to the engine. Draw tension on the hoist just enough to support the weight of the motor without elevating it.

✲✲CAUTION

The truck is supported on stands. Tension the hoist slowly and do not disturb the position of the truck on the supports.

ENGINE AND ENGINE REBUILDING

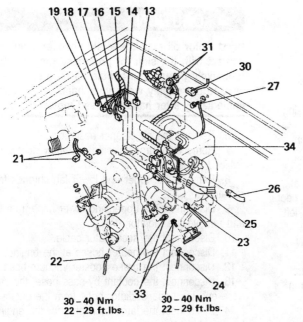

13 Throttle position sensor connector
14 Ignition coil connector
15 Power transistor connector
16 EGR temperature sensor connector
17 Engine coolant temperature gauge unit connector
18 Engine coolant temperaure sensor connector
19 Thermal switch connector <A/T>
20 Oxygen sensor connector
21 Generator connector
22 Oil pressure gauge unit connector
23 A/C engine coolant temperature switch connector <A/C>
24 Ground cable connection
25 Emission control vacuum hose connection
26 Brake booster vacuum hose connection
27 Ground cable connection
28 Idle speed control motor connector
29 Idle speed control motor position sensor connector
30 Idle air control motor connector
31 Control wiring harness connector
32 Heat protector
33 Engine mounting bolt
34 Engine assembly

Fig. 63 Component identification — 2.4L Engines California models

33. Remove the heat shield from the right motor mount.
34. Double check for any remaining cables, hoses, wiring or lines still running to the engine and remove them. Check that any item which was moved out of the way is truly in a safe location.
35. Double check the security of the lifting apparatus. Make certain the cables or chains are tensioned properly. Remove the bolts holding the motor mounts to the vehicle frame.
36. Elevate the hoist slowly and raise the engine assembly out of the truck. Take your time; clearances may be close and the unit is heavy. As soon as the engine is clear, mount it on a stand or support it on wooden blocks. Do not allow it to rest on the oil pan or lie on its side. Never leave an engine hanging from a hoist.

To install:
37. After repairs, make certain the engine is fully reassembled before installation. All components removed when the engine was out of the truck should be in place before reinstallation.
38. Install the engine into the truck and lower it until the bolt holes for the mounts align with the frame rails. Install the bolts and tighten them to 22-29 ft. lbs. (30-40 Nm). Install the heat shield on the right mount.
39. Connect the control wiring harness, the idle air control motor wires, the idle speed control motor position sensor wires, idle speed control motor wires, and the ground cables at the lower left engine block and the firewall above the back of the engine.
40. Fasten the brake booster vacuum hose connection and the emission control vacuum hose connection.
41. Fasten the air conditioning engine coolant temperature switch connector, the oil pressure gauge unit connector, the alternator connector, the oxygen sensor connector, the thermal switch connector (if equipped with an automatic transmission), the engine coolant temperature sensor connector, the engine coolant temperature gauge unit connector, the EGR temperature sensor connector, The power transistor connector, the ignition coil connector and the throttle position sensor connector.
42. Reconnect the water hose, the fuel return hose and the high pressure fuel hose. Check the high pressure fuel hose O-ring for cracks or other damage, replace if any is found. It may be best to replace this O-ring even if no damage is found.

3-50 ENGINE AND ENGINE REBUILDING

43. Fasten the cruise control vacuum hose, if so equipped.
44. Install the power steering pump and the power steering pump drive belt. See Section 1 for adjusting procedures.
45. Install the A/C compressor and drive belt. Refer to Section 1 for adjusting procedures.
46. Fasten the throttle control cable connection, if the vehicle is equipped with an automatic transmission.
47. Fasten the accelerator cable connection.
48. Install the air cleaner duct hose.
49. Install the radiator, tightening the mounting bolts to 6-8 ft. lbs. (8-11 Nm). Install the fan shrouds.
50. Reinstall the transmission adhering to the procedures given in Section 7.
51. Connect the exhaust pipe to the exhaust manifold using a new exhaust gasket.
52. Install the correct amounts and types of oil for the engine and transmission.
53. Fill the cooling system with coolant.
54. Double check all installation items, paying particular attention to loose hoses or hanging wires, untightened nuts, poor routing of hoses and wires (too tight or rubbing) and tools left in the engine area.
55. Lower the vehicle to the ground. Connect the negative battery cable.
56. After making certain that the transmission is in Neutral (manual) or P (automatic), start the engine, allowing it to run at idle. Check carefully for any oil, fuel, vacuum or coolant leaks.
57. Shut the engine off. Top off fluids as necessary.
58. Install the undercovers and splash shields.
59. Install the hood, taking care to align the seams and latch correctly.
60. Perform final adjustments to the belts, throttle cable, idle speed etc. as necessary.
61. Take the vehicle back to the air conditioning mechanic to have the system evacuated and recharged as soon as possible.

Montero (2.6L Engines)
♦ See Figure 64

➡ The transmission, transfer case and front driveshaft must be removed before the engine can be removed. Please refer to Section 7 for detailed instructions on the removal of these components.

1. If the vehicle is equipped with air conditioning, first have the A/C system discharged by a trained and certified mechanic using a recovery/recycling machine.
2. Matchmark and remove the hood. Disconnect the negative battery cable.
3. Elevate and safely support the truck on jackstands. Remove the undercovers and shields below the engine and transmission.
4. Drain the engine coolant.

✱✱CAUTION

When draining the coolant, keep in mind that cats and dogs are attracted by ethylene glycol antifreeze, and are quite likely to drink any that is left in an uncovered container or in puddles on the ground. This will prove fatal in sufficient quantity. Always drain the coolant into a sealable container. Coolant should be reused unless it is contaminated or several years old.

5. Drain the engine oil and the transmission oil. Drain the transfer case and the front axle lubricant.

✱✱CAUTION

Used motor oil may cause skin cancer if repeatedly left in contact with the skin for prolonged periods. Although this is unlikely unless you handle oil on a daily basis, it is wise to thoroughly wash your hands with soap and water immediately after handling used motor oil.

6. Remove the air cleaner assembly.
7. Remove the transmission assembly, following procedures outlined in Section 7.
8. Remove the upper and lower fan shrouds from the radiator.
9. Disconnect the upper and lower hose from the radiator and remove the radiator.
10. Disconnect the accelerator cable.
11. Disconnect the heater hoses at the engine.
12. Disconnect the brake booster vacuum hose.
13. Disconnect the coolant by-pass hose, the small coolant hose and the upper radiator hose from the engine.
14. Disconnect the large multi-plug for the engine control harness.
15. Disconnect the engine coolant temperature sensor wire; if the vehicle is air conditioned, disconnect the coolant temperature switch wire.
16. Disconnect the vacuum hose and the coolant temperature gauge sender.
17. Disconnect the oxygen sensor (carefully) and the oil pressure sending unit wire.
18. Loosen the power steering pump and remove the belt.
19. Remove the power steering pump with the hoses attached and undisturbed; place or hang the pump out of the way. Do not disconnect the hoses.
20. Loosen the air conditioning compressor and remove the belt.
21. Remove the air conditioning compressor with the hoses attached and undisturbed; place or hang the unit out of the way. Do not loosen or disconnect the hoses; do not allow the hoses to become crimped or crushed when the unit is placed aside.

✱✱WARNING

Do not allow the compressor to hang by the hoses.

22. Disconnect the fuel hoses (on fuel injected models release the fuel pressure first!). Label or identify them for correct reinstallation.
23. Disconnect the 3 wiring connectors at the alternator.
24. Remove the ground cable from the engine block.
25. Disconnect the exhaust pipe from the bottom of the exhaust manifold. Use wire or string to support the pipe to the side. Do not let the pipe hang of its own weight.
26. Carefully label and disconnect any breather, vacuum and canister lines running to the firewall or body.

ENGINE AND ENGINE REBUILDING 3-51

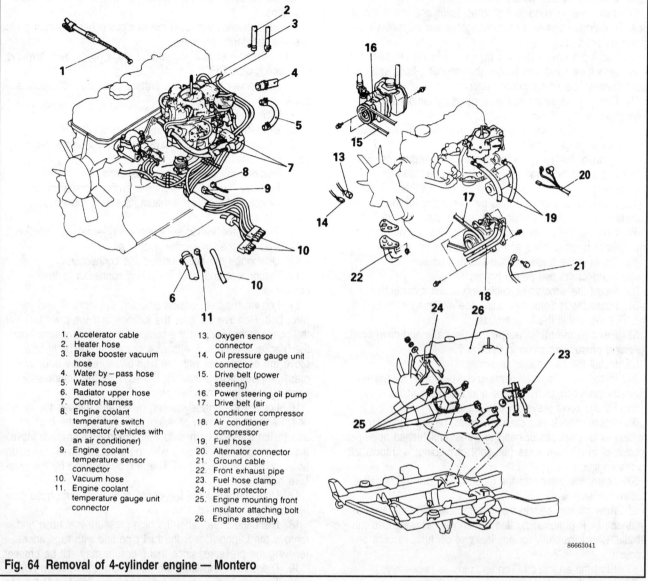

1. Accelerator cable
2. Heater hose
3. Brake booster vacuum hose
4. Water by-pass hose
5. Water hose
6. Radiator upper hose
7. Control harness
8. Engine coolant temperature switch connector (vehicles with an air conditioner)
9. Engine coolant temperature sensor connector
10. Vacuum hose
11. Engine coolant temperature gauge unit connector
13. Oxygen sensor connector
14. Oil pressure gauge unit connector
15. Drive belt (power steering)
16. Power steering oil pump
17. Drive belt (air conditioner compressor
18. Air conditioner compressor
19. Fuel hose
20. Alternator connector
21. Ground cable
22. Front exhaust pipe
23. Fuel hose clamp
24. Heat protector
25. Engine mounting front insulator attaching bolt
26. Engine assembly

Fig. 64 Removal of 4-cylinder engine — Montero

27. Remove the clamp holding the fuel lines to the block.
28. Install the hoisting equipment and make sure it is securely fastened to the engine. Draw tension on the hoist just enough to support the weight of the motor without elevating it.

❊❊CAUTION

The truck is supported on stands. Tension the hoist slowly and do not disturb the position of the truck on the supports.

29. Remove the heat shield from the right motor mount.
30. Double check for any remaining cables, hoses, wiring or lines still running to the engine and remove them. Check that any item which was moved out of the way is truly in a safe location.
31. Double check the security of the lifting apparatus. Make certain the cables or chains are tensioned properly. Remove the bolts holding the motor mounts to the vehicle frame.
32. Elevate the hoist slowly and raise the engine and transmission assembly out of the truck. The assembly may need to be lifted out on an angle. Take your time; clearances may be close and the unit is heavy. As soon as the engine is clear, mount it on a stand or support it on wooden blocks. Do not allow it to rest on the oil pan or lie on its side. Never leave an engine hanging from a hoist.

To install:
33. After repairs, make certain the engine is fully reassembled before installation. All components removed when the engine was out of the truck should be in place before reinstallation.
34. Install the engine into the truck and lower it until the bolt holes for the mounts align with the frame rails. Install the bolts and tighten them to 26 ft. lbs. (35 Nm). Install the heat shield on the right mount.
35. Clamp the fuel lines into place.
36. Connect the various vacuum lines and breather hoses running to the body and firewall.
37. Use a new gasket and new bolts to connect the exhaust pipe to the manifold.
38. Connect the ground cable to the engine.
39. Connect the wiring to the alternator.

3-52 ENGINE AND ENGINE REBUILDING

40. Connect the fuel hoses.
41. Install the air conditioning compressor and install the belt. Tension the belt by hand and tighten the compressor mounting bolts snug.
42. Install the power steering pump and install the belt. Tension the belt by hand and tighten the mounting bolts snug. Both belts will be final tightened later.
43. Connect the wiring to the oil pressure sender and the oxygen sensor.
44. Connect the vacuum hose and the wiring to the coolant temperature gauge sender.
45. Connect the wiring to the coolant temperature sender and the coolant temperature switch if equipped with air conditioning.
46. Carefully connect the multi-pin connector for the engine control harness. Make certain each pin is firmly seated.
47. Connect the upper radiator hose to the engine. Install the coolant by-pass hose and the small coolant hose.
48. Connect the brake booster vacuum hose.
49. Connect the two heater hoses.
50. Install the accelerator cable and adjust it correctly.
51. Install the radiator, tightening the mounting bolts to 5 ft. lbs. (7 Nm). Install the fan shrouds.
52. Reinstall the transmission, transfer case and drive shafts following procedures given in Section 7.
53. Install the air cleaner assembly.
54. Install the correct amounts and types of oil for the engine, transmission, transfer case and axle assemblies.
55. Fill the cooling system with coolant.
56. Double check all installation items, paying particular attention to loose hoses or hanging wires, untightened nuts, poor routing of hoses and wires (too tight or rubbing) and tools left in the engine area.
57. Lower the vehicle to the ground. Connect the negative battery cable.
58. After making certain that the transmission is in Neutral (manual) or P (automatic), start the engine, allowing it to run at idle. Check carefully for any leaks of oil, fuel, vacuum or coolant.
59. Shut the engine off. Top up fluids as necessary.
60. Install the undercovers and splash shields.
61. Install the hood, taking care to align the seams and latch correctly.
62. Recharge the air conditioning system if so equipped.
63. Perform final adjustments to the belts, throttle cable, idle speed etc. as necessary.

4-Wheel Drive Pick-up & Montero (3.0L 12 Valve Engines)

▶ See Figures 65 and 66

➡ When performing this procedure you will need a new fuel line O-ring, two (2) new exhaust manifold to front exhaust pipe gaskets, new oil, transmission fluid, and antifreeze.

1. Matchmark and remove the hood.
2. Elevate and safely support the truck on jackstands. Remove the undercovers and shields below the engine and transmission.

➡ While removing or disconnecting any wires or hoses, make sure to label them. This will make reassembly much easier

3. Depressurize the fuel system by adhering to the following steps:
   a. Disconnect and label the fuel pump harness wiring plug at the fuel tank rear side.
   b. Start the engine and after it stops on its own, turn the ignition switch OFF.
4. Disconnect the negative battery cable from the engine block.
5. Drain the cooling system.
6. Drain the engine and transmission fluid.
7. Disconnect the cooling system overflow, upper radiator, and lower radiator hoses.
8. Remove the one piece radiator shroud.
9. Remove the radiator from the engine compartment.
10. Disconnect the front exhaust pipe from the exhaust manifolds.
11. Remove the transmission assembly — refer to Section 7.
12. Remove the air cleaner duct hose.
13. Disengage the accelerator cable connection.
14. Disconnect and label the wiring connector to the compressor.
15. Loosen the belt tension adjuster, the lockbolt and the pivot bolt. Remove the belt, the lockbolt and the pivot bolt from the compressor. Support the compressor and remove it from the bracket and hook it with wire or rope to the body side. Move the compressor with the high-pressure hose and low-pressure hose still connected. DO NOT let the compressor hang by the high-pressure hoses.
16. Loosen the power steering pump and remove the drive belt. Remove the power steering oil pump from the bracket and fasten it to the body side with strong wire or cord. Move the power steering oil pump with the pressure hose and return hose still connected. DO NOT let the pump hang by the pressure hoses.
17. Unfasten and label the cruise control vacuum hose connection, if so equipped.
18. Disconnect and label the high pressure fuel hose and remove the O-ring. Cover the fuel pipe line with rags after relieving the pressure, since fuel pressure may still be present.
19. Disconnect and label the fuel return and vacuum hoses.
20. Remove both water hose connections located at the top rear of the engine.
21. Remove and label the throttle position sensor, ignition coil and ignition power transistor assembly, EGR temperature sensor, engine coolant temperature gauge unit, engine coolant temperature sensor, alternator, oil pressure gauge unit, and air conditioning engine coolant temperature switch connectors.
22. Unscrew and label the ground cable connections from the left hand side of the intake plenum and the lower left hand side of the engine block.
23. Disengage and label the emission control vacuum hose connection.
24. Remove and label the brake booster vacuum hose connection.
25. Disconnect and label the ground cable from the firewall above the back of the engine.
26. Remove and label the idle air control motor connector.
27. Disconnect and label the control wiring harness.

ENGINE AND ENGINE REBUILDING 3-53

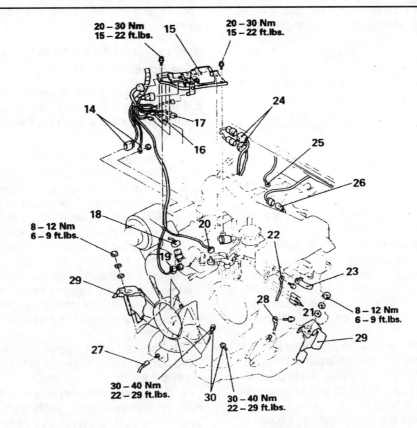

- 14 Generator connector
- 15 Ignition coil and ignition power transistor assembly
- 16 Idle air control motor connector
- 17 Throttle position sensor connector
- 18 Air conditioning engine coolant temperature switch connector <A/C>
- 19 Engine coolant temperature sensor connector
- 20 Engine coolant temperature gauge unit connector
- 21 Emission control vacuum hose connection
- 22 Ground cable connection
- 23 Brake booster vacuum hose connection
- 24 Control wiring harness connector
- 25 Ground cable connection
- 26 EGR temperature sensor connector
- 27 Oil pressure gauge unit connector
- 28 Ground cable connection
- 29 Heat protector
- 30 Engine mounting bolt

Fig. 65 Component identification — 3.0L (12 valve) Engines

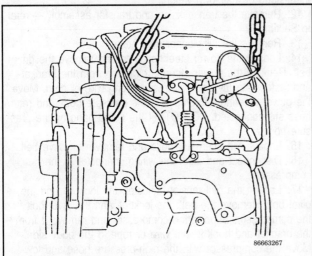

Fig. 66 When hoisting the engine, use the supplied engine chain blocks

28. Install the hoisting equipment and make sure it is securely fastened to the engine. Draw tension on the hoist just enough to support the weight of the motor without elevating it.

✳✳CAUTION

The truck is supported on stands. Tension the hoist slowly and do not disturb the position of the truck on the supports.

29. Remove the heat protector from both motor mounts.
30. Double check for any remaining cables, hoses, wiring or lines still running to the engine and remove them. Check that any item which was moved out of the way is truly in a safe location.
31. Double check the security of the lifting apparatus. Make certain the cables or chains are tensioned properly. Remove the bolts holding the motor mounts to the vehicle frame.
32. Elevate the hoist slowly and raise the engine assembly out of the truck. Take your time; clearances may be close and the unit is heavy. As soon as the engine is clear, mount it on

3-54 ENGINE AND ENGINE REBUILDING

a stand or support it on wooden blocks. Do not allow it to rest on the oil pan or lie on its side. Never leave an engine hanging from a hoist.

To install:

33. After repairs, make certain the engine is fully reassembled before installation. All components removed when the engine was out of the truck should be in place before reinstallation.
34. Install the engine into the truck and lower it until the bolt holes for the mounts align with the frame rails. Install the bolts and tighten them to 22-29 ft. lbs. (30-40 Nm). Install the heat shields on the motor mounts.
35. Connect the control wiring harness, the idle air control motor wires and the three engine ground cables.
36. Fasten the brake booster vacuum hose connection and the emission control vacuum hose connection.
37. Fasten the air conditioning engine coolant temperature switch connector, the oil pressure gauge unit connector, the alternator connector, the engine coolant temperature sensor connector, the engine coolant temperature gauge unit connector, the EGR temperature sensor connector, the ignition coil and ignition power transistor assembly, and the throttle position sensor connector.
38. Reconnect both water hoses at the rear of the intake plenum, the fuel return hose and the high pressure fuel hose. Check the high pressure fuel hose O-ring for cracks or other damage, replace if any is found. It may be best to replace this O-ring even if no damage is found.
39. Reconnect the vacuum hose to the fuel return hose fixture.
40. Fasten the cruise control vacuum hose, if so equipped.
41. Install the power steering pump and the power steering pump drive belt. See Section 1 for adjusting procedures.
42. Install the A/C compressor and drive belt. Refer to Section 1 for adjusting procedures.
43. Fasten the throttle control cable connection, if the vehicle is equipped with an automatic transmission.
44. Fasten the accelerator cable connection.
45. Install the air cleaner duct hose.
46. Install the radiator, tightening the mounting bolts to 6-8 ft. lbs. (8-11 Nm). Install the fan shroud.
47. Reinstall the transmission adhering to the procedures given in Section 7.
48. Connect the front exhaust pipe to the exhaust manifolds using a new gasket.
49. Install the correct amounts and types of oil for the engine and transmission.
50. Fill the cooling system with coolant.
51. Double check all installation items, paying particular attention to loose hoses or hanging wires, untightened nuts, poor routing of hoses and wires (too tight or rubbing) and tools left in the engine area.
52. Lower the vehicle to the ground. Reconnect the negative battery cable.
53. After making certain that the transmission is in Neutral (manual) or **P** (automatic), start the engine, allowing it to run at idle. Check carefully for any oil, fuel, vacuum or coolant leaks.
54. Shut the engine off. Top off all fluids as necessary.
55. Install the undercovers and splash shields.
56. Install the hood, taking care to align the seams and latch correctly.
57. Perform final adjustments to the belts, throttle cable, idle speed etc. as necessary.

Montero (3.0L 24 Valve Engines)
◆ See Figures 67 and 68

➡When performing this procedure you will need a new fuel line O-ring, two (2) new exhaust manifold to front exhaust pipe gaskets, fresh oil, transmission fluid, and antifreeze.

1. Matchmark and remove the hood.
2. Remove the battery and battery tray.
3. Remove the cruise control intermediate link, which is mounted to the firewall of the engine compartment.
    a. Remove the retaining bolt on the link protector and withdrawal the protector from the assembly.
    b. Disconnect the accelerator cable, the throttle cable, and the cruise control cable from the intermediate link. When loosening the adjusting bolts and nuts on the throttle lever, **do not remove** the adjusting bolts or nuts (just loosen).
    c. Remove the four retaining nuts holding the intermediate link to the link bracket. Withdrawal the intermediate link from the bracket.
    d. Remove the link bracket. Make sure to keep all screws and bolts in labeled bags or a similar container to keep from mixing them up.
4. Elevate and safely support the truck on jackstands. Remove the undercovers and shields below the engine and transmission.
5. Disconnect the negative battery cable from the engine block and set aside.
6. Drain the cooling system, refer to Section 1.
7. Drain the engine and transmission fluid, refer to Section 1.
8. Disconnect the cooling system overflow, upper radiator, and lower radiator hoses.
9. Remove the one piece radiator shroud.
10. Remove the radiator from the engine compartment.
11. Disconnect the front exhaust pipes from the exhaust manifolds. Tie them up with strong cord or wire, do not let them hang of their own weight.
12. Remove the transmission and transfer assembly — refer to Section 7.
13. Remove the air cleaner duct hose.
14. Loosen the power steering pump and remove the drive belt. Remove the power steering oil pump from the bracket and fasten it to the body side with strong wire or cord. Move the power steering oil pump with the pressure hose and return hose still connected. DO NOT let the pump hang by the pressure hoses.
15. Loosen the alternator bolts and remove the drive belt.
16. Disconnect and label the wiring connector to the compressor.
17. Loosen the belt tension adjuster, the lockbolt and the pivot bolt. Remove the belt, the lockbolt and the pivot bolt from the compressor. Support the compressor and remove it from the bracket and hook it with wire or rope to the body side. Move the compressor with the high-pressure hose and low-pressure hose still connected. DO NOT let the compressor hang by the high-pressure hoses.
18. Remove the cooling fan and water pump pulley from the front of the engine.

ENGINE AND ENGINE REBUILDING 3-55

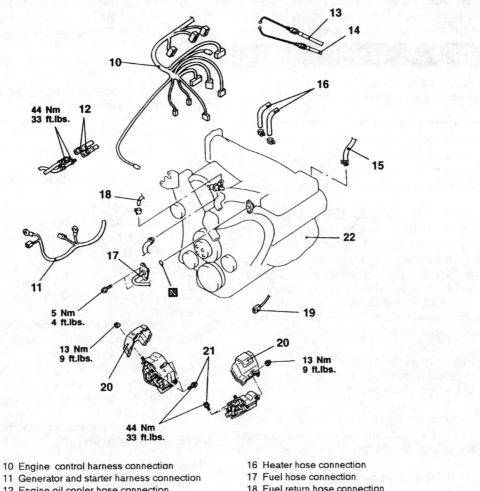

10 Engine control harness connection
11 Generator and starter harness connection
12 Engine oil cooler hose connection
13 Accelerator cable connection
14 Throttle cable connection
15 Brake booster vacuum hose connection
16 Heater hose connection
17 Fuel hose connection
18 Fuel return hose connection
19 Oil pressure switch harness connection
20 Heat protectors
21 Engine mounting bolt

Fig. 67 Component connection identification — 3.0L (24 valve) engines

19. Disconnect and label the engine control wiring harness.
20. Disconnect and label the alternator and starter wiring harness.
21. Unfasten the engine oil cooler hose connection. To loosen these connectors, use two wrenches, usually open end or flare nut, at the same time.
22. Remove the brake booster vacuum hose connection.
23. Remove the two heater hose connections, located at the back of the intake manifold.
24. Disconnect the fuel hose from the intake plenum. The O-ring inside of this connection must be removed and replaced with a new one.
25. Disconnect the fuel return hose.
26. Unfasten the oil pressure switch harness connection.

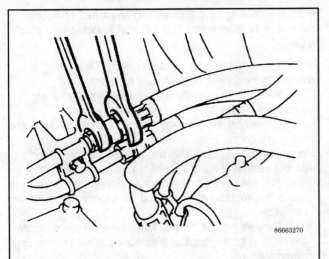

Fig. 68 Use two wrenches to disconnect the oil cooler hose

3-56 ENGINE AND ENGINE REBUILDING

27. Install the hoisting equipment and make sure it is securely fastened to the engine. Draw tension on the hoist just enough to support the weight of the motor without elevating it.

✱✱CAUTION

The truck is supported on stands. Tension the hoist slowly and do not disturb the position of the truck on the supports.

28. Remove the heat protector from both motor mounts.
29. Double check for any remaining cables, hoses, wiring or lines still running to the engine and remove them. Check that any item which was moved out of the way is truly in a safe location.
30. Double check the security of the lifting apparatus. Make certain the cables or chains are tensioned properly. Remove the bolts holding the motor mounts to the vehicle frame.
31. Elevate the hoist slowly and raise the engine assembly out of the truck. Take your time; clearances may be close and the unit is heavy. As soon as the engine is clear, mount it on a stand or support it on wooden blocks. Do not allow it to rest on the oil pan or lie on its side. Never leave an engine hanging from a hoist.

To install:

32. After repairs, make certain the engine is fully reassembled before installation. All components removed when the engine was out of the truck should be in place before reinstallation.
33. Install the engine into the truck and lower it until the bolt holes for the mounts align with the frame rails. Install the bolts and tighten them to 33 ft. lbs. (44 Nm). Install the heat shields on the motor mounts.
34. Reconnect the oil pressure switch harness.
35. Fasten the fuel return hose, the fuel hose, and the two heater hoses. Make sure to replace the O-ring in the fuel hose connection with a new one.
36. Fasten the brake booster vacuum hose connection.
37. Reconnect the oil cooler hose.
38. Reconnect the alternator and starter wiring harness, the engine control harness and the lower, left-hand side engine block ground cable.
39. Install the cooling fan and water pump pulley. Tighten the cooling fan mounting bolts to 7-9 ft. lbs. (10-12 Nm).
40. Install the power steering pump and power steering pump cover.
41. Install the A/C compressor and drive belt. Refer to Section 1 for adjusting procedures.
42. Install the alternator drive belt. Refer to Section 1 for adjusting procedures.
43. Install the power steering pump drive belt. See Section 1 for adjusting procedures.
44. Install the air cleaner duct hose.
45. Install the radiator, tightening the mounting bolts to 6-8 ft. lbs. (8-11 Nm). Install the fan shroud.
46. Reattach the upper and lower radiator hoses, and the overflow hose.
47. Reinstall the transmission and transfer assembly adhering to the procedures given in Section 7.
48. Reattach the front exhaust pipe, or three-way catalytic converter on California models, to the exhaust manifolds. New gaskets are a necessity for this procedure.
49. Install the cruise control intermediate link assembly.
    a. Reattach the link bracket with the center bolt.
    b. Install the intermediate link to the link bracket using the four nuts.
    c. Connect the accelerator cable, the throttle cable, and the cruise control cable to the intermediate link.
50. Adjust the accelerator cable, the throttle cable, and the cruise control cable — refer to Section 5 for this procedure.
51. Install the intermediate link unit's cover.
52. Install the battery and battery tray.
53. Reconnect the negative battery cable to the engine block.
54. Install the correct amounts and types of oil for the engine and transmission.
55. Fill the cooling system with coolant.
56. Double check all installation items, paying particular attention to loose hoses or hanging wires, untightened nuts, poor routing of hoses and wires (too tight or rubbing) and tools left in the engine area.
57. Lower the vehicle to the ground.
58. Reconnect the positive and negative battery cables to the battery.
59. After making certain that the transmission is in Neutral or Park, start the engine, allowing it to run at idle. Check carefully for any oil, fuel, vacuum or coolant leaks.
60. Shut the engine off. Top off all fluids as necessary.
61. Install the undercovers and splash shields.
62. Install the hood, taking care to align the seams and latch correctly.
63. Perform final adjustments to the belts, throttle cable, idle speed etc. as necessary.

Montero (3.5L Engines)

▶ See Figure 69

➡ **When performing this procedure you will need two (2) new exhaust manifold to front exhaust pipe gaskets, fresh oil, transmission fluid, and antifreeze.**

1. Matchmark and remove the hood.
2. Remove the battery and battery tray.

➡ **While removing or disconnecting any wires or hoses, make sure to label them. This will make reassembly much easier.**

3. Remove the cruise control intermediate link, which is mounted to the firewall of the engine compartment.
    a. Remove the retaining bolt on the link protector and withdrawal the protector from the assembly.
    b. Disconnect the accelerator cable, the throttle cable, and the cruise control cable from the intermediate link. When loosening the adjusting bolts and nuts on the throttle lever, **do not remove** the adjusting bolts or nuts (just loosen).
    c. Remove the four retaining nuts holding the intermediate link to the link bracket. Withdrawal the intermediate link from the bracket.
    d. Remove the link bracket. Make sure to keep all screws and bolts in labeled bags or a similar container to keep from mixing them up.
4. Elevate and safely support the truck on jackstands. Remove the undercovers and shields below the engine and transmission.

ENGINE AND ENGINE REBUILDING

5. Disconnect the negative battery cable from the engine block and set aside.
6. Drain the cooling system, refer to Section 1.
7. Drain the engine and transmission fluid, refer to Section 1.
8. Disconnect the cooling system overflow, upper radiator, and lower radiator hoses.
9. Remove the one piece radiator shroud.
10. Remove the radiator from the engine compartment.
11. Disconnect the front exhaust pipes, or warm up three-way catalytic converter if California model, from the exhaust manifolds. Tie them up with strong cord or wire, do not let them hang of their own weight.
12. Remove the transmission and transfer assembly — refer to Section 7.
13. Remove the air cleaner duct hose.
14. Remove the intake manifold plenum cover.
15. Loosen the power steering pump and remove the drive belt. Remove the power steering oil pump from the bracket and fasten it to the body side with strong wire or cord. Move the power steering oil pump with the pressure hose and return hose still connected. DO NOT let the pump hang by the pressure hoses.
16. Loosen the alternator bolts and remove the alternator drive belt.
17. Disconnect and label the wiring connector to the compressor.
18. Loosen the belt tension adjuster, the lockbolt and the pivot bolt. Remove the belt, the lockbolt and the pivot bolt from the compressor. Support the compressor and remove it from the bracket and hook it with wire or rope to the body side. Move the compressor with the high-pressure hose and low-pressure hose still connected. DO NOT let the compressor hang by the high-pressure hoses.
19. Remove the cooling fan and water pump pulley from the front of the engine.
20. Disconnect and label the engine control wiring harness.
21. Disconnect and label the alternator and starter wiring harness.
22. Unfasten the engine oil cooler hose connection. To loosen these connectors, use two wrenches, usually open end or flare nut, at the same time.

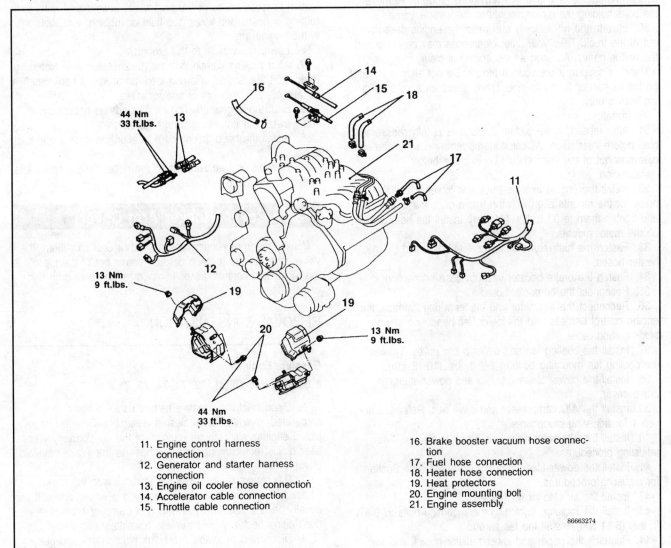

11. Engine control harness connection
12. Generator and starter harness connection
13. Engine oil cooler hose connection
14. Accelerator cable connection
15. Throttle cable connection
16. Brake booster vacuum hose connection
17. Fuel hose connection
18. Heater hose connection
19. Heat protectors
20. Engine mounting bolt
21. Engine assembly

Fig. 69 Component connection identification — 3.5L engines

ENGINE AND ENGINE REBUILDING

23. Remove the brake booster vacuum hose connection.
24. Remove the two heater hose connections, located at the back of the intake manifold.
25. Disconnect the fuel hose and fuel return hose from the left-hand side of the intake plenum.
26. Install the hoisting equipment and make sure it is securely fastened to the engine. Draw tension on the hoist just enough to support the weight of the motor without elevating it.

> **✱✱CAUTION**
>
> **The truck is supported on stands. Tension the hoist slowly and do not disturb the position of the truck on the supports.**

27. Remove the heat protector from both motor mounts.
28. Double check for any remaining cables, hoses, wiring or lines still running to the engine, label and remove them. Check that any item which was moved out of the way is truly in a safe location.
29. Double check the security of the lifting apparatus. Make certain the cables or chains are tensioned properly. Remove the bolts holding the motor mounts to the vehicle frame.
30. Elevate the hoist slowly and raise the engine assembly out of the truck. Take your time; clearances may be close and the unit is heavy. As soon as the engine is clear, mount it on a stand or support it on wooden blocks. Do not allow it to rest on the oil pan or lie on its side. Never leave an engine hanging from a hoist.

To install:

31. After repairs, make certain the engine is fully reassembled before installation. All components removed when the engine was out of the truck should be in place before reinstallation.
32. Install the engine into the truck and lower it until the bolt holes for the mounts align with the frame rails. Install the bolts and tighten them to 33 ft. lbs. (44 Nm). Install the heat shields on the motor mounts.
33. Fasten the fuel return hose, the fuel hose, and the two heater hoses.
34. Fasten the brake booster vacuum hose connection.
35. Reconnect the oil cooler hose.
36. Reconnect the alternator and starter wiring harness, the engine control harness and the lower, left-hand side engine block ground cable.
37. Install the cooling fan and water pump pulley. Tighten the cooling fan mounting bolts to 7-9 ft. lbs. (10-12 Nm).
38. Install the power steering pump and power steering pump cover.
39. Install the A/C compressor and drive belt. Refer to Section 1 for adjusting procedures.
40. Install the alternator drive belt. Refer to Section 1 for adjusting procedures.
41. Install the power steering pump drive belt. See Section 1 for adjusting procedures.
42. Install the air cleaner duct hose.
43. Install the radiator, tightening the mounting bolts to 6-8 ft. lbs. (8-11 Nm). Install the fan shroud.
44. Reattach the upper and lower radiator hoses, and the overflow hose.
45. Reinstall the transmission and transfer assembly adhering to the procedures given in Section 7.
46. Reattach the front exhaust pipe, or three-way catalytic converter on California models, to the exhaust manifolds. New gaskets are a necessity for this procedure.
47. Install the cruise control intermediate link assembly.
    a. Reattach the link bracket with the center bolt.
    b. Install the intermediate link to the link bracket using the four nuts.
    c. Connect the accelerator cable, the throttle cable, and the cruise control cable to the intermediate link.
48. Adjust the accelerator cable, the throttle cable, and the cruise control cable — refer to Section 5 for this procedure.
49. Install the intermediate link unit's cover.
50. Install the battery and battery tray.
51. Reconnect the negative battery cable to the engine block.
52. Install the correct amounts and types of oil for the engine and transmission.
53. Fill the cooling system with coolant.
54. Reconnect the positive and negative battery cables to the battery.
55. Double check all installation items, paying particular attention to loose hoses or hanging wires, untightened nuts, poor routing of hoses and wires (too tight or rubbing) and tools left in the engine area.
56. Lower the vehicle to the ground.
57. After making certain that the transmission is in Neutral or Park, start the engine, allowing it to run at idle. Check carefully for any oil, fuel, vacuum or coolant leaks.
58. Shut the engine off. Top off all fluids as necessary.
59. Install the undercovers and splash shields.
60. Install the hood, taking care to align the seams and latch correctly.
61. Perform final adjustments to the belts, throttle cable, idle speed etc. as necessary.

Valve Cover (Rocker Arm Cover)

Usually made of either pressed steel or cast aluminum, the valve cover serves to keep dirt and debris out of the top of the motor and contain the oil pumped to the valve train and camshafts.

REMOVAL & INSTALLATION

Gasoline Engines
◆ See Figures 70, 71, 72, 73, 74, 75, 76 and 77

1. Disconnect the negative battery cable. If the engine is carbureted, remove the air cleaner assembly. Certain fuel injected engines also require removal of the air cleaner for access or just detachment of the air hose to the intake manifold plenum.
2. Disconnect the breather hoses running to the cover. Check for any other hose or line running over the cover. It can usually be repositioned by removing a clamp or bracket — don't disconnect any hoses unless absolutely necessary.
3. Disconnect and label the spark plug cables and place them to the side. On DOHC engines, remove the plug connectors very carefully — they are easily damaged.

ENGINE AND ENGINE REBUILDING 3-59

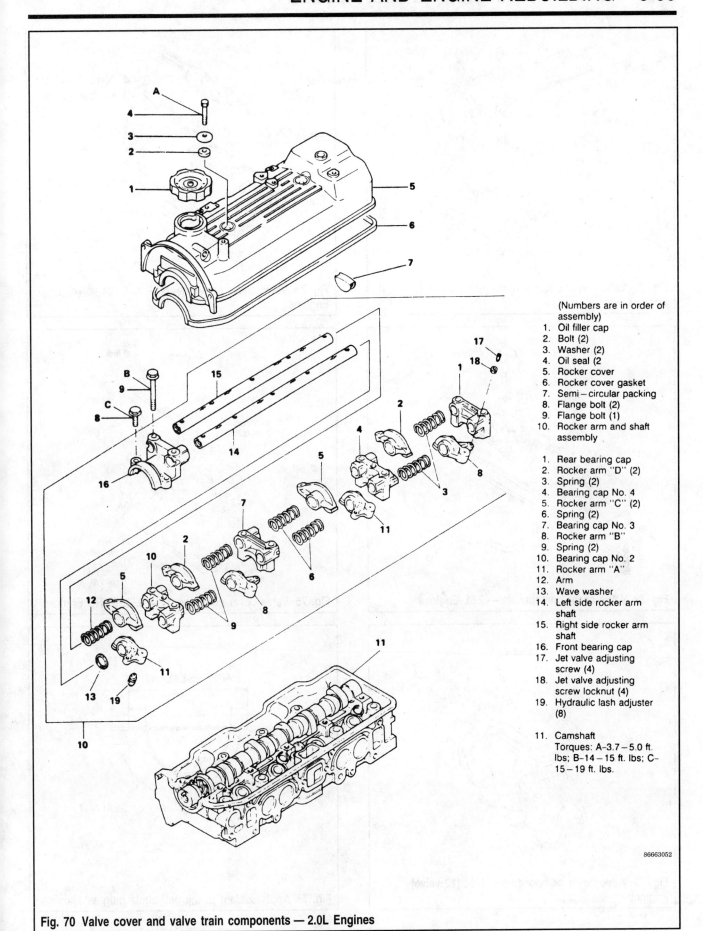

Fig. 70 Valve cover and valve train components — 2.0L Engines

3-60 ENGINE AND ENGINE REBUILDING

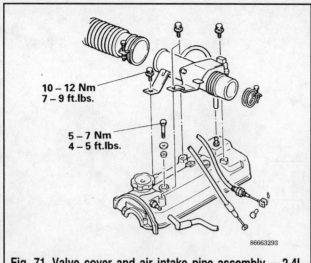

Fig. 71 Valve cover and air intake pipe assembly — 2.4L Engines

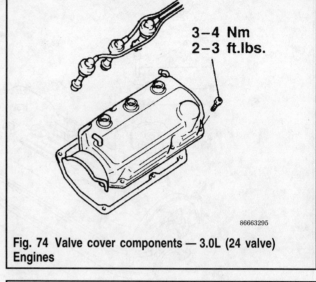

Fig. 74 Valve cover components — 3.0L (24 valve) Engines

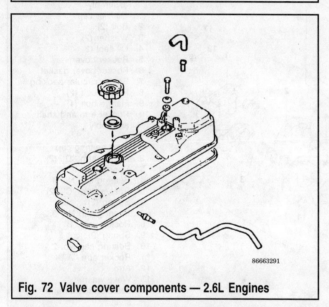

Fig. 72 Valve cover components — 2.6L Engines

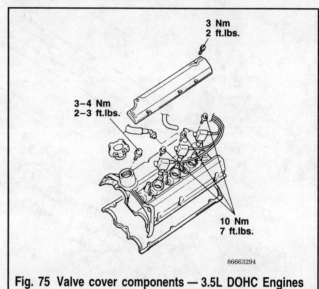

Fig. 75 Valve cover components — 3.5L DOHC Engines

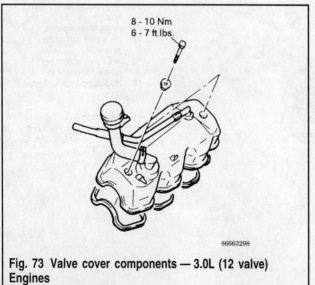

Fig. 73 Valve cover components — 3.0L (12 valve) Engines

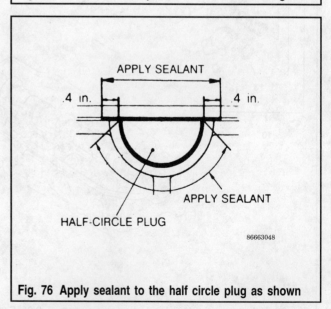

Fig. 76 Apply sealant to the half circle plug as shown

ENGINE AND ENGINE REBUILDING

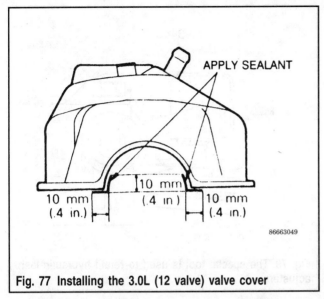

Fig. 77 Installing the 3.0L (12 valve) valve cover

4. Look at the end of the valve cover closest to the pulleys. Some covers overlap the upper timing belt cover. If this is so, remove the retaining bolts holding the valve cover.

5. If the upper timing belt cover overlaps the valve cover, remove the timing belt cover, then remove the valve cover bolts. Cover the belt with a thick rag to protect it.

6. Gently lift the valve cover up and off the head. If it is stuck, release it by tapping it with a wooden tool handle or plastic mallet. Do not ever pry on it; this will distort the cover and guarantee an oil leak after reassembly.

✱✱WARNING

Lift the valve cover and gasket away from the timing belt. Do not allow any oil or foreign matter to fall on the belt. It can be damaged by petroleum based products.

7. At the end farthest from the pulleys, remove the half-circle seal from either the head or the valve cover. The V6 engines do not use this seal.

To install:

8. Using a parts cleaning solvent and a stiff brush, clean the valve cover inside and out. The half-circle seal will have the remains of sealant on it; clean the seal and the head or cover surface to which it mates. If these surfaces are not clean, the new sealant won't adhere properly. Clean the gasket mating surfaces on the head.

9. Check the gasket. The square-section rubber gaskets are generally reusable, but they must be clean and free of distortion. Once the gasket has been on and off a few times, it should be replaced as a preventative measure. Check the groove or channel in the cover, making sure there is no dirt or foreign matter present.

10. Before installing the half-circle plug, coat all of its contact surfaces with a bead of oil-proof gasket sealer. Install the plug in place. Place a very small amount of sealant on the head surface 0.4 in. (10mm) on each side of the seal. This small area will allow the gasket to seal under all temperature conditions as it expands and contracts. For V6 engines, apply a small amount of sealer to the valve cover gasket at the vertical corners.

11. Fit the gasket into the channel in the cover and place it into position on the head. Make sure the gasket stays in place and that the cover lines up properly on all edges.

12. If the timing belt cover was removed, reinstall it, making sure it fits properly and does not rub the belt.

13. Install the rubber bushings, the metal washers and the retaining bolts through the cover. The trick to a leak proof installation is even tightening of the bolts. Bolt torque should not exceed 48-60 inch lbs. (5.4-6.7 Nm); 36-48 inch lbs. (4.0-5.4 Nm) is preferred. The cover must be held just tightly enough to compress the gasket. Overtightening is a leading cause of oil leaks due to deformed valve covers; once distorted, a cover is almost impossible to straighten properly. Cast aluminum covers don't distort: they crack.

14. Reinstall the spark plug wires. Make certain that all wire guides or clips are reinstalled. If an engine ground wire was removed from the valve cover area, reinstall it.

15. Connect the breather hoses and attach the clamps for any other hoses removed. Install the air cleaner assembly as necessary. Reconnect the negative battery cable.

16. It is recommended that the vehicle not be started for 1-2 hours, giving the sealant time to set. While the manufacturer of the sealer may disagree, it seems reasonable to allow a tight bond to develop before subjecting it to a hot, oily environment.

17. After the engine has been run for 20-30 minutes, shut it off and check the valve cover gasket area for oil leaks. Do not attempt to cure a leak by tightening the bolts; it won't work. If leaks are present, the cover must be removed and the cause identified.

Diesel Engines
▶ See Figure 78

1. Disconnect the negative battery cable. Disconnect the breather hose from the valve cover.

2. Remove the two retaining bolts. Catch the washer and seal on each bolt as it is removed.

3. Lift the valve cover up and away from the head. If it does not come loose easily, tap it on a corner with a rubber or plastic mallet.

4. Remove the gasket and check it carefully. If it is not deformed, stretched, nicked or damaged, it may be reused.

5. Clean the inside of the valve cover thoroughly. If your oil and filter changes are frequent, there should be little or no sludge in the valve train.

To install:

6. Position the gasket into the cover, but do not use any sealer or adhesive. Put the cover in place on the engine.

7. Install the bolts, making sure the washer and seal are in place on each.

8. Tighten the bolts to 4 ft. lbs. (5 Nm). This is just tight enough to hold the cover firmly and develop a small amount of compression on the rubber gasket. Overtightening deforms the cover; the gasket cannot seal properly and oil leaks develop.

3-62 ENGINE AND ENGINE REBUILDING

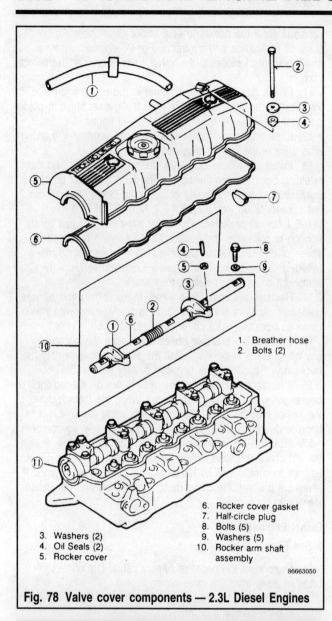

1. Breather hose
2. Bolts (2)
3. Washers (2)
4. Oil Seals (2)
5. Rocker cover
6. Rocker cover gasket
7. Half-circle plug
8. Bolts (5)
9. Washers (5)
10. Rocker arm shaft assembly

Fig. 78 Valve cover components — 2.3L Diesel Engines

Rocker Arms and Shafts

REMOVAL & INSTALLATION

2.0L Engines

▶ See Figures 79 and 80

➡ The Mitsubishi special tool No. MD 998443 (to retain the hydraulic lash adjusters) should be used before removing the rocker arm/shaft assembly.

1. Disconnect the negative battery cable. Remove the rocker cover and the upper timing belt cover. Loosen the camshaft sprocket bolt until you can turn it with your fingers.

2. Turn the engine over until the camshaft sprocket timing mark lines up with the timing mark on the cylinder head. The Top Dead Center mark on the front crankshaft pulley must line up with the timing scale on the front cover.

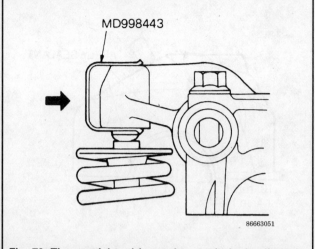

Fig. 79 The special tool is used to retain hydraulic lash adjusters

3. Remove the camshaft sprocket bolt, and without allowing tension on the belt to be lost, place the sprocket in the sprocket holder of the front cover or lower timing belt cover. Make sure you don't lose tension on the belt as this will require a lot of additional labor. Make sure, also, that you do not turn the crankshaft throughout the work. Put the special clips on the eight hydraulic adjusters at the outer ends of all eight rocker arms. Note that these clips go over the lash adjusters that actuate the large intake valves, not on the small adjusting screw for the smaller jet valves.

4. Loosen but do not remove the camshaft bearing cap bolts (not the outer or cylinder head bolts), rotating them outward in steps, a little at a time, moving from bolt to bolt. Remove the bolts and then, holding the ends so the assembly stays together, remove the rocker shaft assembly from the cylinder head.

To install:

5. Place the assembly on a clean work bench. Remove the bolts (which retain the bearing caps in position on the shafts) two at a time along with the associated cap, washers, springs and rockers. Continue until all parts are disassembled. Keep all parts in original order. All parts that will be re-used must be in original positions.

6. Inspection of parts should include verifying that the pads that follow the cam lobes are not worn excessively and that the oil holes are clear. Replace the rockers and/or shafts if rockers are loose on the shafts. If any of the hydraulic lash adjusters have been diagnosed as defective (valves tapping or not closing fully) they can be replaced by removing clips, allowing the adjuster to fall out of the rocker, and replacing the adjuster with another. Install the clip to keep the new adjuster from dropping out of the arm.

7. Assemble the parts of the rocker assembly as follows:

   a. Install the left and right side rocker shafts into the front bearing cap. Notches in the ends of the shaft must be upward.

   b. Install the bolts for the front cap to retain the shafts in place. Note that the left rocker shaft is longer than the right rocker shaft.

   c. Install the wave washer onto the left rocker shaft with the bulge forward.

ENGINE AND ENGINE REBUILDING   3-63

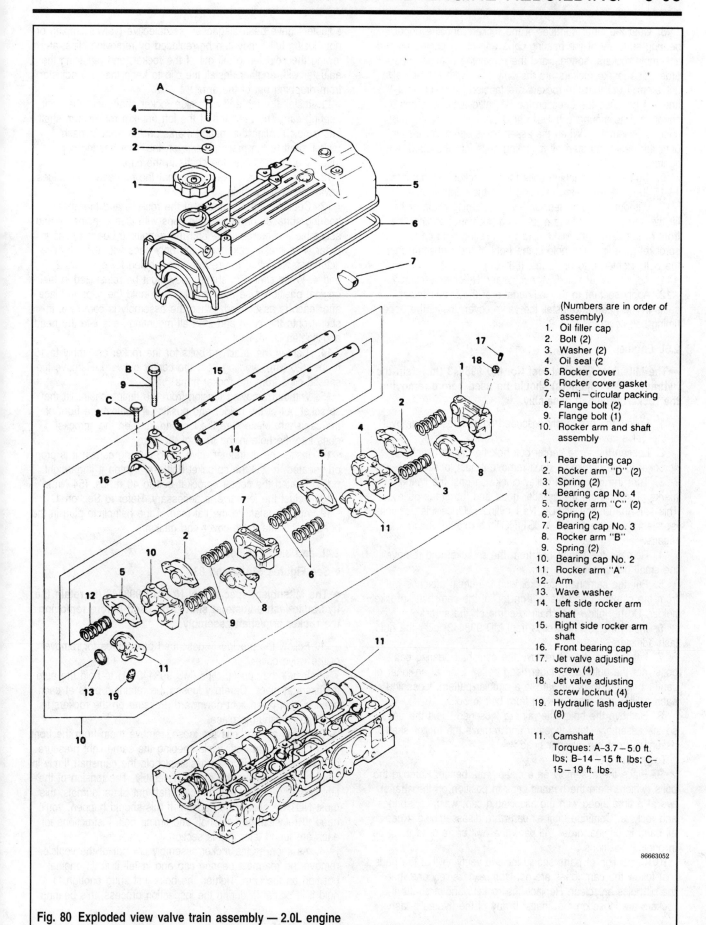

Fig. 80 Exploded view valve train assembly — 2.0L engine

ENGINE AND ENGINE REBUILDING

8. Coat the inner surfaces of the rockers and the upper bearing surfaces of the bearing caps with clean engine oil and assemble rockers, springs, and the remaining bearing caps in order. The intake rockers are the only ones with the jet valve actuators. Note that the rockers are labeled for cylinders 1-3 and 2-4 because the direction the jet valve actuator faces changes. Use mounting bolts to hold the caps in place after each is assembled. When the assembly is complete, install it onto the head and start all mounting bolts into the head and tighten finger-tight.

9. Tighten the attaching bolts for the rocker assembly to 14-15 ft. lbs. (19-20 Nm), working from the center outward.

10. Without removing tension from the timing chain or belt, lift the sprocket out of the holder and position it against the front of the camshaft. Make sure the locating tang on the sprocket goes into the hole in the front of the camshaft. Install the bolt. Tighten it to 65 ft. lbs. (88 Nm).

11. Adjust the jet valves, if necessary. Refer to Section 2.

12. Apply sealant to the top surface of the semi-circular seals in the head and install the valve cover. Install the upper timing belt cover.

2.6L Engines

➡ The Mitsubishi special tool No. MD 998443 (to retain the hydraulic lash adjusters) should be used before removing the rocker arm/shaft assembly.

1. Follow the previously described procedures for removal of the valve cover.

2. Loosen the large center bolt holding the camshaft sprocket to the camshaft; don't remove it, just loosen it.

3. Turn the crankshaft pulley clockwise until the timing marks align on both the cylinder head and the crank pulley. This sets the engine at TDC No.1 cylinder. Once this position is established, the engine MUST NOT be rotated during repairs.

4. Remove the center bolt from the sprocket and remove the small distributor drive gear.

5. Pull the camshaft sprocket with the chain attached out from the camshaft and place it on top of the camshaft sprocket holder. Do not allow the chain to come off the sprocket.

6. Install the retaining clips (tool MD 998443) over the auto lash adjusters.

7. Loosen but do not remove the camshaft bearing cap bolts (not the outer or cylinder head bolts). Start at an inner or center bolt and work outward in a circular pattern, loosening each a little at a time, moving from bolt to bolt.

8. Remove the bolts after all are loosened. Hold the ends so the assembly stays together and remove the rocker shaft assembly from the cylinder head.

To install:

9. Place the assembly on a clean work bench. Remove the bolts (which retain the bearing caps in position on the shafts) two at a time along with the associated cap, washers, springs and rockers. Continue until all parts are disassembled. Keep all parts in original order. All parts that will be re-used must be in original positions.

10. Inspection of parts should include verifying that the pads that follow the cam lobes are not worn excessively and that the oil holes are clear. Replace the rockers and/or shafts if rockers are loose on the shafts. If any of the hydraulic lash adjusters have been diagnosed as defective (valves tapping or not closing fully) they can be replaced by removing clips, allowing the adjuster to fall out of the rocker, and replacing the adjuster with another. Install the clip to keep the new adjuster from dropping out of the arm.

11. Install the left and right side rocker shafts into the front bearing cap. The rear end of the left (intake) rocker arm shaft has a notch. Align the mating marks on the front of each rocker shaft to the mating marks on the front bearing cap. Insert the bolts to hold the shafts in the cap.

12. Install the wave washer so that the rounded side bulges toward the timing chain.

13. Coat the inner surfaces of the rockers and the upper bearing surfaces of the bearing caps with clean engine oil and assemble rockers, springs, and the remaining bearing caps in order. The intake rockers are the only ones with the jet valve actuators. Note that the rockers are labeled for cylinders 2, 3 and 4. While similar in shape, they must be reinstalled in their original position. Use mounting bolts to hold the caps in place after each is assembled. When the assembly is complete, install it onto the head and start all mounting bolts into the head and tighten finger-tight.

14. Tighten the attaching bolts for the rocker assembly to 15 ft. lbs. (20 Nm), working from the center outward. Remove the special clips from the rocker arms.

15. Without removing tension from the timing chain, lift the sprocket out of the holder and position it against the front of the camshaft. Make sure the locating tang on the sprocket goes into the hole in the front of the cam.

16. Install the distributor drive gear, making certain it is properly seated. Install the sprocket bolt and tighten it finger-tight.

17. Tighten the center sprocket bolt to 40 ft. lbs. (54 Nm).

18. Adjust the jet valve, if necessary. Refer to Section 1.

19. Apply sealant to the flat face of the half-circle plug in the head. Install the valve cover and gasket.

2.4L and 3.0L Engines

▸ See Figures 81, 82 and 83

➡ The Mitsubishi special tool No. MD 998443 (to retain the hydraulic lash adjusters) should be used before removing the rocker arm/shaft assembly.

1. Follow the previously described procedures for removal of the valve cover.

2. Install the special clips (MB 998443-01) to hold the auto-adjusters in place. Carefully loosen the retaining bolts at each bearing cap. Keep light downward pressure on the rockers to hold the camshaft in place.

3. When all the bolts are loose, remove them from the front (pulley end) to the rear while keeping the same light pressure on the assembly. Have an assistant hold the camshaft firmly in place as you remove the rocker assembly. The tension of the timing belt will try to pop the camshaft out of its journals; this must NOT be allowed to happen. If this should happen, you'll need to reinstall the camshaft and timing belt. Instructions for each are found later in this section.

4. As soon as the rocker assembly is clear of the vehicle, remove the rearmost bearing cap and install it in its original position on the cam. Tighten the bolts just snug enough to hold it in place. If, during the inspection process, this bearing

ENGINE AND ENGINE REBUILDING 3-65

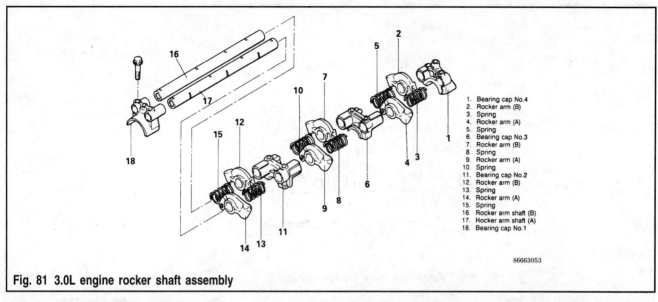

Fig. 81 3.0L engine rocker shaft assembly

1. Bearing cap No.4
2. Rocker arm (B)
3. Spring
4. Rocker arm (A)
5. Spring
6. Bearing cap No.3
7. Rocker arm (B)
8. Spring
9. Rocker arm (A)
10. Spring
11. Bearing cap No.2
12. Rocker arm (B)
13. Spring
14. Rocker arm (A)
15. Spring
16. Rocker arm shaft (B)
17. Rocker arm shaft (A)
18. Bearing cap No.1

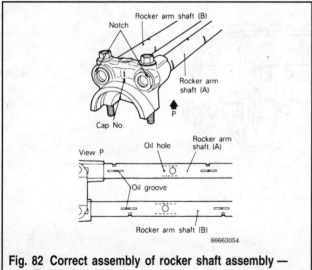

Fig. 82 Correct assembly of rocker shaft assembly — 3.0L Engines

cap must be removed for cleaning or replacement, install the No. 3 bearing cap before removing the No. 4 cap.

➡ Take care during removal that no oil, grease or dirt comes into contact with the timing belt.

To install:

5. Disassemble the rocker assembly, checking each component for wear, scoring, or plugged oil passages. Check the roller for correct and smooth rotation. Inspect the inner diameter of each rocker for any scoring or enlargement. If wear is found inside the rocker, replace it and inspect the shaft for damage.

6. Before reassembly, coat the contact faces liberally with clean motor oil. Observe the numbers on the bearing caps so that they are replaced in the correct location. Reassemble the rocker shafts into the front bearing cap so that the notches face outward. As you continue to assemble the springs, rockers and bearing caps, remember that the arrows on the bearing caps must point in the same direction as the arrow on the head. This is particularly important if the rockers have been removed from both heads.

7. When the rocker arm assembly is complete (except for the rear bearing cap, which is holding the camshaft down), have your assistant hold the camshaft while the rear cap is removed. Assemble the last cap onto the rocker assembly and carefully fit the assembly onto the head. Remember that the camshaft must be held in place during this exchange.

8. Install the retaining bolts finger-tight, making sure each rocker is correctly aligned with the camshaft and valve stem. Once the rockers are securely in place, the special retaining clips may be removed.

9. Starting at the center bearing cap and working outward in two passes, tighten the bearing cap bolts to 15 ft. lbs. (20 Nm).

10. Apply sealer to the correct locations and reinstall the valve cover.

3.5L DOHC Engines

The 3.5L DOHC engines are not equipped with rocker arm shafts like the other engines from Mitsubishi. To remove the rocker arms, the camshafts must first be removed. Refer to the procedures for camshaft removal, located later in this section.

2.3L Diesel Engines

▶ See Figure 78

1. Follow the previously described procedures for removal of the valve cover.
2. Remove the timing belt covers.
3. Using a wrench, turn the crankshaft in a clockwise direction until all the timing marks align. This positions No.1 piston at TDC/compression.
4. Remove the 5 flange bolts and washers holding the rocker rail to the head.
5. Lift the rocker assembly with the rockers attached away from the engine.

To install:

6. If the assembly is to be disassembled, take note that the intake and exhaust rockers are different. Each is labeled **I** or **E**; they must be reassembled in the correct positions.

3-66 ENGINE AND ENGINE REBUILDING

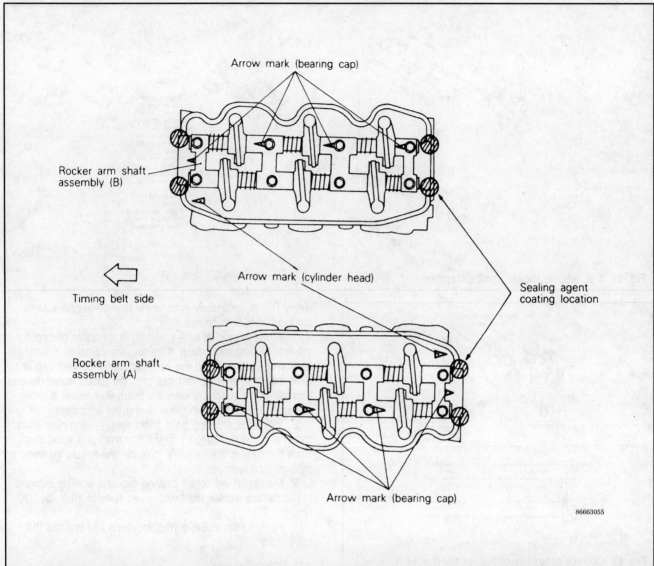

Fig. 83 Directional arrows on the bearing caps must agree with the arrows on the cylinder head assembly

7. Inspect the arms and shafts after a thorough cleaning. The face of the cam follower must be smooth and free of grooves. The end of the adjuster screw must be round (spherical), not mushroomed or flattened. The oil passages in the shaft must be open.

8. Reassemble the rockers onto the shaft, making certain each component is in the correct location. Coat the entire shaft and rockers with a light coat of clean engine oil. Pay particular attention to coating the face of each cam follower.

9. Install the assembly onto the head. Install the 5 retaining bolts and tighten them alternately and evenly to 27 ft. lbs. (37 Nm).

10. Adjust the valves. Refer to Section 1.
11. Install the valve cover.
12. Start the engine and check for oil leaks.

Thermostat

REMOVAL & INSTALLATION

To operate at peak efficiency, an engine must maintain its internal temperatures within certain upper and lower limits. The cooling system circulates fluid around the combustion cylinders and conducts this heated fluid to the radiator, where the heat is exchanged into the airflow created by the fan and the motion of the vehicle.

While most people realize that an engine running too hot (overheated) is a sign of trouble, few know that an engine can run too cool as well. If the proper internal temperatures are not achieved, fuel is not burned efficiently and the lubricating oil does not reach its best working temperature. While a too cold condition is rarely disabling, it can cause a variety of problems which can be mistaken for tune-up or electrical causes.

The thermostat controls the flow of coolant within the system. It reacts to the heat of the coolant and allows more fluid (or less) to circulate. Depending on the amount of fluid being circulated, more or less heat is drawn away from the inside of the engine. While we are beyond the days of having to install different thermostats for summer and winter driving, it is wise to check the function of the thermostat periodically. Special use of the vehicle such as trailer towing or carrying heavy loads may require the installation of a thermostat with different temperature characteristics.

The thermostat requires replacement if the engine runs too cold, as indicated by low operating temperature on the gauge, low heater output, poor gas mileage and a slight lack of performance. If the engine is overheating, there are a number of different possible causes, including not only low coolant due to leakage, but improper engine tuning and/or a dirty cooling system or partially plugged radiator.

If the engine is overheating, it's best to test the thermostat before getting into more complicated diagnosis. You should also test the thermostat if the water temperature gauge is reading low and you're not sure whether it's a thermostat or a gauge problem. Any time the engine has suffered an overheating period, the thermostat should be changed; high temperature can damage it.

✳✳WARNING

Perform this procedure on a cold engine only. If the vehicle has been recently driven, allow it to cool for at least 1-2 hours. Allowing the engine to cool overnight is preferred.

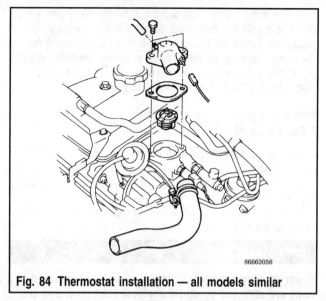

Fig. 84 Thermostat installation — all models similar

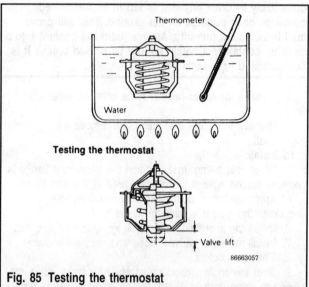

Fig. 85 Testing the thermostat

All Gasoline Engines

▶ See Figures 84 and 85

1. Disconnect the negative battery cable. Remove the air cleaner (if necessary to gain access) and then drain the cooling system below the level of the tubes in the top tank of the radiator.

✳✳CAUTION

When draining the coolant, keep in mind that cats and dogs are attracted by ethylene glycol antifreeze, and are quite likely to drink any that is left in an uncovered container or in puddles on the ground. This will prove fatal in sufficient quantity. Always drain the coolant into a sealable container. Coolant should be reused unless it is contaminated or several years old.

2. You can usually remove the thermostat housing, located at the intake manifold end of the top radiator hose, without disconnecting the hose. It may be necessary to remove other small hoses and/or wiring connectors from the housing. Remove the two mounting bolts and gently lift the housing off the manifold. Once the housing is removed, you may wish to remove it from the end of the hose, making reinstallation easier.

3. If the thermostat is to be tested, suspend it well off the bottom of a pan of water on the stove. Immerse a thermometer that reads up to about 250°F (121°C) in the water. Heat the water. Note the temperature at which the thermostat valve begins to open and continue heating until the water boils. The valve should begin opening around 190°F (88°C), the exact temperature should be stamped on the thermostat, and be wide open as the water boils. The valve must open at least 0.31 in. (8mm) and should start opening within just a few degrees of the specified temperature; otherwise, replace it.

To install:

4. Scrape both gasket surfaces thoroughly, including the indentation in the intake manifold. Install the thermostat with the wax pellet and spring downward. Some models have a ribbed section inside the manifold to keep you from installing the unit upside down. Make sure the unit seats in the indentation so there is a flush surface for the gasket to seal against.

5. Coat a new gasket with sealer on both sides and install it to the manifold with bolt holes matching up. Note that the gasket must be installed above the thermostat.

6. Install the housing and two bolts, turning them gently and going back and forth until they are just snug. You are dealing with cast aluminum which is very brittle — DO NOT overtighten these bolts!

7. Connect the hoses and wiring as necessary and install the air cleaner if it was removed.

3-68 ENGINE AND ENGINE REBUILDING

8. Refill the system with coolant. Operate the engine just above idle until the thermostat opens. Refill the system as necessary. Check for leaks. If any leak is found around the thermostat housing, do not attempt to tighten the bolts. After the engine cools the housing must be removed and scraped, then reinstalled with a fresh gasket and fresh sealer.

Diesel Engine
▶ See Figure 86

The thermostat for the diesel engine is is located on the water intake side of the cooling system, between the lower radiator hose and the block. The thermostat is equipped with a bypass valve.

1. Disconnect the negative battery cable. Drain the coolant to below the level of the thermostat.

> **※※CAUTION**
>
> When draining the coolant, keep in mind that cats and dogs are attracted by ethylene glycol antifreeze, and are quite likely to drink any that is left in an uncovered container or in puddles on the ground. This will prove fatal in sufficient quantity. Always drain the coolant into a sealable container. Coolant should be reused unless it is contaminated or several years old.

2. Remove the lower radiator hose from the water inlet fitting.
3. Remove the water inlet fitting and remove the thermostat.

To install:

4. When reinstalling, make certain the thermostat flange is correctly seated against the dimpled area of the inlet fitting.
5. Apply sealant to both sides of a new inlet fitting gasket and install the gasket against the water pump.
6. Install the inlet fitting and carefully tighten the two bolts.
7. Install the lower radiator hose and secure the clamp.
8. Refill the coolant.
9. Start the engine, checking carefully for leaks. Allow the engine to warm up fully. Watch the temperature gauge for any indication of improper warm-up (underheating or overheating).

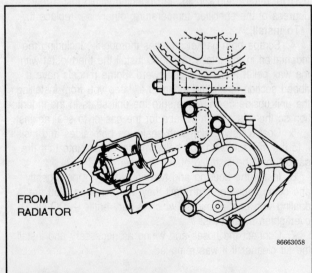

Fig. 86 Location of thermostat — diesel engine

Intake Manifold

REMOVAL & INSTALLATION

2.0L Engine

1. Disconnect the negative battery cable.
2. Drain the cooling system. Disconnect the upper radiator hose at the thermostat housing.

> **※※CAUTION**
>
> When draining the coolant, keep in mind that cats and dogs are attracted by ethylene glycol antifreeze, and are quite likely to drink any that is left in an uncovered container or in puddles on the ground. This will prove fatal in sufficient quantity. Always drain the coolant into a sealable container. Coolant should be reused unless it is contaminated or several years old.

3. Remove the carburetor. This will involve the removal or repositioning of several hoses, lines and wires. Make sure each is properly labeled so it may be reinstalled correctly.
4. Disconnect any remaining vacuum lines and on some models, EGR system piping.
5. Remove all the intake manifold retaining nuts and washers. These may be difficult to break loose. Use rust penetrant freely and keep the wrench square on the bolt to prevent breakage. Gently tap the manifold loose from the gaskets with a rubber or plastic mallet, and remove it.

To install:

6. If a new manifold is being installed, transfer parts such as the thermostat and gasket, and any vacuum fittings. Thoroughly clean the sealing surfaces of the cylinder head and, if it's being re-used, the manifold. Check the cylinder head and manifold sealing surfaces for flatness with a straightedge. Correct a warped surface by replacing the manifold or having the cylinder head surface machined. Install new gasket(s) against the head. Make sure all bolt holes and passages for the cooling water and intake ports align properly.
7. Position the manifold against the gasket and support it there while installing washers and nuts finger-tight. Tighten the bolts evenly and in at least two passes to 15 ft. lbs (22 Nm).
8. Install the carburetor assembly and all hoses, wires and vacuum lines.
9. Refill the coolant. Connect the negative battery cable. Start the engine and check carefully for leaks of either coolant or vacuum.

2.4L Engine
▶ See Figure 87

1. Relieve the fuel pressure by disconnecting the wiring to the electric fuel pump, located back by the fuel pump. Start the vehicle and let it run until it stalls on its own. Turn the vehicle **OFF**.
2. Disconnect the negative battery cable.
3. Drain the engine coolant. Disconnect the upper radiator hose from the thermostat housing.
4. Remove the air intake hoses, breather hose and the air intake pipe.

ENGINE AND ENGINE REBUILDING 3-69

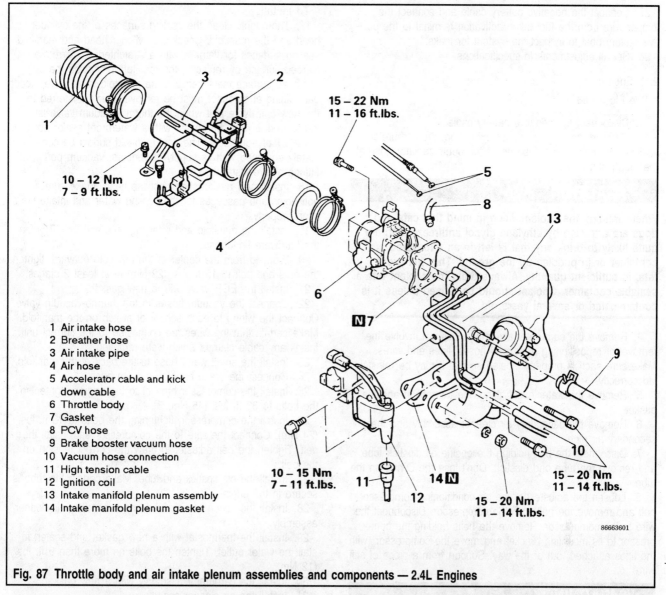

1. Air intake hose
2. Breather hose
3. Air intake pipe
4. Air hose
5. Accelerator cable and kick down cable
6. Throttle body
7. Gasket
8. PCV hose
9. Brake booster vacuum hose
10. Vacuum hose connection
11. High tension cable
12. Ignition coil
13. Intake manifold plenum assembly
14. Intake manifold plenum gasket

Fig. 87 Throttle body and air intake plenum assemblies and components — 2.4L Engines

5. Disconnect all wires, hoses and linkages to the throttle body.
6. Disconnect the ignition coil high tension cable from the distributor.
7. Disconnect the brake booster hose and vacuum hose cluster from the air intake plenum.
8. Unbolt the air intake plenum from the intake manifold and remove the plenum from the engine.
9. Cover the fuel line with a clean shop rag and disconnect the fuel lines from the fuel rail. Keep the line covered or plugged.
10. Remove and label the fuel injector harness connectors. Make sure to retrieve the injector insulator washers, these will have to be replaced upon reinstallation.
11. Remove the fuel rail assembly with injectors intact.
12. Disconnect the heater hose from the manifold.
13. Disconnect the wires to the engine coolant switches.
14. Remove the distributor. Refer to the previously described procedure in this section.
15. Unbolt the intake manifold from the cylinder head and remove from the engine.

16. Clean and dry the mating surfaces of the manifold and cylinder head.

To install:
17. Using a new gasket, install the intake manifold to the head. Starting from the middle and working outward, tighten the retaining nuts to 12 ft. lbs. (16 Nm).
18. Connect the wires to the engine coolant switches. Install the distributor with matchmarks aligned.
19. Install the fuel rail assembly to the manifold and connect the fuel line using a new O-ring.
20. Connect the heater hose to the manifold.
21. Install the air intake plenum with a new gasket. Tighten the retaining bolts to 12 ft. lbs. (16 Nm).
22. Connect the vacuum hoses cluster, brake booster hose and all wires, hoses and linkages to the throttle body.
23. Install the ignition coil.
24. Install the air intake pipe and hoses.
25. Connect the upper radiator hose.
26. Fill the radiator with coolant.

3-70 ENGINE AND ENGINE REBUILDING

27. Connect the negative battery cable and connect the jumper wire from the fuel pump activation terminal to the positive battery post to inspect the system for leaks.
28. Set all adjustments to specifications.

2.6L Engine

▶ See Figure 88

1. Disconnect the negative battery cable.
2. Remove the air cleaner assembly from the carburetor.
3. Drain the coolant. Disconnect the upper radiator hose at the thermostat housing.

> **CAUTION**
>
> When draining the coolant, keep in mind that cats and dogs are attracted by ethylene glycol antifreeze, and are quite likely to drink any that is left in an uncovered container or in puddles on the ground. This will prove fatal in sufficient quantity. Always drain the coolant into a sealable container. Coolant should be reused unless it is contaminated or several years old.

4. Remove the carburetor assembly. This will involve the removal or repositioning of several hoses, lines and wires. Make sure each is properly labeled so that it may be reinstalled correctly.
5. Remove the water outlet fitting, the thermostat and the gasket.
6. Remove the secondary air cleaner assembly and the secondary air pipe.
7. Disconnect the bolt holding the engine oil dipstick tube and remove the tube and dipstick. Don't lose the O-ring on the tube.
8. Loosen the adjuster for the air conditioning compressor belt and remove the belt from the compressor. Disconnect the wire to the compressor. Remove the bolts holding the compressor to its mounting bracket and move the compressor (with the lines attached) out of the way. Support from a piece of stiff wire.

> **WARNING**
>
> Do not loosen any air conditioning lines or discharge the system.

9. Remove the compressor bracket from the engine.
10. Disconnect the heater hose at the manifold.
11. Disconnect the brake booster vacuum hose at the manifold.
12. Disconnect the small water hose at the rear of the manifold.
13. Disconnect the wiring and retainers running to the manifold. Check carefully for concealed wires underneath.
14. Remove the EGR valve. Discard the gasket; it is not reusable.
15. Label and disconnect the vacuum hoses from the thermo valve at the thermostat housing.
16. Loosen the nuts and bolt holding the manifold to the engine. Support the manifold and remove it.

To install:

17. Thoroughly clean the sealing surfaces of the cylinder head and the manifold. Check the cylinder head and manifold sealing surfaces for flatness with a straightedge. Correct a warped surface by replacing the manifold or having the cylinder head surface machined. If the manifold is to be replaced, each fitting and sensor must be removed and transferred to the new unit. Note that when the thermo-vacuum valve is transferred, it must be sealed with a waterproof sealer such as 3M 4171 or equivalent. The new manifold should be completely equipped with its supply of sensors, vacuum ports and fittings before installation.
18. Install new gasket(s) against the head. Make sure all bolt holes and passages for the cooling water and intake ports align properly.
19. Install the manifold and retaining nuts and bolt. Tighten the hardware finger-tight.
20. Working from the center of the manifold outward, tighten the nuts and bolt to 15 ft. lbs. (22 Nm) in at least 2 stages.
21. Install the EGR valve with a new gasket.
22. Connect the vacuum hoses to the thermo-vacuum valve. Connect the wiring to each sensor or switch on the manifold. Make certain that the wires are connected to the correct unit. Install any cable clamps which were removed.
23. Install the small water hose to the back of the manifold.
24. Connect the brake booster vacuum hose.
25. Install the compressor bracket to the engine, tightening the bolts to 33 ft. lbs. (45 Nm).
26. Install the compressor, tightening the bolts to 18 ft. lbs. (24 Nm). Connect the wire to the compressor and install the belt. Tighten the belt adjuster to draw the correct tension on the belt.
27. Install the oil dipstick and tube. Make certain the tube is secure in the engine.
28. Install the secondary air tube and secondary air cleaner assembly.
29. Install the thermostat with a new gasket and sealer; install the water outlet. Tighten the bolts no more than 9 ft. lbs. (12 Nm).
30. Install the carburetor assembly with a new gasket.
31. Install the air cleaner assembly.
32. Refill the coolant and connect the negative battery cable.
33. Start the engine and check for leaks of coolant, oil or vacuum. Confirm the proper operation of all electrical lights and gauges on the dashboard. Turn on the air conditioning and confirm the compressor's operation.

3.0L and 3.5L Engines

▶ See Figures 89, 90, 91 and 92

The basic procedure for removal of the intake system is essentially the same for 3.0L (12 valve), 3.0L (24 valve) and the 3.5L DOHC engines. The only differences may exist in the slightly different location or shape of the same parts. If unsure of the exact part name or location, the diagrams show the parts and their locations for removal.

1. Relieve the pressure in the fuel system. Detach the wiring from the electric fuel pump by the fuel tank. Start the vehicle and let it run until it stalls by itself. Turn the engine **OFF**.
2. Disconnect the negative battery cable.

ENGINE AND ENGINE REBUILDING 3-71

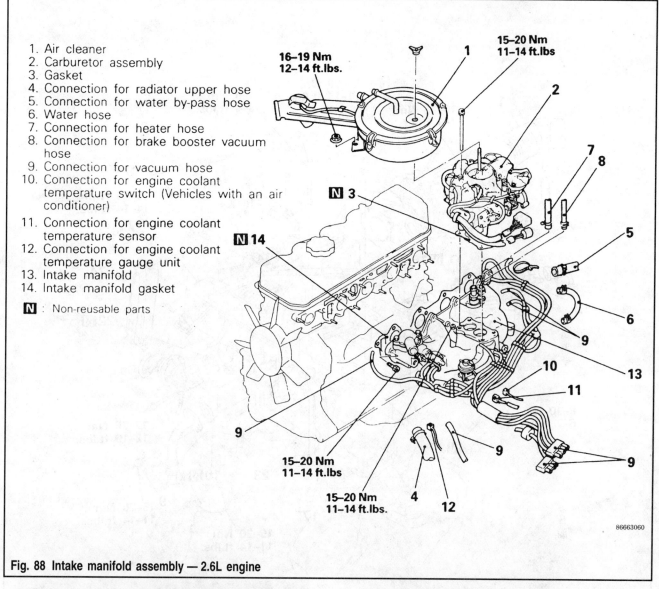

1. Air cleaner
2. Carburetor assembly
3. Gasket
4. Connection for radiator upper hose
5. Connection for water by-pass hose
6. Water hose
7. Connection for heater hose
8. Connection for brake booster vacuum hose
9. Connection for vacuum hose
10. Connection for engine coolant temperature switch (Vehicles with an air conditioner)
11. Connection for engine coolant temperature sensor
12. Connection for engine coolant temperature gauge unit
13. Intake manifold
14. Intake manifold gasket

N : Non-reusable parts

Fig. 88 Intake manifold assembly — 2.6L engine

3. Drain most of the coolant out of the cooling system so the intake manifold passages will be empty.

✳✳CAUTION

When draining the coolant, keep in mind that cats and dogs are attracted by ethylene glycol antifreeze, and are quite likely to drink any that is left in an uncovered container or in puddles on the ground. This will prove fatal in sufficient quantity. Always drain the coolant into a sealable container. Coolant should be reused unless it is contaminated or several years old.

4. Disconnect the air intake hose.
5. Remove the accelerator cable adjusting bolts.
6. On vehicles equipped with an automatic transmission, remove the connection for the throttle control cable.
7. Unfasten the connection for the accelerator cable.
8. Disconnect the vacuum and the brake booster vacuum hoses.
9. Unfasten the EGR temperature sensor connector.

10. Disengage the connection for the three vacuum hoses on the left-hand side of the air intake plenum.
11. Remove the EGR pipe attaching bolts. The gasket in this connection will need to be replaced with a new one during reassembly.
12. Unplug the ignition coil high tension cable from the coil.
13. Remove the ignition coil assembly from the top of the air intake plenum.
14. Remove the mounting bolt for the engine oil filler neck bracket.
15. Disconnect the PVC hose from the intake plenum.
16. Remove the throttle body assembly by removing the throttle body installation bolts, taking care not to disturb the throttle body during the operation. Leave the water hoses attached to the throttle body assembly. Set safely aside.

✳✳WARNING

The bolts holding the throttle body are of two different sizes. Note or diagram their locations for reassembly.

3-72 ENGINE AND ENGINE REBUILDING

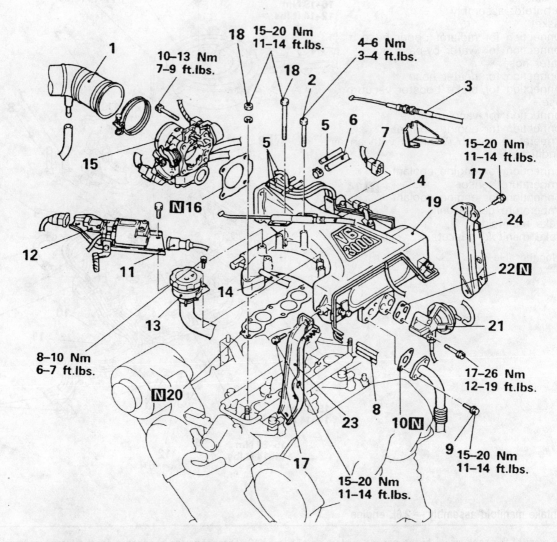

1. Connection for air intake hose
2. Accelerator cable adjusting bolts
3. Connection for throttle control cable (Vehicles with an automatic transmission)
4. Connection for accelerator cable
5. Connection for vacuum hose
6. Connection for brake booster vacuum hose
7. EGR temperature sensor connector
8. Connection for vacuum hose
9. EGR pipe attaching bolts
10. Gasket
11. Connection for high tension cable
12. Ignition coil
13. Engine oil filler neck bracket
14. Connection for PCV hose
15. Throttle body assembly
16. Throttle body gasket
17. Bolts
18. Bolts and nuts
19. Air intake plenum
20. Air intake plenum gasket
21. EGR valve
22. EGR gasket
23. Air intake plenum front stay
24. Air intake plenum rear stay

N : Non-reusable parts

Fig. 89 Air intake plenum removal and installation components — 3.0L Engines

ENGINE AND ENGINE REBUILDING

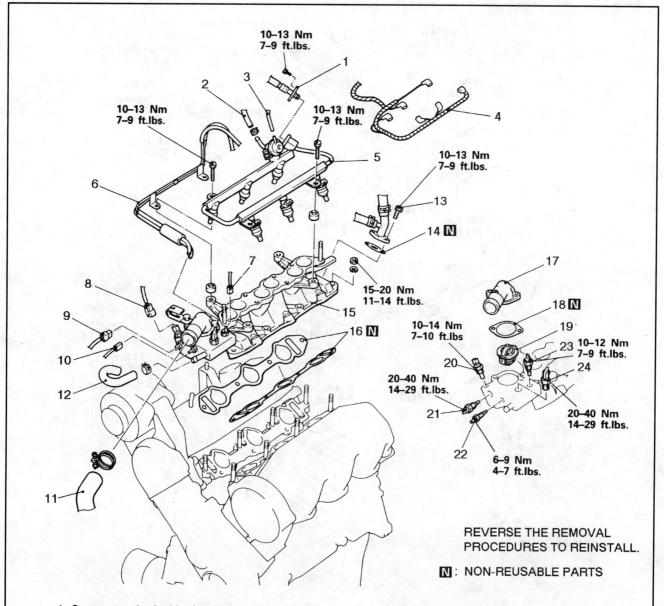

1. Connection for fuel high pressure hose
2. Connection for fuel return hose
3. Vacuum hose
4. Connection for control harness
5. Delivery pipe, fuel injector and pressure regulator
6. Vacuum hose and pipe assembly
7. Connection for engine coolant temperature gauge unit connector
8. Connection for engine coolant temperature switch connector (vehicles with an air conditioner)
9. Connection for engine coolant temperature sensor connector
10. Connection for thermo switch connector (vehicles with an automatic transmission)
11. Radiator upper hose
12. Connection for water by-pass hose
13. Heater pipe attaching bolts
14. Gasket
15. Intake manifold
16. Intake manifold gasket
17. Water outlet fitting assembly
18. Gasket
19. Thermostat
20. Engine coolant temperature switch (vehicles with an air conditioner)
21. Engine coolant temperature sensor
22. Thermo switch (vehicles with an automatic transmission)
23. Engine coolant temperature gauge unit
24. Thermo valve assembly

REVERSE THE REMOVAL PROCEDURES TO REINSTALL.

N : NON-REUSABLE PARTS

Fig. 90 Air intake manifold removal and installation components — 3.0L Engines

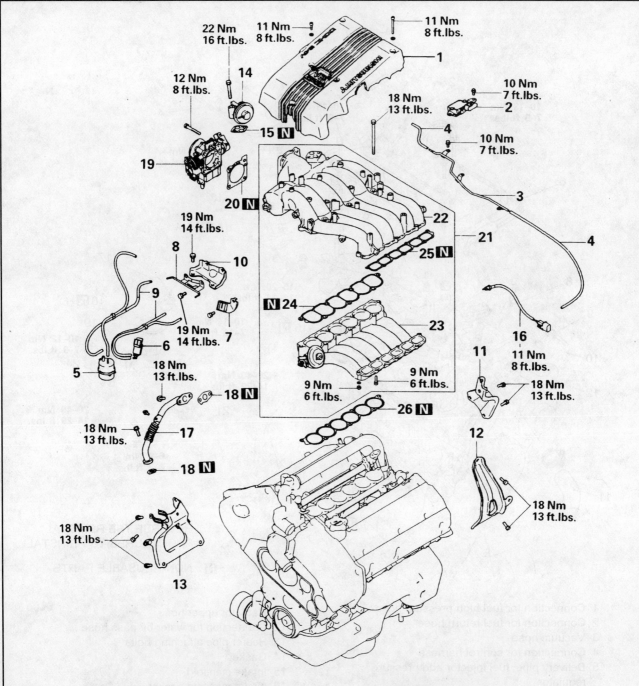

Fig. 91 Air intake plenum removal and installation components — 3.5L Engines

ENGINE AND ENGINE REBUILDING

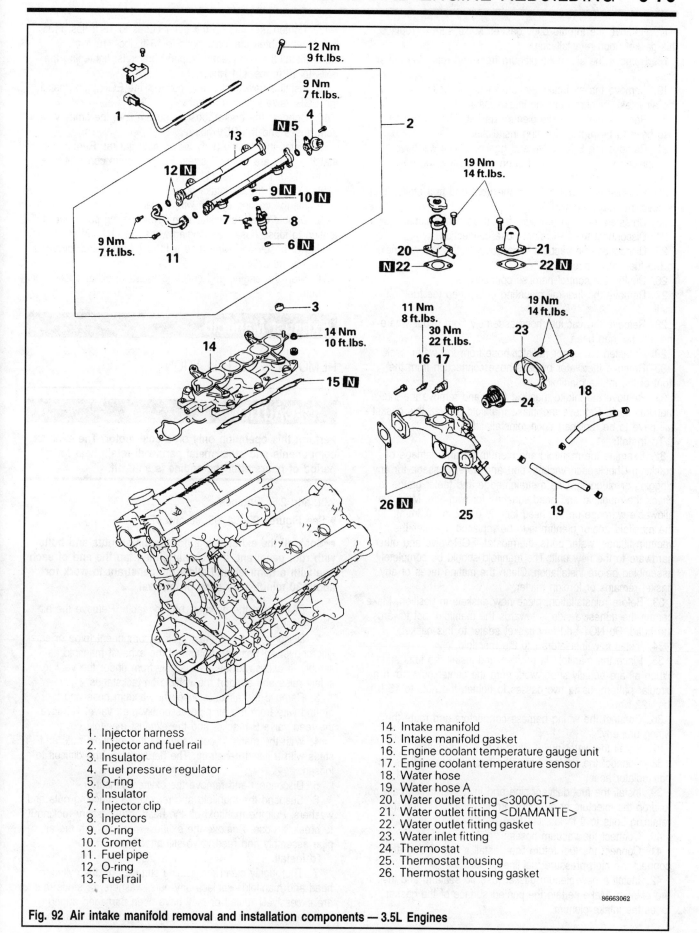

1. Injector harness
2. Injector and fuel rail
3. Insulator
4. Fuel pressure regulator
5. O-ring
6. Insulator
7. Injector clip
8. Injectors
9. O-ring
10. Gromet
11. Fuel pipe
12. O-ring
13. Fuel rail
14. Intake manifold
15. Intake manifold gasket
16. Engine coolant temperature gauge unit
17. Engine coolant temperature sensor
18. Water hose
19. Water hose A
20. Water outlet fitting <3000GT>
21. Water outlet fitting <DIAMANTE>
22. Water outlet fitting gasket
23. Water inlet fitting
24. Thermostat
25. Thermostat housing
26. Thermostat housing gasket

Fig. 92 Air intake manifold removal and installation components — 3.5L Engines

ENGINE AND ENGINE REBUILDING

17. Remove the throttle body gasket. Make sure to replace this gasket upon reinstallation.
18. Remove the air intake plenum front and rear stay upper bolts.
19. Remove the air intake plenum mounting bolts. Remove the air intake plenum from the intake manifold.
20. Remove the air intake plenum gasket. This gasket will also have to be replaced during installation.
21. Remove the EGR valve and gasket, which will have to be replaced with a new gasket upon reinstallation of the EGR valve.
22. Remove the lower bolts on the front and rear stays, then remove the stays themselves.
23. Unfasten the high pressure fuel hose connection.
24. Disconnect the fuel return hose connection.
25. Disconnect the vacuum hose going to the fuel rail, next to the fuel return hose.
26. Unplug the control harness connections.
27. Remove the fuel rail mounting bolts, then the fuel rail itself.
28. Remove the vacuum hose assembly from under where the fuel rail had been.
29. Unfasten the upper radiator hose from the water neck.
30. Remove the water by-pass hose connection from the front of the intake manifold.
31. Remove the intake manifold bolts and remove the intake manifold itself. Retrieve the intake manifold gasket, this gasket will have to be replaced upon reinstallation.

To install:
32. Examine the manifold and plenum for any damage or cracking. Check each vacuum port and water passage for any clogging or plugging. Use a straightedge and feeler gauge to check the manifold and head surfaces for warpage. Maximum allowable warpage on the head face is 0.012 in. (0.3mm). If the manifold and/or plenum is to be replaced, transfer the vacuum fittings, water ports, thermostat, EGR valve and other hardware to the new unit. The manifold should be completely assembled before installation. Clean the mating faces of any gasket remains or foreign matter.
33. Before reinstallation, place new gaskets in position. Make certain the adhesive side is towards the manifold, not towards the head. Do NOT add any gasket sealer to this gasket.
34. Install new insulators into the manifold.
35. Place the manifold in position and install the nuts snug. When all are equally snug, work from the center outward in a circular pattern, using two passes to tighten the nuts to 15 ft. lbs. (22 Nm).
36. Connect the wiring harness connectors and route the wiring properly.
37. Install the water outlet fitting with a new gasket.
38. Connect the coolant hose, the heater hose and the upper radiator hose.
39. Install the fuel delivery pipe and injectors, taking care not to drop the injectors or subject them to impact. Tighten the retaining bolts to 8 ft. lbs. (11 Nm).
40. Connect the vacuum hoses.
41. Connect the fuel return line. Install a new O-ring and connect the high pressure fuel line.
42. Install a new plenum gasket (use no sealant) and install the plenum. Make certain the printed surface of the gasket faces the intake plenum.
43. Tighten the nuts on the outer edges to 13 ft. lbs. (18 Nm), then tighten the inner bolts to 13 ft. lbs. (18 Nm).
44. Install the two plenum support brackets, tightening the bolts to 10 ft. lbs. (14 Nm).
45. Install a new gasket and connect the EGR pipe. Install the EGR valve with a new gasket.
46. Connect the brake booster vacuum line, the small vacuum hoses and the PCV hose.
47. Install the throttle body with a new gasket. Remember that the bolts are different sizes. Position them correctly and tighten them to 8 ft. lbs. (11 Nm).
48. Connect the air intake hose.
49. Fill the cooling system with coolant.
50. Double check all installation items, paying particular attention to loose hoses or hanging wires, untightened nuts, poor routing of hoses and wires (too tight or rubbing) and tools left in the engine area.
51. Start the engine and check for leaks of either coolant or vacuum.

Exhaust Manifold

REMOVAL & INSTALLATION

✱✱CAUTION

Perform this operation only on a cold motor. The exhaust components and other metal parts will retain heat for a period of hours after the engine is shut off.

2.0L and 2.6L Engines
◆ See Figures 93, 94, 95, 96 and 97

➡Soak all the exhaust manifold retaining nuts and bolts with rust penetrant, if necessary, then tap the end of each one with a hammer and allow the penetrant to work for about 30 minutes for ease of removal.

1. Disconnect the negative battery cable. Remove the air cleaner.
2. Disconnect the exhaust system component (pipe or catalytic converter) from the bottom of the exhaust manifold assembly. If access to this part is only from under the vehicle, safely raise and support the vehicle on jackstands.
3. Remove the air duct hose, the vacuum hose and the ground wire from the air pipe assembly reed valve. Remove the reed valve B bracket and the exhaust manifold cover.
4. With the manifold cold, soak all the manifold nuts and studs with a liquid penetrant. The hardware will be difficult to loosen.
5. Disconnect and remove the oxygen sensor.
6. Suspend the manifold and remove all attaching nuts and washers. Pull the manifold off the head, if necessary rocking it to break it loose. Remove the exhaust manifold with the air pipe assembly and reed valve still attached.

To install:
7. Thoroughly clean the sealing surfaces on the cylinder head and manifold. Replace any nuts, washers, or studs that are excessively rusted or may have been damaged during removal. Studs may sometimes be removed by installing two

ENGINE AND ENGINE REBUILDING

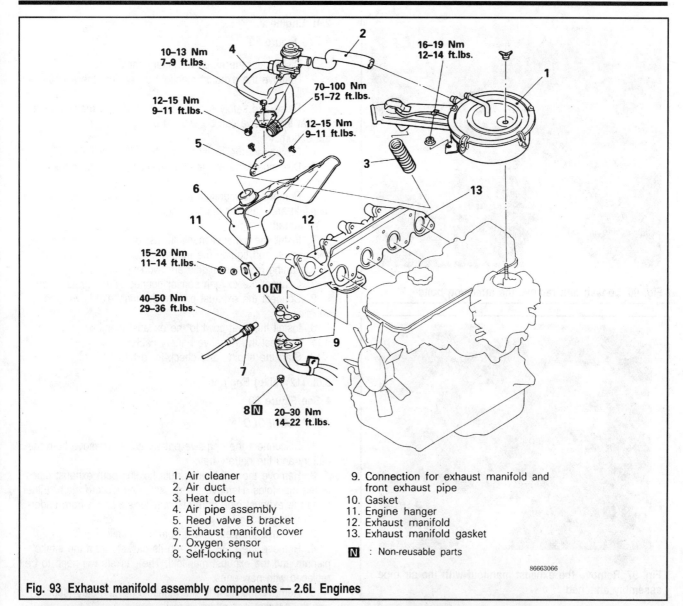

1. Air cleaner
2. Air duct
3. Heat duct
4. Air pipe assembly
5. Reed valve B bracket
6. Exhaust manifold cover
7. Oxygen sensor
8. Self-locking nut
9. Connection for exhaust manifold and front exhaust pipe
10. Gasket
11. Engine hanger
12. Exhaust manifold
13. Exhaust manifold gasket

N : Non-reusable parts

Fig. 93 Exhaust manifold assembly components — 2.6L Engines

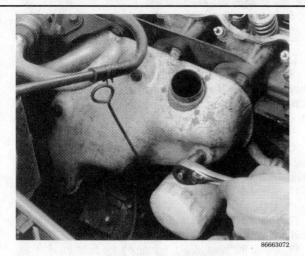

Fig. 94 Loosen and remove the bolts holding the exhaust manifold cover onto the manifold . . .

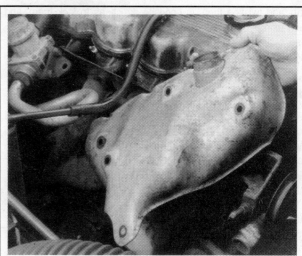

Fig. 95 . . . then remove the exhaust manifold cover

Fig. 96 Loosen and remove the attaching bolts

Fig. 97 Remove the exhaust manifold with the air pipe assembly attached

nuts and twisting them in opposite directions to lock them, and then using your wrench on the inner nut, to remove the stud. Use a straightedge to check manifold and cylinder head sealing surfaces for flatness.

➡ **Maximum acceptable warpage is 0.012 in. (0.3mm). Replace the the manifold assembly if necessary.**

8. Install new gaskets in such a way that all bolt holes and ports are lined up. Make sure all the nuts turn freely, oiling them lightly if necessary. Also make sure all the studs are screwed all the way into the block.
9. Place the manifold in position and support it while you install all the washers and nuts hand-tight.
10. Tighten the nuts and bolts alternately and in several stages to 12 ft. lbs. (16 Nm) working from the center position out.
11. Install the piping, heat stoves, shields and exhaust components. Always use new exhaust gaskets upon installation.
12. Operate the engine and check carefully for exhaust leaks.

2.4L Engine

♦ See Figure 98

1. Disconnect the negative battery cable.
2. Remove the exhaust manifold cover from the exhaust manifold.
3. Raise and safely support the vehicle to a medium height, so that the vehicle can be worked on from both the top and the bottom.
4. Disconnect the exhaust pipe from the manifold.
5. Disconnect the oxygen sensor connector and ground cable.
6. Remove the manifold mounting nuts and remove the manifold and gasket from the engine.

To install:

7. Install the exhaust manifold, using a new exhaust manifold gasket, and tighten the mounting nuts to 13 ft. lbs. (18 Nm), starting from the middle and working outward.
8. Connect the oxygen sensor connector and ground cable.
9. Connect the exhaust pipe to the manifold. Lower the vehicle.
10. Install the heat cowl to the exhaust manifold.
11. Connect the negative battery cable.
12. Start the engine and check for exhaust leaks.

3.0L (12 Valve) Engines

♦ See Figure 99

LEFT MANIFOLD

1. Disconnect the negative battery cable. Remove both the battery and the battery tray.
2. Remove the self-locking nuts holding both exhaust pipes to the manifolds. Remove the gaskets and discard them. Either hang the exhaust pipe from stiff wire or support it from underneath with wooden blocks.
3. Remove the heat shield from the manifold.
4. Remove the EGR pipe and its gaskets from the intake plenum and the exhaust manifold. The gaskets will need to be replaced with new ones.
5. Remove the front intake manifold plenum stay (support) and the bracket it is attached to.
6. Remove the retaining nuts and remove the exhaust manifold.
7. Remove the manifold gasket.

To install:

8. Inspect the manifold and head surfaces for cracks or distortion. Use a straightedge and feeler gauge to check the surfaces for warpage. Maximum acceptable warpage is 0.012 in. (0.3mm).
9. Place a new gasket against the head.
10. Install the manifold. Tighten all the nuts snug. In two passes and working from the inner nuts outward, tighten the retaining nuts to 11-16 ft. lbs. (15-22 Nm).
11. Reinstall the front intake manifold plenum stay and bracket.
12. Reattach the EGR pipe to the intake manifold plenum and the exhaust manifold using new gaskets.
13. Install the heat shield, tightening the bolts to 10 ft. lbs. (13 Nm).
14. Using new exhaust gaskets and new self-locking nuts, connect the exhaust system to both manifolds.

ENGINE AND ENGINE REBUILDING 3-79

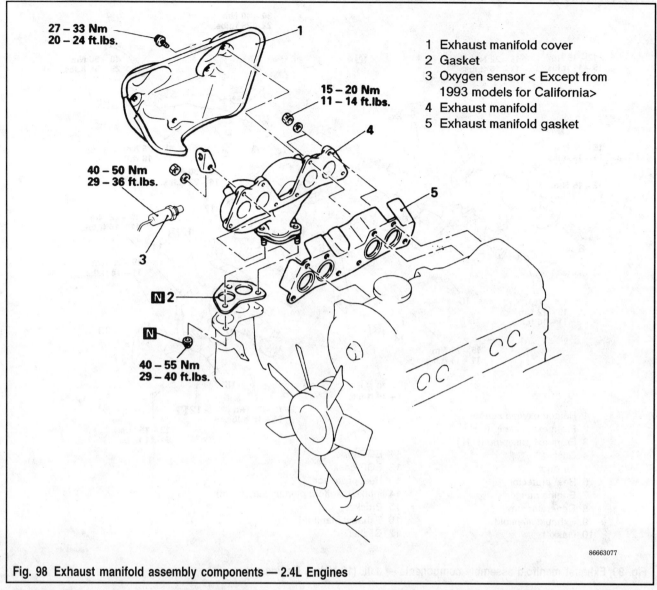

Fig. 98 Exhaust manifold assembly components — 2.4L Engines

1 Exhaust manifold cover
2 Gasket
3 Oxygen sensor < Except from 1993 models for California>
4 Exhaust manifold
5 Exhaust manifold gasket

15. Reinstall the battery tray, then the battery. Reconnect the negative battery cable.
16. Start the engine and check for exhaust leaks.

RIGHT MANIFOLD

1. Disconnect the negative battery cable. Remove the air duct.
2. Remove the self-locking nuts holding both exhaust pipes to both manifolds. Remove the gaskets and discard them. Either hang the exhaust pipe from stiff wire or support it from underneath with wooden blocks.
3. Remove the heat shield, the rear engine hanger and the alternator stay from the exhaust manifold.
4. Remove the retaining nuts and remove the manifold and its gasket.

To install:

5. Inspect the manifold and head surfaces for cracks or distortion. Use a straightedge and feeler gauge to check the surfaces for warpage. Maximum acceptable warpage is 0.012 in. (0.3mm).

6. Install a new gasket and place the manifold in position. Tighten the all the nuts snug. In two passes and working from the inner nuts outward, tighten the retaining nuts to 11-16 ft. lbs. (15-22 Nm).
7. Install the alternator stay and tighten the upper bolt to 9-11 ft. lbs. (12-15 Nm) and the lower bolt to 11-16 ft. lbs. (15-22 Nm). Install the rear engine hanger. Install the heat shield, tightening the bolts to 10 ft. lbs. (14 Nm).
8. Using new gaskets and new self-locking nuts, connect the exhaust system.
9. Connect the negative battery cable and start the engine. Check for exhaust leaks.

3.0L (24 Valve) and 3.5L Engines

▶ See Figures 100 and 101

LEFT MANIFOLD

1. Disconnect the negative battery cable. Remove both the battery and the battery tray.
2. Remove the self-locking nuts holding both exhaust pipes to the manifolds. Remove the gaskets and discard them. Either

3-80 ENGINE AND ENGINE REBUILDING

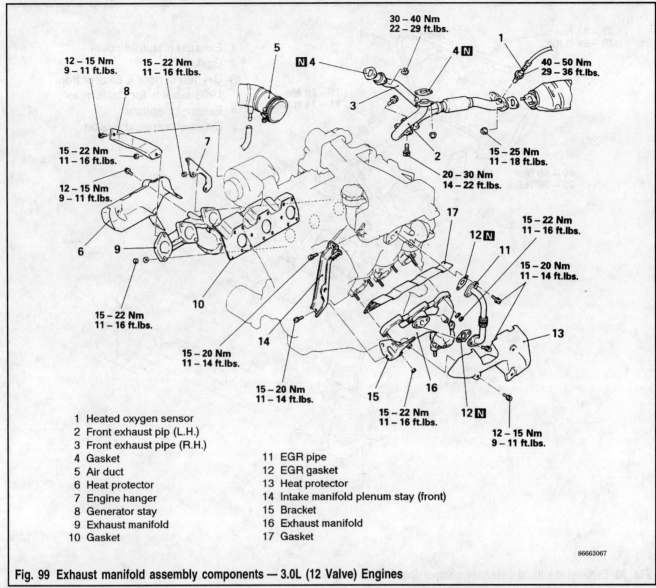

1. Heated oxygen sensor
2. Front exhaust pip (L.H.)
3. Front exhaust pipe (R.H.)
4. Gasket
5. Air duct
6. Heat protector
7. Engine hanger
8. Generator stay
9. Exhaust manifold
10. Gasket
11. EGR pipe
12. EGR gasket
13. Heat protector
14. Intake manifold plenum stay (front)
15. Bracket
16. Exhaust manifold
17. Gasket

Fig. 99 Exhaust manifold assembly components — 3.0L (12 Valve) Engines

hang the exhaust pipe from stiff wire or support it from underneath with wooden blocks.

3. Remove the heat shield from the manifold.
4. Remove the retaining nuts and remove the exhaust manifold.
5. Remove the manifold gasket.

To install:

6. Inspect the manifold and head surfaces for cracks or distortion. Use a straightedge and feeler gauge to check the surfaces for warpage. Maximum acceptable warpage is 0.012 in. (0.3mm).
7. Place a new gasket against the head.
8. Install the manifold. Tighten all the nuts snug. In two passes and working from the inner nuts outward, tighten the retaining nuts to 22 ft. lbs. (29 Nm) on both the 3.0L (24 valve) and 3.5L engines.
9. Install the heat shield, tightening the bolts to 10 ft. lbs. (14 Nm).
10. Using new exhaust gaskets and new self-locking nuts, connect the exhaust system to both manifolds.

11. Reinstall the battery tray, then the battery. Reconnect the negative battery cable.
12. Start the engine and check for exhaust leaks.

RIGHT MANIFOLD

1. Disconnect the negative battery cable. Remove the air duct and air cleaner cover.
2. Remove the self-locking nuts holding both exhaust pipes to both manifolds. Remove the gaskets and discard them. Either hang the exhaust pipe from stiff wire or support it from underneath with wooden blocks.
3. Only on the 3.5L engine remove the EGR pipe. Remove the heat shield from the manifold.
4. Remove the retaining nuts and remove the manifold and its gasket.

To install:

5. Inspect the manifold and head surfaces for cracks or distortion. Use a straightedge and feeler gauge to check the surfaces for warpage. Maximum acceptable warpage is 0.012 in. (0.3mm).

ENGINE AND ENGINE REBUILDING

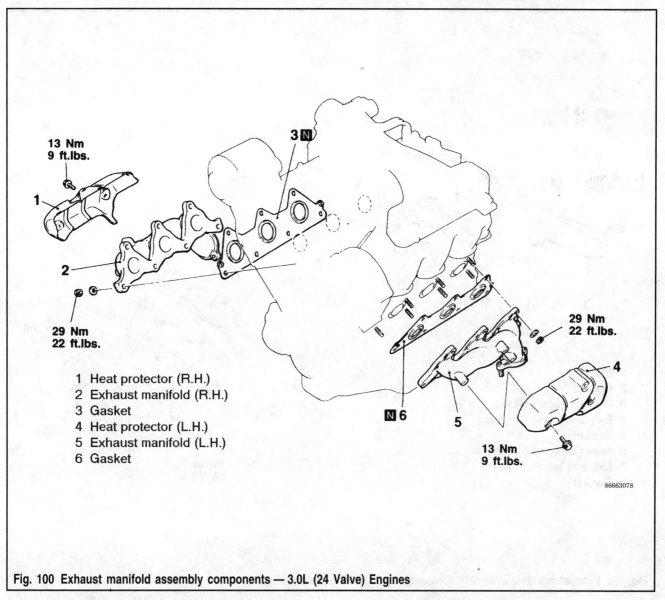

1. Heat protector (R.H.)
2. Exhaust manifold (R.H.)
3. Gasket
4. Heat protector (L.H.)
5. Exhaust manifold (L.H.)
6. Gasket

Fig. 100 Exhaust manifold assembly components — 3.0L (24 Valve) Engines

6. Install a new gasket and place the manifold in position. Tighten the all the nuts snug. In two passes and working from the inner nuts outward, tighten the retaining nuts to 14 ft. lbs. (19 Nm) on the 3.0L (12 valve) engine or to 22 ft. lbs. (29 Nm) on the 3.0L (24 valve) and 3.5L engines.
7. Install the EGR pipe and new gasket, if working on the 3.5L engine. Install the heat shield, tightening the bolts to 10 ft. lbs. (14 Nm).
8. Using new gaskets and new self-locking nuts, connect the exhaust system.
9. Connect the negative battery cable and start the engine. Check for exhaust leaks.

Combination Manifold

➡ Although they are individual pieces, the diesel engine intake and exhaust manifolds share the same gasket. Whenever one is removed, the other must also be removed. The gasket must always be replaced.

REMOVAL & INSTALLATION

▶ See Figure 102

1. Perform this work only on a cold engine. Make sure all surfaces are cool to the touch before beginning. Disconnect the negative battery cable.
2. Disconnect the exhaust pipe from the turbocharger and remove the gasket. Remove the heat shield.
3. Remove the bolts holding the inlet fitting to the inlet manifold. Loosen the hose clamps and remove the fitting; remove the rubber connecting hose.
4. Loosen the hose clamps from the oil return line and remove it.
5. Carefully disconnect the oil supply pipe from the top of the turbocharger.
6. Disconnect the vacuum hose from the top of the wastegate actuator. Remove the wastegate actuator mounting bolts.
7. Remove the bolts holding the turbocharger to the exhaust flange. Remove the turbocharger and discard the gasket.

3-82 ENGINE AND ENGINE REBUILDING

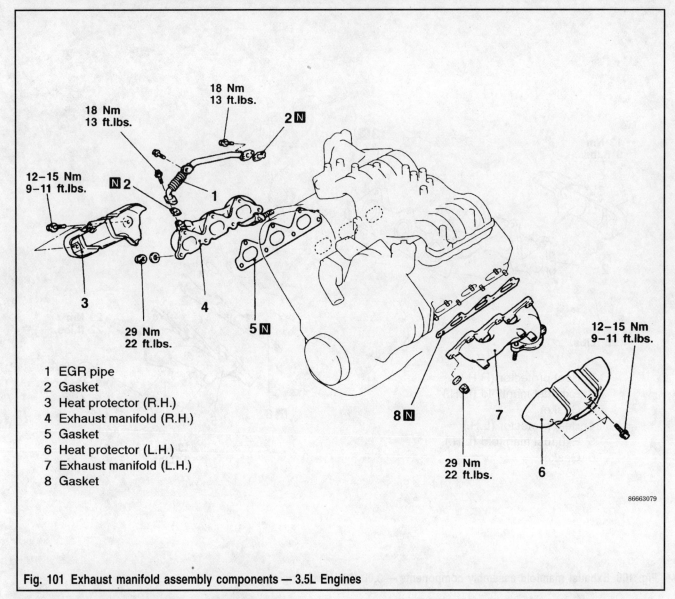

Fig. 101 Exhaust manifold assembly components — 3.5L Engines

1. EGR pipe
2. Gasket
3. Heat protector (R.H.)
4. Exhaust manifold (R.H.)
5. Gasket
6. Heat protector (L.H.)
7. Exhaust manifold (L.H.)
8. Gasket

8. Immediately after removing the turbocharger, plug the intake and exhaust ports and the oil ports with clean, lint-free cloth rags or pieces of crumpled paper. The turbocharger MUST be protected from dirt and grit; place the unit in a protected location away from the work area.
9. Disconnect the intake manifold retaining bolts and remove the manifold.
10. Disconnect the exhaust manifold retaining bolts and remove the manifold. Remove the gasket(s) and discard. Make certain the mating surfaces are completely free of gasket material.

To install:
11. Install the new gasket and install the manifolds (exhaust first, then intake) into position.
12. Start the retaining nuts and bolts by hand; tighten them to 12 ft. lbs. (16 Nm) working from the center to the ends.
13. Remove the plugs from the turbocharger passages and install the turbocharger with a new gasket onto the exhaust manifold. Tighten the nuts evenly.
14. Install the wastegate mounting bolts and tighten to 10 ft. lbs. (14 Nm). Reconnect the wastegate actuator vacuum hose.
15. Connect the oil return line; make certain the clamp is secure.
16. Pour clean engine oil into the oil supply port to pre-lubricate the turbocharger. Install the oil supply line and tighten it carefully.
17. Install the rubber connecting hose on the inlet fitting. Use a new gasket and attach the inlet fitting to the intake manifold.
18. Use a new gasket and install the exhaust pipe to the bottom of the turbocharger. Install the heat shield.
19. Start the engine, listening carefully for any sign of unusual noise from either the manifolds or the turbocharger. Watch the oil pressure indicator for any sign of incorrect pressure.

Turbocharger

A turbocharger is an exhaust-driven turbine which drives a compressor wheel on the other end of the same shaft. The turbine is located in the exhaust flow, generally just below the exhaust manifold. The compressor is located in the intake air

ENGINE AND ENGINE REBUILDING

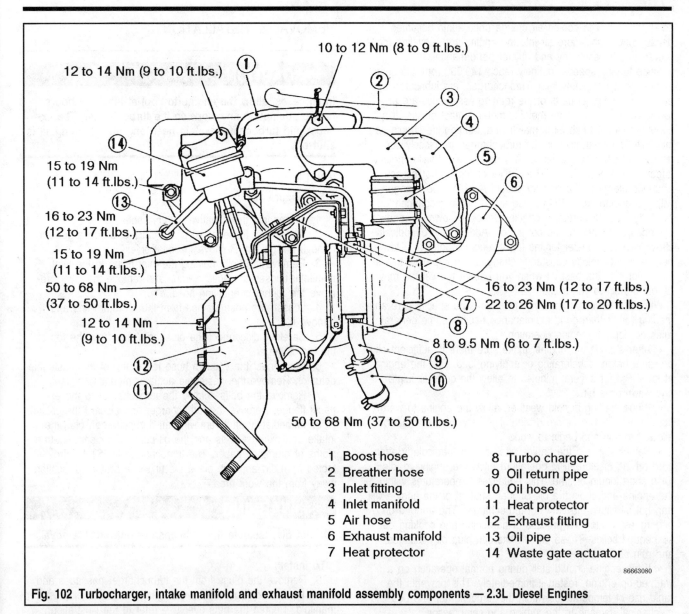

Fig. 102 Turbocharger, intake manifold and exhaust manifold assembly components — 2.3L Diesel Engines

1 Boost hose
2 Breather hose
3 Inlet fitting
4 Inlet manifold
5 Air hose
6 Exhaust manifold
7 Heat protector
8 Turbo charger
9 Oil return pipe
10 Oil hose
11 Heat protector
12 Exhaust fitting
13 Oil pipe
14 Waste gate actuator

path, usually between the air cleaner and the intake manifold. Even though the exhaust and intake air channels are connected by the turbocharger shaft, the exhaust is kept separate from the intake air at all times. (Think of the old water wheel turning the grinding wheel at the mill; the force is transferred but the water never touches the grain.)

By compressing the intake air, more air is squeezed into the cylinder carrying more oxygen for better combustion. Since more air is introduced, more fuel can be introduced, yielding more power. Turbocharging is one way of coaxing more power out of relatively small engines. A larger engine with the same power output would add additional weight, thus reducing overall performance. A turbocharger system is relatively light in comparison, and the additional weight of strengthened components is still well below the weight of a larger motor.

It is possible to get too much of a good thing. Turbocharging is self-perpetuating; that is, as boost (air compression) builds, the exhaust volume builds and the turbine turns faster, providing more boost and so on. If left alone, the turbocharger would build pressure well beyond the operating ability of the engine. To prevent these costly and spectacular failures, boost is held to a reasonable level by a wastegate. Usually located in the output elbow area, the wastegate is a pressure valve which activates at a pre-determined level of pressure. When it opens, it simply allows exhaust flow to bypass the turbine, thus limiting its speed. NEVER attempt to change the setting of the wastegate. If the wastegate or actuator is suspected of faulty operation, replace it.

As the intake air is compressed in the turbocharger, it becomes heated and expands. This expanding air flow is less dense so less air is forced into the engine, partially defeating the purpose of the turbocharger. To overcome this condition, some engines are fitted with an intercooler to remove heat from the air charge. A properly designed intercooler system can reduce air temperature by 90°F (32°C) or more. The intercooler is simply a heat exchanger located between the turbocharger and the intake manifold.

The compressed air charge is directed through ductwork to the intercooler where it is cooled and then on to the intake manifold. The system works in the same fashion as the radiator for the cooling system except that air is being cooled instead of fluid. In some cases the intercooler even looks like a small radiator. The cooled air charge once again becomes dense, introducing more air into the engine and providing more power, greater economy and quieter performance.

Since turbine speeds routinely reach 140,000 rpm, adequate lubrication is absolutely vital. Turbochargers are lubricated by engine oil. Since all parts of the rotating assemblies are protected by a film of oil, no metal-to-metal contact occurs. If a supply of clean, fresh oil is maintained, bearing life should be indefinite. All clearances in the turbocharger are closely controlled and carefully machined. Any dirt in the oil will seriously affect the life of the unit. Oil and filter changes should occur at frequent intervals. The oil filter should ALWAYS be changed with the engine oil. ALWAYS use an oil of the recommended viscosity for your particular engine. Check the owner's manual or underhood label for the correct oil. Additionally, periodically check with your dealer for the latest recommendations. New petroleum technology constantly changes and improves available motor oils; the best oil when you bought the vehicle may be old news two years later.

While the turbocharging system requires no special care and feeding (other than good maintenance habits), some general rules do apply to engine operation.

- After starting the engine, make sure there is sufficient oil pressure before accelerating or applying load. Run the engine at low RPM for several minutes to allow the oil to circulate and warm up a bit.
- When starting in cold weather, allow the engine to warm up a minute or two before driving. The very thick oil must be allowed to thin and begin to work.
- Before stopping the engine after driving (particularly after hard driving or high RPM operation) allow the engine to idle for a short length of time. This equalizes temperatures within the engine and cools the oil. Sudden shut-off of the engine will trap hot oil within the very hot turbocharger. The trapped oil undergoes "coking", a process of hardening and cooking out suspended solids. These carbon bits can plug oil passages and ruin bearings.
- If the engine should stall during normal operation on a warmed-up engine, restart it immediately. This prevents the rapid rise of temperatures within the turbocharger and possible mechanical damage to the turbine and compressor.
- If it is necessary to transport a turbocharged engine removed from the vehicle, plug the air passages with rags or tape them shut. This will prevent entry of dirt and dust which can damage the bearings. Additionally, this will prevent the turbine from turning on its bearings while no oil is being delivered.
- Be sensitive to what the engine tells you as you drive. Impaired performance, noise, vibration or smoking can be a sign of impending failure. Periodically check for loose or restricted hoses and fittings. Replace the air filter at frequent intervals.

➡ After the engine is shut off, the turbocharger may whine as it runs down. Don't confuse this air whine with bearing failure noise, usually a more mechanical high-pitched sound.

REMOVAL & INSTALLATION

✲✲CAUTION

In all cases, allow the engine to cool at least 3-4 hours before attempting any work on the turbocharger. The exhaust and turbocharger retain heat long after the engine is shut off.

2.3L Diesel
▶ See Figure 102

1. Disconnect the negative battery cable.
2. Disconnect the exhaust pipe from the turbocharger and remove the gasket. Remove the heat shield.
3. Remove the bolts holding the inlet fitting to the inlet manifold. Loosen the hose clamps and remove the fitting; remove the rubber connecting hose.
4. Loosen the hose clamps from the oil return line and remove it.
5. Carefully disconnect the oil supply pipe from the top of the turbocharger.
6. Disconnect the vacuum hose to the top of the wastegate actuator. Remove the wastegate actuator mounting bolts.
7. Remove the bolts holding the turbocharger to the exhaust flange. Remove the turbocharger and discard the gasket.
8. Immediately after removing the turbocharger, plug the intake and exhaust ports and the oil ports with clean, lint-free cloths or crumpled paper. The turbocharger MUST be protected from dirt and grit; place the unit in a protected location away from the work area.

✲✲WARNING

Do not disassemble the turbocharger unit after removal.

To install:

9. Remove the plugs from the turbocharger passages and install the turbocharger with a new gasket onto the exhaust manifold. Tighten the nuts evenly — refer to the illustration.
10. Install the wastegate actuator mounting bolts and tighten to 10 ft. lbs. (14 Nm). Reconnect the upper vacuum hose.
11. Connect the oil return line; make certain the clamp is secure.
12. Pour clean engine oil into the oil supply port to pre-lubricate the turbocharger. Install the oil supply line and tighten it carefully.
13. Install the rubber connecting hose on the inlet fitting. Use a new gasket and attach the inlet fitting to the intake manifold.
14. Use a new gasket and install the exhaust pipe to the bottom of the turbocharger. Install the heat shield.
15. Start the engine, listening carefully for any sign of unusual noise from either the manifolds or the turbocharger. Watch the oil pressure indicator for any sign of incorrect pressure.

Radiator

The radiator is nothing more than a large heat exchanger. It is mounted so that the airflow at the front of the vehicle is

ENGINE AND ENGINE REBUILDING 3-85

forced through the fins of the unit, carrying heat away from the engine coolant flowing through the unit. The fan(s) supplement the airflow by drawing in cool air, thus providing cooling even when the vehicle is not moving.

Because of the need for good air flow, modern radiators and the fans have shrouding or ducting to guide the air through the fins. This duct work, including undervehicle covers and shields, must be in place for proper cooling. Leaving the shrouds and covers off can reduce cooling efficiency and reduce driveability.

Periodically, check the radiator surfaces for blockage by leaves, insects, or mud. Most debris can be removed by hand, and the force of a water hose can be useful in dislodging other items. When cleaning the radiator fins, don't use anything metallic or sharp; the fins are very thin and are easily bent or punctured. Generally, the only times a radiator must be removed are either for repair of a leak or to allow access to other components.

It should be noted that most radiators are mounted to rubber bushings rather than directly to the bodywork. This allows the unit to serve as a vibration damper while the engine is running. The mounts and bushings must be properly reinstalled. Replace the rubber bushings if they show signs of wear or lack of flexibility.

REMOVAL & INSTALLATION

CAUTION

Always perform this work on a completely cool engine. The coolant will retain heat and pressure for several hours after use. Never remove the radiator cap or hoses if the radiator feels hot or warm to the touch. Liquid under pressure can turn to steam instantly when released; severe scalding can result. Before beginning any work, make certain the ignition is OFF.

4-Cylinder Engines

♦ See Figures 103 and 104

➡On some models the battery and shroud(s) may have to be removed for additional working access.

1. Disconnect the negative battery cable. Set the heater temperature control to HOT inside the truck.
2. Using a large capacity container, loosen the radiator drain plug and drain the cooling system. It will drain quicker with the radiator cap removed.

CAUTION

When draining the coolant, keep in mind that cats and dogs are attracted by ethylene glycol antifreeze, and are quite likely to drink any that is left in an uncovered container or in puddles on the ground. This will prove fatal in sufficient quantity. Always drain the coolant into a sealable container. Coolant should be reused unless it is contaminated or several years old.

3. Disconnect the upper and lower hoses at the radiator and disconnect the overflow hose to the reserve tank.
4. If equipped with an automatic transmission, disconnect the oil cooler lines at the radiator and at the transmission. Be prepared to contain oil spillage. Remove the hose assembly from the truck; plug the transmission ports and hose ends quickly to keep oil in and dirt out.
5. Remove the radiator mounting bolts and lift out the radiator.

To install:

6. Position the radiator in the truck, making certain all the mounts and bushings are correctly installed. Tighten the mounting bolts enough to hold well but do not overtighten them. Double check the draincock to make sure it is closed.
7. Reassemble the automatic transmission oil cooler lines and install the retaining bracket. Make certain the hoses are properly routed and firmly attached at both ends.
8. Connect the upper and lower radiator hoses and the overflow hose. Check each fitting and the inside of the hose for any corrosion or debris which would prevent a good seal.
9. Fill the system with coolant.
10. Start the engine and allow it to idle with the radiator cap removed. When the engine has warmed enough, the thermostat will open and water flow will be visible within the radiator. Fill the coolant to the bottom of the radiator neck and install the cap.

CAUTION

Do not lean over the radiator neck while the engine is running. Hot coolant may be splashed out by air trapped within the system. Keep hands, tools and clothing away from the fans which may engage at any time as the coolant heats up.

11. Shut the engine off and check the hose connections carefully for leaks. With the engine fully warm, a small leak may be emitting steam rather than liquid. If any leaks are found, allow the engine to cool completely before attempting repairs.

6-Cylinder Engines

♦ See Figures 104 and 105

CAUTION

If vehicle if equipped with SRS airbag system refer to Section 6 for the system precautions. Be extremely careful not to subject the front impact sensor to any shocks during removal and installation of the radiator assembly. The SRS system can be extremely dangerous; heed all system cautions.

1. Set the heater temperature control to HOT inside the vehicle.
2. Disconnect and remove the battery.

3-86 ENGINE AND ENGINE REBUILDING

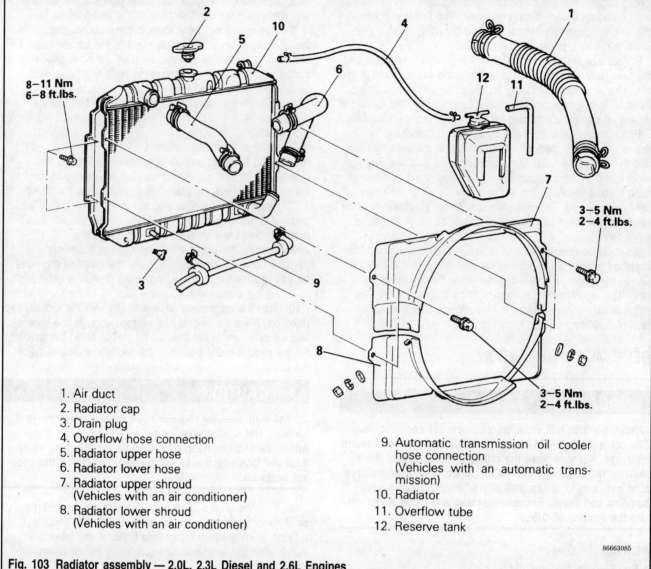

1. Air duct
2. Radiator cap
3. Drain plug
4. Overflow hose connection
5. Radiator upper hose
6. Radiator lower hose
7. Radiator upper shroud (Vehicles with an air conditioner)
8. Radiator lower shroud (Vehicles with an air conditioner)
9. Automatic transmission oil cooler hose connection (Vehicles with an automatic transmission)
10. Radiator
11. Overflow tube
12. Reserve tank

Fig. 103 Radiator assembly — 2.0L, 2.3L Diesel and 2.6L Engines

3. Using a large capacity container, loosen the radiator drain plug and drain the cooling system. It will drain quicker with the radiator cap removed.

❄❄CAUTION

When draining the coolant, keep in mind that cats and dogs are attracted by ethylene glycol antifreeze, and are quite likely to drink any that is left in an uncovered container or in puddles on the ground. This will prove fatal in sufficient quantity. Always drain the coolant into a sealable container. Coolant should be reused unless it is contaminated or several years old.

4. Disconnect the upper and lower hoses and disconnect the overflow hose to the reserve tank.
5. Remove the upper fan shroud, then the lower fan shroud for the 3.0L (12 valve) engines. For the 3.0L (24 valve) and 3.5L engines, the fan shroud is one piece.
6. Disconnect the automatic transmission cooler lines.
7. Remove the radiator retaining bolts and remove the radiator.

To install:

8. Reinstall the radiator in position, making certain all the mounts and bushings are correctly installed. Tighten the mounting bolts enough to hold well but do not overtighten them. Double check the draincock to make sure it is closed.
9. Connect the automatic transmission cooler lines.
10. Install the lower shroud first, then the upper shroud.
11. Connect the upper and lower radiator hoses and the overflow hose. Check the each fitting and the inside of the hose for any corrosion or debris which would prevent a good seal.
12. Install the battery and connect the terminals.
13. Fill the system with coolant. Start the engine and allow it to idle with the radiator cap removed. When the engine has warmed enough, the thermostat will open and water flow will

ENGINE AND ENGINE REBUILDING

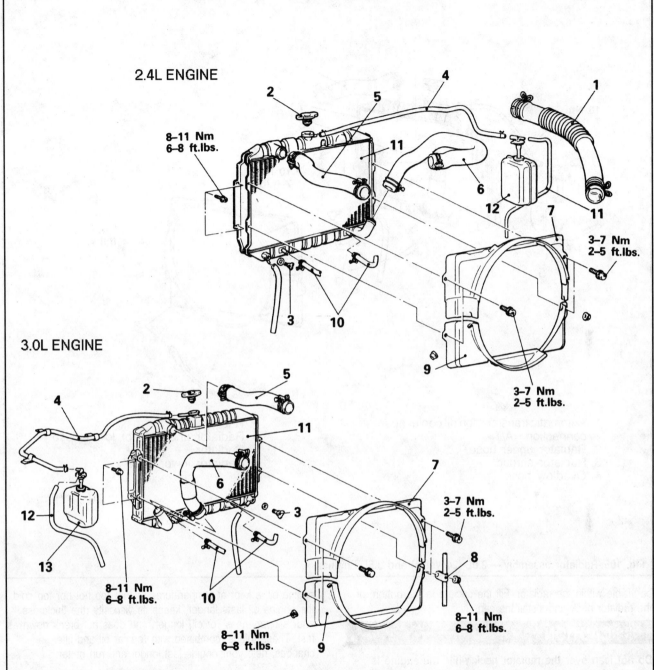

1. Air duct <2.6L Engine>
2. Radiator cap
3. Drain plug
4. Connection for overflow hose
5. Radiator upper hose
6. Radiator lower hose
7. Radiator upper shroud
8. Hose clamp <3.0L Engine>
9. Radiator lower shroud
10. Connection for automatic oil cooler hoses (Vehicles with an automatic transmission)
11. Radiator
12. Overflow tube
13. Reserve tank

Fig. 104 Radiator assembly — 2.4L and 3.0L (12 Valve) Engines

3-88 ENGINE AND ENGINE REBUILDING

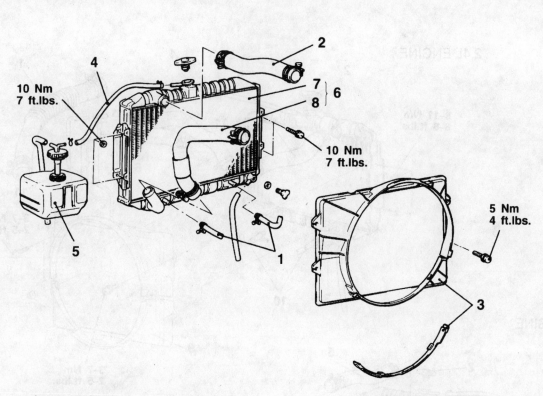

1. Automatic transmission oil cooler hose connection <A/T>
2. Radiator upper hose
3. Radiator shroud
4. Overflow hose
5. Reserve tank
6. Radiator and radiator lower hose
7. Radiator
8. Radiator lower hose

Fig. 105 Radiator assembly — 3.0L (24 Valve) and 3.5L Engines

be visible within the radiator. Fill the coolant to the bottom of the radiator neck and install the cap.

✽✽CAUTION

Do not lean over the radiator neck while the engine is running. Hot coolant may be splashed out by air trapped within the system. Keep hands, tools and clothing away from the fan.

14. Shut the engine off and check the hose connections carefully for leaks. With the engine fully warm, a small leak may be emitting steam rather than liquid. If any leaks are found, allow the engine to cool completely before attempting repairs.

Engine Oil Cooler

▶ See Figures 106, 107, 108 and 109

The engine oil cooler is a heat exchanger whose job is to help keep the temperature of the engine oil controlled. When engine oil is kept at temperatures neither too hot nor too cold, the engine oil lasts longer, keeps its viscosity (the thickness and "slippery-ness" of oil) longer, and does not break down as fast. These things, combined with regular oil and filter changes, help your engine last longer and run better.

REMOVAL & INSTALLATION

1. Disconnect the negative battery cable.
2. Drain the engine oil.
3. Remove the radiator grille from the front of the vehicle.
4. Remove the eye bolts from the engine oil cooler. Be sure to hold the weld nut of the oil cooler with a wrench while loosening the eye bolt. Discard the two eye bolt gaskets (total number of four (4) gaskets in all).
5. Remove the oil cooler mounting bolts and extract the oil cooler from the vehicle.
6. Detach the securing clamps of the hoses and tubes, loosen the flare nuts, and remove the tubes and hoses.

ENGINE AND ENGINE REBUILDING   3-89

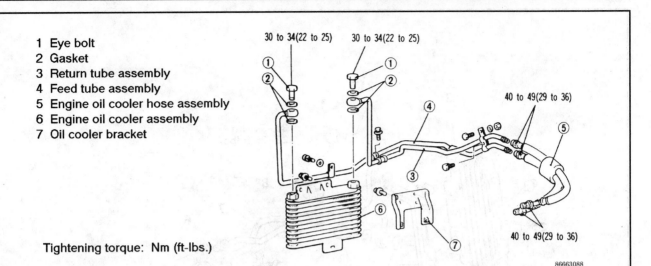

1. Eye bolt
2. Gasket
3. Return tube assembly
4. Feed tube assembly
5. Engine oil cooler hose assembly
6. Engine oil cooler assembly
7. Oil cooler bracket

Tightening torque: Nm (ft-lbs.)

Fig. 106 Engine oil cooler assembly — 2.0L, 2.3L and 2.6L Engines

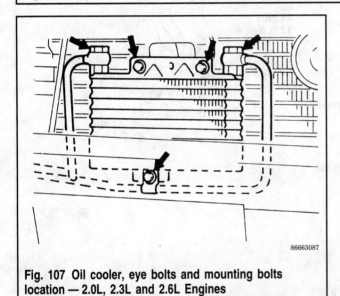

Fig. 107 Oil cooler, eye bolts and mounting bolts location — 2.0L, 2.3L and 2.6L Engines

To Install:

7. Check the oil cooler for bent or damaged fins, and the oil cooler pipes for cracks, damage, clogging or deterioration. If damage exists have the component repaired, or replace if repairing is not possible. Clean the cooling fins of dirt or other foreign matter.

8. Reconnect the oil cooler hoses and securing clamps. On the 2.0L, the 2.3L and the 2.6L engines, when installing the hose assembly, install the long hose (return side) upward and the short hose (feed side) downward. On 2.4L, 3.0L and 3.5L engines, perform the following procedure.

　a. Provisionally tighten the eye bolts, and install the clamp so that it touches the crimps on the hoses.
　b. Fully tighten the eye bolt on the return hose.
　c. Place the feed hose against the stopper, and fully tighten the eye bolt on the feed hose.

9. Mount the oil cooler back into its original position and secure with the mounting bolts.

10. Connect the feed and return oil cooling hoses back to the oil cooler. Once again use a wrench to hold the weld nut in place while tightening the eye bolt into the oil cooler.

11. Place the radiator grille back onto the front of the vehicle and secure with the screws or bolts removed.

12. Fill the engine with oil. Connect the negative battery cable to the battery.

13. Start the vehicle and check for any oil leaks.

14. Shut the vehicle **OFF**. Check the oil level and add if needed.

Engine Fan

REMOVAL & INSTALLATION

▶ See Figure 110

1. Disconnect the negative battery cable.
2. Remove all drive belts.
3. Unfasten the fan shroud securing bolts and remove the fan shroud.

➥The radiator and/or the air conditioning condenser may need to be removed to facilitate fan removal.

4. Loosen the fan-to-fluid coupling bolts and remove the fan.
5. If the fluid coupling needs to be removed, loosen and remove the water pump pulley connecting bolts and pull the fluid coupling device off of the engine.

To install:

6. Install the fluid coupling device and tighten the nuts to 7 ft. lbs. (10 Nm).
7. Install the fan on the fluid coupling and tighten the bolts evenly to 6-9 ft. lbs. (8-12 Nm).
8. Install the fan shroud.
9. Install all drive belts and adjust them as detailed previously.
10. Check the cooling system level, start the engine and check for proper operation.

3-90 ENGINE AND ENGINE REBUILDING

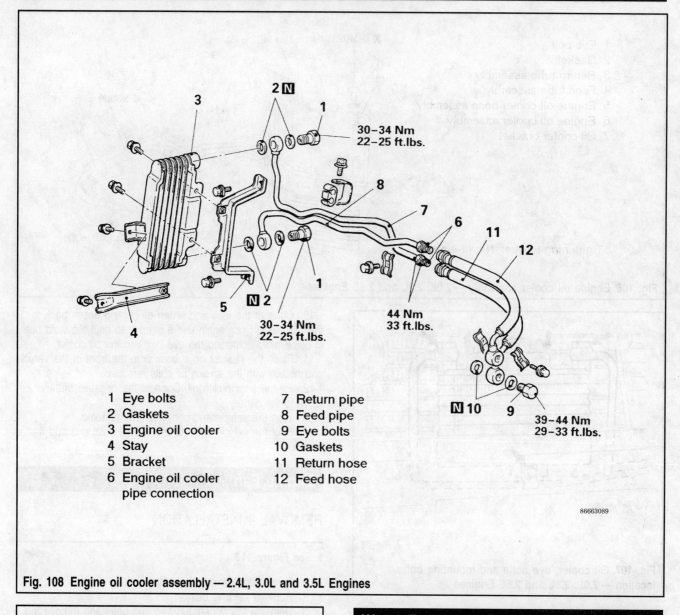

1. Eye bolts
2. Gaskets
3. Engine oil cooler
4. Stay
5. Bracket
6. Engine oil cooler pipe connection
7. Return pipe
8. Feed pipe
9. Eye bolts
10. Gaskets
11. Return hose
12. Feed hose

Fig. 108 Engine oil cooler assembly — 2.4L, 3.0L and 3.5L Engines

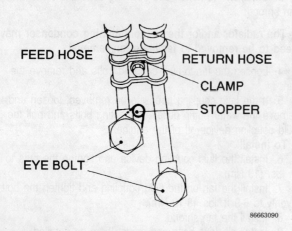

Fig. 109 Correct mounting position of the return and feed pipe to the engine — 2.4L, 3.0L and 3.5L Engines

Water Pump

REMOVAL & INSTALLATION

✱✱CAUTION

When draining the coolant, keep in mind that cats and dogs are attracted by ethylene glycol antifreeze, and are quite likely to drink any that is left in an uncovered container or in puddles on the ground. This will prove fatal in sufficient quantity. Always drain the coolant into a sealable container. Coolant should be reused unless it is contaminated or several years old.

ENGINE AND ENGINE REBUILDING   3-91

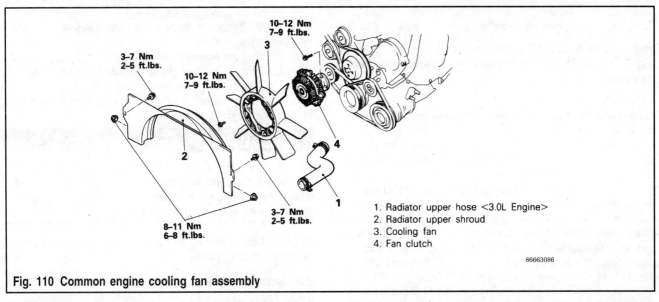

Fig. 110 Common engine cooling fan assembly

2.0L Engine

♦ See Figure 111

➡ This operation requires removal of the timing belt. Do not attempt to replace the water pump if you are not familiar with timing belt procedures.

1. Disconnect the negative battery cable.
2. Turn the crankshaft pulley clockwise until all the timing marks align. This sets No. 1 piston at TDC/compression. Once set in this position, the motor position must not be disturbed during repairs.
3. Drain the cooling system.
4. Remove the upper radiator shroud.
5. Loosen the power steering pump and remove the belt.
6. Remove the air conditioner compressor belt tensioner and remove the belt.
7. Loosen the alternator and remove the drive belt.
8. Remove the cooling fan and clutch assembly.
9. Remove the water pump pulley.
10. Remove the center bolt holding the crankshaft pulleys. The bolt will be tight.
11. Remove the smaller pulley bolt and remove the power steering pulley and the crankshaft pulley from the crankshaft.
12. Remove the upper and lower timing belt covers.
13. Confirm that all the timing marks align and that the engine is at TDC/compression on No. 1 cylinder.
14. Loosen and remove the timing belt adjuster.
15. Use chalk or a crayon to mark the direction of rotation on the timing belt. Carefully remove the timing belt from the pulleys.
16. Disconnect the lower radiator hose from the water pump.
17. Remove the water pump mounting bolts and remove the alternator bracket.

➡ The water pump bolts are of different lengths and must be reinstalled in the correct position. Label or diagram each at the time of removal.

18. Carefully remove the pump from the engine block and separate the pump from the water pipe.
19. Check the pump thoroughly for damage or cracks. Turn the shaft, checking for binding or noise. If the pump was

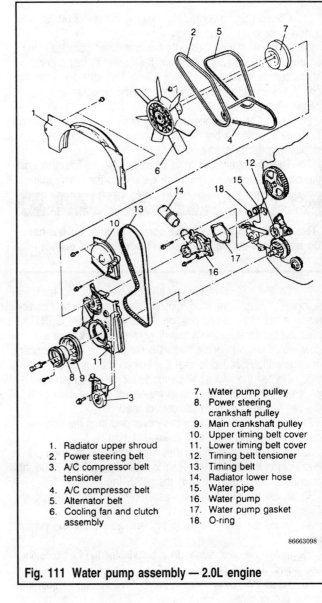

Fig. 111 Water pump assembly — 2.0L engine

3-92 ENGINE AND ENGINE REBUILDING

leaking coolant through the vent hole, the internal seal has failed and the pump must be replaced.

→Do not confuse light moisture accumulation with outright leakage. The vent hole area may be moist in normal operation.

To install:

20. Clean the mating surfaces of the pump and block. Remove all traces of the old gasket.
21. Install a new O-ring on the front end of the water pipe and wet the O-ring with water or coolant. Do not apply grease or oil.
22. Install a new water pump gasket, making sure all the bolt holes line up. Do not apply any sealer to the gasket.
23. Install the water pump retaining bolts in their correct holes (don't forget the alternator brace) and tighten each finger-tight. Tighten the bolts in two steps to their final torque: Bolt heads marked **4** tighten to 10 ft. lbs. (14 Nm) and bolt heads marked **7** tighten to 18 ft. lbs. (24 Nm).
24. Install the water pipe, making sure the O-ring stays in position.
25. Connect the lower radiator hose to the water pump. Use a new hose clamp if necessary.
26. Install the timing belt onto the crankshaft sprocket, the oil pump sprocket and the camshaft sprocket in that order. Make certain there is no slack between the crankshaft sprocket and the oil pump or between the oil pump sprocket and the camshaft sprocket.
27. Loosen the tensioner mounting bolts. Allow the tensioner to move against the belt under its spring tension; do not force or push on the tensioner.
28. Turn the crankshaft in its normal clockwise direction until the camshaft pulley has moved 2 teeth from the timing mark.

✲✲WARNING

This rotation is to create the correct tension on the belt. Do not rotate the engine counterclockwise and do not press on the belt.

29. Carefully observe the belt and the way it fits on each sprocket. It may have lifted from the camshaft sprocket, particularly in the "10 o'clock" position. If this is the case, GENTLY push on the tensioner with your fingers (counterclockwise or downward) to take up the slack. Do not push on the tensioner any more than needed to seat the belt on the sprocket.
30. Tighten the lower bolt on the tensioner first, then the pivot or upper bolt. The order is important; if done incorrectly, the belt will not be under the correct tension.
31. Install the lower timing belt cover and then the upper cover. Tighten the bolts to 8 ft. lbs. (11 Nm).
32. Install the crankshaft pulley and the power steering pulley. Tighten the small pulley bolt to 19 ft. lbs. (25 Nm) and the crankshaft center bolt to 88 ft. lbs. (120 Nm).
33. Install the pulley and fan assembly onto the water pump. Tighten the nuts to 8 ft. lbs. (11 Nm).
34. Install the alternator drive belt and adjust it to the proper tension.
35. Install the compressor drive belt. Install the belt tensioner and adjust the belt tension.
36. Install the power steering pump drive belt and adjust it to the proper tension.
37. Install the upper radiator shroud.
38. Double check the draincock, closing it if necessary, and refill the cooling system.
39. Connect the negative battery cable.
40. Start the engine and check carefully for leaks. Allow the engine to warm up with the radiator cap removed. After the thermostat opens, adjust the coolant level to the bottom of the radiator neck and install the radiator cap.

✲✲CAUTION

Do not lean over the radiator neck while the engine is running. Hot coolant may be splashed out by air trapped within the system. Keep hands, tools and clothing away from the fans which may engage at any time.

41. Shut the engine off and perform any necessary adjustments to the drive belts.

2.3L Diesel Engine

▶ See Figure 112

→Perform this procedure only on a completely cold engine.

1. Disconnect the negative battery cable.
2. Drain the coolant into clean containers so that it may be reused if still in good condition.
3. Disconnect the lower radiator hose from the water pump.
4. Loosen the bolts holding the fan, fan pulley and fan clutch.
5. Loosen the drive belt and remove it.
6. Remove the cooling fan, fan clutch and pulley.
7. Remove all the retaining bolts and remove the water pump. If it is stuck in place, it may be tapped with a plastic or rubber mallet.
8. Inspect the pump parts for any sign of wear or damage. Turn the shaft by hand, checking for binding, noise or uneven rotation.
9. Check the seal area for any sign of leakage. Any fluid leakage from the breather hole requires replacement of the pump.

→Very slight moisture or vapor around the hole is normal. Use common sense to determine outright leakage.

To install:

10. Make certain the mating surfaces are free of any trace of the old gasket.
11. Use a new gasket, install the pump and tighten the bolts evenly and alternately. Do not overtighten.
12. Install the fan clutch, pulley and cooling fan.
13. Install the drive belt and adjust it to the proper tension.
14. Connect the lower radiator hose; make certain the clamp is correctly placed and secure.
15. Check the draincock and make sure it is firmly closed. Refill the cooling system with coolant.
16. Connect the negative battery cable. Start the engine and check for leaks.

2.4L Engine

1. Disconnect the negative battery cable.
2. Drain the cooling system.
3. Remove the upper radiator shroud.

ENGINE AND ENGINE REBUILDING

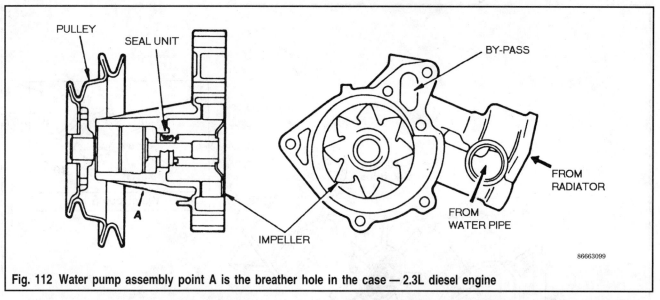

Fig. 112 Water pump assembly point A is the breather hole in the case — 2.3L diesel engine

4. Remove all accessory belts. Remove the air conditioning compressor tensioner pulley, if equipped.
5. Remove the cooling fan and clutch assembly and remove the water pump pulley.
6. Disconnect the radiator hose from the water pump.
7. Remove the crankshaft pulley(s).
8. Remove the timing belt covers.
9. If the same timing belt will be reused, mark the direction of the timing belt's rotation, for installation in the same direction. Make sure the engine is positioned so the No. 1 cylinder is at the TDC of its compression stroke and the sprockets' timing marks are aligned with the engine's timing mark indicators. Remove the timing belt.
10. The water pump bolts are different lengths, note their positions before removing. Remove the water pump mounting bolts and remove the pump from the block and the water pipe connection. Remove the O-ring from the water pipe connection.

To install:

11. Check the water pump for cracks, damage or wear. Check the bearing for damage, abnormal noise and sluggish rotation. Check the seal unit for leaks. Replace the water pump assembly if any of these symptoms is present.
12. Clean and dry the mating surfaces of the block and water pump. Install a new O-ring to the water pipe connection. Coat the new O-ring with water to aid in installation.
13. Install the water pump with a new gasket to the block and tighten the bolts to 10 ft. lbs. (13 Nm). Tighten the alternator bracket bolt to 17 ft. lbs. (23 Nm).
14. Install the timing belt onto the crankshaft sprocket, the oil pump sprocket and the camshaft sprocket in that order. Make certain there is no slack between the crankshaft sprocket and the oil pump or between the oil pump sprocket and the camshaft sprocket.
15. Loosen the tensioner mounting bolts. Allow the tensioner to move against the belt under its spring tension; do not force or push on the tensioner.
16. Turn the crankshaft in its normal clockwise direction until the camshaft pulley has moved 2 teeth from the timing mark.

✱✱WARNING

This rotation is to create the correct tension on the belt. Do not rotate the engine counterclockwise and do not press on the belt.

17. Carefully observe the belt and the way it fits on each sprocket. It may have lifted from the camshaft sprocket, particularly in the "10 o'clock" position. If this is the case, GENTLY push on the tensioner with your fingers (counterclockwise or downward) to take up the slack. Do not push on the tensioner any more than needed to seat the belt on the sprocket.
18. Tighten the lower bolt on the tensioner first, then the pivot or upper bolt. The order is important; if done incorrectly, the belt will not be under the correct tension.
19. Check to see that the clearance between the outside of the belt and the cover are within the standard value by grasping the tension side (between the camshaft sprocket and oil pump sprocket) of the center part of the timing belt between the thumb and index finger.
20. Install the crankshaft pulley(s).
21. Connect the radiator hose to the water pump.
22. Install the water pump pulley. Install the cooling fan and clutch assembly.
23. Install the air conditioning compressor tensioner pulley, if equipped.
24. Install and adjust the accessory belts.
25. Install the upper radiator shroud.
26. Fill the radiator with coolant. This cooling system has a self-bleeding thermostat, so system bleeding is not required.
27. Connect the negative battery cable, run the vehicle until the thermostat opens and fill the overflow tank. Check for leaks.
28. Once the vehicle has cooled, recheck the coolant level.

2.6L Engine

♦ See Figure 113

1. Disconnect and remove the battery.

3-94 ENGINE AND ENGINE REBUILDING

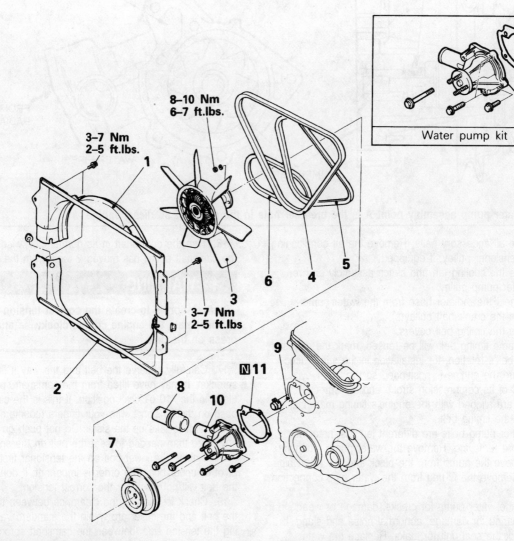

1. Radiator upper shroud
2. Radiator lower shroud
3. Cooling fan clutch assembly
Adjustment of air conditioner compressor drive belt deflection
4. Air conditioner compressor drive belt (Vehicles with an air conditioner)
Adjustment of alternator drive belt deflection
5. Alternator drive belt
Adjustment of power steering oil pump drive belt deflection
6. Power steering oil pump drive belt
7. Water pump pully
8. Connection of radiator lower hose
9. Connection of heater hose
10. Water pump
11. Water pump gasket

NOTE
Reverse the removal procedures to reinstall.
N : Non-reusable parts

Fig. 113 Water pump assembly — 2.6L Engines

ENGINE AND ENGINE REBUILDING 3-95

2. Drain the cooling system.
3. Remove the lower radiator hose from the water pump.
4. Remove the upper fan shroud. Carefully unbolt and remove the fan with the clutch assembly and store the fan in an upright position.
5. Loosen the adjuster and remove the air conditioning compressor drive belt. Loosen the alternator and power steering pump and remove their drive belts.
6. Remove the water pump pulley from the studs.
7. Disconnect the heater hose from the water pump.
8. Remove the mounting bolts holding the pump to the engine. Note that the bolts are of different lengths; label or diagram them as they are removed.
9. Remove the water pump and gasket. Clean the mating surfaces thoroughly and remove all traces of the old gasket.
10. Check the pump thoroughly for damage or cracks. Turn the shaft, checking for binding or noise. If the pump was leaking coolant through the vent hole, the internal seal has failed and the pump must be replaced.

➡ Do not confuse light moisture accumulation with outright leakage. The vent hole area may be moist in normal operation.

11. If the fan was removed, inspect the fan blades for cracking or bending. Check the area of each bolt hole for cracks.
   To install:
12. Install a new gasket to the engine, making sure all the bolt holes are aligned. Install the water pump and tighten the bolts finger-tight. Don't forget to install the alternator bracket. When all the bolts are snug, tighten them to about 11 ft. lbs. (21 Nm) in steps.
13. Connect the heater hose to the water pump.
14. Install the water pump pulley.
15. Install the drive belts beginning with the power steering belt, then the alternator belt and the A/C compressor belt. Seat the belts on all the pulleys but do not tighten them yet.
16. Install the fan and clutch assembly. Install the upper fan shroud.

✻✻WARNING

The shroud retaining bolts must not exceed 0.63 in. (16mm) in length. Longer bolts will interfere with the radiator and cause leaks.

17. Adjust the drive belts to the proper tension and tighten the adjusting fittings for each component.
18. Install the lower radiator hose.
19. Install the battery.
20. Double check the draincock, closing it if necessary, and refill the cooling system.
21. Start the engine and check carefully for leaks. Allow the engine to warm up with the radiator cap removed. After the thermostat opens, adjust the coolant level to the bottom of the radiator neck and install the radiator cap.

3.0L Engines
▸ See Figure 114

➡ Develop a system of identifying bolts by location and component so that reassembly is made easier. Damage can be caused by forcing a long bolt into a shorter passage. The use of the correct special tools or their equivalent is most important for this procedure. The crankshaft sprocket must be removed and installed with the use of tools MB 990767-01 and MB 998715 or their equivalents. Attempts to use other tools or methods may cause damage to the sprocket and/or crankshaft.

1. Release fuel pressure in the system by disconnecting the fuel pump wiring harness. The fuel pump is located back by the fuel tank. Start the engine and let the vehicle idle until it stalls on its own. Shut the engine **OFF**.
2. Disconnect the negative battery cable.
3. Drain the cooling system. Remove the thermostat and housing/case assembly, upper radiator hose, bypass hose and the radiator shroud.
4. Remove the cooling fan clutch assembly.
5. Remove the drive belts from the A/C compressor, the power steering pump and the alternator.
6. Remove the cooling fan pulley.
7. Remove the power steering pump, power steering pump bracket and mounting bracket. When the power steering pump is removed from the bracket, hold it toward the body using wire or similar materials. Move the pump with the pressure hose and return hose still attached.
8. Remove the tension pulley bracket.
9. Disconnect the air conditioning compressor and remove the compressor bracket. Hold the air conditioning compressor toward the body using wire or similar materials. Move the compressor with the high pressure hose and low pressure hose still attached.
10. Remove the cooling fan bracket assembly.
11. Remove the timing belt covers. If the same timing belt will be reused, mark the direction of the timing belt's rotation, for installation in the same direction. Make sure the engine is positioned so the No. 1 cylinder is at the TDC of its compression stroke and the sprockets timing marks are aligned with the engine's timing mark indicators.
12. Loosen the timing belt tensioner bolt and remove the belt. Position the tensioner as far away from the center of the engine as possible and tighten the bolt.
13. Remove the water pump mounting bolts, separate the pump from the water inlet pipe and remove the pump from the engine. Remove the water inlet fitting from the pump.
   To install:
14. Check the pump thoroughly for damage or cracks. Turn the shaft, checking for binding or noise. If the pump was leaking coolant through the vent hole, the internal seal has failed and the pump must be replaced.
15. Clean and dry the mating surfaces of the block and water pump. Install a new O-ring to the water inlet pipe.
16. Install the pump and water inlet fitting, with new gaskets, to the engine and water pipe. Tighten the water pump mounting bolts evenly to 20 ft. lbs. (27 Nm).
17. Align all timing marks. Install the timing belt — refer to the necessary service procedures located in this section.
18. Install the timing belt covers.
19. Install the crankshaft pulley and flange.
20. Install the air conditioning compressor and its bracket.
21. Install the power steering pump bracket and the pump.
22. Install the water pump pulley. Install the cooling fan and clutch assembly.
23. Install the drive belts, adjust as necessary.

3-96 ENGINE AND ENGINE REBUILDING

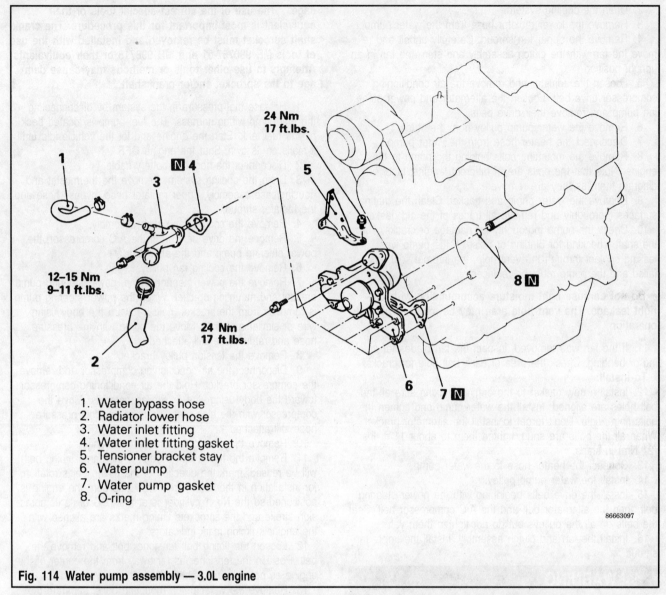

Fig. 114 Water pump assembly — 3.0L engine

1. Water bypass hose
2. Radiator lower hose
3. Water inlet fitting
4. Water inlet fitting gasket
5. Tensioner bracket stay
6. Water pump
7. Water pump gasket
8. O-ring

24. Install thermostat and housing assembly/case. Connect the upper radiator hose, bypass hose and the upper radiator shroud.
25. Fill the cooling system.
26. Connect the negative battery cable, run the vehicle until the thermostat opens and fill the overflow tank. Check for leaks.
27. Once the vehicle has cooled, recheck the coolant level.

3.5L Engines
▶ See Figure 115

For this procedure, you will need a water pump O-ring, a water pump/engine block gasket, a water pump/water pump fitting gasket, a water pump fitting/thermostat case gasket, a water outlet fitting/thermostat case gasket, a water outlet fitting O-ring, and any disposable parts needed for a timing belt removal. Refer to the timing belt procedure in this section.

➡Develop a system of identifying bolts by location and component so that reassembly is made easier. Damage can be caused by forcing a long bolt into a shorter passage. The use of the correct special tools or their equivalent is most important for this procedure. The crankshaft sprocket must be removed and installed with the use of tools MB 990767-01 and MB 998715 or their equivalents. Attempts to use other tools or methods may cause damage to the sprocket and/or crankshaft.

1. Release fuel pressure in the system by disconnecting the fuel pump wiring harness. The fuel pump is located back by the fuel tank. Start the engine and let the vehicle idle until it stalls on its own. Shut the engine **OFF**.
2. Disconnect the negative battery cable.
3. Drain the cooling system.
4. Remove the radiator, the upper radiator hose and the cooling fan shroud. Refer to the previously described procedure in this section.
5. Remove the alternator, refer to the previously described procedure in this section.
6. Remove the battery and battery tray.
7. Remove the under skid plate and the undercover.
8. Remove the cooling fan clutch assembly and the water pump pulley.

ENGINE AND ENGINE REBUILDING 3-97

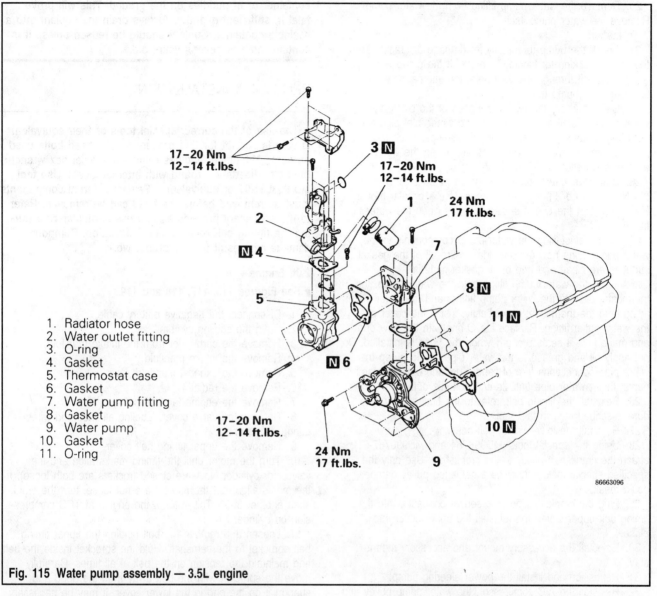

1. Radiator hose
2. Water outlet fitting
3. O-ring
4. Gasket
5. Thermostat case
6. Gasket
7. Water pump fitting
8. Gasket
9. Water pump
10. Gasket
11. O-ring

Fig. 115 Water pump assembly — 3.5L engine

9. Remove the air conditioning compressor and compressor bracket. Remove the air conditioning compressor with the hoses attached. Suspend the removed air conditioning compressor, by using wire or similar material, at a place where no damage will occur to it.
10. Remove the power steering pump, the power steering pump cover and the accessory mount stay. As with the A/C compressor, move the power steering pump with the hoses still attached and secure it to the body with either strong wire or strong cord.
11. Remove the accessory mount from the front of the engine.
12. Remove the timing belt upper cover.
13. Unplug the crankshaft position sensor connector.
14. Remove the crankshaft pulley from the crankshaft. Use only the specified tools (MD 998754 and MB990767-01), or a damaged pulley damper could result.
15. Remove the timing belt lower cover.
16. Remove the timing belt.
   a. Align the timing marks on the pulleys with the marks on the engine heads.
   b. Loosen the center bolt on the tensioner pulley to remove the timing belt. Make a mark on the back of the timing belt indicating the direction of rotation so it may be reassembled in the same direction if it is to be reused.

❋❋CAUTION

The cam of the left bank camshaft lifts the valve by means of the rocker arm, the spring force of the valve will easily turn the sprocket, so be careful not to insert your fingers, etc.

17. Remove the auto tensioner, the tension pulley and the tension arm assembly.
18. Remove the water outlet fitting, by removing the bracket over it first. Discard the old O-ring and gasket, these will need to be replaced for installation.
19. Unbolt the thermostat case from the water pump fitting. A new gasket for this juncture will also be needed before reassembly.

3-98 ENGINE AND ENGINE REBUILDING

20. Remove the water pump fitting from the water pump. Remove the water pump itself.

To Install:

21. Check the pump thoroughly for damage or cracks. Turn the shaft, checking for binding or noise. If the pump was leaking coolant through the vent hole, the internal seal has failed and the pump must be replaced.

22. Clean and dry the mating surfaces of the block and water pump. Install a new O-ring to the water inlet pipe. Do not apply oil and grease to water pipe O-rings. Keep the water pipe connections free of sand, dust, etc. Insert the water pipe until its end bottoms.

23. Install the water pump and tighten its bolts to 17 ft. lbs. (24 Nm) in two or three gradual steps. Use a circular motion. Make sure that the O-ring seated properly when the water pipe fit into it.

24. When reinstalling the water pump components to the water pump using new gaskets. Make sure all of the gasket surfaces are clean and free of all gasket material.

25. Install the water pump fitting onto the water pump, the thermostat case to the water pump fitting, and the water outlet fitting onto the thermostat case. Lastly, install the cover over the water outlet fitting. Replace the O-ring with the new one and make sure it seats properly when the hose is installed. Do not apply oil and grease to the water pipe O-ring. Keep the water pipe connections free of sand, dust and other contaminants. Insert water pipe until its end bottoms out.

26. Reinstall the timing belt, refer to the timing belt procedure described later in this section.

27. Reinstall the lower timing belt cover.

28. Using the special tools (MD998754 and MD990767-01), attach the crankshaft pulley to the crankshaft. Use only the specified special tools, otherwise a damaged pulley damper could result.

29. Plug the crankshaft position sensor connector into the timing belt upper cover and reinstall the upper cover into place.

30. Reinstall the accessory mount and accessory mount stay.

31. Reposition and install the power steering pump.

32. Reinstall the A/C compressor, the water pump pulley and the cooling fan clutch assembly.

33. Install the under skid plate and the undercovers.

34. Install the battery tray and battery.

35. Install the alternator assembly.

36. Install the radiator assembly.

37. Install the drive belts, adjust as necessary.

38. Fill the cooling system.

39. Plug the fuel pump harness back in.

40. Connect the negative battery cable, run the vehicle until the thermostat opens and fill the overflow tank. Check for leaks.

41. Once the vehicle has cooled, recheck the coolant level.

Cylinder Head and Gasket

✱✱CAUTION

When draining the coolant, keep in mind that cats and dogs are attracted by ethylene glycol antifreeze, and are quite likely to drink any that is left in an uncovered container or in puddles on the ground. This will prove fatal in sufficient quantity. Always drain the coolant into a sealable container. Coolant should be reused unless it is contaminated or several years old.

REMOVAL & INSTALLATION

➡The use of the correct special tools or their equivalent is very important for this procedure. The head bolts used on some Mitsubishi engines require a special hex wrench; the bolts' heads are round with internal facets. Use tool MD 998051-01 or equivalent. Several external components must be removed before the head can be removed. Refer to the necessary procedures and use particular care during the timing belt removal and installation. Component damage can result from improper work.

2.0L Engine

▶ See Figures 116, 117, 118 and 119

1. Disconnect the negative battery cable.
2. Drain the cooling system.
3. Remove the carburetor from the intake manifold.
4. Remove the intake manifold.
5. Remove the exhaust manifold.
6. Remove the radiator.
7. Remove the engine fan and pulley.
8. Remove the valve cover, labeling all lines and wires during disassembly.
9. Remove the upper timing belt cover.
10. Turn the motor until the timing marks align and the valves for cylinder No.1 are closed (rockers are both loose). If the marks align, but the rockers are not loose, turn the crankshaft another 360°. This will put the engine at TDC compression on cylinder No. 1.
11. Loosen and remove the bolt holding the upper timing belt sprocket to the camshaft. Hold the sprocket inside the belt and maintain the tension on the belt at all times. Carefully move the sprocket with the belt away from the head and support it on the pad in the lower cover. It may be necessary to use a small shim of soft material to keep the sprocket in position and under tension.

✱✱WARNING

If the belt tension is lost or the sprocket fall out of place, the timing belt must be reinstalled "from scratch", requiring the removal of the lower cover and other items.

12. Double check the head to insure no remaining wires or hoses are connected.
13. Loosen the head bolts in the order shown (this is important) and in three passes until all are finger loose. Remove the bolts.
14. Rock the head gently to break it loose; if tapping is necessary, do so with a rubber or wooden mallet at the corners of the head. DO NOT pry the head up by wedging tools between the head and the block.
15. Lift the head free of the engine. Support it on wooden blocks on the workbench. Refer to cleaning and inspection following this procedure for work to be done before installing

ENGINE AND ENGINE REBUILDING

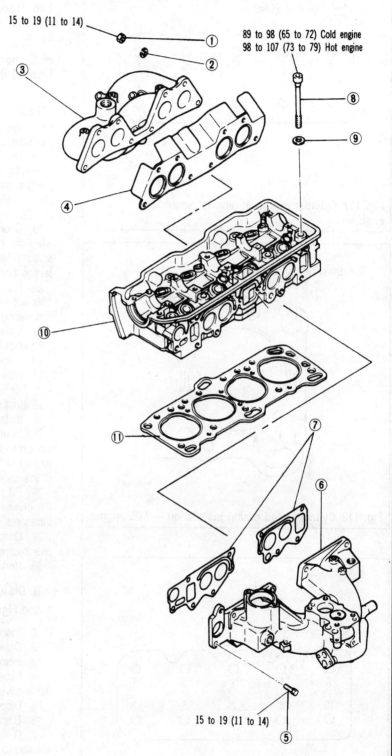

(1) Nut (8)
(2) Spring washer (7)
(3) Exhaust manifold
(4) Exhaust manifold gasket
(5) Bolt (5)
(6) Inlet manifold
(7) Inlet manifold gasket
(8) Cylinder head bolt (10)
(9) Washer (10)
(10) Cylinder head
(11) Cylinder head gasket

Tightening torque: Nm (ft-lbs.)

Fig. 116 Cylinder head assembly — 2.0L engine

3-100 ENGINE AND ENGINE REBUILDING

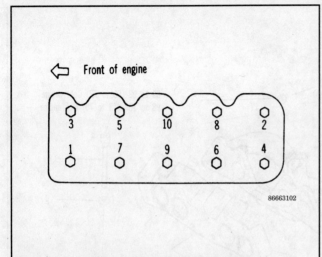

Fig. 117 Cylinder head bolt removal sequence — 2.0L engine

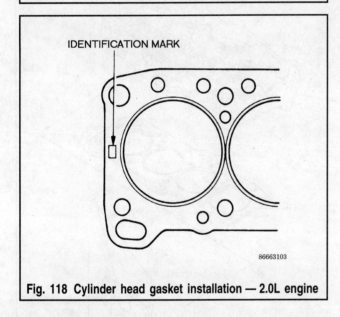

Fig. 118 Cylinder head gasket installation — 2.0L engine

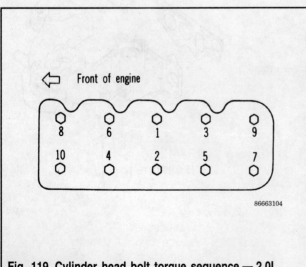

Fig. 119 Cylinder head bolt torque sequence — 2.0L engine

the head. If the head has been removed for work other than gasket replacement, the rocker assembly and the camshaft may be removed.

To install:

16. Place a new gasket onto the engine block. Make certain it is installed with the identifying mark facing up (towards the head).

➡ Do not apply any sealant to the gasket or mating surfaces of the head and block.

17. Install the head straight down onto the block. Try to eliminate most of the side-to-side adjustments as this may move the gasket out of position. Install the bolts by hand and just start each bolt 1 or 2 turns on the threads.
18. The head bolt torque specification is 68 ft. lbs. (92 Nm). The bolts must be tightened in order in two equal steps. On the first pass, tighten all the bolts to 34 ft. lbs. (46 Nm); then proceed through the order again, tightening each bolt to 68 ft. lbs. (92 Nm).
19. Check that the cam and rocker position has not changed while the head was removed. Carefully install the camshaft sprocket with the timing belt to the camshaft. Tighten the retaining bolt to 65 ft. lbs. (88 Nm).
20. Install the upper timing belt cover and the valve cover. This will protect the valve train and timing belt from contamination during reassembly of other components.
21. Install the exhaust manifold with a new gasket.
22. Install the intake manifold with a new gasket.
23. Install the carburetor.
24. Connect all wiring, hoses, and lines to the proper location. Double check all connectors and clamps.
25. Install the engine fan and pulley.
26. Install the radiator.
27. Double check all installation items, paying particular attention to loose hoses or hanging wires, untightened nuts, poor routing of hoses and wires (too tight or rubbing) and tools left in the engine area.
28. Fill the cooling system with the proper coolant. Changing the engine oil and filter is recommended to eliminate any contaminants in the oil.
29. Connect the negative battery cable. Start the engine and check carefully for leaks of oil, vacuum, fuel or coolant.
30. Perform final engine adjustments as necessary.

2.3L Diesel Engine

▶ See Figures 120 and 121

1. Disconnect the negative battery cable.
2. Drain the coolant into clean containers for reuse.
3. Remove the upper radiator hose from the engine.
4. Label and disconnect the breather hoses running to the valve cover and across the head.
5. Disconnect the heater hose at the intake manifold.
6. Disconnect the wiring to the temperature sensor.
7. Disconnect and remove the inlet piping and the oil lines running to the turbocharger. Make certain that both the lines and the turbocharger ports are plugged or covered as soon as they are disconnected. NO foreign matter is permitted in the turbocharger system.
8. Carefully disconnect the fuel injection lines. See Section 5.
9. Disconnect the glow plug connections.

ENGINE AND ENGINE REBUILDING

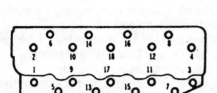

Fig. 120 Cylinder head bolt removal sequence — 2.3L diesel engine

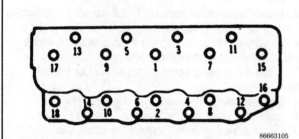

Fig. 121 Cylinder head bolt torque sequence — 2.3L diesel engine

10. Remove the intake and exhaust manifolds.
11. Remove the upper and lower timing belt covers. Turn the crankshaft and align all the timing marks; this positions the motor at TDC/compression for No. 1 cylinder.
12. Mark the direction of rotation on the timing belts with chalk or a felt marker.
13. Remove the center bolt holding the crankshaft pulley and remove the pulley. Do not turn the engine out of position during this removal.
14. Remove the lower timing belt cover.
15. Slightly loosen the retaining (lock) bolts for the timing belt tensioner. Move the tensioner towards the water pump and tighten the bolts; this holds the tensioner in the slack position.
16. Remove the camshaft sprocket and the injection pump sprocket. Do not let either shaft turn out of place during the removal. Remove the timing belt without forcing it or kinking it.

17. Remove the fuel injection pump.

✲✲WARNING

Do not move or disturb the position of the engine once the timing belt is removed!

18. Remove the valve cover and gasket.
19. Loosen the head bolts in the correct sequence. Loosen each bolt gradually and in two or three passes. Sudden loosening of any bolt may cause head damage by warping.
20. Remove the head by lifting it off the engine; do not attempt to slide it to the side.
21. Remove the gasket. Clean the mating surfaces of all gasket remains. If the head is to be off the engine for any length of time, plug the cylinders with clean cloths or crumpled newspaper. Place the head on wood blocks on the work bench for examination or further work. The head may be cleaned, inspected or disassembled following instructions given later in this section.

To install:
22. Before reinstallation, double check the head and block faces for any trace of gasket material. Remove any plugs or paper from the head and/or cylinders.
23. Place a new head gasket in position on the block. Pay close attention to the alignment of the gasket with the holes and passages in the block. The gaskets generally do not have any identifying marks for up or front. An incorrectly installed gasket may block an oil or coolant passage.
24. Place the head onto the block carefully; do not disturb the position of the gasket. Start the head bolts in each hole by hand and turn them in 1 or 2 turns.
25. Following the correct bolt-tightening pattern, tighten each bolt to 40 ft. lbs. (54 Nm). This is one half the final tension.
26. Repeat the bolt tightening sequence, tightening each bolt to 80 ft. lbs. (108 Nm) The acceptable range is 76-83 ft. lbs. (103-113 Nm).
27. Install the crankshaft pulley; tighten the bolt to 52 ft. lbs. (70 Nm).
28. Install the fuel injection pump.
29. Move the timing belt tensioner all the way toward the water pump and temporarily tighten it in this position.
30. Install the timing belt onto the injection pump sprocket and camshaft sprocket, making certain NOT to turn the crankshaft sprocket during the procedure. Make certain there is no slack in the belt between the camshaft sprocket and the injection pump sprocket.

➡The injection pump sprocket tends to rotate during this installation. Hold it in place while installing the timing belt.

31. With the belt in place, check the alignment of the timing marks on both sprockets. Loosen the tensioner mounting bolts which will apply tension to the belt.
32. Tighten the belt tensioner nut and bolt. Tighten the slot side bolt first, then the fulcrum side. If it's done the other way, the adjuster will turn and the belt will be too tight.
33. Check the alignment of the timing marks.
34. Using a wrench on the crankshaft bolt, turn the engine clockwise enough to move the camshaft sprocket 2 teeth past the timing mark. Use the wrench to reverse (counterclockwise) direction and turn the engine back to the original position with the timing marks aligned.

35. Use your finger to press down on the belt midway between the cam sprocket and the injection pump sprocket. Deflection in the belt should be 0.16-0.20 in. (4-5mm).
36. Install the valve cover.
37. Install the upper and lower timing covers.
38. Install the intake and exhaust manifolds. Always use new gaskets.
39. Connect the fuel injection lines.
40. Connect the wiring for the glow plug circuit.
41. Install the turbocharger if it was removed from the manifolds. Remove any plugs from the turbocharger passages; prime the unit by pouring a clean engine oil into the turbocharger oil inlet port.
42. Connect the oil drain hose and the oil inlet pipe to the turbocharger.
43. Connect the turbocharger air inlet fitting and rubber hose (duct).
44. Connect the temperature sensor lead. Connect the heater hose to the intake manifold.
45. Connect the breather and vacuum hoses as necessary.
46. Install the upper radiator hose to the engine and make certain the clamp is properly located and secure.
47. Refill the cooling system with coolant.
48. Connect the negative battery cable; start the engine and check for leaks of fuel, oil or coolant.
49. Check the injection timing, adjusting it if needed.

2.4L Engine

▶ See Figures 122 and 123

1. Disconnect negative battery cable. Properly relieve the fuel system pressure.
2. Drain the cooling system.
3. Remove the upper radiator hose. Disconnect the heater hoses.
4. Remove the air intake hose and pipe.
5. Disconnect the accelerator, and if equipped, kickdown cable.
6. Disconnect and plug the fuel lines.
7. If equipped with power steering, unbolt the power steering pump from its brackets and position it to the side.

➡ **Do not disconnect the power steering lines.**

8. Remove the timing belt upper cover and valve cover.
9. Rotate the crankshaft clockwise and align the timing mark on the camshaft sprocket with the timing mark on the cylinder head.
10. Remove the camshaft bolt.
11. Remove the sprocket from the camshaft (with the timing belt attached) and allow it to rest on the lower cover. Secure the belt to the sprocket so that they do not become disengaged.

✷✷WARNING

Do not rotate the crankshaft after the camshaft sprocket is removed from the camshaft. Secure the sprocket and timing belt so there is no slack in the belt. Make sure the sprocket does not become disengaged from the timing belt. If the engine is disturbed or the timing belt moved, the camshaft timing will have to be reset.

12. Label and disconnect the spark plug wires from the spark plugs. Remove the distributor cap and wires.
13. Mark the position of the rotor and distributor housing in relation to the cylinder head and remove the distributor. If unfamiliar with the procedures for removing the distributor, refer to those directions in Section 2.
14. Label and disconnect all vacuum lines, hoses and wiring connectors from the manifolds and cylinder head.
15. Raise the vehicle and support safely.
16. Remove the exhaust pipe from the exhaust manifold.
17. Lower the vehicle.
18. Remove cylinder head bolts, starting from the outside and working inward. Remove the cylinder head from the engine.
19. If necessary, remove the intake and exhaust manifolds from the cylinder head.
20. Clean the cylinder head gasket mating surfaces.

To install:

21. If removed, install the intake and exhaust manifolds to the cylinder head, using new gaskets.
22. Install a new head gasket to the block and position the cylinder head assembly with all head bolts and washers. tighten the bolts in sequence, in 2 or 3 steps, to 72 ft. lbs. (100 Nm).
23. Install the camshaft sprocket to the camshaft and tighten the bolt to 72 ft. lbs. (100 Nm).
24. Install the distributor, aligning the marks made during removal. Install the distributor cap and spark plug wires.
25. Install the power steering pump and adjust the belt tension.
26. Connect the heater hoses and upper radiator hose.
27. Install the valve cover and upper timing belt cover.
28. Connect the fuel lines. Connect the accelerator, and if equipped, kickdown cables.
29. Connect all remaining wiring connectors, vacuum lines and hoses in their proper locations.
30. Install the air intake pipe and hose.
31. Reconnect the exhaust system to the exhaust manifold.
32. Refill the cooling system.
33. Connect the negative battery cable.
34. Start the engine and check for leaks. Check the ignition timing.

2.6L Engine

▶ See Figures 124, 125, 126, 127, 128, 129, 130, 131, 132 and 133

1. Disconnect the negative battery cable.
2. Drain the coolant.
3. Remove the air conditioner compressor drive belt.
4. Remove the heat shield over the brake master cylinder.
5. Disconnect the oxygen sensor harness connector and remove the heat shield from the exhaust manifold.
6. Remove the air cleaner assembly.
7. Remove the upper radiator hose from the thermostat housing.
8. Remove the air intake duct and the secondary air cleaner.
9. Disconnect the PCV hose and accelerator cable from the valve cover. Remove the valve cover and gasket. Remove the half-circle plug from the head.

ENGINE AND ENGINE REBUILDING 3-103

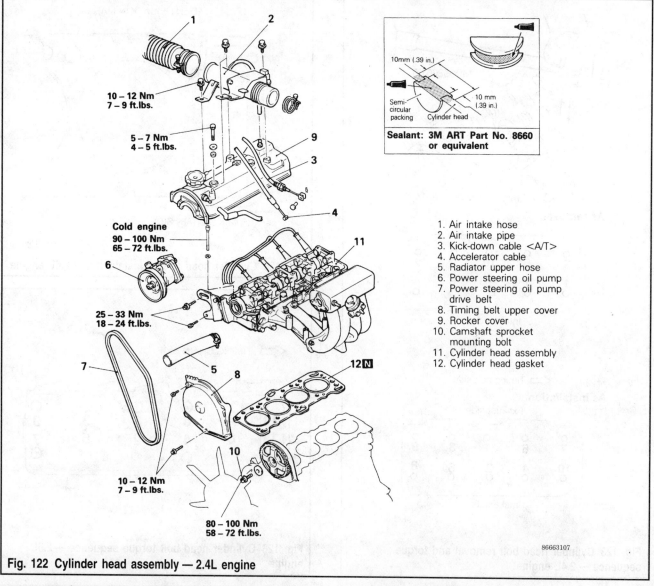

Fig. 122 Cylinder head assembly — 2.4L engine

1. Air intake hose
2. Air intake pipe
3. Kick-down cable <A/T>
4. Accelerator cable
5. Radiator upper hose
6. Power steering oil pump
7. Power steering oil pump drive belt
8. Timing belt upper cover
9. Rocker cover
10. Camshaft sprocket mounting bolt
11. Cylinder head assembly
12. Cylinder head gasket

10. Turn the crankshaft clockwise to align the timing marks and set the engine to TDC/Compression for No. 1 cylinder.
11. Label and disconnect the coil and spark plug wires from the distributor.
12. Disconnect the fuel lines.
13. Disconnect the vacuum hose to the power brake booster from the engine. Disconnect the heater hose and engine coolant hose at the rear of the engine.
14. Remove the dipstick tube and dipstick. Discard the O-ring.
15. Disconnect the rear (bottom) of the catalytic converter from the exhaust system.
16. Remove the distributor.
17. Label and disconnect the vacuum hoses at or near the head.
18. Label and disconnect the electrical connectors at the head and intake manifold. Remember the connectors on the rear of the head, too.
19. Remove the ground cable connection at the engine.
20. Remove the bolt holding the camshaft sprocket to the camshaft. Remove the distributor drive gear.

21. Pull the camshaft sprocket (with the timing chain attached) from the camshaft and place it on top of the camshaft sprocket holder.

✱✱WARNING

The crankshaft must not be rotated after the sprocket is removed from the camshaft. Do not allow the timing chain to come off the camshaft sprocket.

22. Double check the head to insure no remaining wires or hoses are connected.
23. Loosen the head bolts in the order shown (this is important) and in three passes until all are finger loose. Remove the bolts. Note that the two front bolts must each be loosened before bolt No. 2 is loosened.
24. Rock the head gently to break it loose; if tapping is necessary, do so with a rubber or wooden mallet at the corners of the head. DO NOT pry the head up by wedging tools between the head and the block.

3-104 ENGINE AND ENGINE REBUILDING

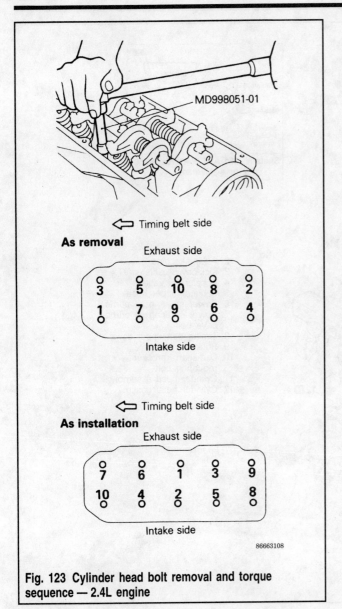

Fig. 123 Cylinder head bolt removal and torque sequence — 2.4L engine

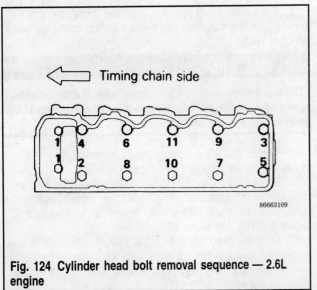

Fig. 124 Cylinder head bolt removal sequence — 2.6L engine

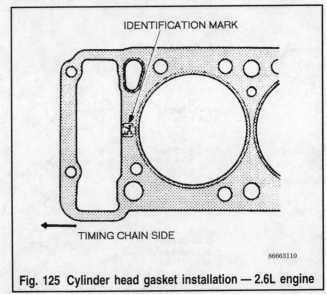

Fig. 125 Cylinder head gasket installation — 2.6L engine

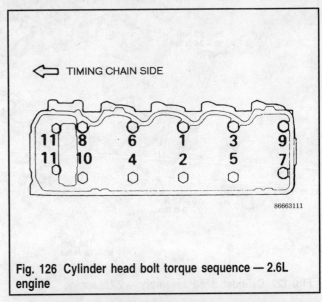

Fig. 126 Cylinder head bolt torque sequence — 2.6L engine

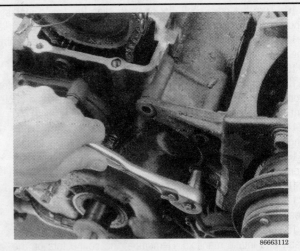

Fig. 127 Detach any brackets which might be in the way — 2.6L engine

ENGINE AND ENGINE REBUILDING

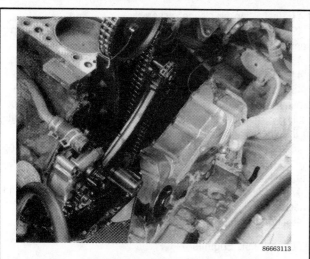

Fig. 128 Remove the timing belt covers to access the timing belt — 2.6L engine

Fig. 131 Loosen and remove the remaining cylinder head bolts in the proper sequence — 2.6L engine

Fig. 129 After removing the camshaft sprocket, secure it to the chain and rest it on the lower pads — 2.6L engine

Fig. 132 If removal of the cylinder head is difficult, tap the corners with a wood or rubber mallet. Do not pry the head off with anything — 2.6L engine

Fig. 130 Remove the cylinder head bolts, starting with the two foremost bolts — 2.6L engine

Fig. 133 Remove the old gasket and dispose of it. Use a new gasket upon installation — 2.6L engine

25. Lift the head free of the engine, remembering that the intake and exhaust manifolds (with the catalytic converter) are still attached to the head. This assembly is heavy; an assistant is helpful when lifting. Support the head assembly on wooden blocks on the workbench. Refer to "Cleaning and Inspection" following this section for work to be done before installing the head. If the head has been removed for work other than gasket replacement, the manifolds, rocker assembly and the camshaft may be removed.

26. Before placing a new gasket onto the engine block, apply 3M® ART sealant No. 8660 or equivalent to the top surface of each joint between the timing case and the cylinder block. Apply just enough sealant to fill the seam and make certain NO sealant enters the oil passages in the block. Place the new gasket on the block with the identifying mark facing up (towards the head).

➡ **Do not apply any sealant to the gasket or mating surfaces of the head and block.**

27. Install the head straight down onto the block. Try to eliminate most of the side-to-side adjustments as this may move the gasket out of position. Install the bolts by hand and just start each bolt 1 or 2 turns on the threads.

28. The head bolt torque specification is 68 ft. lbs. (92 Nm). The bolts must be tightened in the order shown in 3 steps. On the first pass, tighten all the bolts to about 20 ft. lbs. (27 Nm) then proceed through the order tightening each bolt to about 45 ft. lbs. (61 Nm). The final torque is achieved on the third pass.

✱✱WARNING

The two bolts into the timing case are tightened ONLY to 13 ft. lbs. (19 Nm). Achieve this torque in steps of 4 ft. lbs. (5.4 Nm), 9 ft. lbs. (12 Nm) and 13 ft. lbs. (18 Nm). Do not overtighten these bolts.

29. Check that the cam and rocker position has not changed while the head was removed. Carefully install the camshaft sprocket with the timing chain to the camshaft. Install the distributor drive gear and tighten the retaining bolt to 40 ft. lbs. (54 Nm).

30. Connect the engine ground strap and connect each electrical connector to its proper component.
31. Connect the vacuum hoses to the proper ports, making sure each is firmly seated and not crimped or crushed.
32. Install the distributor.
33. Connect the catalytic converter to the exhaust system.
34. Install a new O-ring on the dipstick tube and install the tube to the engine. Make sure the O-ring is not damaged during installation.
35. Connect the small coolant hose and the heater hose to the head or intake manifold. Install the brake booster vacuum hose.
36. Connect the fuel lines.
37. Install the spark plug and coil wires.
38. Install the half-circle plug with sealant. Install the valve cover and gasket.
39. Install the accelerator cable and connect the PCV hose to the valve cover.
40. Install the secondary air cleaner and connect the air intake ducts.
41. Install the air cleaner assembly.
42. Install the manifold heat shield and connect the oxygen sensor wiring.
43. Install the heat shield on the brake master cylinder.
44. Install and adjust the air conditioning compressor belt.
45. Double check all installation items, paying particular attention to loose hoses or hanging wires, untightened nuts, poor routing of hoses and wires (too tight or rubbing) and tools left in the engine area.
46. Fill the cooling system with the proper coolant. Changing the engine oil and filter is recommended to eliminate any contaminants in the oil.
47. Connect the negative battery cable. Start the engine and check carefully for leaks of oil, vacuum, fuel or coolant.
48. Perform final engine adjustments as necessary.

3.0L And 3.5L Engines
▸ See Figures 134 and 135

✱✱WARNING

The use of the correct special tools or their equivalent is most important for this procedure. The camshaft sprocket must be removed and installed with the use of tools MB 990767-01 and MB 998715 or their equivalents. Attempts to use other tools or methods may cause damage to the sprocket and/or camshaft.

1. Relieve the fuel system pressure.
2. Disconnect the negative battery cable.
3. Drain the cooling system. Remove the upper radiator hose.
4. Remove the air intake hose.
5. Remove the accessory drive belts, fan and pulleys.
6. Remove the air conditioning compressor, power steering pump and mounting brackets and position them to the side, without disconnecting the lines.
7. If removing the right cylinder head, disconnect the wiring and remove the alternator cover, alternator and alternator stay.
8. Remove the timing belt covers.

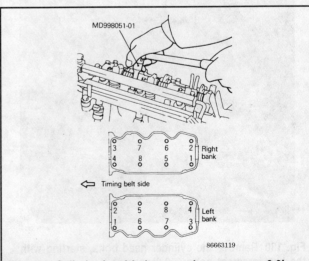

Fig. 134 Cylinder head bolt removal sequence — 3.0L and 3.5L engines

ENGINE AND ENGINE REBUILDING 3-107

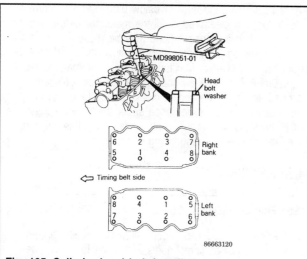

Fig. 135 Cylinder head bolt installation torque sequence — 3.0L and 3.5 engines

9. Remove the timing belt as follows:
   a. Rotate the crankshaft and bring the piston in No. 1 cylinder to TDC on the compression stroke. Align the camshaft and crankshaft sprocket timing marks.
   b. Mark the timing belt in the direction of rotation for reinstallation purposes.
   c. Loosen the timing belt tensioner bolt and turn the tensioner counterclockwise. Remove the timing belt.

✳︎✳︎WARNING

Do not rotate the crankshaft or camshaft sprockets after the timing belt has been removed.

10. Label and disconnect the spark plug wires from the spark plugs.
11. If removing the left cylinder head, remove the distributor cap (if so equipped). Mark the position of the rotor and the distributor housing in relation to the cylinder head, then remove the distributor. If unfamiliar with the distributor removal procedures, refer to these directions in Section 2.
12. Remove the valve cover.
13. Remove the EGR pipe and the air intake plenum stays.
14. Disconnect the accelerator and, if equipped, throttle control cable. Disconnect and plug the fuel lines.
15. Label and disconnect the wiring connectors, vacuum lines and hoses from the air intake plenum, intake manifold and cylinder head.
16. Remove the air intake plenum and intake manifold.
17. Remove the exhaust manifold.
18. If removing the right cylinder head, remove the dipstick tube.
19. If necessary, remove the camshaft sprocket bolt and camshaft sprocket. Remove the alternator bracket and/or timing belt rear cover.
20. Remove the cylinder head bolts starting from the outside and working inward.
21. Remove the cylinder head from the engine.
22. Clean the gasket mounting surfaces.

To install:

23. Install the new cylinder head gasket over the dowels on the engine block, with the identification mark at front top.
24. Install the cylinder head on the engine and tighten the cylinder head bolts in sequence using 3 even steps, to 65-72 ft. lbs. (90-100 Nm) for all vehicles up to 1991. On all 1992-95 engines, tighten cylinder head bolts in sequence using 3 steps to 76-83 ft. lbs. (103-113 Nm).
25. If removed, install the timing belt rear cover and/or alternator bracket. tighten the alternator bracket bolts in the proper sequence.
26. If removed, install the dipstick tube using a new O-ring coated with clean engine oil.
27. Install the exhaust manifold.
28. Install the intake manifold and air intake plenum.
29. Install the EGR pipe and air intake plenum stays.
30. Connect the fuel lines, accelerator and, if equipped, throttle control cable, wiring connectors, vacuum lines and hoses.
31. Install the valve cover.
32. Install the distributor (if so equipped), if removed, aligning the marks made during removal. Install the distributor cap and spark plug wires.
33. Make sure the camshaft and crankshaft sprocket timing marks are aligned.
34. Install the timing belt. Refer to the timing belt procedures later in this section.
35. Install the timing belt covers.
36. If removed, install the alternator, alternator cover and alternator stay.
37. Install the air conditioning compressor and power steering pump with the brackets.
38. Install the pulleys, cooling fan and accessory drive belts. Tighten the crankshaft pulley bolt.
39. Install the upper radiator hose and the air intake hose.
40. Fill the cooling system.
41. Connect the negative battery cable.
42. Start the engine and check for leaks. Check the ignition timing.

CYLINDER HEAD CLEANING AND INSPECTION

▶ See Figures 136 and 137

Immediately after the head is removed from the engine, the old gasket material should be removed from the head and block. Stuff the cylinders with newspaper or rags to prevent accumulation of foreign matter, but don't press down on the pistons too hard. If a piston moves, the crankshaft will turn and the timing relationship will be lost. (This is particularly important if the belt is still attached to the camshaft gear.)

Remove as much of the old gasket as possible from the block with your fingers. The rest must be scraped with a tool which will not damage the metal by scoring or scratching. A hardwood or plastic scraper is highly recommended. Take extreme care not to get any of the scrappings into the water and oil passages or the bolt holes in the head. This common error carries very expensive penalties.

Once the head is on the work bench, the manifolds and other external components should be removed for all but the simplest of cleaning. The head is easier to manipulate and removed items are less likely to be damaged. Unless the head is to be fully stripped, the cam and rockers may stay in place,

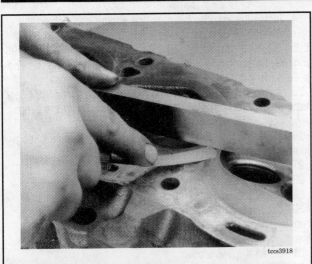

Fig. 136 Check the head surface in all directions and take several measurements

Fig. 137 With the valves removed, a wire wheel may be used to clean the combustion chambers of carbon deposits

but take measures to protect the valve train from dust and impact during repairs.

The head surface should be scraped with the non-marring scraper, again observing the protection of oil and water passages. The combustion chamber and valves will have carbon deposits in them. Attempt to remove as much of the deposit as possible with the scraper. The remaining deposits will need to be removed with a rotary wire brush mounted in a drill. Make sure that the brush is actually removing the deposit and not just polishing it into invisibility.

After the head surface is completely clean and free of any debris, place a quality straightedge across the gasket surface of the cylinder head. Using feeler gauges, determine the clearance at the center of the straightedge and at other locations on either side. Measure along the diagonals of the head, along the longitudinal center line and across the head at several locations. Keep a written record of the thickest feeler which will pass between the straightedge and the head at every location.

Acceptable warpage is at or below 0.0020 in. (0.05mm) at EVERY location measured. If all the measurements are at or below this specification, the head may be reused as is. If any one measurement is greater than this limit, the head must be resurfaced or replaced. Maximum allowable warpage is 0.0080 in. (0.20mm). Warpage exceeding this requires replacement of the head. Measure the engine block surface (deck) in the same fashion. Look for clearance under the straightedge to be 0.0020 in. (0.05mm) or less. Any reading over this value requires replacement or resurfacing of the block.

RESURFACING

▶ See Figures 138 and 139

If the head is to be resurfaced (planed), it must be completely stripped of all components and taken to a machine shop. At this point, it is worthwhile considering having the head crack-checked by the shop. Commercial methods allow much more sophisticated checking than can be done by eye. Additionally, having the head cleaned in a solvent bath will remove deposits from all the passages within the head.

❋❋WARNING

Do not allow the head to be immersed in "hot tank" cleaners designed for iron heads. The chemicals will attack the aluminum alloy in the Mitsubishi head and make the head unusable. Specify to the shop that the head contains aluminum.

Resurfacing the head involves removing metal from the face of the head, much as a wooden surface is leveled with a plane. This can remove the high and low spots and give a true surface to mount onto the engine, clamping the gasket evenly across its surface. The two critical dimensions in resurfacing the head are the amount of metal removed (usually expressed in thousandths of a millimeter) and the overall thickness of the head.

The minimum head thickness must NOT be exceeded under any circumstance. If the amount of resurfacing required would cause the head to become too thin, the head is unusable and must be discarded. The maximum thickness allowed to be removed from a Mitsubishi head is 0.0080 in. (0.20mm). This figure is valid for a head which has never been cut (original thickness) and would reduce it to minimum thickness.

An additional reminder: when resurfacing a V6 head, the profile of the intake manifold mating surface will change (relative to the manifold) after the head is cut. The intake manifold face will have to be resurfaced to compensate for this. Most machine shops will do this automatically; remind them anyway.

After the head is reconditioned, build it back up on the workbench by installing any and all removed components. It is much easier to install a head complete with cam, rockers and manifolds than to attempt final assembly with the head on the block. Keep the head clean and free of grit. Refer to the applicable procedures of this section for details and tighten specifications for individual components (camshaft, manifolds, etc.)

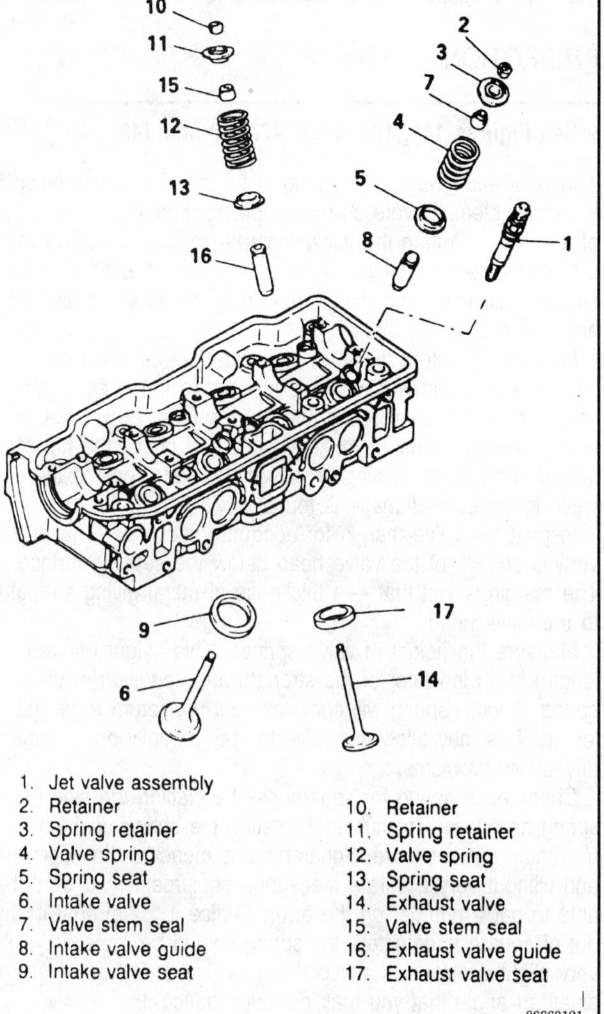

ENGINE AND ENGINE REBUILDING

Valves and Valve Springs

➡ On all engines, the cylinder head must be removed from the vehicle and the camshaft(s) and rockers removed from the head. Valves and components must be reinstalled in their original positions. Develop a system to store and identify parts as they are removed. Identify components by location and application (Example: Cyl. 2 exhaust) to avoid confusion. Keep the work area clean and store small parts immediately. The use of the correct special tools or their equivalents may be required for some procedures.

REMOVAL

▸ See Figures 140, 141, 142 and 143

The early Mitsubishi engine use a small "jet valve" next to the main intake valve. This small additional valve directs a portion of the air/fuel mixture at the spark plug, allowing hotter and more complete combustion. The jet valve must be removed before removing the larger valves.

The jet valve assembly is screwed into the head. The use of special tool (socket) MD 998310 is required to remove and install the valve. When using the tool, make certain that the wrench is not tilted with respect to the center of the valve. If the wrench is tilted, the excess pressure on the spring retainer may bend the valve stem, causing it to bind during operation. Remove the assembly and set it aside.

On all heads, install a standard valve spring compressor and tighten it to compress the spring. Remove the valve spring retainers (collets) from the valve stem. A small magnet can be very helpful for this. Slowly loosen the spring compressor and remove the valve spring retainer, the spring and the lower spring washer (spring seat). The valve can now be withdrawn from the combustion side of the head. Use a pair of pliers to remove the valve stem seal from each position. Discard the seals immediately; they cannot be reused.

Each valve should be cleaned to remove the carbon deposits from the valve head and stem. This may be done either with chemical cleaners (wear eye protection and gloves) or

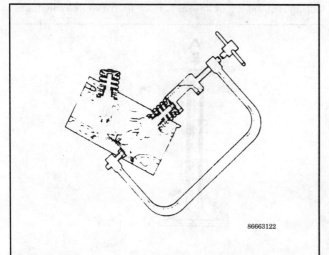

1. Jet valve assembly
2. Retainer
3. Spring retainer
4. Valve spring
5. Spring seat
6. Intake valve
7. Valve stem seal
8. Intake valve guide
9. Intake valve seat
10. Retainer
11. Spring retainer
12. Valve spring
13. Spring seat
14. Exhaust valve
15. Valve stem seal
16. Exhaust valve guide
17. Exhaust valve seat

Fig. 138 Common type of cylinder head assembly — the jet valve is not found on all engines

Fig. 139 Use a spring compressor when removing and installing valve springs and retainers

Fig. 140 The location of one of the jet valves on a Mitsubishi head

3-110 ENGINE AND ENGINE REBUILDING

Fig. 141 Use a spring compressor to compress the valve spring and remove the valve spring retainers

Fig. 142 Once the valve spring retainers are removed, separate the valve spring from the cylinder head

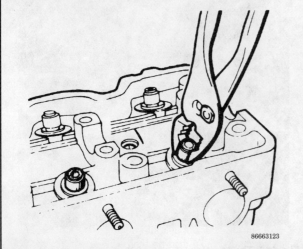

Fig. 143 Remove the valve stem seals with a pair of pliers

with the wire brush and power drill used to clean the head. Do not use sandpaper or other abrasives to clean the valves.

INSPECTION

▶ See Figures 144, 145, 146, 147, 148 and 149

Inspect each valve for burning of the head or deformation of the lower stem. Severe corrosion, pitting, burning, or cracking of these parts mean the valve must be replaced. Inspect the top of the stem for pitting. If the pitting is light and the valve is otherwise in good condition, pitting may be removed by light application of an oil stone.

Measure the stem diameter at several points along its length. A common wear point is just above the head of the valve where the stem enters the valve guide. If this area is worn, the valve must be replaced. Visually check the top of the stem for mushrooming or deformation and replace the valve if any such damage is found.

Inspect the valve margin for adequate thickness. This is the vertical section of the valve head below the seating surface. The margin is just that — a thickness of metal giving strength to the valve head.

Measure the height of valve springs. This height (or free length) is an indicator of the strength and condition of the spring. A worn spring will compress but not return to its full length. This may allow the valve to stay partially open, causing driveability problems.

Check each spring for squareness by placing the side of the spring next to a T-square and rotating the spring until you find maximum out-of-square. Squareness is measured in degrees and without very accurate measuring equipment, you won't be able to put a number on the error. Suffice it to say that if any out of square is detected, the spring should be replaced. A very slight deviation is acceptable (the limit is 2°) but this is so small an angle that you may not even notice it.

Measure stem-to-guide clearance. Depending on the tools available, either measure the valve stem and the valve guide individually, then subtract to find the clearance or install the valve in its guide and locating a securely mounted dial indicator against the top of the stem. Rock the valve away from the

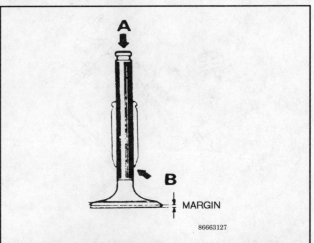

Fig. 144 Check for deformation or pitting at A and wear at B — do not re-use the valve if the margin is at or below the minimum

ENGINE AND ENGINE REBUILDING

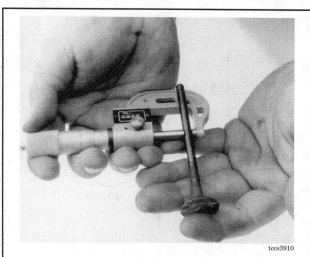

Fig. 145 Use a micrometer to check the valve stem diameter

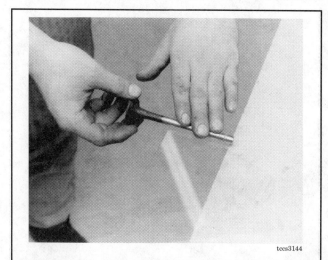

Fig. 146 Valve stems may be rolled on a flat surface to check for bends

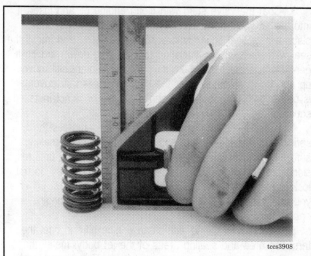

Fig. 147 Check the valve spring for squareness on a flat surface, a carpenters square can be used

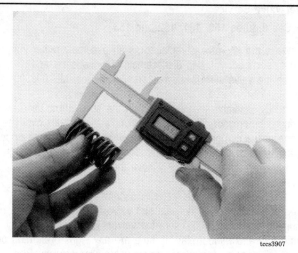

Fig. 148 Use a caliper gauge to check the valve spring free-length

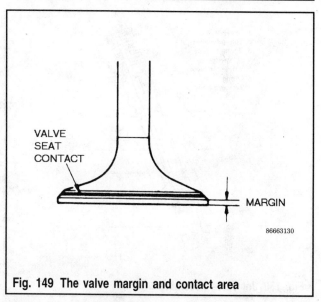

Fig. 149 The valve margin and contact area

indicator and hold it there while you zero the indicator. Then rock the valve all the way toward the indicator and hold it there as you take the reading. If the stem-to-guide clearance exceeds specifications the valve and/or guide should be replaced.

The valve face and seat surfaces must have consistent contact near the center of each. Coat the surface of the seat with a metal dye (such as Prussian blue) or a thin coat of lapping compound. Insert the valve all the way so the marking material is transferred to the valve face. If a consistent mark is found toward the center of the valve face all around, the valve and seat may be re-used as is, assuming there are no other problems with the valve. If the seating is questionable or obviously bad, the valve and seat should be machined to the proper dimensions and angles by an automotive machine shop.

It then must be lapped in (polished) by coating the seating surfaces with a lapping compound and rotating the valve with a special suction cup tool until the surfaces mate perfectly; lapping compound must then be thoroughly cleaned from both surfaces.

3-112 ENGINE AND ENGINE REBUILDING

Jet Valve Service

▶ See Figures 150, 151, 152 and 153

While the jet valve is replaced as an assembly rather than being serviced, you can disassemble it to inspect it and replace the valve stem seal. You should use tool MD 998309 or equivalent.

1. Disassemble the jet valve with the special tool used as a spring compressor. The forked portion of the tool fits over the ridge at the center of the assembly while the hole in the upper section fits over the valve stem. Compress the spring enough to remove the keepers and remove them; then slowly release the pressure.

2. Disassemble each jet valve assembly so all parts are kept together. Check the seat and valve faces for any signs of burning or cracks. Make sure the valve slides in the jet body smoothly and without play. If there are any signs of burning or excessive valve stem wear, the jet valve assembly must be replaced. Inspect the spring for cracks of any kind, and replace defective springs.

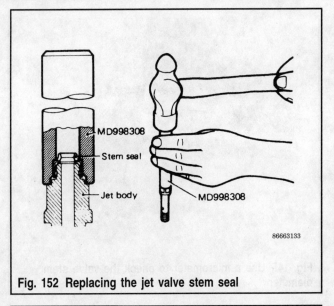

Fig. 152 Replacing the jet valve stem seal

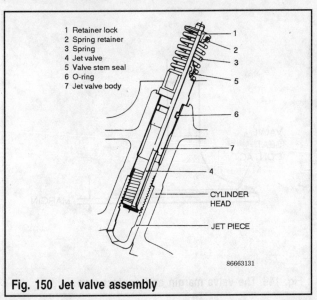

Fig. 150 Jet valve assembly

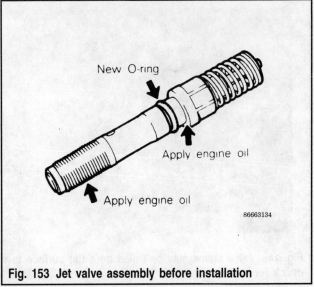

Fig. 153 Jet valve assembly before installation

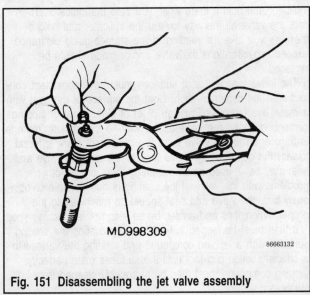

Fig. 151 Disassembling the jet valve assembly

3. Coat the valve stem with clean engine oil and insert it into the jet valve body. Sit the assembly on a flat surface, place a new seal squarely over the valve stem, and install the seal with the special installation tool, tapping downward on it gently with a hammer. Make sure the valve stem still moves up and down smoothly after the seal is installed; if movement is not smooth, the seal has been damaged or is not installed squarely.

4. Install the spring and spring retainer, compress the spring with the special tool and install the keepers, making sure they remain squarely in the valve stem groove until tension is removed and the retainer covers them. If they do not lock in place securely, the valve spring could pop out either as you are working on the engine, or later, when the engine is operating.

5. Coat a new O-ring with clean engine oil and coat the threaded area and sealing areas of the jet body. Install the new O-ring into the groove provided for it. Then, insert each jet valve squarely into the hole in the cylinder head and start the threads carefully. Use the special socket and a torque

ENGINE AND ENGINE REBUILDING 3-113

wrench to tighten the assembly to 13-15 ft. lbs. (18-20 Nm) keeping the wrench square to the valve stem at all times.

INSTALLATION

♦ See Figure 154

Before reassembly, check that all lapping compound has been cleaned from the head. Remember that each piece being reused must be reinstalled in its original position.

1. Position the spring seat on the head. The seat will be forced into its final position in the next step.
2. Install the valve stem seals and insure proper installation of the spring seats.
3. Thoroughly coat the valve stem with clean engine oil. Insert the valve slowly from underneath, guiding it gently through the new seal. Make sure the valve will slide up and down smoothly, indicating the seal is in proper position.
4. Install the valve springs with the color-coded end facing upward. Install the retainer on top of the spring. Compress the retainer with the valve spring compressor until the retainer sits below the groove in the valve stem.
5. Install the keepers, making sure they stay properly positioned in the stem groove, and release the tension on the compressor slowly. Note that if the keepers do not sit squarely in the grooves, the valve spring tension could be released suddenly either as you work, or later as the vehicle is in operation. Once the keepers have locked securely, remove the valve spring compressor.
6. Repeat the sequence for each intake and exhaust valve.

➡ After valve spring assembly installation tap the top of the valve springs to make sure the valve keepers are seated properly.

Valve Seats

♦ See Figures 155 and 156

REMOVAL & INSTALLATION

As in the case of valves, seats may corrected for inadequate seal width and trueness with machining. After the seat is machined the valve must be lapped in. If the valve seat insert shows severe damage or wear it must be replaced.

The seats should be replaced by an experienced machine shop. The procedure requires precise measuring and milling of the head. The new seat is installed through the use of extreme temperatures which require the use of special safety and handling equipment.

Before removal, the seat thickness is removed through drilling or reaming, leaving about 0.5 mm around the outside. The seat can then be collapsed out of the head. After cleaning the area, the opening in the head is precisely enlarged to accept the new, oversize valve seat. The hole in the head must conform exactly to the outer diameter and depth of the seat insert. These dimensions vary for each engine; select your parts carefully.

On most applications, the seat is installed after the head is heated to 482°F (250°C). The room temperature seat is pressed (not hammered) into the heated head. As the head cools, it grips the seat very tightly.

After the seat is installed, the metal must be allowed to return to normal temperatures slowly. Resist any temptation to hurry the heating or cooling; sudden temperature change can cause the metal to crack. Simply let the head sit, usually on the press or workbench for a few hours.

After the head has returned to normal temperature, the new seat insert must be machined to the correct angles. The valve is then inserted and the valve and seat lapped with compound to assure perfect contact.

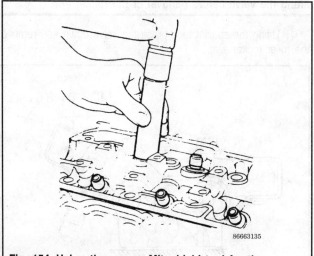

Fig. 154 Using the proper Mitsubishi tool for the installation of the valve seals

Fig. 155 Installing a common valve seal

3-114 ENGINE AND ENGINE REBUILDING

Fig. 156 Mitsubishi recommends using a tool to fit the seals to the guides — but a properly sized socket (neither too large nor too small) can be used

Valve Guides

REMOVAL & INSTALLATION

The valve guides are so named because they guide the stem of the valve during its motion. By properly locating the stem of the valve, the head and face of the valve are kept in proper relationship to the head. Occasionally, the valve guides will wear causing poor valve sealing and poor engine performance. If this problem is suspected, the valve stem-to-guide clearance must be checked.

If the wear is minimal, it may not be necessary to replace the guides. A competent machine shop can knurl the guides. This process raises a spiral ridge on the inside of the guide (without removing it), thereby eliminating the wobble in the stem. This is not a cure-all; it can only be done to correct minimal wear.

Excessive wear of the valve guides is indicated by excessive stem-to-guide clearance. You can try inserting a new valve into the old guide and measuring stem-to-guide clearance again. If this brings tolerances well within specifications, it may be considered satisfactory to just to replace the valve.

If this does not bring the clearance to within specifications, the guide must be replaced. Provided the old valve is in good condition, it may be checked in the new guide. However, if the engine is receiving a complete overhaul, maximum service life will be achieved by replacing both the guide and valve. Valve guides should be replaced by an automotive machine shop equipped with the proper tools and presses.

The guide must be pressed out of the cylinder head using special tools and a hydraulic press. The tool number varies by specific motor (due to the diameter of the guide) and equivalent tools should be available from most suppliers. Use the tool backed up by a large press to press the guide out the bottom of the cylinder head.

Do NOT replace the guide with one of the same size, but use a guide that is one size oversize. The guide bore in the cylinder head must be machined to the exact dimensions of the outside diameter of the new guide. Press the new guide in, from the top, using the same tool. All these operations should be performed with the parts at room temperature. For each motor, the amount of projection of the valve guide is critical. Use of the correct special tool will ensure the guide is pressed in the correct distance.

The new or re-used valves should operate freely inside the new guides. A new guide may require reaming to achieve the correct stem-to-guide clearance. Once the guide is correctly mated to the valve, the valve face and seat must be lapped and then checked for correct seating. This must be done because the new guide may change the exact angle of the valve stem.

Hydraulic Lash Adjusters

REMOVAL & INSTALLATION

▶ See Figures 157 and 158

The hydraulic lash adjusters are designed to keep the lash between the rocker arm and the valve at a constant rate, and also eliminate almost all periodical adjustment of the valve lash. The lash adjuster is a precision part, and it needs to be kept free from dust and other foreign matter. The lash adjusters do not need to be disassembled.

3.5L Engines

▶ See Figures 159 and 160

✻✻CAUTION

In the cylinder from which the lash adjuster is removed, the piston interferes with the valve when the valve is pushed down. Therefore, turn the crankshaft to keep the piston position down. In addition, the rocker arm located at the valve lifted by the camshaft cannot be removed. Therefore, turn the crankshaft to keep the camshaft from lifting the valve before removal of the rocker arm.

1. Using the special tool, press the valve down and remove the roller rocker arm.

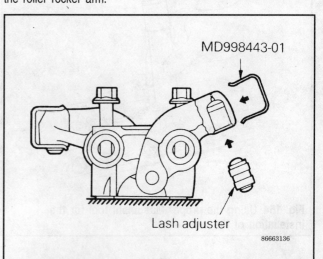

Fig. 157 The location of the lash adjuster and lash adjuster retainer upon installation

ENGINE AND ENGINE REBUILDING

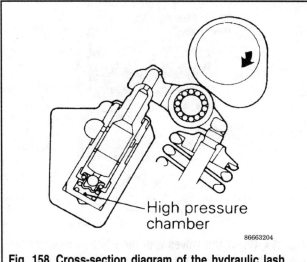

Fig. 158 Cross-section diagram of the hydraulic lash adjuster

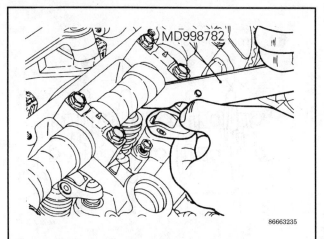

Fig. 160 While the valve is depressed, removal and installation of the roller rocker arm and lash adjuster is facilitated

2. Pull out the lash adjuster from the cylinder head.

To install:

3. Install a new lash adjuster, with all air previously removed, to the cylinder head.
4. With the valve pressed down by the special tool, install the roller rocker arm.

All Other Engines

For the procedures for all other engines refer to the camshaft removal and installation procedures, located in this section.

INSPECTION

On Vehicle Inspection

▶ See Figures 161, 162, 163, 164, 165 and 166

If a clanging noise due to the lash adjuster produced immediately after the engine has started or during operation persists, perform the following checks:

1. Check the engine oil, and add or replace if necessary.
   a. If the engine oil is low, the air drawn in from the oil strainer will be trapped in the oil passage.
   b. If the engine oil level is higher than the specified level, agitation of oil by the crankshaft could cause a large amount of air to enter the oil.
   c. A deteriorated oil contains a large amount of air, because the air, once trapped, is not readily separated from the oil.

If the air trapped due to these causes enters the high pressure chamber in the lash adjuster, the air in the high pressure chamber will be compressed while the valve is in the opened position. The lash adjuster will be drawn too far in, and will produce noise when the valve closes. This is the same phenomenon that occurs when the valve clearance is adjusted to an excessive dimension.

In this case, the normal condition will be restored if the air escapes from the lash adjuster.

2. Start the engine and slowly race (accelerate the engine from idling speed to 3,000 rpm in 30 seconds and then decelerate it to the idling speed in 30 seconds) it several times (less than 10 times). If racing the engine causes the noise to die away, it means that the air has escaped from the high pressure chamber of the lash adjuster and that the lash adjuster has regained its normal functions.
   a. When the vehicle is parked on a slope for a long period of time, the oil in the lash adjuster will decrease. When the engine is started, the air might enter the high pressure chamber.
   b. After a long period of parking during which the oil in the oil passage goes away, it will take some time before the oil is re-supplied to the lash adjuster. Therefore, the air could enter the high pressure chamber.

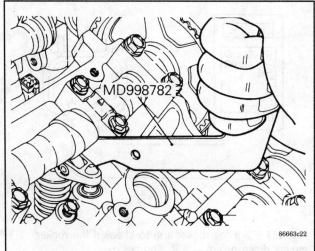

Fig. 159 Press the valve down by using the tool MD998782

3-116 ENGINE AND ENGINE REBUILDING

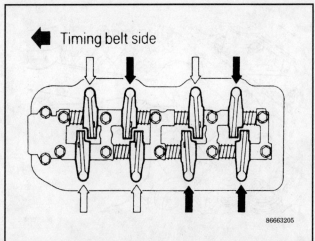

Fig. 161 Check the valves with the white arrows, rotate the crankshaft 360°, then check the valves with the black arrows — 2.0L, 2.3L, 2.4L and 2.6L Engines

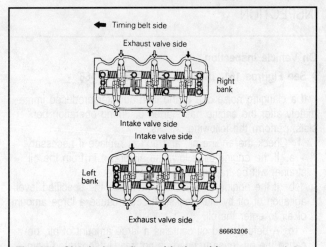

Fig. 162 Check the valves with the white arrows, rotate the crankshaft 360°, then check the valves with the black arrows — 3.0L (12 Valve) Engines

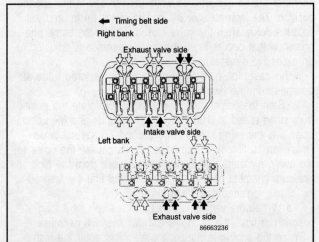

Fig. 163 Check the valves with the white arrows, rotate the crankshaft 360°, then check the valves with the black arrows — 3.0L (24 Valve) Engines

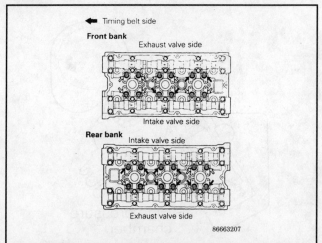

Fig. 164 Check the valves with the white arrows, rotate the crankshaft 360°, then check the valves with the black arrows — 3.5L Engines

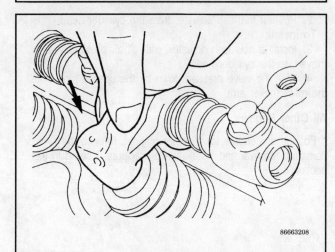

Fig. 165 Push the rocker arm to check if the rocker moves down or not — all engines except for the 3.5L Engines.

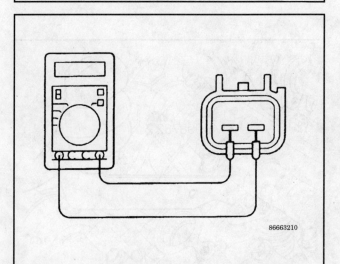

Fig. 166 Push the rocker arm to check if the rocker moves down or not — 3.5L Engines.

ENGINE AND ENGINE REBUILDING   3-117

3. If any abnormal noise is not eliminated by racing, check the lash adjuster.

   a. Stop the engine.

   b. Set the engine so that cylinder No. 1 is positioned at the top dead center of the compression stroke.

   c. Press the rocker arm at the area indicated by the white arrow mark to check whether the rocker arm is lowered or not.

   d. Slowly turn the crankshaft 360° clockwise.

   e. In the same procedure as step "c", check the rocker arm at the area indicated by the black arrow mark.

   f. Push down the rocker arm at a portion located right above the lash adjuster. If the rocker arm goes down readily, the lash adjuster is defective. Replace it with a new one in accordance with Step 4. In addition, when replacing the lash adjuster, be sure to remove air positively from the lash adjuster before installation. Then perform inspection in accordance with steps "a" through "e" to make sure that there is no abnormality. In addition, if the rocker arm is felt to be very stiff or cannot be pushed down when it is pushed, the lash adjuster is in the normal condition. Therefore, check for other causes of noise.

Bleeding and Checking

When cleaning or checking the lash adjusters, use only clean diesel fuel. To test the lash adjusters, follow this procedure.

1. Immerse the lash adjuster in clean diesel fuel.
2. While lightly pushing down the inner steel ball using a small wire, move the plunger up and down four or five times to bleed air. The use of the retainer facilitates the air bleeding of the rocker arm mounted type lash adjuster.
3. Remove the small wire and press the plunger in. If the plunger is hard to be pushed in, the lash adjuster is normal. If the plunger can be pushed in all the way readily, bleed the lash adjuster again and test again. If the plunger is still loose, replace the lash adjuster.

➡Upon completion of bleeding the air out of the lash adjuster, hold the adjuster upright to avoid the diesel fuel from spilling.

4. After air bleeding, set the lash adjuster on the special tool (leak down tester MD998440).
5. After the plunger has gone down somewhat 0.008-0.020 in. (0.2-0.5mm), measure the time needed for it to go down an additional 0.04 in. (1mm). Replace if the measured time is not between 4-20 seconds.

Timing Belt and Covers

REMOVAL & INSTALLATION

➡Elevate and safely support the vehicle as necessary for access to various components. Depending on tools used and part location, some components may be easier to reach from underneath. When working over the fender, always use a fender cover to protect the paint work. If the timing belt is to be reused, extreme care must be taken to protect it from oil, fluids and grease. These chemicals can cause the belt to deteriorate and possibly break during engine operation. When handling the belt, do not crimp or fold it; sharp bends can break the fibers within the belt.

2.0L Engine

▶ See Figures 167, 168, 169, 170, 171, 172, 173, 174, 175, 176, 177, 178 and 179

➡These engines have two timing belts. One connects the crankshaft and camshaft, the second one drives the right silent shaft.

1. With the vehicle in Neutral (manual) or **P** (automatic), make certain the parking brake is set and the wheels are blocked. Disconnect the negative battery cable.
2. Remove the water pump drive belt and water pump pulley. If the vehicle is equipped with air conditioning, loosen the compressor belt tensioner and remove the compressor belt.
3. Remove the bolts holding the crankshaft pulley(s) and remove the pulley(s).
4. Remove the upper and lower timing belt covers and their gaskets.
5. Use a socket wrench on the projecting crankshaft bolt to turn the engine clockwise (only!) and align all the timing marks on the sprockets and cases. It may be necessary to wipe off the area to see the marks clearly. Do not use spray cleaners around the timing belt. When all of the marks align exactly, the engine is set to TDC/compression on No. 1 cylinder. From this point onward, the camshaft and crankshaft positions MUST NOT be changed.
6. If the timing belt is to be reused, make a chalk or crayon arrow on the belt showing the direction of rotation so that it may be reinstalled correctly.
7. Loosen the bolts holding the tensioner and pivot the tensioner towards the water pump. Temporarily tighten the bolts to hold the tensioner in its slack position.
8. Carefully slide the belt off the sprockets. Place the belt in a clean, dry, protected location away from the work area.
9. Remove the camshaft sprocket bolt and remove the sprocket.
10. Remove the crankshaft sprocket retaining bolt. Remove the sprocket and flange.
11. Remove the plug from the left side of the cylinder block and insert a suitable tool (with the shape of a screwdriver) to keep the left silent shaft in place. The suitable tool should have a shaft diameter of 0.3 in. (8mm) and a shaft length of at least 2.4 in. (60mm).
12. Remove the bolt holding the oil pump sprocket and remove the sprocket.
13. Loosen the mounting bolt for the right silent shaft sprocket until it can be turned with your fingers. Do not remove it.
14. Remove the tensioner for the silent shaft belt (tensioner B).
15. Remove the silent shaft belt (timing belt B). Remove the crankshaft pulley for the silent shaft belt if so desired.

✴✴WARNING

After the timing belt has been removed, do not attempt to loosen the silent shaft bolt by holding the sprocket with pliers. If the sprocket is to be removed, hold the pulley with your fingers.

ENGINE AND ENGINE REBUILDING

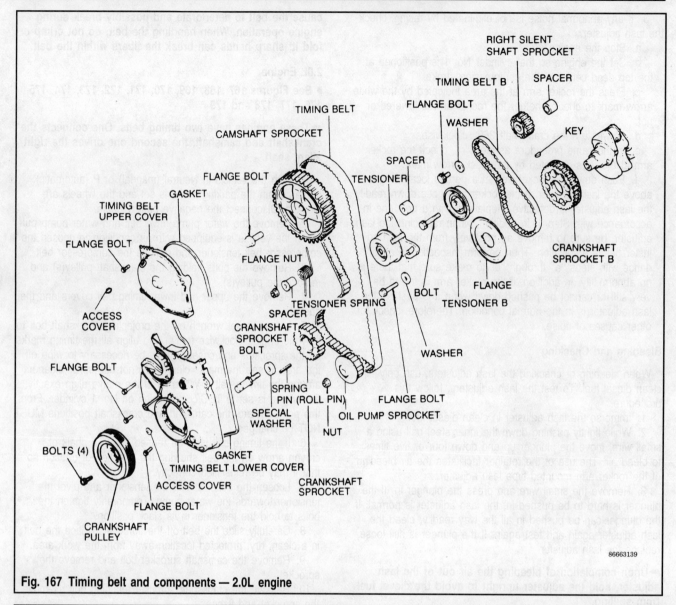

Fig. 167 Timing belt and components — 2.0L engine

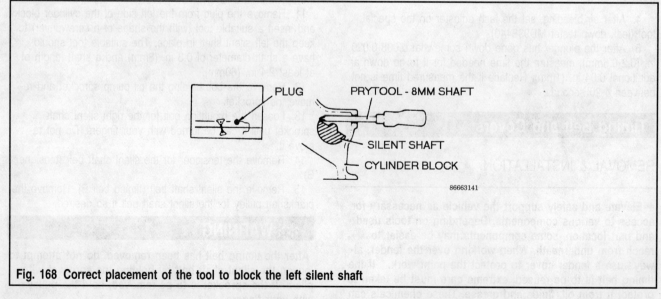

Fig. 168 Correct placement of the tool to block the left silent shaft

ENGINE AND ENGINE REBUILDING 3-119

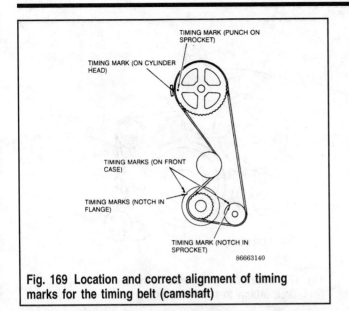

Fig. 169 Location and correct alignment of timing marks for the timing belt (camshaft)

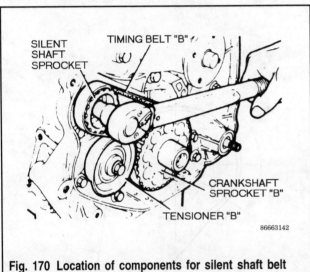

Fig. 170 Location of components for silent shaft belt (timing belt B)

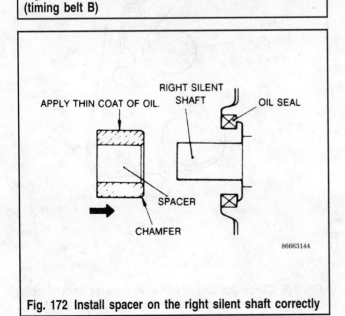

Fig. 172 Install spacer on the right silent shaft correctly

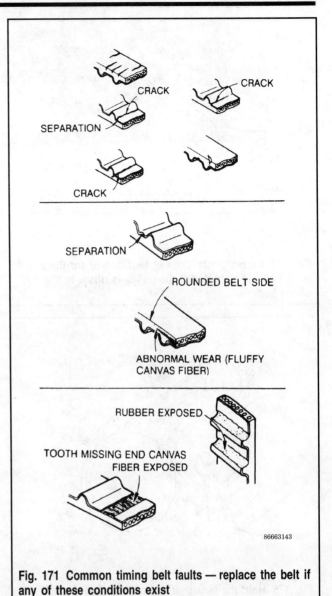

Fig. 171 Common timing belt faults — replace the belt if any of these conditions exist

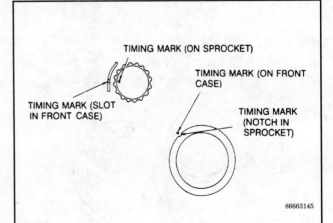

Fig. 173 Align the timing marks before installing the silent shaft

3-120 ENGINE AND ENGINE REBUILDING

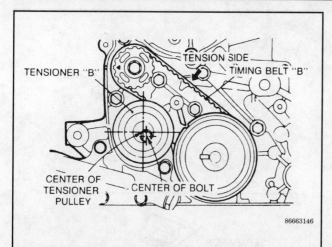

Fig. 174 Correct installation of tensioner B for the silent shaft belt — the center of the pulley is offset to the left of the bolt

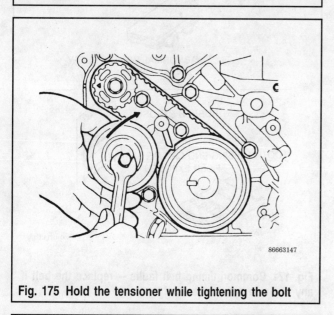

Fig. 175 Hold the tensioner while tightening the bolt

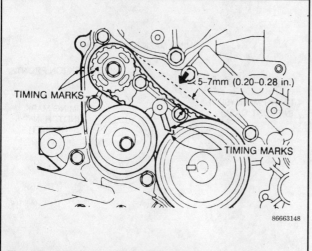

Fig. 176 Check the silent shaft belt for the proper deflection

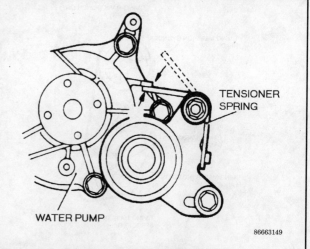

Fig. 177 Install the lower end of the tensioner spring first, then attach to the water pump

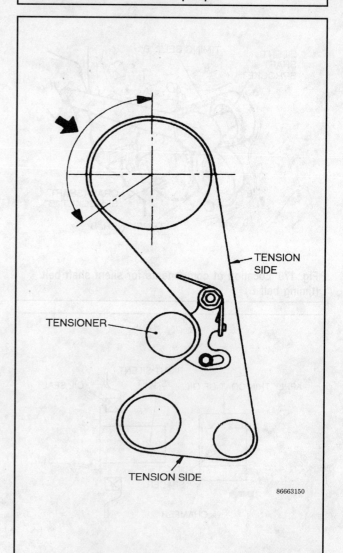

Fig. 178 Check the cam sprocket in the area shown for any sign of the timing belt lifting after it is tensioned

ENGINE AND ENGINE REBUILDING 3-121

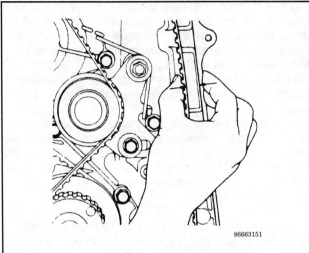

Fig. 179 The pinch test for checking timing belt deflection — refer to the service procedure

16. Inspect the timing belts in detail for any flaw or wear. If the belt is not virtually perfect, replace it. A case can be made for replacing the belt every time it is removed, particularly on high-mileage engines. Some of the conditions to look for are:
   • Hardened back surface; non-elastic and glossy; hard to mark with a fingernail.
   • Cracking on back of belt, bottom of teeth or side of belt.
   • Missing teeth or teeth lifting from belt.
   • Side of belt worn or fuzzy. Normal belt should have clean sides as if cut with a sharp knife.
   • Wear on teeth as shown by distinct color change or worn rubber.
   • Separation of inner coating from backing.
   • Any uneven wear patterns on the teeth of the belt. Wear pattern should be even across each tooth and not differ from one tooth to another.

17. Check the sprockets and tensioner for wear. The sprocket teeth should be well defined, not rounded and the valleys between the teeth should be clean. The tensioners should spin freely with no binding or unusual noise. Replace the tensioner if there is any sign of grease leaking from the seal. Clean everything with a clean, dry cloth.

✱✱WARNING

Do not spray or immerse the sprockets or tensioners in cleaning solvent. The sprocket may absorb the solvent and transfer it to the belt. The tensioners are internally lubricated and the solvent will dilute or dissolve the lubricant.

To install:

18. Install the sprocket for the silent shaft belt onto the crankshaft. Make certain it is installed correctly.
19. If the sprocket for the right silent shaft was removed, coat the spacer with a light coating of clean engine oil and install the spacer to the shaft. Be sure to install it in the correct direction. Install the silent shaft sprocket and tighten the bolt finger-tight.
20. Double check the timing marks on the silent shaft sprocket and crankshaft sprocket. Carefully align the marks if necessary.
21. Install the silent shaft belt, observing the direction of rotation mark made earlier. Handle the belt carefully and do not use metal tools to guide or force the belt into place. When installing the belt, make sure the tension side (opposite from the tensioner) has no slack in it.
22. Install the tensioner (B) with the center of the pulley located to the left side of the mounting bolt and with the pulley flange to the front of the engine.
23. Lift the tensioner with your hand so that the belt becomes taut. Hold the tensioner in this position and tighten its bolt. Use care that only the bolt and not the tensioner shaft is turned during tightening. The bolt should be tightened to 13 ft. lbs. (18 Nm).
24. Tighten the right silent shaft bolt to 26 ft. lbs. (35 Nm).
25. Check that the timing marks are still aligned. Push down on the center of the tension side of the belt with your finger. Correct belt deflection is 0.25 in. (5-7mm). If the deflection is not correct, the tensioner must be released fully and the belt re-tensioned.
26. Install the flange and the crankshaft sprocket onto the crankshaft. Make certain the flange is installed in the correct direction. If it is put on incorrectly, the belt will wear and break.
27. Install the special washer and the sprocket retaining bolt to the crankshaft. Tighten the bolt to 88 ft. lbs. (119 Nm).
28. Install the camshaft sprocket to the camshaft and tighten the bolts to 66 ft. lbs. (89 Nm).
29. Install the spacer, tensioner and tensioner spring if they were removed. Install the lower end of the spring to its position on the tensioner, then place the upper end in position at the water pump. Move the tensioner towards the water pump and temporarily tighten it in this position.
30. Install the oil pump sprocket, tightening the nut to 40 ft. lbs. (54 Nm). Remember that the oil pump drives the left silent shaft, which is still blocked by the tool. Hold the sprocket by hand when tightening the nut.
31. Double check the alignment of the timing marks for the cam, crank and oil pump sprockets. If any adjustment is needed to the oil pump sprocket, remove the tool blocking the silent shaft before adjusting the sprocket. Once everything is aligned, replace the tool in the left side of the block and leave it there until installation of the timing belt is complete.

➡**If the tool can only be inserted about 1 in. (25mm) or less, the shaft is out of position. Turn the oil pump sprocket through one full turn clockwise; the screwdriver should then go in about 2.5 in. (63mm).**

32. Install the timing belt onto the crankshaft sprocket, the oil pump sprocket and the cam sprocket in that order. Keep the belt taut between sprockets. If reusing an old belt, make certain that the direction of rotation arrow is properly oriented.
33. Loosen the tensioner mounting nut and bolt. The spring will move the tensioner against the belt and tension it.
34. Check the belt as it passes over the camshaft sprocket. The belt may tend to lift in the area to the left of the sprocket. (Roughly 7 o'clock to 12 o'clock when viewed from the pulley end.) Make certain the belt is well seated and not rubbing on any flanges or nearby surfaces.

ENGINE AND ENGINE REBUILDING

35. At the tensioner, tighten the bolt in the slotted hole first, then tighten the nut on the pivot. If this order is not followed, the belt will become too tight and break.
36. Once again, check all the timing marks for alignment. Nothing should have changed; check anyway.
37. Remove the tool blocking the left side silent shaft. Using a socket on the crankshaft bolt, turn the crankshaft smoothly one full turn (360°). Do NOT turn the engine backwards.
38. Loosen the tensioner nut and bolt. The spring will allow the tensioner to tighten a little bit more because of the slack picked up during engine rotation. Tighten the bolt, then the nut to 36 ft. lbs. (49 Nm).
39. Check the deflection of the belt. At the middle of the right (tension) side of the belt, deflect the belt outward with your finger, toward the timing case. The distance between the belt and the line of the cover seal should be about 0.5 in. (14mm).
40. Install the lower and upper timing belt covers. Make certain the gaskets are properly seated in the covers and that they don't come loose during installation.
41. Install the crankshaft pulley.
42. Install the water pump pulley, tightening the bolts and install the belt. Adjust the belt to the correct tension.
43. Connect the negative battery cable. Start the engine and let it idle, listening for any unusual noises from the area of the timing belt. Possible causes of noise are the belt rubbing against the covers or a sprocket flange, the belt being too loose and slapping, or a tensioner binding. Do not accelerate the engine if abnormal noises are heard from the timing belt assembly — severe damage can result.

2.3L Diesel Engine

▶ See Figures 180, 181, 182, 183, 184, 185, 186, 187 and 188

1. Disconnect the negative battery cable.
2. Remove all drive belts and any accessories which may block access to the timing belt covers.
3. Remove the upper timing belt cover. Note that the bolts are of different lengths; diagram their location for proper reinstallation.
4. Rotate the engine by hand (clockwise) until all the timing marks align. This positions No. 1 cylinder at TDC/compression.
5. If either or both of the timing belts are to be reused, mark the direction of rotation on the belt with chalk or a felt marker.
6. Remove the center bolt and washer holding the crankshaft pulley and remove the pulley. Do not turn the engine out of position during this removal.
7. Remove the lower timing belt cover.
8. Slightly loosen the retaining (lock) bolts for the timing belt tensioner. Move the tensioner towards the water pump and tighten the bolts; this holds the tensioner in the slack position.

➡ If the silent shaft timing belt is not being removed, DO NOT loosen the tensioner for the silent shaft belt.

9. Remove the camshaft sprocket and the injection pump sprocket. Do not let either shaft turn out of place during the removal. Remove the timing belt without forcing it or kinking it.
10. If the silent shaft belt is to be removed, the silent shafts must be locked in place. On the right side, remove the rubber plug from the silent shaft case and insert a screwdriver-shaped prytool; the shaft should have a diameter of 0.32 in. (8mm). On the left side, remove the cover panel from the silent shaft case and insert a broad round bar; a short extension for a socket wrench is an ideal tool.
11. Remove the silent shaft sprockets; label each sprocket so it may be reinstalled in its original position. Remove the timing belt.
12. Check the belt carefully for damage, wear or deterioration. Check the teeth carefully for any cracking at the base or separation from the backing. Do not use detergents or solvents to clean the belt or the sprockets. Clean them only with dry cloths or brushes. Pay particular attention to the edges of the belt; they should not be frayed or worn.

To install:
13. Install the crankshaft sprocket for the silent shaft belt, the flange and the timing belt sprocket. Make certain the sprockets and flange are installed in the correct positions; if either sprocket is installed backwards, the belt may jump off or become damaged.
14. Install the crankshaft pulley; tighten the bolt to 52 ft. lbs. (70 Nm).
15. Install the injection pump sprocket and flange.
16. Install the spacer on the left silent shaft with its tapered end inward or toward the oil seal. If the spacer is not placed correctly, the oil seal will be damaged.
17. Install the silent shaft sprockets and flanges, tightening the bolts to 28 ft. lbs. (38 Nm). The shafts must be held in place with the tools used during disassembly.
18. The tensioner for the silent shaft belt should still be in the slack position. Align the timing mark of the crankshaft sprocket if necessary.
19. Align the timing marks of each silent shaft sprocket.
20. Install the silent shaft timing belt (belt B), making certain that there is no slack between the crankshaft pulley and the right (upper) silent shaft. This side of the belt is called the tension side.

✱✱WARNING

If the belt is being reused, it must be reinstalled in the same direction of rotation. Observe the marks made during disassembly.

21. Use a finger to push down on the slack side of the belt — between the tensioner and the right side silent shaft — and check that the timing marks on the sprockets remain in alignment.
22. Keep your finger on the belt. Loosen the mounting nut and bolt on the tensioner and allow the tensioner to move against the belt under its own spring tension.

➡ Do not assist the tensioner by pushing on it.

23. Tighten the tensioner nut, then the bolt to hold the tensioner in place. If the bolt is tightened first, the tensioner will turn and the tension will become incorrect.
24. Check the timing marks; all should be aligned. Press on the mid-point of the belt's tension side; belt deflection should be 0.16–0.20 in. (4–5mm).
25. The tensioner for the timing belt (belt A) should be in the slack position. Double check the alignment of all three pulleys for belt A; the timing marks should all align.

ENGINE AND ENGINE REBUILDING

26. Install the timing belt (in the correct direction if reusing the old belt) first onto the crankshaft sprocket, then onto the injection pump and then onto the camshaft sprocket. The tension side (crankshaft to injection pump) must be kept taut.

➡ The sprocket on the injection pump tends to rotate by itself during this procedure. Hold the sprocket in position while installing the belt. An assistant can be useful during this three-handed installation.

27. Loosen the tensioner mounting bolt and allow the spring tension to take up the slack in the belt. Don't push on the tensioner.
28. Tighten the tensioner slot side bolt before tightening the fulcrum bolt. If bolts are tightened out of order, the tensioner turns and the belt is placed under incorrect tension. After the belt is under tension, check the camshaft pulley; the belt may be lifting on the upper, outer edge (10 o'clock position). Make certain the belt is firmly seated on the sprocket.
29. Check the timing marks on all sprockets for correct alignment. Anything not in correct position will require removal of the belt(s) and alignment of the sprockets.
30. Using a wrench on the crankshaft pulley bolt, turn the engine clockwise until the camshaft sprocket has moved 2 teeth. Now turn the engine back (counterclockwise) to its original position.
31. Use a finger to press on the belt midway between the camshaft sprocket and the injection pump sprocket. Belt deflection should be 0.16-0.20 in. (4-5mm).
32. Install the upper and lower timing belt covers.
33. Install the crankshaft pulley and tighten to 47-54 ft. lbs. (64-73 Nm).
34. Install any accessories and drive belts that were removed.
35. Connect the negative battery cable back to the battery.

2.4L Engine

▶ See Figures 189 and 190

1. Disconnect negative battery cable.
2. Remove the radiator upper shroud.
3. Remove the power steering oil pump drive belt.
4. Loosen and remove the air conditioning tension pulley.

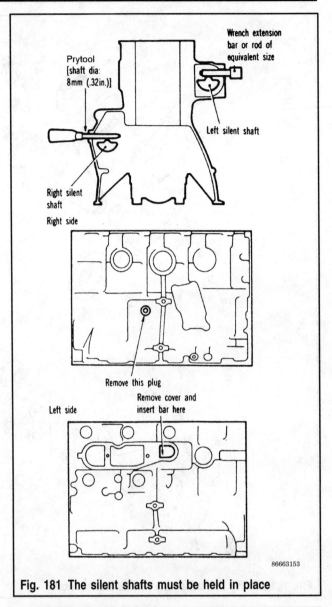

Fig. 181 The silent shafts must be held in place

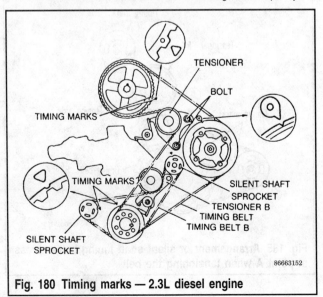

Fig. 180 Timing marks — 2.3L diesel engine

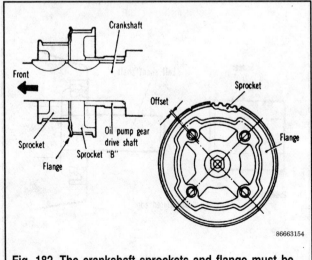

Fig. 182 The crankshaft sprockets and flange must be correctly installed

3-124 ENGINE AND ENGINE REBUILDING

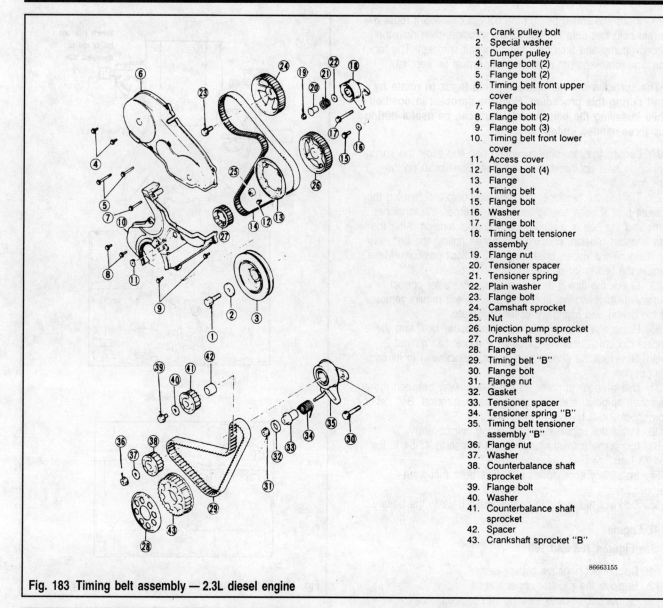

1. Crank pulley bolt
2. Special washer
3. Dumper pulley
4. Flange bolt (2)
5. Flange bolt (2)
6. Timing belt front upper cover
7. Flange bolt
8. Flange bolt (2)
9. Flange bolt (3)
10. Timing belt front lower cover
11. Access cover
12. Flange bolt (4)
13. Flange
14. Timing belt
15. Flange bolt
16. Washer
17. Flange bolt
18. Timing belt tensioner assembly
19. Flange nut
20. Tensioner spacer
21. Tensioner spring
22. Plain washer
23. Flange bolt
24. Camshaft sprocket
25. Nut
26. Injection pump sprocket
27. Crankshaft sprocket
28. Flange
29. Timing belt "B"
30. Flange bolt
31. Flange nut
32. Gasket
33. Tensioner spacer
34. Tensioner spring "B"
35. Timing belt tensioner assembly "B"
36. Flange nut
37. Washer
38. Counterbalance shaft sprocket
39. Flange bolt
40. Washer
41. Counterbalance shaft sprocket
42. Spacer
43. Crankshaft sprocket "B"

Fig. 183 Timing belt assembly — 2.3L diesel engine

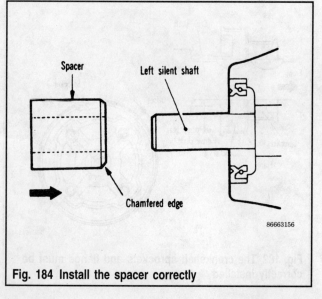

Fig. 184 Install the spacer correctly

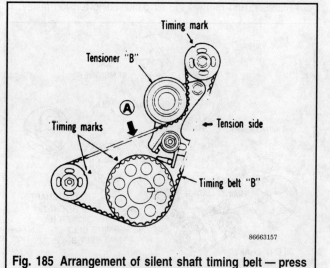

Fig. 185 Arrangement of silent shaft timing belt — press at point A when tensioning the belt

ENGINE AND ENGINE REBUILDING   3-125

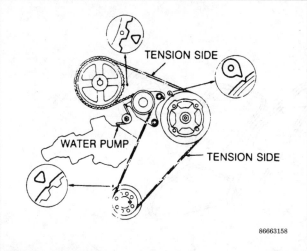

Fig. 186 Timing marks must be aligned before installing the timing belt

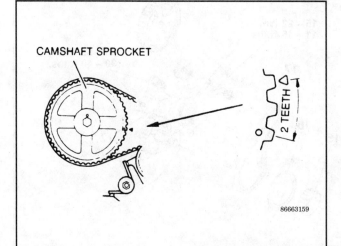

Fig. 187 Turn the crankshaft until the camshaft moves 2 teeth

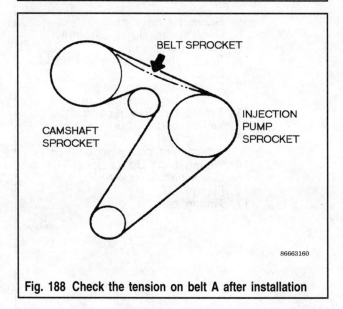

Fig. 188 Check the tension on belt A after installation

5. Remove the air conditioning compressor drive belt.
6. Remove the alternator drive belt.
7. Loosen and remove the four (4) nuts holding the cooling fan assembly on, then remove the cooling fan.
8. Remove the water pump pulley.
9. Remove the power steering pump crankshaft pulley.
10. Remove the crankshaft pulley.
11. Remove the timing belt upper and lower covers.
12. Rotate the crankshaft only clockwise and align the timing mark. Do not rotate the crankshaft counterclockwise. Do not rotate the crankshaft after the timing belt has been removed. If the timing belt is to be re-used, although it should be replaced whenever removed, mark an arrow with chalk or crayon in its direction of travel.
13. Loosen and remove the timing belt tensioner.
14. Make sure the timing marks are still lined up and slide the timing belt off of the sprockets with out pinching or binding. Do not rotate any of the engine shafts after the timing belt has been removed.
15. Mark a directional arrow on the second, smaller timing belt as well. Remove this timing belt's tensioner.
16. Remove the crankshaft sprocket and flange.
17. Remove the second timing belt, and make sure that the sprockets do not turn after removal.

To install:

18. Check the belt in detail. A case can be made to replace the timing belts whenever they are removed. Some conditions to check for are:

- Hardened back surface rubber. If the belt has a glossy back surface, is non-elastic and is so hard that even if a finger nail is forced into it, no mark is produced.
- Cracked back surface rubber.
- Cracked or exfoliated canvas.
- Cracked tooth bottom.
- The side of belt is cracked.
- The side of the belt is badly worn. A normal belt should have clear-cut sides as if cut with a sharp knife.
- The belt has badly worn teeth. The canvas on the load side tooth flank of the belt is worn (fluffy canvas fibers,

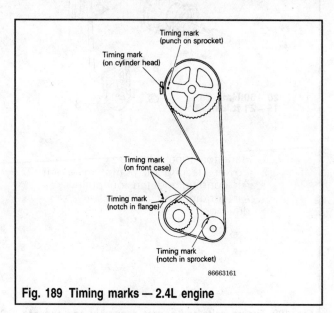

Fig. 189 Timing marks — 2.4L engine

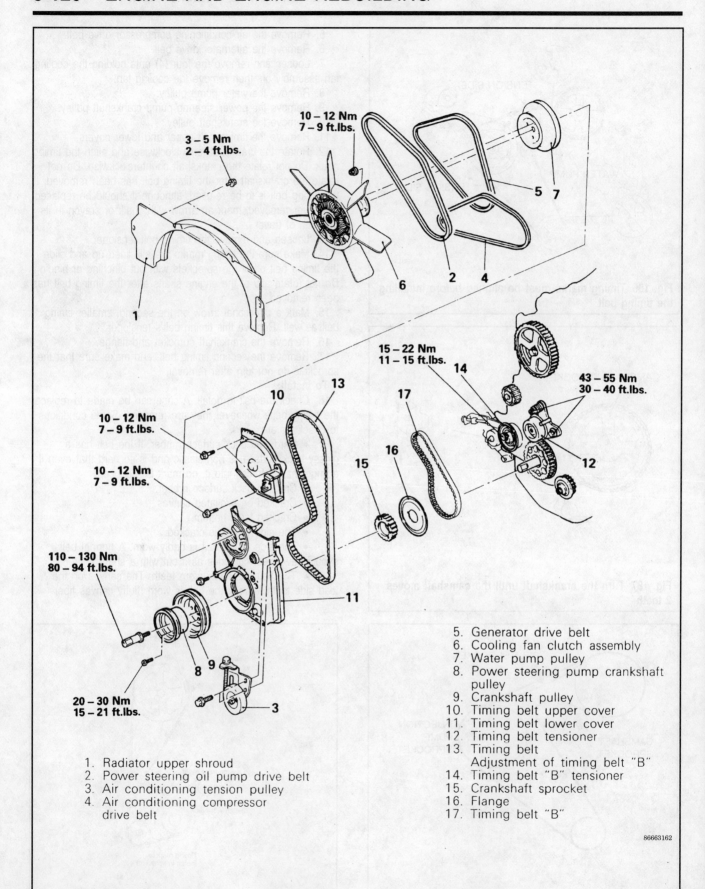

Fig. 190 Timing belts and covers assembly components — 2.4L engine

ENGINE AND ENGINE REBUILDING 3-127

rubber gone and color changed to white, and unclear canvas texture) or the canvas on the load side tooth flank is worn down and the rubber is exposed (tooth width reduced).
• The belt is missing teeth.

19. Check the sprockets and tensioner for wear. The sprocket teeth should be well defined, not rounded and the valleys between the teeth should be clean. The tensioners should spin freely with no binding or unusual noise. Replace the tensioner if there is any sign of grease leaking from the seal. Clean everything with a clean, dry cloth.

✳✳WARNING

Do not spray or immerse the sprockets or tensioners in cleaning solvent. The sprocket may absorb the solvent and transfer it to the belt. The tensioners are internally lubricated and the solvent will dilute or dissolve the lubricant.

20. Align the timing mark of the silent shaft belt sprockets on the crankshaft and silent shaft with the marks on the front case. Wrap the silent shaft belt around the sprockets so there is no slack in the upper span of the belt and the timing marks are still in line.
21. Install the tensioner initially so the actual center of the pulley is above and to the left of the installation bolt, and temporarily attach the tensioner pulley so that the flange is toward the front of the engine.
22. Move the pulley up by hand to create tension in the belt so that the center span of the long side of the belt deflects about 0.25 in. (6mm).
23. Hold the pulley tightly so it does not rotate when the bolt is tightened. Tighten the bolt to 15 ft. lbs. (20 Nm). If the pulley has moved, the belt will be too tight.
24. Install the crankshaft sprocket and flange onto the crankshaft.
25. Install the timing belt tensioner fully toward the water pump and temporarily tighten the bolts. Place the upper end of the spring against the water pump body. Align the timing marks of the cam, crankshaft and oil pump sprockets with the corresponding marks on the front case or head.

➡ **If the following step is not followed exactly, there is a chance that the silent shaft alignment will be 180 degrees off. This will cause a noticeable vibration in the engine and the entire procedure will have to be repeated.**

26. Before installing the timing belt, ensure that the left side silent shaft is in the correct position.

➡ **It is possible to align the timing marks on the camshaft sprocket, crankshaft sprocket and the oil pump sprocket with the left balance shaft out of alignment.**

27. With the timing mark on the oil pump pulley aligned with the mark on the front case, check the alignment of the left balance shaft to assure correct shaft timing.
   a. Remove the plug located on the left side of the block in the area of the starter.
   b. Insert a tool having a shaft diameter of 0.3 in. (8mm) into the hole.
   c. With the timing marks still aligned, the tool must be able to go in at least $2\frac{1}{3}$ in. (59mm). If it can only go in about 1 in. (25.4mm), turn the oil pump sprocket 1 complete revolution.
   d. Recheck the position of the balance shaft with the timing marks realigned. Leave the tool in place to hold the silent shaft while continuing.
28. Install the belt to the crankshaft sprocket, oil pump sprocket and the camshaft sprocket, in that order. While doing so, make sure there is no slack between the sprockets except where the tensioner will take it up when released.
29. Recheck the timing marks' alignment.
30. If all are aligned, loosen the tensioner mounting bolt and allow the tensioner to apply tension to the belt.
31. Remove the tool that is holding the silent shaft in place and turn the crankshaft clockwise a distance equal to 2 teeth of the camshaft sprocket. This will allow the tensioner to automatically tension the belt the proper amount.

➡ **Do not manually apply pressure to the tensioner. This will overtighten the belt and will cause a howling noise.**

32. First tighten the lower mounting bolt and then tighten the upper spacer bolt.
33. To verify that belt tension is correct, check that the deflection of the longest span (between the camshaft and oil pump sprockets) is $\frac{1}{2}$ in. (13mm).
34. Install the timing belt covers.
35. Install the crankshaft pulley and the power steering pump crankshaft pulley.
36. Install the water pump pulley and the cooling fan clutch assembly.
37. Install the alternator, air conditioning and power steering drive belts. Install the air conditioning tension pulley. Adjust the belts to proper tension — refer to Section 1.
38. Install the radiator upper shroud.
39. Connect the negative battery cable and road test the vehicle for proper operation.

3.0L and 3.5L Engines

◆ See Figures 191, 192, 193, 194, 195, 196, 197, 198, 199, 200, 201, 202, 203, 204, 205 and 206

REMOVAL

1. Disconnect the negative battery cable. Remove the battery and battery tray, if necessary for additional working clearance.
2. Drain the cooling system.
3. Disconnect the upper radiator hose at the radiator.
4. Remove the fan shroud from the radiator.
5. Remove the cooling fan and clutch assembly.
6. Loosen the necessary components and remove the drive belts from the air conditioning compressor, the power steering pump and the water pump/alternator.
7. Remove the fan pulley from the water pump.
8. Remove the power steering pump from the engine. Hang it out of the way by using string or stiff wire. Do not disconnect any lines or hoses; just move the whole pump with the lines attached.

3-128 ENGINE AND ENGINE REBUILDING

9. Remove the power steering pump bracket and mount.

✻✻WARNING

The bolts are of different lengths. Label or diagram each bolt and its location; correct reassembly is required.

10. Remove the tension pulley bracket (idler pulley) located just behind the power steering pump bracket.
11. Remove the air conditioning compressor. Hang it out of the way without kinking or twisting the lines.

✻✻WARNING

Do not disconnect any lines from the compressor. Do not allow the compressor to hang by the lines; support it securely.

12. Remove the compressor bracket.
13. Remove the cooling fan bracket assembly or the accessory mount.
14. Remove the upper timing belt covers and their gaskets; keep the gaskets with the covers. Removal of the crankshaft position sensor connector will be necessary on the 3.5L engines.

✻✻WARNING

Bolts are of three different lengths; label or diagram their location during removal.

15. Remove the crankshaft pulley. Use only the specified special tools (MD998754 & MB990767-01), or a damaged pulley damper could result.
16. Remove the lower timing belt cover.
17. Use a wrench on the crankshaft bolt to turn the engine clockwise until all the timing marks align. This positions the engine at TDC/compression for No. 1 piston. Once positioned, the engine must not be moved out of place.
18. Install the special counter-holding tools or equivalent and remove the crankshaft pulley. The large center bolt will be tight; do not turn the motor during removal.
19. Remove the front flange from the crankshaft sprocket.
20. Loosen the timing belt tensioner bolt and turn the timing belt tensioner counterclockwise along the elongated hole; this will relax the belt tension.

➙If vehicle is equipped with automatic type timing belt tensioner, remove the assembly, compress the pushrod and insert a small pin to reset the tensioner.

21. If the timing belt is to be reused, mark the direction of rotation on the belt with chalk or crayon. The belt must be reinstalled in its original position.
22. Carefully slide the belt off the sprockets. Place the belt in a clean, dry, protected location away from the work area. if the tensioner is to be removed, disconnect the spring and remove the retaining bolt.

3.0L ENGINE INSTALLATION

1. Inspect the timing belt in detail for any flaw or wear. If the belt is not virtually perfect, replace it. A case can be made for replacing the belt every time it is removed, particularly on high-mileage engines. Some of the conditions to look for are:
   • Hardened back surface; non-elastic and glossy; hard to mark with a fingernail.
   • Cracking on back of belt, bottom of teeth or side of belt.
   • Missing teeth or teeth lifting from belt.
   • Side of belt worn or fuzzy. Normal belt should have clean sides as if cut with a sharp knife.
   • Wear on teeth as shown by distinct color change or worn rubber.
   • Separation of inner coating from backing.
   • Any uneven wear patterns on the teeth of the belt. Wear pattern should be even across each tooth and not differ from one tooth to another.
2. Check the sprockets and tensioner for wear. The sprocket teeth should be well defined, not rounded and the valleys between the teeth should be clean. The tensioners should spin freely with no binding or unusual noise. Replace the tensioner if there is any sign of grease leaking from the seal. Clean everything with a clean, dry cloth.

✻✻WARNING

Do not spray or immerse the sprockets or tensioners in cleaning solvent. The sprocket may absorb the solvent and transfer it to the belt. The tensioners are internally lubricated and the solvent will dilute or dissolve the lubricant.

3. If the tensioner was removed, it must be reinstalled. After bolting it loosely in place, connect the spring onto the pin. Make certain the spring faces in the correct direction on the tensioner. Turn the tensioner to the extreme counterclockwise position on the elongated hole and tighten the bolt just enough to hold the tensioner in this position.
4. Double check the alignment of the timing marks on the camshaft and crankshaft sprockets.
5. Observing the direction of rotation marks made earlier, install the belt onto the crankshaft sprocket and then onto the left bank camshaft sprocket. Maintain tension on the belt between the sprockets.
6. Continue installing the belt onto the water pump, the right bank cam sprocket and the tensioner.
7. With your fingers, apply gentle counterclockwise force to the left camshaft sprocket. When the belt is taut on the tension side, the timing marks should align perfectly.
8. Install the flange on the crankshaft sprocket.
9. Loosen the bolt holding the tensioner one or two turns and allow the spring tension to draw the tensioner against the belt.
10. Using special tool MB 998716-01 or equivalent adapter, turn the crankshaft two complete revolutions clockwise. Turn the crank smoothly and re-align the timing marks at the end of the second revolution. This allows the tensioner to compensate for the normal amount of slack in the belt.
11. With the timing marks aligned, tighten the tensioner bolt to 18 ft. lbs. (25 Nm).
12. Inspect each gasket for the timing covers. The gaskets should be clean and pliable. Replace any which are distorted, cracked or broken. Coat the channel in each timing cover with

ENGINE AND ENGINE REBUILDING

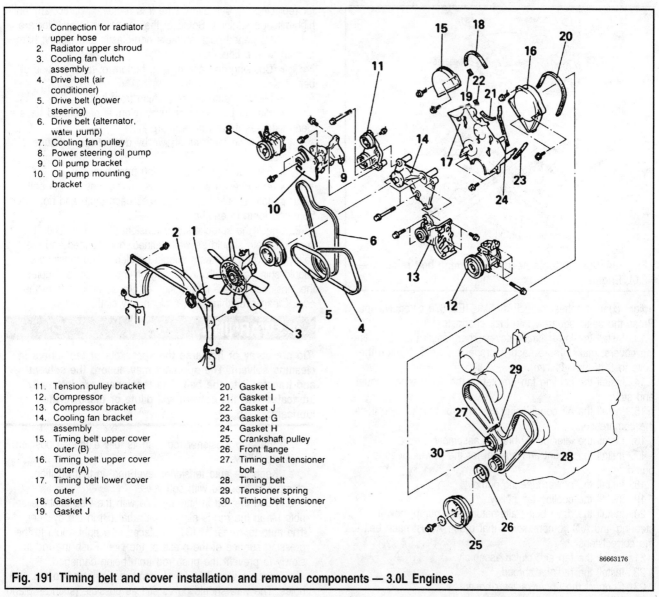

1. Connection for radiator upper hose
2. Radiator upper shroud
3. Cooling fan clutch assembly
4. Drive belt (air conditioner)
5. Drive belt (power steering)
6. Drive belt (alternator, water pump)
7. Cooling fan pulley
8. Power steering oil pump
9. Oil pump bracket
10. Oil pump mounting bracket
11. Tension pulley bracket
12. Compressor
13. Compressor bracket
14. Cooling fan bracket assembly
15. Timing belt upper cover outer (B)
16. Timing belt upper cover outer (A)
17. Timing belt lower cover outer
18. Gasket K
19. Gasket J
20. Gasket L
21. Gasket I
22. Gasket J
23. Gasket G
24. Gasket H
25. Crankshaft pulley
26. Front flange
27. Timing belt tensioner bolt
28. Timing belt
29. Tensioner spring
30. Timing belt tensioner

Fig. 191 Timing belt and cover installation and removal components — 3.0L Engines

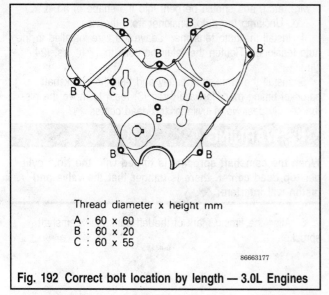

Thread diameter x height mm
A : 60 x 60
B : 60 x 20
C : 60 x 55

Fig. 192 Correct bolt location by length — 3.0L Engines

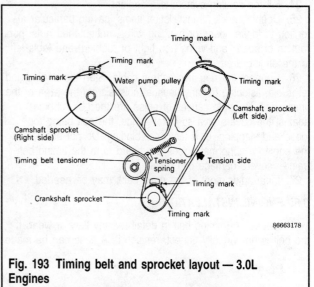

Fig. 193 Timing belt and sprocket layout — 3.0L Engines

3-130 ENGINE AND ENGINE REBUILDING

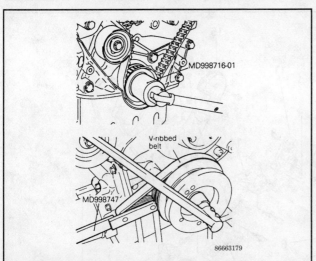

Fig. 194 Special tools needed for timing belt removal — 3.0L Engines

a light coat of adhesive such as 3M® EC 870 or equivalent. Press the seals squarely into their channels.

13. Install the lower timing cover. Then install the upper timing covers, making sure each is properly seated. Tighten the bolts to 8 ft. lbs. (11 Nm).
14. Install the cooling fan bracket or the accessories mount and stay.
15. Install the air conditioning compressor bracket and install the compressor.
16. Install the idler pulley bracket assembly.
17. Install the mount and bracket for the power steering pump.
18. Install the power steering pump.
19. Install the cooling fan pulley.
20. Install the drive belts: alternator/water pump, power steering and A/C compressor in that order. Adjust each belt to the correct tension.
21. Install the fan and clutch assembly.
22. Install the radiator shroud.
23. Connect the upper radiator hose.
24. Fill the cooling system with coolant.
25. Double check all installation items, paying particular attention to loose hoses or hanging wires, untightened nuts, poor routing of hoses and wires (too tight or rubbing) and tools left in the engine area.
26. Connect the negative battery cable. Start the engine and let it idle, listening for any unusual noises from the area of the timing belt. Possible causes of noise are the belt rubbing against the covers or a sprocket flange, the belt being too loose and slapping, or a tensioner binding. Do not accelerate the engine if abnormal noises are heard from the timing belt train — severe damage can result.
27. Final adjustment of the drive belts may be needed.

3.5L ENGINE INSTALLATION

1. Inspect the timing belt in detail for any flaw or wear. If the belt is not virtually perfect, replace it. A case can be made for replacing the belt every time it is removed, particularly on high-mileage engines. Some of the conditions to look for are:
- Hardened back surface; non-elastic and glossy; hard to mark with a fingernail.
- Cracking on back of belt, bottom of teeth or side of belt.
- Missing teeth or teeth lifting from belt.
- Side of belt worn or fuzzy. Normal belt should have clean sides as if cut with a sharp knife.
- Wear on teeth as shown by distinct color change or worn rubber.
- Separation of inner coating from backing.
- Any uneven wear patterns on the teeth of the belt.

Wear pattern should be even across each tooth and not differ from one tooth to another.

2. Check the sprockets and tensioner for wear. The sprocket teeth should be well defined, not rounded and the valleys between the teeth should be clean. The tensioners should spin freely with no binding or unusual noise. Replace the tensioner if there is any sign of grease leaking from the seal. Clean everything with a clean, dry cloth.

✱✱WARNING

Do not spray or immerse the sprockets or tensioners in cleaning solvent. The sprocket may absorb the solvent and transfer it to the belt. The tensioners are internally lubricated and the solvent will dilute or dissolve the lubricant.

3. If the auto tensioner rod is in its fully extended position, reset it as follows.
   a. Keep the auto tensioner level and, in that position, clamp it in the vise with soft jaws.
   b. Push the rod in little by little with the vise until the set hole (A) in the rod is aligned with that (B) in the cylinder. The auto tensioner MUST be placed at a right angle to the pressing surface of the press or the vice. Push the rod in slowly to prevent the push rod from being damaged.
   c. Insert a wire 0.55 in. (1.4mm) in diameter into the set holes. The wire should be as stiff as possible (such as piano wire, etc.), and should be bent into the shape of an "L".
   d. Unclamp the auto tensioner from the vise.
4. Install the auto tensioner. Leave the wire installed in the auto tensioner. Tighten the retaining bolts to 17 ft. lbs. (24 Nm).
5. Install the crankshaft pulley and turn the crankshaft sprocket timing mark forward three (3) teeth to move the piston slightly past No. 1 cylinder top dead center.

✱✱WARNING

When the camshaft sprocket is turned with the No.1 cylinder top dead center, there is danger that the valve and piston will interfere.

6. Align the timing mark of the left bank side camshaft sprocket.

ENGINE AND ENGINE REBUILDING

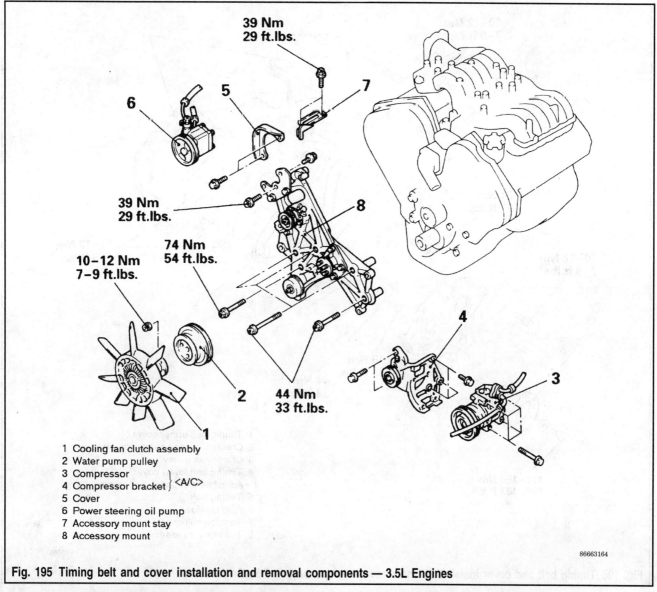

1. Cooling fan clutch assembly
2. Water pump pulley
3. Compressor
4. Compressor bracket } <A/C>
5. Cover
6. Power steering oil pump
7. Accessory mount stay
8. Accessory mount

Fig. 195 Timing belt and cover installation and removal components — 3.5L Engines

7. Align the timing mark of the right bank side camshaft sprocket and support it not to rotate with a box-end wrench.

※※WARNING

The camshaft sprocket will easily turn because of the valve spring force, so be careful not to insert your fingers, etc. If the sprocket on one side of the right bank is turned one full revolution while the sprocket timing marks on the opposite side of the right bank are aligned, the intake and exhaust valves may cause interference.

8. Check that the camshaft sprocket timing mark of the right bank side is aligned and clamp the timing belt in place with double clips. If the timing belt is reused, install so that the arrow marked on it at the time of removal is pointing in the clockwise direction.
9. Set the timing belt onto the water pump pulley.
10. Check that the camshaft sprocket timing mark of the left bank side is aligned and clamp the timing belt in place with double clips.
11. Set the timing belt onto the idler pulley.
12. After aligning the crankshaft sprocket timing marks, turn the crankshaft one touch counter-clockwise.
13. Set the timing belt onto the crankshaft sprocket.
14. Set the timing belt onto the tensioner pulley.
15. Place the tensioner pulley pin hole so that it is towards the top. Press the tensioner pulley onto the timing belt, and provisionally tighten the fixing bolt.
16. Align the crankshaft sprocket timing marks.
17. Check that each of the sprocket timing marks is aligned.
18. Remove the four (4) double clips holding the belt in place.
19. After turning the crankshaft a 1/4 turn counter-clockwise, turn it clockwise to the position where the timing marks are aligned.
20. Loosen the center bolt on the tensioner pulley. Using the special tool and torque wrench, apply tensioning torque to the timing belt and, at the same time, tighten the center bolt to

3-132 ENGINE AND ENGINE REBUILDING

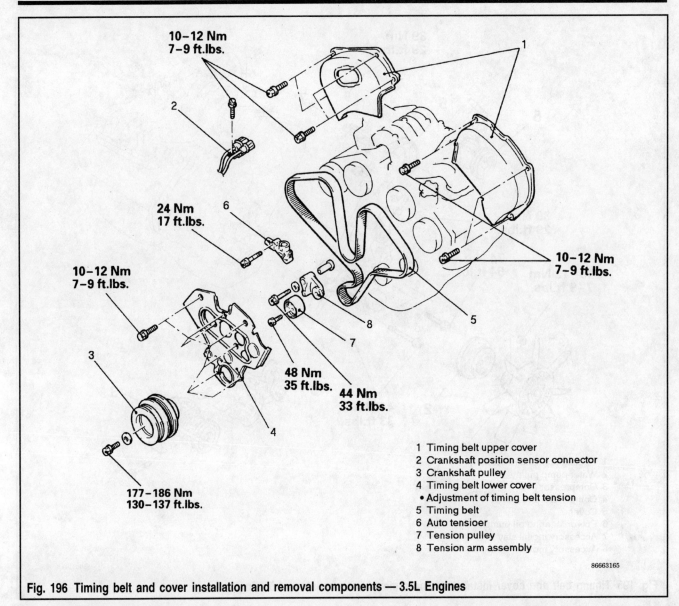

Fig. 196 Timing belt and cover installation and removal components — 3.5L Engines

1. Timing belt upper cover
2. Crankshaft position sensor connector
3. Crankshaft pulley
4. Timing belt lower cover
 • Adjustment of timing belt tension
5. Timing belt
6. Auto tensioer
7. Tension pulley
8. Tension arm assembly

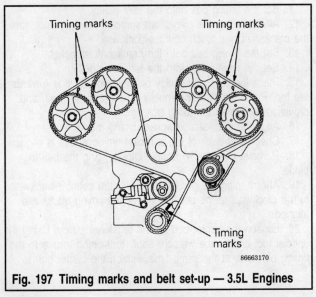

Fig. 197 Timing marks and belt set-up — 3.5L Engines

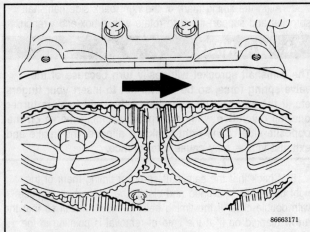

Fig. 198 Make sure to install the belt so that the arrow marked on it during removal points in a clockwise direction — 3.5L Engines

ENGINE AND ENGINE REBUILDING

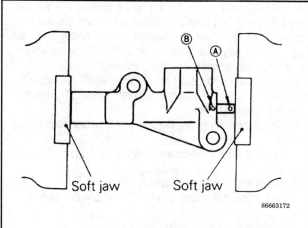

Fig. 199 Push the rod in little by little with a vise until the set hole (A) in the rod is aligned with hole (B) in the cylinder — 3.5L Engines

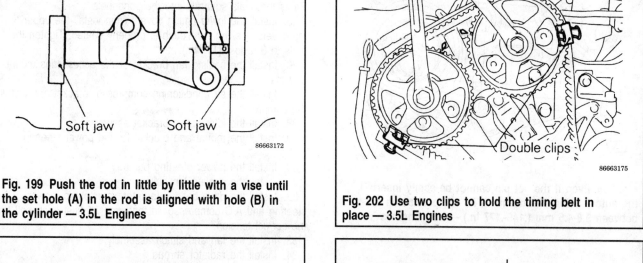

Fig. 202 Use two clips to hold the timing belt in place — 3.5L Engines

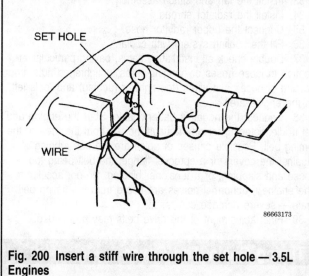

Fig. 200 Insert a stiff wire through the set hole — 3.5L Engines

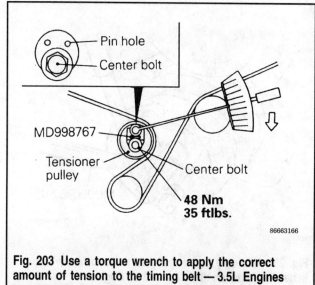

Fig. 203 Use a torque wrench to apply the correct amount of tension to the timing belt — 3.5L Engines

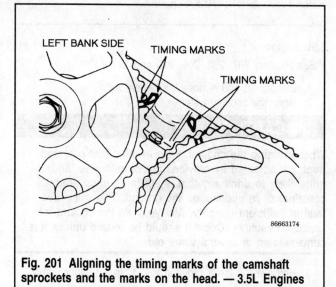

Fig. 201 Aligning the timing marks of the camshaft sprockets and the marks on the head. — 3.5L Engines

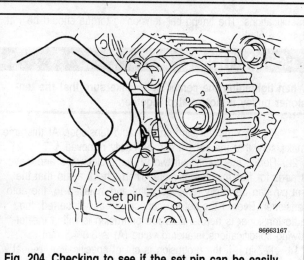

Fig. 204 Checking to see if the set pin can be easily removed and installed into the auto tensioner — 3.5L Engines

3-134 ENGINE AND ENGINE REBUILDING

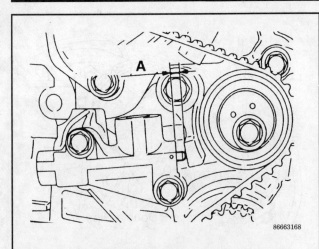

Fig. 205 Even if the set pin cannot be easily inserted, the auto tensioner is normal if its rod protrusion is between 3.8-4.5 mm (.149-.177 in.) — 3.5L Engines

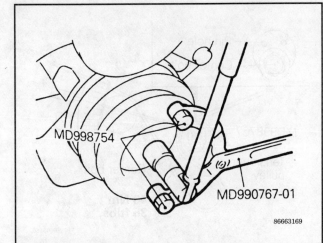

Fig. 206 Use only these specified tools to remove and install the crankshaft pulley — 3.5L Engines

specifications. The timing belt tensioning torque should be 7 ft. lbs. (9.4 Nm).

✱✱CAUTION

When tightening the center bolt, make sure that the tensioner pulley is not rotated together.

21. Remove the set pin from the auto tensioner. At this time, make sure that the set pin can be easily removed.
22. Rotate the crankshaft two (2) turns clockwise and leave it as is for five minutes or more. Then, check again that the set pin can be easily removed from, and installed to, the auto tensioner. Even if the set pin cannot be easily inserted, the auto tensioner is normal if its protrusion is within the following specifications: standard value (A) is 3.8-4.5 mm (.149-.177 in.). If the protrusion is out of specification, repeat steps 46-49.
23. Check again that the timing marks on all sprockets are aligned properly.

24. Inspect each gasket for the timing covers. The gaskets should be clean and pliable. Replace any which are distorted, cracked or broken. Coat the channel in each timing cover with a light coat of adhesive such as 3M® EC 870 or equivalent. Press the seals squarely into their channels.
25. Install the lower timing cover. Then install the upper timing covers, making sure each is properly seated. Tighten the bolts to 8 ft. lbs. (11 Nm).
26. Install the cooling fan bracket or the accessories mount and stay.
27. Install the air conditioning compressor bracket and install the compressor.
28. Install the idler pulley bracket assembly.
29. Install the mount and bracket for the power steering pump.
30. Install the power steering pump.
31. Install the cooling fan pulley.
32. Install the drive belts: alternator/water pump, power steering and A/C compressor in that order. Adjust each belt to the correct tension.
33. Install the fan and clutch assembly.
34. Install the radiator shroud.
35. Connect the upper radiator hose.
36. Fill the cooling system with coolant.
37. Double check all installation items, paying particular attention to loose hoses or hanging wires, untightened nuts, poor routing of hoses and wires (too tight or rubbing) and tools left in the engine area.
38. Connect the negative battery cable. Start the engine and let it idle, listening for any unusual noises from the area of the timing belt. Possible causes of noise are the belt rubbing against the covers or a sprocket flange, the belt being too loose and slapping, or a tensioner binding. Do not accelerate the engine if abnormal noises are heard from the timing belt train — severe damage can result.
39. Final adjustment of the drive belts may be needed.

Timing Chain And Cover

REMOVAL & INSTALLATION

2.6L Engine

▶ See Figures 207, 208, 209, 210 and 211

1. Disconnect and remove the battery.
2. Drain the cooling system.

✱✱CAUTION

When draining the coolant, keep in mind that cats and dogs are attracted by ethylene glycol antifreeze, and are quite likely to drink any that is left in an uncovered container or in puddles on the ground. This will prove fatal in sufficient quantity. Always drain the coolant into a sealable container. Coolant should be reused unless it is contaminated or several years old.

3. Remove the lower radiator hose from the water pump.

ENGINE AND ENGINE REBUILDING 3-135

4. Remove the upper fan shroud. Carefully unbolt and remove the fan with the clutch assembly and store the fan in an upright position.

5. Loosen the adjuster and remove the air conditioning drive belt. Loosen the alternator and power steering pump and remove their drive belts.

6. Remove the water pump pulley.
7. Disconnect the heater hose from the water pump.
8. Remove the mounting bolts holding the water pump to the engine. Note that the bolts are of different lengths; label or diagram them as they are removed.
9. Remove the water pump and gasket.
10. Remove the crankshaft bolt. Use a puller to remove the crankshaft pulley.
11. Remove the valve cover.
12. Remove ONLY the two small front bolts from the cylinder head; these screw into and seal the top of the timing cover.

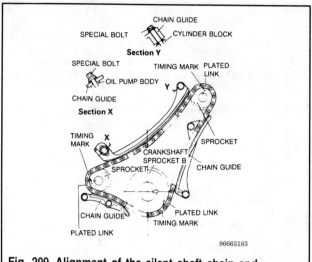

Fig. 209 Alignment of the silent shaft chain and sprockets — 2.6L engine

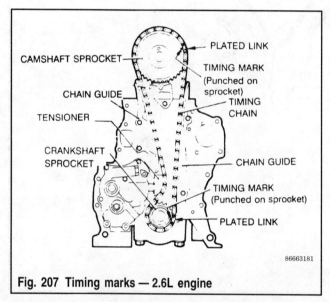

Fig. 207 Timing marks — 2.6L engine

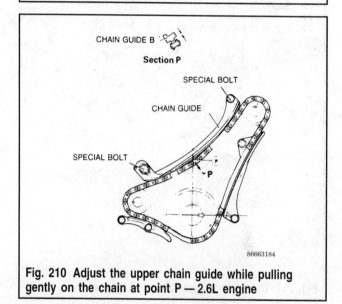

Fig. 210 Adjust the upper chain guide while pulling gently on the chain at point P — 2.6L engine

13. Remove the oil pan bolts (front and side) that screw into the timing cover. Use a moderately sharp instrument to separate the oil pan gasket from the underside of the front cover.
14. Unbolt and remove the timing cover.
15. Clean all the gasket surfaces. If the oil pan gasket was damaged in removing the front cover, carefully cut the oil pan gasket off flush with the front of the block on both sides and remove the cutoff piece of gasket.
16. The crankshaft seal should be replaced any time the timing cover is removed. Carefully pry the oil seal out of the cover without scratching the bore into which the seal fits. Install a new seal with an seal driver or installer such as MD 998376-01 and MB 990938-01 or their equivalents.
17. Turn the crankshaft clockwise until the timing marks align. (Both valves for cylinder No. 1 are closed and the rockers are loose.).
18. Remove the 3 chain guides bearing on the silent shaft chain. Label or diagram each bolt as it is removed. The bolts are of different lengths and styles; they must be replaced in the correct locations.

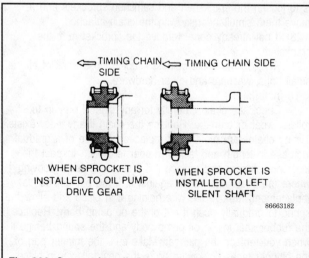

Fig. 208 Correct installation of the silent shaft chain sprockets — 2.6L engine

ENGINE AND ENGINE REBUILDING

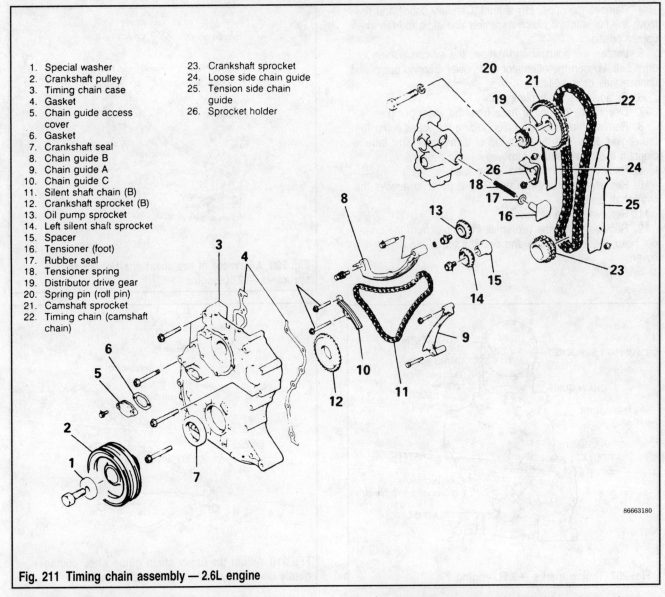

1. Special washer
2. Crankshaft pulley
3. Timing chain case
4. Gasket
5. Chain guide access cover
6. Gasket
7. Crankshaft seal
8. Chain guide B
9. Chain guide A
10. Chain guide C
11. Silent shaft chain (B)
12. Crankshaft sprocket (B)
13. Oil pump sprocket
14. Left silent shaft sprocket
15. Spacer
16. Tensioner (foot)
17. Rubber seal
18. Tensioner spring
19. Distributor drive gear
20. Spring pin (roll pin)
21. Camshaft sprocket
22. Timing chain (camshaft chain)
23. Crankshaft sprocket
24. Loose side chain guide
25. Tension side chain guide
26. Sprocket holder

Fig. 211 Timing chain assembly — 2.6L engine

19. Remove the chain from the silent shaft and oil pump sprockets and then from the crankshaft sprocket. If the chain will not come off the sprockets, don't pry on it. Unbolt the silent shaft and oil pump sprockets and carefully remove them with the chain. If the chain is removed individually, remove the sprockets afterward. Some sprockets use a small key on the shaft; don't lose it.

➡The two sprockets are identical, however the oil pump drive sprocket is installed with the concave side toward the engine while the left silent shaft sprocket has the concave side out.

20. Remove the crankshaft sprocket for the silent shaft chain.
21. The timing chain tensioner maintains constant spring pressure on the chain. To prevent it from popping out of the oil pump it must be fastened in place. Run a piece of wire around the plunger and the left side of the oil pump.
22. Remove the camshaft sprocket bolt and pull the distributor drive gear and the sprocket off the camshaft. Separate the chain from the sprockets and remove it. An alternate method is to loosen the crankshaft and camshaft sprockets and remove them simultaneously with the chain attached.
23. If not already done, remove the sprocket from the crankshaft.
24. Remove the chain guides, keeping close track of all the small bolts, washers and other hardware.

To install:
25. Check the chain carefully for any sign of play in the rollers, wear or damage. Replace the chain if any wear exists. Timing chains are most often replaced because of a gradual increase in length and lack of proper tension. Inspect the tensioner plunger and replace it if it shows a deep grooving where the chain has ridden against it. If the plunger needs replacement, remove the wire holding it in place and allow the spring to gradually push it out of the oil pump body. Replace the rubber seal in the oil pump body and the spring behind it when you replace the plunger. Make sure the thinner part of the plunger faces downward. Wire the new follower in place just as the old one was. If the timing chain guides show heavy grooving, they should be replaced. Sprockets should be replaced if the teeth are rounded, deformed or cracked.

ENGINE AND ENGINE REBUILDING 3-137

26. Install the left and right chain guide. Don't forget the small sprocket holder; it will be needed.
27. Both the crankshaft sprocket and the camshaft sprocket have a small dot on their faces. Assemble the chain and sprockets so that each dot aligns with the plated links on the chain. Both sets of marks and links must align. Install the crankshaft sprocket onto the crankshaft with the chain in place and support the camshaft sprocket on the sprocket holder.
28. Make certain the chain is riding correctly on the guides. Have the camshaft sprocket retaining bolt and the proper wrench at hand. Lift the cam sprocket into place against the cam and install the bolt. Make certain the sprocket engages the guide pin correctly. Tighten the bolt to 40 ft. lbs. (54 Nm).
29. Remove the wire holding the tensioner plunger and allow it to tension the chain.
30. Assemble the silent shaft sprocket, the oil pump sprocket and the silent shaft chain. Again, each sprocket is marked with a dot which must be aligned with the plated links on the chain.

※※WARNING

The sprocket for the silent shaft mounts with the concave side facing out (away from the engine block) and the sprocket for the oil pump mounts with the concave side facing the engine block.

31. Install the silent shaft drive sprocket onto the crankshaft (sprocket B).
32. Holding the two sprockets with the chain installed and aligned, fit the chain over the crankshaft sprocket so that the third plated link aligns with the mark on the crank sprocket. Once the chain is in place, install the other sprockets to the left silent shaft and the oil pump. Install the bolts just snug but not tight.
33. Rotate both silent shaft sprockets (the left side and oil pump sprockets) inward slightly to position the slack in the chain in the center of the longest side of the chain (point P in the figure).
34. Install the chain guides and adjust the upper guide (B) so that when the chain is pulled downward at the center (point P), the clearance between the chain and the guide is 0.008-0.031 in. (0.20-0.80mm). Tighten the guide locking bolt to 14 ft. lbs. (19 Nm).
35. Before replacing the timing chain cover, check all the gasket surfaces for cleanliness and inspect the cover for cracks or damage.
36. Using a new gasket, cut an exact replacement for the section of oil pan gasket damaged during removal. Insert this piece of gasket onto the front of the pan in the exact position of the old piece. Use liquid sealer on the joint between the two sections of gasket on both sides.
37. Using new gaskets carefully positioned, install the chain cover and tighten the bolts to 10 ft. lbs. (14 Nm).
38. Lightly coat the outside diameter of the crankshaft pulley boss with clean engine oil. Install the pulley onto the crankshaft, followed by the washer and bolt. Tighten the bolt to force the pulley all the way on. Tighten the bolt to 88 ft. lbs. (119 Nm).
39. Install the two head bolts through the head and into the timing chain cover. Tighten the bolts to 14 ft. lbs. (19 Nm).
40. Install the oil pan bolts, tightening them evenly to 4 ft. lbs. (5 Nm).

41. Install the water pump with a new gasket. Don't forget to install the alternator bracket.
42. Connect the heater hose to the water pump.
43. Install the water pump pulley.
44. Install the drive belts beginning with the power steering belt, then the alternator belt and the A/C compressor belt.
45. Install the fan and clutch assembly if it was removed. Install the upper fan shroud.
46. Adjust the drive belts to the proper tension and tighten the adjusting fittings for each component.
47. Install the lower radiator hose.
48. Install the battery.
49. Double check the draincock, closing it if necessary, and refill the cooling system.
50. Connect the negative battery cable and start the engine. Check carefully for leaks, particularly around the water pump and oil pan seams.
51. Listen for any unusual noises from the area of the timing chain. Possible causes of noise are the chain rubbing against the cover or a guide being too loose or too tight. Do not accelerate the engine if abnormal noises are heard from the timing chain case — severe damage can result.
52. After making needed adjustments, shut the engine off, allow it to cool and set the coolant to the correct level.

Oil Pan

REMOVAL & INSTALLATION

▶ See Figures 212, 213, 214, 215 and 216

※※CAUTION

Used motor oil may cause skin cancer if repeatedly left in contact with the skin for prolonged periods. Although this is unlikely unless you handle oil on a daily basis, it is wise to thoroughly wash your hands with soap and water immediately after handling used motor oil.

The oil pan must be pulled downward as much as 6 in. (15.2 cm) to clear the oil pickup. In many cases, this requires that the engine mounts be disconnected and the engine raised to clear a crossmember underneath the shallower section of the pan.

If raising the engine is needed, support the vehicle securely high enough to allow easy access to the underside. Refer to the engine removal and installation procedures and disconnect all the hoses and wires that would prevent the engine from being lifted the necessary distance.

※※WARNING

Raising the engine will require the use of overhead or hoist type equipment attached to the top of the engine by chains. Do not attempt to raise the engine with a jack underneath. Damage or injury can result.

Support the engine on the hoist and remove the through-bolts from the mounts. Elevate the engine only enough to gain

ENGINE AND ENGINE REBUILDING

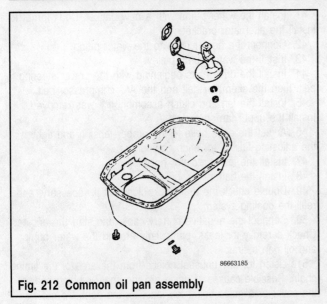

Fig. 212 Common oil pan assembly

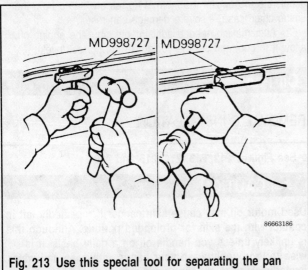

Fig. 213 Use this special tool for separating the pan from the block — other tools may cause damage

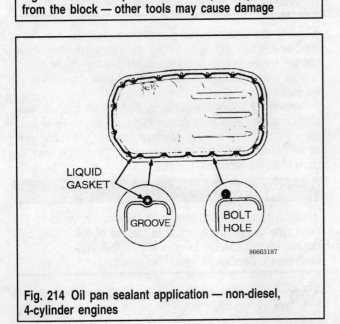

Fig. 214 Oil pan sealant application — non-diesel, 4-cylinder engines

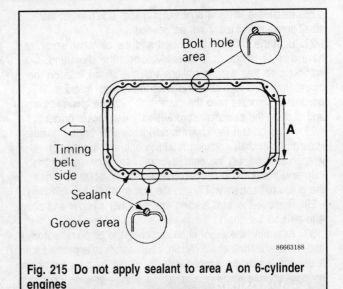

Fig. 215 Do not apply sealant to area A on 6-cylinder engines

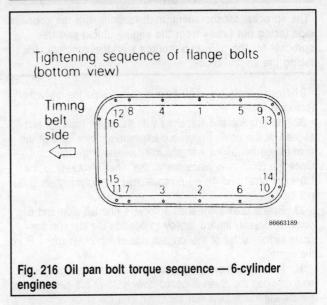

Fig. 216 Oil pan bolt torque sequence — 6-cylinder engines

the needed clearance and keep a close watch for interference with other components as the engine is raised.

➡ The 3.0L (24 Valve) and 3.5L engines are equipped with an upper and lower oil pan assembly. To remove the lower oil pan use a wooden block to tap the oil pan down. Do not use the oil pan remover tool (MD998727) to remove the lower pan which is made of aluminum. The upper oil pan can be removed but use caution on the back bolt holes that go through the transmission assembly.

1. Disconnect the negative battery cable. Drain the engine oil. When the pan is empty, reinstall the drain plug finger-tight.
2. Remove the bolts holding the oil pan. The length may vary by location. Keep the bolts in order or identify them so that they may be correctly installed.
3. Most engines use a bead of sealer on the oil pan rather than a paper gasket. After the bolts are removed, tap the corners of the pan with a rubber or plastic mallet. If the pan pops loose, it may be removed on the spot. In most cases, the pan will remain firmly glued to the block. Insert the special

ENGINE AND ENGINE REBUILDING 3-139

gasket slicing tool (MB 998727 or equivalent) by tapping it into place with a mallet or hammer. Once in place, drive the tool along the pan to release the sealer and the pan.

✲✲WARNING

DO NOT attempt to loosen the pan with a screwdriver or chisel. The flange of the pan will deform and cause oil leaks after reassembly.

4. When the pan is removed, both the flange surface of the pan and the matching surface of the block must be completely cleaned of any remaining gasket or sealer. When cleaning with a scraper, take great care not to gouge or mar the metal surfaces.

5. With the pan removed, the oil pick-up screen and tube may be removed by unbolting it from the block. If removed, it should be thoroughly cleaned in a solvent bath and air dried before reinstallation. The gasket surfaces must be cleaned and the old gasket discarded.

6. Clean the oil pan with solvent, taking great care to remove all particulate matter and sludge before air drying. The pan must be reinstalled free of any matter which could be circulated through the engine. Remove the drain plug, replace the washer and reinstall the plug, tightening it.

7. If the pick-up was removed, reinstall it with a new gasket and tighten the bolts to 13 ft. lbs. (18 Nm).

8. On all engines except for the diesel engines, apply a thin bead of sealant (MZ 100168 or equivalent) all around the pan. Cut the tip of the sealant tube to allow a bead about 0.16 in (4mm) to flow. The sealant goes in the groove in the pan flange. Note that the bead should flow to the inside of the bolt holes or between the bolt and the pan. The V6 engines also receive the bead of sealer but the area between the two rearmost bolt holes must not be sealed. Care must be taken to prevent oozing of sealant into the rear valley of the pan. The diesel truck engines use a paper gasket which does not require sealant. Sealant such as 3M® ART 8660 or its equivalent must be applied to the block at the four seams where the front and rear cases join. Apply the sealer only to these areas.

9. Once the sealant is applied, the pan must be installed within 10 minutes or the sealant becomes unusable. Fit the pan into place and install the bolts finger-tight only, making certain the bolts are replaced in the correct holes as determined by their length.

10. Tighten the pan bolts evenly and alternating from side to side to 5 ft. lbs. (7 Nm) only. The pan bolts for the 6-cylinder engine must be tightened in the pattern shown; the pattern may be used as a guide for the correct tightening of other pans.

✲✲WARNING

The tightness of the bolts does not seal the pan, the gasket or sealer does. DO NOT OVERTIGHTEN.

11. Since the oil has been drained, replacement of the oil filter is recommended before refilling the engine oil.

12. Install the correct amount of engine oil. Make sure the drain plug is tight before pouring the oil.

13. If the engine is supported by a hoist, lower it into position and secure the mounts and through-bolts as necessary.

14. If possible, the sealant should be allowed to cure for a period of hours before the engine is started. Once started, let the engine warm up at idle and then shut it off.

15. Check the pan carefully around the seam for any sign of leakage. If any leakage is found, do not attempt to cure it by tightening the pan bolts. This usually aggravates the leak rather than curing it. The pan will need to be removed and resealed.

Oil Pump and Silent Shafts

➡ 6-cylinder engines do not use a silent shaft assembly.

REMOVAL & INSTALLATION

2.0L And 2.4L Engines

◆ See Figures 217, 218, 219, 220, 221 and 222

1. Disconnect the negative battery cable.
2. Remove the timing belts by following the directions for removal of the timing belts found in this section.
3. Remove the oil filter. Drain the engine oil.

✲✲CAUTION

Used motor oil may cause skin cancer if repeatedly left in contact with the skin for prolonged periods. Although this is unlikely unless you handle oil on a daily basis, it is wise to thoroughly wash your hands with soap and water immediately after handling used motor oil.

4. Remove the oil pan by following the previously outlined procedures.
5. Remove the oil pick up screen and gasket.
6. Remove the plug from the left side of the cylinder block and insert a screwdriver-sized prytool to keep the left silent shaft in place. The prytool should have a shaft diameter of 0.3 in. (8mm) and a shaft length of at least 2.4 in. (60mm).
7. Remove the bolt holding the oil pump sprocket and remove the sprocket.
8. Loosen the mounting bolt for the right silent shaft sprocket until it can be turned with your fingers. Do not remove it.
9. If not already done, remove the tensioner for the silent shaft belt (tensioner B).
10. If not already done, remove the silent shaft belt (timing belt B). Remove the crankshaft pulley for the silent shaft belt, if so desired.

✲✲WARNING

After the timing belt has been removed, do not attempt to loosen the silent shaft bolt by holding the sprocket with pliers. If the sprocket is to be removed, hold the pulley with your fingers.

3-140 ENGINE AND ENGINE REBUILDING

11. Remove the front case mounting bolts; remove the front case assembly and its gasket.

※※WARNING

The front case bolts are of several different lengths and must be replaced exactly as removed. Labeling or diagramming their placement during removal is very important.

12. Remove the prytool from the left silent shaft. Remove the silent shafts from the engine.
13. Remove the oil pump cover from the front case.
14. Remove the oil pump gears.

To install:

15. Visually check the contact surfaces of the oil pump cover and the front case for wear. If excessive wear is present, replace the components. Inspect the oil passages for clogging and clean them as needed. The silent shafts should be inspected for any of wear or seizure. Closely inspect the oil

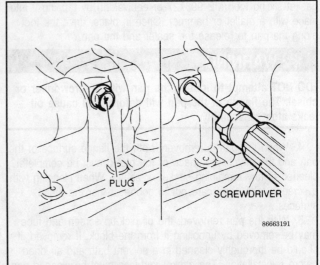

Fig. 218 Use a prytool to block the left silent shaft

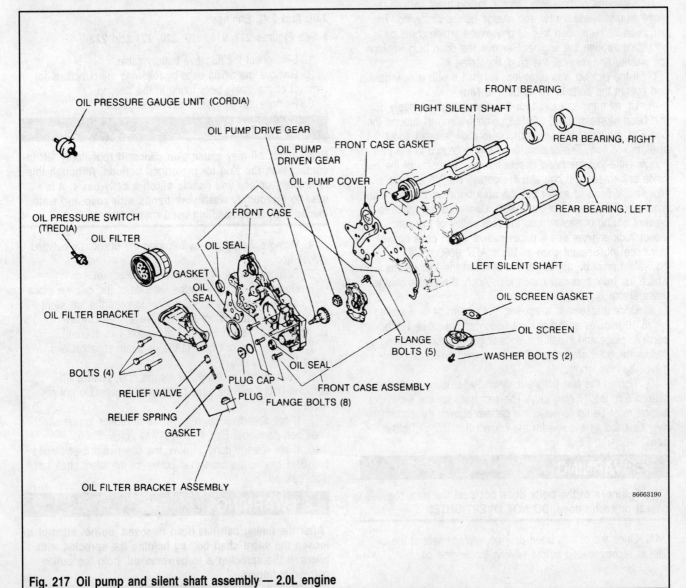

Fig. 217 Oil pump and silent shaft assembly — 2.0L engine

ENGINE AND ENGINE REBUILDING 3-141

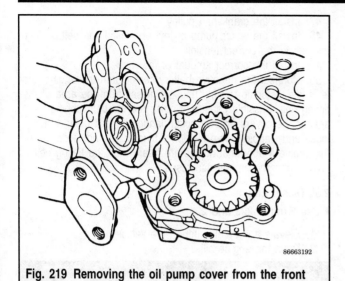

Fig. 219 Removing the oil pump cover from the front case

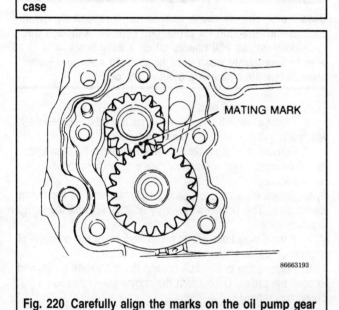

Fig. 220 Carefully align the marks on the oil pump gear

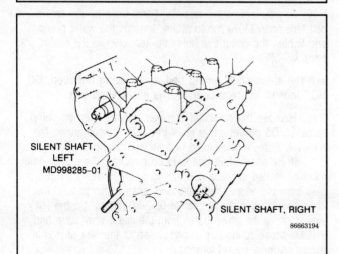

Fig. 221 Seal guide tool in place — note the engine is removed and inverted

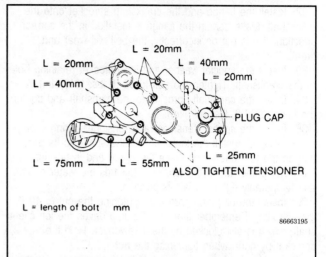

Fig. 222 Correct bolt placement by length — 2.0L and 2.4L engines

seals on the front case, replacing any with worn or damaged lips.

16. Insert the relief plunger into the oil filter bracket and check for smooth operation.
17. Apply engine clean engine oil to the oil pump gears and install them in the front case. Align the mating marks.
18. Install the oil pump cover to the front case and tighten the bolts to 16 ft. lbs. (21 Nm).
19. Carefully install the left and right silent shafts into the engine; the bearings can be damaged by careless installation.
20. Install a seal guide tool such as MD 998285-01 or its equivalent to the crankshaft. Note that the tapered end should face away from the block. Coat the surface of the guide with clean engine oil. This guide will help in installing a new seal in the front case. If the seal is already in place, the guide will protect the seal during installation.
21. Install a new front case gasket, carefully aligning all the bolt holes.
22. Install the front case to the block and lightly tighten the eight bolts. When the case is properly seated, the seal guide may be removed.
23. Insert a prytool through the left side access hole and block the left silent shaft.
24. Install the silent shaft end bolt and tighten it to 26 ft. lbs. (35 Nm).
25. Install the oil filter bracket and gasket. Tighten the front case mounting bolts to 17 ft. lbs. (23 Nm) and the oil filter bracket bolts.
26. Install the plug cap at the end of the left silent shaft.
27. Coat the relief plunger and spring in clean engine oil and install them into the oil filter bracket. Install the relief plug and gasket, tightening the plug to 34 ft. lbs. (46 Nm).
28. Install the oil screen and gasket.
29. Install the oil pan.
30. Install a new oil filter.
31. Check that the timing marks are still aligned. Push down on the center of the tension side of the belt with your finger. Correct belt deflection is approximately 1/4 in. (5-7mm). If the deflection is not correct, the tensioner must be released fully and the belt re-tensioned.

3-142 ENGINE AND ENGINE REBUILDING

32. Install the flange and the crankshaft sprocket onto the crankshaft. Make certain the flange is installed in the correct direction. If it is put on incorrectly, the belt will wear and break.
33. Install the special washer and the sprocket retaining bolt to the crankshaft. Tighten the bolt to 88 ft. lbs. (119 Nm).
34. Install the camshaft sprocket to the camshaft and tighten the bolts to 66 ft. lbs. (89 Nm).
35. Install the spacer, tensioner and tensioner spring if they were removed. Install the lower end of the spring to its position on the tensioner, then place the upper end in position at the water pump. Move the tensioner towards the water pump and temporarily tighten it in this position.
36. Install the oil pump sprocket, tightening the nut to 40 ft. lbs. (54 Nm). Remember that the oil pump drives the left silent shaft, which is still blocked by the screwdriver. Hold the sprocket by hand when tightening the nut.
37. Double check the alignment of the timing marks for the cam, crank and oil pump sprockets. If any adjustment is needed to the oil pump sprocket, remove the screwdriver blocking the silent shaft before adjusting the sprocket. Once everything is aligned, replace the screwdriver in the left side of the block and leave it there until installation of the timing belt is complete.

➡ If the screwdriver can only be inserted about 1 in. (25mm) or less, the shaft is out of position. Turn the oil pump sprocket through one full turn clockwise; the screwdriver should then go in about 2½ in. (63mm).

38. Install the timing belt onto the crankshaft sprocket, the oil pump sprocket and the cam sprocket in that order. Keep the belt taut between sprockets. If reusing an old belt, make certain that the direction of rotation arrow is properly oriented.
39. Loosen the tensioner mounting nut and bolt. The spring will move the tensioner against the belt and tension it.
40. Check the belt as it passes over the camshaft sprocket. The belt may tend to lift in the area to the left of the sprocket (roughly 7 o'clock to 12 o'clock when viewed from the pulley end.) Make certain the belt is well seated and not rubbing on any flanges or nearby surfaces.
41. At the tensioner, tighten the bolt in the slotted hole first, then tighten the nut on the pivot. If this order is not followed, the belt will become too tight and break.
42. Once again, check all the timing marks for alignment. Nothing should have changed; check anyway.
43. Remove the prytool blocking the left side silent shaft. Using a socket on the crankshaft bolt, turn the crankshaft clockwise smoothly one full turn (360°). Do NOT turn the engine backwards.
44. Loosen the tensioner nut and bolt. The spring will allow the tensioner to tighten a little bit more because of the slack picked up during engine rotation. Tighten the bolt, then the nut to 36 ft. lbs. (48 Nm).
45. Check the deflection of the belt. At the middle of the right (tension) side of the belt, deflect the belt outward with your finger, toward the timing case. The distance between the belt and the line of the cover seal should be about 9/16 in. (14mm).
46. Install the lower and upper timing belt covers. Make certain the gaskets are properly seated in the covers and that they don't come loose during installation. Tighten the bolts to 8 ft. lbs. (11 Nm).
47. Install the crankshaft pulley.
48. Install the water pump pulley, and install the belt. Adjust the belt to the correct tension.
49. Install the correct amount of motor oil.
50. Connect the negative battery cable. Start the engine and let it idle, listening for any unusual noises from the area of the timing belt. Possible causes of noise are the belt rubbing against the covers or a sprocket flange, the belt being too loose and slapping, or a tensioner binding. Do not accelerate the engine if abnormal noises are heard from the timing belt train — severe damage can result.

2.3L Diesel Engine

▶ See Figures 223 and 224

1. Elevate and safely support the vehicle.
2. Drain the engine oil.

✱✱CAUTION

Used motor oil may cause skin cancer if repeatedly left in contact with the skin for prolonged periods. Although this is unlikely unless you handle oil on a daily basis, it is wise to thoroughly wash your hands with soap and water immediately after handling used motor oil.

3. Remove the oil pan.
4. Remove all of the drive belts and any other accessories which are in the way of the timing belt covers.
5. Remove the upper timing belt cover. Note that the bolts are of different lengths; diagram their location for proper reinstallation.
6. Rotate the engine by hand (clockwise) until all the timing marks align. This positions No. 1 cylinder at TDC on the compression stroke.
7. If the timing belts are to be reused, mark the direction of rotation on the belt with chalk or a felt marker.
8. Remove the center bolt holding the crankshaft pulley and remove the pulley. Do not turn the engine out of position during this removal.
9. Remove the lower timing belt cover.
10. Slightly loosen the retaining (lock) bolts for the timing belt tensioner. Move the tensioner towards the water pump and tighten the bolts; this holds the tensioner in the slack position.

➡ If the silent shaft timing belt is not being removed, DO NOT loosen the tensioner for the silent shaft belt.

11. Remove the camshaft sprocket and the injection pump sprocket. Do not let either shaft turn out of place during the removal. Remove the timing belt without forcing or kinking it.
12. If the silent shaft belt is to be removed, the silent shafts must be locked in place. On the right side, remove the rubber plug from the silent shaft case and insert a thin prytool; the shaft should have a diameter of 0.32 in. (8mm). On the left side, remove the cover panel from the silent shaft case and insert a broad round bar or prytool, about the size of a short extension for a socket wrench.
13. Remove the silent shaft sprockets; label each sprocket so it may be reinstalled in its original position. Remove the timing belt.

ENGINE AND ENGINE REBUILDING   3-143

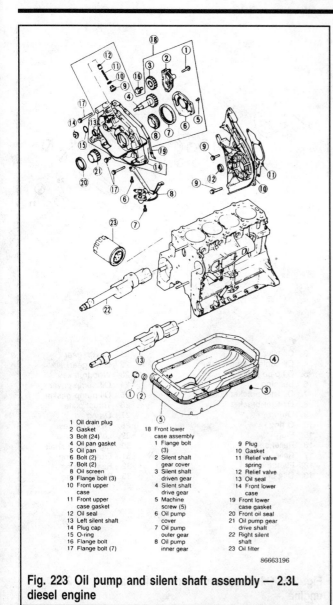

1. Oil drain plug
2. Gasket
3. Bolt (24)
4. Oil pan gasket
5. Oil pan
6. Bolt (2)
7. Bolt (2)
8. Oil screen
9. Flange bolt (3)
10. Front upper case
11. Front upper case gasket
12. Oil seal
13. Left silent shaft
14. Plug cap
15. O-ring
16. Flange bolt
17. Flange bolt (7)
18. Front lower case assembly
    1. Flange bolt (3)
    2. Silent shaft gear cover
    3. Silent shaft driven gear
    4. Silent shaft drive gear
    5. Machine screw (5)
    6. Oil pump cover
    7. Oil pump outer gear
    8. Oil pump inner gear
    9. Plug
    10. Gasket
    11. Relief valve spring
    12. Relief valve
    13. Oil seal
    14. Front lower case
    19. Front lower case gasket
    20. Front oil seal
    21. Oil pump gear drive shaft
    22. Right silent shaft
    23. Oil filter

Fig. 223 Oil pump and silent shaft assembly — 2.3L diesel engine

14. Check the belt carefully for damage, wear or deterioration. Check the teeth carefully for any cracking at the base or separation from the backing. Do not use detergents or solvents to clean the belt or the sprockets. Clean them only with dry cloths or brushes. Pay particular attention to the edges of the belt; they should not be frayed or worn.
15. Remove the upper front case. The oil seal in the case should be carefully removed and discarded. This seal must be replaced any time the case is removed.
16. Remove the lower front case.
17. Before removing the inner and outer drive gears, make alignment marks on both gears for reference during reassembly.
18. Remove the oil pump.
19. Remove the oil pump cover from the front case.
20. Remove the oil pump gears.
21. Visually check the contact surfaces of the oil pump cover and the front case for wear. If excessive wear is present, replace the components. Inspect the oil passages for clogging and clean them as needed. The silent shafts should be inspected for any wear or seizure. Closely inspect the oil seals on the front case, replacing any which have worn or damaged lips.
22. If the silent shafts are to be removed, do so by lifting each out of its housing. Handle the shafts carefully and do not damage the bearing surfaces.

To install:
23. When reinstalling the silent shafts, coat each one liberally with clean engine oil. Tighten the chamber cover bolts to 3.5 ft. lbs (5 Nm).
24. Install the oil pump gears in the same direction as before. Take great care to align the the marks.
25. Coat the gears with a light coat of clean engine oil before installing the oil pump cover.
26. Install the oil pump cover to the front case and tighten the bolts to 16 ft. lbs. (22 Nm).
27. Install the oil pump.
28. Install a seal guide tool such as MD 998393 or its equivalent to the crankshaft. Note that the tapered end should face away from the block. Coat the surface of the guide with clean engine oil. This guide will help in installing a new seal in the front case. If the seal is already in place, the guide will protect the seal during installation.
29. Install a new front case gasket, carefully aligning all the bolt holes.
30. Install the upper and lower front cases to the block and lightly tighten the bolts. When the case is properly seated, the seal guide may be removed.
31. Install the crankshaft sprocket for the silent shaft belt, the flange and the timing belt sprocket. Make certain the sprockets and flange are installed in the correct positions; if either sprocket is installed backwards, the belt may jump off or become damaged.
32. Install the crankshaft pulley.
33. Install the injection pump sprocket and flange.
34. Install the spacer on the left silent shaft with its tapered end inward or toward the oil seal. If the spacer is not placed correctly, the oil seal will be damaged.
35. Install the silent shaft sprockets and flanges, tightening the bolts to 28 ft. lbs. (38 Nm). The shafts must be held in place with the tools used during disassembly.

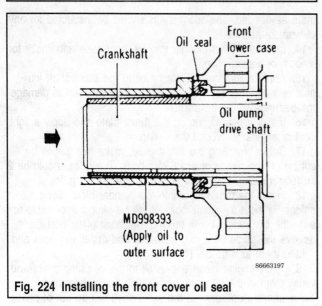

Fig. 224 Installing the front cover oil seal

ENGINE AND ENGINE REBUILDING

36. The tensioner for the silent shaft belt should still be in the slack position. Align the timing mark of the crankshaft sprocket if necessary.
37. Align the timing marks of each silent shaft sprocket.
38. Install the silent shaft timing belt (belt B), making certain that there is no slack between the crankshaft pulley and the right (upper) silent shaft.
39. The tensioner for the timing belt (belt A) should be in the slack position. Double check the alignment of all three pulleys for belt A; the timing marks should all align.
40. Install the timing belt.
41. Install the upper and lower timing belt covers.
42. Install the oil pan.
43. Install the drive belts and other accessories removed for the procedure.
44. Lower the vehicle back to the ground.
45. Fill the engine fluids and double check the engine compartment for loose wires, etc.
46. Perform any engine adjustments needed.

2.6L Engine

▶ See Figures 225, 226, 227, 228, 229 and 230

1. Remove the timing chains and sprockets, following procedures outlined previously in this Section.
2. Drain the engine oil and remove the oil filter.

✱✱CAUTION

Used motor oil may cause skin cancer if repeatedly left in contact with the skin for prolonged periods. Although this is unlikely unless you handle oil on a daily basis, it is wise to thoroughly wash your hands with soap and water immediately after handling used motor oil.

3. Disconnect the oil cooler bypass valve and the oil pipe joint at the filter bracket.
4. Remove the bracket bolt and nut. Remove the bushing and remove the oil filter bracket from the engine. Remove both O-rings from the filter bracket.
5. Remove the oil pan and gasket, following procedures outlined previously in this Section.
6. Remove the oil pick-up screen and gasket.
7. Remove the plug, relief spring and plunger from the side of the oil pump body.
8. Remove the flange bolts and remove the oil pump body. Don't lose the small pins in the case.
9. Remove the driven gear and drive gear. Don't lose the woodruff key in the end of the silent shaft.
10. Remove the oil pump cover and its gasket.
11. If the left silent shaft is to be removed, the thrust plate must be extracted first. Install two 8 mm diameter bolts into the threaded holes of the flange and turn the bolts to remove the thrust plate by forcing it outward.
12. If desired, remove the right silent shaft.

To install:

13. Visually check the contact surfaces of the oil pump cover and the front case for wear. Use a straight edge and a feeler gauge to measure the clearance between the gears and the case. Maximum allowable clearance is 0.0060 in. (0.15mm). If excessive wear is present, replace the gears or case as needed. Inspect the oil passages for clogging and clean

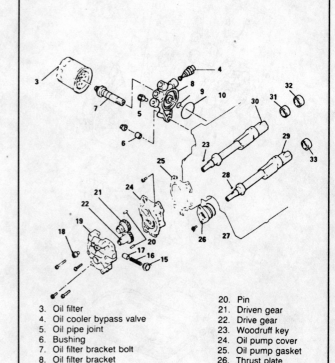

3. Oil filter
4. Oil cooler bypass valve
5. Oil pipe joint
6. Bushing
7. Oil filter bracket bolt
8. Oil filter bracket
9. O-ring
10. O-ring
15. Plug
16. Relief spring
17. Relief plunger
18. Flange bolt
19. Oil pump body
20. Pin
21. Driven gear
22. Drive gear
23. Woodruff key
24. Oil pump cover
25. Oil pump gasket
26. Thrust plate
27. O-ring
28. Woodruff key
29. Left silent shaft
30. Right silent shaft
31. Front bearing
32. Rear bearing
33. Rear bearing

Fig. 225 Oil pump and silent shaft assembly — 2.6L engine

them as needed. The silent shafts should be inspected for any of wear or seizure.

14. Insert the relief plunger into the front case and check for smooth operation.
15. Apply a light coat of engine oil to the silent shaft journals and carefully reinstall the shafts. Use care not to damage the bearings during installation.
16. Install a new O-ring on the thrust plate and apply a light coat of engine oil around the O-ring.
17. Before installing the thrust plate, make two guides by cutting off the heads of two 6 mm bolts. The bolts should be 2 in. (50mm) long. Install the bolts in the threaded holes.
18. Install the thrust plate into the cylinder block along the guides. Without the guide bolts, it will be almost impossible to align the bolt holes. While holding the thrust plate in place, remove the guide bolts one at a time and install the bolts and tighten them to 8 ft. lbs. (11 Nm).
19. Apply engine clean engine oil to the oil pump gears and install them in the oil pump body. Align the mating marks. If the marks are not aligned, the silent shaft will be out of phase and engine vibration will occur.

ENGINE AND ENGINE REBUILDING

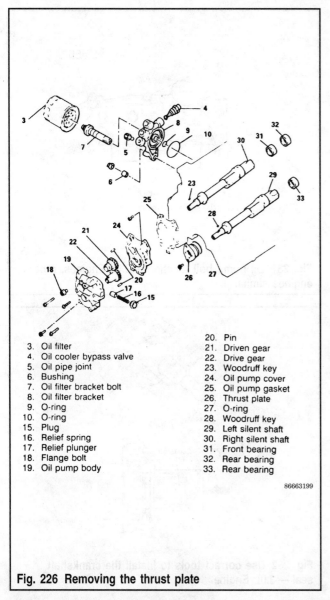

3. Oil filter
4. Oil cooler bypass valve
5. Oil pipe joint
6. Bushing
7. Oil filter bracket bolt
8. Oil filter bracket
9. O-ring
10. O-ring
15. Plug
16. Relief spring
17. Relief plunger
18. Flange bolt
19. Oil pump body
20. Pin
21. Driven gear
22. Drive gear
23. Woodruff key
24. Oil pump cover
25. Oil pump gasket
26. Thrust plate
27. O-ring
28. Woodruff key
29. Left silent shaft
30. Right silent shaft
31. Front bearing
32. Rear bearing
33. Rear bearing

Fig. 226 Removing the thrust plate

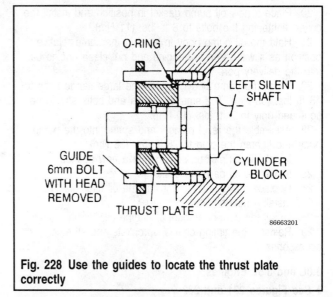

Fig. 228 Use the guides to locate the thrust plate correctly

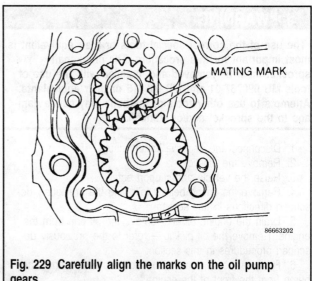

Fig. 229 Carefully align the marks on the oil pump gears

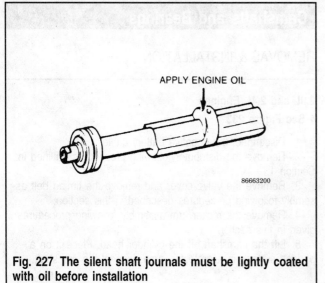

Fig. 227 The silent shaft journals must be lightly coated with oil before installation

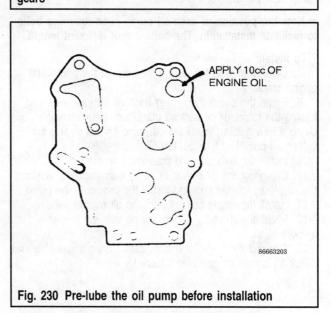

Fig. 230 Pre-lube the oil pump before installation

3-146 ENGINE AND ENGINE REBUILDING

20. Place a new oil pump gasket in position and install the cover, tightening the bolts to 8 ft. lbs. (11 Nm).
21. Hold the oil pump body in hand in the same relative position as it will be on the engine and put clean engine oil into the delivery port.
22. Install the pump body. Tighten the large flange bolts to 48 ft. lbs. (65 Nm). The smaller screws and bolts should be tightened only to 7 ft. lbs. (10 Nm).
23. Assemble the relief plunger and spring into the pump body and tighten the plug to 28 ft. lbs. (38 Nm).
24. Install the oil pick-up screen with a new gasket.
25. Install the oil pan and gasket.
26. Replace the O-rings and install the oil filter bracket.
27. Install a new oil filter.
28. Install the correct amount of engine oil.
29. Reinstall the timing chains, sprockets and all other accessories.

3.0L and 3.5L Engines
▶ See Figures 231 and 232

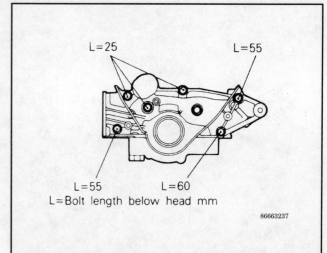

Fig. 231 Oil pump bolt location — 3.0L Engines, other engines similar

> ✳✳ WARNING
>
> The use of the correct special tools or their equivalent is most important for this procedure. The crankshaft sprocket must be removed and installed with the use of tools MB 990767-01 and MB 998715 or their equivalents. Attempts to use other tools or methods may cause damage to the sprocket and/or crankshaft.

1. Disconnect the negative battery cable.
2. Remove the dipstick.
3. Raise the vehicle and support safely.
4. Remove the timing belt by following the previously described directions in this section.
5. Drain the engine oil and remove the oil pan from the engine. Remove the oil pickup — refer to the previously described procedures in this section.
6. Remove the oil pump mounting bolts and remove the pump from the front of the engine.

➡ Note the position of each oil pump case retaining bolts to facilitate installation. The bolts are of different length.

To install:

7. Clean the gasket mounting surfaces of the pump and engine block.
8. Prime the pump by pouring fresh oil into the inlet and turning the rotors or by packing pump with petroleum jelly. Using a new gasket, install the oil pump on the engine and tighten all bolts to 11 ft. lbs. (15 Nm).
9. Install the balancer and crankshaft sprockets.
10. Clean out the oil pickup or replace as required. Replace the oil pickup gasket ring and install the pickup to the pump.
11. Install the timing belt, oil pan and all related parts.
12. Install the dipstick. Fill the engine with the proper amount of oil.
13. Connect the negative battery cable. Start the engine and check for proper oil pressure. Check for leaks.

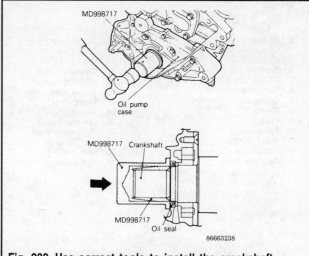

Fig. 232 Use correct tools to install the crankshaft seal — 3.0L Engine

Camshafts and Bearings

REMOVAL & INSTALLATION

2.0L and 2.4L Engines
▶ See Figure 233

1. Disconnect the negative battery cable.
2. Remove the distributor following procedures outlined in Section 1.
3. Remove the valve cover and remove the timing belt assembly following procedures described in this section.
4. Remove the rocker arm assembly following procedures given in this section.
5. Lift the camshaft off the cylinder head. Place it on a padded surface in a protected area. Inspect the camshaft.

ENGINE AND ENGINE REBUILDING

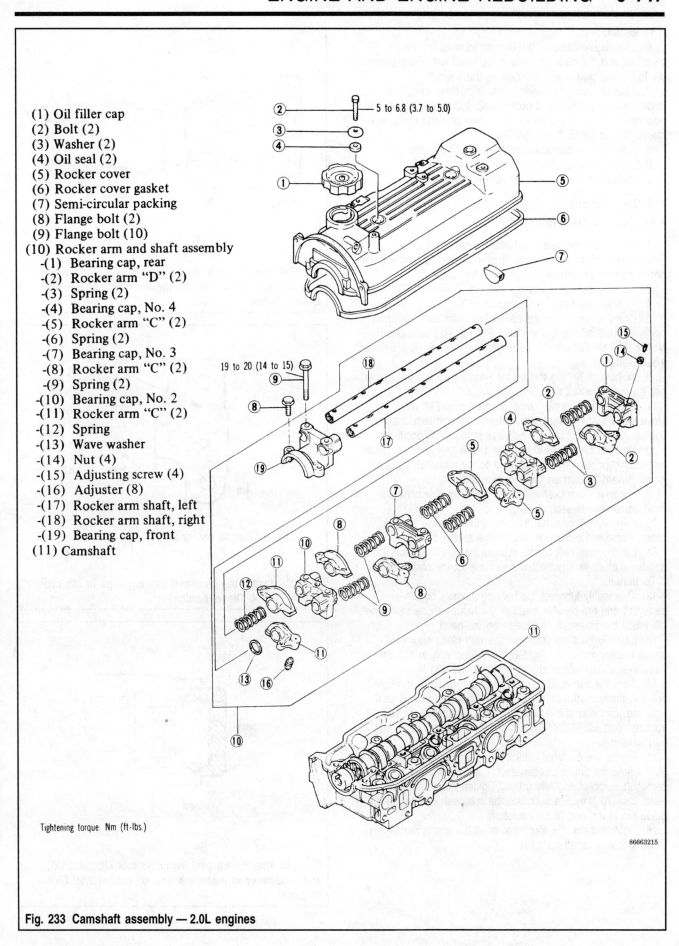

Fig. 233 Camshaft assembly — 2.0L engines

3-148 ENGINE AND ENGINE REBUILDING

To install:

6. Thoroughly lubricate the bearing journals on the camshaft and the bearing seats in the head with clean engine oil. Place the camshaft in position on the head.

7. Liberally coat the bearing caps with clean engine oil. Install the rocker shaft and bearing cap assembly. Tighten the bearing cap assembly bolts evenly in two or three steps; final torque should be 15 ft. lbs. (20 Nm).

8. Reinstall the camshaft sprocket and timing belt.

9. Install the valve cover.

10. Install the distributor. Connect the negative battery cable.

2.3L Diesel Engine

♦ See Figures 234, 235 and 236

1. Disconnect the negative battery cable.
2. Label and disconnect the breather hose and any vacuum hoses running around or across the valve cover.
3. Remove the valve cover.
4. Remove the upper timing cover.
5. Using a wrench on the center bolt, turn the crankshaft clockwise until all the timing marks line up. This positions the No.1 piston on TDC/compression. Both valves for No.1 cylinder are closed.
6. Without disturbing the engine position, loosen the center bolt on the camshaft sprocket.
7. Loosen the retaining bolts on the timing belt adjuster. Move the adjuster towards the water pump and temporarily tighten the bolts, holding the adjustor in the slack position.
8. Remove the sprocket with the timing belt attached. Note the relative position of the camshaft to the camshaft sprocket. Keep upward tension on the sprocket and belt.
9. Remove the rocker assembly, following directions outlined earlier in this section.
10. Remove the camshaft bearing caps, keeping them in order. Remove the front camshaft oil seal and discard.
11. Lift the camshaft off the cylinder head. Place it on a padded surface in a protected area. Inspect the camshaft.

To install:

12. Thoroughly lubricate the bearing journals on the camshaft and the bearing seats in the head with clean engine oil. Place the camshaft in position on the head.
13. Liberally coat the bearing caps with clean engine oil. Install the bearing caps. Tighten the bolts evenly in two or three steps; final torque should be 15 ft. lbs. (20 Nm).
14. After the camshaft bearing caps are in place and tightened to the correct amount of torque, drive the new oil seal onto the front end of the camshaft with a camshaft oil seal installer (part MD998381-01) with a hammer until the oil seal is fully seated.
15. Rotate the camshaft until it is in its original position (do NOT rotate the timing belt assembly) and reinstall the camshaft sprocket and timing belt. Tighten the center bolt to 52 ft. lbs. (70 Nm). Make certain the sprocket lines up on the guide pin in the end of the camshaft.
16. Double check the alignment of all the timing marks, particularly the camshaft sprocket.

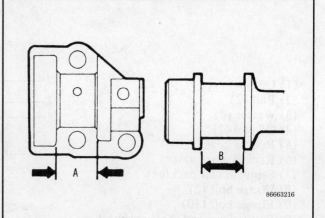

Fig. 234 Find the camshaft end-play by subtracting the measurement of A from the measurement of B — 2.3L Diesel Engines

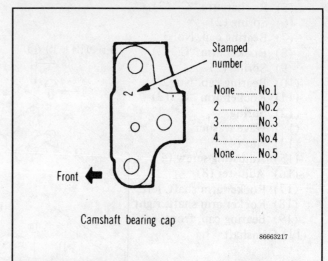

Fig. 235 Install the camshaft bearing caps in the order shown — 2.3L Diesel Engines

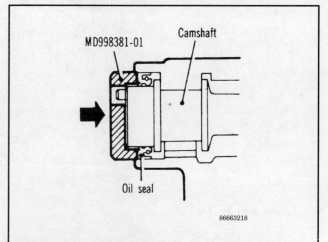

Fig. 236 Use the oil seal installing tool MD998381-01 and a hammer to insert the new oil seal — 2.3L Diesel Engines

ENGINE AND ENGINE REBUILDING 3-149

17. Loosen the bolts on the tensioner, allowing the tensioner to move against the belt under its spring tension.

✱✱WARNING

Do not assist the tensioner by pushing on it — let the spring do the work.

18. Use a wrench on the crankshaft bolt to turn the engine clockwise until the camshaft sprocket has moved two teeth out of position. Note that this is done with the belt adjustor loose. Be certain to turn the engine smoothly and only by the specified amount; the motion is necessary to correctly tension the belt.
19. Tighten the upper bolt on the tensioner, then tighten the lower bolt. Correct torque for each bolt is 18 ft. lbs. (25 Nm). If the order of tightening is not followed the tensioner will turn, allowing the belt to receive incorrect tension.
20. Turn the engine in the reverse (counterclockwise) direction until all the timing mark align.
21. Push on the belt midway between the cam sprocket and the injection pump; correct deflection is 0.16-0.20 in. (4-5mm).
22. Install the valve cover and gasket.
23. Reconnect the vacuum and breather hoses.

2.6L Engine

♦ See Figure 237

1. Disconnect the negative battery cable.
2. Remove the distributor.
3. Remove the valve cover by adhering to the previously described directions in this section.
4. Remove the rocker arm assembly following procedures given in this section.
5. Remove the camshaft sprocket along with the timing chain. Secure these two pieces together with wire or string. Set these two down on the lower chain stays.
6. Remove the rear bearing cap bolts and the cap. These are not associated with the rocker shaft assembly.
7. Lift the camshaft off the cylinder head. Place it on a padded surface in a protected area. Refer to the inspection section following for correct procedures.

To install:

8. Thoroughly lubricate the bearing journals on the camshaft and the bearing seats in the head with clean engine oil. Place the camshaft in position on the head. Be careful not to damage the camshaft journals during installation.
9. Apply a sealer to the outside diameter of the circular seal for the rear bearing and install it in the head with one side directly in contact with the rear of the camshaft. The packing will end up under the rearmost portion of the rear bearing cap.
10. Install the rocker shaft/bearing cap assembly. tighten retaining bolts in steps working from the center out to 15 ft. lbs. (20 Nm).
11. Reinstall the cam sprocket and chain to the camshaft.
12. Inspect the semi-circular seal that goes in the front of the timing chain cover. Seal the top with an adhesive such as 3M® Adhesive 8001 or equivalent.
13. Install the rocker cover.

14. Install the distributor. Connect the negative battery cable.
15. Start the engine and allow it to idle until the temperature gauge indicates normal operating temperature. Check to make sure that it is running correctly and not leaking.

3.0L Engines

♦ See Figures 238, 239, 240, 241, 242, 243 and 244

1. Disconnect the negative battery cable.
2. Follow the directions for timing belt removal, located in this section.
3. Remove the two camshaft sprockets using the special Mitsubishi tools (MB990767-01 and MIT308239).
4. Remove the cooling fan stay, the alternator bracket stay and the alternator bracket.
5. Remove the oil filler hose filter and hose.
6. Remove the rocker arm covers and the old gaskets. Discard the gaskets as new ones will be needed for installation.
7. Remove the circular packing from the rear of the camshaft bore, and the oil seal from the front of the camshaft bore. Discard both, as new ones will be needed for installation.
8. Remove the distributor adapter and O-ring. Discard the old O-ring, a new one will be needed for installation.
9. Remove the rocker arm shaft assemblies.
    a. Install the special tools (MD998443-01) to the rocker arms to hold the lash adjusters.
    b. Loosen the camshaft bearing cap bolt. Do not remove the bolts from the cap.
    c. Remove the rocker arms, shaft and bearing cap as an assembly.
10. Lift the camshaft off the cylinder head. Place it on a padded surface in a protected area. Refer to the inspection section following for correct procedures.

To install:

11. Thoroughly lubricate the bearing journals on the camshaft and the bearing seats in the head with clean engine oil. Place the camshaft in position on the head. Be careful not to damage the camshaft journals during installation.
12. Install the rocker arm shaft assembly (if the rocker arm shaft assembly was disassembled, refer to the rocker arm removal directions, outlined previously in this section).
    a. Apply a minimum amount of the specified sealant (3M® Nut Locking No. 4171 or equivalent) on the four places of the cylinder head.

➡**Be sure the sealing agent does not swell out onto the camshaft journal surface of the cylinder heads. If it swells out, immediately wipe it off before it can dry.**

    b. Install the rocker arms, shafts and bearing caps such that the arrow mark on the bearing cap faces in the same direction as the arrow mark on the cylinder heads.
    c. Tighten the bearing cap bolts to the specified torque.
    d. Remove the special tools from all of the rocker arms.
13. Install the distributor adapter with a new O-ring.
14. Install the new camshaft oil seal. Use the special Mitsubishi tool (MD998713-01 or equivalent) to install the new camshaft oil seal. Coat the rubber seal edge with a layer of clean engine oil before installation.

3-150 ENGINE AND ENGINE REBUILDING

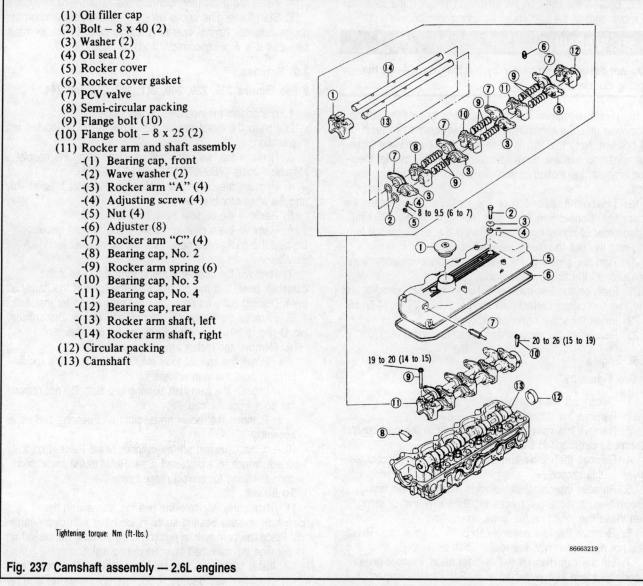

(1) Oil filler cap
(2) Bolt – 8 x 40 (2)
(3) Washer (2)
(4) Oil seal (2)
(5) Rocker cover
(6) Rocker cover gasket
(7) PCV valve
(8) Semi-circular packing
(9) Flange bolt (10)
(10) Flange bolt – 8 x 25 (2)
(11) Rocker arm and shaft assembly
 -(1) Bearing cap, front
 -(2) Wave washer (2)
 -(3) Rocker arm "A" (4)
 -(4) Adjusting screw (4)
 -(5) Nut (4)
 -(6) Adjuster (8)
 -(7) Rocker arm "C" (4)
 -(8) Bearing cap, No. 2
 -(9) Rocker arm spring (6)
 -(10) Bearing cap, No. 3
 -(11) Bearing cap, No. 4
 -(12) Bearing cap, rear
 -(13) Rocker arm shaft, left
 -(14) Rocker arm shaft, right
(12) Circular packing
(13) Camshaft

Tightening torque: Nm (ft-lbs.)

Fig. 237 Camshaft assembly — 2.6L engines

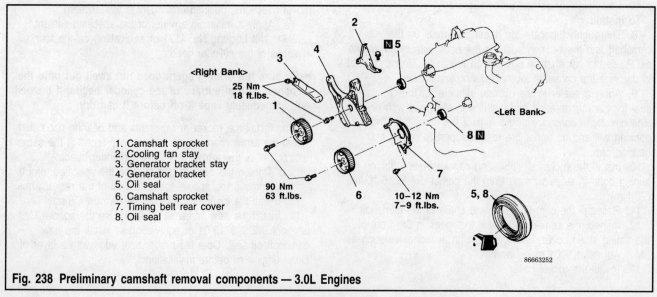

1. Camshaft sprocket
2. Cooling fan stay
3. Generator bracket stay
4. Generator bracket
5. Oil seal
6. Camshaft sprocket
7. Timing belt rear cover
8. Oil seal

Fig. 238 Preliminary camshaft removal components — 3.0L Engines

ENGINE AND ENGINE REBUILDING 3-151

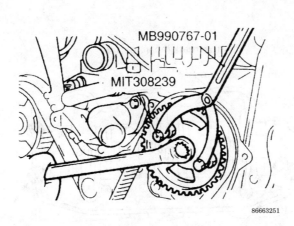

Fig. 239 Use these two special tools to remove the camshaft sprockets — 3.0L Engines

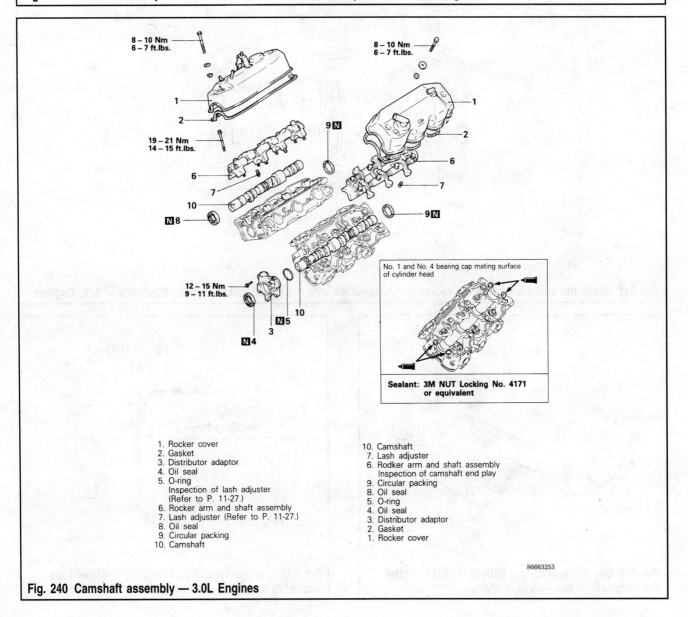

1. Rocker cover
2. Gasket
3. Distributor adaptor
4. Oil seal
5. O-ring
   Inspection of lash adjuster
   (Refer to P. 11-27.)
6. Rocker arm and shaft assembly
7. Lash adjuster (Refer to P. 11-27.)
8. Oil seal
9. Circular packing
10. Camshaft

10. Camshaft
7. Lash adjuster
6. Rodker arm and shaft assembly
   Inspection of camshaft end play
9. Circular packing
8. Oil seal
5. O-ring
4. Oil seal
3. Distributor adaptor
2. Gasket
1. Rocker cover

Fig. 240 Camshaft assembly — 3.0L Engines

3-152 ENGINE AND ENGINE REBUILDING

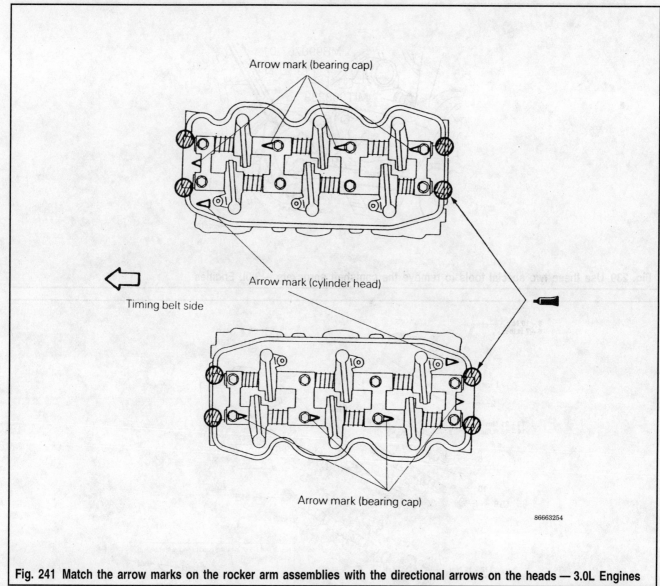

Fig. 241 Match the arrow marks on the rocker arm assemblies with the directional arrows on the heads — 3.0L Engines

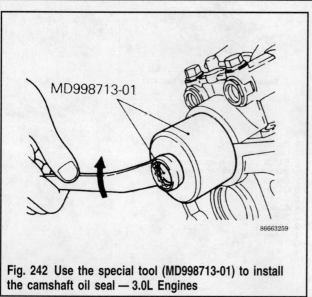

Fig. 242 Use the special tool (MD998713-01) to install the camshaft oil seal — 3.0L Engines

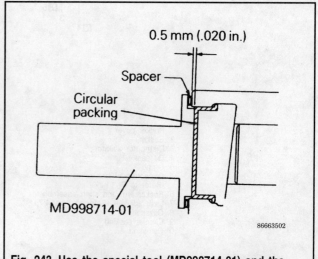

Fig. 243 Use the special tool (MD998714-01) and the washer to install the circular packing — 3.0L Engines

ENGINE AND ENGINE REBUILDING 3-153

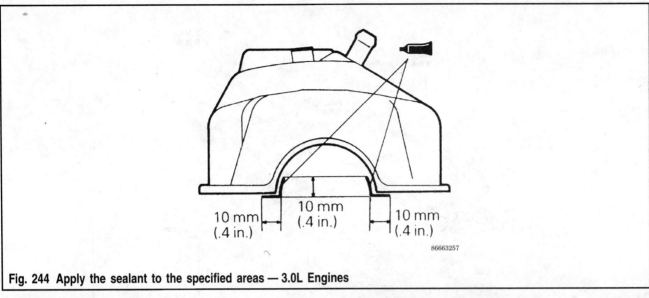

Fig. 244 Apply the sealant to the specified areas — 3.0L Engines

15. Install the rear circular packing.
   a. Install a 0.052-0.059 in. (1.3-1.5mm) thick spacer to the special tool (MD998714-01) and drive the circular packing in.

➡ Use of the tool (MD724328) spacer for transmissions is recommended.

※※CAUTION

The packing is overdriven if no spacer is fitted to the special tool.

16. Install the rocker arm cover with the old gasket, if the old gasket is not damaged (for more information — refer to the removal instructions for rocker arm covers in this section). Apply specified sealant (3M® ATD Part No. 8660 or equivalent) on the areas specified in the illustration.
17. Install the oil filler tube and oil filler filter.
18. Install the alternator bracket, the alternator bracket stay and the cooling fan stay.
19. Install the camshaft sprockets using the special tools that were used upon removal.
20. Follow the timing belt installation instructions and install all equipment taken off the engine during removal.

3.5L DOHC Engines

▶ See Figures 245, 246, 247, 248, 249, 250 and 251

1. Follow the directions for intake manifold removal and installation outlined previously in this section.
2. Remove the timing belts by following the procedures described previously in this section.
3. Remove the center covers on the rocker arm covers.
4. Remove the three (3) ignition coils and their respective spark plug wires. Label the wires for ease of installation.
5. Remove the rocker arm covers from the heads.
6. Remove the four (4) camshaft sprockets. Use a wrench to hold the hexagonal part of the camshaft to prevent the crankshaft from turning, and then loosen the camshaft sprocket bolt.

※※CAUTION

Do not hold the camshaft sprocket with a tool, or a damaged sprocket could be the result.

7. Remove the camshaft bearing caps in this order: front, rear, No. 2, No. 4, No. 3.
8. Lift the camshafts off the cylinder head. Place them on a padded surface in a protected area. Refer to the inspection section following for correct procedures.
9. Remove the rocker arms and lash adjusters, for service of these two parts refer to the procedures described previously in this section.

To install:

10. Install the rocker arms on the valve stems and the lash adjusters.
11. Thoroughly lubricate the bearing journals on the camshaft and the bearing seats in the head with clean engine oil. Place the camshaft carefully in position on the head. Be careful not to damage the camshaft journals during installation.
   a. Make sure the crankshaft is still at top dead center on No. 1 cylinder.
   b. Check to make sure that the rocker arm is installed correctly on the lash adjuster and valve.
   c. Install the camshaft while noting the identification mark (stamped on the hexagonal section).
   d. Install the camshafts with their dowel pins positioned as shown in the illustration.
12. Loosely install the bearing caps onto the head. Install the bearing caps according to the identification mark and cap number. No. 2, 3 and 4 bearing caps have the front mark stamped on them. Install these caps with the front mark pointing in the same direction as that on the cylinder head. Each cap will have either an "I", for intake, or an "E", for exhaust.
13. Gradually tighten the bearing caps in two or three steps. In the final step, tighten to the specified torque.
14. Install the camshaft oil seals into the front camshaft bearing cap. Use the special tool (MD998761-01) to correctly

3-154 ENGINE AND ENGINE REBUILDING

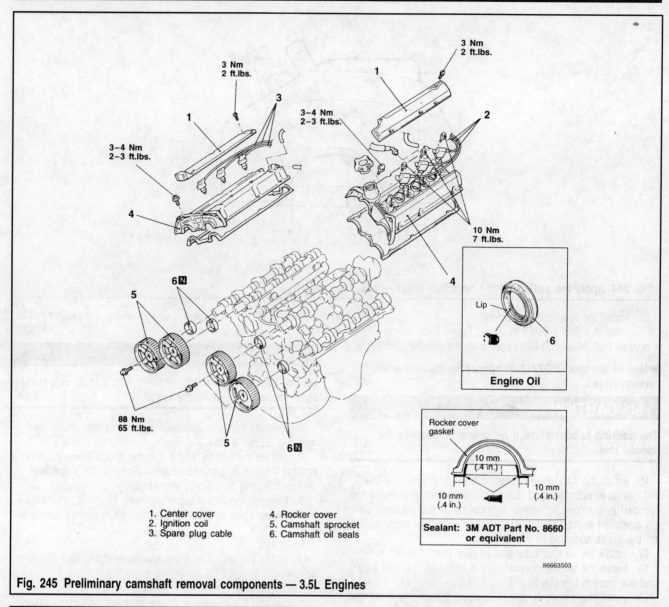

Fig. 245 Preliminary camshaft removal components — 3.5L Engines

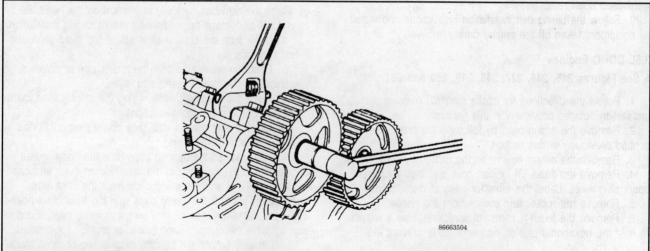

Fig. 246 Use these two wrenches to remove the camshaft sprockets, make sure to grip the camshaft hexagonal section — 3.5L Engines

ENGINE AND ENGINE REBUILDING

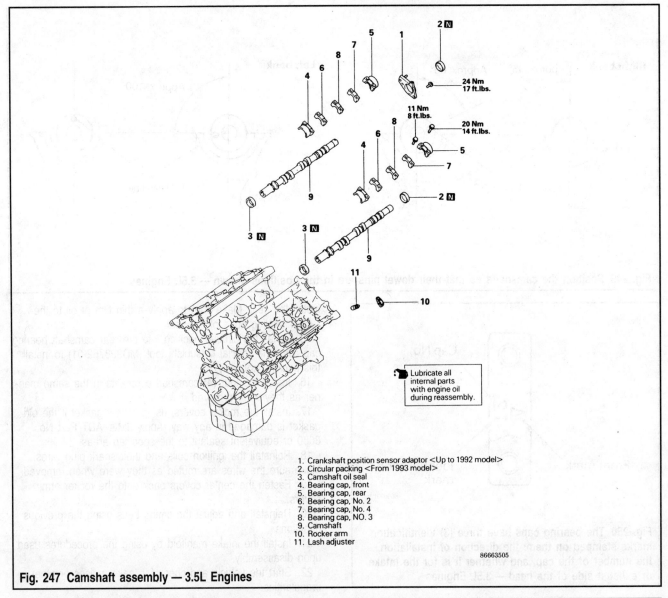

1. Crankshaft position sensor adaptor <Up to 1992 model>
2. Circular packing <From 1993 model>
3. Camshaft oil seal
4. Bearing cap, front
5. Bearing cap, rear
6. Bearing cap, No. 2
7. Bearing cap, No. 4
8. Bearing cap, NO. 3
9. Camshaft
10. Rocker arm
11. Lash adjuster

Fig. 247 Camshaft assembly — 3.5L Engines

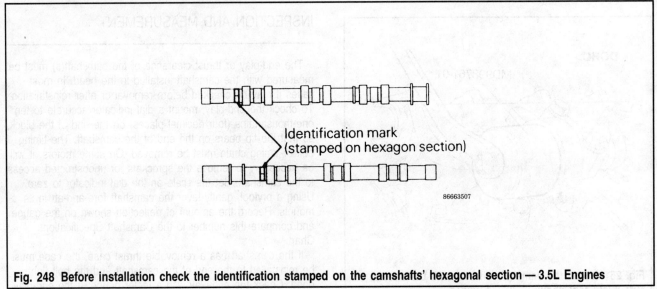

Fig. 248 Before installation check the identification stamped on the camshafts' hexagonal section — 3.5L Engines

3-156 ENGINE AND ENGINE REBUILDING

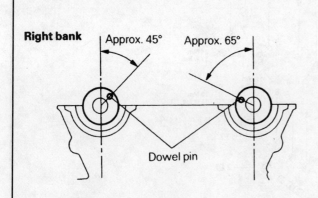
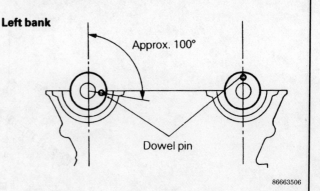

Fig. 249 Position the camshafts so that their dowel pins are in the positions shown — 3.5L Engines

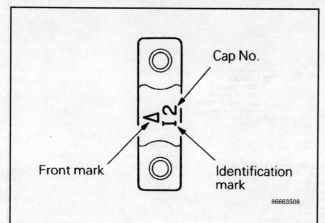

Fig. 250 The bearing caps have three (3) identification marks stamped on them: the direction of installation, the number of the cap, and whether it is for the intake or exhaust side of the head — 3.5L Engines

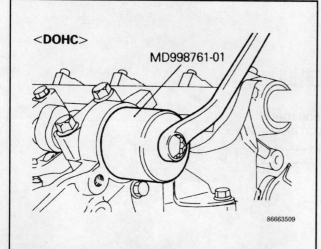

Fig. 251 Install the camshaft oil seal with the special Mitsubishi tool shown — 3.5L Engines

install the seal. Make sure to apply a thin film of oil to the rubber sealing edge of the seal.

15. Install the circular packing into the rear camshaft bearing cap. Use the special Mitsubishi tool (MD998762-01) to install this packing correctly.
16. Install the four (4) camshaft sprockets in the same manner as they were removed.
17. Install the rocker covers, using a new gasket if the old gasket is damaged in any way. Apply 3M® ADT Part No. 8660 or equivalent sealant to the specified areas.
18. Reinstall the ignition coils and their spark plug wires. Make sure the wires are routed as they were when removed.
19. Fasten the center covers back onto the rocker arm covers.
20. Reinstall and adjust the timing belts using the previous directions.
21. Install the intake manifold by using the procedures used upon disassembly.
22. Start the vehicle and check for leaks, noises or other irregularities.

INSPECTION AND MEASUREMENT

The end-play or thrust clearance of the camshaft(s) must be measured with the camshaft installed in the head. In most cases it may be checked before removal or after reinstallation. To check the end-play, mount a dial indicator accurate to ten one-thousandths (four decimal places) on the end of the block, so that the tip bears on the end of the camshaft. The timing belt or timing chain must be removed. On some motors, it will be necessary to remove the sprockets for unobstructed access to the camshaft. Set the scale on the dial indicator to zero. Using a prytool, gently lever the camshaft fore-and-aft in its mounts. Record the amount of deflection shown on the gauge and compare this number to the Camshaft Specifications Chart.

If the camshaft has a removable thrust case, the case must be reinstalled to the end of the camshaft and the bolt tightened. Check the end-play with a feeler gauge. If the end-play is excessive, replace the thrust case with a new one and

ENGINE AND ENGINE REBUILDING 3-157

recheck the clearance. If the end-play is still excessive, replace the camshaft.

Excessive end-play may indicate either a worn camshaft or a worn head; the worn camshaft is most likely and much cheaper to replace. Chances are good that if the camshaft is worn in this dimension (axial), substantial wear will show up in other measurements.

Using a micrometer or Vernier caliper, measure the diameter of all the journals and the height of all the lobes. Record the readings and compare them to the Camshaft Specifications Chart in the beginning of this Section. Any measurement beyond the stated limits indicates wear and the camshaft must be replaced.

Lobe wear is generally accompanied by scoring or visible metal damage on the lobes. Overhead camshaft engines are very sensitive to proper lubrication with clean, fresh oil.

If a new camshaft is required, order new rockers to accompany it so that there are two new surfaces in contact.

The clearance between the camshaft and its journals (bearings) must also be measured. The amount of space between the camshaft and its bearings is the oil clearance; less space means less oil and more wear. Excessive play will allow the camshaft to hammer the bearing and cause wear. Clean the camshaft, the journals and the bearing caps of any remaining oil and place the camshaft in position on the head. Lay a piece of compressible gauging material (Plastigage® or similar) on top of each journal on the cam.

Install the bearing caps in their correct order and install the rocker assembly. Tighten the bolts in three passes to the correct torque.

✸✸WARNING

Do not turn the camshaft with the gauging material installed.

Remove the rocker assembly and measure the gauging material at its widest point by comparing it to the scale provided with the package. Compare these measurements to the Camshaft Specifications Chart. Any measurement beyond specifications indicates wear. If you have already measured the cam (or replaced it) and determined it to be usable, excess bearing clearance indicates the need for a new head due to wear of the journals.

Remove the camshaft from the head and remove all traces of the gauging material. Check carefully for any small pieces clinging to contact faces.

After a careful cleaning, coat the camshaft and bearing surfaces with clean engine oil and continue reassembly.

Pistons and Connecting Rods

➡This procedure requires removal of the cylinder head and oil pan. It is much easier to perform this work with the engine removed from the vehicle and mounted on a stand. These procedures require certain hand tools which may not be in your tool box. A cylinder ridge reamer, a numbered punch set, piston ring expander, snapring tools and piston installation tool (ring compressor) are all necessary for correct piston and rod repair.

REMOVAL

▶ See Figures 252, 253, 254, 255, 256, 257, 258, 259 and 260

1. Remove the cylinder head, following correct procedures listed earlier in this section.
2. Remove the oil pan, following correct procedures listed earlier in this section.
3. The connecting rods are marked to indicate which surface faces front, but the bearing caps should be matchmarked with numbers (front to rear) before disassembly. Use a marking punch and a small hammer; install the number over the seam so that each piece will be re-used in its original location.
4. Remove the connecting rod cap bolts, pull the caps off the rods, and place them on a bench in order.
5. Inspect the upper portions of the cylinder (near the head) for a ridge formed by ring wear. If there is a ridge, it must be removed by first shifting the piston down in the cylinder and then covering the piston top completely with a clean rag. Use a ridge reamer to remove metal at the lip until the

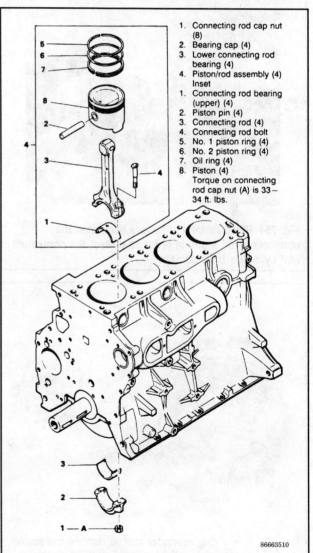

Fig. 252 Exploded view of pistons, connecting rods and related parts — 2.6L engine

3-158 ENGINE AND ENGINE REBUILDING

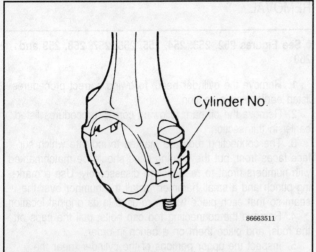

Fig. 253 Mark the side of the piston rod and rod cap with the cylinder number

Fig. 254 Place lengths of rubber hose over the connecting rod studs in order to protect the crankshaft and cylinders from damage

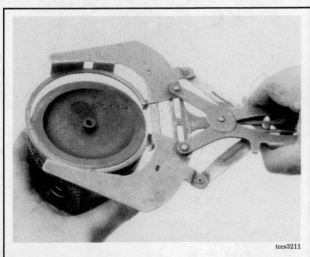

Fig. 256 Use a ring expander tool to remove the piston rings

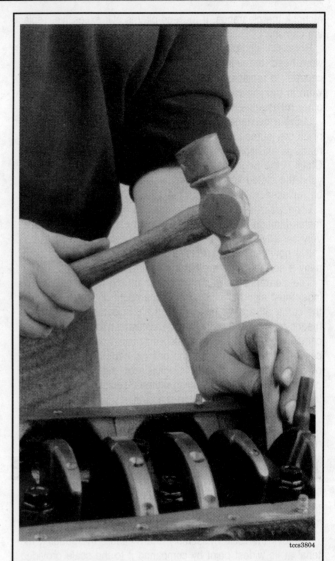

Fig. 255 Carefully tap the piston out of the bore using a wooden dowel

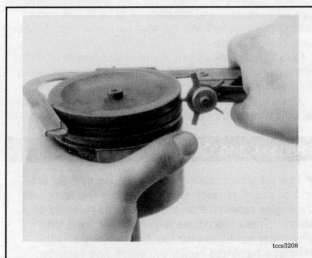

Fig. 257 Clean the piston grooves using a ring groove cleaner

ENGINE AND ENGINE REBUILDING 3-159

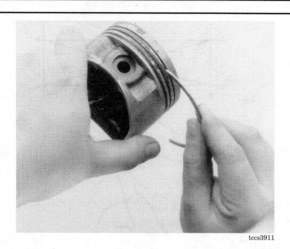

Fig. 258 A piece of an old ring can be used to clean out the grooves, BUT be careful — the ring is very sharp

Fig. 259 Remove the glazing on the cylinder walls with a flexible drill hone

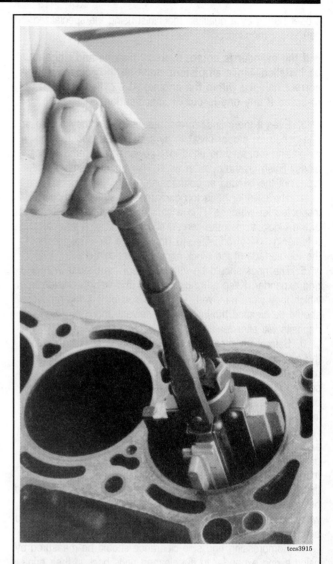

Fig. 260 Remove the ridge from the cylinder bore using a ridge cutter

cylinder is smooth. If this is not done, the rings will be damaged during removal of the piston.

6. Once the ridges have been removed, the pistons and rods may be pushed upward and out of the cylinders. Place pieces of rubber tubing over the rod bolts to protect the cylinder walls. Use a piece of wood or a hammer handle under the piston to tap it upward. If you're working under an engine that's still installed in the vehicle with the crankshaft still in position, turn the crankshaft until the crankpin for each cylinder is in a convenient position. Be careful not to subject the piston and/or rod to heavy impact and do not allow the piston rod to damage the cylinder wall on the way out. The slightest nick in the metal can cause problems after reassembly.

7. Clean the pistons, rings and rods in parts solvent with a bristle brush. Do not use a wire brush, even to remove heavy carbon. The metal may be damaged. Use a piston groove cleaner to clean the lands (grooves) in the piston.

INSPECTION

▶ See Figures 261, 262, 263 and 264

1. Measure the bore of the cylinder at three levels and in two dimensions (fore-and-aft and side-to-side). That's six measurements for each cylinder. By comparing the three vertical readings, the taper of the cylinder can be determined and by comparing the front-rear and left-right readings the out-of-round can be determined. The block should be measured: at the level of the top piston ring at the top of piston travel; in the center of the cylinder; and at the bottom. Compare your readings with the specifications in the chart.

2. If the cylinder bore is within specifications for taper and out-of-round, and the wall is not scored or scuffed, it need not be bored. If not, it should be bored oversize as necessary to ensure elimination of out-of-round and taper. Under these circumstances, the block should be taken to a machine shop for

proper boring by a qualified machinist using the specialized equipment required.

→ If the cylinder is bored, oversize pistons and rings must be installed. Since all pistons must be the same size (for correct balance within the engine) ALL cylinders must be re-bored if any one is out of specification.

3. Even if the cylinders need not be bored, they should be fine honed for proper break-in by a qualified machine shop. A de-glazing tool may be used in a power drill to remove the glossy finish on the cylinder walls. Use only the smooth stone type, not the beaded or bottle-brush type.

4. The cylinder head top deck (gasket surface) should be inspected for warpage. Run a straightedge along all four edges of the block, across the center, and diagonally. If you can pass a feeler gauge of 0.004 in. (0.1mm) under the straightedge, the top surface of the block should be machined.

5. The rings should be removed from the pistons with a ring expander. Keep all rings in order and with the piston from which they were removed. The rings and piston ring grooves should be cleaned thoroughly with solvent and a brush as deposits will alter readings of ring wear.

6. Before any measurements are begun, visually examine the piston for any signs of cracks, particularly in the skirt area or scratches in the metal. Anything other than light surface scoring disqualifies the piston from further use. The metal will become unevenly heated and the piston may break apart during use.

7. Piston diameter should be measured at the skirt, at right angles to the piston pin. Compare either with specified piston diameter or subtract the diameter from the cylinder bore dimension to get clearance, depending upon the information in the specifications. If clearance is excessive, the piston should be replaced. If a new piston still does not produce piston-to-wall clearance within specifications, use an oversize piston and bore out the cylinder accordingly.

8. Compression ring side clearance should be measured by using a ring expander to put cleaned rings back in their original positions on the pistons. Measure side clearance on one side by attempting to slide a feeler gauge of the thickness specified between the ring and the edge of the ring groove. If

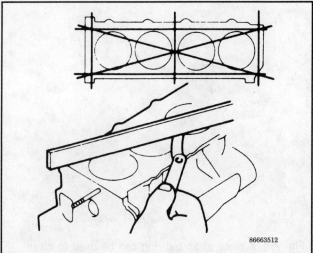

Fig. 262 Inspect the top deck of the block in the directions shown

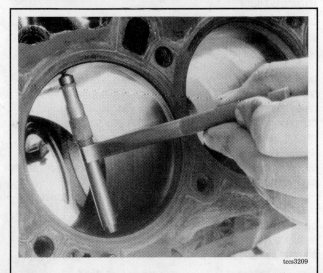

Fig. 261 Measure the inside bore of the cylinders

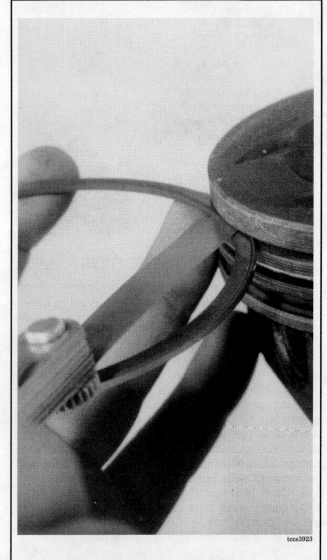

Fig. 263 Measuring piston ring side clearance

ENGINE AND ENGINE REBUILDING 3-161

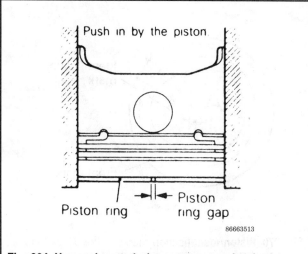

Fig. 264 Use an inverted piston to seat a ring in the cylinder — measure the end-gap with a feeler gauge

the gauge will not pass into the groove, the ring may be re-used. If the gauge will pass, but a gauge of slightly greater thickness representing the wear limit will not, the piston may be re-used, but new rings must be installed.

9. Ring end-gap must be measured for all three rings in the cylinder by using a piston top (upside down) to press the ring squarely into the top of the cylinder. The rings must be at least 0.59 in. (15mm) from the bottom of the bore. Use a feeler gauge to measure the end-gap and compare it with specifications. If cylinder bore wear is very slight, you may use new rings to bring the end-gap to specification without boring the cylinder. Measure the gap with the ring located near the minimum dimension at the bottom of the cylinder, not nearer the top where wear is greatest.

10. The connecting rods must be free from wear, cracking and bending. Visually examine the rod, particularly at its upper and lower ends. Look for any sign of metal stretching or wear. The piston pin should fit cleanly and tightly through the upper end, allowing no side-play or wobble. The bottom end should also be an exact half-circle, with no deformity of shape. The bolts must be firmly mounted and parallel. The rods may be taken to a machine shop for exact measurement of twist or bend.

INSTALLATION

▶ See Figures 265, 266, 267, 268, 269, 270 and 271

1. Remember that if you are installing an oversize piston, you must also use new rings that are also of the correct oversize.

2. Install the rings on the piston, lowest ring first. Generally the side rails for the oil control ring can be installed by hand with care. The other rings require the use of the ring expander. There is a high risk of ring breakage or piston damage if the rings are installed without the expander. The correct spacing of the ring end-gaps is critical to oil control. No two gaps should align; they should be evenly spaced around the piston with the gap in the oil ring expander facing the front of the piston (aligned with the mark on the top of the piston).

Once the rings are installed, the pistons must be handled carefully and protected from dirt and impact.

3. Install the number two compression ring next and then the top compression ring using a ring expander. Note that these rings have the same thickness but different cross-sections; make sure positioning is correct. Make sure all markings face upward and that the gaps are all staggered. Gaps must also not be in line with either the piston pin or thrust faces of the piston.

4. All the pistons, rods and caps must be reinstalled in the correct cylinder. Make certain that all labels and stamped numbers are present and legible. Double check the piston rings; make certain that the ring gaps DO NOT line up, but are evenly spaced around the piston at about 120° intervals. Double check the bearing insert at the bottom of the rod for proper mounting. Reinstall the protective rubber hose pieces on the bolts.

5. Liberally coat the cylinder walls and the crankshaft journals with clean, fresh engine oil. Also apply oil to the bearing surfaces on the connecting rod and the cap.

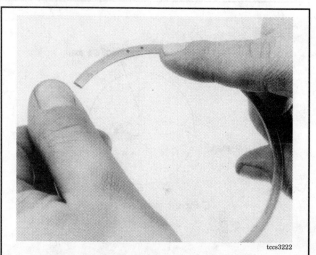

Fig. 265 Most rings are marked to show which side should face upward

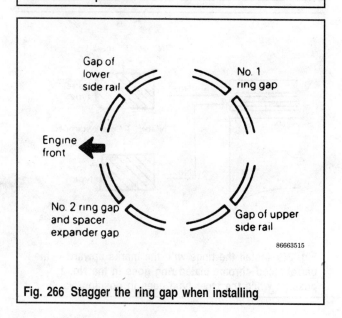

Fig. 266 Stagger the ring gap when installing

3-162 ENGINE AND ENGINE REBUILDING

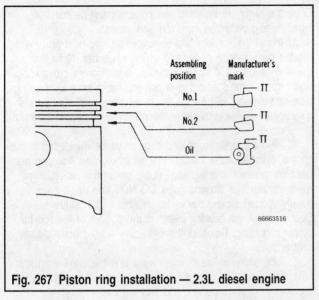

Fig. 267 Piston ring installation — 2.3L diesel engine

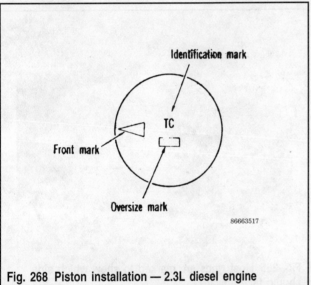

Fig. 268 Piston installation — 2.3L diesel engine

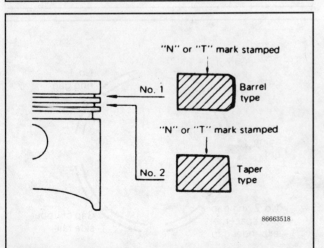

Fig. 269 Install the rings with the marks upward — the barrel faced chrome plated ring goes in the No. 1 position while the taper type goes in position No. 2

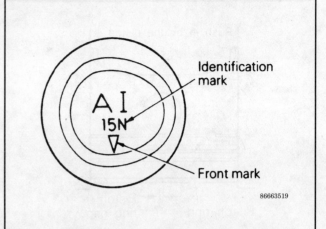

Fig. 270 Piston identification marks — the V6 pistons use L and R depending on which bank contained the piston

Fig. 271 Installing the piston into the block using a ring compressor and the handle of a hammer

6. Identify the front mark on each piston and rod and position the piston loosely in its cylinder with the marks facing the front (pulley end) of the motor.

WARNING

Failure to observe the "Front" marking and its correct placement can lead to sudden engine failure.

7. Install the ring compressor (piston installation tool) around one piston and tighten it gently until the rings are compressed almost completely.
8. Gently push down on the piston top with a wooden hammer handle or similar soft-faced tool and drive the piston into

ENGINE AND ENGINE REBUILDING

the cylinder bore. Once all three rings are within the bore, the piston will move with some ease.

※※WARNING

If any resistance or binding is encountered during the installation, DO NOT apply force. Tighten or adjust the ring compressor and/or reposition the piston. Brute force will break the ring(s) or damage the piston.

9. From underneath, pull the connecting rod into place on the crankshaft. Remove the rubber hoses from the bolts. Check the rod cap to confirm that the bearing is present and correctly mounted, then install the rod cap (observing the correct number and position) and its nuts. Leaving the nuts finger-tight will make installation of the remaining pistons and rods easier.

10. Assemble the remaining pistons in the same fashion, repeating steps 7, 8 and 9.

11. With all the pistons installed and the bearing caps secured finger-tight, the retaining nuts may be tightened to their final setting. Refer to the torque specifications chart for the correct torque for the engine in your vehicle. For each pair of nuts, make three passes alternating between the two nuts on any given rod cap. The three tightening steps should each be about one third of the final torque. The intent is to draw each cap up to the crank straight and under even pressure at the nuts.

12. Turn the crankshaft through several clockwise rotations, making sure everything moves smoothly and there is no binding. With the piston rods connected, the crank may be stiff to turn. Try to turn it in a smooth continuous motion so that any binding or stiff spots may be felt.

13. Reinstall the oil pan. Even if the engine is to remain apart for other repairs, install the oil pan to protect the bottom end and tighten the bolts to the correct specification; this eliminates one easily overlooked mistake during future reassembly.

14. If the engine is to remain apart for other repairs, pack the cylinders with crumpled newspaper or clean rags (to keep out dust and grit) and cover the top of the motor with a large rag. If the engine is on a stand, the whole block can be protected with a large plastic trash bag.

15. If no further work is to be performed, continue reassembly by installing the head, timing belt, etc.

PISTON PIN (WRIST PIN) REPLACEMENT

♦ See Figures 272, 273, 274, 275 and 276

➡The piston and pin are a matched set and must be kept together. Label everything and store parts in identified containers. The use of the correct special tools or their equivalent is very important for this procedure. You will need tool set MD 998184-01 for removing and installing the piston pin. You will also need access to a hydraulic press, capable of delivering up to 4000 lbs. (17,792 N) of force.

1. Remove the pistons from the engine and remove the rings from the pistons.

2. Set up the tool set with the pin pusher on the press bed. Make certain the press plates are securely fastened to the press frame.

3. Fit the piston and rod assembly onto the body of the special tool, with the front marks on the rod and piston facing up. The numbers or letters on the side of the rod serve as the front mark for the rod; the piston usually has an arrowhead pointing to the front. The V6 pistons are marked "L" and "R"; if the piston came from the left (firewall) side, the "L" designates the front. If from the right bank, use the "R" as the front marker.

4. Position the piston so that the lip of the tool insert fits between the connecting rod boss and the inside of the piston. The boss should contact as much of the tool insert as possible to support it.

5. Adjust the support at the rear of the connecting rod until the rod is horizontal (level) to the press bed. Misalignment of the pin and receiving tube may occur if the rod is crooked.

6. Position the pin pusher on the pin and use the press to drive the pin out. As the pin is removed, it must pass through the receiving tube. Adjust the alignment as necessary.

7. Remove the rod and piston from the press. Check the pin for obvious wear or damage. A quick check of piston pin wear is to push the pin into the piston (without the rod) with your thumb. The pin should fit with slight resistance. If the pin is too loose or has any play, either the pin or piston is worn; both should be replaced.

8. To reinstall, set up the special press tools with the stop plug threaded about halfway into the bottom of the receiving tube.

9. Select the largest piston pin guide that will pass through the piston and rod. Install the spring, spacer and guide into the receiving tube.

10. With the connecting rod removed from the piston, insert the piston pin into the piston so that it is centered in the piston. Accurately measure the amount of protrusion (or recess) between the pin and piston and make certain the measurement is the same on the opposite side. Record the measurement.

11. Place the connecting rod and piston over the special tool on the press with the front marks facing up. The spring loaded

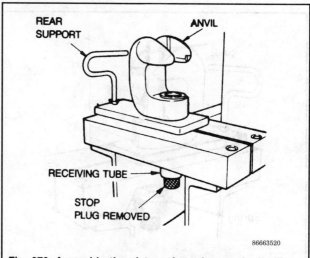

Fig. 272 Assemble the piston pin tools onto hydraulic press

ENGINE AND ENGINE REBUILDING

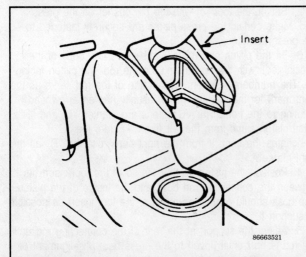

Fig. 273 The insert will support the rod during removal and installation

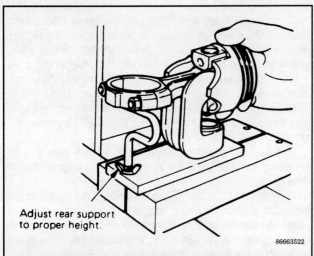

Fig. 274 Adjust the rear support so the rod is horizontal and parallel to the press bed

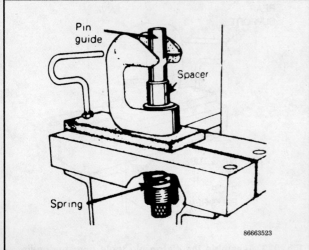

Fig. 275 Install the spring, spacer and guide into the receiving tube

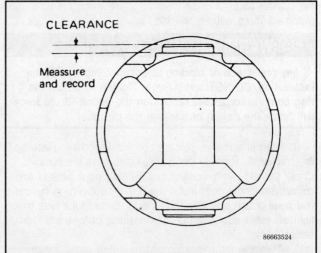

Fig. 276 The clearance measurement is important — the pin must be centered after installation

piston guide will pass through the piston and rod and align them.

12. Coat the piston pin with fresh motor oil and install it into the piston. Place the pin pusher on the piston pin and use the press to insert it through the connecting rod and opposite side of the piston.

13. Continue pressing until the pin projection (or recess) is equal to that measured in Step 10. apply pressure to the pin and adjust the stop plug until it comes into contact with the spacer.

14. Remove the piston assembly from the press and check the piston pin for centering. If it is not centered, adjust the stop plug up or down to obtain the correct position. The pin stop is now set for any remaining pistons.

✳✳WARNING

Installation pressure on the press ranges from about 1100 lbs. (4,893 N) to about 3600 lbs. (16,013 N). If the pin cannot be installed within this range, replace the rod and/or pin. Pressure extremes will break the piston or rod.

ROD BEARING REPLACEMENT

▸ See Figures 277, 278, 279 and 280

1. Crankshaft crankpins (the part of the crank on which the bearing rides) should be inspected for clogged oil holes and scoring. If scoring is evident, the crankpins should be machined undersize and the proper undersize (thicker) bearings installed in all the rods.

2. If the crankpins and rod bearing shells appear to show minimal wear and normal wear patterns, they still must be checked for proper fit with Plastigage®. Clean all the surfaces of oil and then cut plastic measuring media (Plastigage® or similar) inserts to a length that will fit the width of the bearing. Apply the insert to an area of the crankpin away from an oil hole and install the rod cap. Tighten the bolts to specification. Make sure you don't turn the crankshaft.

3. Remove the connecting rod bolts and then remove the caps. Read the width of the Plastigage® at its widest part by

ENGINE AND ENGINE REBUILDING 3-165

Fig. 277 Apply a strip of gauging material to the bearing journal, then install and tighten the cap

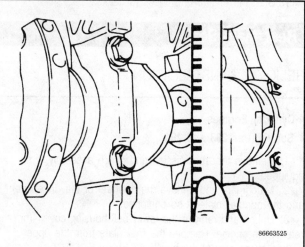

Fig. 278 Reading the width of the mark left by Plastigage®

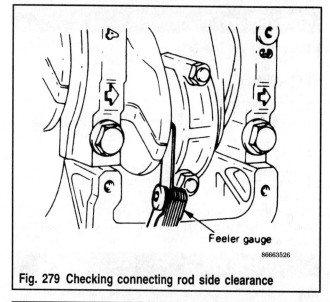

Fig. 279 Checking connecting rod side clearance

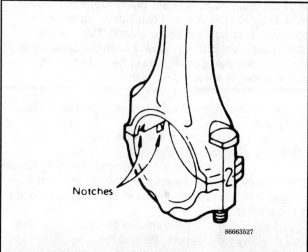

Fig. 280 Notches in the connecting rod and connecting rod cap should be on the same side

comparing it with the scale on the package to get actual bearing clearance. If the bearing clearance is within specification, the bearings may be reused.

4. You must also measure side clearance between the connecting rod side surface and the crank with the rod assembled and tightened to the specified torque. Excessive wear here indicates the rod and cap must be replaced to restore clearance.

5. When the bearing shells are replaced, make sure bearing backs and inner surfaces of the rod and cap are dry and that bearings and crankpins are well lubricated with clean engine oil. It is important to insure that all bearings are installed with the notches (in the bearings, rod and cap) aligned; this will keep the bearing from spinning out of position. Notches in the rod and cap should be on the same side. This is a way to make sure rod and cap are properly assembled.

6. tighten the rod bolts to specification, going back and forth in several steps.

ENGINE AND ENGINE REBUILDING

Core (Freeze) Plugs

♦ See Figures 281 and 282

REPLACEMENT

Core plugs need replacement only if they are found to be leaking, are excessively rusty, have popped due to freezing or, if the engine is being overhauled.

If the plugs are accessible with the engine in the truck, they can be removed as-is. If not, the engine will have to be removed.

1. If necessary, remove the engine and mount it on a work stand. If the engine is being left in the truck, drain the engine coolant and engine oil

※※CAUTION

When draining the coolant, keep in mind that cats and dogs are attracted by ethylene glycol antifreeze, and are quite likely to drink any that is left in an uncovered container or in puddles on the ground. This will prove fatal in sufficient quantity. Always drain the coolant into a sealable container. Coolant should be reused unless it is contaminated or several years old.

2. Remove anything blocking access to the plug or plugs to be replaced.
3. Drill or center-punch a hole in the plug. For large plugs, drill a ½ in. hole; for small plugs, drill a ¼ in. hole.
4. For large plugs, using a slide-hammer, thread a machine screw adapter or insert 2-jawed puller adapter into the hole in the plug. Pull the plug from the block; for small plugs, pry the plug out with a pin punch.
5. Thoroughly clean the opening in the block, using steel wool or emery paper to polish the hole rim.
6. Coat the outer diameter of the new plug with sealer and place it in the hole.

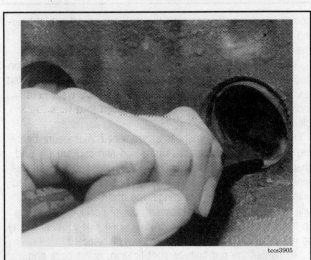

Fig. 281 Using a punch and hammer, the freeze plug can be loosened in the block

Fig. 282 A freeze plug removed from the block

7. For cup-type core plugs: these plugs are installed with the flanged end outward. The maximum diameter of this type of plug is located at the outer edge of the flange. Carefully and evenly, drive the new plug into place with a large socket-shaped tool, which fits snugly into the freeze plug.
8. For expansion-type plugs: these plugs are installed with the flanged end inward. The maximum diameter of this type of plug is located at the base of the flange. It is imperative that the correct type of installation tool is used with this type of plug. Under no circumstances is this type of plug to be driven in using a tool that contacts the crowned portion of the plug. Driving in this plug incorrectly will cause the plug to expand prior to installation. When installed, the trailing (maximum) diameter of the plug MUST be below the chamfered edge of the bore to create an effective seal. If the core plug replacing tool has a depth seating surface, do not seat the tool against a non-machined (casting) surface.
9. Install any removed parts and, if necessary, install the engine in the truck.
10. Refill the cooling system and crankcase.
11. Start the engine and check for leaks.

Rear Main Seal

REMOVAL & INSTALLATION

4-Cylinder Engines

♦ See Figures 283 and 284

1. Remove the transmission and clutch assembly, if so equipped.
2. Matchmark and remove the flywheel or the driveplate and the adapter plate, if so equipped.
3. Unbolt and remove the lower bell housing cover from the rear of the engine. Remove the rear plate from the upper portion of the rear of the block.
4. The lower surface of the oil seal casing presses against the oil pan gasket (or seal) at the rear of the pan. On engines

ENGINE AND ENGINE REBUILDING

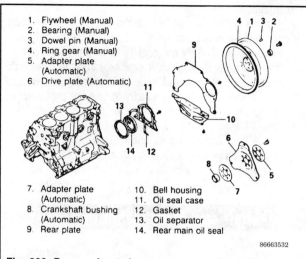

Fig. 283 Rear main seal components — 4-cylinder engines

1. Flywheel (Manual)
2. Bearing (Manual)
3. Dowel pin (Manual)
4. Ring gear (Manual)
5. Adapter plate (Automatic)
6. Drive plate (Automatic)
7. Adapter plate (Automatic)
8. Crankshaft bushing (Automatic)
9. Rear plate
10. Bell housing
11. Oil seal case
12. Gasket
13. Oil separator
14. Rear main oil seal

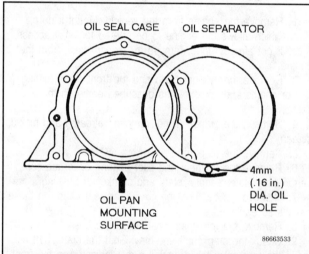

Fig. 284 When pushing the oil separator into the case make sure the oil hole is in the bottom position

with a gasket (2.3L diesel engines), carefully separate the gasket from the bottom of the seal case.

➡ You may want to loosen the oil pan bolts slightly at the rear to make it easier to separate the two surfaces. If the gasket is damaged, the oil pan will have to be removed and the gasket replaced.

5. On vehicles employing sealant around the oil pan, unbolt and remove the pan. Clean the pan and block surfaces of all traces of sealant.
6. Remove the oil seal case bolts, and pull the case straight off the rear of the crankshaft. Remove the case gasket.
7. Remove the seal retainer or oil separator (not all engines use an oil separator) from the case and pry out the seal. Take care not to gouge or damage the metal surrounding the seal. Inspect the sealing surface at the rear of the crankshaft. If a deep groove is worn into the surface, the crankshaft will have to be replaced. Lubricate the sealing surface with clean engine oil.

To install:

8. Using a seal installer of the correct size, install the new seal into the bore of rear oil seal case. Make certain that the flat side of the seal will face outward when the case is installed on the engine. The inside of the seal must be flush with the inside surface of the seal case.
9. Install the retainer or oil separator over the seal with the small hole located at the bottom. Install a new gasket onto the block surface and install the seal case to the rear of the block. Tighten the bolts to 8 ft. lbs. (11 Nm).
10. Reinstall the oil pan if it was removed, making certain the sealant is correctly applied. If the pan was only loosened, retighten the bolts to the correct torque. If the oil was drained, install the correct amount of motor oil.
11. Install the rear plate and bell housing cover.
12. Observing the matchmarks made earlier, install the flywheel or drive plate.
13. Install the transmission and related components as necessary.

6-Cylinder Engines

▶ See Figures 285, 286, 287 and 288

1. Remove the transmission and clutch assembly, if so equipped.
2. Matchmark and remove the flywheel or driveplate and adapter plate. For the 3.0L engines, use the Mitsubishi tools (MB990767-01 and MIT308239) to hold the crankshaft and flywheel stationary while loosening the flywheel bolts. For the 3.5L engines, use the Mitsubishi tool (MD998781) to hold the flywheel in position.
3. Remove the rear oil seal.
   a. Cut out a portion in the crankshaft oil seal lip.
   b. Cover the tip of a small prytool with a cloth and apply it to the cutout in the oil seal to pry the oil seal out.

✹✹CAUTION

Take care not to damage the crankshaft and oil seal case.

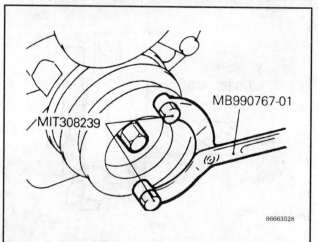

Fig. 285 Use these tools to hold the crankshaft in position while loosening or tightening the flywheel bolts — 3.0L Engines

3-168 ENGINE AND ENGINE REBUILDING

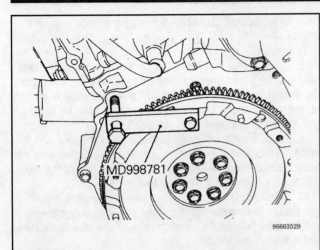

Fig. 286 Use this bracket to keep the flywheel from spinning while loosening or tightening the flywheel bolts

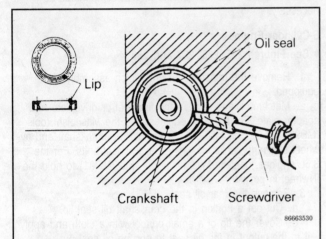

Fig. 287 Cut a little piece out of the seal, so that it can be pried out with a prytool — upon installation, apply engine oil to the sealing lip

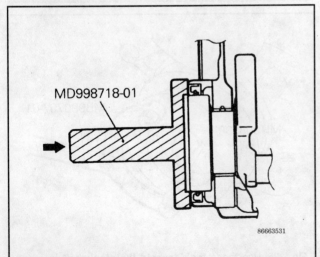

Fig. 288 Use this tool to install the rear oil seal straight and flush in the rear oil seal cover

To install:

4. Inspect the sealing surface at the rear of the crankshaft. If a deep groove is worn into the surface, the crankshaft will have to be replaced. Coat the sealing lip of the seal with fresh, clean engine oil. Press the new seal into the case with a seal installing tool. The seal must be pressed in square until it bottoms in the case. It is necessary to use the proper tool (MD998718-01) to fit the seal into place.
5. Install the rear plate.
6. Install the transmission mounting plate.
7. Install the flywheel or drive plate and adapter.
8. Install the transmission and related components as necessary.

Crankshaft and Main Bearings

REMOVAL

◆ See Figures 289 and 290

1. Remove the engine from the vehicle. Mount it on a stand which allows the engine to be rotated for easy access.
2. If not already done, remove the transmission from the engine.
3. Remove the cylinder head and the front cover, timing belt or chains and sprockets as described earlier in this section.
4. Remove the oil pan (if not removed already) and pick-up screen.
5. Remove the oil pump.
6. Remove the flywheel and pilot bearing or torque converter drive plate, adapter plate and crankshaft bushing.
7. Remove the bell housing cover and rear plate, on those models so equipped.
8. Remove the rear main seal housing and seal.
9. Check the crankshaft end-play. Shift the crankshaft as far to the rear as possible. Install a dial indicator on the front of the engine with the stem directly on the front of the crankshaft. Zero the indicator and use a small pry bar to shift the crankshaft forward as far as it will go. The amount of motion will show on the dial indicator; read the figure and compare with specifications. Excess end-play is a sign of wear.

➡ It's best to check the end-play with the engine in its upright (normal) position.

10. Invert the engine, and remove the connecting rod caps, keeping them in exact order. Loosen the bolts slowly and evenly. It is recommended that each cap be marked with its cylinder number before disassembly.
11. Remove the main bearing caps, keeping them in order. The caps are marked with arrows indicating the front of the engine, and are numbered in order from front to rear.

ENGINE AND ENGINE REBUILDING 3-169

Fig. 289 A dial gauge may be used to check crankshaft end-play

Fig. 290 Carefully pry the shaft back and forth while reading the dial gauge for play

12. Once all the bearing caps are removed, the crankshaft may be removed if necessary.

✽✽CAUTION

The crankshaft is a heavy component and can cause damage or injury if dropped or banged into other components during removal. The polished surfaces must be protected from dirt and abrasion during repairs.

13. Clean the crankshaft and bearing surfaces with a parts solvent. If necessary, remove the bearing shells from the caps, block and connecting rods in order to soak them. Note or diagram their positions and keep them in exact order at all times. Once used, bearing shells are NOT interchangeable and must be reinstalled in their original location.

INSPECTION AND CLEANING

▶ See Figures 291, 292 and 293

1. Inspect the crankshaft journals for scuffing, grooving or scoring. If crankshaft wear is uneven, the crankshaft should be turned on a lathe by a competent machine shop. Each journal will be undersize and new, undersize (thicker) bearings must be installed.
2. Make sure all the oil passages are clean, if necessary cleaning them with solvent and a stiff bristle brush.
3. It is recommended that crankpins and main bearing journals be measured with a micrometer to check for wear, out-of-roundness, and taper. Out-of-round is checked by measuring the journal across 2 diameters which are 90° apart (north-south and east-west) and comparing the two readings. The difference between the two readings gives the out-of-round specification. Taper is determined by measuring each journal at each end and comparing the numbers. If one reading is larger than the other, the journal tapers from large to small.
4. Inspect the bearing surfaces for burning, grooving, or scoring of any kind. Any but the slightest scratches disqualify a bearing from reuse; if there is any doubt, replace the bearings. (A good rule of thumb is to replace the bearing every time they are disassembled.) If the end-play, as measured above, is excessive, new bearings must be installed.

➡ If new bearings fail to correct end-play problems, the crankshaft will have to be machined undersize and different bearings installed.

INSTALLATION

▶ See Figures 294, 295, 296, 297, 298 and 299

1. Once the crankshaft has been cleaned, inspected or machined, the bearing clearance must be checked. this must be done in all cases, including reuse of old bearings or installation of thicker bearings. The bearing clearance (or oil clearance) is a critical dimension and must be proper before final assembly.

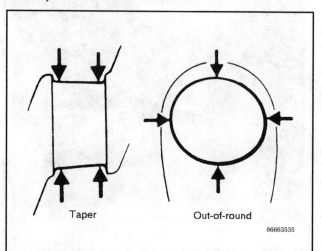

Fig. 291 Checking the crankshaft for out-of-round and taper is extremely important

3-170 ENGINE AND ENGINE REBUILDING

Fig. 292 A dial gauge can also be used to check crankshaft runout — mount the gauge so that it reads off of the crankshaft main journals

Fig. 293 Mount the gauge on all the different main bearing journals

2. Install the crankshaft into the block (the block is inverted) and make sure all bearing surfaces and journals are clean and dry. Cut pieces of plastic measuring media (such as Plastigage® or similar product) to fit the width of the bearing. Lay the piece widthwise across the journal. The insert must not rest on top of oil holes and the bearing shell that clamps against it must not be a grooved one.

3. Install and tighten all rod and main bearing caps to specifications. Do NOT turn the crankshaft. Remove the caps and read the width of the measuring media at the widest point. This is done by comparing the width at its widest point to lines on the scale included with the package. If working on a V6 engine, install the bearing cap to the engine block as shown in the illustration. Tighten the bearing cap bolts to the specified torque in the sequence shown.

4. If the bearing clearance is to specification, old bearings may be re-used if otherwise in good condition. In cases of moderate, regular wear, it may be feasible to replace the bearings with new ones of original size and avoid machining the crankshaft, provided this gives the specified clearance.

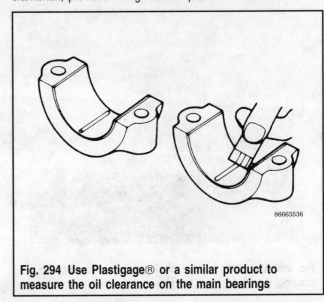

Fig. 294 Use Plastigage® or a similar product to measure the oil clearance on the main bearings

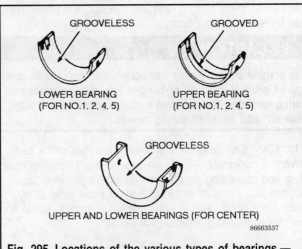

Fig. 295 Locations of the various types of bearings — notches in the bearing caps correspond with tabs on the bearing

ENGINE AND ENGINE REBUILDING

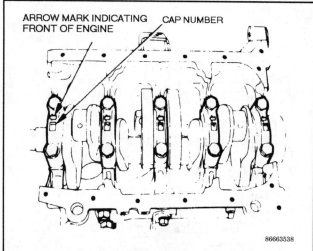

Fig. 296 The bearing caps must be installed with the arrow facing the front of the engine and in proper order

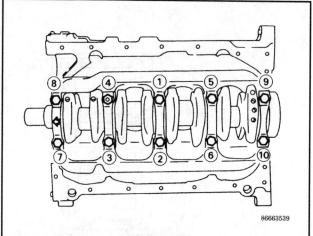

Fig. 297 Tighten the bearing caps in the order shown to the specified torque — 4-cylinder Engines with one-piece or separate bearing caps

5. If there is any question about the wear pattern or the ability of the new bearing shells to provide correct clearance, it is advisable to have the crankshaft machined and install matching bearings.

6. Before final assembly, all bearing surfaces must be thoroughly lubricated with clean engine oil. Make sure the main bearings have the oil holes lined up and that the upper shells for bearings 1, 2, 4, and 5 have grooves so that oil will pass to the connecting rod journals. The center main bearings, both top and bottom, are groove-less but include thrust bearing surfaces (flanges). Make sure connecting rod bearings have the notches in the bearing and cap together. Make sure all main bearing caps are installed in their original positions and directions. Main bearing caps are numbered in order from front to rear and have an arrow indicating the front (crankshaft pulley) end of the engine. Bearing cap bolts should be tightened evenly in 3 stages to specification. Work in sequence, starting with the center main bearing cap, then moving to No. 2, No. 4, front, and rear caps.

7. Reinstall the crankshaft seals, the oil pick-up screen and oil pan as soon as possible. Even if other work is to be performed on the engine, the seals and pan will serve to protect the crankshaft, rods and bearings from dirt and impact damage.

Flywheel and Ring Gear

REMOVAL & INSTALLATION

♦ See Figure 300

All Engines

➡The clutch cover and the pressure plate are balanced as an assembly; if replacement of either part becomes necessary, replace both parts as an assembly.

1. Remove the transmission by using the procedures from Section 7.

✱✱CAUTION

The clutch disc contains asbestos, which has been determined to be a cancer causing agent. Never clean the clutch assembly with compressed air! Avoid inhaling dust from the clutch during disassembly. When cleaning components, use commercially available brake cleaning fluids.

2. Remove the clutch assembly, if equipped — refer to Section 7.
3. Most trucks equipped with automatic transmissions have separate drive plates, but the 2.0L engines have the ring gear attached to the torque converter. For all trucks except for the 2.0L engine trucks, remove the retaining bolts holding the adapter plate, the drive plate, the crankshaft adapter and the crankshaft bushing to the crankshaft. For trucks equipped with the 2.0L engine, remove the torque converter and examine the teeth on the ring gear.
4. For trucks with manual transmissions, matchmark the flywheel and crankshaft. Loosen the retaining bolts evenly and in a crisscross pattern. Support the flywheel during removal of the last bolts and remove the flywheel.

To install:

5. Carefully inspect the teeth on the flywheel, torque converter, or drive plate for any signs of wearing or chipping. If anything beyond minimal contact wear is found, replace the unit.

➡Since the flywheel is driven by the starter gear, it would be wise to inspect the starter drive if any wear is found on the flywheel teeth. A worn starter can cause damage to the flywheel.

6. When reassembling, place the flywheel or drive plate in position on the crankshaft and make sure the matchmarks align. Install the retaining bolts finger-tight. The torque converter should be reinstalled on the transmission, not on the engine.

3-172 ENGINE AND ENGINE REBUILDING

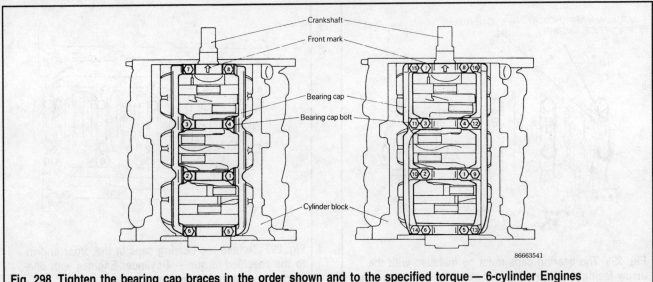

Fig. 298 Tighten the bearing cap braces in the order shown and to the specified torque — 6-cylinder Engines

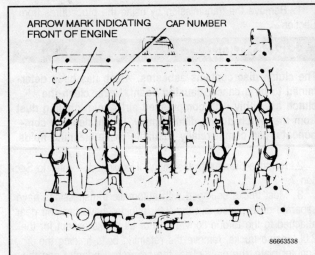

Fig. 299 A common newer 4-cylinder engine crankshaft assembly with a one-piece bearing cap brace

7. Tighten the flywheel or drive plate bolts in a diagonal pattern and in three passes to 98 ft. lbs. (133 Nm) for 4-cylinder engines or 54 ft. lbs. (74 Nm) for 6-cylinder engines.
8. Install and adjust the clutch assembly.
9. Install the transmission.

RING GEAR REPLACEMENT

If the the ring gear teeth on the tighten converter are damaged, the torque converter must be replaced. The ring gear cannot be separated or reinstalled individually.

If a ring gear is damaged on the flywheel of a manual transmission car, it is usually cheaper and easier to buy a new flywheel assembly than to change the gear. If you possess the proper equipment for heating and handling the ring gear, the procedure is as follows:

1. With the flywheel removed from the car, tap around the ring gear to loosen it and remove it from the flywheel.

An alternate method is to drill a 0.39 in. (10mm) hole between any two teeth on the gear, being careful not to drill into the flywheel. Mount the flywheel in a vise and split the ring with a hammer and sharp chisel.

✱✱WARNING

Do NOT heat the ring gear or flywheel during removal. The gear cannot be removed when heated.

2. Heat the new ring gear to 572°F (300°C). Use heavy gloves and tongs to manipulate the gear. Since the required temperature is more than twice the boiling point of water and well above the flame point of paper, take special care to keep anything flammable out of the area.
3. Position the ring gear onto the flywheel. It should align easily. If it does not, tap it lightly with a brass punch to get it in position.
4. Allow the ring gear and flywheel to air cool for a period of hours. Do not attempt to cool the metal with water, oil or other fluids.
5. Install the flywheel and tighten the bolts to specification.

ENGINE AND ENGINE REBUILDING 3-173

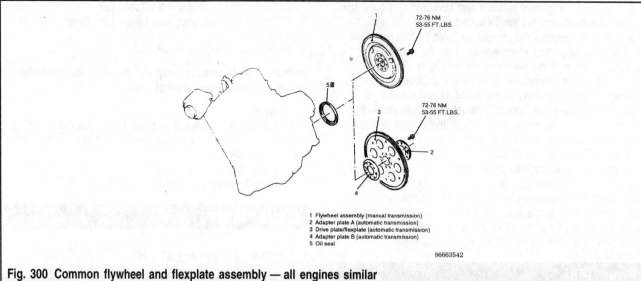

Fig. 300 Common flywheel and flexplate assembly — all engines similar

EXHAUST SYSTEM

Safety Precautions

For a number of reasons, exhaust system work can be dangerous. Always observe the following precautions:
• Support the vehicle securely by using jackstands, or equivalent, under the frame of the vehicle.
• Wear safety goggles to protect your eyes from metal chips that may fly free while working on the exhaust system.
• When using a torch, be careful not to come close to any fuel lines.
• Always use the proper tool for the job.
• NEVER WORK ON A HOT EXHAUST SYSTEM!!

Special Tools

A number of special exhaust tools can be rented or bought from a local auto parts store. A common one is a tail pipe expander, designed to enable you to join pipes of identical diameter.

It may also be quite helpful to use solvents designed to loosen rusted nuts or bolts. Soaking rusted hardware the night before you do the job can speed the work of freeing rusted parts considerably. Remember that these products are often flammable, apply only to parts after they are cool.

The Mitsubishi exhaust system generally consists of four pieces. At the front of the vehicle, the front section of pipe connects the exhaust manifold or turbocharger to the catalytic converter. The front pipe on some models contains a section of flexible, braided pipe which allows the system to move without breaking. The catalytic converter, a sealed, non-serviceable unit, can be unbolted from the system and replaced, if necessary.

The diesel engine does not require a catalyst and no converter is found on diesel powered trucks. The gasoline engines in the truck have two catalytic converters. Although they are referred to as front and rear converters, they are attached to each other and are located immediately below the exhaust manifold. Replacement of the front (upper) catalyst requires removal of the exhaust manifold before removal of the converter.

Inspection

An intermediate or center pipe runs from the catalytic converter to the muffler at the rear of the vehicle. Depending on the engine in the vehicle, this pipe may contain a resonator or pre-muffler which serves to quiet and smooth out the exhaust flow before it enters the rear muffler. The resonator is welded into the pipe and is not removable.

The muffler, with its entry pipe and tail pipe, complete the system and serve to further quiet and cool the exhaust. Mitsubishi mufflers have lead-in and exhaust pipes welded to the body of the muffler. If one of the pipes is damaged, the complete muffler assembly should be replaced.

The exhaust system is attached to the body by several welded hooks and flexible rubber hangers; these hangers absorb exhaust vibrations and isolate the system from the body of the car. A series of metal heat shields runs along the exhaust piping, protecting the underbody from excess heat. These heat shields should be the first place to look when chasing a light metallic rattle or buzz under the vehicle. Because the shields are exposed under the body work, they may become bent, loose or packed with road debris.

When inspecting or replacing exhaust system parts, make sure there is adequate clearance from all points on the body to avoid possible overheating of the floor pan. Check the complete system for broken, damaged, missing or poorly positioned parts. Rattles and vibrations in the exhaust system may be caused by misalignment of parts. When aligning the system, leave all the nuts and bolts loose until everything is in its proper place, then tighten the hardware working from the front to the rear. Remember that what appears to be proper clearance during repair may change as the vehicle moves down the road. The motion of the engine, body and suspension must be considered when replacing parts.

3-174 ENGINE AND ENGINE REBUILDING

For exhaust component removal and installation, a good rule of thumb is to disconnect the forward-most joint first, then the rear joint, then remove the component from its hangers. This is particularly important when working with the front pipe or the complete system; if the weight of the pipe(s) is allowed to rest on the ground, the length of the system can develop enough leverage to break or damage the manifold or turbocharger.

In general, if any component is attached at both ends, it's easier to remove the entire system from the vehicle and do the repair where you can see it. Remember to use new gaskets at every disassembled joint.

When disassembling the rubber and/or metal hangers, take close note of the style and location of each, labeling or diagraming them if necessary to insure correct reinstallation. The manufacturer selects the hangers and bushings to deliver the maximum support and isolation. If any of the hangers or hardware look to be in poor condition, replace the piece.

Front Pipe

REMOVAL & INSTALLATION

▶ See Figures 301, 302, 303 and 304

1. Support the vehicle securely by using jackstands or equivalent under the frame of the vehicle.
2. Remove the exhaust pipe clamps and any front exhaust pipe shield.
3. Soak the exhaust manifold front pipe mounting studs with penetrating oil. Remove attaching nuts and gasket from the manifold.

➡ If these studs snap off, while removing the front pipe the manifold will have to be removed and the stud will have to be drilled out and the hole tapped.

4. Remove front pipe from the muffler/catalytic converter.
5. Remove any exhaust pipe mounting hanger or bracket.

To install:

6. Upon installation use a new gasket at every disassembled junction.
7. Connect the exhaust pipe loosely to the hangers.
8. Install the front pipe on the manifold with seal if so equipped.
9. Connect the pipe to the muffler/catalytic converter. Assemble all parts loosely and position the pipe to insure proper clearance from body of vehicle.
10. Tighten mounting studs, bracket bolts and exhaust clamps.
11. Install exhaust pipe shield.
12. Start engine and check for exhaust leaks.

Catalytic Converter

REMOVAL & INSTALLATION

▶ See Figures 301, 302, 303 and 304

1. Remove the converter lower shield.
2. Disconnect converter from front pipe.
3. Disconnect converter from center pipe or tail pipe assembly.
4. Remove catalytic converter.

➡ Assemble all parts loosely and position the converter before tightening the exhaust clamps.

To install:

5. Position the converter in place and install the mounting bolts to the front and tail pipes. Always use new clamps and exhaust gaskets.
6. Install the converter lower shield.
7. Start the engine and check for leaks.

Center Pipe

REMOVAL & INSTALLATION

▶ See Figures 301, 302, 303 and 304

1. Raise the vehicle and support it with safety stands.
2. Soak all nuts and bolts at the converter and muffler connections with penetrating oil.
3. Disconnect the pipe at the converter.
4. While supporting the pipe, disconnect it at the muffler and remove it.

To install:

5. Position the pipe and connect it to the muffler, using new exhaust gaskets at all disassembled junctions.
6. Connect the pipe to the converter and then install any other mounting brackets or hardware.

Tailpipe And Muffler

REMOVAL & INSTALLATION

▶ See Figures 301, 302, 303 and 304

1. Disconnect the center pipe at the catalytic converter assembly.
2. Remove all brackets and exhaust clamps.
3. Remove the tail pipe from the muffler. On some models the tail pipe and muffler are one piece. Remove the center pipe from the muffler.

To install:

4. Reassemble the tail and center pipes to the muffler. Always use new clamps and exhaust seals, start engine and check for leaks.
5. Connect all brackets and hangers. Connect the center pipe to the catalytic converter.

ENGINE AND ENGINE REBUILDING 3-175

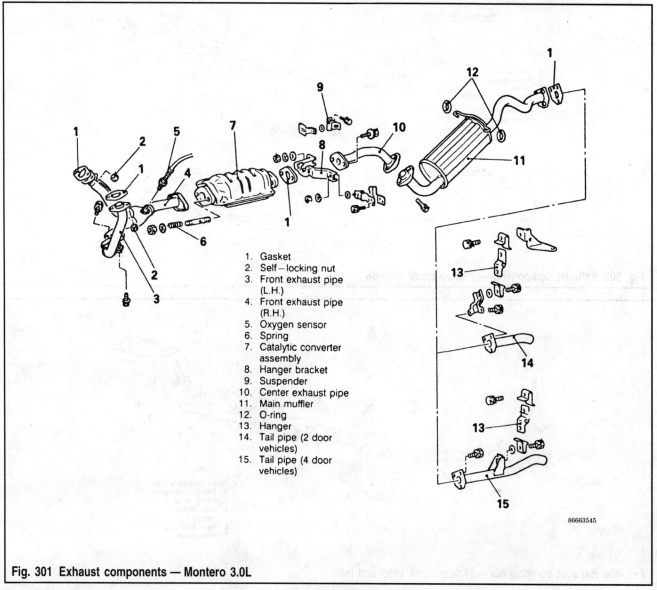

Fig. 301 Exhaust components — Montero 3.0L

1. Gasket
2. Self-locking nut
3. Front exhaust pipe (L.H.)
4. Front exhaust pipe (R.H.)
5. Oxygen sensor
6. Spring
7. Catalytic converter assembly
8. Hanger bracket
9. Suspender
10. Center exhaust pipe
11. Main muffler
12. O-ring
13. Hanger
14. Tail pipe (2 door vehicles)
15. Tail pipe (4 door vehicles)

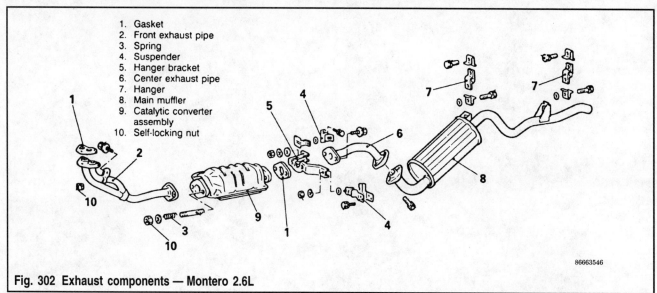

Fig. 302 Exhaust components — Montero 2.6L

1. Gasket
2. Front exhaust pipe
3. Spring
4. Suspender
5. Hanger bracket
6. Center exhaust pipe
7. Hanger
8. Main muffler
9. Catalytic converter assembly
10. Self-locking nut

3-176 ENGINE AND ENGINE REBUILDING

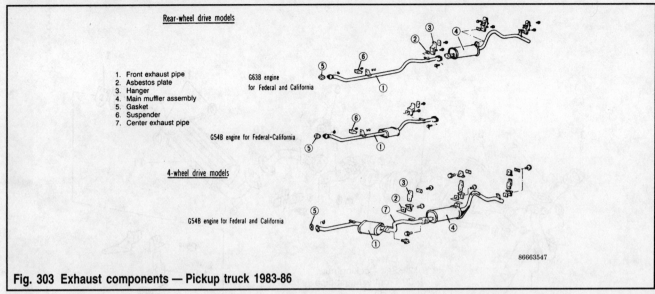

Fig. 303 Exhaust components — Pickup truck 1983-86

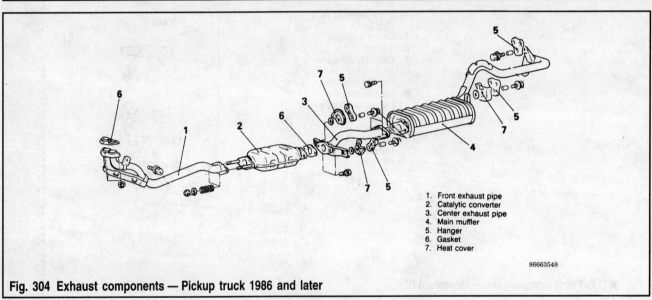

Fig. 304 Exhaust components — Pickup truck 1986 and later

CARBURETED ENGINE CONTROLS
  FEEDBACK CARBURETOR SYSTEM
    (FBC) 4-21
**COMPONENT LOCATION AND
VACUUM DIAGRAMS 4-80**
DIAGNOSTIC TROUBLE CODES
  CLEARING DIAGNOSTIC TROUBLE
    CODES 4-69
  GENERAL INFORMATION 4-68
  READING DIAGNOSTIC TROUBLE
    CODES 4-68
**DIESEL ENGINES EMISSION
CONTROLS**
  AIR CONDITIONING IDLE SPEED
    COMPENSATOR 4-21
  CRANKCASE VENTILATION
    SYSTEM 4-20
  EVAPORATIVE EMISSION
    CONTROLS 4-20
  EXHAUST GAS RECIRCULATION
    (EGR) 4-20
  FUEL INJECTION 4-21
  HIGH ALTITUDE COMPENSATION
    SYSTEM 4-21
FUEL INJECTION ENGINE CONTROLS
  BAROMETRIC PRESSURE
    SENSOR 4-31
  CAMSHAFT POSITION SENSOR 4-48
  CHECK ENGINE/MALFUNCTION
    INDICATOR LAMP 4-22
  CLOSED THROTTLE POSITION
    SWITCH 4-39
  COMPONENT INSPECTION
    PROCEDURES 4-22
  CRANKSHAFT POSITION
    SENSOR 4-50
  EGR TEMPERATURE SENSOR 4-52
  ENGINE CONTROL MODULE
    (ECM) 4-57
  ENGINE CONTROL MODULE (ECM)
    POWER GROUND 4-27
  ENGINE COOLANT TEMPERATURE
    SENSOR 4-32
  EVAPORATIVE EMISSION PURGE
    SOLENOID 4-54
  HARNESS CHECKING
    PRECAUTIONS 4-22
  IDLE AIR CONTROL MOTOR (DC
    MOTOR) 4-42
  IDLE AIR CONTROL MOTOR
    (STEPPER MOTOR) 4-43
  IDLE SPEED CONTROL MOTOR 4-40
  IDLE SPEED CONTROL MOTOR
    POSITION SENSOR 4-41
  INTAKE AIR TEMPERATURE
    SENSOR 4-30
  KNOCK SENSOR 4-55
  MULTI-PORT FUEL INJECTION
    RELAY 4-23
  OXYGEN SENSOR 4-45
  THROTTLE POSITION SENSOR 4-34

  TROUBLESHOOTING
    PROCEDURES 4-22
  VARIABLE INDUCTION CONTROL
    SOLENOID 4-56
  VOLUME AIR FLOW SENSOR 4-28
**GASOLINE ENGINE EMISSION
CONTROLS**
  EVAPORATIVE EMISSION CONTROL
    SYSTEM 4-3
  EXHAUST GAS RECIRCULATION
    (EGR) SYSTEM 4-7
  HIGH ALTITUDE COMPENSATION
    SYSTEM 4-13
  INTAKE AIR TEMPERATURE
    CONTROL SYSTEM 4-17
  JET VALVE 4-12
  MAINTENANCE REMINDER
    LIGHTS 4-12
  MIXTURE CONTROL VALVE
    (MCV) 4-17
  POSITIVE CRANKCASE VENTILATION
    (PCV) SYSTEM 4-2
  SECONDARY AIR SUPPLY
    SYSTEM 4-10
  VACUUM REGULATOR VALVE 4-13
  VARIABLE INDUCTION SYSTEM
    (VIS) 4-18

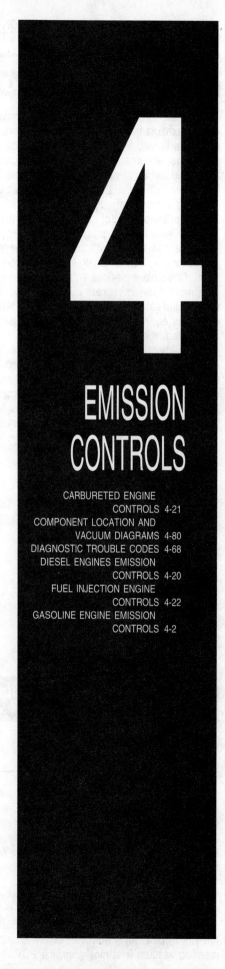

4
EMISSION CONTROLS

CARBURETED ENGINE
CONTROLS 4-21
COMPONENT LOCATION AND
VACUUM DIAGRAMS 4-80
DIAGNOSTIC TROUBLE CODES 4-68
DIESEL ENGINES EMISSION
CONTROLS 4-20
FUEL INJECTION ENGINE
CONTROLS 4-22
GASOLINE ENGINE EMISSION
CONTROLS 4-2

4-2 EMISSION CONTROLS

GASOLINE ENGINE EMISSION CONTROLS

There are three sources of automotive pollutants; crankcase fumes, exhaust gases, and gasoline evaporation. The pollutants formed from these substances fall into three categories: unburned hydrocarbons (HC), carbon monoxide (CO), and oxides of nitrogen (NOx). The equipment used to limit these pollutants is called emission control equipment.

Due to varying state, federal, and provincial regulations, specific emission control equipment may vary by area of sale. The U.S. emission equipment is divided into two categories: California and 49 State. In this section, the term "California" applies only to cars originally built to be sold in California. Some California emissions equipment is not shared with equipment installed on cars built to be sold in the other 49 states. Models built to be sold in Canada also have specific emissions equipment, although in many cases the 49 State and Canadian equipment is the same.

Both carbureted and fuel injected cars require an assortment of systems and devices to control emissions. Newer cars rely more heavily on computer management of many of the engine controls. This eliminates the many of the vacuum hoses and linkages around the engine. In the lists that follow, remember that not every component is found on every car.

ECM CONTROLLED SYSTEMS
- Fuel Evaporative Control
- Carburetor Feedback System
- Deceleration Fuel Cutoff
- Ignition Timing Controls
- Cold Mixture Heater

NON-ECM CONTROLLED SYSTEMS
- Positive Crankcase Ventilation (PCV)
- Throttle Positioner
- Exhaust Gas Recirculation (EGR)
- Air Suction
- High Altitude Compensation
- Automatic Hot Air Intake
- Automatic Choke
- Choke Breaker
- Choke Opener

Positive Crankcase Ventilation (PCV) System

SYSTEM OPERATION

A closed positive crankcase ventilation system is used on all Mitsubishi models. This system cycles incompletely burned fuel (which works its way past the piston rings into the crankcase) back into the intake manifold for reburning with the fuel/air mixture. The oil filler cap is sealed and the air is drawn from the top of the crankcase into the intake manifold through a valve with a variable orifice.

This valve (commonly known as the PCV valve) regulates the flow of air into the manifold according to the amount of manifold vacuum. When the throttle plates are open fairly wide, the valve opens fully. However, at idle speed, when the manifold vacuum is at maximum, the PCV valve reduces the flow in order not to unnecessarily affect the small volume of mixture passing into the engine.

During most driving conditions, manifold vacuum is high and all of the vapor from the crankcase, plus a small amount of fresh air, is drawn into the manifold via the PCV valve. At full throttle, the increase in the volume of blow-by and the decrease in manifold vacuum make the flow via the PCV valve inadequate. Under these conditions, excess vapors are drawn into the air cleaner and pass into the engine along with the fresh air.

A plugged valve or hose may cause a rough idle, stalling or low idle speed, oil leaks in the engine and/or sludging and oil deposits within the engine and air cleaner. A leaking valve or hose could cause an erratic idle or stalling.

TESTING

If any of the following operating problems occur, inspect the PCV system:
- Rough idle, not explained by an ordinary vacuum leak, or fuel delivery problem.
- Oil leaks past the valve cover, oil pan seals or even front and rear crankshaft seals not explainable by age, high mileage or lack of basic maintenance.
- Excessive dirtiness of the air cleaner cartridge at low mileage.
- Noticeable dirtiness in the engine oil due to fuel dilution well before normal oil change interval.

➡**An engine with badly worn piston rings and/or valve seals may produce so much blow-by that even a normally functioning PCV system cannot deal with it. A compression test should be performed if extreme wear is suspected.**

The PCV system is easily checked with the engine running at normal idle speed (warmed up). Remove the PCV valve from the valve cover, but leave it connected to its hose. Place your thumb over the end of the valve to check for vacuum. If there is no vacuum, check for plugged hoses or ports. If these are open, the valve is faulty. With the engine off, remove the PCV valve completely. Shake it end to end, listening for the rattle of the needle inside the valve. Generally, if no rattle is heard, the needle is jammed (probably with oil sludge) and the valve should be replaced. If the valve is a threaded type, it may be necessary to use a thin probe inserted in the threaded end to check for plunger motion. If no motion is felt, replace the valve.

An engine which is operated without crankcase ventilation can be damaged very quickly. It is important to check and change the PCV valve at regular maintenance intervals.

REMOVAL & INSTALLATION

Remove the PCV valve from the valve cover. Most valves are pressed into the valve cover but some are threaded and screw into the cover. Remove the hose from the valve. Take

EMISSION CONTROLS 4-3

note of which end of the valve was in the manifold. This one-way valve must be reinstalled correctly or it will not function. While the valve is removed, the hoses should be checked for splits, kinks and blockages. Check the vacuum port (that the hoses connect to) for any clogging.

Remember that the correct function of the PCV system is based on a sealed engine. An air leak at the oil filler cap and/or around the oil pan can defeat the design of the system.

Evaporative Emission Control System

♦ See Figures 1, 2, 3 and 4

※※CAUTION

Fuel vapors are EXTREMELY explosive. Observe no smoking/no open flame precautions. Have a "Type B-C" (dry chemical) fire extinguisher within arm's reach at all times and know how to use it.

OPERATION

The heart of this system is a charcoal canister located in the engine compartment. Fuel vapor that collects in the carburetor float bowl and/or gas tank is stored in the canister instead of being released into the atmosphere.

In order to restore the ability of the charcoal to hold fuel, fresh air is drawn through the canister under certain operating conditions, drawing the fuel vapors back out and introducing them into the combustion chambers.

At idle speed, or when the engine is cold, the addition of any fuel vapor to the correct mixture would cause excessive tailpipe emissions. For this reason, a port in the carburetor or fuel injection system throttle body allows the fuel vapors to be drawn out of the canister only after the throttle has been opened past the normal idle position. If there is no vacuum, the canister purge valve remains closed.

The flow of air and fuel are further restricted by a thermal valve when the engine is cold. This valve prevents the vacuum signal from going to the canister purge valve until the engine reaches a pre-determined temperature.

When the canister purge valve opens, air is drawn under slight vacuum from the air intake. If the engine exhibits operating problems during warm-up and basic fuel system and engine tune-up adjustments are correct, check the thermo valve for proper operation.

TESTING

Purge Control System
♦ See Figure 5

The purge control system controls the flow of the fuel fumes from the evaporative canister. Check the entire system as a whole first, then, if the system exhibits signs of incorrect functioning, check each of its components to find the exact problem with the system.

1. Disconnect the black vacuum hose (2.4L, 3.0L and 3.5L engines' vacuum hose has two red stripes) from the intake manifold and plug the nipple.
2. Connect a hand vacuum pump to the disconnected vacuum hose.
3. Check the following figures both when the engine is cold (engine coolant temperature is 122°F/50°C or less) and when it is hot (185-205°F/85-95°C).
4. On 2.0L and 2.6L engines:
   a. When the engine is cold, the engine is run at 2,500 rpm and the amount of applied vacuum is 15.7 in.Hg. (53 kPa), the vacuum should not leak.
   b. When the engine is hot, the engine is idling and the vacuum is the same as above, the vacuum should also not leak.
   c. When the engine is hot, the engine rpm is 2,500 and the same amount of vacuum is applied, the vacuum should leak down. If the system does not do this, check each component to find the faulty part.
5. On 2.4L, 3.0L and 3.5L engines:
   a. When the engine is cold, the engine is idling and 15.7 in.Hg. (53 kPa) of vacuum is applied, the vacuum should not leak. Run the engine at 3,000 rpm and check again. The vacuum should still not leak.
   b. Run the engine until hot and then shut the engine OFF.
   c. Restart the engine. With the engine hot, idling and the same amount of vacuum applied as above, the vacuum should not leak. Within 3 minutes of starting the engine, run the engine at 3,000 rpm and try to apply vacuum. The vacuum should leak. Then, after three minutes have passed, once again apply the vacuum at 3,000 rpm. The vacuum should be maintained momentarily, after which it should start to leak. The vacuum will leak continuously if the altitude is 2,200 m (7,200 ft.) or higher, or the intake air temperature is 50°C (122°F) or higher.
6. Reconnect the vacuum hose.

Purge Control Valve
♦ See Figure 6

The 2.0L and 2.6L engines utilize a vacuum controlled purge control valve. First check all the hoses and connections for

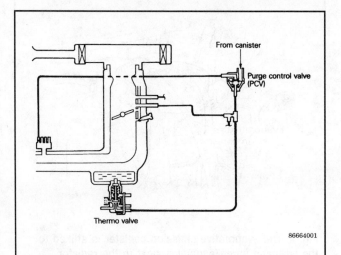

Fig. 1 The purge control system components — 2.0L and 2.6L engines

4-4 EMISSION CONTROLS

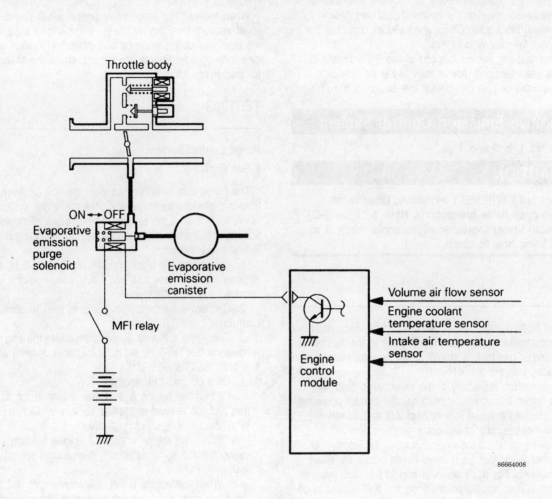

Fig. 2 The purge control system components — 2.4L, 3.0L and 3.5L engines

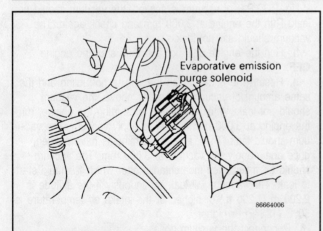

Fig. 3 The evaporative emission purge solenoid is located on the left-hand side of the engine compartment next to the radiator overflow reservoir — 2.4L, 3.0L and 3.5L engines

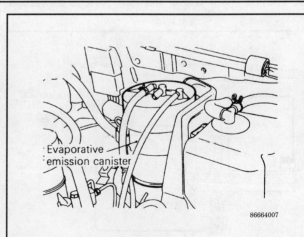

Fig. 4 The evaporative emission canister is affixed to the left-hand inner fenderwell, next to the radiator overflow reservoir

EMISSION CONTROLS 4-5

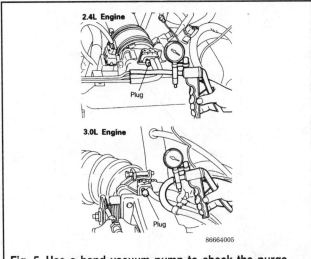

Fig. 5 Use a hand vacuum pump to check the purge control system

proper attachment, cracks, bends and leaks. Many problems relate simply to poor mechanical connections within the system or restricted hoses.

Four cylinder Pick-up and Montero engines require the removal of the valve to test it. Label each hose before removal; correct re-connection is essential. With the valve off the car, connect a hand vacuum pump to the bottom vacuum nipple of the valve. Apply a vacuum of 15.7 in. Hg. (53 kPa) for to check for air tightness. The valve should hold the vacuum. Release the vacuum and blow gently into the canister side hose port. With no vacuum applied, no air should pass through the valve. Now draw a vacuum of at least 8 in. Hg. (27 kPa) and blow into the port again. Air should flow through the valve.

Evaporative Emission Purge Solenoid
♦ See Figures 7 and 8

The 2.4L, 3.0L and 3.5L engines control the canister through the Evaporative Emission Purge Solenoid (EEPS) valve. This can be checked in place on the firewall. With the engine off, label and disconnect the two vacuum hoses running to the valve. One hose will have a red stripe on it; take note of which port it was connected to. Remove the electrical connector from the valve. Connect a hand vacuum pump to the port which contained the hose with the red stripe. Draw a light vacuum on the pump; no vacuum should flow (the system holds vacuum).

Using jumper wires, connect battery voltage to the terminals on the solenoid valve. Make absolutely CERTAIN that the polarity of the wiring is correct. The (+) side of the battery must be connected to the terminal which forms the top of the "T" in the pattern of the terminals. Once the solenoid is energized, vacuum should leak (or flow) when the hand pump is used. Disconnect the jumpers and the vacuum pump. Use an ohmmeter to check the resistance across the terminals of the solenoid valve. Correct resistance is 36-44 ohms at 20°C (68°F).

➡ Resistance will change with temperature. Make common sense allowances for temperature variation.

Purge Port Vacuum
♦ See Figures 9, 10 and 11

On 2.4L, 3.0L and 3.5L engines it is necessary to check the purge port with a hand vacuum pump to check for obstructions in the port itself. Make sure the vehicle is warmed up. Disconnect the vacuum hose from the throttle body purge hose nipple and connect the hand vacuum pump to the nipple. Start the engine. For all vehicles except those equipped with a pre-1993 2.4L engine, check to see that, after raising the engine speed by racing the engine, vacuum rises proportionately with the rise in engine speed. On pre-1993 2.4L engines, check to see that, after raising the engine speed by racing the engine, vacuum remains fairly constant. If there is a problem with the outcome of this test, it is possible that the throttle body port may be clogged and require cleaning.

Thermo Valve
♦ See Figure 12

The thermo valve found on all pre-1990 4-cylinder Pick-up and Montero engines is located on the intake manifold or cylinder head where its bottom will be immersed in coolant. The 2.4L, 3.0L and 3.5L engines control the purge system through the ECM. The thermo valve blocks or passes vacuum to the

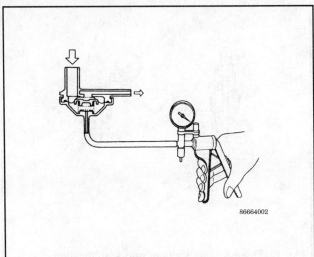

Fig. 6 Checking the purge control valve with a hand vacuum pump — 2.0L and 2.6L engines

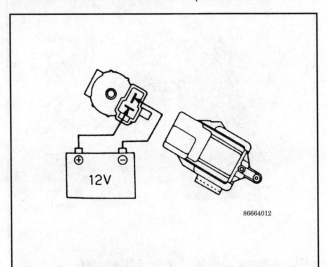

Fig. 7 Be certain that the power source is connected to the correct terminals

4-6 EMISSION CONTROLS

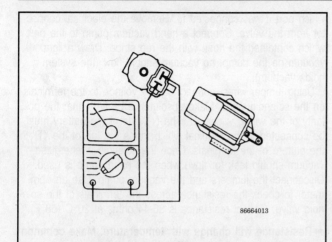

Fig. 8 Check the resistance across the two terminals of the evaporative emission control solenoid — 2.4L, 3.0L and 3.5L engines

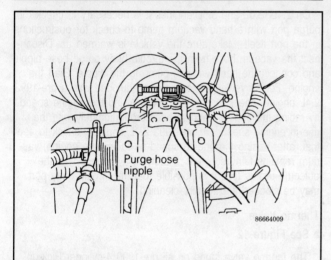

Fig. 9 Connect the hand vacuum pump to the throttle body purge hose nipple — 3.0L engines

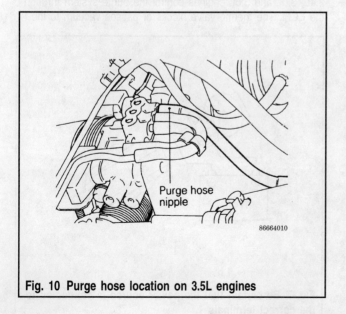

Fig. 10 Purge hose location on 3.5L engines

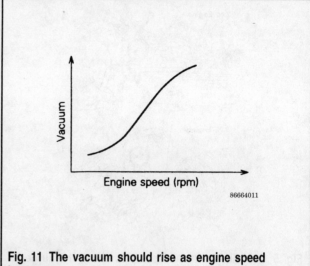

Fig. 11 The vacuum should rise as engine speed increases

purge control valve depending on the coolant temperature. Test after the engine has cooled overnight.

Label and disconnect the two vacuum hoses on the thermo valve. With the engine turned **OFF**, connect a hand-held vacuum pump to the upper port on the two port valves. Apply vacuum and confirm that the valve leaks or does NOT hold vacuum. Start the engine and allow it to warm up. When the coolant has reached normal operating temperature, disconnect the appropriate hose as before and repeat the test; the valve should hold vacuum and not leak.

Disconnect all of the vacuum hoses from the thermo valve. Connect the hand vacuum pump to nipples (B), (C) and (D) one at a time; plug the nipples not being tested at the time. Apply vacuum to check the valves condition. The first check should be done with the coolant 104°F (40°C) or less. In this case the valve should not hold the vacuum. On the second test of each nipple the engine coolant should be at least 176°F (80°C) and the valve should hold vacuum. If the valve failed either of these tests it needs to be replaced with a new unit.

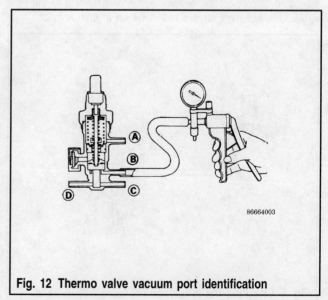

Fig. 12 Thermo valve vacuum port identification

EMISSION CONTROLS 4-7

If it becomes necessary to replace the thermo valve, do so only on a cold motor. Fit the wrench only onto the faceted base of the valve, never on the plastic parts. When installing the new unit, apply sealer such as 3M®No. 4171 or equivalent to the threads and tighten the new unit to 22 ft. lbs. (30 Nm).

2-Way Overfill Limiter

This device, sometimes mistaken for a fuel filter, is found in the vapor line running from the tank to the canister. Usually located at or near the tank, this valve is both a pressure and suction-sensitive unit. Its purpose is to compensate for the pressure changes within the fuel tank. Since the filler cap is tight enough to be considered sealed, the pressure must be equalized somewhere within the system.

When the pressure builds within the tank, such as on a very hot day or after a long period of driving, the valve releases the pressure and vapor into the charcoal canister, thereby venting the tank without raising emissions. Conversely, should the tank develop a vacuum, the valve will bleed some air (and vapor) from the canister into the tank.

The control pressures within the valve are pre-set and not adjustable, but a quick check can be performed as follows:

1. Look at the valve and observe which end is toward the tank. Label or diagram the correct position.
2. Remove the valve from the vapor line. It may be necessary to remove other obvious components such as a parking brake cable bracket for access.
3. Lightly blow through either end of the valve. If air passes after some resistance, the valve is in good condition.
4. Install the valve into the line in the correct direction and secure the clamps. Make certain the lines are firmly seated on the ports before installing the clamps.

Fuel Check Valve (Roll-Over Valve)

Usually mounted near the fuel tank, this simple valve is found in the vapor line coming from the tank to the charcoal canister. Normally the line carries vapor which is easily absorbed and held by the charcoal in the canister. If the car rolls over, the line would fill with liquid gasoline and exceed the canister's ability to absorb fuel. Once saturated, the canister would allow the liquid fuel to run out, possibly onto the hot surfaces of the engine. Since an engine fire is the last thing you want when your car is on its head (or any other time), the roll-over valve will block the vapor line and keep the fuel out of the canister.

This valve rarely, if ever, fails. If it must be checked, unbolt it and remove the hoses. Shake it — if it rattles, it's OK. Since even the simplest job can be done incorrectly, make sure the valve is reinstalled right side up. Connect the hoses firmly and install the clamps.

Charcoal Canister

Since the canister cannot be tested on the workbench, it should be checked periodically for cracks, obstructions and proper hose connections. Check the maintenance schedule for your car; many models require replacement of the canister after a period of years or miles/kilometers. Additionally, the canister should be considered suspect on a carbureted car after any incident of severe engine flooding. It is possible to deliver enough fuel vapor from the carburetor to overcome the capacity of the charcoal. This can result in the air/fuel mixture becoming too rich and causing driveability problems. If a high mileage carburetor is overhauled or replaced, a new charcoal canister should be installed as well.

When working around the canister, remember to label or diagram every hose before removal. Vacuum and vapor must flow correctly if the system is to work properly.

Exhaust Gas Recirculation (EGR) System

OPERATION

▶ See Figure 13

Exhaust Gas Recirculation (EGR) is used to reduce peak flame temperatures in the combustion chamber. A small amount of exhaust gas is diverted from the exhaust manifold and re-entered into the intake manifold, where it mixes with the air/fuel charge and enters the cylinder to be burned. Cooler combustion reduces the formation of Nitrogen Oxide (NO_2) emissions.

The system consists of the EGR valve, controlling the flow of exhaust gas and various vacuum and/or electric controls to keep the EGR from working at the incorrect time.

No EGR is required when the engine is cold due to lower flame temperatures in the engine. EGR under these conditions would produce rough running so EGR function is cut off either by a thermo valve or by the ECM (which is monitoring coolant temperature.) Additionally, EGR flow is cut off at warm idle to eliminate any roughness or stumble on initial acceleration.

Cooler combustion temperatures also result in slightly reduced power output. This isn't felt during normal, part-throttle driving and the emission benefits outweigh the slight loss. However, in a wide-open throttle situation a power reduction is not desirable; full power could be the margin of success in a passing or accident avoidance situation. For this reason, EGR function is eliminated when the engine goes on wide-open throttle. Normally, the vacuum to the EGR valve can overcome the spring tension within the valve and hold it open. When the throttle opens fully, vacuum to the EGR is reduced and the spring closes the valve.

A common symptom of EGR malfunction is light engine ping at part throttle, particularly noticeable under load such as going uphill or carrying several passengers. An EGR valve which fails to close properly can also cause a rough or uneven idle. If the engine is correctly tuned and other common causes (vacuum leaks, bad plug wires, etc.) are eliminated, EGR function should be considered as a potential cause when troubleshooting a rough idle.

Since the majority of EGR components do not require routine maintenance and should not clog or corrode if unleaded gas is used, you should check all other reasonable causes of a problem before checking this system.

TESTING

➡ A vacuum pump capable of producing more than 10 in. Hg. (33.8 kPa) of vacuum will be needed to perform this test.

4-8 EMISSION CONTROLS

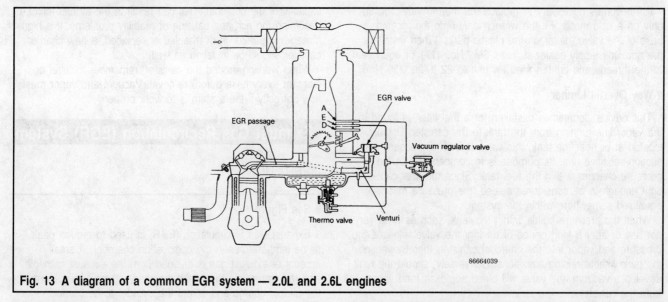

Fig. 13 A diagram of a common EGR system — 2.0L and 2.6L engines

EGR System

▶ See Figures 14 and 15

2.0L AND 2.6L ENGINES

1. Allow the engine to cool overnight. Since the EGR system works differently for warm and cold engines, a completely cold engine is required for testing.
2. Disconnect the vacuum hose with the green stripe from either the throttle body (fuel injected) or the base of the carburetor. Attach the end of the hose to vacuum pump.
3. Plug the port from which the hose was removed. Start the engine and attempt to draw a vacuum with the hand pump. The system should NOT hold vacuum with the engine cold and running at idle.
4. Allow the engine to warm up to normal operating temperature. The coolant must be 80-85°C (175-185°F) before testing. Using the pump, draw a vacuum of 1.2-1.7 in. Hg (4-5.75 kPa). The carbureted system will leak vacuum at warm idle.
5. For carbureted engines, increase the engine speed to 3500 rpm. Slowly draw vacuum with the hand pump and observe the vacuum gauge on the pump. The system should leak vacuum until the pump reaches about 1.5 in. Hg, (5.75 kPa) at which time the vacuum should be held.

2.4L, 3.0L AND 3.5L ENGINES

▶ See Figures 14 and 15

1. Allow the engine to cool overnight.
2. Disconnect the vacuum hose with the green stripe from the EGR valve. Use a "T" connector to connect the hand vacuum pump into the system and connect the hose back to the EGR.
3. Start the engine and observe the vacuum gauge on the hand pump. Press the accelerator suddenly to race the engine. On a cold engine, there should be no change in the vacuum; normal (atmospheric) pressure is maintained.
4. Allow the engine to warm up to normal operating temperature, generally 158-170°F (68-77°C). Repeat the sudden rpm test while watching the gauge on the pump. The vacuum should rise temporarily to about 3.9 in. Hg (13.2 kPa).

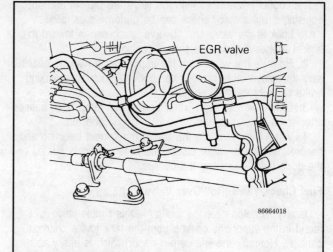

Fig. 14 Check the EGR system vacuum by using a hand vacuum pump and a "T" connector

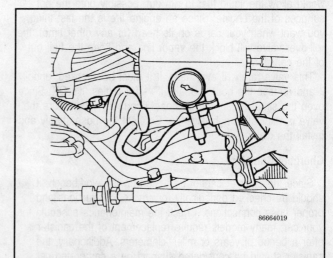

Fig. 15 Disconnect the three-way terminal (the "T"), and connect the hand vacuum pump to the EGR valve

EMISSION CONTROLS 4-9

5. Disconnect the "T" from the system and connect the vacuum pump directly to the EGR valve.

6. Draw a vacuum of at least 9.4 in. Hg (31.75 kPa) for the 2.4L and the 3.0L (12 valve) engines or 9.1 in. Hg (30.7 kPa) for the 3.0L (24 valve) and 3.5L engines while the engine is at warm idle. The quality of the idle should change noticeably, becoming rough or even stalling as the EGR valve opens. The exact vacuum level at which this occurs varies by engine family.

EGR Valve Control Vacuum

Coolant temperature should be 185-203°F (85-95°C).

1. Disconnect the vacuum hose from the throttle body EGR vacuum nipple and connect a hand vacuum pump to the nipple.

2. Start the engine and check to see that, after raising the engine speed by racing the engine, vacuum raises proportionately with the rise in engine speed.

➡If there is a problem with the change in vacuum, it is possible that the throttle body port may be clogged and requires cleaning.

3. Reconnect the vacuum hoses.

EGR Valve

▶ See Figure 16

1. Label and disconnect the hoses from the valve. Carefully loosen and remove the retaining bolts, remembering that they are probably heat-seized and rusty. Use penetrating oil freely.

2. Remove the valve and clean the gasket remains from both mating surfaces.

3. Inspect the valve for any sign of carbon deposits or other cause of binding or sticking. The valve must close and seal properly; the pintle area may be cleaned with solvent to remove soot and carbon.

4. Attach the vacuum pump to the vacuum port on the EGR valve. If the valve has two vacuum ports, pick one and plug the other.

5. Perform the vacuum holding test. Refer to the following information and draw the correct amount of vacuum, making sure it is held. If the correct vacuum cannot be maintained, the valve is leaking internally.

- 2.0L and 2.6L engines: holds vacuum at 10 in. Hg (33.75 kPa), will not pass air at 1.2 in. Hg (4 kPa), air passes through at 3.3 in. Hg (11.2 kPa).
- 3.0L 1989 engines: holds vacuum at 9.7 in. Hg (32.75 kPa), will not pass air at 0.8 in. Hg (2.7 kPa), air passes through at 3.3 in. Hg (11.2 kPa).
- 2.4L, 3.0L and 3.5L post 1990-95 engines: will not pass air at 1.2 in. Hg (4 kPa) or less, air passes through at 9.1-9.4 in. Hg (30.7-31.75 kPa) or more.

6. Release the vacuum but keep the pump attached to the valve. Devise a way to blow into the valve while drawing a vacuum and reading the gauge on the pump. Draw a slight vacuum and make sure that the valve is closed (your breath does not pass) at the specified vacuum.

7. Now increase the vacuum and check that the valve passes air at the specified vacuum. If the valve is sticky or worn, it may not open properly. A weakened valve will open too soon. If either condition is encountered, replace the valve.

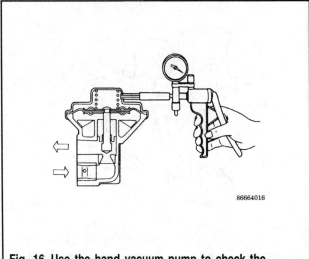

Fig. 16 Use the hand vacuum pump to check the functioning of the EGR valve

8. Install the valve with a new gasket. (Don't forget to remove the plug from the second vacuum port). Tighten the bolts.

9. Connect the hoses and lines to their proper ports.

EGR Control Solenoid

▶ See Figures 17, 18 and 19

1. Label and disconnect the vacuum hoses, taking note of the position of each.

2. Remove the wiring harness connector.

3. Connect the hand vacuum pump to the port which contained the vacuum hose with the red stripe (green stripe on 1990-95 2.4L, 3.0L and 1994-95 3.5L engines).

4. Use jumper wires to bridge battery voltage to the terminals of the solenoid. Draw a vacuum with the pump. When battery voltage is present, the solenoid should hold vacuum. When the voltage is removed, it should not be possible to draw and hold a vacuum. If either condition is not met, the unit must be replaced.

Thermo Valve

➡A vacuum pump capable of producing more than 10 in. Hg (33.8 kPa) of vacuum will be needed to perform this test.

The engine families which contain the secondary air system use the purge control temperature thermovalve to enable or trigger the EGR system. Later engines (G64B, 4G15, 4G61, 4G63 and 6G72) use a separate sensor devoted just to the EGR system.

1990-95 Pick-up trucks and Monteros use an additional sensor which measures the temperature of the exhaust gas at the EGR. The electrical signal generated by this EGR temperature sensor is used in conjunction with the signal from the oxygen sensor to fine tune the air/fuel mixture very accurately under all driving conditions.

The vacuum valve is checked by removing the hoses (label them!) and installing the vacuum pump on one port. Draw a vacuum with the pump: if the coolant temperature is below 122°F (50°C), vacuum should leak. Once the engine is warmed up to its normal operating temperature 175-185°F

4-10 EMISSION CONTROLS

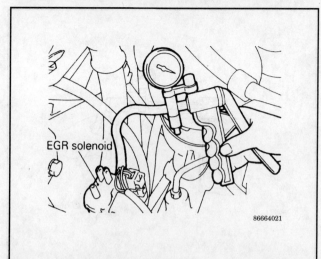

Fig. 17 Connect the hand vacuum pump to the nipple where the green-stripped vacuum hose was attached

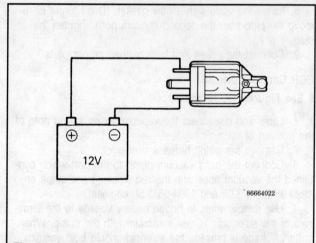

Fig. 18 Apply a vacuum and check to see if vacuum is maintained when a 12 volt current is supplied — be certain to connect the wires to the correct terminals

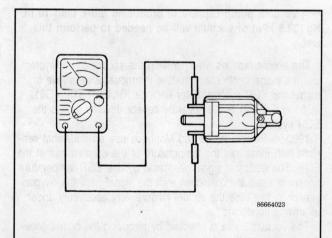

Fig. 19 Measure the resistance between the terminals of the solenoid valve — the resistance should be 36-44 ohms at 20°C (68°F)

(80-85°C) the vacuum should be held. If it is necessary to remove the valve, partially drain the coolant until it is below the level of the sensor. Perform this work only on a cold engine. Carefully unscrew the unit, applying wrench force only to the faceted part, never on the plastic. Before reinstalling, coat the threads with sealant (3M®No. 4171 or equivalent); tighten the valve to 22 ft. lbs. (30 Nm) and refill the coolant.

The California electric EGR temperature sensor must be removed from the car before testing. With the motor cold, carefully disconnect the wiring connector and unscrew the sensor from the EGR valve.

Place the sensor in a pan of water. Use a thermometer to measure the water temperature as you heat the pan. Use an ohmmeter to measure the resistance at the terminals of the sensor as the temperature increases. At 50°C (122°F) the resistance should be 60,000-83,000 ohms. When the water reaches 212°F (100°C), the resistance should be 11,000-14,000 ohms. The sensor should be replaced if there is significant deviation in the resistance. When reinstalling the sensor, tighten it to 8 ft. lbs. (11 Nm).

Secondary Air Supply System

OPERATION

▶ See Figure 20

The 2.0L and 2.6L engines use a second system to supply fresh air into the exhaust stream. This extra air is rich with oxygen and enhances the conversion process within the catalytic converter. At certain times, the exhaust system pressure is less than the pressure in the secondary air supply system. This is due to the natural power pulses of an engine. When the exhaust pressure is low, the higher pressure of the system pushes the reed open and admits fresh air into the exhaust stream. As the high pressure exhaust pulse passes the reed valve, the high pressure stops the fresh air from entering the exhaust stream and closes the valve itself. The fresh air may be either from the air cleaner or a separate intake system containing its own air cleaner. The entire system is controlled by the secondary air control solenoid, an electrically operated vacuum switch capable of disabling the system when it is not needed.

TESTING

▶ See Figures 21, 22, 23 and 24

➡A vacuum pump capable of producing more than 10 in. Hg (33.8 kPa) of vacuum will be needed to perform this test.

Although this system rarely fails, to check the secondary air control valve, remove the valve from the pipes connecting it to the right-hand exhaust manifold. Blow air into the valve from the air filter side to ensure that air does not blow through. Connect a hand vacuum pump to the secondary air control valve nipple. Apply a vacuum of 19.75 in. Hg (67 kPa) and check for air tightness (air does not blow through). Apply a vacuum of 40.75 in. Hg (20 kPa) and blow air into the valve from the air filter side; air should blow through. Still with the

EMISSION CONTROLS 4-11

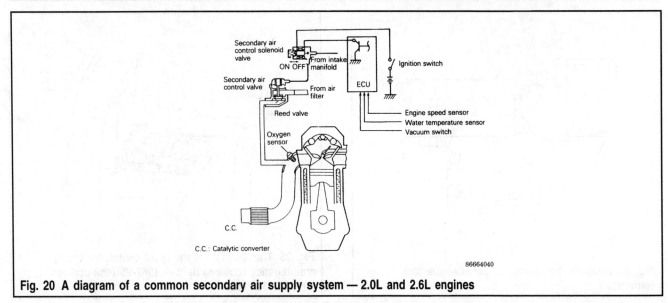

Fig. 20 A diagram of a common secondary air supply system — 2.0L and 2.6L engines

vacuum applied, blow through the valve from the exhaust manifold side; air should not blow through. If any fault is found in the above checks, replace the secondary air control valve. Install the secondary air control valve to 37-44 ft. lbs. (50-60 Nm).

To test the secondary air control solenoid valve, label and remove the vacuum hoses, taking note of the location of each. Generally, the hoses are marked with a blue, yellow and white stripe OR a green stripe and a white stripe or a red stripe. Attach the vacuum pump to the port which contained the red striped hose. Use jumper wires to bridge battery voltage to the terminals. The 4-cylinder Pick-up and Montero use a four conductor plug; apply the voltage to terminals 2 and 3.

Draw vacuum with the pump and check that the unit holds vacuum when power is applied and the other hose nipples are blocked or plugged. Remove the other nipple's plug and the vacuum should leak. When the power is removed and the other nipple is unplugged, the vacuum should leak. Use an ohmmeter to check the resistance across the terminals; correct resistance is 36-44 ohms at 68°F (10°C).

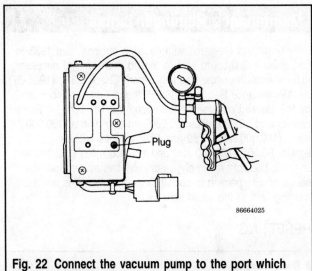

Fig. 22 Connect the vacuum pump to the port which contained the red stripe and apply a vacuum

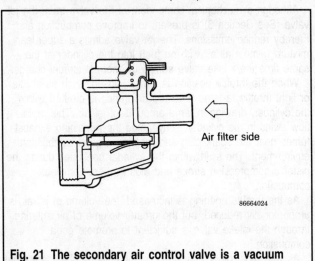

Fig. 21 The secondary air control valve is a vacuum assisted reed type valve — attach the hand vacuum pump to the top vacuum port to test

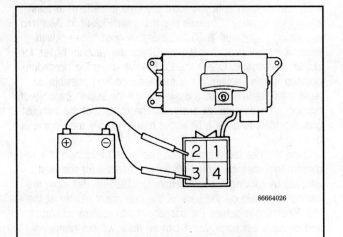

Fig. 23 Connect a 12 volt direct current power source to the No. 2 (+) and No. 3 (-) terminals — check to see if the vacuum leaks or is held

4-12 EMISSION CONTROLS

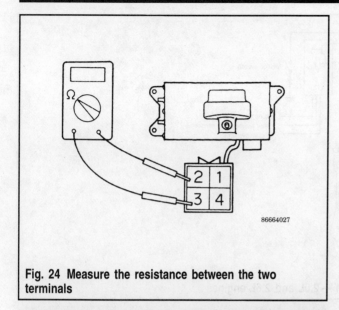

Fig. 24 Measure the resistance between the two terminals

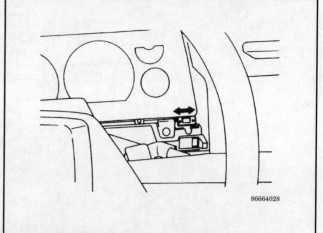

Fig. 25 The location of the reset switch for the maintenance required light — 1987-95 Pick-ups and 1990-95 Monteros

Maintenance Reminder Lights

Pick-ups and Monteros with gasoline engines from 1985 incorporate a dashboard light to remind the owner of necessary maintenance intervals. Reading either "Check EGR" (1985-87) or "Maintenance Reqd" (1988-95), these lights are controlled by the elapsed mileage on the odometer. The light will illuminate first at 50,000 miles (80,000 km) and then at 100,000 miles (160,000 km). 1989 and later vehicles will energize the light at 80,000 (128,000 km) and 120,000 miles (192,000 km) as well. The light is simply a reminder that system maintenance and inspection is due at these intervals; it is not a warning light in the usual sense.

RESETTING

▶ See Figure 25

When the light comes on, resist the temptation to simply turn it off. Leave it on until you have the time to perform at least basic maintenance. Chances are that your Pick-up or Montero is running pretty well at 50,000 miles. A good tune up with fresh filters will keep it that way. Check the vacuum hoses for cracks or damage. Check the EGR valve using the procedure outlined in this section. If this maintenance isn't possible, at least take the vehicle to a dealer or service station capable of running it on an exhaust analyzer. The content of the exhaust gas can tell you quite a bit about how efficiently the engine is burning fuel.

Once the needed maintenance has been performed, the reminder light may be turned off (which resets it for the next interval) by moving a small switch. On 1985-86 Pick-ups, the switch is located on the back of the instrument cluster at the top. Reaching it behind the dashboard can require contortions and creative language, but it can be done without removing the instrument cluster. On 1987-95 Pick-ups and 1990-95 Monteros, the switch is on the front of the instrument cluster, at the bottom right; the clear plastic window over the cluster must be removed to gain access. Carefully remove the four screws and the lens, reset the switch and reinstall the lens.

All 1985-89 Monteros have the reset switch located on the back of the speedometer, roughly behind the 80 mph (128 km/h) marking. Again, reaching it can be difficult, but it can be done. When reaching up behind the dashboard, be careful not to dislodge any wiring harnesses or connectors; don't force or push any components out of the way.

Mitsubishi recommends that after the 100,000 mile (160,000 km), and on 1989 and later models, 120,000 mile (192,000 km), service is performed, the light bulb be removed from the indicator. This does not relieve the owner from performing needed maintenance; it just turns the light off permanently. Removing the bulb or disabling the system before the specified mileage is a violation of the Federal Automobile Emissions Act.

Jet Valve

OPERATION

Although a mechanical component of the valve train, the jet valve (See Section 3) is present to improve combustion and thereby reduce emissions. The jet valve admits a super-lean mixture (almost all air with no fuel) into the cylinder at the same time the intake valve admits the normal air/fuel charge.

When the throttle position is almost closed, such as at idle or light throttle, a large pressure difference is created within the cylinder, drawing in large amounts of jet air. This rapid flow swirls in the cylinder, scavenging the remaining exhaust fumes near the spark plug and creating a good combustion environment. The swirl within the cylinder continues during the piston's compression stroke and aids in more complete combustion.

As the throttle opening is increased, the volume of jet air is proportionally reduced, but the greater volume of air entering through the intake valve is sufficient to promote good combustion.

As with most other emission control systems, there are times when the leaning effect of the jet valve is undesirable. A cold start, with the choke engaged, requires a rich mixture to allow the engine to run smoothly. The jet valve still opens with each intake stroke (it's driven by the camshaft) but the air passage is blocked by the jet air volume control valve. The valve is controlled by coolant temperature; when the engine warms to a pre-determined point, the temperature sensor activates the jet air volume control valve and the air flows into the cylinder on each intake stroke. The temperature sensor controlling this system is the same one used to control the EGR valve.

Vacuum Regulator Valve

The vacuum regulator valve is located in the device box, which is mounted on the left-hand inner fenderwell of the engine compartment. This system is found on 2.0L and 2.6L engines.

INSPECTION

➡ A vacuum pump capable of producing more than 10 in. Hg (33.8 kPa) of vacuum will be needed to perform these tests.

Valve

1. Disconnect the vacuum hose (white stripe) from the device box and connect a hand vacuum pump up to the port.
2. Apply a vacuum of 15.70 in. Hg (53 kPa) and check the VRV condition as follows:
   a. With the engine not running, the vacuum should leak.
   b. With the engine running at 3,500 rpm, the vacuum should not leak.
3. Remove the hand vacuum pump and reconnect the vacuum hose to the device box.

Control Vacuum

▶ See Figures 26 and 27

1. Locate the three vacuum hoses plugged into the back of the carburetor together. One leads to the EGR valve, one leads to the distributor and the last leads to the VRV. Disconnect the VRV vacuum hose.
2. Connect a hand vacuum pump to the VRV carburetor nipple.
3. Start and race the engine to make sure that the VRV vacuum increases gradually with the engine speed. If an abnormality is found in the change of vacuum, the VRV carburetor port could be blocked. Clean the port as necessary.

High Altitude Compensation System

OPERATION

▶ See Figure 28

➡ This system is only found on the 2.0L and 2.6L engines.

The carburetor meters fuel according to the volumetric flow rate of air and supplies the resultant mixture to the engine. Therefore, even if the carburetor is set for optimum air-fuel rate at low altitude, the mixture becomes too rich at high altitude since the air is less dense at high altitudes.

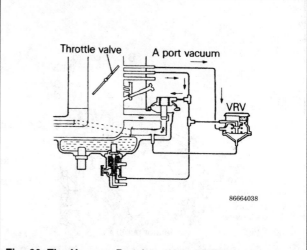

Fig. 26 The Vacuum Regulator Valve (VRV) system vacuum hose configuration

At high altitude, this high altitude compensation system supplies additional bleed air to the carburetor emulsion well and jet air passage to dilute the fuel, preventing an over-rich air-fuel mixture caused by the drop of air density at high altitudes.

The system also advances the ignition timing by a fixed amount to reduce CO and HC emission and to secure driveability at high altitudes.

At low altitudes, the HAC opens to allow the intake manifold vacuum to escape from the HAC into the atmosphere. Therefore, the vacuum switching valve and HAC's additional bleed air passage remain closed and bleed air is not supplied to the carburetor. At high altitudes, the HAC closes and the intake manifold vacuum is applied to the vacuum switching valve, HAC's additional bleed air passage and distributor. As a result, the vacuum switching valve and HAC's additional bleed air passage are opened to supply bleed air to the carburetor. At the same time, the distributor advances the ignition timing.

INSPECTION — NON-CALIFORNIA MODELS

▶ See Figures 29 and 30

➡ A vacuum pump capable of producing more than 10 in. Hg (33.8 kPa) of vacuum will be needed to perform this test.

System Inspection

ALTITUDES BELOW 3,900 FT. (1,200M)

When disconnecting the vacuum hose, put a mark on the hose so that it may be reconnected to the original position. The engine coolant temperature should be 185-205°F (85-95°C).

1. Connect a timing light to the engine.
2. Remove the air filter.
3. Disconnect the vacuum hoses (black, red stripe, black) from the carburetor primary emulsion well bleed nipple, secondary emulsion well bleed nipple and the jet air nipple; plug the nipples.

4-14 EMISSION CONTROLS

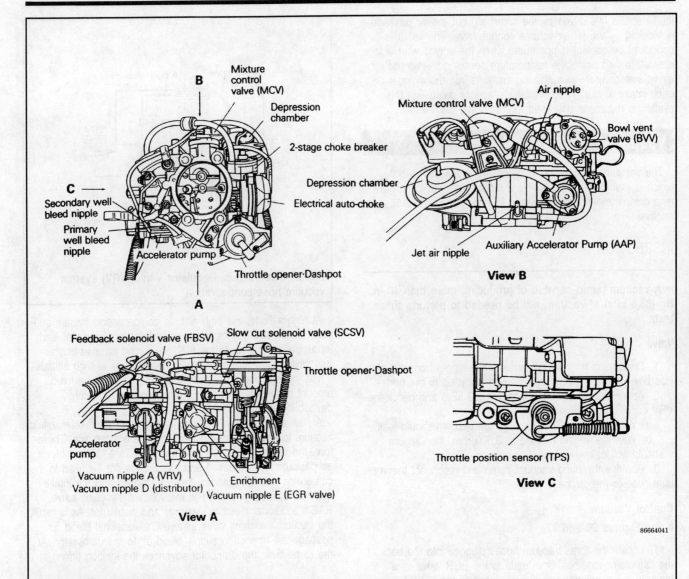

Fig. 27 The external vacuum connections and components of the Mitsubishi carburetor

4. Connect a hand vacuum pump to the vacuum hoses, one hose at a time, and check air tightness (the vacuum should NOT leak) while running the engine at idle.
5. Connect the vacuum hoses back to their original positions.
6. Run the engine at idle and check ignition timing. The timing should be 6-10° BTDC for the 2.0L engines and 5-9° BTDC for the 2.6L engines.
7. While the engine is idling, disconnect the vacuum hose (yellow stripe) from the HAC and put a finger at the hose end to check that vacuum is felt.
8. Connect everything back to its original position and remove the timing light.

ALTITUDES ABOVE 3,900 FT. (1,200M)

When disconnecting the vacuum hose, put a mark on the hose so that it may be reconnected to the original position. The engine coolant temperature should be 185-205°F (85-95°C).

1. Connect a timing light to the engine.

2. Remove the air filter.
3. Disconnect the vacuum hoses (black, red stripe, black) from the carburetor primary emulsion well bleed nipple, secondary emulsion well bleed nipple and the jet air nipple; plug the nipples.
4. Connect a hand vacuum pump to the vacuum hoses, one hose at a time, and check the air tightness (the vacuum SHOULD leak) while running the engine at idle.
5. Connect the vacuum hoses back to their original positions.
6. Run the engine at idle and check ignition timing. The timing should be 13° BTDC for the 2.0L engines and 12° BTDC for the 2.6L engines.
7. Connect everything back to its original position and remove the timing light.

High Altitude Compensator (HAC)

➡A vacuum pump capable of producing more than 10 in. Hg (33.8 kPa) of vacuum will be needed to perform this test.

EMISSION CONTROLS 4-15

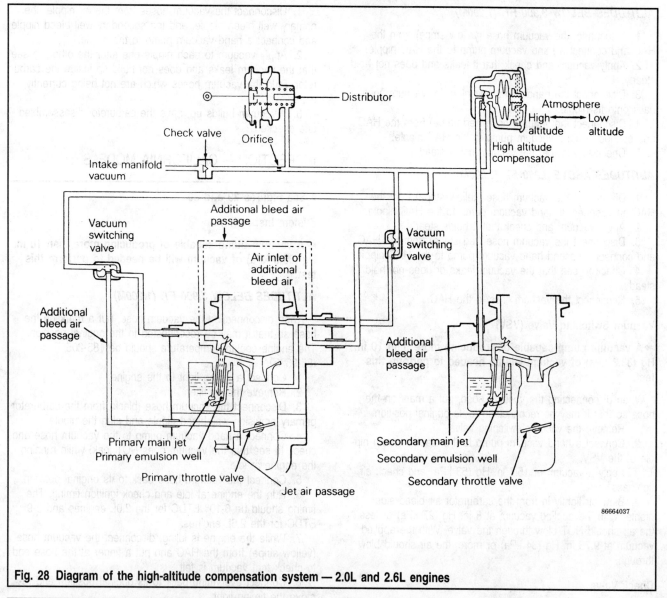

Fig. 28 Diagram of the high-altitude compensation system — 2.0L and 2.6L engines

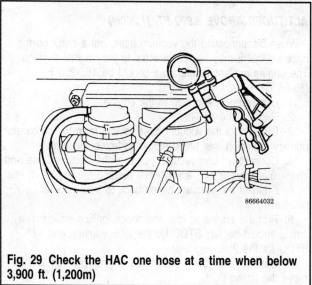

Fig. 29 Check the HAC one hose at a time when below 3,900 ft. (1,200m)

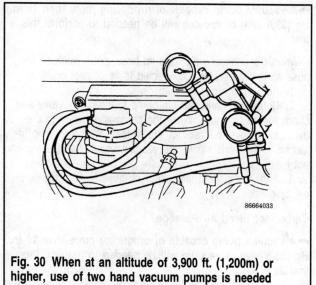

Fig. 30 When at an altitude of 3,900 ft. (1,200m) or higher, use of two hand vacuum pumps is needed

4-16 EMISSION CONTROLS

ALTITUDES BELOW 3,900 FT. (1,200M)

1. Disconnect the vacuum hose (yellow stripe) from the HAC and connect a hand vacuum pump to the HAC nipple.
2. Apply vacuum and check that it leaks and does not hold steady.
3. Disconnect the hand pump and plug the vacuum hose back onto its nipple.
4. Disconnect the vacuum hose (red stripe) from the HAC and connect a hand vacuum pump to the HAC nipple.
5. Check to see that the vacuum holds steady.

ALTITUDES ABOVE 3,900 FT. (1,200M)

1. Disconnect the vacuum hose (yellow stripe) from the HAC and connect a hand vacuum pump to the HAC nipple.
2. Apply vacuum and check that it holds steady.
3. Disconnect the vacuum hose (red stripe) from the HAC and connect a second hand vacuum pump to the HAC nipple.
4. Check to see that the vacuum leaks or does not hold steady.
5. Reconnect the vacuum lines to the HAC.

Vacuum Switching Valve (VSV)

➡ A vacuum pump capable of producing more than 10 in. Hg (33.8 kPa) of vacuum will be needed to perform this test.

When disconnecting the vacuum hose, put a mark on the hose so that it may be reconnected to its original position.

1. Remove the vacuum switching valve.
2. Connect a hand vacuum pump to the black vacuum nipple of the VSV.
3. Apply a vacuum of 15.7 in. Hg (53 kPa) and check air tightness.
4. Blow air lightly in from the carburetor air bleed side nipple. With the applied vacuum at 8 in. Hg (27 kPa) or less, the air should NOT blow through the valve. With the applied vacuum at 9.75 in. Hg (34 kPa) or more, the air should blow through.

Check Valve

➡ A vacuum pump capable of producing more than 10 in. Hg (33.8 kPa) of vacuum will be needed to perform this test.

When disconnecting the vacuum hose, put a mark on the hose so that it may be reconnected to its original position.

1. Remove the check valve.
2. Connect a hand vacuum pump to the check valve and check the air tightness. When the pump is attached to the dark blue nipple, the vacuum should leak. With the pump attached to the white nipple, the vacuum should hold steady and not leak.
3. If the component fails either of these tests, it needs to be replaced.

Carburetor Bleed Air Passage

➡ A vacuum pump capable of producing more than 10 in. Hg (33.8 kPa) of vacuum will be needed to perform this test.

1. Disconnect the vacuum hoses from the air nipple, the primary well bleed nipple, and the secondary well bleed nipple and connect a hand vacuum pump to the nipple.
2. Apply vacuum to each nipple one after the other, to see that the vacuum leaks and does not build up inside the carburetor. Plug the vacuum hoses which are not being currently tested.
3. If vacuum builds up, have the carburetor disassembled and checked.

INSPECTION — CALIFORNIA MODELS

◆ See Figures 29 and 30

System Inspection

➡ A vacuum pump capable of producing more than 10 in. Hg (33.8 kPa) of vacuum will be needed to perform this test.

ALTITUDES BELOW 3,900 FT. (1,200M)

When disconnecting the vacuum hose, put a mark on the hose so that it may be reconnected to the original position. The engine coolant temperature should be 185-205°F (85-95°C).

1. Connect a timing light to the engine.
2. Remove the air filter.
3. Disconnect the vacuum hose (black) from the carburetor primary emulsion well bleed nipple and plug the nipple.
4. Connect a hand vacuum pump to the vacuum hose and check to see that the vacuum does NOT leak, while running the engine at idle.
5. Connect the vacuum hose back to its original position.
6. Run the engine at idle and check ignition timing. The timing should be 6-10° BTDC for the 2.0L engines and 5-9° BTDC for the 2.6L engines.
7. While the engine is idling, disconnect the vacuum hose (yellow stripe) from the HAC and put a finger at the hose end to check that vacuum is felt.
8. Connect everything back to its original position and remove the timing light.

ALTITUDES ABOVE 3,900 FT. (1,200M)

When disconnecting the vacuum hose, put a mark on the hose so that it may be reconnected to the original position. The engine coolant temperature should be 185-205°F (85-95°C).

1. Connect a timing light to the engine.
2. Remove the air filter.
3. Disconnect the vacuum hose (black) from the carburetor primary emulsion well bleed nipple and plug the nipple.
4. Connect a hand vacuum pump to the vacuum hose and check that the vacuum leaks, while running the engine at idle.
5. Connect the vacuum hoses back to their original positions.
6. Run the engine at idle and check ignition timing. The timing should be 13° BTDC for the 2.0L engines and 12° BTDC for the 2.6L engines.
7. Connect everything back to its original position and remove the timing light.

EMISSION CONTROLS 4-17

High Altitude Compensator (HAC)

➡ A vacuum pump capable of producing more than 10 in. Hg (33.8 kPa) of vacuum will be needed to perform this test.

ALTITUDES BELOW 3,900 FT. (1,200M)

1. Disconnect the vacuum hose (white stripe, two nipples side) from the HAC and connect a hand vacuum pump to the HAC nipple.
2. Apply vacuum and check that it leaks and does not hold steady.
3. Disconnect the hand pump and plug the vacuum hose back onto its nipple.
4. Disconnect the vacuum hose (black) from the HAC and connect a hand vacuum pump to the HAC nipple.
5. Check to see that the vacuum holds steady.

ALTITUDES ABOVE 3,900 FT. (1,200M)

1. Disconnect the vacuum hose (white stripe, two nipples side) from the HAC and connect a hand vacuum pump to the HAC nipple.
2. Apply vacuum and check that it holds steady.
3. Disconnect the vacuum hose (black) from the HAC and connect a second hand vacuum pump to the HAC nipple.
4. Check to see that the vacuum leaks or does not hold steady.
5. Reconnect the vacuum lines to the HAC.

Vacuum Switching Valve (VSV)

These procedures are the same as the non-California procedures described earlier.

Check Valve

These procedures are the same as the non-California procedures described earlier.

Carburetor Bleed Air Passage

These procedures are the same as the non-California procedures described earlier.

Intake Air Temperature Control System

This system is only found on the 2.0L and 2.6L gasoline engines.

INSPECTION

▶ See Figure 31

System Inspection

1. Remove the air filter cover and air duct.
2. Run the engine at idle and check the air control valve condition. At 86°F (30°C) or less, the cold air side inlet should close. At 113°F (45°C) or more, the cold air side inlet should open.

If necessary, apply compressed air to cool or apply hot air using a hair dryer, etc. to heat.

Air Control Valve

▶ See Figure 32

➡ A vacuum pump capable of producing more than 10 in. Hg (33.8 kPa) of vacuum will be needed to perform this test.

1. Remove the air filter.
2. Disconnect the vacuum hose from the air control valve and connect a hand vacuum pump to the valve nipple.
3. Apply a vacuum of 19.75 in. Hg (67 kPa) and check to make sure the vacuum holds steady and does not leak down.
4. Check the air control valve operation. With a vacuum of 2.85 in. Hg (9 kPa) or less the cold air side inlet opens. With a vacuum of 7.53 in. Hg (25 kPa) or more the cold air side inlet closes.
5. Connect the vacuum hose back to their original position.

Thermo Sensor

▶ See Figure 33

➡ A vacuum pump capable of producing more than 10 in. Hg (33.8 kPa) of vacuum will be needed to perform this test.

1. Connect a hand vacuum pump to the thermo sensor nipple and check air tightness. With the thermo sensor temperature at 86°F (30°C) or less the vacuum should not leak. With the thermo sensor temperature at 113°F (45°C) or more the vacuum should leak.
2. If any fault is found in the above checks, replace the air filter body.

Mixture Control Valve (MCV)

This procedure is only relevant to 2.0L and 2.6L engines equipped with manual transmissions.

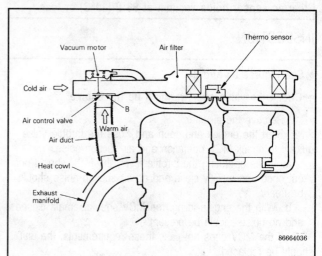

Fig. 31 A diagram of a common intake air temperature control system — 2.0L and 2.6L engines

4-18 EMISSION CONTROLS

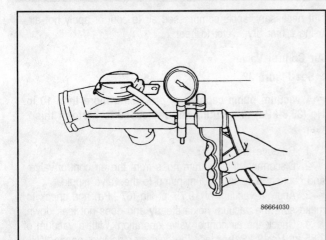

Fig. 32 Use a hand vacuum pump to check the air control valve

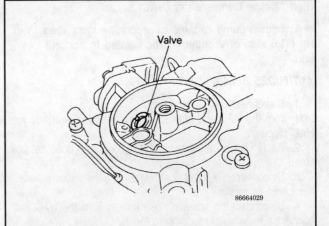

Fig. 34 The location of the MCV on the carburetor — when the throttle lever is operated, this valve pops out once and then quickly closes

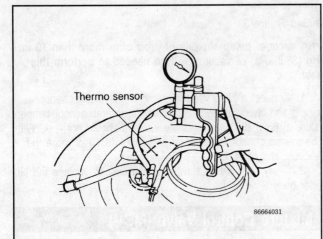

Fig. 33 Using a vacuum pump, check to make sure the thermo sensor holds vacuum at 86°F (30°C) or less.

INSPECTION

▶ See Figures 34 and 35

➡Check the valve after warming up the engine

1. Remove the air filter.
2. Start the engine and open and close the throttle valve quickly to check MCV operation and air suction noise.
   a. While operating the throttle lever, the MCV should pop out once and quickly close and an air suction noise should be heard.
   b. With the engine idling, the MCV should remain closed and no noise should be present.
3. If the MCV does not pass these requirements, the unit should be replaced.

Variable Induction System (VIS)

➡This system is only found on the 3.5L engines.

OPERATION

The Variable Induction System (VIS) controls the dual-stage intake manifold through vacuum pressure. The dual-stage intake manifold is equipped with two sets of intake runners: a short pair and a long pair. The different lengths of the runners makes the intake manifold more versatile at various speeds and allows for better performance. The system components include the vacuum tank and the variable induction control solenoid.

INSPECTION

▶ See Figure 36

System Testing

1. Allow the engine to reach normal operating temperature.
2. Connect a tachometer to the engine.
3. Connect a vacuum gauge, by using a T-joint, between the variable induction control solenoid and the variable induction control vacuum actuator.
4. Start the engine and check to be sure that the negative pressure is applied to the vacuum gauge according to the following items:
   a. At the engine rpm of 3,200 or less, negative pressure should be maintained and the control valve should be closed.
   b. When the engine is raced from 3,200 rpm or less, the negative pressure should not change and the control valve should be closed.
   c. With the engine rpm at or above 3,400, the negative pressure should leak and the control valve should be open.
5. Check to be sure that the rod of the variable induction control vacuum actuator moves during inspection.

EMISSION CONTROLS 4-19

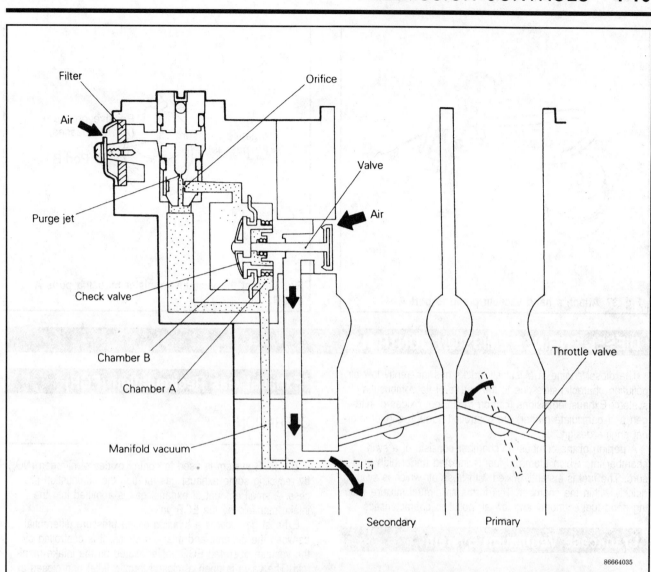

Fig. 35 A diagram of a common mixture control valve system — 2.0L and 2.6L engines

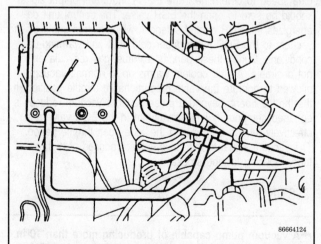

Fig. 36 Attach a vacuum gauge, using a T-joint, to the hose running between the vacuum tank and the variable induction control solenoid

Vacuum Tank Testing

▶ See Figures 37 and 38

➡ A vacuum pump capable of producing more than 10 in. Hg (33.8 kPa) of vacuum will be needed to perform this test.

1. Install a hand vacuum pump to port A and check to be sure that air-tightness is maintained when 19.7 in. Hg (66.5 kPa) of vacuum is applied. After checking, remove the hand vacuum pump.
2. Check to be sure that air passes through when negative pressure is applied to port B, and that air does not pass through when positive pressure is applied to port B.

Variable Induction Control (VIC) Solenoid

For electrical testing of the VIC solenoid and harness, refer to the procedures presented later in the electrical engine controls portion of this section.

4-20 EMISSION CONTROLS

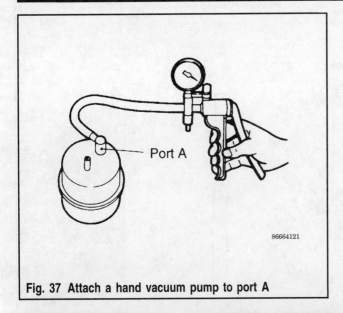

Fig. 37 Attach a hand vacuum pump to port A

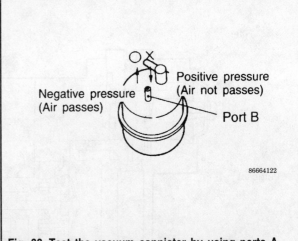

Fig. 38 Test the vacuum cannister by using ports A and B

DIESEL ENGINES EMISSION CONTROLS

The diesel engine in Mitsubishi trucks has inherently low air pollution characteristics due to the nature of its combustion system. Exhaust emissions (carbon monoxide, oxides of nitrogen and particulate matter) are controlled through careful internal engine design.

A portion of the combustion chamber consists of a swirl chamber into which atomized fuel is injected under high pressure. The fuel is instantly mixed with fresh air which is swirling quickly within the chamber. This forms an air/fuel mixture yielding good fuel economy and low air pollution characteristics.

Crankcase Ventilation System

GENERAL INFORMATION

A sealed crankcase ventilation system is used to prevent the blow-by gasses from escaping into the atmosphere. This system consists of a breather hose, an outlet pipe at the rocker cover, and an inlet pipe at the air intake hose.

Blow-by gasses generated in the crankcase are drawn into the air intake hose through the breather hose in the valve cover. Aside from keeping the hose and passages clean and open, the system requires no maintenance.

Evaporative Emission Controls

GENERAL INFORMATION

Diesel fuel does not pose an evaporant hazard; therefore no serviceable evaporative controls are installed on the diesel truck.

Exhaust Gas Recirculation (EGR)

OPERATION

The EGR system is used to control oxides of nitrogen (NO_x) by recycling some exhaust gas through the combustion process. A small amount of exhaust gas is admitted into the intake manifold by the EGR valve.

Exhaust gas flow is a function of the pressure differential between the exhaust and intake systems. It is controlled by the vacuum operated EGR valve located on the intake manifold. The valve is open at closed throttle (idle) and closed at wide open throttle.

The EGR control unit monitors coolant temperature and engine speed to determine if the EGR control unit output signal should be sent to open the EGR valve. The control unit de-energizes the EGR control solenoid, which in turn shuts off the vacuum to the EGR valve under cold start and warming-up conditions. During these periods, exhaust gas recirculation is not desirable. As the coolant warms up to normal, vacuum is allowed to operate the valve proportional to throttle position. If the throttle opens beyond a preset level, a vacuum reducer is engaged, temporarily limiting vacuum to the EGR valve and effectively turning it off during the wide open throttle period. The vacuum signal to the valve is also reduced at high altitude, reducing tailpipe smoke.

TESTING

➡ **A vacuum pump capable of producing more than 10 in. Hg (33.8 kPa) of vacuum will be needed to perform this test.**

1. Start the engine and warm it to full operating temperature. This can be achieved quicker by driving the vehicle than by allowing it to idle.

EMISSION CONTROLS 4-21

2. Disconnect the vacuum hose from the EGR valve and install a vacuum pump and short piece of hose. Alternately, disconnect the EGR hose at the control solenoid and connect the hose directly to the vacuum pump.

3. With the engine running at warm idle, draw a vacuum with the hand pump; the EGR valve should open. The engine sound should change, possibly even stalling, depending on how far the valve is opened.

4. Draw and release vacuum several times and at different speeds. The response of the valve (and therefore the idle speed/quality) should be proportional the the vacuum signal from the pump.

5. Disconnect the test equipment and reconnect the vacuum hose.

REMOVAL & INSTALLATION

1. Label and disconnect the hoses from the valve. Carefully loosen and remove the retaining bolts, remembering that they are probably heat-seized and rusty. Use penetrating oil freely.

2. Remove the valve and clean the gasket remains from both mating surfaces.

3. Inspect the valve for any sign of carbon deposits or other cause of binding or sticking. The valve must close and seal properly; the pintle area may be cleaned with solvent to remove soot and carbon.

4. After cleaning and inspection, recheck the motion of the valve with a hand vacuum pump. Watch the shaft area closely. A stuck or binding valve will not open properly; a weakened valve will open too soon. If either condition is encountered, replace the valve.

To install:

5. Install the valve with a new gasket. Tighten the bolts evenly to 5 ft. lbs. (7 Nm).

6. Connect the hoses and lines to their proper ports.

High Altitude Compensation System

The diesel trucks sold as Federal specification low altitude vehicles will meet the Federal high altitude requirements with no adjustments required. Trucks sold for principal use in high altitude areas are adjusted by the dealer in order to reduce levels of visible smoke and soot. The adjustment involves re-calibrating the injection timing from 5° ATDC to 3° ATDC.

Air Conditioning Idle Speed Compensator

When the optional air conditioning unit is installed, the idle speed must be increased when the air conditioner compressor is engaged. The compressor adds sufficient load to the engine that uneven idle or stalling may occur if compensation is not provided.

This system consists of a vacuum actuator and a vacuum solenoid valve. When the compressor is switched on, the solenoid valve is energized, allowing some vacuum from the power brake booster assembly to be bled into the actuator. As a result, the control lever is slightly pulled by the actuator, admitting more fuel to the engine. Consequently, the engine runs at a faster idle speed.

Fuel Injection

The fuel injection system, the timing of fuel delivery into the engine and the quality of fuel used all contribute significantly to emission reduction. Correct diesel fuel — either 1-D, 2-D or winterized 2-D is to be used. Do not use heating fuel oil or diesel fuels intended for industrial or marine use. Buy fuel from a reputable dealer and avoid the use of additives. The diesel fuel system is discussed in detail in Section 5 of this manual.

CARBURETED ENGINE CONTROLS

Feedback Carburetor System (FBC)

All Mitsubishi feedback carburetors employ an oxygen sensor as part of the feedback system controlling the air/fuel mixture within the carburetor. Although this technology is now rather dated in view of the sophisticated fuel injection controls available today, the FBC system represents the first real attempt to control fuel delivery by means other than gravity and air pressure.

Reduced to its simplest form, the oxygen sensor reads the content of the exhaust and communicates electrically with the Engine Control Module (ECM). This micro computer evaluates the signal and adjusts the amount of fuel entering the engine to achieve the best combustion and therefore the lowest emissions. Although the oxygen sensor plays a critical role in this control process, the ECM also interprets signals from other sensors reading throttle position, engine temperature and other variables.

In ordinary operating conditions after engine warm up, the air/fuel ratio is within the usable range of the oxygen sensor. By attempting to keep the air and fuel mixed at the proportionally perfect (stoichiometric) ratio, the exhaust will be almost completely processed by the catalytic converter. If the exhaust changes drastically, pollutants may exceed the capacity of the converter. This constant process of checking the exhaust and adjusting the mixture is called closed loop operation.

There are times during normal operation, however, when the exhaust is outside these usable limits. Conditions of engine start-up, partially warmed driving, high load operation or sudden deceleration can each cause sufficient exhaust conditions to exceed the range of the oxygen sensor. For this reason, when the ECM is advised by the sensors of one of these conditions, the system enters open loop operation. The ECM then controls the carburetor based on pre-programmed values (often called default values) and disregards the signal from the oxygen sensor. These pre-programmed values are in the ROM (Read Only Memory) of the ECM and, since they are installed at manufacture, cannot be altered by the oxygen sensor. They control the carburetor to allow the quickest warm up of a cold engine or the best compromise of performance and emissions under the existing condition.

The feedback carburetor is covered in Section 5.

4-22 EMISSION CONTROLS

FUEL INJECTION ENGINE CONTROLS

Troubleshooting Procedures

➡ **The 1993-95, rear-wheel drive 2.4L (California) Pick-ups have the same engine controls as the 3.0L engines. All other 2.4L engines are listed as 2.4L engines.**

To effectively troubleshoot the Multi-port Fuel Injection (MFI) system, distinct steps or procedures should be systematically followed. Haphazard work in this area will not effectively lead one to the true problem in the MFI system. Adhere to the following procedures for constantly precise troubleshooting.

1. Confirm the malfunction. Cause the malfunction to reoccur and check the details of the malfunction and the conditions (engine condition, driving conditions, etc.) when the malfunction reoccurs.
2. Read the diagnostic trouble codes out. When the diagnostic trouble code is read out and the diagnostic trouble code is output, refer to the diagnostic procedures and repair the indicated malfunction.
3. Assume the probable cause of the malfunction and determine the appropriate check points. Refer to the procedures by their problem symptoms and confirm the check point and the sequence of checking for the indicated malfunction.
4. Check the engine control module's input and output signals. Using the scan tool or an oscilloscope, check the input and output signals of the engine control module. If the input and output signals are normal, sensor input/actuator control can be judged to be normal, so check the input and output signals of the subsequent check point.
5. Check the MFI component's harness. If an abnormal condition of the input and output signals of the engine control unit is discovered, check, and repair if necessary, the body harness of the MFI component. After making this repair, again check the input and output signals of the engine control module; if they are normal proceed to the checking of the input and output signals of the subsequent check point.
6. Check the MFI component itself. If the input and output signals of the engine control module are abnormal even though the body harness is normal, check the MFI component itself, and repair or replace as necessary. After making this repair or replacement, again check the input and output signals of the engine control module; if they are normal, proceed to the checking of the input and output signals of the subsequent check point.

Harness Checking Precautions

The following precautions will help make the procedures for checking the wiring harnesses easier and safer. Damage can be done to a harness if wires are not hooked up correctly, and the directions for checking the harnesses are written with certain descriptions which stay consistent throughout all of the procedures.
- Connector symbols are described as seen from the end of the terminal for the connector.
- The abbreviation "B+" used for the normal judgement value when checking the voltage is the abbreviation for battery positive voltage.
- Be sure to use the special tool (test harness) when, for a waterproof connector, checking while the circuit is conductive. If a probe is inserted from the harness side, the waterproof capability will be lowered, thereby causing corrosion. Never do so.
- Also, if there is no test harness with an applicable connector, the test harness set (MB991348) can be used to make connections directly between the terminals.
- When a connector is disconnected in order to check terminal voltage, etc., never insert a probe if the terminal to be checked is a female pin, because the forceful insertion of a probe will cause improper or incomplete contact.
- When checking for damaged or disconnected wiring of a harness (open circuit) and if both ends of the harness are unconnected, use a jumper wire to ground one end of the harness, and then check for continuity between the other end and ground. By doing this, you can check for damaged or disconnected wiring, and, if there is no continuity, the harness should be repaired.
- When checking for a harness short-circuit (short-circuit to ground), open one end of the harness and then check for continuity between the other end and ground. If there is continuity, the harness is short-circuited to ground and should be repaired.
- If the voltage (power-supply voltage) supplied to a sensor is not normal, repair the harness. If the voltage to the sensor is still not normal after the harness has been repaired, replace the engine control unit and check again.

Check Engine/Malfunction Indicator Lamp

Among the on-board diagnostic items, a check engine/malfunction indicator lamp comes on to notify the driver of the emission control items when an irregularity is detected. However, when an irregular signal returns to normal and the engine control module judges that it has returned to normal, the check engine/malfunction indicator lamp goes out. Moreover, when the ignition switch is turned **OFF**, the light goes out. Even if the ignition switch is turned on again, the light does not come on until the irregularity is detected. Here, immediately after the ignition switch is turned on, the check engine/malfunction indicator lamp is lit for 5 seconds to indicate that the check engine/malfunction indicator lamp operates normally.

Component Inspection Procedures

Some components on the MFI system need to be inspected using a special Mitsubishi tool (Multi Use Tester MUT/MUT II), this tool is expensive and, unless extensive testing is foreseen, it might be more financially feasible to test the harness yourself and have the component tested by an automotive mechanic familiar with Mitsubishi vehicles.

EMISSION CONTROLS 4-23

Multi-port Fuel Injection Relay

OPERATION

▶ See Figures 39 and 40

While the ignition switch is **ON**, battery power is supplied to the engine control module, injectors, volume air-flow sensor, etc. Turning the ignition switch to the **ON** position causes current to flow from the ignition switch through the MFI relay coil to ground. This turns the MFI relay switch on and supplies power from the battery to the engine control module via the MFI relay switch.

HARNESS TESTING

2.4L Engines
▶ See Figure 41

➡ The 1993-95, rear-wheel drive 2.4L (California) Pick-ups have the same engine controls as the 3.0L engines. All other 2.4L engines are listed as 2.4L engines.

1. With the MFI relay harness unplugged from the relay and the ignition switch on, measure the power supply voltage to terminal 8 of the harness connector. If there is no voltage present, the problem lies in the ignition switch or in the harness between the ignition switch and the MFI relay connector, these should be repaired or replaced as needed.
2. Check for continuity of the ground circuit on terminal 6 of the MFI relay harness connector. The ignition should be **OFF** and the MFI relay still unplugged. If continuity is present, stop the check, the problem lies elsewhere. If there is no continuity, repair or replace the harness between the MFI relay and ground.
3. With the MFI connector still unplugged, measure the power supply voltage of terminal 4 of the MFI relay. If the power is not sufficient (the power is the voltage of the battery, therefore around 10-12 volts), repair the harness between terminal 4 and the battery.
4. Disconnect the Engine Control Module (ECM) connector, and the MFI relay connector should still be unplugged. Check for an open-circuit, or short-circuit to ground, between the ECM (terminals 102 and 107) and the MFI relay (terminal 3). If an open-circuit is found, repair the harness between the ECM and terminal 3 of the MFI relay. If no short-circuit is found, the component is OK.

3.0L (12 Valve) Engines
▶ See Figure 42

When performing these checks, use a harness side connector to fasten onto the different terminals.

1. Disconnect the ECM harness. Measure the ignition switch-IG terminal (terminal 110) input voltage. With the ignition

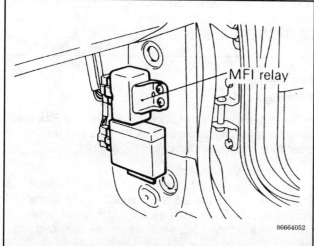

Fig. 39 Location of the Multi-port Fuel Injection (MFI) relay on 2.4L equipped vehicles

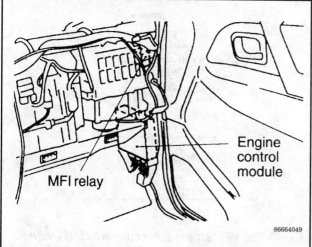

Fig. 40 Location of the Multi-port Fuel Injection (MFI) relay on 3.0L and 3.5L equipped vehicles

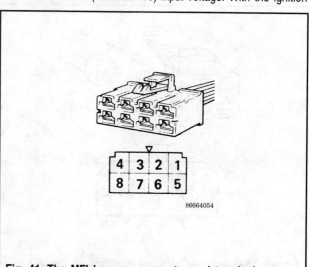

Fig. 41 The MFI harness connector and terminal identification — 2.4L engines

EMISSION CONTROLS

switch **OFF**, the voltage should be 0-1 volts. When the ignition switch is **ON**, the voltage should be that of the battery (around 10-12 volts). If the circuit shows these voltages, go on to Step 2, otherwise repair the harness between terminal 110 and the ignition switch.

  2. Turn the ignition switch **OFF** and unplug the MFI relay connector. Measure the power supply voltage of the MFI relay (terminal 10 to ground). If the voltage is that of the battery, continue on to Step 3, otherwise repair the harness between terminal 10 and the battery.

  3. With both the ECM and the MFI relay connectors unplugged, check for an open-circuit or a short-circuit to ground between the MFI relay (terminal 8) and the ECM (terminals 66 and 63). If no short-circuit is found, move on to Step 4, otherwise repair the harnesses between terminal 8 and terminals 66 and 63.

  4. Keep both the ECM and the MFI relay connectors unplugged and check for an open-circuit between terminal 4 of the MFI and terminals 102 and 107 of the ECM. If a short-circuit is found, repair the harnesses between the terminals, otherwise go on to Step 5.

  5. Connect the ECM and the MFI relay. While cranking the engine to start it, measure the power supply voltage of the actuator (terminal 5 of the MFI to ground). The voltage should be 8 volts or higher. Once the engine is running, race the engine and measure the voltage again. The voltage should be at the battery level. If the voltage is not as indicated, either the ECM or the MFI relay is defective.

3.0L (24 Valve) and 3.5L Engines
▶ See Figure 43

When performing these checks, use a harness side connector to fasten onto the different terminals.

  1. Disconnect the ECM harness. Measure the ignition switch-IG terminal (terminal 62) input voltage. With the ignition switch **OFF**, the voltage should be 0-1 volts. When the ignition switch is **ON**, the voltage should be that of the battery (around 10-12 volts). If the circuit shows these voltages, go on to Step 2, otherwise repair the harness between terminal 62 and the ignition switch.

  2. Turn the ignition switch **OFF** and unplug the MFI relay connector. Measure the power supply voltage of the MFI relay (terminals 4 and 8 to ground). If the voltage is that of the battery, continue on to Step 3, otherwise repair the harness between terminals 4 and 8 and the battery.

  3. With both the ECM and the MFI relay connectors unplugged, check for an open-circuit or a short-circuit to ground between the MFI relay (terminal 6) and the ECM (terminal 38). If no short-circuit is found, move on to Step 4, otherwise repair the harnesses between terminal 6 and terminal 38.

  4. Keep both the ECM and the MFI relay connectors unplugged and check for an open-circuit between terminal 2 of the MFI and terminals 12 and 25 of the ECM. If a short-circuit is found, repair the harnesses between the terminals, otherwise go on to Step 5.

  5. Connect the ECM and the MFI relay. While cranking the engine to start it, measure the power supply voltage of the actuator (terminal 3 of the MFI to ground). The voltage should be 8 volts or higher. Once the engine is running, race the engine and measure the voltage again. The voltage should be at the battery level. If the voltage is not as indicated, either the ECM or the MFI relay is defective.

COMPONENT INSPECTION

2.4L Engines
▶ See Figure 44

➡ The 1993-95, rear-wheel drive 2.4L (California) Pick-ups have the same engine controls as the 3.0L engines. All other 2.4L engines are listed as 2.4L engines.

  1. Remove the MFI relay.
  2. Check the continuity between the MFI relay terminals. Check terminals 3-5 and 2-5, which should register approximately 95 ohms. Check terminals 6-7, which should register 35 ohms. Check terminals 6-8, which will only be connected one way.

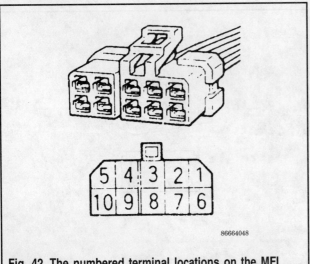

Fig. 42 The numbered terminal locations on the MFI harness connector — 3.0L (12 valve) engines

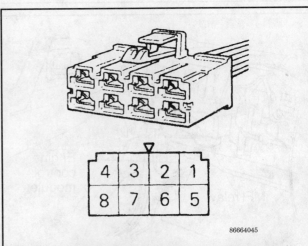

Fig. 43 The MFI harness connector on 3.0L (24 valve) and 3.5L engines — make certain that the wires for testing are hooked to the correct terminals

EMISSION CONTROLS 4-25

3. By using the jumper wire, connect terminal 7 of the MFI relay to the (+) terminal of the battery and terminal 6 to the (-) terminal of the battery.

※※WARNING

If the jumper leads are not connected to the proper terminals, the relay will be damaged.

4. While connecting the jumper wire on the (-) side of the battery, check continuity between the terminals 1 and 4 of the MFI relay. While connected, the resistance should be 0 ohms, and infinite resistance when disconnected.

5. By using the jumper wire, connect terminal 2 of the MFI relay to the (+) terminal of the battery and terminal 5 to the (-) terminal of the battery.

6. While connecting the jumper wire on the (-) side of the battery, check continuity between the terminals 1 and 4 of the MFI relay. While connected, the resistance should be 0 ohms, and infinite resistance when disconnected.

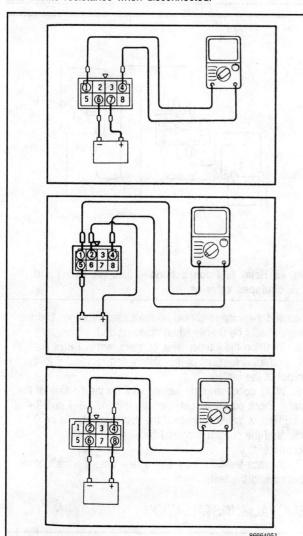

Fig. 44 Testing connections for the MFI relay — if the relay does not pass all of the tests, it needs to be replaced with a new unit

7. By using the jumper wire, connect terminal 8 of the MFI relay to the (+) terminal of the battery and terminal 6 to the (-) terminal of the battery.

8. While connecting the jumper wire on the (-) side of the battery, check continuity between the terminals 2 and 4 of the MFI relay. While connected, the resistance should be 0 ohms, and infinite resistance when disconnected.

9. Replace the MFI relay with a new one, if a malfunction occurred with the tests.

3.0L (12 Valve) Engines
▶ See Figure 45

1. Remove the MFI relay.
2. By using the jumper wire, connect terminal 10 of the MFI relay to the (+) terminal of the battery and terminal 8 to the (-) terminal of the battery.

※※WARNING

If the jumper leads are not connected to the proper terminals, the relay will be damaged.

3. While connecting the jumper wire on the (-) side of the battery, check the voltage at terminals 4 and 5 of the MFI relay. While connected the voltage should be the battery voltage, and the voltage should be 0 volts when disconnected.

4. By using the jumper wire, connect terminal 9 of the MFI relay to the (+) terminal of the battery and terminal 6 to the (-) terminal of the battery.

5. While connecting the jumper wire on the (-) side of the battery, check continuity between the terminals 2 and 3 of the MFI relay. While connected the continuity should be 0 ohms, and the continuity should be ∞ ohms when disconnected.

6. By using the jumper wire, connect terminal 3 of the MFI relay to the (+) terminal of the battery and terminal 7 to the (-) terminal of the battery.

7. While connecting the jumper wire on the (-) side of the battery, check the voltage between at terminal 2 of the MFI relay. While connected the voltage should be the battery voltage, and the voltage should be 0 volts when disconnected.

8. Replace the MFI relay with a new one, if a malfunction occurred with the tests.

3.0L (24 Valve) and 3.5L Engines
▶ See Figures 46, 47 and 48

1. Remove the MFI relay.
2. Check the continuity between the MFI relay terminals. Check terminals 5-7, which should register continuity. Check terminals 6-8, which will only show continuity in one direction.

3. By using the jumper wire, connect terminal 7 of the MFI relay to the (+) terminal of the battery and terminal 5 to the (-) terminal of the battery.

※※WARNING

If the jumper leads are not connected to the proper terminals, the relay will be damaged.

4. While connecting the jumper wire on the (-) side of the battery, check the voltage at terminal 1 of the MFI relay. While

4-26 EMISSION CONTROLS

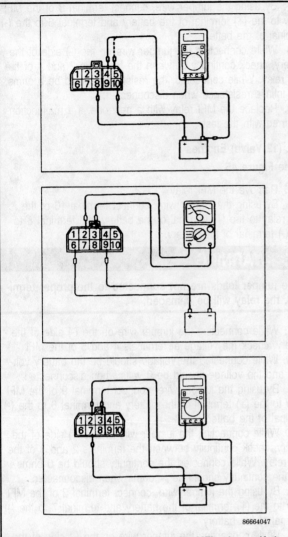

Fig. 45 Testing connections for the MFI relay — if the relay does not pass all of the tests, it needs to be replaced with a new unit

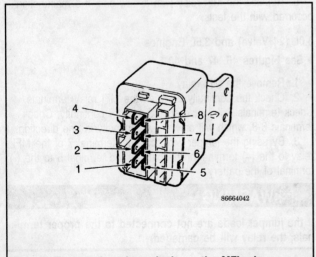

Fig. 46 The numbered terminals on the MFI relay — 3.0L (24 valve) and 3.5L engines

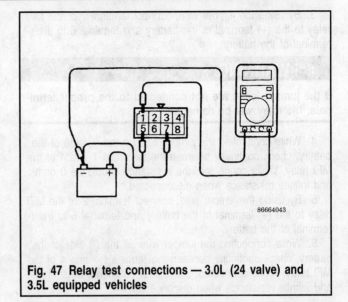

Fig. 47 Relay test connections — 3.0L (24 valve) and 3.5L equipped vehicles

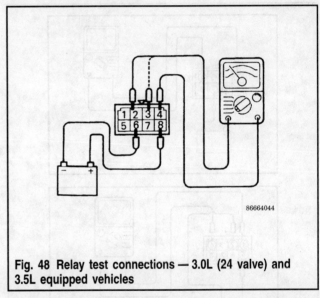

Fig. 48 Relay test connections — 3.0L (24 valve) and 3.5L equipped vehicles

connected the voltage should be the battery voltage, and the voltage should be 0 volts when disconnected.

5. By using the jumper wire, connect terminal 8 of the MFI relay to the (+) terminal of the battery and terminal 6 to the (-) terminal of the battery.

6. While connecting the jumper wire on the (-) side of the battery, check continuity between the terminals 1-4 and 3-4 of the MFI relay. While connected the continuity should be 0 ohms, and the continuity should be ∞ ohms when disconnected.

7. Replace the MFI relay with a new one, if a malfunction occurred with the tests.

REMOVAL & INSTALLATION

The MFI relay is located on the right-hand side kick-panel, or up under the dashboard. It is the uppermost of the two relays located there. It is mounted to the kick-panel with two screws and has one 8-terminal (for 2.4L, 3.0L 24 valve, and

EMISSION CONTROLS

3.5L engines) or a 10-terminal (for 3.0L 12 valve engines) connector plugged into it. Remove the screws and unplug the connector to remove.

Engine Control Module (ECM) Power Ground

OPERATION

If there is incorrect or incomplete contact of the engine control module's ground line, the engine will not function correctly. Therefore, checking before further diagnosis can save a lot of aggravating time.

✱✱CAUTION

To prevent the possibility of permanent ECM damage, the ignition switch must always be OFF when disconnecting power from or reconnecting power to the module. This includes unplugging the ECM connector, disconnecting the negative battery cable, removing the unit fuse or even attempting to jump your dead battery using jumper cables.

TESTING

▶ See Figures 49 and 50

With the ECM connector unplugged, check for continuity of the ground circuit. Connect terminals 101 and 106 (for 2.4L and 3.0L 12 valve engines) or terminals 13 and 26 (for 3.0L 24 valve or 3.5L engines) to a grounded ohmmeter. There should be continuity between these terminals and the ground, if there is not, the circuit (harness) needs to be repaired or replaced from terminals 101 and 106 to ground or 13 and 26 to ground, depending which applies.

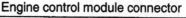
Engine control module connector

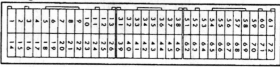

Fig. 50 The engine control module connector terminals — 3.0L (24 valve) and 3.5L engines

REMOVAL & INSTALLATION

▶ See Figure 51

✱✱CAUTION

The ECM is extremely sensitive to electrical and mechanical damage. NEVER touch the connector pins or soldered components on the circuit board in order to prevent possible electrostatic discharge damage to the components. Always wear a wrist grounding strap when handling the ECM.

Although this particular circuit cannot be removed on its own, the ECM can be removed from the vehicle for easier testing. If any repairs need to be done for this procedure, however, the repairs will be in the harness and not the ECM. The ECM is located under the dashboard on the far right-hand side of the vehicle. It is attached with either 2 or 4 screws, so make sure all screws have been removed before trying to do so, and a harness connector.

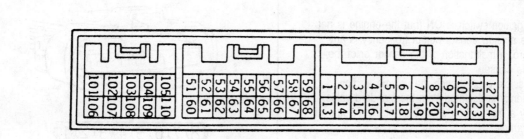

Fig. 49 The engine control module connector terminals — 2.4L and 3.0L (12 valve) engines

4-28 EMISSION CONTROLS

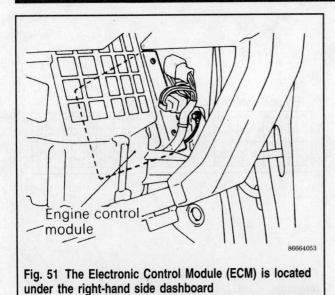

Fig. 51 The Electronic Control Module (ECM) is located under the right-hand side dashboard

Volume Air Flow Sensor

OPERATION

The volume air flow sensor is incorporated within the air cleaner; it functions to convert the amount of engine air intake to pulse signals of a frequency proportional to the amount of engine air intake, and to input those signals to the ECM. The ECM then, based upon those signals, calculates the amount of fuel injection and ignition timing.

The power for the volume air flow sensor is supplied from the MFI relay to the volume air flow sensor, and is grounded at the ECM. The volume air flow sensor, by intermitting the flow of the 5 volts applied from the ECM, produces pulse signals.

TROUBLESHOOTING HINTS

1. If the engine sometimes stalls, try starting the engine and shaking the volume air flow sensor harness. If the engine, as a result of shaking the harness, stalls, incorrect or improper contact of the volume air flow sensor connector is the probable cause.

2. If, when the ignition switch is ON (but the engine is not started), the volume air flow sensor output frequency is any value other than zero, a malfunction of the sensor or of the ECM is the probable cause.

3. If idling is possible even though the volume air flow sensor output frequency is deviated from the standard value, the cause is usually a malfunction other than of the volume air flow sensor. For example:

   a. The flow of air within the sensor is disturbed. (An air duct is disconnected or the air cleaner element is clogged.)

   b. Incomplete combustion within a cylinder is happening. (Malfunction of sparkplugs, ignition coil, injectors, compression pressure, etc.)

   c. Air is taken into the intake manifold through a leaking gasket, etc.

   d. There is an improper adhesion of the valve sheet on the EGR valve.

HARNESS TESTING

When performing these checks, use a harness side connector to fasten onto the different terminals.

2.4L Engines
▶ See Figures 52 and 53

➡ The 1993-95, rear-wheel drive 2.4L (California) Pick-ups have the same engine controls as the 3.0L engines. All other 2.4L engines are listed as 2.4L engines.

1. Unplug the volume air flow sensor harness connector from the sensor itself. With the ignition switch ON, measure the power supply voltage to terminal 4 of the harness connector. If the voltage is not that of the battery (around 10-12 volts), repair or replace the harness between terminal 4 and the MFI relay, or check the MFI relay, otherwise go on to Step 2.

2. Measure the terminal voltage at terminal 3. If the voltage is 4.8-5.2 volts continue on to Step 3, otherwise repair or replace the harness from terminal 3 to the ECM terminal 10.

3. Turn the ignition switch OFF, and check for continuity of the ground circuit (terminal 5 on the volume air flow sensor harness connector). If there is continuity, continue on to Step 4, otherwise repair the harness between the volume air flow sensor terminal 5 and the ECM terminals 14 and 24.

4. Disconnect the ECM harness plug from the ECM. Check for continuity between the volume air flow sensor (terminal 7) and the ECM (terminal 57). If there is continuity, then the part is OK, otherwise repair or replace the harness between the two terminals.

3.0L and 3.5L Engines
▶ See Figures 52 and 54

1. With the the MFI harness connector and the volume air flow sensor harness connector unplugged, check for continuity between the volume air flow sensor connector (terminal 4) and

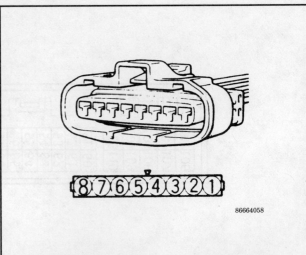

Fig. 52 The volume air flow sensor harness connector terminal arrangement

EMISSION CONTROLS 4-29

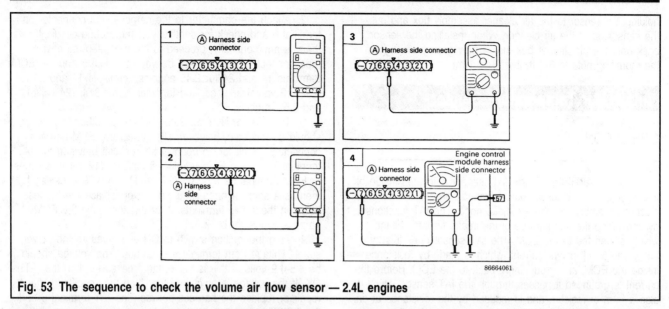

Fig. 53 The sequence to check the volume air flow sensor — 2.4L engines

the MFI relay connector (terminal 4 for 3.0L 12 valve engines, otherwise terminal 2). Touch the ohmmeter probes to both ends of the harness. If there is no continuity, repair or replace the harness between the two components, otherwise go on to Step 2.

2. Check for continuity from terminal 5 of the volume air flow sensor harness connector to ground. If continuity is present, go on to Step 3, otherwise repair or replace the harnesses between terminal 5 of the volume air flow sensor and the terminals (17 and 24 for 3.0L 12 valve engines, otherwise 72) of the ECM.

3. Unplug the ECM harness connector. Check for an open or short-circuit between the volume air flow sensor harness connector terminals 7 and 3 and ECM terminals 57 and 10 (70 and 19 on 3.0L 24 valve and 3.5L engines). If a short or open-circuit is found, repair or replace the harnesses between the two connectors, otherwise go on to Step 4.

4. Turn the ignition switch **ON** and measure the applied voltage from terminal 3 of the volume air flow sensor harness connector to ground. The voltage should be 4.8-5.2 volts, if it is not, then replace the ECM with a new unit, otherwise the volume air flow sensor is good. Plug all connectors back in and turn the ignition **OFF**.

COMPONENT TESTING

Mitsubishi does not give testing procedures of the volume air flow sensor itself with an ohmmeter. To test this component an oscilloscope is needed. Due to the high cost, and special skills needed to operate the tool, have the system diagnosed by a dealer or reputable service facility.

REMOVAL & INSTALLATION

▶ See Figure 55

The volume air flow sensor is located between the air cleaner and the air inlet hose. To remove the sensor, remove the air intake hose by loosening the screw-type hose clamp. Pull the air intake hose off of the sensor. Loosen the four nuts

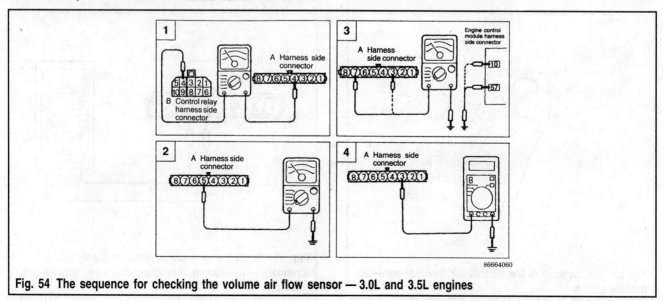

Fig. 54 The sequence for checking the volume air flow sensor — 3.0L and 3.5L engines

4-30 EMISSION CONTROLS

holding the sensor to the air cleaner assembly box and pull the sensor off of the air cleaner. When installing the sensor back onto the air cleaner box, use a new gasket and tighten the mounting nuts to 6-7 ft. lbs. (8-10 Nm).

Intake Air Temperature Sensor

OPERATION

The Intake Air Temperature (IAT) sensor is located inside of the volume air flow sensor box, attached between the air cleaner assembly and the air intake hose. The IAT functions by converting the temperature of the intake air into voltage, which is then fed to the ECM. The IAT sensor does this by taking the 5 volt power current, which is made by resistance inside the ECM, and grounding it back to the ECM. Before the current is grounded it passes through the IAT sensor, which is also a form of resistor, and is reduced by the resistance at an inverse proportion to the intake air temperature. In other words, the higher the intake air temperature, the lower the resistance of the IAT sensor. The ECM, based upon these signals from the IAT sensor, then corrects the amount of fuel injection and ignition timing.

➡**Since the air temperature is measured after the air cleaner, the air has some chance to warm up. Therefore, the IAT sensor indicates a different temperature than the outside temperature.**

HARNESS TESTING

➡**The 1993-95, rear-wheel drive 2.4L (California) Pick-ups have the same engine controls as the 3.0L engines. All other 2.4L engines are listed as 2.4L engines.**

1. Unplug the wiring harness connector from the side of the Volume air flow sensor box, this harness connector also services the IAT sensor.

2. Attach the ohmmeter to the harness side connector on terminal 5 and check for continuity of the ground circuit. If there is any continuity, proceed to the next step, otherwise repair or replace the harness between terminal 5 and the ECM terminals 14 and 24 for 2.4L engines, terminals 17 and 24 for 3.0L (12 valve) engines, and terminal 72 for 3.0L (24 valve) and 3.5L engines.

3. This step is NOT for 2.4L engines, all other engines should do this step, however. Disconnect the ECM harness connector. Check for an open or short-circuit between the IAT terminal 6 and the ECM terminal 8 for 3.0L (12 valve) engines, or ECM terminal 52 for 3.0L (24 valve) and 3.5L engines. If there is a short or open-circuit, repair or replace the harness between these two terminals. After this test, plug the ECM harness connector back in.

4. Turn the ignition switch to **ON** and measure the power supply voltage from terminal 6 to ground. The voltage should be 4.5-4.9 volts, if this is the voltage measured then the IAT is OK. If the voltage is anything other than the specified voltage, repair or replace the harness between the IAT terminal 6 and the ECM terminal 8.

5. Plug all connectors back into place.

COMPONENT TESTING

▶ See Figure 56

1. Unplug the volume air flow sensor connectors.
2. Measure the resistance between the volume air flow sensor/IAT sensor harness connector terminals 5 and 6. The resistance between these two terminals depends on their temperature. If the temperature is around 32°F (0°C) the resistance should be 6.0 kilohms, at 68°F (20°C) the resistance should be 2.7 kilohms, and with the temperature at 176°F (80°C) the resistance should be 0.4 kilohms.
3. Measure the resistance while heating the sensor with a hair drier. The higher the temperature (the hotter the IAT sensor), the smaller the resistance should become.
4. If resistance does not decrease as temperature increases or the resistance remains unchanged, replace the volume air flow sensor assembly.

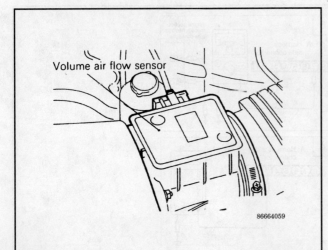

Fig. 55 The location of the volume air flow sensor — all models similar

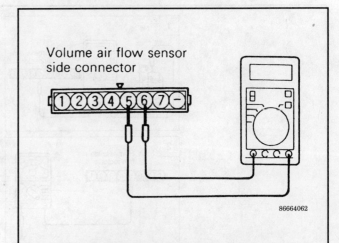

Fig. 56 Check the resistance between these two terminals — resistance should go down as temperature goes up

EMISSION CONTROLS 4-31

REMOVAL & INSTALLATION

♦ See Figure 57

Since the IAT sensor is integral to the volume air flow sensor assembly, refer to the previously described removal and installation procedures for the volume air flow sensor.

Barometric Pressure Sensor

OPERATION

The Barometric Pressure Sensor (BPS), located in the volume air flow sensor assembly, converts the barometric pressure into voltage, and then inputs that voltage (as signals) to the ECM. The ECM, based upon those signals, then corrects the amount of fuel injection, etc.

If there is a malfunction of the BPS, driveability of the vehicle will become worse particularly at higher altitudes. When it appears as though the BPS is malfunctioning, first check the air cleaner for a clog or restriction. A restriction of the air cleaner sometimes causes a reactions as though the BPS were malfunctioning.

HARNESS TESTING

♦ See Figure 58

➥ The 1993-95, rear-wheel drive 2.4L (California) Pick-ups have the same engine controls as the 3.0L engines. All other 2.4L engines are listed as 2.4L engines.

1. Since the BPS is located in the volume air flow sensor box, the harness connector (HC) to this box will be the wiring tested throughout this procedure. Unplug the volume air flow sensor HC from the volume air flow sensor assembly box.

2. Check terminal 5 of the HC for continuity to ground. If there is no continuity in the circuit, repair or replace the harness between terminal 5 of the HC to terminals 14 and 24 of the ECM.

3. Turn the ignition switch to **ON** and measure the power supply voltage of terminal 1 of the HC. If the voltage is not between 4.8-5.2 volts, repair or replace the harness between terminal 1 (HC) and terminals 13 and 23 (ECM).

4. Turn the ignition switch **OFF** and unplug the ECM harness connector. Check for an open or short-circuit to ground, between terminal 16 (ECM) and terminal 2 (HC) for 2.4L and 3.0L (12 valve) engines, terminal 61 (ECM) and terminal 2 (HC) for 3.0L (24 valve) and 3.5L engines. For all engines except for the 2.4L, check also terminal 23 (ECM) and terminal 1 (HC) for 3.0L (12 valve) engines, terminal 65 (ECM) and terminal 1 (HC) for 3.0L (24 valve) and 3.5L engines. If there is a short or open-circuit, repair or replace the harness between these terminals, otherwise the harness is good.

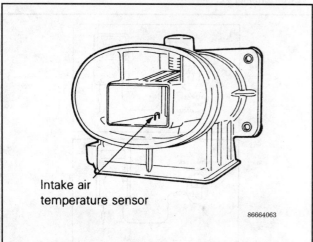

Fig. 57 The Intake Air Temperature (IAT) sensor is located inside the volume air flow sensor assembly — all models similar

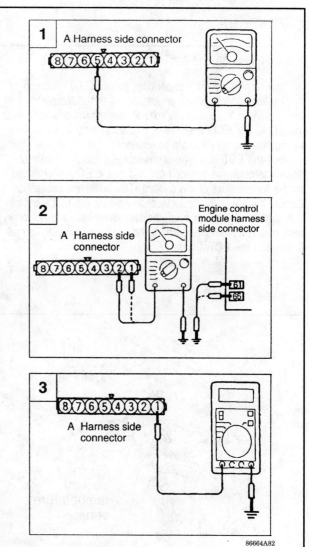

Fig. 58 The BPS harness testing procedure — 3.0L (24 valve) and 3.5L engines

4-32 EMISSION CONTROLS

COMPONENT TESTING

To correctly test this component a Mitsubishi MUT or MUT II tool is needed. Mitsubishi does not give testing procedures of the volume air flow sensor itself with an ohmmeter. Due to the high cost, and special skills needed to operate the tool, have the system diagnosed by a dealer or reputable service facility.

REMOVAL & INSTALLATION

Since the BPS sensor is integral to the volume air flow sensor assembly, refer to the previously described removal and installation procedures for the volume air flow sensor.

Engine Coolant Temperature Sensor

OPERATION

▶ See Figures 59 and 60

The Engine Coolant Temperature sensor (ECT), located on the intake manifold of the engine, converts the temperature of the coolant in the engine to voltage, and inputs this voltage to the ECM. The ECM uses these signals to help it decide how to regulate the MFI system components.

Since the ECT is a type of resistor, the electrical current flowing through the sensor from and to the ECM is changed by the temperature of the coolant. The hotter the coolant becomes, the lower the resistance created by the ECT sensor.

If, during engine warm up, the fast idling speed is not correct, or black smoke is emitted, the problem is usually a malfunction of the ECT sensor.

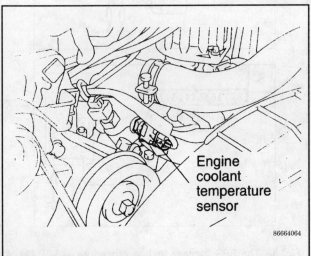

Fig. 59 The location of the Engine Coolant Temperature (ECT) sensor — 3.0L and 3.5L engines

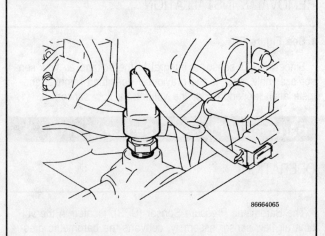

Fig. 60 The location of the Engine Coolant Temperature (ECT) sensor — 2.4L engines

HARNESS TESTING

▶ See Figures 61, 62 and 63

➡The 1993-95, rear-wheel drive 2.4L (California) Pick-ups have the same engine controls as the 3.0L engines. All other 2.4L engines are listed as 2.4L engines.

1. Unplug the ECT sensor wire connector form the ECT sensor. Check terminal 1, for 2.4L engines, or terminal 2, for all 3.0L and 3.5L engines, of the connector for continuity to ground. If there is no continuity the problem lies in the wire harness, repair or replace the harness between the following terminals:

    a. 2.4L engines: ECT terminal 1 and ECM terminal 14.

    b. 3.0L (12 valve) engines: ECT terminal 2 and ECM terminals 17 and 24.

    c. 3.0L (24 valve) and 3.5L engines: ECT terminal 2 and ECM terminal 72.

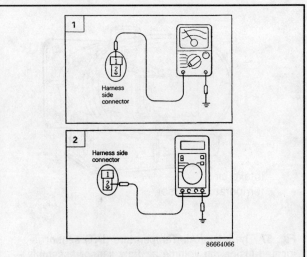

Fig. 61 The ECT harness testing procedure — 2.4L engines

EMISSION CONTROLS 4-33

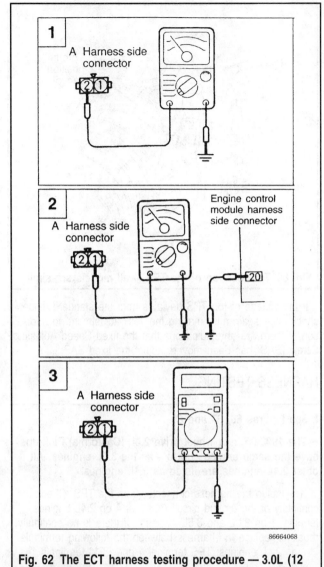

Fig. 62 The ECT harness testing procedure — 3.0L (12 valve) engines

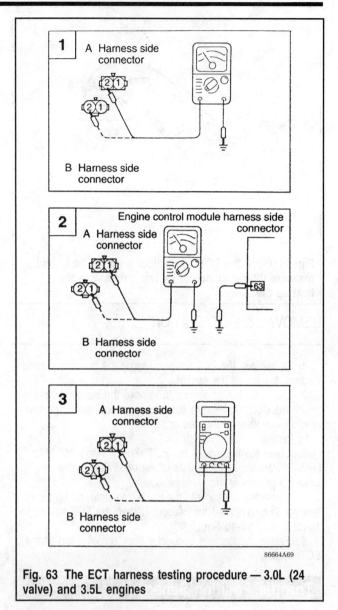

Fig. 63 The ECT harness testing procedure — 3.0L (24 valve) and 3.5L engines

2. This step is for all engines except for the 2.4L. Unplug the ECM harness connector. Check for an open or short circuit between the following ECT sensor and ECM terminals:

   a. 3.0L (12 valve) engines: ECT terminal 1 and ECM terminal 20.
   b. 3.0L (24 valve) and 3.5L engines: ECT terminal 1 and ECM terminal 63.

3. Make sure the ECM harness connector is plugged in. Turn the ignition switch to **ON** and measure the power supply from terminal 2 (2.4L) or terminal 1 (3.0L and 3.5L) of the ECT harness connector. If the voltage is not 4.5-4.9 volts, replace the ECM itself, otherwise the harness and ECM are good.

➡ Before replacing the ECM or other computer module, either check or have the component in question checked.

COMPONENT TESTING

▶ See Figure 64

This procedure is the same for all engine models.

1. Remove the ECT sensor from the intake manifold. Be careful not to damage the plastic portion or the electrodes of the ECT sensor when unscrewing the sensor from the manifold.

2. With the temperature sensing portion of the ECT sensor immersed in water, check the resistance between the two terminals. Slowly heat the water up and watch to see the change in resistance. With the water at 32°F (0°C) the resistance should be 5.9 kilohms, at 68°F (20°C) the resistance should be 2.5 kilohms, at 104°F (40°C) the resistance should be 1.1 kilohms, and at 176°F (80°C) the resistance should be 0.3 kilohms. If the resistance deviates from the standard value greatly, replace the sensor with a new one.

4-34 EMISSION CONTROLS

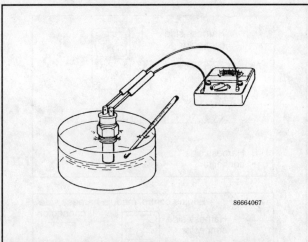

Fig. 64 Place the ECT sensor into a pot of water and measure the resistance of its two terminals while heating the water

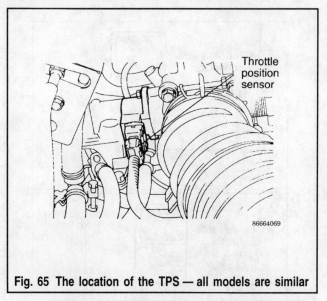

Fig. 65 The location of the TPS — all models are similar

REMOVAL & INSTALLATION

1. To remove the ECT sensor, unplug the harness connector from the top of the sensor.
2. Use a socket or wrench to remove the sensor from the intake manifold. Make sure not to damage the plastic portion or the electrodes on the sensor.

To Install:

3. Clean the threads of the ECT sensor. Apply 3M®Nut Locking Part No. 4171 sealant or the equivalent to the threaded portion of the sensor.
4. Screw the sensor into the intake manifold, making certain not to crossthread the sensor. Tighten the ECT sensor to 15-29 ft. lbs. (20-40 Nm).
5. Fasten the harness connector back onto the top of the ECT sensor.

Throttle Position Sensor

OPERATION

▶ See Figure 65

The Throttle Position Sensor (TPS), located at the entrance to the intake manifold plenum, sends voltage signals to the ECM depending on the position of the throttle. The electrical current the ECM sends into the TPS is monitored when it returns back to the ECM. Since the TPS is a resistor, the 5 volt current is changed depending on the TPS position. The further open the throttle is rotated, the greater the resistance in the TPS. Therefore, the more open the throttle, the less of the 5 volt current returns to the ECM.

The ECM can measure this voltage and knows how far open the throttle is. It uses this knowledge to help decide how to control the other MFI system components.

The signals of the TPS are more important for control of the automatic transmission than for control of the engine; shifting "impact shocks" are produced if there is a malfunction of the TPS.

If the voltage of the TPS deviates from the standard value, check once again after making the TPS adjustment. In addition, if there are any indications that the fixed Speed Adjusting Screw (SAS) has been moved, adjust the fixed SAS.

HARNESS TESTING

▶ See Figures 66, 67 and 68

➡The 1993-95, rear-wheel drive 2.4L (California) Pick-ups have the same engine controls as the 3.0L engines. All other 2.4L engines are listed as 2.4L engines.

1. Unplug the harness connector from the TPS. Check for continuity of the ground circuit (terminal 4 on 2.4L engines, terminal 1 on 3.0L and 3.5L engines). If there is no continuity, repair or replace the harness between the following terminals:
   a. 2.4L engines: TPS terminal 4 and ECM terminals 14 and 24.
   b. 3.0L (12 valve) engines: TPS terminal 1 and ECM terminals 17 and 24.
   c. 3.0L (24 valve) and 3.5L engines: TPS terminal 1 and ECM terminal 72.
2. Unplug the ECM harness connector and all other control module connectors which use TPS output. Check for an open or short-circuit between the TPS and the ECM. If a short or open-circuit is found, repair or replace the harnesses between the following terminals (use these terminals also for the initial test):
   a. 2.4L engines: TPS terminal 2 and ECM terminals 19.
   b. 3.0L (12 valve) engines: TPS terminal 3 and 4 and ECM terminals 19 and 23, respectively.
   c. 3.0L (24 valve) and 3.5L engines: TPS terminal 3 and 4 and ECM terminal 61 and 64, respectively.
3. Unplug the harness connector from the TPS and turn the ignition switch to **ON**. Measure the power supply voltage of the TPS terminal 1 (2.4L) or terminal 4 (3.0L and 3.5L). If the voltage is not 4.8-5.2 volts, replace the ECM with a new unit.

➡Before replacing the ECM or other computer module, either check or have the component in question checked.

EMISSION CONTROLS 4-35

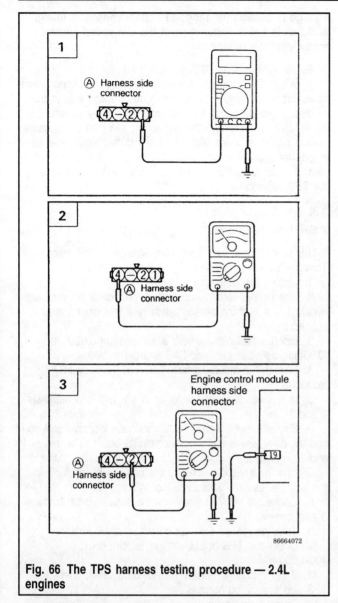

Fig. 66 The TPS harness testing procedure — 2.4L engines

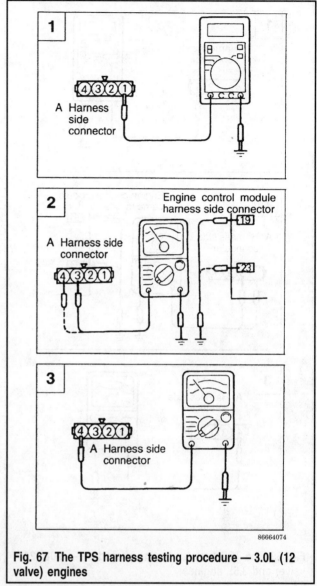

Fig. 67 The TPS harness testing procedure — 3.0L (12 valve) engines

COMPONENT TESTING

➡ The 1993-95, rear-wheel drive 2.4L (California) Pick-ups have the same engine controls as the 3.0L engines. All other 2.4L engines are listed as 2.4L engines.

1. Unplug the TPS harness connector.
2. Measure the resistance between terminals 1 and 4. The resistance should be between 3.5-6.5 kilohms.
3. Connect a pointer type ohmmeter between terminal 4 and 2 (2.4L) or terminal 1 and 3 (3.0L and 3.5L).
4. Operate the throttle valve slowly from the idle position to the full open position and check that the resistance changes smoothly in proportion with the throttle valve opening angle.

5. If the resistance fails to change smoothly, or is grossly out of specification, replace the TPS with a new unit.

ADJUSTMENT

2.4L Engines

➡ The 1993-95, rear-wheel drive 2.4L (California) Pick-ups have the same engine controls as the 3.0L engines. All other 2.4L engines are listed as 2.4L engines.

1. Unplug the TPS harness connector and link the two unplugged connector halves using the test harness set (Mitsubishi special tool).
2. Connect a voltmeter between terminal 2 and 4 of the TPS itself.

4-36 EMISSION CONTROLS

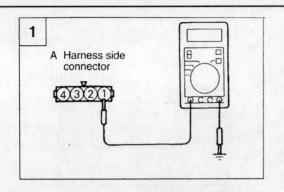

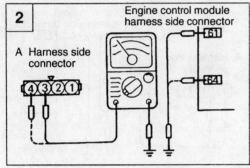

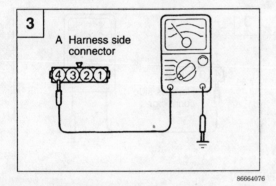

Fig. 68 The TPS harness testing procedure — 3.0L (24 valve) and 3.5L engines

3. Turn the ignition switch **ON** and wait for 15 seconds or longer.

→When the ignition switch is turned ON, the idle speed control plunger extends to the fast idle opening angle. After 15 seconds has passed, the plunger contracts and stops at the initial position, which is the position corresponding to the idle position where the output voltage of the idle speed control motor position sensor is 0.9 volts.

4. Check that the output voltage of the TPS is 0.48-0.52 volts.
5. If the voltage deviates from the standard value, loosen the TPS mounting bolts and turn the TPS body to adjust. Tighten the bolts securely after the adjustment.

→Turning the TPS body clockwise raises the output voltage. When the TPS is adjusted, there may be an output of diagnostic trouble code of idle speed control motor position sensor malfunction. This occurs when the TPS output voltage is outside the range of normal voltage at idling, and this is not a malfunction of the idle speed control motor position sensor.

6. Turn the ignition **OFF**.
7. When the diagnostic trouble code is output during adjustment, erase the diagnostic trouble code by using a scan tool or disconnect the negative cable from the battery terminal for 10 seconds or more and then reconnect it. When the negative battery cable has been removed and installed again, idle the engine for approximately 15 minutes to warm it up. This will clear the memory of all diagnostic trouble codes output during the TPS adjustment.

3.0L and 3.5L Engines
▶ See Figures 69 and 70

This procedure is also for the adjustment of the Closed Throttle Position (CTP) switch.

1. Unplug the connector.
2. Use jumper wires to connect an ohmmeter between terminals 1 and 2 of the sensor/switch itself (not the harness connector).
3. Insert a feeler gauge with a thickness of 0.0256 in. (0.65mm) between the fixed SAS and the throttle lever.
4. Loosen the mounting bolts and turn the body fully clockwise.
5. In this condition, check for continuity between terminals 1 and 2.
6. Slowly turn the sensor/switch counter-clockwise until you find the point where there is no continuity between terminals 1 and 2. Tighten the mounting bolts to hold it in this position.
7. Link the sensor/switch and the harness connector with the test harness set (Mitsubishi tool MB991348).
8. Connect a voltmeter between the sensor/switch terminal 3 and terminal 1.
9. Turn the ignition **ON**, but do not start the engine.
10. Check the TPS output voltage, which should be 400-1,000 mV.
11. If the voltage is outside the standard value, check the TPS and associated harness.
12. Remove the feeler gauge.

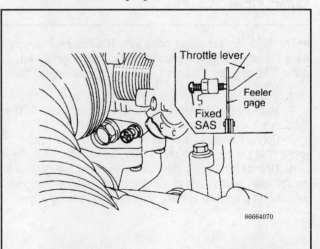

Fig. 69 Insert a feeler gauge between the fixed SAS and the throttle lever

EMISSION CONTROLS 4-37

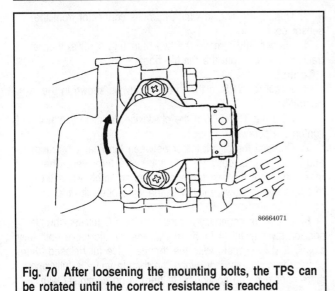

Fig. 70 After loosening the mounting bolts, the TPS can be rotated until the correct resistance is reached

13. Turn the ignition switch to **OFF**.

REMOVAL & INSTALLATION

2.4L Engines

♦ See Figures 71, 72 and 73

1. Disconnect the negative battery cable.
2. Remove any components obstructing access to the TPS.
3. Unplug the engine wiring harness connector from the sensor connector.
4. Loosen and remove the two mounting screws, then extract the sensor from the throttle body.

To install:

5. Install the TPS to the throttle body as shown in the illustration.
6. Turn the TPS 90° in the clockwise direction to set it and tighten the screws.
7. Connect the circuit tester between terminal 4 (ground) and 2 (output), or between 2 (output) and 1 (power). Then,

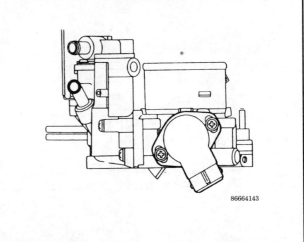

Fig. 72 Rotate the TPS 90° clockwise to set it — 2.4L engines

make sure that the resistance changes smoothly when the throttle valve is slowly moved to the wide open throttle position.

8. If the resistance does not change smoothly, adjust the TPS following the previously described procedures.

3.0L Engines

♦ See Figures 74 and 75

1. Disconnect the negative battery cable.
2. Remove any components obstructing access to the TPS.
3. Unplug the engine wiring harness connector from the sensor connector.
4. Loosen and remove the two mounting screws, then extract the sensor from the throttle body.

To install:

5. Install the TPS to the throttle body as shown in the illustration.
6. Turn the TPS 90° counter-clockwise to set it, and tighten the screws.
7. Connect the circuit tester between terminal 4 (ground) and 2 (output), or between 2 (output) and 1 (power). Then,

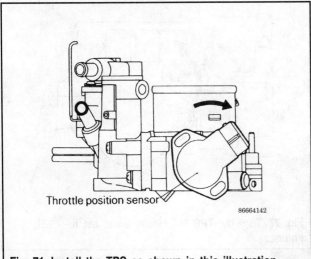

Fig. 71 Install the TPS as shown in this illustration — 2.4L engines

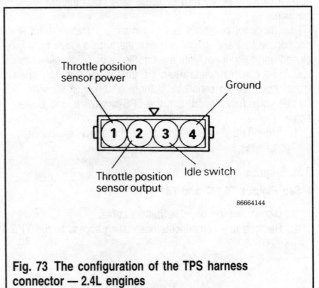

Fig. 73 The configuration of the TPS harness connector — 2.4L engines

4-38 EMISSION CONTROLS

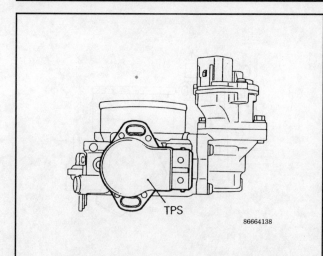

Fig. 74 Install the TPS as shown in the illustration — 3.0L engines

3. Unplug the engine wiring harness connector from the sensor connector.
4. Loosen and remove the two mounting screws, then extract the sensor from the throttle body.

To install:

5. Install the TPS to the throttle body as shown in the illustration.
6. Turn the TPS 90° in the clockwise direction to set it and tighten the screws.
7. Connect the circuit tester between terminal 4 (ground) and 2 (output), or between 2 (output) and 1 (power). Then, make sure that the resistance changes smoothly when the throttle valve is slowly moved to the wide open throttle position.
8. Check for continuity across terminals 3 (closed throttle position switch) and 4 (ground) with the throttle valve both fully closed and fully open. With the throttle valve fully closed, there should be continuity, and when the throttle valve is fully open there should be no continuity. If there is no continuity with

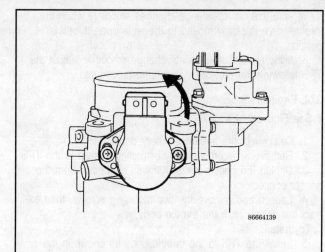

Fig. 75 Turn the TPS 90° counter-clockwise to set it — 3.0L engines

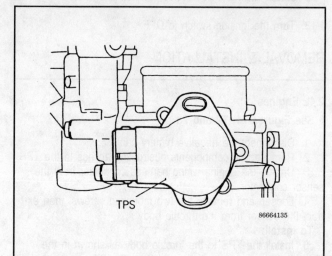

Fig. 76 Install the TPS as shown in the illustration — 3.5L engines

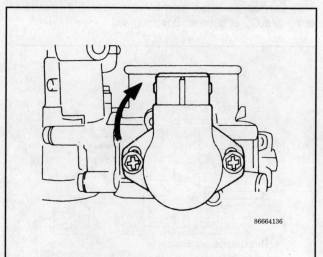

Fig. 77 Turn the TPS 90° clockwise to set it — 3.5L engines

make sure that the resistance changes smoothly when the throttle valve is slowly moved to the wide open throttle position.

8. Check for continuity across terminals 2 (closed throttle position switch) and 1 (ground) with the throttle valve both fully closed and fully open. With the throttle valve fully closed, there should be continuity, and when the throttle valve is fully open there should be no continuity. If there is no continuity with throttle valve fully closed, turn the TPS clockwise, and check again.
9. If the TPS does not conform to the above standards, this unit needs to be replaced with a new one.

3.5L Engines

▶ See Figures 76, 77 and 78

1. Disconnect the negative battery cable.
2. Remove any components obstructing access to the TPS.

EMISSION CONTROLS 4-39

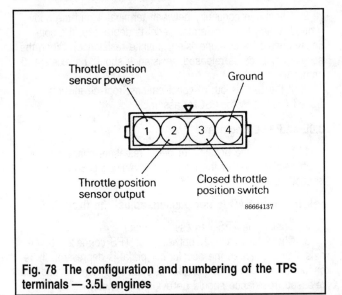

Fig. 78 The configuration and numbering of the TPS terminals — 3.5L engines

Fig. 79 The location of the CTPS on the 3.0L and 3.5L engines is inside the Idle Speed Control Motor (ISCM)

throttle valve fully closed, turn the TPS counter-clockwise, and check again.

➡ Some TPS units are not provided with the position switch. In that case, the check described in Step 4 cannot be accomplished.

9. If the previous specifications are not met, replace the TPS unit.

Closed Throttle Position Switch

OPERATION

♦ See Figure 79

The Closed Throttle Position switch (CTP) is located either in the TPS (3.0L and 3.5L engines) or in the Idle Speed Control Motor (ISCM) assembly (2.4L engines). The CTP sends electrical signals to the ECM, telling it whether the accelerator is depressed or released. The ECM regulates the idle-speed control motor based upon those signals. Voltage within the ECM is applied, by way of the resistance, to the CTP. When the foot is taken off of the accelerator, the CTP is switched on and the current is grounded. As a result, the CTP voltage changes from a high to a low level.

If there is an abnormal condition of the CTP output even though the results of the check of the CTP and harness, the cause may be either improper adjustment of the accelerator cable or the cruise control cable, or improper adjustment of the fixed Speed Adjusting Screw (SAS).

HARNESS TESTING

♦ See Figure 80

2.4L Engines

➡ The 1993-95, rear-wheel drive 2.4L (California) Pick-ups have the same engine controls as the 3.0L engines. All other 2.4L engines are listed as 2.4L engines.

Unplug the harness connector going to the CTP and turn the ignition **ON**. Measure the power supply voltage of the CTP terminal 4. If the voltage is not 4 volts or higher, repair or replace the harness between the CTP terminal 4 and the ECM terminal 6, otherwise the harness is fine.

3.0L and 3.5L Engines

♦ See Figure 81

1. Unplug the ECM harness connector and the TPS connector. Check for an open or short-circuit between the CTP terminal 2 and the ECM terminal 14 or 67. If there is an open or short-circuit, repair or replace the harness between the two terminals.

   a. 3.0L (12 valve) engines: TPS terminal 2 and ECM terminal 14.

   b. 3.0L (24 valve) and 3.5L engines: TPS terminal 2 and ECM terminal 67.

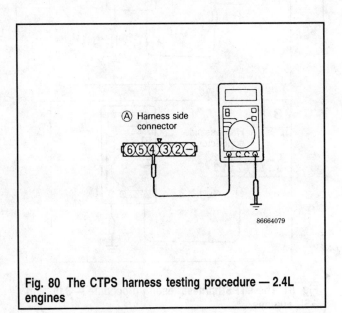

Fig. 80 The CTPS harness testing procedure — 2.4L engines

4-40 EMISSION CONTROLS

2. Plug the ECM harness connector back into the ECM. Check for continuity from the TPS terminal 1 to ground. If no continuity is found, repair the harness.

   a. 3.0L (12 valve) engines: TPS terminal 1 and ECM terminals 17 and 24.

   b. 3.0L (24 valve) and 3.5L engines: TPS terminal 1 and ECM terminal 72.

3. Turn the ignition switch **ON** and measure the power supply voltage of the CTP terminal 2. If the voltage is not 4 volts or higher, replace the ECM with a new unit.

COMPONENT TESTING

2.4L Engines

➡ The 1993-95, rear-wheel drive 2.4L (California) Pick-ups have the same engine controls as the 3.0L engines. All other 2.4L engines are listed as 2.4L engines.

1. Unplug the idle speed control motor position sensor harness connector.

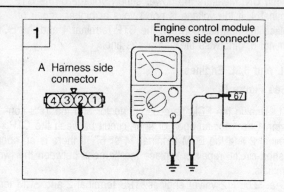

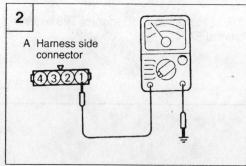

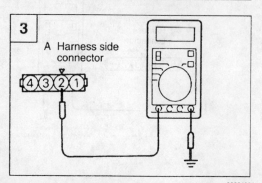

Fig. 81 The CTPS harness testing procedure — 3.0L and 3.5L engines

2. Check the continuity between terminal 4 and the body ground. When the accelerator pedal is depressed, the continuity should be be nonexistent (infinite resistance). When the accelerator pedal is released, resistance should be existent (0 ohms).

3. If the CTP is out of specifications, replace the Idle Speed Control Motor (ISCM) assembly.

3.0L and 3.5L Engines

1. With the the accelerator pedal released, check that the throttle valve lever or the fixed Speed Adjusting Screw (SAS) is pushed.

➡ If the fixed SAS is not pushed, adjust the fixed SAS.

2. Unplug the TPS harness connector.

3. Check for continuity between the TPS connector terminals 1 and 2. When the accelerator pedal is depressed, there should be no continuity (infinite resistance). When the pedal is released, resistance should be evident (0 ohms).

➡ If there is no continuity when the accelerator pedal is released, loosen the TPS installation screw; then, after turning the TPS all the way clockwise, check again.

4. Replace the TPS (with the built-in CTP) if there is a malfunction.

COMPONENT ADJUSTING

Since the CTP switches are located in other components, refer to those components for adjusting procedures. That is, for the 2.4L engine, refer to the idle speed control motor position sensor, and for the other engines, refer to the TPS procedures.

REMOVAL & INSTALLATION

Since the CTP switches are located in other components, refer to those components for removal and installation procedures. That is, for the 2.4L engine, refer to the idle speed control motor position sensor, and for the other engines, refer to the TPS procedures.

Idle Speed Control Motor

OPERATION

➡ This procedure is only for 2.4L engines.

The intake air volume during idling is controlled by extending or retracting the servo plunger and by opening and closing the throttle valve. The servo plunger is extended or retracted by the DC motor inside of the ISCM being driven in either a clockwise or counter-clockwise direction. The switching of the direction of the DC motor's direction is controlled by the motor drive IC in the ECM.

If idling is unstable or the engine stalls and it is difficult to determine the cause, turn the ignition switch **ON** and wait at least 15 seconds, then disconnect the servo connector. Check-

EMISSION CONTROLS 4-41

ing in this state makes it much easier to ascertain where the cause of the trouble is located. Also adjust the engine by turning the engine speed adjusting screw during this check, if needed.

HARNESS TESTING

▶ See Figure 82

Check for an open-circuit, or a short-circuit to ground, between the ECM (terminals 58 and 59) and the ISCM (terminals 1 and 2). If a short or open-circuit is found, repair the harness between ECM terminal 58 and ISCM terminal 1, and ECM terminal 59 and ISCM terminal 2. Otherwise the harness is good.

COMPONENT TESTING

1. Unplug the ISCM harness connector.
2. Check for continuity of the ISCM coil, that is, between terminals 1 and 2 of the ISCM. The continuity should be 5-35 ohms at 68°F (20°C) from terminal 1 to terminal 2.
3. Connect a 6V DC power source between terminal 1 and 2 of the ISCM harness plug, and check to be sure that the ISCM operates.

✳✳WARNING

Apply only a 6V DC or lower voltage power source. Application of a higher voltage could cause locking of the servo gears.

4. If the ISCM fails any of these tests, replace the whole ISCM assembly with a new unit.

REMOVAL & INSTALLATION

To remove the ISCM simply unplug the harness connector and remove the retaining screws. Make certain to remember which side was up. To install the ISCM, put the ISCM into place and tighten the retaining screws securely. Plug the harness connector onto the ISCM.

Idle Speed Control Motor Position Sensor

OPERATION

➡This procedure is only for 2.4L engines

The Idle Speed Control Motor (ISCM) position sensor is located at the mouth of the intake manifold plenum, near the TPS. The ISCM position sensor converts the position of the plunger in the ISCM to a voltage, which it then inputs to the ECM. The ECM takes the signal and controls the ISCM.

If the plunger in the ISCM moves from the contracted to the extended position, the resistance between the variable resistor terminal and the ground terminal of the ISCM position sensor increases in proportion to the amount of extension. Thus the voltage at the variable resistor terminal of the ISCM position sensor increases as the plunger is extended.

The ISCM position sensor is the most important sensor in control of the idle speed. In many cases where failures occur due to changes in the engine load, as when the air conditioning switch is turned on while the engine is idling, the fault lies with this sensor.

If the output voltage of the ISCM position sensor deviates from the standard value even though the results of the checks of the ISCM position sensor wire harness and individual components are normal, the problems may be that the basic idle speed adjustment is faulty, there are deposits on the throttle valve, air is being sucked into the intake manifold through gaps in a gasket, the EGR valve seat is sealed poorly, or the combustion in the cylinders is poor (spark plugs, ignition coil, injectors, compression pressure, etc.).

HARNESS TESTING

▶ See Figure 83

1. Unplug the ISCM position sensor harness connector from the sensor. Turn the ignition **ON**, and measure the power supply voltage from ISCM harness terminals 2 and 6. If the voltage is not 4.8-5.2 volts, repair the harness from ISCM terminals 2 and 6 to ECM terminals 13 and 23.
2. Turn the ignition switch to **OFF** and check for continuity from ISCM harness terminal 3 to ground. If no continuity is detected, repair the harness from ISCM terminal 3 to ECM terminals 14 and 24.
3. Unplug the ECM harness from the ECM, and check for an open-circuit (or short-circuit) between the ECM and the ISCM position sensor. Check from ISCM terminal 5 to ECM terminal 17, if a short or open-circuit is found, repair the harness. If no short or open-circuit is found, the harness is good.

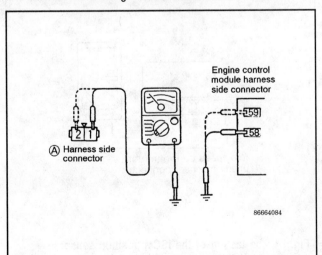

Fig. 82 The ISCM harness testing procedure — 2.4L engines

4-42 EMISSION CONTROLS

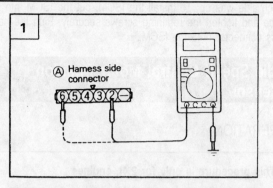
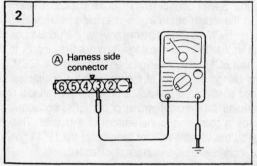
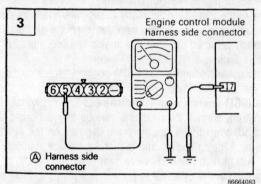

Fig. 83 The ISCM position sensor harness testing procedure — 2.4L engines

COMPONENT TESTING

♦ See Figure 84

1. Unplug the ISCM position sensor connector.
2. Measure the resistance between terminals 2 and 3, which should be between 4-6 kilohms.
3. Hook a 6V DC power source up between terminals 1 and 2 of the ISCM harness connector, and then measure the resistance between terminals 3 and 6 of the ISCM position sensor connector when the ISCM is activated (caused to expand and contract). The resistance should decrease smoothly as the ISCM plunger contracts.

※※WARNING

Apply only a 6V DC or lower voltage source. Application of higher voltages could cause a locking of the servo gears.

4. If there is a deviation from the standard value, or if the change is not smooth, replace the ISCM assembly.

REMOVAL & INSTALLATION

To remove the ISCM simply unplug the harness connector and remove the retaining screws. Make certain to remember which side was up. To install the ISCM, put the ISCM into place and tighten the retaining screws securely. Plug the harness connector onto the ISCM.

Idle Air Control Motor (DC Motor)

➡This procedure is for 2.4L engines (California — RWD, 1993 and later models)

OPERATION

♦ See Figure 85

The volume of intake air during engine idling is controlled by the opening and closing of the Idle Air Control (IAC) valve for bypassing the throttle valve, located at the air intake port. The DC motor turns clockwise or counter-clockwise according to the change in the direction of current in the motor drive IC inside the ECM. The IAC valve opens and closes depending on which direction the DC motor inside the IAC motor is turning.

While the engine is idling, if the idle speed and IAC valve position (step) change when the air conditioning switch is turned to On and Off, it can be assumed that the idle air control motor and the IAC valve position sensor are operating normally.

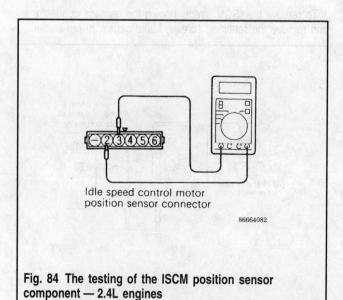

Fig. 84 The testing of the ISCM position sensor component — 2.4L engines

EMISSION CONTROLS 4-43

If the IAC valve position (step) is outside the standard position, the malfunction is probably one of the following:
- Basic idle speed adjustment is wrong.
- Some deposit is adhering to the throttle valve.
- Air is being drawn into the air intake manifold through a defective gasket seal.
- The EGR valve sheet adhesion is defective.
- Combustion malfunction inside a cylinder (spark plug, ignition coil, injector or compression pressure is defective.)

HARNESS TESTING

▶ See Figure 86

Unplug the IAC motor harness connector and the ECM harness connector. Check for an open or short-circuit to ground, between the IAC motor terminal 5 and the ECM terminal 4, and the IAC motor terminal 6 and the ECM terminal 17. If a short or open-circuit is present, repair the harness between the respective terminals.

COMPONENT TESTING

➡ For this procedure you will need a sound scope.

This test checks the actuator. Use a sound scope to check the sound of the idle air control servo. Listen to the servo while the ignition switch is turned **ON**. Shortly after the ignition is turned **ON**, the sound of the servo should be able to be heard.

REMOVAL & INSTALLATION

The idle air control motor (DC motor) is attached to the throttle body. The air intake hose must be removed, and the throttle body removed from the intake manifold plenum. Label any hoses or cables removed and save all bolts and nuts for re-installation. For further instructions on the removal and installation of the throttle body, refer to Section 5.

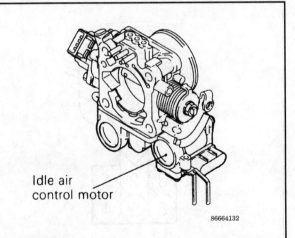

Fig. 85 The location of the idle air control motor, attached to the throttle body — 2.4L (California, 1993 and later) engines

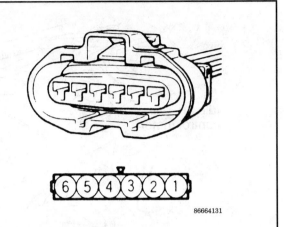

Fig. 86 The harness side connector and terminal numbering — 2.4L (California, 1993 RWD and later) engines

Idle Air Control Motor (Stepper Motor)

➡ This motor is found on 3.0L and 3.5L engines

OPERATION

▶ See Figures 87 and 88

The amount of air taken in during idling is regulated by the opening and closing of the servo valve located in the air passage that bypasses the throttle valve. The servo valve is opened or closed by the activation of the stepper motor (incorporated within the idle air control motor) in the forward or reverse (clockwise or counter-clockwise) direction. Battery positive (+) voltage is supplied, by way of the MFI relay, to the coil of the stepper motor. The ECM switches the power transistors on (located within the ECM) in sequential order, and, when current floes to the stepper motor coil, the stepper motor is activated in the forward or reverse direction.

If the number of stepper motor steps increases to 100-200 steps or decreases to 0 steps, the cause is probably a malfunction of the stepper motor or damaged or disconnected wiring of the harness.

If the number of stepper motor steps is outside the standard value, even though the results of checking the harness of the idle air control motor and of the component itself indicates no abnormal condition, the cause is probably one of the following: incorrect adjustment of the standard idling speed, deposits adhering to the throttle valve, air drawn into the intake manifold from a leaking gasket, incomplete combustion inside a cylinder (malfunction of spark plugs, ignition coil, injectors, compression pressure, etc.).

HARNESS TESTING

▶ See Figure 89

1. Unplug the IAC motor harness connector and the MFI relay connector. By touching the probes of the ohmmeter to both ends of the wiring harness, check for continuity between

4-44 EMISSION CONTROLS

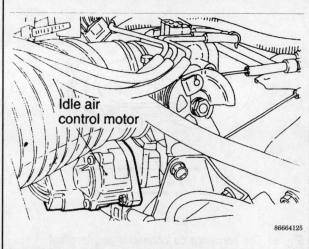

Fig. 87 The location of the idle air control motor (stepper motor) — 3.0L and 3.5L engines

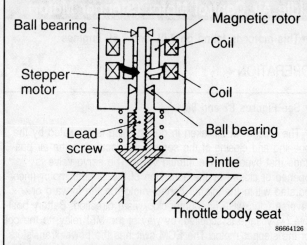

Fig. 88 Internal components of the idle air control motor — 3.0L and 3.5L engines

the IAC motor terminal 2 and 5 to the MFI relay terminal 5 (3.0L 12 valve engines) or MFI relay terminal 3 (3.0L 24 valve and 3.5L engines). If no continuity exists, repair the harness between the three terminals.

2. Unplug the ECM connector, and check for an open or short-circuit to ground between the ECM terminal and the IAC motor terminals:

3.0L (12 valve) engines:
- ECM terminal 58 and IAC motor terminal 1.
- ECM terminal 59 and IAC motor terminal 3.
- ECM terminal 67 and IAC motor terminal 4.
- ECM terminal 68 and IAC motor terminal 6.

3.0L (24 valve) and 3.5L Engines:
- ECM terminal 4 and IAC motor terminal 1.
- ECM terminal 17 and IAC motor terminal 3.
- ECM terminal 5 and IAC motor terminal 4.
- ECM terminal 18 and IAC motor terminal 6.

3. If there is a short or open-circuit, repair the above harnesses, otherwise the harness is good.

4. Plug all connectors back into their respective components.

COMPONENT TESTING

Actuator Testing

1. Check that the operating sound of the stepper motor can be heard over the idle air control motor when the ignition switch is turned to the **ON** position (without starting the engine).
2. If no operating sound can be heard, check the stepper motor drive circuit. (If the circuit is good, a defective stepper motor or ECM is suspected.)

Coil Resistance Testing

1. Unplug the idle air control motor connector and fasten the test harness (Mitsubishi tool MD998463).
2. Measure the resistance between terminal 2 (white clip on test harness) of the connector at the idle air control motor side and terminal 1 (red clip) or terminal 3 (blue clip). The resistance should be 28-33 ohms at 68°F (20°C).
3. Measure the resistance between terminal 5 (green clip) of the connector at the idle air control motor side and terminal 6 (yellow clip) or terminal 4 (black clip). The resistance should be the same as above.
4. If the coil failed either of these tests, replace the idle air control motor.

Idle Air Control Motor

▶ See Figures 90, 91 and 92

1. Remove the throttle body. Refer to section 5 for instructions.
2. Remove the stepper motor.
3. Fasten the test harness to the idle air control motor harness connector.
4. Connect the positive (+) terminal of a power source (approximately 6 V) to the white or the green clip.
5. While holding the idle air control motor as shown in the illustration, connect a negative (-) power source terminal to

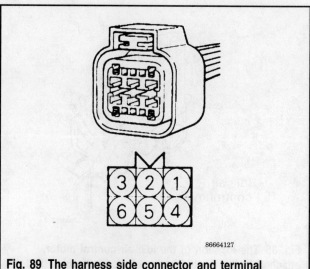

Fig. 89 The harness side connector and terminal numbering — 3.0L and 3.5L engines

EMISSION CONTROLS 4-45

each clip in the following sequence, and check whether there is vibration of the stepper motor as a result of activation of the stepper motor.

    a. Connect the negative (-) power source terminal to the red and black clips.

    b. Connect the negative (-) power source terminal to the blue and black clips.

    c. Connect the negative (-) power source terminal to the blue and yellow clips.

    d. Connect the negative (-) power source terminal to the red and yellow clips.

    e. Connect the negative (-) power source terminal to the red and black clips.

    f. Repeat the test in the reverse (e through a) sequence.

6. If vibration is felt as a result of this test, the stepper motor can be considered to be normal.

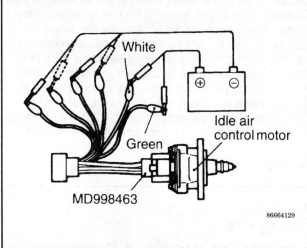

Fig. 92 Use Mitsubishi tool MD998463 to test the idle air control motor — 3.0L and 3.5L engines

REMOVAL & INSTALLATION

To remove the idle air control motor, the throttle body must first be removed from the intake manifold plenum. Once the throttle body has been removed, the stepper motor unbolts from the it. Label any hoses or cables removed and save the bolts and nuts for re-installation. Attach the new stepper motor to the throttle body. Tighten to 8-10 ft. lbs. (12-14 Nm). Install any parts removed for access to the throttle body.

Oxygen Sensor

OPERATION

▶ See Figures 93, 94 and 95

With the exception of the 1983-84 Pick-up and the 1984 49-state Montero, all Mitsubishi engines use an oxygen sensor to aid in the control of the air/fuel mixture. The ideal mixture within the engine is 14.7 parts of air to one part of fuel. If this ratio can be maintained under all conditions, emissions will be kept to an absolute minimum. The trick is to inform the Engine Control Module (ECM) of any change in conditions so that it can react and make necessary changes. The oxygen sensor is one of many sensors which detect changes during driving.

Located in either the exhaust manifold or the exhaust pipe ahead of the catalytic converter, the oxygen sensor reads the amount of oxygen in the exhaust flow and generates a proportional electrical voltage. This voltage is transmitted to ECM which interprets it and sends necessary messages to fuel and air control components. Remember that the oxygen sensor is reading the result of combustion and reacting to it. If there is a problem in the air/fuel mixture entering the engine, the combustion will be imperfect and the oxygen sensor will generate a signal which shows the error. The signal does not necessarily indicate that the sensor has failed, only that it has detected a different oxygen concentration.

Since the oxygen sensor is the furthest "downstream" in the combustion process, it essential to check all other sensors and

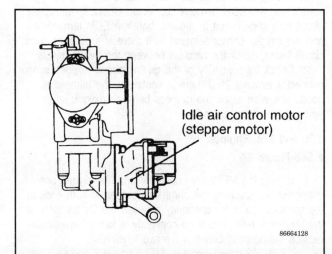

Fig. 90 The location of the idle air control motor — 3.0L and 3.5L engines

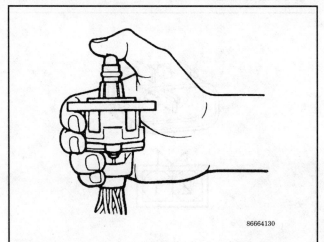

Fig. 91 Test the idle air control motor for vibration when the harness is connected to it — 3.0L and 3.5L engines

4-46 EMISSION CONTROLS

controls on the engine before assuming this sensor to be bad. Obviously, if the engine is running inefficiently, replacing the oxygen sensor won't cure the problem; the new sensor will continue to correctly read the imperfect exhaust content. About the only failure common to all oxygen sensors is loose or corroded connectors in the electrical wires. If a trouble code indicates an oxygen sensor malfunction, the first place to look is at the connector, making sure the pins are clean and fit tightly together. The low voltages flowing in this system can be changed or blocked by a high resistance (poor) connection.

HARNESS TESTING

2.4L Engines
♦ See Figure 96

➠ The 1993-95, rear-wheel drive 2.4L (California) Pick-ups have the same engine controls as the 3.0L engines. All other 2.4L engines are listed as 2.4L engines.

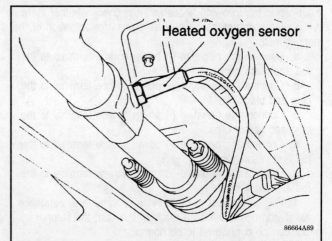

Fig. 95 The O_2 sensor is located in the front exhaust pipe, before the catalytic converter — 3.0L and 3.5L engines

1. Unplug both the harness connector going to the oxygen sensor and the ECM harness connector. Check for an open-circuit or a short-circuit to ground, between ECM terminal 4 and the oxygen sensor terminal 1. If there is a short or open-circuit found, repair the harness between the two terminals.

2. Check for continuity of the ground circuit (oxygen sensor connector terminal 2). If there is continuity, the harness is good, otherwise repair the harness between oxygen sensor terminal 2 and the ground.

3.0L and 3.5L Engines
♦ See Figure 97

1. Unplug the MFI relay connector and the oxygen sensor connector. By touching the ohmmeter probes to both ends of the harness, check for continuity between the O_2 sensor and the MFI relay terminals. If no continuity is found, repair or replace the harness between the two terminals.

  a. 3.0L (12 valve) engines: MFI terminal 5 and O_2 sensor terminal 1.
  b. 3.0L (24 valve) and 3.5L engines: MFI terminal 3 and O_2 sensor terminal 1.

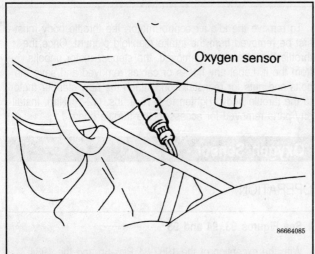

Fig. 93 The location of the oxygen (O_2) sensor on 2.4L equipped Pick-ups — right-hand side of the engine

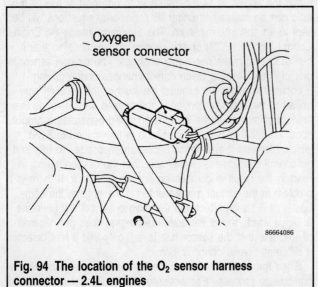

Fig. 94 The location of the O_2 sensor harness connector — 2.4L engines

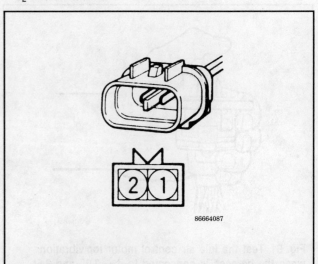

Fig. 96 The O_2 sensor harness connector terminals — 2.4L engines

EMISSION CONTROLS 4-47

2. Unplug the ECM harness connector and plug the MFI relay connector back in. Check for an open-circuit or a short-circuit to ground between the O_2 sensor and the ECM. If there is a short or open-circuit, repair or replace the harness between the two terminals.

    a. 3.0L (12 valve) engines: ECM terminal 4 and O_2 sensor terminal 4.

    b. 3.0L (24 valve) and 3.5L engines: ECM terminal 56 and O_2 sensor terminal 4.

3. Check for continuity in the ground circuit. If no continuity is found, repair or replace the harness between the terminals.

    a. 3.0L (12 valve) engines: ECM terminal 17 and 24 and O_2 sensor terminal 2, Ground circuit and O_2 sensor terminal 3.

    b. 3.0L (24 valve) and 3.5L engines: ECM terminal 72 and O_2 sensor terminal 2, Ground circuit and O_2 sensor terminal 3.

COMPONENT TESTING

2.0L, 2.4L and 2.6L Engines

➡ The 1993-95, rear-wheel drive 2.4L (California) Pick-ups have the same engine controls as the 3.0L engines. All other 2.4L engines are listed as 2.4L engines.

1. Before testing, warm the engine to normal operating temperature. Coolant temperature must be 80-85°C (175-185°F) or more.

➡ An accurate digital voltmeter is required for this test.

2. Shut the engine off. Disconnect the oxygen sensor connector and connect the positive probe of the voltmeter to the sensor connector.

3. Ground the negative probe of the meter to the body or the engine as convenient but do not ground it back to the sensor or connect it to the second terminal.

4. Place the meter where it can be seen from the driver's seat. Start the engine.

5. Race the engine to about 4000 rpm and observe the meter; it should show about 1 volt (600-1000 mV).

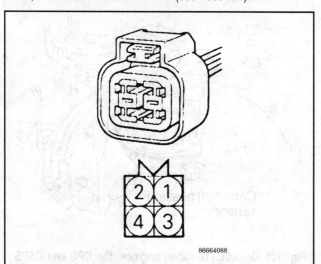

Fig. 97 The O_2 sensor harness connector terminals — 3.0L and 3.5L engines

6. Shut the engine off, remove the test equipment and reconnect the sensor harness.

7. If the sensor does not show this value, replace it with a new one.

3.0L and 3.5L Engines

1. Unplug the connector to the O_2 sensor, then use the test harness (Mitsubishi MD998464) to link the connector and the sensor together.

2. Check that there is continuity between terminal 1 (black clip of the test harness) and terminal 3 (white clip) of the O_2 sensor connector. There should be approximately 20 ohms at 68°F (20°C).

3. If there is no continuity, replace the O_2 sensor.

4. Warm the engine until the coolant temperature is 176°F (80°C) or higher.

5. Using jumper wires, connect terminal 1 (black clip) and terminal 3 (white clip) of the O_2 sensor with the positive battery terminal and negative battery terminal respectively.

✱✱WARNING

When connecting the jumper wires, be careful not to connect them to the wrong terminals, since this could damage the O_2 sensor. The terminals should be terminal 1 (+) and terminal 3 (-).

6. Connect a digital voltmeter to terminal 2 (red clip) and terminal 4 (blue clip).

7. While repeatedly racing the engine, measure the O_2 sensor's output voltage. The voltage should measure between 0.6-1.0 volts.

8. If the measurements are not as specified, the cause is probably a malfunction of the O_2 sensor.

REMOVAL & INSTALLATION

✱✱CAUTION

Perform this work only after the exhaust system has cooled enough to avoid burns.

It is more common to remove the oxygen sensor for protection or access during other repairs than to replace it because of failure. Once the sensor is removed, it must be protected from impact and/or chemical contact. Never attempt to clean the tip with solvent and never allow the tip to contact grease, oil or other chemicals. The zirconia element in the tip will be polluted and the sensor will function poorly, if at all.

1. Locate the oxygen sensor. On 4-cylinder engines, it will be located in the exhaust manifold, usually mounted either underneath or from the side. The 6-cylinder engines have the oxygen sensor mounted in the exhaust pipe, just beyond the Y where the left and right pipes connect into one.

2. Follow the wiring from the sensor to the first connector and disconnect it. Do not attempt to disconnect the wiring at the sensor.

3. Particularly on non-turbocharged, 4-cylinder engines, the sensor may be obstructed by heat shields on the exhaust manifold. Remove them as necessary.

4-48 EMISSION CONTROLS

4. Install the proper size wrench on the flats of the sensor. Place the socket on the sensor and use a box wrench to turn the socket.

5. Keeping the wrench (or socket) square to the sensor while removing it. Do not allow the wrench to become crooked or to come off the flats. Remember that the sensor has been exposed to extreme temperature and corrosive exhaust gasses. It may be difficult to remove.

6. Once the sensor is removed, place it in a clean, protected location. For reinstallation, the threads of the sensor may be lightly coated with an anti-seize compound but extreme care must be taken to protect the tip and shield area of the sensor from even the slightest contamination.

7. Handle the oxygen sensor carefully, protecting it from impact, and install it in place. Start the threads by hand and hand-tighten it as far as possible.

8. Finish by tightening the sensor to 33 ft. lbs. (45 Nm).

9. Install the heat shields if any were removed. Tighten the bolts to 12 ft. lbs. (16 Nm).

10. Connect the sensor wiring to the harness connector. Make certain the wiring is correctly run and out of the way of hot or moving components.

Camshaft Position Sensor

OPERATION

♦ See Figures 98, 99, 100, 101, 102 and 103

The Camshaft Position Sensor (CPS) is located integral to the distributor on 2.4L and 3.0L (12 valve) engines, and is located on the front-upper timing belt cover of 3.0L (24 valve) and 3.5L engines. The CPS detects the top dead center position of the No. 1 cylinder. It sends a signal to the ECM when this position is detected. The ECM uses this information to help it decide how to run the fuel injection sequence, among other functions. The power for the CPS is supplied from the MFI relay and is grounded to the vehicle body. The CPS, by intermitting the flow (to ground) of the 5 volts applied from the ECM, produces the pulse signals.

If there is a malfunction of the CPS, the sequential multiport fuel injection will not be correct. This will result in such problems as engine stalling, unstable idling, and poor acceleration.

HARNESS TESTING

2.4L Engines

♦ See Figure 104

➡The 1993-95, rear-wheel drive 2.4L (California) Pick-ups have the same engine controls as the 3.0L engines. All other 2.4L engines are listed as 2.4L engines.

1. Disconnect the CPS harness connector and turn the ignition **ON**. Measure the power supply voltage of CPS harness terminal 3 (to ground). The voltage should be the battery voltage (i.e. 10-12 volts). If this is not the case, repair or replace the harness between CPS terminal 3 and the ignition switch.

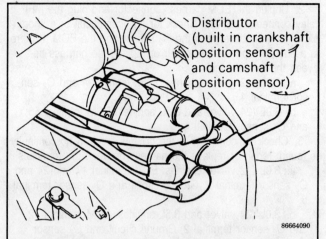

Fig. 98 The CPS, along with the crankshaft position sensor, is located inside of the distributor — 2.4L engines

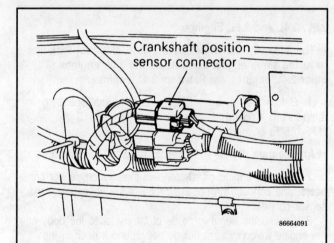

Fig. 99 The CPS and CSPS harness connector is mounted on the firewall on Pick-up trucks equipped with the 2.4L engine

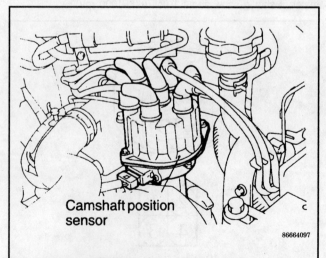

Fig. 100 On 3.0L (12 valve) engines, the CPS and CSPS are located inside of the distributor

EMISSION CONTROLS 4-49

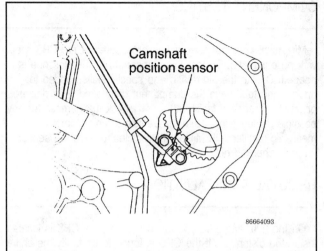

Fig. 101 The CPS location under the upper left-hand timing belt cover — 3.0L (24 valve) and 3.5L engines

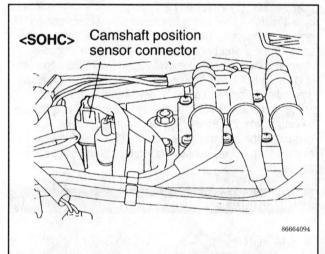

Fig. 102 The location of the CPS harness connector — 3.0L (24 valve) engines

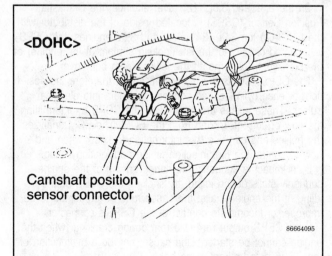

Fig. 103 The location of the CPS harness connector — 3.5L engines

2. Turn the ignition switch **OFF** and check for continuity of the ground circuit, which is CPS harness terminal 2. If no continuity is found, repair or replace the harness from CPS terminal 2 to ground.

3. Turn the ignition switch back **ON** and check the voltage of the CPS harness terminal 4 (output circuit). The voltage should be between 4.8-5.2 volts, if it is not, repair the harness from CPS terminal 4 to ECM terminal 22.

4. If the harness passed all of the tests, it is good.

5. Plug all connectors back into their original components.

3.0L (12 Valve) Engines
▶ See Figure 105

1. Unplug the MFI relay harness connector and the distributor harness connector. While touching the ohmmeter probes to both ends of the harness, check for continuity between the CPS (distributor) harness terminal 3 and the MFI relay harness terminal 5. If no continuity is found, repair the harness between these two terminals.

2. Plug the MFI relay harness back in, and check for continuity in the ground circuit, that is terminal 4 of the CPS (distributor) harness to ground. If no continuity exists, repair the harness between CPS harness terminal 4 and the ground.

3. Unplug the ECM harness connector. Check for an open or short-circuit between the CPS harness terminal 1 and ECM harness terminal 22, and CPS harness terminal 2 and ECM harness terminal 21. If a short or open-circuit is found, repair the harness between the two terminals.

4. Turn the ignition switch **ON**. Check the output circuit voltage. This is CPS harness terminal 1 to ground. If the output voltage is not 4.8-5.2V, replace the ECM with a new unit.

5. Plug all connectors back into their original components.

3.0L (24 Valve) and 3.5L Engines
▶ See Figure 106

1. Unplug the MFI relay harness connector and the CPS harness connector. Insert the probes of the circuit tester into both ends of the harness. Check for continuity between the

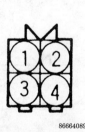

Fig. 104 The numbering of the terminals on the component (sensor) side of the CPS connector — 2.4L engines

EMISSION CONTROLS

COMPONENT TESTING

Mitsubishi does not give testing procedures of the CPS itself with an ohmmeter. To test this component an oscilloscope is needed. Due to the high cost, and special skills needed to operate the tool, have the component diagnosed by a dealer or reputable service facility after you check the harness for any possible malfunctions. Take the vehicle to an automotive mechanic familiar with Mitsubishi vehicles to have the sensor properly checked out.

REMOVAL & INSTALLATION

On the 2.4L and 3.0L (12 valve) engines, the CPS is located inside the distributor. If the CPS is found to be faulty the entire distributor must be removed and either rebuilt or replaced. Refer to Section 3 for distributor removal and installation procedures.

On the 3.0L (24 valve) and 3.5L engines, the CPS is located under the front-upper timing belt cover. The sensor can be removed and replaced with a new one, however refer to section 3 for the timing belt removal and installation procedures for preliminary accessory removal. Only remove the upper left-hand timing belt cover, though, and do not remove the timing belts, other timing belt covers or anything else that is not directly in the way. Unbolt the two mounting bolts and remove the sensor from the front of the engine head.

To install the new sensor, secure it in place with the bolts removed from the old sensor. Reinstall all of the components removed for access.

Crankshaft Position Sensor

OPERATION

▶ See Figures 98, 99, 100, 107, 108 and 109

On the 2.4L and 3.0L (12 valve) engines, the Crankshaft Position Sensor (CSPS) is located inside of the distributor with the CPS. On the 3.0L (24 valve) and 3.5L engines, the CSPS is mounted directly to the left (looking straight at the engine) of the crankshaft, underneath the lower timing belt cover.

The CSPS detects the crankshaft angle, therefore the position, of each cylinder, and converts this data into electrical pulses. These pulses are created by the CSPS by intermitting the flow of current applied from the ECM. The ECM analyzes the pulses and can use this information to know when and how to control the fuel injection timing and ignition timing.

If an impact is suddenly felt during driving or the engine suddenly stalls during idling, try shaking the CSPS during idling. If the engine stalls, the cause may be presumed to be improper or incomplete contact of the CSPS's connector.

If the CSPS output rpm is 0 rpm during cranking (when the engine cannot be started), the cause may be a malfunction of the CSPS or a broken timing belt.

If the indicated value of the CSPS output is 0 rpm during cranking (the engine will not start), the ignition coil's primary current may be malfunctioning to intermittently pulse correctly,

Fig. 105 The numbering of the terminals on the component (sensor) side of the CPS connector — 3.0L (12 valve) engines

CPS harness terminal 3 and MFI relay terminal 3. If there is no continuity, repair the harness between these two terminals.

2. Plug the MFI relay connector back in. Check for continuity in the ground circuit, which is CPS terminal 1 to ground. If no continuity in the circuit, repair the harness between CPS harness terminal 1 to ground.

3. Unplug the ECM harness connector. Check the CPS harness terminal 2 and ECM terminal 68 for an open or short-circuit to ground. If there is a short or open-circuit, repair the harness between the two terminals.

4. Turn the ignition switch ON. Measure the applied voltage of CPS harness terminal 2 to ground. The voltage should be between 4.8-5.2 volts, if it is not, replace the ECM with a new unit. If the harness passed all of the tests, the harness is good.

5. Plug all connectors back into their original components.

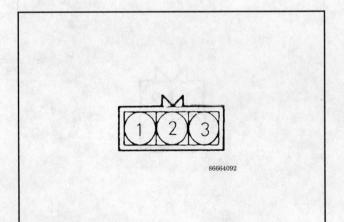

Fig. 106 The numbering of the terminals on the component (sensor) side of the CPS connector — 3.0L (24 valve) and 3.5L Engines

EMISSION CONTROLS 4-51

so a malfunction of the ignition system circuitry, the ignition coil and/or the ignition power transistor is probably the cause.

If idling is possible even though the CSPS indicated rpm is a deviation from the standard value, the cause is usually a malfunction of something other than the CSPS. It could be a malfunction of the ECT sensor, the ISCM, or an improper adjustment of the standard idling speed.

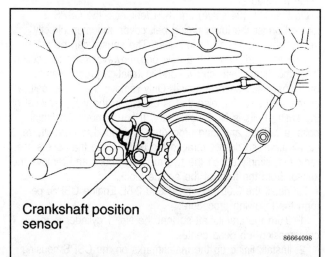

Fig. 107 The CSPS location under the lower timing belt cover — 3.0L (24 valve) and 3.5L engines

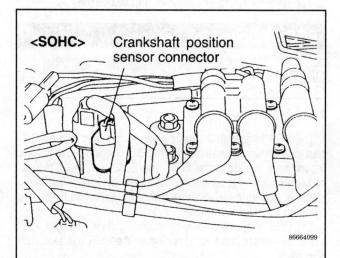

Fig. 108 The location of the CSPS harness connector — 3.0L (24 valve) engines

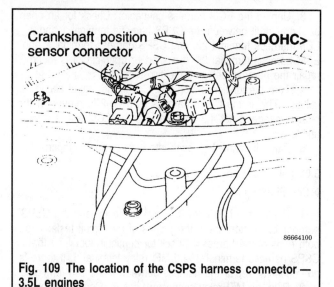

Fig. 109 The location of the CSPS harness connector — 3.5L engines

HARNESS TESTING

2.4L Engines

♦ See Figure 104

➥The 1993-95, rear-wheel drive 2.4L (California) Pick-ups have the same engine controls as the 3.0L engines. All other 2.4L engines are listed as 2.4L engines.

1. Disconnect the CSPS harness connector (the same as the CPS) and turn the ignition **ON**. Measure the power supply voltage of CPS harness terminal 3 (to ground). The voltage should be the battery voltage (i.e. 10-12 volts). If this is not the case, repair or replace the harness between CSPS terminal 3 and the ignition switch.

2. Turn the ignition switch **OFF** and check for continuity of the ground circuit, which is CSPS harness terminal 2. If no continuity is found, repair or replace the harness from CSPS terminal 2 to ground.

3. Turn the ignition switch back **ON** and check the voltage of the CSPS harness terminal 4 (output circuit). The voltage should be between 4.8-5.2 volts, if it is not, repair the harness from CSPS terminal 1 to ECM terminal 21.

4. If the harness passed all of the tests, it is good.

5. Plug all connectors back into their original components.

3.0L (12 Valve) Engines

♦ See Figure 105

1. Unplug the MFI relay harness connector and the distributor harness connector. While touching the ohmmeter probes to both ends of the harness, check for continuity between the CSPS (CPS/distributor) harness terminal 3 and the MFI relay harness terminal 5. If no continuity is found, repair the harness between these two terminals.

2. Plug the MFI relay harness back in, and check for continuity in the ground circuit, that is terminal 4 of the CSPS (distributor) harness to ground. If no continuity exists, repair the harness between CSPS harness terminal 4 and the ground.

4-52 EMISSION CONTROLS

3. Unplug the ECM harness connector. Check for an open or short-circuit between the CSPS harness terminal 1 and ECM harness terminal 22, and CSPS harness terminal 2 and ECM harness terminal 21. If a short or open-circuit is found, repair the harness between the two terminals.

4. Turn the ignition switch **ON**. Check the output circuit voltage. This is the CSPS harness terminal 2 to ground. If the output voltage is not 4.8-5.2V, replace the ECM with a new unit.

5. Plug all connectors back into their original components.

3.0L (24 Valve) and 3.5L Engines

♦ See Figure 110

1. Unplug the MFI relay harness connector and the CSPS harness connector. Insert the probes of the circuit tester into both ends of the harness. Check for continuity between the CSPS harness terminal 3 and MFI relay terminal 3. If there is no continuity, repair the harness between these two terminals.

2. Plug the MFI relay connector back in. Check for continuity in the ground circuit, which is the CSPS terminal 1 to ground. If no continuity in the circuit, repair the harness between CSPS harness terminal 1 to ground.

3. Unplug the ECM harness connector. Check the CSPS harness terminal 2 and ECM terminal 69 for an open or short-circuit to ground. If there is a short or open-circuit, repair the harness between the two terminals.

4. Turn the ignition switch **ON**. Measure the applied voltage of CSPS harness terminal 2 to ground. The voltage should be between 4.8-5.2 volts, if it is not, replace the ECM with a new unit. If the harness passed all of the tests, the harness is good.

5. Plug all connectors back into their original components.

COMPONENT TESTING

Mitsubishi does not give testing procedures of the CSPS itself with an ohmmeter. To test this component an oscilloscope is needed. Due to the high cost, and special skills needed to operate the tool, have the system diagnosed by a dealer or reputable service facility.

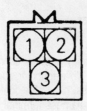

Fig. 110 The numbering of the terminals on the component (sensor) side of the CSPS connector — 3.0L (24 valve) and 3.5L engines

REMOVAL & INSTALLATION

On the 2.4L and 3.0L (12 valve) engines, the CSPS is located inside the distributor. If the CSPS is found to be faulty the entire distributor must be removed and either rebuilt or replaced. Refer to Section 3 for distributor removal and installation procedures.

On the 3.0L (24 valve) and 3.5L engines, the CSPS is located under the lower timing belt cover, mounted directly next to the crankshaft timing gear. The sensor can be removed and replaced with a new one, however refer to Section 3 for the timing belt removal and installation procedures for preliminary accessory removal. Only remove the timing belt covers and anything else that is directly in the way; removal of the timing belts is NOT necessary. If you remove the timing belts, a tedious and long procedure is needed to correctly re-install them, so do not remove the belts. Once the covers are removed, simply unbolt the two mounting bolts and remove the sensor from the front of the engine block.

To install the 3.0L (24 valve) and 3.5L Engine CSPS, perform the following procedure:

1. Turn the crankshaft so that the No. 1 cylinder is at compression top dead center.
2. Install, lining up the matchmarks on the CSPS housing and the coupling.
3. Install the mounting bolts securely.

Reinstall all of the components removed for access — refer to section 3 for timing belt removal and installation.

EGR Temperature Sensor

OPERATION

♦ See Figure 111

The EGR temperature sensor converts changes in the EGR gas temperature downstream of the EGR valve to voltages and inputs these signals to the ECM. The ECM uses these pulses to keep track of the condition of the EGR system. When the signal is judged to be abnormal, the engine warning lamp lights up to inform the driver.

The EGR temperature sensor terminal voltage changes with the EGR gas temperature, dropping as the EGR gas temperature rises.

The EGR temperature sensor is located on the top right-hand (driver's) side of the engine intake manifold plenum.

HARNESS TESTING

♦ See Figure 112

2.4L Engines

➡The 1993-95, rear-wheel drive 2.4L (California) Pick-ups have the same engine controls as the 3.0L engines. All other 2.4L engines are listed as 2.4L engines.

1. Unplug the EGR temperature sensor connector and turn the ignition switch **ON**. Measure the power supply between the EGR temperature sensor harness connector terminal 2 and a

EMISSION CONTROLS 4-53

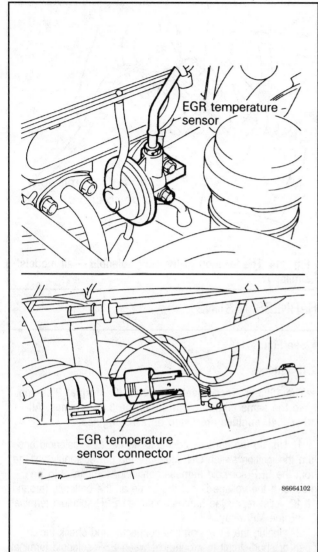

Fig. 111 The location of the EGR temperature sensor and its harness connector — all models similar

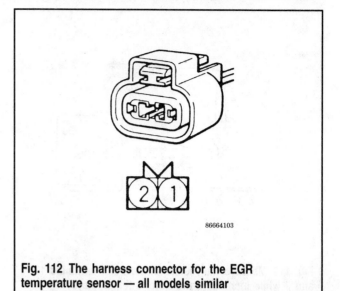

Fig. 112 The harness connector for the EGR temperature sensor — all models similar

ground. If the voltage measured is not between 4.3-4.7, repair the harness between the terminal 2 and the ECM terminal 15.

2. Turn the ignition switch **OFF**. Check for continuity from the EGR temperature sensor connector terminal 1 to ground. If no continuity is found, repair the harness between the EGR temperature sensor harness connector terminal 1 to the ECM terminals 14 and 24, otherwise the harness is good.

3. Plug all connectors back into their original components.

3.0L Engines

Only the California models of the 3.0L engines have the EGR temperature sensor.

1. Unplug the EGR temperature sensor connector and turn the ignition switch **ON**. Measure the power supply between the EGR temperature sensor harness connector terminal 1 and a ground. If the voltage measured is not between 4.3-4.7, repair the harness between the terminal 1 and the ECM terminal 15.

2. Turn the ignition switch **OFF**. Check for continuity from the EGR temperature sensor connector terminal 2 to ground. If no continuity is found, repair the harness between the EGR temperature sensor harness connector terminal 2 to the ECM terminals 17 and 24, otherwise the harness is good.

3. Plug all connectors back into their original components.

3.5L Engines

1. Unplug the EGR temperature sensor harness connector. Check for continuity from the EGR temperature sensor connector terminal 1 to ground. If no continuity is found, repair the harness between the EGR temperature sensor harness connector terminal 1 to the ECM terminal 72, otherwise the harness is good.

2. Unplug the ECM module harness connector and check for an open or short-circuit to ground between the EGR temperature sensor terminal 2 and the ECM terminal 53. If there is a short or open-circuit, repair the harness between these two terminals.

3. Plug the ECM harness connector back in and turn the ignition switch **ON**. Measure the power supply between the EGR temperature sensor harness connector terminal 2 and a ground. If the voltage measured is not between 4.3-4.7, replace the ECM with a new unit.

4. Plug all connectors back into their original components.

COMPONENT TESTING

▶ See Figure 113

1. Remove the EGR temperature sensor.
2. Place the EGR temperature sensor in water, and then measure the resistance value between terminals 1 and 2 while increasing the temperature of the water.
   a. At 122°F (50°C), the resistance should be 60-83 kilohms.
   b. At 212°F (100°C), the resistance should be 11-14 kilohms.
3. Replace the EGR temperature sensor if there is a significant deviation from the standard value.
4. Install the EGR temperature sensor and tighten to 7.3-8.6 ft. lbs. (10-12 Nm).
5. Plug all connectors back into their original components.

4-54 EMISSION CONTROLS

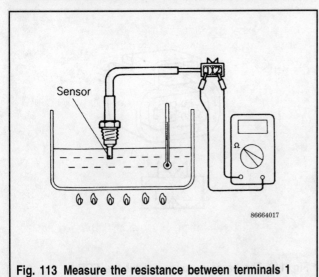

Fig. 113 Measure the resistance between terminals 1 and 2 while increasing the water's temperature

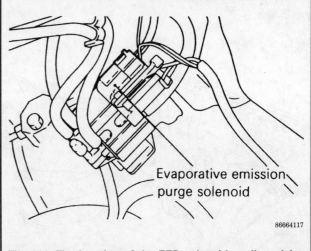

Fig. 114 The location of the EEP solenoid — all models similar

REMOVAL & INSTALLATION

To remove the EGR temperature sensor, unplug the EGR temperature sensor harness connector and, using a wrench, loosen the EGR temperature sensor from its boss. Make sure the wrench is correctly applied to the faces of the EGR temperature sensor. Remove the sensor along with its harness.

To replace, thread the harness back into its original position. Apply some thread sealant to the threads and screw the sensor into the mounting boss. Make certain not to crossthread the sensor. Tighten the sensor to 7-9 ft. lbs. (10-12 Nm).

Evaporative Emission Purge Solenoid

OPERATION

▶ See Figure 114

While the 2.0L and 2.6L engines use a vacuum actuated Evaporative Emission Purge (EEP) solenoid, the 2.4L, 3.0L and 3.5L engines use an electrically controlled EEP solenoid. The EEP solenoid is an ON/OFF type of solenoid valve. It regulates the introduction of purge air from the evaporative emission canister to the intake air plenum.

The EEP solenoid is powered by the battery, which is channeled through the MFI relay. When the ECM switches the ignition power transistor within the unit, current floes to the coil and purge air is allowed to enter the manifold.

This solenoid is located on the left-hand side of the engine compartment, and is integral with the EGR solenoid.

HARNESS TESTING

▶ See Figure 115

2.4L Engines

➡The 1993-95, rear-wheel drive 2.4L (California) Pick-ups have the same engine controls as the 3.0L engines. All other 2.4L engines are listed as 2.4L engines.

1. Unplug the harness connector to the EEP solenoid and turn the ignition switch **ON**. Measure the power supply voltage from the EEP solenoid harness connector terminal 2 to a ground. If the voltage is not the same as the battery's (around 10-12 volts) repair the harness between EEP solenoid terminal 2 and the MFI relay.

2. Unplug the ECM harness connector and check for an open or short-circuit to ground, between EEP solenoid terminal 1 and the ECM terminal 62. If a short or open-circuit is found, repair the harness between these two terminals.

3. Plug all connectors back into their original components.

2.4L (California), 3.0L and 3.5L Engines

1. Unplug the EEP solenoid harness connector and the MFI relay connector. By touching the circuit tester probes to both ends of the harness, check for continuity between the EEP solenoid terminal and the MFI relay. If no continuity is found, repair the circuit between these terminals.

   a. 2.4L (California) engines: EEP terminal 2 and MFI relay terminal 2.

   b. 3.0L (12 valve) engines: EEP terminal 1 and MFI relay terminal 5.

   c. 3.0L (24 valve) and 3.5L engines: EEP terminal 1 and MFI relay terminal 3.

2. Unplug the ECM harness connector and check for an open or short-circuit to ground between the ECM terminal and the EEP solenoid terminal. If a short or open-circuit is found, repair the harness between these terminals.

   a. 2.4L (California) engines: EEP terminal 1 and ECM terminal 9.

   b. 3.0L (12 valve) engines: EEP terminal 2 and ECM terminal 62.

EMISSION CONTROLS 4-55

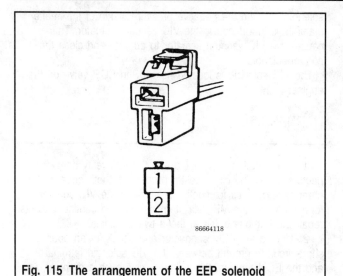

Fig. 115 The arrangement of the EEP solenoid terminals — all models

c. 3.0L (24 valve) and 3.5L engines: EEP terminal 2 and ECM terminal 9.
3. Plug all connectors back into their original components.

COMPONENT TESTING

Although the EEP valve can be tested with a hand vacuum pump (see the earlier described direction for this procedure), the EEP solenoid cannot be tested without an expensive scan tool from Mitsubishi. If the harness checks out OK and the diagnostic code still calls for this component to be checked, take the vehicle to an automotive mechanic familiar with Mitsubishi vehicles.

REMOVAL & INSTALLATION

The EEP solenoid is a part of the EEP valve, and is located next to the EGR valve assembly. Removal of this entire component is necessary. Disconnect and label the vacuum hoses going to the assembly. Unplug and label the harness connectors plugged into the EEP assembly. Remove the mounting bolts holding the assembly in place.
To install the new component, mount the assembly back in place with the mounting bolts. Tighten the bolts until they are snug, plug the harness connectors and the vacuum hoses back into the assembly.

Knock Sensor

➡The knock sensor is only found on 3.5L engines.

OPERATION

♦ See Figure 116

The knock sensor converts cylinder block vibrations, due to knocking, into voltage signals. These signals are various voltages according to the strength of the knocking. These signals are input into the ECM. The ECM controls the delay in the spark timing according to these signals.
When knocking occurs when driving at maximum load, the following problems, other than the knock sensor, can be the cause: incorrect spark plug heat rating, incorrect gasoline, incorrect standard spark timing adjustment.
The knock sensor is located in the intake valley, under the intake.

HARNESS TESTING

♦ See Figure 117

1. Unplug the knock sensor connector and the ECM harness connector. Check for an open or short-circuit to ground between the knock sensor terminal 1 and the ECM terminal 58. If a short or open-circuit does exist, repair the harness between the two terminals.
2. Check for continuity from the knock sensor terminal 2 and ground. If there is no continuity, repair the terminal between the knock sensor terminal 2 and the ground.
3. Plug all connectors back into their original components.

COMPONENT TESTING

The knock sensor cannot be tested without an expensive scan tool from Mitsubishi. If the harness turns out all right and all other possible reasons for the malfunction have been investigated, take the vehicle to an automotive mechanic familiar with Mitsubishi vehicles.

REMOVAL & INSTALLATION

To remove the knock sensor, the intake manifold plenum must be removed (refer to Section 3 for these procedures). When all components are removed, simply unscrew the knock sensor with an open or box wrench. Make sure not to damage the sensor.

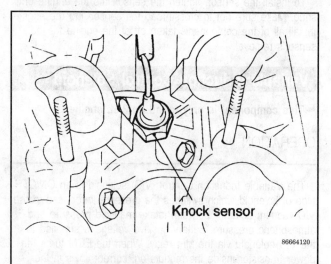

Fig. 116 The location of the knock sensor in the intake manifold valley — 3.5L engine

4-56 EMISSION CONTROLS

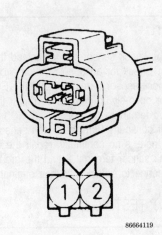

Fig. 117 The arrangement of the knock sensor connector terminals — 3.5L engines

To install the sensor, tighten the sensor into the engine until snug. Make sure not to crossthread the sensor into the engine. Install all of the components taken off of the engine for the sensor's removal.

Variable Induction Control Solenoid

➥This component is only found on 3.5L engines

OPERATION

The Variable Induction Control (VIC) solenoid is an ON/OFF type of solenoid which switches the pressure introduced to the VIC vacuum actuator between intake manifold pressure and atmospheric pressure. Battery positive voltage is supplied to the VIC solenoid via the MFI relay. When the ECM turns the power transistor inside the module on, current flows in the solenoid coil and the negative pressure inside of the intake manifold is introduced to the VIC vacuum actuator. This causes the VIC vacuum actuator to operate and close the control solenoid.

The VIC solenoid is located next to the EGR valve on the engine.

HARNESS TESTING

1. Unplug the VIC solenoid and the MFI relay harness connectors. Insert the circuit tester probes into both ends of the harness and check for continuity between the VIC solenoid terminal 1 and the MFI relay terminal 3. If no continuity exists, repair the harness between these two terminals.

2. Unplug the ECM connector and check for an open or short-circuit to ground between the VIC solenoid terminal 2 and the ECM terminal 20. If there exists such a problem, repair the harness between the two terminals, otherwise the harness is good.

COMPONENT TESTING

Actuator

➥A vacuum pump capable of producing more than 10 in. Hg (33.8 kPa) of vacuum will be needed to perform this test.

1. Apply negative pressure to the vacuum tank side nipple (connect to the white vacuum hose) of the VIC solenoid with a hand vacuum pump. Check for air-tightness when voltage is applied and removed from the VIC solenoid terminals.

   a. When the battery positive (+) voltage is applied, the other nipple of the VIC solenoid should be open and the negative pressure should leak.

   b. With the battery positive (+) voltage applied, cover the other nipple with your finger and the negative pressure should be maintained.

   c. When the battery positive (+) voltage is not applied and the other nipple is left open, the negative pressure should be maintained.

Coil

▶ See Figure 118

Use an ohmmeter to measure the resistance between the two terminals (coil) on the VIC solenoid. The resistance should be 36-44 ohms at 68°F (20°C). If the VIC solenoid fails any of these tests, replace it with a new one.

REMOVAL & INSTALLATION

To remove the VIC solenoid, unplug and label the harness connector and vacuum lines and remove the mounting bolt. Remove the component from the engine. To install, securely attach the component to the engine with the mounting bolt. Connect all of the vacuum lines and the harness plug.

EMISSION CONTROLS 4-57

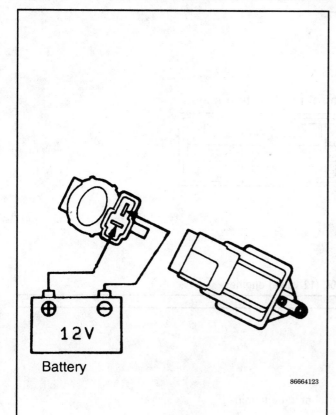

Fig. 118 Test the resistance between the two terminals on the VIC solenoid coil — the resistance should be between 36-44 ohms at 68°F (20°C)

Engine Control Module (ECM)

TERMINAL VOLTAGE TESTING

▶ See Figures 119, 120, 121, 122, 123, 124, 125, 126, 127, 128, 129, 130 and 131

1. Connect extra fine wire probes to the voltmeter probes (paper clip, etc.)
2. Insert the extra fine wire probes from the wire side to each terminal of the ECM connector and measure the voltage while referring to the check chart.

➡Measure the voltage with the ECM connector plugged in. Measure the voltage between each terminal and the grounding terminal. You may find it convenient to pull the ECM out to make it easier to reach the connector terminals. It is not necessary to do inspection in the order of the chart.

- 2.4L Engines: terminal 106.
- 2.4L (California RWD 1993-95) Engines: terminal 26.
- 3.0L (12 valve) Engines: terminal 106.
- 3.0L (24 valve) and 3.5L Engines: terminal 26.

✱✱WARNING

Short-circuiting the positive (+) probe between a connector terminal and any ground other than the grounding terminal could damage the vehicle wiring, the sensor or the ECM. Use care to prevent this from happening.

3. If the voltmeter shows any deviation from the standard value, check the corresponding sensor, actuator and related electrical wiring, then repair or replace as needed.
4. After the repair or replacement, recheck the ECM with the voltmeter to confirm that the repair has corrected the malfunction.

4-58 EMISSION CONTROLS

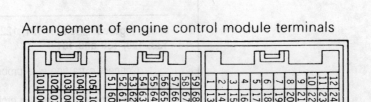

Fig. 119 The arrangement of the ECM terminals — 2.4L and 3.0L (12 valve) engines

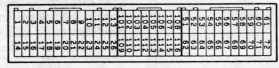

Fig. 120 The arrangement of the ECM terminals — 2.4L (California RWD 1993-), 3.0L (24 valve) and 3.5L engines

Terminal No.	Item	Check condition (engine condition)	Standard value	Note
103	Back-up power	Ignition switch: OFF	B+	
102	Power	Ignition switch: ON	B+	
107				
109	Fuel pump driving signal	Engine: cranking	More than 8V	
		Engine: idling	B+	

Fig. 121 The ECM terminal voltage check chart — 2.4L engines

EMISSION CONTROLS 4-59

Terminal No.	Item	Check condition (engine condition)		Standard value	Note
56	MFI relay (fuel pump)	Ignition switch: ON		B+	
		Engine: idling		0 – 3V	
23	Sensor applied voltage	Ignition switch: ON		4.5 – 5.5V	
10	Volume air flow sensor	Engine: idling		2.2 – 3.2V	
		Engine: 2000rpm			
57	Volume air flow sensor reset signal	Engine: idling		0 – 1V	
		Engine: 3000rpm		6 – 9V	
8	Intake air temperature	Ignition switch: ON	When intake air temperature is 0°C (32°F)	3.2 – 3.8V	
			When intake air temperature is 20°C (68°F)	2.3 – 2.9V	
			When intake air temperature is 40°C (104°F)	1.5 – 2.1V	
			When intake air temperature is 80°C (176°F)	0.4 – 1.0V	
16	Barometric pressure sensor	Ignition switch: ON	When altitude is 0m (0ft.)	3.7 – 4.3V	
			When altitude is 1200 m (3937 ft.)	3.2 – 3.8V	
20	Engine coolant temperature sensor	Ignition switch: ON	When engine coolante temperature is 0°C (32°)	3.2 – 3.8V	
			When engine coolant temperature is 20°C (68°F)	2.3 – 2.9V	
			When engine coolant temperature is 40°C (104°F)	1.3 – 1.9V	
			When engine coolant temperature is 80°C (176°F)	0.3 – 0.9V	
19	Throttle position sensor	Ignition switch: Maintain more than 15 seconds woth "ON" condition.	Idle	0.4 – 0.6V	
			wide open throttle	4.5 – 5.5V	
6	Closed throttle position switch	Ignition switch: ON	Set throttle valve to idling position.	0 – 1V	
			Open throttle valve slightly	More than 4V	
22	Camshaft position sensor	Engine: creakning		0.1 – 3.0V	
		Engine: idling			
21	Crankhsaft position sensor	Engine: cranking		0.2 – 3.0V	
		Engine: idling			
108	Ignition switch-ST	Engine: cranking		More than 8V	
104	Park/Neutral position switch	Ignition switch: ON	Set selector lever to P or N	0 – 3V	<A/T>
			Set selector lever to D, 2, L or R	8 – 14V	
18	Vehicle speed sensor	• Ignition switch: ON • Slowly drive vehicle forward		0↔5V (repeat changes)	
7	Air-conditioning switch	Engile: idling	Turn off air-conditioning switch	0 – 3V	
			Turn on air-conditioning switch (Air-conditioning compressor is in driving condition.)	B+	
65	Air conditioning compressor clutch relay	• Engine: idling • Air-conditioning switch: OFF → ON (Air-conditioning compressor is in driving condition.)		B+ or momentarily over 6V ↓ 0 – 3V as A/C clutch cycles	

Fig. 122 The ECM terminal voltage check chart — 2.4L engines

4-60 EMISSION CONTROLS

Terminal No.	Item	Check condition (engine condition)		Standard value	Note
4	Oxygen sensor	Engine: warm, 2000 rpm. (Check with digital of voltmeter.)		0↔0.8V (Repeat changes)	
51	No.1 injector	Engine: Step on the accelerator pedal sharply from idling condition after warm-up		Temporarily drops slightly from 11 – 14V	
52	No.2 injector				
60	No.3 injector				
61	No.4 injector				
54	Ignition power transistor unit	Engine: 3000 rpm		0.3 – 3.0V	
62	Evaporative emission purge solenoid	Ignition switch: ON		B+	
		Engine: warm, 3000rpm		0 – 3V	
12	Ignition timing adjustment terminal	Ignition switch: ON	Ground ignition timing adjustment terminal	0 – 1V	
			Release ignition timing adjustment terminal from ground	4.0 – 5.5V	
64	Check engine/ malfunction indicator lamp	Ignition switch: OFF → ON		0 – 3V ↓ 9 – 13V (after passing several seconds)	
53	EGR solenoid	ignition switch: ON		B+	
		Engine: idle suddenly depress the accelerator pedal.		Temporaily drops slightly fron B+	
15	EGR temperature sensor	Ignition switch: ON	When sensor temperature is 50°C (122°F)	3.6 – 4.4V	<Federal (From 1994 models), California>
			When senosr temperature is 100°C (212°F)	2.2 – 3.0V	
58	Idle speed control motor (connection)	Ignition switch: When 15 seconds has passed immediately after ON		More than 4V (momentarily) ↓ 0 – 1V	
59	Idle speed control motor (contraction)	Ignition switch: immediately after ON		More than 4V (momentarily) ↓ 0 – 1V	
17	Idle speed control motor position sensor	Ignition switch: ON → After 15 seconds have passed		More than 1.2V ↓ 0.7 – 1.1V	

Fig. 123 The ECM terminal voltage check chart — 2.4L engines

Terminal No.	Check Item	Check Condition (Engine Condition)	Standard value
60	Backup power supply	Ignition: OFF	B+
12	Power supply	Ignition: ON	B+
25			
8	MFI relay (Fuel pump)	Ignition switch: ON	B+
		Engine: Idle	0 – 3V
61	Sensor impressed voltage	Ignition switch: ON	4.5 – 5.5V

Fig. 124 The ECM terminal voltage check chart — 2.4L (California RWD 1993-95) engines

EMISSION CONTROLS 4-61

Terminal No.	Check Item	Check Condition (Engine Condition)			Standard value
70	Volume air flow sensor	Engine: Idle			2.2 – 3.2V
		Engine: 2,000 rpm			
19	Volume air flow sensor reset signal	Engine: Idle			0 – 1V
		Engine: 3,000 rpm			6 – 9V
52	Intake air temperature sensor	Ignition switch: ON	When intake air temperature is 0°C (32°F)		3.2 – 3.8V
			When intake air temperature is 20°C (68°F)		2.3 – 2.9V
			When intake air temperature is 40°C (104°F)		1.5 – 2.1V
			When intake air temperature is 80°C (176°F)		0.4 – 1.0V
65	Barometric pressure sensor	Ignition switch: ON	When altitude is 0 m		3.7 – 4.3V
			When altitude is 1,200 m (3,937 ft.)		3.2 – 3.8V
63	Engine coolant temperature sensor	Ignition switch: ON	When engine coolant temperature is 0°C (32°F)		3.2 – 3.8V
			When engine coolant temperature is 20°C (68°F)		2.3 – 2.9V
			When engine coolant temperature is 40°C (104°F)		1.3 – 1.9V
			When engine coolant temperature is 80°C (176°F)		0.3 – 0.9V
64	Throttle position sensor	Ignition switch: ON		Idle	0.3 – 0.6V
				Wide open throttle.	4.5 – 5.5V
67	Closed throttle position switch	Ignition switch: ON		Set throttle valve to idle position.	0 – 1V
				Slightly open throttle valve	4V or more
68	Camshaft position sensor	Engine: Cranking			0.2 – 3.0V
		Engine: Idle			
69	Crankshaft position sensor	Engine: Cranking			0.2 – 3.0V
		Engine: Idle			
51	Ignition switch – ST	Engine: Cranking			8V or more
71	Park/neutral position switch	Ignition switch: ON		Set selector lever to P or N.	0 – 3V
				Set selector lever to D, 2, L or R.	8 – 14V
66	Vehicle speed sensor	Ignition switch: ON Move the vehicle slowly forward.			0 ↔ 5V (Changes repeatedly)
115	Air conditioning switch	Engine: Idle		Turn the air conditioning switch OFF.	0 – 3V
				Turn the air conditioning switch ON. (Air conditioning compressor is operating)	B+
22	Air conditioning compressor relay	Engine: Idle Air conditioning switch: OFF → ON Turn the air conditioning switch ON. (Air conditioning compressor is operating)			B+ or temporarily 6V or more ↓ 0 – 3V as A/C clutch cycles

Fig. 125 The ECM terminal voltage check chart — 2.4L engines

EMISSION CONTROLS

Terminal No.	Check Item	Check Condition (Engine Condition)		Standard value
56	Heated oxygen sensor (front)	Engine: warm, 2000 rpm. (Check using a digital type voltmeter.)		0 ↔ 0.8V (Changes repeatedly)
55	Heated oxygen sensor (rear)	• Transmission: 2nd gear <M/T> "L" range <A/T> • Accelerate the vehicle with wide open throttle. • Engine: Over 3,500 rpm		0.6 – 1.0V
1	No.1 injector	While engine is idling after having warmed up, suddenly depress the accelerator pedal.		From 11–14V, momentarily drops slightly
14	No.2 injector			
2	No.3 injector			
15	No.4 injector			
10	Ignition power transistor unit	Engine: 3,000 rpm		0.3 – 3.0V
9	Evaporative emission purge solenoid	Ignition switch: ON		B+
		Engine: warm, 3000 rpm		0 – 3V
104	Ignition timing adjustment terminal	Ignition switch: ON	Earth the ignition timing adjustment terminal	0 – 1V
			Remove the earth connection from the ignition timing adjustment terminal.	4.0 – 5.5V
106	Check engine/malfunction indicator lamp	Air conditioning switch: OFF → ON		0 – 3V ↓ 9 – 13V (After several seconds have elapsed)
6	EGR solenoid	Ignition switch: ON		B+
		Engine: idle, warm		From B+ momentarily drops
53	EGR temperature sensor	Ignition switch: ON	When the sensor temperature is 50°C (122°F).	3.6 – 4.4V
			When the sensor temperature is 180°C (212°F).	2.2 – 3.0V
105	Oxygen sensor heater	Engine: idle, warm		0 – 3V
		Engine: 5,000 rpm		B+
5	Idle air control valve position sensor No.1	Ignition switch: Immediately after turning ON		1.5 – 4V (Momentarily) ↓ 0 – 1V or 4.5 – 5.5V
18	Idle air control valve position sensor No.2	Ignition switch: Immediately after turning ON		1.5 – 4V (Momentarily) ↓ 0 – 1V or 4.5 – 5.5V

Fig. 126 The ECM terminal voltage check chart — 2.4L engines

EMISSION CONTROLS 4-63

Terminal No.	Check Item	Check Condition (Engine Condition)	Standard value
4	Idle air control motor (closed)	Ignition switch: Immediately after turning ON	2V or more (Momentarily) ↓ 0–1V
17	Idle air control motor (open)	Ignition switch: Immediately after turning ON	4V or more (Momentarily) ↓ 0–1V

Fig. 127 The ECM terminal voltage check chart — 2.4L engines

4-64 EMISSION CONTROLS

Terminal No.	Check Item	Check Condition (Engine Condition)		Standard value	Remarks
10	Volume air flow sensor	Engine: Idling		2.2–3.2 V	
		Engine: 2,000 rpm			
57	Volume air flow sensor reset signal	Engine: Idling		0–1 V	
		Engine: 3,000 rpm		6–9 V	
8	Intake air temperature sensor	Ignition switch: ON	Air intake temperature of 0°C (32°F)	3.2–3.8 V	
			Air intake temperature of 20°C (68°F)	2.3–2.9 V	
			Air intake temperature of 40°C (104°F)	1.5–2.1 V	
			Air intake temperature of 80°C (176°F)	0.4–1.0 V	
16	Barometric pressure sensor	Ignition switch: ON	Altitude of 0 m (0 ft.)	3.7–4.3 V	
			Altitude of 1,200 m (3,937 ft.)	3.2–3.8 V	
20	Engine coolant temperature sensor	Ignition switch: ON	Coolant temperature of 0°C (32°F)	3.2–3.8 V	
			Coolant temperature of 20°C (68°F)	2.3–2.9 V	
			Coolant temperature of 40°C (104°F)	1.3–1.9 V	
			Coolant temperature of 80°C (176°F)	0.3–0.9 V	
19	Throttle position sensor	Ignition switch: ON for 15 seconds or more	Idling	0.3–1.0 V	
			Wide open throttle	4.5–5.5 V	
14	Closed throttle position switch	Ignition switch: ON	Set the throttle valve to the idle position.	0–1 V	
			Slightly open the throttle valve.	4 V or higher	
22	Camshaft position sensor	Engine: Cranking		0.2–3.0 V	
		Engine: Idling			
21	Crankshaft position sensor	Engine: Cranking		0.2–3.0 V	
		Engine: Idling			
108	Ignition switch-ST	Engine: Cranking		8 V or higher	M/T
104	Park/neutral position switch	Ignition switch: ON	Set the selector lever to P or N.	0–3 V	A/T
			Set the selector lever to D, 2, L or R.	8–14 V	
18	Vehicle speed sensor	• Ignition switch: ON • Move the vehicle slowly forward.		0 ↔ 5 V (Changes repeatedly)	
5	Power steering pressure switch	Engine: Idling after having warmed up	Set the steering wheel to the straight-forward.	B+	
			Half turn the steering wheel.	0–3 V	

Fig. 128 The ECM terminal voltage check chart — 3.0L (12 valve) engines

EMISSION CONTROLS 4-65

Terminal No.	Check Item	Check Condition (Engine Condition)		Standard value	Remarks
7	Air conditioning switch	Engine: Idling	Turn the air conditioning switch to OFF.	0–3 V	
			Turn the air conditioning switch to ON (air conditioning compressor is operating).	B+	
65	Air conditioning compressor clutch relay	• Engine: Idling • Air conditioning switch: OFF → ON (Air compressor is operating)		B+ or temporarily 6 V or more → 0–3 V	
4	Heated oxygen sensor	Engine: Warmed up, 2,000 rpm (Check using a digital type voltmeter.)		0 ↔ 0.8 V (Changes repeatedly)	
51	No. 1 injector	Engine: Idling after having warmed up, rapidly depress the accelerator pedal		From 11–14 V, momentarily drops slightly	
52	No. 2 injector				
60	No. 3 injector				
61	No. 4 injector				
105	No. 5 injector				
109	No. 6 injector				
58	Stepper motor coil <A1>	• Engine: 3,000 rpm • Check immediately after hot restart		B+ ↑↓ 0–3 V (Changes repeatedly)	
59	Stepper motor coil <A2>				
67	Stepper motor coil <B1>				
68	Stepper motor coil <B2>				
54	Ignition power transistor unit	Engine: 3,000 rpm		0.3–3 V	
62	Evaporative emission purge solenoid	Ignition switch: ON		B+	
		Engine: Warmed up, 3,000 rpm		0–3 V	
12	Ignition timing adjustment terminal	Ignition switch: ON	Ground the ignition timing adjustment terminal.	0–1 V	
			Disconnect the ground from the ignition timing adjustment terminal.	4.0–5.5 V	
64	Check engine/malfunction indicator lamp	Ignition switch: OFF → ON		0–3 V ↓ 9–13 V (After several seconds have elapsed)	
9	Anti-lock braking signal	Engine: Idling		B+	
		• When vehicle first starts to move after turning the ignition switch to ON • Vehicle speed: 0 → 10 km/h (0–6 mph)		B+ ↓ 0–3 V (temporarily)	

Fig. 129 The ECM terminal voltage check chart — 3.0L (12 valve) engines

EMISSION CONTROLS

Terminal No.	Check item	Check condition (Engine condition)		Standard value	Remarks
45	Air conditioning switch	Engine: Idling	Turn the air conditioning switch to OFF.	0–3 V	
			Turn the air conditioning switch to ON (Air conditioning compressor is operating)	B+	
22	Air conditioning compressor clutch relay	• Engine: Idling • Air conditioning switch: OFF → ON (Air compressor is operating)		Changes from B+ or temporarily 6 V higher to 0–3 V as A/C clutch cycles	
24	Electrical load switch	Engine: Idling	Turn the lighting switch off.	0–3 V	DOHC
			Turn the lighting switch on.	B+	
55 <California–SOHC> 56 <Federal, California>	Heated oxygen sensor	Engine: Warmed up, 2,000 rpm (Check using a digital type voltmeter)		0 ↔ 0.8 V (changes repeatedly)	
32<California> 35<California–SOHC>	Heated oxygen sensor (rear)	• Transaxle: second<M/T> L range<A/T> • Drive with wide open throttle • Engine: 3,500 rpm or more		0.6–1.0 V	
1	No. 1 Injector	Engine: While engine is idling after having warmed up, rapidly depress the accelerator pedal		From 11–14 V, momentarily drops slightly	
14	No. 2 Injector				
2	No. 3 Injector				
15	No. 4 Injector				
3	No. 5 Injector				
16	No. 6 Injector				
4	Stepper motor coil <A1>	Engine: Warmed up Check immediately after hot restart		Changes repeatedly between B+ and 0–3 V	
17	Stepper motor coil <A2>				
5	Stepper motor coil <B1>				
18	Stepper motor coil <B2>				
10	Ignition power transistor unit A	Engine: 3,000 rpm		0.3–3 V	
23	Ignition power transistor unit B				
11	Ignition power transistor unit C				

Fig. 130 The ECM terminal voltage check chart — 3.0L (24 valve) and 3.5L engines

EMISSION CONTROLS 4-67

Terminal No.	Check item	Check condition (Engine condition)		Standard value	Remarks
9	Evaporative emission purge solenoid <No. 1>	Ignition switch: ON		B+	
		Engine: Warmed up, 3,000 rpm		0–3 V	
20	Evaporative emission purge solenoid <No. 2>	Ignition switch: ON		B+	California –SOHC
		Engine: Warmed up, 3,500 rpm		0–10 V	
31	Engine ignition signal	Engine: 3,000 rpm		0.3–3 V	
34	Ignition timing adjustment terminal	Ignition switch: ON	Ground the ignition timing adjustment terminal.	0–1 V	
			Disconnect the ground from the ignition timing adjustment terminal.	4.0–5.5 V	
36	Check engine/ malfunction indicator lamp	Ignition switch: OFF → ON		Changes from 0–3 V to 9–13 V (after several seconds have elapsed)	
6	EGR solenoid	Ignition switch: ON		B+	DOHC
		Engine: While engine is idling after having warmed up, rapidly depress the accelerator pedal		Temporarily drops slightly from B+	
53	EGR temperature sensor	Ignition switch: ON	Sensor temperature of 50°C (122°F)	3.6–4.4 V	DOHC
			Sensor temperature of 100°C (212°F)	2.2–3.0 V	
20	Variable induction control solenoid	Engine: Idling		0–3 V	DOHC
		Engine: 5,000 rpm		B+	
44	Anti-lock braking signal	Engine: Idling		B+	
		• When vehicle first starts to move after turning the ignition switch to ON • Vehicle speed: 0 → 10 km/h (0 → 0.6 mph)		Changes B+ to 0–3 V (temporarily)	

Fig. 131 The ECM terminal voltage check chart — 3.0L (24 valve) and 3.5L engines

4-68 EMISSION CONTROLS

DIAGNOSTIC TROUBLE CODES

General Information

The Engine Control Module (ECM) monitors the output signals (some signals at all times and the others under specified conditions) from various sensors. When it is noticed that a signal irregularity has continued for a specified time or longer, the ECM judges that an irregularity has occurred, memorizes the diagnostic trouble code, and outputs the signal to the diagnostic output terminals.

There are 16 (for Pick-up trucks) or 18 (for Monteros) diagnostic items. Since memorization of the diagnostic trouble codes is backed up directly by the battery, the diagnostic results are memorized even if the ignition key is turned OFF. The diagnostic trouble codes will, however, be erased when the battery terminal or the ECM connector is disconnected. The diagnostic trouble codes are also erased by setting the ignition switch to the ON position and then disconnecting the negative (-) terminal of the battery for ten seconds or longer.

The 16 or 18 diagnostic items are listed in the Diagnostic Trouble Charts and are all indicated sequentially from the smallest code number to the largest.

➡The ignition timing adjustment signal diagnostic trouble code is output when the ignition timing adjustment terminal line is short-circuited to ground. This is not a malfunction.

To read out the trouble codes, a voltmeter will be needed. The trouble codes are output as a series of voltage spikes of two different lengths. The different lengths are 1.5 seconds long for the large spike and 0.5 seconds long for the short spike. Each long spike is equal to ten and each short spike is equal to one, therefore if a trouble code reads out as two long spikes followed by four short spikes, this would be trouble code No. 24.

Reading Diagnostic Trouble Codes

- When the battery voltage is low, no detection of component failure is possible. Be sure to check the battery for voltage and other conditions before starting the test.
- The diagnostic item is erased if the battery or the ECM connector is disconnected. Do not disconnect the battery before the diagnostic results are completely read.

PICK-UP TRUCKS

▶ See Figure 132

1. Connect the analog voltmeter to the diagnostic output terminal 1 and the ground terminal 12 of the data link link connector (white). The data link connector is located under the far left-hand side dashboard console, underneath of the fuse board.
2. Turn the ignition switch to ON.
3. Check the swing of the voltmeter pointer and read the diagnostic output.
4. Repair the problem location, referring to the diagnostic chart.

MONTEROS

▶ See Figures 133 and 134

1983-93 Models

1. Connect the analog voltmeter to the diagnostic output terminal 1 and the ground terminal 12 of the data link link connector (white). The data link connector is located under the far left-hand side dashboard console, underneath of the fuse board.
2. Turn the ignition switch to ON.
3. Check the swing of the voltmeter pointer and read the diagnostic output.
4. Repair the problem location, referring to the diagnostic chart.

1994-95 Models

SOHC ENGINES

1. Use the special tool (diagnostic trouble code check harness) to connect an analog-type voltmeter to the diagnostic output terminal 25 and to the ground terminals 4 or 5 of the data link connector.
2. Turn the ignition switch ON.
3. Read the diagnostic trouble code pattern from the voltmeter and record it.
4. Referring to the diagnostic chart, repair the faulty part.

DOHC ENGINES

➡This procedure uses the check engine/malfunction indicator lamp to read the trouble codes; a voltmeter is not needed.

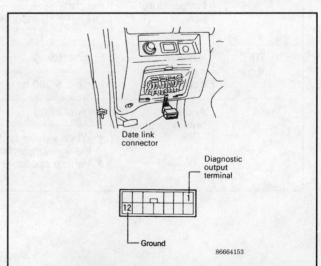

Fig. 132 The location of the data link connector and its terminals — Pick-up trucks

EMISSION CONTROLS 4-69

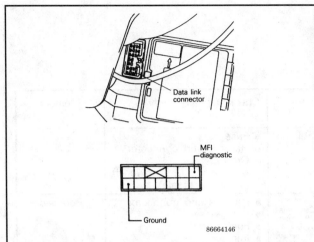

Fig. 133 The location of the data link connector and the specific terminals needed for testing — models up to 1993

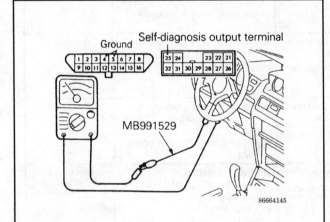

Fig. 134 Connect the data link terminal 4 or 5 with the self-diagnosis output terminal 25 — SOHC 1994 and later models

1. Use the special tool (diagnostic trouble code check harness) to connect the diagnostic test mode control terminal (terminal 1) of the data link connector to the ground.
2. Turn the ignition switch **ON**.
3. Read the diagnostic trouble code pattern from the flashing of the check engine/malfunction indicator lamp.
4. Referring to the diagnostic chart, repair the faulty part.

Clearing Diagnostic Trouble Codes

ALL MODELS

To clear the diagnostic trouble codes from the computer's memory, do the following procedure:
1. Turn the ignition **OFF**.
2. After removing the battery cable from the battery terminals for 10 seconds or more, reconnect the cable.
3. Turn the ignition **ON** and take a reading of the diagnostic output to check if a normal code is output.
4. Let the engine warm up and then run it at idle for about 10 minutes. Recheck for any diagnostic codes.

4-70 EMISSION CONTROLS

No.	Diagnostic trouble code / Output signal pattern	Diagnostic item	Check item (Remedy)	Memory
—	12A0104	Engine control module	• Fuse • Harness and connector • Ground (Replace ECM if power ⊕ ground available)	—
11	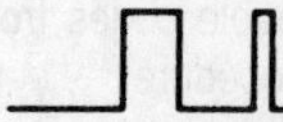 12A0104	Oxygen sensor <Federal, California-RWD (Up to 1992 models), California-4WD> Heated oxygen sensor (front) <California-RWD (From 1993 models)>	• Harness and connector • Oxygen sensor • Fuel pressure • Injectors (Replace if defective.) • Intake air leaks	Retained
12	 12A0104	Volume air flow sensor	• Harness and connector (If harness and connector are normal, replace volume air flow sensor assembly.)	Retained
13	 12A0104	Intake air temperature sensor	• Harness and connector • Intake air temperature sensor	Retained
14	 12A0104	Throttle position sensor	• Harness and connector • Throttle position sensor • Closed throttle position switch	Retained
15	 12A0104	Idle speed control motor position sensor <Federal, California-RWD (Up to 1992 models), California-4WD>	• Harness and connector • Idle speed control motor position sensor • Throttle position sensor	Retained
21	 12A0107	Engine coolant termperature sensor	• Harness and connector • Engine coolant temperature sensor	Retained
22	 12A0107	Crankshaft position sensor	• Harness and connector (If harness and connector are normal, replace distributor assembly.)	Retained
23	 12A0107	Camshaft position sensor	• Herness and connector (If harness and connector are normal, replace distributor assembly.)	Retained
24	 12A0107	Vehicle speed sensor	• Harness and connector • Vehicle speed sensor	Retained

Fig. 135 Diagnostic trouble code chart — Pick-up trucks

Diagnostic trouble code: No.	Diagnostic trouble code: Output signal pattern	Diagnostic item	Check item (Remedy)	Memory
25	H 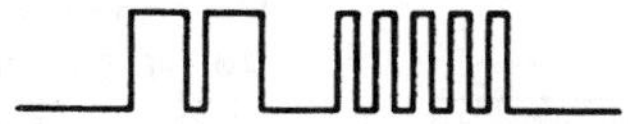 L 12A0107	Barometric pressure sensor	• Harness and connector (If harness and connector are normal, replace barometric pressure sensor assembly.)	Retained
36	H 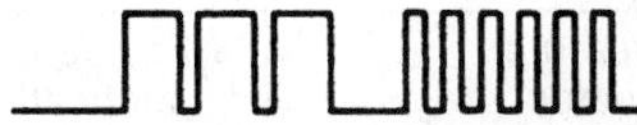 L 12A0104	Ignition timing adjustment signal	• Harness and connector	–
41	H 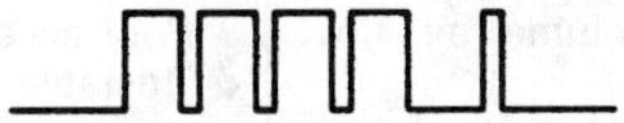 L 12A0105	Injector	• Harness and connector • Injector coil resistance	Retained
42	H 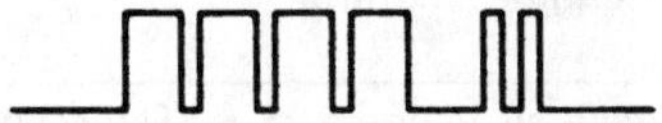 L 12A0105	Fuel pump <Federal, California-RWD (Up to 1992 models), California-4WD>	• Harness and connector • MFI relay	Retained
43	H 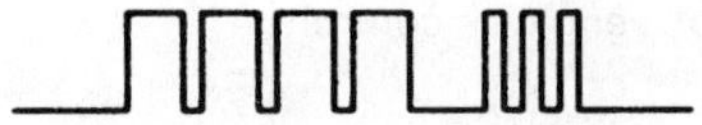 L 12A0105	EGR <Federal (From 1994 models), California>	• Harness and connector • EGR temperature sensor • EGR valve • EGR solenoid • EGR valve control vacuum	Retained
55	H 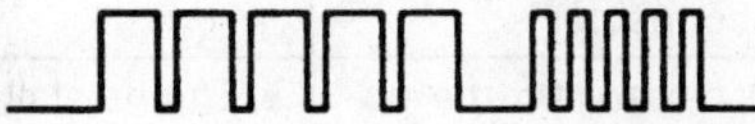 L 12A0107	Idle air control valve position sensor <California-RWD (From 1993 models)>	• Harness and connector • IAC valve movement (If harness, connector and IAC valve movement are normal, replace Idle air control motor assembly)	Retained
59	H 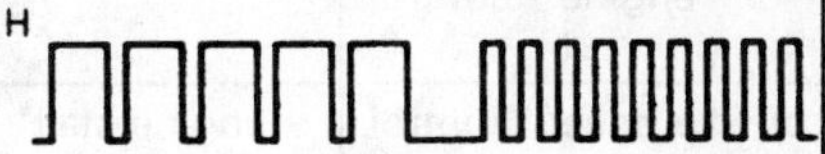 L 12A0107	Heated oxygen sensor (rear) <California-RWD (From 1993 models)>	• Harness and connector • Heated oxygen sensor	Retained
–	H (Continuous) 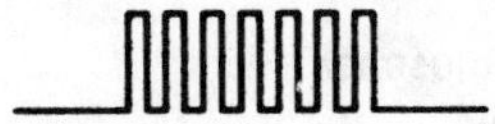 L 12A0104	Normal state	–	–

NOTE
Do not replace the ECM until a thorough terminal check reveals there are no short/open circuit.

86664155

Fig. 136 Diagnostic trouble code chart — Pick-up trucks

EMISSION CONTROLS

NOTE
*: The fail safe/back up funcftions are operating

Diagnostic trouble code No.	Diagnostic item	Trouble symptom	Main cause	Note (malfunction symptom, etc.)
–	Engine control module	Malfunction of engine control module itself	–	• Engine stall • No starting
11	Oxygen sensor < Federal, California-RWD (Up to 1992 models), California-4WD > Heated oxygen sensor (rear) < California-RWD (From 1993 models) >	Although mixture feedback control (closed loop control) is in progress, the oxygen sensor signal voltage does not change (lean/rich) for 30 seconds.	(1) Malfunction of oxygen sensor (2) Open-circuit, shortcircuit of oxygen sensor circuit or poor connection of connector	• Failure of exhaust gas purification function*
			(3) Inadequate fuel pressure (4) Malfunction of injector (5) Air absorption from gasket space (6) Malfunction of engine control module	• Failure of exhaust gas purification function* • Poor starting • Unstable idling • Poor acceleration
12	Volume air flow sensor	Although the engine is operating, the volume air flow sensor signal frequency is 10 Hz or lower for 4 seconds.	(1) Malfunction of volume air flow sensor (2) Open-circuit, shortcircuit of volume air flow sensor circuit or poor connection of connector (3) Malfunction of engine control module	• Poor acceleration* • Unstable idling*
13	Intake air temperature sensor	(1) Intake air temperature sensor signal voltage is 4.5V or more for 4 seconds. (2) Intake air temperature sensor signal voltage is 0.27V or lower for 4 seconds.	(1) Malfunction of intake air temperature sensor (2) Open-circuit, shortcircuit of intake air temperature sensor circuit of poor connection of connector (3) Malfunction of engine control module	• Slightly poor drivability* • In high temperature: (a) Poor starting* (b) Unstable idling*
14	Throttle position sensor	(1) Throttle position sensor signal voltage is 0.2V or lower for 4 seconds. (2) Although the closed throttle position switch turns on, the throttle position sensor signal voltage is 2V or more.	(1) Malfunction or maladjustment of throttle position sensor (2) Open-circuit, shortcircuit of throttle position sensor circuit or poor connection of connector	• Slightly poor acceleration < MT > • Poor drivability < A/T > • Engine stall
			(3) Malfunction at closed throttle position switch "ON" (4) Shortcircuit of closed throttle position switch signal line (5) Malfunction of engine control module	• Engine stall • No racing
15	Idle speed control motor position sensor < Federal, California-RWD (Up to 1992 models), California-4WD >	(1) The idle speed control motor position sensor signal voltage is more than 4.8V. (2) The idle speed control motor position sensor signal voltage is less than 0.2V. (3) During idling, the idle speed control motor position sensor signal voltage is too low for the opening of the throttle valve.	(1) Malfunction of idle speed control motor position sensor (2) Open-circuit, shortcircuit of idle speed control position sensor circuit or poor connection of connector (3) Malfunction ormaladjustment of throttle position sensor (4) Open-circuit, shortcircuit of throttle position sensor circuit or poor connection of connector (5) Malfunction of engine control module.	• Engine stall*

Fig. 137 Troubleshooting chart — Pick-up trucks

EMISSION CONTROLS 4-73

Diagnostic trouble code No.	Diagnostic item	Trouble symptom	Main cause	Note (malfunction symptom, etc.)
21	Engine coolant temperature sensor	(1) Engine coolant temperature sensor signal voltage is 4.6V or more for 4 seconds. (2) Engine coolant temperature sensor signal voltage is 0.11V or lower for 4 seconds. (3) Engine coolant temperature sensor signal indicated a drop of engine coolant temperature during the engine warming operation.	(1) Malfunction of engine coolant temperature sensor (2) Open circuit, shortcircuit of engine coolant temperature sensor circuit of poor connection of connector (3) Malfunction of engine control module	In cold engine condition; • No starting* • Unstable idling* • Poor acceleration*
22	Crankshaft position sensor	The crankshaft position sensor signal voltage does not change (High/Low), even though the engine is cranking more than 4 seconds	(1) Malfunction of crankshaft position sensor (2) Opening circuit, shortcircuit of crankshaft position sensor circuit or poor connection of connector (3) Malfunction of engine control module	• Engine stall • No starting
23	Camshaft position sensor	Although the engine is operating, the camshaft position sensor signal voltage does not change (High/Low) for 4 seconds.	(1) Malfunction of camshaft position sensor (2) Open circuit, shortcircuit of camshaft position sensor circuit or poor connection of connector (3) Malfunction of engine control module	• Unstable idling* • Poor acceleration*
24	Vehicle speed sensor	The vehicle speed sensor signal voltage does not change (High/Low), when the vehicle is accelerated with engine speed more than 3000 rpm.	(1) Malfunction of vehicle speed sensor (2) Open circuit, shortcircuit of vehicle speed sensor or poor connection of connector (3) Malfunction of engine control module	Engine may stall when the vehicle is stopped from reduced speed driving.
25	Barometric pressure sensor	(1) Barometric pressure sensor signal voltage is 4.5V or more for 4 seconds. (2) Barometric pressure sensor signal voltage is 0.2V or lower for 4 seconds.	(1) Malfunction of barometric pressure sensor (2) Open circuit, shortcircuit of barometric pressure sensor or poor connection of connector (3) Malfunction of engine control module	• Unstable idling* • Poor acceleration* • Poor starting*
36	Ignition timing adjustment signal	The ignition timing adjustment signal line to ground is shortcircuited.	(1) Shortcircuit of ignition timing adjustment signal line to ground (2) Malfunction of engine control module	• Poor acceleration • Engine overheat
41	Injector	During engine cranking or idling, the injector does not continue to drive more than 4 seconds.	(1) Malfunction of injector (2) Open circuit, shortcircuit of injector circuit or poor connection of connector (3) Malfunction of engine control module	• Unstable idling • Poor acceleration • Poor starting
42	Fuel pump <Feberal, California-RWD (Up to 1992 models), California-4WD>	The fuel pump is not provided with motor driving power, even though the engine is operating.	(1) Open-circuit, shortcircuit of fuel pump power circuit or poor connection of connector (2) Malfunction of MFI relay (3) Malfunction of engine control module	• No starting • Engine stall

Fig. 138 Troubleshooting chart — Pick-up trucks

4-74 EMISSION CONTROLS

Diagnostic trouble code No.	Diagnostic item	Trouble symptom	Main cause	Note (malfunction symptom, etc.)
43	EGR <Federal (From 1994 models) California>	During operation after engine warm-up: The EGR volume is scarce. (EGR temperature sensor signal voltage is too high.)	(1) EGR valve does not open (2) EGR valve control negative pressure is too low (3) Malfunction of EGR solenoid (4) Malfunction of EGR temperature sensor (5) Open-circuit, short circuit of EGR temperature sensor circuit or poor connection of connector (6) Malfunction of engine control module	• Failure of exhaust gas purification function
55	Idle air control valve position sensor <California-RWD (From 1993 models)>	IAC valve does not move to the intended position (opening angle), even though idle air control motor operates many times.	(1) Idle air control valve position sensor malfunction (2) Open circuit or short circuit in idle air control valve position sensor, or connector contact, is defective (3) Idle air control motor (DC motor) malfunction (4) Open circuit or short circuit in idle air control motor (DC motor), or connector contact is defective (5) Engine control module malfunction	• Inappropriate idle speed* • Engine stops* • Unstable idling*
59	Heated oxygen sensor (rear) <California-RWD (From 1993 models)>	The heated oxygen sensor signal voltage does not go over 0.1V even though the engine is warmed up.	(1) Heated oxygen sensor malfunction. (2) Open circuit or short circuit in heated oxygen sensor circuit, or connector contact is defective (3) Engine control module malfunction.	• Reduction in exhaust gas purification efficiency*

Fig. 139 Troubleshooting chart — Pick-up trucks

Diagnostic trouble code		On-board diagnostic item	Check item (Remedy)	Memory
No.	Diagnostic code pattern			
—	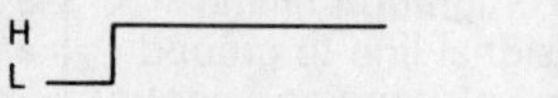 H⎯⎽ L	Engine control module	• Fuse • Harness and connector • Ground (Replace ECM if power + ground available)	—
11	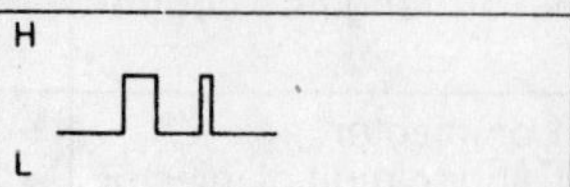 H L	Heated oxygen sensor	• Harness and connector • Heated oxygen sensor • Fuel pressure • Injectors (Replace it defective.) • Intake air leaks	Retained
12	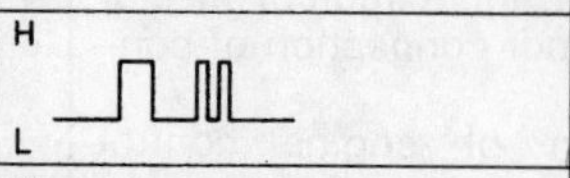 H L	Volume air flow sensor	• Harness and connector (If harness and connector are normal, replace volume air flow sensor assembly.)	Retained
13	 H L	Intake air temperature sensor	• Harness and connector • Intake air temperature sensor	Retained
14	 H L	Throttle position sensor	• Harness and connector • Throttle position sensor • Closed throttle position switch	Retained
21	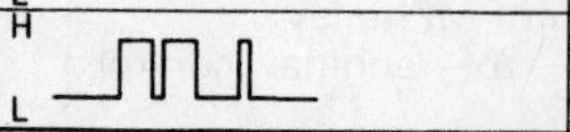 H L	Engine coolant temperature sensor	• Harness and connector • Engine coolant temperature sensor	Retained

Fig. 140 Diagnostic trouble code chart — Monteros

EMISSION CONTROLS 4-75

No.	Diagnostic trouble code Diagnostic code pattern	On-board diagnostic item	Check item (Remedy)	Memory
22	H 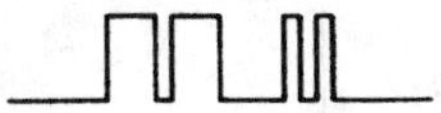 / L	Crankshaft position sensor	• Harness and connector (If harness and a connector are normal, replace distributor assembly.)	Retained
23	H  / L	Camshaft position sensor	• Harness and connector (If harness and a connector are normal, replace distributor assembly.)	Retained
24	H  / L	Vehicle speed sensor (reed switch)	• Harness and connector • Vehicle speed sensor (reed switch)	Retained
25	H  / L	Barometric pressure sensor	• Harness and connector (If harness and a connector are normal, replace barometric sensor assembly.)	Retained
31	H 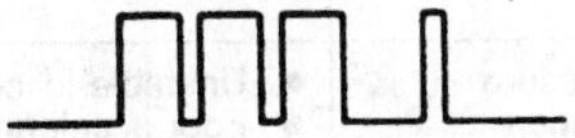 / L	Knock sensor <DOHC>	• Harness and connector (If harness and a connector are normal, replace knock sensor.)	Retained
36	H  / L	Ignition timing adjustment signal	• Harness and connector	—
41	H  / L	Injector	• Harness and connector • Injector oil resistance	Retained
43	H 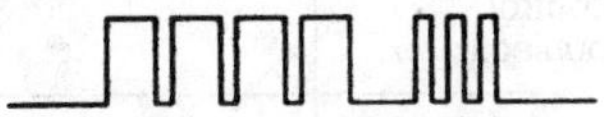 / L	EGR <DOHC>	• Harness and connector • EGR temperature sensor • EGR valve • EGR solenoid • EGR valve control vacuum	Retained
44	H 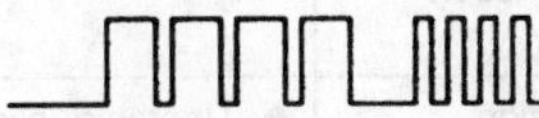 / L	Ignition coil, Ignition power transistor unit (No. 1 - 4 cylinder) <DOHC>	• Harness and connector • Ignition coil • Ignition power transistor unit	Retained
52	H  / L	Ignition coil, Ignition power transistor unit (No. 2 - 5 cylinder) <DOHC>	• Harness and connector • Ignition coil • Ignition power transistor unit	Retained
53	H 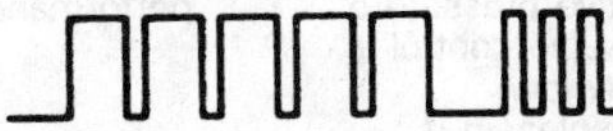 / L	Ignition coil, Ignition power transistor unit (No. 3 - 6 cylinder) <DOHC>	• Harness and connector • Ignition coil • Ignition power transistor unit	Retained
—	H  / L	Normal state	—	—

NOTE
Do not replace the ECM until a thorough terminal check reveals there are no short/open circuits.

Fig. 141 Diagnostic trouble code chart — Monteros

4-76 EMISSION CONTROLS

Diagnostic trouble code No.	Diagnostic item	Diagnostic contents	Probable cause	Remark (Trouble symptom, etc.)
22	Crankshaft position sensor	(1) Voltage of crankshaft position sensor signal does not change (high/low) even though engine has been cranking for 4 seconds or more. (2) Crankshaft position sensor signal pattern is abnormal.	(1) Crankshaft position sensor malfunction (2) Open-circuit or short in crankshaft position sensor circuit, or connector contact is defective (3) Engine control module malfunction	• Engine stalls • Starting is impossible
23	Camshaft position sensor	(1) Camshaft position sensor signal voltage does not change (high/low) for 4 seconds even though the engine is running. (2) Camshaft position sensor signal pattern is abnormal.	(1) Camshaft position sensor malfunction (2) Open-circuit or short in camshaft position sensor circuit, or connector contact is defective (3) Engine control module malfunction	• Unstable idling* • Poor acceleration*
24	Vehicle speed sensor (Reed switch)	Voltage of vehicle speed sensor does not change (high/low), even though vehicle is accelerating at an engine speed of 3000 rpm or more.	(1) Vehicle speed sensor malfunction (2) Open-circuit or short in vehicle speed sensor circuit, or connector contact is defective (3) Engine control module malfunction	Engine stalls when vehicle stops after decelerating
25	Barometric pressure sensor	(1) Barometric pressure sensor signal voltage is 4.5 V or more for 4 seconds. (2) Barometric pressure sensor signal voltage is 0.2 V or less for 4 seconds.	(1) Barometric pressure sensor signal malfunction (2) Open-circuit or short in barometric pressure sensor circuit, or connector contact is defective (3) Engine control module malfunction	• Unstable idling* • Poor acceleration* • Poor starting
31	Knock sensor <DOHC>	Signal voltage is abnormal.	(1) The knock sensor is troubled. (2) Open-circuit, short circuit or improper connector contact occurs in the knock sensor circuit. (3) The engine control module is troubled.	• Improper acceleration*
36	Ignition timing adjustment signal	Ignition timing adjustment signal wire is short-circuited to the ground.	(1) Ignition timing adjustment signal wire is short-circuited to the ground (2) Engine control module malfunction	• Poor acceleration • Engine overheats
41	Injectors	Injectors do not operate for a continuous 4 second period while engine is cranking or idling.	(1) Injector malfunction (2) Open-circuit or short in injector circuit, or connector contact is defective (3) Engine control module malfunction	• Unstable idling* • Poor acceleration* • Poor starting
43	EGR <DOHC>	During engine running after warming-up. (1) The EGR amount is small. (The signal voltage of the EGR temperature sensor is excessively high.) (2) The signal voltage of the EGR temperature sensor is 0.1V or less.	(1) The EGR valve is not opened. (2) The negative pressure of the EGR valve control is excessively low. (3) The EGR solenoid is troubled. (4) The EGR temperature sensor is troubled. (5) Open-circuit, short-circuit or improper connector contact occurs in the EGR temperature sensor circuit. (6) The engine control module is troubled.	• Deterioration of exhaust gas purifying performance

NOTE
*: Fail-safe/backup function is operating.

Fig. 142 Diagnostic code troubleshooting chart — Monteros

EMISSION CONTROLS 4-77

Diagnostic trouble code No.	Diagnostic item	Diagnostic contents	Probable cause	Remark (Trouble symptom, etc.)
—	Engine control module	Abnormality in engine control module	—	• Engine stalls • Starting is impossible
11	Heated oxygen sensor	Oxygen sensor signal voltage does not change (lean/rich) for 30 seconds even though air/fuel ratio feedback control (closed-loop control) is being carried out.	(1) Heated oxygen sensor malfunction (2) Broken circuit wire or short in heated oxygen sensor circuit, or connector contact is defective	• Reduction in exhaust gas purification efficiency*
			(3) Inappropriate fuel pressure (4) Injector malfunction (5) Air leaking in through clearance in gasket (6) Engine control module malfunction	• Reduction in exhaust gas purification efficiency* • Poor starting • Unstable idling • Poor acceleration
12	Volume air flow sensor	Frequency of volume air flow sensor signal is 10 Hz or below for 4 seconds even though engine is running.	(1) Volume air flow sensor malfunction (2) Open-circuit or short in volume air flow sensor circuit, or connector contact is defective (3) Engine control module malfunction	• Poor acceleration* • Inappropriate idle speed* • Unstable idling*
13	Intake air temperature sensor	(1) Intake air temperature sensor signal voltage is 4.5 V or more for 4 seconds. (2) Intake air temperature sensor signal voltage is 0.27 V or less for 4 seconds.	(1) Intake air temperature sensor malfunction (2) Open-circuit or short in the intake air temperature sensor circuit, or connector contact is defective (3) Engine control module malfunction	• Slightly poor driveability* • At high temperatures: (a) Poor starting* (b) Unstable idling*
14	Throttle position sensor	(1) Throttle position sensor signal voltage is 0.2 V or less for 4 seconds. (2) Throttle position sensor signal voltage is 2 V or more even though the closed throttle position switch is ON.	(1) Throttle position sensor malfunction, or adjustment is defective (2) Open-circuit or short in throttle position sensor circuit, or connector contact is defective	• Slightly poor acceleration <M/T> • Poor driveability <A/T> • Engine stalls
			(3) Closed throttle position switch malfunctio when ON (4) Short circuit in closed throttle position witch signal wire (5) Engine control module malfunction	• Engine stalls • Racing is impossible
21	Engine coolant temperature sensor	(1) Engine coolant temperature sensor signal voltage is 4.6 V or more for 4 seconds. (2) Engine coolant temperature sensor signal voltage is 0.11 V or less for 4 seconds. (3) Engine coolant temperature sensor signal indicates that the engine coolant temperature drops while the engine is warming up.	(1) Engine coolant temperature sensor (2) Open-circuit or short in engine coolant temperature sensor circuit, or connector contact is defective (3) Engine control module malfunction	When engine is cold: • Poor starting* • Unstable idling* • Poor acceleration*

NOTE
*: Fail-safe/backup function is operating.

Fig. 143 Diagnostic code troubleshooting chart — Monteros

EMISSION CONTROLS

Diagnostic trouble code No.	Diagnostic item	Diagnostic contents	Probable cause	Remark (Trouble symptom, etc.)
44	Ignition coil/ignition power transistor unit for cylinders 1 and 4 <DOHC>	During engine running, the ignition signal is not input. However, cases in which ignition signals are input to all cylinders are excluded.	(1) The ignition coil is troubled. (2) Open-circuit, short-circuit or improper connector contact occurs in the ignition primary circuit. (3) The ignition power transistor unit is troubled. (4) The engine control module is troubled.	• Unstable idling* • Improper acceleration* • Improper start*
52	Ignition coil/ignition power transistor unit for cylinders 2 and 5 <DOHC>	During engine running, the ignition signal is not input. However, cases in which ignition signals are input to all cylinders are excluded.	(1) The ignition coil is troubled. (2) Open-circuit, short-circuit or improper connector contact occurs in the ignition primary circuit. (3) The ignition power transistor unit is troubled. (4) The engine control module is troubled.	• Unstable idling* • Improper acceleration* • Improper start*
53	Ignition coil/ignition power transistor unit for cylinders 3 and 6 <DOHC>	During engine running, the ignition signal is not input. However, cases in which ignition signals are input to all cylinders are excluded.	(1) The ignition coil is troubled. (2) Open-circuit, short-circuit or improper connector contact occurs in the ignition primary circuit. (3) The ignition power transistor unit is troubled. (4) The engine control module is troubled.	• Unstable idling* • Improper acceleration* • Improper start*

NOTE
*: Fail-safe/backup function is operating.

Fig. 144 Diagnostic code troubleshooting chart — Monteros

EMISSION CONTROLS 4-79

When the main sensor malfunctions are detected by the on-board diagnostic, the vehicle is controlled by means of the pre-set control logic to maintain safe conditions for driving.

Malfunctioning item	Control contents during malfunction
Volume air flow sensor	(1) Uses the throttle position sensor (TPS) signal and engine speed signal (crankshaft position sensor signal) to take readings of the basic injector drive timing and basic ignition timing from the pre-set mapping. (2) Fixes the idle air control motor in the appointed position so idle air control is not performed.
Intake air temperature sensor	Controls the intake air temperature at 25°C (77°F).
Throttle position sensor (TPS)	No increase in fuel injection amount during acceleration due to the throttle position sensor signal.
Engine coolant temperature sensor	Controls the engine coolant temperature to 80°C (176°F). (This control will be continued until the ignition switch turns to OFF even though the sensor signal returns to normal.)
Camshaft position sensor	Injects fuel simultaneously into all cylinders. (However, When the No. 1 cylinder top dead center is not detected at all after the ignition switch is turned to ON.)
Barometric pressure sensor	Controls the barometric pressure to 760 mmHg (30 in.Hg).
Knock sensor <DOHC>	Switches from ignition timing for premium gasoline to ignition timing for regular gasoline.
Ignition power transistor unit, ignition coil <DOHC>	Fuel is cut in the cylinder where the ignition signal is abnormal.
Heated oxygen sensor	Air/fuel ratio closed loop control is not performed.

Fig. 145 Fail-safe/backup function quick reference table

4-80 EMISSION CONTROLS

COMPONENT LOCATION AND VACUUM DIAGRAMS

For any model not shown in these diagrams or in case of conflict between the actual engine arrangement and a diagram, refer to the underhood label on the vehicle. Changes in equipment and vacuum routing may be made throughout the production of the vehicle; the underhood label will show the correct data for the car.

In the following diagrams, the symbol **C** within a circle indicates that the wire so noted connects to the computer (ECM).

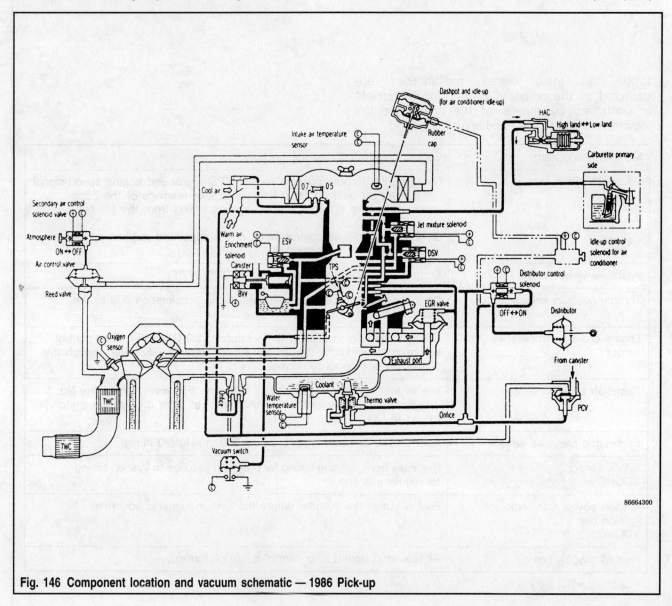

Fig. 146 Component location and vacuum schematic — 1986 Pick-up

EMISSION CONTROLS 4-81

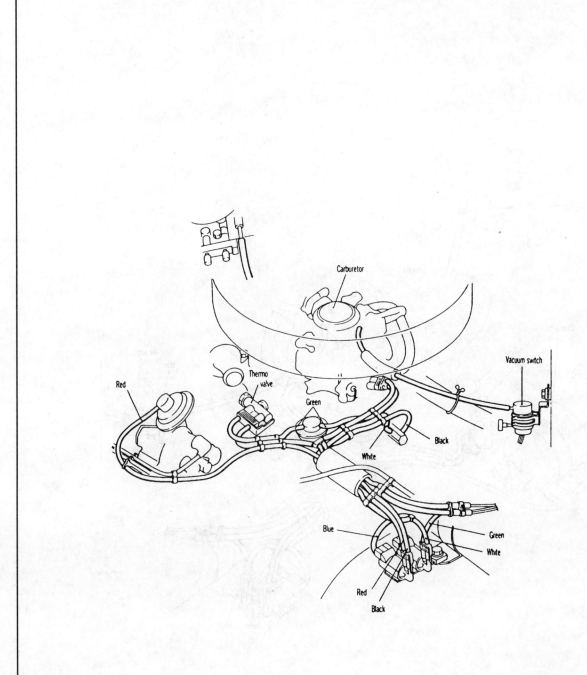

Fig. 147 Emission control vacuum hose routing — 1986 Pick-up with 2.6L engine

4-82 EMISSION CONTROLS

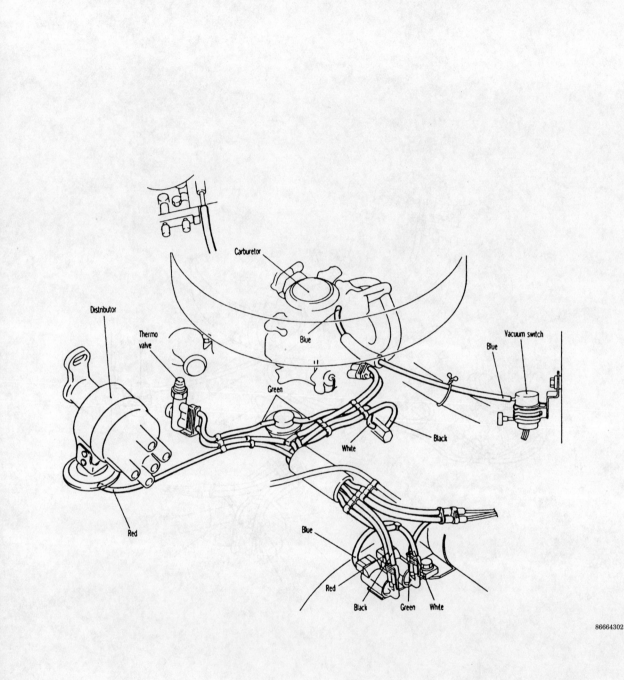

Fig. 148 Emission control vacuum hose routing — 1986 Pick-up with 2.0L engine

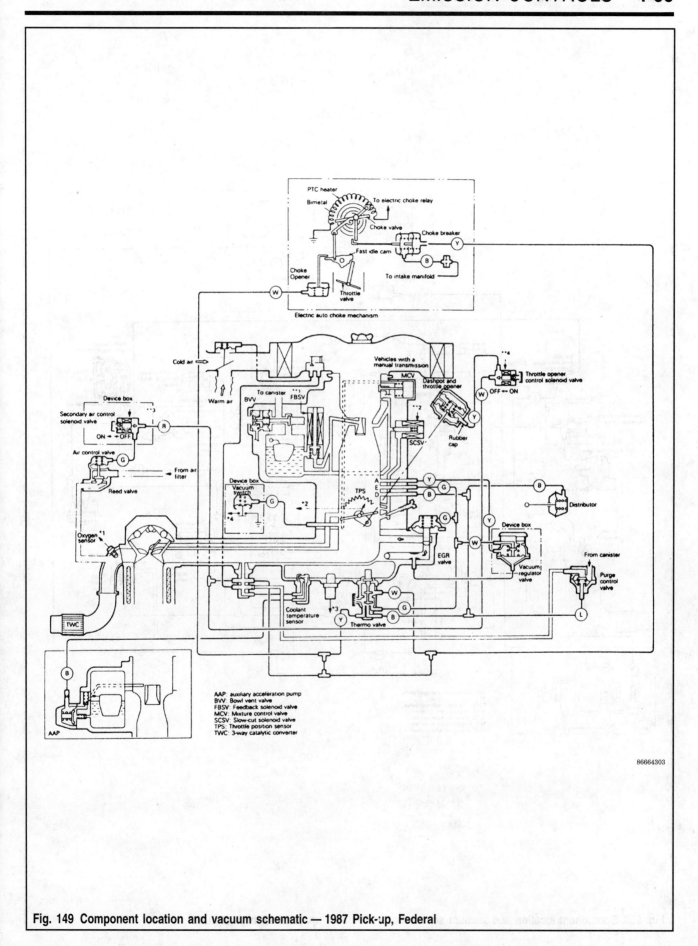

Fig. 149 Component location and vacuum schematic — 1987 Pick-up, Federal

4-84 EMISSION CONTROLS

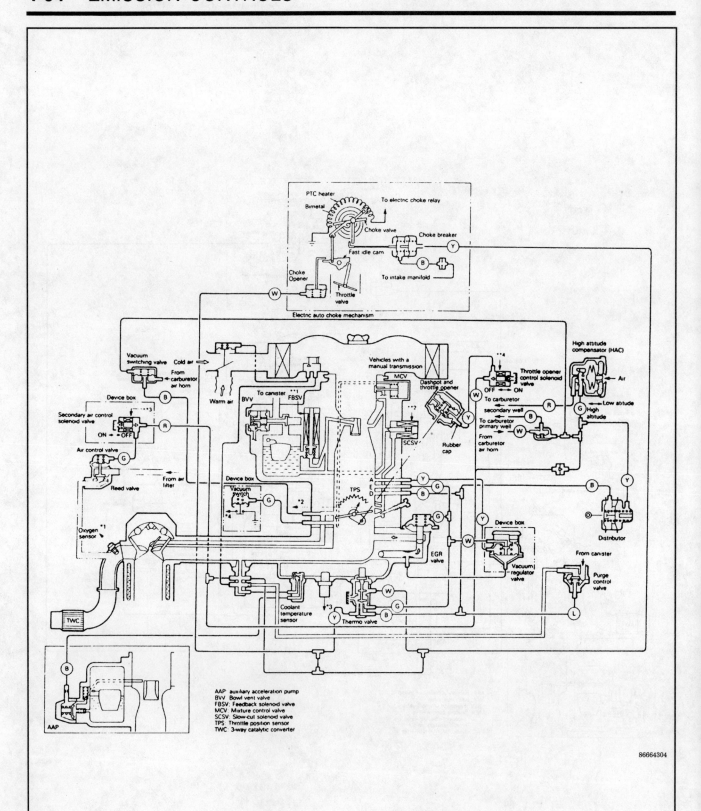

Fig. 150 Component location and vacuum schematic — 1987 Pick-up, Federal, high altitude

EMISSION CONTROLS 4-85

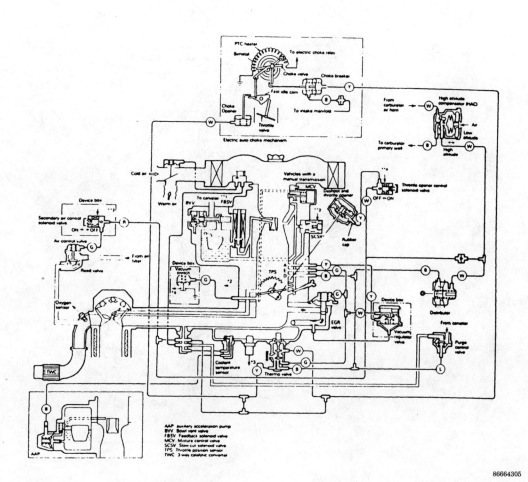

Fig. 151 Component location and vacuum schematic — 1987 Pick-up, California

4-86 EMISSION CONTROLS

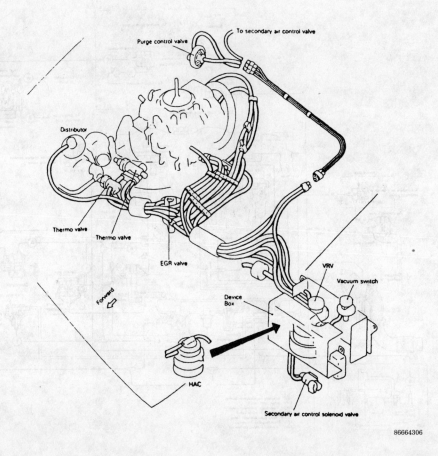

Fig. 152 Emission control vacuum hose routing — 1987 Pick-up

EMISSION CONTROLS 4-87

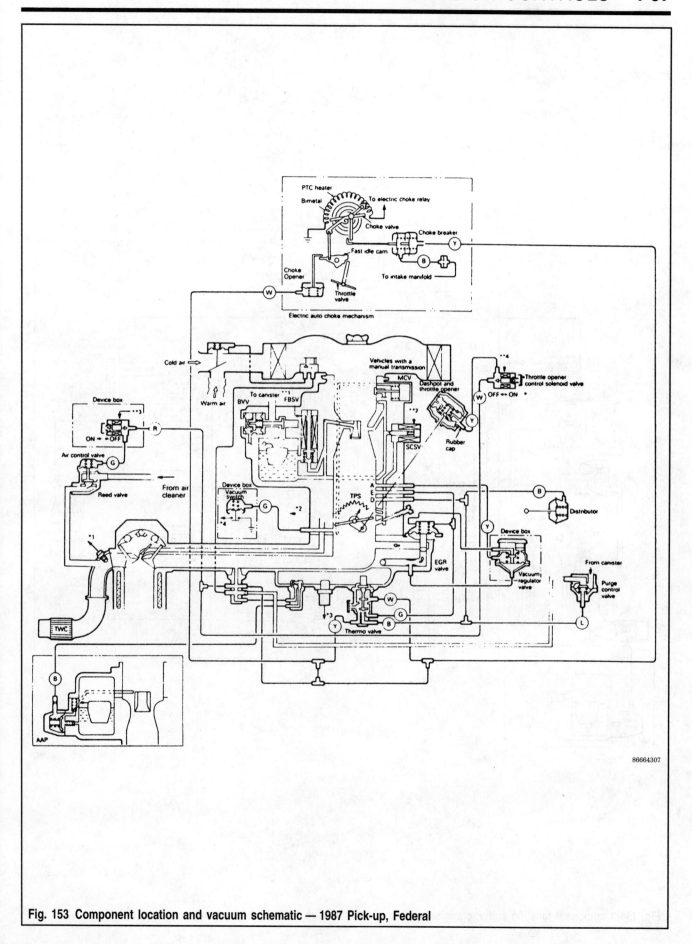

Fig. 153 Component location and vacuum schematic — 1987 Pick-up, Federal

4-88 EMISSION CONTROLS

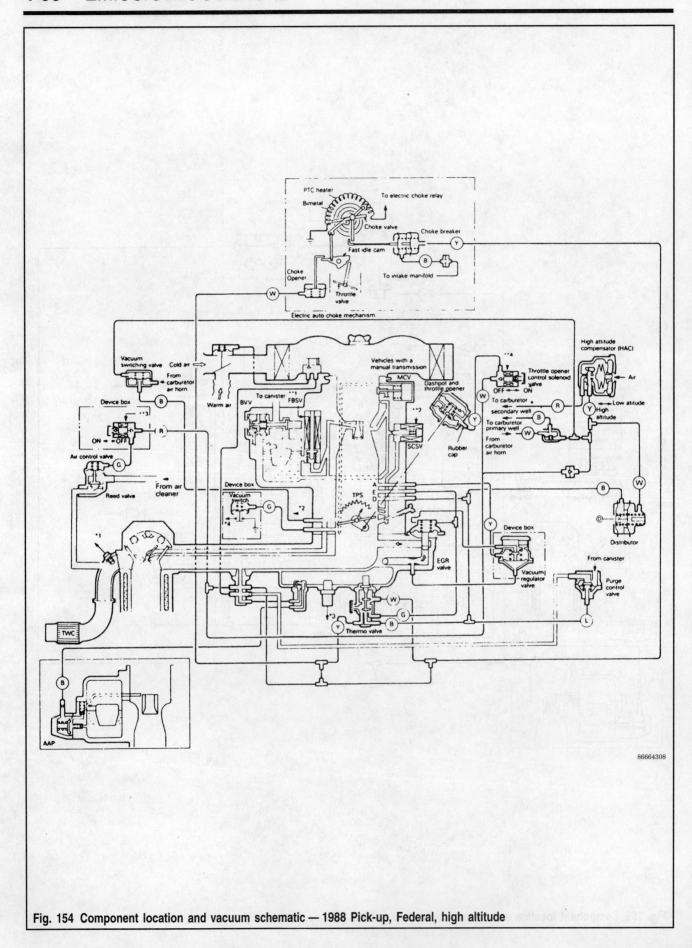

Fig. 154 Component location and vacuum schematic — 1988 Pick-up, Federal, high altitude

EMISSION CONTROLS 4-89

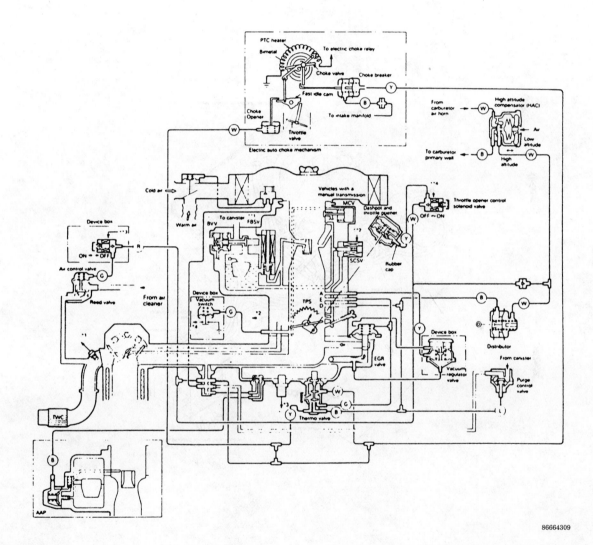

Fig. 155 Component location and vacuum schematic — 1988 Pick-up, California

4-90 EMISSION CONTROLS

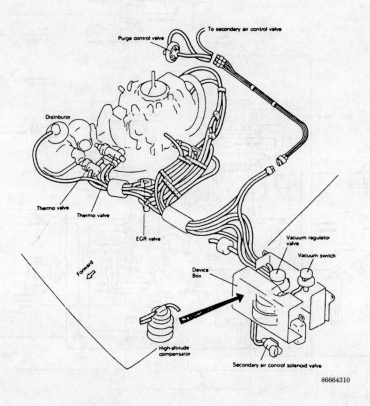

Fig. 156 Emission control vacuum hose routing — 1988 Pick-up

EMISSION CONTROLS 4-91

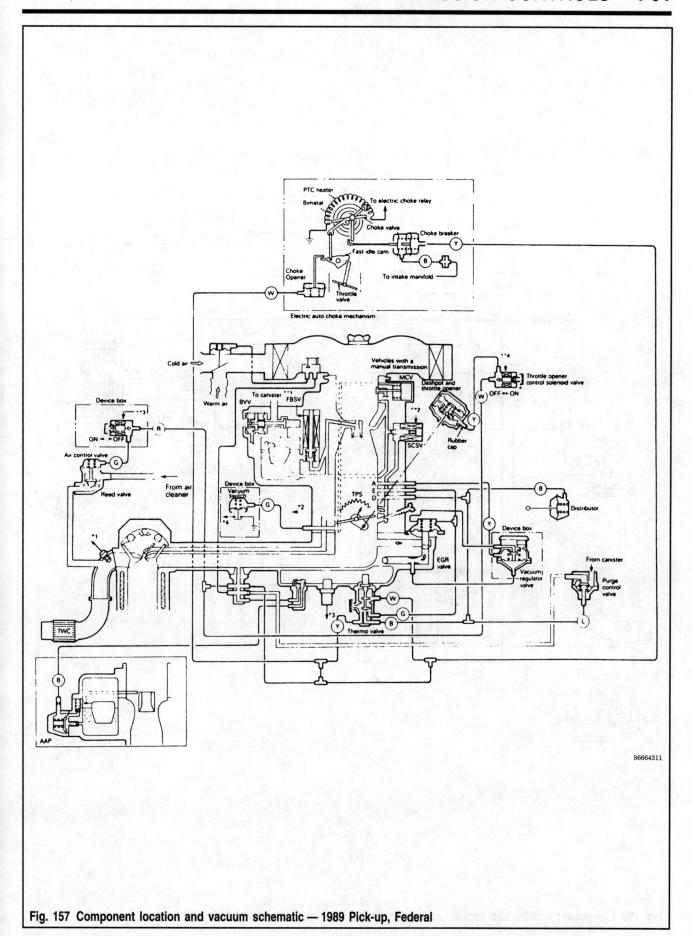

Fig. 157 Component location and vacuum schematic — 1989 Pick-up, Federal

4-92 EMISSION CONTROLS

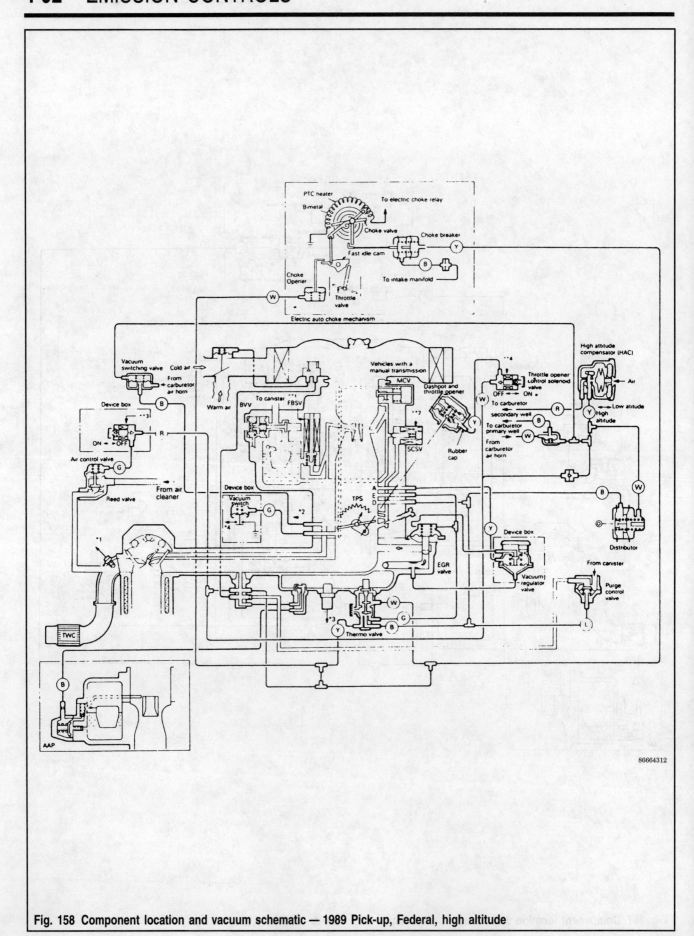

Fig. 158 Component location and vacuum schematic — 1989 Pick-up, Federal, high altitude

EMISSION CONTROLS 4-93

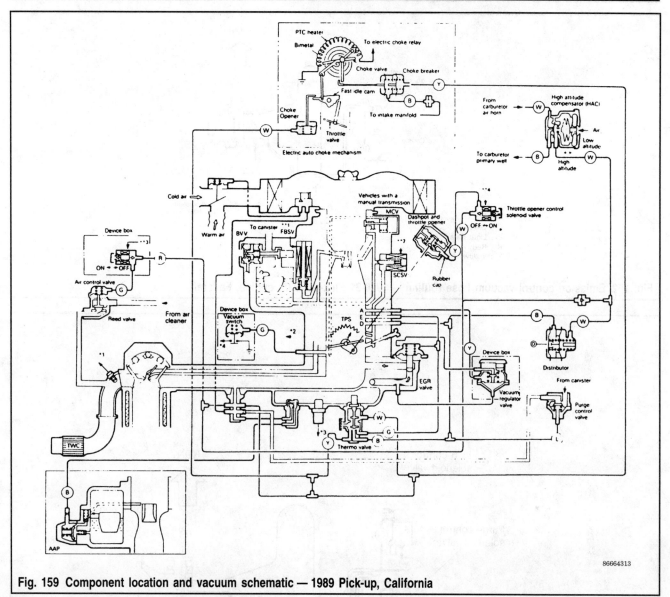

Fig. 159 Component location and vacuum schematic — 1989 Pick-up, California

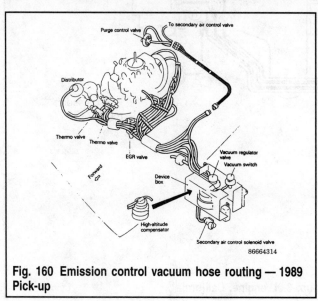

Fig. 160 Emission control vacuum hose routing — 1989 Pick-up

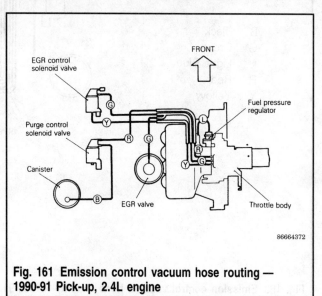

Fig. 161 Emission control vacuum hose routing — 1990-91 Pick-up, 2.4L engine

4-94 EMISSION CONTROLS

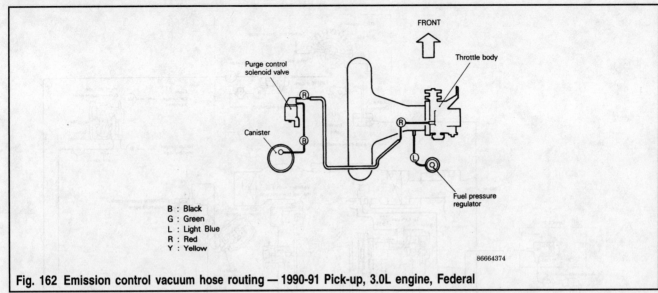

Fig. 162 Emission control vacuum hose routing — 1990-91 Pick-up, 3.0L engine, Federal

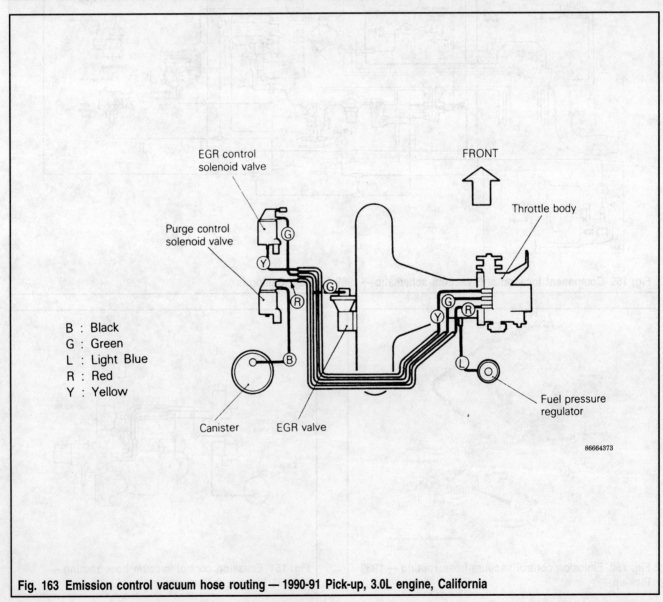

Fig. 163 Emission control vacuum hose routing — 1990-91 Pick-up, 3.0L engine, California

EMISSION CONTROLS 4-95

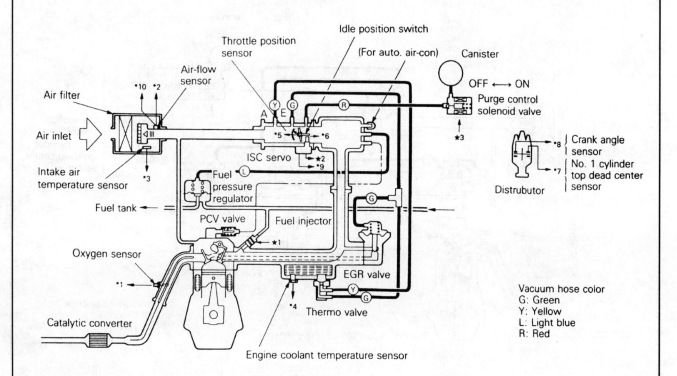

Fig. 164 Multi-port fuel injection component location and vacuum schematic — 1990-91 Pick-up, 2.4L engine, Federal

4-96 EMISSION CONTROLS

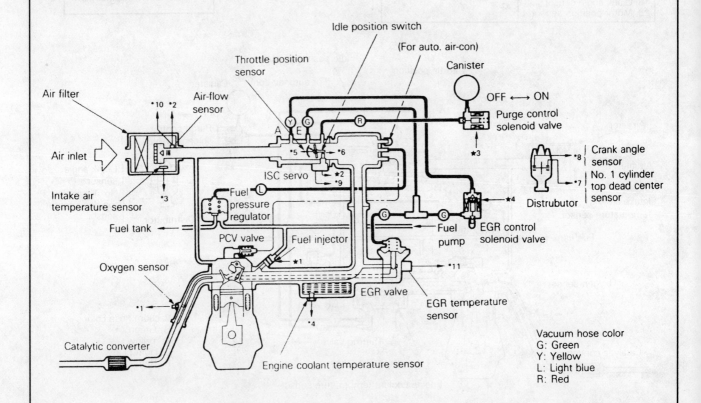

Fig. 165 Multi-port fuel injection component location and vacuum schematic — 1990-91 Pick-up, 2.4L engine, California

EMISSION CONTROLS 4-97

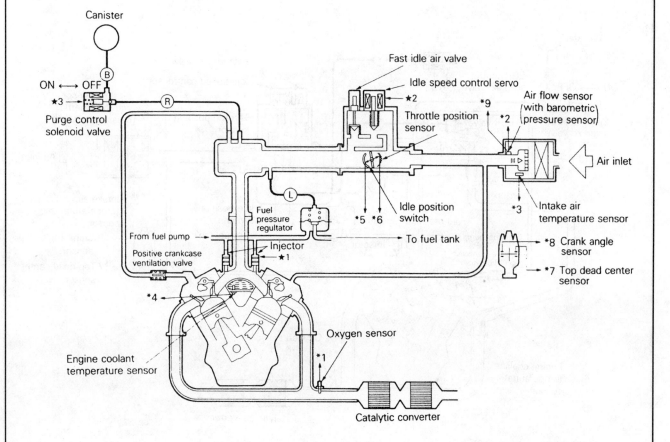

Fig. 166 Multi-port fuel injection component location and vacuum schematic — 1990-91 Pick-up, 3.0L engine, Federal

4-98 EMISSION CONTROLS

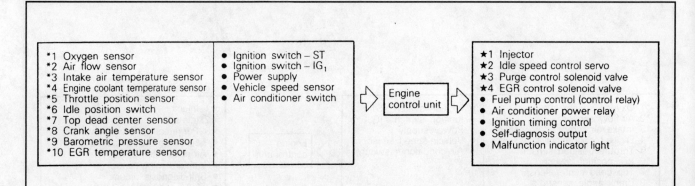
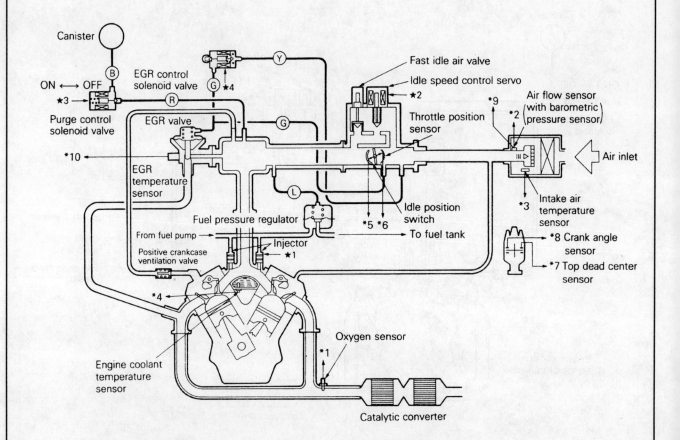

Fig. 167 Multi-port fuel injection component location and vacuum schematic — 1990-91 Pick-up, 2.4L engine, California

EMISSION CONTROLS 4-99

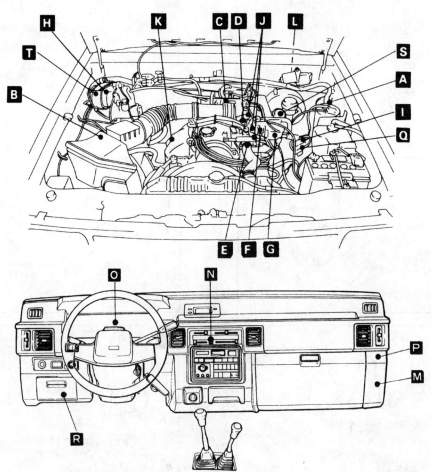

Name	Symbol	Name	Symbol
Air conditioning compressor clutch relay	A	Ignition coil (ignition power transistor)	F
Air conditioning switch	N	Ignition timing adjustment terminal	T
Volume air-flow sensor (incorporating intake air temperature sensor and barometric pressure sensor)	B	Park/neutral position switch <A/T>	L
		Injector	J
Crankshaft position sensor and camshaft position sensor	G	Idle speed control motor (closed throttle position switch, idle speed control motor position sensor)	C
EGR solenoid	Q	Oxygen sensor	K
EGR temperature sensor <Federal (From 1994 models), California>	S	Evaporative emission purge solenoid	I
Engine control module	P	Data link connector	R
MFI relay	M	Throttle position sensor	D
Engine coolant temperature sensor	E	Vehicle-speed sensor (reed switch)	O
Fuel pump check terminal	H		

Fig. 168 Electronic engine control component location — 1992-95 Pick-up, 2.4L engine

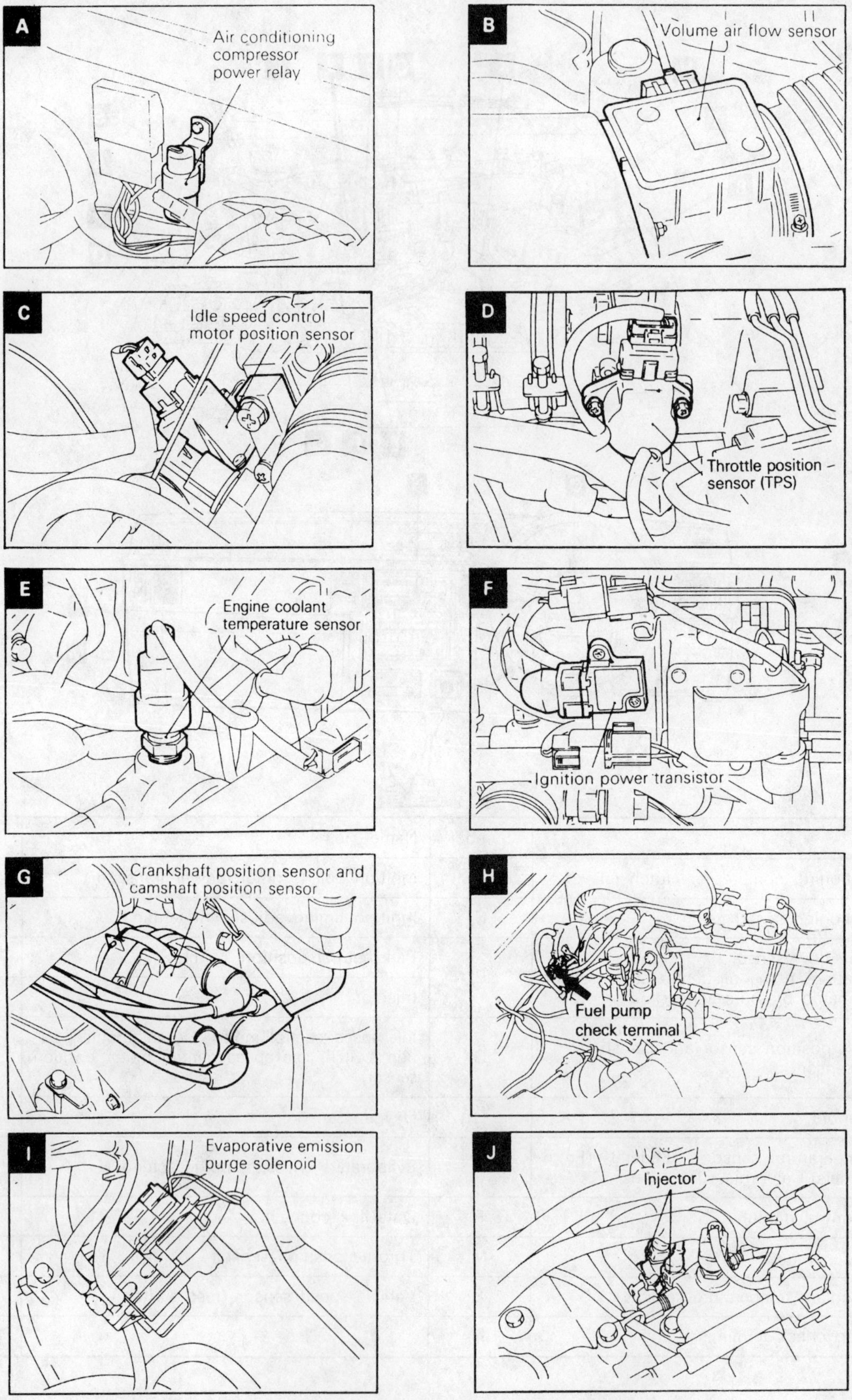

4-100 EMISSION CONTROLS

A — Air conditioning compressor power relay

B — Volume air flow sensor

C — Idle speed control motor position sensor

D — Throttle position sensor (TPS)

E — Engine coolant temperature sensor

F — Ignition power transistor

G — Crankshaft position sensor and camshaft position sensor

H — Fuel pump check terminal

I — Evaporative emission purge solenoid

J — Injector

Fig. 169 Electronic engine control component location — 1992-95 Pick-up, 2.4L engine

EMISSION CONTROLS 4-101

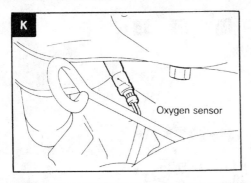
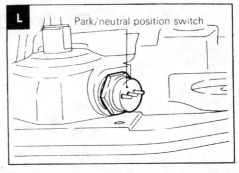
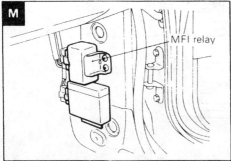
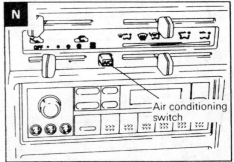
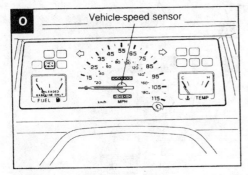
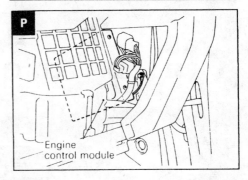
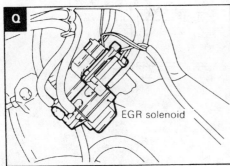

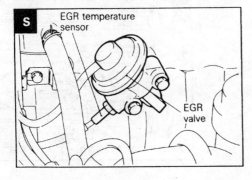
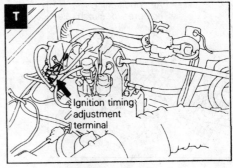

Fig. 170 Electronic engine control component location — 1992-95 Pick-up, 2.4L engine

4-102 EMISSION CONTROLS

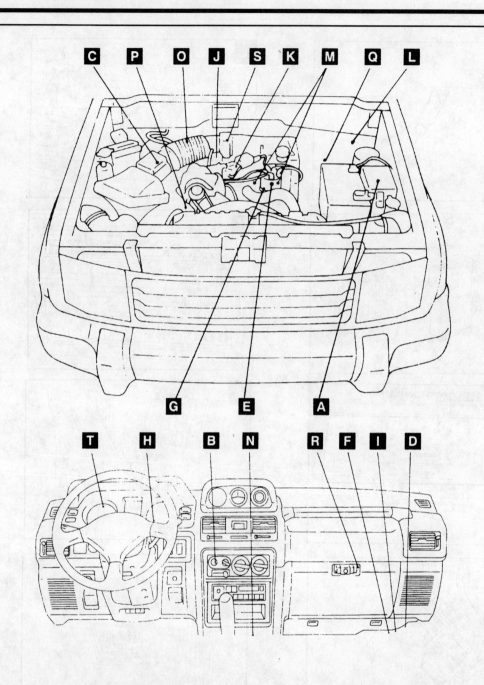

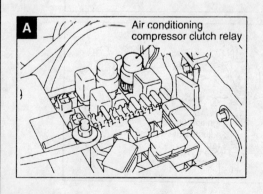

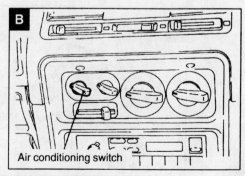

Fig. 171 Electronic engine control component location — 1992-95 Pick-up, 3.0L (12 valve) engine

EMISSION CONTROLS 4-103

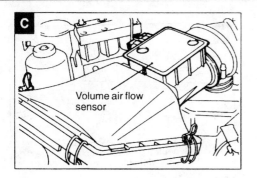
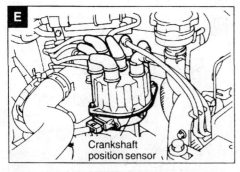
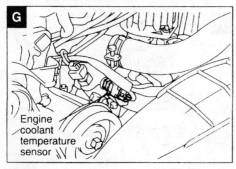
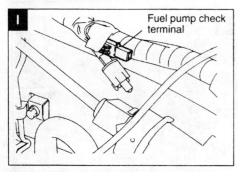
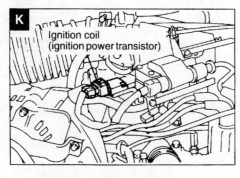
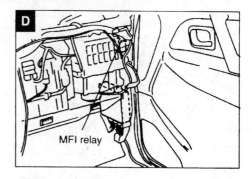
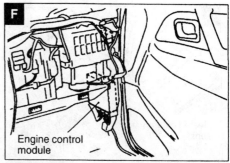
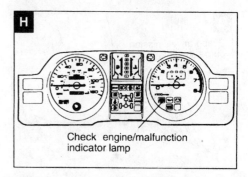
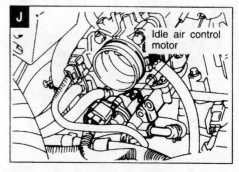
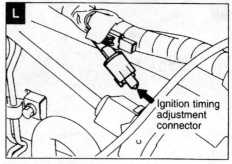

Fig. 172 Electronic engine control component location — 1992-95 Pick-up, 3.0L (12 valve) engine

4-104 EMISSION CONTROLS

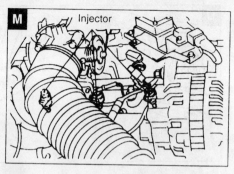

M — Injector

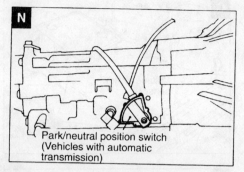

N — Park/neutral position switch (Vehicles with automatic transmission)

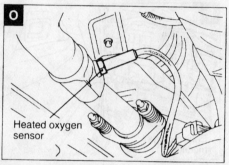

O — Heated oxygen sensor

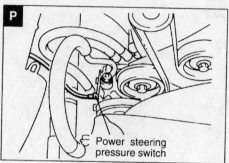

P — Power steering pressure switch

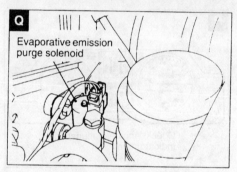

Q — Evaporative emission purge solenoid

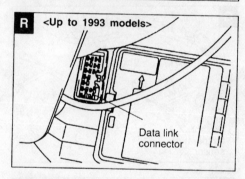

R <Up to 1993 models> — Data link connector

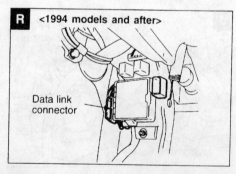

R <1994 models and after> — Data link connector

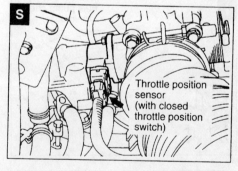

S — Throttle position sensor (with closed throttle position switch)

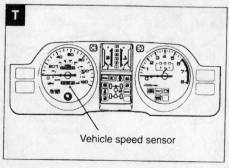

T — Vehicle speed sensor

Fig. 173 Electronic engine control component location — 1992-95 Pick-up, 3.0L (12 valve) engine

EMISSION CONTROLS 4-105

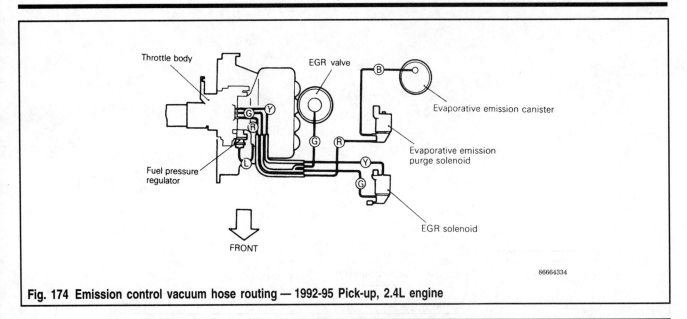

Fig. 174 Emission control vacuum hose routing — 1992-95 Pick-up, 2.4L engine

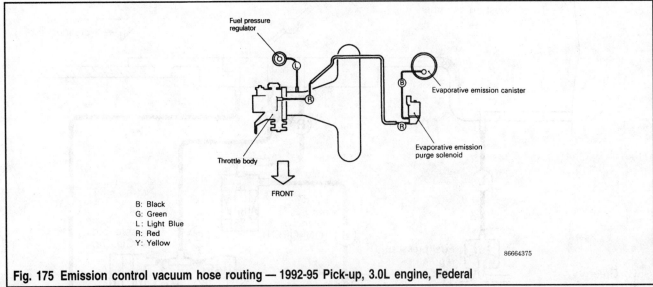

Fig. 175 Emission control vacuum hose routing — 1992-95 Pick-up, 3.0L engine, Federal

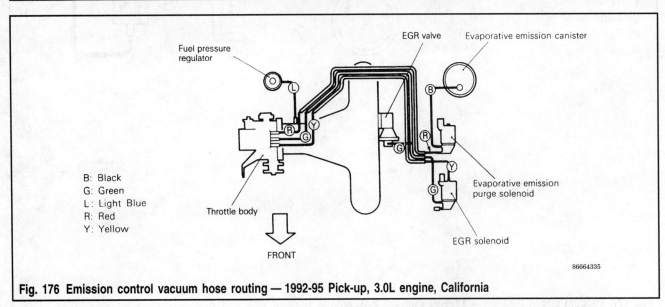

Fig. 176 Emission control vacuum hose routing — 1992-95 Pick-up, 3.0L engine, California

4-106 EMISSION CONTROLS

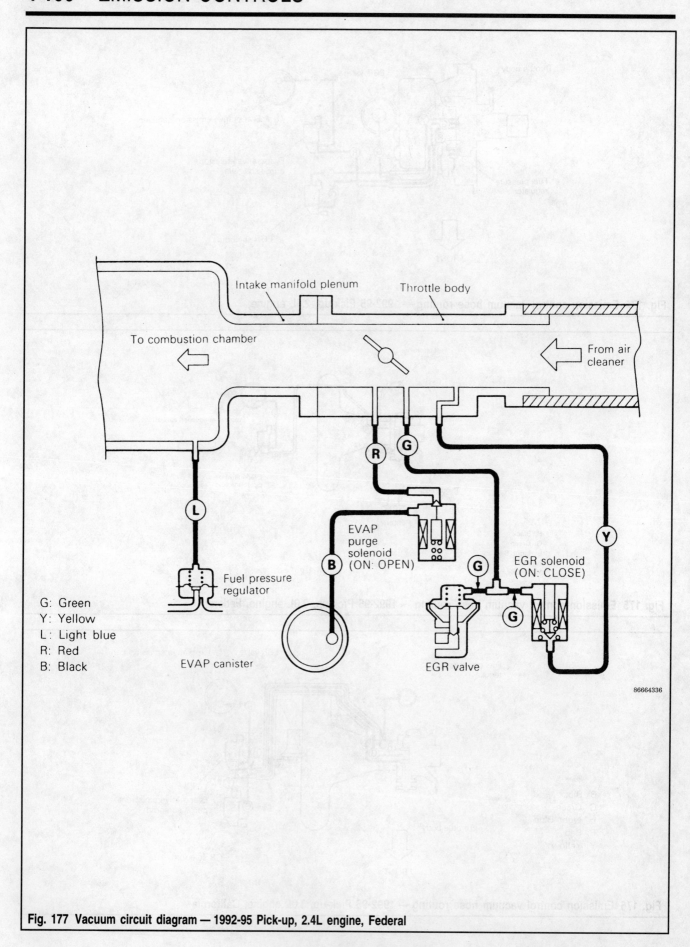

Fig. 177 Vacuum circuit diagram — 1992-95 Pick-up, 2.4L engine, Federal

EMISSION CONTROLS 4-107

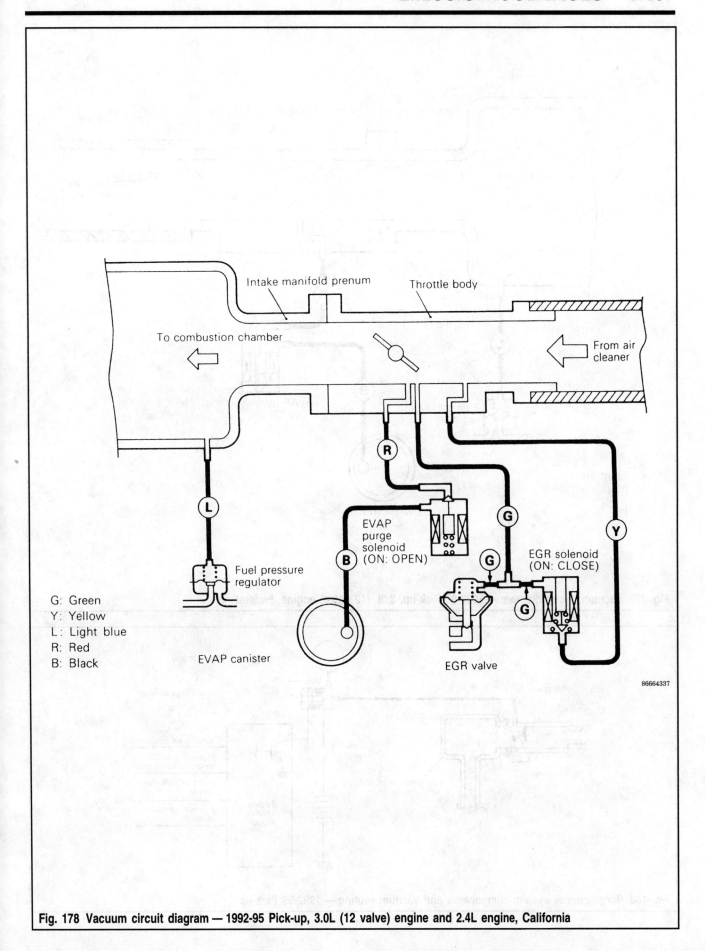

Fig. 178 Vacuum circuit diagram — 1992-95 Pick-up, 3.0L (12 valve) engine and 2.4L engine, California

4-108 EMISSION CONTROLS

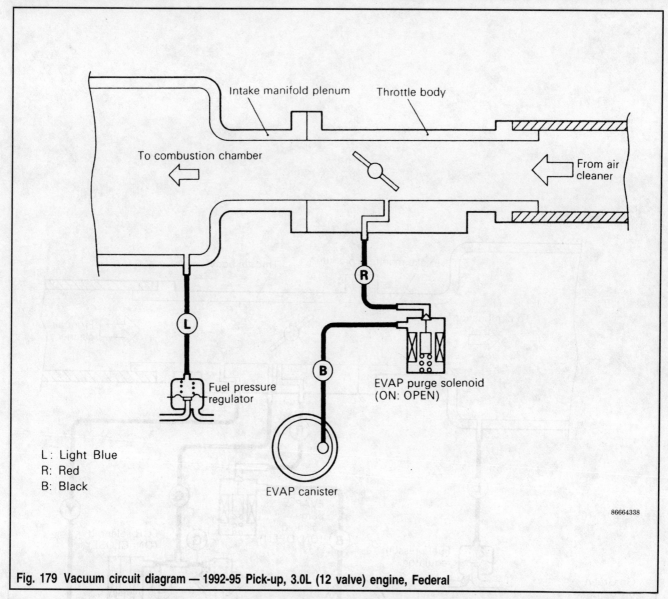

Fig. 179 Vacuum circuit diagram — 1992-95 Pick-up, 3.0L (12 valve) engine, Federal

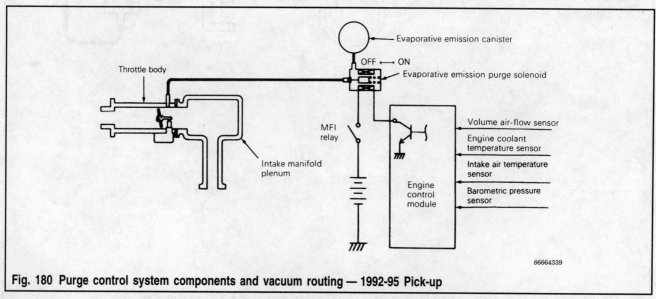

Fig. 180 Purge control system components and vacuum routing — 1992-95 Pick-up

EMISSION CONTROLS 4-109

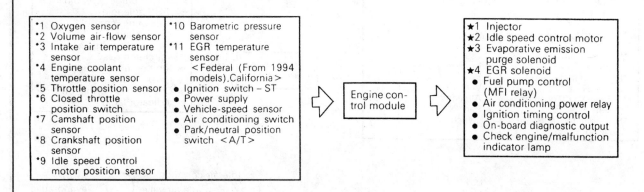

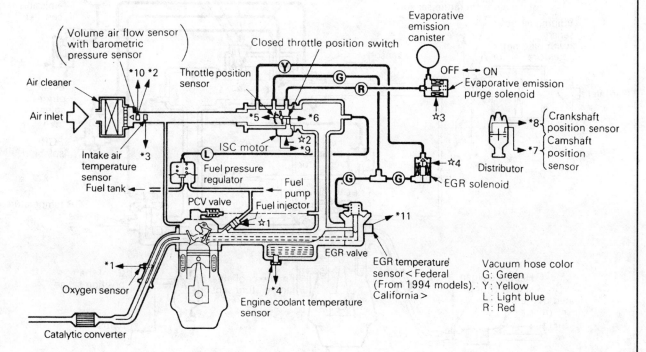

Fig. 181 Multi-port fuel injection system diagram — 1992-95 Pick-up, 2.4L engine, Federal, California-RWD (1992) and 4WD

4-110 EMISSION CONTROLS

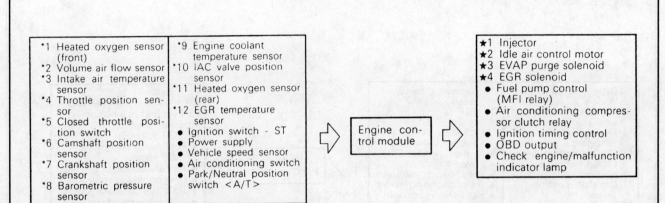

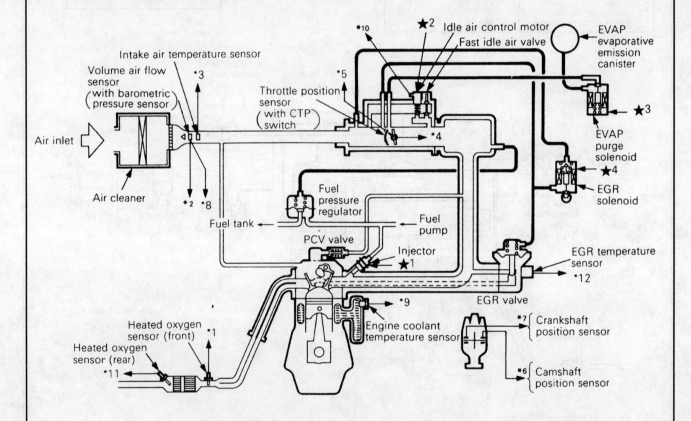

Fig. 182 Multi-port fuel injection system diagram — 1992-95 Pick-up, 2.4L engine, California-RWD (from 1993 on)

EMISSION CONTROLS 4-111

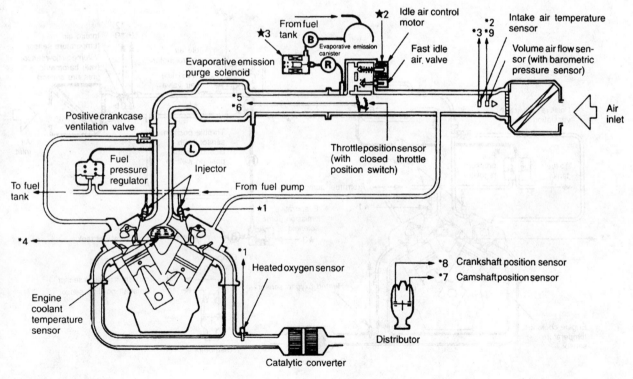

Fig. 183 Multi-port fuel injection system diagram — 1992 Pick-up, 3.0L (12 valve) engine

4-112 EMISSION CONTROLS

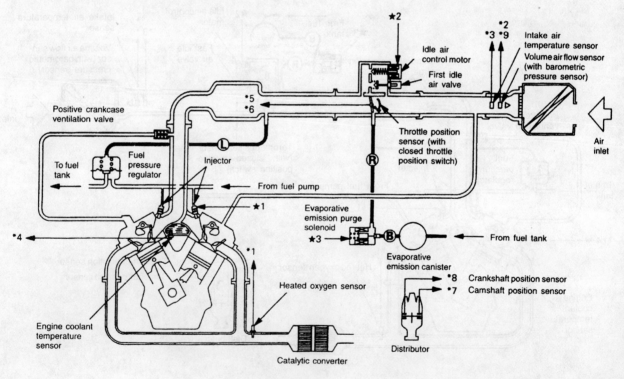

Fig. 184 Multi-port fuel injection system diagram — 1993-95 Pick-up, 3.0L (12 valve) engine

EMISSION CONTROLS 4-113

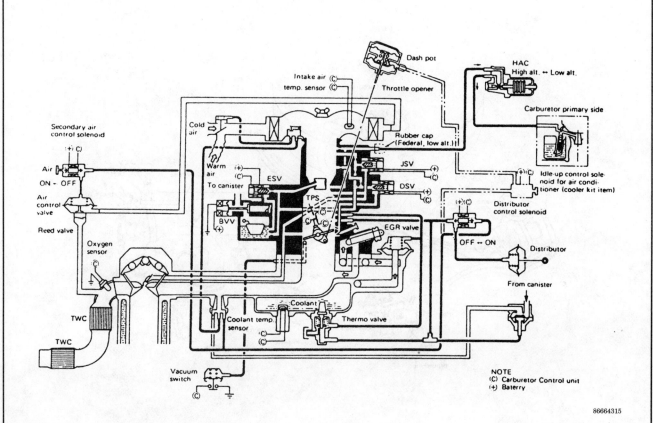

Fig. 185 Component location and vacuum schematic — 1986 Montero

4-114 EMISSION CONTROLS

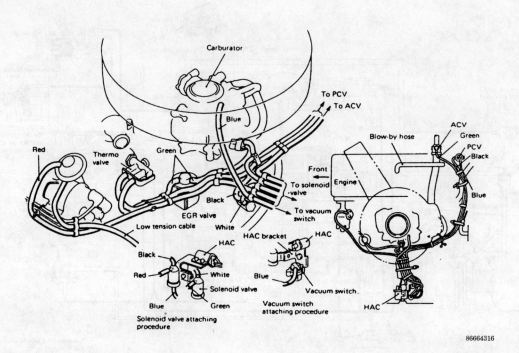

Fig. 186 Emission control vacuum hose routing — 1986 Montero

EMISSION CONTROLS 4-115

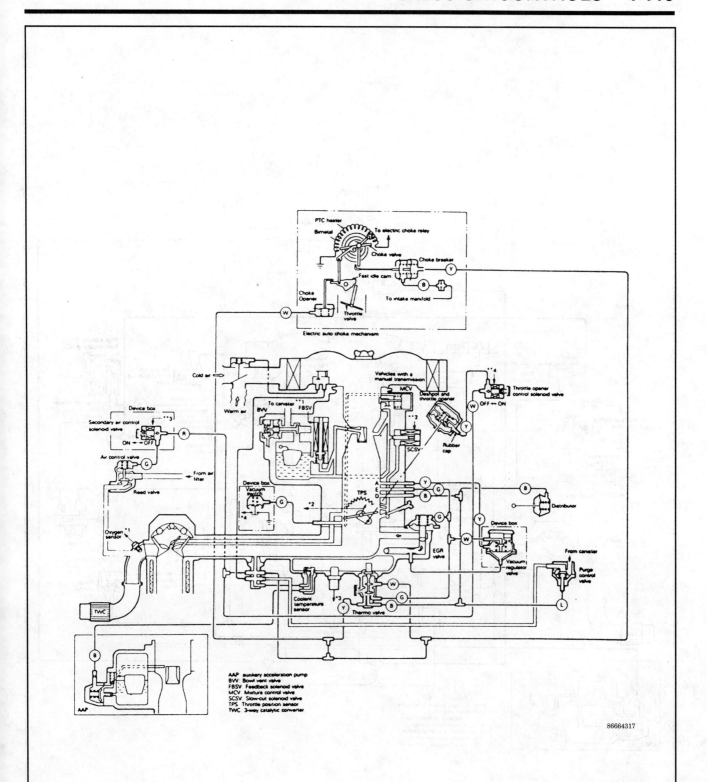

Fig. 187 Component location and vacuum schematic — 1987 Montero, Federal

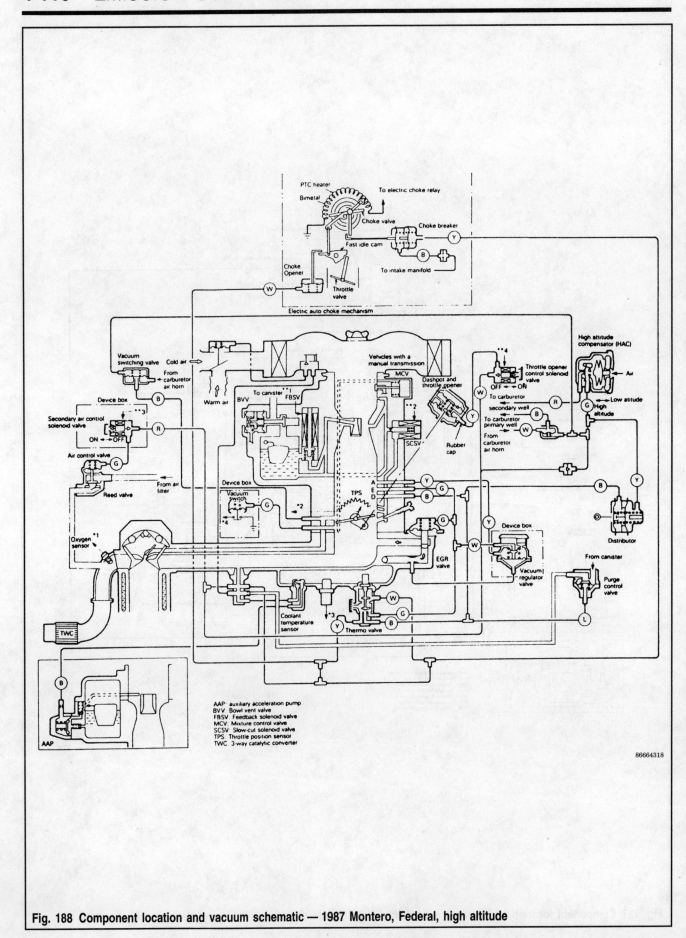

Fig. 188 Component location and vacuum schematic — 1987 Montero, Federal, high altitude

EMISSION CONTROLS 4-117

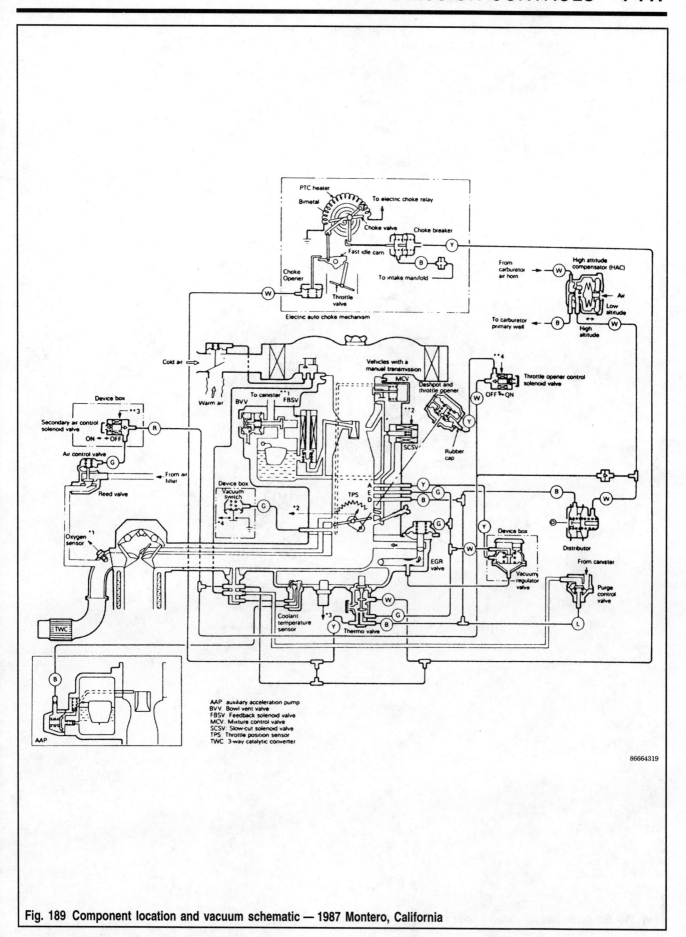

Fig. 189 Component location and vacuum schematic — 1987 Montero, California

4-118 EMISSION CONTROLS

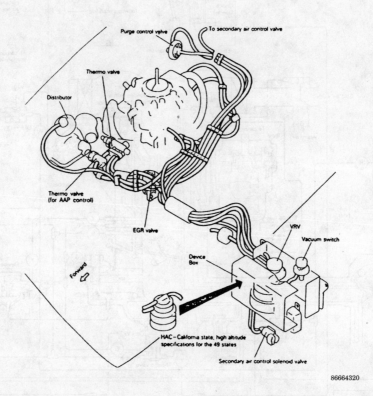

Fig. 190 Emission control vacuum hose routing — 1987 Montero

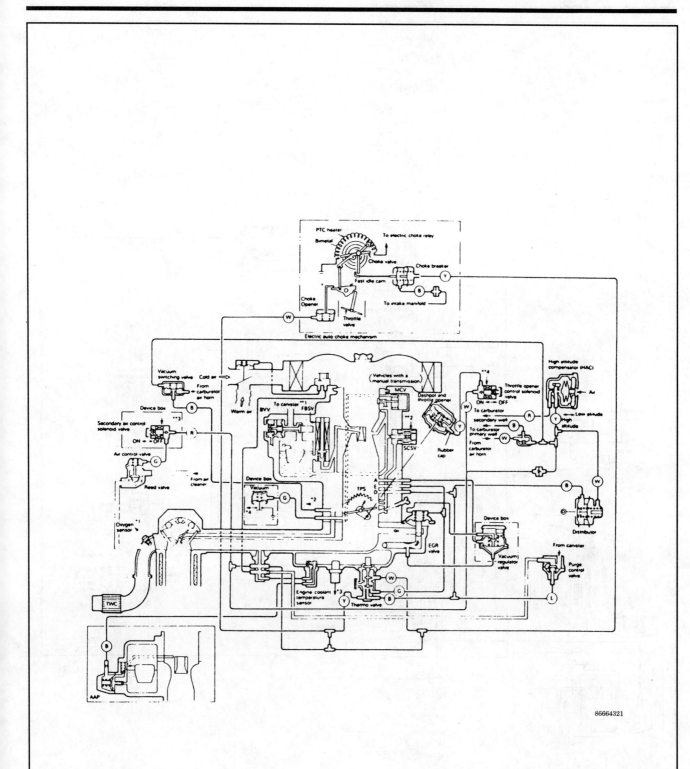

Fig. 191 Component location and vacuum schematic — 1988 Montero, Federal

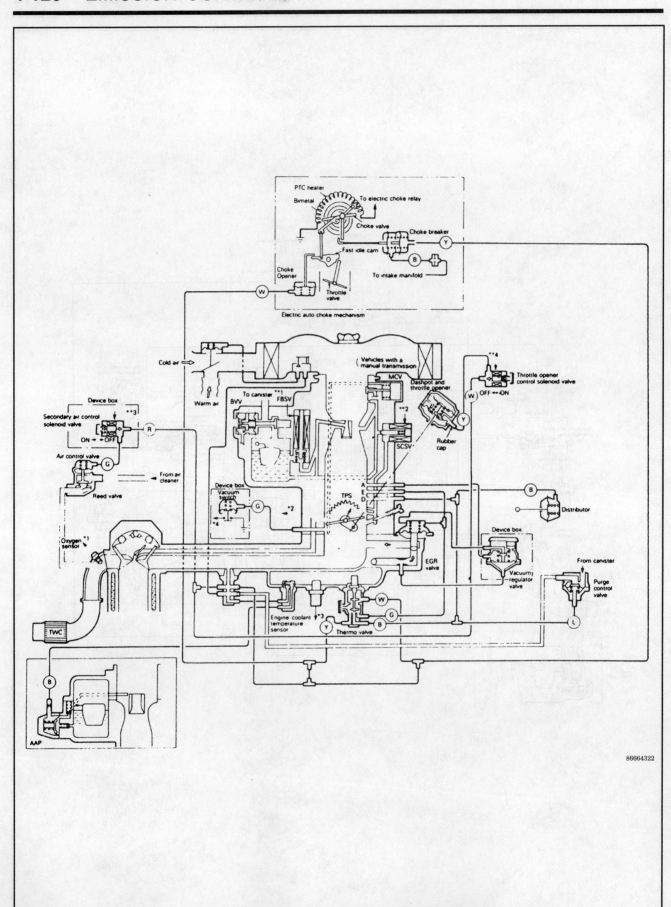

Fig. 192 Component location and vacuum schematic — 1988 Montero, Federal, high altitude

EMISSION CONTROLS 4-121

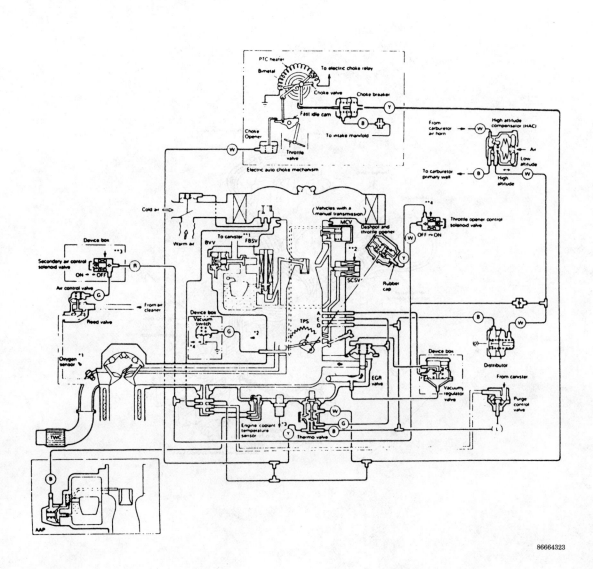

Fig. 193 Component location and vacuum schematic — 1989 Montero, California

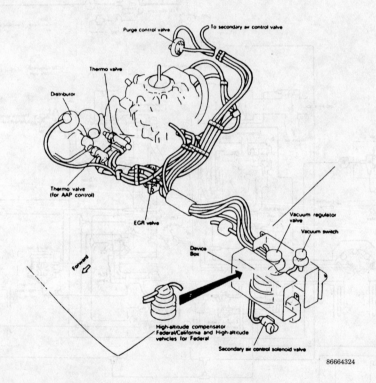

Fig. 194 Emission control vacuum hose routing — 1988 Montero

EMISSION CONTROLS 4-123

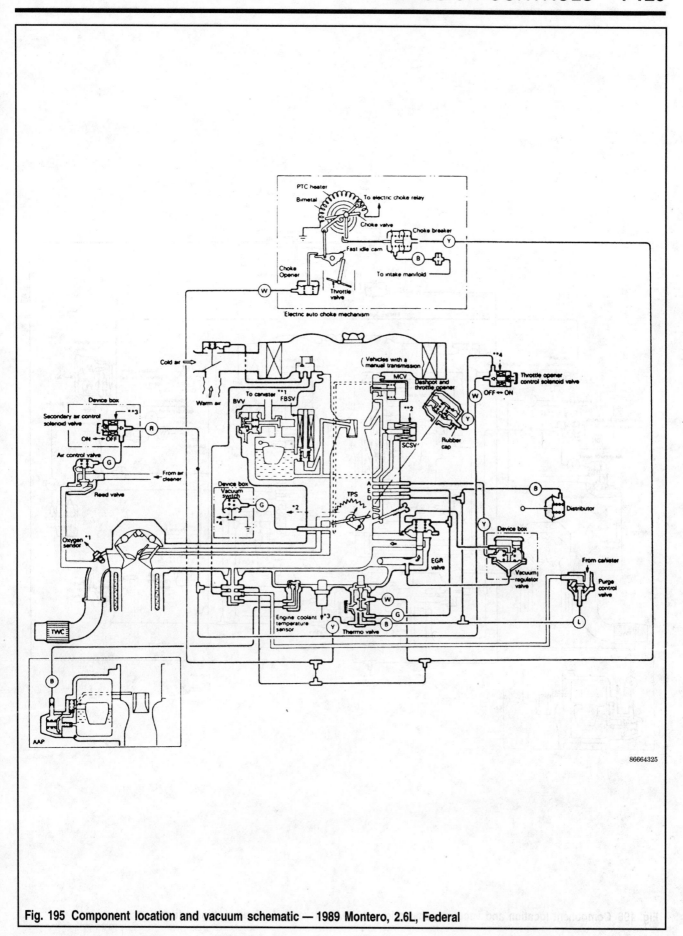

Fig. 195 Component location and vacuum schematic — 1989 Montero, 2.6L, Federal

4-124 EMISSION CONTROLS

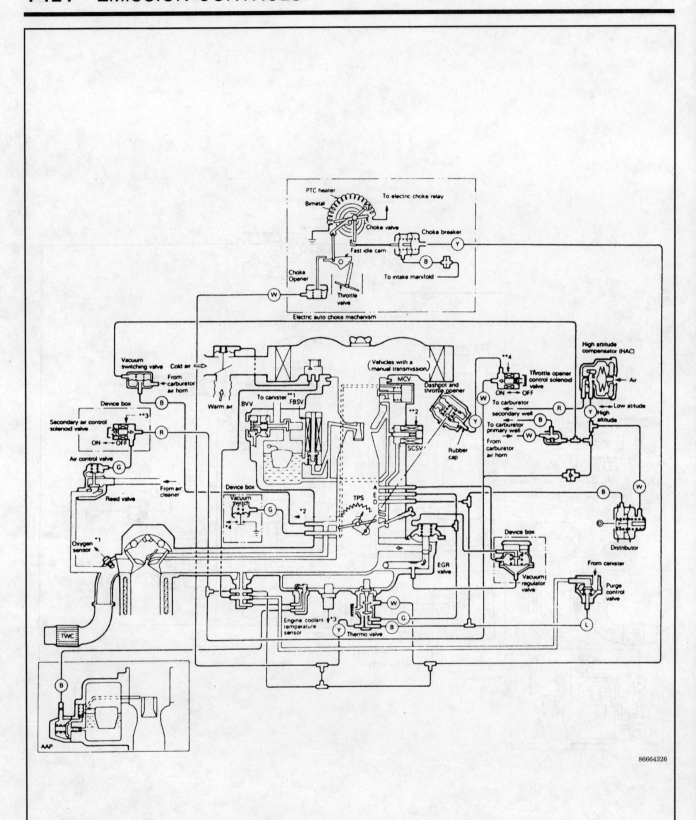

Fig. 196 Component location and vacuum schematic — 1989 Montero, 2.6L, Federal, high altitude

EMISSION CONTROLS 4-125

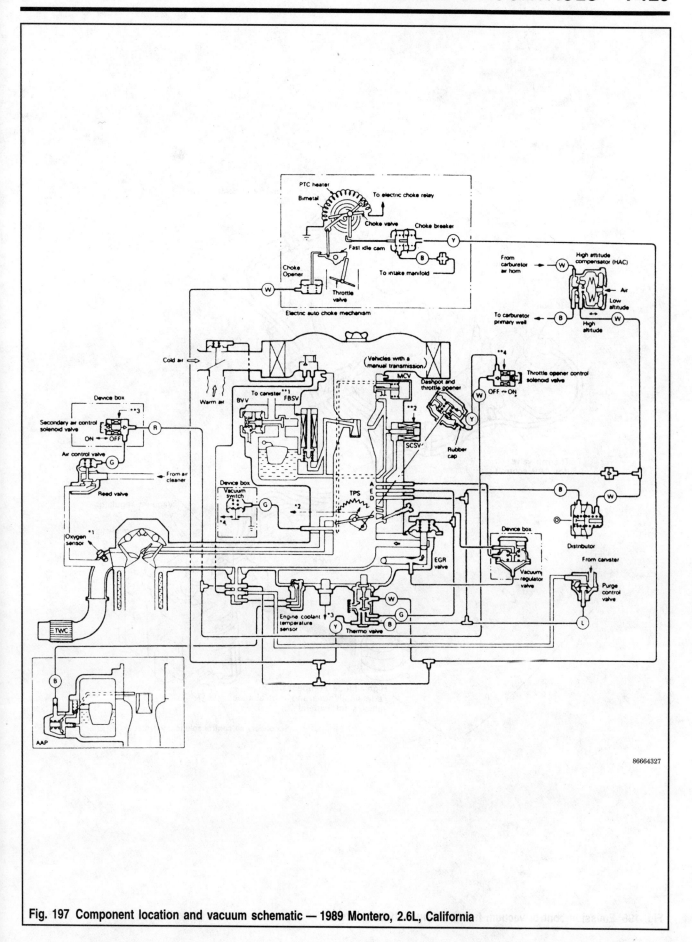

Fig. 197 Component location and vacuum schematic — 1989 Montero, 2.6L, California

4-126 EMISSION CONTROLS

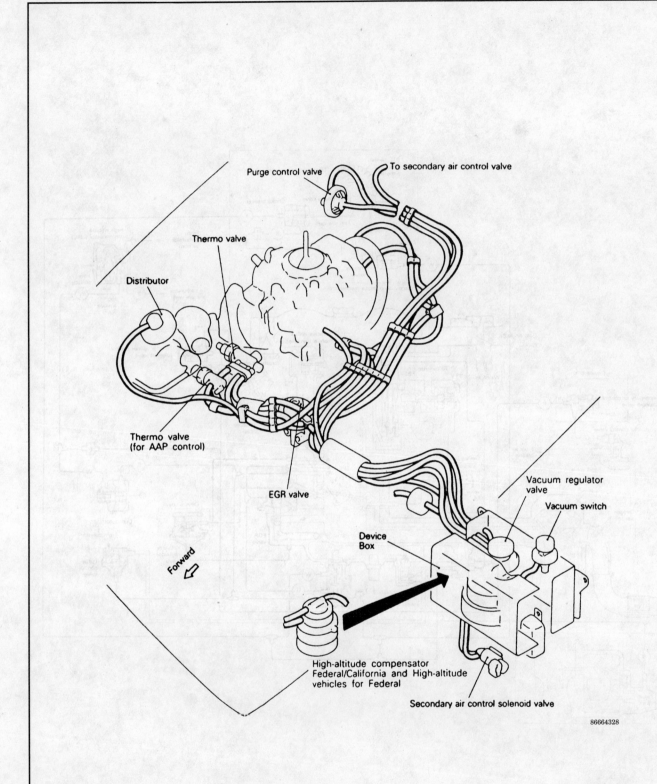

Fig. 198 Emission control vacuum hose routing — 1989 Montero

EMISSION CONTROLS 4-127

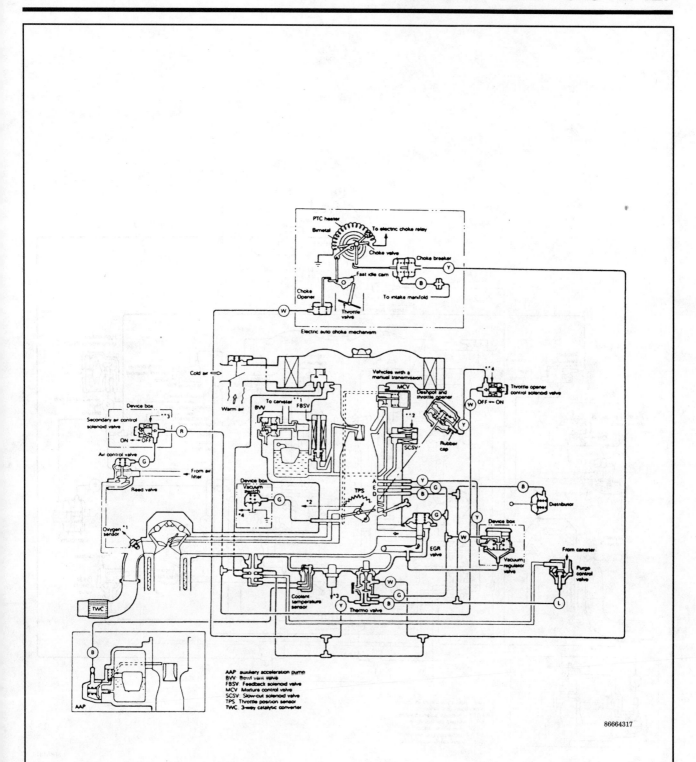

Fig. 199 Component location and vacuum schematic — 1987 Montero, Federal

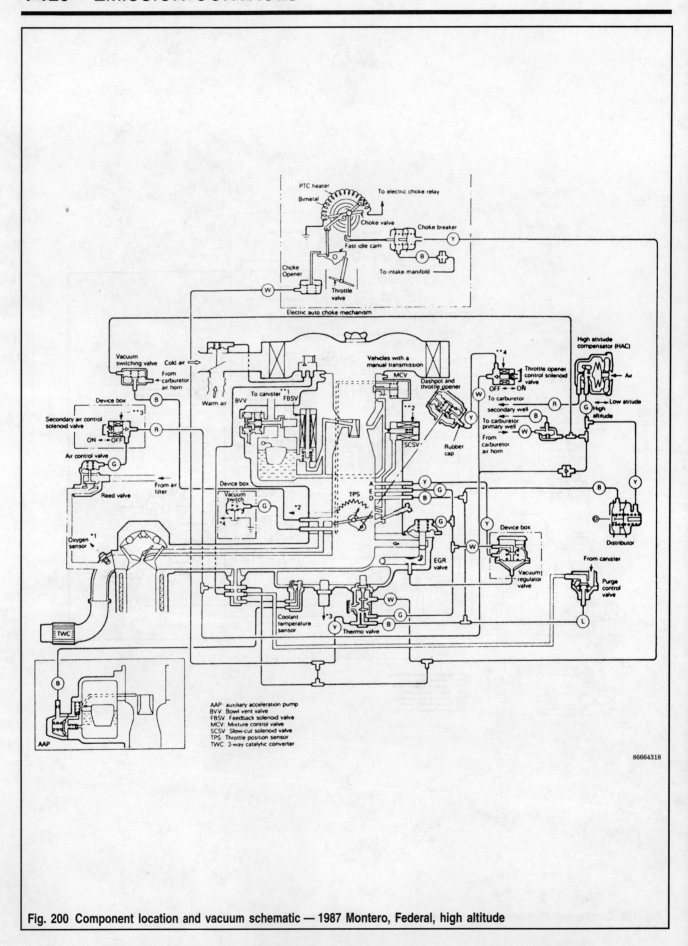

Fig. 200 Component location and vacuum schematic — 1987 Montero, Federal, high altitude

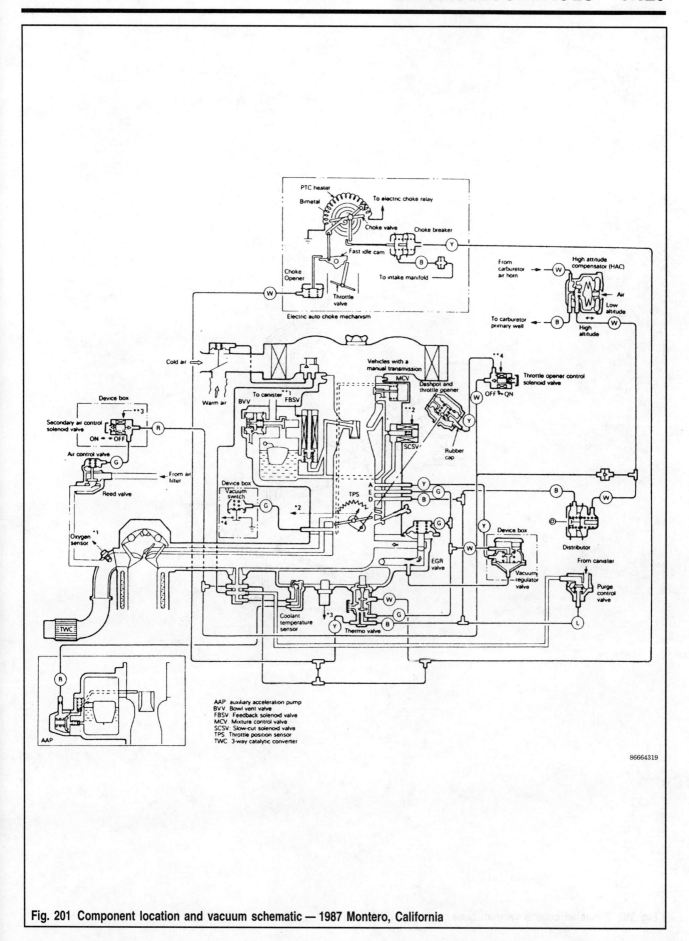

Fig. 201 Component location and vacuum schematic — 1987 Montero, California

4-130 EMISSION CONTROLS

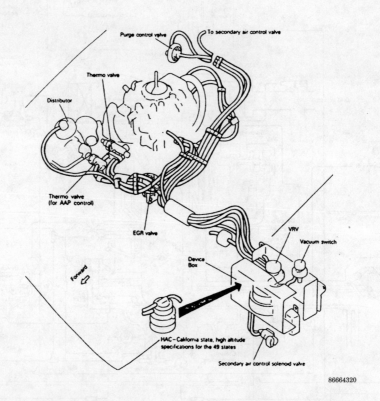

Fig. 202 Emission control vacuum hose routing — 1987 Montero

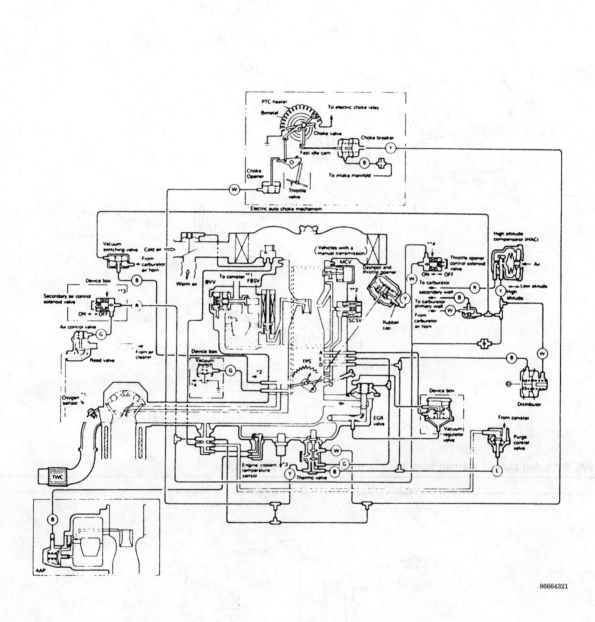

Fig. 203 Component location and vacuum schematic — 1988 Montero, Federal

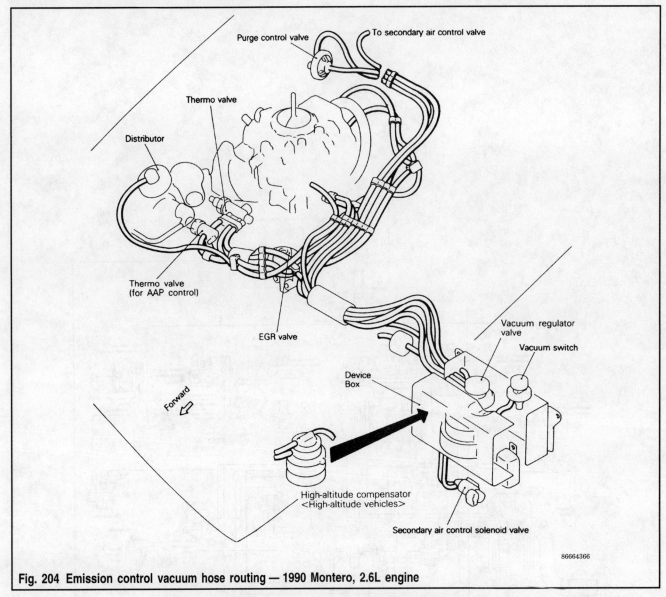

Fig. 204 Emission control vacuum hose routing — 1990 Montero, 2.6L engine

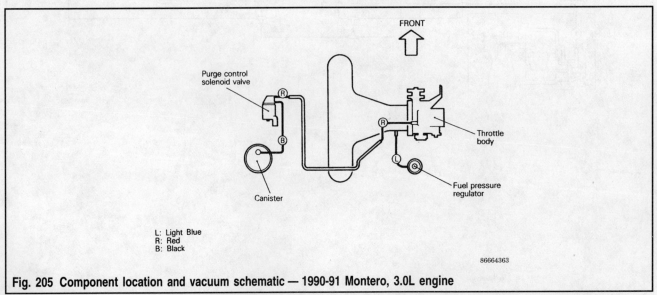

L: Light Blue
R: Red
B: Black

Fig. 205 Component location and vacuum schematic — 1990-91 Montero, 3.0L engine

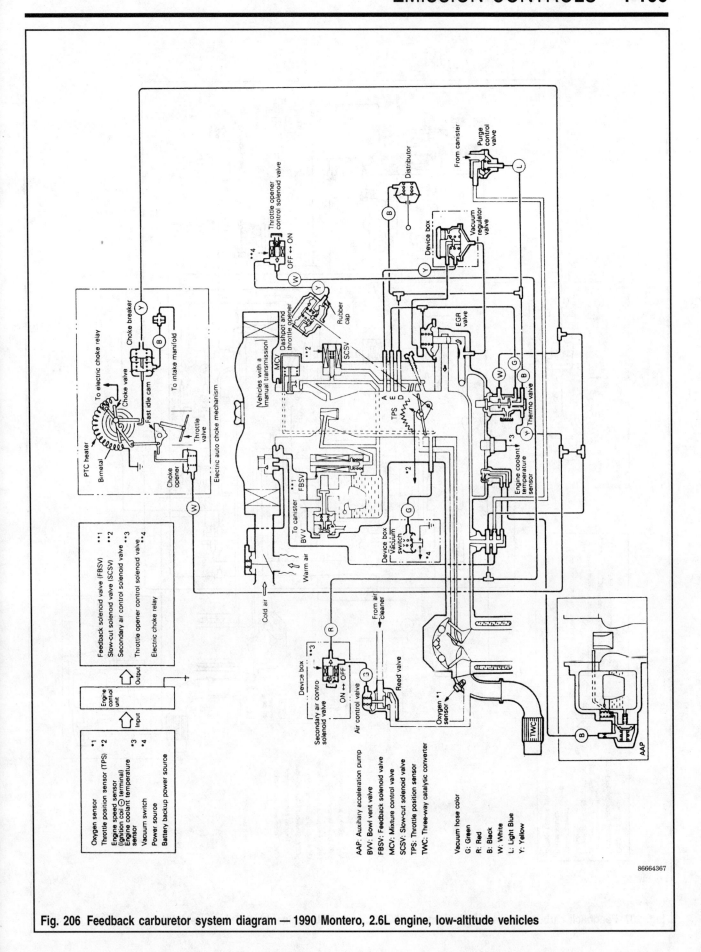

Fig. 206 Feedback carburetor system diagram — 1990 Montero, 2.6L engine, low-altitude vehicles

4-134 EMISSION CONTROLS

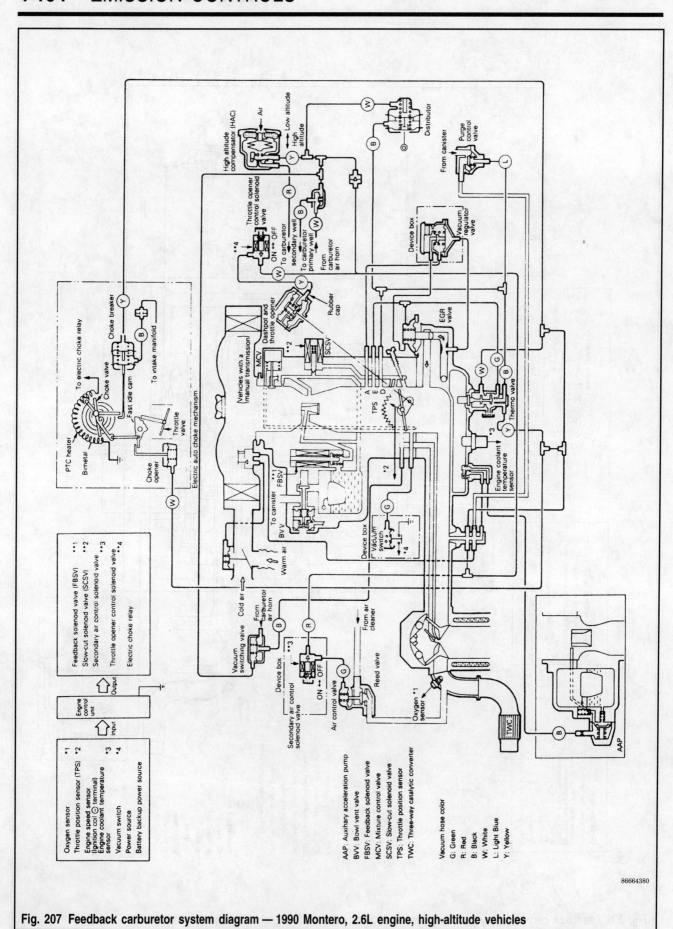

Fig. 207 Feedback carburetor system diagram — 1990 Montero, 2.6L engine, high-altitude vehicles

EMISSION CONTROLS 4-135

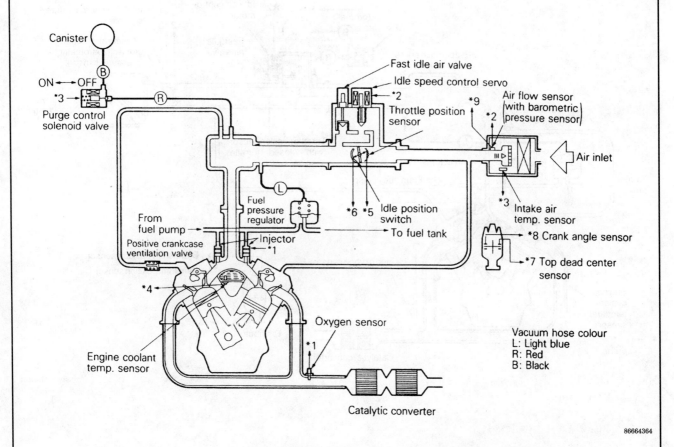

Fig. 208 Multi-port fuel injection system component location and vacuum schematic — 1990-91 Montero, 3.0L engine

4-136 EMISSION CONTROLS

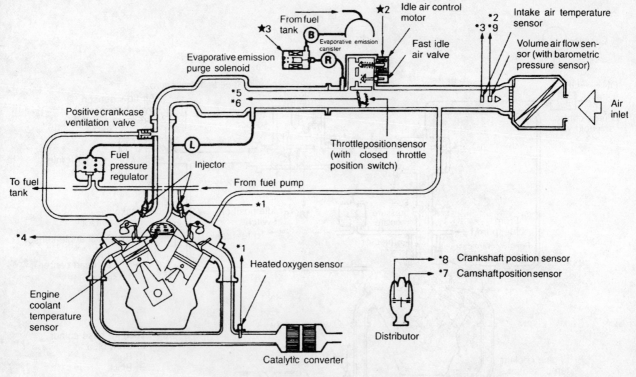

Fig. 209 Multi-port fuel injection system component location and vacuum schematic — 1992 Montero, 3.0L (12 valve) engine

EMISSION CONTROLS 4-137

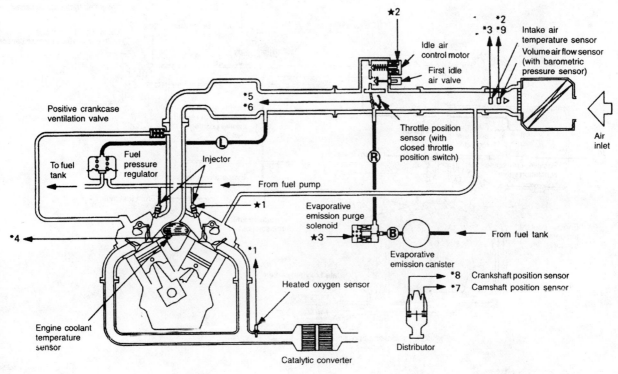

Fig. 210 Multi-port fuel injection system component location and vacuum schematic — 1993-94 Montero, 3.0L (12 valve) engine

4-138 EMISSION CONTROLS

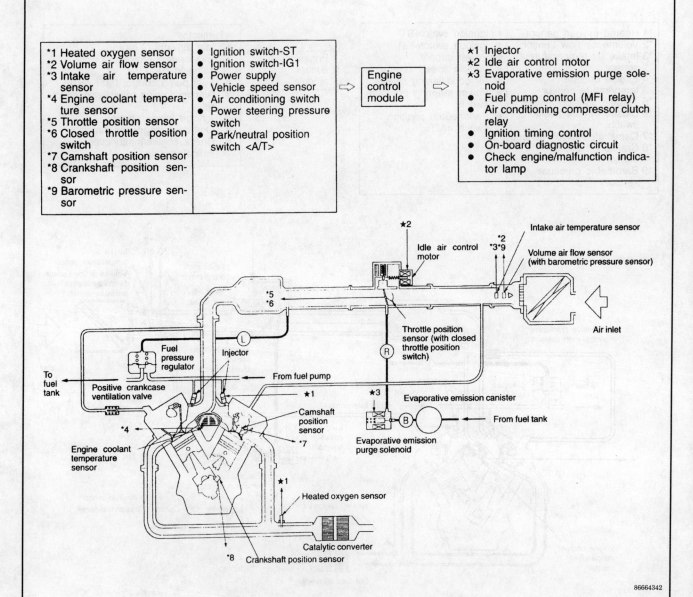

Fig. 211 Multi-port fuel injection system component location and vacuum schematic — 1995 Montero, 3.0L (24 valve) engine, Federal

EMISSION CONTROLS 4-139

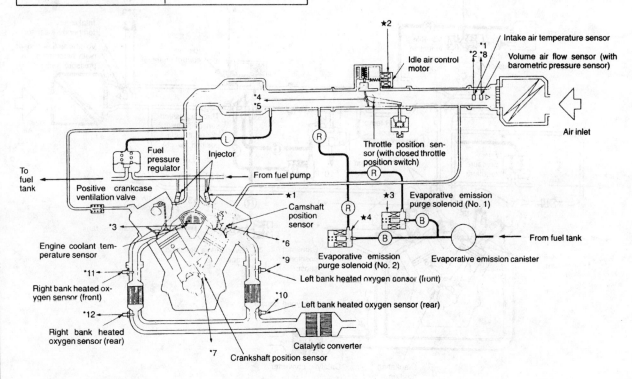

Fig. 212 Multi-port fuel injection system component location and vacuum schematic — 1995 Montero, 3.0L (24 valve) engine, California

4-140 EMISSION CONTROLS

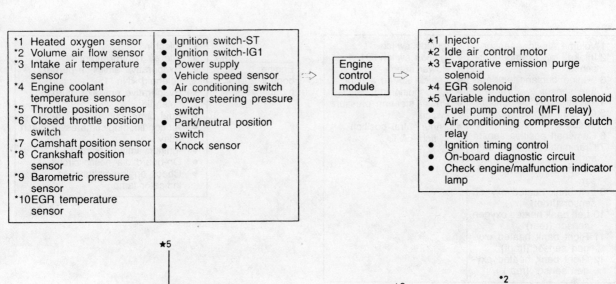

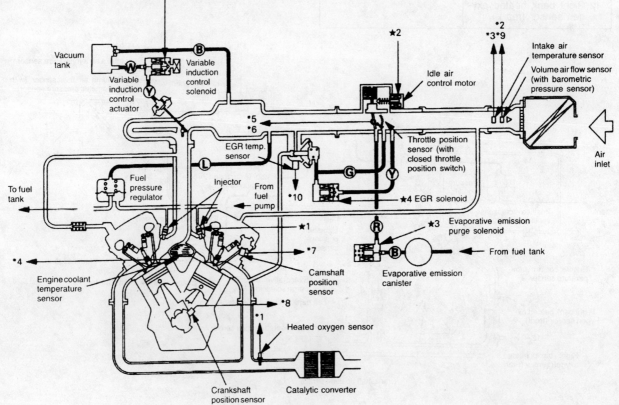

Fig. 213 Multi-port fuel injection system component location and vacuum schematic — 1994-95 Montero, 3.5L DOHC engine, Federal and 1994 California

EMISSION CONTROLS 4-141

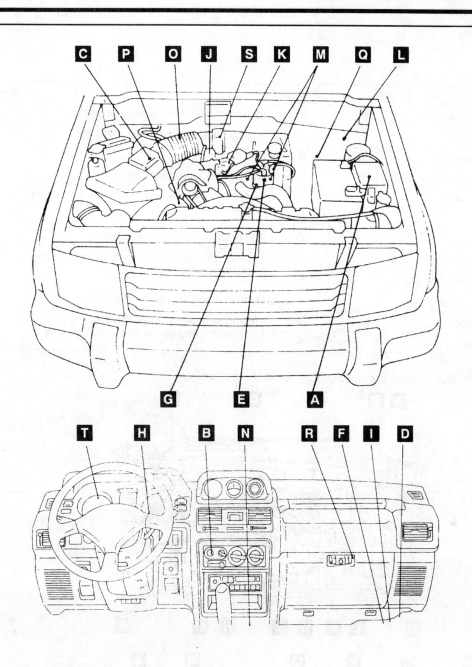

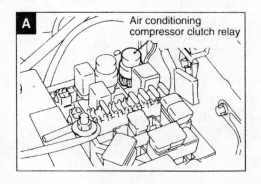

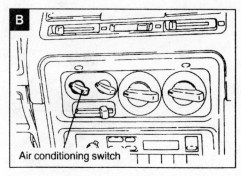

Fig. 214 Multi-port fuel injection system component location and vacuum schematic — 1995 Montero, 3.5L DOHC engine, California

4-142 EMISSION CONTROLS

Name	Symbol	Name	Symbol
Air conditioning compressor clutch relay	F	Ignition coil (Ignition power transistor)	H
Air conditioning switch	Q	Ignition timing adjustment connector	B
Camshaft position sensor	G	Injector	D
Check engine / Malfunction indicator lamp	P	Multiport fuel injection (MFI) relay	S
Crankshaft position sensor	I	Park / Neutral position switch (Vehicles with automatic transmission)	U
Data link connector	N		
Engine control module	R	Power steering pressure switch	L
Engine coolant temperature sensor	J	Throttle position sensor (with closed throttle position switch)	C
Evaporative emission purge solenoid	E		
Fuel pump check connector	A	Vehicle speed sensor (reed switch)	O
Heated oxygen sensor	T	Volume air flow sensor (with intake air temperature sensor and barometric pressure sensor)	M
Idle air control motor	K		

NOTE
The entries in the "Name" column are arranged in alphabetical order.

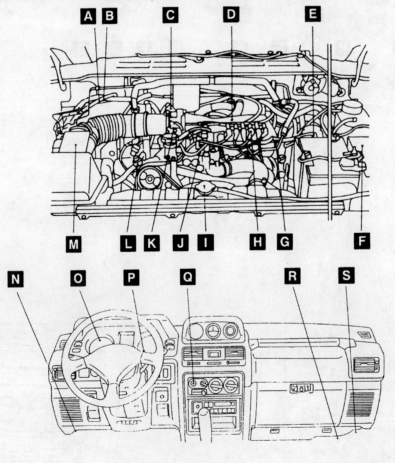

Fig. 215 Component location diagram — 1995 Montero, 3.0L (24 valve) engine

EMISSION CONTROLS 4-143

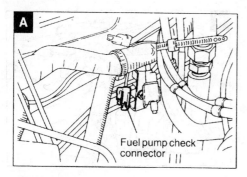

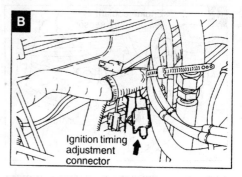

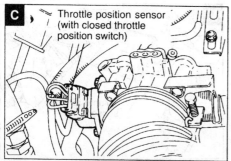

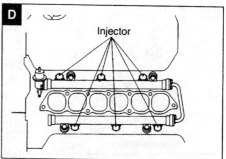

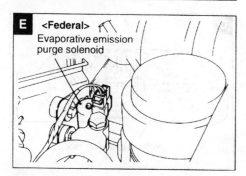

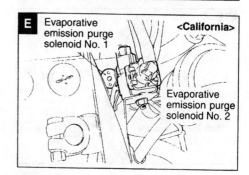

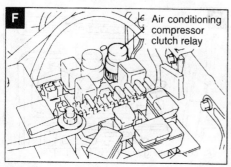

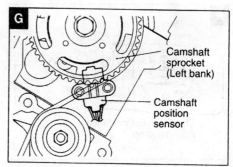

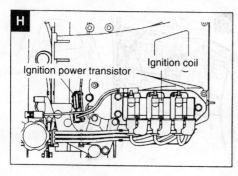

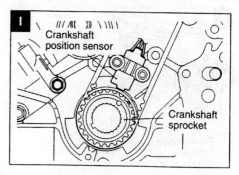

Fig. 216 Component location diagram — 1995 Montero, 3.0L (24 valve) engine

4-144 EMISSION CONTROLS

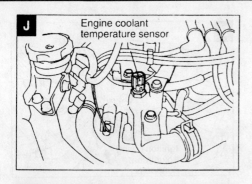

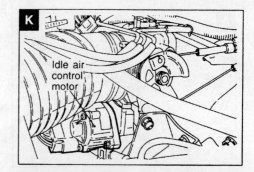

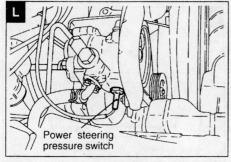

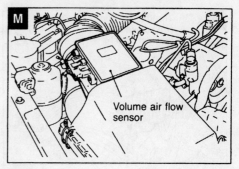

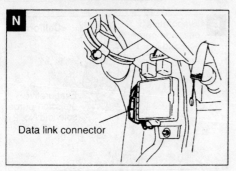

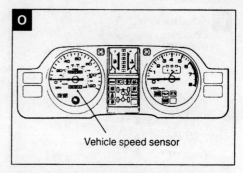

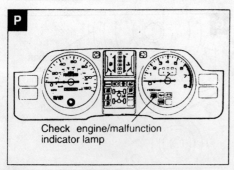

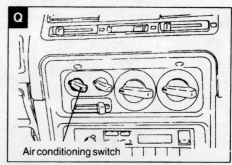

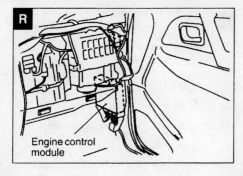

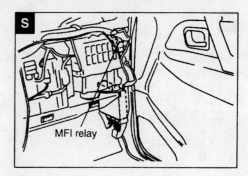

Fig. 217 Component location diagram — 1995 Montero, 3.0L (24 valve) engine

EMISSION CONTROLS 4-145

Name	Symbol	Name	Symbol
Air conditioning compressor clutch relay	A	Ignition coil (ignition power transistor)	O
Air conditioning switch	B	Ignition timing adjustment connector	P
Camshaft position sensor	C	Injector	Q
Check engine/malfunction indicator lamp	D	Multiport fuel injection (MFI) relay	R
Crankshaft position sensor	E	Park/neutral position switch (Vehicles with automatic transmission)	S
Data link connector	F		
EGR solenoid	G	Power steering pressure switch	T
EGR temperature sensor	H	Throttle position sensor (with closed throttle position switch)	U
Engine control module	I		
Engine coolant temperature sensor	J	Variable induction control solenoid	V
Evaporative emission purge solenoid	K	Vehicle speed sensor (reed switch)	W
Fuel pump check connector	L	Volume air flow sensor (with intake air temperature sensor and barometric pressure sensor)	X
Heated oxygen sensor	M		
Idle air control motor	N		

NOTE
The entries in the "Name" column are arranged in alphabetical order.

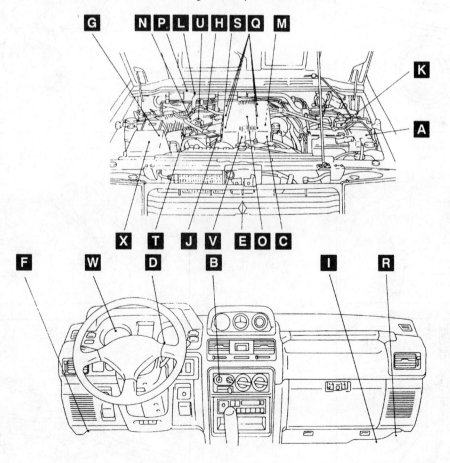

Fig. 218 Component location diagram — 1994-95 Montero, 3.5L DOHC engine

4-146 EMISSION CONTROLS

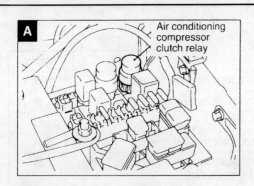
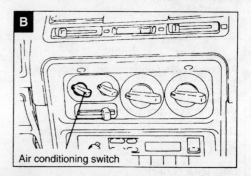
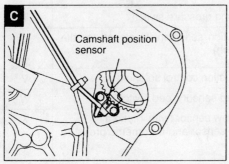
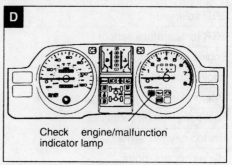
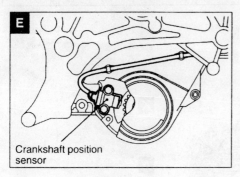
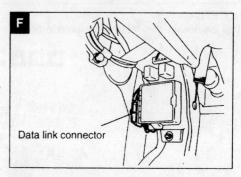
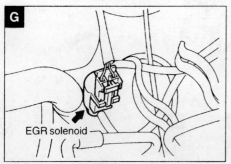
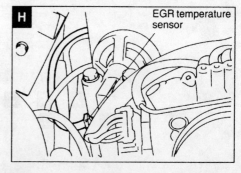
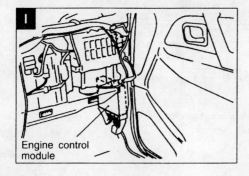
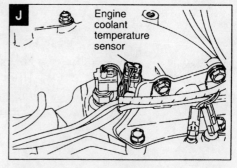

Fig. 219 Component location diagram — 1994-95 Montero, 3.5L DOHC engine

EMISSION CONTROLS 4-147

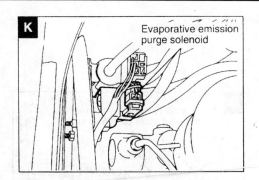
K — Evaporative emission purge solenoid

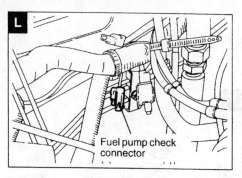
L — Fuel pump check connector

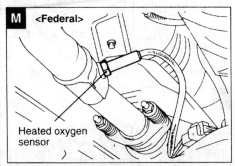
M <Federal> — Heated oxygen sensor

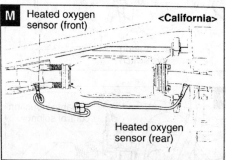
M <California> — Heated oxygen sensor (front), Heated oxygen sensor (rear)

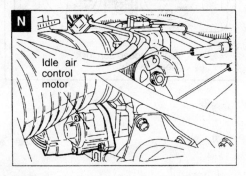
N — Idle air control motor

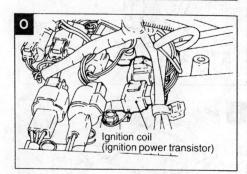
O — Ignition coil (ignition power transistor)

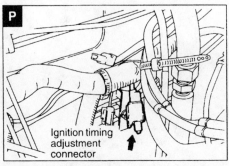
P — Ignition timing adjustment connector

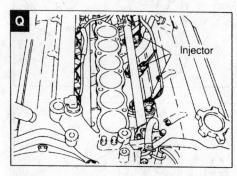
Q — Injector

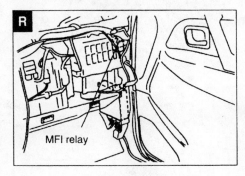
R — MFI relay

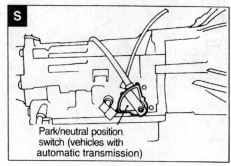
S — Park/neutral position switch (vehicles with automatic transmission)

Fig. 220 Component location diagram — 1994-95 Montero, 3.5L DOHC engine

4-148 EMISSION CONTROLS

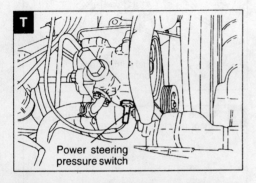

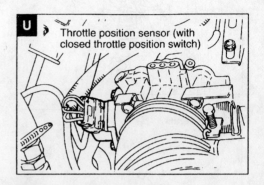

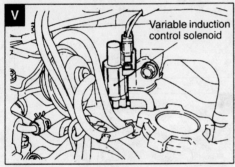

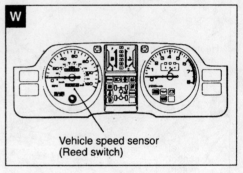

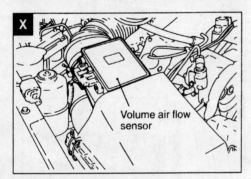

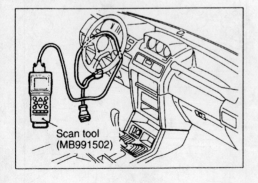

Fig. 221 Component location diagram — 1994-95 Montero, 3.5L DOHC engine

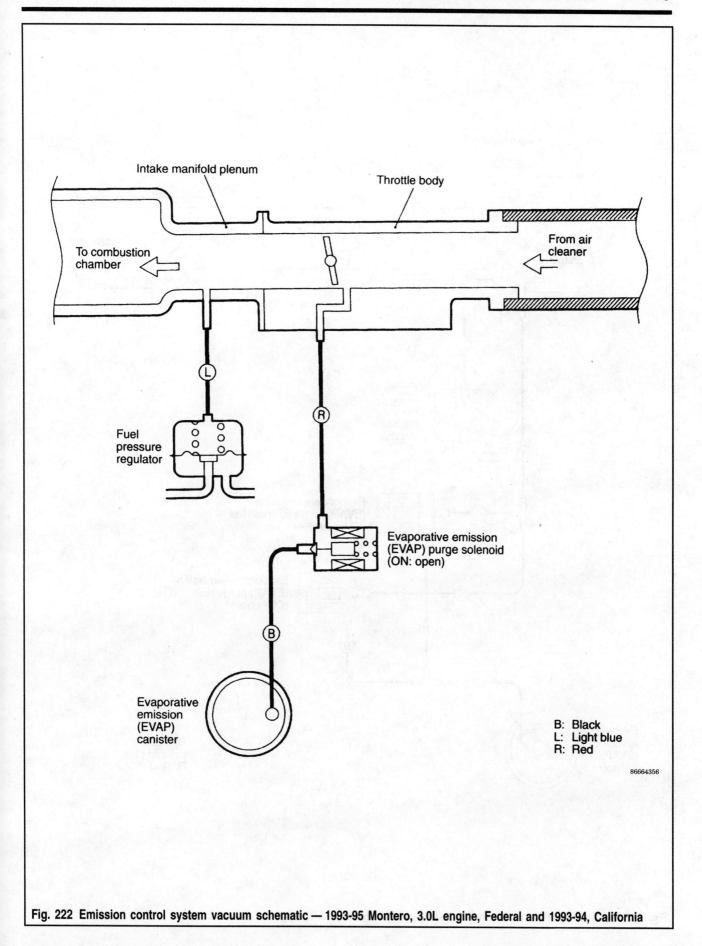

Fig. 222 Emission control system vacuum schematic — 1993-95 Montero, 3.0L engine, Federal and 1993-94, California

4-150 EMISSION CONTROLS

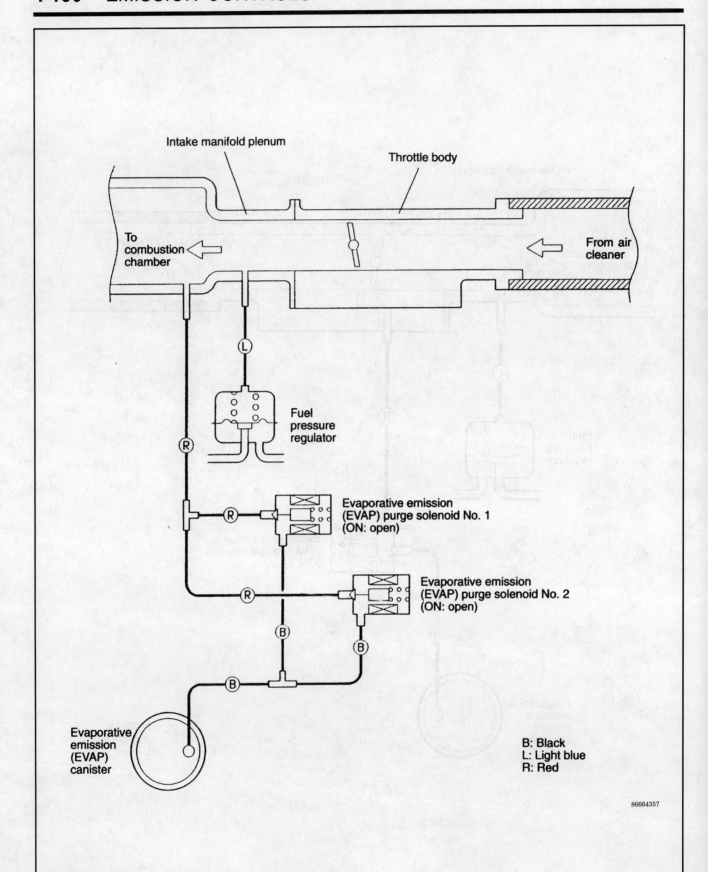

Fig. 223 Emission control vacuum hose routing — 1995 Montero, 3.0L engine, California

EMISSION CONTROLS 4-151

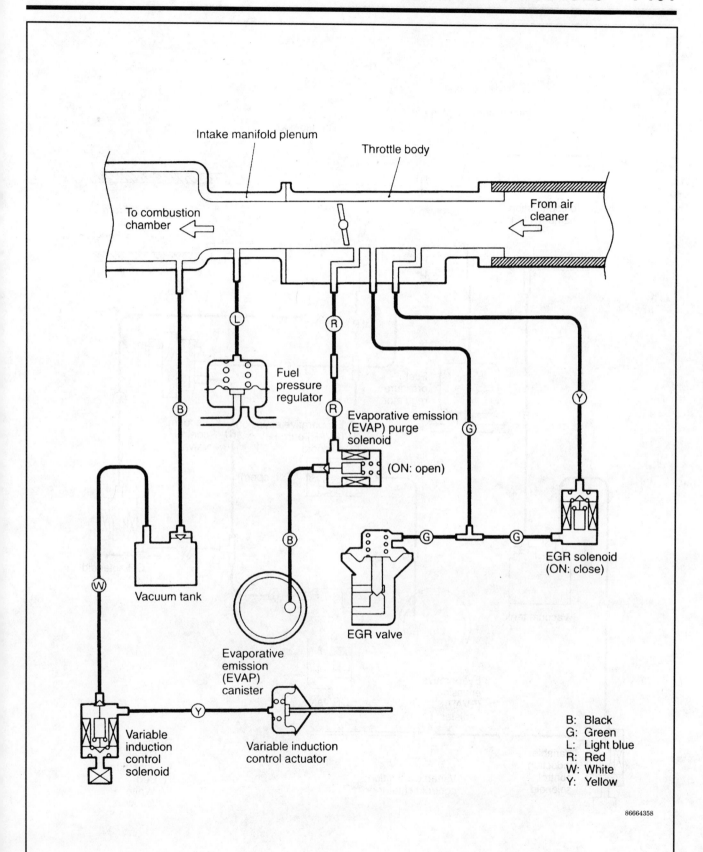

Fig. 224 Emission control vacuum hose routing — 1994 Montero, 3.5L DOHC engine

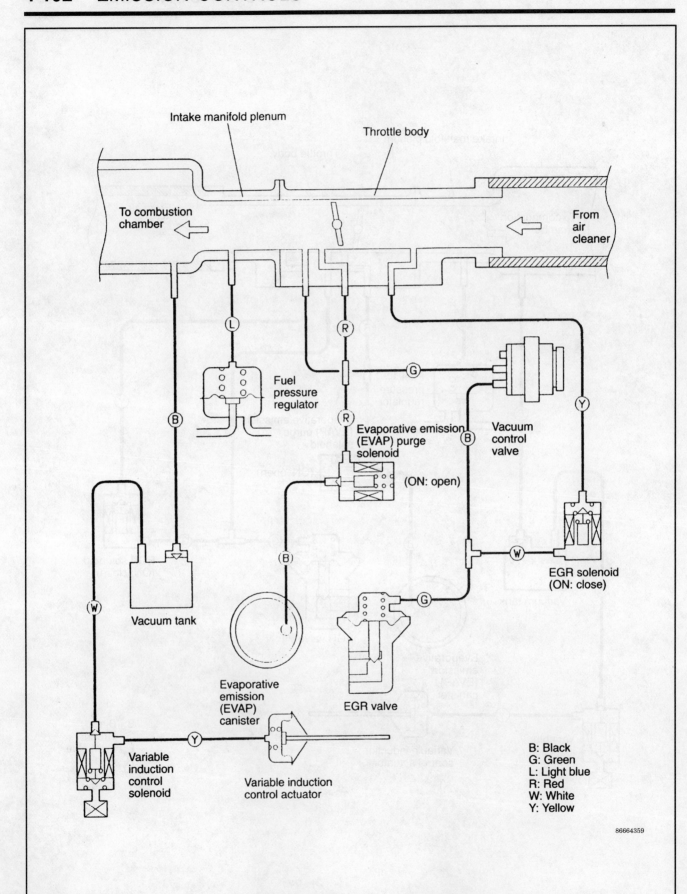

Fig. 225 Emission control vacuum hose routing — 1995 Montero, 3.5L DOHC engine

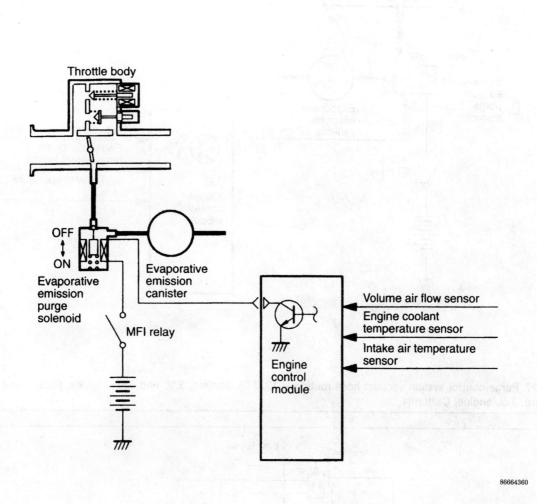

Fig. 226 Purge control system vacuum hose routing — 1992 Montero, 3.0L engine

4-154 EMISSION CONTROLS

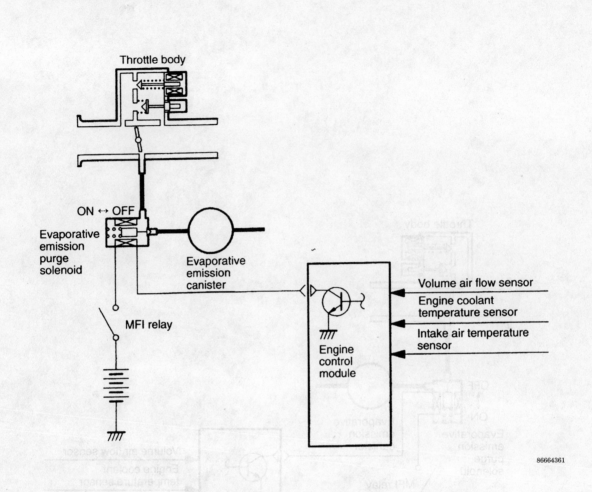

Fig. 227 Purge control system vacuum hose routing — 1993-95 Montero, 3.0L and 3.5L engines, Federal and 1993-94 Montero, 3.0L engine, California

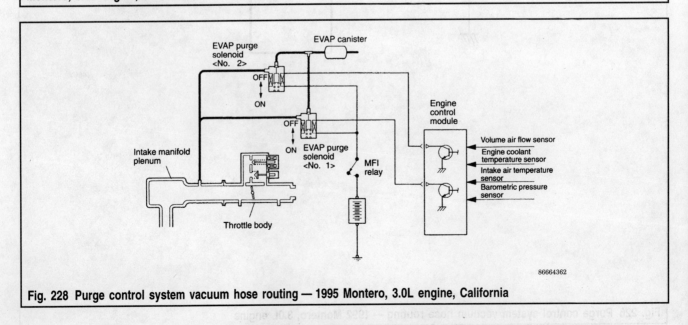

Fig. 228 Purge control system vacuum hose routing — 1995 Montero, 3.0L engine, California

CARBURETED FUEL SYSTEM
  CARBURETORS 5-3
  MECHANICAL FUEL PUMP 5-2
DIESEL FUEL SYSTEM
  GLOW PLUGS 5-45
  IDLE SPEED ADJUSTMENT 5-45
  IDLE-UP ADJUSTMENT 5-45
  INJECTION LINES 5-40
  INJECTION PUMP 5-40
  INJECTORS 5-40
FUEL TANK
  TANK ASSEMBLY 5-46
MULTI-PORT FUEL INJECTION SYSTEM (MFI)
  ELECTRIC FUEL PUMP 5-19
  FUEL FILTER 5-33
  FUEL INJECTORS 5-26
  FUEL PRESSURE REGULATOR 5-30
  PRESSURE RELIEF VALVE 5-33
  RELIEVING FUEL SYSTEM
    PRESSURE 5-19
  THROTTLE BODY 5-20
SPECIFICATIONS CHARTS
  CARBURETOR
    SPECIFICATIONS 5-18

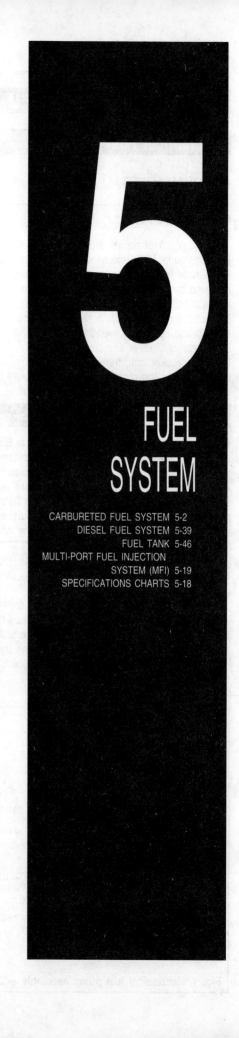

5

FUEL SYSTEM

CARBURETED FUEL SYSTEM 5-2
DIESEL FUEL SYSTEM 5-39
FUEL TANK 5-46
MULTI-PORT FUEL INJECTION
SYSTEM (MFI) 5-19
SPECIFICATIONS CHARTS 5-18

5-2 FUEL SYSTEM

CARBURETED FUEL SYSTEM

Mechanical Fuel Pump

REMOVAL & INSTALLATION

♦ See Figures 1 and 2

Mechanical fuel pumps are found on all non-diesel Pick-up trucks through 1989 and all 4-cylinder Monteros. The mechanical fuel pump is mounted on the side of the head, centered between the runners of the intake manifold. The intake manifold does not require removal to remove the pump, but access can be tricky. A selection of socket extensions, swivels and open end wrenches can be helpful.

A pushrod or lever arm runs from a special camshaft lobe to the pump rocker arm, transferring the motion to the pump diaphragm. As the diaphragm moves, fuel is alternately drawn into the pump and then sent to the carburetor.

✱✱CAUTION

Gasoline in either liquid or vapor state is EXTREMELY explosive. Take great care to contain spillage. Work in an open or well-ventilated area. Do not connect or disconnect electrical connectors while fuel hoses are removed or loosened. Observe no smoking/no open flame rules during repairs. Have a dry-chemical fire extinguisher (type B-C) within arm's reach at all times and know how to use it.

1. Disconnect the negative battery cable. Depending on the vehicle, it may be necessary to remove the air cleaner assembly.
2. Set the engine to TDC/compression for No. 1 cylinder. This is generally done by turning the crankshaft pulley bolt clockwise (never counterclockwise) until the timing marks on the case align. Double check the positioning by removing the distributor cap and making sure the rotor points to the No. 1 terminal.

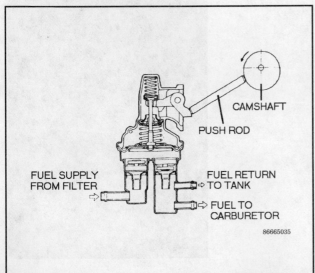

Fig. 1 Mechanical fuel pump assembly — 2.0L engines

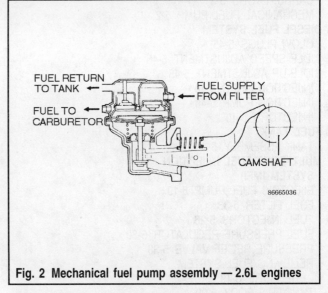

Fig. 2 Mechanical fuel pump assembly — 2.6L engines

3. Label or diagram the fuel lines running to the pump. Disconnect the three fuel lines by using a pair of pliers to shift clamps away from the nipples on the pump and then pulling the lines off with a twisting motion.
4. Remove the two mounting bolts from the head. Remove the pump, spacer, and two gaskets from the head. As you pull the pump off the head, catch the pushrod which is located just behind the pump. If the pump is the lever type, it will come off with the pump.
5. Clean the gasket surfaces of the insulator, pump and cylinder head. Insert the two bolts through the pump's mounting base. Slide a new gasket, the insulator, and a second new gasket into position over the two bolts. Turn the pump so its mounting surface faces the cylinder head.
6. Apply a light coat of clean engine oil to the pushrod. Place the pump pushrod against the cupped surface of the operating lever and angle it upward in the position it was in during removal. Hold the pushrod at that angle as you insert it into the bore in the head. Once the pushrod is in the bore in the cylinder head, you can release your fingers and move the pump toward the head along the line of the pushrod. If the pump is the lever-arm type, make certain the arm is installed above the camshaft. The cam will move the lever as it turns.
7. Start the two bolts into the bores in the head and tighten them finger tight. Tighten the mounting bolts alternately and evenly to 12 ft. lbs. (16 Nm).
8. Inspect the hoses for cracks (even hairline cracks can leak) and replace if necessary. Reconnect the fuel hoses. Make sure the hoses are installed all the way onto the nipples.
9. Work the clamps into position. Make sure the clamps are located well past the bulged portion of the nipples but do not sit at the extreme ends of the hoses.
10. Replace the distributor cap.

➡**If the fuel pump was replaced because of a ruptured diaphragm, there is a strong possibility of fuel contamination in the engine oil. Changing the oil and oil filter is strongly recommended after the fuel pump is replaced.**

11. Install the air cleaner assembly if it was removed.

FUEL SYSTEM 5-3

12. Connect the negative battery cable. Start the engine and check for leaks. Because the fuel lines have been partially emptied, the engine may crank for a longer than normal period before starting.

TESTING

On Vehicle Testing

1. Disconnect the inlet line (coming from the filter) at the pump. Connect a vacuum gauge to the pump nipple. Remove the coil-to-distributor high tension cable at the coil.
2. Have an assistant crank the engine with the key as you watch the gauge. A vacuum should be produced in a regular cycle as the pump turns over. There should be no blowback of pressure (abrupt drop in vacuum or even positive pressure for a short time). If there is blowback of pressure, the inlet valve on the pump is leaking and the unit must be replaced.
3. The vacuum shown with each pump stroke should be strong and constant. If the vacuum is low (or none at all), the diaphragm is leaking and the pump must be replaced.

Off Vehicle Testing

1. Inspect the small breather hole or tube (above the diaphragm) which vents the pump's upper chamber. Leakage of fuel or oil here confirms that the diaphragm or oil seal is leaking.
2. Inspect the end of the pushrod and the contact surface on the pump operating lever. Replace the pushrod or pump if there is obvious wear. If the camshaft end of the pushrod is badly worn, remove the valve cover and inspect the camshaft eccentric (which operates the fuel pump) for excessive wear.
3. If the pump has a lever arm, check the contact pad (face) of the lever for wear or scoring. Inspect the camshaft if heavy scoring is present. Check the motion of the arm and the tension of the spring.

Carburetors

ADJUSTMENTS

Accelerator Cable

1. Warm the engine up completely and make certain the high idle cam is not engaged.
2. Check the action of the accelerator pedal. There should be little to no free-play between the normal "at rest" position and the point that engine speed begins to increase. Correct free-play measurement is 0-0.8 in. (0-20mm).
3. If there is excessive free-play, loosen the cable adjusting nuts located on the cable mount on the carburetor. With both nuts loosened, the cable may be adjusted to remove slack. Do not adjust the cable beyond the point of no free-play; the engine idle speed will be changed.

Float and Fuel Level

♦ See Figures 3 and 4

1. Remove the top of the carburetor and remove the gasket. The carburetor does not need to be removed for this procedure. Follow the directions within the carburetor overhaul procedure later in this section.
2. Hold the carburetor top and float upside down. Use a float gauge or depth gauge to measure the distance from the bottom of the float (now the top, since it's upside down) to the inside of the carburetor top. The correct distance is 0.76-0.84 in. (19-21mm).
3. If the dimension is not correct, the shim below the needle seat must be changed. Use a thicker shim to increase the measurement (or lower the float level) or a thinner shim to decrease the measurement (raise the float level).
4. The shim kit contains 3 shims: 0.3mm, 0.4mm and 0.5mm. The float measurement will change by 3 times the measurement of the shim installed or removed. Shims may be combined as necessary. Some arithmetic after measuring should provide the correct shim thickness on the first try.
5. To replace the shim, remove the float pin and the float. Remove the needle valve.
6. Use a pair of pliers to gently unscrew the seat at its widest point. Take great care not to damage the seat or the surrounding metal.
7. Slip the shim(s) over the narrow portion of the seat and then reinstall the seat. Gently tighten it without damage.
8. Reassemble the needle valve and float.
9. Re-measure the float height and replace the shim(s) if necessary to get the correct height.
10. Reinstall the carburetor top.

Throttle Opener

1983-87 PICK-UP AND 1983-86 MONTERO

♦ See Figures 5, 6 and 7

1. Disconnect the vacuum hose running to the throttle opener port.
2. Connect a hand vacuum pump to the nipple of the throttle opener. Install a tachometer to read engine speed. Make certain the idle speed for the engine is set correctly.
3. Run the engine at curb idle (fully warm). Apply 11.8 in. Hg (39.8 kPa) with the hand pump; idle speed should increase

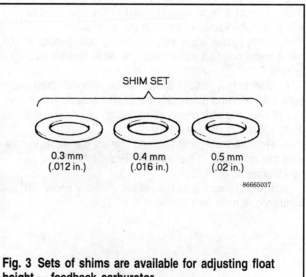

Fig. 3 Sets of shims are available for adjusting float height — feedback carburetor

5-4 FUEL SYSTEM

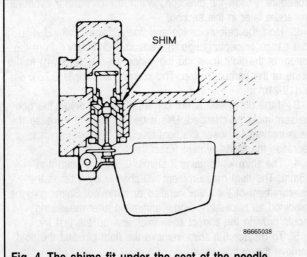

Fig. 4 The shims fit under the seat of the needle valve — feedback carburetor

to 900-950 rpm. If the idle does not increase, replace the dashpot/throttle opener assembly.

 a. Remove the throttle return spring from the throttle lever.

 b. Disconnect the throttle opener rod from the free lever.

 c. Remove the two attaching screws and remove the throttle opener/dashpot assembly.

 d. Install the new unit, connect the rod and install the spring.

4. Reconnect the vacuum hose to the nipple. Start the engine, run it at curb idle and turn the air conditioning on. Idle speed should increase to 900-950 rpm.

5. If the throttle opener needs adjustment, do so by turning the screw on the throttle opener/dashpot assembly. Don't turn any other screws to adjust the throttle opener; the curb idle or other important settings may be affected.

1988-89 PICK-UP AND 1987-90 MONTERO

1. Identify the vacuum control solenoid on the firewall. Label and carefully remove the two vacuum hoses running to the unit.

2. Disconnect the electrical connector from the solenoid.

3. Connect a hand vacuum pump to the solenoid port which held the vacuum hose with the white stripe.

4. Using jumper wires, connect battery voltage to one terminal of the solenoid and connect the other terminal to a good ground.

5. With battery voltage applied to the solenoid, draw vacuum with the hand pump. Use the chart to check the properties of the valve with proper combinations of vacuum and electricity applied or removed.

6. Remove the 12 volt jumpers. Use an ohmmeter to measure the resistance of the solenoid. Resistance should be 40-46 ohms at 68°F (20°C).

7. If the solenoid does not behave correctly under ALL test conditions, it must be replaced.

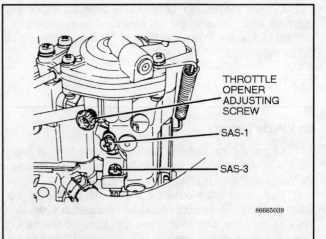

Fig. 5 The location of the throttle opener adjusting screw. Do not adjust SAS 1 or 3 — 1983-87 Pick-ups and 1983-86 Monteros

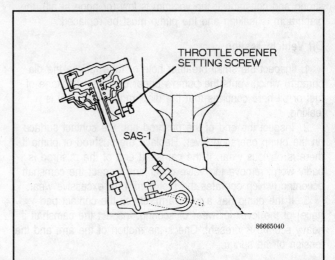

Fig. 6 Adjusting the throttle opener — 1983-87 Pick-ups and 1983-86 Monteros

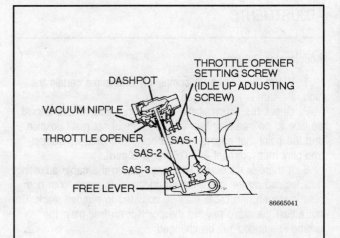

Fig. 7 The location of the throttle opener setting screw — 1983-87 Pick-ups and 1983-86 Monteros

FUEL SYSTEM 5-5

REMOVAL & INSTALLATION

✳︎✳︎CAUTION

Gasoline in either liquid or vapor state is EXTREMELY explosive. Take great care to contain spillage. Work in an open or well-ventilated area. Do not connect or disconnect electrical connectors while fuel hoses are removed or loosened. Observe no smoking/no open flame rules during repairs. Have a dry-chemical fire extinguisher (type B-C) within arm's reach at all times and know how to use it.

Non-Feedback Models

➡ All 1983-84 non-diesel Pick-ups, as well as 1983 Monteros and 1984 non-California Monteros, were equipped with a non-feedback carburetor.

1. Disconnect the negative battery cable.
2. Remove the air cleaner.
3. Place a container under the fuel line inlet fitting to contain spillage. Disconnect the fuel inlet line from the carburetor nipple.
4. Label or diagram each vacuum hose connection; disconnect the hoses from the carburetor ports.
5. Disconnect the throttle cable at the carburetor.
6. Remove the mounting bolts. Lift the carburetor off the engine and remove it to a workbench. Keep it level to avoid spilling fuel from the float bowl.
7. Carefully drain the carburetor into a container with an airtight lid.
8. Before installation, inspect the mating surfaces of the carburetor and manifold. They should be clean and free of nicks, burrs or any pieces of gasket material. Clean the surfaces as necessary and remove any slight imperfections with crocus or emery cloth.
9. Put a new carburetor gasket on the surface of the manifold.
10. Carefully locate the carburetor on top of the gasket with all holes lined up. Install the carburetor bolts, tightening them alternately and evenly in small increments. The gasket must be compressed evenly to prevent leakage.
11. Connect the throttle linkage. Have someone depress the accelerator pedal and make sure the throttle blade opens all the way; it should be exactly vertical at full throttle. Adjust the cable as needed to obtain full opening of the throttle plate.
12. Connect the vacuum hoses according to your drawing or labeling. Make sure all are soft and free of cracks to make a good seal. Replace hoses that are hard or cracked.
13. Reconnect the fuel hoses.
14. Reinstall the air cleaner assembly. Check the filter element and replace it if needed.
15. Connect the negative battery cable.
16. Double check all installation items, paying particular attention to loose hoses or hanging wires, untightened nuts, poor routing of hoses and wires (too tight or rubbing) and tools left in the engine area.
17. Start the engine. It will require a longer cranking period and two or three pumps of the accelerator pedal before it starts.

✳︎✳︎CAUTION

Do NOT prime the engine by pouring fuel into the air horn of the carburetor. This outdated and foolish practice can result in severe injury or damage.

18. While the engine is warming up, check the work area carefully for any sign of fuel or vacuum leaks and attend to them immediately. Set the idle speed and make other necessary adjustments after the engine is running smoothly and is fully warmed up.

Feedback Models

➡ All 1985-89 Pick-ups and Monteros, as well as 1984 California Monteros and 1990 4-cylinder Monteros, were equipped with a feedback carburetor.

1. Disconnect the negative battery cable.

➡ All wires and hoses should be labeled at the time of removal. The amount of time saved during reassembly makes the extra effort well worthwhile.

2. Drain the engine coolant to a point below the level of the intake manifold.

✳︎✳︎CAUTION

When draining the coolant, keep in mind that cats and dogs are attracted by ethylene glycol antifreeze, and are quite likely to drink any that is left in an uncovered container or in puddles on the ground. This will prove fatal in sufficient quantity. Always drain the coolant into a sealable container. Coolant should be reused unless it is contaminated or several years old.

3. Remove the air cleaner assembly, disconnecting the various hoses and ducts as necessary.
4. Disconnect the accelerator cable from the carburetor.
5. Disconnect the coolant hose connection from the back of the carburetor.
6. Disconnect the vacuum hoses.
7. For vehicles with automatic transmissions, remove the lock pin and disconnect the throttle control cable linkage.
8. Disconnect the wiring to the throttle position sensor.
9. Disconnect the wiring to the solenoid valve.
10. Unplug the connector for the carburetor heater. Take care to separate the connectors without pulling on the wiring.
11. Use a small pan or jar to hold under the fuel hose connections. Disconnect them one at a time and catch any spilled fuel.
12. Remove the carburetor retaining bolts and lift the carburetor off the engine. Keep the carburetor level to avoid spillage. Once the carburetor is clear of the car, drain the remaining fuel into a container with an airtight lid.
13. Remove the carburetor base gasket (insulator). If desired, remove the heater element. Handle the heater carefully, protecting it from impact or pulling on the wire.
14. Before installation, inspect the mating surfaces of the carburetor and manifold. They should be clean and free of

5-6 FUEL SYSTEM

nicks, burrs or any pieces of gasket material. Clean the surfaces as necessary and remove any slight imperfections with crocus or emery cloth.

15. Reinstall the heater element and put a new carburetor gasket on the surface of the manifold.

16. Carefully locate the carburetor on top of the gasket with all holes lined up. Install the carburetor bolts, tightening them alternately and evenly in small increments. The gasket must be compressed evenly to prevent leakage. Tighten the bolts to 13 ft. lbs. (18 Nm).

17. Connect the wiring harness to the heater element, the solenoid valve and the throttle position sensor.

18. If equipped with an automatic transmission, connect the throttle control cable and install the locking pin securely.

19. Connect the vacuum hoses, inspecting each one for any sign of cracking or hardening. If a hose is suspect, replace it.

20. Install the coolant hose and secure the clamp.

21. Connect the accelerator cable.

22. Reinstall the air cleaner, connecting the hoses and ductwork as necessary. Make certain the air cleaner body is properly seated on the carburetor and that each hose is firmly attached.

23. Refill the cooling system with coolant.

24. Connect the negative battery cable.

25. Double check all installation items, paying particular attention to loose hoses or hanging wires, untightened nuts, poor routing of hoses and wires (too tight or rubbing) and tools left in the engine area.

26. Start the engine. It will require a longer cranking period and two or three pumps of the accelerator pedal before it starts.

※※CAUTION

Do NOT prime the engine by pouring fuel into the air horn of the carburetor. This outdated and foolish practice can result in severe injury or damage.

27. While the engine is warming up, check the work area carefully for any sign of fuel, coolant or vacuum leaks and attend to them immediately. Set the idle speed and make other necessary adjustments to the accelerator or throttle control cables after the engine is running smoothly and is fully warmed up.

OVERHAUL

Non-Feedback Models

➡The carburetor is comprised of three main sections: the top or float bowl cover, the main body and the throttle body or base. Separating these sections requires the removal and installation of many small parts and fittings. Do not disassemble anything unnecessarily. Some important components are not removed or adjusted during an overhaul.

1. Remove the carburetor from the car, following the instructions earlier in this section. Drain any remaining fuel into a container with an airtight lid. Place the carburetor on the workbench in a clean, dry area. Placing it on a large, lint-free cloth will help prevent parts from getting lost or rolling around.

2. Remove the coolant hose from the throttle body and from the wax element.

3. Using a small hand grinder or similar tool, remove the heads from the two lock screws in the choke cover.

4. Disconnect the fuel cut-off solenoid ground wire from the top of the carburetor.

5. Remove the throttle return spring and the damper spring.

6. Disconnect the vacuum hose running from the depression chamber to the throttle body.

7. Remove the accelerator pump rod from the throttle lever.

8. Remove the dashpot rod (for manual transmission) or the throttle opener rod (automatic transmission) from the free lever.

9. Remove the depression chamber rod from the secondary throttle lever.

10. Remove the six screws from the carburetor top. The four outer ones connect to the main body of the carburetor; the two bolts within the air passage connect to the throttle body.

➡**Many of the screws use Phillips-type heads. Use a screwdriver which fits the head exactly. An improper tool can damage the head of the screw and cause problems during reassembly.**

11. Remove the main body with the top attached (the top cannot come free yet) by lifting straight up. Do not turn the carburetor upside down during the removal; if it is inverted, the accelerator pump check weight, ball and steel ball of the anti-overfill device will fall out.

12. Remove the E-clip from the lower end of the choke unloader rod and disconnect the rod from the lever.

13. Separate the top from the main body.

14. At this point, the carburetor is disassembled enough to perform common overhaul replacements. Do not disassemble any further components without good diagnostic reasons. In particular, do not disassemble the automatic choke system or attempt to remove the throttle plates; both systems require very precise alignment which is beyond the ability of the home mechanic.

15. Remove the float from the float arm by removing the pivot pin.

16. Inspect the float bowl for any sign of particulate dirt or solid matter. Carefully wipe the bowl clean. Shake the float, listening for any sign of liquid fuel inside. If the float has absorbed fuel, it must be replaced.

17. Remove the retaining screw and bracket holding the needle valve. Carefully remove the needle valve and inspect it for uneven wear or pitting. (A magnifying glass is very helpful for checking the tip.) Check the seat for signs of pitting. Don't remove the seat without planning to replace it; if it looks OK, leave it alone. If the seat is to be removed, it must be carefully unscrewed with pliers. It will be difficult to loosen and care must be take not to damage or deform the seat. When the seat is removed, the spacing shim below it must be recovered and reinstalled. This shim determines float level adjustment.

18. Remove the accelerator pump and the fuel cut-off solenoid. Remove the check weight and ball.

19. Wearing eye protection and gloves, carefully clean the fuel and air passages with a spray cleaner and, if available, compressed air. A majority of carburetor problems are caused by very small bits of dirt lodging in the air or fuel passages. Clean everything thoroughly.

FUEL SYSTEM 5-7

20. Inspect the motion of both the choke and throttle plates. They must move smoothly with absolutely no sign of binding or notching. Clean the linkages and plates as necessary, then apply a small amount of lubricant to the pivot points.

21. If any of the fuel jets are to be replaced due to wear or etching, they must be replaced with the identical item. Each jet has a number on the side of it to aid in identification. (The jets are selected based on precise airflow measurements during assembly. Installation of the wrong jet will send the wrong fuel mixture to the engine under almost all conditions).

22. Install the accelerator pump, the fuel cut-off solenoid and the check weight and ball.

23. Install the needle valve and its retainer.

24. Hold the float in position and install the pivot pin.

25. Carefully place the carburetor top onto the main body and install the four retaining screws holding the top to the body.

26. Install the choke unloader rod to the lever and install the E-ring to hold the rod in place.

➡ Be careful that the E-ring does not spring out of place during installation.

27. Install the two screws through the air horn and tighten them.

28. Install the depression chamber rod to the secondary throttle lever.

29. Connect the dashpot or throttle opener rod to the free lever.

30. Install the accelerator pump rod to the throttle lever.

31. Install the vacuum hose between the depression chamber and the throttle body.

32. Install the throttle return spring and the damper spring.

33. Connect the ground wire for the fuel cut-off solenoid.

34. Install new screws to hold the choke cover in place.

35. Install the coolant hose from the throttle body to the wax element.

36. Move the carburetor linkage by hand, checking that motions are smooth and there is no binding in any of the mechanisms.

37. Reinstall the carburetor.

Feedback Models

▶ See Figures 8, 9, 10, 11, 12, 13, 14, 15, 16, 17, 18, 19, 20, 21, 22, 23, 24, 25, 26, 27, 28, 29, 30, 31, 32, 33, 34, 35, 36, 37, 38, 39, 40 and 41

✲✲WARNING

Certain parts or assemblies must not be disassembled or altered during overhaul. The choke plate and shaft, automatic choke linkage, inner venturi, throttle plate and shaft, and fuel inlet nipple must be left alone. Damage and or reduced performance may result from tampering with these components. Many of the screws have Phillips-type heads. Use a screwdriver which fits the head exactly. An improper tool can damage the head of the screw and cause problems during reassembly.

1. Remove the carburetor from the vehicle, following the instructions earlier in this section. Drain any remaining fuel into a container with an airtight lid. Place the carburetor on the workbench in a clean, dry area. Placing it on a large, lint-free cloth will help prevent parts from getting lost or rolling around.

2. Remove the throttle return spring and the damper spring.

3. Remove the throttle opener (automatic transmission) or the dashpot (manual transmission) rod from the free lever and remove the opener or dashpot unit from the top of the carburetor.

4. If equipped with automatic transmission, remove the Idle Speed Control (ISC) servo by removing the bracket screws. Put the ISC servo out of the way until reassembly.

✲✲WARNING

Do not attempt to test the servo with battery voltage. It runs on a lower voltage sent from the ECM. Applying battery voltage to this unit will destroy it.

5. Remove the connector bracket.

6. Remove the vacuum hose running from the base to the choke breaker. This vacuum line will have a delay valve in it.

7. Remove the five screws holding the top of the carburetor to the body and base.

8. The top of the carburetor will be firmly held to the mixing body by the gasket. Do not attempt to lift the top by hand. Use a screwdriver blade or similar thin, flat tool inserted between the top and the enrichment cover. Lightly pry the top upwards and lift the top slowly. Do not apply excessive force and don't try to rush the job.

9. Remove the float pivot pin and remove the float. Carefully remove the needle valve.

✲✲WARNING

Do not let the float drop and do not apply any force to the float. The needle valve controlled by the float will be damaged.

10. The valve seat may be removed by using two small, flat-bladed screwdrivers to gently lever the seat upwards and out of position. Use care not to damage the surrounding area or the seat mounts during this process.

11. Find the electrical connector for the Feedback Solenoid Valve (FBSV). The terminals must be removed from the connector housing before the valve can be removed. Use a very thin, flat tool (such as a jeweler's screwdriver) inserted into the connector to loosen the stopper and remove each terminal.

12. Remove the grommet from the top of the carburetor. Remove the retainer and remove the FBSV attaching screw; remove the feedback solenoid valve.

13. Remove the retainer and remove the Slow Cut Solenoid Valve (SCSV) from the carburetor top. Hold the solenoid by the body and avoid pulling on the wiring.

14. Using the same small screwdriver technique as in Step 11, disconnect the SCSV terminals from the plastic connector body and remove the SCSV from the carburetor.

15. Use a hand grinder or similar tool to remove the heads of the rivets holding the cover of the choke assembly. Remove the small screw in the bottom of the cover.

16. Remove the packing (gasket), bimetal assembly and plate.

5-8 FUEL SYSTEM

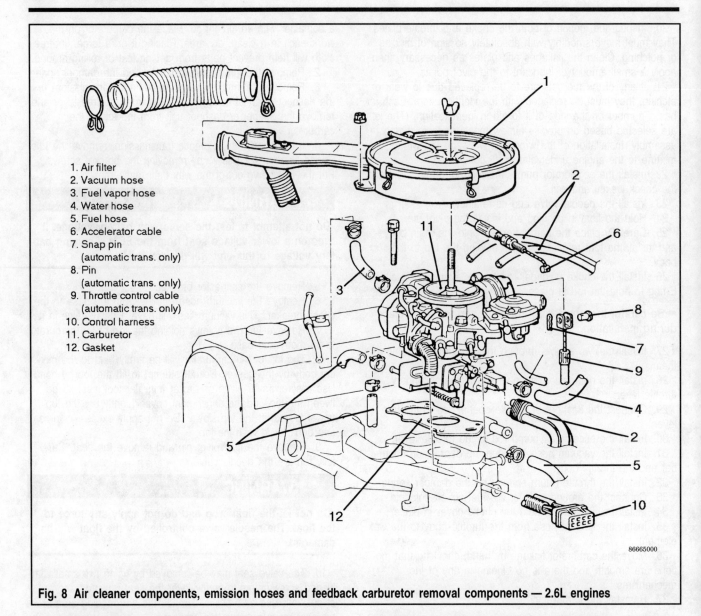

Fig. 8 Air cleaner components, emission hoses and feedback carburetor removal components — 2.6L engines

17. Using a pin punch or similar tool, remove the remainder of the rivets from each hole. Take care not to damage the surrounding material.
18. Remove the bimetal terminal from the wiring connector.
19. Remove the bowl vent valve.

➡There are small springs within this unit. Take note of their location and placement, and be careful not to lose them.

20. Remove the choke breaker cover.
21. Remove the check weight and its ball and remove the steel ball from the anti-overfill device.
22. Remove the accelerator pump rod from the throttle shaft lever.
23. Remove the screw from the throttle body assembly, taking care not to raise burrs on the head of the screw. Any deformation will prevent the base from mating to the manifold properly.
24. Using a screwdriver which exactly matches the groove, remove the main jets.

25. Remove the accelerator pump mounting screws and remove the pump cover link assembly, the diaphragm, spring, pump body and gasket from the carburetor body.
26. Remove the three attaching screws from the enrichment valve and remove the cover, spring, and diaphragm assembly from the main body of the carburetor.
27. Remove the vacuum hose running between the depression chamber and the throttle body.
28. Disconnect the depression chamber rod from the secondary throttle lever. Unbolt and remove the depression chamber.
29. Using a screwdriver that exactly matches the screw heads, remove the throttle position sensor from the throttle body (base) of the carburetor.
30. Wearing eye protection and gloves, carefully clean the fuel and air passages with a spray cleaner and, if available, compressed air. A majority of carburetor problems are caused by very small bits of dirt lodging in the air or fuel passages. Do NOT use metal wire or similar to clean the passages.
31. Check the diaphragms carefully for any sign of damage or cracking.

FUEL SYSTEM 5-9

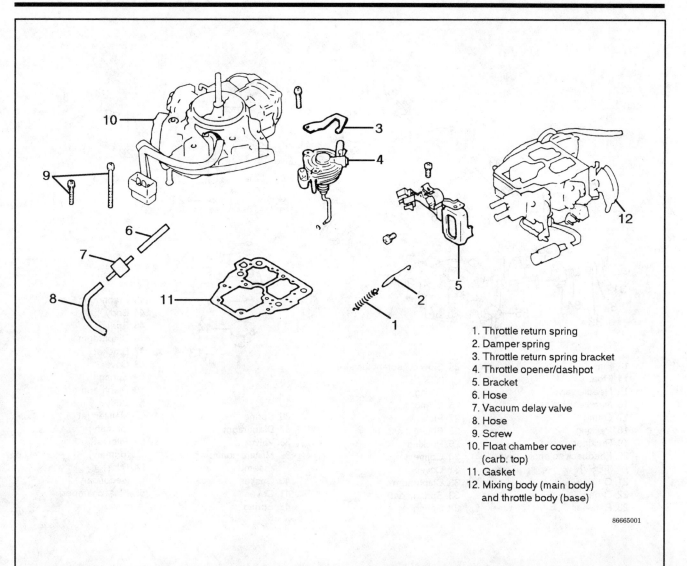

1. Throttle return spring
2. Damper spring
3. Throttle return spring bracket
4. Throttle opener/dashpot
5. Bracket
6. Hose
7. Vacuum delay valve
8. Hose
9. Screw
10. Float chamber cover (carb. top)
11. Gasket
12. Mixing body (main body) and throttle body (base)

Fig. 9 Components removed when separating the float chamber from the mixing body of a feedback carburetor — 2.6L engines

32. Check the operation of the needle valve; it should move lightly and smoothly. If any binding is felt, replace it.
33. Check the fuel inlet filter (above the needle valve) for clogging.
34. Check the float for cracks, deformation or internal leakage.
35. Inspect the motion of the various linkages and pivots. If any binding is felt, clean the system thoroughly and apply a light coat of lubricant.
36. Check the operation of both solenoid valves. Apply battery voltage to the terminals; the solenoid should operate with a distinct click each time power is applied or removed. Inspect the tip of the FBSV to insure the jet is open and clean.
37. Use an ohmmeter to check the solenoids' resistance. The SCSV should have a resistance of 48-60 ohms at 68°F (20°C); the FBSV should have 54-66 ohms resistance at the same temperature.

➡**Resistance will increase or decrease as the temperature rises or falls. Make common sense allowances for the temperature in which you are testing the units.**

38. Hold one lead of the ohmmeter against the case of each solenoid and touch the other lead to each terminal. There should NOT be continuity between the case and the terminals.
39. Use the ohmmeter to check the bimetal assembly. Connect one lead to the wire terminal and the other lead to the body of the assembly. Correct resistance is approximately 6 ohms at 68°F (20°C).
40. The dashpot should be inspected by pulling outward on the rod. Resistance should be felt; when released, the lever should return quickly to its original position.
41. The depression chamber is checked by pushing the rod all the way into the unit and then blocking the vacuum port firmly with a finger. Release the rod; if it stays in place (with the vacuum port blocked), the unit is good. If the rod returns to its original extended position, the diaphragm inside has failed.

5-10 FUEL SYSTEM

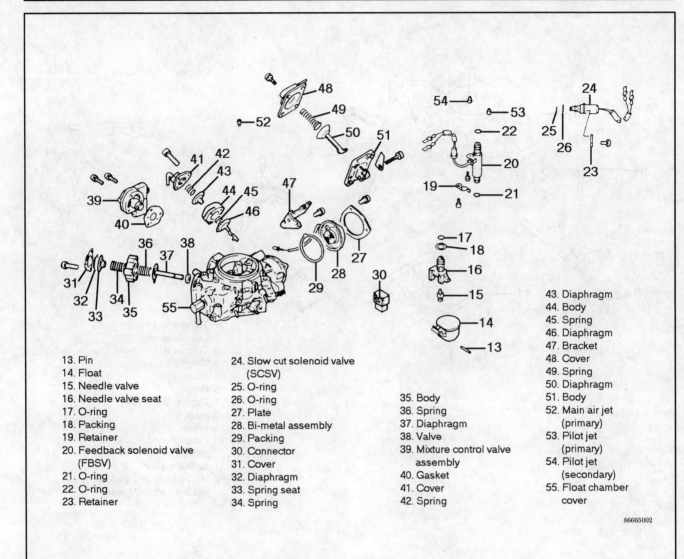

Fig. 10 Feedback carburetor float chamber disassembly components — 2.6L engines

13. Pin
14. Float
15. Needle valve
16. Needle valve seat
17. O-ring
18. Packing
19. Retainer
20. Feedback solenoid valve (FBSV)
21. O-ring
22. O-ring
23. Retainer
24. Slow cut solenoid valve (SCSV)
25. O-ring
26. O-ring
27. Plate
28. Bi-metal assembly
29. Packing
30. Connector
31. Cover
32. Diaphragm
33. Spring seat
34. Spring
35. Body
36. Spring
37. Diaphragm
38. Valve
39. Mixture control valve assembly
40. Gasket
41. Cover
42. Spring
43. Diaphragm
44. Body
45. Spring
46. Diaphragm
47. Bracket
48. Cover
49. Spring
50. Diaphragm
51. Body
52. Main air jet (primary)
53. Pilot jet (primary)
54. Pilot jet (secondary)
55. Float chamber cover

42. Before reassembly, make certain that all parts are clean and dry. Any gasket or O-ring which was removed MUST be replaced with a new one during reinstallation.

➡ If the jets are to be replaced, the new ones must be exact replacements. The jets have a number stamped on them for identification. Jet size is selected based on sophisticated air flow measurements during assembly of the carburetor; changing the jets will lead to extreme driveability and emission problems.

To assemble:

43. Install the throttle position sensor onto the throttle body and tighten the screws without causing damage.
44. Install the depression chamber. Connect the chamber rod to the secondary throttle lever.
45. Install the vacuum hose from the depression chamber to the throttle body.
46. Assemble the diaphragm, spring and cover for the enrichment valve. Install the valve and cover assembly onto the main body.
47. Assemble the accelerator pump and install it to the body with a new gasket.
48. Install the main jets without damaging them.
49. Install the screw into the throttle body assembly without damaging the head of the screw.
50. Install the accelerator pump rod to the throttle shaft lever.
51. Install the check weight and ball and the steel ball for the anti-overfill device.
52. Install the choke breaker cover.
53. Install the bowl vent valve, taking care to assemble the small springs and diaphragm correctly.
54. Loosely hold the bimetal choke assembly in place. Route the terminal and wiring correctly to the plastic connector. Install the terminal in the connector by pushing it into the correct location. The stopper pin will engage the terminal automatically. A slight click may be heard or felt when the terminal is in position.

FUEL SYSTEM 5-11

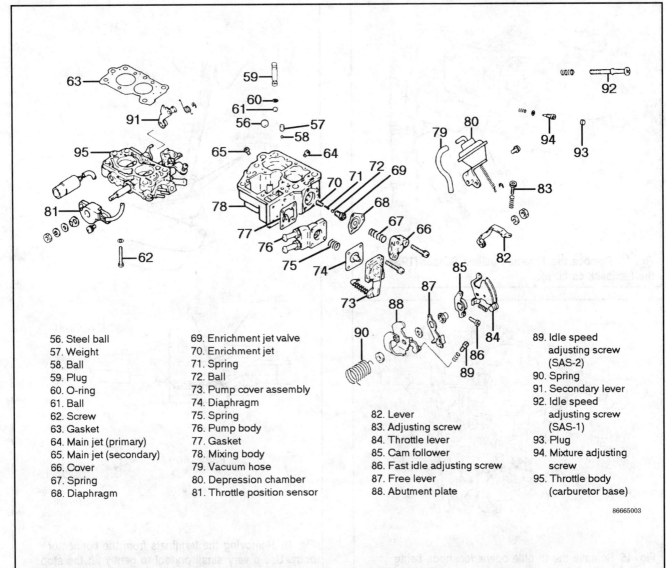

56. Steel ball
57. Weight
58. Ball
59. Plug
60. O-ring
61. Ball
62. Screw
63. Gasket
64. Main jet (primary)
65. Main jet (secondary)
66. Cover
67. Spring
68. Diaphragm
69. Enrichment jet valve
70. Enrichment jet
71. Spring
72. Ball
73. Pump cover assembly
74. Diaphragm
75. Spring
76. Pump body
77. Gasket
78. Mixing body
79. Vacuum hose
80. Depression chamber
81. Throttle position sensor
82. Lever
83. Adjusting screw
84. Throttle lever
85. Cam follower
86. Fast idle adjusting screw
87. Free lever
88. Abutment plate
89. Idle speed adjusting screw (SAS-2)
90. Spring
91. Secondary lever
92. Idle speed adjusting screw (SAS-1)
93. Plug
94. Mixture adjusting screw
95. Throttle body (carburetor base)

Fig. 11 Feedback carburetor mixing body and throttle body disassembly components — 2.6L engines

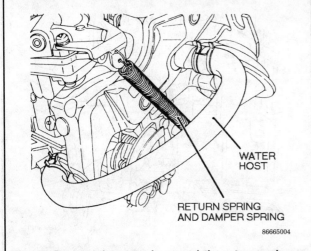

Fig. 12 Remove the water hose and the return and damper springs — feedback carburetor

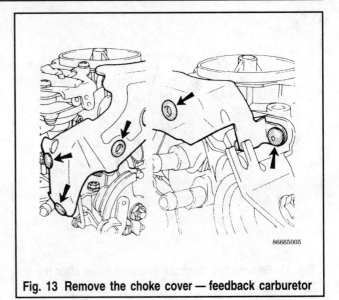

Fig. 13 Remove the choke cover — feedback carburetor

5-12 FUEL SYSTEM

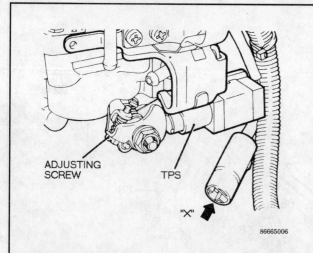

Fig. 14 Remove the Throttle Position Sensor (TPS) from the feedback carburetor

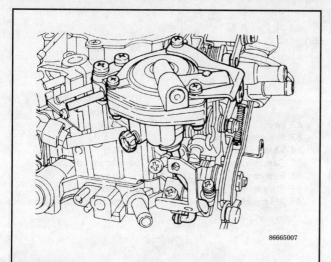

Fig. 15 Remove the throttle opener/dashpot, being careful not to bend the actuating rods

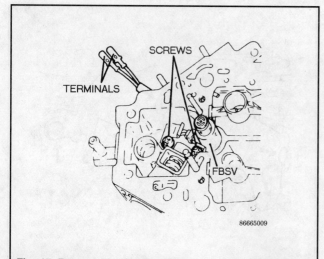

Fig. 17 Remove the feedback solenoid valve after the terminals are free

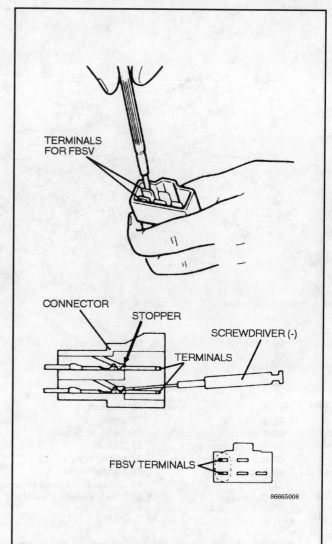

Fig. 16 Removing the terminals from the connector body. Use a very small prytool to gently lift the stop tab within the plastic shell

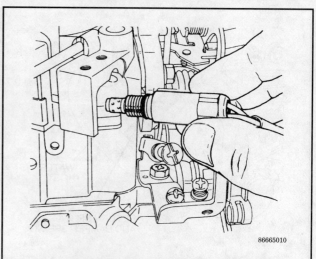

Fig. 18 Removing the solenoid valve — feedback carburetor

FUEL SYSTEM 5-13

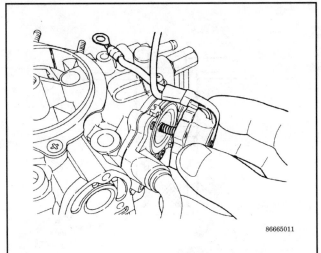

Fig. 19 Removing the bowl vent solenoid — feedback carburetor

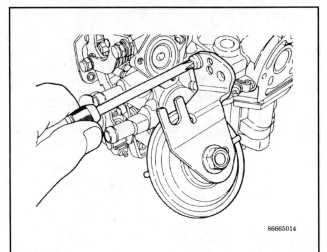

Fig. 22 Removing the depression chamber — feedback carburetor

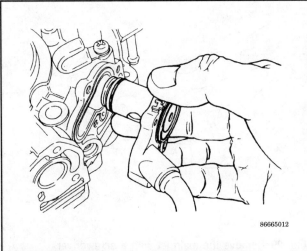

Fig. 20 Removing the Bowl Vent Valve (BVV) — feedback carburetor

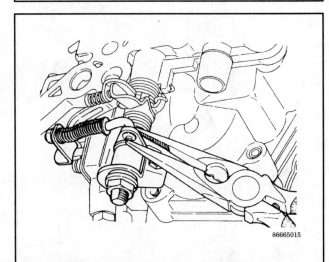

Fig. 23 Removing the accelerator pump rod — feedback carburetor

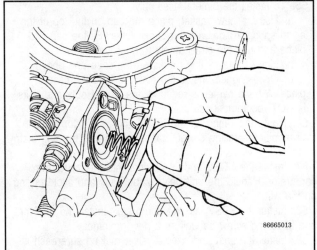

Fig. 21 Removing the choke breaker cover — feedback carburetor

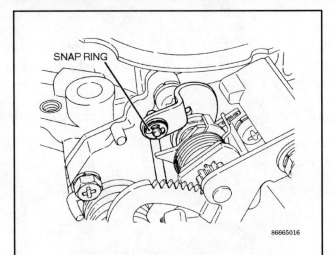

Fig. 24 Remove the snapring and then the choke rod — feedback carburetor

5-14 FUEL SYSTEM

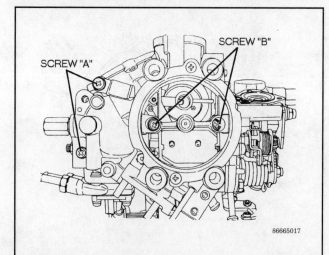

Fig. 25 Remove the float chamber cover — feedback carburetor

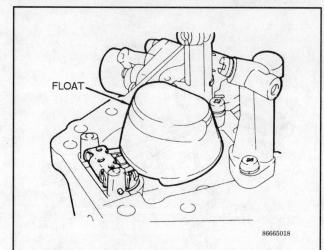

Fig. 26 Remove the pin, then remove the float and needle — feedback carburetor

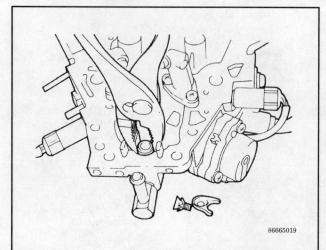

Fig. 27 Use a pair of pliers to remove the needle seat — feedback carburetor

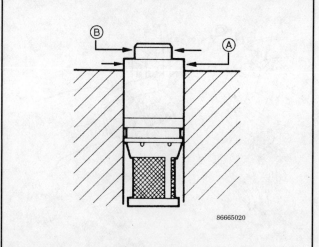

Fig. 28 When removing the needle seat, grasp it by area A, not area B — feedback carburetor

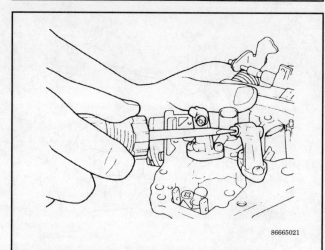

Fig. 29 Remove the main jet with a screwdriver equipped with the proper-sized blade for the main jet slot — feedback carburetor

55. To install the bimetal assembly:
   a. Using a new gasket, place the cup on the top of the spiral spring in line with the choke lever. Fit the cap into place and use the small screw to hold it in place.
   b. Line up the mating marks on the case and body.
   c. Once aligned, install the rivets to hold the case in place. A hand riveter is required; the use of nuts and bolts is NOT recommended.
56. Route the wiring for the FBSV and SCSV correctly. Install terminals into the correct connector port by pushing them in firmly.
57. Install the SCSV and its retainer into the top of the carburetor. Handle the unit only by the body and avoid pulling on the wire.
58. Install the FBSV into the carburetor. Install the retainer and attaching screw as well as a new grommet.
59. Install the seat and needle valve, making sure each is correctly placed and securely installed.

FUEL SYSTEM 5-15

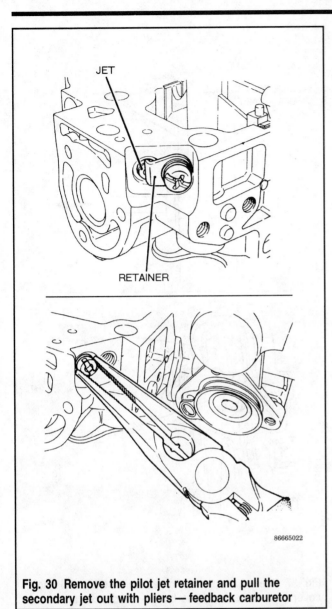

Fig. 30 Remove the pilot jet retainer and pull the secondary jet out with pliers — feedback carburetor

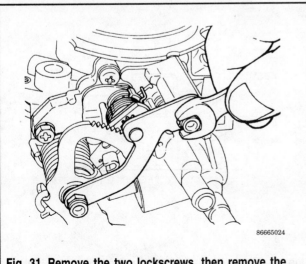

Fig. 31 Remove the two lockscrews, then remove the choke pinion assembly — feedback carburetor

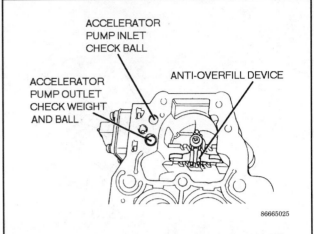

Fig. 32 Remove the check weight and ball, as well as the steel ball of the anti-overfill device — feedback carburetor

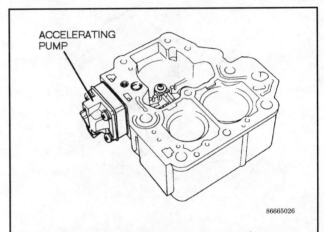

Fig. 33 Remove the accelerator pump mounting screws, then remove the pump cover link assembly, diaphragm, spring, body and gasket from the main body — feedback carburetor

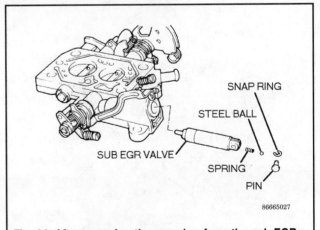

Fig. 34 After removing the snapring from the sub-EGR control valve pin, remove the pin and the link from the valve. Take the little steel ball and spring out of the sub-EGR control valve — feedback carburetor

5-16 FUEL SYSTEM

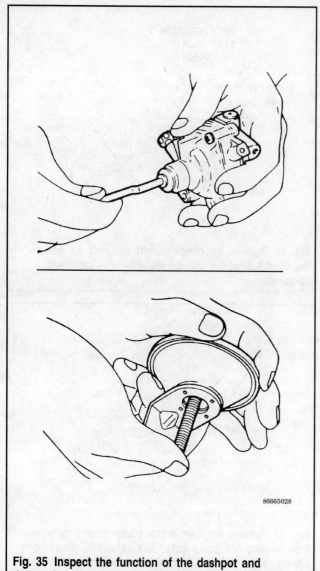

Fig. 35 Inspect the function of the dashpot and depression chamber — feedback carburetor

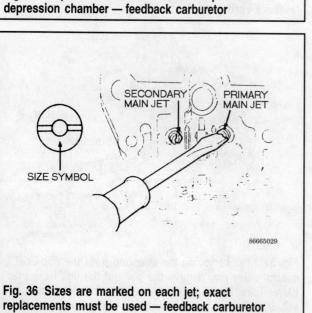

Fig. 36 Sizes are marked on each jet; exact replacements must be used — feedback carburetor

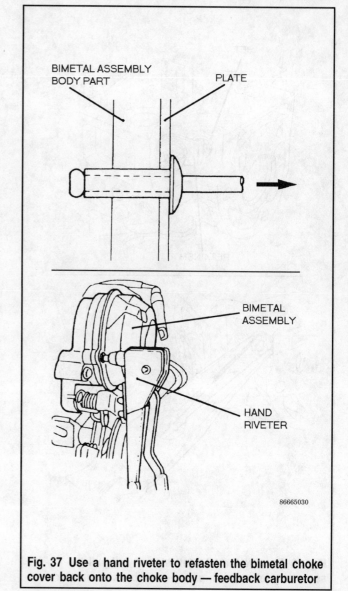

Fig. 37 Use a hand riveter to refasten the bimetal choke cover back onto the choke body — feedback carburetor

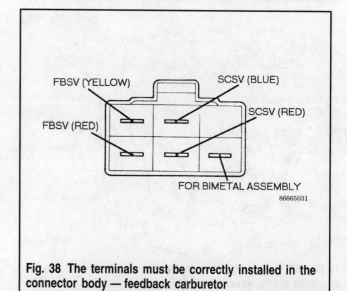

Fig. 38 The terminals must be correctly installed in the connector body — feedback carburetor

FUEL SYSTEM 5-17

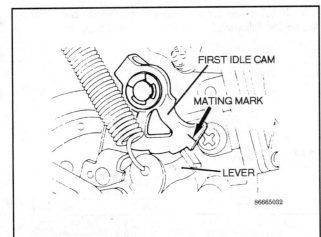

Fig. 39 The throttle plate clearance must be checked after the high idle cam is correctly set — feedback carburetor

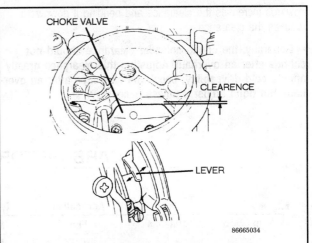

Fig. 41 Check the choke plate clearance with the throttle wide open; make needed adjustments by carefully bending the lever — feedback carburetor

60. Install the float and pivot pin, taking great care not to put any undue force on the float. Refer to the carburetor adjustment section for float level adjustment procedure.

61. Install the carburetor top to the main body (with a new gasket) and install the five screws. Make sure each screw is tight without deforming the head.

62. Install the vacuum hose and delay valve running from the base of the carburetor to the choke breaker.

63. Install the connector bracket.

64. Install the ISC servo if it was removed (automatic transmission only).

65. Install either the dashpot or throttle opener unit and connect the rod to the free lever.

66. Install the throttle return spring and damper spring.

67. Move the carburetor linkages by hand, checking that motions are smooth and there is no binding in any of the mechanisms.

68. With the carburetor correctly assembled, some adjustments must be made on the bench. Set the high idle cam to the second highest position and turn the carburetor upside down. Using a drill bit of known diameter or a clearance tool, check the clearance between the primary throttle plate and the throttle bore. Correct clearances are:

- 2.0L engine with manual transmission: 0.025 in. (0.63mm)
- 2.0L engine with automatic transmission: 0.028 in. (0.71mm)
- 2.6L engine with manual transmission: 0.028 in. (0.71mm)
- 2.6L engine with automatic transmission: 0.031 in. (0.80mm)

69. Check the choke unloader clearance by using your finger to lightly press and set the choke plate. When it is fully closed, move the throttle linkage to open the throttle plate(s) all the way; the throttle plates should be vertical in their bores. Measure the clearance between the choke plate and the choke bore. Correct clearance is 0.079 in. (2mm) for Pick-ups and Monteros. If adjustment is needed, gently bend the throttle lever to achieve the correct clearance. Bending the lever

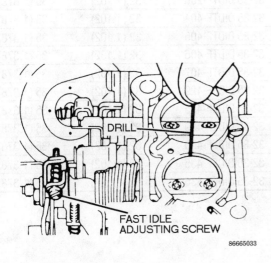

Fig. 40 Use a drill or feeler gauge to check the throttle plate clearance — feedback carburetor

5-18 FUEL SYSTEM

upwards increases the clearance and bending it downward reduces the clearance.

➥ Generally, the choke unloader clearance should not change after an overhaul. Adjusting this clearance greatly affects cold driveability; check the clearance after an overhaul, but don't adjust it unless necessary.

70. Install a new base gasket and reinstall the carburetor on the engine, following instructions given previously in this section.

CARBURETOR CHART–PICK-UP

Engine	Year	Application	Type	Model No.	Throttle Bore mm (in.) Primary	Secondary
Pick-up G63B	1983–84	Fed	N	—	32 (1.260)	35 (1.378)
		Cal	N	—	32 (1.260)	35 (1.378)
	1985–86	M/T	F	32-35 DIDTF-205	32 (1.260)	35 (1.378)
		A/T	F	32-35 DIDTF-206	32 (1.260)	35 (1.378)
	1987–89	Fed-M/T	F	32-35 DIDEF-400	32 (1.260)	35 (1.378)
		Fed-A/T	F	32-35 DIDEF-401	32 (1.260)	35 (1.378)
		Cal-M/T	F	32-35 DIDEF-402	32 (1.260)	35 (1.378)
		Cal-A/T	F	32-35 DIDEF-403	32 (1.260)	35 (1.378)
Pick-up G54B	1983–84	Fed	N	—	32 (1.102)	35 (1.378)
		Cal	N	—	32 (1.102)	35 (1.378)
	1985–86	M/T	F	32-35 DIDTF-207	32 (1.102)	35 (1.378)
		A/T	F	32-35 DIDTF-208	32 (1.102)	35 (1.378)
	1987	Fed-M/T	F	32-35 DIDTF-404	32 (1.102)	35 (1.378)
		Cal-A/T	F	32-35 DIDTF-405	32 (1.102)	35 (1.378)
		Fed-M/T	F	32-35 DIDTF-406	32 (1.102)	35 (1.378)
		Cal-A/T	F	32-35 DIDTF-407	32 (1.102)	35 (1.378)
	1988	Fed	F	32-35 DIDEF-429	32 (1.102)	35 (1.378)
		Cal/Fed	F	32-35 DIDEF-435	32 (1.102)	35 (1.378)
	1989	Fed M/T	F	32-35 DIDEF-429	32 (1.102)	35 (1.378)
		Fed A/T	F	32-35 DIDEF-430	32 (1.102)	35 (1.378)
		Cal M/T	F	32-35 DIDEF-435	32 (1.102)	35 (1.378)
		Cal A/T	F	32-35 DIDEF-436	32 (1.102)	35 (1.378)

N NON - FEEDBACK CARBURETOR
F FEEDBACK CARBURETOR

CARBURETOR CHART – MONTERO

Engine	Year	Application	Type	Model Number	Throttle Bore mm (in.) Primary	Secondary
Montero G54B	1983	All	N	32-35 DIDTA-106	32 (1.260)	35 (1.378)
	1984	Fed HA, M/T	N	32-35 DIDTA-170	32 (1.260)	35 (1.378)
		Fed HA, A/T	N	32-35 DIDTA-171	32 (1.260)	35 (1.378)
		Fed M/T	N	32-35 DIDTA-186	32 (1.260)	35 (1.378)
		Fed A/T	N	32-35 DIDTA-187	32 (1.260)	35 (1.378)
		Cal M/T	F	32-35 DIDTA-184	32 (1.260)	35 (1.378)
		Cal A/T	F	32-35 DIDTA-185	32 (1.260)	35 (1.378)
	1985-86	M/T	F	32-35 DIDTF-209	32 (1.260)	35 (1.378)
		A/T	F	32-35 DIDTF-210	32 (1.260)	35 (1.378)
	1987	Fed M/T	F	32-35 DIDEF-410	32 (1.260)	35 (1.378)
		Fed A/T	F	32-35 DIDEF-411	32 (1.260)	35 (1.378)
		Cal M/T	F	32-35 DIDEF-412	32 (1.260)	35 (1.378)
		Cal A/T	F	32-35 DIDEF-413	32 (1.260)	35 (1.378)
	1988	Fed M/T	F	32-35 DIDEF-431	32 (1.260)	35 (1.378)
		Fed A/T	F	32-35 DIDEF-432	32 (1.260)	35 (1.378)
		Cal M/T	F	32-35 DIDEF-433	32 (1.260)	35 (1.378)
		Cal A/T	F	32-35 DIDEF-434	32 (1.260)	35 (1.378)
	1989-90	Fed	F	32-35 DIDEF-431	32 (1.260)	35 (1.378)
		Cal	F	32-35 DIDEF-441	32 (1.260)	35 (1.378)

N NON - FEEDBACK CARBURETOR
F FEEDBACK CARBURETOR

MULTI-PORT FUEL INJECTION SYSTEM (MFI)

Relieving Fuel System Pressure

The Multi-port Fuel Injection (MFI) system incorporate an electric fuel pump to create high pressures in the fuel lines. This pressure is needed to adequately feed the fuel through the injectors, which sufficiently vaporize the fuel in the runners of the intake manifold. Since the fuel pump creates such high pressures, the MFI system incorporates special fuel hoses designed to withstand these pressures.

Because of the systems high operating pressures, whenever the system is opened to the outside (loosening fuel lines or similar fuel components) the high pressure must first be relieved. The following procedure will lessen the fuel system pressure so that work can be accomplished:

1. Remove the cover plate from over the fuel pump. The fuel pump cover plate is located under the rear carpeting in the trunk area of the truck or Montero.

2. Label and unplug the harness connectors to the electric fuel pump.
3. Start the vehicle and allow it to idle until it stalls on its own.
4. Turn the ignition switch to **OFF**.
5. Remove the negative (-) battery cable form the negative (-) terminal of the battery.
6. The fuel system pressure should now be lowered to normal, however, when loosening fuel lines or related components have rags close at hand to soak up any residual gasoline in the system.

Electric Fuel Pump

Mechanical fuel pumps are generally not capable of delivering the high pressures required for fuel injection. Because of alternating motion of the pump arm, the mechanical pump is prone to wear and failure. An electric motor, driving a rotary or

5-20 FUEL SYSTEM

vane-type pump, provides a constant speed and pressure in the fuel line.

REMOVAL & INSTALLATION

▶ See Figures 42 and 43

✶✶WARNING

In almost every case, replacing the Mitsubishi electric fuel pump requires removal of the fuel tank. Please refer to the fuel tank removal and installation procedure at the end of this section.

1. Lift or move aside the carpet in the rear cargo area.
2. Remove the oval cover plate from the access hole.
3. Disconnect the fuel pump harness connector at the rear of the fuel tank.
4. Start the engine, allowing it to run until it runs out of fuel. This relieves pressure within the fuel system.
5. Turn the ignition switch **OFF** and disconnect the negative battery cable.
6. Remove the fuel filler cap.
7. Drain the fuel from the fuel tank into suitable container(s), each with an airtight lid.
8. Carefully, and using two wrenches on the fittings, disconnect the high pressure fuel line and the main fuel line from the pump. Remove the high pressure fuel hose at the body-side main pipe connection, unfasten the pump-side connection. Wrap each joint in a rag before loosening; some pressure may remain within the system.
9. Remove the fuel pump retaining nuts and carefully remove the pump from the tank.

➡ The fuel pump may still contain liquid fuel. Drain it into a suitable container with an airtight lid before performing any tests or inspections.

To install:

10. Carefully install the pump into the tank and install the 6 retaining nuts onto the studs. Tighten the nuts evenly and in a crisscross pattern to 12 ft. lbs. (16 Nm).
11. Reinstall the high pressure and supply line to the pump.

✶✶WARNING

Take great care when connecting the threaded high pressure lines. Start the threads by hand, then, while holding the hose side to keep it from turning, tighten the flare nut to 27 ft. lbs. (37 Nm).

12. Connect the pump wiring harness.
13. Connect the negative battery cable. Turn the ignition key to **ON** but do not start the engine. You should hear the pump running. Allow it to run for about 10 seconds to build line pressure, then start the engine.
14. After the engine has run smoothly for a minute or two, shut the ignition off and check the work area for any trace of leakage. Attend to any fuel leaks immediately, remembering that the system has repressurized and must be handled safely.
15. Reinstall the cover on the access hole and replace the carpet in the cargo area.

TESTING

When diagnosing engine problems, particularly a "no start" condition, the fuel pump function should be checked. Because of the difficulty in reaching the fuel pump connections, Mitsubishi automobiles have a conveniently located test connector under the hood. V6 Monteros have the connector taped to a wiring harness inside the right front kick panel, near the ECM and relays.

This connector bypasses all of the controls in the system (ignition switch, pump relay, etc.) and sends voltage directly to the pump. When power is applied to the pump you should be able to hear it running, although the in-tank pumps require you remove the fuel filler cap and listen at the filler.

To use the test connector, make certain the ignition is **OFF**. Great damage may be caused if the system is tested with the key on. Locate the test connector in the engine compartment (on later Monteros it is located under the right-hand side of the dashboard) — it is generally on a short lead out of a wiring harness and looks like a plug that isn't connected to anything. Once found, use a jumper wire to connect the test terminal to the positive battery terminal. Listen for the fuel pump; if it runs, you know the pump motor is good. While the pump is running, gently squeeze one of the fuel lines and confirm that the pump is delivering fuel pressure. If it does not run, the pump itself is most likely defective OR the connector to the pump is faulty.

When the pump is removed from the tank, it may be checked by applying battery voltage (+) to the connector and grounding the housing. Only run the pump for one or two seconds during this test; the pump can be damaged by running without liquid.

✶✶CAUTION

Make certain the pump has been drained of residual fuel and is free of fuel vapors. Connecting and removing the test leads will cause sparks which can ignite any vapor in the area.

Additionally, the pump motor may be checked with an ohmmeter. Some resistance should be seen on the meter scale. If the meter shows no motion (infinite), the motor has an internal fault and the pump must be replaced.

Throttle Body

On the Multi-port Fuel Injection (MFI) system, each injector is mounted directly into the head at the cylinder. On this system, the throttle body controls the amount of air entering the intake system, but does not meter fuel in any way.

In this section, the term throttle body will be used to indicate the MFI system.

FUEL SYSTEM   5-21

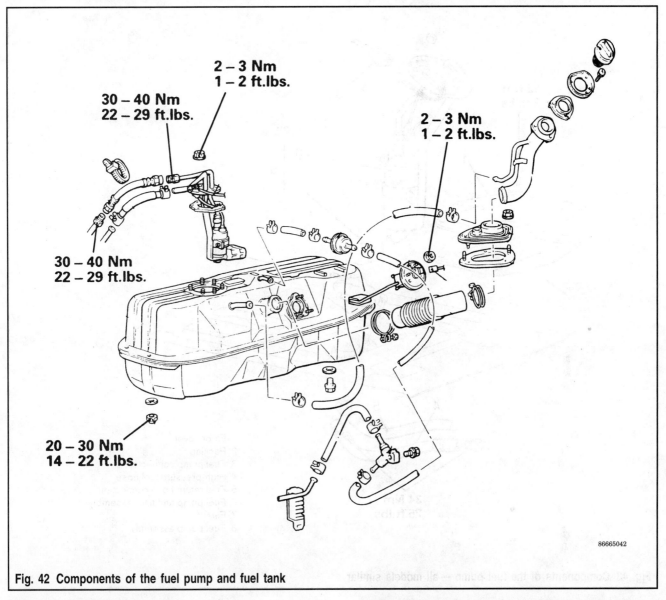

Fig. 42 Components of the fuel pump and fuel tank

REMOVAL & INSTALLATION

▶ See Figures 44 and 45

> ※※**CAUTION**
>
> Gasoline in either liquid or vapor state is EXTREMELY explosive. Take great care to contain spillage. Work in an open or well-ventilated area. Do not connect or disconnect electrical connectors while fuel hoses are removed or loosened. Observe smoking/no open flame rules during repairs. Have a dry-chemical fire extinguisher (type B-C) within arm's reach at all times and know how to use it.

➡ The throttle body for each of the MFI engines is slightly different. The procedure below is general and may require slight alteration of sequence depending on the engine.

1. Disconnect the negative battery cable.
2. Drain the coolant, at least to a level below the intake manifold.

> ※※**CAUTION**
>
> When draining the coolant, keep in mind that cats and dogs are attracted by ethylene glycol antifreeze, and are quite likely to drink any that is left in an uncovered container or in puddles on the ground. This will prove fatal in sufficient quantity. Always drain the coolant into a sealable container. Coolant should be reused unless it is contaminated or several years old.

3. Disconnect the main air intake duct from the throttle body.
4. Label and disconnect the vacuum and breather hoses running to the throttle body. Take care that clamps are not bent or distorted during removal.
5. Disconnect the accelerator cable and the cruise control cable if so equipped.

5-22 FUEL SYSTEM

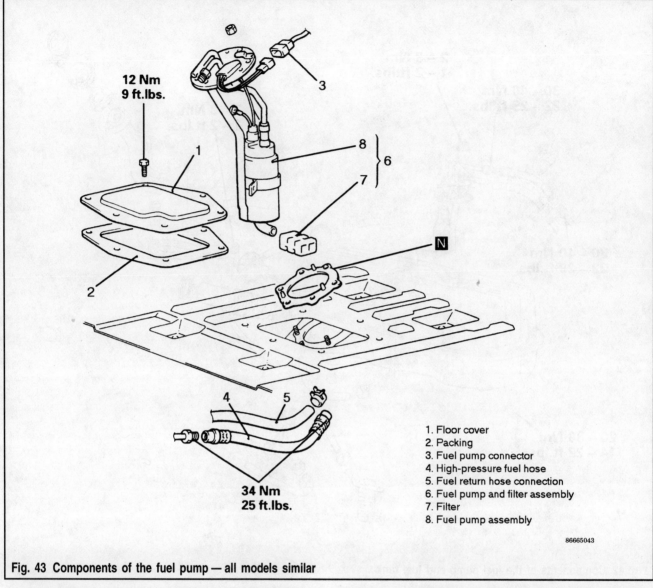

1. Floor cover
2. Packing
3. Fuel pump connector
4. High-pressure fuel hose
5. Fuel return hose connection
6. Fuel pump and filter assembly
7. Filter
8. Fuel pump assembly

Fig. 43 Components of the fuel pump — all models similar

6. Label and disconnect the coolant hoses running to the throttle body.

7. Follow each wire running from the throttle body. Label and disconnect each at its connector. Most of the wiring harnesses have connectors at some distance from the throttle body. Loosen, release or remove any clips or brackets holding the throttle body harnesses in place.

8. Carefully remove the four nuts and bolts holding the throttle body to the manifold. The throttle body may use bolts of two different lengths; take careful note of each bolt's position.

9. Lift the throttle body away from the manifold and handle it carefully. Place it in a protected location.

10. If the unit is being replaced, compare the old and the new ones. Look for any components which need to be transferred from the old unit. The throttle plate area may be cleaned with a spray cleaner, but must be completely dry before installation. Disassembly of the throttle body is not recommended.

11. Before reinstalling, make sure that all remains of the old gasket are removed from both the manifold flange and the base of the throttle body. Place a new gasket on the manifold and hold the throttle body in place. Don't forget the support bracket on the twin cam engines.

12. Install the four nuts and bolts finger tight. Make certain the bolts are in the correct holes by length. Tighten the bolts evenly, alternating from bolt to bolt, in small increments. The bolts must draw down evenly, creating an airtight seal against the gasket.

13. For all engines, tighten the nuts and bolts to the following torque figures. Do NOT overtighten these bolts. The figures are:

- 1990-91 2.4L Engines — 11-16 ft. lbs. (15-22 Nm)
- 1992 2.4L Engines — 9 ft. lbs. (12 Nm)
- 1993-95 2.4L Engines — 14 ft. lbs. (19 Nm)
- 1989-91 3.0L Engines — 7-9 ft. lbs. (10-13 Nm)
- 1992-95 3.0L Engines — 10 ft. lbs. (14 Nm)
- 1994-95 3.5L DOHC Engines — 8 ft. lbs. (12 Nm)

14. Reconnect the wiring connectors to the harnesses. Secure, tighten or reinstall any wire clips or retainers. The harnesses must be kept clear of moving or hot surfaces.

FUEL SYSTEM 5-23

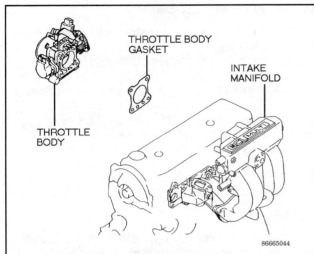

Fig. 44 Orientation of the throttle body to the intake manifold plenum — 2.4L engines

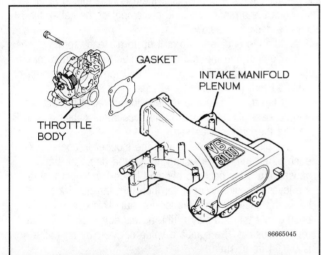

Fig. 45 Orientation of the throttle body to the intake manifold plenum — 3.0L engines

15. Install the coolant hoses, making sure each is properly clamped to its port.
16. Install the accelerator cable and cruise control cable if so equipped. Adjust the cable to the correct tension and make certain the lock nuts or bracket bolts are secure.
17. Connect the vacuum hoses. Examine the end of each and replace any showing signs of cracking or hardening.
18. Install the main air duct and connect the breather tubes.
19. Refill the coolant to the proper level.
20. Connect the negative battery cable.

ADJUSTMENTS

The throttle body itself does not need any adjustments, however, the throttle body does hold or contain many items which do need occasional adjusting, such as: the throttle position sensor, the idle speed control motor, the idle air control motor, the accelerator position sensor, etc. For adjusting procedures regarding these components, please refer to Section 4 in this manual.

The throttle body also holds the accelerator cable, which needs to be adjusted occasionally upon reinstallation. For accelerator cable adjustment, use the following procedures:

Models Without Cruise Control

♦ See Figures 46, 47 and 48

2.4L ENGINE

➡Adjust the accelerator cable with no load on the engine or electrical system

1. Turn the air conditioning and lamps off.
2. Warm the engine until stabilized at idle.
3. Confirm that idle rpm is at the prescribed rpm — refer to Section 2 for more extensive procedures regarding idle speed.
4. Stop the engine (turn the ignition switch **OFF**.
5. Confirm that there are no sharp bends in the accelerator cable.
6. Check the inner cable for the correct amount of slack. The slack should be 0.04-0.08 in. (1-2mm) for a manual transmission, and 0.12-0.02 in. (3-5mm) for an automatic transmission.
7. If there is too much or not enough slack, adjust the play by the following procedure:
    a. Turn the ignition switch to the **ON** position and hold the switch in that position for 15 seconds (with the engine not running).
    b. Loosen the adjusting nut so that the throttle lever is free. Turn the accelerator adjusting nut to the point where the throttle lever just starts moving, then back off one turn and lock the adjusting nut with the lock nut.
    c. The accelerator cable free-play should now be the standard setting.
8. Confirm that the throttle cable stopper touches the air idle control actuator.

✱✱CAUTION

The above inspection should be done after turning the ignition switch ON (engine stopped) and leaving it in that condition for 15 seconds.

3.0L (12 VALVE) ENGINE

1. Check that there are no sharp bends in the routing of the accelerator cable.
2. Check that the throttle link is touching the fixed SAS (stopper).
3. Loosen the adjusting bolts on the intake manifold plenum, then secure the outer cable so that the free-play of the inner cable will be the standard value of 0.04-0.08 in. (1-2mm).

➡If there is excessive play in the accelerator cable, the vehicle speed drop ("undershoot") when climbing a slope will be large. If there is no play (excessive tension) in the accelerator cable, the idling speed will increase.

4. After adjustment confirm that the throttle valve fully opens and closes by operating the pedal.

5-24 FUEL SYSTEM

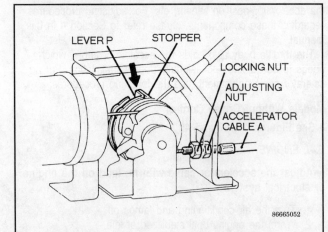

Fig. 46 Make certain that lever P is against the stopper while adjusting the slack in the cable — 2.4L and 3.0L Pick-up engines

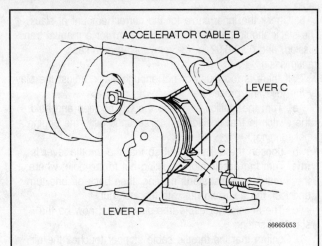

Fig. 47 The gap (C), between lever C and lever P should be 0-0.08 in. (0-2mm) — 2.4L and 3.0L Pick-up engines

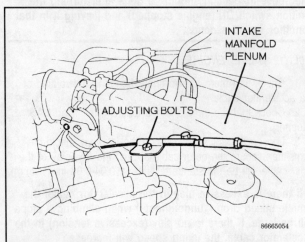

Fig. 48 Loosen the adjusting bolts on the intake manifold plenum to adjust the free-play in the cable — 3.0L engines

3.0L (24 VALVE) AND 3.5L ENGINES

1. Check to see if there is any sharp bend in the wiring of the accelerator cable.
2. Confirm that the throttle lever is touched by the fixed SAS.
3. Make certain that the inner cable play is at the standard value: 0.04-0.08 in. (1-2mm).
4. If the play is outside of the standard value, adjust by turning the adjusting nut so that the inner cable play is brought to the standard value and then tighten the lock nut.

Models With Cruise Control
▶ See Figures 49, 50, 51 and 52

PICK-UPS

1. Depress the accelerator pedal and check to be sure that the throttle lever moves smoothly from fully open to fully closed.
2. If there is excessive sagging of the inner cable of the accelerator cables A and B, or if there is no sagging at all, first check the arrangement and placement of the cable, then adjust as described below.
    a. Idle the engine to warm it up to a steady rpm.
    b. Confirm that the idle rpm matches the specified rpm. In this case the specified rpm is 650-850 rpm (2.4L engines) or 600-800 rpm (3.0L Engines) at idle.
    c. Stop the engine and remove the cover of the actuator.
    d. Loosen the lock nuts of the accelerator cables A and B and let the inner cables sag.
    e. While keeping lever P and the stopper in contact with each other, turn the adjusting nut (in the direction that causes the outer cable to become longer) until just before the lever P begins to operate, then turn the nut 1/2 turn back and secure it with the locking nut. This operation will ensure that the free-play of the accelerator cable A is correctly adjusted. The standard value of free-play for cable A is 0-0.04 in. (0-1mm).
    f. In order to get the play of accelerator cable B (the distance C, between levers C and P) to the standard value, alter the adjusting nuts of the throttle body and adjust. The standard value for accelerator cable B is 0-0.08 in. (0-2mm).

➡If there is excessive play of the accelerator cables, the vehicle speed drop ("undershoot") when climbing a slope will be large. If there is no play (excessive tension) of the cables, the idling speed will increase.

3. After adjusting the cables, operate the accelerator pedal, turn the lever steadily and make sure that the throttle lever is being activated. Also move the actuator, turn the lever steadily and make sure that the throttle lever is being activated.
4. Install the actuator cover.

MONTEROS

1. Remove the link protector plate.
2. Check if there is any deflection in the inner cables of the the accelerator cable, cruise control cable and the throttle cable. If there is excessive deflection or no play in the inner cables, loosen the adjusting bolts and nuts to release each link from the throttle lever. Do not remove the adjusting nuts or bolts.

FUEL SYSTEM 5-25

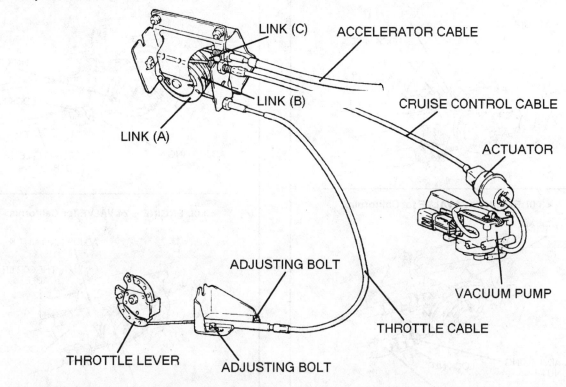

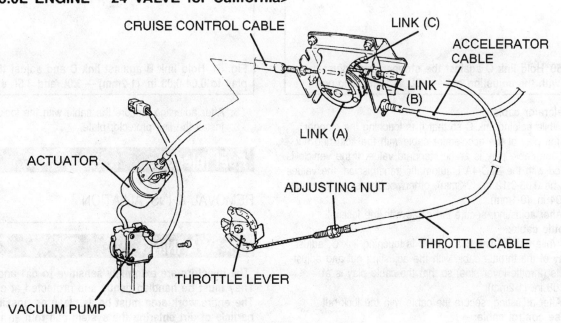

Fig. 49 Accelerator cable location and assembly components — 3.0L and 3.5L engines

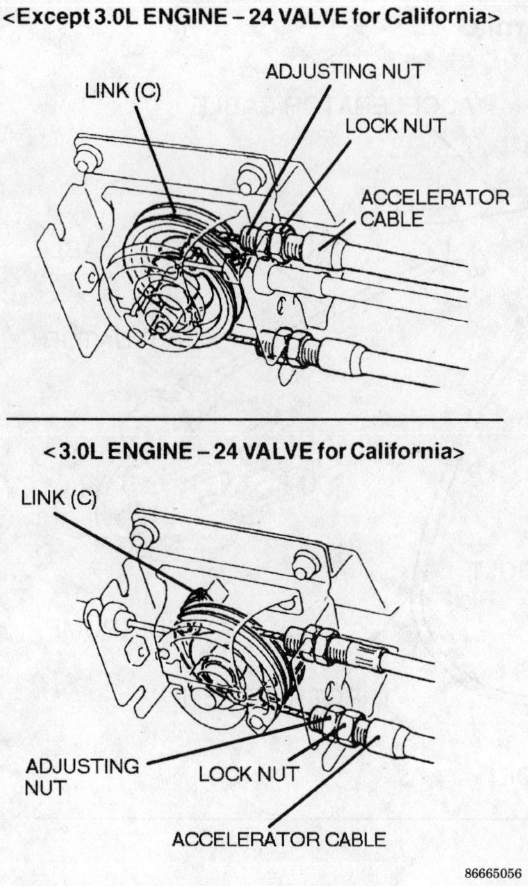

5-26 FUEL SYSTEM

Fig. 50 Hold link C against the stopper and adjust the play with the adjusting nut — 3.0L and 3.5L engines

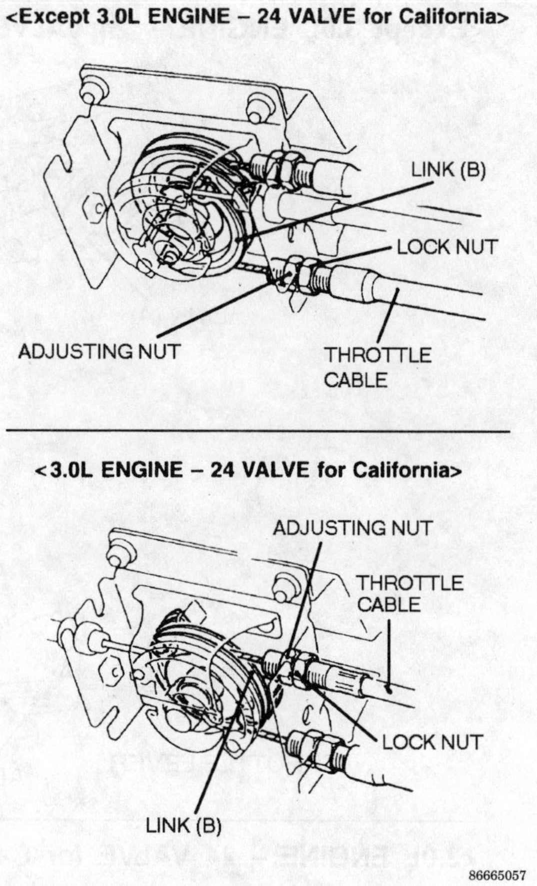

Fig. 51 Hold link B against link C and adjust the free-play to 0.04-0.08 in. (1-2mm) — 3.0L and 3.5L engines

Accelerator cable:

3. While holding link C so that it is touching the stopper, adjust the play of the accelerator cable with the adjusting nut so that the cable play is at the standard value. If the vehicle is equipped with the ELC-4A/T automatic transmission, the values should be 0.08-0.12 in. (2-3mm), otherwise the standard value is 0-0.04 in. (0-1mm).

4. After adjusting, secure the cable with the locknut.

Throttle cable:

5. While holding link B so that it is touching link C, adjust the play of the throttle cable with the adjusting nut and adjusting bolts (throttle lever side) so that the cable play is at 0.04-0.08 in. (1-2mm).

6. After adjusting, secure the cable with the lock nut.

Cruise control cable:

7. While holding link A so that it is touching link B, adjust the play of the cruise control cable with the adjusting nut so that the cable play is at the standard value of 0.04-0.08 in. (1-2mm).

8. After adjusting, secure the cable with the locknut.
9. Install the link protector plate.

Fuel Injectors

REMOVAL & INSTALLATION

✲✲WARNING

The injectors are extremely sensitive to dirt and impact. They must be handled gently and protected at all times. The entire work area must be as clean as possible. Any particle of dirt entering the system can foul an injector or change its operation. Any gaskets or O-rings removed with the injector MUST be replaced with new ones at reassembly. Do not attempt to reuse these seals; high pressure fuel leaks may result.

FUEL SYSTEM 5-27

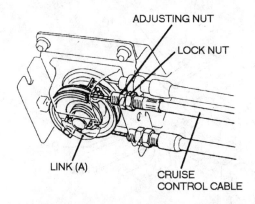

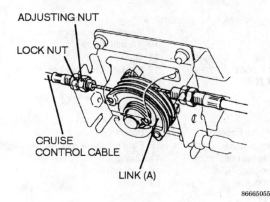

Fig. 52 Adjust the free-play in the cruise control cable while holding link A against link B — 3.0L and 3.5L engines

2.4L Engines

▶ See Figures 53 and 54

1. Safely relieve the pressure within the fuel system — refer to the procedures described previously in this section.

✱✱CAUTION

The fuel system is under pressure. Release pressure slowly and contain spillage. Observe no smoking/no open flame precautions. Have a Class B-C (dry powder) fire extinguisher within arm's reach at all times.

2. Disconnect the negative battery cable.
3. Drain the engine coolant to a level below that of the intake manifold.
4. Remove the air intake hose, the breather hose, the wiring harness connector, the air intake pipe and the air hose.
5. Disconnect the accelerator and kickdown cables from the throttle body.
6. Remove the water and vacuum hoses from the throttle body.
7. Remove the throttle body from the intake manifold plenum — refer to the procedures described previously in this section.
8. Extract the gasket from the throttle body/intake manifold plenum mounting surface; a new gasket will be needed upon assembly.
9. Unplug the fuel injector wiring harness from the fuel injectors.
10. Disconnect the high pressure fuel line from the fuel rail.

✱✱WARNING

There may still be residual pressure in the fuel system; release the fuel line slowly to allow pressure to escape. Have rags handy to wipe or soak up any spilled fuel.

11. Remove the fuel return hose, the mounting bolt and the fuel pressure regulator from the fuel rail. The O-ring between the fuel rail and the fuel pressure regulator will need to be replaced with a new one upon assembly.
12. Remove the fuel rail with the fuel injectors installed.

➡**The fuel injectors are extremely sensitive to contaminants, impact or wrong voltages. Do not allow any of the injectors to fall when removing the fuel rail. If an injector does drop to the floor or other hard surface, it should be replaced.**

13. Remove the fuel rail to intake manifold insulators.
14. Remove the injectors (with a slight pull), the O-rings, the grommets and the insulators from the fuel rail. The O-rings, the grommets and the insulators must all be replaced with new ones upon assembly.

To Install:

15. Reassembly begins by installing a new grommet and O-ring (in that order) onto the injector. Coat the O-ring with a light coating of oil.
16. Install each injector into the rail by turning it to the left and right. Make sure that the injector turns freely when in place. If it does not turn under finger pressure, remove it, inspect the O-ring and reinsert the injector. While the injector does not turn during its operation, its ability to turn is an indicator of correct installation.
17. Replace the seats (insulator) in the intake manifold. Install the delivery pipe and the injectors onto the manifold without dropping an injector. Make certain the rubber bushings are correctly seated in the installation holes.
18. Install a new O-ring on the fuel pressure regulator and coat it lightly with clean gasoline. Install and tighten the fuel pressure regulator to fuel rail mounting bolts to 5-8 ft. lbs. (7-11 Nm).
19. Install the fuel rail mounting bolts through the fuel return hose mounting bracket, through the fuel rail and into the intake manifold. Tighten them to 6-9 ft. lbs. (8-12 Nm).
20. Connect the fuel return hose to the fuel return rail. Fasten it in place with the spring clip.
21. Connect the high pressure fuel line to the fuel rail, using as new O-ring. Coat the O-ring with a little gasoline to help it seal better. Tighten the two retaining bolts to 1-2 ft. lbs. (2-3 Nm).
22. Plug the fuel injector harness back onto the fuel injectors.

5-28 FUEL SYSTEM

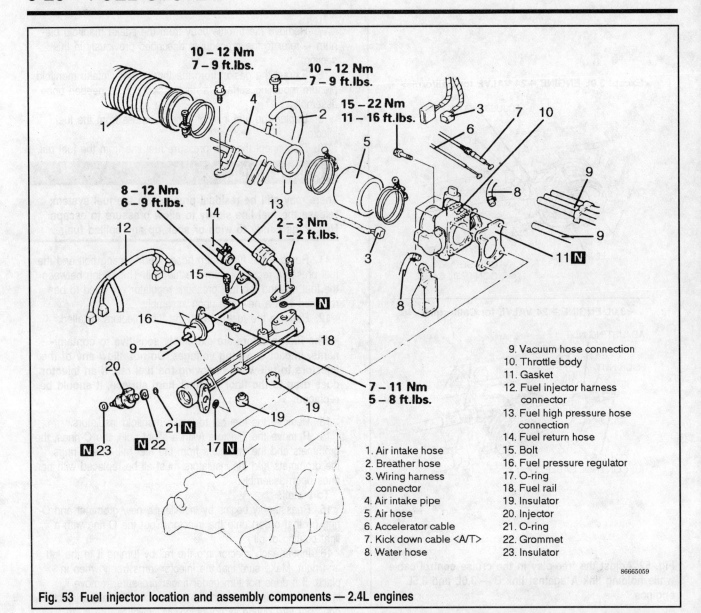

Fig. 53 Fuel injector location and assembly components — 2.4L engines

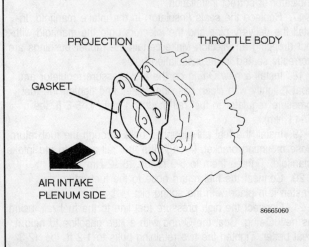

Fig. 54 Make certain that the intake manifold-to-throttle body gasket is in the correct position — 2.4L engines

23. Install the throttle body back onto the intake manifold plenum with a new gasket. Make sure the gasket goes onto the intake manifold plenum to throttle body opening with the gasket projection up and to the left (look straight into the intake manifold plenum opening, in other words from the passenger-side of the engine compartment). Install the throttle body mounting bolts with 11-16 ft. lbs. (15-22 Nm) of torque.

24. Connect the vacuum hose, the water hose, the automatic transmission kickdown cable (if so equipped) and the accelerator cable. Refer to the accelerator cable adjustment procedure described earlier in this section.

25. Connect the air hose, the air intake pipe, the breather hose and the air intake hose.

26. Plug the wiring harness connector back into the throttle body.

27. Fill the engine coolant system back up.

28. Connect the negative (-) battery cable back to the battery and start the engine. The engine may turn over a few times until the fuel system pressure can be brought back up to normal.

29. Check for any coolant or fuel leaks, repair if necessary.

FUEL SYSTEM 5-29

3.0L and 3.5L Engines

▶ See Figures 55, 56, 57, 58, 59 and 60

1. Safely relieve the pressure within the fuel system.

❈❈CAUTION

The fuel system is under pressure. Release pressure slowly and contain spillage. Observe no smoking/no open flame precautions. Have a Class B-C (dry powder) fire extinguisher within arm's reach at all times.

2. Disconnect the negative battery cable.
3. Remove the air intake plenum, following procedures listed in Section 3.
4. Disconnect the high pressure fuel line at the delivery pipe (rail). The O-ring inside the fitting is not reusable.

❈❈WARNING

Have a clean rag or towel handy to catch the residual gasoline in the fuel system. Some pressure will remain within the system.

5. Disconnect the fuel return hose and remove its O-ring. Disconnect the vacuum hose from the fuel pressure regulator.
6. Remove the fuel pressure regulator and its O-ring.
7. Remove the cover piece from the fuel rail, if so equipped.
8. Unplug the electrical connector from each injector.
9. Remove the bolts holding the delivery pipe to the engine. Note that the fuel rail is one continuous piece serving both banks of cylinders.
10. Lift the rail with the injectors attached up and away from the engine. Take great care not to drop any of the injectors during this removal.

➡ **The fuel injectors are extremely sensitive to contaminants, impact or wrong voltages. Do not allow any of the injectors to fall when removing the fuel rail. If an injector does drop to the floor or other hard surface, it should be replaced.**

11. The injectors may be removed from the rail with a gentle pull. On the 3.5L engines, the injectors are held in place by injector clips; slide these off to remove the injectors. Both the grommet (insulator) and O-ring on the top of the injector must be discarded and replaced. The lower insulator or seat must also be removed and replaced.

❈❈WARNING

Some fuel will flow out of the delivery pipe when the injectors are removed.

12. Reassembly begins by installing a new grommet and O-ring (in that order) onto the injector. Coat the O-ring with a light coating of clean engine oil.

➡ **Use care not to let the engine oil enter the delivery pipe.**

13. Install each injector into the rail, making sure that the injector turns freely when in place. If it does not turn under finger pressure, remove it, inspect the O-ring and reinsert the injector. (While the injector does not turn during its operation, its ability to turn is an indicator of correct installation). On 3.5L engines, install the injector clips by sliding the open ends onto both injector and fuel rail.
14. Replace the seats (insulators) in the intake manifold. Install new rubber bushings onto the mounting points of the fuel rail. Install the delivery pipe and the injectors onto the manifold without dropping an injector. Make certain the rubber bushings are in place under the delivery pipe brackets.
15. Tighten the fuel rail bolts to 9 ft. lbs. (12 Nm).
16. Plug each electrical connector onto its proper injector.
17. Install the cover on the fuel rail (if so equipped); tighten the bolts to 6 ft. lbs. (8 Nm).
18. Replace the O-ring, coat it lightly with gasoline and install the fuel pressure regulator. Tighten the fasteners to 7 ft. lbs. (9 Nm).
19. Connect the fuel return hose and any vacuum hoses.
20. Install the new O-ring, coat it lightly with gasoline and install the high pressure fuel line. Make certain the O-ring is not damaged during installation. Tighten the bolts to 1.5-2 ft. lbs. (2-3 Nm).
21. Reinstall the air intake plenum; refer to Section 3 of this manual for the removal and installation procedure.
22. Reattach any component removed earlier.
23. Connect the negative battery cable.

TESTING

The simplest way to test the injectors is simply to listen to them with the engine running. Use either a stethoscope-type tool or the blade of a long screwdriver to touch each injector while the engine is idling. You should hear a distinct clicking as each injector opens and closes. Check that the operating sound increases as the engine speed is increased.

➡ **The sounds of the other injector(s) may be heard, even though the one being checked is not operating. Listen to each injector to get a feel for normal sounds; the one with the abnormal sound is the problem.**

Additionally, the resistance of the injector can be easily checked. Disconnect the negative battery cable and remove the electrical connector from the injector to be tested. Use an ohmmeter to check the resistance across the terminals of the injector. Correct resistance at 68°F (20°C) is 13-16 ohms.

Slight variations are acceptable due to temperature conditions.

Bench testing of the injectors can only be done using expensive special equipment. Generally this equipment can be found at a dealership and sometimes at a well-equipped machine shop or performance shop. There is no provision for field testing the injectors by the owner/mechanic. DO NOT attempt to test the injector by removing it from the engine and making it spray into a jar.

Never attempt to check a removed injector by hooking it directly to the battery. The injector runs on a much smaller voltage and the 12 volts from the battery will destroy it inter-

5-30 FUEL SYSTEM

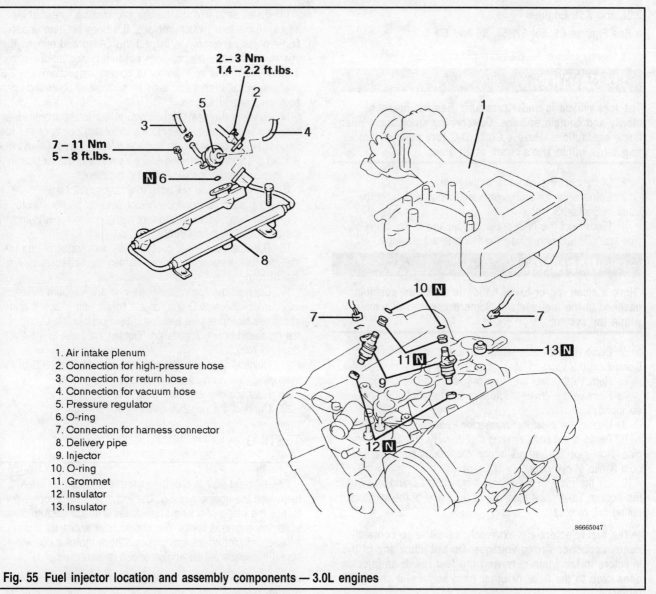

1. Air intake plenum
2. Connection for high-pressure hose
3. Connection for return hose
4. Connection for vacuum hose
5. Pressure regulator
6. O-ring
7. Connection for harness connector
8. Delivery pipe
9. Injector
10. O-ring
11. Grommet
12. Insulator
13. Insulator

Fig. 55 Fuel injector location and assembly components — 3.0L engines

nally. Since this happens at the speed of electricity, you don't get a second chance.

➡ For diagnostic procedures and trouble codes regarding the Electronic Control Module (ECM) and MFI system, refer to Section 4 of this manual.

Fuel Pressure Regulator

On the MFI system, since the fuel injectors injector the fuel into the intake manifold, the system needs to operate at a high pressure to inject the fuel at a good velocity and to help atomize the fuel charge as it enters the manifold. The fuel system pressure is extremely important for the function of the fuel injectors. The electric fuel pump provides more than enough pressure and the fuel pressure regulator has the job of bringing that fuel pressure down to a level that the injectors need and help to stabilize any surges the fuel pump might create.

REMOVAL & INSTALLATION

2.4L Engines

▶ See Figures 53 and 54

1. Safely relieve the pressure within the fuel system — refer to the procedures described previously in this section.

✱✱CAUTION

The fuel system is under pressure. Release pressure slowly and contain spillage. Observe no smoking/no open flame precautions. Have a Class B-C (dry powder) fire extinguisher within arm's reach at all times.

2. Disconnect the negative battery cable.
3. Drain the engine coolant to a level below that of the intake manifold.
4. Remove the air intake hose, the breather hose, the wiring harness connector, the air intake pipe and the air hose.

FUEL SYSTEM 5-31

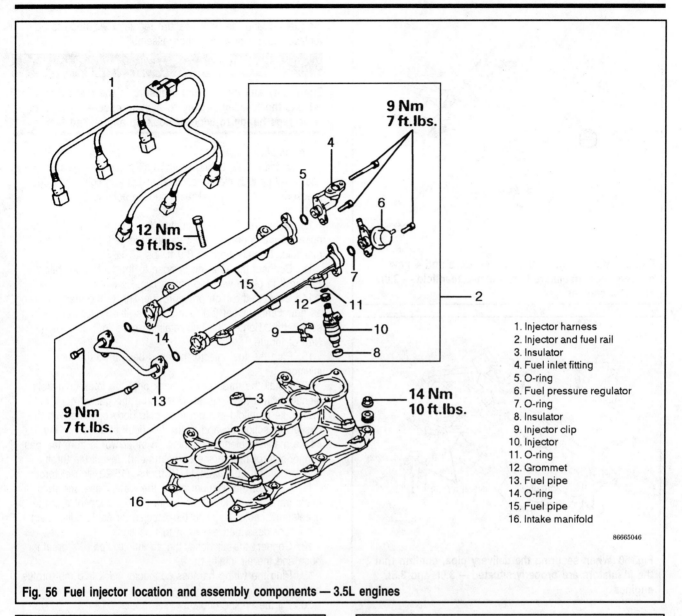

Fig. 56 Fuel injector location and assembly components — 3.5L engines

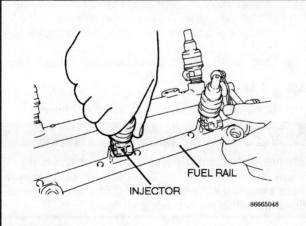

Fig. 57 When installing the injectors on the fuel rails, make certain that the injector clip openings face toward the center of the two rails — 3.5L engines

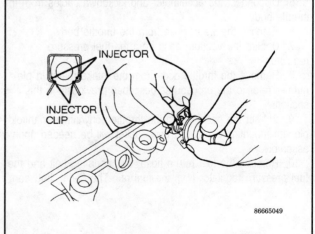

Fig. 58 The injector clips MUST be used to secure the injectors onto the fuel rails — 3.5L engines

5-32 FUEL SYSTEM

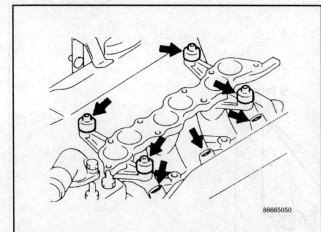

Fig. 59 Attach 6 new injector insulators and 4 new delivery pipe insulators to the intake manifold — 3.0L and 3.5L engines

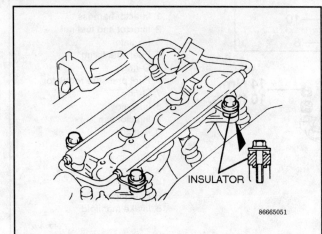

Fig. 60 When securing the delivery pipe, confirm that the insulators are properly situated — 3.0L and 3.5L engines

5. Disconnect the accelerator and kickdown cables from the throttle body.
6. Remove the water hose from the throttle body.
7. Unplug the vacuum hose from the fuel pressure regulator.
8. Remove the throttle body from the intake manifold plenum — refer to the procedures described previously in this section.
9. Extract the gasket from the throttle body/intake manifold plenum mounting surface; a new gasket will be needed upon assembly.
10. Remove the fuel return hose, the mounting bolt and the fuel pressure regulator from the fuel rail. The O-ring between the fuel rail and the fuel pressure regulator will need to be replaced with a new one upon assembly.

**WARNING

There may still be residual pressure in the fuel system; release the fuel line slowly to allow pressure to escape. Have rags handy to wipe or soak up any spilled fuel.

To Install:
11. Install a new O-ring on the fuel pressure regulator and coat it lightly with clean gasoline. Install and tighten the fuel pressure regulator to fuel rail mounting bolts to 5-8 ft. lbs. (7-11 Nm).
12. Install the fuel rail mounting bolts through the fuel return hose mounting bracket, through the fuel rail and into the intake manifold. Tighten them to 6-9 ft. lbs. (8-12 Nm).
13. Connect the fuel return hose to the fuel return rail. Fasten it in place with the spring clip.
14. Connect the high pressure fuel line to the fuel rail, using as new O-ring. Coat the O-ring with a little gasoline to help it seal better. Tighten the two retaining bolts to 1-2 ft. lbs. (2-3 Nm) of torque.
15. Plug the fuel injector harness back onto the fuel injectors.
16. Install the throttle body back onto the intake manifold plenum with a new gasket. Make sure the gasket goes onto the intake manifold plenum to throttle body opening with the gasket projection up and to the left (look straight into the intake manifold plenum opening, in other words from the passenger-side of the engine compartment). Install the throttle body mounting bolts with 11-16 ft. lbs. (15-22 Nm) of torque.
17. Connect the vacuum hose, the water hose, the automatic transmission kickdown cable (if so equipped) and the accelerator cable. Refer to the accelerator cable adjustment procedure described earlier in this section.
18. Connect the air hose, the air intake pipe, the breather hose and the air intake hose.
19. Plug the wiring harness connector back into the throttle body.
20. Fill the engine coolant system back up.
21. Connect the negative (-) battery cable back to the battery and start the engine. The engine may turn over a few times until the fuel system pressure can be brought back up to normal.
22. Check for any coolant or fuel leaks, repair if necessary.

3.0L and 3.5L Engines

1. Safely relieve the pressure within the fuel system.

**CAUTION

The fuel system is under pressure. Release pressure slowly and contain spillage. Observe no smoking/no open flame precautions. Have a Class B-C (dry powder) fire extinguisher within arm's reach at all times.

2. Disconnect the negative battery cable.
3. Remove the air intake plenum, following procedures listed in Section 3.

FUEL SYSTEM 5-33

4. Disconnect the fuel pressure regulator from the fuel rail and remove its O-ring. Disconnect the vacuum and fuel return hoses from the fuel pressure regulator.

✱✱WARNING

Have a clean rag or towel handy to catch the residual gasoline in the fuel system. Some pressure will remain within the system.

To Install:

5. Replace the O-ring, coat it lightly with gasoline and install the fuel pressure regulator. Tighten the fasteners to 7 ft. lbs. (9 Nm).
6. Connect the fuel return hose and any vacuum hoses.
7. Reinstall the air intake plenum; refer to Section 3 in this manual for more in-depth directions on the installation or removal of the intake manifold plenum.
8. Install the throttle body onto the intake manifold plenum; use the earlier described instructions in this section.
9. Fill the coolant system back up to normal operating level.
10. Connect the negative battery cable.
11. Start the engine and check for fuel and coolant leaks; repair if needed.

Pressure Relief Valve

The pressure relief valve keeps the pressure inside of the fuel tank below a specified maximum. If the fuel tank becomes over-pressurized, the pressure relief valve opens and allows the excess pressure to exit the fuel tank. The pressure relief valve is attached to the vapor hose, located between the fuel tank nipple and the fuel tank rollover valve. The vapor hose leads the over-pressurized air/fuel vapor mixture to the charcoal canister, which stores the vapors for later use.

REMOVAL & INSTALLATION

▶ See Figure 61

The 2.4L, 3.0L and 3.5L Pick-up trucks and Monteros have similar pressure relief valve locations and use the same procedures for removal and installation.

➡Access to the pressure relief valve is much easier with the fuel tank removed; refer to the procedures for fuel tank removal, found later in this section. However, access to the pressure relief valve may be possible on some models without removing the fuel tank.

1. Compress the spring retaining clips holding the vapor return hose and the pressure relief valve-to-fuel tank hose onto the relief valve with a pair of pliers and slide them to the side.
2. Work the valve free of both hoses. Make certain to not allow any contaminants (dirt, rust, etc.) into either hose.
3. Insert the new pressure relief valve into the hoses, making certain that it is installed in the correct direction (the pressure relief valve will only flow in one direction — refer to the illustration). The inlet side goes toward the fuel tank and the outlet side goes toward the rollover check valve.

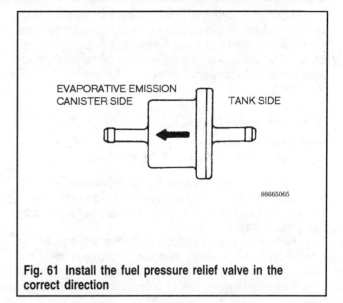

Fig. 61 Install the fuel pressure relief valve in the correct direction

4. Slide the spring retaining clips up past the nipple bulge of the relief valve and clamp the hoses in place.

TESTING

▶ See Figure 62

1. Attach a clean rubber hose to both sides of the pressure relief valve.
2. Blow lightly from the inlet side of the valve and the air should pass through after a slight resistance is felt.
3. Blow lightly from the outlet side and the air should not pass through.
4. If the valve fails either of these tests, replace it with a new unit.

Fuel Filter

The fuel filter purifies the fuel coming from the fuel tank before it reaches the injectors. With any gasoline system, car-

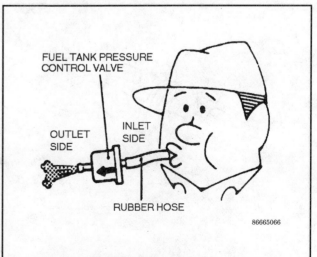

Fig. 62 Check the fuel pressure relief valve by blowing through it in each direction

5-34 FUEL SYSTEM

bureted or fuel injected, it is extremely important for the fuel to be clean and free of debris. Dirt, rust or any other contaminant can create extensive and costly damage to the system. Periodical fuel filter changes are a good way to insure against dirt or other unwanted contaminants entering the fuel delivery components.

REMOVAL & INSTALLATION

▶ See Figures 63, 64, 65, 66, 67, 68, 69 and 70

The fuel filter is located on the firewall in the engine compartment, or mounted on the vehicle's frame right below the engine compartment, and is connected to the fuel hard-lines by high pressure flexible hoses.

1. Reduce the internal pressure of the fuel delivery system — refer to the previously described procedure in this section.
2. Disconnect the negative (-) battery cable from the battery.
3. Remove the fuel filter protector plate, which is usually affixed to the filter with two mounting bolts.
4. Disconnect the fuel filter mounting bolt.
5. Using two wrenches, disconnect the high pressure fuel line and the fuel hard-line from the filter. The Monteros are equipped with two high pressure hoses, rather than one high pressure hose and one hard-line.

To Install:

6. Connect the high pressure line(s) to the fuel filter. First temporarily tighten the flare nut by hand, then tighten it to the specified torque of 22 ft. lbs. (30 Nm). Make certain that the fuel hose does not become twisted. Connect the hard-line to the fuel filter with the same amount of torque.
7. Mount the fuel filter with the mounting bolt and install the protector plate.
8. Connect the negative (-) battery cable to the battery.
9. Start the engine (it may take a while for the system to pressurize and start the vehicle).
10. Check for any gasoline leaks from the fuel filter fittings, repair if needed.

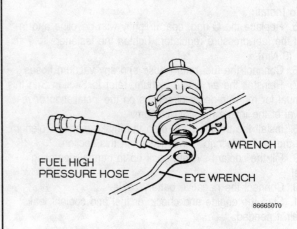

Fig. 64 Use two wrenches when loosening and tightening the high pressure fuel lines to keep them from being damaged

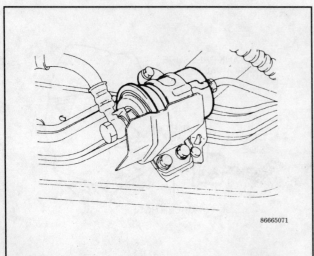

Fig. 63 On some models, the fuel filter is mounted to the frame of the vehicle in this position

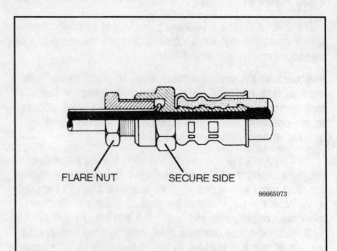

Fig. 65 The high pressure fuel line fitting can be damaged if the wrong side is turned, or only one wrench is used

FUEL SYSTEM 5-35

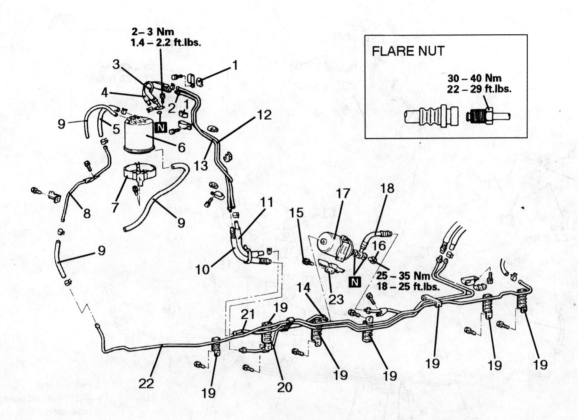

1. Clip
2. High pressure hose connection
3. High pressure hose
4. Fuel return hose
5. Vapor hose
6. Evaporative emission canister
7. Evaporative emission canister holder
8. Pipe
9. Vapor hose
10. High pressure hose
11. Fuel return hose
12. Main pipe
13. Return pipe
14. Fuel filter and main pipe connection
15. Bolt
16. Eye bolt
17. Fuel filter
18. High pressure hose
19. Clip
20. Main pipe
21. Return pipe
22. Vapor pipe
23. Fuel filter protector

Fig. 66 The fuel and vapor line assembly and component locations — 2.4L Pick-ups

5-36 FUEL SYSTEM

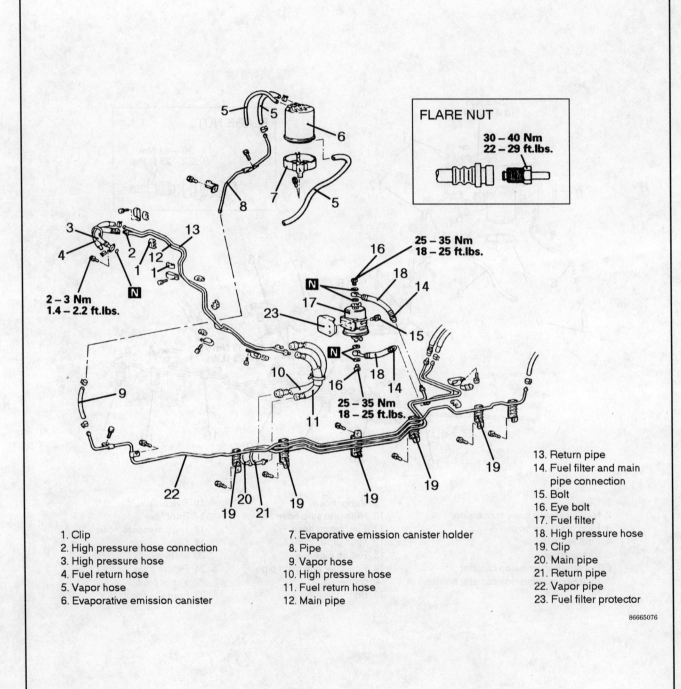

Fig. 67 The fuel and vapor line assembly and component locations — 3.0L Pick-ups

FUEL SYSTEM 5-37

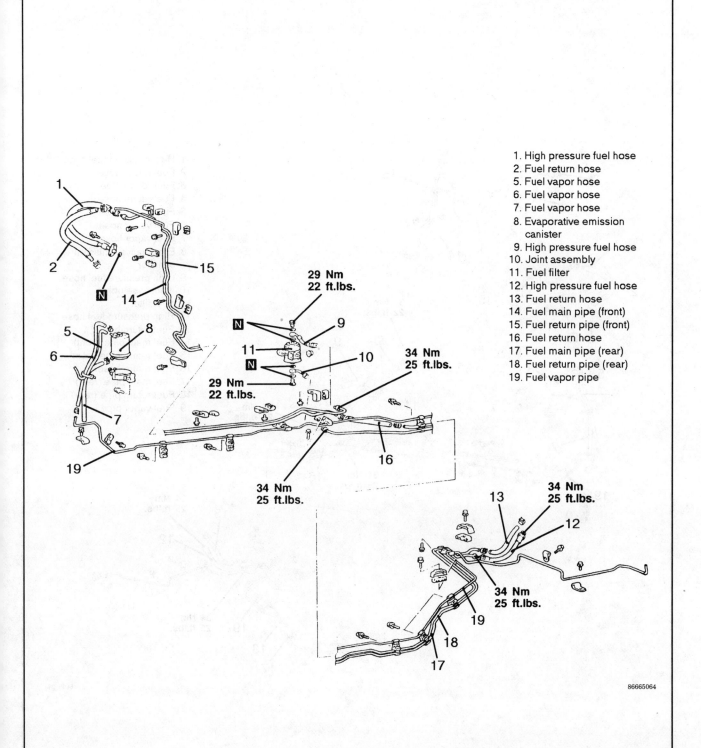

Fig. 68 The fuel and vapor line assembly and component locations — pre-1994 Monteros

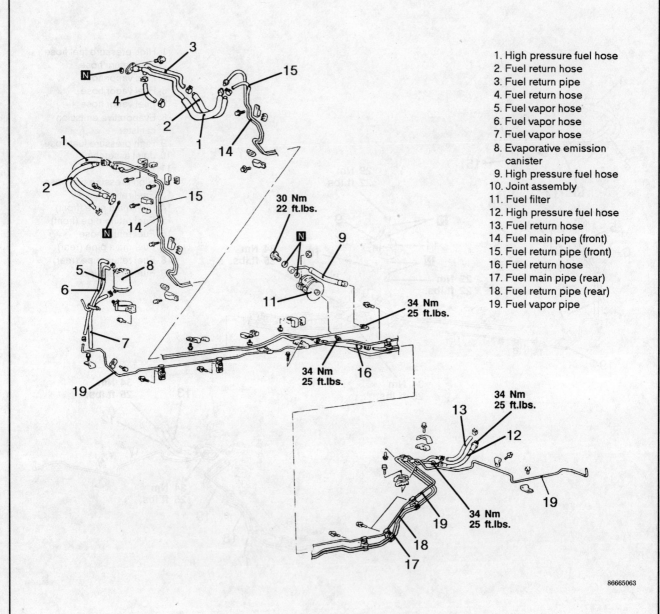

Fig. 69 The fuel and vapor line assembly and component locations — 1994-95 Monteros (3.0L 12 valve and 3.5L engines)

FUEL SYSTEM 5-39

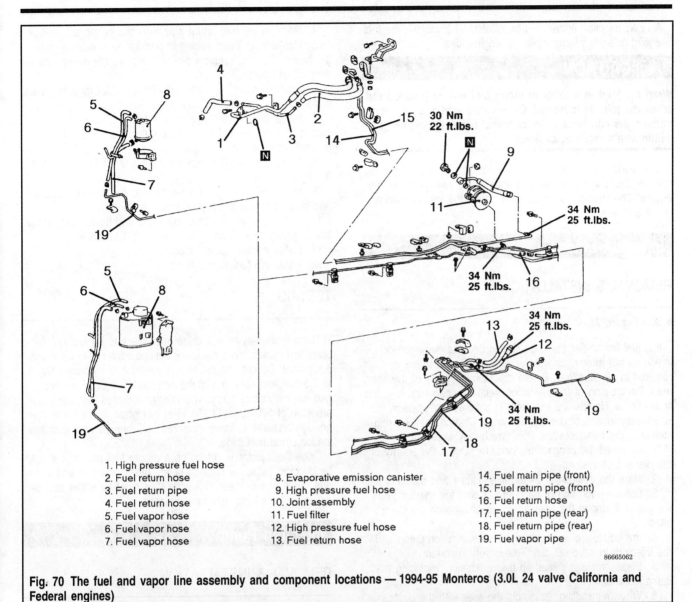

1. High pressure fuel hose
2. Fuel return hose
3. Fuel return pipe
4. Fuel return hose
5. Fuel vapor hose
6. Fuel vapor hose
7. Fuel vapor hose
8. Evaporative emission canister
9. High pressure fuel hose
10. Joint assembly
11. Fuel filter
12. High pressure fuel hose
13. Fuel return hose
14. Fuel main pipe (front)
15. Fuel return pipe (front)
16. Fuel return hose
17. Fuel main pipe (rear)
18. Fuel return pipe (rear)
19. Fuel vapor pipe

Fig. 70 The fuel and vapor line assembly and component locations — 1994-95 Monteros (3.0L 24 valve California and Federal engines)

DIESEL FUEL SYSTEM

The diesel combustion system works on different principals than a gasoline engine's combustion system. In a gasoline engine, the air/fuel charge is moderately compressed by the stroke of the piston. The spark plug fires, burning the fuel in the cylinder. As it burns, the air/fuel charge expands, driving the piston downward.

The diesel is compression fired; that is, no external source of heat or spark is used to ignite the fuel charge. As any substance is compressed, it develops heat. If a flammable substance — fuel — is compressed sufficiently, it will burn or explode. The diesel engine uses compression within the cylinders to ignite the fuel charge, which again drives the piston downward as it expands. For reference, normal compression within a gasoline engine is about 8.5:1 with fully developed racing engines reaching around 13.0:1. The Mitsubishi diesel engine uses a compression ratio of about 21.0:1. The act of compressing the air/fuel mixture to this level generates tremendous heat (about 1700°F) and it is this heat which ignites the fuel.

The fuel injection pump is the heart of the system, drawing fuel through the filter and delivering it to the injectors at the proper time in the combustion cycle. This mechanically driven pump must be kept in perfect synchronization with the engine.

The first signs of diesel trouble usually show up at the injection nozzles. An injector may fail or it may become blocked or stuck from dirt in the fuel. Some signs of injector trouble are:
• Heavy knocking noises from the injectors of one or more cylinders.
• Engine overheating.
• Loss of power, particularly under load or acceleration.
• Smokey black exhaust
• Increased fuel consumption.

5-40 FUEL SYSTEM

A faulty injection nozzle can be located by loosening the fuel line joint at each injector while the engine idles.

✱✱CAUTION

Wrap the joint in a towel or cloth; fuel will be pumped out when the joint is loosened. Observe no smoking/no open flame rules and have a dry chemical fire extinguisher within arm's reach at all times.

If an injector is working properly, changing its fuel supply (by opening the line) will cause a change in the idle quality of the engine. The injector that does NOT cause the idle to change is the problem.

Injection Lines

REMOVAL & INSTALLATION

▶ See Figure 71

If a fuel line under the hood should become damaged or leaky, do not attempt to weld or repair it. Replacement is required in all situations. Before disassembling any of the lines, mark the location of the plastic or metal clamps which hold the lines. These clamps are important in suppressing vibration which may loosen or damage the lines. Clamps must be reinstalled in the exact position from which they were removed.

1. Clean all the connections carefully. Dirt is the enemy of the diesel fuel system.
2. Mark the position of all the line clamps and brackets.
3. Loosen each end of each line at both the injector and the pump. Remove the lines as an assembly with the clamps intact.
4. Immediately plug the fittings on the injection pump and the injectors to keep out dirt. This is very important.
5. Disassemble the lines on the workbench, replacing the damaged line(s).
6. When reinstalling, assemble the lines with the clamps correctly placed on the workbench.
7. Remove the plugs from the pump and injector ports. Install the fuel line assembly and start each threaded fitting by hand. Tighten each fitting to 20 ft. lbs. (27 Nm).
8. Double check that the line clamps are in place and secure. Bleed the fuel system.
9. Start the engine and check each joint for leaks. Because the system was open, it may require a few seconds until it runs smoothly.

Injectors

REMOVAL & INSTALLATION

▶ See Figures 72, 73 and 74

1. Remove the injection lines as previously described.
2. Disconnect the fuel return hose.
3. Remove the nut which holds the fuel return pipe to each injector. Use a second wrench to counterhold the injector.
4. Remove the fuel return pipe from the injectors.
5. Use a long, deep socket to remove each injector from the head. After the injectors are out, remove the gasket from each nozzle hole.
6. When reinstalling, clean the nozzle holder mounting area of the cylinder head.
7. Install a new nozzle tip gasket and a new nozzle holder gasket into the cylinder head. New gaskets are required to prevent high pressure fuel leaks.
8. Install the injectors into the head and tighten each to 41 ft. lbs. (55 Nm).
9. Install the return pipe gaskets onto the injector nozzles and install the fuel return pipe onto the injectors.
10. Use two wrenches to counterhold and tighten the return line retaining nuts. Tighten them to 26 ft. lbs. (35 Nm).
11. Install the fuel return hose.
12. Install the fuel supply lines.

TESTING

There is no way to test diesel fuel injectors without an elaborate test bench. Your dealer or service center may have this equipment. Do not attempt to test the injector by connecting it to the fuel line while out of the engine. You'll bend the line and run the risk of injury. The injector releases fuel under a pressure of 1707 psi (11,761 kPa) or higher; this is more than enough pressure to force atomized fuel through your skin and induce blood poisoning.

Once the injector is set up on a proper test bench, it should be checked for spray pattern, noise, break pressure and leakage. Each one of these characteristics must be within specification or the injector must be replaced.

Injection Pump

REMOVAL & INSTALLATION

▶ See Figure 75

✱✱WARNING

After the pump is reinstalled, the injection timing must be reset. Adjusting the injection timing requires special tools for measuring the prestroke of the pump. If these tools are not available, do NOT remove the injection pump or attempt to work on it.

1. Disconnect the battery ground cable.
2. Remove the upper cover from the timing belt.
3. Remove the nut and washer holding the injection pump sprocket to its shaft.

➡Be careful not to drop the nut and/or washer into the lower timing cover.

4. Turn the crankshaft to bring the No.1 piston to TDC/compression. When the piston is in this position, all the timing marks align.

FUEL SYSTEM 5-41

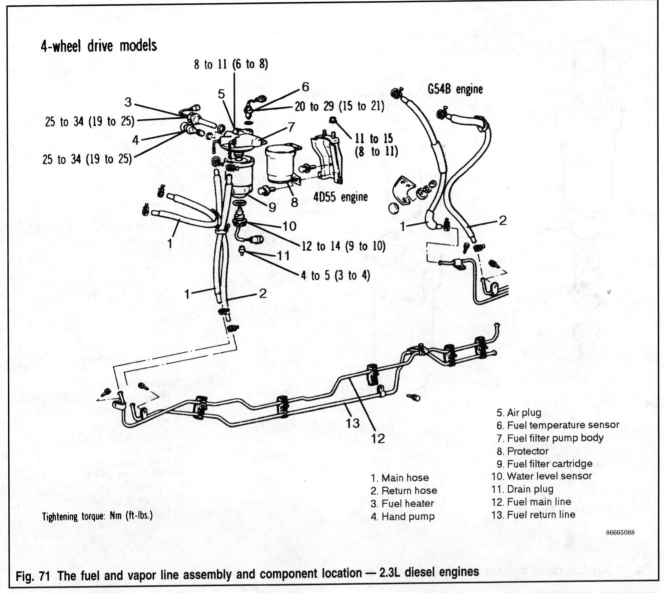

Fig. 71 The fuel and vapor line assembly and component location — 2.3L diesel engines

5. Use a pulley puller or extractor to carefully remove the injection pump sprocket. Catch the small Woodruff key and don't lose it.

✽✽WARNING

Do not subject the shaft to impact. Protect the timing belt from any stress, twisting or fluids. After the sprocket is removed, do not turn the crankshaft.

6. Disconnect the boost compensator hose at the injector pump.
7. Remove the vacuum hose from the vacuum regulating valve and the injection pump.
8. Disconnect the electrical harness running to the pump.
9. Wrap the joints with rags or cloth and remove the fuel supply hose and the return hose from the injection pump and plug the ports immediately. Fuel will be released when these joints are opened.
10. Loosen the union nuts and remove the injection lines. Use a second wrench to counterhold fittings where possible.

11. Remove the two mounting bolts, then remove the mounting nuts and washers.

➡ The pump is heavy and awkward to hold; make sure you or a helper have a firm grip before releasing the mounts.

✽✽WARNING

Do not hold the pump by either the accelerator lever or the fast idle lever. Additionally, these levers must not be removed from the pump. The levers' adjustment and placement is critical to the correct operation of the pump.

12. Remove the injection pump mounting bracket from the engine.
13. Before reinstalling the pump, make certain the entire pump is clean and the lines have been checked for dirt and foreign matter.
14. Install the mounting bracket on the engine and tighten the bolts to 26 ft. lbs. (35 Nm).

5-42 FUEL SYSTEM

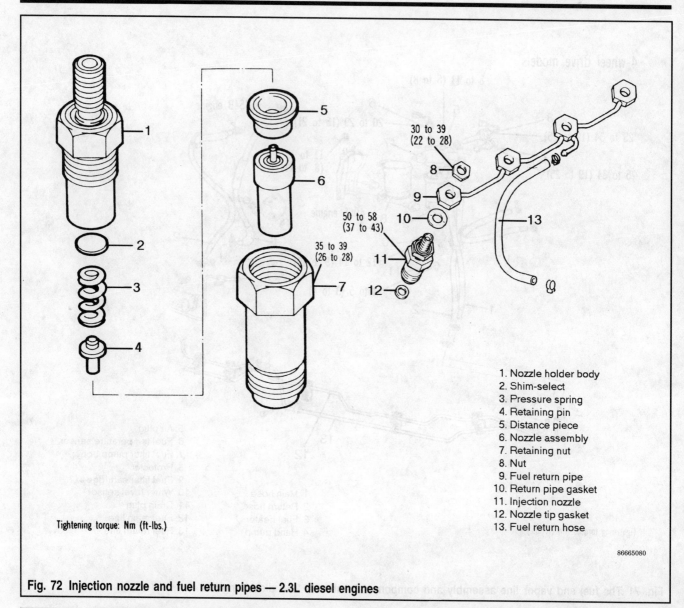

Fig. 72 Injection nozzle and fuel return pipes — 2.3L diesel engines

1. Nozzle holder body
2. Shim-select
3. Pressure spring
4. Retaining pin
5. Distance piece
6. Nozzle assembly
7. Retaining nut
8. Nut
9. Fuel return pipe
10. Return pipe gasket
11. Injection nozzle
12. Nozzle tip gasket
13. Fuel return hose

Tightening torque: Nm (ft-lbs.)

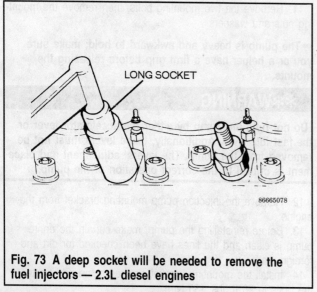

Fig. 73 A deep socket will be needed to remove the fuel injectors — 2.3L diesel engines

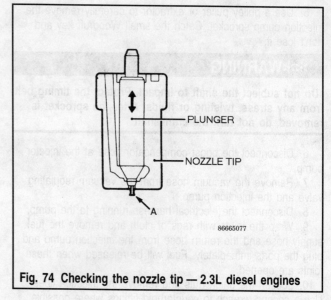

Fig. 74 Checking the nozzle tip — 2.3L diesel engines

FUEL SYSTEM 5-43

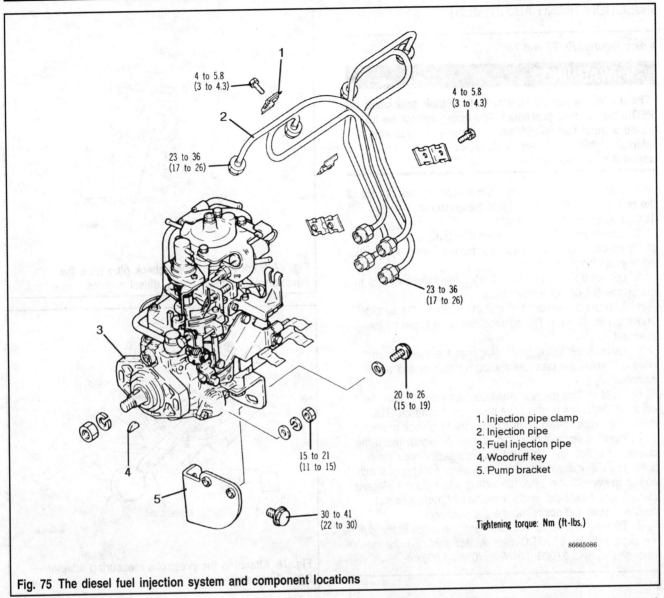

1. Injection pipe clamp
2. Injection pipe
3. Fuel injection pipe
4. Woodruff key
5. Pump bracket

Tightening torque: Nm (ft-lbs.)

Fig. 75 The diesel fuel injection system and component locations

15. Carefully mount the pump and tighten the nuts and bolts to 19 ft. lbs. (26 Nm).
16. Install the fuel delivery lines. Counterhold the joint while tightening the line fittings to 22 ft. lbs. (29 Nm).
17. Carefully connect the the fuel supply hose and the return hose to the pump.
18. Connect the wiring harness to the pump.
19. Install the vacuum hoses between the regulating valve and the injection pump.
20. Connect the boost compensator hose and connect the negative battery cable.
21. Carefully install the sprocket on the shaft of the injection pump. Do not subject the shaft to impact and do not damage the timing belt. Make sure the Woodruff key is correctly seated in its groove.
22. Perform the ignition timing adjustment procedure outlined in this section.

✱✱WARNING

This adjustment MUST be performed each time the injection pump is removed.

23. Bleed the fuel system.
24. Install the upper timing belt cover.

5-44 FUEL SYSTEM

INJECTION TIMING ADJUSTMENT

♦ See Figures 76, 77 and 78

> **✱✱WARNING**
>
> The use of the correct special tools or their equivalent is REQUIRED for this procedure. The timing cannot be adjusted without tool MD998384, the prestroke measuring adaptor. If this tool is not available, do not attempt to adjust the injection timing.

1. With the engine off, turn the crankshaft clockwise until all the timing marks align. This places the engine at TDC/compression for No.1 piston.
2. Loosen but do not remove the injection pipe union nuts at the injection pump. Always use a second wrench to counterhold the pump fitting.
3. Loosen the nuts and bolts holding the injection pump to the engine but do not remove them.
4. Examine the special tool and make certain the tip protrudes 0.4 in. (10mm). The pushrod can be adjusted by the inner nut.
5. Remove the timing check plug from the injector pump head and install the prestroke measuring adapter and a dial indicator.
6. Carefully turn the crankshaft counterclockwise until the notch on the pulley is at a point about 30° before the TDC mark. The notch should be in roughly the 11 o'clock position.
7. Zero the dial indicator. With a wrench, slightly move the crankshaft in both the clockwise and counterclockwise directions. The dial indicator should stay at zero; if it changes with engine movement, the initial 30° setting was incorrect. Repeat steps 6 and 7 until the gentle movement of the crankshaft does not move the pointer on the dial indicator.
8. Turn the crankshaft clockwise until the notch in the pulley aligns with the 5° ATDC mark. At this point, the dial indicator should read 0.0383–0.0405 in. (0.997–1.003mm).

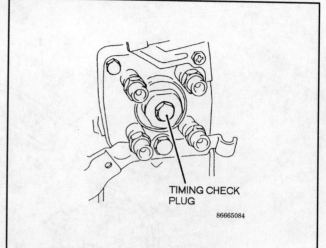

Fig. 77 Remove the timing check plug from the injection pump head — 2.3L diesel engines

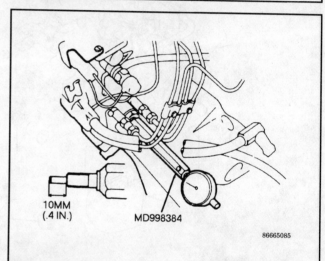

Fig. 78 Attaching the prestroke measuring adapter — 2.3L diesel engines

9. If the gauge does not show the specified value, move the injection pump body right or left on the mounts until the dial indicator reads correctly. Tighten the mounting nuts and bolts to hold the pump in place.
10. After the pump is locked in place, repeat the check procedure. Set the engine 30° before, zero the indicator, move the engine to 5° ATDC and check for the correct gauge reading. Each time the pump is moved, the entire check should be repeated.
11. Remove the special tools. Install a new copper gasket and the timing check plug. Tighten it to 6 ft. lbs. (8 Nm). Do not overtighten this bolt.
12. Tighten the fuel delivery lines. Counterhold the joint while tightening the line fittings to 22 ft. lbs. (29 Nm).
13. Since the fuel lines were loosened, bleeding the system is recommended.

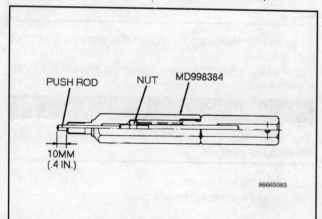

Fig. 76 Mitsubishi tool MD998384 or equivalent is needed to check the injection timing — 2.3L diesel engines

FUEL SYSTEM 5-45

Idle Speed Adjustment

▶ See Figure 79

1. The idle must be set with the engine fully warmed up, the lights and all electrical accessories off, the transmission in NEUTRAL and the parking brake applied.
2. Run the engine for more than 5 seconds at a speed between 2000 and 3000 rpm.
3. Allow the engine to idle for at least 2 minutes.
4. Using a diesel tachometer, check the idle speed. Correct speed is 700-800 rpm.
5. Adjust the speed to specification as necessary using the idle speed adjusting screw. Loosen the locknut and turn the screw to attain correct idle. Retighten the locknut.

✻✻WARNING

Do not adjust any other screws in the area.

Idle-up Adjustment

▶ See Figure 80

➡ This procedure is only for Pick-up trucks equipped with air conditioning.

When doing this procedure, the following conditions should be met:
- Cooling water temperature: 185-205°F (85-95°C)
- All lights and accessories: OFF
- Parking brake: applied
- Vehicles with power steering: set tires in the straight ahead position to prevent the power steering pump from being loaded. Set the steering wheel in the stationary position.

1. Make sure that the curb idle speed is within the specified speed and reset it by turning the idle speed adjusting screw, if necessary.
2. Loosen nuts 1 and 2. Turn the adjuster to set the length of the vacuum actuator rod so that the end of the rod is positioned as shown in the illustration in relation to the accelerator lever. Retighten the two nuts.

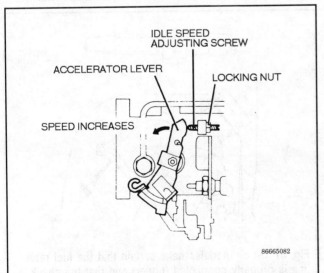

Fig. 79 Adjusting the idle speed — 2.3L diesel engines

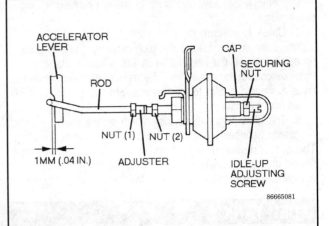

Fig. 80 Nuts 1 and 2 must be loosened before adjusting the length of the vacuum actuator rod when adjusting the idle-up — A/C equipped 2.3L diesel Pick-ups

3. Turn the air conditioner switch on and off several times, then set it to the Off position and confirm that the rod and accelerator lever are not in contact with each other.
4. Loosen the nut which secures the idle-up adjusting screw, adjust the engine speed to the specified speed by using the idle-up adjusting screw, and then retighten the nut. The idle-up engine speed should be 850-900 rpm.
5. Turn the air conditioner on and off several times to check the vacuum actuator for operation (lever goes up and down).

Glow Plugs

The glow plug system is a preheat system which rapidly warms the combustion chamber during a cold start. This initial heating allows the fuel to burn quicker, resulting in better cold driveability.

The earliest diesels required the driver to wait while the glow plugs worked; when the correct temperature was achieved, the engine could be started. Mitsubishi eliminated this long wait with what they call the Super Quick Glow System. By using both relays and a dropping resistor, the plugs are brought to their operating temperature almost instantly. The engine may be started within 5-10 seconds. Additionally, when the engine is running and the coolant temperature is below 130°F (55°C), the glow system will remain engaged, resulting in stable heat generation and reduced engine noise.

The glow plugs themselves are nothing more than resistance heating elements. Because they heat quickly, they consume heavy amounts of electricity. The high amperage required is conducted through a buss-bar (solid conductor) connected to each plug. It is important that the battery in a diesel vehicle be kept fully charged; otherwise the draw of the glow plugs combined with the draw of the starter will drain the battery.

REMOVAL & INSTALLATION

1. Disconnect the negative battery cable.
2. Loosen the small nut on the top of each glow plug.

5-46 FUEL SYSTEM

3. Remove the glow plug plate (buss-bar) running among the 4 plugs.
4. Carefully unscrew the plug from the head, using a 12mm, deep socket. Handle the plug carefully, protecting the tip at all times. If the unit falls or strikes a hard surface, it must be considered unusable. (The heating element inside the tip is a winding similar to the filament within a light bulb; it will break under impact.).
5. The glow plug may be tested with an ohmmeter. Check resistance between the top terminal and the body of the plug. The correct resistance is 0.23 ohms; make certain the meter is set on the correct scale to read this value. If there is no resistance or excessive resistance, the plug is unusable.

6. Inspect the buss-bar for any damage to the protective coating.
7. Install each glow plug with your fingers. Take great care to install the plug straight into the port, protecting the tip from impact.
8. Once each plug is finger-tight, use a wrench to tighten the plug to 13 ft. lbs. (17 Nm).
9. Install the buss-bar to the glow plugs. Make certain the retaining nuts make firm contact and tighten them to 1 ft. lb. (1.4 Nm).
10. Connect the negative battery cable.

FUEL TANK

Tank Assembly

REMOVAL & INSTALLATION

▶ See Figure 81

✱✱CAUTION

Gasoline in either liquid or vapor form is EXTREMELY explosive. An empty tank can sometimes be more hazardous than a full one due to vapor accumulation. Observe no smoking/no open flame rules during this procedure. Take extreme care to avoid sparks from any source, including electric switches and dropped tools. Have a dry powder (type B-C) fire extinguisher within arm's reach at all times and know how to use it. Always store drained fuel in metal containers with an airtight lid. Liquid gasoline may cause skin irritation or allergic reaction; avoid splashing fuel on clothing.

➡ Common sense dictates that the tank be very close to empty before draining it. Have a funnel and a supply of rags on hand before beginning.

Pick-up

1983-90 MODELS

▶ See Figure 82

1. Disconnect the negative battery cable.
2. Elevate and safely support the vehicle on jackstands. For 1986-89 trucks, remove the small side skirt panel on the left side.
3. Remove the filler cap to equalize pressure within the tank.
4. Remove the drain plug and in-tank filter. Drain the fuel into a suitable metal container with an airtight lid. Later models may not have the in-tank filter.
5. Loosen the fuel hose clamps (main, return and vapor) at the fuel tank and disconnect the hoses. Label or identify each hose and port.
6. Remove the electrical harness from the fuel gauge unit.
7. Remove the filler neck retaining bolts from the body. Remove the filler hose protector.

8. Use a floor jack and a broad piece of wood positioned under the fuel tank to support the tank.
9. Remove the tank mounting nuts.
10. Lower the tank to the ground.
11. Install the new tank by raising it into place with the jack.
12. Install the retaining nuts and tighten them to 32 Nm (24 ft. lbs.). When the tank is secure, the jack may be removed.
13. Connect the electrical harness to the gauge unit.
14. Install the filler hose protector and install the neck-to-body retaining screws.
15. Install the hoses to their ports, making certain that each clamp is secure.
16. If not already done, install the in-tank filter and drain bolt. Tighten the bolt to 44 ft. lbs. (59 Nm).
17. Using a funnel, carefully refill the tank with the drained fuel. Avoid splashing fuel onto the painted bodywork. Install the filler cap.
18. Lower the vehicle to the ground and connect the negative battery cable.
19. Start the engine. It will take a period of cranking before the engine fires and runs smoothly due to the lines being emptied.
20. Check the entire work area carefully for any sign of leakage. A small leak will not drip but will only show up a slight moisture on the hose. Attend to all leaks immediately.

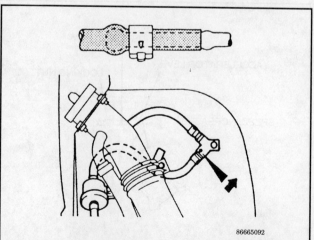

Fig. 81 On all models, make certain that the fuel return line is properly connected (upper) and that the check valve (arrow) is not upside down

FUEL SYSTEM   5-47

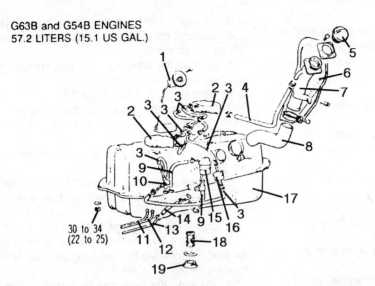

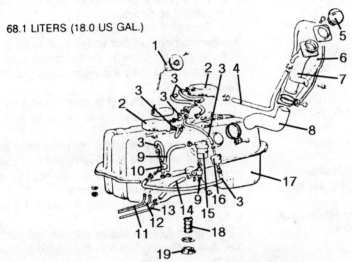

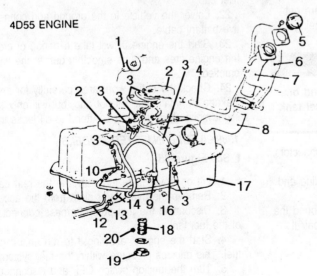

1. Fuel gauge unit
2. Separator tank
3. Vapor hose
4. Breather hose
5. Fuel filler cap
6. Filler hose protector
7. Filler neck
8. Connecting hose
9. Main hose
10. Check valve
11. Fuel vapor pipe
12. Fuel main pipe
13. Fuel return pipe
14. Return hose
15. Fuel filter
16. Overfill limiter
17. Fuel tank
18. Fuel filter (in tank)
19. Drain plug
20. Valve

Tightening torque: Nm (ft-lbs.)

Fig. 82 Fuel tank assembly and component location — 1983-90 Pick-ups

FUEL SYSTEM

1991-95 MODELS
▶ See Figure 83

Although some slight differences may appear between different model years or makes, the procedures for all 1991-95 Mitsubishi trucks is essentially the same. Refer to the illustrations if any discrepancies arise.

1. Disconnect the negative battery cable.
2. Unplug all wiring harnesses from the fuel tank (fuel pump and fuel gauge unit connectors).
3. Elevate and safely support the vehicle on jackstands.
4. Remove the filler cap to equalize pressure within the tank.
5. Remove the drain plug and drain the fuel into a clean, suitably-sized container with a sealable top.
6. Remove the side skirt panel stay.
7. Remove the main hose, return hose, vapor hose, breather hose and fuel filler hose connections.
8. On vehicles equipped with a 3.0L engine, remove the fuel tank protector.
9. Position a floor jack under the fuel tank and support the tank with it.
10. Remove the fuel tank mounting nuts and lower the fuel tank with the floor jack.
11. At this time, the fuel tank pressure relief valve, fuel pump assembly, or fuel gauge sending unit can be removed from the fuel tank.

To install:
12. Lift the fuel tank back up into its position with the floor jack.
13. Secure the tank in place with the fuel tank mounting nuts. Tighten them to 18-22 ft. lbs. (25-30 Nm).
14. Install the fuel tank protector plate, if so equipped.
15. Install the fuel filler hose until its end contacts the fuel tank.
16. Install the breather and return hoses to the nipples, making sure that they are pushed at the least 1-1.20 in. (25-30mm) on.
17. Reattach the vapor hose connection.
18. Install the main hose. Temporarily tighten the flare nut by hand, then tighten it to 22-29 ft. lbs. (30-40 Nm), being careful that the fuel hose does not become twisted.

✲✲CAUTION

When tightening the flare nut, be careful not to bend or twist the line, which could cause damage to the fuel tank connection.

19. Attach the fuel pump and the fuel gauge unit connectors.
20. Install the side skirt panel stay.
21. Fill the tank with approximately 1 gallon of gasoline and connect the negative battery cable.
22. Start the engine and check for any fuel leaks around the hoses going to the fuel tank and their connections; repair if necessary.

Montero

1983-90 CARBURETED MODELS
▶ See Figure 84

1. Disconnect the negative battery cable.
2. Remove the gas filler cap to equalize the pressure within the tank.
3. Elevate and safely support the vehicle on jackstands.
4. Although not required, the job is easier if the rear wheels are removed.
5. Remove the drain plug and drain the fuel into a suitable metal container with an airtight lid.
6. Remove the fuel filler protector.
7. Disconnect the vapor hose, the check valve and the overfill limiter. Make sure each line is labeled for correct reassembly.
8. Loosen the clamp and remove the large filler hose from the tank.
9. Disconnect the breather hose from the tank.
10. Label and disconnect the main hose and the return hose from the tank.
11. Unplug the fuel gauge electrical connector.
12. Position a floor jack and a broad piece of wood under the tank. Remove the tank retaining nuts and lower the tank from the vehicle.
13. The separator tanks and the protective stone shield may be removed if so desired.
14. Reinstall the tank by elevating it with the floor jack. Install the retaining nuts and tighten them to 20 ft. lbs. (27 Nm).
15. Connect the wiring to the fuel gauge sender.
16. Connect the main and return hoses and insure the clamps are tight.
17. Fit the filler neck and breather hoses onto the tank and secure the clamps. Make certain the hoses are properly seated on the ports.
18. Install the overfill limiter, the check valve and the vapor hose. Make certain the check valve (roll over valve) is mounted so that the arrow on its case point upwards.
19. Install the filler neck protector.
20. If not already done, install the drain plug and tighten it to 12 ft. lbs. (16 Nm).
21. Using a funnel, carefully refill the tank with the drained fuel. Avoid splashing fuel onto the painted bodywork. Install the filler cap.
22. Lower the vehicle to the ground and connect the negative battery cable.
23. Start the engine. It will take a period of cranking before the engine fires and runs smoothly due to the lines being emptied.
24. Check the entire work area carefully for any sign of leakage. A small leak will not drip, but will only show up as slight moisture on the hose. Attend to all leaks immediately.

1989-90 FUEL INJECTED MODELS
▶ See Figure 85

1. Lift or move aside the carpet in the rear cargo area.
2. Remove the oval cover plate from the access hole.
3. Disconnect the fuel pump harness connector at the rear of the fuel tank.
4. Start the engine, allowing it to run until it runs out of fuel. This relieves pressure within the fuel system.
5. Turn the ignition switch **OFF** and disconnect the negative battery cable.
6. Remove the fuel filler cap.
7. Safely elevate and support the vehicle on jackstands.

FUEL SYSTEM   5-49

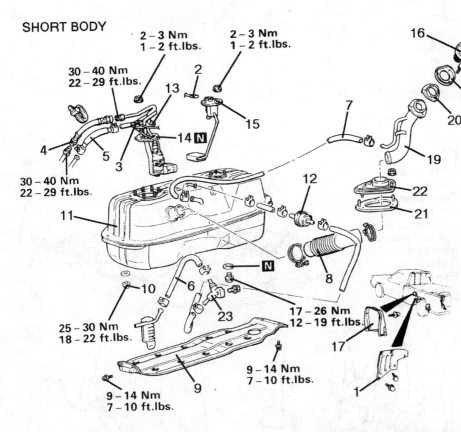

1. Side skirt panel stay
2. Fuel gauge unit connector
3. Fuel pump connector
4. Main hose connection
5. Return hose connection
6. Vapor hose connection
7. Breather hose connection
8. Fuel filler hose connection
9. Fuel tank protector <3.0L engine>
10. Fuel tank mounting nut
11. Fuel tank
12. Fuel tank pressure relief valve
13. Fuel pump assembly
14. Packing
15. Fuel gauge unit

Fuel tank filler tube - Removal steps

7. Breather hose connection
8. Fuel filler hose connection
16. Fuel tank filler tube cap
17. Fuel tank filler tube cover
18. Reinforcement
19. Fuel tank filler tube
20. Packing
21. Retainer
22. Grommet

Check valve and fuel tank pressure control valve - Removal steps

6. Vapor hose connection
23. Fuel tank rollover valve
12. Fuel tank pressure relief valve

Fig. 83 Fuel tank assembly and component location — 1991-95 Pick-ups

5-50 FUEL SYSTEM

1. Drain plug
2. Fuel filler cap
3. Fuel filler hose protector
4. Vapor hose
5. Check valve
6. Overfill limiter (two-way valve)
7. Clamp assembly
8. Fuel filler hose
9. Breather hose
10. Packing
11. Fuel filler neck
12. Main hose
13. Return hose
14. Fuel gauge unit connector
15. Fuel tank assembly mounting nuts
16. Fuel tank
17. Pipe assembly
18. Separator tanks
19. Fuel tank protector

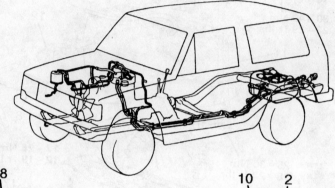
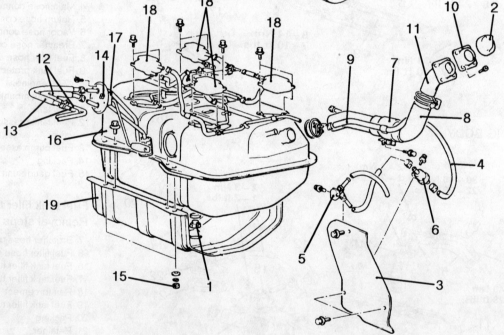

Fig. 84 Fuel tank assembly and component location — 1983-90 carbureted Monteros

FUEL SYSTEM 5-51

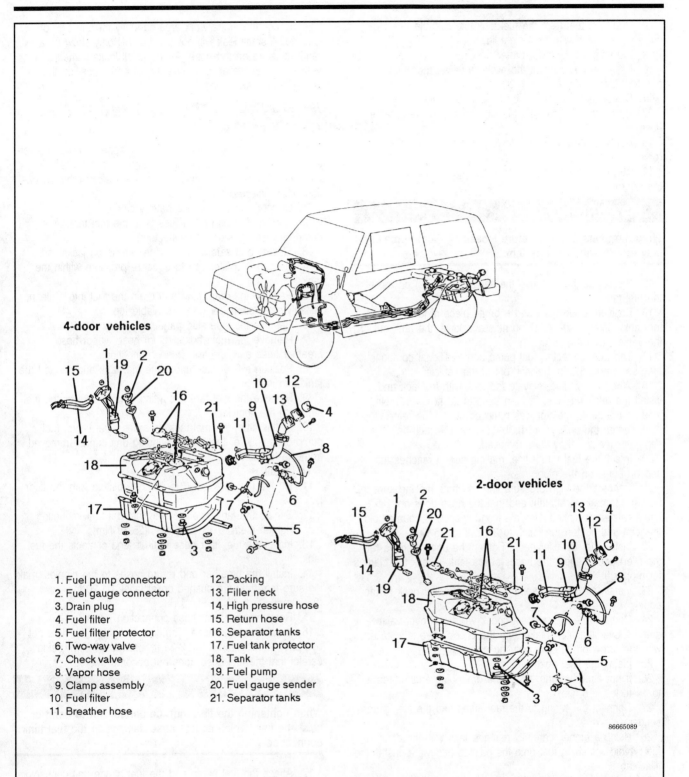

Fig. 85 Fuel tank assembly and component location — 1989-90 fuel injected Monteros

5-52 FUEL SYSTEM

8. Drain the fuel from the fuel tank into suitable container(s), each with an airtight lid.
9. Remove the fuel filler protector.
10. Label and disconnect the two-way valve, the check valve and the vapor hose.
11. Disconnect the clamp holding the filler neck to the body.
12. At the tank, disconnect the filler hose and the breather hose.
13. Remove the retaining screws at the top of the filler and remove the filler neck from the car.
14. At the fuel pump, disconnect the high pressure line and the return line.

✲✲WARNING

Wrap the joints in a rag before loosening. Some pressure may remain within the system.

15. Disconnect and remove the separator tank(s) on top of the fuel tank.
16. Position a floor jack and a broad piece of lumber under the tank. Remove the retaining nuts and lower the tank out of the vehicle with the jack.
17. The tank protector, fuel pump and additional components may be removed with the tank out of the vehicle.
18. When reinstalling, elevate the tank with the jack and install the retaining nuts. Tighten them to 20 ft. lbs. (27 Nm).
19. If the tank protector was removed, it may be reinstalled now; tighten the bolts to 10 ft. lbs. (14 Nm). Install the separator tanks if they were removed.
20. Install the fuel return line, making sure it reaches the second bulge on the pipe.
21. Connect the high pressure line by hand, making sure the flare nut threads in straight and that the hose does not become twisted or kinked. Use a second wrench to counterhold the joint and tighten the fitting to 26 ft. lbs. (35 Nm).
22. Install the fuel filler neck. Tighten the upper screws (at the filler) and connect the hoses at the tank. Check the clamps for proper fit.
23. Install the clamp assembly holding the filler neck to the body.
24. Connect the vapor hose, check valve (roll over valve) and the overfill limiter. Make sure the check valve is installed with the arrow on its case pointing up.
25. Install the filler hose protector.
26. If not already done, install the drain plug and tighten it to 14 ft. lbs. (18 Nm).
27. Connect the wiring to the fuel pump and the fuel gauge sender.
28. Using a funnel, carefully refill the tank with the drained fuel. Avoid splashing fuel onto the painted bodywork. Install the filler cap.
29. Lower the vehicle to the ground and connect the negative battery cable.
30. Start the engine. It will take a period of cranking before the engine fires and runs smoothly due to the lines being emptied.

31. Check the entire work area carefully for any sign of leakage. A small leak will not drip but will only show up a slight moisture on the hose. Attend to all leaks immediately.
32. Reinstall the access cover panel and reposition the carpeting in the cargo area.

1991-95 MODELS
▶ See Figures 86 and 87

Although some slight differences may appear between different model years or makes, the procedures for all 1991-95 Mitsubishi trucks is essentially the same. Refer to the illustrations if any discrepancies arise.

1. Disconnect the negative battery cable.
2. Unplug all wiring harnesses from the fuel tank (fuel pump and fuel gauge unit connectors).
3. Elevate and safely support the vehicle on jackstands.
4. Remove the filler cap to equalize pressure within the tank.
5. Remove the drain plug and drain the fuel into a clean, suitably-sized container with a sealable top.
6. Remove the side skirt panel stay.
7. Remove the main hose, return hose, vapor hose, breather hose and fuel filler hose connections.
8. Position a floor jack under the fuel tank and support the tank with it.
9. Remove the fuel tank mounting nuts and lower the fuel tank with the floor jack.
10. At this time, the fuel tank pressure relief valve, fuel pump assembly, or fuel gauge sending unit can be removed from the fuel tank.

To install:
11. Lift the fuel tank back up into its position with the floor jack.
12. Secure the tank in place with the fuel tank mounting nuts. Tighten them to 18-22 ft. lbs. (25-30 Nm).
13. Install the fuel filler hose until its end contacts the fuel tank.
14. Install the breather and return hoses to the nipples, making sure that they are pushed on at least 1-1.20 in. (25-30mm).
15. Reattach the vapor hose connection.
16. Install the main hose. Temporarily tighten the flare nut by hand, then tighten it to 22-29 ft. lbs. (30-40 Nm), being careful that the fuel hose does not become twisted.

✲✲CAUTION

When tightening the flare nut, be careful not to bend or twist the line, which could cause damage to the fuel tank connection.

17. Attach the fuel pump and the fuel gauge unit connectors.
18. Install the side skirt panel stay.
19. Fill the tank with approximately 1 gallon of gasoline and connect the negative battery cable.
20. Start the engine and check for any fuel leaks around the hoses going to the fuel tank and their connections; repair if necessary.

FUEL SYSTEM 5-53

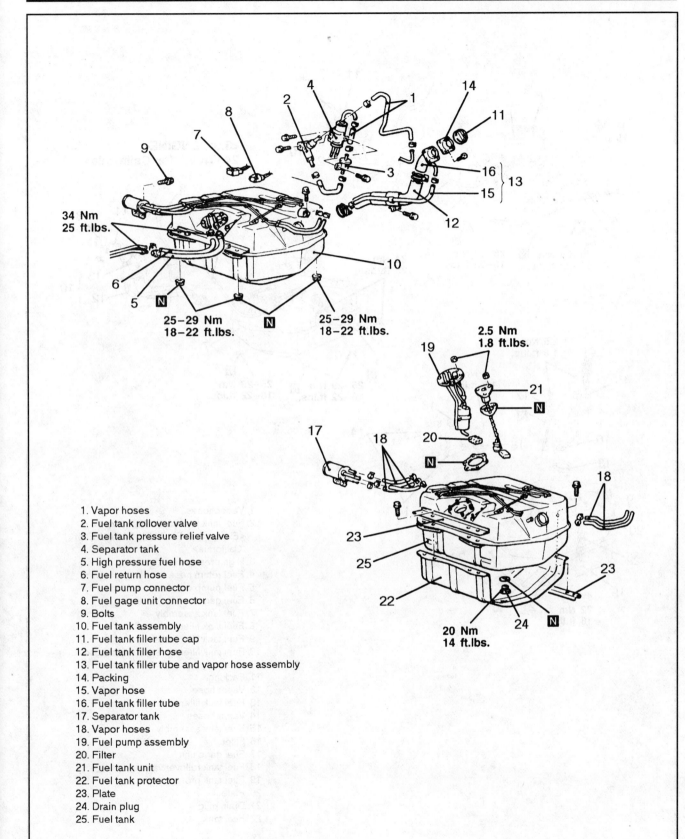

1. Vapor hoses
2. Fuel tank rollover valve
3. Fuel tank pressure relief valve
4. Separator tank
5. High pressure fuel hose
6. Fuel return hose
7. Fuel pump connector
8. Fuel gage unit connector
9. Bolts
10. Fuel tank assembly
11. Fuel tank filler tube cap
12. Fuel tank filler hose
13. Fuel tank filler tube and vapor hose assembly
14. Packing
15. Vapor hose
16. Fuel tank filler tube
17. Separator tank
18. Vapor hoses
19. Fuel pump assembly
20. Filter
21. Fuel tank unit
22. Fuel tank protector
23. Plate
24. Drain plug
25. Fuel tank

Fig. 86 Fuel tank assembly and component location — 1991-94 Monteros

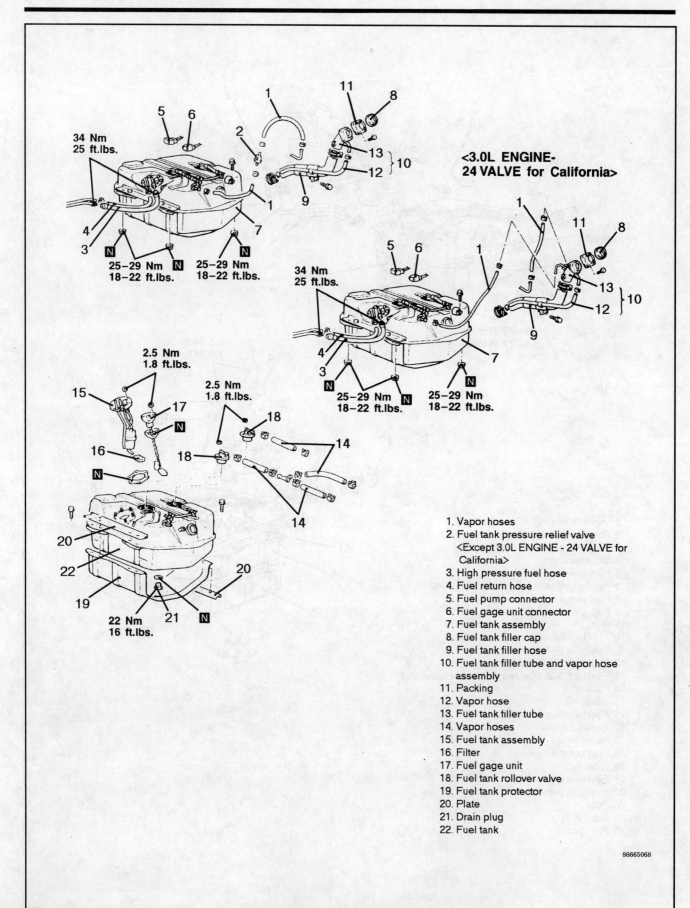

Fig. 87 Fuel tank assembly and component location — 1995 Monteros

AIR CONDITIONER
  AIR CONDITIONING CONTROL
    UNIT 6-50
  AIR CONDITIONING SWITCH 6-46
  COMPRESSOR 6-34
  CONDENSER 6-36
  CONDENSER FAN MOTOR 6-40
  EVAPORATOR CORE 6-41
  EXPANSION VALVE 6-50
  REAR EVAPORATOR
    ASSEMBLY 6-44
  RECEIVER/DRIER 6-50
  REFRIGERANT LINES 6-50
  SUB-CONDENSER AND FAN
    MOTOR 6-40
AUDIO SYSTEM
  RADIO, TAPE PLAYER AND CD
    PLAYER 6-63
  SPEAKERS 6-64
CIRCUIT PROTECTION
  COMPONENT LOCATIONS 6-97
  FUSES 6-97
  FUSIBLE LINKS 6-96
CRUISE CONTROL
  CLUTCH PEDAL POSITION
    SWITCH 6-55
  COLUMN SWITCH 6-54
  MAIN SWITCH 6-54
  PARK/NEUTRAL POSITION
    (INHIBITOR) SWITCH 6-56
  SPEED SENSOR 6-55
  STOP LIGHT/BRAKE SWITCH 6-55
  VACUUM SYSTEM
    COMPONENTS 6-56
HEATER
  BLOWER MOTOR 6-13
  CONTROL CABLES 6-34
  CONTROL PANEL AND BLOWER
    SWITCH 6-30
  HEATER CORE 6-14
INSTRUMENTS AND SWITCHES
  CLOCK 6-85
  COMBINATION SWITCH
    (HEADLIGHT/TURN
    SIGNAL/WINDSHIELD WIPER
    SWITCH) 6-85
  FUEL GAUGE 6-82
  INSTRUMENT
    CLUSTER/COMBINATION
    METER 6-78
  MULTI-METER 6-82
  OIL PRESSURE GAUGE 6-82
  PRINTED CIRCUIT BOARD 6-82
  SPEEDOMETER CABLE 6-80
  TEMPERATURE GAUGE 6-82
  VOLTMETER 6-82
  WINDSHIELD AND REAR WINDOW
    WIPER SWITCHES 6-83
LIGHTING
  HEADLIGHTS 6-87
  SIGNAL AND MARKER LIGHTS 6-92

**SUPPLEMENTAL RESTRAINT SYSTEM
(AIR BAG)**
  AIR BAG MODULE AND CLOCK
    SPRING 6-12
  GENERAL INFORMATION 6-11
TRAILER WIRING 6-96
**UNDERSTANDING AND
TROUBLESHOOTING ELECTRICAL
SYSTEMS**
  MECHANICAL TEST
    EQUIPMENT 6-10
  SAFETY PRECAUTIONS 6-3
  TROUBLESHOOTING AND
    DIAGNOSIS 6-3
  WIRING HARNESSES 6-8
**UNDERSTANDING BASIC
ELECTRICITY 6-2**
WINDSHIELD WIPERS
  INTERMITTENT WIPER RELAY 6-76
  REAR WIPER AND WASHER
    MOTOR 6-74
  WINDSHIELD WASHER MOTOR AND
    FLUID RESERVOIR 6-71
  WINDSHIELD WIPER MOTOR AND
    LINKAGE 6-70
  WIPER BLADE AND ARM 6-70
WIRING DIAGRAMS 6-103

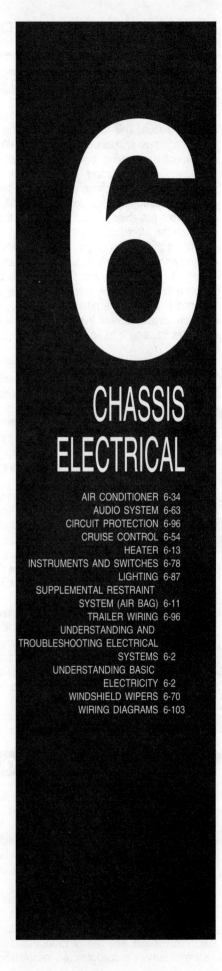

6
CHASSIS ELECTRICAL

AIR CONDITIONER 6-34
AUDIO SYSTEM 6-63
CIRCUIT PROTECTION 6-96
CRUISE CONTROL 6-54
HEATER 6-13
INSTRUMENTS AND SWITCHES 6-78
LIGHTING 6-87
SUPPLEMENTAL RESTRAINT
  SYSTEM (AIR BAG) 6-11
TRAILER WIRING 6-96
UNDERSTANDING AND
TROUBLESHOOTING ELECTRICAL
  SYSTEMS 6-2
UNDERSTANDING BASIC
  ELECTRICITY 6-2
WINDSHIELD WIPERS 6-70
WIRING DIAGRAMS 6-103

6-2 CHASSIS ELECTRICAL

UNDERSTANDING BASIC ELECTRICITY

For any electrical system to operate, it must make a complete circuit. This simply means that the power flow from the battery must make a complete circle. When an electrical component is operating, power flows from the battery to the component, passes through the component causing it to perform its function (lighting a light bulb), and then returns to the battery through the ground of the circuit. This ground is usually (but not always) the metal part of the vehicle or truck on which the electrical component is mounted.

Perhaps the easiest way to visualize this is to think of connecting a light bulb with two wires attached to it to the battery. If one of the two wires attached to the light bulb were attached to the negative post of the battery and the other were attached to the positive post of the battery, you would have a complete circuit. Current from the battery would flow to the light bulb, causing it to light, and return to the negative post of the battery.

The normal automotive circuit differs from this simple example in two ways. First, instead of having a return wire from the bulb to the battery, the light bulb often returns the current to the battery through the chassis of the vehicle. Since the negative battery cable is attached to the chassis and the chassis is made of electrically conductive metal, the chassis of the vehicle can serve as ground wire to complete the circuit. Secondly, most automotive circuits contain switches to turn components on and off as required.

Every complete circuit must include a component or load which is using the power from the power source. If you were to disconnect the light bulb from the wires and touch the two wires together (don't do this), the power source, in this case the battery, would attempt to deliver all its power instantly to the opposite pole. This tremendous current flow causes extreme heat to build within the system, often melting the insulation on the wiring.

Because grounding a wire from a power source makes a complete circuit, but without a load to use it, this phenomenon is called a short circuit. Common causes are: broken insulation (exposing the metal wire to a metal part of the vehicle), or an internally shorted switch. Water leaking into normally sealed components can also conduct electricity to undesired locations.

Some electrical components which require a large amount of current to operate (blower motor, rear defogger, headlights, etc.) have a relay in their circuit. Since these circuits carry a large amount of current, the thickness of the wire in the circuit (gauge size) is also greater. If this large wire were connected from the component to the control switch on the instrument panel, and then back to the component, a voltage drop would occur in the circuit. To prevent this potential drop in voltage, an electromagnetic switch (relay) is used.

The large wires in the circuit are connected from the battery to one side of the relay, and from the opposite side of the relay to the component. The relay is normally open, preventing current from passing through the circuit. An additional, smaller, wire is connected from the relay to the control switch for the circuit. When the control switch is turned on, it grounds the smaller wire from the relay and completes the circuit. This closes the relay and allows current to flow from the battery to the component. The horn, headlight, and starter circuits are three which use relays.

It is possible for larger surges of current to pass through the electrical system of your vehicle or truck. If this surge of current were to reach an electrical component, it could burn it out. To prevent this, fuses, circuit breakers and fusible links are connected into the current supply wires. These are nothing more than pre-planned weak spots within the system. Too much current WILL damage something; if the something is easy to find and easy to replace, major components and wiring will be protected. Fuses are designed to pass the amount of current (amperes) shown on the fuse. If a current flow develops which exceeds the rating, the fuse element separates (blows), causing an open circuit and stopping the current flow.

A circuit breaker is essentially a self-repairing fuse. The circuit breaker opens the circuit the same way a fuse does. However, when either the short is removed from the circuit or the surge subsides, the circuit breaker resets itself and does not have to be replaced as a fuse does.

A fuse link or main link is a wire that acts as a fuse. It is normally connected between the starter relay and the main wiring harness or the positive battery cable and the main wiring harness under the hood. The fuse link (if installed) protects all the chassis electrical components by cutting off the main power supply if necessary. If the vehicle shows absolutely no electrical response when the key is turned, the main link is suspect. Check it first; it's rare for a battery to get so low that something won't work.

UNDERSTANDING AND TROUBLESHOOTING ELECTRICAL SYSTEMS

At the rate which both import and domestic manufacturers are incorporating electronic control systems into their production lines, every new vehicle already is equipped with one or more on-board computer. These electronic components (with no moving parts) should theoretically last the life of the vehicle, provided nothing external happens to damage the circuits or memory chips.

While it is true that electronic components should never wear out, in the real world malfunctions do occur. It is also true that any computer-based system is extremely sensitive to electrical voltages and cannot tolerate careless or haphazard testing or service procedures. An inexperienced individual can literally do major damage looking for a minor problem by using the wrong kind of test equipment or connecting test leads or connectors with the ignition switch ON. When selecting test equipment, make sure the manufacturers instructions state that the tester is compatible with whatever type of electronic control system is being serviced. Read all instructions carefully and double check all test points before installing probes or making any test connections.

The following section outlines basic diagnosis techniques for dealing with computerized automotive control systems. Along with a general explanation of the various types of test equipment available to aid in servicing modern electronic

CHASSIS ELECTRICAL 6-3

automotive systems, basic repair techniques for wiring harnesses and connectors is given. Read the basic information before attempting any repairs or testing on any computerized system, to provide the background of information necessary to avoid the most common and obvious mistakes that can cost both time and money. Although the replacement and testing procedures are simple in themselves, the systems are not, and unless one has a thorough understanding of all components and their function within a particular computerized control system, the logical test sequence these systems demand cannot be followed. Minor malfunctions can make a big difference, so it is important to know how each component affects the operation of the overall electronic system to find the ultimate e cause of a problem without replacing good components unnecessarily. It is not enough to use the correct test equipment; the test equipment must be used correctly.

Safety Precautions

※※CAUTION

Whenever working on or around any computer based microprocessor control system, always observe these general precautions to reduce the possibility of personal injury or damage to electronic components.

- Never install or remove battery cables with the key **ON** or the engine running.
- Jumper cables should be connected with the key **OFF** to avoid power surges that can damage electronic control units.
- Engines equipped with computer controlled systems should avoid both giving and getting jump starts due to the possibility of serious damage to components from arcing in the engine compartment when connections are made with the ignition **ON**.
- Always remove the battery cables before charging the battery.
- Never use a high output charger on an installed battery or attempt to use any type of "hot shot" (24 volt) starting aid.
- Exercise care when inserting test probes into connectors to insure good connections without damaging the connector or spreading the pins. Always probe connectors from the rear (wire) side, NOT the pin side, to avoid accidental shorting of terminals during test procedures.
- Never remove or attach wiring harness connectors with the ignition switch **ON**, especially to an electronic control unit.
- Do not drop any components during service procedures and never apply 12 volts directly to any component (like a solenoid or relay) unless instructed specifically to do so. Some component electrical windings are designed to safely handle only 4 or 5 volts and can be destroyed in seconds if 12 volts are applied directly to the connector.
- Remove the electronic control unit (module) if the vehicle is to be placed in an environment where temperatures exceed approximately 176°F (80°C), such as a paint spray booth or when arc or gas welding near the control unit location in the truck.

Troubleshooting and Diagnosis

ORGANIZED TROUBLESHOOTING

When diagnosing a specific problem, organized troubleshooting is a must. The complexity of a modern automobile demands that you approach any problem in a logical, organized manner. There are certain troubleshooting techniques that are standard:

1. Establish when the problem occurs. Does the problem appear only under certain conditions? Were there any noises, odors, or other unusual symptoms?
2. Isolate the problem area. To do this, make some simple tests and observations; then eliminate the systems that are working properly. Check for obvious problems such as broken wires, dirty connections or split or disconnected vacuum hoses. Always check the obvious before assuming something complicated is the cause.
3. Test for problems systematically to determine the cause once the problem area is isolated. Are all the components functioning properly? Is there power going to electrical switches and motors? Is there vacuum at vacuum switches and/or actuators? Is there a mechanical problem such as bent linkage or loose mounting screws? Doing careful, systematic checks will often turn up most causes on the first inspection without wasting time checking components that have little or no relationship to the problem.
4. Test all repairs after the work is done to make sure that the problem is fixed. Some causes can be traced to more than one component, so a careful verification of repair work is important to pick up additional malfunctions that may cause a problem to reappear or a different problem to arise. A blown fuse, for example, is a simple problem that may require more than another fuse to repair. If you don't look for a problem that caused a fuse to blow, for example, a shorted wire may go undetected.

Experience has shown that most problems tend to be the result of a fairly simple and obvious cause, such as loose or corroded connectors or air leaks in the intake system; making careful inspection of components during testing essential to quick and accurate troubleshooting. Special, hand held computerized testers designed specifically for diagnosing the system are available from a variety of aftermarket sources, as well as from the vehicle manufacturer, but care should be taken that any test equipment being used is designed to diagnose that particular computer controlled system accurately without damaging the control unit (ECU) or components being tested.

➡**Pinpointing the exact cause of trouble in an electrical system can sometimes only be accomplished by the use of special test equipment. The following describes commonly used test equipment and explains how to put it to best use in diagnosis. In addition to the information covered below, the manufacturer's instructions booklet provided with the tester should be read and clearly understood before attempting any test procedures.**

6-4 CHASSIS ELECTRICAL

SPECIFIC TROUBLESHOOTING

Electrical problems generally fall into one of three areas:
- The component is not receiving current.
- The component has failed and cannot use the current properly, if at all.
- The component is not properly grounded.

The electrical system can be checked with a test light and a jumper wire. A test light is a device that looks like a pointed screwdriver with a wire attached to it and has a light bulb in its handle. A jumper wire is a piece of insulated wire with an alligator clip attached to each end.

It should be noted that a test light will only show that voltage is present; it will not indicate the amount of voltage. Certain components will only function if a minimum voltage is supplied. If a more than a "yes or no" answer is required during diagnosis, use a voltmeter. If this must be purchased, purchase a multimeter or volt/ohmmeter (VOM). These reasonably inexpensive tools include several scales for both AC (household) and DC volts as well as an ohmmeter for checking resistance and continuity.

If a component is not working, you must follow a systematic plan to determine which of the three causes is the villain.

1. Turn on the switch that controls the inoperable component.
2. Disconnect the power supply wire from the component.
3. Attach the ground wire on the test light to a good metal ground.
4. Touch the probe end of the test light to the end of the power supply wire that was disconnected from the component. If the component is receiving current, the test light will go on.

➡ **Some components work only when the ignition switch is turned ON.**

If the test light does not go on, then the problem is in the circuit between the battery and the component. This includes all the switches, fuses and relays in the system. Follow the wire that runs back to the battery. The problem is an open circuit between the battery and the component. If the fuse is blown and, when replaced, immediately blows again, there is a short circuit in the system which must be located and repaired. If there is a switch in the system, bypass it with a jumper wire. This is done by connecting one end of the jumper wire to the power supply wire into the switch and the other end of the jumper wire to the wire coming out of the switch. If the test light lights with the jumper wire installed, the switch or whatever was bypassed is defective.

➡ **Never substitute the jumper wire for the component, since a load is required to use the power from the battery.**

5. If the bulb in the test light goes on, then the current is getting to the component that is not working. This eliminates the first of the three possible causes. Connect the power supply wire and connect a jumper wire from the component to a good metal ground. Do this with the switch which controls the component turned on, and also the ignition switch turned on if it is required for the component to work. If the component works with the jumper wire installed, then it has a bad ground. This is usually caused by the metal area on which the component mounts to the chassis being coated with some type of foreign matter.

6. If neither test located the source of the trouble, then the component itself is defective. Remember that for any electrical system to work, all connections must be clean and tight.

TEST EQUIPMENT

Jumper Wires

Jumper wires are simple, yet extremely valuable, pieces of test equipment. Jumper wires are merely wires that are used to bypass sections of a circuit. The simplest type of jumper wire is merely a length of multi-strand wire with an alligator clip at each end. Jumper wires are usually fabricated from lengths of standard automotive wire and whatever type of connector (alligator clip, spade connector or pin connector) that is required for the particular vehicle being tested. The well equipped tool box will have several different styles of jumper wires in several different lengths. Some jumper wires are made with three or more terminals coming from a common splice for special purpose testing. In cramped, hard-to-reach areas it is advisable to have insulated boots over the jumper wire terminals in order to prevent accidental grounding, sparks, and possible fire, especially when testing fuel system components.

Jumper wires are used primarily to locate open electrical circuits, on either the ground (-) side of the circuit or on the hot (+) side. If an electrical component fails to operate, connect the jumper wire between the component and a good ground. If the component operates only with the jumper installed, the ground circuit is open. If the ground circuit is good, but the component does not operate, the circuit between the power feed and component is open. You can sometimes connect the jumper wire directly from the battery to the hot terminal of the component, but first make sure the component uses 12 volts in operation. Some electrical components, such as fuel injectors, are designed to operate on about 4 volts and running 12 volts directly to the injector terminals can burn out the wiring. By inserting an inline fuseholder between a set of test leads, a fused jumper wire can be used for bypassing open circuits. Use a 5 amp fuse to provide protection against voltage spikes. When in doubt, use a voltmeter to check the voltage input to the component and measure how much voltage is being applied normally. By moving the jumper wire successively back from the lamp toward the power source, you can isolate the area of the circuit where the open is located. When the component stops functioning, or the power is cut off, the open is in the segment of wire between the jumper and the point previously tested.

✲✲CAUTION

Never use jumpers made from wire that is of lighter gauge than used in the circuit under test. If the jumper wire is of too small gauge, it may overheat and possibly melt. Never use jumpers to bypass high resistance loads (such as motors) in a circuit. Bypassing resistances, in effect, creates a short circuit which may, in turn, cause damage and fire. Never use a jumper for anything other than temporary bypassing of components in a circuit.

CHASSIS ELECTRICAL 6-5

12 Volt Test Light

The 12 volt test light is used to check circuits and components while electrical current is flowing through them. It is used for voltage and ground tests. Twelve volt test lights come in different styles but all have three main parts; a ground clip, a probe, and a light. The most commonly used 12 volt test lights have pick-type probes. To use a 12 volt test light, connect the ground clip to a good ground and probe wherever necessary with the pick. The pick should be sharp so that it can penetrate wire insulation to make contact with the wire, without making a large hole in the insulation. The wrap-around light is handy in hard to reach areas or where it is difficult to support a wire to push a probe pick into it. To use the wrap around light, hook the wire to probed with the hook and pull the trigger. A small pick will be forced through the wire insulation into the wire core.

✳✳CAUTION

Do not use a test light to probe electronic ignition spark plug or coil wires. Never use a pick-type test light to probe wiring on computer controlled systems unless specifically instructed to do so. Any wire insulation that is pierced by the test light probe should be taped and sealed with silicone after testing.

Like the jumper wire, the 12 volt test light is used to isolate opens in circuits. But, whereas the jumper wire is used to bypass the open to operate the load, the 12 volt test light is used to locate the presence of voltage in a circuit. If the test light glows, you know that there is power up to that point; if the 12 volt test light does not glow when its probe is inserted into the wire or connector, you know that there is an open circuit (no power). Move the test light in successive steps back toward the power source until the light in the handle does glow. When it does glow, the open is between the probe and point previously probed.

➡**The test light does not detect that 12 volts (or any particular amount of voltage) is present; it only detects that some voltage is present. It is advisable before using the test light to touch its terminals across the battery posts to make sure the light is operating properly.**

Self-Powered Test Light

The self-powered test light usually contains a 1.5 volt penlight battery. One type of self-powered test light is similar in design to the 12 volt test light. This type has both the battery and the light in the handle and pick-type probe tip. The second type has the light toward the open tip, so that the light illuminates the contact point. The self-powered test light is dual purpose piece of test equipment. It can be used to test for either open or short circuits when power is isolated from the circuit (continuity test). A powered test light should not be used on any computer controlled system or component unless specifically instructed to do so. Many engine sensors can be destroyed by even this small amount of voltage applied directly to the terminals.

Open Circuit Testing

To use the self-powered test light to check for open circuits, first isolate the circuit from the vehicle's 12 volt power source by disconnecting the battery or wiring harness connector. Connect the test light ground clip to a good ground and probe sections of the circuit sequentially with the test light. (start from either end of the circuit). If the light is out, the open is between the probe and the circuit ground. If the light is on, the open is between the probe and end of the circuit toward the power source.

Short Circuit Testing

By isolating the circuit both from power and from ground, and using a self-powered test light, you can check for shorts to ground in the circuit. Isolate the circuit from power and ground. Connect the test light ground clip to a good ground and probe any easy-to-reach test point in the circuit. If the light comes on, there is a short somewhere in the circuit. To isolate the short, probe a test point at either end of the isolated circuit (the light should be on). Leave the test light probe connected and open connectors, switches, remove parts, etc., sequentially, until the light goes out. When the light goes out, the short is between the last circuit component opened and the previous circuit opened.

➡**The 1.5 volt battery in the test light does not provide much current. A weak battery may not provide enough power to illuminate the test light even when a complete circuit is made (especially if there are high resistances in the circuit). Always make sure that the test battery is strong. To check the battery, briefly touch the ground clip to the probe; if the light glows brightly the battery is strong enough for testing. Never use a self-powered test light to perform checks for opens or shorts when power is applied to the electrical system under test. The 12 volt vehicle power will quickly burn out the 1.5 volt light bulb in the test light.**

Voltmeter

A voltmeter is used to measure voltage at any point in a circuit, or to measure the voltage drop across any part of a circuit. It can also be used to check continuity in a wire or circuit by indicating current flow from one end to the other. Voltmeters usually have various scales on the meter dial and a selector switch to allow the selection of different voltages. The voltmeter has a positive and a negative lead. To avoid damage to the meter, always connect the negative lead to the negative (-) side of circuit (to ground or nearest the ground side of the circuit) and connect the positive lead to the positive (+) side of the circuit (to the power source or the nearest power source). Note that the negative voltmeter lead will always be black and that the positive voltmeter will always be some color other than black (usually red). Depending on how the voltmeter is connected into the circuit, it has several uses.

A voltmeter can be connected either in parallel or in series with a circuit and it has a very high resistance to current flow. When connected in parallel, only a small amount of current will flow through the voltmeter current path; the rest will flow through the normal circuit current path and the circuit will work normally. When the voltmeter is connected in series with a circuit, only a small amount of current can flow through the

6-6 CHASSIS ELECTRICAL

circuit. The circuit will not work properly, but the voltmeter reading will show if the circuit is complete or not.

Available Voltage Measurement

Set the voltmeter selector switch to the 20V position and connect the meter negative lead to the negative post of the battery. Connect the positive meter lead to the positive post of the battery and turn the ignition switch **ON** to provide a load. Read the voltage on the meter or digital display. A well charged battery should register over 12 volts. If the meter reads below 11.5 volts, the battery power may be insufficient to operate the electrical system properly. This test determines voltage available from the battery and should be the first step in any electrical trouble diagnosis procedure. Many electrical problems, especially on computer controlled systems, can be caused by a low state of charge in the battery. Excessive corrosion at the battery cable terminals can cause a poor contact that will prevent proper charging and full battery current flow.

Normal battery voltage is 12 volts when fully charged. When the battery is supplying current to one or more circuits it is said to be "under load". When everything is off the electrical system is under a "no-load" condition. A fully charged battery may show about 12.5 volts at no load; will drop to 12 volts under medium load; and will drop even lower under heavy load. If the battery is partially discharged the voltage decrease under heavy load may be excessive, even though the battery shows 12 volts or more at no load. When allowed to discharge further, the battery's available voltage under load will decrease more severely. For this reason, it is important that the battery be fully charged during all testing procedures to avoid errors in diagnosis and incorrect test results.

Voltage Drop

When current flows through a resistance, the voltage beyond the resistance is reduced (the larger the current, the greater the reduction in voltage). When no current is flowing, there is no voltage drop because there is no current flow. All points in the circuit which are connected to the power source are at the same voltage as the power source. The total voltage drop always equals the total source voltage. In a long circuit with many connectors, a series of small, unwanted voltage drops due to corrosion at the connectors can add up to a total loss of voltage which impairs the operation of the normal loads in the circuit.

INDIRECT COMPUTATION OF VOLTAGE DROPS

1. Set the voltmeter selector switch to the 20 volt position.
2. Connect the meter negative lead to a good ground.
3. Probe all resistances in the circuit with the positive meter lead.
4. Operate the circuit in all modes and observe the voltage readings.

DIRECT MEASUREMENT OF VOLTAGE DROPS

1. Set the voltmeter switch to the 20 volt position.
2. Connect the voltmeter negative lead to the ground side of the resistance load to be measured.
3. Connect the positive lead to the positive side of the resistance or load to be measured.

4. Read the voltage drop directly on the 20 volt scale.

Too high a voltage indicates too high a resistance. If, for example, a blower motor runs too slowly, you can determine if there is too high a resistance in the resistor pack. By taking voltage drop readings in all parts of the circuit, you can isolate the problem. Too low a voltage drop indicates too low a resistance. If, for example, a blower motor runs too fast in the **MED** and/or **LOW** position, the problem can be isolated in the resistor pack by taking voltage drop readings in all parts of the circuit to locate a possibly shorted resistor. The maximum allowable voltage drop under load is critical, especially if there is more than one high resistance problem in a circuit because all voltage drops are cumulative. A small drop is normal due to the resistance of the conductors.

HIGH RESISTANCE TESTING

1. Set the voltmeter selector switch to the 4 volt position.
2. Connect the voltmeter positive lead to the positive post of the battery.
3. Turn on the headlights and heater blower to provide a load.
4. Probe various points in the circuit with the negative voltmeter lead.
5. Read the voltage drop on the 4 volt scale. Some average maximum allowable voltage drops are:
   - FUSE PANEL — 0.7 volts
   - IGNITION SWITCH — 0.5 volts
   - HEADLIGHT SWITCH — 0.7 volts
   - IGNITION COIL (+) — 0.5 volts
   - ANY OTHER LOAD — 1.3 volts

➥Voltage drops are all measured while a load is operating; without current flow, there will be no voltage drop.

Ohmmeter

The ohmmeter is designed to read resistance (ohms) in a circuit or component. Although there are several different styles of ohmmeters, all will usually have a selector switch which permits the measurement of different ranges of resistance (usually the selector switch allows the multiplication of the meter reading by 10, 100, 1000, and 10,000). A calibration knob allows the meter to be set at zero for accurate measurement. Since all ohmmeters are powered by an internal battery (usually 9 volts), the ohmmeter can be used as a self-powered test light. When the ohmmeter is connected, current from the ohmmeter flows through the circuit or component being tested. Since the ohmmeter's internal resistance and voltage are known values, the amount of current flow through the meter depends on the resistance of the circuit or component being tested.

The ohmmeter can be used to perform continuity test for opens or shorts (either by observation of the meter needle or as a self-powered test light), and to read actual resistance in a circuit. It should be noted that the ohmmeter is used to check the resistance of a component or wire while there is no voltage applied to the circuit. Current flow from an outside voltage source (such as the vehicle battery) can damage the ohmmeter, so the circuit or component should be isolated from the vehicle electrical system before any testing is done. Since

CHASSIS ELECTRICAL 6-7

the ohmmeter uses its own voltage source, either lead can be connected to any test point.

➡ **When checking diodes or other solid state components, the ohmmeter leads can only be connected one way in order to measure current flow in a single direction. Make sure the positive (+) and negative (-) terminal connections are as described in the test procedures to verify the one-way diode operation.**

In using the meter for making continuity checks, do not be concerned with the actual resistance readings. Zero resistance, or any resistance readings, indicate continuity in the circuit. Infinite resistance indicates an open in the circuit. A high resistance reading where there should be none indicates a problem in the circuit. Checks for short circuits are made in the same manner as checks for open circuits except that the circuit must be isolated from both power and normal ground. Infinite resistance indicates no continuity to ground, while zero resistance indicates a dead short to ground.

RESISTANCE MEASUREMENT

The batteries in an ohmmeter will weaken with age and temperature, so the ohmmeter must be calibrated or "zeroed" before taking measurements. To zero the meter, place the selector switch in its lowest range and touch the two ohmmeter leads together. Turn the calibration knob until the meter needle is exactly on zero.

➡ **All analog (needle) type ohmmeters must be zeroed before use, but some digital ohmmeter models are automatically calibrated when the switch is turned on. Self-calibrating digital ohmmeters do not have an adjusting knob, but it's a good idea to check for a zero readout before use by touching the leads together. All computer controlled systems require the use of a digital ohmmeter with at least 10 megohms impedance for testing. Before any test procedures are attempted, make sure the ohmmeter used is compatible with the electrical system or damage to the on-board computer could result.**

To measure resistance, first isolate the circuit from the vehicle power source by disconnecting the battery cables or the harness connector. Make sure the key is **OFF** when disconnecting any components or the battery. Where necessary, also isolate at least one side of the circuit to be checked to avoid reading parallel resistances. Parallel circuit resistances will always give a lower reading than the actual resistance of either of the branches. When measuring the resistance of parallel circuits, the total resistance will always be lower than the smallest resistance in the circuit. Connect the meter leads to both sides of the circuit (wire or component) and read the actual measured ohms on the meter scale. Make sure the selector switch is set to the proper ohm scale for the circuit being tested to avoid misreading the ohmmeter test value.

✲✲WARNING

Never use an ohmmeter with power applied to the circuit. Like the self-powered test light, the ohmmeter is designed to operate on its own power supply. The normal 12 volt automotive electrical system current could damage the meter!

Ammeters

An ammeter measures the amount of current flowing through a circuit in units called amperes or amps. Amperes are units of electron flow which indicate how fast the electrons are flowing through the circuit. Since Ohms Law dictates that current flow in a circuit is equal to the circuit voltage divided by the total circuit resistance, increasing voltage also increases the current level (amps). Likewise, any decrease in resistance will increase the amount of amps in a circuit. At normal operating voltage, most circuits have a characteristic amount of amperes, called "current draw" which can be measured using an ammeter. By referring to a specified current draw rating, measuring the amperes, and comparing the two values, one can determine what is happening within the circuit to aid in diagnosis. An open circuit, for example, will not allow any current to flow so the ammeter reading will be zero. More current flows through a heavily loaded circuit or when the charging system is operating.

An ammeter is always connected in series with the circuit being tested. All of the current that normally flows through the circuit must also flow through the ammeter; if there is any other path for the current to follow, the ammeter reading will not be accurate. The ammeter itself has very little resistance to current flow and therefore will not affect the circuit, but it will measure current draw only when the circuit is closed and electricity is flowing. Excessive current draw can blow fuses and drain the battery, while a reduced current draw can cause motors to run slowly, lights to dim and other components to not operate properly. The ammeter can help diagnose these conditions by locating the cause of the high or low reading.

Multimeters

Different combinations of test meters can be built into a single unit designed for specific tests. Some of the more common combination test devices are known as Volt/Amp testers, Tach/Dwell meters, or Digital Multimeters. The Volt/Amp tester is used for charging system, starting system or battery tests and consists of a voltmeter, an ammeter and a variable resistance carbon pile. The voltmeter will usually have at least two ranges for use with 6, 12 and 24 volt systems. The ammeter also has more than one range for testing various levels of battery loads and starter current draw and the carbon pile can be adjusted to offer different amounts of resistance. The Volt/Amp tester has heavy leads to carry large amounts of current and many later models have an inductive ammeter Pick-up that clamps around the wire to simplify test connections. On some models, the ammeter also has a zero-center scale to allow testing of charging and starting systems without switching leads or polarity. A digital multimeter is a voltmeter, ammeter and ohmmeter combined in an instrument

6-8 CHASSIS ELECTRICAL

which gives a digital readout. These are often used when testing solid state circuits because of their high input impedance (usually 10 megohms or more).

The tach/dwell meter combines a tachometer and a dwell (cam angle) meter and is a specialized kind of voltmeter. The tachometer scale is marked to show engine speed in rpm and the dwell scale is marked to show degrees of distributor shaft rotation. In most electronic ignition systems, dwell is determined by the control unit, but the dwell meter can also be used to check the duty cycle (operation) of some electronic engine control systems. Some tach/dwell meters are powered by an internal battery, while others take their power from the truck battery in use. The battery powered testers usually require calibration much like an ohmmeter before testing.

Special Test Equipment

A variety of diagnostic tools are available to help troubleshoot and repair computerized engine control systems. The most sophisticated of these devices are the console type engine analyzers that usually occupy a garage service bay, but there are several types of aftermarket electronic testers available that will allow quick circuit tests of the engine control system by plugging directly into a special connector located in the engine compartment or under the dashboard. Several tool and equipment manufacturers offer simple, hand held testers that measure various circuit voltage levels on command to check all system components for proper operation. Although these testers often cost about $300-500, consider that the average computer control unit (or ECM) can cost just as much and the money saved by not replacing perfectly good sensors or components in an attempt to correct a problem could justify the purchase price of a special diagnostic tester the first time it's used.

These computerized testers can allow quick and easy test measurements while the engine is operating or while the truck is being driven. In addition, the on-board computer memory can be read to access any stored trouble codes; in effect allowing the computer to tell you where it hurts and aid trouble diagnosis by pinpointing exactly which circuit or component is malfunctioning. In the same manner, repairs can be tested to make sure the problem has been corrected. The biggest advantage these special testers have is their relatively easy hookups that minimize or eliminate the chances of making the wrong connections and getting false voltage readings or damaging the computer accidentally.

➥**It should be remembered that these testers check voltage levels in circuits; they don't detect mechanical problems or failed components if the circuit voltage falls within the preprogrammed limits stored in the tester PROM unit. Also, most of the hand held testers are designed to work only on one or two systems made by a specific manufacturer.**

A variety of aftermarket testers are available to help diagnose different computerized control systems. Owatonna Tool Company (OTC), for example, markets a device called the OTC Monitor which plugs directly into the Assembly Line Diagnostic Link (ALDL). The OTC tester makes diagnosis a simple matter of pressing the correct buttons and, by changing the internal PROM or inserting a different diagnosis cartridge, it will work on any model from full size to subcompact, over a wide range of years. An adapter is supplied with the tester to allow connection to all types of ALDL links, regardless of the number of pin terminals used. By inserting an updated PROM into the OTC tester, it can be easily updated to diagnose any new modifications of computerized control systems.

Wiring Harnesses

The average automobile contains about ½ mile (⅘ km) of wiring, with hundreds of individual connections. To protect the many wires from damage and to keep them from becoming a confusing tangle, they are organized into bundles, enclosed in plastic or taped together and called wire harnesses. Different wiring harnesses serve different parts of the vehicle. Individual wires are color coded to help trace them through a harness where sections are hidden from view.

A loose or corroded connection or a replacement wire that is too small for the circuit will add extra resistance and an additional voltage drop to the circuit. A ten percent voltage drop can result in slow or erratic motor operation, for example, even though the circuit is complete. Automotive wiring or circuit conductors can be in any one of three forms:

1. Single strand wire
2. Multi-strand wire
3. Printed circuitry

Single strand wire has a solid metal core and is usually used inside such components as alternators, motors, relays and other devices. Multi-strand wire has a core made of many small strands of wire twisted together into a single conductor. Most of the wiring in an automotive electrical system is made up of multi-strand wire, either as a single conductor or grouped together in a harness. All wiring is color coded on the insulator, either as a solid color or as a colored wire with an identification stripe. A printed circuit is a thin film of copper or other conductor that is printed on an insulator backing. Occasionally, a printed circuit is sandwiched between two sheets of plastic for more protection and flexibility. A complete printed circuit, consisting of conductors, insulating material and connectors for lamps or other components is called a printed circuit board. Printed circuitry is used in place of individual wires or harnesses in places where space is limited, such as behind instrument panels.

WIRE GAUGE

Since computer controlled automotive electrical systems are very sensitive to changes in resistance, the selection of properly sized wires is critical when systems are repaired. The wire gauge number is an expression of the cross section area of the conductor. The most common system for expressing wire size is the American Wire Gauge (AWG) system.

Gauge numbers are assigned to conductors of various cross section areas. As gauge number increases, area decreases and the conductor becomes smaller. A 5 gauge conductor is smaller than a 1 gauge conductor and a 10 gauge is smaller than a 5 gauge. As the cross section area of a conductor decreases, resistance increases and so does the gauge

CHASSIS ELECTRICAL 6-9

number. A conductor with a higher gauge number will carry less current than a conductor with a lower gauge number.

➡Gauge wire size refers to the size of the conductor, not the size of the complete wire. It is possible to have two wires of the same gauge with different diameters because one may have thicker insulation than the other.

12 volt automotive electrical systems generally use 10, 12, 14, 16 and 18 gauge wire. Main power distribution circuits and larger accessories usually use 10 and 12 gauge wire. Battery cables are usually 4 or 6 gauge, although 1 and 2 gauge wires are occasionally used. Wire length must also be considered when making repairs to a circuit. As conductor length increases, so does resistance. An 18 gauge wire, for example, can carry a 10 amp load for 10 feet without excessive voltage drop; however if a 15 foot wire is required for the same 10 amp load, it must be a 16 gauge wire.

An electrical schematic shows the electrical current paths when a circuit is operating properly. It is essential to understand how a circuit works before trying to figure out why it doesn't. Schematics break the entire electrical system down into individual circuits and show only one particular circuit. In a schematic, no attempt is made to represent wiring and components as they physically appear on the vehicle; switches and other components are shown as simply as possible. Face views of harness connectors show the cavity or terminal locations in all multi-pin connectors to help locate test points.

If you need to backprobe a connector while it is on the component, the order of the terminals must be mentally reversed. The wire color code can help in this situation, as well as a keyway, lock tab or other reference mark.

WIRING REPAIR

Soldering is a quick, efficient method of joining metals permanently. Everyone who has the occasion to make wiring repairs should know how to solder. Electrical connections that are soldered are far less likely to come apart and will conduct electricity much better than connections that are only "pig-tailed" together. The most popular (and preferred) method of soldering is with an electrical soldering gun. Soldering irons are available in many sizes and wattage ratings. Irons with higher wattage ratings deliver higher temperatures and recover lost heat faster. A small soldering iron rated for no more than 50 watts is recommended, especially on electrical systems where excess heat can damage the components being soldered.

There are three ingredients necessary for successful soldering; proper flux, good solder and sufficient heat. A soldering flux is necessary to clean the metal of tarnish, prepare it for soldering and to enable the solder to spread into tiny crevices. When soldering, always use a resin flux or resin core solder which is non-corrosive and will not attract moisture once the job is finished. Other types of flux (acid core) will leave a residue that will attract moisture and cause the wires to corrode. Tin is a unique metal with a low melting point. In a molten state, it dissolves and alloys easily with many metals. Solder is made by mixing tin with lead. The most common proportions are 40/60, 50/50 and 60/40, with the percentage of tin listed first. Low priced solders usually contain less tin, making them very difficult for a beginner to use because more heat is required to melt the solder. A common solder is 40/60 which is well suited for all-around general use, but 60/40 melts easier, has more tin for a better joint and is preferred for electrical work.

Soldering Techniques

Successful soldering requires that the metals to be joined be heated to a temperature that will melt the solder usually 360-460°F (182-238°C). Contrary to popular belief, the purpose of the soldering iron is not to melt the solder itself, but to heat the parts being soldered to a temperature high enough to melt the solder when it is touched to the work. Melting flux-cored solder on the soldering iron will usually destroy the effectiveness of the flux.

➡Soldering tips are made of copper for good heat conductivity, but must be "tinned" regularly for quick transference of heat to the project and to prevent the solder from sticking to the iron. To "tin" the iron, simply heat it and touch the flux-cored solder to the tip; the solder will flow over the hot tip. Wipe the excess off with a clean rag, but be careful as the iron will be hot.

After some use, the tip may become pitted. If so, simply dress the tip smooth with a smooth file and "tin" the tip again. An old saying holds that "metals well cleaned are half soldered." Flux-cored solder will remove oxides but rust, bits of insulation and oil or grease must be removed with a wire brush or emery cloth. For maximum strength in soldered parts, the joint must start off clean and tight. Weak joints will result in gaps too wide for the solder to bridge.

If a separate soldering flux is used, it should be brushed or swabbed on only those areas that are to be soldered. Most solders contain a core of flux and separate fluxing is unnecessary. Hold the work to be soldered firmly. It is best to solder on a wooden board, because a metal vise will only rob the piece to be soldered of heat and make it difficult to melt the solder. Hold the soldering tip with the broadest face against the work to be soldered. Apply solder under the tip close to the work, using enough solder to give a heavy film between the iron and the piece being soldered, while moving slowly and making sure the solder melts properly. Keep the work level or the solder will run to the lowest part and favor the thicker parts, because these require more heat to melt the solder. If the soldering tip overheats (the solder coating on the face of the tip burns up), it should be retinned. Once the soldering is completed, let the soldered joint stand until cool. Tape and seal all soldered wire splices after the repair has cooled.

Wire Harness and Connectors

The on-board computer (ECM) wire harness electrically connects the control unit to the various solenoids, switches and sensors used by the control system. Most connectors in the engine compartment or otherwise exposed to the elements are protected against moisture and dirt which could create oxidation and deposits on the terminals. This protection is important because of the very low voltage and current levels used by the computer and sensors. All connectors have a lock which secures the male and female terminals together, with a secondary lock holding the seal and terminal into the connector. Both terminal locks must be released when disconnecting ECM connectors.

These special connectors are weather-proof and all repairs require the use of a special terminal and the tool required to service it. This tool is used to remove the pin and sleeve terminals. If removal is attempted with an ordinary pick, there is a good chance that the terminal will be bent or deformed. Unlike standard blade type terminals, these terminals cannot be straightened once they are bent. Make certain that the connectors are properly seated and all of the sealing rings in place when connecting leads. On some models, a hinge-type flap provides a backup or secondary locking feature for the terminals. Most secondary locks are used to improve the connector reliability by retaining the terminals if the small terminal lock tongs are not positioned properly.

Molded-on connectors require complete replacement of the connection. This means splicing a new connector assembly into the harness. All splices in on-board computer systems should be soldered to insure proper contact. Use care when probing the connections or replacing terminals in them as it is possible to short between opposite terminals. If this happens to the wrong terminal pair, it is possible to damage certain components. Always use jumper wires between connectors for circuit checking and never probe through weatherproof seals.

Open circuits are often difficult to locate by sight because corrosion or terminal misalignment are hidden by the connectors. Merely wiggling a connector on a sensor or in the wiring harness may correct the open circuit condition. This should always be considered when an open circuit or a failed sensor is indicated. Intermittent problems may also be caused by oxidized or loose connections. When using a circuit tester for diagnosis, always probe connections from the wire side. Be careful not to damage sealed connectors with test probes.

All wiring harnesses should be replaced with identical parts, using the same gauge wire and connectors. When signal wires are spliced into a harness, use wire with high temperature insulation only. With the low voltage and current levels found in the system, it is important that the best possible connection at all wire splices be made by soldering the splices together. It is seldom necessary to replace a complete harness. If replacement is necessary, pay close attention to insure proper harness routing. Secure the harness with suitable plastic wire clamps to prevent vibrations from causing the harness to wear in spots or contact any hot components.

➡ **Weatherproof connectors cannot be replaced with standard connectors. Instructions are provided with replacement connector and terminal packages. Some wire harnesses have mounting indicators (usually pieces of colored tape) to mark where the harness is to be secured.**

In making wiring repairs, it's important that you always replace damaged wires with wires that are the same gauge as the wire being replaced. The heavier the wire, the smaller the gauge number. Wires are color-coded to aid in identification and whenever possible the same color coded wire should be used for replacement. A wire stripping and crimping tool is necessary to install solderless terminal connectors. Test all crimps by pulling on the wires; it should not be possible to pull the wires out of a good crimp.

Wires which are open, exposed or otherwise damaged are repaired by simple splicing. Where possible, if the wiring harness is accessible and the damaged place in the wire can be located, it is best to open the harness and check for all possible damage. In an inaccessible harness, the wire must be bypassed with a new insert, usually taped to the outside of the old harness.

When replacing fusible links, be sure to use fusible link wire, NOT ordinary automotive wire. Make sure the fusible segment is of the same gauge and construction as the one being replaced and double the stripped end when crimping the terminal connector for a good contact. The melted (open) fusible link segment of the wiring harness should be cut off as close to the harness as possible, then a new segment spliced in as described. In the case of a damaged fusible link that feeds two harness wires, the harness connections should be replaced with two fusible link wires so that each circuit will have its own separate protection.

➡ **Most of the problems caused in the wiring harness are due to bad ground connections. Always check all vehicle ground connections for corrosion or looseness before performing any power feed checks to eliminate the chance of a bad ground affecting the circuit.**

Repairing Hard Shell Connectors

Unlike molded connectors, the terminal contacts in hard shell connectors can be replaced. Weatherproof hard-shell connectors with the leads molded into the shell have non-replaceable terminal ends. Replacement usually involves the use of a special terminal removal tool that depress the locking tongs (barbs) on the connector terminal and allow the connector to be removed from the rear of the shell. The connector shell should be replaced if it shows any evidence of burning, melting, cracks, or breaks. Replace individual terminals that are burnt, corroded, distorted or loose.

➡ **The insulation crimp must be tight to prevent the insulation from sliding back on the wire when the wire is pulled. The insulation must be visibly compressed under the crimp tabs, and the ends of the crimp should be turned in for a firm grip on the insulation.**

The wire crimp must be made with all wire strands inside the crimp. The terminal must be fully compressed on the wire strands with the ends of the crimp tabs turned in to make a firm grip on the wire. Check all connections with an ohmmeter to insure a good contact. There should be no measurable resistance between the wire and the terminal when connected.

Mechanical Test Equipment

VACUUM GAUGE

Most gauges are graduated in inches of mercury (in. Hg), although a device called a manometer reads vacuum in inches of water (in. H_2O). The normal vacuum reading usually varies between 18-22 in. Hg (61-74 kPa) at sea level. To test engine vacuum, the vacuum gauge must be connected to a source of manifold vacuum. Many engines have a plug in the intake manifold which can be removed and replaced with an adapter fitting. Connect the vacuum gauge to the fitting with a suitable rubber hose or, if no manifold plug is available, connect the vacuum gauge to any device using manifold vacuum, such as

CHASSIS ELECTRICAL 6-11

EGR valves, etc. The vacuum gauge can be used to determine if enough vacuum is reaching a component to allow its actuation.

HAND VACUUM PUMP

Small, hand-held vacuum pumps come in a variety of designs. Most have a built-in vacuum gauge and allow the component to be tested without removing it from the vehicle. Operate the pump lever or plunger to apply the correct amount of vacuum required for the test specified in the diagnosis routines. The level of vacuum in inches of Mercury (in. Hg) is indicated on the pump gauge. For some testing, an additional vacuum gauge may be necessary.

Intake manifold vacuum is used to operate various systems and devices on late model vehicles. To correctly diagnose and solve problems in vacuum control systems, a vacuum source is necessary for testing. In some cases, vacuum can be taken from the intake manifold when the engine is running, but vacuum is normally provided by a hand vacuum pump. These hand vacuum pumps have a built-in vacuum gauge that allow testing while the device is still attached to the component. For some tests, an additional vacuum gauge may be necessary.

SUPPLEMENTAL RESTRAINT SYSTEM (AIR BAG)

General Information

SYSTEM OPERATION AND COMPONENTS

The Supplemental Restraint System (SRS) is designed to supplement the front seat belts to help reduce the risk or severity of injury to the driver by activating and deploying a driver's-side air bag in certain frontal collisions.

The SRS consists of: left front and right front impact sensors (located on the right and left radiator support panels); air bag module (located in the center of the steering wheel). The air bag module contains a folded air bag and an inflator unit. The SRS also contains: a SRS diagnosis unit with a safing impact sensor (located in front of the shift lever); a SRS warning light to indicate operational status of the SRS (located on the instrument panel); a clock spring (mounted behind the steering wheel); and wiring.

The SRS is designed so that the air bags will deploy (inflate) when the safing sensor, plus either or both of the left front and right front impact sensors simultaneously activate while the ignition switch is in the **ON** position. These sensors are designed to be activated in frontal or near-frontal impacts of moderate to severe force.

✱✱WARNING

Only authorized service personnel should work on or around SRS components. Extreme care must be used when servicing the SRS to avoid injury to the mechanic (by inadvertent deployment of the air bag) or vehicle operator (by rendering the SRS inoperative).

The diagnosis unit monitors the SRS system and stores data concerning any detected faults in the system. When the ignition key is in the **ON** or the **START** position, the SRS warning light should illuminate for about 7 seconds and then turn off. That indicates that the SRS system is in operational order. If the SRS warning light does any of the following, immediate inspection by an authorized dealer is needed:
- The SRS warning light does not illuminate as described previously.
- The SRS warning light stays on for more than 7 seconds.
- The SRS warning light illuminates while driving.

If a vehicle's SRS warning light is in any of these three conditions when brought in for inspection, the SRS system must be inspected, diagnosed and serviced by a qualified automotive mechanic.

SERVICE PRECAUTIONS

1. In order to avoid injury to yourself or others from accidental deployment of the air bag during servicing, have a qualified automotive mechanic work on the system.
2. Do not use any electrical test equipment on or near SRS components, except for the following:
   - MB991502 Scan tool (MUT-II)
   - ROM Pack
   - MB991349 SRS Check Harness
   - MB686560 SRS Air Bag Adapter Harness A
   - MB628919 SRS Air Bag Adapter Harness B
3. Never attempt to repair the following components:
   - Front impact sensors
   - SRS Diagnosis Unit (SDU)
   - Clock spring
   - Air bag module

If any of these components are thought to be faulty, take the vehicle to a qualified mechanic in order to have them replaced.

4. Do not attempt to repair the wiring harness connectors of the SRS. If any of the connectors are found to be faulty, replace the wiring harness.
5. After disconnecting the battery cable, wait 60 seconds or more before proceeding with any work on the SRS system. The system is designed to retain enough voltage to deploy the air bag for a short time even after the battery has been disconnected, so serious injury may result from unintended air bag deployment if work is done on the SRS system immediately after the battery cables are disconnected.
6. SRS components should not be subjected to heat over 200°F (93°C), so remove the front impact sensors, SRS diagnosis unit, air bag module and clock spring before drying or baking the vehicle after painting.
7. Whenever you finish servicing the SRS, check the SRS warning light operation to make sure that the system functions properly.
8. Make certain that the ignition switch is **OFF** when any scan tool is connected or disconnected.
9. If the SRS components are removed for the purpose of inspection, sheet metal repair, painting or other work, they

6-12 CHASSIS ELECTRICAL

should be stored in a clean, dry place until they are reinstalled.

➡ **Serious injury can result from unintentional air bag deployment, so it is best to have the SRS components and system worked on by a qualified mechanic.**

DISARMING AND ENABLING THE SYSTEM

To disarm the SRS system, disconnect the negative battery from the battery and wait at least 60 seconds. Waiting for 60 seconds after disconnecting the cable is imperative because the SRS control module holds voltage. The module is designed to hold voltage, so that if, during a frontal collision, one or more of the battery cables is severed, the module will still have the voltage needed to deploy the air bag.

✽✽WARNING

Wait at least 60 seconds after disconnecting the battery cable before doing any further work.

To arm the SRS system, reconnect the negative battery cable to the battery. Turn the ignition key **ON**. Make sure the SRS warning light illuminates for about 7 seconds, turns off and then remains off for at least 45 seconds. If the warning light does anything other than this, the SRS system is not functioning correctly and should immediately be taken to a qualified automotive mechanic.

Air Bag Module and Clock Spring

SYSTEM PRECAUTIONS

When working on the Supplemental Restraint System (SRS) adhere to the following system precautions, otherwise bodily injury or component damage may result.

• Wait at least 60 seconds after disconnecting the battery cable before doing any further work.
• Never attempt to disassemble or repair the air bag module or clock spring. If faulty, have it replaced. Do not drop the air bag module or clock spring or allow contact with water, grease or oil. Replace it if a dent, crack, deformation or rust is detected.
• The air bag module should be stored on a flat surface and placed so that the pad surface is facing upward. Do not place anything on top of the air bag modules.
• After deployment of an air bag, replace the clock spring with a new one. Wear gloves and safety glasses when handling an air bag that has deployed.
• The connection between the air bag module and the clock spring does not take much force to disconnect it, do not apply excessive force to it. The removed air bag module should be stored in a clean, dry place with the pad cover face up.
• Do not hammer on the steering wheel. Doing so may damage the collapsible column mechanism.
• If the clock spring's mating mark is not properly aligned, the steering wheel may not be completely rotational during a turn, or the flat cable within the clock spring may be severed, obstructing normal operation of the SRS.
• Be sure when installing the steering wheel, that the harness of the clock spring does not become caught or tangled.
• If the an undeployed air bag module was replaced with a new unit, take the undeployed air bag module to a Mitsubishi dealer to have it correctly disposed of.

REMOVAL & INSTALLATION

✽✽WARNING

Both Mitsubishi and Chilton recommend strongly against working on the SRS system, unless you have experience and training for working on these types of systems. SRS system and personal damage could result from inexperienced people working on these systems.

1. Set the steering wheel and the front wheels to the straight ahead position.
2. Remove the ignition key from the ignition switch.
3. Disconnect the negative battery cable from the battery. Wrap tape around the terminal end of the cable to prevent accidental grounding of the vehicle.
4. Remove the air bag module mounting nut using a socket wrench from the back side.
5. When removing the connector of the clock spring from the air bag module, press the air bag's lock toward the outer side to spread it open. Use a flat-tipped, skinny prytool to remove the connector gently.
6. Remove the cap from the underside of the assembly.
7. Using a steering wheel puller, remove the steering wheel from the steering column.
8. Remove the six lower column cover retaining screws and pull the cover off of the column.
9. Unplug the clock spring and body wiring harness connectors.
10. Remove the retaining screws and lift the clock spring off of the steering column.
11. To install, align the mating mark and the Neutral position indicator of the clock spring, then slide the clock spring onto the steering column.
12. Secure the clock spring to the column switch with the mounting screws.
13. Plug the clock spring and body wiring harness connectors back together.
14. Install the column lower cover and tighten the six mounting screws until snug.
15. Before installing make sure that the front wheels are still pointing straight ahead and that the Neutral mark and the mating mark on the clock spring are still aligned.
16. Tighten the steering wheel retaining nut to 29 ft. lbs. (39 Nm) of torque, then turn the steering wheel all the way in both directions to confirm that steering is normal.
17. Install the lower-rear cap onto the steering wheel.
18. When installing the air bag module, route the horn switch harnesses as shown in the illustration. Install the air bag module without clamping the harnesses. Secure the air

CHASSIS ELECTRICAL 6-13

bag module onto the steering wheel with the retaining nuts, which should be tightened to 2.5 ft. lbs. (3.4 Nm).

19. Arm the SRS system — follow the previously described procedure.

HEATER

Blower Motor

REMOVAL & INSTALLATION

1983-86 Pick-ups

1. Remove the dashboard, following instruction given later in this section.
2. Disconnect the electric harness to the fan motor.
3. Remove the 3 bolts holding the fan assembly.
4. Remove the fan and motor assembly from the blower housing.
5. Remove the fan from the motor shaft.
6. Reassemble fan on the motor shaft and install it in the blower housing.
7. Tighten the mounting screws.
8. Connect the wiring to the motor.
9. Reinstall the dashboard.

1987-95 Pick-ups

♦ See Figures 1 and 2

1. Remove the glove box stopper and unbolt the glove box frame.
2. Remove the glove box from the dashboard.
3. Disconnect the air selection control wire.
4. Remove the large center duct running from the blower to the heater unit.
5. Disconnect the electrical harness from the motor. If necessary (for access), disconnect the ground connector near the motor.
6. Remove the three retaining bolts and remove the blower assembly. The fan may be removed from the shaft when the motor is clear of the housing.
7. When reinstalling, attach the fan to the shaft and install the assembly into the housing.
8. Tighten the mounting bolts and connect the electrical connector. Reconnect the ground harness if it was removed.
9. Install the large center duct.
10. Move the dashboard air selection lever to the "FRESH" position. Push the air selection damper lever against its stop (clockwise motion) and connect the air selection cable to the lever. Secure the outer part of the cable in place with the cable clip.
11. Install the glove box and stoppers.

1983-89 Monteros

♦ See Figure 3

➡ The order of steps may vary slightly for 1983-86 vehicles.

1. Remove the lap heater duct under the right-side dashboard.
2. Remove the glove box stoppers and the glove box assembly.
3. Disconnect the air selection control wire.
4. Remove the large center duct running from the blower to the heater unit. (Except on 1983-86 Monteros)
5. Disconnect the electrical harness from the motor. If necessary for access, disconnect the ground connector near the motor. The resistor may be removed from the motor as necessary.
6. Disconnect the small breather hose from the motor. Remove the three retaining bolts and remove the blower assembly. The fan may be removed from the shaft when the motor is clear of the housing.
7. When reinstalling, attach the fan to the shaft and install the assembly into the housing. Make sure the gasket is seated correctly.
8. Tighten the mounting bolts and connect the electrical connector. Reconnect the ground harness if it was removed and secure the resistor if needed.
9. Install the large center duct if it was removed.
10. Move the dashboard air selection lever to the "RECIRCULATE" position. Push the air selection damper lever against its stop (counterclockwise motion) and connect the air selection cable to the lever. Secure the outer part of the cable in place with the cable clip.
11. Install the glove box and stoppers.
12. Install the lap heater duct.

1990-95 Monteros

♦ See Figures 4 and 5

➡ If only the blower assembly (resistor, blower motor, blower case) needs to be removed, it can be done simply by removing the foot shower duct first, then the blower assembly; the other components need not be removed.

Prior to this procedure, if the evaporator is going to be removed, the air conditioning system must be discharged by an automotive mechanic trained and certified to work on air conditioning systems. If the evaporator is not going to be removed, the A/C system need not be discharged.

1. On vehicles with air conditioning, disconnect the drain hose, the liquid pipe and the suction hose.
2. Extract the glove box assembly from the dashboard.
3. The speaker grille cover should be removed, then the speaker itself.
4. Unscrew the four mounting bolts and then remove the lower frame.
5. Remove the right-hand side foot shower duct.
6. Unplug the engine control relay assembly and remove its bracket.
7. Disconnect the air selection control wire from the air selection damper lever.
8. On vehicles with air conditioning, the evaporator will need to be removed. On vehicles without air conditioning, remove the duct joint instead.
9. Loosen and remove the blower assembly mounting bolts, then drop the assembly down and out of the dashboard.

6-14 CHASSIS ELECTRICAL

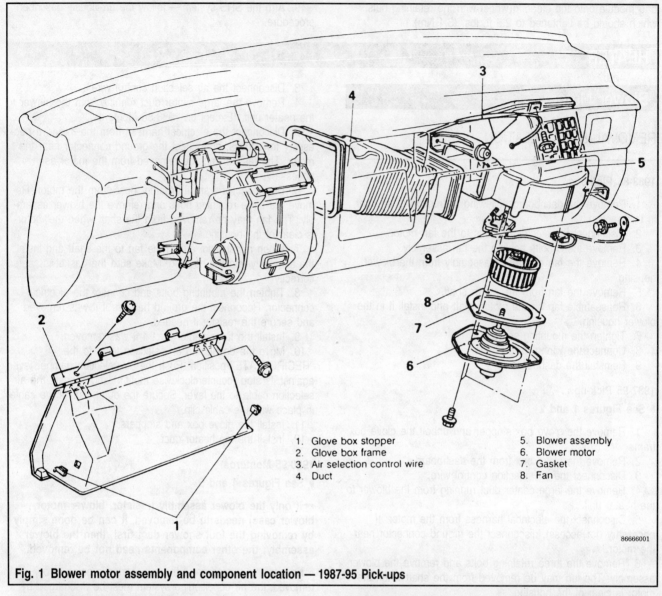

1. Glove box stopper
2. Glove box frame
3. Air selection control wire
4. Duct
5. Blower assembly
6. Blower motor
7. Gasket
8. Fan

Fig. 1 Blower motor assembly and component location — 1987-95 Pick-ups

To install:
10. Secure the blower assembly in its position with the mounting bolts.
11. Install the duct joint (vehicles with no A/C) or the evaporator (vehicles with A/C).
12. Connect the air selection control wire to the air selection damper lever by following the following steps:

  a. Move the air selection control lever to the recirculation position.

  b. With the air selection damper lever pressed inward in the direction indicated by the arrow, connect the inner cable of the air selection lever, then use a clip to secure the outer cable.

13. Install the engine control relay assembly and its bracket.
14. Secure the right-hand side foot shower duct in place and tighten the retaining bolts.
15. Install the lower frame, the speaker and grille cover, and the glove box.
16. Reattach the liquid pipe, the suction hose, and the drain hose, if removed earlier.
17. Have the A/C system evacuated and charged by an automotive mechanic.

Heater Core

The heater core is a small heat exchanger located inside the vehicle, similar to the radiator at the front of the vehicle. Coolant is circulated from the engine through the heater core and back to the engine. The heater fan blows fresh, outside air through the heater core; the air is heated and sent on to the interior of the vehicle.

About the only time the heater core will need removal is for replacement due to clogging or leaking. Thankfully, this doesn't happen too often; removing the heater core is a major task. Some are easier than others, but all require working in unusual positions inside the vehicle and fitting tools into very cramped quarters behind the dashboard. In many cases, the dashboard must be removed during the procedure — another major project.

CHASSIS ELECTRICAL 6-15

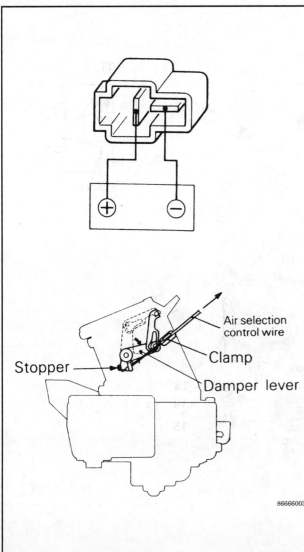

Fig. 2 The air selection control wire and damper lever assembly — 1987-95 Pick-ups

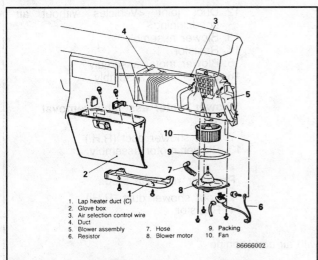

1. Lap heater duct (C)
2. Glove box
3. Air selection control wire
4. Duct
5. Blower assembly
6. Resistor
7. Hose
8. Blower motor
9. Packing
10. Fan

Fig. 3 Blower motor assembly and component location — 1983-89 Monteros

REMOVAL & INSTALLATION

1983-86 Pick-ups
▶ See Figure 6

1. Disconnect the battery ground cable.
2. Move the temperature selector lever to the off position.
3. Drain the coolant.

✻✻CAUTION

When draining the coolant, keep in mind that cats and dogs are attracted by ethylene glycol antifreeze, and are quite likely to drink any that is left in an uncovered container or in puddles on the ground. This will prove fatal in sufficient quantity. Always drain the coolant into a sealable container. Coolant should be reused unless it is contaminated or several years old.

4. In the cab, remove the parcel tray, the center vent grille and duct and the defroster duct.
5. Label and disconnect the control cables at the heater box.
6. Disconnect the coolant hoses running to the heater core. There will be some coolant within the hoses; don't spill it in the cab.
7. Disconnect the fan wiring. Remove the upper and center mounting nuts holding the heater assembly. Remove the assembly.

✻✻WARNING

The heater core still contains coolant; be careful not to spill it in the cab.

8. With the heater case removed, the core may be easily removed on the workbench.
9. Place the new core in the heater case. Install the case in position and tighten the retaining nuts and bolts. Make sure the case is properly seated before tightening the mounting nuts.
10. Connect the wiring to the fan motor.
11. Install the water hoses, making sure each is pushed onto its pipe at least 1 in. (25mm). Use new clamps if needed.
12. Connect the cables to the heater unit. Pay attention to the labels.
13. Install the defroster duct, the center ventilator duct and grille and the parcel tray.
14. Double check the draincock, closing it securely if not already done. Fill the cooling system with coolant.
15. Connect the negative battery cable.
16. Start the engine. Move the temperature selector to HOT and allow the engine to warm up. Check the function of each heater control and inspect the hose connections for leaks.

6-16 CHASSIS ELECTRICAL

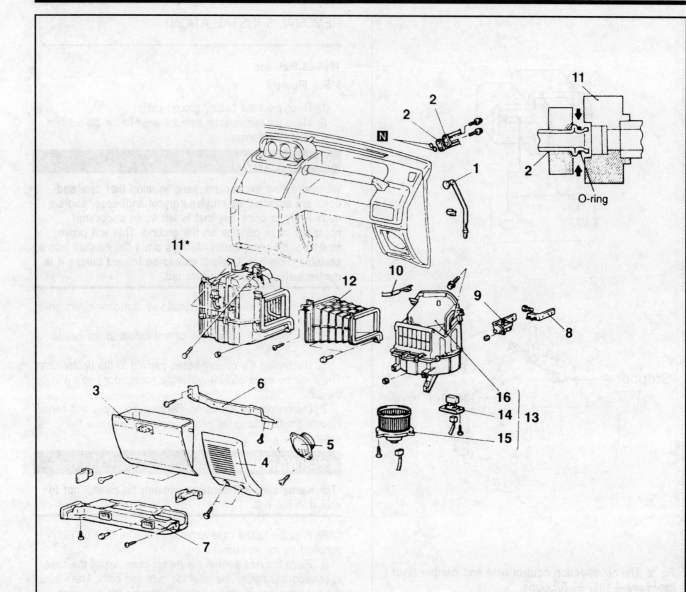

Fig. 4 Blower motor assembly and component location — 1990-95 Monteros

CHASSIS ELECTRICAL 6-17

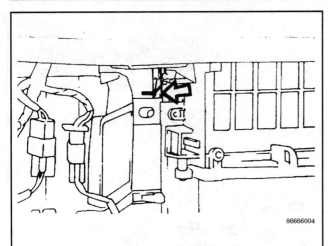

Fig. 5 Press the air selection damper lever in the direction of the arrow, and connect the inner cable of the air selection control wire — 1990-95 Monteros

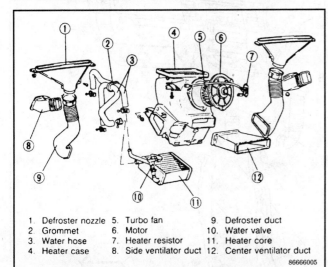

1. Defroster nozzle
2. Grommet
3. Water hose
4. Heater case
5. Turbo fan
6. Motor
7. Heater resistor
8. Side ventilator duct
9. Defroster duct
10. Water valve
11. Heater core
12. Center ventilator duct

Fig. 6 Heater system components — 1983-86 Pick-ups

1987-95 Pick-ups

♦ See Figures 7, 8, 9, 10, 11 and 12

1. With the engine cold, set the temperature control lever to the extreme hot position. Drain the engine coolant through the radiator drain plug.

✱✱CAUTION

When draining the coolant, keep in mind that cats and dogs are attracted by ethylene glycol antifreeze, and are quite likely to drink any that is left in an uncovered container or in puddles on the ground. This will prove fatal in sufficient quantity. Always drain the coolant into a sealable container. Coolant should be reused unless it is contaminated or several years old.

2. Disconnect the coolant hoses running to the heater pipes at the firewall. Take note of hose location and placement; they are formed to fit exactly as needed.
3. Disconnect the negative battery cable.
4. Remove the hazard flasher switch.
5. Pop out the hole cover, or starter unlock switch, on the opposite side (from the hazard flasher switch) of the instrument hood.
6. Remove the instrument hood.
7. Remove the instrument cluster, remembering to disconnect the wiring and speedometer cable at the rear.
8. Remove the fusebox cover. Remove the fusebox from the dashboard and let it hang.
9. Remove the glove box. Move the stopper out of the way before lowering the door.
10. Remove the two defroster ducts.
11. Label and disconnect the 3 control wires running to the heater unit.
12. Remove the speaker cover panels on each side.
13. Remove the clock or small compartment from the dash.
14. Pop out the small square cover at the upper center of the dash. Its easiest to reach in the clock hole and release the clips with your fingers.
15. Remove the lower center faceplate or cover from the center of the dash.
16. Remove the shifter knob.
17. Remove the floor console assembly.
18. Disconnect the left side and front harness connectors running to the dashboard.
19. Disconnect the air conditioner and heater wiring harnesses. Disconnect the instrument cluster harness and the radio connectors, including the antenna wire.
20. Remove the retaining nuts and bolts holding the dash. They are located across the top, along the sides and at the bottom.
21. Use the tilt release lever and move the steering column all the way downward.
22. Remove the dash carefully; several components are still connected to it.
23. Remove the large duct running between the heater case and the fan case.
24. Remove the center ventilator duct.
25. Remove the defroster duct.
26. Remove the two vertical reinforcing bars.
27. Remove the retaining nuts and bolts and remove the heater case.

✱✱WARNING

The heater core still contains coolant; be careful not to spill it inside the cab.

28. Remove the hose cover on the side of the heater case.
29. Disconnect the hose clamps on the small hoses.
30. Use a sharp knife or razor to cut away the small hoses. Attempting to remove the hoses may damage the tubing.
31. Remove the retaining plate.
32. Remove the heater core from the case.

To install:

33. Install the core into the case, making sure all the padding and insulation is in place.
34. Install the plate holding the two pipes. Install new small hoses and new clamps. Make certain the joints are correctly fitted and that the clamps are tightened. These hoses are not accessible after installation; any leak will require removal again.

6-18 CHASSIS ELECTRICAL

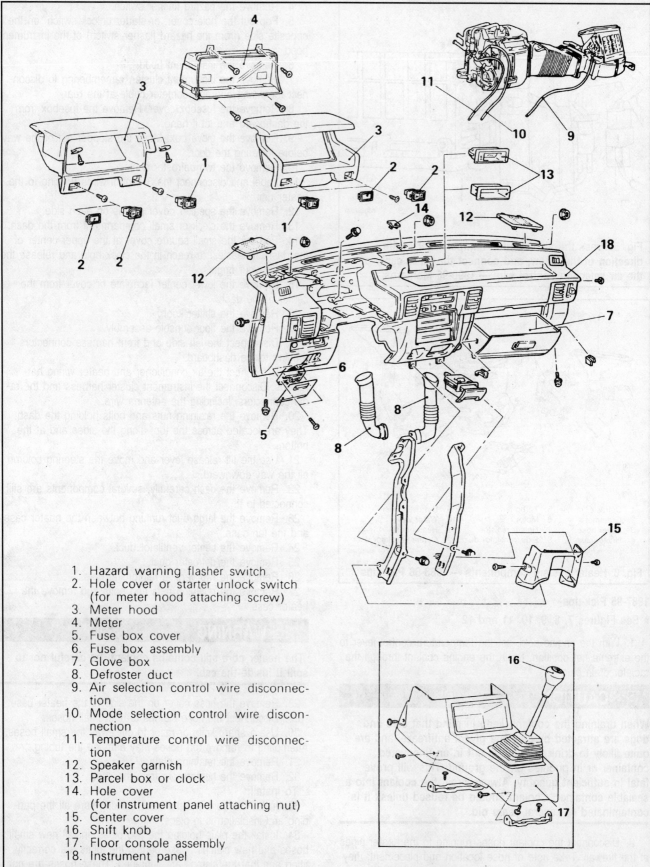

1. Hazard warning flasher switch
2. Hole cover or starter unlock switch (for meter hood attaching screw)
3. Meter hood
4. Meter
5. Fuse box cover
6. Fuse box assembly
7. Glove box
8. Defroster duct
9. Air selection control wire disconnection
10. Mode selection control wire disconnection
11. Temperature control wire disconnection
12. Speaker garnish
13. Parcel box or clock
14. Hole cover (for instrument panel attaching nut)
15. Center cover
16. Shift knob
17. Floor console assembly
18. Instrument panel

Fig. 7 Instrument panel (dashboard) removal and installation components — 1987-95 Pick-ups

CHASSIS ELECTRICAL 6-19

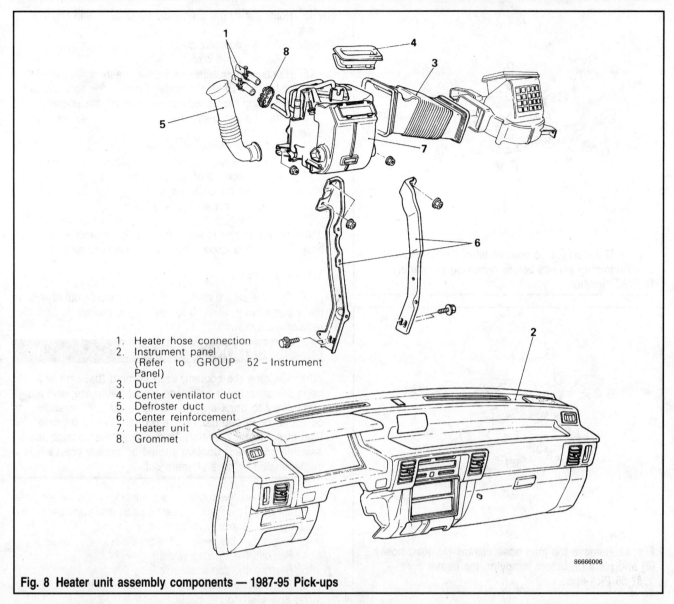

1. Heater hose connection
2. Instrument panel (Refer to GROUP 52 – Instrument Panel)
3. Duct
4. Center ventilator duct
5. Defroster duct
6. Center reinforcement
7. Heater unit
8. Grommet

Fig. 8 Heater unit assembly components — 1987-95 Pick-ups

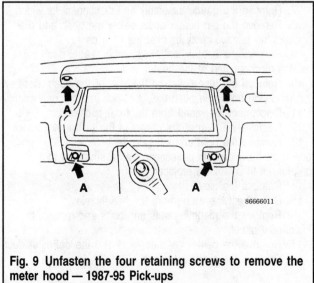

Fig. 9 Unfasten the four retaining screws to remove the meter hood — 1987-95 Pick-ups

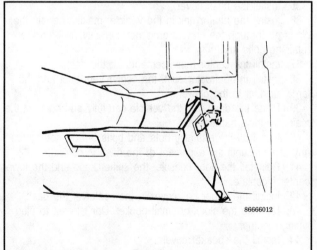

Fig. 10 To remove the glove box assembly, the glove box stopper must first be released — 1987-95 Pick-ups

6-20 CHASSIS ELECTRICAL

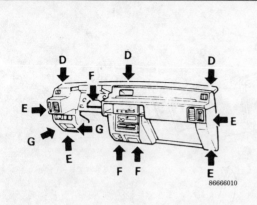

Fig. 11 Make certain to remove all of the instrument panel retaining screws before removing the panel — 1987-95 Pick-ups

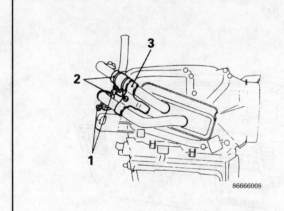

Fig. 12 Remove the joint hose clamps (1), joint hoses (2) and plate (3) before removing the heater core — 1987-95 Pick-ups

35. Install the hose cover.
36. Install the heater unit in the vehicle, making certain the pipes go through the firewall grommet correctly. Tighten the retaining bolts and nuts.
37. Install the center reinforcements for the dash.
38. Install the defroster duct, the center vent duct and the large air duct to the blower.
39. Place the dash into the vehicle carefully and connect the main electrical harnesses.
40. Install all the retaining nuts and bolts but do not tighten any of them until the alignment of the dash is correct.
41. Reinstall the floor console, the shifter knob and the lower center faceplate.
42. Install the upper bolt cover in the center of the dash.
43. Replace the clock or small pocket. Don't forget to plug wiring together.
44. Install the speaker covers.
45. Route the heater control cables properly and connect them.
46. Install the defroster ducts.
47. Install the glove box.
48. Remount the fusebox on the dash and install the lid.
49. Install the instrument cluster. Connect the speedometer cable and wiring harnesses before tightening the screws.
50. Install the instrument cluster hood and install the bolt cover.
51. Install the hazard warning switch.
52. Connect the heater hoses to the pipes at the firewall.
53. Double check the draincock, closing it securely if not already done. Fill the cooling system with coolant.
54. Connect the negative battery cable.
55. Start the engine. Move the temperature selector to HOT and allow the engine to warm up. Check the function of each heater control and inspect the hose connections for leaks.

1983-86 Monteros

1. With the engine cold, set the temperature control lever to the extreme hot position. Drain the engine coolant through the radiator drain plug.

※※CAUTION

When draining the coolant, keep in mind that cats and dogs are attracted by ethylene glycol antifreeze, and are quite likely to drink any that is left in an uncovered container or in puddles on the ground. This will prove fatal in sufficient quantity. Always drain the coolant into a sealable container. Coolant should be reused unless it is contaminated or several years old.

2. Disconnect the coolant hoses running to the heater pipes at the firewall. Take note of hose location and placement; they are formed to fit exactly as needed.
3. Disconnect the negative battery cable.
4. Remove the steering wheel.
5. Remove the center console.
6. Remove the meter assembly. Carefully disconnect the wiring and speedometer cable.
7. Remove the gauge assembly and disconnect its wiring.
8. Remove the lap heater ducts below the dash and disconnect the release cable bracket from the dash.
9. Remove the heater control assembly. Label each cable as it is disconnected at the heater end.
10. Remove the fuse cover on the side of the dash. unscrew the mounting bolts and push the fuseblock into the dashboard.
11. Disconnect the wiring from the front speakers.
12. Remove the plug at the center of the instrument panel.
13. Remove the right and left side defroster grilles by gently prying on the mounting projections with a small flat tool. Be careful not to break the projections off.
14. Remove the glove box.
15. Disconnect the wiring from the heater relay.
16. Remove the mounting nuts and bolts and remove the instrument panel.
17. Remove the center ventilator duct and the defroster duct.
18. Remove the rear heater duct.

19. Remove the upper and lower retaining nuts and remove the heater case.

> ✱✱**WARNING**
>
> **The heater core still contains coolant; be careful not to spill it in the cab.**

20. Remove the heater control lever arm and remove the water valve cover.
21. Remove the heater pipe and the water valve.
22. Disconnect the control arm linkage and remove the control arm.
23. Remove the heater core by sliding it out of the case sideways. Do not remove the protective felt liner when removing the core. It insulates the core within the case.

To install:
24. Install the new core into the case with the felt insulator positioned properly.
25. Install the control arm on the linkage.
26. Install the water control valve and heater pipe
27. Move the center ventilator damper (door) to the closed position. Turn the arm fully clockwise and connect it to the link.
28. Move the defroster/heater damper to the defroster position. Turn the arm fully counterclockwise and connect it to the link.
29. Close the water valve completely and close the air intake damper completely. Connect the arm to the link.
30. Install the heater hoses, pushing them onto the pipes as far as the second bulge. Tighten the clamps. Make sure the clamps are positioned so as not to interfere with other hoses.
31. Reinstall the heater unit into the vehicle. Tighten the retaining nuts.
32. Connect the rear heater duct, the defroster duct and the center ventilator duct.
33. Position the dashboard and secure the mounting bolts.
34. Connect the heater relay wiring and install the glove box.
35. Install the side defroster grilles and the plug at the center of the dash.
36. Connect the front speaker wiring.
37. Position the fusebox and install its screws; install the cover.
38. Install the heater control assembly and connect the control cables.
39. Install the two lap heater ducts and install the hood release cable bracket.
40. Install the combination gauge unit and connect the wiring.
41. Install the instrument cluster. Connect the wiring and the speedometer cable.
42. Install the center console.
43. Install the steering wheel.
44. Connect the heater hoses to the pipes at the firewall.
45. Double check the draincock, closing it securely if not already done. Fill the cooling system with coolant.
46. Connect the negative battery cable.
47. Start the engine. Move the temperature selector to HOT and allow the engine to warm up. Check the function of each heater control and inspect the hose connections for leaks.

1987-91 Montero

▶ See Figures 13, 14, 15, 16, 17, 18, 19, 20, 21, 22, 23, 24, 25, 26, 27, 28, 29, 30 and 31

1. With the engine cold, set the temperature control lever to the extreme hot position. Drain the engine coolant through the radiator drain plug.

> ✱✱**CAUTION**
>
> **When draining the coolant, keep in mind that cats and dogs are attracted by ethylene glycol antifreeze, and are quite likely to drink any that is left in an uncovered container or in puddles on the ground. This will prove fatal in sufficient quantity. Always drain the coolant into a sealable container. Coolant should be reused unless it is contaminated or several years old.**

2. Disconnect the coolant hoses running to the heater pipes at the firewall. Take note of hose location and placement; they are formed to fit exactly as needed.
3. Disconnect the negative battery cable.
4. Remove the lap heater air ducts under the left and right side dash.
5. Remove the hood release cable bracket from the dashboard.
6. Remove the left and right side defroster grilles. Use a small flat tool to raise the attaching projections. Don't break the tabs.
7. Remove the glove box.
8. Use a flat, padded tool to remove the rear cover from the instrument cluster.
9. Remove the instrument cluster (meter case). Disconnect the speedometer cable and the wiring harnesses to the back of the cluster.
10. Disconnect the wiring to each switch on the cluster. Label each as it is removed.
11. Remove the screws from the side of the combination meter unit and remove the cover.
12. Remove the 4 screws at the base of the combination meter and gently remove the meter. Disconnect the wire harnesses.

> ✱✱**WARNING**
>
> **The inclinometer can be damaged by dropping or bumping it. Do not tilt the unit so far as to exceed the maximum indication on the scale. Resist the temptation to play with the unit.**

13. Remove the center panel or lower console.
14. Disconnect the 3 heater control cables running to the heater unit. Label each one.
15. Remove the center reinforcement and bring it out as a unit with the radio or stereo. Disconnect the wiring when the unit is free.
16. Remove the horn pad and the steering wheel.
17. Remove the fuse box cover and release the fusebox from the dashboard.
18. Unfasten the dashboard mounting nuts and bolts, and remove the dashboard.
19. Remove the center ventilator duct and the defroster duct.
20. Remove the rear heater duct.

6-22 CHASSIS ELECTRICAL

Fig. 13 Use a prytool and a piece of rubber or rag to remove the meter cover — 1987-91 Monteros

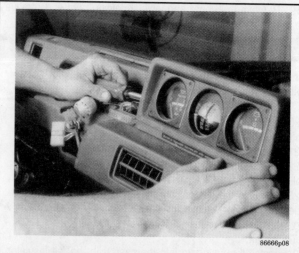

Fig. 16 Removing the combination meter case from the instrument panel — 1987-91 Monteros

Fig. 14 Remove the fuse box cover and the fuse box itself — 1987-91 Monteros

Fig. 17 The center panel and its controls must also be removed — 1987-91 Monteros

Fig. 15 Removing the retaining screws and the meter case from the instrument panel — 1987-91 Monteros

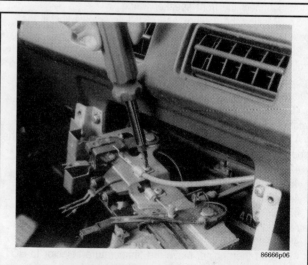

Fig. 18 Unhook the heater control assembly cables from the heater control levers — 1987-91 Monteros

CHASSIS ELECTRICAL 6-23

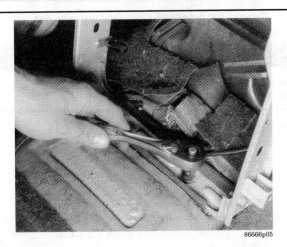

Fig. 19 Remove all of the mounting bolts from the center reinforcement, then remove the reinforcement itself — 1987-91 Monteros

Fig. 20 After removing all of its mounting screws, remove the instrument panel from the vehicle — 1987-91 Monteros

Fig. 21 With the instrument panel removed from the vehicle, access to the heater system is easy — 1987-91 Monteros

Fig. 22 Disconnect the heater unit-to-duct brackets — 1987-91 Monteros

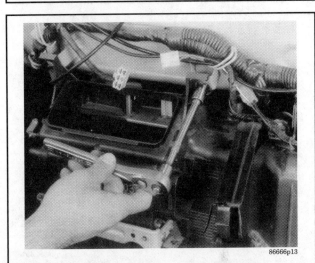

Fig. 23 Unscrew the heater unit mounting bolts — 1987-91 Monteros

Fig. 24 Extract the heater unit, being careful not to damage the water pipes — 1987-91 Monteros

6-24 CHASSIS ELECTRICAL

Fig. 25 Unfasten the heater core cover mounting screws — 1987-91 Monteros

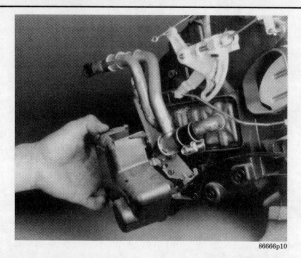

Fig. 26 Remove the heater core cover from the heater unit — 1987-91 Monteros

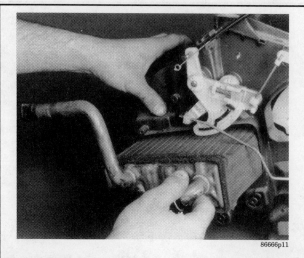

Fig. 27 Slide the heater core out of the heater unit — 1987-91 Monteros

21. Remove the upper and lower retaining nuts and remove the heater case.

✲✲WARNING

The heater core still contains coolant; be careful not to spill it in the cab.

22. Remove the heater control lever arm and remove the water valve cover.
23. Remove the heater pipe and the water valve.
24. Disconnect the control arm linkage and remove the control arm.
25. Remove the heater core by sliding it out of the case sideways. Do not remove the protective felt liner when removing the core. It insulates the core within the case.

To install:

26. Install the new core into the case with the felt insulator positioned properly.
27. Install the control arm on the linkage.
28. Install the water control valve and heater pipe
29. Move the center ventilator damper (door) to the closed position. Turn the arm fully clockwise and connect it to the link.
30. Move the defroster/heater damper to the defroster position. Turn the arm fully counterclockwise and connect it to the link.
31. Close the water valve completely and close the air intake damper completely. Connect the arm to the link.
32. Install the heater hoses, pushing them onto the pipes as far as the second bulge. Tighten the clamps. Make sure the clamps are positioned so as not to interfere with other hoses.
33. Reinstall the heater unit into the vehicle. Tighten the retaining nuts.
34. Connect the rear heater duct, the defroster duct and the center ventilator duct.
35. Position the dashboard and start each nut and bolt. Make certain the dashboard is aligned, then tighten the fittings.
36. Install the fusebox assembly and its cover.
37. Install the steering wheel and horn pad.
38. Install the radio and center reinforcement. Connect the wiring before tightening the bolts.
39. Carefully route and connect the heater control cables.
40. Install center panel and secure the heater controls.
41. Install the combination gauges and connect the wiring.
42. Install the cover for the combination gauges.
43. Install the instrument cluster, connecting the speedometer cable and the many wiring connectors.
44. Install the cluster cover; make sure it is firmly seated in place.
45. Replace the glove box
46. Reinstall the side defroster grilles.
47. Connect the hood release cable bracket.
48. Install the two lap heater ducts.
49. Connect the heater hoses to the pipes at the firewall.
50. Double check the draincock, closing it securely if not already done. Fill the cooling system with coolant.
51. Connect the negative battery cable.
52. Start the engine. Move the temperature selector to HOT and allow the engine to warm up. Check the function of each heater control and inspect the hose connections for leaks.

CHASSIS ELECTRICAL 6-25

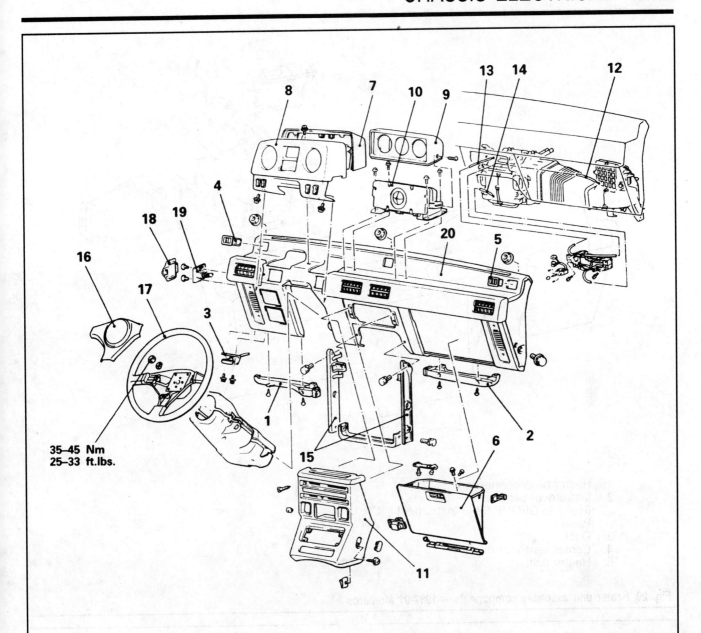

1. Lap heater duct (B)
2. Lap heater duct (C)
3. Hood release cable bracket
4. Demister grille (L.H.)
5. Demister grille (R.H.)
6. Glove box
7. Meter cover
8. Meter case
9. Combination meter pad
10. Combination meter case
11. Center panel
12. Connection of recirculation/fresh air changeover control wire
13. Connection of mode selection control wire
14. Connection of water valve control wire
15. Center reinforcement
16. Horn pad
17. Steering wheel
18. Fuse box cover
19. Fuse box assembly
20. Instrument panel

Fig. 28 The instrument panel removal and installation components — 1987-91 Monteros

6-26 CHASSIS ELECTRICAL

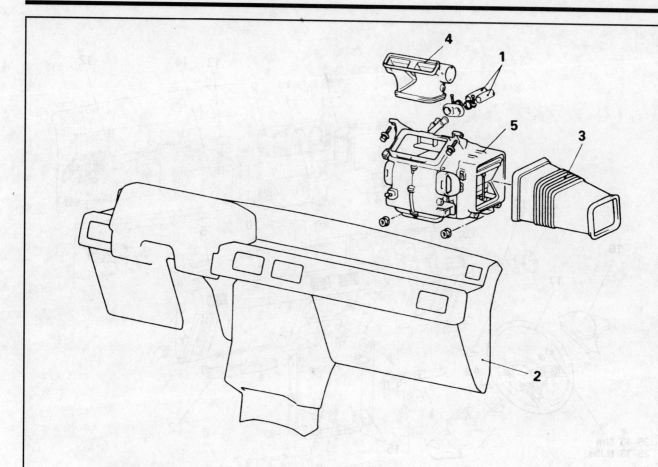

1. Heater hose connection
2. Instrument panel
   (Refer to GROUP 52 – Instrument Panel.)
3. Duct
4. Center ventilator duct
5. Heater unit

Fig. 29 Heater unit assembly components — 1987-91 Monteros

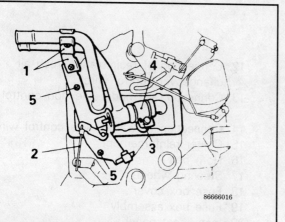

Fig. 30 Remove the piping clamp (1), water valve link (2), joint hose clamp (3), joint hose (4) and screws (5) to remove the water valve from the heater core — 1987-91 Monteros

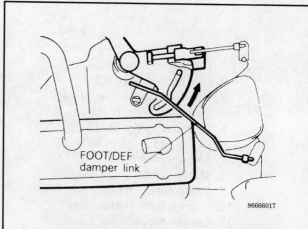

Fig. 31 Remove the FOOT/DEF damper link from the mode selection damper lever to remove the heater core — 1987-91 Monteros

CHASSIS ELECTRICAL 6-27

1992-95 Montero

▶ See Figures 32, 33, 34 and 35

1. Drain the coolant system into a clean, large container so that reuse of the coolant is possible.

> **✱✱CAUTION**
>
> When draining the coolant, keep in mind that cats and dogs are attracted by ethylene glycol antifreeze, and are quite likely to drink any that is left in an uncovered container or in puddles on the ground. This will prove fatal in sufficient quantity. Always drain the coolant into a sealable container. Coolant should be reused unless it is contaminated or several years old.

2. Disconnect the negative (-) battery cable from the battery.
3. Remove the floor console inside the vehicle by following these steps:
   a. Remove the switch panel and the suspension control switch or hole cover.
   b. Unplug the rear console harness connector.
   c. Remove side panel A.
   d. Unscrew the four retaining screws and remove the rear console assembly.
   e. If so equipped, remove the manual transmission shift lever knob.
   f. Remove the transfer shift lever knob.
   g. Disconnect the floor console harness.
   h. Unscrew the three retaining screws and remove the front console assembly. When removing the automatic transmission front console assembly, set the automatic transmission selector lever to **L**.

> **✱✱WARNING**
>
> When installing or removing the instrument panel, do not allow any impact or shock to the SRS diagnosis unit.

4. Remove the instrument panel and dashboard. Use the following steps:
   a. Remove the hood lock release and the fuel filler door lock release handles.
   b. Loosen and remove the retaining screws, then the instrument under and corner covers.
   c. Remove the glove box stopper and then the glove box.
   d. Unscrew the mounting screws and remove the center panel A.
   e. Extract the heater control assembly.
   f. Take the radio and tape player out.
   g. Pop the meter hood plug out, then remove the meter bezel assembly by loosening and removing the mounting screws.
   h. Remove the screws holding the combination meter (instrument cluster) in place and extract the meter.
   i. Disconnect and remove the speedometer cable adapter. First disconnect the speedometer cable at the transmission end of the cable, then remove the lock of the speedometer cable adapter from the instrument panel. Pull the speedometer cable slightly toward the vehicle interior, then remove the speedometer cable adapter.
   j. Remove the four screws holding the column cover in place, then remove the cover itself.
   k. Remove the clock or clock plug.
   l. Remove the side defroster garnish, the door mirror control switch, the front speaker, the rheostat, the rear wiper and washer switch and the door lock switch.
   m. Disconnect the ventilation control wire and the wiring harness.
   n. Unscrew the steering column installation bolts.
   o. Unscrew all the retaining screws holding the dashboard (instrument panel assembly) in place and remove the assembly.

5. Disconnect the two water hoses from the heater core, located in the engine compartment. Make certain not to damage the metal pipes.
6. Remove the left-hand and right-hand foot shower ducts.
7. Take the lap cooler duct A out.
8. On vehicles with A/C, re,ove the evaporator mounting bolts and nut. On vehicles with no A/C, remove the joint duct instead.
9. Take the center duct assembly and the two center reinforcements out.
10. Remove the heater unit. Be careful, there is still coolant in the heater core, which is in the heater unit. Set the heater unit outside of the vehicle and either pour the coolant out of the heater core while still in the heater unit, or wait until the heater core has been removed from the heater unit.
11. Remove the foot distribution duct.
12. Loosen and remove the heater core retaining bolts, and remove the heater core from the heater unit.

To install:

13. While the system is apart, check the operation of the dampers and link mechanism. Also check the heater core for clogging and water leakage.
14. Set the heater core back into the heater unit and secure it in place with the retaining bolts.
15. With the two mounting bolts, mount the foot distribution duct back in place.
16. Install the heater unit back into its original position, and use the two bolts and the two nuts to secure it there.
17. Install the two center reinforcements and affix the center duct assembly to the top of the heater unit.
18. Install either the joint duct, or the evaporator mounting bolts and nut, depending on whichever was removed.
19. Put the lap cooler duct and the right-hand and left-hand foot shower ducts back in place. Secure them with their mounting screws.
20. Attach the two water cooling hoses to the heater core. Make sure the hoses are on all the way, the clamps are properly installed and the rubber firewall gasket/grommet is correctly installed. It is important to make certain that the hoses are correctly secured to the pipes, because if a coolant leak is created, the whole system must once again be disassembled to repair it.
21. Install the instrument panel assembly:
    a. Hold the assembly in position and secure it in place with its mounting screws.
    b. Tighten the steering column installation bolts and plug the harness connector back together.
    c. To install the ventilation control wire, set the cool air bypass dial to the closed position. Close the cool air bypass lever at the heater unit side (lever is lightly hit against the

6-28 CHASSIS ELECTRICAL

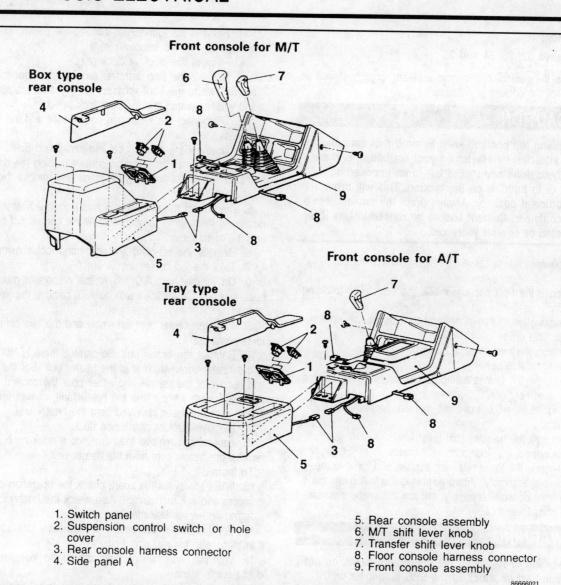

1. Switch panel
2. Suspension control switch or hole cover
3. Rear console harness connector
4. Side panel A
5. Rear console assembly
6. M/T shift lever knob
7. Transfer shift lever knob
8. Floor console harness connector
9. Front console assembly

Fig. 32 Floor console removal and installation component locations — 1992-95 Monteros

stopper). Install the ventilation control wire and secure it with the clip.

   d. Install the door lock switch, the rear wiper and washer switch, the rheostat, the front speaker, the door mirror control switch, the side defroster garnish, the clock or clock plug and the column cover.

   e. Hook the speedometer cable adapter back up the back of the combination meter (instrument cluster).

   f. Install the combination meter, the meter bezel assembly and the meter hood plug.

   g. hook the radio and tape player back up and install them in the dashboard.

   h. Install the heater control assembly, center panel A, the glove box assembly and the glove box stopper.

   i. Install the instrument under and corner covers, the fuel filler door lock release handle and the hood lock release handle.

22. Install the floor console:

   a. Attach the front console assembly and plug the floor console harness connector back together.

   b. Install the transfer shift lever knob, and the manual transmission shift lever knob, if so equipped.

   c. Attach the rear console assembly and side panel A.

   d. Plug the rear console harness connector back together.

   e. Install the suspension control switch or the hole cover to the switch panel, then install the switch panel to the rear console assembly.

23. Fill the cooling system back up to full and connect the negative (-) battery cable to the battery.

24. Start the engine and check for any coolant system leaks at the heater core cooling hose connection.

CHASSIS ELECTRICAL 6-29

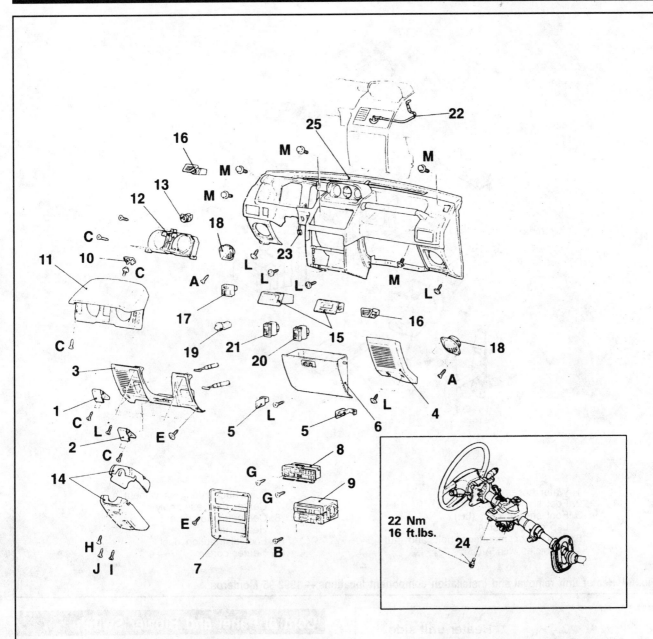

1. Hood lock release handle
2. Fuel filler door lock release handle
3. Instrument under cover
4. Instrument corner cover
5. Glove box stopper
6. Grove box assembly
7. Center panel A
8. Heater control assembly
9. Radio and tape player
10. Meter hood plug
11. Meter bezel assembly
12. Combination meter
13. Speedometer cable adapter <Up to 1993 models>
14. Column cover
15. Clock or clock plug
16. Side defroster garnish
17. Door mirror control switch
18. Front speaker
19. Rheostat
20. Rear wiper and washer switch
21. Door lock switch
22. Ventilation control wire
23. Harness connector
24. Steering column installation bolts
25. Instrument panel assembly

Fig. 33 Instrument panel removal and installation component locations — 1992-95 Monteros

6-30 CHASSIS ELECTRICAL

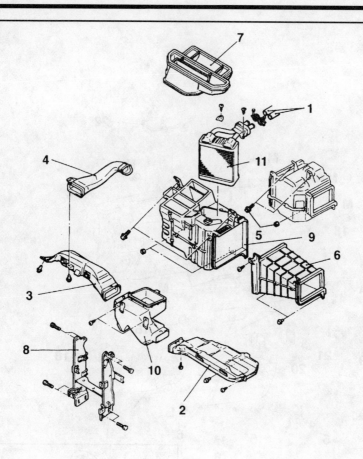

1. Water hoses connection
2. Foot shower duct (RH)
3. Foot shower duct (LH)
4. Lap cooler duct A
5. Evaporator mounting bolt and nut <Vehicles with A/C>
6. Joint duct <Vehicles without A/C>
7. Center duct assembly
8. Center reinforcement
9. Heater unit
10. Foot distribution duct
11. Heater core

Fig. 34 Heater unit removal and installation component locations — 1992-95 Monteros

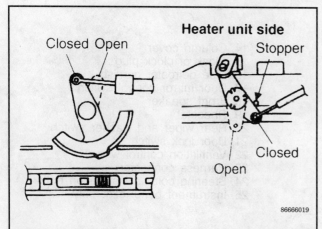

Fig. 35 To install the ventilation control wire, set the cool air bypass dial to the closed position and close the cool air bypass lever at the heater unit side — 1992-95 Monteros

Control Panel and Blower Switch

REMOVAL & INSTALLATION

1983-86 Pick-up

▶ See Figure 36

1. Remove the knobs from the heater control sliders and from the radio.
2. Remove the instrument cluster, referring to the detailed procedure later in this section.
3. Label and disconnect each control cable at the heater case. Once the cables are disconnected, do not change the positions of the levers on the case.
4. Remove the 4 screws holding the heater control panel to the dashboard. Remove the panel carefully. Remember that the cables are still attached and must be routed through the dash.

CHASSIS ELECTRICAL 6-31

5. Begin reinstallation by routing the cables through the opening in the dash. Make sure the cables are correctly routed to the levers on the case.
6. Install the retaining screws for the panel.
7. Connect the cables to the proper levers. Double check any retainers or locking clips.
8. Check the operation of each heater control; it should move smoothly and without binding.
9. Install the instrument cluster assembly.
10. Install the knobs for the radio and heater controls.

1987-95 Pick-up

▶ See Figure 37

1. Remove the glove box stopper.
2. Remove the glove box.
3. Disconnect the air selection control cable.
4. Remove the knobs from the heater control panel.
5. Carefully remove the center trim panel. The retaining screws are concealed.
6. Disconnect the left defroster duct.
7. Remove the mode selection control cable.
8. Disconnect the temperature control cable.
9. Remove the screws holding the heater control panel to the dashboard. Remove the panel carefully. Remember that the cables are still attached and must be routed through the dash.
10. Begin reinstallation by routing the cables through the opening in the dash. Make sure the cables are correctly routed to the levers on the case.
11. Install the retaining screws for the panel.
12. Connect the temperature control cable and the mode selection cable.
13. Install the left defroster duct.
14. Install the center panel and tighten the fittings.
15. Install the knobs on the heater control panel.
16. Connect the air selection cable.
17. Install the glove box and stoppers.
18. Test the motion of each heater control; the movement should be smooth, with no binding or jerkiness.

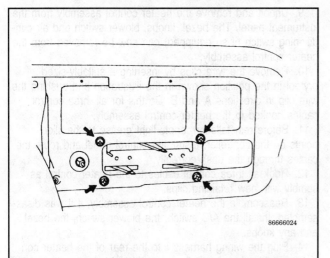

Fig. 36 Removing the heater controls requires removal of the instrument cluster — 1983-86 Pick-ups

1983-91 Montero

▶ See Figure 38

1. Remove the glove box stoppers and remove the glove box.
2. Disconnect the air selection control cable.

➡ Each control cable has a spring clamp holding the outer sheath of the cable in place. Before disconnecting the inner part of the cable from its lever, use a small screwdriver to lift the spring tab up and release the outer cable.

3. Remove the knobs from the heater control panel.
4. Remove the small plugs from the side of the lower dash or console. Check the upper section of the console (near the levers). Remove any small concealment caps to reveal the attaching screws.
5. Remove the center panel by removing the upper and lower mounting screws.
6. As the center panel comes loose, reach behind it and disconnect the wiring connector.
7. Disconnect the lap heater duct under the dash.
8. Remove the defroster duct.
9. Disconnect the mode selector control cable.
10. Disconnect the temperature control cable to the water valve.
11. Unbolt and remove the heater control assembly.
12. Remove and disconnect the blower switch if so desired.
13. Before reinstallation, apply light grease to the slide points on the controller. Install the control panel and route the cables through the dash correctly.
14. Tighten the mounting screws.
15. Move the temperature control lever to the extreme left position.
16. With the water valve lever turned fully in the counterclockwise direction, connect the inner cable to the lever and then secure the sheath in the retaining clip.
17. Move the mode selection lever to the extreme left position.
18. With the mode selection damper lever (on the heater case) moved fully in a counterclockwise direction, connect the inner cable to the lever and secure the outer cable in the spring clip.
19. Install the defroster duct and install the lap heater duct.
20. Carefully connect the center panel wiring harness to the dash harness and install the center panel. Tighten the screws evenly and install the cover plugs.
21. Install the knobs on the heater control levers.
22. On the dash, move the air selection control lever to the "RECIRCULATE" position.
23. Move the air selection damper lever fully in the counterclockwise direction and connect the inner cable to the lever. Secure the outer cable in the spring clip.
24. Install the glove box and its stoppers.
25. Test the motion of each heater control; the movement should be smooth, with no binding or jerkiness.

1992-95 Montero

▶ See Figures 39, 40, 41, 42 and 43

1. Remove the glove box stoppers and remove the glove box.

6-32 CHASSIS ELECTRICAL

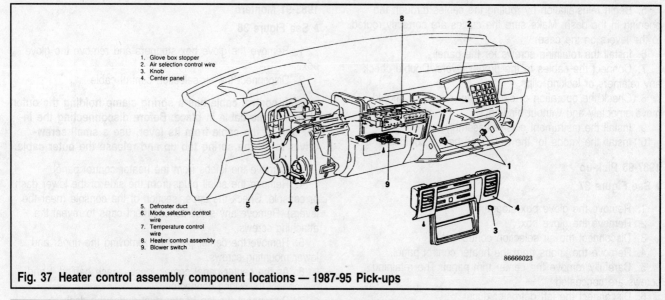

Fig. 37 Heater control assembly component locations — 1987-95 Pick-ups

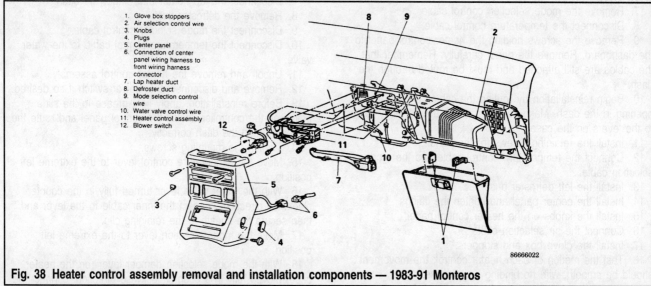

Fig. 38 Heater control assembly removal and installation components — 1983-91 Monteros

2. Disconnect the air selection control cable and the temperature control cable from the heater unit.

➡ Each control cable has a spring clamp holding the outer sheath of the cable in place. Before disconnecting the inner part of the cable from its lever, use a small screwdriver to lift the spring tab up and release the outer cable.

3. Remove the knobs from the heater control panel.
4. Remove the small plugs from the side of the lower dash or console. Check the upper section of the console (near the levers). Remove any small concealment caps to reveal the attaching screws.
5. Remove the instrument under cover, then remove the left-hand side cooler and foot shower ducts.
6. Remove the center panel by removing the upper and lower mounting screws.
7. As the center panel comes loose, reach behind it and disconnect the wiring connector.
8. Disconnect the mode selector control cable.

9. Unbolt and remove the heater control assembly from the instrument panel. The bezel, knobs, blower switch and air conditioning switch (if so equipped) can now be removed from the heater control assembly.
10. Remove the wire clips by inserting a suitably-sized prytool in the position shown in the illustration and pushing the wire clip in directions A and B. Do this for all three control cables hooked to the heater control assembly.
11. Before reinstallation, apply light grease to the slide points on the controller. Install the control panel and route the cables through the dash correctly.
12. Hook all three control cables to the heater control assembly with new retaining clips.
13. Reassemble the heater control assembly, if it was disassembled. Install the A/C switch, the blower switch, the bezel and any knobs.
14. Plug the wiring harness into the rear of the heater control assembly, then insert the assembly into its mounting bracket and secure it in place with the retaining screws.
15. Install the center panel.

CHASSIS ELECTRICAL 6-33

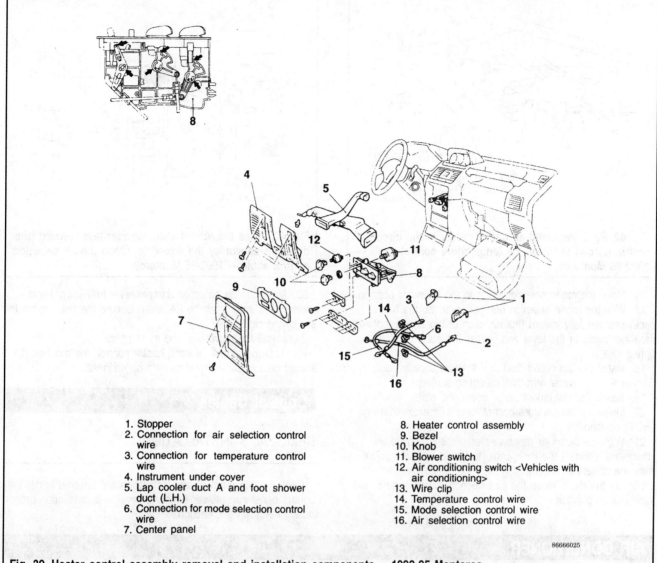

1. Stopper
2. Connection for air selection control wire
3. Connection for temperature control wire
4. Instrument under cover
5. Lap cooler duct A and foot shower duct (L.H.)
6. Connection for mode selection control wire
7. Center panel
8. Heater control assembly
9. Bezel
10. Knob
11. Blower switch
12. Air conditioning switch <Vehicles with air conditioning>
13. Wire clip
14. Temperature control wire
15. Mode selection control wire
16. Air selection control wire

Fig. 39 Heater control assembly removal and installation components — 1992-95 Monteros

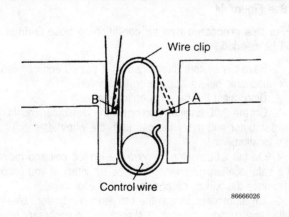

Fig. 40 Use a prytool to push the wire clip in directions A and B to remove the three control cables — 1992-95 Monteros

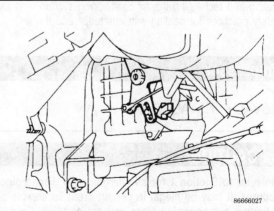

Fig. 41 Push the air selection damper lever in the direction shown by the arrow to connect the mode selection control cable — 1992-95 Monteros

6-34 CHASSIS ELECTRICAL

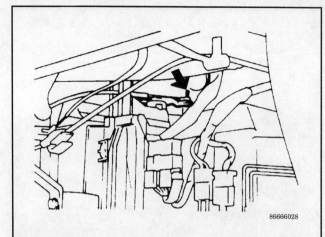

Fig. 42 Push the blend air damper lever in the direction shown (arrow) to install the temperature control cable — 1992-95 Monteros

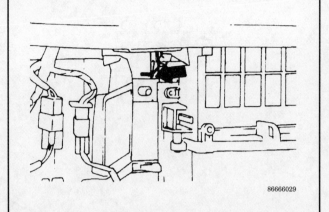

Fig. 43 Press the air selection damper lever inward (the direction shown by the arrow) to attach the air selection control cable — 1992-95 Monteros

16. Move the mode selection lever to the defroster position.
17. With the mode selection damper lever (on the heater case) pressed fully inward (the direction of the arrow), connect the inner cable to the lever and secure the outer cable in the spring clip.
18. Install the lap cooler duct and the foot shower duct. Secure them in place with their mounting screws.
19. Screw the instrument under cover into place.
20. Move the temperature control lever to the extreme right (HOT) position.
21. With the blend air damper lever pressed completely downward, connect the inner cable to the lever and then secure the sheath in the retaining clip.
22. On the dash, move the air selection control lever to the recirculation position.

23. Move the air selection damper lever fully inward and connect the inner cable to the lever. Secure the outer cable in the spring clip.
24. Install the glove box and its stoppers.
25. Test the motion of each heater control; the movement should be smooth, with no binding or jerkiness.

Control Cables

REMOVAL & INSTALLATION

Control cable removal and installation are covered under the control panel and blower switch removal and installation procedures, earlier in this section.

AIR CONDITIONER

Refer to Section 1 for the correct procedures for discharging, evacuating and recharging the air conditioning system, and for the safety practices for dealing with antifreeze and the A/C system.

Compressor

REMOVAL & INSTALLATION

✱✱CAUTION

Please refer to Section 1 for all charging and discharging procedures. It may be illegal in certain areas to service air conditioning components unless you are certified. Consult with your local authorities. Always wear eye protection. Don't smoke when handling refrigerant.

4-Cylinder Engines
▶ See Figure 44

➡For this procedure new air conditioning hose O-rings will be needed.

1. Have the system discharged by a trained and certified auto mechanic before working on this system.
2. Disconnect the negative battery cable.
3. On the 2.0L engine, disconnect the distributor cap from the distributor and move the cap (with the wires attached) to a safe location.
4. On the 2.6L engine, remove the ignition coil and move it to a safe location. The wiring may be left intact. If you choose to remove the wiring, be certain to label each wire.
5. On all models, loosen the belt tension adjuster (idler pulley) and remove the belt. It should not be necessary to remove this pulley if only the compressor is to be removed.
6. Label and disconnect the wiring running from the compressor. Some models have more than one wire connector. Check carefully.
7. Disconnect the discharge hose and the suction hose from the compressor. As soon as the lines are free, remove

CHASSIS ELECTRICAL 6-35

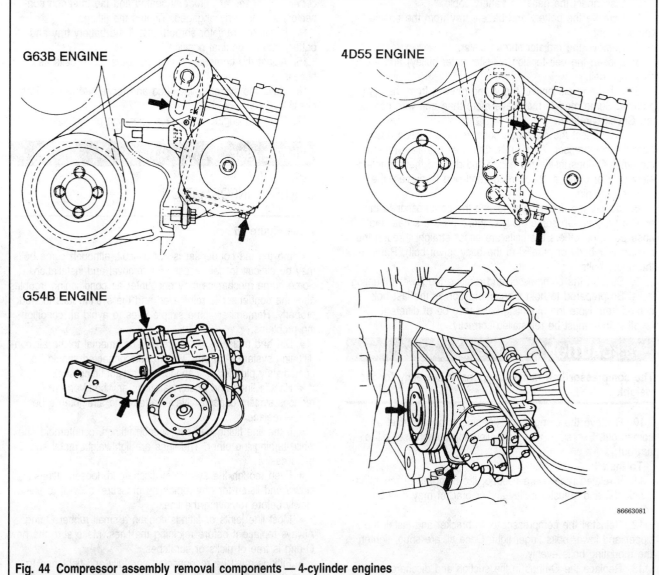

Fig. 44 Compressor assembly removal components — 4-cylinder engines

the small O-rings inside the hoses and discard them. Immediately plug or cover each open line and port to prevent the entry of any dirt.

8. Identify the upper bolts holding the compressor to the engine bracket. On some models, these bolts are near the hose ports; on others the bolts are either straight through the compressor body or parallel to the body (pivot bolt). Remove the upper bolts.

9. Support the compressor and remove the lower mounting bolts. Be prepared to hold the compressor as the last bolt comes free. Note the mounting bolts may be of different lengths; they must be reinstalled correctly.

✱✱CAUTION

The compressor is a heavy component; be ready for the weight.

10. Remove the compressor, keeping it roughly level as it comes out. Excessive tilting will allow the compressor oil to run out.

To install:

11. If related repairs are needed, the compressor mounting bracket(s) and adjuster pulley and/or bracket may be removed.

12. Reinstall the compressor to its bracket and install the upper and lower bolts finger tight. Once all are snug, tighten all mounting bolts.

13. Replace the O-rings in the suction and discharge hoses with new ones and install the hoses to the compressor. Make certain the hoses are properly seated and the threaded connectors are properly engaged. Tighten the fittings.

14. Proceed with reassembly following the procedure for your particular vehicle model in reverse order.

15. Have the system evacuated and recharged.

16. Adjust the compressor belt and other drive belts as necessary.

17. Start the engine and check operation of A/C system.

6-Cylinder Engines

▶ See Figure 45

1. Have the air conditioning system discharged by a trained and certified automotive mechanic.

6-36 CHASSIS ELECTRICAL

2. Disconnect the negative battery cable.
3. Remove the battery and battery tray from the engine compartment.
4. Remove the radiator shroud cover.
5. Loosen the belt tension adjuster (idler pulley) and remove the belt.
6. Label and disconnect the wiring running from the compressor. Some models may have more than one wire connector. Check carefully.
7. Disconnect the discharge hose and the suction hose from the compressor. As soon as the lines are free, remove the small O-rings inside the hoses and discard them. Immediately plug or cover each open line and port to prevent the entry of any dirt.
8. Identify the upper bolts holding the compressor to the engine bracket. On some models, these bolts are near the hose ports; on others the bolts are either straight through the compressor body or parallel to the body (pivot bolt). Remove the upper bolts.
9. Support the compressor and remove the lower mounting bolts. Be prepared to hold the compressor as the last bolt comes free. Note the mounting bolts may be of different lengths; they must be reinstalled correctly.

✱✱CAUTION

The compressor is a heavy component; be ready for the weight.

10. Remove the compressor, keeping it roughly level as it comes out. Excessive tilting will allow the compressor oil to run out.

To install:

11. If related repairs are needed, the compressor mounting bracket(s) and adjuster pulley and/or bracket may now be removed.
12. Reinstall the compressor to its bracket and install the upper and lower bolts finger tight. Once all are snug, tighten the mounting bolts evenly.
13. Replace the O-rings in the suction and discharge hoses with new ones and install the hoses to the compressor. Make certain the hoses are properly seated and the threaded connectors are properly engaged. Tighten the fittings.
14. Install the radiator shroud. Install the battery tray and battery into its original position.
15. Adjust the compressor belt and other drive belts as necessary.
16. Have the system evacuated and recharged as soon as possible.
17. Start the engine and check operation of A/C system.

Condenser

GENERAL INFORMATION

▶ See Figure 46

Removing the condenser is not difficult, although some bolts may be difficult to gain access for removal and installation. Some home mechanics may encounter air conditioning troubles after the condenser is reinstalled and the system fails to work properly. Remember some simple rules to avoid air conditioning problems:

• Dirt and moisture are extremely detrimental to the air conditioning system. Any line which is disconnected should be immediately plugged or capped with a tight-fitting seal.
• Handle lines and hoses carefully; any bend or kink reduces system capacity. Never attempt to straighten a bent line — replace it.
• If the line fitting has two wrench fittings, counterhold one while turning the other. The lines are lightweight metal and bend easily.
• Even though the system is discharged, loosen fittings slowly and listen for any remaining pressure; allow it to bleed slowly before removing the line.
• Most line joints or fittings contain a small rubber O-ring. Always replace it before rejoining the lines. Make sure the new O-ring is free of nicks or scratches.
• Fittings must be reconnected carefully and tightened to the correct torque. Too tight damages threads, too loose will leak.

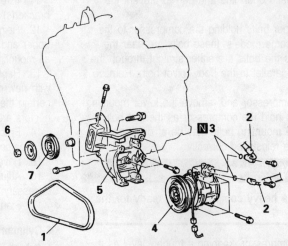

1. Compressor drive belt
2. Connection of high-pressure hose and low-pressure hose
3. O-ring
4. Compressor
5. Compressor bracket
6. Tension pulley mounting nut
7. Tension pulley

Fig. 45 Air conditioning compressor and tension pulley assembly — 6-cylinder engines

CHASSIS ELECTRICAL 6-37

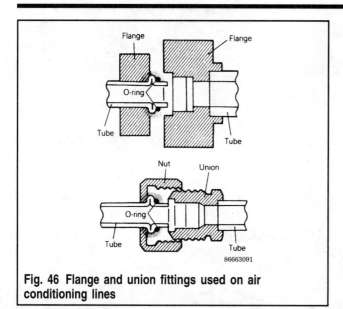

Fig. 46 Flange and union fittings used on air conditioning lines

REMOVAL & INSTALLATION

Except Montero
♦ See Figure 47

✴✴CAUTION

The compressed refrigerant used in the air conditioning system expands into the atmosphere at a temperature of 2°F (-16.5°C) or lower. This will freeze any surface, including your eyes, that it contacts. In addition, the refrigerant decomposes into a poisonous gas in the presence of a flame. Do not open or disconnect any part of the air conditioning system until you have read the safety warnings in Section 1.

1. Have the air conditioning system discharged by a trained and certified mechanic using a recovery/recycling machine.
2. Disconnect the negative battery cable.

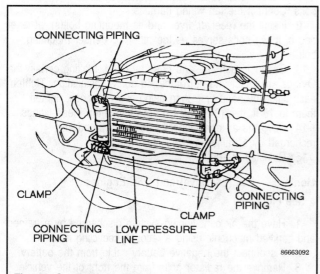

Fig. 47 Condenser assembly and related components

3. Remove the grille and, if needed for access to the condenser mounting bolts, the front bumper. Remove anything else that is restricting access to the condenser mounting bolts.
4. Remove the condenser refrigerant lines and plug them.

➡On the some models the receiver/drier assembly need to be removed first to allow easier access to the condenser.

5. Remove the condenser retaining bolts. Remove the condenser from the vehicle.

To install:
6. Install the condenser in the vehicle and evenly tighten all the mounting bolts equally.

➡Always use new O-rings in all refrigerant lines.

7. Reconnect all the refrigerant lines.
8. Install all the equipment which was removed to gain access to the condenser mounting bolts.
9. Connect the negative battery cable.
10. Have the system evacuated and recharged as soon as possible.

Montero — 2.6L and 3.0L (12 Valve) Engines

SINGLE AIR CONDITIONER
♦ See Figure 48

1. Have the air conditioning system discharged by a trained and certified mechanic using a recovery/recycling machine.
2. Remove the grille. Use care not to damage the plastic.
3. Carefully, and with the use of two wrenches, disconnect the lines in front of and running to the condenser. On the V6 there are three line junctions on the right side of the condenser and and four junctions on the left side. On the 4-cylinder engine, disconnect the lines at the condenser ports and disconnect the receiver/dryer.
4. Remove the receiver/dryer mounting bolts(s) and remove the receiver/drier.
5. Unplug the electric condenser fan, remove the mounting bolts for the fan and remove the fan.
6. Remove the condenser mounting bolt on the side opposite the receiver/dryer and carefully lift the condenser out of the vehicle.

To install:
7. Install the unit and tighten the right retaining bolt.
8. Install the fan assembly and tighten the retaining bolts. Connect the fan wiring harness.
9. Install the receiver/dryer and its mounting bolt(s); at least one of these bolts serves as the other condenser mounting bolt.
10. Connect each line and hose by hand, making certain the O-rings are intact and not damaged during installation. Tighten each fitting by hand first, then using two wrenches, tighten them to 18 ft. lbs. (24 Nm). Do not overtighten these fittings.
11. Evacuate and recharge the system; check carefully for leaks.
12. Reinstall the grille.

DUAL AIR CONDITIONERS
♦ See Figure 49

1. Have the air conditioner system discharged by a trained and qualified mechanic using a recovery/recycling machine.
2. Remove the grille. Use care not to damage the plastic.

6-38 CHASSIS ELECTRICAL

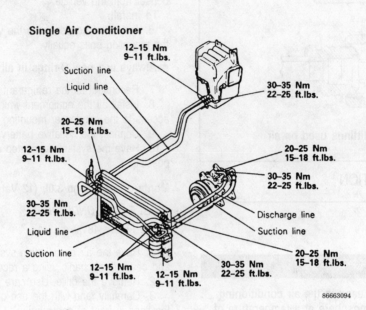

Fig. 48 Montero single air conditioner system line routing and torque specifications — 2.6L and 3.0L (12 valve) engines

3. Carefully, and with the use of two wrenches, disconnect the lines in front of and running to the condenser. The presence of the second condenser adds two more lines to the maze; disconnect the lines at the joints closest to each condenser. Disconnect both ends of any line in the way; do not bend or force the lines out of the way.

4. Remove the receiver/dryer mounting bolts(s) and remove the receiver/drier.

5. Unplug the large electric condenser fan, remove the mounting bolts for the fan and remove the fan.

6. Remove the condenser mounting bolt on the side opposite the receiver/dryer and carefully lift the main condenser out of the vehicle.

7. Unplug the connector for the smaller auxiliary fan. Unbolt the sub-condenser from its brackets and remove it.

8. Remove the fan assembly from the sub-condenser.

To install:

9. Attach the small fan to the condenser before placing the unit in the vehicle. Install the bracket bolts.

10. Install the main condenser and tighten the right retaining bolt.

11. Install the large fan assembly and tighten the retaining bolts. Connect the fan wiring harness.

12. Install the receiver/dryer and its mounting bolt(s); at least one of these bolts serves as the other condenser mounting bolt.

13. Connect each line and hose by hand, making certain the O-rings are intact and not damaged during installation. Tighten each fitting by hand first, then using two wrenches, tighten them to 18 ft. lbs. (24 Nm). Do not overtighten these fittings.

14. Have the air conditioner system evacuated and recharged by a trained and qualified mechanic.

15. Reinstall the grille.

Montero — 3.0L (24 Valve) and 3.5L Engines
♦ See Figure 50

1. Have the air conditioning system discharged by a trained and certified mechanic using a recovery/recycling machine.

2. Disconnect the negative battery cable from the battery.

3. Remove the radiator grille from the front of the vehicle.

4. Remove the hood latch stay and bracket assembly.

5. Remove the transmission cooler mounting bolts.

CHASSIS ELECTRICAL 6-39

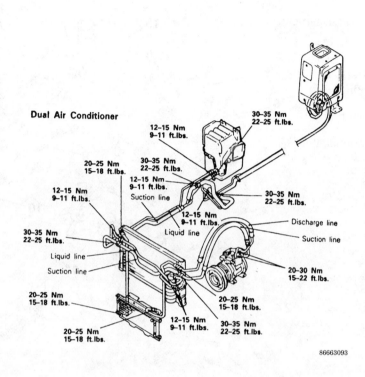

Fig. 49 Montero dual air conditioner system line routing and torque specifications — 2.6L and 3.0L (12 valve) engines

6. Remove the engine oil cooler mounting bolts and bracket.
7. Remove the condenser mounting bolts.
8. Disconnect the high pressure pipe A from the condenser and the receiver. Remove and replace the O-rings in these fittings with new upon reinstallation.
9. Remove the receiver bracket and receiver.
10. Remove the condenser electric fan assembly from the front of the condenser.
11. Disconnect the upper high pressure hose from the condenser. This hose junction will also need a new O-ring upon reassembly.
12. Extract the condenser out of the engine compartment. After the condenser has been removed from the vehicle, the three side seals can be removed from it.

To Install:

13. Before installation, check the condenser fan for crushing and other damage. Check the condenser's high pressure hose and pipe installation parts for damage or deformation. Check the condenser fan shroud for damage. Clean any dirt or debris from the condenser vanes.
14. Affix the side seals to the condenser, position the condenser back into its place, and reattach the upper high pressure hose to the condenser. Reinstall any brackets for this hose.
15. Install the fan assembly back to the front of the condenser. Reattach the receiver assembly to the condenser.
16. Install the lower high pressure hose, then secure the condenser in place, using the condenser mounting bolts.
17. Remount the transmission cooler and the engine oil cooler into their respective places.
18. Remount the hood latch and assembly bracket.
19. Install the radiator grille back in place. Reconnect the negative battery cable.
20. Have the air conditioning system evacuated and recharged as soon as possible.

6-40 CHASSIS ELECTRICAL

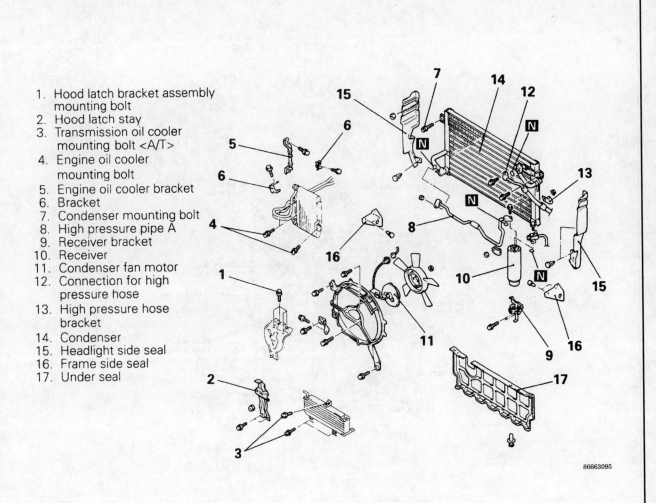

1. Hood latch bracket assembly mounting bolt
2. Hood latch stay
3. Transmission oil cooler mounting bolt <A/T>
4. Engine oil cooler mounting bolt
5. Engine oil cooler bracket
6. Bracket
7. Condenser mounting bolt
8. High pressure pipe A
9. Receiver bracket
10. Receiver
11. Condenser fan motor
12. Connection for high pressure hose
13. High pressure hose bracket
14. Condenser
15. Headlight side seal
16. Frame side seal
17. Under seal

Fig. 50 Condenser and condenser fan motor assembly — 3.0L (24 valve) and 3.5L engines

Condenser Fan Motor

REMOVAL & INSTALLATION

Please refer to the A/C condenser removal and installation procedures earlier in this section.

Sub-Condenser and Fan Motor

➡ These components are found on 1983-91 Monteros equipped with the dual air conditioner system.

REMOVAL & INSTALLATION

◆ See Figure 51

1. Have the A/C system discharged by a qualified automotive mechanic.
2. Disconnect the negative (-) battery cable from the battery.
3. Remove the radiator grille from the front of the vehicle — refer to Section 10 for this procedure.
4. Using two wrenches, disconnect the two lines going to the sub-condenser assembly. Once the lines are removed, cap the hoses or pipes with a blank plug to prevent entry of dust, dirt and water.
5. Discard the old O-rings; new O-rings will be needed for re-assembly.
6. Unplug the sub-condenser fan motor assembly harness connector.
7. Remove the four sub-condenser to mounting bracket bolts and extract the sub-condenser/fan motor assembly from the vehicle.
8. Unscrew the four fan motor to sub-condenser retaining bolts and separate the two components.
9. At this time, the mounting brackets can be removed, if needed for repairs.
10. To install the sub-condenser/fan motor assembly, bolt the mounting brackets back in place.

CHASSIS ELECTRICAL   6-41

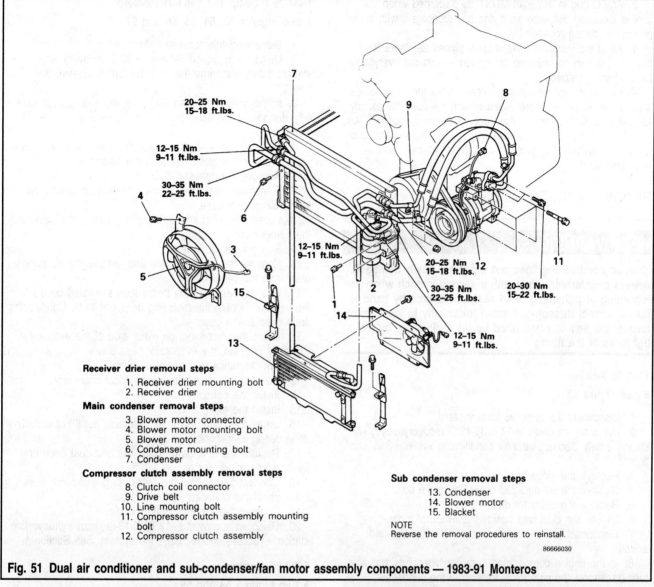

Receiver drier removal steps
1. Receiver drier mounting bolt
2. Receiver drier

Main condenser removal steps
3. Blower motor connector
4. Blower motor mounting bolt
5. Blower motor
6. Condenser mounting bolt
7. Condenser

Compressor clutch assembly removal steps
8. Clutch coil connector
9. Drive belt
10. Line mounting bolt
11. Compressor clutch assembly mounting bolt
12. Compressor clutch assembly

Sub condenser removal steps
13. Condenser
14. Blower motor
15. Blacket

NOTE
Reverse the removal procedures to reinstall.

Fig. 51 Dual air conditioner and sub-condenser/fan motor assembly components — 1983-91 Monteros

11. Mount the fan motor assembly to the sub-condenser with the four mounting bolts.
12. Bolt the sub-condenser/fan motor assembly to the mounting brackets on the vehicle.
13. Plug the fan motor wiring harness connector back together.
14. Apply a light coating of compressor oil to the new O-rings and install them onto the A/C lines.
15. Attach the A/C lines to the sub-condenser and tighten the fittings, using two wrenches, to 15-18 ft. lbs. (20-25 Nm).
16. Install the radiator grille and attach the negative (-) battery cable to the battery.
17. Have the A/C system evacuated and charged by a qualified mechanic.

Evaporator Core

PRECAUTIONS

The evaporator core and other components are located within the casing. In all cases, the casing must be opened for access to the various parts. Use a broad tool wrapped in a cloth to remove the clips. Use of an unpadded tool may damage or break the case. Check the case carefully for any hidden screws holding the halves together.

The best theory to apply during disassembly is removing the case from the evaporator, not the evaporator from the case. Once the components are exposed, the air-flow sensor, expansion valve, thermostat and/or insulating panels will need removal. Use a few simple rules to guide your work:

1. Don't force anything. All components are delicate and easily broken or stripped.
2. Use two wrenches whenever a joint is loosened or tightened.

CHASSIS ELECTRICAL

3. Any O-ring in any joint MUST be discarded when the joint is loosened. Replace each ring and lubricate it with compressor oil during reassembly.

4. All of the insulation and support pieces are there for a reason. Don't try to redesign the system — just put everything back where it belongs.

After components are replaced, reassemble the case halves around the evaporator core. Make certain the case fits exactly as it should; the correct air flow around the core is crucial. An air leak from a poorly fitted case could upset the system function. Install all the clips properly and the small case screws if any were used.

REMOVAL & INSTALLATION

✱✱WARNING

The air conditioning lines and pipes are easily damaged. Always counterhold joints with a second wrench when loosening or tightening. Start each connection by hand to insure correct threading. Tighten joints only to the correct torque; the seal is established by the O-ring, not by the tightness of the fitting.

1983-86 Pick-up

▶ See Figure 52

1. Disconnect the negative battery cable.
2. Using an approved R-12 or R-134a recovery/recycling station, safely discharge the air conditioning system. See Section 1.
3. Remove the glove box.
4. Remove the air duct just below the glove box.
5. Remove the defroster duct.
6. Loosen the duct joint beside the evaporator case.
7. Disconnect the electrical harness for the relays and switch.
8. In the engine compartment, use two wrenches and carefully disconnect the pipe joints running to the condenser.
9. Disconnect the drain hose from the case.
10. Remove the retaining nuts and bolts and remove the cooling unit from the truck.
11. After repairs, carefully install the assembled case into the vehicle. Tighten the mounting bolts and nuts. Connect the drain hose to the case.
12. Replace the O-rings and connect the lines in the engine compartment.
13. Connect the wiring to the relays and switch behind the dash to their wiring harnesses.
14. Secure the duct joint.
15. Install the defroster duct and the small lap heater duct (behind the glove box).
16. Install the glove box.
17. Connect the negative battery cable.
18. Using an approved R-12 or R-134a recovery/recycling station, evacuate and recharge the air conditioning system. See Section 1.

1987-95 Pick-up and 1983-91 Montero

▶ See Figures 53, 54, 55, 56 and 57

1. Disconnect the negative battery cable.
2. Using an approved R-12 or R-134a recovery/recycling station, safely discharge the air conditioning system. See Section 1.
3. In the engine compartment, use two wrenches to carefully disconnect the liquid line and the suction line at the firewall fittings.
4. Remove the retaining nut, located just above the hoses.
5. Remove the glove box from the dashboard.
6. Remove the defroster duct.
7. Disconnect the main electrical harness connector running to the evaporator case.
8. Loosen the duct joint between the evaporator case and the heater case.
9. Remove the drain hose.
10. Remove the retaining bolts and remove the evaporator case from the vehicle.
11. After repairs, carefully install the assembled case into the vehicle. Tighten the mounting bolts and nuts. Connect the drain hose to the case.
12. Install the duct joints on either side of the evaporator. Correctly installed, the evaporator case should have about 0.12 in. (3mm) clearance to the joint on each side.
13. Connect the main harness electrical connectors.
14. Install the defroster duct.
15. Install the glove box.
16. Under the hood, install the retaining nut if not already done during installation.
17. Replace the O-rings on each line and coat each ring lightly with compressor oil.
18. Connect and tighten the lines, using a second wrench to counterhold the opposite side of the joint.
19. Connect the negative battery cable.
20. Using an approved R-12 or R-134a recovery/recycling station, evacuate and recharge the system. See Section 1.

1992-95 Montero

▶ See Figures 58 and 59

1. Disconnect the negative battery cable.
2. Using an approved R-12 or R-134a recovery/recycling station, safely discharge the air conditioning system. See Section 1.
3. In the engine compartment, use two wrenches to carefully disconnect the high pressure pipe and the low pressure hose at the firewall fittings. Two new O-rings will be needed upon installation.
4. Remove the drain hose.
5. Remove the glove box stopper from the instrument panel, then remove the glove box assembly.
6. Unscrew the speaker grille retaining screws and remove the grille.
7. Unbolt the four bolts holding the lower frame in place and remove the frame from under the instrument panel (dashboard).
8. Extract the right-hand side foot shower duct and unplug the A/C wiring harness from the A/C control unit.
9. Remove the retaining bolts and remove the evaporator case from the vehicle.

CHASSIS ELECTRICAL 6-43

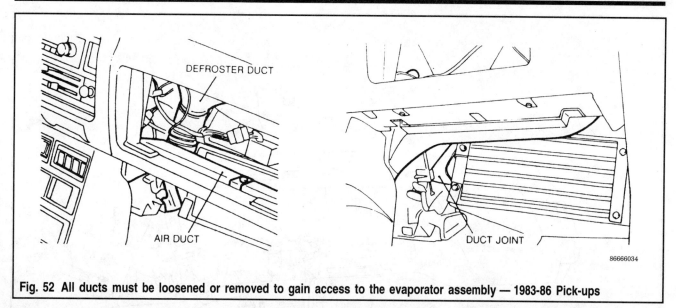

Fig. 52 All ducts must be loosened or removed to gain access to the evaporator assembly — 1983-86 Pick-ups

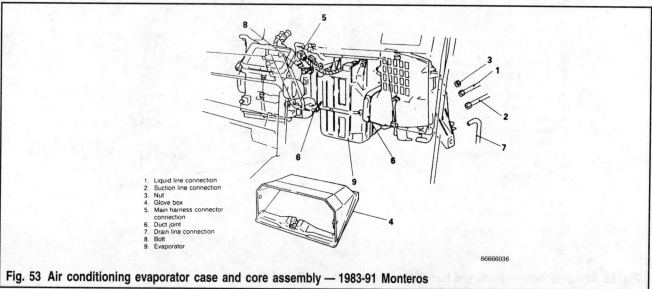

1. Liquid line connection
2. Suction line connection
3. Nut
4. Glove box
5. Main harness connector connection
6. Duct joint
7. Drain line connection
8. Bolt
9. Evaporator

Fig. 53 Air conditioning evaporator case and core assembly — 1983-91 Monteros

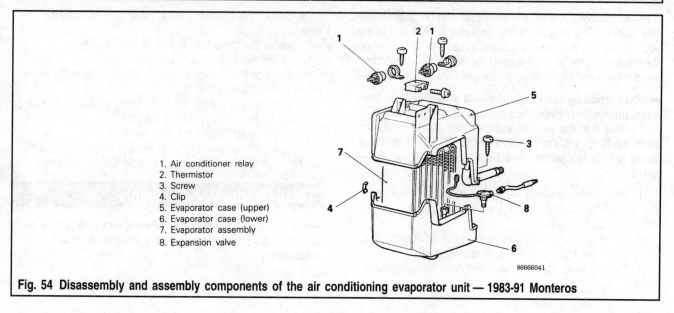

1. Air conditioner relay
2. Thermistor
3. Screw
4. Clip
5. Evaporator case (upper)
6. Evaporator case (lower)
7. Evaporator assembly
8. Expansion valve

Fig. 54 Disassembly and assembly components of the air conditioning evaporator unit — 1983-91 Monteros

6-44 CHASSIS ELECTRICAL

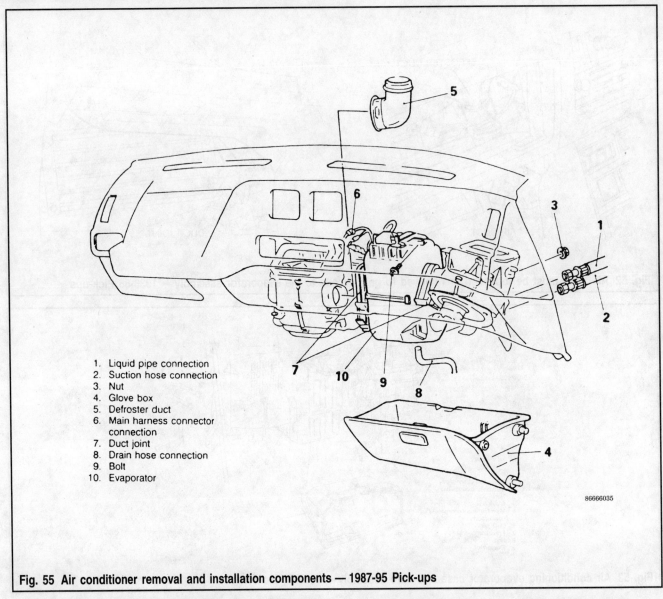

1. Liquid pipe connection
2. Suction hose connection
3. Nut
4. Glove box
5. Defroster duct
6. Main harness connector connection
7. Duct joint
8. Drain hose connection
9. Bolt
10. Evaporator

Fig. 55 Air conditioner removal and installation components — 1987-95 Pick-ups

10. At this point in time, the evaporator can be disassembled and repaired. The evaporator unit assembly is made up of the A/C control unit, the upper and lower evaporator case, the air thermo sensor, the air inlet sensor, the expansion valve, the low pressure and high pressure pipes and the evaporator itself.

➡When replacing the evaporator with a new unit, fill the evaporator with the specified amount of compressor oil and install it in the vehicle. Vehicles using R-12 refrigerant need 2.0 fl. oz. (60 cm^3) of NIPPONDENSO OIL 6, vehicles using R-134a refrigerant need 1.4 fl. oz. (40 cm^3) of ND — OIL 8.

11. After repairs, carefully install the assembled case into the vehicle. Tighten the mounting bolts and nuts. Connect the drain hose to the case.
12. Connect the A/C wiring harness to the A/C control unit.
13. Install the right-hand side foot shower duct.
14. Bolt the lower frame in place with the four mounting bolts.
15. Secure the speaker grille in place with its retaining screws.
16. Install the glove box and glove box stopper.
17. Under the hood, attach the drain hose.
18. Replace the O-rings on each line and coat each ring lightly with compressor oil.
19. Connect and tighten the lines, using a second wrench to counterhold the opposite side of the joint.
20. Connect the negative battery cable.
21. Using an approved R-12 or R-134a recovery/recycling station, evacuate and recharge the system. See Section 1.

Rear Evaporator Assembly

➡Some 1983-91 Monteros were equipped with a dual air conditioning system, which has a secondary, rear evaporator.

CHASSIS ELECTRICAL 6-45

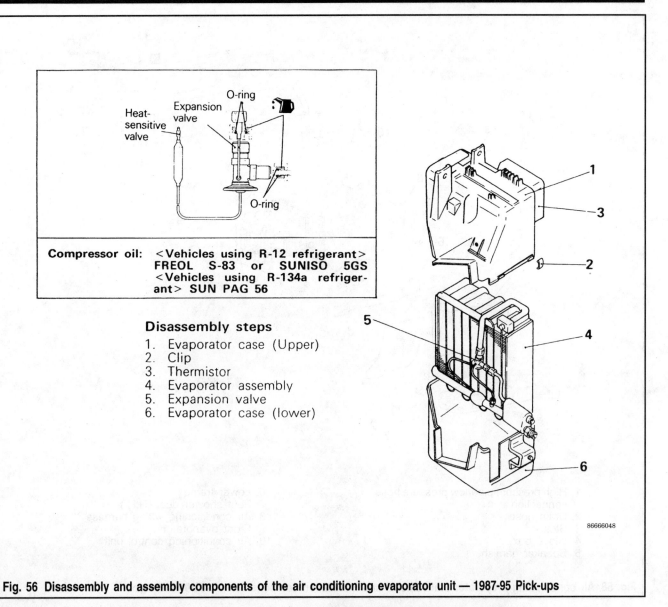

Fig. 56 Disassembly and assembly components of the air conditioning evaporator unit — 1987-95 Pick-ups

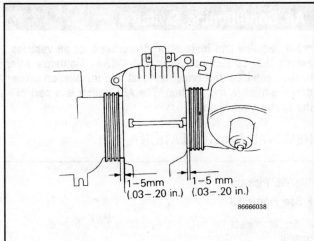

Fig. 57 Upon installation, adjust the clearance between the evaporator and the duct joints — 1987-95 Pick-ups

REMOVAL & INSTALLATION

♦ See Figures 60, 61 and 62

1. Have the A/C system discharged by a qualified automotive mechanic.
2. Disconnect the negative (-) battery cable from the battery.
3. Remove the passenger-side (right-hand side) rear pillar trim. The evaporator is located under this trim piece.
4. Remove the liquid line and the suction line connections from the evaporator assembly. Once the lines are removed, cap the hoses or pipes with a blank plug to prevent entry of dust, dirt and water.
5. Make certain to discard the old O-rings from the A/C lines and acquire new ones for reassembly.
6. Disconnect the upper duct from the top of the rear evaporator assembly.
7. Unplug the wiring harness connector.

6-46 CHASSIS ELECTRICAL

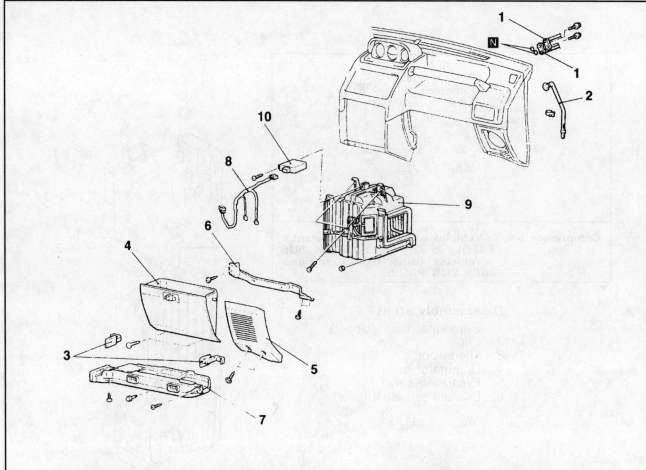

1. High pressure pipe/low pressure hose connection
2. Drain hose
3. Stopper
4. Glove box
5. Speaker garnish
6. Lower frame
7. Foot shower duct (R.H.)
8. Air conditioning wiring harness
9. Front evaporator
10. Air conditioning control unit

Fig. 58 Air conditioning evaporator unit removal and installation components — 1992-95 Monteros

8. Unbolt the mounting bolts and remove the evaporator assembly from the vehicle. Be careful, the evaporator will still have some coolant inside; do not spill it in the vehicle.
9. To install the rear evaporator, set it in place in the vehicle.
10. Secure it in place with the mounting bolts and plug the wiring harness connector back together.
11. Install the upper duct into the top of the evaporator.
12. Apply a coating of compressor oil (DENSO OIL 6 or SUNISO 5GS) to the new O-rings and place them in position in the liquid and suction lines. Refer to the refrigerant line procedure in this section for specifics.
13. Bolt the lines tightly to the evaporator.
14. Reinstall the rear pillar trim and connect the negative (-) battery cable to the battery.
15. Have the A/C system evacuated and charged by a certified mechanic.

Air Conditioning Switch

➡For removal and installation procedures for all vehicles except 1987-95 Pick-up trucks and 1983-91 Monteros, refer to the heater control panel removal and installation procedures, earlier in this section. The A/C switch is a part of the heater control panel.

REMOVAL & INSTALLATION

1987-95 Pick-up
♦ See Figure 63

1. Disconnect the negative (-) battery cable from the battery.
2. Pull all of the knobs off of the heater control panel levers. Note which knob goes where.

CHASSIS ELECTRICAL 6-47

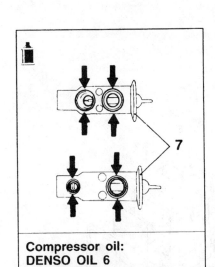
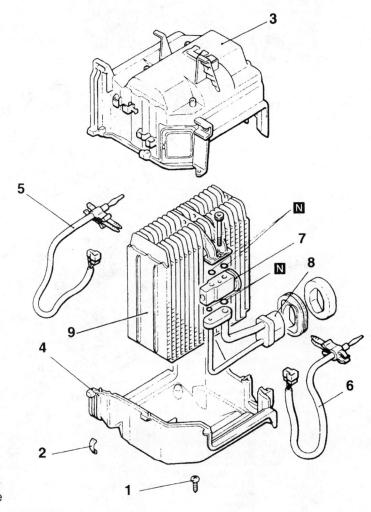

Compressor oil:
DENSO OIL 6
Vehicles using R-12
ND-OIL 8
Vehicles using R-134a

1. Screw
2. Clip
3. Evaporator case (upper)
4. Evaporator case (lower)
5. Air thermo sensor
6. Air inlet sensor
7. Expansion valve
8. Low pressure/high pressure pipe
9. Evaporator

Fig. 59 Evaporator unit disassembly and reassembly components — 1992-95 Monteros

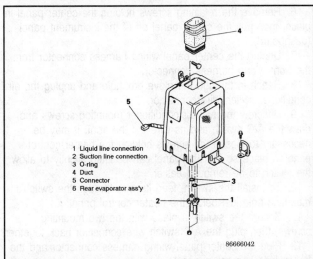

1. Liquid line connection
2. Suction line connection
3. O-ring
4. Duct
5. Connector
6. Rear evaporator ass'y

Fig. 60 Air conditioning rear evaporator unit removal and installation components — 1983-91 Monteros

3. Remove the glove box stopper and either remove the glove box assembly totally or just swing it down and out of the way.
4. Remove the retaining screws holding the center panel in place, then pull the center panel off of the instrument panel (dashboard).
5. Reach in through the glove box hole and unplug the air conditioner switch wire connector.
6. Unscrew the two air conditioner mounting screws and draw the A/C switch and its wire out the front. It may be necessary to loosen the screws holding the heater control panel in place so that the panel can be lifted slightly to allow the switch to be removed easier.
7. To install the switch, feed its wire through the switch's mounting hole just below the heater control panel.
8. Secure the switch in place with the two mounting screws, then plug the A/C switch wire connector back together.
9. Place the center panel in position and affix it there with the retaining screws.
10. Install the glove box assembly and install the glove box stopper in place.

6-48 CHASSIS ELECTRICAL

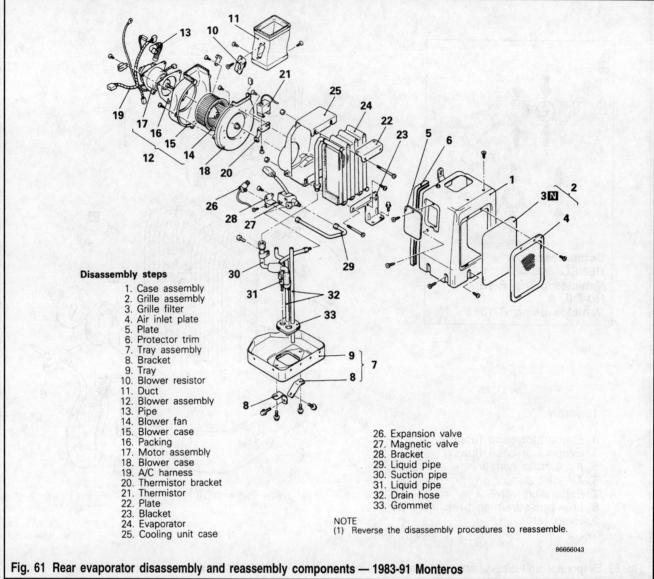

Disassembly steps

1. Case assembly
2. Grille assembly
3. Grille filter
4. Air inlet plate
5. Plate
6. Protector trim
7. Tray assembly
8. Bracket
9. Tray
10. Blower resistor
11. Duct
12. Blower assembly
13. Pipe
14. Blower fan
15. Blower case
16. Packing
17. Motor assembly
18. Blower case
19. A/C harness
20. Thermistor bracket
21. Thermistor
22. Plate
23. Blacket
24. Evaporator
25. Cooling unit case
26. Expansion valve
27. Magnetic valve
28. Bracket
29. Liquid pipe
30. Suction pipe
31. Liquid pipe
32. Drain hose
33. Grommet

NOTE
(1) Reverse the disassembly procedures to reassemble.

Fig. 61 Rear evaporator disassembly and reassembly components — 1983-91 Monteros

11. Push all of the heater control panel knobs back onto their respective levers.
12. Connect the negative (-) battery cable back to the battery.

1983-91 Montero

▶ See Figure 64

1. Disconnect the negative (-) battery cable from the battery.
2. Remove the lap heater duct from underneath of the glove box area, by removing the two retaining screws holding it in place.
3. Pull all of the knobs off of the heater control panel levers. Note which knob goes where.
4. Remove the glove box stopper and either remove the glove box assembly totally or just swing it down and out of the way.
5. Pop the retaining screw cover plugs out of the center cover, there should be one on each side.
6. Remove the retaining screws holding the center panel in place, then pull the center panel off of the instrument panel (dashboard).
7. Unplug the center panel wiring harness connector from the front wiring harness connector.
8. Reach in through the glove box hole and unplug the air conditioner switch wire connector.
9. Unscrew the two air conditioner mounting screws and draw the A/C switch and its wire out the front. It may be necessary to loosen the screws holding the heater control panel in place so that the panel can be lifted slightly to allow the switch to be removed easier.
10. To install the switch, feed its wire through the switch's mounting hole just below the heater control panel.
11. Secure the switch in place with the two mounting screws, then plug the A/C switch wire connector back together.
12. Plug the center panel wiring harness connector and the front wiring harness connector back together.
13. Place the center panel in position and affix it there with the retaining screws.
14. Push the retaining screw covers back in place.

CHASSIS ELECTRICAL   6-49

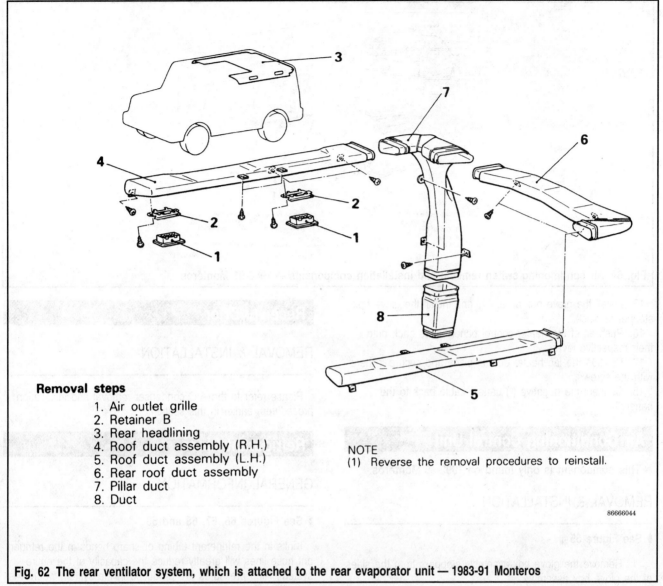

Removal steps
1. Air outlet grille
2. Retainer B
3. Rear headlining
4. Roof duct assembly (R.H.)
5. Roof duct assembly (L.H.)
6. Rear roof duct assembly
7. Pillar duct
8. Duct

NOTE
(1) Reverse the removal procedures to reinstall.

Fig. 62 The rear ventilator system, which is attached to the rear evaporator unit — 1983-91 Monteros

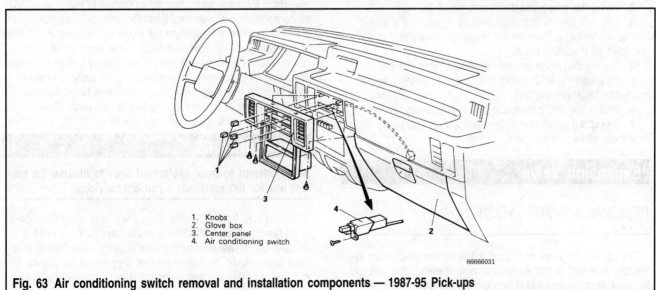

1. Knobs
2. Glove box
3. Center panel
4. Air conditioning switch

Fig. 63 Air conditioning switch removal and installation components — 1987-95 Pick-ups

6-50 CHASSIS ELECTRICAL

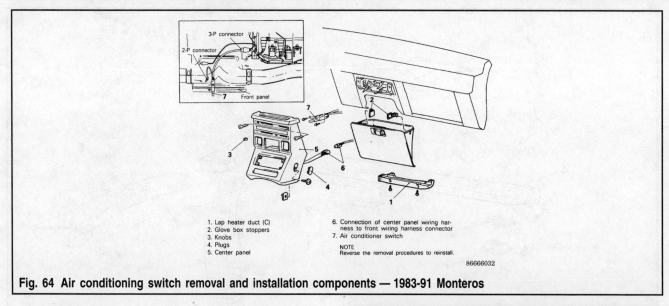

Fig. 64 Air conditioning switch removal and installation components — 1983-91 Monteros

15. Install the glove box assembly and install the glove box stopper in place.
16. Push all of the heater control panel knobs back onto their respective levers.
17. Position the lap heater duct in place and secure it there with the screws.
18. Connect the negative (-) battery cable back to the battery.

Air Conditioning Control Unit

➥This component is only found on 1992-95 Monteros.

REMOVAL & INSTALLATION

▸ See Figure 65

1. Remove the glove box assembly stoppers from the sides of the glove box assembly.
2. Swing the glove box down and out of the way.
3. The A/C control unit is located up inside of the glove box assembly hole. Remove the retaining screws and pull the unit from its mounting place.
4. Unplug the wiring harness connector from the unit.
5. To install the A/C control unit, plug the wiring harness connector back into the unit.
6. Attach the unit in its place with the retaining screws.
7. Swing the glove box assembly back up into place and install the glove box stoppers.

Expansion Valve

REMOVAL & INSTALLATION

The expansion valve is located inside of the evaporator assembly, attached to the evaporator core. Refer to the evaporator core removal and installation procedure, earlier in this section.

Receiver/Drier

REMOVAL & INSTALLATION

Please refer to the A/C condenser removal and installation procedures, earlier in this section.

Refrigerant Lines

GENERAL INFORMATION

▸ See Figures 66, 67, 68 and 69

Kinks in the refrigerant tubing or sharp bends in the refrigerant hose lines will greatly reduce the capacity of the entire system. High pressures are produced in the system when it is operating. Extreme care must be exercised to make sure that all connections are pressure tight. Dirt and moisture can enter the system when it is opened for repair or replacement of lines or components. The following precautions must be observed.

The system must be completely discharged before opening any fitting or connection in the refrigeration system. Open fittings with caution even after the system has been discharged. If any pressure is noticed as a fitting is loosened, allow trapped pressure to bleed off very slowly.

✻✻WARNING

Never attempt to bend preformed lines to fit. Use the correct line for the installation you are servicing.

A good rule for the flexible hose lines is to keep the radius of all bends at least 10 times the diameter of the hose. Sharper bends will reduce the flow of refrigerant. The flexible hose lines should be routed so that they are at least 3 in. (80mm) from the exhaust manifold. It is good practice to inspect all flexible hose lines at least once a year to make sure they are in good condition and properly routed.

CHASSIS ELECTRICAL 6-51

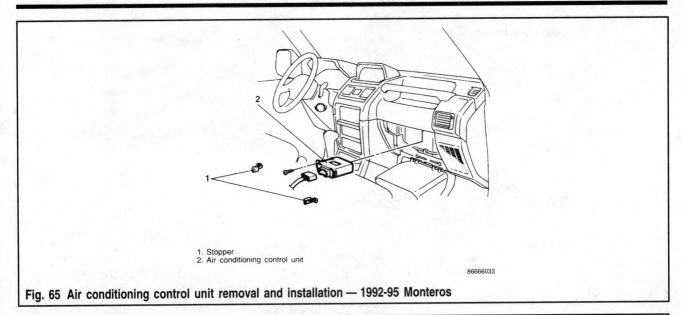

Fig. 65 Air conditioning control unit removal and installation — 1992-95 Monteros

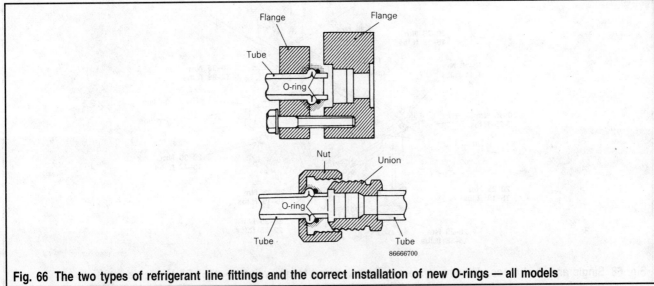

Fig. 66 The two types of refrigerant line fittings and the correct installation of new O-rings — all models

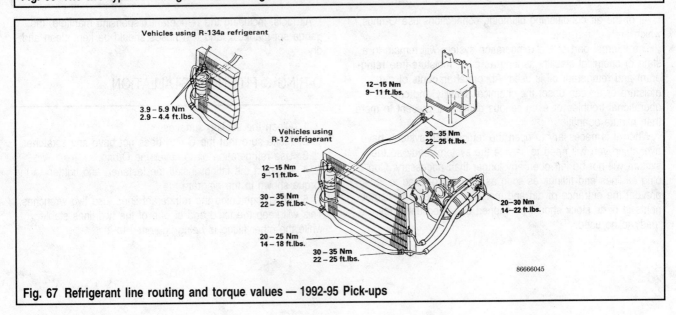

Fig. 67 Refrigerant line routing and torque values — 1992-95 Pick-ups

6-52 CHASSIS ELECTRICAL

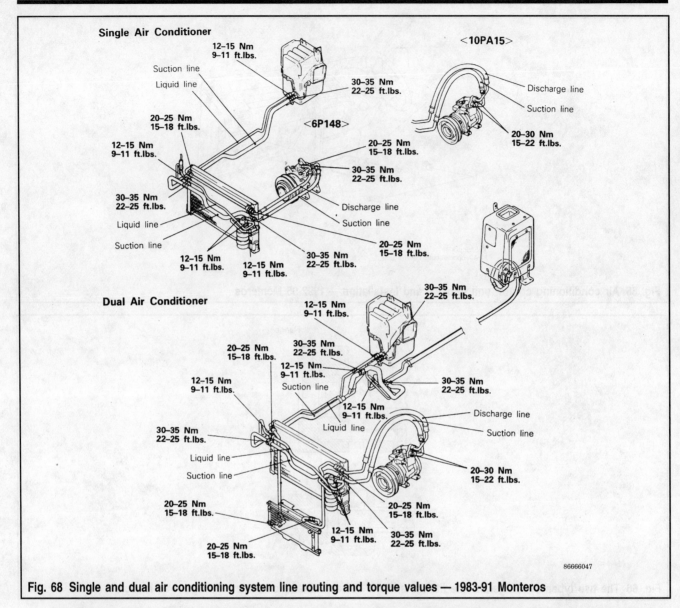

Fig. 68 Single and dual air conditioning system line routing and torque values — 1983-91 Monteros

All of the air conditioning plumbing connections use O-rings, which are not reusable.

The internal part of the refrigeration system will remain in a state of chemical stability as long as pure, moisture-free refrigerant and refrigerant oil is used. Abnormal amounts of dirt, moisture or air can upset the chemical stability and cause operational troubles or even serious damage if present in more than minute quantities.

When it is necessary to open the refrigeration system, have everything you will need to service the system ready so the system will not be left open any longer than necessary. Cap or plug all lines and fittings as soon as they are opened to prevent the entrance of dirt and moisture. All lines and components in parts stock should be capped or sealed until they are ready to be used.

All tools, including the refrigerant dispensing manifold, the gauge set manifold, and test hoses should be kept clean and dry.

O-RING & FITTING INSTALLATION

1. Clean the sealing surfaces.
2. Make sure that the O-ring does not have any scratches.
3. Use refrigeration oil to cover the O-ring.
4. Connect the fitting, install the fastener, and tighten to the torque shown in the diagram.
5. When tightening the refrigerant lines, use two wrenches. This will keep the fixed end of one of the two lines stable while the other fitting is being tightened to it.

CHASSIS ELECTRICAL   6-53

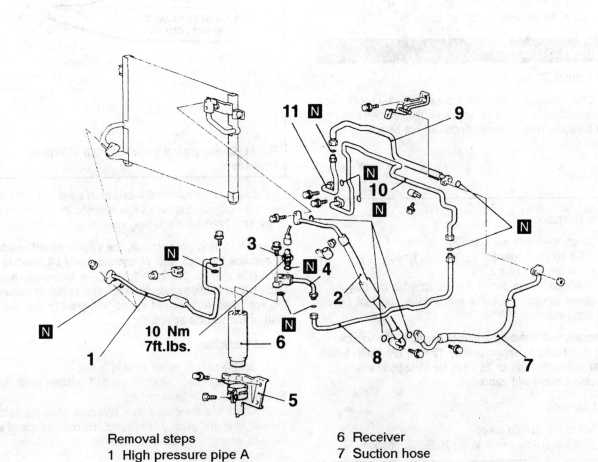

Removal steps
1 High pressure pipe A
2 High pressure hose
3 Dual pressure switch
4 High pressure pipe B
5 Receiver bracket
6 Receiver
7 Suction hose
8 High pressure pipe C
9 Suction pipe A
10 High pressure pipe D
11 Suction pipe B

Fig. 69 Air conditioning system refrigerant line routing — 1992-95 Monteros

CHASSIS ELECTRICAL

CRUISE CONTROL

Main Switch

INSPECTION

1. Turn the ignition switch to **ON**.
2. Check to be sure that the indicator light within the switch illuminates when the switch is turned On.

Column Switch

▶ See Figures 70 and 71

All of the following cruise control column switch checks must be performed while driving the vehicle. If the vehicle cannot be driven for some reason, these checks will not be able to be performed.

TESTING

System Operation

1. Switch the MAIN switch to On.
2. Drive above a speed of 25 mph (40 km/h).
3. Press the SET button.
4. Check and make certain that the speed is not only the same speed as when the button was released, but that the speed stays constant.

➡ If, because of climbing a hill for example, the vehicle speed decreases to approximately 12 mph (20 km/h) below the set speed, (9 mph or 15 km/h for 1992-95 Monteros), the cruise control will cancel.

Speed Increase

1. Set to the desired speed.
2. Turn the control switch to the RESUME position.

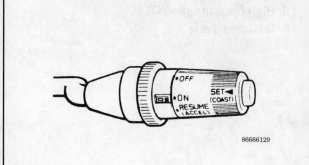

Fig. 70 Cruise control column switch — 1988-95 Pickups and 1988-91 Monteros

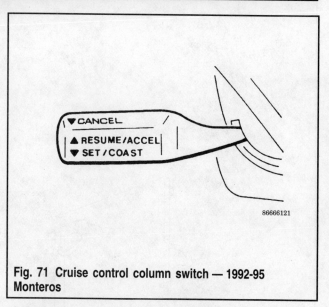

Fig. 71 Cruise control column switch — 1992-95 Monteros

3. Check to be sure that acceleration continues while the switch is held, and that when it is released the constant speed at the time becomes the driving speed.

➡ Even if, during acceleration, the vehicle speed reaches or exceeds the high limit of approximately 90 mph/145 km/h (124 mph or 200 km/h for 1992-95 Monteros), acceleration will continue; however, when the switch is released, the set speed (memorized speed) will slow to the high limit of the vehicle speed.

Speed Reduction

1. Set the vehicle to the desired speed.
2. Turn the control switch to the SET position (push the SET button in on older models).
3. Check that the deceleration continues while the switch is pressed, and that when it is released the constant speed at the time when it was released becomes the new driving speed.

➡ When the vehicle speed reaches the low limit (approximately 25 mph or 40 km/h) during deceleration, the cruise control will cancel.

Cancellation and Resume

1. Set the cruise speed control.
2. Check that there is a return to ordinary driving (cruise control is cancelled) when either of the operations below is performed.
   - The cruise control switch is turned to the CANCEL position.
   - The brake pedal is depressed.
   - The clutch pedal is depressed (manual transmission).
   - The transmission selector lever is moved to the "N" position (automatic transmission).
3. Turn the control switch to the RESUME position while driving at a vehicle speed of approximately 25 mph (40 km/h) or higher, and check that there is a return to the cruise control

CHASSIS ELECTRICAL 6-55

speed before it was cancelled. The vehicle should also maintain this speed.

4. When driving at a constant speed and the main switch is turned OFF, check that the vehicle returns to normal driving conditions.

Switch Continuity

PICK-UP

♦ See Figure 72

1. Unplug the column switch harness connector.
2. Operate the switch and check the continuity between the switch-side terminals.
   - Terminals 17 and 4 should show continuity when the SET switch is on.
   - Terminals 5 and 6 should show continuity when the Main switch is on.
   - Terminals 17 and 16 and terminals 5 and 6 should show continuity when the Main switch is on RESUME.
3. If the switch does not show the correct continuity, the switch needs replacement with a new switch.

Speed Sensor

REMOVAL & INSTALLATION

In order to access the speed sensor, the combination meter/instrument cluster must be removed from the dashboard. Refer to the instrument cluster removal and installation procedure, later in this section.

TESTING

Pick-up

♦ See Figure 73

1. Remove the combination meter/instrument cluster from the dashboard. Refer to the procedure later in this section for removal and installation instructions.
2. Connect an ohmmeter (resistance X 1 ohm range) between reed switch (+) terminal 1 and GND terminal 10 on the printed circuit board of the combination meter, then turn the speedometer cable shaft slowly, using a thin rod or similar tool.
3. It is normal if continuity and discontinuity between terminals alternate four times for every revolution of the shaft. If it does not do this, replace the speedometer assembly.

Stop Light/Brake Switch

INSPECTION

♦ See Figure 74

1. Unplug the stop light/brake switch harness connector.
2. Check the switch for continuity between the following terminals:
   - When the brake pedal is depressed, terminals 2 and 3 should be continuous.
   - When the brake pedal is not depressed, terminals 1 and 4 should be continuous.
3. If the switch does not show these continuities, replace the switch with a new unit.

Clutch Pedal Position Switch

This procedure is only for vehicles equipped with a manual transmission.

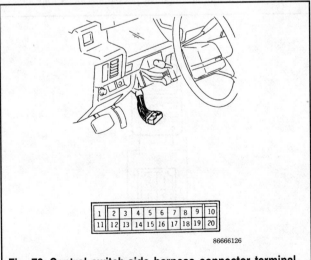

Fig. 72 Control switch-side harness connector terminal configuration — 1988-95 Pick-ups

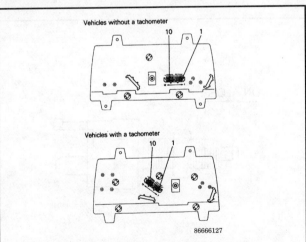

Fig. 73 Vehicle-speed sensor terminals on the printed circuit board of the combination meter — 1988-95 Pick-ups

6-56 CHASSIS ELECTRICAL

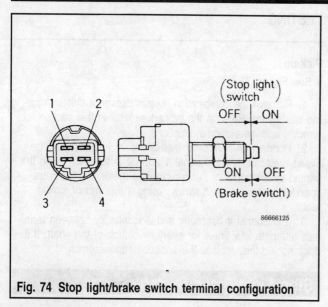

Fig. 74 Stop light/brake switch terminal configuration

INSPECTION

▶ See Figure 75

1. Unplug the clutch pedal position switch wiring harness connector.
2. Check for continuity between the connector terminals when the clutch pedal is depressed.
3. If there is no continuity when the pedal is depressed, replace the switch with a new switch.

Park/Neutral Position (Inhibitor) Switch

This procedure is only for vehicles equipped with an automatic transmission.

INSPECTION

Montero

▶ See Figures 76 and 77

1. Unplug the park/neutral position switch wiring harness connector.
2. Check for continuity between the following connector terminals when the transmission selector lever is moved to the **N** position:
   - 1983-91 Monteros — terminals 3 and 4
   - 1992-93 Monteros — terminals 7 and 12
   - 1994-95 Monteros — terminals 5 and 6
3. If no continuity is found in the preceding tests, replace the switch with a new unit.

Vacuum System Components

TESTING

Pick-up

GENERAL OPERATION

▶ See Figure 78

1. Check the vacuum circuit for leaks or clogging.
2. Unplug the vacuum pump wiring harness connector.
3. Disconnect the check valve vacuum hose at the actuator and vacuum pump side, and connect a vacuum gauge in its place.
4. Connect a 12 volt battery's positive (+) terminal to the connector L terminal (power side) and the battery's negative (-) terminal to connector B terminal (GND side) and operate the motor to check the generated negative pressure (vacuum).
5. The standard value of the negative pressure produced is 5.9 in. Hg/min. (9.3 kPa/min.) or more.
6. After the release of the negative pressure, connect the battery again to generate negative pressure. Check that the

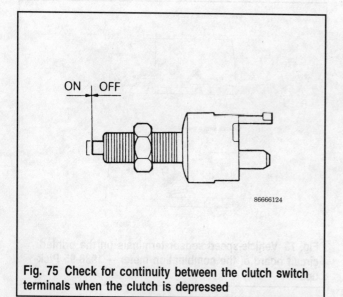

Fig. 75 Check for continuity between the clutch switch terminals when the clutch is depressed

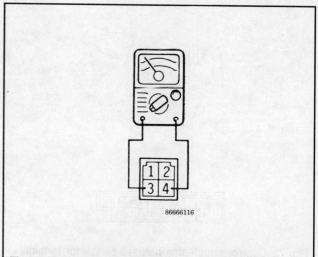

Fig. 76 Park/neutral position switch terminal configuration — 1988-91 Monteros

CHASSIS ELECTRICAL 6-57

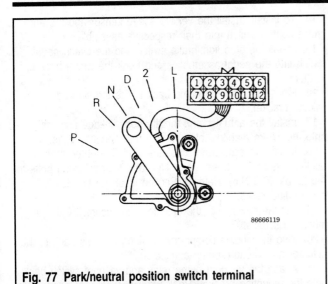

Fig. 77 Park/neutral position switch terminal configuration — 1992-95 Monteros

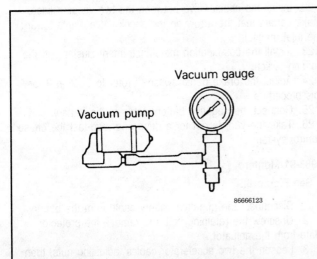

Fig. 78 Attach a vacuum gauge to the pump to check the vacuum created — 1988-95 Pick-up

disconnection of the battery to stop the motor does not cause a sudden loss of vacuum.

7. If the pump does not meet the previously stated standards, it will need to be replaced by a new unit.

VACUUM SWITCH

1. Unplug the wiring connector from the vacuum switch.
2. Use a hand vacuum pump, connected to the negative pressure port of the switch, to apply a vacuum.
3. Check for continuity between the switch terminals.
4. Check that there is no sudden drop of vacuum.
5. If the switch fails either of these tests, replace it with a new one.

VACUUM PUMP RELAY

♦ See Figure 79

1. Unplug the wiring harness connector from the vacuum pump relay.

2. Hook the battery (12 volt) positive terminal to terminal 1 (green wire) of the relay and the negative terminal to terminal 2 (blue and red wire). This will energize the relay coil.
3. Check for continuity between terminals 3 and 4 when the coil is energized. There should be continuity present (approximately. 0 ohms).
4. Disconnect the battery from the relay (this will de-energize the coil) and check terminals 3 and 4 again. There should be no continuity present.
5. Check the continuity across terminals 1 and 2 with the coil de-energized. There should be approximately 70 ohms present.
6. If the relay fails any of the preceding tests, replace it with a new unit.

VACUUM CHECK VALVE

1. Remove the check valve from its vacuum lines.
2. Blow into nipple B and check that air blows out from nipple A.
3. If the air does not blow through, replace the check valve with a new unit.
4. Install the check valve back between the vacuum lines. Make certain that the arrow on the check valve points toward the inlet manifold side.

1992-95 Montero

SOLENOID VALVE

♦ See Figure 80

➡A control valve and release valve are contained within the solenoid valve.

1. Unplug the wiring harness connector from the cruise vacuum pump.
2. Measure the resistance value between terminals 1 and 2, and also between terminals 1 and 3. The standard value should be 50-60 ohms.
3. Check that the solenoid valve makes an operating noise when battery voltage is applied to terminals 1 and 2, as well as to terminals 1 and 3.
4. If there is a malfunction of the solenoid valve, replace the cruise vacuum pump assembly.

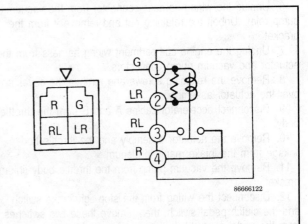

Fig. 79 Connect a 12 volt battery's positive terminal to terminal 1 (green) and its negative terminal to terminal 2 (blue with red) to energize the relay coil — 1988-95 Pick-ups

6-58 CHASSIS ELECTRICAL

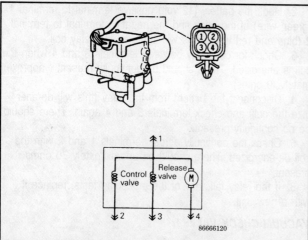

Fig. 80 Measure the solenoid valve resistance between terminals 1 and 2, and also between terminals 1 and 3 — 1992-95 Monteros

MOTOR

1. Unplug the wiring harness to the cruise vacuum pump.
2. Check that the motor revolves when battery positive voltage is applied between terminals 1 and 4.
3. If the motor does not operate, replace it with a new unit.

REMOVAL & INSTALLATION

Pick-up

▶ See Figure 81

1. Disconnect the negative battery cable from the battery.
2. Remove the column switch — refer to the procedures in this section.
3. Remove the instrument cluster/combination meter — refer to the procedures in this section.
4. Disconnect the vacuum hoses and remove the check valve.
5. Disconnect the wiring harness plug and the vacuum hose from the vacuum switch and remove it.
6. Unplug the wiring harness connector from the vacuum pump relay. Unbolt the retaining nut and remove it from the bracket.
7. Unplug the engine compartment wiring harness from the actuator and vacuum pump connectors.
8. Remove the retaining screws and remove the cover from over the actuator assembly.
9. Disconnect accelerator cables A and B from the throttle body.
10. Remove the actuator assembly and the throttle body linkage from the intake manifold plenum.
11. Remove the vacuum pump from the throttle body linkage bracket.
12. Disconnect the wiring from the stop light/brake switch and the clutch pedal switch, then remove these two switches from the pedal assemblies.
13. Remove the vehicle speed sensor (incorporated into the speedometer) and the auto-cruise control switch (integrated into the column switch) from their respective assemblies.

14. To install, install the vehicle speed sensor and the auto-cruise control switch into their respective assemblies.
15. Install the stop light/brake switch and the clutch pedal switch into the pedal assemblies and hook the wiring back up to them.
16. Mount the vacuum pump back to the throttle body linkage bracket.
17. Install the actuator and throttle body linkage assembly onto the intake manifold plenum with the mounting bolts.
18. Install the accelerator cables A and B. Tighten their retaining nuts to 6-8 ft. lbs. (8-11 Nm) and their retaining bolts to 4-6 ft. lbs. (6-8 Nm). Refer to Section 5 for the accelerator cable adjustments.
19. Set the cover over the actuator and secure it with the mounting screws.
20. Plug the engine compartment wiring harness back to the actuator and vacuum pump connectors.
21. Attach the vacuum pump relay and vacuum switch back onto the mounting bracket. Hook the wiring harness connectors back together and install the vacuum hoses.
22. Install the check valve between the two vacuum lines. Make certain that the arrow on the check valve points toward the inlet manifold.
23. Install the combination meter/instrument cluster and the column switch.
24. Adjust the brake pedal switch — refer to Section 9 for this procedure.
25. Connect the negative battery cable to the battery.
26. Take the vehicle out for a drive and road-test the cruise control system.

1988-91 Montero

▶ See Figure 82

1. Disconnect the negative battery cable from the battery.
2. Unscrew the retaining bolt and remove the protector plate from the actuator.
3. Loosen the two accelerator cables' adjusting nuts, then unfasten the actuator side inner cables from the actuator.
4. Unplug the actuator wiring harness connector.
5. Remove the three mounting nuts, then remove the actuator assembly itself from the mounting bracket. Make sure to retain the mounting nuts, washers, spacers, and rubber grommets for reassembly.
6. Unplug the wiring harness connectors to the accelerator, stop light, clutch and inhibitor switches. Remove these switches from their respective assemblies.
7. Remove the combination meter, by using the procedures found in this section. Remove the vehicle speed sensor, which is attached to the speedometer.
8. The auto-cruise control switch is an integral part of the column switch, remove this switch — refer to the instructions in this section.
9. To remove the control unit, remove the front rail cover and the front door opening trim.
10. Remove the cowl side trim, then remove the auto-cruise control unit.
11. To install, mount the auto-cruise control unit and install the cowl side trim.
12. Install the door opening trim and secure the front rail cover in place with the mounting screws and retainers.

CHASSIS ELECTRICAL 6-59

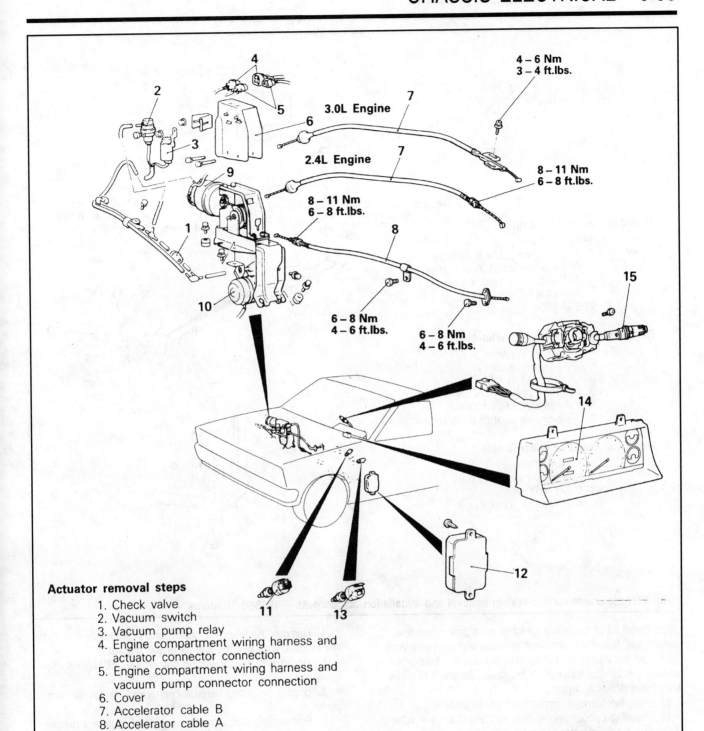

Actuator removal steps
1. Check valve
2. Vacuum switch
3. Vacuum pump relay
4. Engine compartment wiring harness and actuator connector connection
5. Engine compartment wiring harness and vacuum pump connector connection
6. Cover
7. Accelerator cable B
8. Accelerator cable A
9. Actuator assembly
10. Vacuum pump

Removal of switches and control unit
11. Stop light switch/Brake switch
12. Electronic control unit (ECU)
13. Clutch switch
14. Vehicle speed sensor (incorporated in speedometer)
15. Auto-cruise control switch (integrated into column switch)

Fig. 81 Auto-cruise control system removal and installation components — 1988-95 Pick-ups

6-60 CHASSIS ELECTRICAL

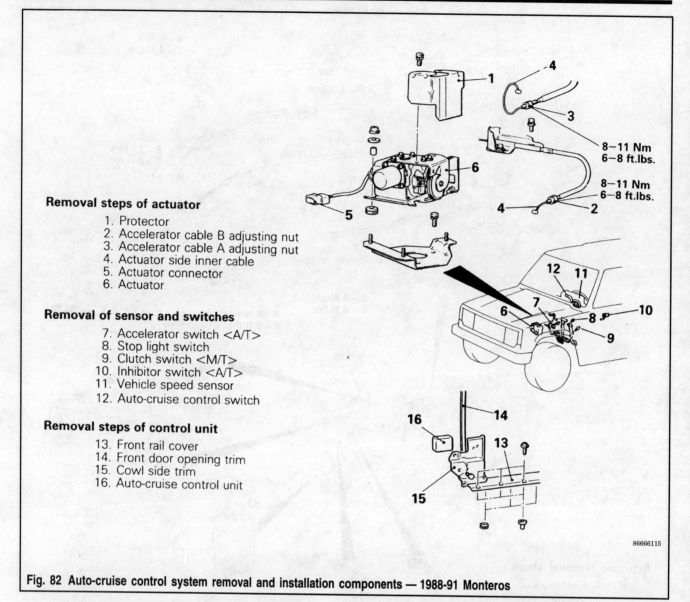

Fig. 82 Auto-cruise control system removal and installation components — 1988-91 Monteros

13. Install all of the inside switches onto their respective assemblies. Install the combination meter and column switch.
14. Set the rubber grommets onto the actuator bracket's studs, then set the actuator on the studs. Secure it in place with the washers and nuts.
15. Plug the harness connectors back together.
16. Install the accelerator cables and adjust them — refer to Section 5 for accelerator cable adjustment procedures.
17. Tighten the accelerator cable adjuster-nut locking nuts to 6-8 ft. lbs. (8-11 Nm).
18. Mount the protector plate on the actuator with the retaining bolt.
19. Connect the negative battery cable to the battery.

1992-95 Montero

▶ See Figures 83 and 84

1. Disconnect the negative battery cable from the battery.
2. Unbolt the retaining bolt and remove the link protector from the intermediate link.
3. Disconnect the accelerator cable, the throttle cable and the cruise control cable from the intermediate link.
4. Remove the four nuts holding the intermediate link to the link bracket, then pull the link off of the bracket.
5. Unscrew the one retaining bolt on the link bracket. Remove the bracket.
6. Unplug the wiring harness connector from the vacuum pump.
7. Remove the vacuum pump vacuum hoses and remove the pump from the pump bracket.
8. Remove the three nuts holding the vacuum pump bracket onto the actuator bracket, then pull the bracket off.
9. Disconnect the vacuum hose to the actuator and remove the actuator from the actuator bracket.
10. To install, mount the actuator onto the actuator bracket.
11. Mount the vacuum pump bracket onto the actuator bracket with the three nuts and washers.
12. Install the vacuum pump to the vacuum pump bracket. Connect the actuator and the vacuum pump with the vacuum hose.
13. Plug the wiring harness connectors back together.
14. Mount the link bracket with the one retaining bolt.

CHASSIS ELECTRICAL   6-61

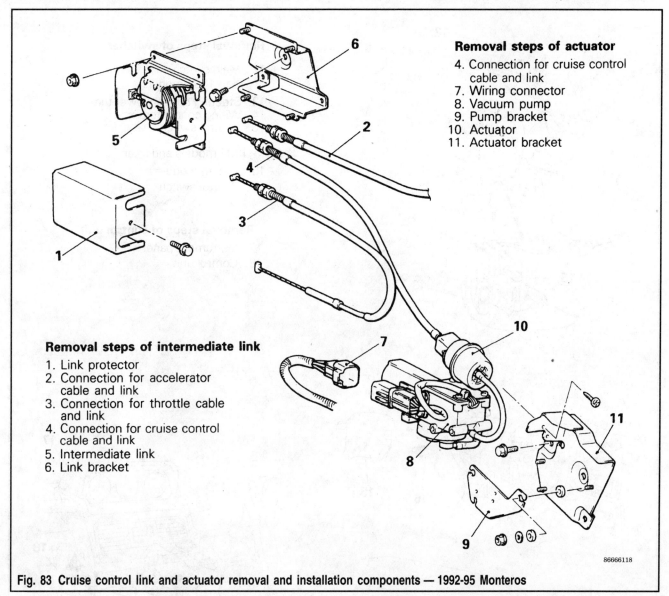

Fig. 83 Cruise control link and actuator removal and installation components — 1992-95 Monteros

15. Using the four nuts, secure the intermediate link to the link bracket.
16. Attach the accelerator cable, the throttle cable, and the cruise control cable to the intermediate link.
17. Adjust the accelerator cable, the throttle cable, and the cruise control cable. Check to see if there is any deflection in the inner cables of the accelerator cable, cruise control cable and throttle cable. If there is excessive deflection or no play in an inner cable, loosen the adjusting bolts and nuts to release each link from the throttle lever (do not remove the adjusting bolts or nuts).

   a. Accelerator cable: while holding link (C) so that it is touching the stopper, adjust the play of the accelerator cable with the adjusting nut so that the cable play is 0-0.04 in. (0-1mm) for non-California vehicles or 0.08-0.12 in. (2-3mm) for California models. After adjusting, secure the cable with the lock nut.

   b. Throttle cable: while holding link (B) so that it is touching link (C), adjust the play of the accelerator cable with the adjusting nut and adjusting bolts (throttle lever side) so that the cable play is 0.04-0.08 in. (1-2mm). After adjusting, secure the cable with the lock nut.

   c. Cruise control cable: while holding link (A) so that it is touching link (B), adjust the play of the accelerator cable with the adjusting nut so that the cable play is 0.04-0.08 in. (1-2mm). After adjusting, secure the cable with the lock nut.
18. Replace the intermediate link unit's cover.
19. Connect the negative battery cable to the battery.

6-62 CHASSIS ELECTRICAL

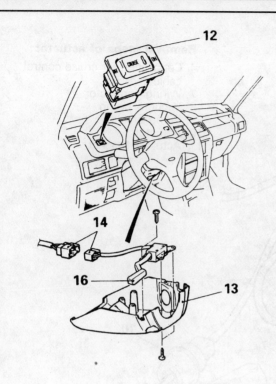

Removal steps of switches

12. Main switch
    Up to 1993 models
13. Steering column lower trim
14. Wiring connectors
16. Control switch

   1994 models and after
15. Air bag module
16. Control switch

Removal steps of control unit

17. Instrument panel
18. Control unit

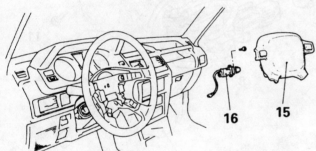

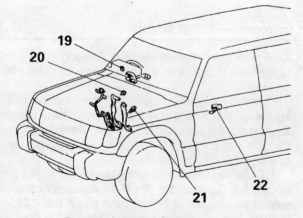

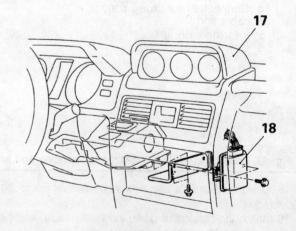

Removal steps of sensors

19. Vehicle speed sensor (reed switch)
20. Stop light switch
21. Clutch pedal position switch <M/T>
22. Park/Neutral position switch <A/T>
23. TPS (Throttle position sensor)

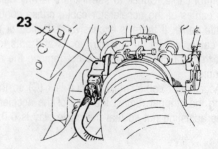

Fig. 84 Cruise control system sensor and switch removal and installation components — 1992-95 Monteros

CHASSIS ELECTRICAL 6-63

AUDIO SYSTEM

Radio, Tape Player and CD Player

REMOVAL & INSTALLATION

1983-86 Pick-up

➡ The radio fuses are mounted on the right rear corner of the radio. The radio need not be removed to change a fuse; just remove the glove box and reach through the opening.

1. Disconnect the negative (-) battery cable from the battery.
2. Remove the instrument cluster bezel (trim piece).
3. Remove the radio bracket attaching screws from the instrument panel and remove the radio bracket.
4. Pull the radio outward slightly and disconnect the speaker wires, the power wire and the antenna cable.
5. Remove the radio.
6. When reinstalling, note that the mounting support attaches to the radio differently for AM radios or AM/FM radios.
7. Hold the radio in position and connect the wiring and antenna leads.
8. Before setting the radio into the dash, make certain that the mounting stay fits into the slot in the reinforcement.
9. Install the radio and secure it in place with the mounting screws.
10. Mount the instrument bezel with its retaining screws.
11. Connect the negative (-) battery cable to the battery.

1987-95 Pick-up

▶ See Figure 85

1. Disconnect the negative (-) battery cable from the battery.
2. Remove the heater control lever knobs.
3. Remove the screws holding the center trim panel. Use a plastic or padded stick or flat tool to remove the panel without marring it.
4. Remove the radio bracket mounting screws.
5. Pull the radio unit out of the dash (with the brackets attached) a short distance.
6. Disconnect the wiring and antenna lead at the rear of the radio.
7. Remove the radio from the truck and remove the brackets from the radio.
8. When reinstalling, mount the brackets to the radio and install them as a unit.
9. Before fitting the radio completely into the dash, connect the wiring and antenna cable.
10. Tighten the bracket mounting screws.
11. Install the face plate and the heater control lever knob.
12. Connect the negative (-) battery cable to the battery.

1983-86 Montero

1. Disconnect the negative (-) battery cable from the battery.
2. Remove the radio faceplate. Use a flat plastic tool or protected metal tool so as not to mar the plastic
3. Remove the plug on the each side of the center console. Reach through the holes and remove the radio mounting screws.
4. Disconnect the antenna lead wire, the speaker connector and the power wire from the back of the radio.
5. Remove the radio.
6. To install, insert the radio into its hole.
7. Connect the antenna lead wire, the speaker wires and the power wire to the back of the radio.
8. Insert and tighten the radio mounting screws through the side holes of the center console.
9. Push the radio faceplate back into position.
10. Connect the negative (-) battery cable to the battery.

1987-91 Montero

▶ See Figure 86

1. Disconnect the negative (-) battery cable from the battery.
2. Remove the knobs from the heater controls.
3. Remove the plugs from the side of the console.
4. Remove the center console from the dash. As it comes clear, disconnect the wiring harness.
5. Remove the radio faceplates.
6. Disconnect the radio brackets from the console.
7. Remove the radio with the brackets attached. Once clear of the console, the brackets may be removed.
8. When reinstalling, attach the brackets to the radio and the radio to the console.
9. As the console is moved into place, connect the wiring harness.
10. Secure the center console, install the hole plugs and install the heater control knobs.
11. Connect the negative (-) battery cable to the battery.

1992-95 Montero

▶ See Figure 87

1. Disconnect the negative (-) battery cable from the battery.
2. Remove the screws holding the center panel in place and pull it off of the dashboard.
3. Unscrew the radio/tape player mounting screws from the side brackets.
4. Pull the radio assembly out of its hole just far enough to disconnect and label any wiring attached to it. It may be necessary to remove the glove box assembly to get at the back of the radio before it is pulled all the way out.
   a. To remove the glove box, pop the glove box stoppers out and swing the glove box down and out of the way.
5. Extract the radio from the dashboard.
6. To install the radio/tape player unit, attach the wiring to the back of the radio and slide it into its hole (the wiring may need to be attached through the glove box hole after it has been slid in).
7. Install the glove box and stoppers to the dashboard.

6-64 CHASSIS ELECTRICAL

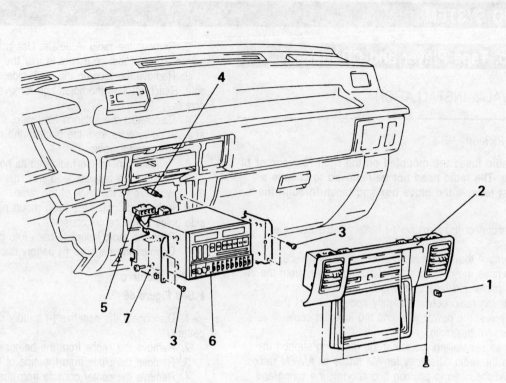

1. Heater control lever knob
2. Center panel
3. Radio bracket mounting screws
4. Feeder wire plug connection
5. Connector connection
6. Radio with tape player
7. Bracket

Fig. 85 Installation and removal components for the radio/tape player — 1987-95 Pick-ups

8. Secure the radio in place with the retaining screws and place the center panel in place over the radio.
9. Insert and tighten the center panel screws.
10. Connect the negative (-) battery cable to the battery.

Speakers

REMOVAL & INSTALLATION

1983-86 Pick-up
▶ See Figure 88

The speaker for the 1983-86 Pick-ups is located under the center of the defroster garnish by the windshield.

1. Disconnect the battery negative (-) cable from the battery.
2. Pull off the radio knobs and unscrew the radio nuts on the two adjuster stems. Note which side the knobs go on and in which order they should be installed.
3. Pull off the heater fan switch and heater control lever knobs.
4. Remove the cluster panel retaining screws and pull the panel out of its hole in the instrument panel (dashboard). Pull it out just far enough to gain access to the wires attached to it in the rear.
5. Detach any electrical wires and the speedometer cable from the cluster panel. Make sure to label the wires. If the cluster will not pull out far enough to get at the back of it, go up from under the dashboard to remove the wires and cable.
6. Remove the cluster panel totally form the dashboard.
7. Reach through the cluster panel opening, and disconnect and label any wires going to the speaker.
8. Loosen and remove the speaker retaining screws, then pull the speaker out through the cluster hole. If too many wires or ductwork is in the way, lower the speaker and remove it from under the dashboard.
9. To install the speaker, position it up under the defroster garnish, then secure it in its original position with the mounting screws.
10. Attach the electrical wires to the speaker.

CHASSIS ELECTRICAL 6-65

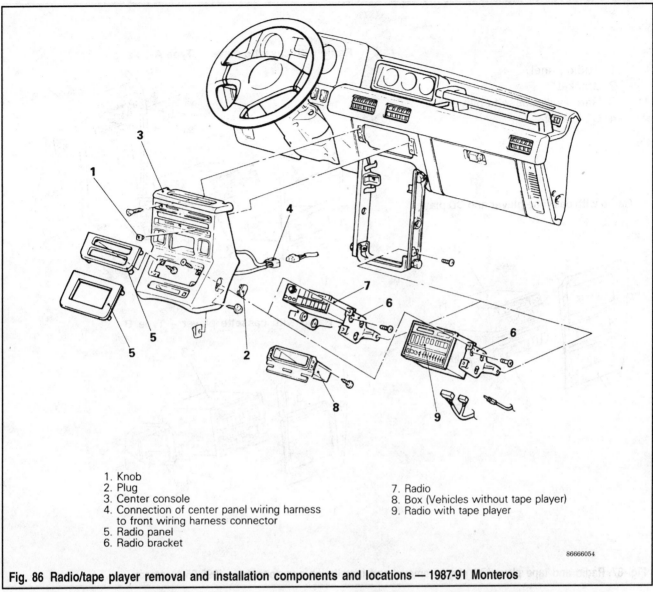

1. Knob
2. Plug
3. Center console
4. Connection of center panel wiring harness to front wiring harness connector
5. Radio panel
6. Radio bracket
7. Radio
8. Box (Vehicles without tape player)
9. Radio with tape player

Fig. 86 Radio/tape player removal and installation components and locations — 1987-91 Monteros

11. Place the cluster panel close enough to its hole to attach the speedometer and wires to it.
12. Mount the cluster in place with its retaining screws.
13. Push all of the heater control knobs back on their respective levers.
14. Tighten the radio nuts only until they are snug and push the knobs back onto the adjuster stems.
15. Connect the negative (-) battery cable to the battery.

1987-95 Pick-up

FRONT SPEAKERS

▶ See Figure 89

The front speakers on the 1987-95 Pick-ups are located on the upper far corners of the instrument panel (dashboard), by the windshield.

1. Disconnect the negative (-) battery cable from the battery.

2. For the driver's side speaker follow these next procedures, for the passenger's side skip to step 3.
   a. Remove the hazard warning switch, the starter unlock switch or the hole cover from the meter hood.
   b. Unplug the wiring going to these two switches.
   c. Loosen and remove the four meter hood mounting screws, then pull the meter hood off of the instrument panel.
3. Insert a trim stick into the speaker garnish and pry the speaker garnish to remove it from the instrument panel.
4. Remove the two mounting screws holding the speaker in place.
5. Lift the speaker out of its hole and disconnect the electrical wiring.
6. To install the speaker, attach the electrical radio wiring to the speaker.
7. Set the speaker into its hole so that the mounting holes line up with each other.
8. Secure the speaker in place with the mounting screws and push the speaker garnish back into place over the speaker.

6-66 CHASSIS ELECTRICAL

1 Audio panel
2 Bracket
3 Radio with cassette player
4 CD player

Radio with cassette player and CD player

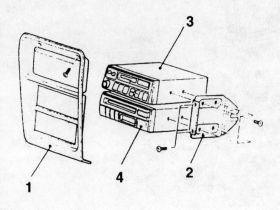

Radio with cassette player – Type A

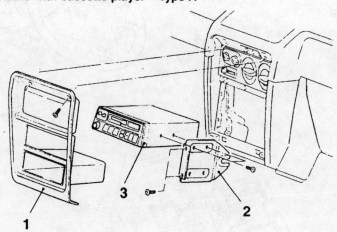

Radio with cassette player – Type B

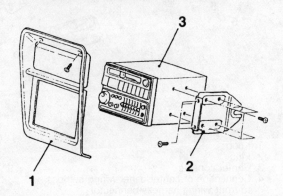

Fig. 87 Radio and tape player/CD player removal and installation components — 1992-95 Monteros

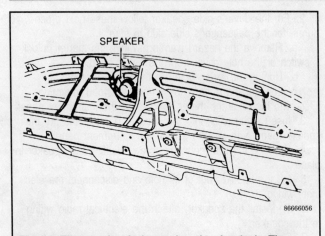

Fig. 88 The speaker is located under the dash. The speaker will have to be loosened from underneath the dashboard and pulled down to remove — 1983-86 Pick-ups

9. For the driver's side:
 a. Set the meter hood in place and install the four retaining screws snugly.
 b. Connect the hazard switch and starter unlock switch wiring to the switches.
 c. Pop the hazard switch, the starter unlock switch or the hole cover back in place.
10. Connect the negative (-) battery cable to the battery.

REAR SPEAKERS
▶ See Figure 90

The rear speakers are located in the door rear pillar molding on both sides of the Pick-up.

1. Disconnect the negative (-) battery cable from the battery.
2. Remove the four retaining screws holding the rear speaker garnish in place.
3. Pull the garnish off of the speaker, then loosen and remove the four nuts holding the speaker in place.
4. Pull the speaker out far enough to gain access to the wiring and detach the wiring from it.

CHASSIS ELECTRICAL 6-67

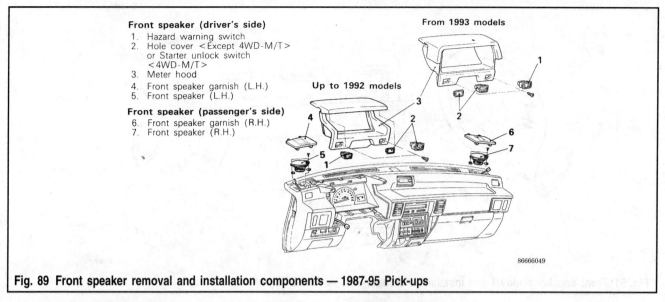

Fig. 89 Front speaker removal and installation components — 1987-95 Pick-ups

5. Once the speaker is removed, pull the speaker waterproof cover out of the speaker hole.
6. To install the speaker, insert the waterproof cover with the arrow marks pointing up.
7. Attach the wires to the speakers.
8. Install the speaker into the waterproof cover. Install the right (passenger's side) speaker so that the terminal faces the rear, and the left speaker so that the terminal faces forward.
9. Install the four retaining nuts onto the studs and tighten until they are snug.
10. Install the speaker garnishes with the arrows pointing up, then secure them in place with the retaining screws.
11. Attach the negative (-) battery cable to the battery.

1983-91 Montero

FRONT SPEAKERS

♦ See Figure 91

The front speakers on the 1983-91 Monteros are located on the extreme right and left-hand sides of the lower instrument panel, behind the oblong grilles.

1. Disconnect the negative (-) battery cable from the battery.
2. Unscrew the mounting screws holding the speakers in place.
3. Reach up under the dashboard and extract the speakers from behind the grilles.
4. Disconnect the wiring to the speakers.
5. To install the speakers, attach the wiring to them.
6. Hold the speaker up, behind the speaker grilles of the dashboard while screwing the mounting screws in place. Tighten the screws only until they are snug.
7. Attach the negative (-) battery cable to the battery.

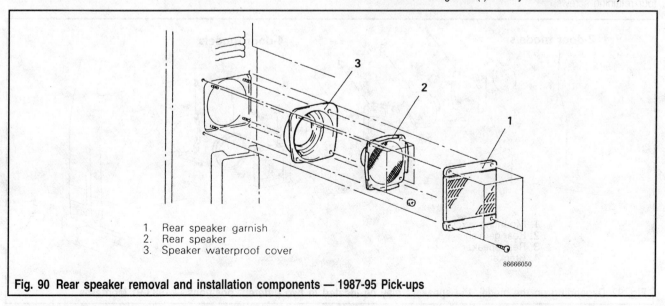

Fig. 90 Rear speaker removal and installation components — 1987-95 Pick-ups

6-68 CHASSIS ELECTRICAL

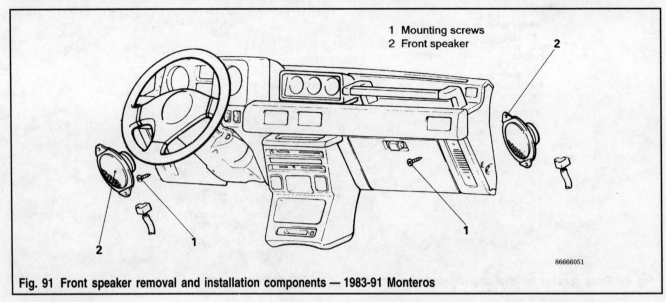

Fig. 91 Front speaker removal and installation components — 1983-91 Monteros

REAR SPEAKERS
▶ See Figure 92

The position of the rear speakers depends on whether the Montero is a two-door or four-door version. The two-door version has the speakers mounted to the wall, in front of the rear wheel-well. The four-door version has the speaker mounted above and to the rear of the rear wheel well on the wall.

1. Disconnect the negative (-) battery cable from the battery.
2. Loosen and remove the four speaker garnish retaining screws and pull the garnish off of the speaker.
3. Unscrew the four speaker to bracket retaining screws and pull the speaker out far enough to disconnect the wiring attached to it.
4. Remove the speaker.
5. Remove the four bracket mounting screws holding the bracket to the wall of the Montero.
6. To install, hold the bracket in place and secure with the four retaining screws.
7. Plug the speaker wiring back into the speaker and install the speaker with the four retaining screws to the bracket.
8. Fasten the speaker garnish over the speaker with the four garnish retaining screws.
9. Attach the negative (-) battery cable to the battery.

1992-95 Montero
▶ See Figure 93

The 1992-95 Monteros are equipped with six speakers, two on the lower corners of the instrument panel, two on the bottom corner of the front doors, and two located by the rear wheel well.

FRONT SPEAKERS

1. Disconnect the negative (-) battery cable from the battery.
2. Unscrew the retaining screws holding the grille cover in place over the speaker.

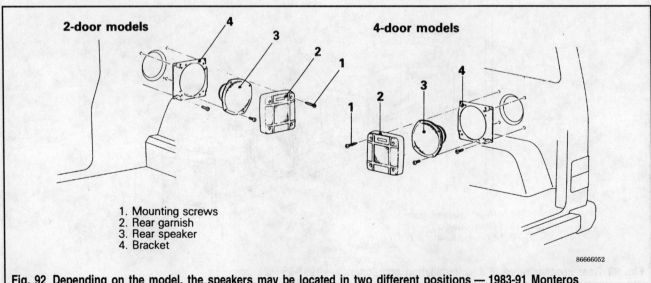

Fig. 92 Depending on the model, the speakers may be located in two different positions — 1983-91 Monteros

CHASSIS ELECTRICAL 6-69

3. Remove the two mounting screws from the speaker and pull the speaker out of its mounting hole far enough until the wiring can be removed from the back of the speaker.

4. Remove the speaker totally.

5. To install the speaker, first attach the wiring to the speaker.

6. Set the speaker in its mounting hole so that the screw holes line up, then secure it in place with the retaining screws.

7. Install the grille cover over the speaker and install the grille cover's screws.

8. Connect the negative (-) battery cable back to the battery.

REAR SPEAKERS

The four rear speakers have essentially the same removal procedure, the only difference being the removal and installation of the trim pieces.

1. Disconnect the negative (-) battery cable from the battery.

2. For the door speakers, remove the door trim and for the rear speakers, remove the quarter trim. Refer to Section 10 for the procedures of these trim pieces.

3. Loosen and remove the four screws holding the speaker in place.

4. Pull the speaker out far enough to gain access to the wiring and detach the wiring from it.

5. Once the speaker is removed, remove the speaker waterproof cover out of the speaker hole. The speaker waterproof cover is held in place with three screws on the door and four screws by the rear wheel-well.

6. To install the speaker, insert the waterproof cover and secure in place with the retaining screws.

7. Attach the wires to the speakers.

8. Install the speaker into the waterproof cover. Install the speakers in the same position as they were when removed.

9. Install the four retaining screws and tighten until they are snug.

10. Install the trim pieces removed earlier. Once again, refer to Section 10 for more detailed instructions.

11. Attach the negative (-) battery cable to the battery.

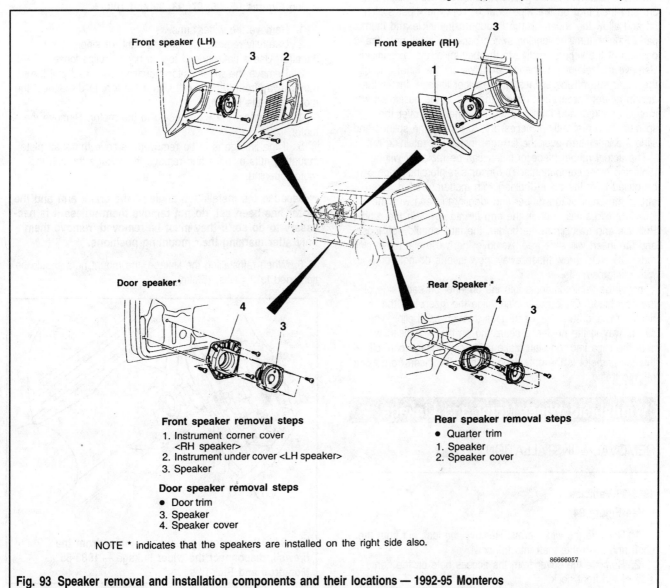

Fig. 93 Speaker removal and installation components and their locations — 1992-95 Monteros

WINDSHIELD WIPERS

Wiper Blade and Arm

REPLACEMENT

The wiper blade assembly may be replaced by simply compressing the locking tab and pulling the mounting prong out of the arm. On a few models, you may have to unscrew a Phillips screw.

If you need to replace the entire blade assembly, simply bend the retaining clip down and pull the assembly away from the arm. Make sure to pull the arm away from the windshield until it passes the detent before releasing it. Otherwise, it may spring against the windshield and scratch it. Replace the blade assembly.

If you have to replace an arm assembly, make sure to start out with the wipers in the park position. Turn the ignition key on, and then turn on the wiper switch. Turn the wiper switch off and allow the wipers to run through a full cycle and then park. Finally, turn the ignition switch back off. Now, note the position of the wipers relative to the bottom of the windshield. Remove the retaining nut and carefully work the inner end of the wiper arm off the splines on the wiper linkage. Install the arm in reverse order, first putting the arm in the position it was in when parked, and then pushing the inner arm over the splines. Turn it slightly, if necessary, to line up the splines and slide the inner arm over the linkage. Install the retaining nut.

The actual rubber blade or insert can be replaced without replacing other components. The insert's reinforcing strips can be pulled from the blade (holder) with moderate effort. Examine the ends of the rubber and identify the end with the lock tab; two small teeth in the arm fit into a slot in the rubber. Pull this end away from the holder; the tabs should pop loose and the insert will slide free. Remove the two metal reinforcing rods. DO NOT throw these away; new inserts do not come with reinforcements.

Install the reinforcements into the channel on each side of the new blade. Once assembled, slide the insert into the holder, making sure the insert goes through each clip. When all the way in the holder, a sharp tug on the new insert will seat the clips into the slot. Make sure they engage correctly; this tab-and-slot is the only thing keeping the insert from flying off into traffic.

Windshield Wiper Motor and Linkage

REMOVAL & INSTALLATION

1983-86 Vehicles
► See Figure 94

1. Remove the wiper arms. Remove the lock nut from the shaft and push the shaft into the cowl.
2. Remove the cover from the access hole on the right side of the front deck.
3. Remove the bolts that hold the motor bracket to the body.
4. Pull the wiper motor towards you.
5. Matchmark the motor and linkage arm.
6. Disconnect the wiper motor and linkage so that the motor shaft and linkage are at right angles. Hold the linkage with your right hand when disconnecting.
7. The linkage maybe removed through the access hole as necessary.
8. Reinstall and position the matchmarks correctly. (If not done, the wipers will not park correctly.)
9. Connect the linkage to the motor and install the motor. Install the motor bracket bolts.
10. Lift the linkage so the shafts project through the holes and install their locknuts.
11. Install the access hole cover. Install the wiper arms.

1987-95 Vehicles
► See Figures 95, 96, 97, 98, 99 and 100

1. Remove the wiper arms.
2. Carefully remove the left and right, or single, cowl panel(s). Use a flat, padded tool to pry the clips loose.
3. Remove the wiper motor mounting bolts and pull the motor away from the firewall. Use a flat tool to disconnect the motor from the linkage.
4. Disconnect the wiring going to the motor. Remove the motor.
5. If the linkage is to be removed, remove the shaft plate mounting nuts or bolts and remove the linkage through the motor opening.

➡ Because the installation angle of the crank arm and the motor has been set, do not remove them unless it is necessary to do so. If they must be removed, remove them only after marking their mounting positions.

6. When reinstalling the linkage, the mount nuts should be tightened to 3 ft. lbs. (5 Nm).

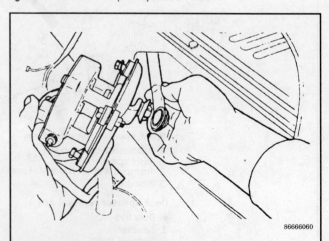

Fig. 94 After removing the wiper motor from the firewall, disconnect the wiper linkage — 1983-86 Monteros and Pick-ups

CHASSIS ELECTRICAL 6-71

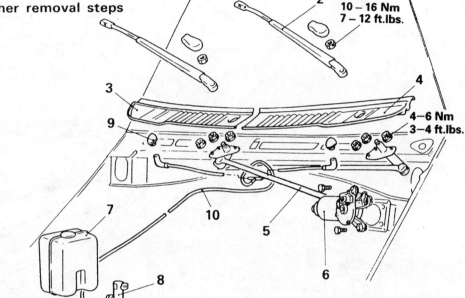

Windshield wipers removal steps
1. Wiper blade
2. Wiper arm
3. Front deck garnish (Passenger's side)
4. Front deck garnish (Driver's side)
5. Wiper link
6. Wiper motor

Windshield washer removal steps
7. Washer tank
8. Washer motor
9. Washer nozzle
10. Washer tube

Fig. 95 Windshield wiper and washer system removal and installation components — 1987-95 Pick-ups

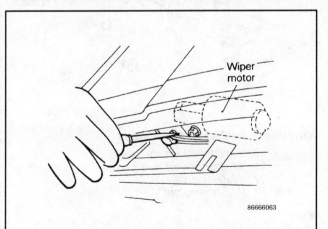

Fig. 96 Reach through the front deck with a prytool to disconnect the wiper linkage from the motor — 1987-95 Pick-ups

7. Connect the motor to the linkage and install the motor mount bolts.
8. Connect the wiring to the wiper motor.
9. install the left and right cowl panels, if they were removed.
10. Install the wiper arms.

Windshield Washer Motor and Fluid Reservoir

REMAVAL & INSTALLATION

▶ See Figures 95, 97 and 98

Although the mounting positions may be different, the windshield washer motor and reservoir on all Mitsubishi Pick-ups and Monteros is essentially the same. Each vehicle has a washer tank or fluid reservoir, a washer motor or pump, washer nozzles and the washer system fluid tubes.

6-72 CHASSIS ELECTRICAL

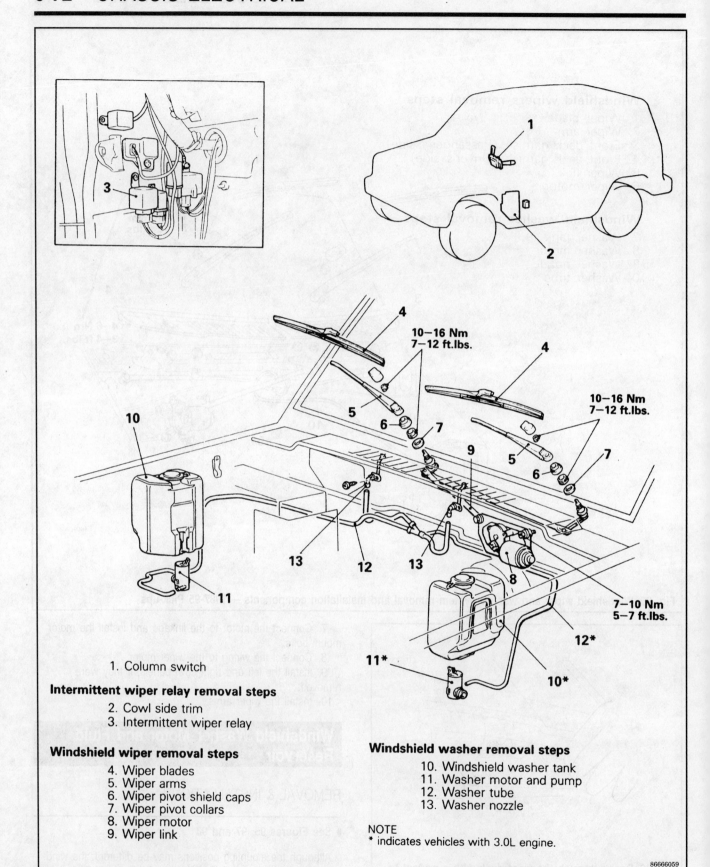

1. Column switch

Intermittent wiper relay removal steps
2. Cowl side trim
3. Intermittent wiper relay

Windshield wiper removal steps
4. Wiper blades
5. Wiper arms
6. Wiper pivot shield caps
7. Wiper pivot collars
8. Wiper motor
9. Wiper link

Windshield washer removal steps
10. Windshield washer tank
11. Washer motor and pump
12. Washer tube
13. Washer nozzle

NOTE
* indicates vehicles with 3.0L engine.

Fig. 97 Windshield wiper and washer system components and their locations — 1987-91 Monteros

CHASSIS ELECTRICAL 6-73

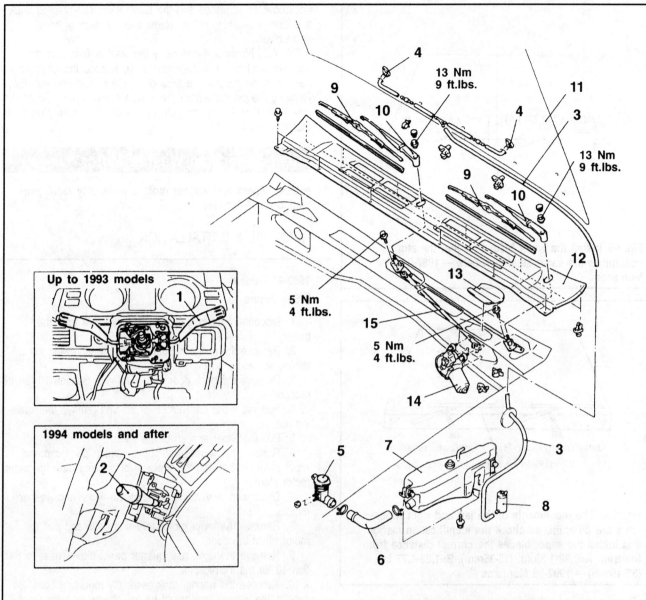

1. Column switch (with built-in wiper and washer switch, and wiper relay)
2. Wiper and washer switch
3. Washer tube
4. Washer nozzle
9. Wiper blade
10. Wiper arm
14. Wiper motor

Washer tank removal steps
- Splash shield
- Washer fluid draining
3. Washer tube
5. Cap
6. Hose
7. Washer tank assembly
8. Washer motor

Linkage removal steps
10. Wiper arm
11. Hood
12. Front deck garnish
13. Hole cover
14. Wiper motor
15. Linkage

Fig. 98 Windshield wiper and washer system components and their locations — 1992-95 Monteros

6-74 CHASSIS ELECTRICAL

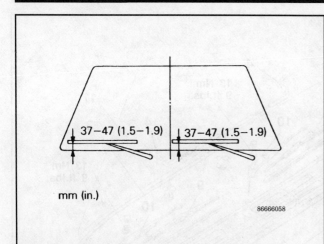

Fig. 99 Install the wiper blades so that the stop position is the same as shown here — 1987-91 Monteros

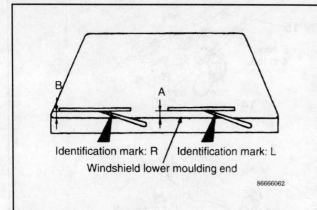

Fig. 100 The movements of the left and right wiper arms are different, so check the identification marks and install the wiper blades the correct distance from the trim: A=0.98-1.38 in. (25-35mm), B=1.38-1.77 in. (35-45mm) — 1992-95 Monteros

The windshield washer fluid reservoir is mounted in the engine compartment, on either the right-hand or left-hand inner fender-well. It is mounted to the fender-well with a few mounting screws or bolts. Attached to the fluid reservoir are the fluid tubes and the washer motor. To remove the reservoir, disconnect the fluid tube and the wiring harness from the washer motor and remove the reservoir mounting bolts. The motor will come off with the reservoir.

The washer motor is mounted to the fluid tank, is retained there by a mounting bracket. The motor also has wires connected to it, which send to the motor the signals from the wiper switch inside the vehicle. To remove the washer motor, disconnect the fluid tube and wiring harness from it. Then unbolt the mounting bracket or screws and remove the motor from the reservoir.

The washer nozzles are mounted underneath of the deck cowls or garnishes or in the hood. They are hooked to the fluid tubes and are the components which aim the washer fluid spray onto the windshield. To remove the nozzles, remove the deck cowl or garnishes and the fluid tubes. The nozzles usually have a retaining nut or screw into the deck or hood themselves.

The fluid hoses or tubes carry the washer fluid from the reservoir and motor to the nozzles. To replace the tubes with new ones, simply pull the tube off of the nozzle and the motor. Remove the old hose from the engine compartment. Route the new tube along the same path as the old one, then plug it into the nozzle and motor.

Rear Wiper and Washer Motor

➡ Rear wipers and washer motors were only factory-installed on Monteros.

REMOVAL & INstallATION

1983-91 Models
◆ See Figure 101

1. Disconnect the negative (-) battery cable from the battery.
2. Remove the inside handle cover by unscrewing the retaining screws.
3. Remove the back door trim — refer to Section 10 for the procedure.
4. Pop the wiper arm nut cover off and remove the retaining nut.
5. Pull the wiper arm off of the stem.
6. Remove the wiper pivot nut, the wiper pivot cap, the wiper pivot washer and the wiper pivot packing from the wiper motor stem.
7. Disconnect any wiring going to the wiper and washer motors.
8. Remove the wiper motor retaining bolts and pull the motor off of the door.
9. Remove the right, rear quarter panel trim — refer to Section 10 for the instructions.
10. Remove the washer tank assembly mounting bolts. Be careful, the washer tank might be full of washer fluid. Empty the washer tank assembly into a clean container.
11. Disconnect the wiring from the washer motor and pump assembly.
12. Remove the washer motor and pump from its the washer fluid tank.
13. To install, place the washer motor in its mounting bracket and tighten the bracket to hold it.
14. Connect the washer tube to the motor, and mount the washer tank assembly to the wall.
15. Hook the wiring back up to the washer motor and pump assembly.
16. Install the rear quarter trim panel.
17. Mount the wiper motor onto the door and torque the mounting bolts to 5-7 ft. lbs. (7-10 Nm).
18. Connect the electrical wiring to the wiper motor.
19. Slide the wiper pivot packing, washer, pivot cap and nut onto the stem. Tighten the nut to 6-9 ft. lbs. (8-12 Nm).
20. Install the wiper arm onto the wiper motor stem so that the wiper blade is parallel to the lower edge of the window glass.

CHASSIS ELECTRICAL 6-75

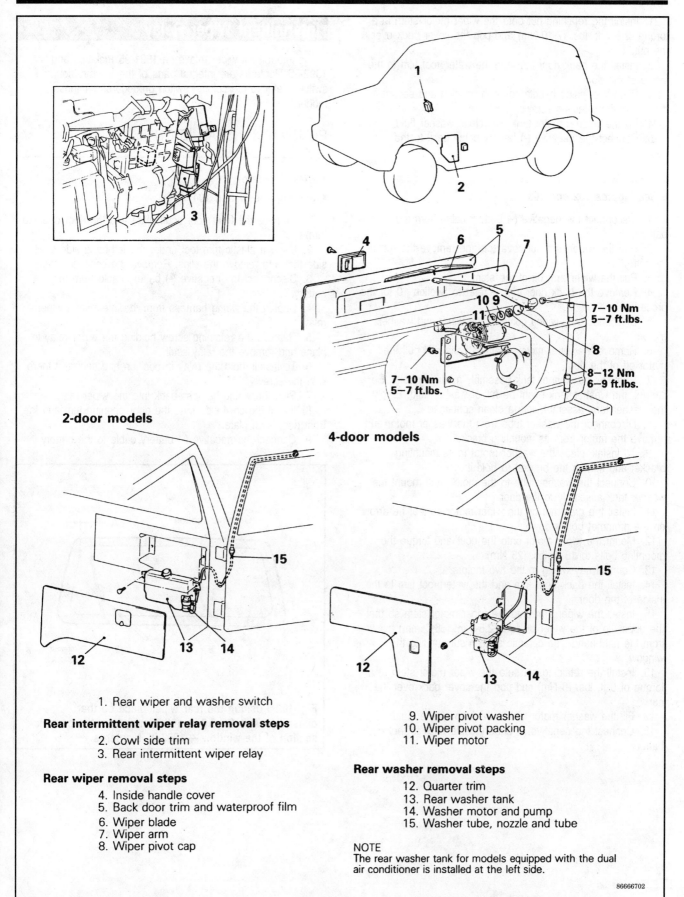

Fig. 101 Rear window wiper and washer system components and their locations — 1983-91 Monteros

6-76 CHASSIS ELECTRICAL

21. Install the retaining nut onto the wiper motor stem at a torque of 5-7 ft. lbs. (7-10 Nm) and pop the cover back over the nut.
22. Install the back door trim and the waterproof film to the inside of the door.
23. Place the inside handle cover in position and secure it there with the retaining screws.
24. Fill the washer motor tank with clean washer fluid.
25. Connect the negative (-) battery cable back to the battery.

1992-95 Models

▶ See Figures 102 and 103

1. Disconnect the negative (-) battery cable from the battery.
2. Pop the wiper arm nut cover off and remove the retaining nut.
3. Pull the wiper arm off of the stem.
4. Remove the back door trim — refer to Section 10 for the procedure.
5. Disconnect any wiring going to the wiper and washer motors.
6. Remove the wiper motor retaining bolts and pull the motor off of the door.
7. Remove the washer tank assembly mounting bolts. Be careful, the washer tank might be full of washer fluid. Empty the washer tank assembly into a clean container.
8. Disconnect the washer tube from the washer motor and remove the motor from its mounting bracket.
9. To install, place the washer motor in its mounting bracket and tighten the bracket to hold it.
10. Connect the washer tube to the motor, and mount the washer tank assembly to the door.
11. Install the grommet on the wiper motor so that he arrow on the grommet points down.
12. Mount the wiper motor onto the door and torque the mounting bolts to 18 ft. lbs. (25 Nm).
13. Connect any wiring to the two motors.
14. Install the back door trim and the waterproof film to the inside of the door.
15. Install the wiper arm onto the wiper motor stem so that the upper tip of the wiper blade is 2.5-3 in. (65-75mm) away from the right-hand side ceramic section surrounding the window.
16. Install the retaining nut onto the wiper motor stem at a torque of 6 ft. lbs. (8 Nm) and pop the cover back over the nut.
17. Fill the washer motor tank with clean washer fluid.
18. Connect the negative (-) battery cable back to the battery.

Intermittent Wiper Relay

The intermittent wiper relays on 1991-95 Pick-ups and 1992-95 Monteros are integral parts of the column-mounted switch. Refer to the column switch procedures for those models.

REMOVAL & INSTALLATION

1987-91 Montero

▶ See Figures 97 and 104

1. Remove the front door scuff plate and the door opening trim.
2. Insert a plastic trim tool under the left-hand side cowl side trim and remove the clips. Remove the cowl side trim.
3. Disconnect the negative (-) battery cable from the battery.
4. Unplug the wiring harness from the intermittent wiper relay.
5. Remove the retaining screw holding the wiper relay in place and remove the relay itself.
6. To install, hold the relay in position and secure it there with the screw.
7. Plug the wiring harness back into the wiper relay.
8. Install the cowl side trim, the door opening trim and the front door scuff plate.
9. Connect the negative (-) battery cable to the battery.

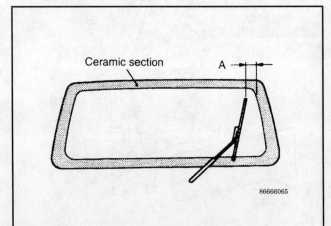

Fig. 102 Position the rear wiper blade so that dimension A is 2.5-3.0 in. (65-75mm) from the ceramic section of the window — 1992-95 Monteros

CHASSIS ELECTRICAL 6-77

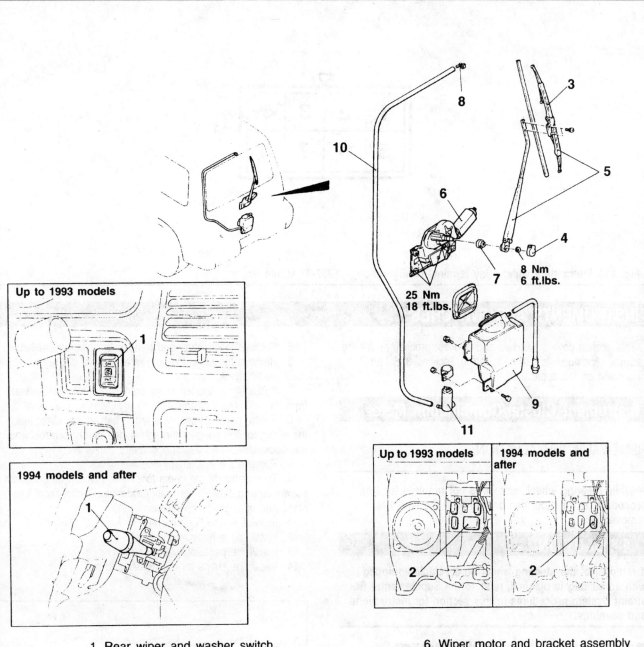

1. Rear wiper and washer switch 1994 models and after
3. Wiper blade
8. Washer nozzle

Rear intermittent wiper relay removal steps
- Instrument under cover
2. Rear intermittent wiper relay

Wiper motor removal steps
4. Cover
5. Wiper arm and blade assembly
- Back door trim

6. Wiper motor and bracket assembly
7. Grommet

Washer tank and motor removal steps
- Back door trim
9. Washer tank assembly
- Washer fluid draining
10. Washer tube
11. Washer motor

Fig. 103 Rear window wiper and washer system components and their locations — 1992-95 Monteros

6-78 CHASSIS ELECTRICAL

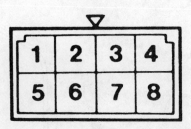

Fig. 104 Intermittent wiper relay terminal configuration — 1987-91 Monteros

INSTRUMENTS AND SWITCHES

This section covers switches and gauges; procedures for the gauges' sending units can be found in Sections 3, 5 and 7, depending on the application.

Instrument Cluster/Combination Meter

REMOVAL & INSTALLATION

➡ While steering wheel removal is not required for this procedure, extra room can be gained if the wheel is removed.

✴✴CAUTION

If removal of the steering wheel on a vehicle equipped with an air bag is planned, refer to the Supplemental Restraint System procedures in this section for instructions and warnings.

➡ In many cases, it will be easier to disconnect the speedometer cable at the transmission first. Push some slack in the cable into the vehicle; this will allow the instrument cluster to come out of the dash far enough to reach the upper connector.

1983-86 Pick-up

▶ See Figure 105

➡ The instrument cluster and radio/heater faceplate are all one piece.

1. Remove the radio knobs.
2. Remove the fan control knob and the heater control slider knobs.
3. Remove the instrument cluster panel attaching screws.
4. Remove the locking nuts from the radio shafts and the fan control switch.
5. Remove the ashtray and remove the ashtray bracket.
6. Remove the attaching screws at the top of the instrument assembly.
7. Beginning at the left lower corner, remove the faceplate from the instrument cluster.
8. As the panel comes free, reach behind and disconnect the wiring for the gauges, the lighter, the heater controls and the clock. Disconnect the speedometer cable.
9. Remove the instrument cluster.
10. To reinstall, fit the panel loosely in position and connect each wire and cable to its proper location. Fit the panel into place and engage the spring tabs.
11. Install the retaining screws.
12. Install the ashtray and bracket.
13. Install the panel attaching screws.
14. Install the knobs for the heater and radio.

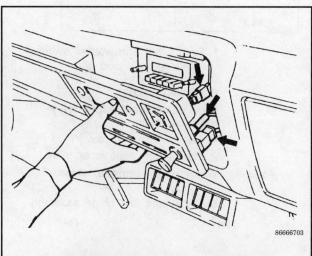

Fig. 105 Disconnect the wiring as the panel comes loose — 1983-86 Pick-ups

CHASSIS ELECTRICAL 6-79

1987-95 Pick-up
▶ See Figure 106

1. Disconnect the negative (-) battery cable from the battery.
2. Use a small protected flat tool to pry the hazard flasher switch, the starter unlock switch or the hole cover out of the dash.
3. Disconnect the wiring to these switches.
4. Remove the 4 screws (two need to be accessed through the switch holes) holding the hood in place. Remove the hood.
5. Remove the 4 screws holding the instrument cluster. Pull the cluster towards you.
6. Disconnect the electrical harnesses and the speedometer cable from the instrument cluster.
7. To install, connect the electrical wiring to the back of the instrument cluster.
8. Insert the speedometer cable into the back of the instrument cluster until its stopper properly fits to the meter-side groove. After installing the speedometer, pull the speedometer cable through the grommet in the toe-board (fire wall) until the cable marking is visible from the engine compartment side.
9. Secure the instrument cluster in place with the four retaining screws.
10. Install the meter hood and mount in place with the four mounting screws.
11. Connect the hazard switch and the starter unlock switch, if so equipped, to their respective wiring.
12. Push the two switches back into their holes until they click in place.
13. Connect the negative (-) battery cable back to the battery.

1983-86 Montero

1. Disconnect the speedometer cable at the transmission.
2. Using a screwdriver or similar tool wrapped in a rag, gently pry the instrument cover loose. Note that this is the rear panel on top of the cluster, between the instrument pod and the defroster grille.
3. Remove the screws from the bottom of the instrument cluster and remove the bolt from the top of the cluster.

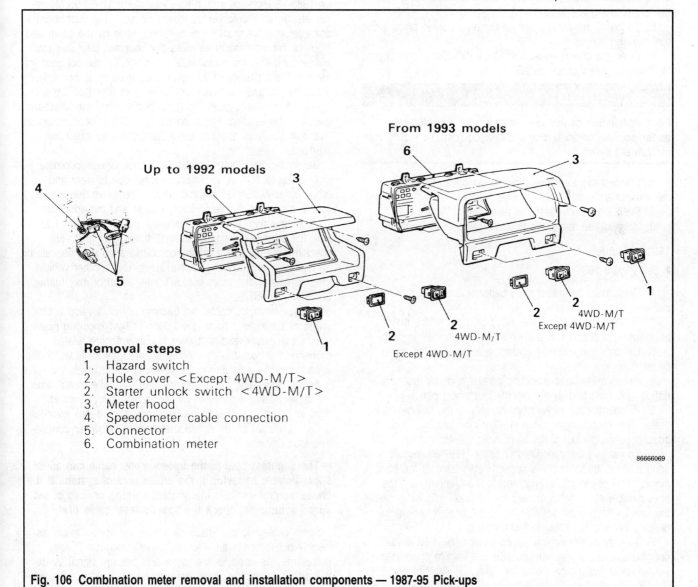

Fig. 106 Combination meter removal and installation components — 1987-95 Pick-ups

6-80 CHASSIS ELECTRICAL

4. Work the cluster out from the dash and disconnect the speedometer cable.
5. Carefully disconnect the wiring harnesses from the cluster and remove the cluster.
6. Reinstall in reverse order, taking care that each cable and harness is securely installed.
7. When installing the cluster cover, make certain it snaps into place and is correctly positioned.
8. Connect the speedometer cable at the transmission.

1987-91 Montero
♦ See Figure 107

1. Disconnect the negative (-) battery cable from the battery.
2. Using a plastic trim tool, pry the meter cover off of the meter assembly. If no plastic trim tool is available, use a metal prytool wrapped in cloth.
3. Remove the speedometer and any wiring (make sure to label the wires) from the back of the meter assembly.
4. Remove the lower and upper retaining screws holding the meter assembly in place and remove the assembly from the instrument panel (dashboard).
5. To install, mount and secure the assembly in place with the retaining screws.
6. Insert the speedometer cable until its stopper properly fits into the speedometer groove.

✳✳WARNING

Poor installation of the cable may cause a fluctuating meter pointer, or noise and a damaged harness inside the instrument panel.

7. Connect the electrical wiring to the back of the instrument meter assembly.
8. Push the meter cover into place until it "clicks".
9. Connect the negative (-) battery cable to the battery.

1992-95 Montero
♦ See Figure 108

1. Disconnect the negative (-) battery cable from the battery.
2. Pop the upper-rear meter hood plug out of the meter hood, using a plastic trim tool or a prytool wrapped in cloth.
3. Unscrew the retaining screws, then remove the meter bezel.
4. Remove the three mounting screws and pull the combination meter out and away from the instrument panel.
5. Disconnect the wiring from the rear of the combination meter. The speedometer cable (1992-93 models only) and adapter pulls right out of the combination meter.
6. To remove the speedometer cable (1992-93 models only) adapter, disconnect the speedometer cable at the transmission end of the cable. From inside the Montero, pull the speedometer able slightly toward the vehicle's interior, release the lock by turning the adapter to the left or right, and then remove the adapter from the instrument panel.
7. To remove the vehicle speed sensor (1994-95 models only), unplug the wiring harness from the vehicle speed sensor, which is located on the driver's side, tail-shaft of the transmission. Unscrew the sensor from the tail-shaft of the transmission.
8. To install, screw the vehicle speed sensor back into the transmission and hook up the wiring harness to it, or insert the speedometer cable adapter back into the instrument panel.
9. Connect the electrical wiring and the speedometer cable adapter to the combination meter, then set the meter in place.
10. Secure the meter with the three retaining screws.
11. Place the meter bezel over the combination meter and fasten it there with the mounting screws.
12. Pop the meter hood plug back into its hole and connect the negative (-) battery cable to the battery.

Speedometer Cable

REMOVAL & INSTALLATION

The speedometer cable connects a rotating gear within the transmission to the dashboard speedometer/odometer assembly on 1983-95 Pick-ups and 1983-93 Monteros (1994-95 Monteros use an electronic vehicle speed sensor). The dashboard unit interprets the number of turns the made by the cable and displays the information as miles per hour and total mileage.

Assuming that the transmission contains the correct gear for the car, the accuracy of the speedometer depends primarily on tire condition and tire diameter. Badly worn tires (too small in diameter) or overinflation (too large in diameter) can affect the speedometer reading. Replacement tires of the incorrect overall diameter (such as oversize snow tires) can also affect the readings.

Generally, manufacturers state that speedometer/odometer error of up to 10% is considered normal due to wear and other variables. Stated another way, if you drove the vehicle over a measured 1 mile (1.6 km) course and the odometer showed anything between 0.9 and 1.1 miles (1.45 and 1.77 km), the error is considered normal. If you plan to do any checking, always use a measured course such as mileposts on an interstate highway or turnpike. Never use another vehicle for comparison; the other vehicle's inherent error may further cloud your readings.

The speedometer cable can become dry or develop a kink within its case. As it turns, the ticking or light knocking noise it makes can easily lead an owner to chase engine related problems in error. If such a noise is heard, carefully watch the speedometer needle during the speed range in which the noise is heard. Generally, the needle will jump or deflect each time the cable binds. The needle motion may be very small and hard to notice; a helper in the back seat should look over the driver's shoulder at the speedometer while the driver concentrates on driving.

➡The slightest bind in the speedometer cable can cause unpredictable behavior in the cruise control system. If the cruise control exhibits intermittent surging or loss of set speed symptoms, check the speedometer cable first.

Some cables do not attach directly to the speedometer assembly but rather to an electrical pulse generator. These pulses may be used for the meter and mileage signal. Additionally, the electric signals representing the speed of the vehi-

CHASSIS ELECTRICAL 6-81

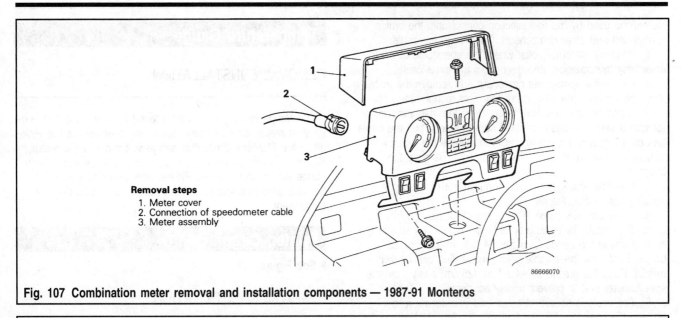

Fig. 107 Combination meter removal and installation components — 1987-91 Monteros

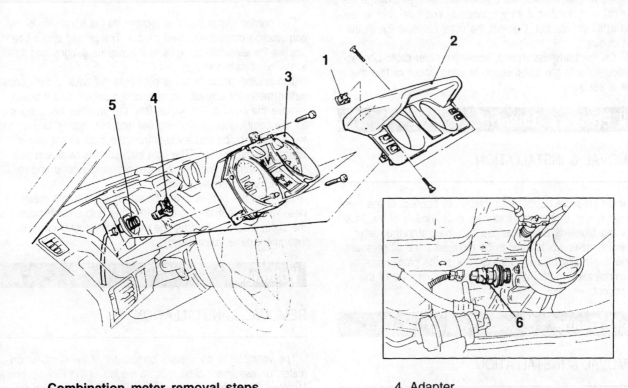

Fig. 108 Combination meter removal and installation components — 1992-95 Monteros

cle can be used by the fuel injection control unit, the cruise control unit and other components. To change the cable:

1. Unscrew the outer collar where the speedometer cable enters the transmission, and disconnect the inner cable.
2. Follow the appropriate procedure and remove the instrument cluster from the dashboard. On most models, the cable housing can be released from the back of the instrument cluster with a simple release tang. On some models, pull the cable toward you (out of the dash) and then release the lock by turning the adaptor to either the left or right. Remove the adaptor.
3. The inner cable may now be pulled out of the cable housing from inside the car.
4. Install the new cable into the sheath. It's a good idea to check the sheath for binding or kinking. The routing should not contain any sharp bends or angles. A little lithium grease brushed onto the inner cable before installation can be very helpful. Keep the grease at least 1 in. (25mm) away from the speedometer end to prevent messy accidents.
5. The cable is square at either end in order for it to engage the speedometer and transmission fittings. Engage the inner cable by rotating it as you insert it. You'll be able to feel when it fits into its slot. Connect the outer cable at the instrument cluster.
6. On the transmission end, screw the outer cable connector securely onto the transmission housing. Make certain the cable is secure.

Oil Pressure Gauge

REMOVAL & INSTALLATION

The oil pressure gauge is an integral component of the combination meter, or instrument cluster, in all models of the Pick-up or the Montero. Since this gauge is integral to the meter, the entire meter would need to be replaced if the oil pressure gauge was malfunctioning. Please refer to the procedures for the combination meter/instrument cluster for removal of this component.

Fuel Gauge

REMOVAL & INSTALLATION

The fuel gauge is an integral component of the combination meter, or instrument cluster, in all models of the Pick-up or the Montero. Since this gauge is integral to the meter, the entire meter would need to be replaced if the fuel gauge was malfunctioning. Please refer to the procedures for the combination meter/instrument cluster for removal of this component.

Temperature Gauge

REMOVAL & INSTALLATION

The temperature gauge is an integral component of the combination meter, or instrument cluster, in all models of the Pick-up or the Montero. Since this gauge is integral to the meter, the entire meter would need to be replaced if the temperature gauge was malfunctioning. Please refer to the procedures for the combination meter/instrument cluster for removal of this component.

Printed Circuit Board

▶ See Figure 109

REMOVAL & INSTALLATION

The printed circuit board is located on the back side of the combination meter/instrument cluster. The printed circuit board applies the electrical currents to the various gauges and lights in the combination meter.

The printed circuit board is held onto the back of the combination meter by screws. To remove the printed circuit board, remove the combination meter (see the previous instructions for the combination meter removal and installation). Unplug the various light bulbs and sockets from the back of the circuit board and meter case. The light bulb sockets twist and pull out. Then remove the circuit board retaining screws and remove it from the meter case.

To install the printed circuit board, mount it to the meter case with the retaining screws. Plug the bulb sockets back into the circuit board and install the combination meter into the instrument panel (dashboard).

Voltmeter

REMOVAL & INSTALLATION

The voltmeter is an integral component of the combination meter, or instrument cluster, in all models of the Pick-up or the Montero. Since this gauge is integral to the combination meter, the entire combination meter would need to be replaced if the voltmeter was malfunctioning. Please refer to the procedures for the combination meter/instrument cluster for removal of this component.

Multi-Meter

➡ This gauge set is only found on 1987-95 Monteros. It includes the voltmeter, the inclinometer, and the oil pressure gauge.

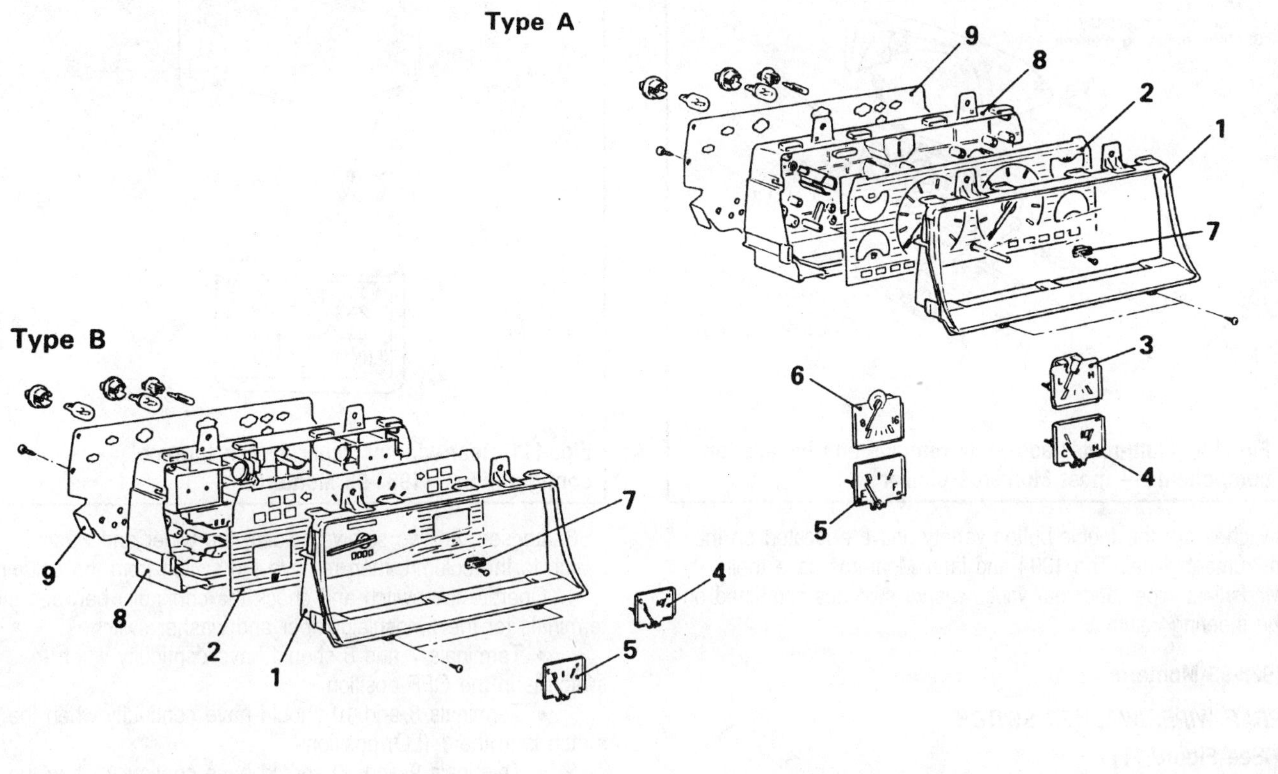

CHASSIS ELECTRICAL 6-83

1. Meter glass
2. Speedometer and Tachometer (with meter panel) – Type A
   Speedometer (with meter panel) – Type B
3. Oil pressure gauge
4. Engine coolant temperature gauge
5. Fuel gauge
6. Voltage meter
7. Maintenance required warning indicator reset switch
8. Meter case
9. Printed circuit board

Fig. 109 On most models, the printed circuit board can be unscrewed from the back of the meter case

REMOVAL & INSTALLATION

▶ See Figure 110

1. Disconnect the negative (-) battery cable from the battery.
2. Remove the two side-mounting screws.
3. Pull the meter hood off of the multi-meter.
4. Remove the multi-meter mounting screws and pull the multi-meter up high enough to disconnect the wiring hooked to the bottom of it.
5. Remove the multi-meter assembly.
6. Remove the meter mounting bracket from the instrument panel.
7. To install, screw the meter mounting bracket to the instrument panel.
8. Hook the electrical wiring to the under-side of the multi-meter and set the meter in place.
9. Secure the meter onto the instrument panel with the retaining screws.
10. Fasten the meter hood to the multi-meter with the two side-mounting screws.
11. Hook the negative (-) battery cable back to the battery.

Windshield and Rear Window Wiper Switches

REMOVAL & INSTALLATION

On 1983-89 Pick-ups and Monteros, the windshield wiper switch is integral with the headlight switch. See the procedure for headlight switch removal and installation. On 1990-95 Pick-ups and 1990-93 Monteros, the windshield wiper switch is a part of the column-mounted combination switch. See the procedure for combination switch removal and installation, later in this section. On 1994 and later Monteros, the windshield wiper and washer switch is separate from the column switch.

The rear wiper and washer switches found on Monteros are also covered in this section. Pre-1994 rear wiper and washer

6-84 CHASSIS ELECTRICAL

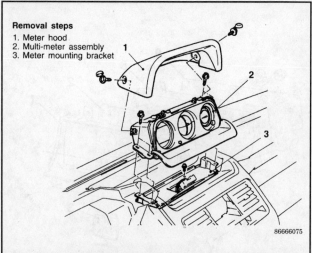

Fig. 110 Multi-meter assembly removal and installation components — most Monteros similar

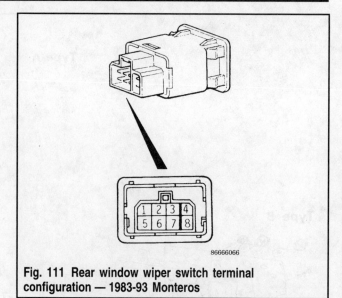

Fig. 111 Rear window wiper switch terminal configuration — 1983-93 Monteros

switches are the toggle-button variety and are located on the instrument panel. The 1994 and later Monteros have their windshield wiper and rear wiper/washer switches combined on the steering column.

1983-93 Montero

REAR WIPER/WASHER SWITCH

▶ See Figure 111

1. Insert a plastic trim tool under the edge of the switch and pry the switch to remove it from the instrument panel.
2. Using an ohmmeter with side-mount terminals, check for continuity in the switch as follows:
   - Terminals 7 and 2 should be continuous when the wiper switch is in the ON position.
   - Terminals 5 and 2 should be continuous when the wiper switch is in the OFF position.
   - Terminals 5 and 2, and terminals 3 and 8 should be continuous when the wiper switch is in the INT position.
   - Terminals 7 and 6 should be continuous when the washer switch is in the ON position.
3. If the switch failed any of the tests, replace it with a new unit.
4. Disconnect the negative (-) battery cable from the battery.
5. Disconnect the electrical wiring from the back of the switch and remove the switch.
6. To install the switch, attach any electrical wiring removed earlier.
7. Push the switch back into its hole on the instrument panel until it clicks in place.
8. Attach the negative (-) battery cable back to the battery.

1994-95 Montero

FRONT AND REAR WIPER/WASHER SWITCH

▶ See Figure 112

1. Remove the column lower cover.
2. Remove the column upper cover.

3. Loosen the two screws holding the wiper and washer switch to the column, then remove the switch from the column.
4. Operate the switch and check the continuity between the terminals for the windshield wiper and washer switch.
   - Terminals 7 and 8 should have continuity when the switch is in the OFF position.
   - Terminals 8 and 10 should have continuity when the switch is in the 1 (LO) position.
   - Terminals 9 and 10 should have continuity when the switch is in the 2 (HI) position.
   - Terminals 6 and 10 should have continuity when the washer switch is in the ON position.
5. Operate the switch and check for continuity between the terminals for the rear window wiper and washer switch.
   - Terminals 2 and 7 should have continuity when the wiper switch is in the ON position.
   - Terminals 4 and 7 should have continuity when the wiper switch is in the OFF position.
   - Terminals 1 and 6, and terminals 4 and 7 should have continuity when the wiper switch is in the INT position.

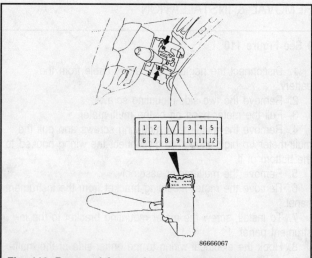

Fig. 112 Rear and front wiper and washer switch terminal configuration — 1994-95 Monteros

- Terminals 2 and 3 should have continuity when the washer switch is in the ON position.
- Terminals 5 and 8 should have continuity when the illumination light is on.

6. If the switch failed any of these tests, replace the old switch with a new unit.
7. Disconnect the negative (-) battery switch from the battery.
8. Unplug the harness connector to the switch and discard the old switch.
9. Attach the wiper and washer switch to the column with the two retaining screws.
10. Plug any harness connectors back together.
11. Install the upper and lower column covers.
12. Connect the negative (-) battery cable to the battery.

Combination Switch (Headlight/Turn Signal/Windshield Wiper Switch)

→Although the stalk on the column controls various functions, the switch behind the steering wheel is actually one integrated unit. In most cases, any problem with a stalk or switch function will require replacement of the entire switch assembly.

REMOVAL & INSTALLATION

▶ See Figures 113, 114 and 115

❋❋CAUTION

If removal of the steering wheel on a vehicle equipped with an air bag is planned, refer to the Supplemental Restraint System procedures in this section for instructions and warnings.

1. Disconnect the negative battery cable.
2. Remove the steering wheel. Use a steering wheel puller to remove the steering wheel.

❋❋WARNING

Do not hammer on the steering wheel to remove it; doing so may damage the collapsible mechanism.

3. On some late model Monteros, it may be necessary to remove the instrument under cover to gain access to the lower column cover.
4. Remove the upper and lower steering column covers. On some models it may be necessary to remove the lower dash panel or knee protector for access.
5. Follow the combination switch wiring harness down the column to the connector. Detach the connector and release any wire clips holding the harness to the switch.
6. Remove the retaining screws holding the switch to the column. Work the switch off the column.

→Some models have the turn signal and hazard switches mounted on a separate base plate. Removing the mounting screws will allow separate removal or replacement.

7. Install the switch, tighten the mounting screws and route the wire harness to the connector. Make sure the wire does not interfere with the moving parts of the tilt column mechanism.

❋❋WARNING

The switch must be centered in the column. If it is installed on an angle, the self-cancelling aspect of the turn signals will be affected.

8. Connect the harness and install any wire clips or retainers.
9. Install the upper and lower column covers.
10. Install the instrument under cover, if removed earlier.
11. Install the steering wheel.
12. Connect the battery cable back to the battery.

Clock

REMOVAL & INSTALLATION

1983-86 Vehicles
▶ See Figure 116

On these vehicles, the clock is mounted behind the instrument cluster. Refer to the instructions regarding the removal and installation of the combination meter/instrument cluster and remove the cluster from the vehicle. Remove the clock from the back of the instrument cluster.

To install the clock, simply attach the clock to the back of the instrument cluster and connect the wires. Install the instrument cluster in the instrument panel.

1987-91 Montero

The 1987-91 Monteros have their clocks mounted behind the center panel. Disconnect the negative battery cable from the battery. Remove the center panel from the instrument panel and the floor. Disconnect the wiring from the back of the clock and unscrew the mounting screws. Pull the clock off of the back of the center panel.

To install the clock, mount it in its place with the mounting screws. Plug the wires back into it and install the center panel to the instrument panel and floor. Connect the battery cable back to the battery. Refer to the procedures regarding removal and installation of the center panel in Section 10 for more instructions.

1987-95 Pick-up and 1992-95 Montero

For these models, the clock is simply an assembly which is plugged into the dashboard. To remove the clock, use a plastic trim tool and pry the clock assembly out towards the front of the dashboard. Disconnect the negative battery cable from the battery. The clock should pop right out of its mounting hole. Disconnect the wiring to the back of it.

To install the clock, attach the wiring to the back of it and push it back into its mounting hole until it pops in place. Reconnect the negative battery cable.

6-86 CHASSIS ELECTRICAL

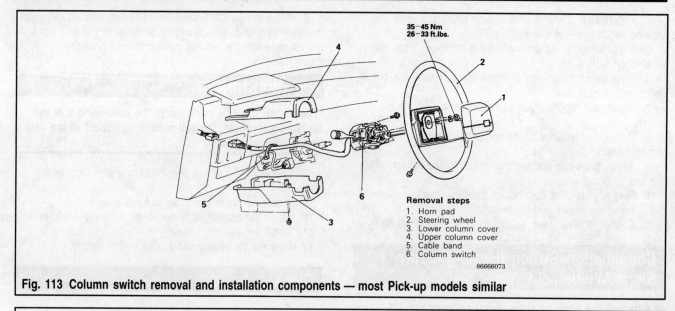

Fig. 113 Column switch removal and installation components — most Pick-up models similar

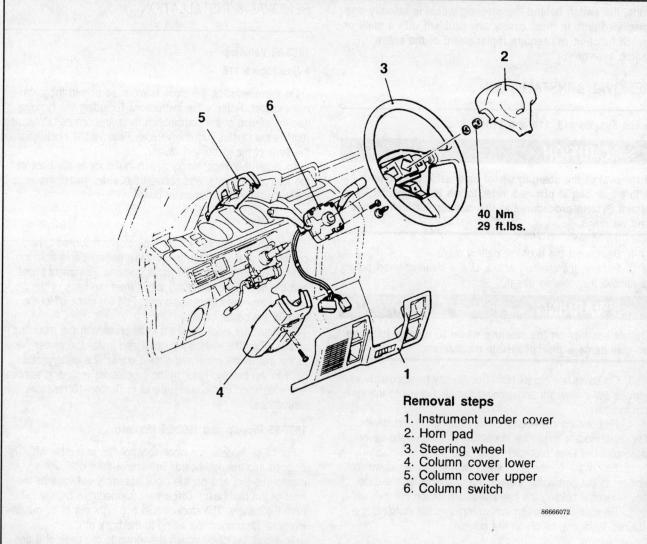

Fig. 114 Column switch removal and installation components — most Monteros similar

CHASSIS ELECTRICAL 6-87

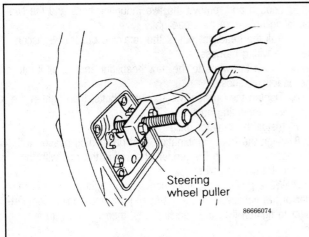

Fig. 115 Use a steering wheel puller to remove the steering wheel — do NOT hammer on the steering wheel to remove it

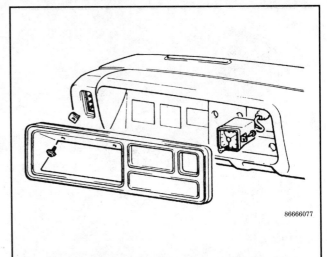

Fig. 116 The clock is mounted behind, and attached to, the instrument cluster — 1983-86 Montero and Pick-up

LIGHTING

Headlights

Headlights, like any other lighting device, can fail due to broken filaments. The front of any vehicle is the worst possible location for a lighting device since it is subject to impact, extensive temperature change and severe vibration, all of which shorten the life of the light. The front of the vehicle is also where good lighting is needed the most, so it's not uncommon to have to replace a headlight during the life of the vehicle.

There are two general styles of headlamps, the sealed beam and replaceable bulb type. The sealed beam is by far the most common and includes almost all of the circular and rectangular lamps found on cars built through the early 1980's. The sealed beam is so named because it includes the lamp (filament), the reflector and the lens in one sealed unit. Sealed beams are available in several sizes and shapes.

The replaceable bulb is the newer technology. Using a small halogen bulb, only the lamp is replaced, while the lens and reflector are part of the body of the car. This is generally the style found on wrap-around or "European" lighting systems. While the replaceable bulbs are more expensive than sealed beams, they generally produce more and better light. The fixed lenses and reflectors can be engineered to allow better frontal styling and better light distribution for a particular vehicle.

It is quite possible to replace a headlight of either type without affecting the alignment (aim) of the light. Sealed beams mount into a bracket (bucket) to which springs are attached. The adjusting screws control the position of the bucket which in turn aims the light. Replaceable bulbs simply fit into the back of the reflector. The lens and reflector unit are aimed by separate adjusting screws.

Take a moment before disassembly to identify the large adjusting screws (generally two for each lamp, one above and one at the side) and don't change their settings.

With the exception of the oldest cars, sealed beams are removed from the outside of the car. Start with the outer trim pieces and work your way in to the lamp and its retainer. Bulb type units are almost always replaced from under the hood.

REMOVAL & INSTALLATION

✱✱CAUTION

Most headlight retaining rings and trim bezels have very sharp edges. Wear gloves. Never pry or push on a headlamp; it can shatter suddenly. When changing a replaceable bulb, never touch the new bulb with your fingers; always hold it by the base. The natural oils in your skin will cling to the glass and create hot spots on the bulb envelope. These hot spots will shorten the life of the bulb substantially. If the bulb has been handled, clean the glass with alcohol and dry it thoroughly before installation.

Pick-up

▶ See Figures 117 and 118

1. Remove the grille.
2. Remove the four small screws holding the metal retaining ring or frame.
3. Pull the lamp forward and out; disconnect the wiring harness. Install the new lamp and connect the harness.
4. Fit the lamp into place and install the retaining ring.
5. Install the grille.

1983-91 Montero

▶ See Figure 119

1. Remove the radiator grille.
2. Remove the front sidemarker lamp.
3. Remove the plastic headlamp bezel.
4. Loosen or remove the three screws holding the headlight retaining ring. Once all three screws are loosened, the ring may be turned slightly and removed. This trick eliminates removing very small screws.

6-88 CHASSIS ELECTRICAL

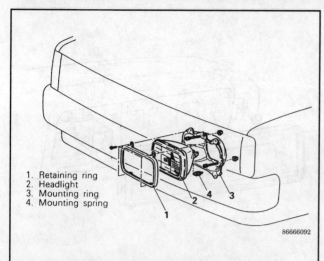

Fig. 117 Headlight assembly removal and installation components — all Pick-ups similar

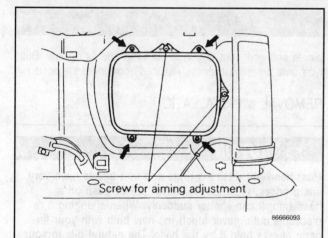

Fig. 118 Only remove the four screws around the retaining ring — do not loosen or remove the aiming screws

5. Remove the headlamp and disconnect the wiring harness. Install the new lamp and connect the wiring. Fit the lamp in place and install the retaining ring.

6. Install the plastic bezel, the sidemarker assembly and the grille.

1992-95 Montero

♦ See Figures 120, 121, 122 and 123

The 1992-95 Monteros have a replaceable bulb rather than a sealed bulb. This procedure tells how to remove the headlight assembly and the small replaceable bulb.

HEADLIGHT ASSEMBLY

1. Remove the front combination light (turn signal & side marker light) mounting screws and set spring. Remove the combination light by pulling it towards the front of the vehicle.

➡ For the left side, before removing the front combination light, remove the engine coolant reserve tank.

2. Remove the radiator grille.

3. Loosen and remove the two mounting bolts and the two mounting nuts from the headlight.
4. Pull the headlight out of the retaining bezel and disconnect the wiring harness.
5. Attach the wires to the new headlight and install it into the headlight recess.
6. Tighten the outside two mounting bolts, then tighten the two inside mounting nuts.
7. Install the radiator grille.
8. Align the front combination light positioning bosses with the insertion holes in the fender, then align the ribs with the headlight insertion holes.
9. While pushing the front combination light in towards the rear of the vehicle, pull the set spring into the engine compartment to tighten it to the vehicle body, then tighten with the screw.
10. Install the engine coolant reserve tank, if removed.

REPLACEABLE BULB

1. Remove the engine coolant reserve tank (left-hand side only).
2. Disconnect the negative (-) battery cable from the battery.
3. Unplug the harness connector, then pull the socket cover out.
4. Remove the locking cap by rotating it counterclockwise, then remove the socket and bulb together.

➡ If the socket cover is not securely installed, the lens will be out of focus, or water will get inside the light unit, so the cover should be securely installed.

✱✱WARNING

Never hold the halogen light bulb with a bare hand, dirty glove, etc. If the glass surface is dirty, be sure to clean it with alcohol or paint thinner and install it after drying it thoroughly.

5. Insert the bulb and socket together into the back of the headlight assembly.
6. Turn the locking cap clockwise to lock it in place.
7. Install the socket cover and wiring harness to the headlight assembly.
8. Connect the battery cable to the battery.
9. Install the engine coolant reserve tank.

AIMING

The 1992-95 Monteros are the only models covered by this manual on which the headlights can be aimed without special tools. To aim the headlights of Pick-ups and 1983-91 Monteros, you will need to get (or make) a wall screen. If, after adjusting the headlights on the wall screen, they are still out of alignment, take the vehicle to an automotive mechanic to have them adjusted.

1992-95 Montero

♦ See Figures 124 and 125

The 1992-95 Monteros have adjusting screws with scales to aim the headlight assembly. Since the headlight bulbs are the

CHASSIS ELECTRICAL 6-89

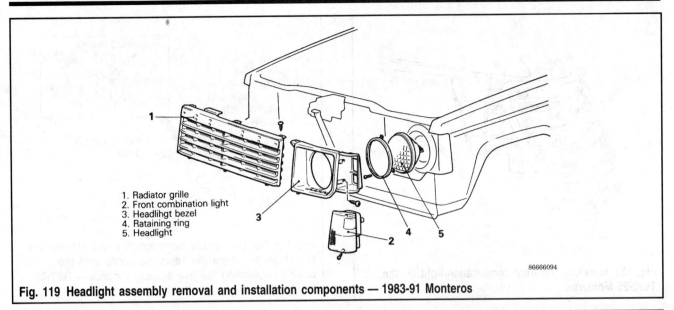

Fig. 119 Headlight assembly removal and installation components — 1983-91 Monteros

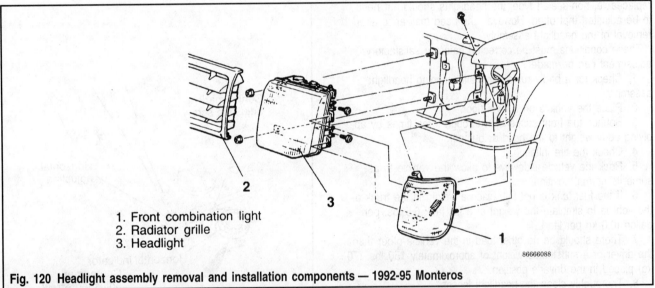

Fig. 120 Headlight assembly removal and installation components — 1992-95 Monteros

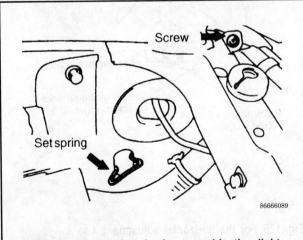

Fig. 121 When removing the front combination light, make sure to disconnect the set spring from the fender — 1992-95 Monteros

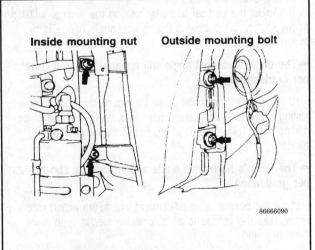

Fig. 122 Tighten the outside mounting bolts, then the inside nuts — 1992-95 Monteros

6-90 CHASSIS ELECTRICAL

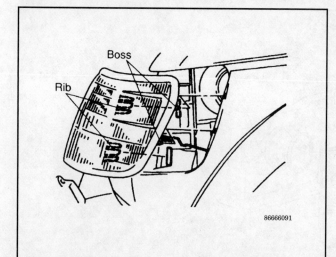

Fig. 123 Installing the front combination light on the 1992-95 Monteros

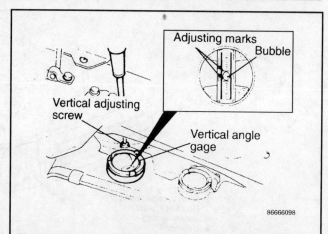

Fig. 124 For the vertical adjustment, use a screwdriver to tighten or loosen the adjusting screw until the bubble is between the two adjusting marks — 1992-95 Monteros

replaceable, non-sealed type, the headlights should not need to be adjusted that often. However, the need may arise after removal of the headlight assembly.

These conditions must be corrected before a satisfactory adjustment can be made:

1. Check for a badly rusted or malfunctioning headlight assembly.
2. Place the vehicle on a level floor.
3. Bounce the front suspension through three times by applying body weight to the hood or bumper.
4. Check the tire inflation.
5. Rock the vehicle sideways to allow the vehicle to assume its normal position.
6. If the fuel tank is not full, place a weight in the trunk of the vehicle to simulate the weight of a full tank 6.5 lbs. per gallon (0.8 kg per liter).
7. There should be no other load in the vehicle other than the driver or a substituted weight of approximately 150 lbs. (70 kg) placed in the driver's position.
8. Thoroughly clean the headlight lenses.

To aim the headlights:

9. Adjust the vertical angle by rotating the vertical adjusting screw so that the bubble in the vertical angle gauge is between the two adjusting marks.

➥The beam's vertical angle will change by about 0°12' per graduation.

10. Adjust the horizontal angle by turning the horizontal adjusting screw until the center mark (red mark) and the center mating mark (black mark) of the horizontal indicator are aligned.

➥The beam's horizontal angle will change by about 0°23' per graduation.

11. If the beams, after adjustment, still seem aimed incorrectly, take the vehicle to an automotive mechanic to have them adjusted.

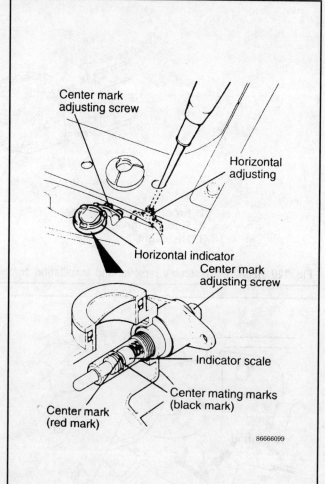

Fig. 125 For the horizontal adjustment, use a screwdriver to tighten or loosen the adjusting screw until the red and black marks are aligned — 1992-95 Monteros

CHASSIS ELECTRICAL 6-91

Except 1992-95 Montero

▶ See Figures 126, 127, 128 and 129

These models of the Montero and the Pick-up use two aiming screws to adjust the headlight beams. Tightening the screw makes the light point more toward the screw being tightened. Loosening the aiming screw points the headlight beam more away from the screw.

These conditions must be corrected before a satisfactory adjustment can be made:

1. Check for a badly rusted or malfunctioning headlight assembly.
2. Place the vehicle on a level floor.
3. Bounce the front suspension through three times by applying body weight to the hood or bumper.
4. Check the tire inflation.
5. Rock the vehicle sideways to allow the vehicle to assume its normal position.
6. If the fuel tank is not full, place a weight in the trunk of the vehicle to simulate the weight of a full tank 6.5 lbs. per gallon (0.80 kg per liter).
7. There should be no other load in the vehicle other than the driver or a substituted weight of approximately 150 lbs. (70 kg) placed in the driver's position.
8. Thoroughly clean the headlight lenses.

Headlight aim preparation:

9. Measure the center of the headlight bulb as shown in the illustration.
10. Position the vehicle on a known level floor 9.8 ft. (3 m) from an aiming screen or brightly colored wall.

➡ Four lines of adhesive tape or similar are needed on the screen or wall.

11. Position a vertical piece of tape so that it is aligned with the vehicle center line.
12. Position a horizontal piece of tape with reference to the center line of each of the headlight bulbs.
13. Position a vertical piece of tape on the screen with reference to the center line of each of the headlight bulbs.

Headlight adjustment:

14. A properly aimed lower beam on the aiming screen will appear 25 ft. (7.6 m) in front of the vehicle on the road. Adjust

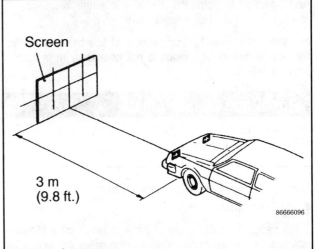

Fig. 127 Position the vehicle on a level floor and about 9.8 ft. (3 m) from the screen or a brightly colored wall

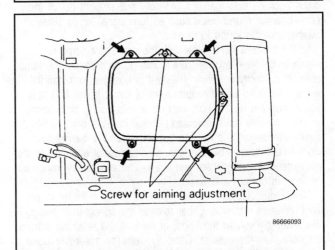

Fig. 128 Do not confuse the two aiming screws with the retaining ring screws (indicated by the arrows); round headlights similar

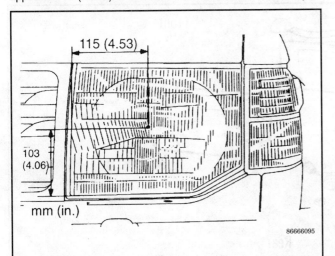

Fig. 126 Measure the center of the headlights so that an aiming screen can be set up for the vehicle

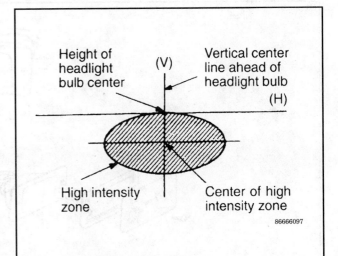

Fig. 129 Aim each headlight so that it illuminates the high intensity zone on the aiming screen

CHASSIS ELECTRICAL

each headlight to match the shaded area in the illustration, which indicates the high intensity zone.

➡ If the visual headlight adjustment at low beam is made, the adjustment at high beam is not necessary (it is also now adjusted).

Signal and Marker Lights

REMOVAL & INSTALLATION

Front Turn Signal and Side Marker Lamps
♦ See Figure 130

To change the bulb on any of the front lamps on a Mitsubishi, remove the mounting screws holding either the lens to the reflector or the lighting unit to the bodywork. If the lens comes off separately, wear gloves to remove the bulb. Very small bulbs such as side markers generally pull straight out of the socket; larger round bulbs such as turn signal lamps usually twist counterclockwise to remove.

Install the new lamp. Note that the socket for some turn signal bulbs have two different length guide grooves; the lamp only fits correctly one way. (While it is possible to install it wrong, doing so would require a lot of force. If you find that you're pushing on the bulb with no results during installation, you've probably got it reversed.) These bulbs should install easily with a gentle push into the socket and a quarter turn clockwise. Reassemble the lens, paying careful attention to the placement of any gaskets and seals. These are very important in keeping out dirt and water.

Newer models have the lens bonded to the reflector to prevent moisture intrusion. Either release the spring from behind the assembly (under the hood) or remove the retaining screws if visible from the outside. Carefully work the assembly loose from the body work, always being wary of the hidden clip or bracket. If you break one of these clips, the lamp will not stay put when you reinstall it.

When the assembly is free of the car, the socket may be removed with a twist counterclockwise. Install the bulb and insert the assembly back into the housing. Fit the whole thing back onto the bodywork, engaging any clips or hooks. Install the retaining screws or the tension spring as necessary.

Rear Lighting and Side Marker Lamps
♦ See Figures 130, 131 and 132

The older Pick-ups and Monteros use separate side marker assemblies, whose lenses may be removed with two screws, and tail light assemblies. The newer (1990 and newer) vehicles use one wrap-around combination light.

The tail, brake and turn lamps usually all fit into one integrated assembly. It is wise to identify the position of the light being changed from the outside of the vehicle before confronting 5 or 6 sockets in the back of the unit. Simply knowing the bad bulb is third from the left can save a large amount of time.

1. Remove the retaining screws holding the lamp assembly to the bodywork and pull the assembly free.
2. Remove the socket from the rear of the reflector (or simply change the bulb if only the lens came off).
3. Examine the holder; usually they are quarter-turn sockets, but a variety of retainers may be found.
4. Carefully remove the socket from the reflector and change the bulb. Almost all rear lamps are the twist out type.
5. Fit the bulb and socket into the housing. Normally, the socket will only fit one way due to plastic guides or keys.
6. Connect the locking device or tab if one was released and reinstall the cover or trim panel.
7. After replacing the lightbulb, carefully fit the unit back in place with the gaskets correctly positioned.
8. Install and tighten the screws.

High-Mount Brake Light (Third Brake Light)

1992-95 PICK-UP

♦ See Figure 133

➡ The high mounted stop light on the Pick-ups does not have individual replaceable bulbs; the whole assembly will need to be replaced.

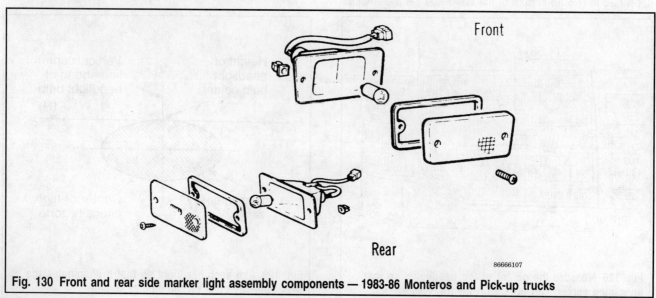

Fig. 130 Front and rear side marker light assembly components — 1983-86 Monteros and Pick-up trucks

CHASSIS ELECTRICAL 6-93

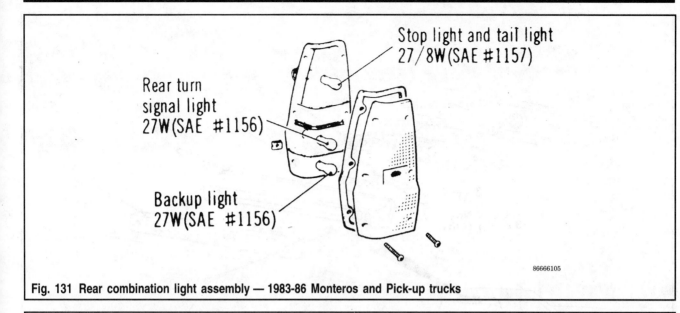

Fig. 131 Rear combination light assembly — 1983-86 Monteros and Pick-up trucks

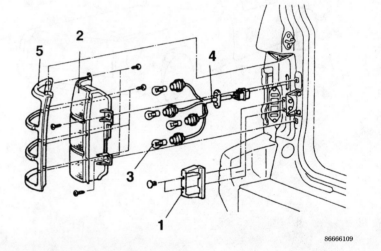

1. Back door bumper cover
2. Rear combination light unit
3. Bulb
• Quarter trim
4. Socket assembly
5. Rear combination light bezel

Fig. 132 Rear wrap-around combination light assembly — 1992-95 Monteros

1. Disconnect the negative (-) battery cable from the battery.
2. Remove the two mounting screws holding the stop light assembly to the bracket.
3. Pull the high-mount brake light assembly, the cover and the screw grommets off of the bracket.
4. Disconnect the wiring harness.
5. Set the cover in place on the bracket and pull the wiring harness through it.
6. Plug the harness to the back connector of the stop light assembly and set the assembly in place.
7. Secure the entire unit in place with the two retaining screws.
8. Connect the negative (-) battery cable to the battery.

1992-95 MONTERO
▶ See Figure 134

The high-mount brake light on the Monteros is mounted to the inside of the back door, right above the window.

1. Disconnect the negative (-) battery cable from the battery.
2. Loosen and remove the two mounting screws.
3. Pull the cover and the socket assembly out of the stop light unit.
4. Pull the bulbs out of the socket assembly.
5. Push the new bulb into the socket assembly, then push the socket assembly into the stop light unit.
6. Place the cover over the socket assembly and secure it in place with the two mounting screws.
7. Reconnect the battery cable.

6-94 CHASSIS ELECTRICAL

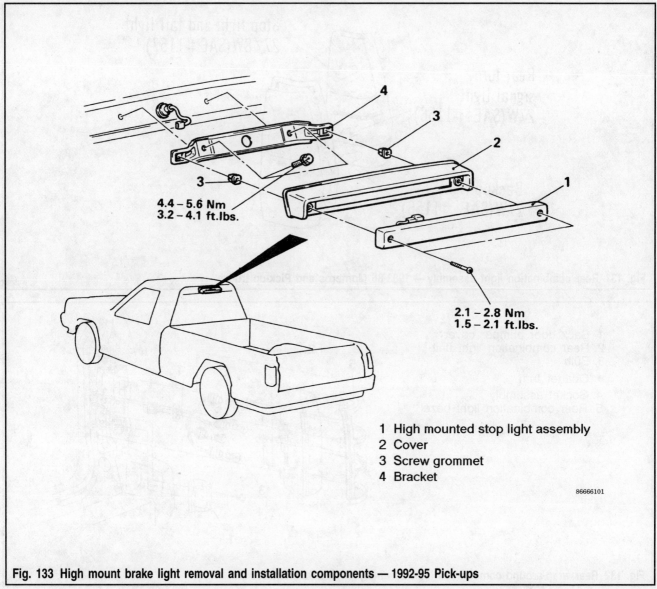

Fig. 133 High mount brake light removal and installation components — 1992-95 Pick-ups

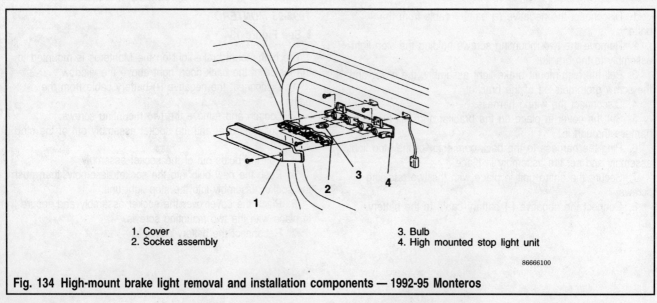

Fig. 134 High-mount brake light removal and installation components — 1992-95 Monteros

CHASSIS ELECTRICAL   6-95

Interior Lights

♦ See Figures 135 and 136

The dome lights, room lights, cargo lights and ceiling mounted map lights on Monteros and Pick-up trucks are of the essentially same design. The light assemblies are comprised of three components: the cover lens, light bulb, and socket assembly.

1. To change the bulb or remove the fixture, pop the cover lens off of the fixture. On most cases the cover lens simply snaps in place on the fixture, but check to make certain that there are no mounting screws on the larger light assemblies.
2. Once the cover lens has been removed, the bulb can be extracted from the light fixture.
3. A new bulb can be inserted, or the fixture itself can be removed from the headliner by removing the mounting screws holding it in place. Some ceiling lights, however, are not held in place with screws, these type snap into the headliner. Check for any mounting screws, and if none are found, the light should be pulled out of the headliner. Make absolutely certain, that there are no mounting screws holding the fixture in place before deciding to "yank" it out of the headliner.
4. After the fixture is loose, disconnect the electrical wiring to the fixture.

To install:

5. Attach the wiring to the fixture, then either push the fixture in place until it is held there by the retaining tabs or clips or mount it in place with the mounting screws, depending on which type of light fixture it is.
6. Install the new bulb and push the cover lens back onto the fixture.

License Plate Lights

♦ See Figures 137 and 138

The license plate lights on the Pick-ups and Monteros have basically the same procedures for removal and installation, although they may look different. The license plate light bulb is inside of the license plate garnish and is held in place by the socket assembly.

1. To change the bulb, remove the cover lenses retaining screws and remove the lenses from the garnish.

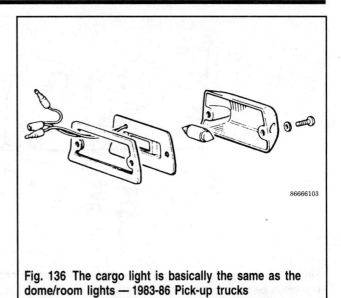

Fig. 136 The cargo light is basically the same as the dome/room lights — 1983-86 Pick-up trucks

2. Pull the old bulbs out of the socket assembly and install the new bulbs.
3. Secure the light lenses in place with their screws.

To remove the whole license light fixture assembly:

4. Remove the lenses and old bulbs.
5. After the lenses and bulbs are removed, remove the lens gaskets, which, as long as they are still in good shape, can be reused upon installation.
6. Before removing the license plate garnish, it may be necessary to remove the rear door inside trim to get at the garnish's mounting screws.
7. Unplug the socket assembly's wiring harness and remove the license plate garnish mounting nuts. Some models, such as the 1992-95 Monteros, also have clips holding the garnish in place, use a flat-tipped prytool to remove these.
8. Remove the garnish and socket assembly together from the back of the vehicle.

To install the license plate light fixture:

9. Mount the garnish and socket assembly together onto the back of the vehicle's door (make certain to route the wiring through the appropriate hole before mounting the assembly),

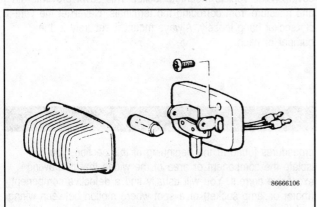

Fig. 135 A common style dome/room light found on Monteros and Pick-up trucks — although some lights may not look the same, the basic components function similarly

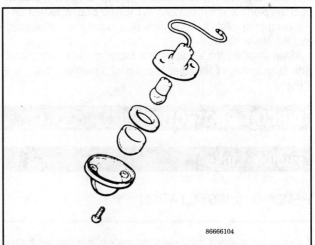

Fig. 137 The single license plate bulb found on 1983-86 Pick-up trucks is a much simpler assembly than those on later model Monteros

6-96 CHASSIS ELECTRICAL

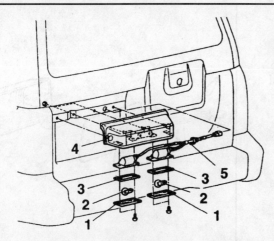

1. Lens
2. Bulb
3. Lens gasket
• Back door trim
  – Door Trim and Waterproof Film.
4. License plate light garnish
5. Socket assembly

Fig. 138 Before removal of the license plate light garnish can be completed, the inside rear door trim and waterproof film must be removed — 1992-95 Monteros

10. Tighten the mounting nuts until they are snug, then plug the wiring harness connectors together.
11. Install the rear door inside trim, if needed.

12. Plug the new light bulbs into the socket assembly and, after placing the gaskets in position, mount the light lenses with their retaining screws.

TRAILER WIRING

Wiring the truck for towing is fairly easy. There are a number of good wiring kits available and these should be used, rather than trying to design your own. All trailers will need brake lights and turn signals as well as tail lights and side marker lights. Most states require extra marker lights for overly wide trailers. Also, most states have recently required back-up lights for trailers, and most trailer manufacturers have been building trailers with back-up lights for several years. Additionally, some Class I, most Class II and just about all Class III trailers will have electric brakes. Add to this number an accessories wire, to operate trailer internal equipment or to charge the trailer's battery, and you can have as many as seven wires in the harness.

Determine the equipment on your trailer and buy the wiring kit necessary. The kit will contain all the wires needed, plus a plug adapter set which included the female plug, mounted on the bumper or hitch, and the male plug, wired into, or plugged into the trailer harness.

When installing the kit, follow the manufacturer's instructions. The color coding of the wires is standard throughout the industry.

One point to note, some domestic vehicles, and most imported vehicles, have separate turn signals. On most domestic vehicles, the brake lights and rear turn signals operate with the same bulb. For those vehicles without separate turn signals, you can purchase an isolation unit so that the brake lights won't blink whenever the turn signals are operated, or, you can go to your local electronics supply house and buy four diodes to wire in series with the brake and turn signal bulbs. Diodes will isolate the brake and turn signals. The choice is yours. The isolation units are simple and quick to install, but far more expensive than the diodes. The diodes, however, require more work to install properly, since they require the cutting of each bulb's wire and soldering in place of the diode.

One final point, the best kits are those with a spring loaded cover on the vehicle mounted socket. This cover prevents dirt and moisture from corroding the terminals. Never let the vehicle socket hang loosely. Always mount it securely to the bumper or hitch.

CIRCUIT PROTECTION

Fusible Links

REMOVAL & INSTALLATION

Every circuit in the car's electrical system is protected by a fuse or fusible link except the starter circuit. A melted fusible link or blown fuse indicates not a bad link, but a short circuit in a component, a grounded wire or an overload. Using the procedures found at the beginning of this section, you can isolate the component or area of the wiring that is drawing excessive current. You will usually find a defective component (motor or lamp socket) or a spot where motion between wiring and the body has rubbed insulation through and caused bare wiring to touch the body or a metal bracket or major drivetrain component.

Mitsubishi vehicles may have one or more fusible links. They appear to be short wires and are always located at the positive battery terminal. Additionally, there may be sub-links within

CHASSIS ELECTRICAL 6-97

particular high amperage circuits. They are simply unplugged and re-plugged to replace them, but you MUST be sure you are using a link of the same capacity. Excessive capacity in a link could result in a vehicle fire.

Fuses

REMOVAL & INSTALLATION

The fuse box on most models is located under the instrument panel on the left (driver's side) above the cowl side trim, other models have it located on the left-hand end of the instrument panel. Almost all vehicles after 1986 have an additional fuse and relay board under the hood near the battery. The most common circuits are on the interior fuseboard for ease of changing the fuse by the driver; the fuses under the hood tend to control peripheral systems used occasionally such as the rear window defogger.

The underhood relay board may have additional fuse boards mounted to it as other circuits run from it. Further additional fuses may be mounted directly at the positive battery terminal, monitoring circuits which originate there. It is in your best interest to read the owner's manual carefully and identify the location of certain fuses which could affect the safety and operation of the vehicle. You could drive home without the heater fan working, but could you make the same drive without headlights?

Component Locations

1983-89 MONTERO

- Main Fusible Links: 1983-87 models at the positive battery terminal; 1988-89 models at bottom of small terminal box near the battery.
- Sub-links: at bottom of small terminal box near battery.
- Fusebox: mounted in left-hand side end panel of dashboard.
- Special fuses not in the fusebox: 1983-89 4-cylinder models' headlight high beam indicator fuse located in harness behind left headlight. 1987-88 models' air conditioner fuse mounted on evaporator case; 1989 model's rear defogger link in harness near sub-link, V6 headlight circuit in wiring harness at driver's side firewall, sunroof fuse above right dash speaker, power door lock fuse in wire harness running just above steering column behind dashboard, rear air conditioner (if so equipped) fuse on blower assembly under right dashboard.
- Turn Signal and Hazard Flasher Relay: separate units mounted under the dash and above the left kick panel. (The hazard flasher is round.)

1990-95 MONTERO

▶ See Figures 139, 140 and 141

- Main Fusible Links: one at positive battery terminal, one in small terminal box near the battery.
- Fusebox: mounted under the left-hand side dashboard, next to the lower speaker.
- Special fuses not in the fusebox: dedicated fuse No. 9 located on relay board, just outside of the fuse box.
- Turn Signal and Hazard Flasher Relay, Overdrive Relay, Power Window Relay, Defogger Relay, and Rear Intermittent Wiper Relay: separate units mounted under the dash, next to the fuse box and the left-hand, lower speaker.
- Tail Light Relay, Condenser Fan Motor Relay, A/C Compressor Relay, Generator Relay, and Headlight Relay: separate units mounted in the small terminal box next to the battery.

PICK-UP

▶ See Figures 142 and 143

- Main Fusible Links: 1983-89 models — two at the positive battery terminal; 1990-95 models — one at the positive battery terminal.
- Sub-links: 1983-86 models have none; 1987-95 models — one on the left inner fender panel.
- Fusebox: under left dashboard.
- Special fuses not in the fusebox: none
- Turn Signal and Hazard Flasher Relay (1983-95 models), Seat Belt Warning Time Relay (1992-95 models), Power Window Relay (1992-95 models), Starter Relay (1992-95 models), Starter Unlock Relay (1992-95 models): combined unit on relay board near the fusebox.
- Vacuum Pump Relay (1992-95 models): mounted in the back, right-hand corner of the engine compartment.
- A/C Compressor Clutch Relay (1992-95 models): mounted on the left, inner fender, next to the sub-fusible link.
- Central Door Lock Control Relay (1992-95 models): mounted in the front of the right-hand door, under the door panel trim.
- Power Window Control Relay (1992-95 models): mounted in the front left-hand door, under the door panel trim.

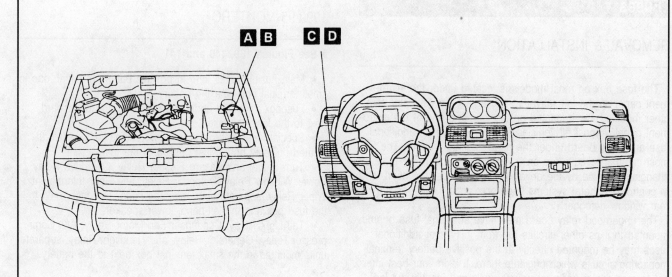
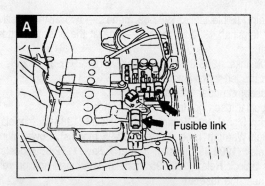
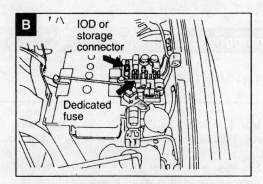
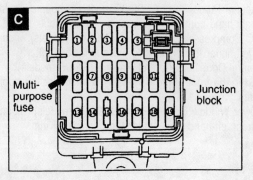
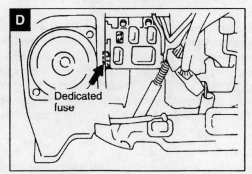

Fig. 139 Fusible link, fuse and Ignition Off Draw (IOD)/storage connector locations — 1992-95 Monteros

CHASSIS ELECTRICAL 6-99

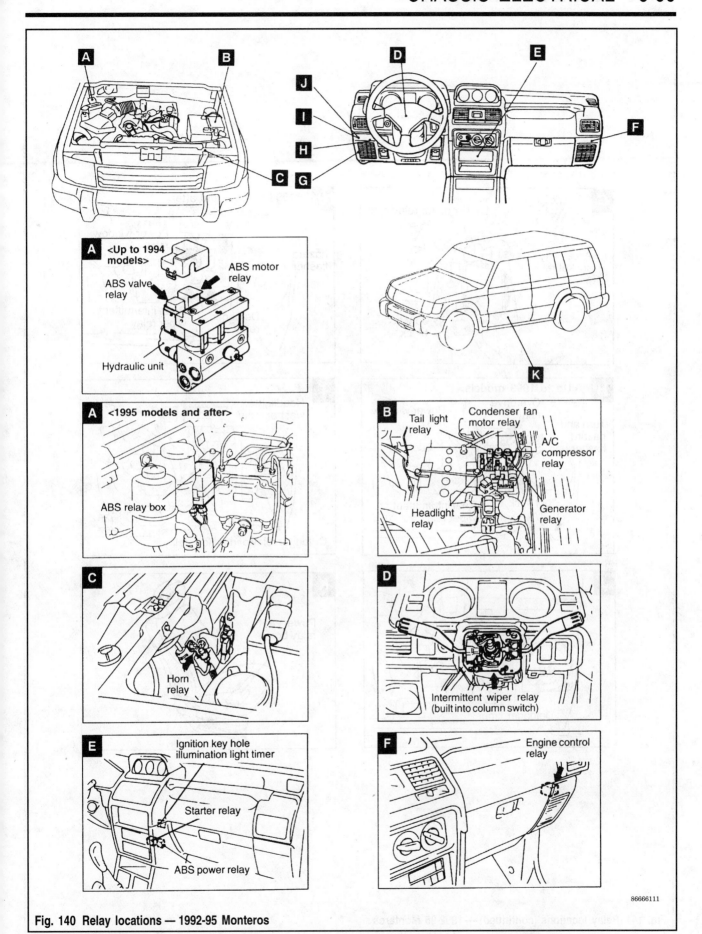

Fig. 140 Relay locations — 1992-95 Monteros

6-100 CHASSIS ELECTRICAL

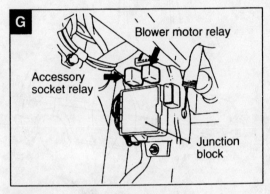
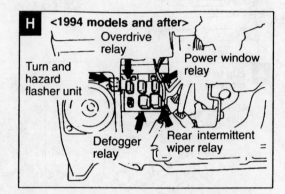
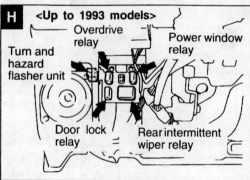
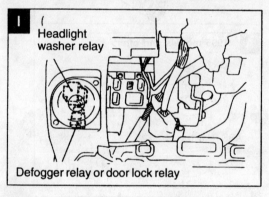
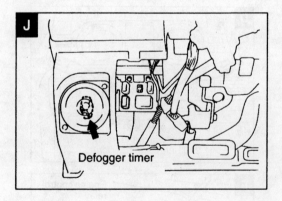
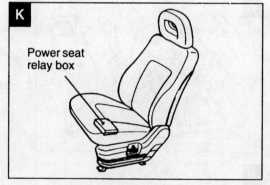

Fig. 141 Relay locations (continued) — 1992-95 Monteros

CHASSIS ELECTRICAL 6-101

ENGINE COMPARTMENT

INSTRUMENT PANEL

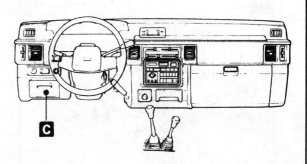

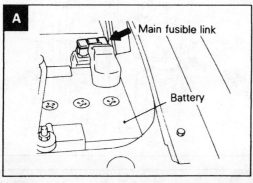

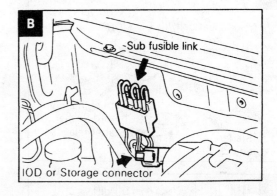

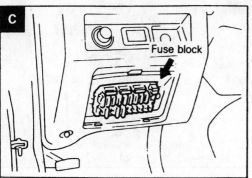

Fig. 142 Fusible link, fuse and Ignition Off Draw (IOD)/storage connector locations — 1987-95 Pick-up trucks

6-102 CHASSIS ELECTRICAL

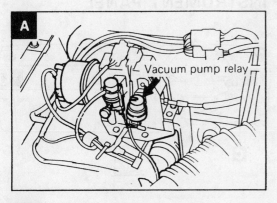
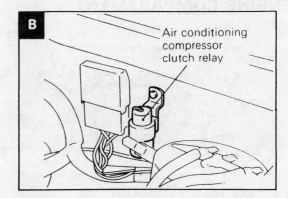
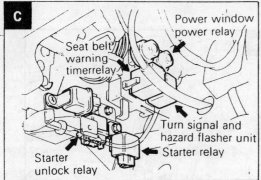
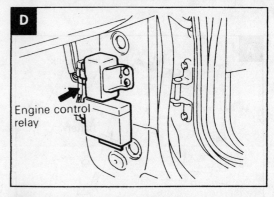
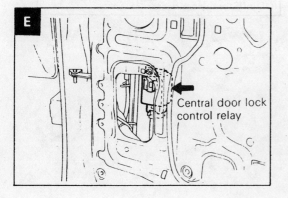
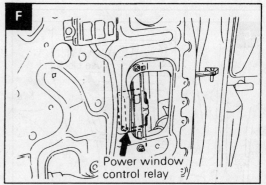

Fig. 143 Relay locations — 1987-95 Pick-up trucks

CHASSIS ELECTRICAL 6-103

WIRING DIAGRAMS

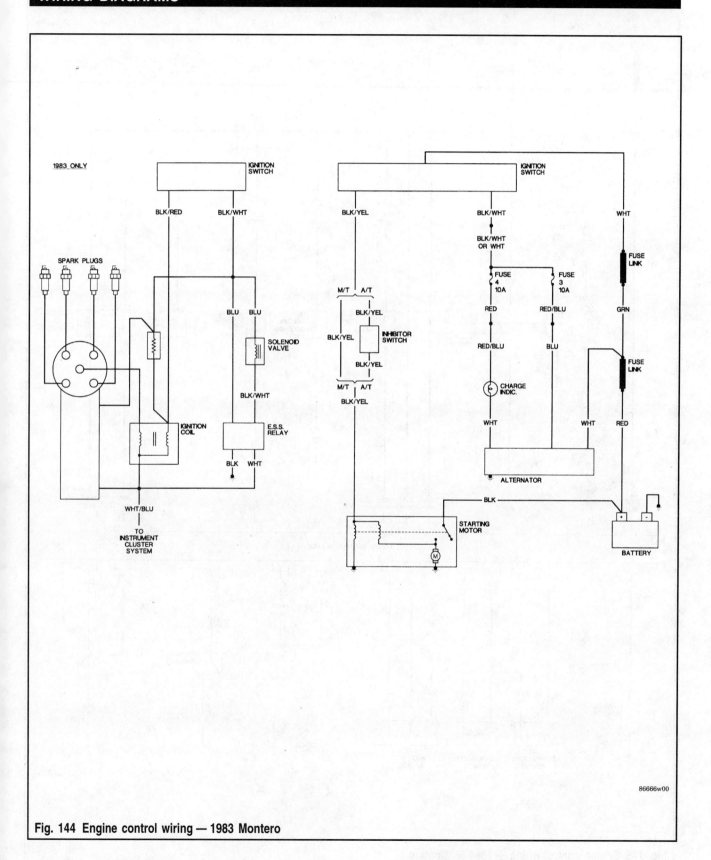

Fig. 144 Engine control wiring — 1983 Montero

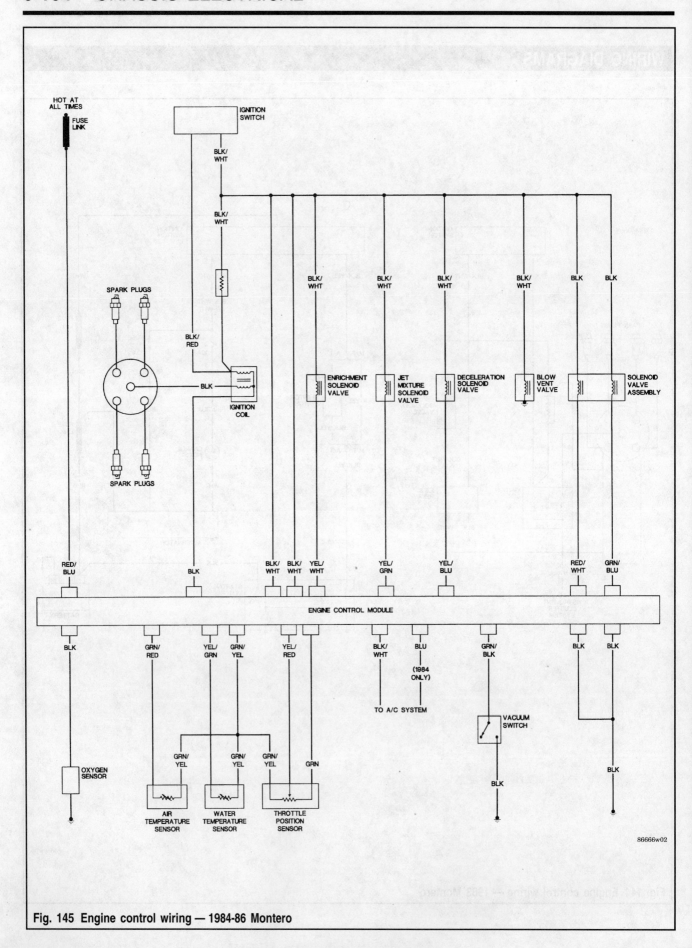

Fig. 145 Engine control wiring — 1984-86 Montero

CHASSIS ELECTRICAL 6-105

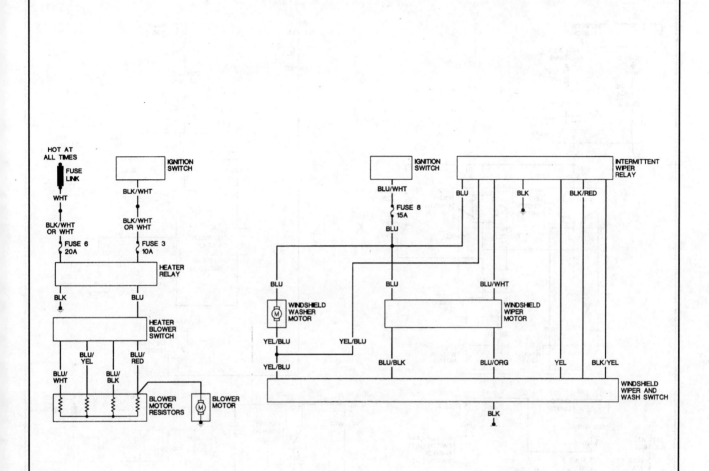

Fig. 146 Body wiring — 1983-86 Montero

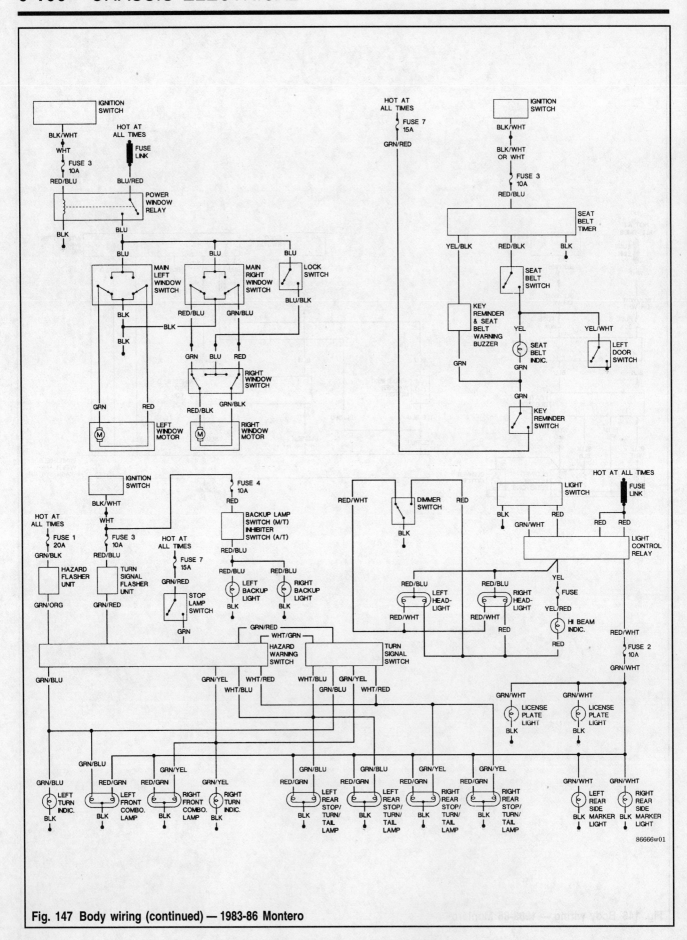

Fig. 147 Body wiring (continued) — 1983-86 Montero

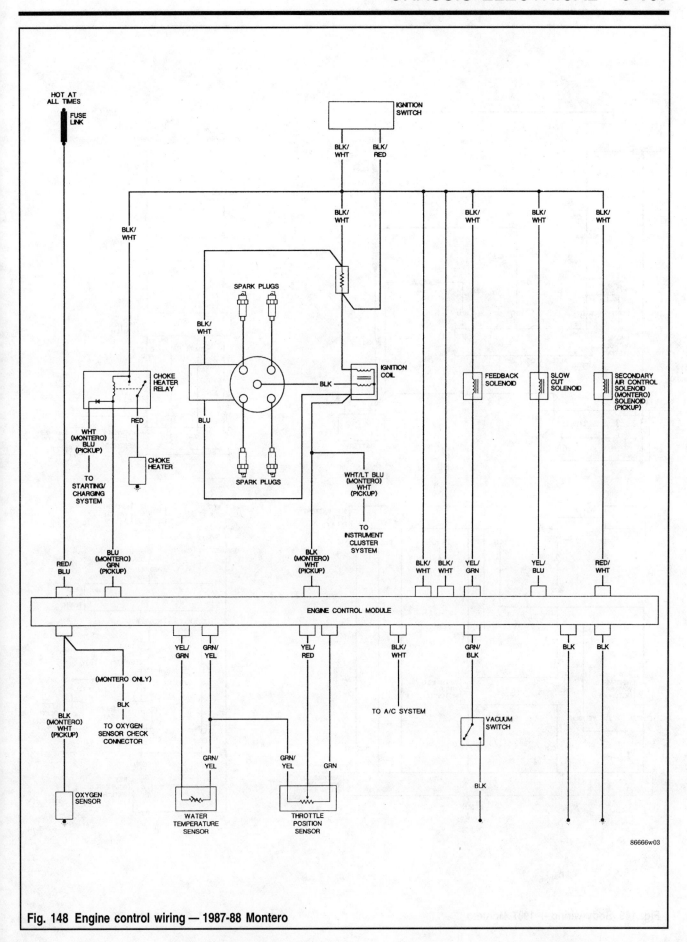

Fig. 148 Engine control wiring — 1987-88 Montero

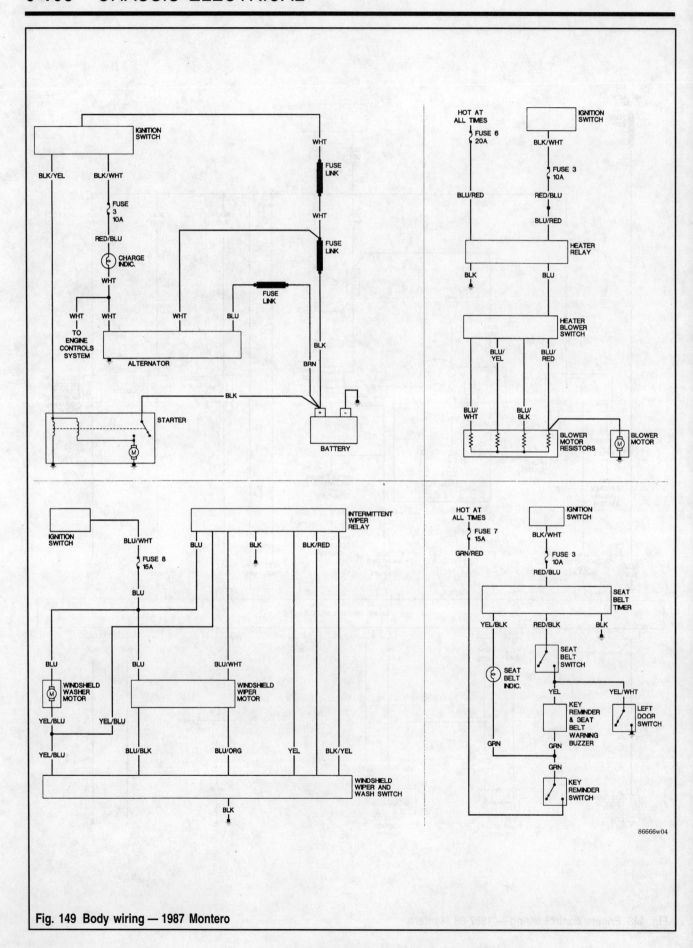

Fig. 149 Body wiring—1987 Montero

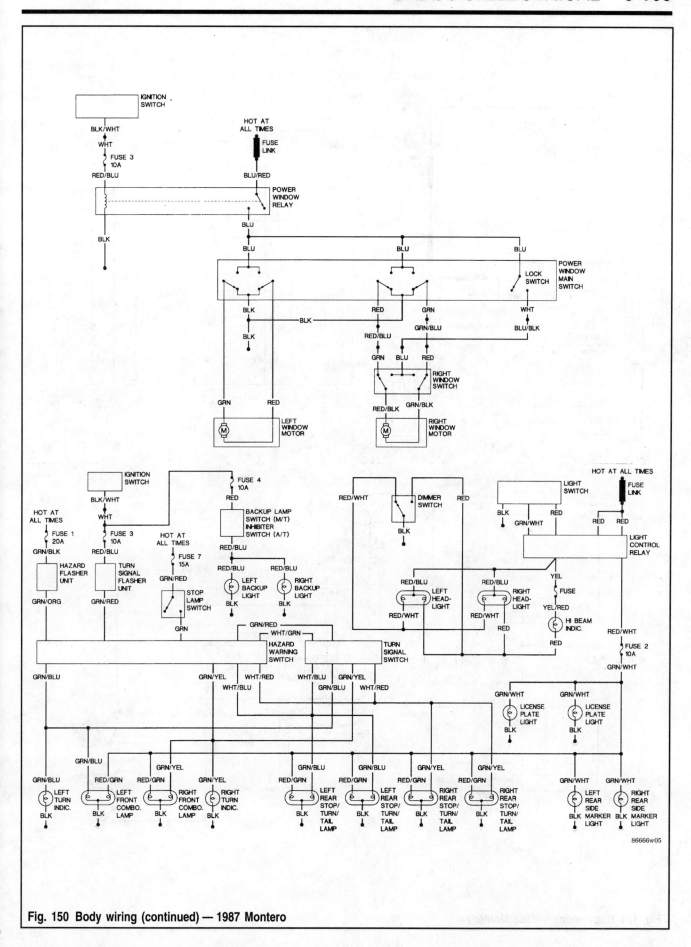

Fig. 150 Body wiring (continued) — 1987 Montero

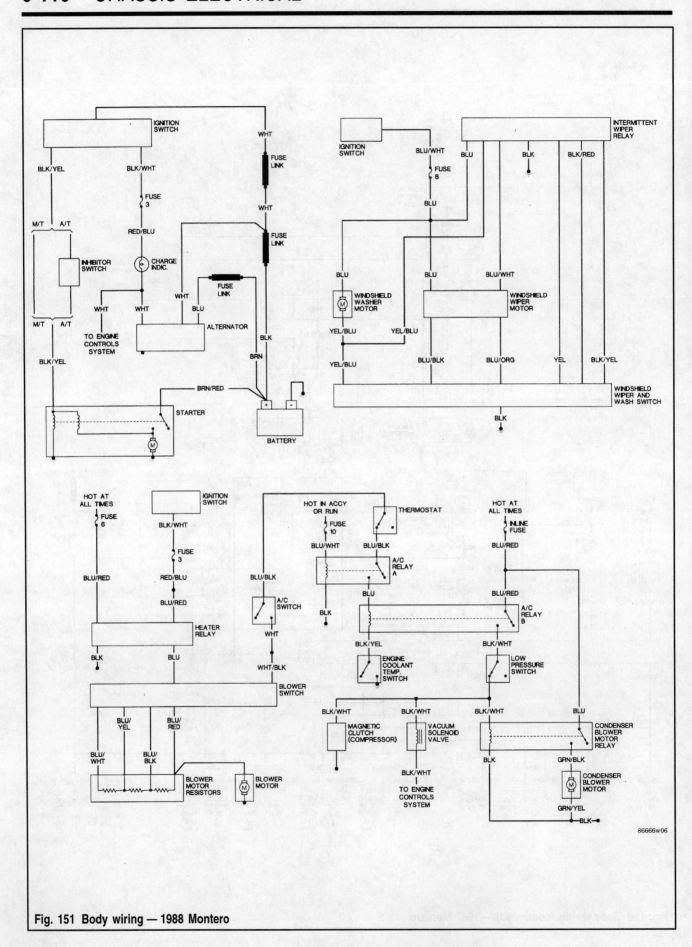

Fig. 151 Body wiring — 1988 Montero

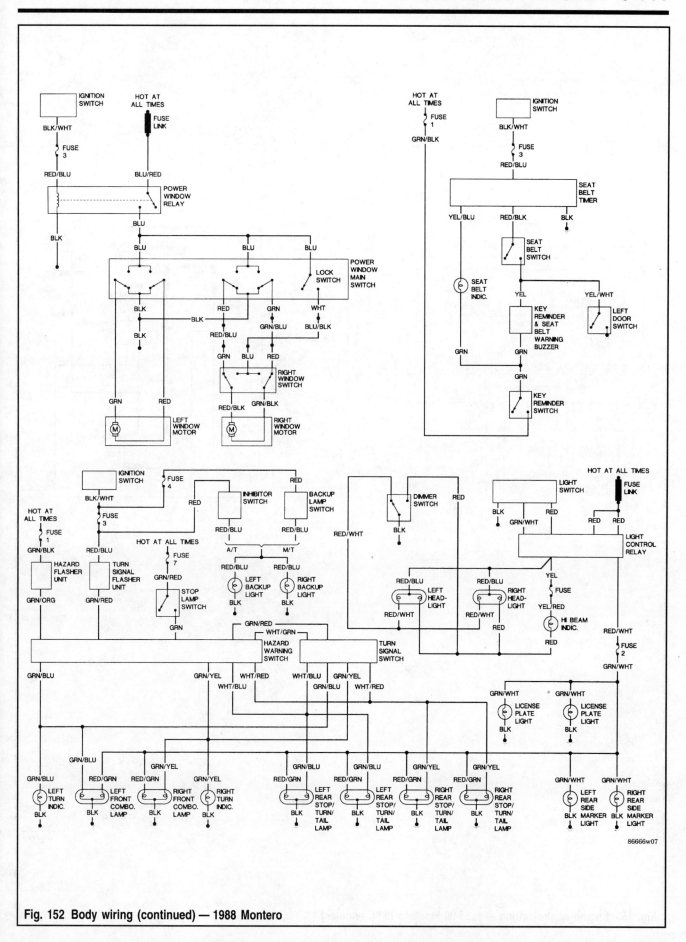

Fig. 152 Body wiring (continued) — 1988 Montero

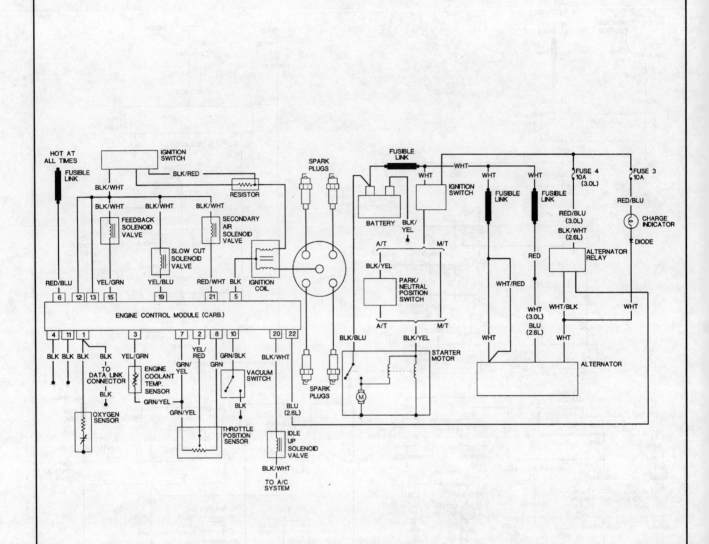

Fig. 153 Engine control wiring — 1989-90 Montero (2.6L engines)

CHASSIS ELECTRICAL 6-113

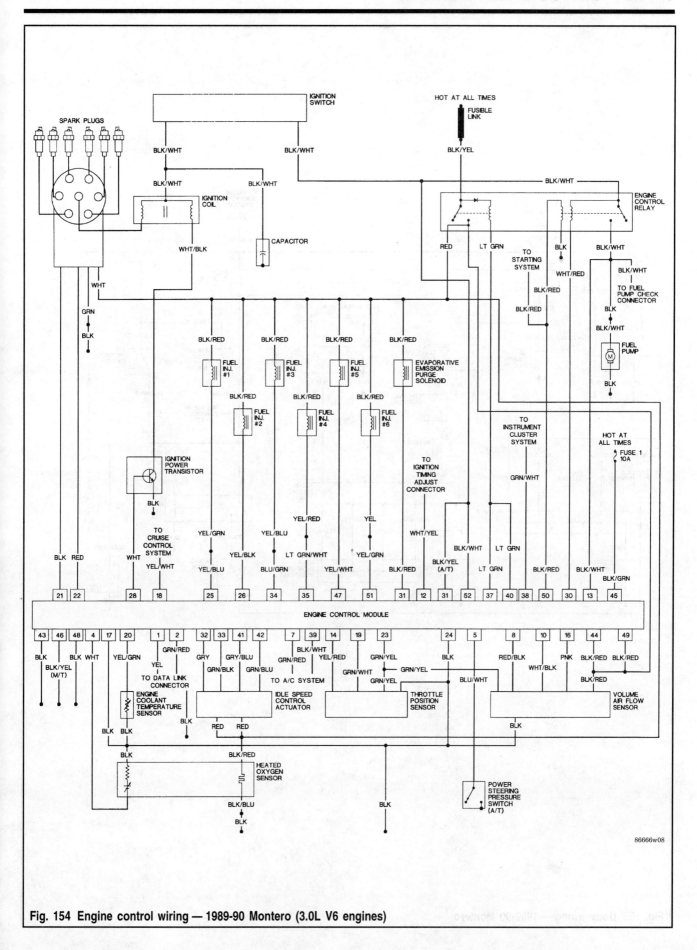

Fig. 154 Engine control wiring — 1989-90 Montero (3.0L V6 engines)

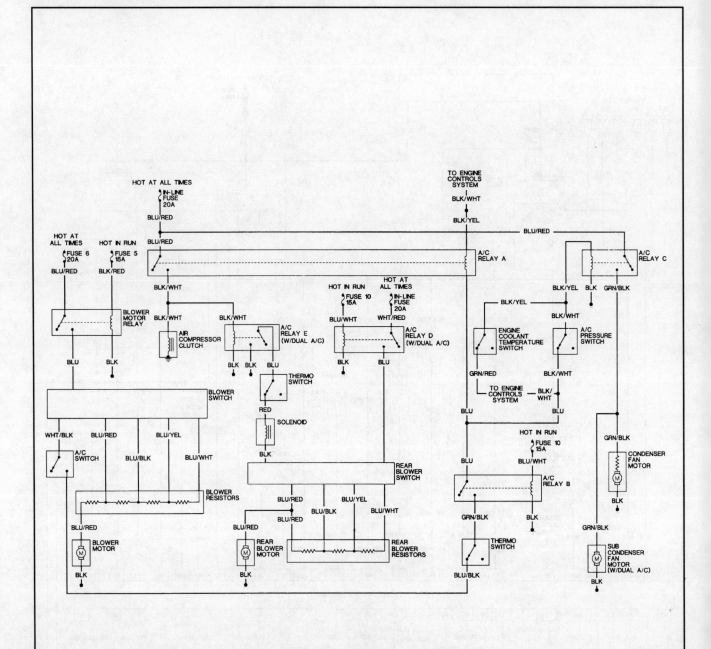

Fig. 155 Body wiring — 1989-90 Montero

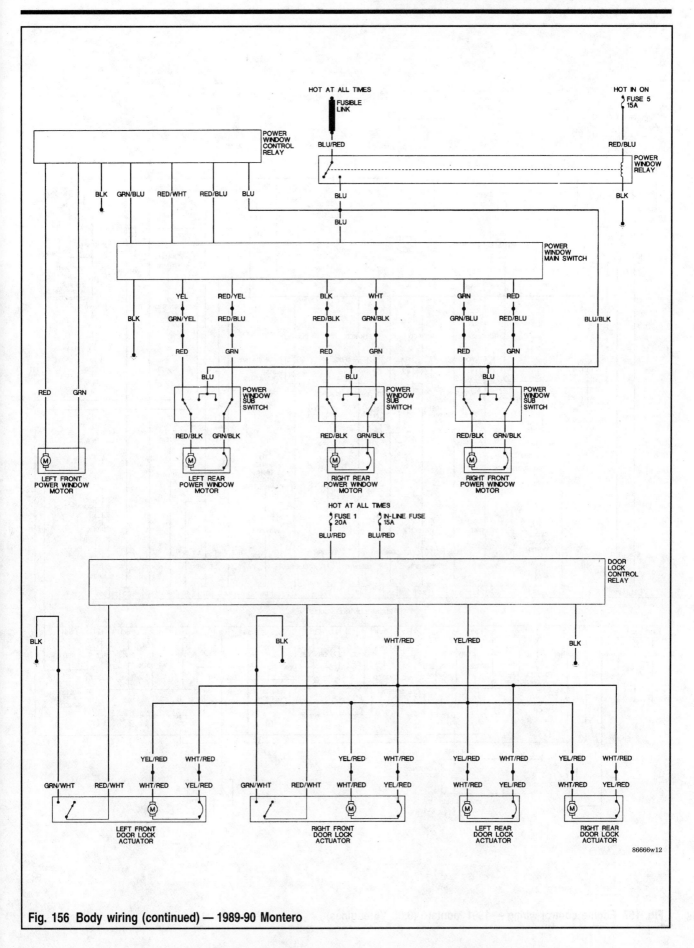

Fig. 156 Body wiring (continued) — 1989-90 Montero

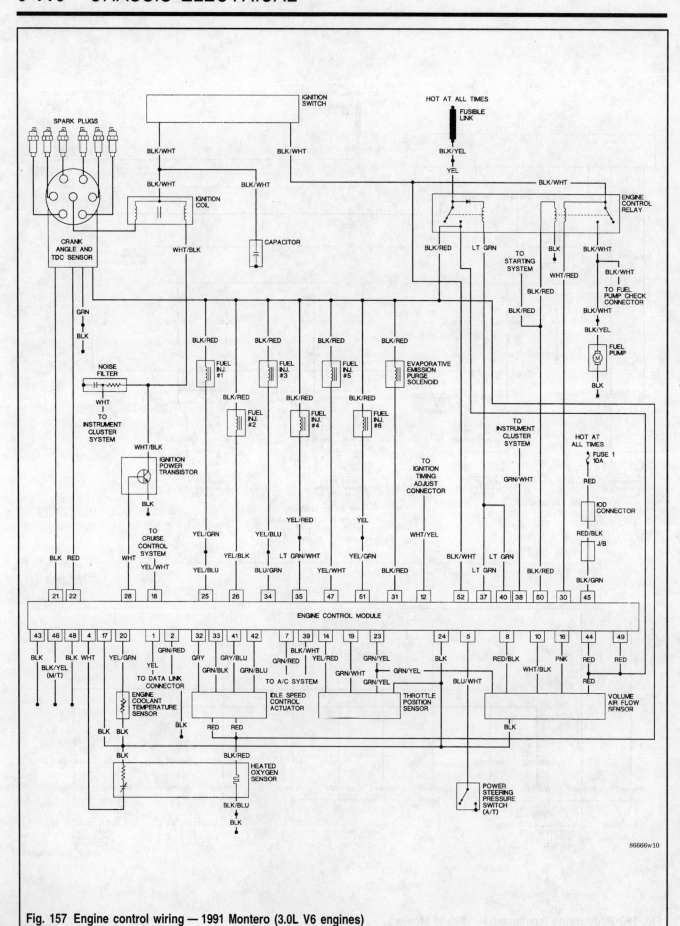

Fig. 157 Engine control wiring — 1991 Montero (3.0L V6 engines)

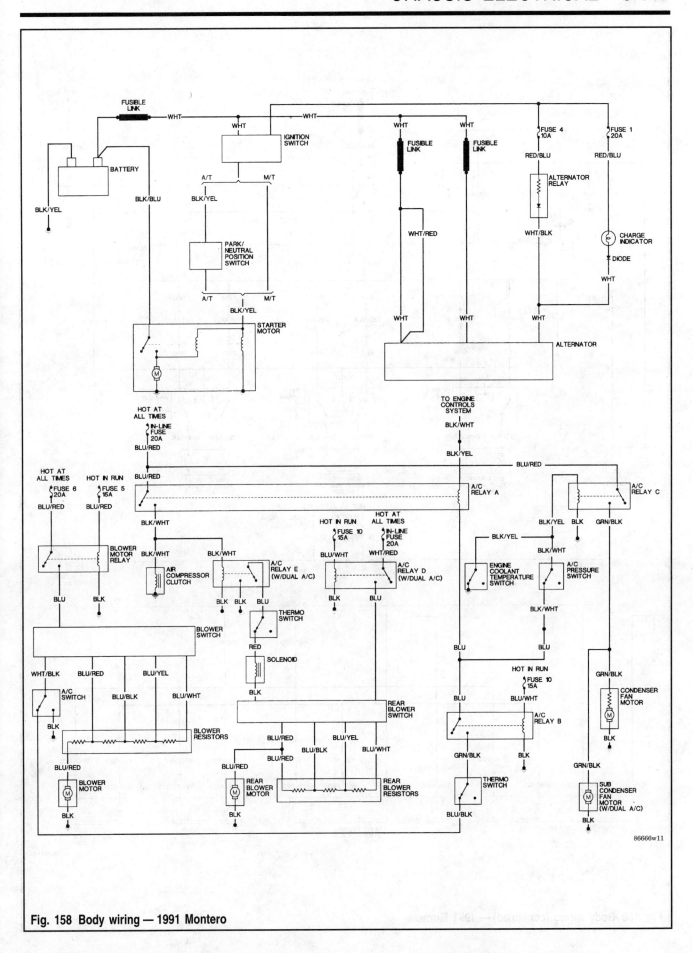

Fig. 158 Body wiring — 1991 Montero

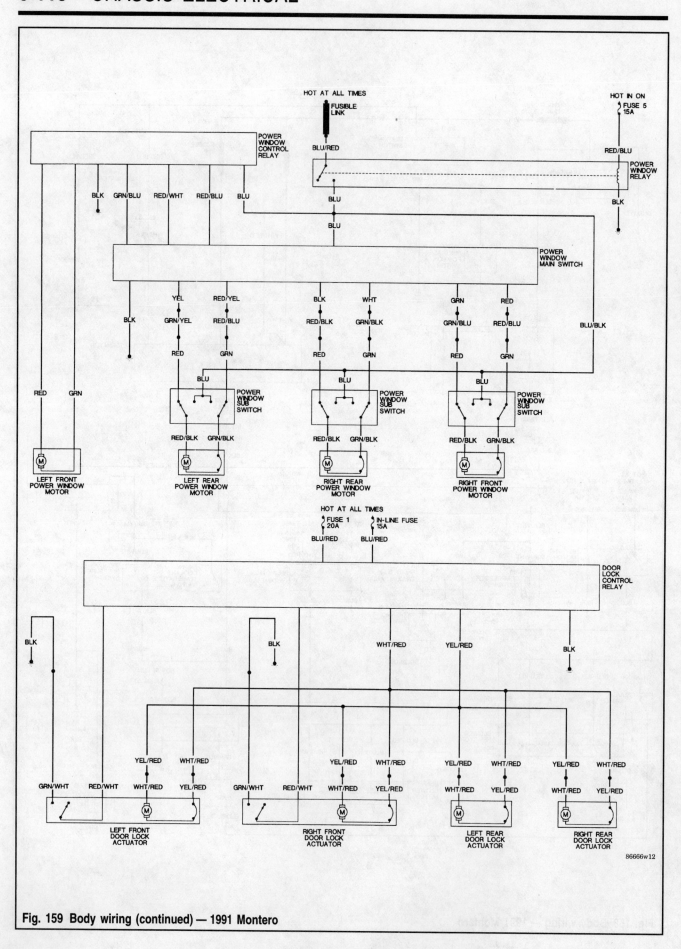

Fig. 159 Body wiring (continued) — 1991 Montero

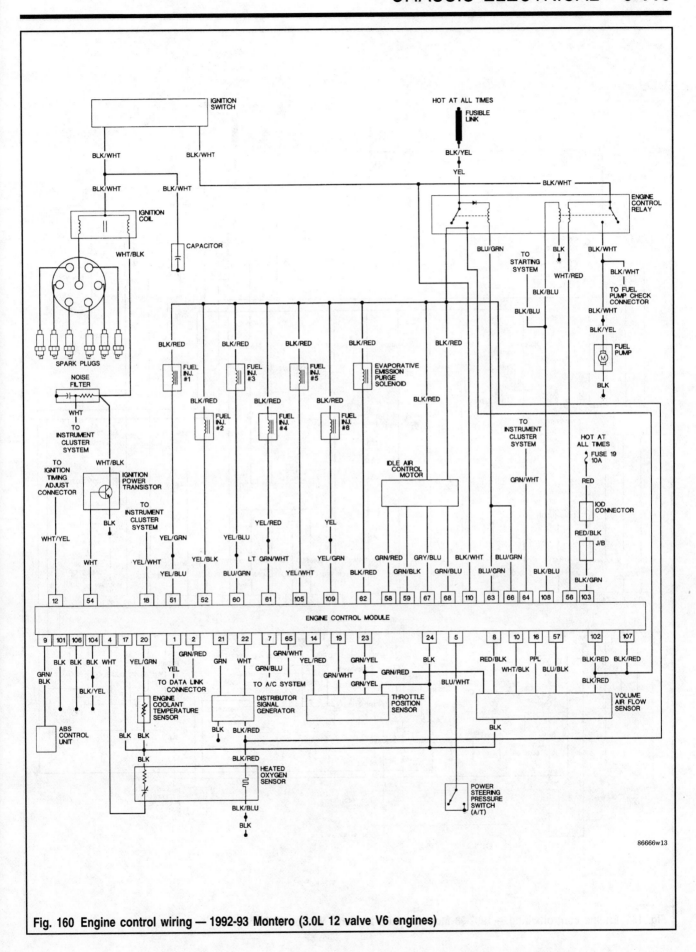

Fig. 160 Engine control wiring — 1992-93 Montero (3.0L 12 valve V6 engines)

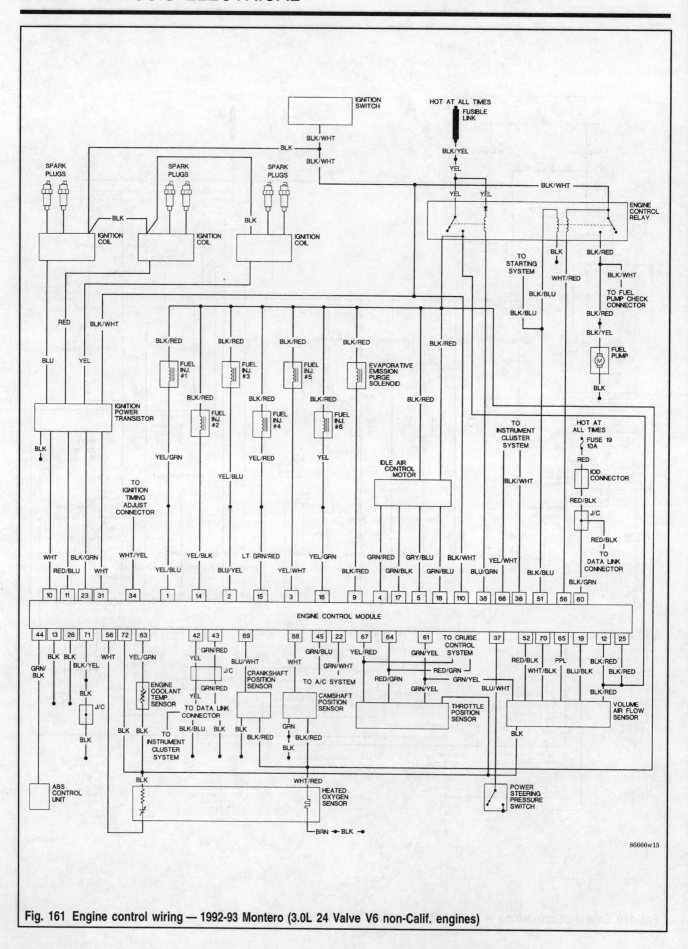

Fig. 161 Engine control wiring — 1992-93 Montero (3.0L 24 Valve V6 non-Calif. engines)

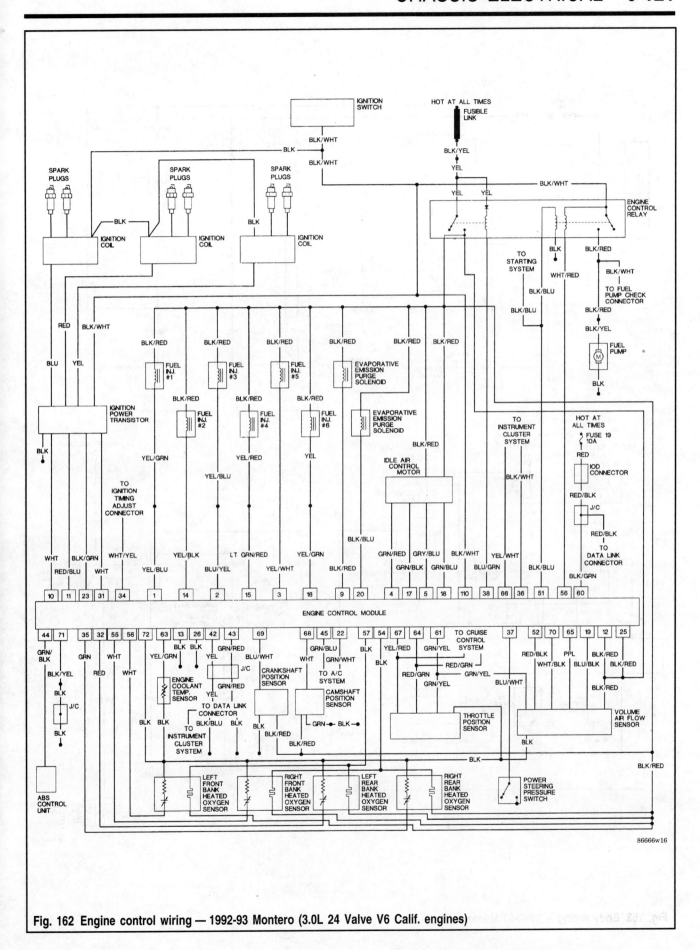

Fig. 162 Engine control wiring — 1992-93 Montero (3.0L 24 Valve V6 Calif. engines)

6-122 CHASSIS ELECTRICAL

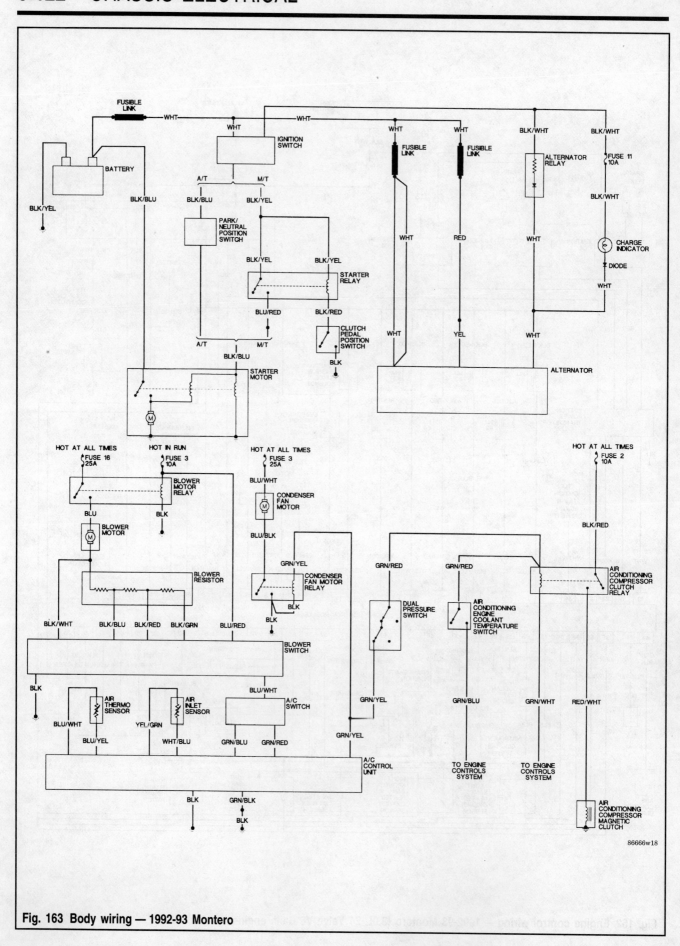

Fig. 163 Body wiring — 1992-93 Montero

CHASSIS ELECTRICAL 6-123

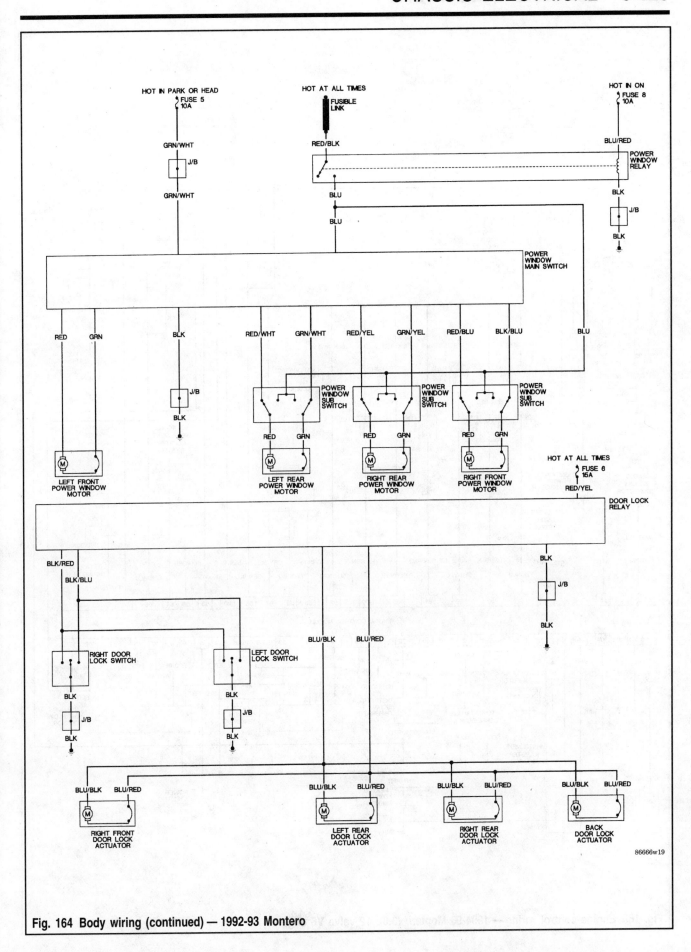

Fig. 164 Body wiring (continued) — 1992-93 Montero

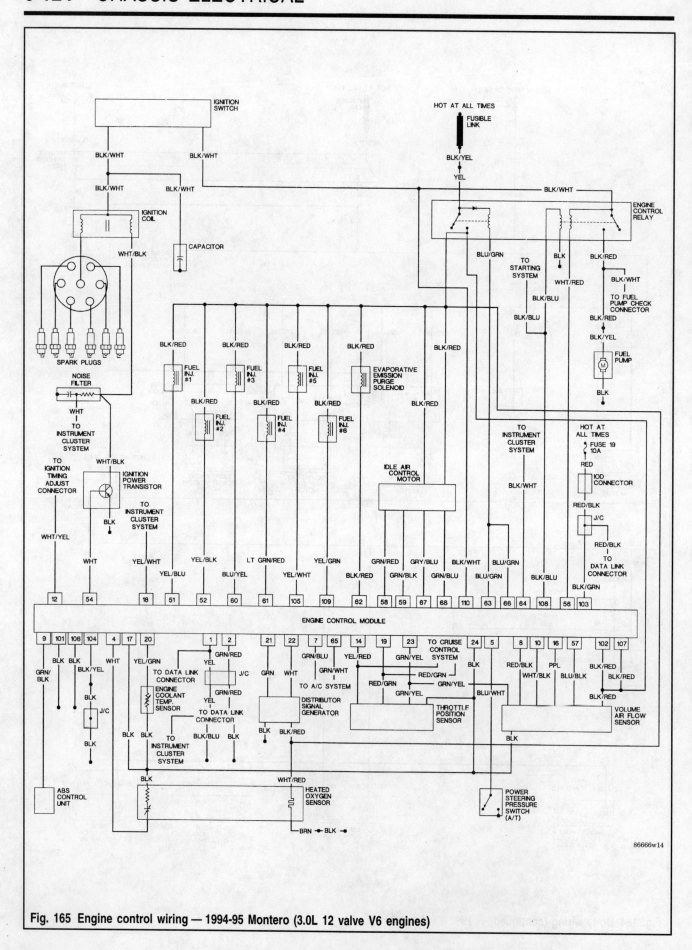

Fig. 165 Engine control wiring — 1994-95 Montero (3.0L 12 valve V6 engines)

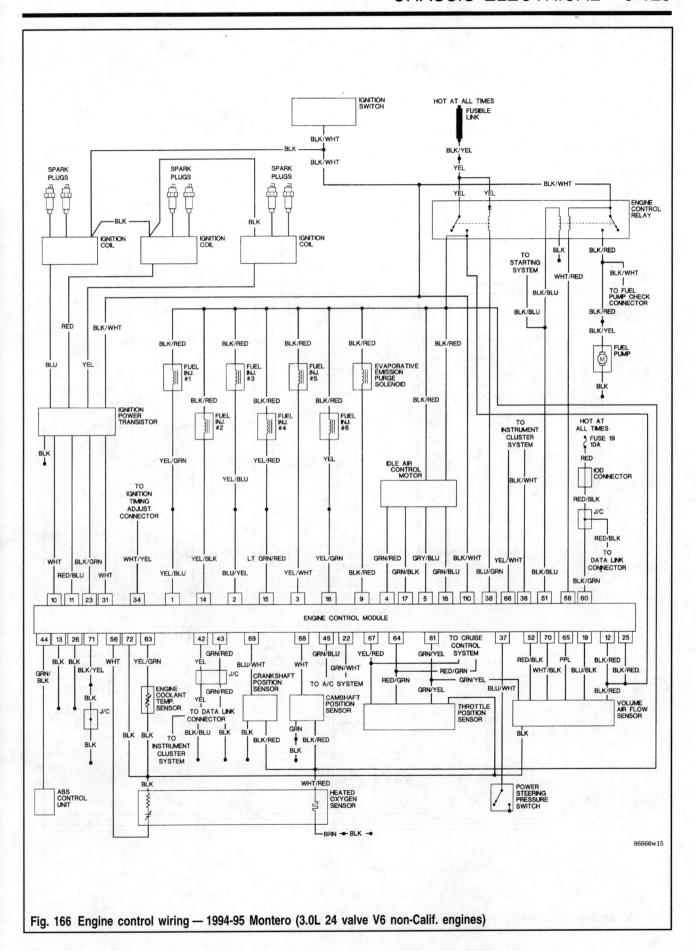

Fig. 166 Engine control wiring — 1994-95 Montero (3.0L 24 valve V6 non-Calif. engines)

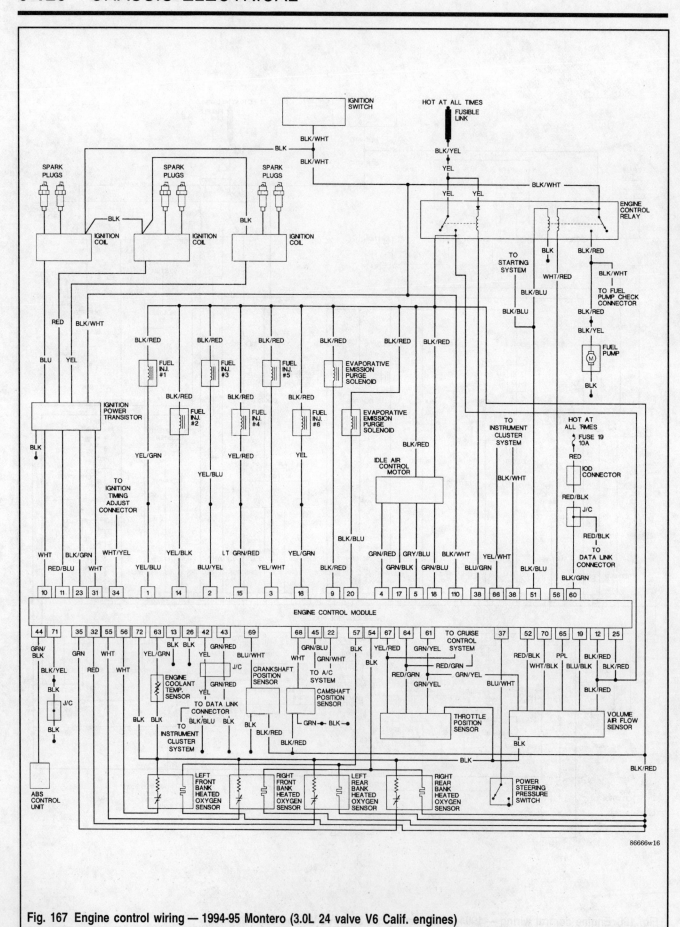

Fig. 167 Engine control wiring — 1994-95 Montero (3.0L 24 valve V6 Calif. engines)

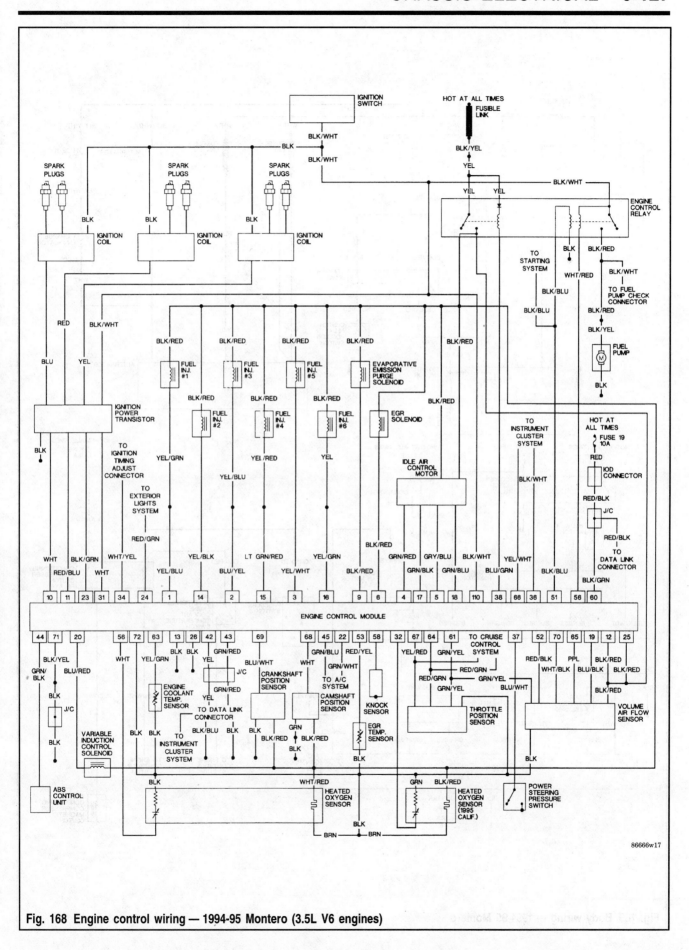

Fig. 168 Engine control wiring — 1994-95 Montero (3.5L V6 engines)

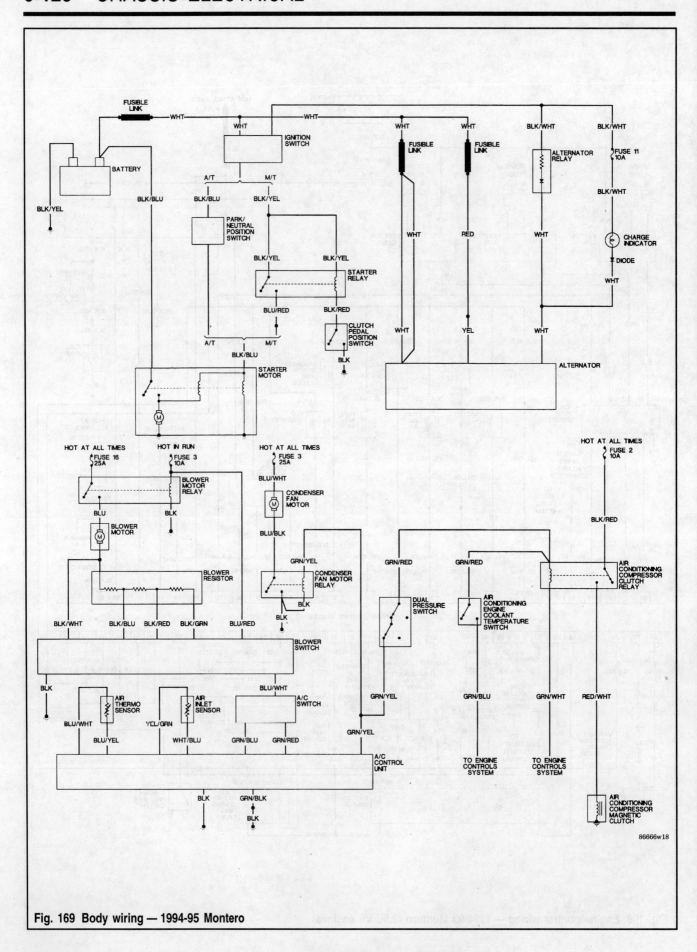

Fig. 169 Body wiring — 1994-95 Montero

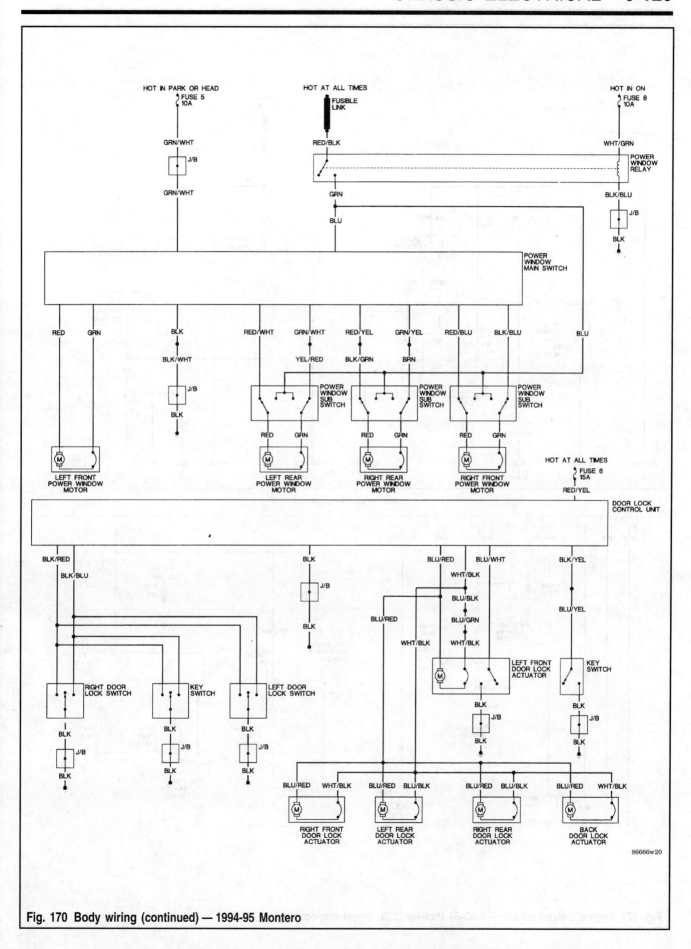

Fig. 170 Body wiring (continued) — 1994-95 Montero

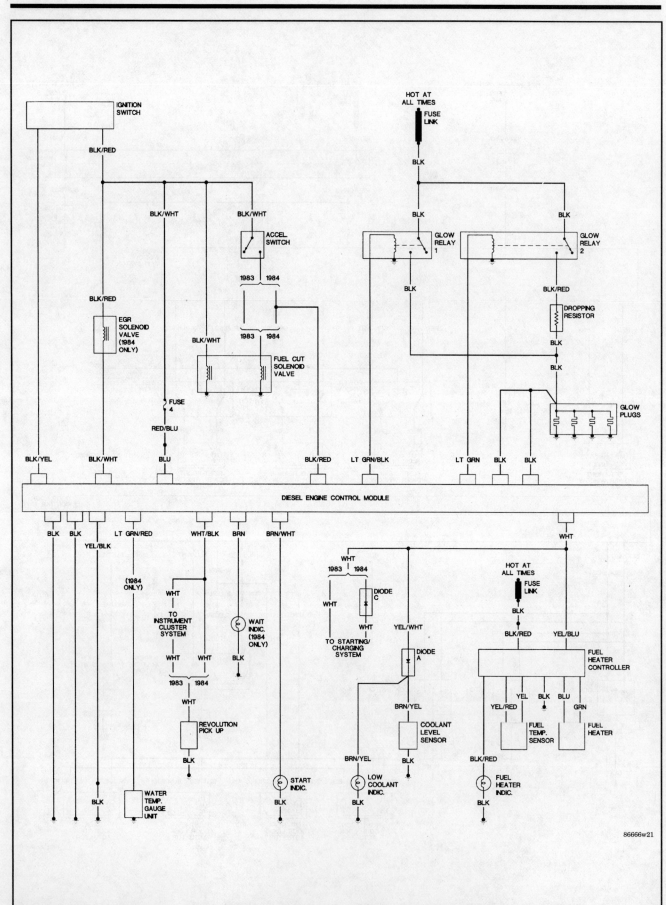

Fig. 171 Engine control wiring — 1983-84 Pick-up (2.3L diesel engines)

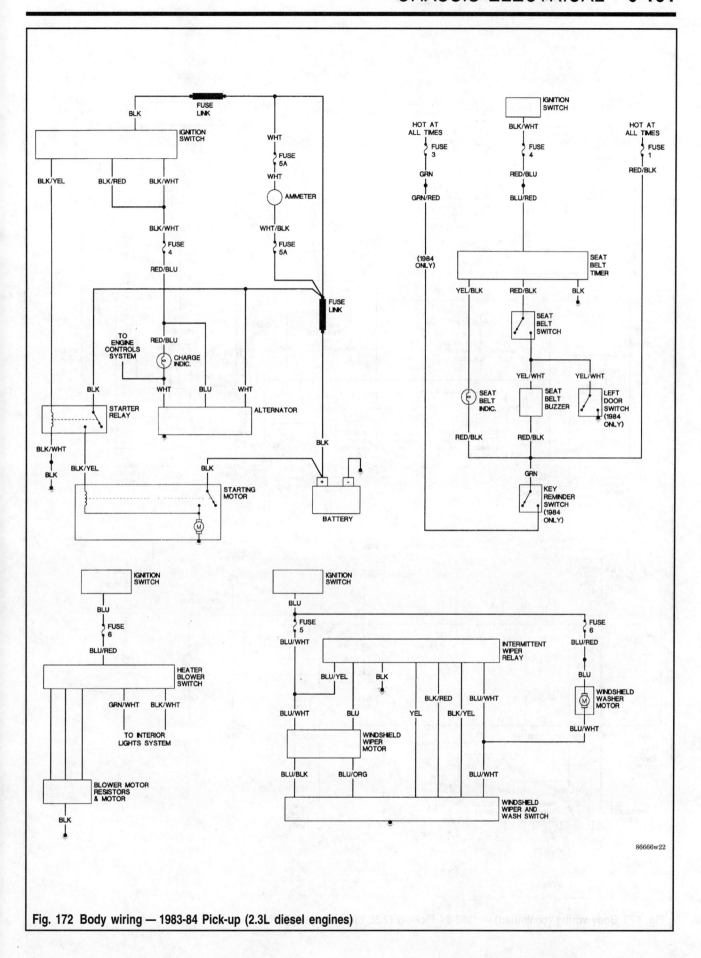

Fig. 172 Body wiring — 1983-84 Pick-up (2.3L diesel engines)

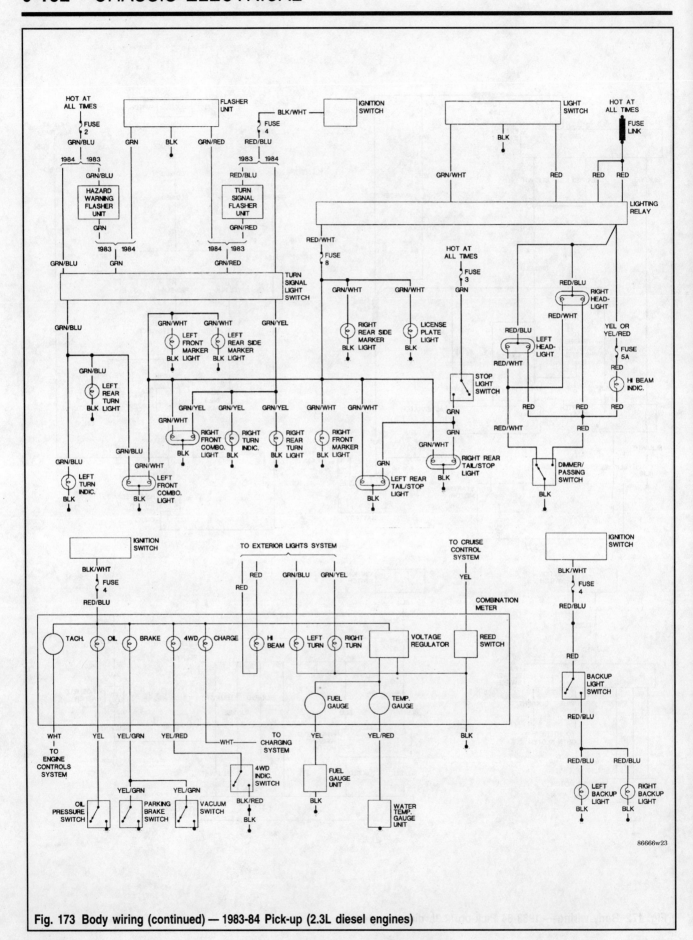

Fig. 173 Body wiring (continued) — 1983-84 Pick-up (2.3L diesel engines)

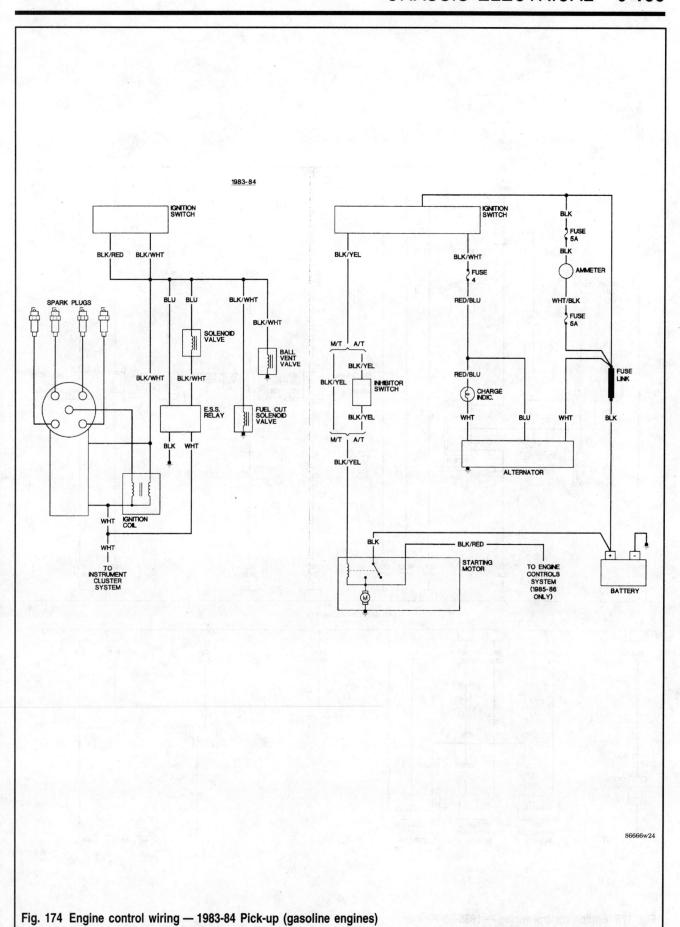

Fig. 174 Engine control wiring — 1983-84 Pick-up (gasoline engines)

6-134 CHASSIS ELECTRICAL

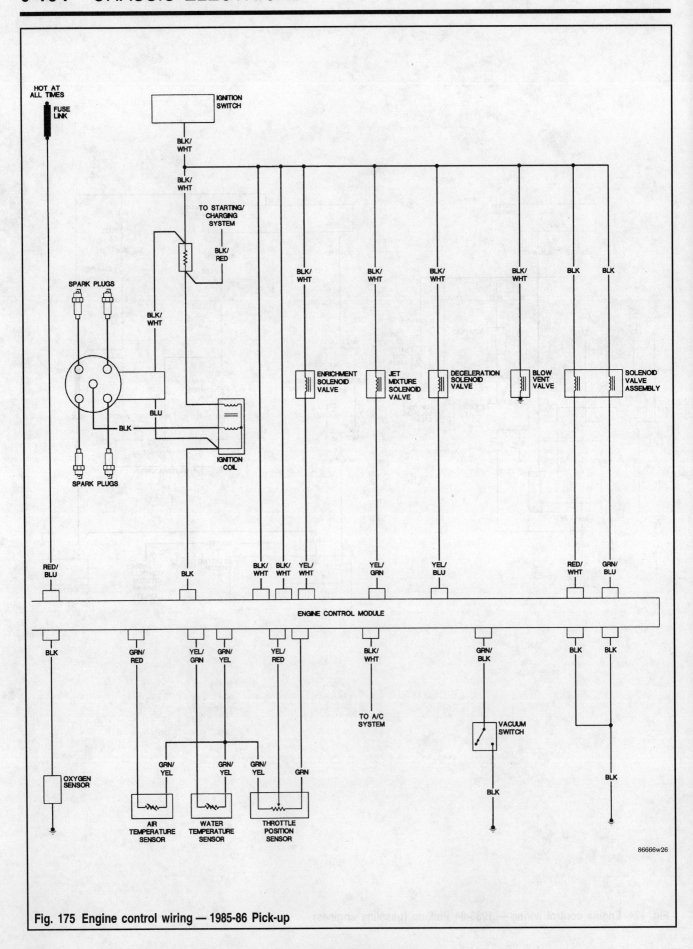

Fig. 175 Engine control wiring — 1985-86 Pick-up

CHASSIS ELECTRICAL 6-135

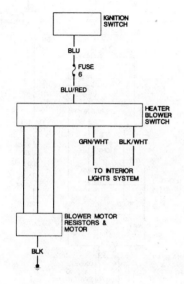
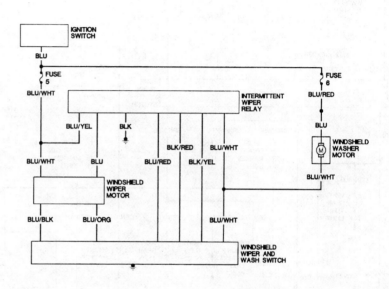

Fig. 176 Body wiring — 1983-86 Pick-up (gasoline engines)

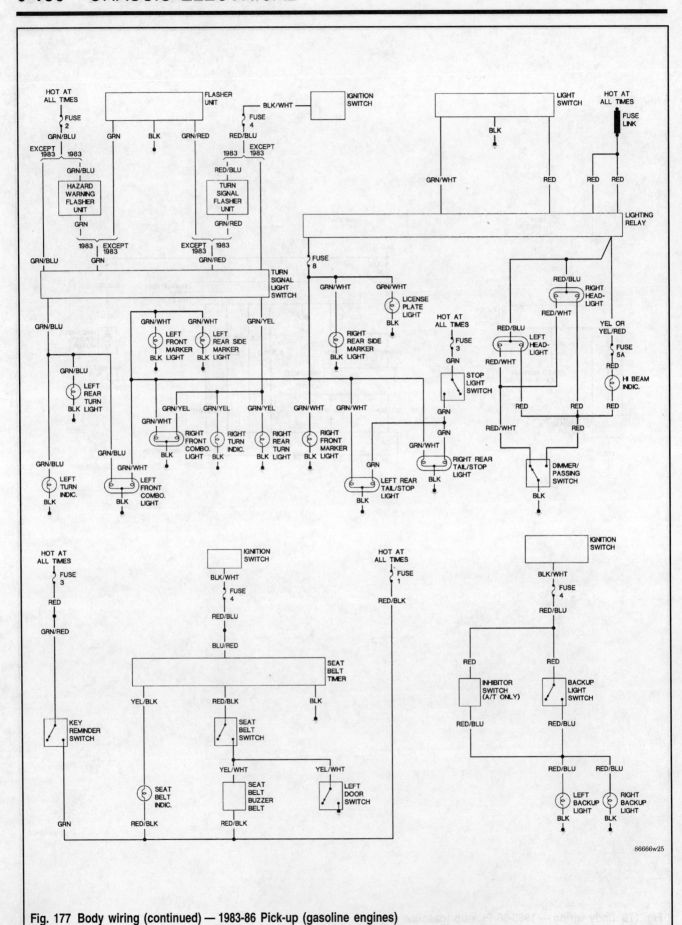

Fig. 177 Body wiring (continued) — 1983-86 Pick-up (gasoline engines)

CHASSIS ELECTRICAL 6-137

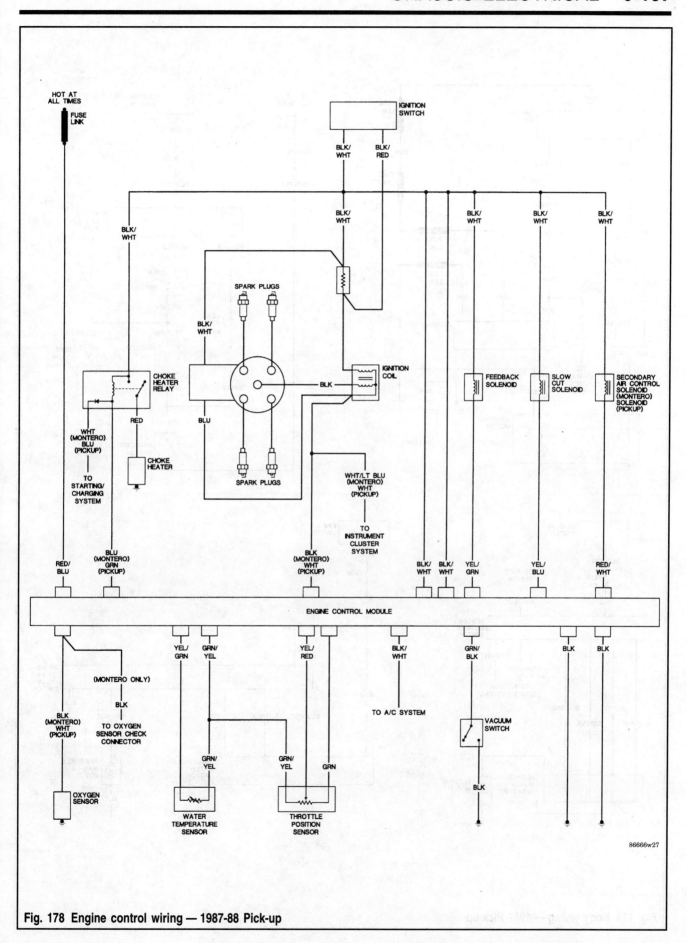

Fig. 178 Engine control wiring — 1987-88 Pick-up

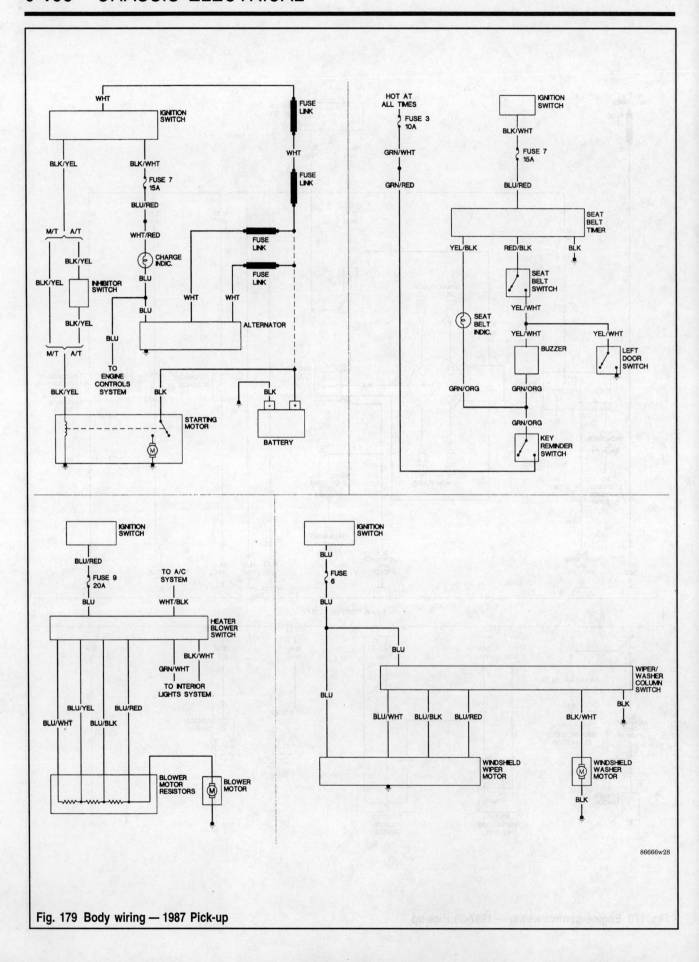

Fig. 179 Body wiring — 1987 Pick-up

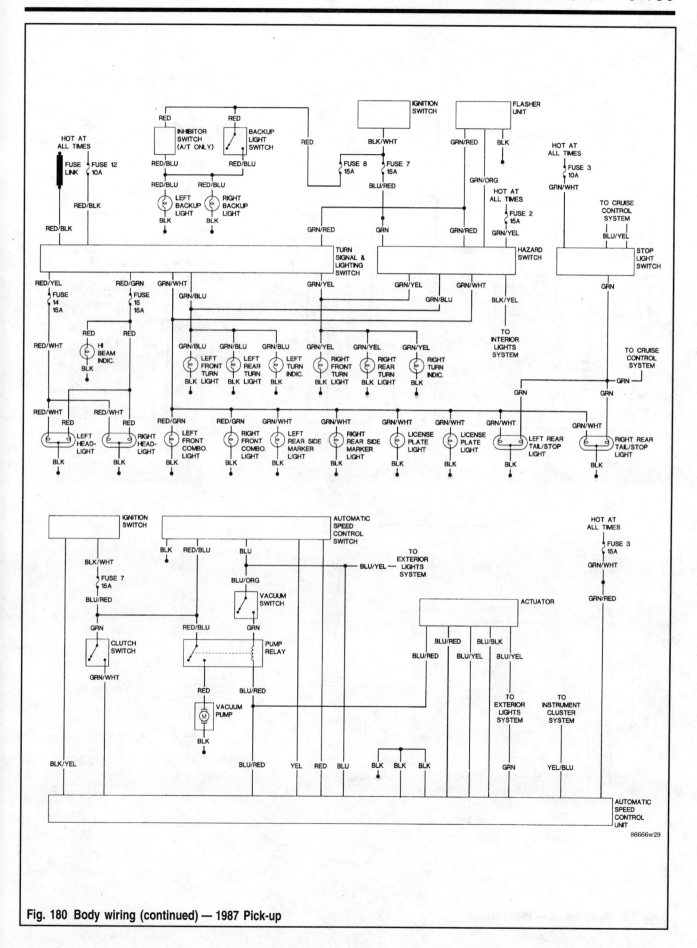

Fig. 180 Body wiring (continued) — 1987 Pick-up

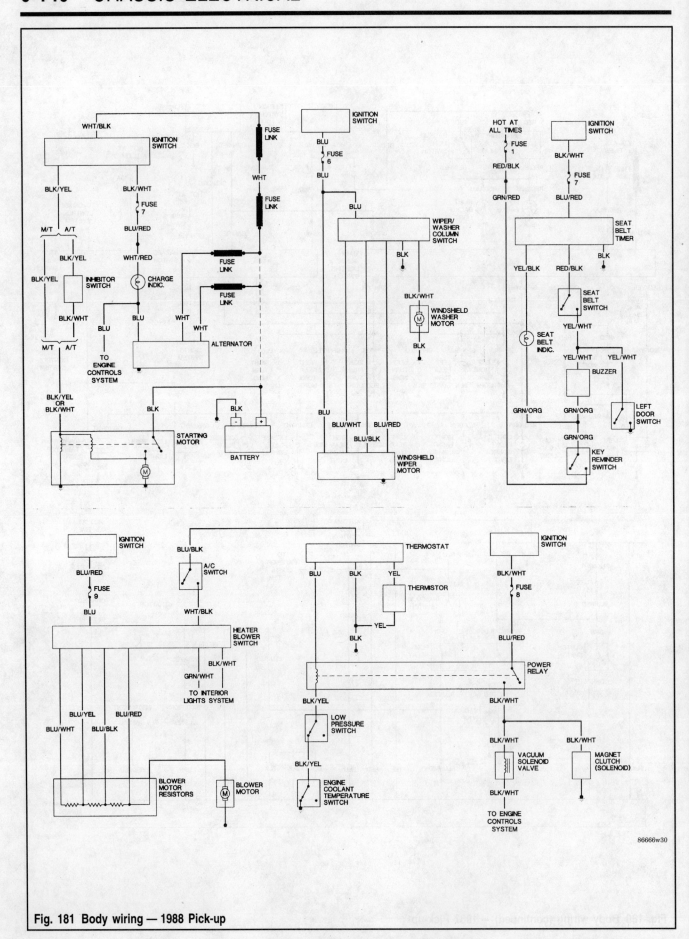

Fig. 181 Body wiring — 1988 Pick-up

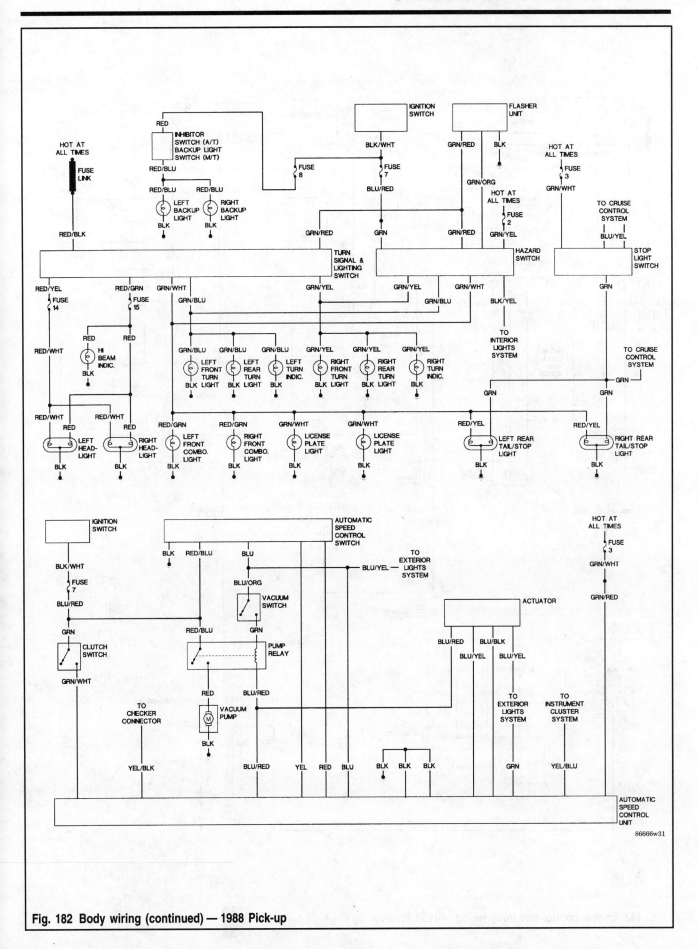

Fig. 182 Body wiring (continued) — 1988 Pick-up

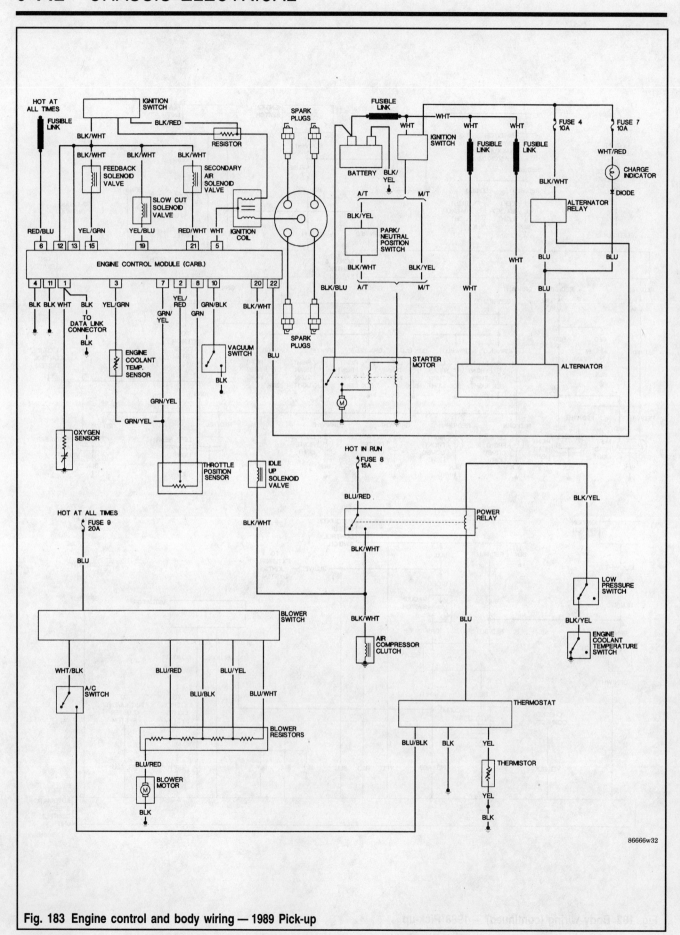

Fig. 183 Engine control and body wiring — 1989 Pick-up

CHASSIS ELECTRICAL 6-143

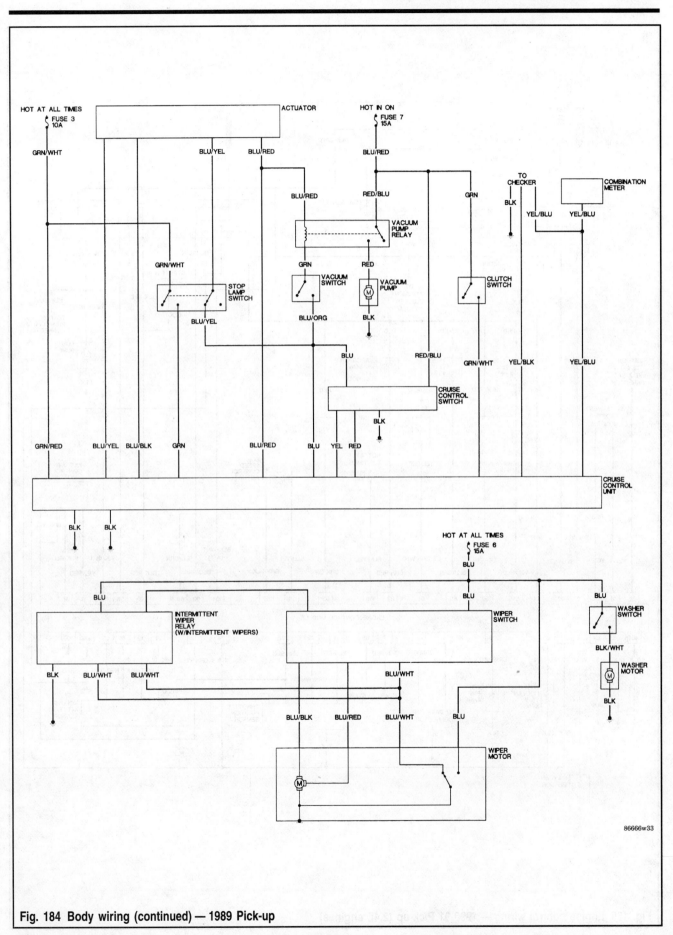

Fig. 184 Body wiring (continued) — 1989 Pick-up

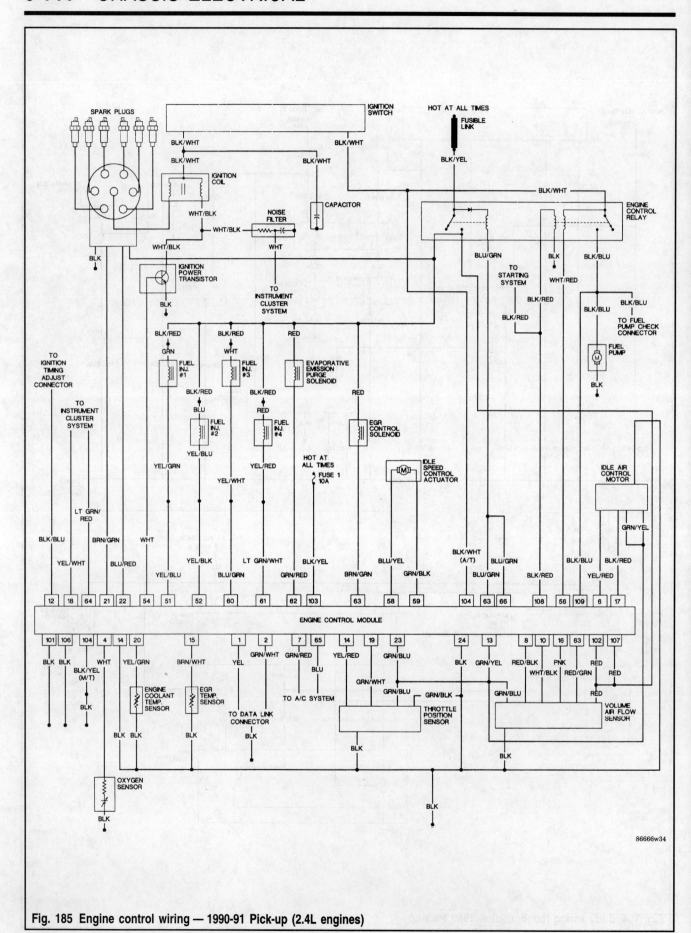

Fig. 185 Engine control wiring — 1990-91 Pick-up (2.4L engines)

CHASSIS ELECTRICAL 6-145

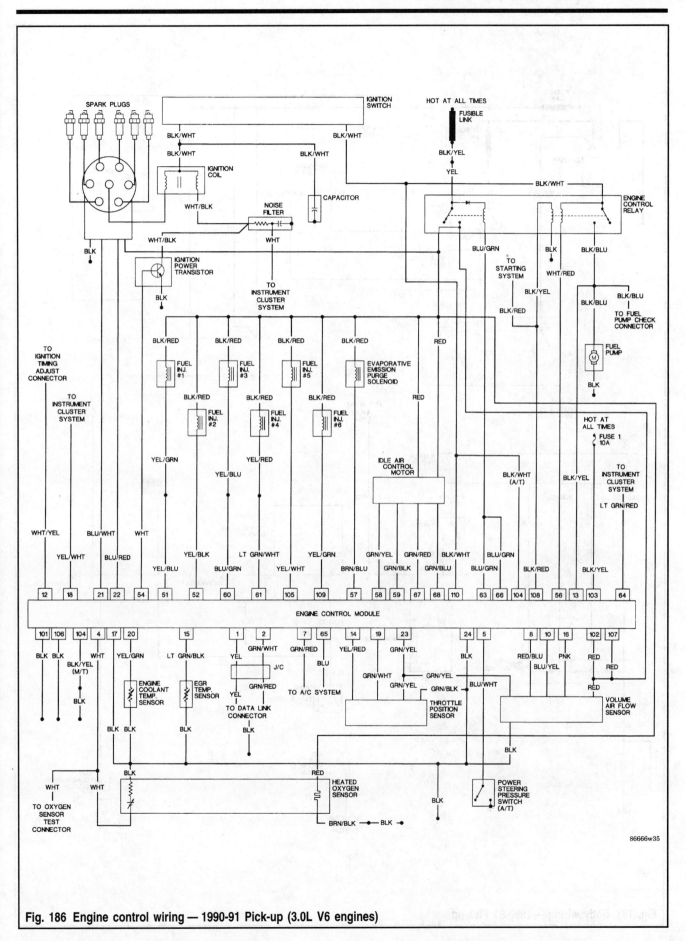

Fig. 186 Engine control wiring — 1990-91 Pick-up (3.0L V6 engines)

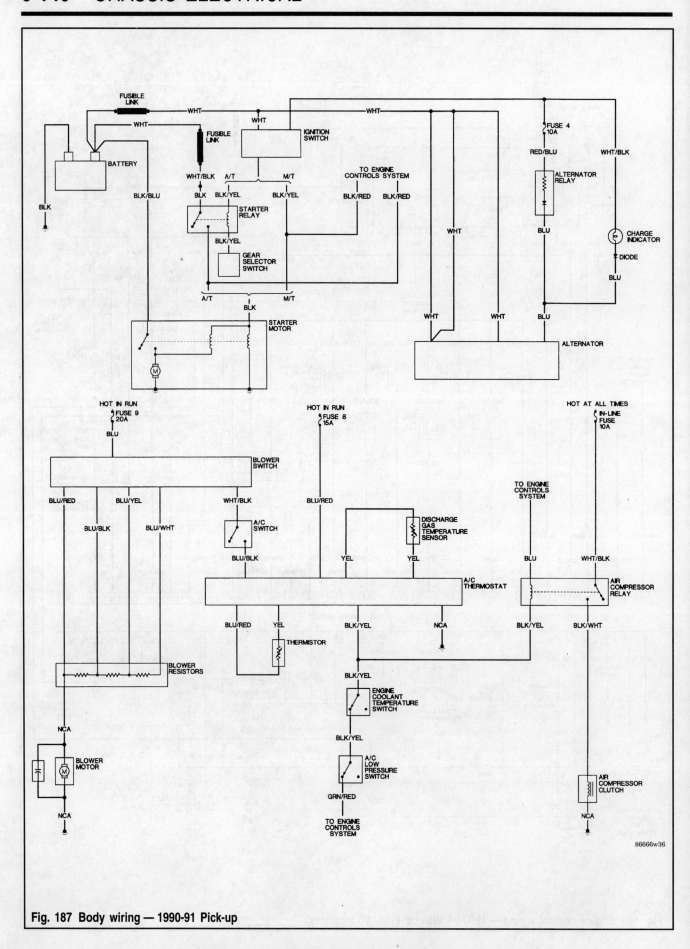

Fig. 187 Body wiring — 1990-91 Pick-up

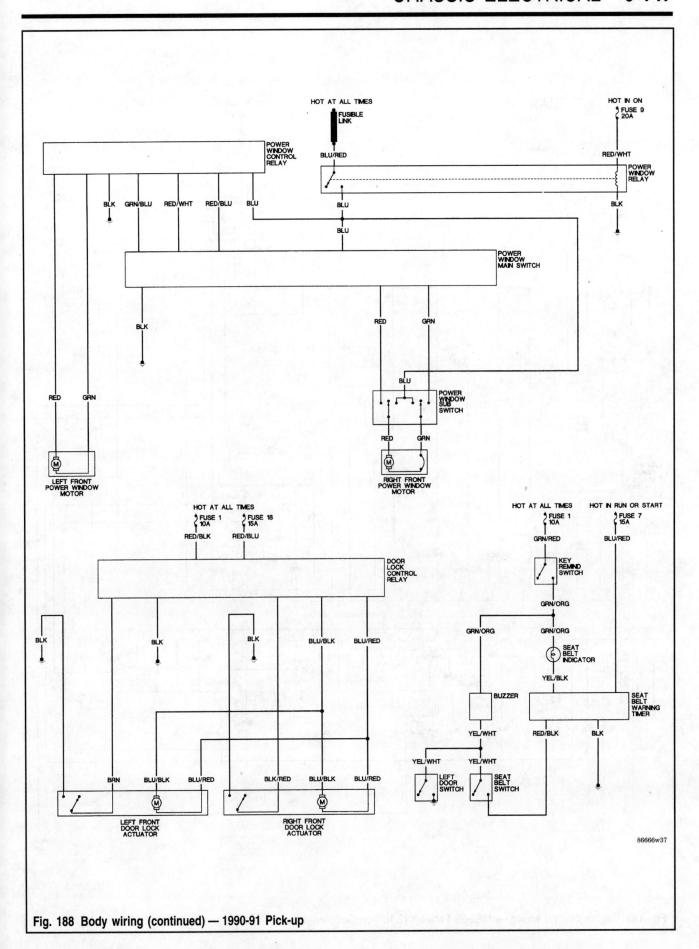

Fig. 188 Body wiring (continued) — 1990-91 Pick-up

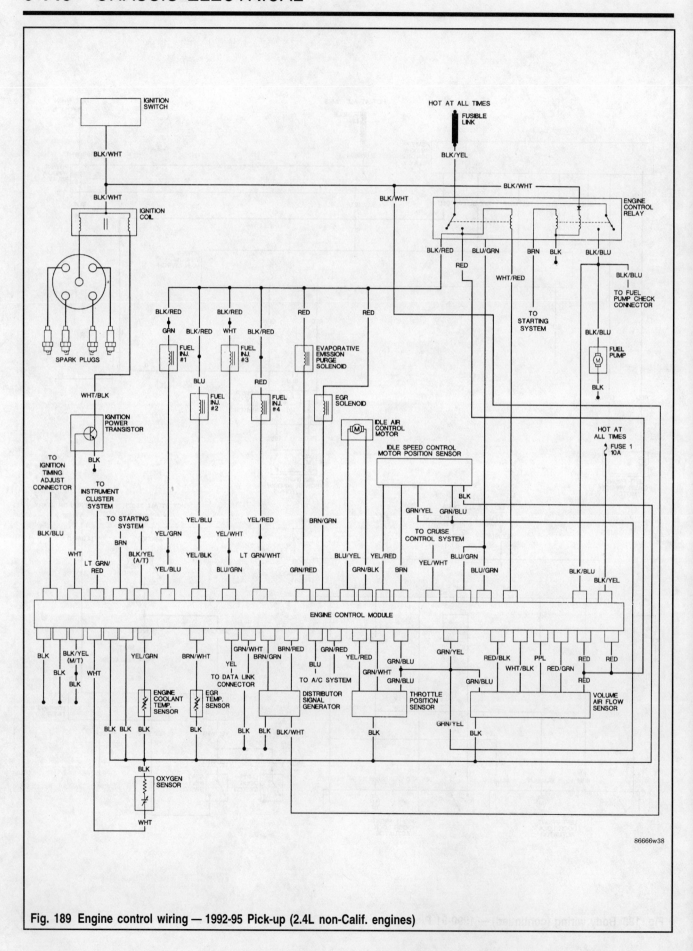

Fig. 189 Engine control wiring — 1992-95 Pick-up (2.4L non-Calif. engines)

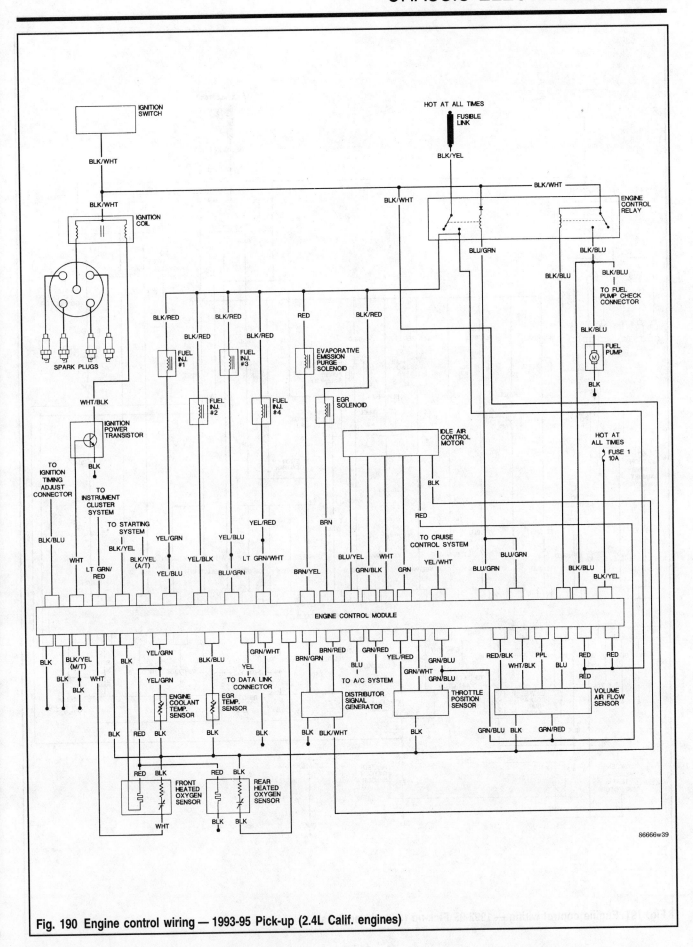

Fig. 190 Engine control wiring — 1993-95 Pick-up (2.4L Calif. engines)

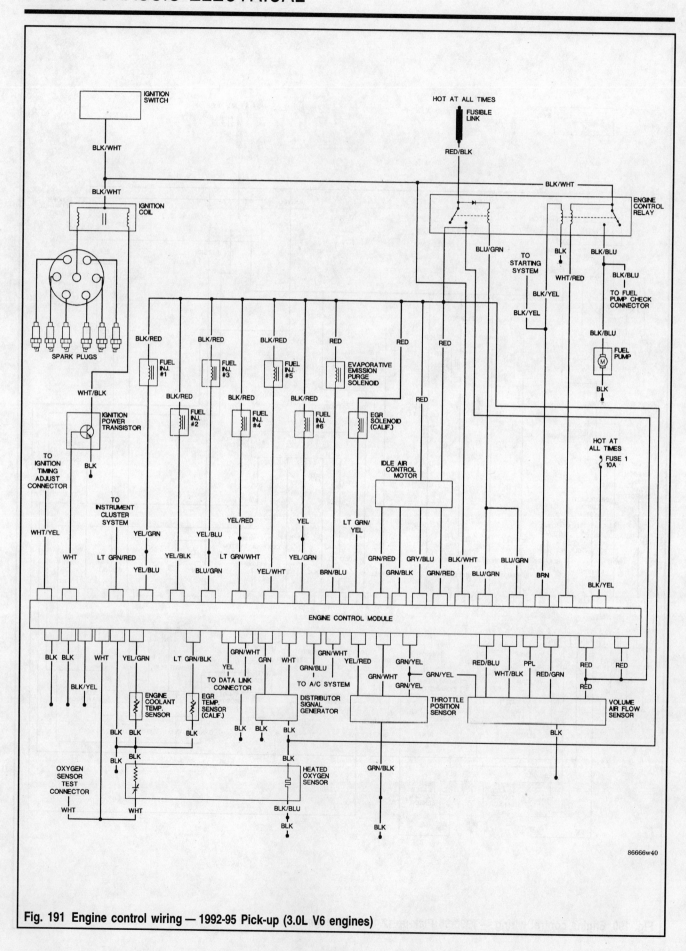

Fig. 191 Engine control wiring — 1992-95 Pick-up (3.0L V6 engines)

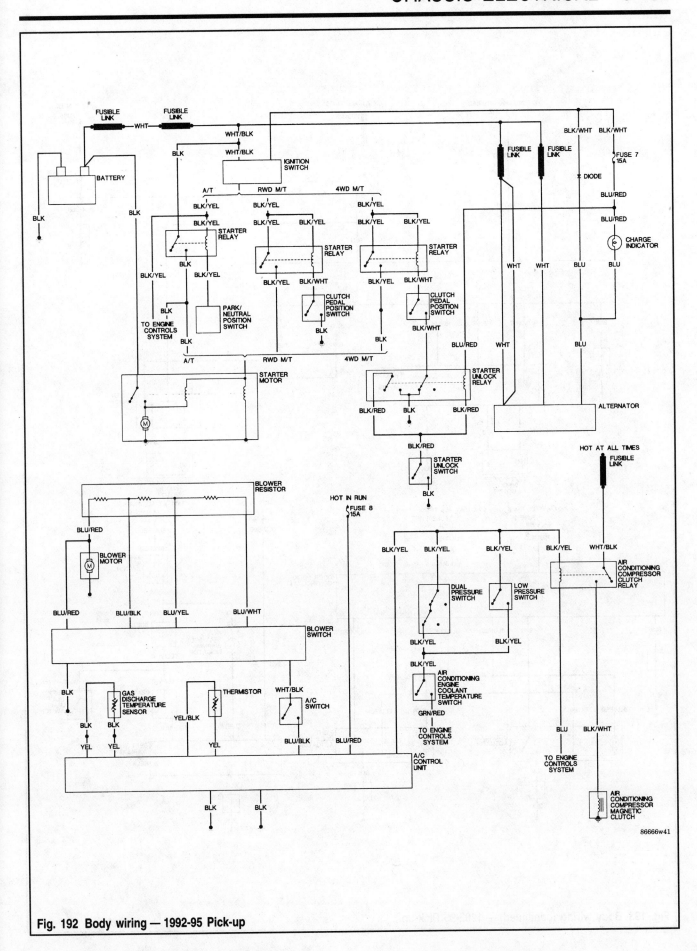

Fig. 192 Body wiring — 1992-95 Pick-up

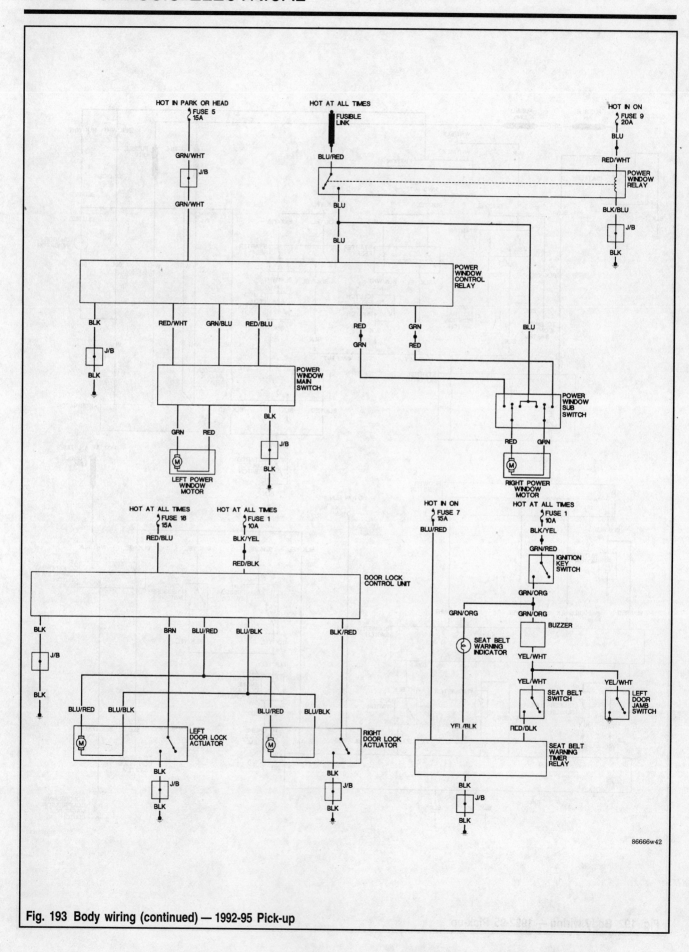

Fig. 193 Body wiring (continued) — 1992-95 Pick-up

AUTOMATIC TRANSMISSION
  ADJUSTMENTS 7-42
  EXTENSION HOUSING SEAL 7-55
  FLUID, PAN AND FILTER
    SERVICE 7-36
  IDENTIFICATION 7-35
  PARK/NEUTRAL POSITION
    (INHIBITOR) SWITCH 7-46
  SHIFTER LINKAGE 7-48
  TRANSMISSION 7-57
CLUTCH
  ADJUSTMENTS 7-21
  CLUTCH LINKAGE 7-26
  CLUTCH MASTER CYLINDER 7-33
  CLUTCH PEDAL 7-21
  CLUTCH RELEASE CYLINDER 7-34
  DISC AND PRESSURE PLATE 7-32
DRIVELINE
  CENTER BEARING 7-73
  FRONT DRIVESHAFT AND U-
    JOINTS 7-70
  REAR DRIVESHAFT AND U-
    JOINTS 7-70
**FRONT DRIVE AXLE — 4-WHEEL
  DRIVE VEHICLES**
  AUTOMATIC LOCKING HUBS 7-76
  CV-JOINTS 7-86
  FRONT AXLE SHAFT, BEARING AND
    SEAL 7-82
  FRONT DIFFERENTIAL
    CARRIER 7-86
  IDENTIFICATION 7-75
  MANUAL LOCKING HUBS 7-75
  PINION SEAL 7-91
MANUAL TRANSMISSION
  ADJUSTMENTS 7-10
  BACK-UP LIGHT SWITCH 7-14
  EXTENSION HOUSING SEAL 7-14
  GEARSHIFT LEVER ASSEMBLY 7-12
  IDENTIFICATION 7-2
  TRANSMISSION 7-15
REAR AXLE
  DETERMINING AXLE RATIO 7-91
  DIFFERENTIAL CARRIER 7-98
  IDENTIFICATION 7-91
  PINION SEAL 7-99
  REAR AXLE HOUSING 7-101
  REAR AXLE SHAFTS, BEARINGS
    AND SEALS 7-92
  UNDERSTANDING DRIVE
    AXLES 7-91
SPECIFICATIONS CHARTS
  AUTOMATIC TRANSMISSION
    IDENTIFICATION 7-35
  MANUAL TRANSMISSION
    IDENTIFICATION 7-2
  TORQUE SPECIFICATIONS 7-105
TRANSFER CASE
  ADJUSTMENTS 7-68
  FRONT AND REAR OUTPUT SHAFT
    SEALS 7-68

IDENTIFICATION 7-68
TRANSFER CASE 7-68

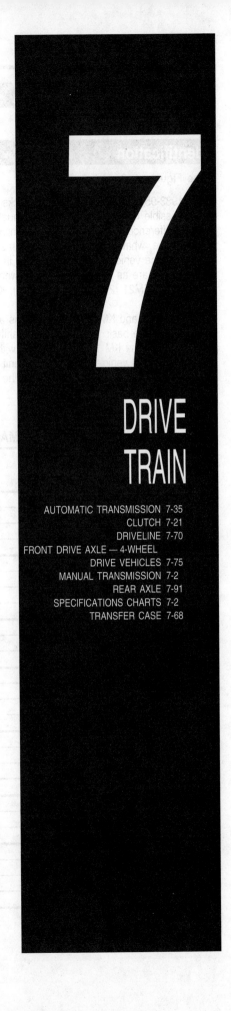

7
DRIVE TRAIN

AUTOMATIC TRANSMISSION 7-35
CLUTCH 7-21
DRIVELINE 7-70
FRONT DRIVE AXLE — 4-WHEEL
DRIVE VEHICLES 7-75
MANUAL TRANSMISSION 7-2
REAR AXLE 7-91
SPECIFICATIONS CHARTS 7-2
TRANSFER CASE 7-68

7-2 DRIVE TRAIN

MANUAL TRANSMISSION

Identification

▶ See Figures 1, 2, 3, 4, 5, 6, 7 and 8

The 1983-95 Monteros and Pick-up trucks came equipped with a possible number of 6 different manual transmissions, with the differences being whether the vehicle has 2- or 4-wheel drive, what engine the vehicle came equipped with, or what years the vehicle was manufactured in. The six transmissions are as follows: KM 130 (2-wheel drive/2WD), KM 132 (2WD), R5M21 (2WD), V5M21 (4WD), KM 145 (4WD), V5MT1 (4WD).

The KM 130 and KM 132 transmissions are internally identical, with the basic difference in the shifter assemblies. The KM 145 is the KM 132 transmission with a transfer case attached for 4WD vehicles. The transfer unit is removable. These transmissions incorporate 5 (4 for the KM 130) forward gears (the fifth gear being an overdrive gear) and a reverse gear. They are fully synchronized in all forward gears and reverse. All gear changes are accomplished with the assistance of synchronizer sleeves. Reverse gear uses a reverse idler gear which is in constant mesh with the countershaft gear.

The top mounted shifter operates the shift rails through a set of shift forks. Shift forks mounted on the rails operate the synchronizer sleeves that allow shifting of the 1/2, 3/4 and overdrive/reverse.

A shift interlock system, located in the side of the transmission case, prevents the shift rails from engaging 2 gears at the same time.

The transmission assembly is composed of 5 major components: a front bearing retainer, a transmission case, a transfer case adapter and a control housing, all of which are cast aluminum. The fifth component is the under cover which is made from stamped aluminum.

MANUAL TRANSMISSION IDENTIFICATION

Year	Model	Engine	Transmission Identification	Drive Train
1983-85	Montero	2.6L	KM145	4WD
	Pick-up	2.0L	KM130	RWD
	Pick-up	2.0L/2.6L	KM132	RWD
	Pick-up	2.3L/2.6L	KM145	4WD
1986	Montero	2.6L	KM145	4WD
	Pick-up	2.0L	KM130	RWD
	Pick-up	2.0L/2.6L	KM132	RWD
	Pick-up	2.6L	KM145	4WD
1987	Montero	2.6L	KM145	4WD
	Pick-up	2.0L/2.6L	KM132	RWD
	Pick-up	2.6L	Km145	4WD
1988-89	Montero	2.6L	KM145	4WD
	Montero	3.0L	V5MT1	4WD
	Pick-up	2.0L/2.6L	KM132	RWD
	Pick-up	2.6L	Km145	4WD
1990	Montero	2.6L	V5M21	4WD
	Montero	3.0L	V5MT1	4WD
	Pick-up	2.4L	R5M21	RWD
	Pick-up	3.0L	V5MT1	4WD
1991-93	Montero	3.0L	V5MT1	4WD
	Pick-up	2.4L	R5M21	RWD
	Pick-up	3.0L	V5MT1	4WD
1994-95	Montero	3.0L	V5MT1	4WD
	Montero	3.5L	V5MT1	4WD
	Pick-up	2.4L	R5M21	2WD
	Pick-up	3.0L	V5MT1	4WD

RWD - Rear Wheel Drive
4WD - Four Wheel Drive

DRIVE TRAIN 7-3

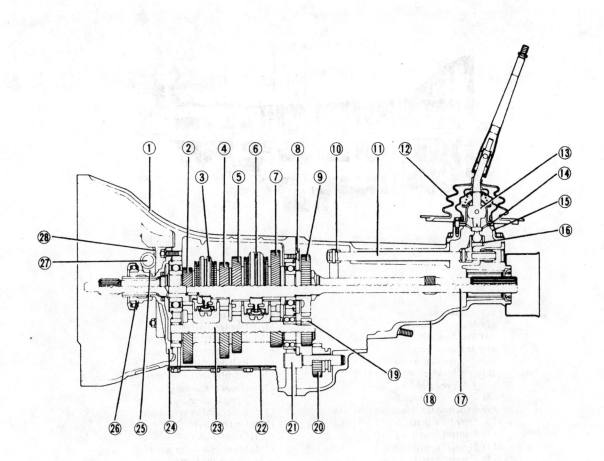

(1) Transmission case
(2) Main drive pinion
(3) Synchronizer assembly (3-4)
(4) 3rd speed gear
(5) 2nd speed gear
(6) Synchronizer assembly (1-2)
(7) 1st speed gear
(8) Rear bearing retainer
(9) Reverse gear
(10) Control finger
(11) Control shaft
(12) Control lever cover
(13) Control lever assembly
(14) Stopper plate
(15) Control housing
(16) Change shifter
(17) Mainshaft
(18) Extension housing
(19) Counter reverse gear
(20) Reverse idler gear
(21) Reverse idler gear shaft
(22) Under cover
(23) Counter gear
(24) Front bearing retainer
(25) Clutch shift arm
(26) Release bearing carrier
(27) Clutch control shaft
(28) Return spring

Fig. 1 Schematic side view of the four speed KM 130 — 1983-86 Monteros and Pick-ups

7-4 DRIVE TRAIN

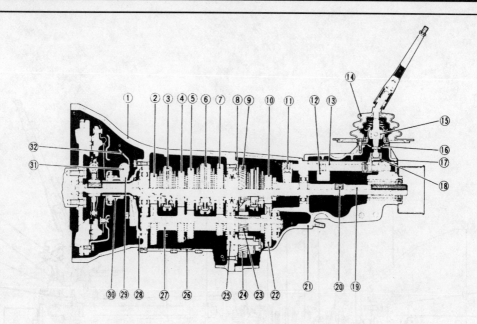

(1) Transmission case
(2) Main drive pinion
(3) Synchronizer assembly (3-4)
(4) 3rd speed gear
(5) 2nd speed gear
(6) Synchronizer assembly (1-2)
(7) 1st speed gear
(8) Rear bearing retainer
(9) Synchronizer assembly (overdrive)
(10) Overdrive gear
(11) Control finger
(12) Neutral return finger
(13) Control shaft
(14) Control lever cover
(15) Control lever assembly
(16) Stopper plate
(17) Control housing
(18) Change shifter
(19) Mainshaft
(20) Speedometer drive gear
(21) Extension housing
(22) Counter overdrive gear
(23) Counter reverse gear
(24) Reverse idler gear
(25) Reverse idler gear shaft
(26) Under cover
(27) Counter gear
(28) Front bearing retainer
(29) Clutch shift arm
(30) Release bearing carrier
(31) Clutch control shaft
(32) Return spring

Fig. 2 Internal component locations of the five speed KM 132 — 1983-89 Monteros and Pick-ups

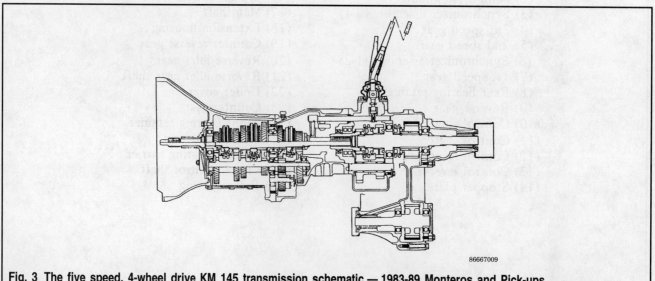

Fig. 3 The five speed, 4-wheel drive KM 145 transmission schematic — 1983-89 Monteros and Pick-ups

DRIVE TRAIN 7-5

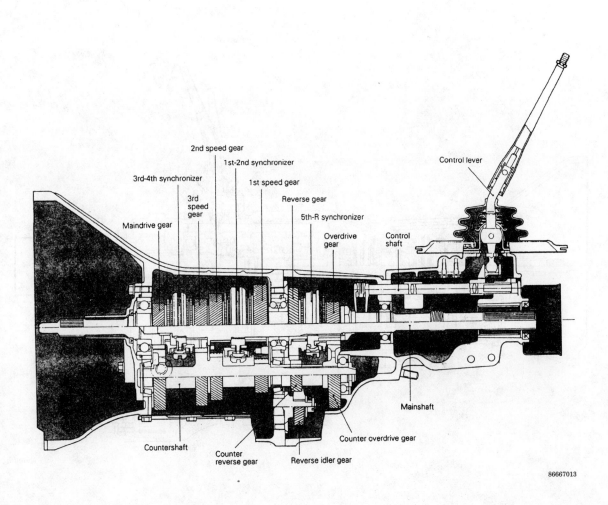

Fig. 4 An internal schematic of the R5M21, which was used on 1990-95 2-wheel drive Pick-ups

7-6 DRIVE TRAIN

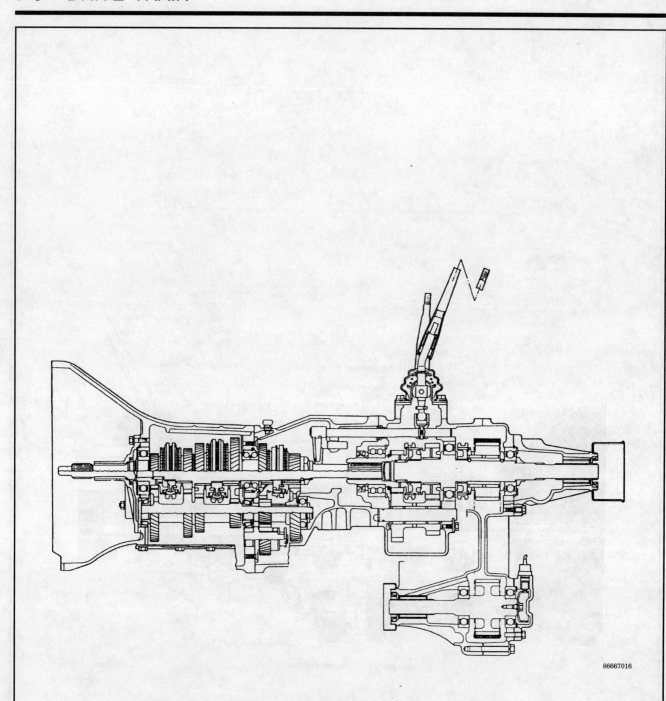

Fig. 5 The 4-wheel drive, 5 speed V5M21 manual transmission was only used on the 1990 2.6L engine-equipped Monteros

DRIVE TRAIN 7-7

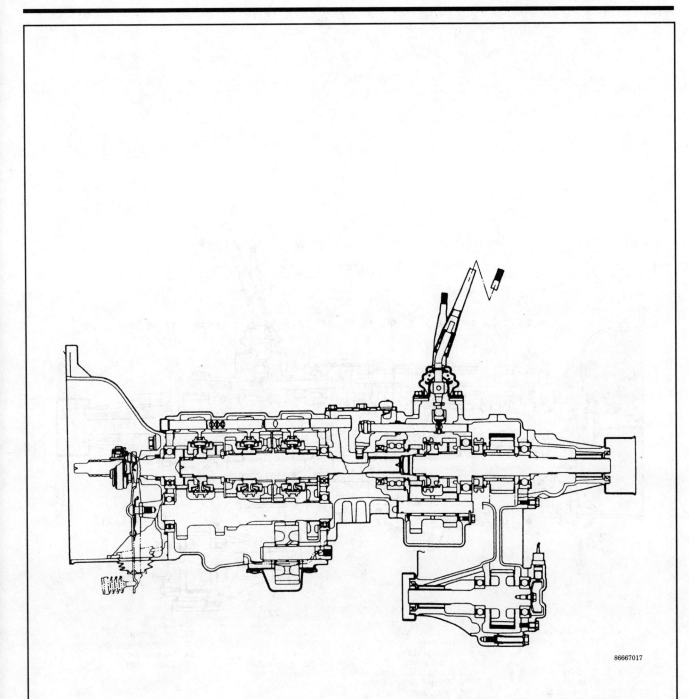

Fig. 6 The V5MT1 series manual transmissions were used on 4-wheel drive Pick-ups and Monteros equipped with 3.0L engines from 1989-95 — this model was used on 1989-91 vehicles

7-8 DRIVE TRAIN

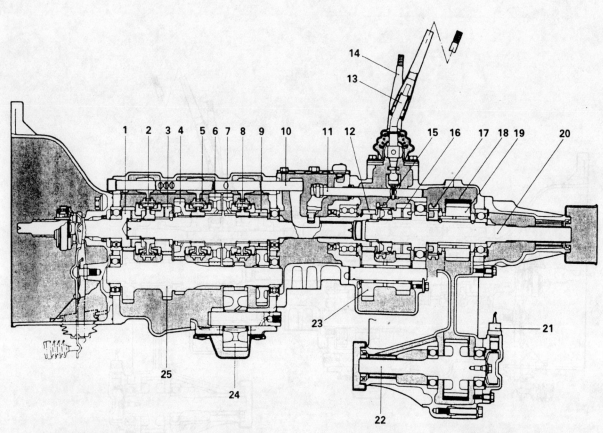

1. Drive pinion
2. 3rd-4th synchronizer
3. 3rd gear
4. 2nd gear
5. 1st-2nd synchronizer
6. 1st gear
7. Reverse gear
8. O.D.-R synchronizer
9. Overdrive gear
10. Shift rail
11. Control shaft
12. Input gear
13. Transmission control lever
14. Transfer control lever
15. H-L clutch
16. Low speed gear
17. 2-4WD clutch
18. Chain
19. Drive sprocket
20. Rear output shaft
21. Pulse generator
22. Front output shaft
23. Transfer counter gear
24. Reverse idler gear
25. Counter shaft

Fig. 7 In 1992 the manual transmission used by the 4-wheel drive Pick-ups and Monteros changed to the V5MT1-2

DRIVE TRAIN 7-9

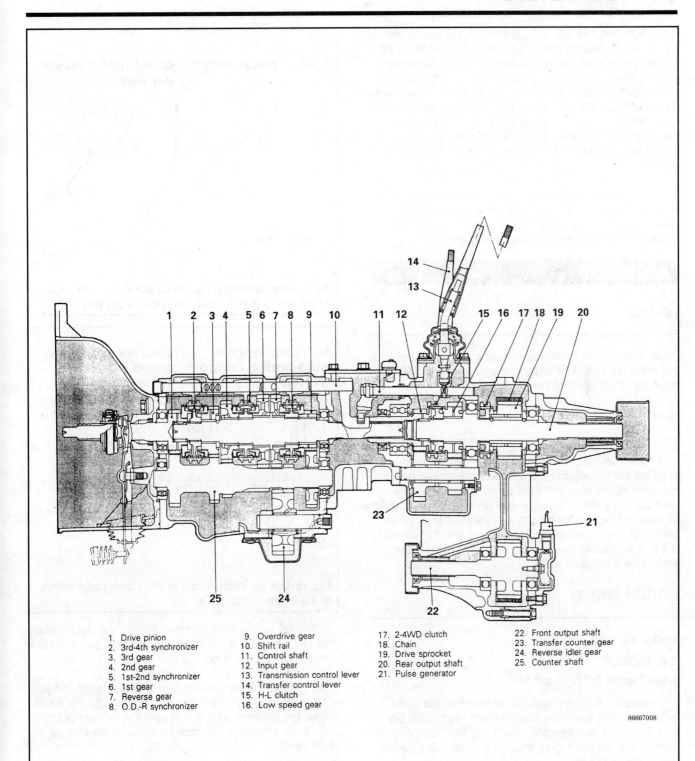

Fig. 8 An internal schematic of the 1993-95 4-wheel drive V5MT1-2 manual transmission

7-10 DRIVE TRAIN

The Mitsubishi KM 130, KM 132 and KM 145 can be identified by an identification code. This code can be found on the transmission identification tag located on the side of the transmission case.

The R5M21, V5MT1 and V5M21 are the newer replacements for the KM 130, KM 132 and the KM 145. These transmissions are in the Monteros and Pick-up trucks from 1990-95. They are essentially the same, also being five speed transmissions. The V5M21 and the V5MT1 are 4-wheel drive transmissions and, therefore, have a transfer case attached to them.

Due to the large number of alloy parts used in the transmissions, torque specifications should be strictly observed. Before installing cap screws into aluminum parts, dip the bolts into clean transmission fluid as this will prevent the screws from galling the aluminum threads, thus causing damage.

Adjustments

LINKAGE

Mitsubishi manual transmission linkages are not adjustable. If the shifter does not work properly, a problem with the size, mounting or lubrication of the linkage is indicated.

SHIFTER

The shifter assemblies on the Mitsubishi Pick-up trucks and Monteros are not adjustable. If the shifter shows play between the upper control lever and lower control lever, replace the lever assembly with a new assembly. Check the lever bushing for wear and replace the bushing if any wear is found. Push the shifter lever in and check that it moves smoothly up and down. If the rubber cover has any damage to it (cracks or rips), replace the cover.

CLUTCH SWITCH

1990-95 Pick-Ups

2.4L ENGINES

▶ See Figures 9, 10, 11 and 12

1. Measure the clutch pedal height from the face of the pedal pad to the floorboard, and the clutch pedal clevis pin play (measured from the upper most pedal face position to the lowest pedal position without moving the clutch). The standard values are as follows:
   - Pedal height — 6.5-6.7 in. (166-171mm).
   - Clevis pin play — 0.8-1.4 in. (20-35mm).
2. If the clutch pedal height is not within the standard value range, turn pedal stopper bolt or clutch pedal position switch until the height is adjusted to the standard value.
3. If the clutch pedal free-play is not within the standard value range, adjust it by the following procedure:
   a. Pull the clutch cable lightly at the toeboard (firewall) and turn the adjusting nut until the adjusting nut-to-insulator clearance is adjusted to 0.12-0.16 in. (3-4mm).

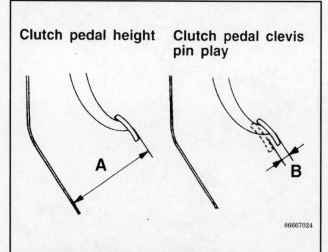

Fig. 9 When adjusting the clutch pedal, A is the pedal height and B is the clutch pedal clevis pin play

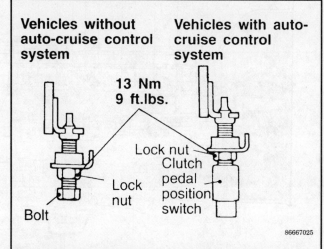

Fig. 10 Use the stopper bolt or the clutch pedal switch to adjust the clutch pedal

   b. After making the adjustment, depress the clutch pedal several times and check that the clutch pedal play is within the standard value range.
4. Measure the distance between the clutch pedal (the face of the pedal pad) and the floorboard when the clutch is disengaged. The standard value should be 2.4 in. (60mm) or more.
5. If the distance is less than the standard value, check the clutch itself.

3.0L ENGINES

▶ See Figures 9, 10, 12 and 13

1. Measure the clutch pedal height (A) from the face of the pedal pad to the floorboard, and the clutch pedal clevis pin play (B), which is measured at the face of the pedal pad. The standard values are as follows:
   - Pedal height — 6.5-6.7 in. (166-171mm).
   - Clevis pin play — 0.31-0.67 in. (8-17mm).

DRIVE TRAIN 7-11

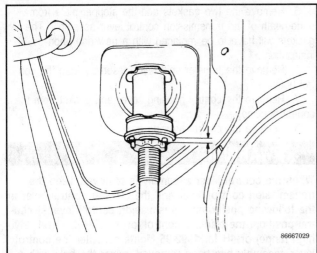

Fig. 11 Adjust the adjusting nut-to-insulator clearance to 0.12-0.16 in. (3-4mm) — cable-actuated clutches only

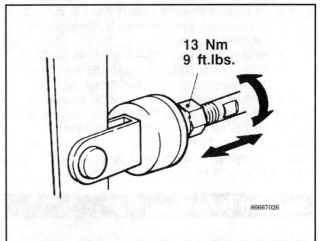

Fig. 13 Turn the clutch master cylinder actuating rod to adjust the clutch pedal height, use the locking nut to hold it at the specified adjustment

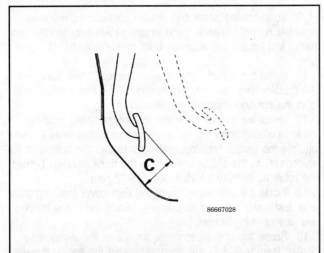

Fig. 12 With the clutch depressed, measure distance C from the floor to the pedal pad

2. If either the clutch pedal height or the clutch pedal clevis pin play are not within the standard value range, adjust as follows:

   a. For vehicles without the cruise control system, turn the stopper bolt to adjust the clutch pedal height to agree with the standard value and then secure the bolt with the locknut.

   b. For vehicles with the cruise control system, unplug the clutch pedal position switch connector and turn the switch for standard clutch pedal height. Lock the switch in place with the locknut.

   c. Turn the pushrod to adjust the clutch pedal clevis pin play to agree with the standard value and then secure the pushrod with the locknut.

✱✱WARNING

When adjusting the clutch pedal clevis pin play, be careful not to push the pushrod toward the master cylinder.

3. After completing the adjustments, confirm that the clutch pedal free-play (measured at the face of the pedal pad) and

the distance between the clutch pedal (the face of the pedal pad) and the floorboard when the clutch is disengaged are within the standard value ranges.
- Pedal free-play — 0.31-0.67 in. (8-17mm)
- Clutch pedal distance — 2.4 in. (60mm) or more.

4. If the clutch pedal free-play (B) and the distance between the clutch pedal and the floorboard (C) when the clutch is disengaged do not agree with the standard values, it is probably the result of either air in the hydraulic system, or a faulty master cylinder or clutch. Bleed the air, or inspect the master cylinder or clutch.

1990-95 Monteros
▶ See Figures 9, 10, 13 and 14

1. Measure the clutch pedal height (A) from the face of the pedal pad to the floorboard, and the clutch pedal clevis pin play (B), which is measured at the face of the pedal pad. The standard values are as follows:
- Pedal height — 7.30-7.50 in. (185.5-190.5mm).
- Clevis pin play — 0.04-0.12 in. (1-3mm).

2. If either the clutch pedal height or the clutch pedal clevis pin play are not within the standard value range, adjust as follows:

   a. For vehicles without the cruise control system, turn the stopper bolt to adjust the clutch pedal height to agree with the standard value and then secure the bolt with the locknut.

   b. For vehicles with the cruise control system, unplug the clutch pedal position switch connector and turn the switch for standard clutch pedal height. Lock the switch in place with the locknut.

   c. Turn the pushrod to adjust the clutch pedal clevis pin play to agree with the standard value and then secure the pushrod with the locknut.

✱✱WARNING

When adjusting the clutch pedal clevis pin play, be careful not to push the pushrod toward the master cylinder.

3. After completing the adjustments, confirm that the clutch pedal free-play (measured at the face of the pedal pad) and

7-12 DRIVE TRAIN

the distance between the clutch pedal (the face of the pedal pad) and the floorboard when the clutch is disengaged are within the standard value ranges.
- Pedal free-play (1992-95) — 0.24-0.51 in. (6-13mm).
- Pedal free-play (1990-91) — 0.31-0.63 in. (8-16mm).
- Clutch pedal distance — 1.4 in. (35mm) or more.

4. If the clutch pedal free-play (B) and the distance between the clutch pedal and the floorboard (C) when the clutch is disengaged do not agree with the standard values, it is probably the result of either air in the hydraulic system, or a faulty master cylinder or clutch. Bleed the air, or inspect the master cylinder or clutch.

Gearshift Lever Assembly

REMOVAL & INSTALLATION

4-Wheel Drive Transmissions
▶ See Figures 15, 16, 17 and 18

For this procedure you will need the following new gaskets:
- Transmission control lever assembly gasket.
- Stopper plate gasket.
- Control housing cover gasket.
- Control housing gasket.

Some differences may exist from year to year, however the following procedures can be applied to all 4-wheel drive Monteros and trucks.

1. Unscrew the two shifter knobs off of the transmission and transfer control lever assemblies.
2. Remove the dust cover retaining plate mounting screws.
3. Pull the dust cover retaining plate and the dust cover up off of the two levers.
4. Remove the transmission control lever assembly. Remove the control housing mounting bolts and transmission control lever assembly mounting bolts with the transfer shift lever in the "2H" ("4H" for 1992-95 Monteros) position and remove the transmission control lever assembly with the control housing.

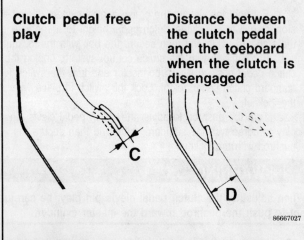

Fig. 14 Measure the clutch pedal free-play C and the clutch's depressed height D

5. Remove the two gaskets and the stopper plate from underneath of the transmission control lever assembly. The gaskets will have to be replaced with a new one upon installation.
6. Remove the transfer shift lever assembly from the control housing.
7. Remove the control housing cover and gasket from the control housing.
8. Remove the control housing gasket.

※※WARNING

When the control lever assembly is removed, keep the transmission control lever and the transfer control lever in the following positions: Transmission control lever — neutral position, the transfer control lever — 2H (2-wheel drive, high range) or 4H for 1992-95 Monteros. After the control lever assembly has been removed, cover the hole with a cloth to prevent entry of foreign substances into the extension housing.

9. To install the assembly, set the control housing cover onto the control housing, using a new gasket between the two parts, and tighten the retaining bolts to 11-15 ft. lbs. (15-22 Nm).
10. Install the transfer shift lever assembly. Make sure that the transfer lever assembly mounting on the transmission side is in the position shown in the illustration.
11. Install the two gaskets and the stopper plate onto the control housing, then set the transmission control lever assembly into the control housing, making sure that the bottom of the shifter sets in the shifter bushing (in the transmission). Tighten the retaining bolts to 11-15 ft. lbs. (15-22 Nm).
12. Install the shift lever cover, the dust cover retaining plate and dust cover over the shift levers. Mount the to the bracket with the retaining screws.
13. Screw the shift lever knobs back onto the levers, only tighten them until they are snug, otherwise the knobs' threads might become stripped.

2-Wheel Drive
▶ See Figure 19

For this procedure, new control lever and stopper plate gaskets will be needed. There may be some slight differences from year to year, however the following procedure is basically the same for all rear-wheel drive trucks and Monteros.

1. Unscrew the shift knob from the control lever.
2. Loosen and remove the dust cover retaining plate screws, then pull the dust cover and retaining plate off of the lever.
3. Remove the three mounting bolts from the shift lever assembly.
4. Pull the control lever assembly with the stopper plate off of the transmission.
5. Remove the two gaskets and the stopper plate off of the under-side of the control lever assembly.
6. To Install the shifter assembly, apply the two gaskets and the stopper plate to the under-side of the control lever assembly.
7. Insert the control lever assembly into the transmission, taking care that the lower arm-ball of the control lever fits into the bushing "cup" inside of the transmission.

DRIVE TRAIN 7-13

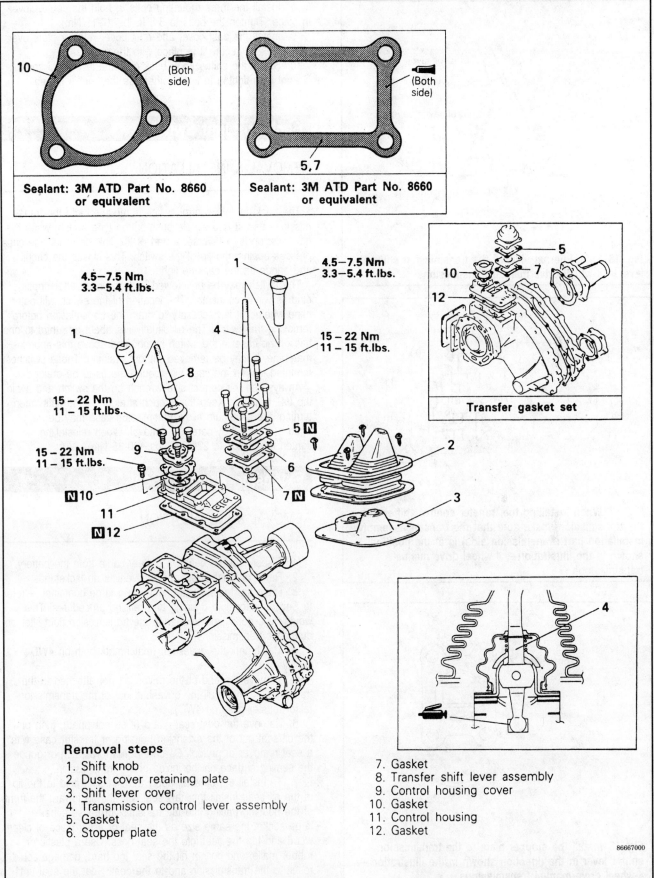

Fig. 15 Transmission and transfer assembly removal and installation — most 4-wheel drive manual transmissions similar

7-14 DRIVE TRAIN

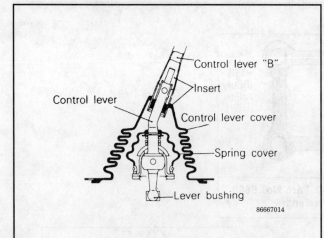

Fig. 16 Side, internal view of the transmission shifter lever — 4-wheel drive manual transmissions

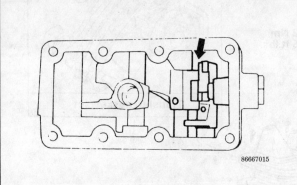

Fig. 17 When installing the transfer control shifter into the transmission, make sure that the transfer assembly installation part (transmission side) is at the position shown in the illustration — 4-wheel drive manual transmissions

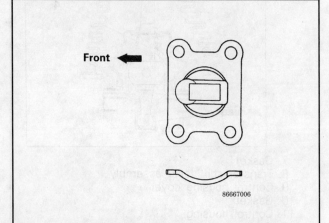

Fig. 18 Install the stopper plate to the transmission shifter lever in the direction shown in the illustration — 4-wheel drive manual transmissions

8. Install the three mounting bolts to hold the control lever in place. Tighten the bolts to 5-8 ft. lbs. (7-11 Nm).

9. Install the dust cover and dust cover retainer plate over the lever and secure it in place with the screws.

10. Screw the shifter knob back onto the lever until snug. Do not over-tighten to knob, otherwise the knob's threads might strip.

Back-up Light Switch

REMOVAL & INSTALLATION

The back-up light control switch is mounted into the transmission case. It is a simple push-button type switch; when the transmission is in reverse, a part of the linkage inside the case presses against the tip of the switch. This closes the circuit and brings on the back-up lights.

The switch may be unscrewed from the case after unplugging its wiring connector. The location of the switch will determine whether it is necessary to drain the transmission before removing the switch. The oil usually fills about one third to one half of the case; if the switch is mounted above this approximate line, it may be removed without draining. Those switches mounted low on the case will require the oil to be drained.

Always use the correct size wrench on the switch and avoid the use of pliers. Keep the wrench straight on the flats when turning; the aluminum alloy transmission case may be damaged or stripped by careless removal. When reinstalling, tighten the switch to 22-25 ft. lbs. (30-35 Nm).

Extension Housing Seal

REMOVAL & INSTALLATION

1. Disconnect the negative battery cable from the battery.
2. Raise and safely support the vehicle on jackstands.
3. Drain the transmission fluid into a large container — refer to Section 1 for fluid draining and refilling procedures. This would be a good time to change the transmission fluid filter on automatic transmissions.
4. Matchmark the driveshaft (matchmark both on 4WD vehicles).
5. Remove the end of the driveshaft not attached to the transmission, then pull the driveshaft out of the transmission (also transfer case on 4WD).
6. Remove the dust seal guard (if so equipped), then pry the oil seal out of the extension housing or transfer case with a seal remover or prytool. Be careful not to scratch or gouge the sealing surface on the inside of the transmission.
7. To install the oil seal, apply transmission fluid to the lip of the oil seal. Install the new seal with the lip toward the front of the housing (facing into the transmission). Use either a large socket the same size as the seal or a flat piece of clean wood and tap the seal into the seal boss. Use a plastic or rubber mallet and do not hit the seal too hard, damage could result to the transmission and to the seal. Seat the seal until the seal is flush with the transmission seal boss.

DRIVE TRAIN 7-15

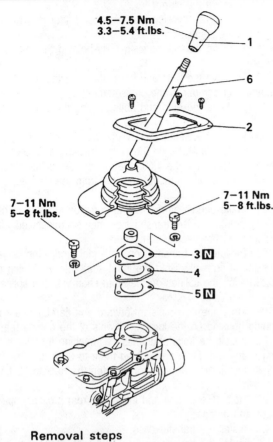

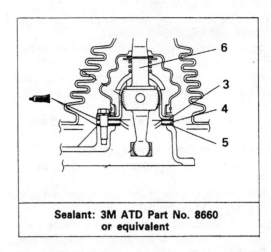

Sealant: 3M ATD Part No. 8660 or equivalent

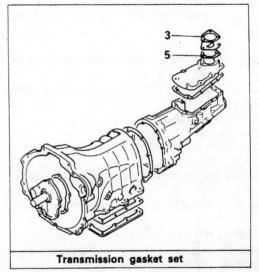

Transmission gasket set

Removal steps
1. Shift knob
2. Dust cover retaining plate
3. Gasket
4. Stopper plate
5. Gasket
6. Control lever assembly

Fig. 19 Transmission shifter lever removal and installation — most 4-wheel drive manual transmissions

8. Tap the dust guard back onto the transmission extension.
9. Slide the driveshafts back into the transmission and transfer case so that the matchmarks line up, then attach the other ends to the differential units or center link.
10. Fill the transmission fluid as per Section 1 in this book.
11. Make certain everything is in place and tightened securely, then lower the vehicle.
12. Connect the negative battery cable back to the battery.
13. Start the vehicle and allow it to warm up. Check for any transmission fluid leaks from the fluid drain or fill holes and from the oil seals.

Transmission

REMOVAL & INSTALLATION

1983-95 2-Wheel Drive Pick-Ups
▶ See Figure 20

1. Elevate and safely support the vehicle on stands.
2. Drain the transmission fluid.
3. Remove the shifter assembly.
4. Disconnect the driveshaft. Matchmark the flanges at the rear axle and remove the bolts. Lower the shaft and slide it to the rear to remove it from the transmission.
5. Disconnect the reverse light wiring and disconnect the speedometer cable.
6. Disconnect the mounting bracket for the exhaust system. While not required, it may be helpful the disconnect the exhaust system at the front joint. Use wire to suspend the pipe out of the way.

7. Remove the bell housing cover.
8. Disconnect the clutch cable.
9. Support the transmission with a floor jack. Use a broad piece of wood to distribute the load evenly.
10. Remove the rear engine support insulator and transmission mount nuts; separate them from the transmission.
11. Remove the mounting bolts for the rear (No. 2) crossmember. Remove the crossmember and the mount.
12. Remove the bolts holding the bell housing to the engine. Take note of the placement and size of each bolt; correct reinstallation is required.
13. Work the transmission rearward and off the engine. Once free, lower the jack and remove the transmission.

To install:

14. When reinstalling, position the transmission on the jack. Before installing, make certain the transmission case is in line with the two locating dowels or guides on the rear of the engine. Slide the transmission straight onto the engine and engage the shaft. Do not allow the transmission to hang on the shaft. Install the engine to bell housing retaining bolts. Make sure the bolts are correctly placed.
15. Install the rear crossmember, tightening the bolts to 33 ft. lbs. (45 Nm).
16. Connect the transmission mount and install the nuts and bolts. Correct torque is 15 ft. lbs. (20 Nm).
17. Connect the clutch cable.
18. Install the bell housing cover.
19. Connect the exhaust system if detached and secure the bracket at the transmission case.
20. Connect the speedometer cable and the backup light wiring harness.
21. Install the driveshaft observing the matchmarks made earlier.
22. Install the shift lever assembly.
23. Install the proper amount of transmission lubricant.
24. Lower the car to the ground.

1983-89 Monteros and 4-Wheel Drive Pick-Ups

1. Place the gear selector in Neutral.
2. Remove the boot retaining screws around the shifter boot; remove the boot and the shift knob.
3. Pull the gearshift lever assembly out of the control housing.
4. Cover the opening in the control housing with a cloth to prevent dirt from entering the transmission.
5. Disconnect the negative battery cable.
6. Raise and safely support the vehicle on stands.
7. Matchmark and disconnect the rear driveshaft at both the rear axle and the transfer case flanges.
8. Disconnect the forward driveshaft from the front axle and remove it by sliding it forward. Install a plug in the transfer case to prevent fluid loss.
9. Disconnect the hydraulic fluid line from the transfer case.
10. Disconnect the starter motor cable, the back up lamp harness and the neutral safety switch wire.
11. Disconnect the speedometer cable.
12. Place a floor jack and a block of wood below the engine oil pan.
13. Support the transfer case and remove it from the vehicle.
14. Remove the starter.
15. Use a transmission jack or second floor jack to place under the transmission.
16. Remove the bell housing bolts holding the unit to the engine.
17. Remove the nuts and bolts attaching the transmission mount and damper to the crossmember.
18. Remove the nuts and bolts holding the crossmember to the frame rails and remove the crossmember.
19. Lower the engine jack. Work the clutch housing off the locating dowels and slide the clutch housing and the transmission rearward until the input shaft clears the clutch disc.
20. Remove the transmission unit from the vehicle.

To install:

21. Make certain all the mating surfaces are free of nicks, burrs, paint, etc. The transmission may be reinstalled with or without the transfer case attached.
22. Elevate the transmission on a jack under the car and position it correctly. Start the input shaft into the clutch disc.
23. Align the splines on the input shaft with the splines in the clutch disc.
24. Move the transmission forward and carefully seat the clutch housing on the locating dowels of the engine rear plate.
25. Install the bolts and washers attaching the clutch housing to the rear plate and tighten them to 36 ft. lbs (49 Nm).
26. Install the starter motor. Tighten the nuts to 20 ft. lbs (27 Nm).
27. Raise the engine and install the rear crossmember, insulator and damper.
28. Install the transfer case.
29. Install the driveshafts front and rear. Make sure the marks made during disassembly are aligned.
30. Connect the starter cable, the backup light wiring and other wires to the transmission.
31. Connect the hydraulic line to the slave cylinder. Bleed the clutch system.
32. Install the speedometer cable.
33. Remove the fill plug and fill the transmission unit to the correct level.
34. Lower the vehicle to the ground.
35. Remove the cloth over the transfer case adapter. Install the gear shift lever assembly in the transfer case. Make sure the ball on the lever is in the socket in the adapter. Install the attaching bolts and tighten to 10 ft. lbs. (12 Nm).
36. Install the boot cover, shift knob and boot retaining screws.

1990-95 Monteros and 4-Wheel Drive Pick-Ups

▶ See Figures 21 and 22

1. Disconnect the negative battery cable from the battery.
2. Remove the transmission and transfer shift lever assembly — refer to the previously described procedures.
3. Raise and support the vehicle on jackstands. Before crawling under the vehicle, make absolute certain that the vehicle is safely supported by the jackstands.
4. Remove the transfer case protector.
5. Unfasten the front exhaust pipe from the two exhaust manifolds, then disconnect it from the intermediate pipe/catalytic convertor (make certain to retain the bolts and nuts for reassembly). Remove the entire pipe from under the vehicle.

DRIVE TRAIN 7-17

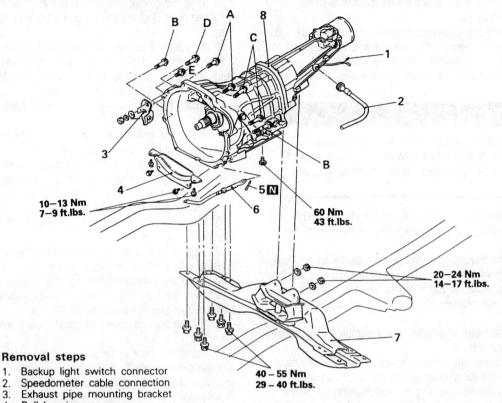

Removal steps
1. Backup light switch connector
2. Speedometer cable connection
3. Exhaust pipe mounting bracket
4. Bell housing cover
5. Cotter pin
6. Clutch cable connection
7. No. 2 crossmember
8. Transmission assembly

Items	Nm	ft.lbs.	O.D. × Length mm (in.)	Bolt identification
A	43 – 55	31 – 40	7 10 × 40 (.4 × 1.6)	D × L
B	43 – 55	31 – 40	7 10 × 65 (.4 × 2.6)	
C	22 – 32	16 – 23	7 10 × 60 (.4 × 2.4)	
D	20 – 27	15 – 20	7 8 × 55 (.3 × 2.2)	
E	20 – 27	15 – 20	7 8 × 25 (.3 × 1.0)	

Fig. 20 Removal and installation for rear-wheel drive manual transmissions — all years similar, some slight differences may exist

6. Drain the transmission fluid into a clean, large pan. With the transmission out for any reason, it is a good time to change the transmission fluid filter on automatic transmissions.
7. Matchmark and disconnect the rear driveshaft at both the rear axle and the transfer case flanges.
8. On 4-wheel drive vehicles, disconnect the forward driveshaft from the front axle and remove it by sliding it forward. Install a plug in the transfer case to prevent residual fluid leakage.
9. Remove the dust seal from the rear extension housing.
10. Label and unplug all wires leading to the transmission, some are: Ground cables, 4WD indicator light switch connector, pulse generator connector, oxygen sensor connector, back-up light switch connector, HI/LO detection switch connector, center differential lock detection switch connection. There may be others depending on the particular year, model and engine the vehicle came equipped with.
11. Unscrew the retaining ring holding the speedometer in the transmission. Pull the speedometer cable out of the transmission and label it.
12. Remove the clutch cylinder heat protector, if so equipped.
13. Remove the retaining bolts, then the clutch release cylinder (with the clutch hose connected to it) from the transmission. Suspend it from the body by using a piece of wire or a similarly safe method.
14. Unbolt the starter motor from the front of the transmission and remove it. Remove the heat shield, if so equipped.
15. Place a floor jack and a block of wood below the engine oil pan.
16. Lift the floor jack under the engine just until the weight of the engine is taken onto the jack — the engine should only barely be lifted by the jack.
17. Use a transmission jack or second floor jack to place under the transmission. Don't support the transmission yet, only lift the jack until it is slightly below the transmission.
18. Remove the left-hand and right-hand side transmission stays from the front of the transmission.
19. Unbolt the bell housing lower cover.
20. Remove the transfer case mounting bracket.

7-18 DRIVE TRAIN

21. Lift the floor jack up until the transmission is being slightly supported by it.
22. Remove the two transmission-to-crossmember bolts. Lift the jack about ¼ in. (6mm) off of the crossmember support. Remove the entire crossmember from the vehicle.
23. Remove the transmission and transfer case assembly together.

✱✱CAUTION

Do not jolt the transmission when it is being removed from the engine. If it is jolted strongly, the front end of the main drive gear, pilot bearing or clutch disk could be damaged.

a. Disconnect the transmission and transfer assembly from the engine by pulling it slowly toward the rear of the vehicle.
b. When the transmission and transfer assembly are lowered, tilt the front of the transmission downward and slowly lower forward, while making sure that the rear of the transmission does not interfere with the rear-most crossmember.

To install:
24. Lift the transmission and transfer assembly close to position with the floor jack.
25. On the engine side, there are two centering locations. Make sure that the transmission mounting bolt holes are aligned with them before mounting the transmission and transfer assembly to the engine. Lowering the rear of the engine SLIGHTLY may help align the two assemblies.
26. Slide the transmission assembly onto the engine making sure the aligning areas stay aligned.
27. Install and tighten the bell housing bolts to 54 ft. lbs. (75 Nm).
28. Lift the transmission/transfer assembly with the floor jack. Since the engine is now attached to the transmission, it also will rise slightly — adjust its jack to keep only slight support.
29. Hold the crossmember in place and secure with the mounting bolts. Tighten the mounting bolts to 47 ft. lbs. (65 Nm).
30. Lower the transmission and transfer case assembly onto the crossmember. Install the two crossmember-to-transmission bolts and tighten them to 13-18 ft. lbs. (18-25 Nm) on Pick-ups and to 29 ft. lbs. (39 Nm) on Monteros.
31. Install the transfer mounting bracket back onto the transfer case.
32. Install the bell housing lower cover, then mount the left-hand and right-hand side transmission stays.
33. Mount the starter motor and clutch release cylinder (and heat shield, if so equipped) to the transmission.
34. Insert the speedometer cable into the transmission and secure it there with the retaining ring.
35. Plug all of the electrical harness connectors back together.
36. Tap the dust seal guard back onto the rear extension housing with a rubber or plastic mallet.
37. Slide the front driveshaft into the transfer case, then attach it to the front differential. Install the rear driveshaft also. Make certain that the matchmarks line up.
38. Fill the transmission and transfer case with oil. Refer to Section 1.
39. Install the front exhaust pipe to the catalytic convertor and the exhaust manifolds. Use new gaskets if it was equipped with them. Make sure that the wiring does not lay on or very close to the exhaust pipes or exhaust manifolds, if they do, reroute them away from the exhaust components.
40. Install the transfer case protector.
41. Lower the vehicle to the ground. Install the transmission and transfer shift lever assembly.
42. Connect the negative battery cable to the battery.
43. Start the vehicle and check for any transmission or transfer case fluid leaks.

DRIVE TRAIN 7-19

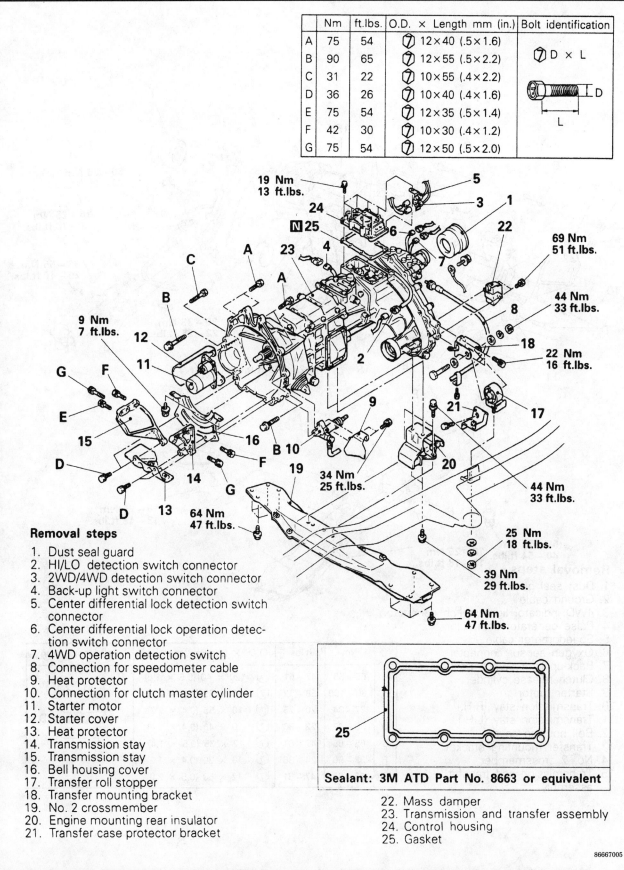

Fig. 21 Removal and installation for the 1992-95 Montero manual transmission — other years are similar, some differences may be present

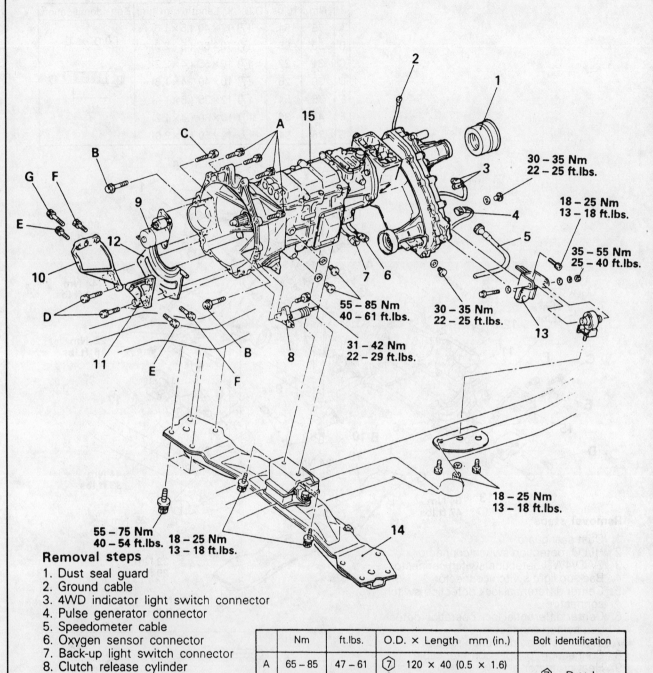

Fig. 22 Although mostly similar with the Montero, the 4-wheel drive manual transmission for the 1990-95 Pick-ups has different torque values and components

DRIVE TRAIN   7-21

CLUTCH

✳✳CAUTION

The clutch driven disc may contain asbestos, which has been determined to be a cancer causing agent. Never clean clutch surfaces with compressed air. Avoid inhaling any dust from the clutch area. When cleaning clutch surfaces, use commercially available brake cleaning solvents.

Adjustments

PEDAL HEIGHT

For 1990-95 Pick-ups and Monteros, refer to the clutch switch adjustment procedures (described earlier in this section).

1. Measure clutch pedal height from the floor to the top of the pedal pad, along the line of pedal travel. The height should be:

- Montero: 7.28-7.48 in. (18.5-19.0cm)
- Pick-up: 6.5-6.7 in. (16.5-17.0cm)

2. Adjust the pedal height, if necessary, by loosening the locknut and turning the adjusting bolt or rotating clutch switch, and retighten the locknut. Now adjust the pedal play as described below.

PEDAL PLAY (FREE-PLAY)

➡When checking pedal play you must not move the pedal far enough to disturb the position of the hydraulic cylinder actuating rod. Move the pedal to a point where resistance stiffens but no more.

1. For hydraulic clutches used in Monteros and later-modeled Pick-ups, loosen the locknut, and turn the pushrod to bring the play within specification. On all models, this is 0.039-0.12 in. (1-3mm). Retighten the locknut.
2. Check the total free-play, which should be 0.24-0.47 in. (6-12mm). Check the pedal height, also. If play is not correct, bleed the system. If that does not correct play, the master cylinder or the clutch itself is faulty.
3. On Pick-ups with the clutch cable, the clutch pedal free-play should be 0.79-1.18 in. (20-30mm). If it is outside specification, pull up gently on the cable at the floorboard and turn the outer cable adjusting nut until the nut-to-insulator clearance is about 0.098 in. (2.5mm). After making the adjustment, depress the clutch pedal to the floor several times and recheck pedal free-play.

Clutch Pedal

REMOVAL & INSTALLATION

Pick-Up

CABLE ACTUATED CLUTCH

♦ See Figure 23

1. Pull the cable out and turn the adjusting nut counterclockwise to increase the cable play.
2. Disconnect the cable from the clutch lever.
3. Unscrew the nut on the right-hand end of the clutch pedal pivot bolt. Remove also the lockwasher and the flat washer.
4. Pull the clutch lever off the end of the pivot bolt. Make sure to note the position of the lever before removal.
5. Slide the clutch pedal out to the left. The pivot bolt will come out with the pedal. Make sure to save the pedal-side clutch pedal bushing.
6. To install, slide the clutch pedal bushing onto the pivot bolt, then slide the pivot bolt through the pedal support member.
7. Insert the right-hand side clutch pedal bushing into the pivot bolt tube, then slide one of the flat washers onto the pivot bolt.
8. Hook the clutch lever into the end of the clutch cable, then slide onto the clutch pedal pivot bolt.
9. Install the flat washer, the lockwasher and the end nut onto the pivot bolt. Tighten until snug.
10. Adjust the clutch and brake pedals — refer to the previous procedures in this section.

HYDRAULIC CLUTCH

♦ See Figure 24

1. Loosen and remove the bracket retaining bolt, then remove the bracket itself.
2. Remove the turnover spring from the clutch pedal.
3. Unscrew the stopper bolt (on vehicles without auto-cruise control system) or the clutch switch (vehicles with the auto-cruise control system). The stopper bolt/ clutch switch is located just above the small bracket on the pedal assembly main bracket.
4. Remove the cotter pin, which is holding the clutch master cylinder rod to the clutch pedal. A new cotter pin will be needed upon installation.
5. Pull the clevis pin out of the clutch pedal, thereby disconnecting the clutch master cylinder from the pedal.
6. Remove the clutch pedal mounting bolt, then pull the clutch pedal down and out of the pedal assembly main bracket. Make sure to retain the two bushings and the spacer.
7. For installation, insert the two bushings and the spacer onto the clutch pedal.
8. Insert the clutch pedal up into the pedal assembly main bracket, then slide the clutch pedal mounting bolt through the main bracket from the right-hand side.

7-22 DRIVE TRAIN

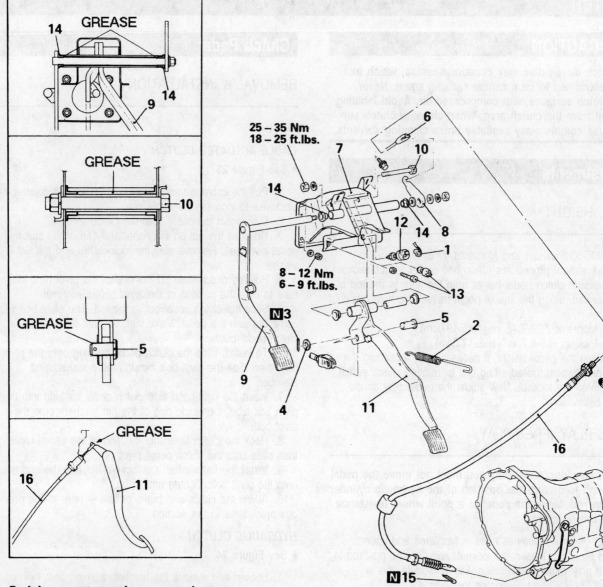

Pedal support member removal steps
1. Stop light switch connector connection
2. Brake pedal return spring
3. Cotter pin
4. Washer
5. Clevis pin
6. Clutch cable connection
7. Pedal support member
8. Clutch lever
9. Clutch pedal
10. Brake pedal shaft
11. Brake pedal
12. Stop light switch
13. Stopper bolt <Vehicles without auto-cruise control system> or Clutch switch <Vehicles with auto-cruise control system>
14. Clutch pedal bushing

Clutch pedal removal steps
6. Clutch cable connection
8. Clutch lever
9. Clutch pedal

Clutch cable removal steps
6. Clutch cable connection
15. Cotter pin
16. Clutch cable
17. Insulator

Fig. 23 Clutch pedal removal and installation — cable-actuated clutches on earlier Monteros and Pick-ups

DRIVE TRAIN 7-23

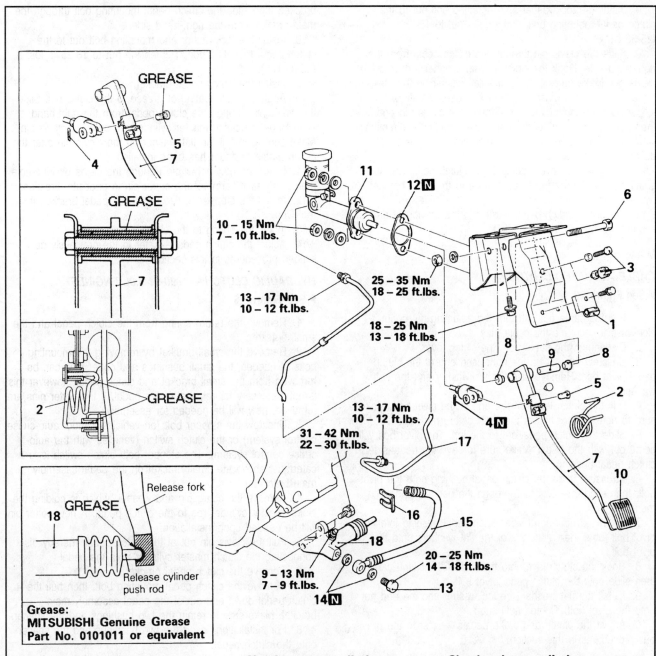

Fig. 24 Clutch pedal removal and installation — 1990-95 Pick-ups equipped with 3.0L engines

7-24 DRIVE TRAIN

9. Install the lockwasher and mounting bolt nut to the clutch pedal mounting bolt. Tighten the nut to 18-25 ft. lbs. (25-35 Nm).

10. Slide the clevis pin through the clutch pedal from the right-hand side. Hook the clutch master cylinder rod onto the clevis pin. Make sure to install the flat washer on the clevis pin after the clutch master cylinder has been installed.

11. Insert the new cotter pin through the clevis pin end-hole, then bend each "arm" of the cotter pin in opposite directions.

12. Install the stopper bolt or clutch switch to the pedal assembly main bracket.

13. Install the turnover spring to the clutch pedal and the small bracket. Mount the small bracket to the main bracket with the retaining bolt.

14. Adjust the clutch pedal — refer to the previously described procedures in this section.

Montero

CABLE ACTUATED CLUTCH

▶ See Figure 23

1. Pull the cable out and turn the adjusting nut counterclockwise to increase the cable play.
2. Disconnect the cable from the clutch lever.
3. Unscrew the nut on the right-hand end of the clutch pedal pivot bolt. Remove also the lockwasher and the flat washer.
4. Pull the clutch lever off the end of the pivot bolt. Make sure to note the position of the lever before removal.
5. Slide the clutch pedal out to the left. The pivot bolt will come out with the pedal. Make sure to save the pedal-side clutch pedal bushing.
6. To install, slide the clutch pedal bushing onto the pivot bolt, then slide the pivot bolt through the pedal support member.
7. Insert the right-hand side clutch pedal bushing into the pivot bolt tube, then slide one of the flat washers onto the pivot bolt.
8. Hook the clutch lever into the end of the clutch cable, then slide onto the clutch pedal pivot bolt.
9. Install the flat washer, the lockwasher and the end nut onto the pivot bolt. Tighten until snug.
10. Adjust the clutch and brake pedals — refer to the previous procedures in this section.

HYDRAULIC CLUTCH — 1989-90 2.6L ENGINES

▶ See Figure 25

1. Remove the spring from the clutch pedal.
2. Remove the cotter pin and washer, which is holding the clutch master cylinder rod to the clutch pedal. A new cotter pin will be needed upon installation.
3. Pull the clevis pin out of the clutch pedal, thereby disconnecting the clutch master cylinder from the pedal.
4. Remove the clutch pedal bracket.
5. Remove the clutch pedal mounting bolt, then pull the clutch pedal down and out of the pedal assembly main bracket. Make sure to retain the two bushings and the spacer.
6. For installation, insert the two bushings and the spacer onto the clutch pedal.
7. Apply a thin coating of grease to the pedal mounting bolt. Insert the clutch pedal up into the pedal assembly main bracket, then slide the clutch pedal mounting bolt through the main bracket from the right-hand side.
8. Install the lockwasher and mounting bolt nut to the clutch pedal mounting bolt. Tighten the nut to 18-25 ft. lbs. (25-35 Nm).
9. Install the clutch pedal bracket.
10. Apply a small coating of grease to the clevis pin. Slide the clevis pin through the clutch pedal from the right-hand side. Hook the clutch master cylinder rod onto the clevis pin. Make sure to install the flat washer on the clevis pin after the clutch master cylinder has been installed.
11. Insert the new cotter pin through the clevis pin end-hole, then bend each "arm" of the cotter pin in opposite directions.
12. Install the stopper bolt to the clutch pedal bracket, if removed.
13. Install the spring to the clutch pedal.
14. Adjust the clutch pedal — refer to the previously described procedures in this section.

HYDRAULIC CLUTCH — 1990-91 3.0L ENGINES

▶ See Figure 26

1. Remove the return spring from the clutch pedal and the small bracket.
2. Remove the small bracket by removing the mounting bolts. If needed, the small bushings and cotter pins can be removed from the small bracket and clutch pedal, however this is not necessary for removal of the pedal. If the cotter pins are removed, new will be needed for assembly.
3. Unscrew the stopper bolt (on vehicles without auto-cruise control system) or the clutch switch (vehicles with the auto-cruise control system). The stopper bolt/ clutch switch is located just above the small bracket on the pedal assembly main bracket.
4. Remove the cotter pin and washer, which is holding the clutch master cylinder rod to the clutch pedal. A new cotter pin will be needed upon installation.
5. Pull the clevis pin out of the clutch pedal, thereby disconnecting the clutch master cylinder from the pedal.
6. Remove the clutch pedal bracket.
7. Remove the clutch pedal mounting bolt, then pull the clutch pedal down and out of the pedal assembly main bracket. Make sure to retain the two bushings and the spacer.
8. For installation, insert the two bushings and the spacer onto the clutch pedal.
9. Apply a thin coating of grease to the pedal mounting bolt. Insert the clutch pedal up into the pedal assembly main bracket, then slide the clutch pedal mounting bolt through the main bracket from the right-hand side.
10. Install the lockwasher and mounting bolt nut to the clutch pedal mounting bolt. Tighten the nut to 18-25 ft. lbs. (25-35 Nm).
11. Install the clutch pedal bracket.
12. Apply a small coating of grease to the clevis pin. Slide the clevis pin through the clutch pedal from the right-hand side. Hook the clutch master cylinder rod onto the clevis pin. Make sure to install the flat washer on the clevis pin after the clutch master cylinder has been installed.
13. Insert the new cotter pin through the clevis pin end-hole, then bend each "arm" of the cotter pin in opposite directions.
14. Install the stopper bolt or clutch switch to the clutch pedal bracket.

DRIVE TRAIN 7-25

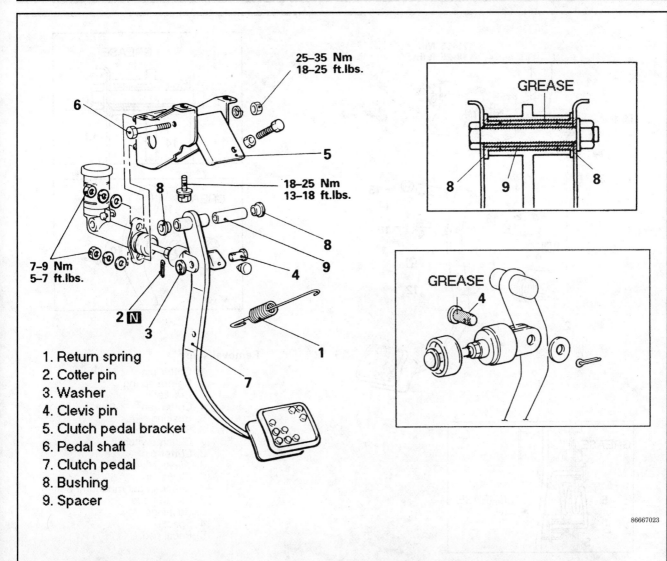

Fig. 25 Clutch pedal removal and installation — 1989-90 Monteros equipped with 2.6L engines (earlier years with hydraulic clutches are similar)

1. Return spring
2. Cotter pin
3. Washer
4. Clevis pin
5. Clutch pedal bracket
6. Pedal shaft
7. Clutch pedal
8. Bushing
9. Spacer

15. Mount the small bracket to the main bracket with the retaining bolts until snug.
16. Install the spring to the clutch pedal. Install the bushings and cotter pins if they were removed upon disassembly.
17. Adjust the clutch pedal — refer to the previously described procedures in this section.

HYDRAULIC CLUTCH — 1992-95

▶ See Figure 27

1. Remove the cotter pin and washer, which is holding the clutch master cylinder rod to the clutch pedal. A new cotter pin will be needed upon installation.
2. Pull the clevis pin out of the clutch pedal, thereby disconnecting the clutch master cylinder from the pedal.
3. Unplug the interlock switch harness connector.
4. Disconnect the clutch master cylinder from the clutch pedal assembly by removing the retaining nuts (one is inside the vehicle in the clutch pedal bracket, the other one is in the engine compartment near the clutch master cylinder).

5. Remove the entire clutch pedal assembly from the firewall. Once the entire unit has been removed, further disassembly can be done.
6. Remove the upper turnover spring retaining clevis pin (a new cotter pin will be needed). Disassemble the turnover spring assembly, which consists of rod A, the spring, rod B and rod B's bushing. Remove the lower retaining clevis pin from the clutch pedal.
7. Unscrew the stopper bolt (on vehicles without auto-cruise control system) or the clutch switch (vehicles with the auto-cruise control system). The stopper bolt/ clutch switch is located on the pedal assembly main bracket.
8. Remove the clutch pedal mounting bolt, then pull the clutch pedal down and out of the pedal assembly main bracket. Make sure to retain the two bushings and the spacer.
9. For installation, insert the two bushings and the spacer onto the clutch pedal.
10. Apply a thin coating of grease to the pedal mounting bolt. Insert the clutch pedal up into the pedal assembly main bracket, then slide the clutch pedal mounting bolt through the main bracket from the right-hand side.

7-26 DRIVE TRAIN

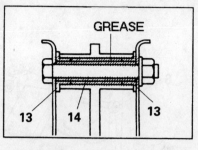
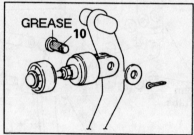
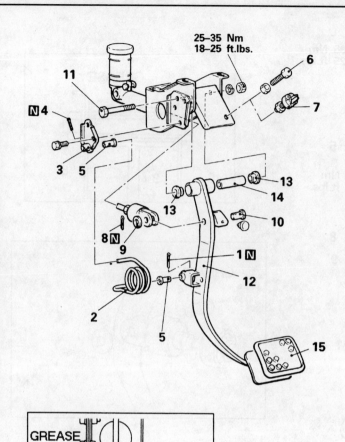

Removal steps

1. Cotter pin
2. Return spring
3. Bracket
4. Cotter pin
5. Bushing
6. Stopper bolt
7. Clutch switch
8. Cotter pin
9. Washer
10. Clevis pin
11. Clutch pedal mounting bolt
12. Clutch pedal
13. Bushings
14. Spacer
15. Pedal pad

Fig. 26 Clutch pedal removal and installation — 1990-91 Monteros equipped with 3.0L engines

11. Install the lockwasher and mounting bolt nut to the clutch pedal mounting bolt. Tighten the nut to 18-25 ft. lbs. (25-35 Nm).

12. Install the stopper bolt or the clutch switch.

13. Assemble the turnover spring (rod A-turnover spring-rod B) and install the lower end onto the clutch pedal with the bushing (oiled), the flat washer and a new cotter pin.

14. Install the upper end of the turnover spring assembly into the main bracket, using the clevis pin, the flat washer and a new cotter pin. Apply a thin film of grease to the clevis pin.

15. Install the clutch pedal assembly onto the firewall and tighten the two mounting bolts (one inside, one outside) to 9 ft. lbs. (13 Nm).

16. Apply a small coating of grease to the clevis pin. Slide the clevis pin through the clutch pedal from the right-hand side. Hook the clutch master cylinder rod onto the clevis pin. Make sure to install the flat washer on the clevis pin after the clutch master cylinder has been installed.

17. Insert the new cotter pin through the clevis pin end-hole, then bend each "arm" of the cotter pin in opposite directions.

18. Adjust the clutch pedal — refer to the previously described procedures in this section.

Clutch Linkage

REMOVAL & INSTALLATION

Cable-Actuated Clutches

CLUTCH CABLE

▶ See Figures 28 and 29

1. From the engine compartment, pull the clutch cable out and turn the adjusting nut counter-clockwise to increase the cable play (which will allow the cable to be easier to remove from the clutch lever.

2. From inside of the vehicle, unhook the cable from the clutch lever.

3. From back in the engine bay, pull the cable out through the firewall.

DRIVE TRAIN 7-27

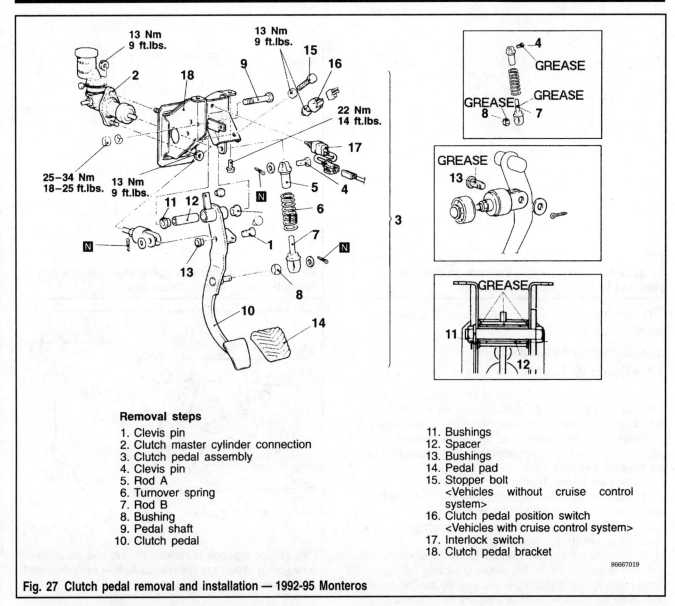

Removal steps
1. Clevis pin
2. Clutch master cylinder connection
3. Clutch pedal assembly
4. Clevis pin
5. Rod A
6. Turnover spring
7. Rod B
8. Bushing
9. Pedal shaft
10. Clutch pedal
11. Bushings
12. Spacer
13. Bushings
14. Pedal pad
15. Stopper bolt
    <Vehicles without cruise control system>
16. Clutch pedal position switch
    <Vehicles with cruise control system>
17. Interlock switch
18. Clutch pedal bracket

Fig. 27 Clutch pedal removal and installation — 1992-95 Monteros

4. Raise the vehicle and safely support it on jackstands.
5. The other end of the clutch cable attaches to the left-hand side of the front of the transmission. Remove the cotter pin holding it in place.
6. Unhook the cable end from the transmission lever, then pull the cable out of the transmission mounting hole.
7. Pull the cable out of the engine mount insulator bracket. Retain the insulator pad for assembly.
8. To install the cable, route it through the engine mount insulator bracket, making sure that the insulator pad is pushed into the bracket (this protects the cable from rubbing against the bracket).
9. Insert the transmission end of the cable through the transmission mounting hole, then hook the cable to the lever.
10. Install a new cotter pin to retain the cable in the transmission.
11. Lower the vehicle back to the ground.
12. Slide the cable through the firewall hole as far as it will go.
13. From inside the vehicle, hook the cable to the clutch lever.

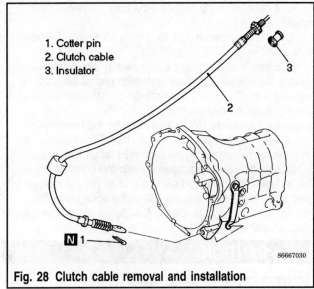

1. Cotter pin
2. Clutch cable
3. Insulator

Fig. 28 Clutch cable removal and installation

7-28 DRIVE TRAIN

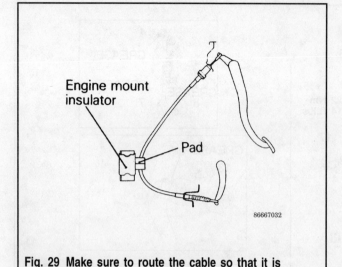

Fig. 29 Make sure to route the cable so that it is retained by the engine mount insulator

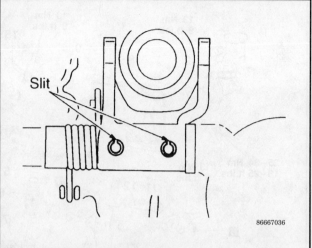

Fig. 30 Install the two new spring clips with the slits facing the tongs on the release fork

14. Adjust the clutch pedal and cable — refer to the clutch switch procedures in this section.

TRANSMISSION INNER-LINKAGE

▶ See Figures 30, 31, 32 and 33

1. To remove the inner linkage of the clutch system, the transmission will have to be removed from the vehicle. To do this, follow the previously outlined instructions in this section.
2. With the transmission out of the vehicle, remove the retaining clips from the clutch release bearing (located on the transmission end shaft.
3. Slide the release bearing off of the shaft.
4. Using a punch with the same outside diameter as the spring clips (two are located in the release fork), tap the spring clips out of the release fork. Discard the old spring clips, new ones will be needed for reassembly.
5. Slide the release fork shaft out of the transmission, making sure to account for the two pieces of packing, the right-hand return spring, the left-hand return spring and the release fork. Note the positions of these components for reassembly.
6. If need be, remove the weight from the release fork shaft.
7. To reassemble, mount the weight back onto the release fork shaft.
8. Before reassembling the release fork and components, grease the release fork shaft from each end of the shaft for about 2-3 in. (25-37mm). Also grease the two front surfaces of the tongs on the release fork.
9. As the release fork shaft is slid into the transmission, install the inner components on the shaft. The order is: packing, left return spring, release fork, right return spring, packing.
10. Line the spring clip holes up with the holes on the release fork shaft, then tap new spring clips into the release fork and shaft. Make sure that the slits in the spring clips are toward the tongs on the release fork (facing toward the transmission end shaft).

✱✱CAUTION

Do not reuse the old spring clips.

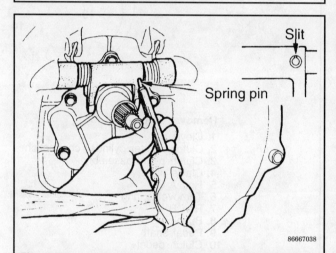

Fig. 31 Use a punch to remove the old and install the new spring clips from the release fork — cable-actuated clutch systems

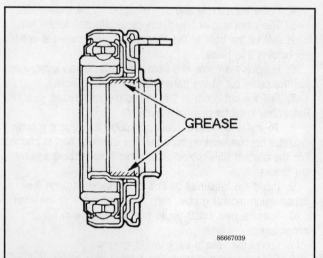

Fig. 32 Grease the areas shown on the inner-surface of the release bearing — all clutch systems are similar

DRIVE TRAIN

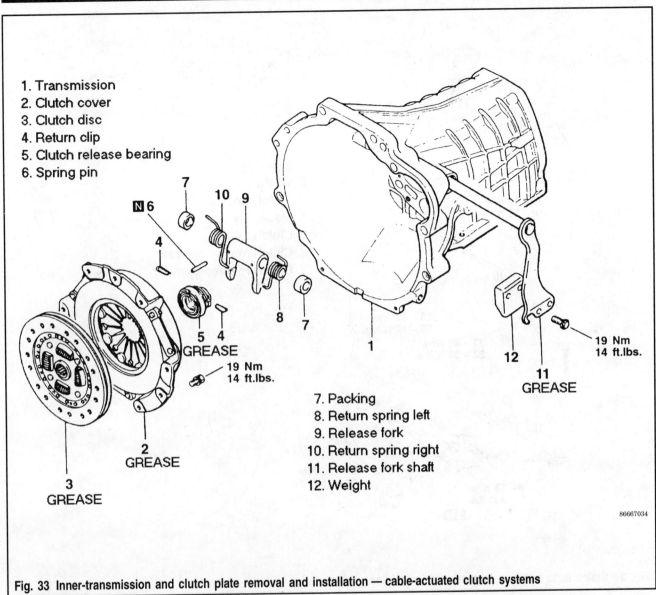

Fig. 33 Inner-transmission and clutch plate removal and installation — cable-actuated clutch systems

1. Transmission
2. Clutch cover
3. Clutch disc
4. Return clip
5. Clutch release bearing
6. Spring pin
7. Packing
8. Return spring left
9. Release fork
10. Return spring right
11. Release fork shaft
12. Weight

11. Grease the inside surface of the clutch release bearing, then slide the bearing onto the transmission end shaft.
12. Secure the release bearing on the shaft with new retaining clips.
13. Install the transmission into the vehicle.

Hydraulic Clutches

CLUTCH HOSE AND TUBE

▶ See Figure 34

This procedure is for all hydraulic clutches with or without ABS. If the vehicle does have ABS, the ABS connection can be removed from the pipe once the entire pipe has been removed from the vehicle.

1. Raise and safely support the vehicle on jackstands.
2. Place a pan or similar container under the clutch hose connection to the clutch release cylinder to catch the clutch fluid.
3. Loosen and remove the eye bolt holding the clutch hose to the clutch release cylinder. Drain the fluid into the container.
4. Discard the two old eye bolt gaskets, new ones will be needed for reassembly.
5. Holding the nut at the clutch hose side, loosen the flare nut of the clutch tube.
6. Remove the clutch hose from the vehicle.
7. Loosen the clutch tube from the clutch master cylinder and extract the tube from the vehicle.
8. At this point the ABS connector can be removed from the tube if need be.
9. To install, carefully thread the upper flare nut of the clutch tube into the clutch master cylinder, then tighten it to 10-12 ft. lbs. (13-17 Nm).
10. Carefully thread the lower flare nut into the upper end of the clutch hose. Once the threads are started, tighten the connection to 10-12 ft. lbs. (13-17 Nm). Take care to not let the clutch hose twist.
11. Connect the clutch hose to the release cylinder at the stepped portion. Make sure to use two new gaskets on the clutch hose connection. Tighten the eye bolt to 14-18 ft. lbs. (20-25 Nm).

7-30 DRIVE TRAIN

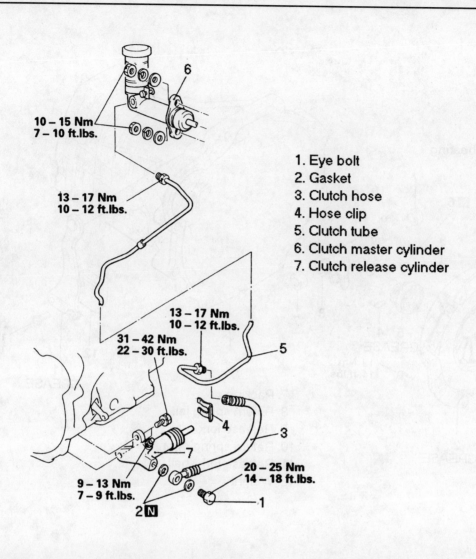

Fig. 34 Clutch hose and tube removal and installation

12. Fill the master cylinder with brake fluid meeting SAE J1703 (DOT3) specifications.

✶✶WARNING

Use the specified brake fluid. Avoid using a mixture of the specified fluid and another fluid.

13. Bleed the system until no air remains in it.
14. Check for clutch fluid leaks at all connections in the system.

TRANSMISSION INNER LINKAGE

▶ See Figures 35, 36 and 37

To service the inner linkage, the transmission must first be removed from the vehicle.

1. With the transmission removed from the vehicle, remove the return spring.
2. Slide the clutch release bearing off of the transmission end shaft.
3. Remove the release cylinder from the side of the transmission.
4. Pop the release fork boot out of the transmission hole.
5. Slide the release fork toward the boot hole in the transmission. This will disengage the fulcrum from the retaining clip. Be careful not to cause any damage to the clip by pushing the release fork in the direction other than that of toward the transmission boot hole or by removing it with force.
6. If needed, unscrew the fulcrum from the transmission.
7. Before reassembly the following areas must be greased with Mitsubishi genuine grease Part No. 0101011 or its equivalent:
   - The fulcrum pivot-point on the release fork.
   - The release cylinder pivot-point on the release fork.
   - The release bearing pivot-point of the release fork.
   - The inner-surface of the release bearing.
8. Screw the fulcrum back into the transmission and tighten to 26 ft. lbs. (36 Nm).
9. Insert the thin end of the release fork through the transmission boot hole from the inside. Slide the fork onto the

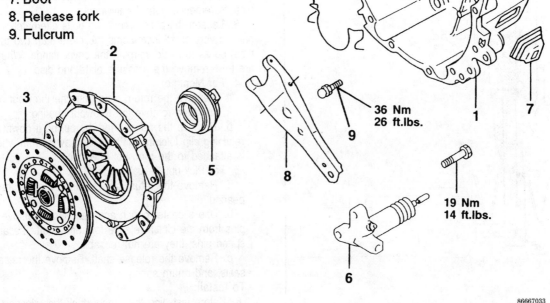

Fig. 35 Inner-transmission clutch linkage and clutch plate removal and installation — hydraulic clutch systems

1. Transmission
2. Clutch cover
3. Clutch disc
4. Return spring
5. Clutch release bearing
6. Release cylinder
7. Boot
8. Release fork
9. Fulcrum

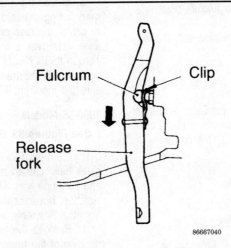

Fig. 36 Slide the release fork toward the release fork boot hole in the transmission to unhook the retainer clip — hydraulic clutch systems

7-32 DRIVE TRAIN

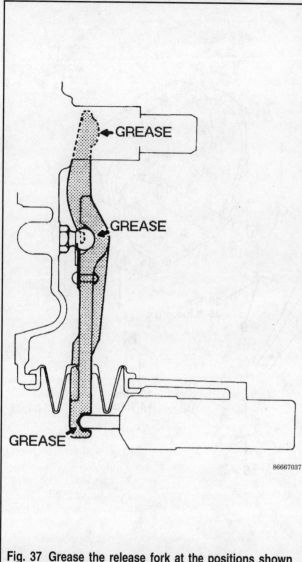

Fig. 37 Grease the release fork at the positions shown prior to installation — hydraulic clutch systems

fulcrum and make sure the clip engages the fulcrum pivot-point.

10. Slide the release fork boot over the outside end of the fork and install it to the transmission hole.
11. Mount the release cylinder onto the outside of the transmission, making sure that the actuating rod fits into the indent on the release fork.
12. Slide the release bearing onto the transmission end shaft (make sure the bearing has already been greased). Make certain that the release bearing is installed facing the correct direction.
13. Install the return spring.
14. Install the transmission into the vehicle.

Disc and Pressure Plate

REMOVAL & INSTALLATION

1983-89 Models
♦ See Figure 38

1. Remove the transmission.
2. Insert the forward end of an old transmission input shaft or a clutch disc guide tool into the splined center of the clutch disc. This will keep the disc from dropping when the pressure plate is removed from the flywheel.
3. Loosen the clutch mounting bolts alternately and diagonally in very small increments (no more than two turns at a time) so as to avoid warping the cover flange. When all bolts are free, remove the pressure plate and disc.
4. For Montero:
    a. Remove the return clips from the throwout bearing and fork, and then remove the throwout bearing.
    b. Remove the release fork by pulling it downward, so the retaining clip fulcrum will be pulled away from where the clip is attached to the fork, not toward it.
5. For Pick-up:
    a. Remove the return clip and then remove the release bearing.
    b. Use a center punch and hammer to remove the spring pins from the clutch release fork and shaft. Discard the spring pins; they are not reusable.
    c. Remove the release shaft. Remove the release fork, seals, and return spring.

To install:

6. Before installing the clutch, pack the release fork fulcrum hole and slave cylinder pushrod hole with grease. Also pack grease into the groove on the inside diameter of the throwout bearing.
7. Wipe all friction surfaces of the clutch disc, pressure plate, and flywheel to remove any grease or oil. Lightly grease the clutch disc splines and transmission input shaft splines.
8. Locate the clutch disc on the flywheel with the stamped mark facing outward. Use a clutch disc guide or old input shaft to center the disc on the flywheel, and then install the pressure plate over it. Install the bolts and tighten them evenly. Torque to 18 Nm (13 ft. lbs.).
9. Install the transmission in reverse of removal, making sure the input shaft is lined up properly to avoid bending it.

1990-95 Models
♦ See Figures 33, 35 and 38

The 1990-95 Pick-ups equipped with the 2.4L 4-cylinder engines have cable-actuated clutches and the 1990-95 Monteros and Pick-ups with 3.0L engines are equipped with hydraulic clutches, however the procedures for removing the clutches from the flywheels is the same.

1. Remove the transmission. For the procedures for the removal of the transmission, refer to the previously described instructions in this section.
2. Remove the clutch cover from the flywheel by unbolting the mounting bolts. Do not allow the clutch disc to fall. If the

DRIVE TRAIN 7-33

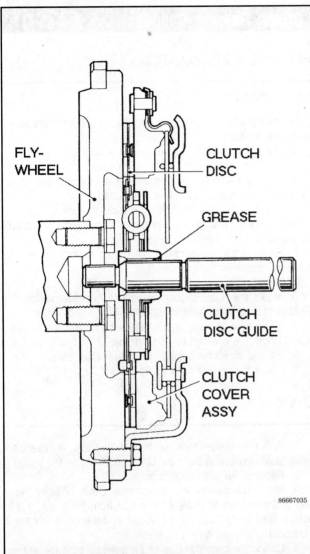

Fig. 38 Side-view internal schematic of the common type of the clutch pressure plate and cover

disc is to be reused, install the clutch disc guide tool before removing the clutch disc.

To install:

3. Apply the specified grease to the clutch disc splines and squeeze it in place with a brush. The specified grease is Mitsubishi genuine grease Part No. 0101011 or equivalent.
4. Use a clutch disc guide to position the disc on the flywheel.
5. Slide the clutch cover over the disc guide and install the mounting bolts until finger tight. Tighten the mounting bolts to 14 ft. lbs. (19 Nm) in a crisscross fashion.
6. Install the transmission.

Clutch Master Cylinder

REMOVAL & INSTALLATION

▶ See Figure 27

1. Loosen the bleeder screw and drain the clutch fluid.

2. Disconnect the pushrod at the clutch pedal by removing the cotter pin, and pulling out the clevis pin.
3. Disconnect the hydraulic tube at the master cylinder. Remove the master cylinder.
4. Discard the old master cylinder gasket, a new one will be needed for reassembly.
5. When reinstalling, mount the cylinder (with a new gasket) and tighten the bolts to 10 ft. lbs (14 Nm).
6. Connect the pushrod to the pedal, using a new cotter pin.
7. Connect the hydraulic tube to the cylinder and tighten it to 11 ft. lbs. (15 Nm).
8. Refill and bleed the system.

OVERHAUL

▶ See Figures 39, 40 and 41

1. Remove the master cylinder.
2. Remove the piston stop ring from the rear of the unit with snapring pliers.
3. Remove the damper and pushrod, if equipped.
4. Pull out the piston assembly.
5. Unscrew the clamp holding the reservoir onto the master cylinder, then pull the reservoir off of the cylinder.
6. Inspect the cylinder bore for rust or scoring. Inspect the piston cup for wear or damage. Inspect the inner bore of the hydraulic line connection for clogging. Clean a clogged bore or replace scored or damaged parts as necessary.
7. Use an inside micrometer to measure the cylinder bore in two directions at right angles to each other in the beginning, middle, and end of the bore. Measure the outer diameter of the piston in a similar way. Compare the figures. If clearance exceeds 0.0059 in. (0.15mm) at any point, replace the cylinder and the piston. This is a critical dimension; measure and compute carefully.

✱✱CAUTION

Do not disassemble the piston assembly.

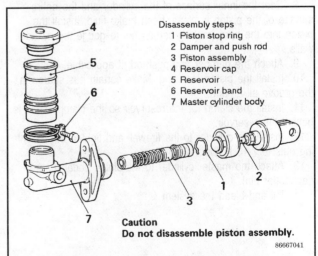

Fig. 39 Clutch master cylinder disassembly and assembly components — hydraulic clutch systems

DRIVE TRAIN

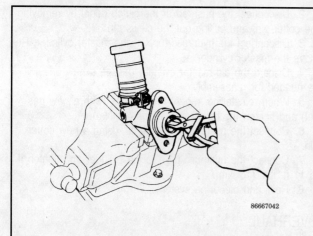

Fig. 40 Using a vice will make disassembly much easier — use snapring pliers to remove the snapring from the master cylinder body

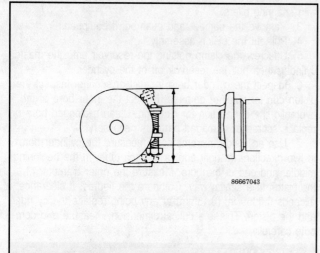

Fig. 41 Install the reservoir retaining clamp so that the adjusting screw falls within the specified area

To assemble:

8. Coat the inner surface of the cylinder and the entire surface of the piston in clean, fresh brake fluid. Install the piston into the bore, making certain not to gouge the cylinder walls.
9. Attach the damper and pushrod, if applicable.
10. Install the piston stop ring. Make certain it is seated in the groove all around.
11. Install the clamp for the reservoir so the clamp screw is behind the reservoir.
12. Install the cylinder to the firewall and tighten the mounting nuts to 7-10 ft. lbs. (10-15 Nm).
13. Attach the master cylinder to the clutch pedal (use a new cotter pin).
14. Fill and bleed the system.

Clutch Release Cylinder

REMOVAL & INSTALLATION

▶ See Figure 27

The clutch release cylinder is mounted to the left-hand side of the transmission.

1. Raise and support the vehicle on jackstands.
2. Position a drain pan under the release cylinder.
3. Loosen the bleeding valve and let the system drain into the pan. Used fluid should never be reused.
4. Remove the eye bolt holding the clutch hose to the release cylinder. Discard the two gaskets from this connection, new ones will be needed for reassembly.
5. Remove the two mounting bolts, then remove the release cylinder from the transmission.

To install:

6. Mount the clutch release cylinder to the transmission. Tighten the mounting bolts to 22-30 ft. lbs. (31-42 Nm).
7. Using new gaskets, attach the clutch hose with the eye bolt. Tighten the eye bolt to 14-18 ft. lbs. (20-25 Nm).
8. Fill and bleed the system. Check for fluid leaks.
9. Lower the vehicle back to the ground.

OVERHAUL

1. With the cylinder removed from the vehicle, remove the valve plate and spring from the fluid port.
2. Remove the pushrod and boots.
3. Mount the cylinder in a bench vise. Cover or block the end of the cylinder to prevent the piston from flying out. Use compressed air to force the piston out of the bore. Apply the air slowly to prevent fluid from splashing.
4. Check the inner surfaces of the cylinder bore for any sign of scratching or rust. Check the piston surface for the same condition and replace it if needed. Inspect the edges of the piston cup; replace if excessive wear or fatigue is found.

To assemble:

5. Coat the internal components is clean fresh brake fluid. Coat the inside of the cylinder. Install the spring and the piston and cup into the cylinder.
6. Install the boots and pushrod. Install the spring and valve plate.
7. Install the unit to the vehicle. Tighten the mounting bolts to 26 ft. lbs. (35 Nm). The end of the pushrod should be greased before installation as should its hole in the clutch release arm. If the bleeder screw was removed, it should be tightened to 8 ft. lbs. (11 Nm).
8. Connect the clutch fluid tube, tightening the hose fitting to 16 ft. lbs. (22 Nm).

SYSTEM BLEEDING

Make sure the clutch master cylinder is filled with the correct fluid (DOT 3 brake fluid). Loosen the bleeder screw at the slave cylinder. Push the clutch pedal down slowly until all air is expelled (bubbles in the fluid); do not release the pedal but

DRIVE TRAIN 7-35

hold it depressed. Retighten the bleeder screw. Release the clutch pedal. Refill the master cylinder to the correct level.

This process may need repetition several times; the key point to remember is that the clutch pedal MUST be held down until the bleeder screw is tightened. If the pedal is released with the bleeder open, air is sucked into the system.

Continue bleeding the system until no air is expelled with the fluid. When a steady stream of fluid is obtained, tighten the bleeder and top off the clutch fluid reservoir.

AUTOMATIC TRANSMISSION

Identification

▶ See Figures 42, 43, 44, 45, 46, 47, 48 and 49

On most models, the Vehicle Information Code Plate mounted on the firewall in the engine compartment holds the information as to which transmission the vehicle is equipped with. On the third line, labeled "Transmission", model is listed followed by a space, then either the axle ratio or the serial number is stamped following the space.

There are five different possible automatic transmissions available for the Pick-ups and Monteros from 1983-95. The transmissions are the MA 904A, the AW 372, the KM148 (4WD), the R4AC1 and the V4AW2 or 3 (4WD). The MA 904A is the oldest transmission available, and was stock on the 1983-86 Monteros and Pick-ups. This transmission is actually a Chrysler Torqueflite transmission. The MA 904A transmission combines a torque converter and a fully automatic 3-speed gear system. The converter housing and transmission case are an integral aluminum alloy die casting. The transmission consists of two multiple disc clutches, an overrunning clutch, two servos and bands, and two planetary gear sets to provide three forward ratios and a reverse ratio. The common sun gear of the planetary gear sets is connected to the front clutch by a driving shell which is splined tot he sun gear and to the

AUTOMATIC TRANSMISSION IDENTIFICATION

Year	Model	Engine	Transmission Identification	Drive Train
1983-86	Montero	2.6L	MA904A	RWD
	Pick-up	2.0L	MA904A	RWD
	Pick-up	2.6L	MA904A	RWD
1987-88	Montero	2.6L	KM148	4WD
	Pick-up	2.0L	AW372	RWD
	Pick-up	2.6L	AW372	RWD
	Pick-up	2.6L	KM148	4WD
1989	Montero	2.6L	KM148	4WD
	Montero	3.0L	V4AW2	4WD
	Pick-up	2.0L	AW372	RWD
	Pick-up	2.6L	AW372	RWD
	Pick-up	2.6L	KM148	4WD
1990-93	Montero	3.0L	V4AW2/ V4AW3	4WD
	Pick-up	2.4L	R4AC1	RWD
	Pick-up	3.0L	R4AC1	4WD
1994-95	Montero	3.0L	V4AW3	4WD
	Montero	3.5L	V4AW3	4WD
	Pick-up	2.4L	R4AC1	RWD
	Pick-up	3.0L	R4AC1	4WD

RWD - Rear Wheel Drive
4WD - Four Wheel Drive

86667c02

7-36 DRIVE TRAIN

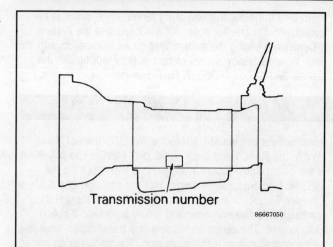

Fig. 42 The transmission identification tag is located on the left side of the transmission — MA 904A, AW 372 and KM 148 transmissions

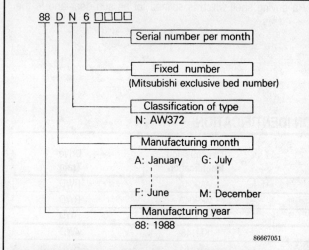

Fig. 43 The AW 372 transmission identification tag breaks down into 5 categories of numbers

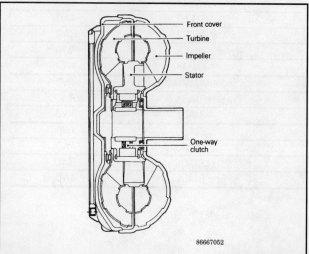

Fig. 44 All of the Mitsubishi trucks and Monteros utilize the same type converter

front clutch retainer. The hydraulic system consists of an oil pump, and a single valve body which contains all of the valves except the governor valve.

The AW 372, KM148 (4WD), R4AC1 and V4AW2 (and V4AW3) (4WD) are 4-speed units, consisting of a gear type oil pump, 3 multiple disc clutches, 3 one-way clutches, 4 multiple disc brakes and 2 planetary gear sets to provide 4 forward ratios and a reverse ratio. Fourth gear is a 0.668:1 overdrive range. The hydraulic control system provides fully automatic operation based on vehicle speed and throttle opening.

The KM148 and the V4AW2 (and V4AW3) incorporate a constant mesh type transfer case for use in 4WD vehicles. The transfer case has a high/low and a 2WD/4WD position. By operating the transfer control lever, running at 2WD high (2H), 4WD high (4H) or 4WD low (4L) can be selected freely. The KM148 transmission is the same as the AW372 from the governor assembly forward, and the V4AW2 (and V4AW3) is the same as the R4AC1 from the governor assembly forward.

Transmission part and identification numbers for the MA 904A, the AW372, and the KM148 are stamped on the left side of the case just above the oil pan gasket surface. The 1st 2 numbers are the manufacturing year. The 3rd number is the manufacturing month with "A" meaning January and "M" meaning December. The 4th number is the transmission classification. The 5th number is a check digit and the last set of numbers is the serial number.

The transmission identification numbers for the R4AC1 and the V4AW2 (and V4AW3) are stamped on the vehicle information plate located on the firewall in the engine compartment. The 1st 5 numbers indicate the transmission model. The last 4 numbers indicate the final gear ratio.

The recommended and preferred fluid is Mopar ATF Plus Type 7176 or Dexron® II (or its preceding fluid) automatic transmission fluid.

All of the automatic transmissions utilize a torque converter that consists of an impeller turbine stator, a oneway clutch and front and rear covers. It is a non-serviceable, sealed constructed unit, in which the surfaces of the outer and inner covers are welded together.

Torque convertors are coded by the manufacturer and sold through varied parts networks. Specific part numbers are used to identify the convertors and to allow proper alignment to the transmission. When replacing a converter, verify the replacement unit is the same as the unit originally used with the engine/transmission combination.

Fluid, Pan and Filter Service

For pan and filter service and fluid change procedures, please refer to Section 1.

DRIVE TRAIN 7-37

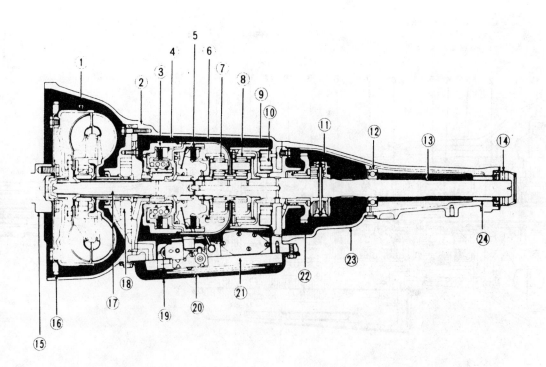

(1) Torque converter
(2) Transmission case
(3) Front clutch
(4) Kickdown band
(5) Rear clutch
(6) Driving shell
(7) Front planetary gear set
(8) Rear planetary gear set
(9) Low-reverse band
(10) Overrunning clutch
(11) Governor
(12) Bearing
(13) Output shaft
(14) Seal
(15) Crankshaft
(16) Ring gear (Drive plate)
(17) Input shaft
(18) Oil pump
(19) Oil pan
(20) Oil filter
(21) Valve body
(22) Governor support (Parking gear)
(23) Extension housing
(24) Bushing

Fig. 45 Internal schematic of the MA 904A automatic transmission components — 1983-86 Pick-ups and Monteros

7-38 DRIVE TRAIN

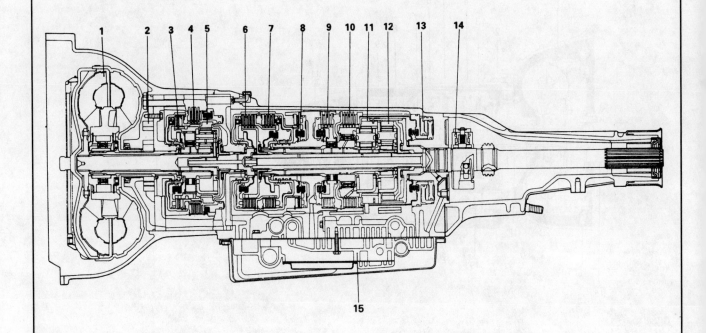

1. Torque coverter
2. Oil pump
3. Over drive clutch
4. Over drive brake
5. Over drive planetary gear
6. Forward clutch
7. Direct clutch
8. Brake No.1
9. Brake No.2
10. Brake No.3
11. Front planetary gear
12. Rear planetary gear
13. Brake No.3 piston
14. Governor
15. Valve body

Fig. 46 Internal schematic of the AW372 automatic transmission components — 1987-89 Pick-ups

DRIVE TRAIN 7-39

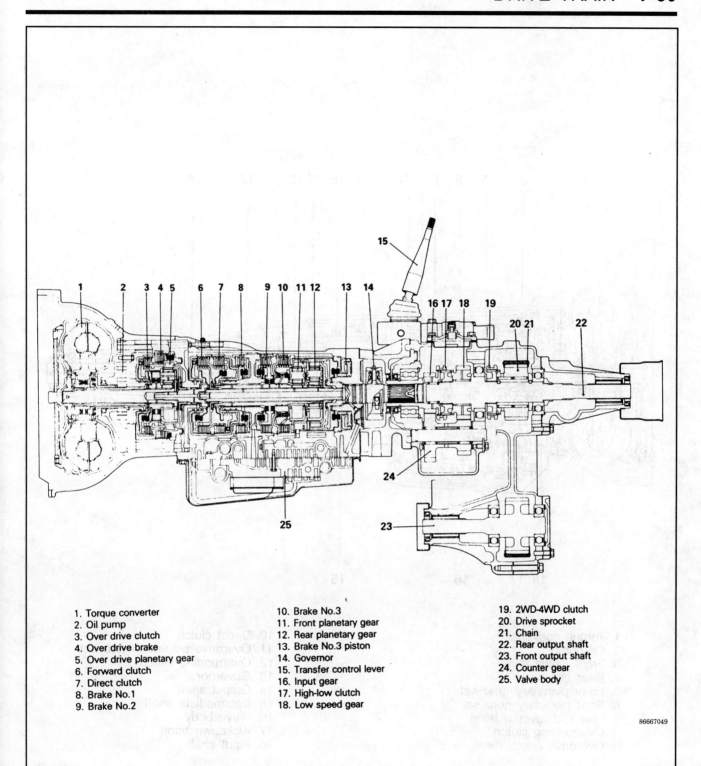

1. Torque converter
2. Oil pump
3. Over drive clutch
4. Over drive brake
5. Over drive planetary gear
6. Forward clutch
7. Direct clutch
8. Brake No.1
9. Brake No.2
10. Brake No.3
11. Front planetary gear
12. Rear planetary gear
13. Brake No.3 piston
14. Governor
15. Transfer control lever
16. Input gear
17. High-low clutch
18. Low speed gear
19. 2WD-4WD clutch
20. Drive sprocket
21. Chain
22. Rear output shaft
23. Front output shaft
24. Counter gear
25. Valve body

Fig. 47 Internal schematic of the KM 148 (4WD) automatic transmission components — 1987-89 Pick-ups and Monteros

7-40 DRIVE TRAIN

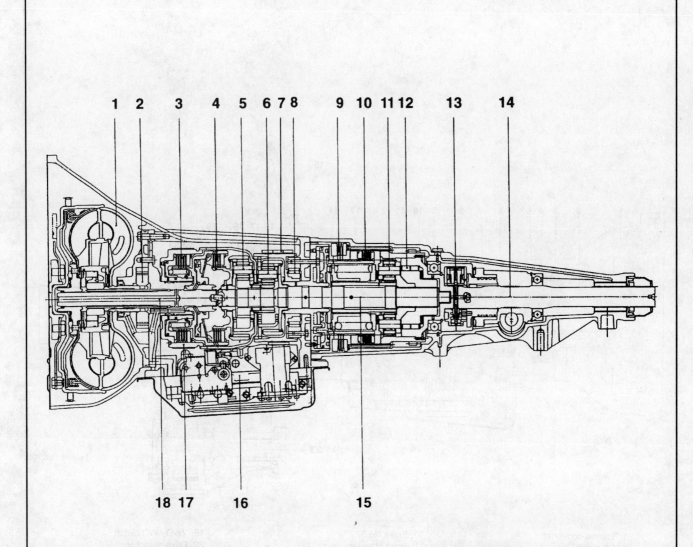

1. Torque converter
2. Oil pump
3. Front clutch
4. Rear clutch
5. Front planetary gear set
6. Rear planetary gear set
7. Low and reverse band
8. Overrunning clutch
9. Overdrive clutch
10. Direct clutch
11. Overdrive planetary gear set
12. Overrunning clutch
13. Governor
14. Output shaft
15. Intermediate shaft
16. Valve body
17. Kickdown band
18. Input shaft

Fig. 48 Internal schematic of the R4AC1 automatic transmission components — 1990-95 Pick-ups

DRIVE TRAIN 7-41

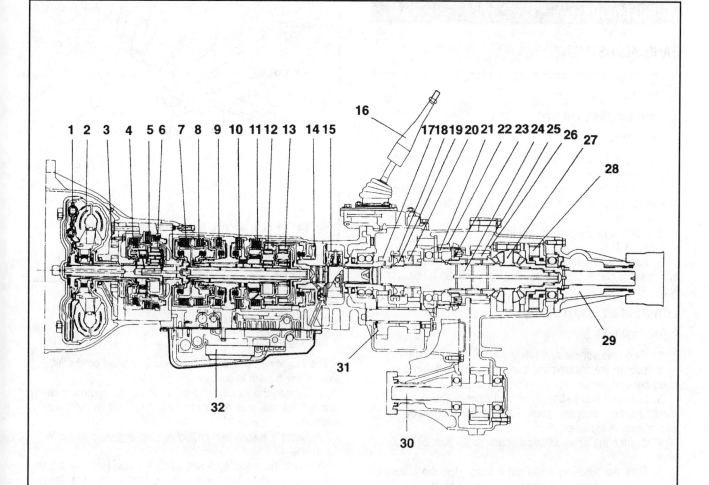

1. Lockup clutch
2. Torque converter
3. Oil pump
4. Overdrive clutch
5. Overdrive brake
6. Overdrive planetary gear
7. Forward clutch
8. Direct clutch
9. Brake No. 1
10. Brake No. 2
11. Brake No. 3
12. Front planetary gear
13. Rear planetary gear
14. Brake No. 3 piston
15. Governor
16. Transfer control lever

17. Input gear
18. High-low sleeve
19. High-low hub
20. Low speed gear
21. Differential lock hub
22. 2WD-4WD synchronizer sleeve
23. 2WD-4WD hub
24. Transfer drive shaft
25. Drive sprocket
26. Chain
27. Center differential
28. Viscous coupling
29. Rear output shaft
30. Front output shaft
31. Counter gear
32. Valve body

Fig. 49 Internal schematic of the V4AW2 (and V4AW3) automatic transmission components — 1990-95 Monteros

7-42 DRIVE TRAIN

Adjustments

BAND ADJUSTMENT

1990-95 Pick-Ups

KICKDOWN BAND (FRONT)

▶ See Figure 50

The kickdown band adjusting screw is located on the left side of the transmission case.

1. Loosen the locknut and back the nut off approximately five turns. Make sure the screw turns freely within the transmission case.
2. Using a torque wrench, tighten the adjusting screw to 72 inch lbs. (8 Nm).
3. Back the adjusting screw out $2^{7}/_{8}$ turns. Hold the adjusting screw in place, then tighten the locknut to 30 ft. lbs. (41 Nm).

LOW-REVERSE BAND (REAR)

▶ See Figure 51

1. Raise the vehicle and safely support it on jackstands.
2. Loosen the transmission pan, drain the fluid into a container and remove the pan from the transmission.
3. Loosen the adjusting screw and locknut, then back the nut off approximately five turns. Test the adjusting screw for free turning in the lever.
4. Tighten the band adjusting screw to 30 inch lbs. (3.5 Nm).
5. Back the adjusting screw out 6 turns. Hold the adjusting screw in this position, and tighten the locknut to 25 ft. lbs. (34 Nm).
6. Reinstall the transmission pan using a new gasket. Tighten the transmission pan bolts to 12.5 ft. lbs. (17 Nm).
7. Fill the transmission with the specified fluid.

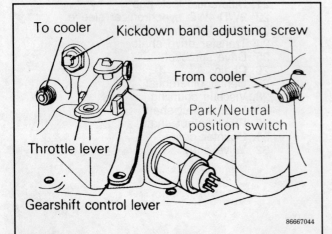

Fig. 50 The kickdown band adjusting screw is in the same location for both the R4AC1 and the MA 904A transmissions

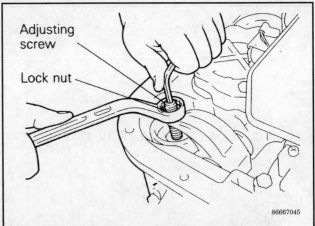

Fig. 51 To gain access to the low-reverse band adjusting screw, the transmission pan must be removed — make sure to have new gaskets for reassembly

1983-86 Pick-Ups and Monteros

KICKDOWN BAND (FRONT)

▶ See Figure 50

The kickdown band adjusting screw is located on the left side of the transmission case.

1. Loosen the locknut and back the nut off approximately five turns. Make sure the screw turns freely within the transmission case.
2. Using a torque wrench, tighten the adjusting screw to 5.8 ft. lbs. (7.8 Nm).
3. Back the adjusting screw out $3^{1}/_{2}$ turns. Hold the adjusting screw in place, then tighten the locknut to 30-41 ft. lbs. (40-55 Nm).

LOW-REVERSE BAND (REAR)

▶ See Figure 51

➡ For this procedure the Mitsubishi tool MD998358 will be needed.

1. Raise the vehicle and safely support it on jackstands.
2. Loosen the transmission pan, drain the fluid into a container and remove the pan from the transmission.
3. Loosen the adjusting screw and locknut, then back the nut off approximately five turns. Test the adjusting screw for free turning in the lever.
4. Tighten the band adjusting screw with the tool MD998358 to 3.6 ft. lbs. (4.9 Nm).
5. Back the adjusting screw out 7 turns. Hold the adjusting screw in this position, and tighten the locknut to 25-35 ft. lbs. (34-47 Nm).
6. Reinstall the transmission pan using a new gasket. Tighten the transmission pan bolts to 11-14 ft. lbs. (15-20 Nm).
7. Fill the transmission with the specified fluid.

SHIFTER CONTROL LINKAGE

1983-86 Pick-Ups and Monteros

1. Set the transmission shifter to the N position.

DRIVE TRAIN 7-43

2. Turn the rod adjusting cam to adjust the distance from the end of the selector lever to the end of the rod adjusting cam as indicated in the illustration.

3. After the adjustment, move the lever through all the ranges to confirm that it moves smoothly and operates correctly.

KICKDOWN LINKAGE ADJUSTMENT

1983-86 Pick-Ups and Monteros

♦ See Figures 52 and 53

➡When the idle speed is adjusted, the throttle linkage will need adjusting.

1. Loosen the bolts so that rods B and C can slide properly.
2. Allow the engine to warm up until the engine coolant temperature reaches the average operating temperature of 170°F (80°C). Confirm complete release of the fast idle. This confirmation can be made by checking to see if the cam surface of the choke lever of the carburetor is completely off the cam follower of the throttle lever.
3. Lightly pushrod A or the transmission throttle lever toward the idle stopper (toward the back of the transmission), and set the rods to the idle position. Tighten the bolt securely to connect rods B and C.
4. Make sure that, when the carburetor throttle valve is wide open, the transmission throttle lever moves with an operating angle of 47.5-54° of movement. Make sure that, when the throttle linkage alone is slowly returned from the fully opened position, the transmission throttle lever completely returns to idle by return spring force.

1987-89 Pick-Ups and Monteros

♦ See Figure 54

1. Check and adjust the idle speed.

➡When the engine idle adjustment has been performed, always adjust the throttle control cable.

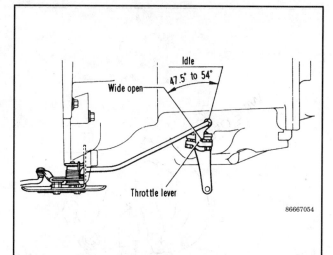

Fig. 53 Check the movement range of the throttle lever on the transmission when finished adjusting the rods

2. Make certain no bend or deformation is present in either the carburetor throttle lever or the cable mounting bracket.
3. Measure the length between the inner cable stopper and the cover end with the throttle wide open.
4. Correct measurement is 2.05-2.09 in. (52-53mm). If the cable is out of specification, adjust the inner cable bracket by moving it up or down as needed.

1990-95 Pick-Ups

♦ See Figures 55, 56, 57 and 58

1. Make sure that the kickdown cable is correctly attached to the bell crank lever.
2. Pull on the inner kickdown cable (transmission end) while the throttle lever is held in the idle position.
3. While the inner cable is being pulled, adjust the inner cable stopper so that the space between the stopper and the outer cable becomes 0.031-0.059 in. (0.8-1.5mm).
4. Pull the inner cable (transmission end) and hold the throttle lever to the full open position. With this position being maintained, loosen and adjust the length of the bell crank lever arm so that the space between the inner cable stopper and the outer cable becomes 1.46-1.50 in. (37-38mm).
5. After adjusting the engine to the regular idling position, fix the inner cable to the throttle lever (engine end of cable). At this time, fasten the outer cable with the adjusting nuts so that the space between the inner cable stopper and the outer cable becomes 0.031-0.05 in. (0.8-1.5mm).
6. With the throttle lever (engine end) being fully open, confirm that the space between the outer cable and the inner cable is 1.30-1.38 in. (33-35mm).

1990-95 Monteros

3.5L AND 1990-94 3.0L ENGINES

♦ See Figure 59

1. Check for a defective or bent throttle lever or throttle cable bracket.
2. With the accelerator depressed, check that the distance between the inner cable stopper and dust cover surface is within 0-0.4 in. (0-1mm).

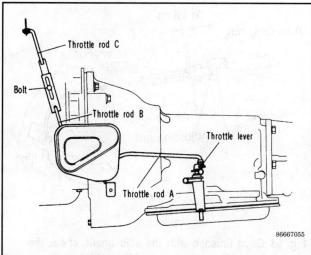

Fig. 52 Throttle rod (kickdown) system — MA904A transmissions

7-44 DRIVE TRAIN

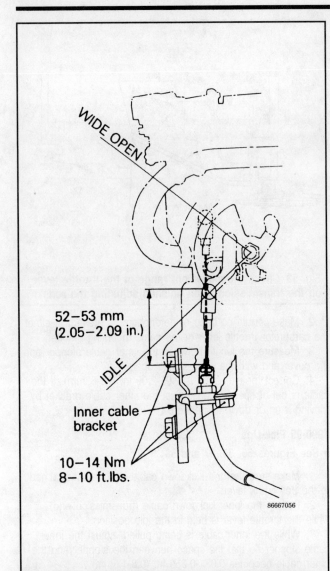

Fig. 54 The kickdown adjustment set up on the AW372 and KM148 transmissions

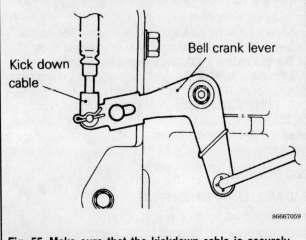

Fig. 55 Make sure that the kickdown cable is securely attached to the bell crank lever before adjusting the cable

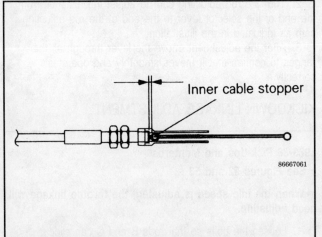

Fig. 56 Move the inner cable stopper to change the clearance at idle — the clearance should be 0.8-1.5mm (0.031-0.05 in.)

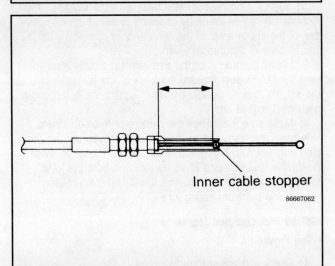

Fig. 57 Measure the clearance when the cable is pulled all the way to the wide open position

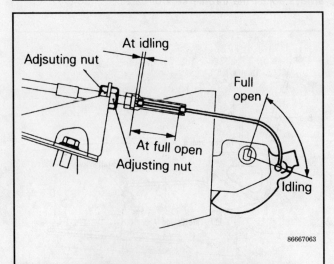

Fig. 58 Once finished with the adjustment, check the clearances on the engine end of the cable

DRIVE TRAIN 7-45

3. If the distance is outside of the standard value, adjust using the adjusting nut on the bracket.

1995 3.0L ENGINES

♦ See Figure 60

1. Make sure that the throttle lever and the bracket have no deformations or damage.
2. Remove the boot at the outer cable so that the inner cable stopper can be seen.
3. Measure the dimension between the end of the inner cable stopper and that of the outer cable with the throttle lever fully open. The standard value is 1.34-1.38 in. (34-35mm).
4. If the distance is outside the standard value, adjust using the adjusting nut on the bracket.

THROTTLE PRESSURE ADJUSTMENTS

♦ See Figure 61

1983-86 Pick-Ups And Monteros

Throttle pressures cannot be tested accurately; therefore, the adjustment should be measured if a malfunction is evident. For this procedure you will need the Mitsubishi tool MIT3763 Throttle Pressure Adjust Tool.

1. Insert the special tool between the throttle lever cam and the kickdown valve.
2. By pushing in on the tool, compress the kickdown valve against its spring so that the throttle valve is completely bottomed inside the valve body.
3. As force is being exerted to compress the spring, turn the throttle lever stop screw with an allen wrench until the head of the screw touches the throttle lever tang with the throttle lever cam touching the tool and the throttle valve bottomed. Be sure the adjustment is made with the spring fully compressed and the valve bottomed in the valve body.

1990-95 Pick-Ups

Throttle pressures cannot be tested accurately; therefore, the adjustment should be measured if a malfunction is evident. For this procedure you will need the Mitsubishi Tool C — 3763.

1. Insert the special tool between the throttle lever cam and the kickdown valve.
2. By pushing in on the tool, compress the kickdown valve against its spring so that the throttle valve is completely bottomed inside the valve body.
3. As force is being exerted to compress the spring, turn the throttle lever stop screw with an allen wrench until the head of the screw touches the throttle lever tang with the throttle lever cam touching the tool and the throttle valve bottomed. Be sure the adjustment is made with the spring fully compressed and the valve bottomed in the valve body.

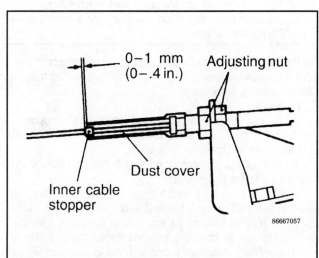

Fig. 59 Adjust the clearance by turning the adjusting nut — 3.0L and 3.5L engines 1990-94

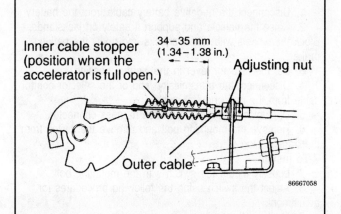

Fig. 60 To correctly adjust the cable, the boot must first be removed — adjust the cable by turning the adjusting nut

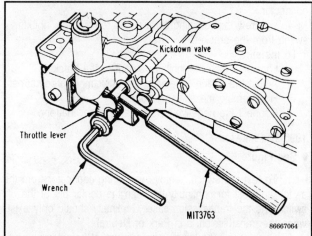

Fig. 61 Use the special tool (either MIT3763 or C — 3763) to depress the kickdown valve while tightening the throttle lever — MA 904A and R4AC1 transmissions

7-46 DRIVE TRAIN

Park/Neutral Position (Inhibitor) Switch

The inhibitor switch on the Mitsubishi trucks and Monteros also houses the back up light switch.

REMOVAL & INSTALLATION

1983-86 Pick-Ups and Monteros

1. Disconnect the negative battery cable from the battery.
2. While pushing the button, remove the selector handle assembly from the selector lever by loosening the set screw.
3. Remove the console box, then remove the position indicator assembly upward by loosening the attaching screws.
4. Disconnect the inhibitor (park/neutral switch) switch wiring.
5. Loosen and remove the attaching screws of the inhibitor switch. Remove the switch from the selector assembly.

To install:

6. Loosely secure the inhibitor switch to the selector assembly.
7. Connect the inhibitor switch wiring.
8. Adjust the inhibitor switch — refer to the following procedures for adjustment.
9. Set the position indicator back in place, then secure it in place with the mounting screws.
10. Install the console box.
11. Mount the selector handle to the lever and secure in place by tightening the set screw.

1987-89 Pick-Ups and Monteros

1. Disconnect the negative battery cable from the battery.
2. Remove the floor console, the selector handle and the indicator panel — refer to the shift linkage removal procedures in this section.
3. Unplug the inhibitor switch (park/neutral switch) harness connector.
4. Unscrew the retaining screws and remove the inhibitor switch from the bracket.

To install:

5. Loosely attach the inhibitor switch to the bracket.
6. Adjust the switch — refer to the adjustment procedures.
7. Reassemble the shifter assembly.
8. Connect the negative battery cable to the battery.

1990-95 Pick-Ups

▶ See Figure 62

1. To test the switch, remove the wiring connector from the switch and test for continuity between the center pin of the switch and the transmission case. Continuity should only exist when the transmission is in Park or Neutral.
2. Check the gearshift linkage adjustment before replacing a switch which tests bad.
3. Unscrew the switch from the transmission case and allow the fluid to drain into a container. Move the selector lever

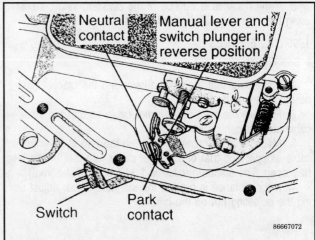

Fig. 62 Unscrew the park/neutral switch from the transmission case and drain the fluid into a container — R4AC1 transmissions

to the Park and then to Neutral positions, and inspect to see that the switch operating lever fingers are centered in the switch opening in the case.

To install:

4. Screw the switch and new seal into the transmission case and tighten to 25 ft. lbs. (34 Nm). Retest the switch with the test lamp or ohmmeter.
5. Add automatic transmission fluid to the transmission to bring it up to the proper level.
6. The back up lamp switch circuit is through the two outside terminals of the 3 terminal switch.
7. To test the back up switch, remove the wiring from the switch and test for continuity between the two outside pins.
8. Continuity should exist only with the transmission in reverse.
9. No continuity should exist from either pin to the case.

1990-95 Monteros

1. Disconnect the negative battery cable from the battery.
2. Raise the vehicle and support it safely on jackstands. Block the wheels with wheel chocks and apply the parking brake.
3. Set the selector lever in the Neutral position.
4. Disconnect the transmission end of the selector control cable from the park/ neutral switch.
5. Unplug the park/neutral switch harness connector.
6. Remove the mounting bolt and remove the switch from the transmission.

To install:

7. Loosely mount the switch with the mounting bolt.
8. Adjust the switch using the following procedures for adjustment.
9. Hook up the harness wiring.
10. Lower the vehicle to the ground. Set the selector lever in the Park position, then remove the wheel chocks.
11. Connect the negative battery cable, if not done already.

DRIVE TRAIN 7-47

ADJUSTMENT

⁂⁂WARNING

If the switch is faulty, the engine may start with the vehicle "in gear". If this happens the vehicle will accelerate when the engine starts. Keep the brakes on and be ready to switch the key off instantly.

1983-86 Pick-Ups and Monteros
▶ See Figures 63 and 64

1. Place the selector lever so that it is in the N position, then temporarily fasten the inhibitor switch (park/neutral switch) to the selector lever bracket. By using an ohmmeter, measure the continuity between the switch terminals BY and BY; move the switch forward and backward, and set it to the position of first continuity (where it first shows continuity).
2. Slide the inhibitor switch (park/neutral switch) so that there is 0.06 in. (1.5mm) clearance between the switch and the shifter lever.
3. Tighten the two attaching screws until snug.

1987-89 Pick-Ups (4WD) and Monteros
▶ See Figures 65, 66 and 67

1. Make sure the harness connector is securely attached.
2. Loosen the mounting screws.
3. Set the shift lever so that the pin at the end of the rod is at the neutral position — see the illustration.
4. Using a circuit tester between terminals 2-BY and 2-BY of the inhibitor switch connector, check the continuity when the inhibitor switch is moved back and forth, and mark the bracket.
5. Tighten the inhibitor switch mounting screws so that the switch edge is 0.1 in. (2.5mm) from the close edge of the lever.

1990-95 Monteros
▶ See Figures 68 and 69

1. Move the transmission selector lever to the N position.

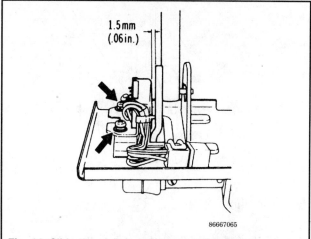
Fig. 64 Slide the switch sideways until the proper clearance of 1.5mm (0.06 in.) exists — MA 904A transmissions

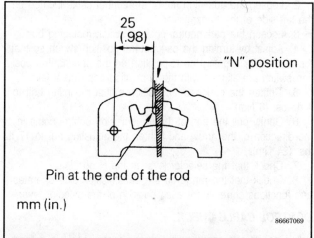
Fig. 65 To adjust the park/neutral switch, move the selector lever until the selector rod is in the position shown — KM 148 transmissions

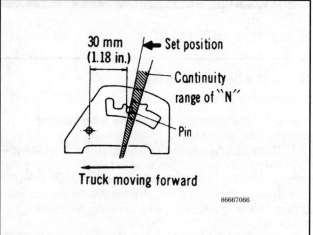
Fig. 63 Set the selector lever in the position shown, then adjust the switch so that it shows continuity — MA 904A transmissions

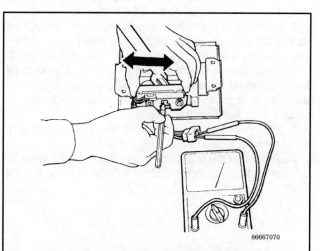
Fig. 66 Once the selector lever is in the correct position, use an ohmmeter to set the park/neutral switch — KM 148 transmissions

7-48 DRIVE TRAIN

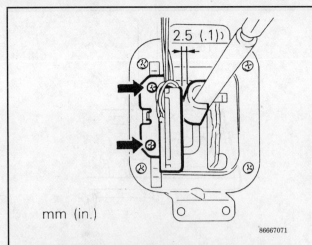

Fig. 67 Move the park/neutral switch to the side until the correct clearance has been reached — KM 148 transmissions

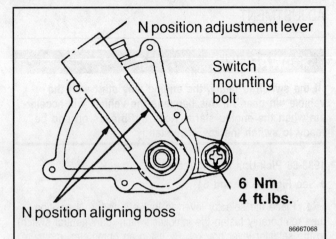

Fig. 68 Position the adjustment lever until it is aligned with the two N position aligning bosses — V4AW2 (V4AW3) transmissions

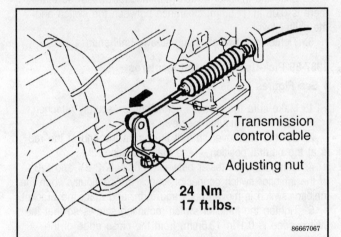

Fig. 69 When adjustments have been made, tighten the adjusting nut to 17 ft. lbs. (24 Nm) — V4AW2 (V4AW3) transmissions

2. Loosen the adjusting nut of the control cable located on the left side of the transmission.
3. Loosen the park/neutral position switch mounting bolt.
4. Adjust by turning the park/neutral position switch so that the bosses for aligning the N position on the park/neutral position switch are aligned with the N position adjustment lever.
5. Tighten the park/neutral position switch mounting bolt to 4 ft. lbs. (6 Nm).
6. Gently pull the end of the transmission control cable in the direction of the arrow and tighten the adjusting nut to 17 ft. lbs. (24 Nm).
7. Check that the selector lever to the N position.
8. Check that the range on the transmission side operates and functions correctly for each position of the selector lever.

CONTROL CABLE CHECK

It is possible to confirm whether the control cable is properly adjusted by checking whether the park/neutral position switch is performing well.
1. Apply the parking and service brakes fully.
2. Set the selector lever to the R position.
3. Turn the ignition key to the **START** position.
4. Slowly move the selector lever upward until it clicks as it fits into the notch of the P position. If the starter motor operates when the lever makes a click, the P position is correct.
5. Then slowly move the selector lever to the N position by the same procedure as in the preceding paragraph. If the starter motor operates when the selector lever is at the N position, then the N position is correct.
6. Also check that the vehicle does not begin to move and the lever does not stop between P — R — N — D.
7. The control cable is properly adjusted if, as described previously, the starter motor starts in both the P range and the N range.

Shifter Linkage

REMOVAL & INSTALLATION

1983-86 Pick-Ups and Monteros
▶ See Figures 70 and 71

Access to the underside of the vehicle will be needed throughout the procedure.
1. Disconnect the negative battery cable from the battery.
2. Remove the selector handle assembly while pushing the button in from the selector lever by loosening the set screw.
3. Remove the console box, then remove the position indicator assembly upward and out by loosening the attaching screws.
4. Disconnect the indicator light and inhibitor switch wiring.
5. Disconnect the control rod from the control arm by loosening the nut from under the floor.

DRIVE TRAIN 7-49

6. Remove the lever bracket assembly by loosening the attaching screws.

To install:

7. Install the lever bracket assembly and tighten the mounting screws until snug.

8. Before installing the transmission control arm to the selector lever, apply a suitable quantity of multipurpose grease (SAE J310a, NLGI grade #3) to the sliding part.

9. Tighten the attaching nut.

10. Check that the control arm moves smoothly without looseness. Replace the bushing if the arm is loose.

11. Attach the inhibitor switch and position indicator light wiring.

12. Adjust the inhibitor switch — refer to the previous procedures for this.

13. Set the position indicator assembly into position, then secure it in place with the mounting screws.

14. Install the console box.

15. Push the selector handle back onto the selector lever and secure it in place with the set screw.

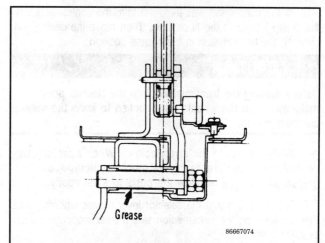

Fig. 71 Grease the transmission control arm in the position shown before installing it to the selector lever — MA 904A transmissions

(1) Selector handle
(2) Push button
(3) Set screw
(4) Rod adjusting cam
(5) Selector lever rod
(6) Selector lever
(7) Detent plate
(8) Position indicator cover (for sports)
(9) Inhibitor switch
(10) Position indicator cover
(11) Lever bracket cover
(12) Control arm
(13) Control rod

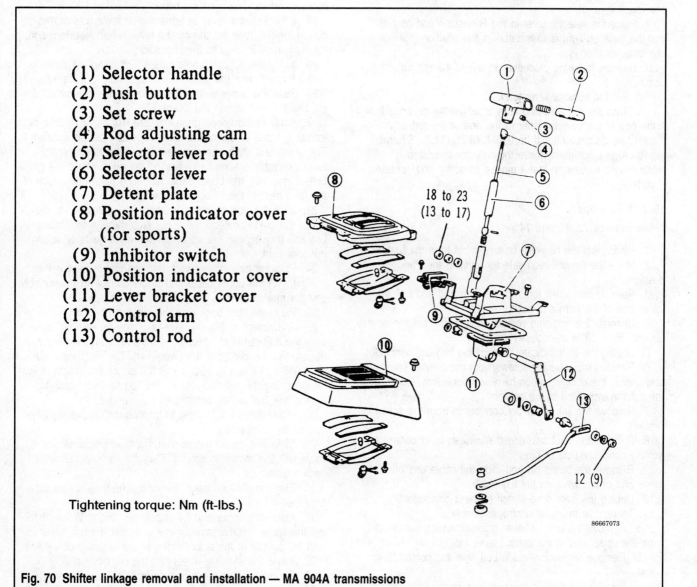

Tightening torque: Nm (ft-lbs.)

Fig. 70 Shifter linkage removal and installation — MA 904A transmissions

7-50 DRIVE TRAIN

16. If the control rod was removed from the control arm, set the selector lever to the N position. Then move the control rod to place the transmission in the neutral position.

✱✱CAUTION

Before moving the transmission into the Neutral position make sure that the wheels are chocked to keep the vehicle from rolling.

➡ Tighten the rod to the lever locknut. When attaching the control rod to the arm, apply as much multipurpose grease as required to allow the parts to slide easily.

17. Make certain that the selector lever moves smoothly and that the lever on the transmission side moves properly to each selector position.
18. Pull the parking brake lever. Fasten the seat belt. Depress the brake pedal, then confirm that the engine does not start while the selector lever is in the D, L, 2 or R positions. Make sure the engine starts when the lever is in the P or N position.
19. Place the selector lever in the R position and confirm that the back up light is illuminated in this position (but not in any other position).
20. Use the following procedure to adjust the rod adjusting cam.
    a. Set the selector lever to N.
    b. Turn the rod adjusting cam to adjust the distance from the end of the selector lever to the end of the rod adjusting cam (the distance should be 0.45-0.49 in. (11.3-12.7mm).
    c. After adjustment, move the selector lever through all the ranges to confirm that it moves smoothly and operates correctly.

1987-89 Pick-Ups

▶ See Figures 72, 73 and 74

1. Disconnect the negative battery cable from the battery.
2. Move the transmission selector lever to the Park position.
3. Remove the screw fitted to the horn pad and pull the lower part of the horn pad out to remove.
4. Unscrew the retaining screws, then lift the column covers up and down off of the column assembly.
5. Unplug the shift indicator illumination harness connector.
6. Remove the retaining screws from the column switch and unplug the column switch harness connectors, then slide the column switch off of the column.
7. Remove the link assembly connection from the column bracket.
8. Remove the boot covering the cable-to-lever control assembly connecting cotter pin.
9. Remove the cotter pin and slide the cable end off of the lower end of the lever control assembly.
10. Unplug the over drive switch harness connector.
11. To remove the lever control assembly:
    a. Remove the change lever guide attaching nut (located on the upper end of the control lever rod).
    b. Remove the bottom guide bolt from the control lever rod.
    c. Push the change lever guide and remove the control rod assembly so that the control rod assembly does not catch the detent plate portion. Make sure to account for the steel ball, the ball support and the spring.
12. Loosen the locking nut from the upper end of the control cable, then slide the cable out of the column mounting bracket.
13. Using a pair of needle-nose pliers, remove the cotter pin from the transmission lever. Disconnect the lower end of the selector cable.
14. Remove the retaining clip near the bottom of the cable.
15. Remove the selector cable from the vehicle.
16. If neither the selector lever nor the transmission lever is moved, the selector cable linkage should not need to be adjusted. However, the linkage should be checked and adjusted if not correct.

To install:

17. To install, route the cable in the vehicle the same way it was before it was removed.
18. Install the retaining clip to the retaining bracket in such a way so that the opening in the retaining clip is facing the opposite direction than the opening in the retaining bracket. Make sure before installing the clip, that the indent on the cable is set into the retaining bracket.
19. If the selector lever or transmission lever was moved during repairs, then set the control lever (when installed) and the transmission lever to the N position.
20. Install the adjusting nuts loosely; adjusting will need to be performed once the control lever assembly is installed.
21. Slide the upper end of the cable into the column retaining bracket and tighten the locknut.
22. Apply Multipurpose grease SAE J310, NLGI No.2 to the following areas of the control lever assembly before installation: The guide bolt shaft, the outside edge of the lower control lever assembly bracket (right around the edge where the guide bolt is inserted), the upper cable mounting pin, bottom face of the tang next to the steel ball assembly.
23. Slide the control lever assembly back into the vehicle, install the guide bolt, set the upper stud through the column bracket, then tighten the upper change lever guide attaching nut until tight.
24. Plug the over drive switch harness connector together.
25. Install the upper cable end to the control lever assembly, using a new cotter pin.
26. Push the boot back over the cable connection.
27. Adjustment of the link assembly may be needed. Measure the adjustable rod (not the short one) from the center of the stud to the center of the pivot shaft. The distance should be 2.098-2.114 in. (53.3-53.7mm). If it is not this length, adjust by loosening the nut and rotating the rod so that the dimension of the link assembly matches the standard.
28. Install the link assembly to the column bracket and the control lever assembly.
29. Slide and mount the column switch in place; secure it there with the mounting screws. Plug the harness connectors back together.
30. Plug the shift indicator illumination harness connector together.
31. Install the upper and lower column covers. When installing the upper column cover, make sure that the link assembly and the pointer of the select indicator are securely connected.
32. Install the steering wheel and the horn cover.
33. Connect the negative battery cable to the battery.
34. If the transmission lever was moved during repairs, set the control lever assembly and the control lever sub assembly

DRIVE TRAIN 7-51

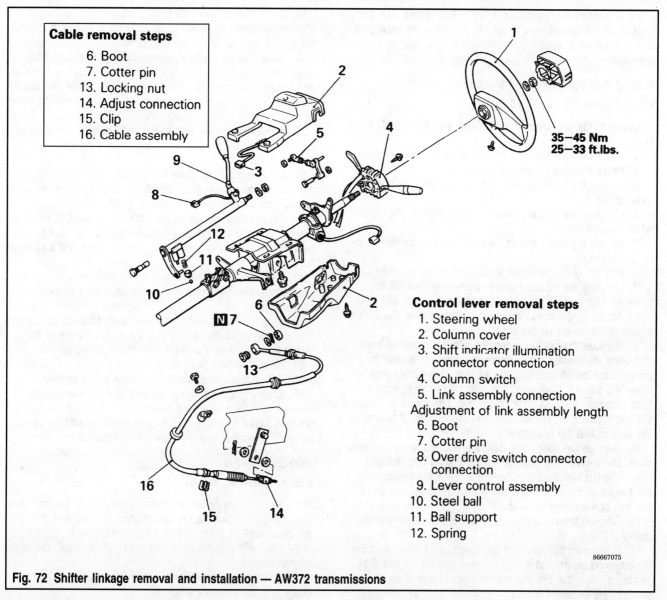

Fig. 72 Shifter linkage removal and installation — AW372 transmissions

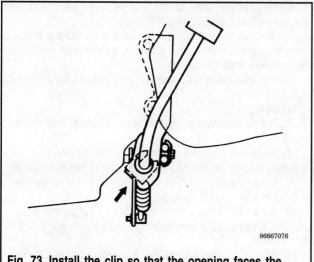

Fig. 73 Install the clip so that the opening faces the direction shown — AW372 transmissions

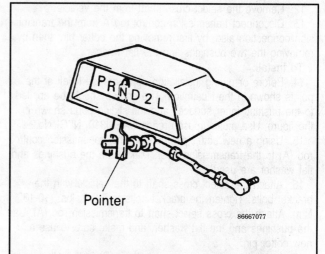

Fig. 74 When installing the upper column cover, make sure that the pointer engages the linkage correctly

in the N position and install the adjuster locknuts so that each position of the control lever assembly matches with each position of the control lever sub assembly and they operate normally.

1987-89 Pick-Ups (4WD) and Monteros
▶ See Figures 75, 76, 77, 78 and 79

Access to the underside of the vehicle will be needed throughout this procedure.

1. Disconnect the negative battery cable from the battery.
2. Press the selector lever cover downward.
3. Unplug the overdrive switch harness connector from the selector handle.
4. Unscrew the handle retaining screw, then lift the handle off of the shift lever.
5. Unplug the lower end of the shift handle over drive harness connector.
6. Pull the shift lever cover up and off of the lever.
7. Unscrew the indicator panel retaining screws, lift the panel up high enough to unplug the inhibitor switch and position indicator light harness connectors, then remove the indicator panel from the shift assembly.
8. Remove the bracket assembly, which consists of the shift lever mechanism. From under the vehicle, remove the transmission control rod from the transmission control arm by loosening the nut. Disconnect the select cross shaft from the heat protector, then from the control rod. Make sure to retain the nut, lockwasher, flat washer, connecting pin and bushings, flat washer and the heat shield.
9. Remove the dust cover from the end of the select cross shaft. Using snap ring pliers remove the snap ring (located under the just removed dust cap), then slide the two springs, the two cross shaft bushings and the cross shaft boot off of the end of the select cross shaft.
10. Remove the cross shaft bracket bolts from the transfer case.
11. Disconnect both ends of the select cross shaft from the transmission control rod A and transmission control rod B by removing the cotter pin. Remove the two bushings and the flat washer and remove the rods.
12. Remove the select cross shaft from the vehicle.
13. Disconnect transmission control rod A from the transmission connection also, by first removing the cotter pin, then by removing the two bushings and the flat washer.

To install:

14. Before or during installation apply grease to all of the points shown in the illustration. The grease should be applied to the bushing inner surfaces and the sliding parts shown in the figure. Use multipurpose grease SAE J310, NLGI No.2.
15. Using a new cotter pin, connect the transmission control rod (A) to the transmission lever. Make sure the bushings and flat washer are used.
16. Attach the select cross shaft to the vehicle with the bracket bolts. Tighten the bracket bolts to 7-9 ft. lbs. (10-13 Nm). Attach the cross select shaft to transmission rod (A). Use the bushings and the flat washer, and make sure to use a new cotter pin.
17. Install the transmission control rod (B) to the other end of the select cross shaft.
18. Install the components to the end of the select cross shaft in the following order: cross shaft boot, spring, bushing, spring, flat washer, snapring, dust cover.
19. Install the lever pin to the removed bracket and shift lever assembly. Insert the pin into the lever, slide the bushing and washer over the pin. Secure with a new cotter pin. Mount the heat shield over the pin end and secure with the mounting screw.
20. Install the bracket and shift lever assembly, then secure it in place with the mounting brackets. Move the shift lever to the Neutral position.
21. Once again from under the vehicle, move the transmission to the Neutral position, and then connect the transmission control arm and transmission control rod B together.
22. Set the indicator panel over the shifter lever, then plug the inhibitor and the indicator light harness connectors together. Mount the indicator panel to the bracket and shift lever assembly.
23. Slide the shifter cover over the shifter lever.
24. Hook the lower end of the over drive wire to the connector.
25. Plug the upper end of the over drive harness connector into the shifter handle, then install the shifter handle to the shifter lever. Secure it in place with the retaining screw.
26. Install the floor console.
27. Connect the negative battery cable to the battery.
28. Check, while driving, to be sure that the transmission is set to each range when the selector lever is shifted to each position. Make sure the over drive is activated and cancelled correctly when the over drive switch is activated.

1990-95 Pick-Ups
▶ See Figures 80 and 81

1. Remove the cotter pin at the upper end of the transmission cable. Remove the flat washer, the bushing and then pull the cable end off of the selector lever attachment.
2. Loosen the locking nut and then pull the cable out of the column bracket.
3. Pull the retaining clip off of the lower retaining bracket, located by the transmission.
4. Remove the cotter pin holding the lower cable end onto the transmission control arm.
5. Remove the flat washer, the cable end and the bushing from the control arm. At this point the transmission cable can be removed from the vehicle.
6. The transmission control arm and rod can be removed at this time.

To install:

7. Before reassembly, grease the pivot points and sliding parts.
8. Install the control arm and rod to the transmission. Install the transmission control rod with the marked position upward.
9. Route the transmission control cable in the same position as it was when removed.
10. Attach the lower end of the cable to the control arm, making sure to use a new cotter pin. Use the bushing and the flat washer.
11. Slide the cable into the lower retaining bracket, then install the clip to hold it in place.

DRIVE TRAIN 7-53

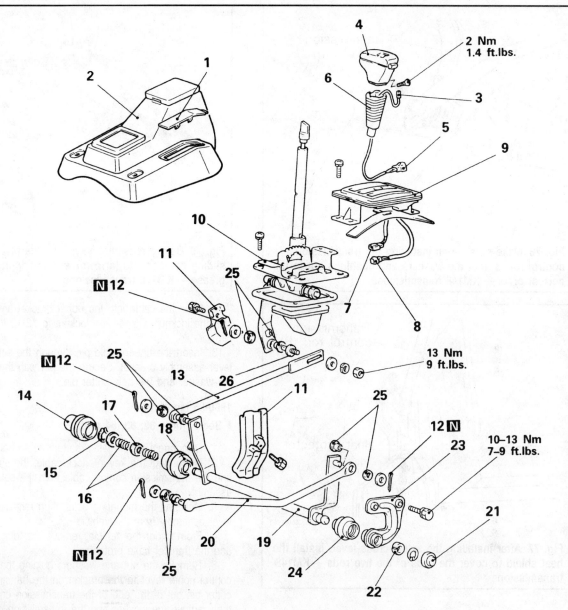

Removal steps

1. Plate B
2. Floor console
3. Overdrive switch connection
4. Selector handle
5. Overdrive switch harness and front wiring harness connection
6. Cover
7. Inhibitor switch and front wiring harness connection
8. Position indicator light and front wiring harness connection
9. Indicator panel
10. Bracket assembly
11. Heat protector
12. Cotter pin
13. Transmission control rod (B)
14. Dust cover
15. Snap ring
16. Spring
17. Cross shaft bushing
18. Cross shaft boot (B)
19. Select cross shaft
20. Transmission control rod (A)
21. Cap
22. Bushing
23. Cross shaft bracket (A)
24. Cross shaft boot
25. Bushing
26. Pin

Fig. 75 Shifter linkage removal and installation — KM148 transmissions

7-54 DRIVE TRAIN

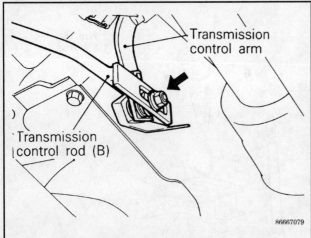

Fig. 76 Make sure, when installing the transmission control rod B, that the washers and bushings are in the correct order — KM148 transmissions

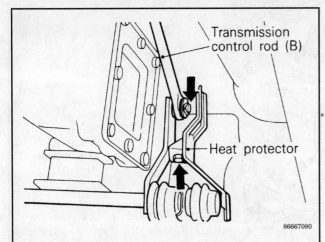

Fig. 77 After installing the select cross lever, install the heat shield to cover the ends of the two rods — KM148 transmissions

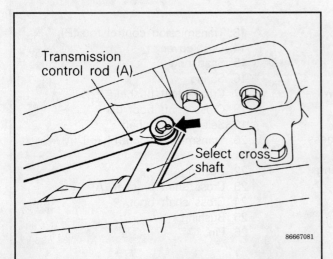

Fig. 78 Always use new cotter pins when reassembling the rod junctions — all transmissions

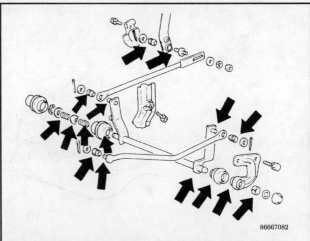

Fig. 79 Before reassembly, grease the pivot points and sliding components (arrows) with multipurpose grease — KM148 transmissions

12. Slide the cable into the upper bracket and retain it there with the locknut. Tighten the locknut to 12-19 ft. lbs. (17-26 Nm).

13. Insert the upper cable pin through the selector control lever assembly bracket. Secure the pin with the bushing, the flat washer and the new cotter pin.

1990-95 Monteros
▶ See Figures 82, 83, 84 and 85

1. Disconnect the negative battery cable from the battery.
2. Disconnect the key-interlock cable, the shift-lock cable and the transmission control cable from the selector lever assembly.
3. Unscrew the mounting bolts, then remove the selector lever assembly from the vehicle.
4. From under the vehicle, remove the four retaining bolts and the transfer case protector.
5. Remove the nut and washers holding the transmission control upper lever to the bracket on the transmission. Pull the cotter pin out of the end of the transmission end of the cable, then remove the washer and the transmission control upper lever from the cable.
6. Remove the transmission cable bracket nuts/bolts. This usually consists of 6-8 nuts/bolts along the length of the cable and at the end bracket (shifter end of cable). Remove the cable bracket and the cable end bracket, both located near the transmission end of the cable.
7. Remove the cable from the vehicle.

To install:

8. To install, install the cable end bracket and the cable brackets to the body of the vehicle.
9. Route the transmission control cable in the vehicle, then attach to the body with the brackets and various mounting screws along its length.
10. Attach the transmission end of the cable to the transmission control upper lever. Make sure to use the flat washer and a new cotter pin. Grease any pivot points with multipurpose grease.
11. Connect the transmission upper control lever to the bracket, using the flat washer, the lockwasher and the nut. Tighten the nut to 17 ft. lbs. (24 Nm).

DRIVE TRAIN 7-55

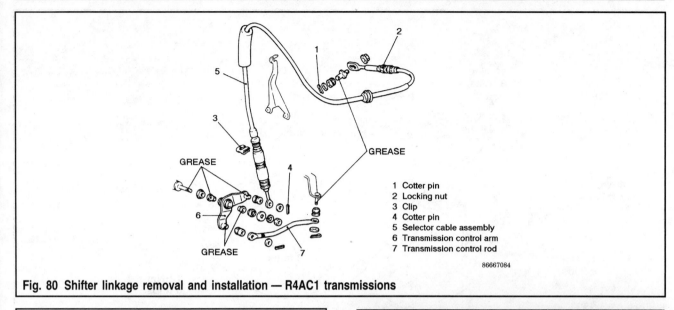

Fig. 80 Shifter linkage removal and installation — R4AC1 transmissions

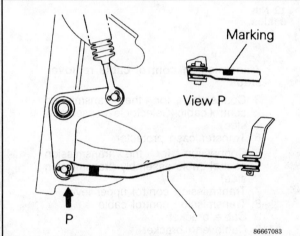

Fig. 81 Install the transmission control rod with the marked section as shown — R4AC1 transmissions

12. Install the transfer case protector.
13. Install the selector lever assembly with the retaining bolts. Tighten the bolts to 8 ft. lbs. (12 Nm).
14. Install the transmission control cable. Move the selector lever to the N position. Loosen the adjusting nut (from under the vehicle), gently pull the end of the transmission control cable in the direction toward the front of the transmission. Tighten the adjusting nut until tight.
15. Install the shift-lock cable. Move the selector lever to the P position. Adjust the shift-lock cable so that the end of the cable (red mark) is at the position shown in the illustration, and then tighten the nut until tight to clamp the shift lock cable.
16. Install the key interlock cable. With the selector lever still in the P position, install the spring and washer that are inserted onto the key interlock cable as shown in the illustration. Gently push the lock cam until the pin stops in the direction of the arrow, and then tighten the nut to clamp the key interlock cable.

Extension Housing Seal

1. Disconnect the negative battery cable from the battery.
2. Raise and safely support the vehicle on jackstands.
3. Drain the transmission fluid into a large container — refer to Section 1 for fluid draining and refilling procedures. This would be a good time to change the transmission fluid filter.
4. Matchmark the driveshaft (matchmark both on 4WD vehicles).
5. Remove the end of the driveshaft not attached to the transmission, then pull the driveshaft out of the transmission (also transfer case on 4WD).
6. Remove the dust seal guard (if so equipped), then pry the oil seal out of the extension housing or transfer case with a seal remover or prytool. Be careful not to scratch or gouge the sealing surface on the inside of the transmission.

To install:
7. To install the oil seal, apply transmission fluid to the lip of the oil seal. Install the new seal with the lip toward the front of the housing (facing into the transmission). Use the Mitsubishi tool MIT3995 and MB990938-01 to install the seal. If these tools are not available, a large socket the same size as the seal can be used. Use a plastic or rubber mallet and do not hit the seal too hard, damage could result to the transmission and to the seal. Seat the seal until the seal is flush with the transmission seal boss.
8. Tap the dust guard back onto the transmission extension.
9. Slide the driveshafts back into the transmission and transfer case so that the matchmarks line up, then attach the other ends to the differential units or center link.
10. Fill the transmission fluid as per Section 1 in this book.
11. Make certain everything is in place and tightened securely, then lower the vehicle.
12. Connect the negative battery cable back to the battery.
13. Start the vehicle and allow it to warm up. Check for any transmission fluid leaks from the fluid drain or fill holes and from the oil seals.

7-56 DRIVE TRAIN

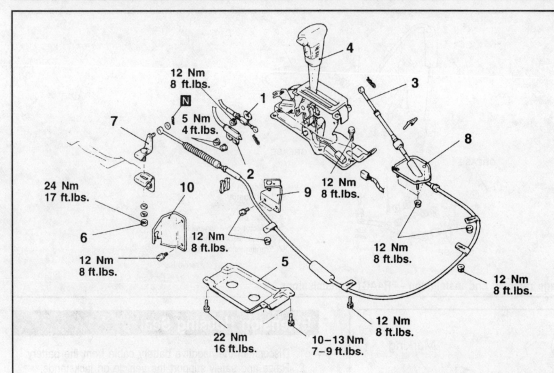

Selector lever assembly removal steps

1. Connection for key-interlock cable (Selector lever assembly side)
2. Connection for shift-lock cable (Selector lever assembly side)
3. Connection for transmission control cable (Selector lever assembly side)
4. Selector lever assembly

Transmission control cable removal steps

3. Connection for the transmission control cable (Selector lever assembly side)
5. Transfer case protector
6. Connection for the transmission control cable assembly (transmission side)
7. Transmission control upper lever
8. Transmission control cable
9. Cable bracket
10. Cable end bracket

Fig. 82 Shifter linkage removal and installation — V4AW2 (V4AW3) transmissions

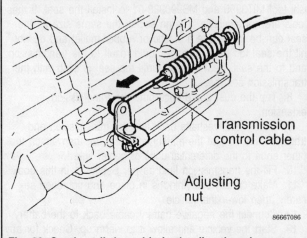

Fig. 83 Gently pull the cable in the direction shown to adjust the shifter cable — V4AW2 (V4AW3) transmissions

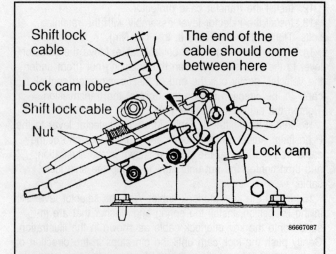

Fig. 84 The shift lock cable adjusting mechanism — V4AW2 (V4AW3) transmissions

DRIVE TRAIN 7-57

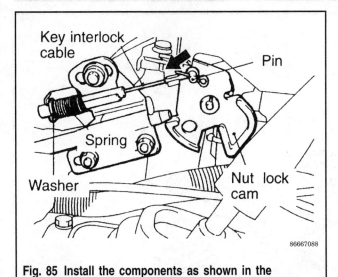

Fig. 85 Install the components as shown in the illustration — V4AW2 (V4AW3) transmissions

Transmission

REMOVAL & INSTALLATION

1983-89 Monteros and Pick-Ups
◆ See Figures 86, 87 and 88

✱✱WARNING

The transmission and the torque converter must always be connected when removed from the engine, otherwise damage to the drive plate, oil pump bushing and oil seal could be the result. The drive plate cannot support any load. Make sure that the weight of the transmission does not act on the drive plate during removal.

1. Disconnect the negative battery cable.
2. It may be necessary to remove the front exhaust pipe (AW372 and KM148 transmissions), do so now. Disconnect the front exhaust pipe from the exhaust manifold and the catalytic converter.
3. Disconnect the kickdown linkage.
4. Remove the selector lever cover and the console. Remove the selector lever.
5. Raise and safely support the vehicle on stands.
6. If 4WD, remove the skid plate from below the transfer case.
7. Remove the shifter control linkage from the transmission. Remove the transfer shifter (if 4WD) and the transmission shifter.
8. Drain the transmission fluid into a suitable container.
9. Disconnect the oil cooler lines and hose from the transmission. Immediately plug or cap the open lines and ports. Suspend the lines vertically to reduce leakage into the work. Disconnect the front and rear driveshafts from the transmission and transfer case.
10. Disconnect the speedometer cable, the back up light wiring and the four wheel drive indicator wiring if so equipped.
11. Remove the converter cover. Remove the torque converter to flex plate (drive plate) retaining bolts. The torque converter will have to be turned as the bolts are loosened and removed, having a friend turn the crankshaft with a large wrench is extremely helpful.
12. Remove the starter mounting bolts and remove the starter. If you're careful, the wiring can remain attached; use stiff wire to tie the starter to a frame rail or other solid mount. Allow it to hang from the wire (NOT the electrical wires — the stiff wire).
13. Support the transmission with a jack or transmission cradle. Distribute the weight of the transmission over a wide area with a piece of lumber.
14. Remove the rear insulator from the crossmember.
15. Remove the crossmember. It is heavy and bulky; a second jack can be helpful in removing it.
16. Remove the transfer case support (if 4WD). Remove the transmission retaining (bell housing) bolts and remove the transmission unit out of the car. Slide the transmission backward and then down, the transmission needs to clear the alignment tangs.

To install:
17. To reinstall, raise the transmission into place and install the retaining (bell housing) bolts. Install the transfer case support if so equipped.
18. Install the rear insulator on the crossmember.
19. Install the starter.
20. Have that friend of yours turn the engine until the flywheel (drive plate) holes line up with the torque converter holes. Install the first bolt finger tight, then have the engine turned and install the next bolt only finger tight. Continue to do this until all of the converter bolts are installed finger tight in the flywheel. Tighten the flywheel to torque converter bolts in two stages up to 25-30 ft. lbs. (35-42 Nm). If the bolts aren't tightened in stages, the flywheel may become tightened down on an angle.
21. Install the converter cover.
22. Connect the speedometer cable, backup light switch wiring and 4WD indicator wiring harness.
23. Connect the front and rear driveshafts to the transmission and transfer case.
24. Connect the oil cooler tubes and hoses from the transmission.
25. Install the shift control arm, cross select shaft and control rod to the transmission.
26. Install the skid plate if so equipped.
27. Lower the vehicle the the ground.
28. Install the shifter linkage — refer to the previous procedures in this section.
29. Install the transfer and transmission shifter assemblies.
30. Connect the kickdown link.
31. Install the front exhaust pipe.
32. Install the proper amount of transmission fluid.
33. Connect the negative battery cable.
34. After installation, refill the transmission with the approved fluid. Check that the transmission will start only in N and P positions and that the back-up lights function only in the R position, otherwise adjusted the shifter linkage.

7-58 DRIVE TRAIN

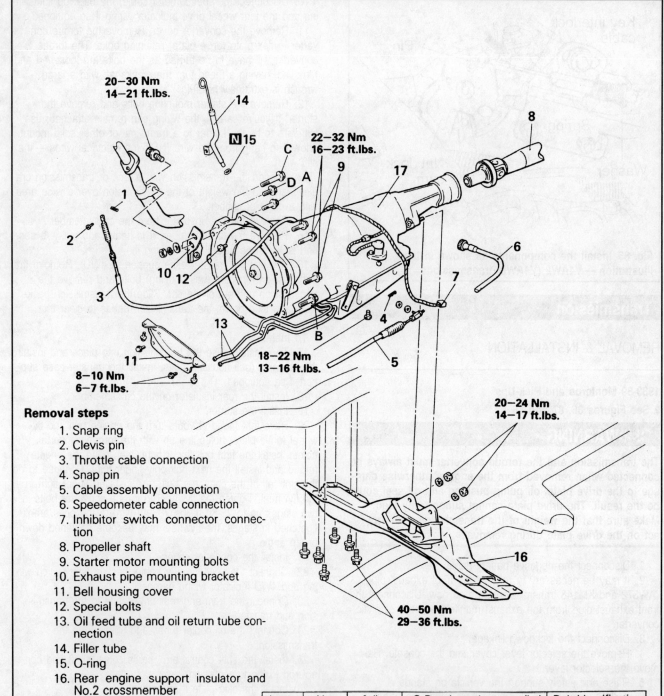

Fig. 86 Automatic transmission removal and installation — AW372 transmissions/ 1987-89 2WD Pick-ups

DRIVE TRAIN 7-59

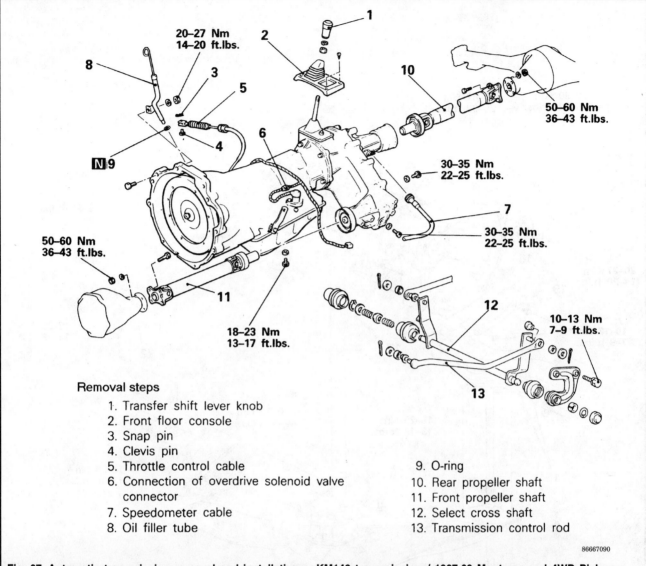

Removal steps

1. Transfer shift lever knob
2. Front floor console
3. Snap pin
4. Clevis pin
5. Throttle control cable
6. Connection of overdrive solenoid valve connector
7. Speedometer cable
8. Oil filler tube
9. O-ring
10. Rear propeller shaft
11. Front propeller shaft
12. Select cross shaft
13. Transmission control rod

Fig. 87 Automatic transmission removal and installation — KM148 transmissions/ 1987-89 Monteros and 4WD Pick-ups

1990-95 Pick-Ups

▶ See Figure 89

1. Disconnect the negative battery cable from the battery.
2. Raise and safely support the vehicle on jackstands. Chock the wheels on the ground to keep the vehicle from rolling when the driveshafts are removed.
3. Remove the under covers.
4. Remove the front exhaust pipe.
5. Matchmark the driveshaft to the rear differential unit and the transmission. Detach the driveshaft from the rear differential and slide the yoke out of the tail of the transmission. Have a pan handy to catch the draining transmission fluid. Try not to spill any transmission fluid because it will stain any clothing and even pavement.
6. Remove the transmission fluid filler tube. Dispose of the O-ring on the end of the tube (it will need to be replaced with a new on upon reassembly).
7. Unplug the transmission harness connector.
8. Loosen the speedometer cable retaining ring, then pull the cable out of the transmission.
9. Disconnect the transmission control rod at the transmission.
10. Disconnect the fluid cooler hoses from the transmission.
11. Disconnect the kickdown rod from the cable.
12. Remove the exhaust pipe mounting bracket from the right side of the transmission.
13. Remove the lower torque converter cover.
14. Remove the torque converter to flywheel bolts. Having a friend turn the engine with a large wrench is a lot of help for this procedure.
15. Support the transmission with a floor jack or a transmission cradle.
16. Remove the two mounting bolts holding the tail of the transmission to the crossmember, then remove the crossmember.
17. With the transmission tilted so that the rear is lowered slightly, remove the flange bolt which fastens the starter motor and bell crank bracket.

➡**Suspend the starter motor from the body by using a piece of wire or a similar method.**

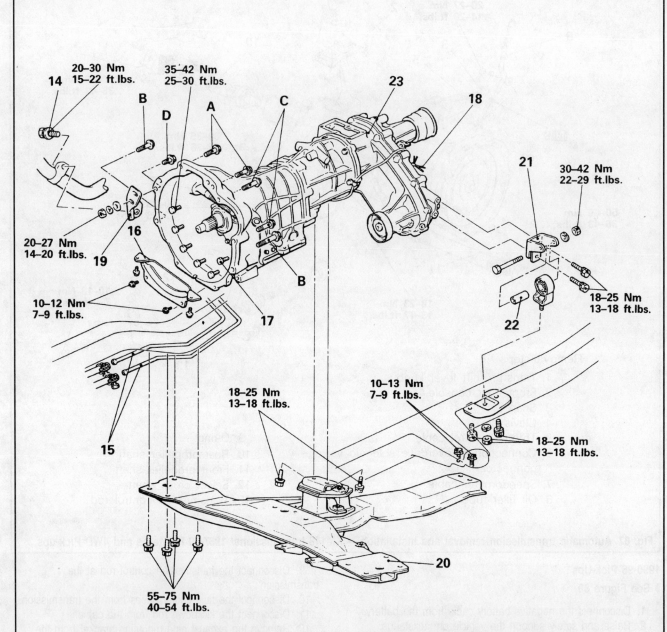

Fig. 88 Automatic transmission removal and installation continued — KM148 transmissions/ 1987-89 Monteros and 4WD Pick-ups

18. Disconnect the transmission assembly from the engine by pulling it slowly toward the rear of the vehicle.

➡ Detach so that the torque converter does not remain at the engine side.

To install:

19. On the engine side, there are two centering locations. Make sure that the transmission mounting bolt holes are aligned with them before mounting the transmission assembly to the engine.

➡ The engine with the transmission assembly removed has its rear slightly raised up due to imbalance caused by such removal. Therefore, disconnect the front exhaust pipe from the catalytic converter to lower the engine.

20. Install the bell housing-to-engine block bolts (refer to the illustration for the torque figures). While installing these bolts also install the bell crank bracket assembly and the starter motor.
21. Install the crossmember, then secure the rear of the transmission to the crossmember with the two mounting nuts. Tighten the rear nuts to 14-20 ft. lbs. (20-27 Nm).
22. Install the torque converter-to-flywheel bolts. Before installing the bolts, make sure that the torque converter paint color can be seen through the check hole in the drive plate. Have that friend of yours turn the engine until the flywheel (drive plate) holes line up (and color can be seen) with the torque converter holes. Install the first bolt finger tight, then have the engine turned and install the next bolt only finger tight. Continue to do this until all of the converter bolts are installed finger tight in the flywheel. Tighten the flywheel-to-torque converter bolts in two stages up to 33-38 ft. lbs. (46-53 Nm). If the bolts aren't tightened in stages, the flywheel may become tightened down on an angle.
23. Install the bell housing lower cover.
24. Install the exhaust pipe mounting bracket to the right side of the transmission. Tighten the mounting nuts to 14-20 ft. lbs. (20-27 Nm).
25. Install the transmission throttle (kickdown) lever.
26. Connect the transmission cooler tubes to the transmission. Make sure to install the cooler tube retainer to the transmission.
27. Connect the transmission control rod to the transmission. Both the kickdown and the control rod will have to be adjusted once the transmission is completely installed — refer to the procedures in this section.
28. Insert the speedometer cable into the transmission, then tighten the retaining ring to hold it in place.
29. Plug the transmission harness connectors together.
30. Insert the fluid filler tube into the transmission with a new O-ring.
31. Fill the transmission up with automatic transmission fluid.
32. Install the driveshaft, making sure that the matchmarks line up.
33. Install the front exhaust pipe.
34. Adjust the kickdown and shifter linkages — refer to the procedures in this section.
35. Install the under covers, then lower the vehicle to the ground.
36. Connect the negative battery cable to the battery.
37. Remove the wheel chocks, start the vehicle up and allow it to warm up. While it is warming up, check for any fluid leaks. Once it is warmed up, check the fluid level and add if needed.

1990-95 Monteros

▶ See Figures 90, 91, 92 and 93

1. Disconnect the negative battery cable from the battery.
2. Remove the transfer shift lever assembly — refer to the manual transmission procedures for the transfer assembly.
3. Raise and support the vehicle on jackstands. Before crawling under the vehicle, make absolute certain that the vehicle is safely supported by the jackstands.
4. Remove the transfer case protector.
5. Unfasten the front exhaust pipe from the two exhaust manifolds, then disconnect it from the intermediate pipe/ catalytic convertor (make certain to retain the bolts and nuts for reassembly). Remove the entire pipe from under the vehicle.
6. Drain the transmission fluid into a clean, large pan. With the transmission out for any reason, it is a good time to change the transmission fluid filter. Drain the transfer case as well.
7. Matchmark and disconnect the rear driveshaft at both the rear axle and the transfer case flanges.
8. On 4-wheel drive vehicles, disconnect the forward driveshaft from the front axle and remove it by sliding it forward. Install a plug in the transfer case to prevent residual fluid leakage.
9. Remove the fluid filler tube from the transmission. Replace the O-ring with a new O-ring upon reassembly.
10. Disconnect the kickdown linkage from the transmission.
11. Remove the dust seal from the rear extension housing.
12. Disconnect the shifter linkage from the transmission.
13. Label and unplug all wires leading to the transmission, some are: Ground cables, 4WD indicator light switch connector, park/ neutral position switch connector, back-up light switch connector, HI/LO detection switch connector, center differential lock detection switch connection. There may be others depending on the particular year, model and engine the vehicle came equipped with.
14. Unscrew the retaining ring holding the speedometer in the transmission. Pull the speedometer cable out of the transmission and label it.
15. Disconnect the transmission cooling lines from the transmission.
16. Unbolt the starter motor from the front of the transmission and remove it. Remove the heat shield, if so equipped.
17. Place a floor jack, with a block of wood on the jacking area, below the engine oil pan.
18. Lift the floor jack under the engine just until the weight of the engine is taken onto the jack — the engine should only barely be lifted by the jack.
19. Use a transmission jack or second floor jack to place under the transmission. Don't support the transmission yet, only lift the jack until it is slightly below the transmission.
20. Remove the left-hand and right-hand side transmission stays from the front of the transmission.
21. Unbolt the bell housing lower cover.
22. Remove the transfer case mounting bracket, roll stopper, mass damper and transfer case protector bracket.
23. Lift the second floor jack up until the transmission is being slightly supported by it.

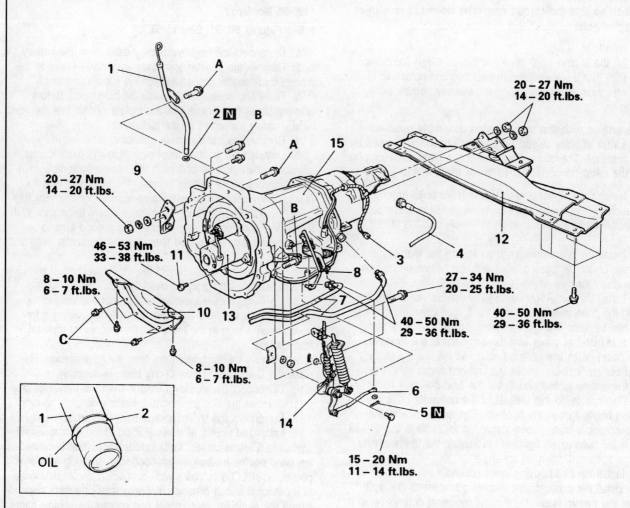

Removal steps
1. Oil filler tube
2. O-ring
3. Transmission harness connector
4. Speedometer cable
5. Cotter pin
6. Transmission control rod (Transmission side)
7. Automatic transmission cooler tube
8. Transmission throttle lever (Bell crank bracket side)
9. Exhaust pipe mounting bracket
10. Bell housing cover
11. Special bolt
12. No.2 crossmember
13. Starter motor
14. Bell crank bracket assembly
15. Transmission assembly

	Nm	ft.lbs.	O.D.×Length mm (in.)	Bolt identification
A	43 – 55	31 – 40	10×50 (.4×2.0)	
B	43 – 55	31 – 40	10×70 (.4×2.8)	
C	30 – 42	21 – 30	10×16 (.4×.6)	

Fig. 89 Automatic transmission removal and installation — R4AC1 transmissions/ 1990-95 Pick-ups

DRIVE TRAIN

24. Remove the two transmission-to-crossmember bolts. Lift the jack about 1/4 of an inch off of the crossmember support. Remove the entire crossmember from the vehicle.
25. Remove the transmission and transfer case together from the vehicle.
   a. Remove the cover from the oil pan upper.
   b. Remove the connecting bolts (6 places) while turning the crankshaft. Having a friend to help with this makes this stage much easier.
   c. Gently lower the rear section of the transmission and transfer assembly to remove the assembly from the engine.

❊❊CAUTION

When removing the transmission and transfer assembly, push the torque converter over to the transmission and transfer assembly side so it does not remain on the engine side.

   d. Next, tilt the front section of the transmission and transfer assembly downward and gently lower it, being careful that the rear section of the transfer does not touch the rear crossmember.

To install:

26. Lift the transmission and transfer assembly close to position with the floor jack.
27. On the engine side, there are two centering locations. Make sure that the transmission mounting bolt holes are aligned with them before mounting the transmission and transfer assembly to the engine. Lowering the rear of the engine SLIGHTLY may help align the two assemblies.
28. Slide the transmission assembly onto the engine making sure the aligning areas stay aligned.
29. Install and tighten the bell housing bolts to 54 ft. lbs. (75 Nm).
30. Lift the transmission/transfer assembly with the floor jack (since the engine is now attached to the transmission, it also will rise slightly — adjust its jack to keep only slight support).
31. Hold the crossmember in place and secure with the mounting bolts. Tighten the mounting bolts to 47 ft. lbs. (65 Nm).
32. Lower the transmission and transfer case assembly onto the crossmember. Install the two crossmember-to-transmission bolts and tighten them to 29 ft. lbs. (39 Nm).
33. Install the torque converter-to-flywheel bolts. Have that friend of yours turn the engine until the flywheel (drive plate) holes line up with the torque converter holes. Install the first bolt finger tight, then have the engine turned and install the next bolt only finger tight. Continue to do this until all of the converter bolts are installed finger tight in the flywheel. Tighten the flywheel-to-torque converter bolts in two stages up to 25-30 ft. lbs. (34-41 Nm). If the bolts aren't tightened in stages, the flywheel may become tightened down on an angle.
34. Install the transfer mounting bracket, the mass damper and the transfer case protector bracket back onto the transfer case.
35. Install the bell housing lower cover, then mount the left-hand and right-hand side transmission stays.
36. Mount the starter motor and heat shield to the transmission.
37. Connect the transmission cooling lines back to the transmission.
38. Insert the speedometer cable into the transmission and secure it there with the retaining ring.
39. Plug all of the electrical harness connectors back together.
40. Install and adjust (refer to previous adjustment procedures) the kickdown and shifter linkages.
41. Insert the fluid filler tube (with a new O-ring installed) into the transmission.
42. Tap the dust seal guard back onto the rear extension housing with a rubber or plastic mallet.
43. Slide the front driveshaft into the transfer case, then attach it to the front differential. Install the rear driveshaft also. Make certain that the matchmarks line up.
44. Fill the transmission and transfer case with oil — refer to Section 1.
45. Install the front exhaust pipe to the catalytic convertor and the exhaust manifolds. Use new gaskets if it was equipped with them. Make sure that the wiring does not lay on or very close to the exhaust pipes or exhaust manifolds, if they do, reroute them away from the exhaust components.
46. Install the transfer case protector.
47. Lower the vehicle to the ground. Install the transmission and transfer shift lever assemblies.
48. Connect the negative battery cable to the battery.
49. Start the vehicle and check for any transmission or transfer case fluid leaks.

7-64 DRIVE TRAIN

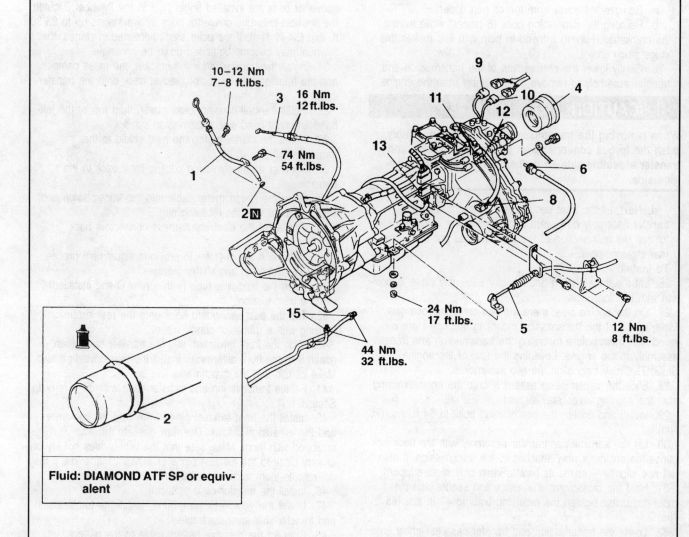

Removal steps
1. Fluid filler pipe
2. O-ring
3. Connection for throttle control cable
4. Dust seal guard
5. Connection for transmission control cable
6. Connection for speedometer cable
8. HI/LO detection switch connector
9. 4WD operation detection switch connector
10. Center differential lock operation detection switch connector
11. Center differential lock detection switch connector
12. 2WD/4WD detection switch connector
13. Park/Neutral position switch connector
15. Connection for fluid cooler pipe

Fig. 90 Automatic transmission removal and installation — V4AW2 transmissions/ 1990-95 Monteros

DRIVE TRAIN 7-65

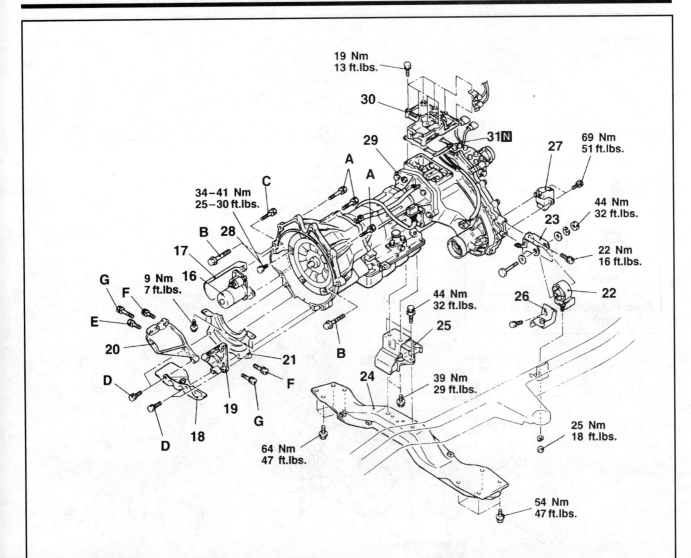

	Nm	ft.lbs.	O.D.×Length mm (in.)	Bolt identification
A	74	54	"7" 12×40 (.5×1.6)	
B	88	65	"7" 12×55 (.5×2.2)	
C	30	22	"7" 10×55 (.4×2.2)	"7" D × L
D	35	26	"7" 10×40 (.4×1.6)	
E	74	54	"7" 12×35 (.5×1.4)	
F	41	31	"7" 10×30 (.4×1.2)	
G	74	54	"7" 12×30 (.5×2.0)	

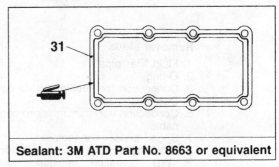

Sealant: 3M ATD Part No. 8663 or equivalent

16. Starter motor
17. Starter cover
18. Heat protector
19. Transmission stay (L.H.)
20. Transmission stay (R.H.)
21. Bell housing cover
22. Transfer roll stopper
23. Transfer mounting bracket

24. No. 2 crossmember
25. Engine mount rear insulator
26. Transfer case protector bracket
27. Mass damper
28. Torque converter connecting bolt
29. Transmission and transfer assembly
31. Gasket
31. Gasket

Fig. 91 Automatic transmission removal and installation continued — V4AW2 transmissions/ 1990-95 Monteros

7-66 DRIVE TRAIN

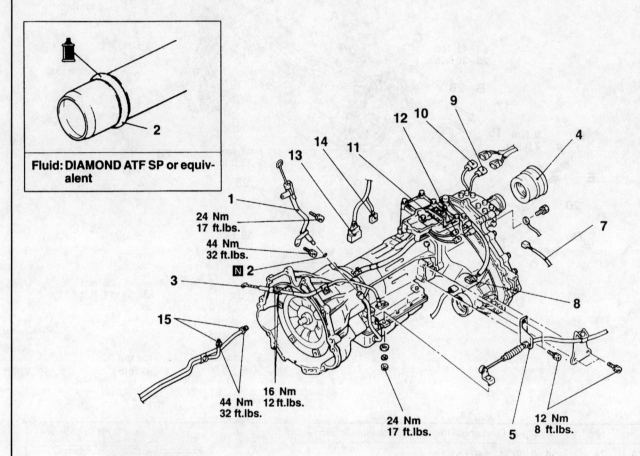

Removal steps
1. Fluid filler pipe
2. O-ring
3. Connection for throttle control cable
4. Dust seal guard
5. Connection for transmission control cable
7. Speed sensor connector
8. HI/LO detection switch connector
9. 4WD operation detection switch connector
10. Center differential lock operation detection switch connector
11. Center differential lock detection switch connector
12. 2WD/4WD detection switch connector
13. Park/Neutral position switch connector
14. Solenoid valve connector
15. Connection for fluid cooler pipe

Fig. 92 Automatic transmission removal and installation — V4AW3 transmissions/ 1994-95 Monteros

DRIVE TRAIN 7-67

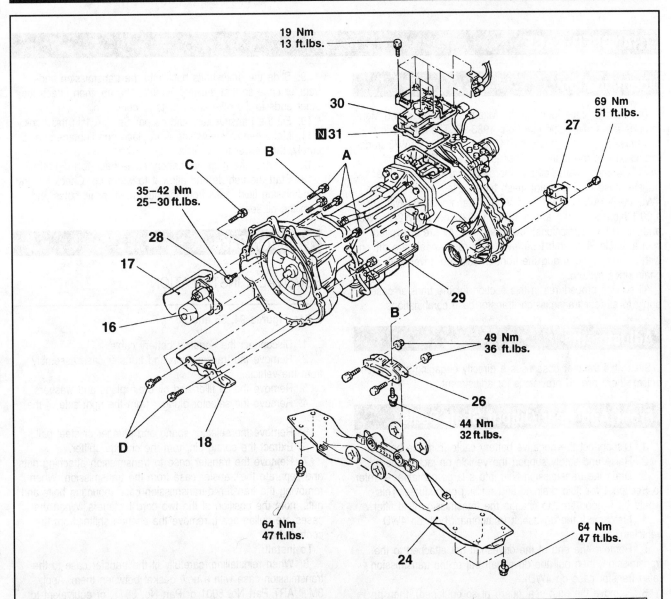

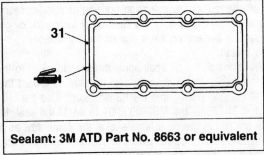

16. Starter motor
17. Starter cover
18. Heat protector
24. No. 2 crossmember
26. Engine rear mount bracket
27. Mass damper
28. Torque converter connecting bolt
29. Transmission and transfer assembly
30. Control housing
31. Gasket

Fig. 93 Automatic transmission removal and installation — V4AW3 transmissions/ 1994-95 Monteros

DRIVE TRAIN

TRANSFER CASE

Identification

The transfer case found in the Mitsubishi Pick-ups and Monteros is the same basic unit from 1983-95. The automatic transmission transfer case (KM148 and V4AW2/ V4AW3 transmissions) and the manual transmission case (KM145 and V5MT1 transmissions) all use the same transfer case. The transfer case is a constant mesh type (called Active Traction 4WD the V4AW2/V4AW3 and V5MT1 transmissions) with a 1.00:1 high gear ratio and a 1.94:1 (1.92:1 on V4AW2/V4AW3 and V5MT1 transmissions) low gear ratio. All are controlled with a single, floor-shift type lever. The rear wheels are driven with a direct system and the front wheels are driven with a chain drive system.

All service procedures in this section for the transfer case apply to all four transmission/ transfer case combinations.

Adjustments

Since this transfer case uses a directly engaging shift mechanism, there are no provisions for adjustment.

Front And Rear Output Shaft Seals

1. Disconnect the negative battery cable from the battery.
2. Raise and safely support the vehicle on jackstands.
3. Drain the transmission fluid into a large container — refer to Section 1 for fluid draining and refilling procedures. This would be a good time to change the transmission fluid filter.
4. Matchmark the driveshaft (matchmark both on 4WD vehicles).
5. Remove the end of the driveshaft not attached to the transmission, then pull the driveshaft out of the transmission (also transfer case on 4WD).
6. Remove the dust seal guard (if so equipped), then pry the oil seal out of the extension housing or transfer case with a seal remover or prytool. Be careful not to scratch or gouge the sealing surface on the inside of the transmission.

To install:

7. To install the oil seal, apply transmission fluid to the lip of the oil seal. Install the new seal with the lip toward the front of the housing (facing into the transmission). Use the Mitsubishi tool MIT3995 and MB990938-01 to install the seal. If these tools are not available, a large socket the same size as the seal can be used. Use a plastic or rubber mallet and do not hit the seal too hard, damage could result to the transmission and to the seal. Seat the seal until the seal is flush with the transmission seal boss.
8. Tap the dust guard back onto the transmission extension.

9. Slide the driveshafts back into the transmission and transfer case so that the matchmarks line up, then attach the other ends to the differential units or center link.
10. Fill the transmission fluid as per Section 1 in this book.
11. Make certain everything is in place and tightened securely, then lower the vehicle.
12. Connect the negative battery cable back to the battery.
13. Start the vehicle and allow it to warm up. Check for any transmission fluid leaks from the fluid drain or fill holes and from the oil seals.

Transfer Case

REMOVAL & INSTALLATION

▶ See Figures 94, 95 and 96

1. Disconnect the negative battery cable.
2. Remove the transmission and transfer case assembly from the vehicle.
3. Remove the oil filler and oil drain plugs and washers.
4. Remove the selector plunger from the right side of the case.
5. Remove the selector spring and plunger or steel ball.
6. Extract the spring pin from the change shifter.
7. Remove the transfer case to transmission attaching nuts and separate the transfer case from the transmission. When removing the transfer-to-transmission case mounting bolts and nuts, note the position of the two cord fasteners. When the cases are pulled apart, remove the change shifter from the control shaft.

To install:

8. When reinstalling, carefully fit the transfer case to the transmission case with a new gasket between them. Apply 3M® ART Part No. 8001 or Part No. 8011, or equivalent to both sides of the gasket and affix it to the rear surface of the adapter. Make sure the end of the transmission output shaft is lightly greased to ease the fit to the transfer case. As the two cases are slid together, install the change shifter to the control shaft.
9. Connect the two units and install the case bolts, making sure to install the two cord fasteners. Tighten the bolts to 22-30 ft. lbs (30-42 Nm).
10. Install the selector spring and plunger or steel ball.
11. Carefully install the spring pin holding the change shifter to the control shaft, using the Mitsubishi tool MD998245-01. Drive the the spring pin in with the slit in the spring pin parallel to the shaft center of the shift rail, so that the dimensions are shown in the illustration.
12. Reinstall the transmission/transfer case as a unit.

DRIVE TRAIN 7-69

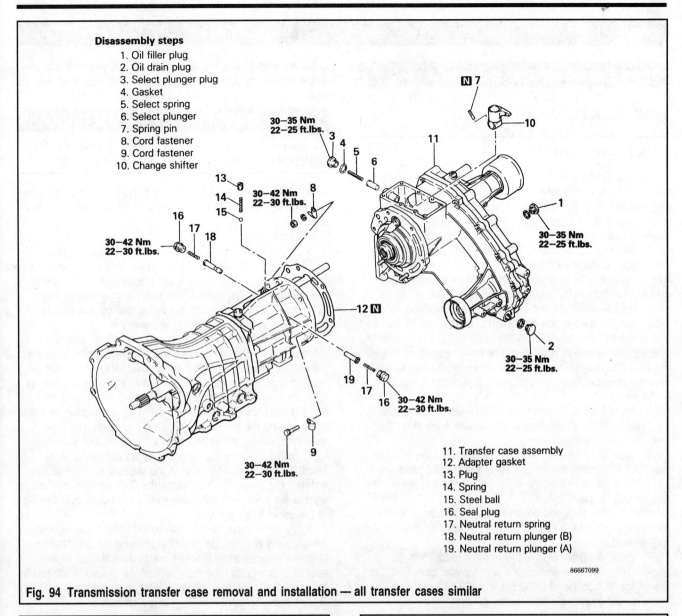

Fig. 94 Transmission transfer case removal and installation — all transfer cases similar

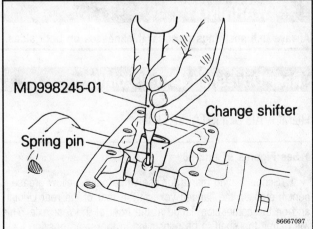

Fig. 95 Use the special tool (or a punch if the tool is not available) MD998245-01 to install the spring clip into the change shifter — all transfer cases similar

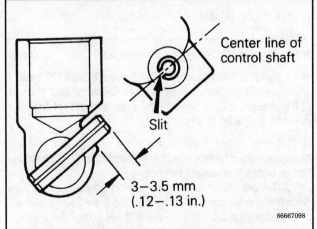

Fig. 96 Install the spring clip so that the slit is facing the direction shown in the illustration — All transfer cases similar

7-70 DRIVE TRAIN

DRIVELINE

Front Driveshaft And U-Joints

REMOVAL & INSTALLATION

▶ See Figures 97 and 98

1. Raise and support the vehicle with jackstands.
2. Remove the transfer case oil fill plug from the back of the transfer case. Set a pan or similar container under the drain plug of the transfer case. Remove the drain plug and drain the transfer case.
3. Matchmark the front flange of the front driveshaft and the flange on the front differential with white paint, a yellow grease pencil or by a similar method. This will permit the shaft to be reinstalled in the same position for balance purposes.
4. Remove the nuts and bolts from the companion flange, and lower the front of the driveshaft. Slide the rear of the shaft out of the transfer case. Cover the opening left at the front of the transfer case with a clean rag or similar means so dirt will be kept out. Have a pan or similar container at hand to catch any transmission fluid which may leak out of the flange.

To install:

5. Apply SAE 80 hypoid gear oil (API GL-4) to the inner and outer surface of the sleeve yoke where it will enter the transmission or center bearing.
6. Slide the driveshaft end into the transfer hole, making sure not to damage the oil seal lip of the transfer case.
7. Align the marks on the yoke and companion flange; line up the bolt holes.
8. Install the nuts and bolts with the lockwashers on the back (nut) side of the flange. Tighten the bolts to 36-43 ft. lbs. (49-58 Nm).
9. Install the drain plug of the transfer to 22-25 ft. lbs. (29-34 Nm). Fill the transfer case with transfer case fluid, then install the fill hole plug to 22-25 ft. lbs. (29-34 Nm).
10. Lower the vehicle back to the ground.

UNIVERSAL JOINT REPLACEMENT

▶ See Figures 99, 100, 101, 102 and 103

➡ To perform this operation, you'll need Mitsubishi Tool MB990840-01 (or an equivalent large C-clamp) and a small collar designed to fit against the yoke and accept the pressed out bearing cups.

1. Scribe mating marks on the yokes of each joint that is to be disassembled. Joints must be assembled in the same position to maintain driveshaft balance.
2. Remove the snaprings from all four positions of the joint. Situate the collar against one side of the yoke so it is perfectly lined up with the hole in the yoke and will accept the bearing cup from that side. Place the flat bottom of the C-clamp under the collar and then screw the upper portion downward against the top of that same yoke cross-shaft.
3. Turn the C-clamp screw handle to press the cross-shaft out. Remove the bearing cups and bearings from that cross-shaft. It's best to turn the unit so the closed end of the cup is at the bottom for retention of the needle bearings when you do this.

✳✳WARNING

Do not tap the journal bearings to remove them, as this will upset the balance of the propeller (drive) shaft.

4. Repeat this step for the other cross-shaft. You can then separate the two yokes of the joint and remove the cross-shaft assembly.

To assemble:

5. Apply multipurpose grease (SAE J310a NLGI grade #2 EP) to the journal outer diameters, ends, and reservoirs on all four arms of the cross-shaft. Apply grease also to the lips of the dust seals in the ends of the bearing cups and to the needles. All parts should be thoroughly coated, but excess grease may make the unit hard to assemble.
6. Place a bearing cup, open end inward, on opposite sides of one of the forks. Place the C-clamp around the cups without the collar. Center the cross-shaft assembly so the bearing cups will be assembled around two of the journals as they are pressed into the yoke. Turn the C-clamp until the bearing cups are pressed in with room for a snapring on either side. Line up the mating mark for the two yokes and then repeat the procedure for the other cross-shaft and yoke.
7. Install the snapring on one side with snapring pliers. Install the collar over the side of the yoke onto which you've installed the snapring. Then install the C-clamp with the bottom against the collar and the lower end of the screw over the top of the upper bearing cup.
8. Turn the clamp to seat the outside of the cup against the snapring on one side. Install a snapring on the other side. Measure the clearance between the outside surface of the second snapring and the outer wall of the groove which retains the snapring in the yoke. If the clearance is more than 0.0024 in. (0.06mm), replace both snaprings with new ones.

✳✳WARNING

Always use snaprings of equal thicknesses on both sides.

Rear Driveshaft And U-Joints

REMOVAL & INSTALLATION

▶ See Figures 97 and 98

1. Scribe mating marks (or use white paint, yellow grease pencil, etc.) on the flanged yoke at the rear of the rear U-joint and on the companion flange at the front of the rear axle. This will permit the shaft to be reinstalled in the same position for balance purposes.
2. Remove the nuts and bolts from the companion flange, and lower the rear of the driveshaft. Slide the front of the shaft out of the transmission. Cover the opening left at the rear of the transmission with a clean rag or similar means so dirt will

DRIVE TRAIN 7-71

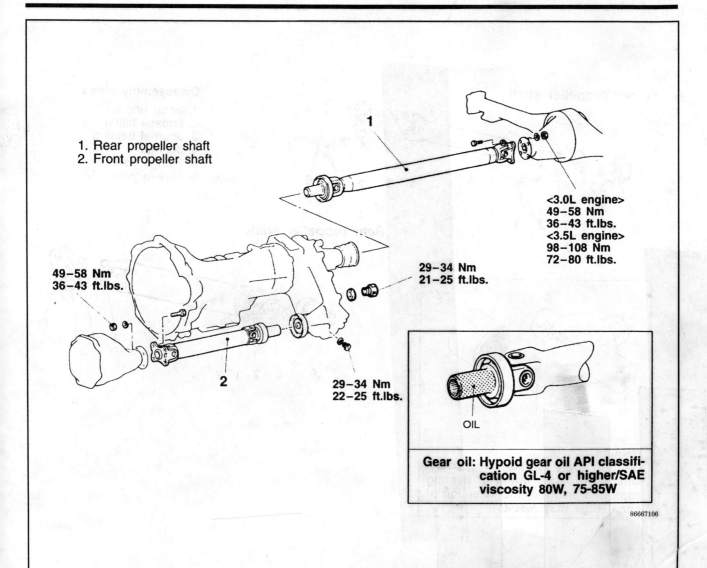

Fig. 97 Driveshaft removal and installation — rear-wheel drive vehicles refer only to the rear driveshaft in the illustration

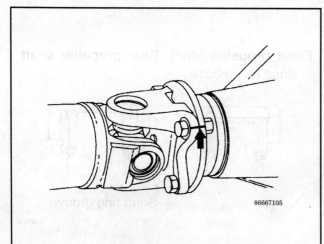

Fig. 98 Matchmark the driveshaft and differential flanges with white paint, a yellow grease pencil or a scribe

be kept out. Have a pan or similar container at hand to catch any transmission fluid which may leak out of the flange.

3. In the case of a driveshaft having a center bearing, the rear shaft must be removed first. Unbolt the bearing or carrier mount from the vehicle and remove it from the front driveshaft. Remove the front driveshaft.

4. When reinstalling, apply SAE 80 hypoid gear oil (API GL-4) to the inner and outer surface of the sleeve yoke where it will enter the transmission or center bearing.

5. Align the marks on the yoke and companion flange; line up the bolt holes.

6. Install the nuts and bolts with the lockwashers on the back (nut) side of the flange. Tighten the bolts to 36-43 ft. lbs. (49-58 Nm) for all vehicles except for the 3.5L V6 equipped Monteros, for these vehicles tighten the bolts and nuts to 72-80 ft. lbs. (98-108 Nm).

7-72 DRIVE TRAIN

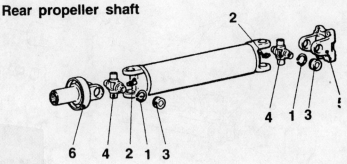

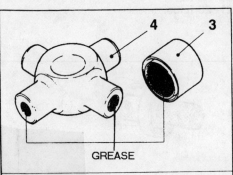

Front propeller shaft

Rear propeller shaft

Disassembly steps
1. Snap ring
2. Grease fitting
3. Journal bearing
4. Journal
5. Flange yoke
6. Sleeve yoke

GREASE

Caution
Do not apply grease excessively. Otherwise, faulty fitting of bearing caps and errors in the selection of snap rings may result.

Fig. 99 Universal joint disassembly and reassembly components — all Pick-up trucks and Monteros similar

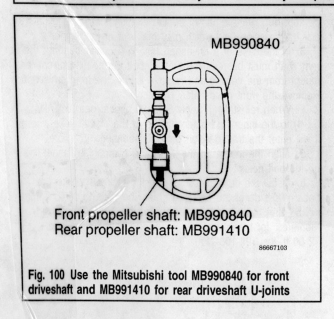

MB990840

Front propeller shaft: MB990840
Rear propeller shaft: MB991410

Fig. 100 Use the Mitsubishi tool MB990840 for front driveshaft and MB991410 for rear driveshaft U-joints

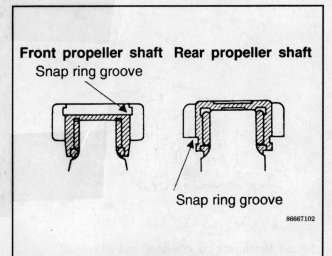

Front propeller shaft **Rear propeller shaft**
Snap ring groove

Snap ring groove

Fig. 101 Remove the snaprings from the universal joints during removal — all models similar

DRIVE TRAIN 7-73

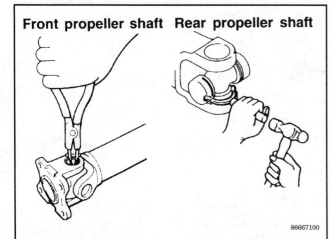

Fig. 102 Depending on which driveshaft the U-joints are being replaced, depends on the location of the snaprings

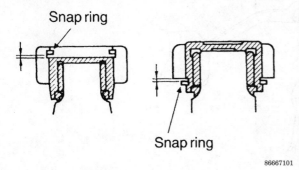

Fig. 103 Measure the clearances of the snaprings — if the clearance is larger than 0.0024 in. (0.06mm) replace the snaprings with new ones

UNIVERSAL JOINT REPLACEMENT

♦ See Figures 99, 100, 101, 102 and 103

This rear driveshaft universal joint replacement procedure is the same as the universal replacement procedures for the front driveshaft. Refer to the previous directions for this procedure.

Center Bearing

REMOVAL & INSTALLATION

♦ See Figures 104, 105, 106 and 107

1. After making mating marks on the front propeller (drive) shaft, center yoke and rear propeller shaft, remove the needle bearing of the center yoke. Then disconnect the front and rear propeller shaft.
2. Remove the nut holding the center yoke and remove the yoke from the center bearing.
3. Gently pry the bearing bracket from the bearing assembly.
4. Use a gear puller or similar tool to remove the bearing.

➡The center bracket and mounting rubber attaching band are spot-welded on the circumference and therefore can not be disassembled.

To install:

5. Fill the bearing grease cavity with multi-purpose grease.
6. Partially insert the center bearing into the shaft and install the bracket to the bearing.
7. Check that the bracket mounting rubber is correctly seated in the bearing groove.
8. Refit the center yoke, making sure to align the notch on the yoke with the notch on the front driveshaft. Use a new attaching nut and tighten it to 157-215 Nm (116-159 ft. lbs).

※※WARNING

The center yoke and center bearing bracket locknuts are self-locking nuts, which must be replaced with new parts.

9. Reinstall the center universal joint, making sure to align the matchmarks of the rear driveshaft with the notch in the yoke.

7-74 DRIVE TRAIN

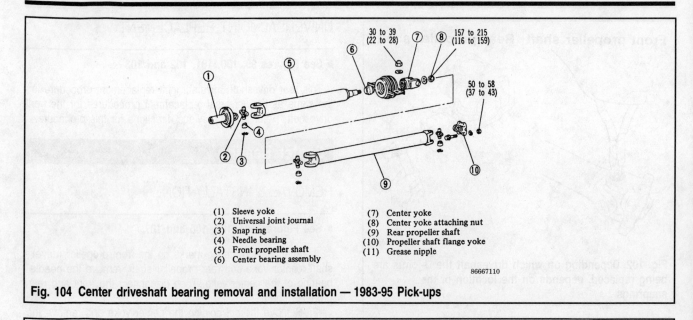

(1) Sleeve yoke
(2) Universal joint journal
(3) Snap ring
(4) Needle bearing
(5) Front propeller shaft
(6) Center bearing assembly
(7) Center yoke
(8) Center yoke attaching nut
(9) Rear propeller shaft
(10) Propeller shaft flange yoke
(11) Grease nipple

Fig. 104 Center driveshaft bearing removal and installation — 1983-95 Pick-ups

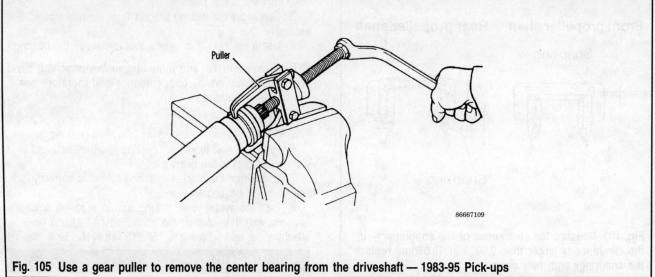

Fig. 105 Use a gear puller to remove the center bearing from the driveshaft — 1983-95 Pick-ups

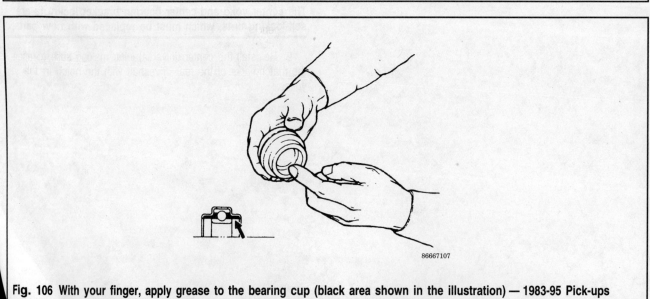

Fig. 106 With your finger, apply grease to the bearing cup (black area shown in the illustration) — 1983-95 Pick-ups

DRIVE TRAIN 7-75

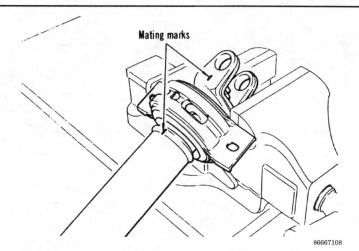

Fig. 107 When reinstalling the universal joint yoke to the driveshaft, make sure that the matchmarks are aligned — 1983-95 Pick-ups

FRONT DRIVE AXLE — 4-WHEEL DRIVE VEHICLES

Identification

▶ See Figure 108

The front axle assembly consists of a front differential, a housing tube, an inner shaft and driveshafts. For better serviceability of the differential, the spacer for backlash adjustment of the final drive gear is placed between the side bearing outer race and the gear carrier.

The double offset joint which can slide in the axial direction, is used at the differential carrier side; the Birfield joint, with a large operation angle, is used at the axial hub side.

To reduce vibration, noise, and fuel consumption when 2WD is applied, manual or automatic free-wheeling hubs are equipped. With the automatic system, one can switch between "LOCK" and "FREE" without having to leave one's seat.

Manual Locking Hubs

▶ See Figure 109

When the control handle is set to the LOCK position, the follower moves along the oblique groove in the control handle and causes the clutch (which is always in mesh with the free-wheeling hub body) to engage the splines of the inner hub, thus coupling the free-wheeling hub body with the driveshaft.

When the control handle is set to the FREE position, the follower moves along the oblique groove in the control handle and uses the tension spring to disengage the clutch from the splines of the ineer hub, thus separating the free-wheeling hub body from the driveshaft.

DISASSEMBLY & ASSEMBLY

▶ See Figures 110, 111 and 112

1983-88 Monteros

1. Remove the front wheels of the vehicle, and support the raised front end of the vehicle with jackstands.
2. Set the control handles on the wheels to the FREE position, then remove the six hub cover retaining bolts. Remove the hub covers from the hubs.
3. Remove and discard the used hub cover gasket. A new gasket will be needed for reassembly.
4. Use snapring pliers to remove the snap ring on the end of the driveshaft. Slide the shim off of the end of the driveshaft.
5. Remove the front brake assembly, by unscrewing the two rear mounting bolts, with the hose connected. Use a wire to suspend the front brake assembly from the upper control arm so that the front brake assembly won't fall.

✱✱WARNING

Do not twist the brake hose.

6. Remove the six manual free-wheeling hub assembly mounting bolts, then slide the hub assembly off of the driveshaft.
7. Pull the free-wheeling hub clutch from the hub assembly.
8. Remove the compression spring, the follower and the tension spring.
9. Using a flat-tipped prytool, remove the snapring from the rear of the free-wheeling hub body. Remove the inner hub.
10. Remove the snapring from the inner hub with a pair of snapring pliers. Separate the free-wheeling hub ring, the spacer and the inner hub.

To install:

11. Apply multipurpose grease (SAE J310, NLGI No. 2) to the entire periphery of the free-wheeling hub ring, the inner

DRIVE TRAIN

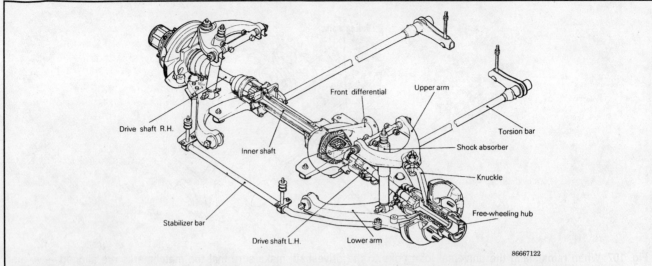

Fig. 108 The front axle system components of the 1988 Pick-ups and Monteros — other models are similar

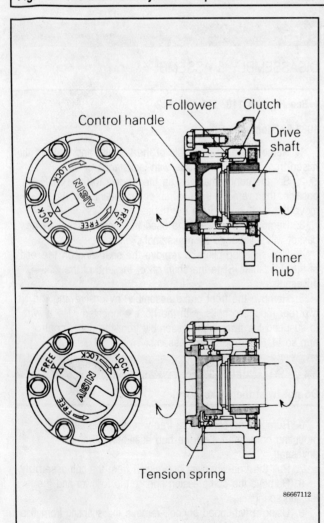

Fig. 109 The internal workings of the manual-change, free-wheeling hub assembly — some 1983-88 Monteros

hub and the free-wheeling hub clutch, the free-wheeling hub cover and the inside of the free-wheeling hub body.

12. Check to be sure that the hub body assembly and hub cover assembly are coated with a sufficient coating of grease; add more grease if necessary.

➡A liberal amount of grease should be applied, especially when grease is wiped away or a new free-wheeling hub is installed.

13. Reassemble the spacer, the free-wheeling hub ring, the inner hub and the snapring. Slide the assembly into the rear of the free-wheeling hub body and secure it in place with the rear snapring.
14. To the rear of the free-wheeling hub cover, install the tension spring, the follower and the compression spring.
15. Slide the free-wheeling hub clutch into the hub body with any spacers that were removed.
16. Install the hub assembly onto the driveshaft. Apply a coating of 3M® ART Part No. 8661, No. 8663, or the equivalent, equally all around, to the free-wheeling hub body assembly and front hub contact surfaces. Tighten the mounting bolts to 36-43 ft. lbs. (50-60 Nm).

➡Make sure that there is no excess specified sealant on the hub outside surface.

17. Install the front shim and snap ring to the hub assembly.
18. Install the free-wheeling hub cover, with a new gasket, to the assembly. Tighten the mounting bolts to 7-10 ft. lbs. (10-14 Nm).
19. Install the wheels to the vehicle, then lower the vehicle to the ground. Tighten the lug nuts again.

Automatic Locking Hubs

▶ See Figure 113

When the transfer is shifted from 2WD to 4WD and riving is begun, rotation of the driveshaft is transmitted from the drive gear to the slide gear to the cam to retainer (A) to brake (A). When this happens, brake (A) is pressed against brake (B) by the function of the cam of retainer (A), and friction force is generated.

DRIVE TRAIN 7-77

Removal steps

1. Free-wheeling hub cover
2. Gasket
   Adjustment of drive shaft end play
3. Snap ring
4. Shim
5. Front brake assembly
6. Manual free-wheeling hub assembly
7. Lock washer
   Adjustment of wheel bearing preload
8. Lock nut
9. Front hub assembly

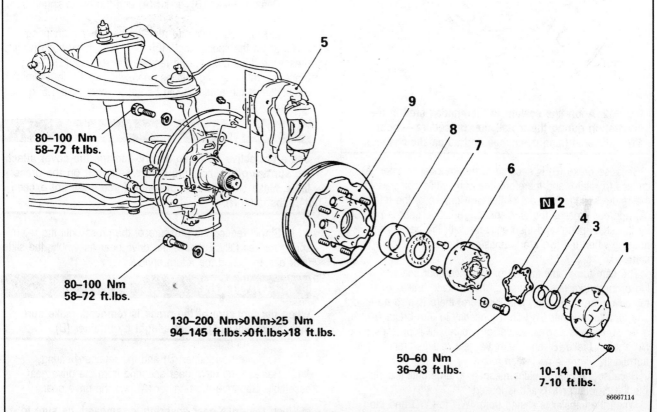

Fig. 110 Front hub and manual free-wheeling hub assemblies' removal and installation — manual-change free-wheeling 4WD systems

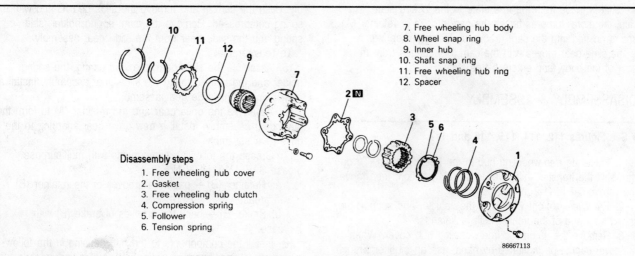

7. Free wheeling hub body
8. Wheel snap ring
9. Inner hub
10. Shaft snap ring
11. Free wheeling hub ring
12. Spacer

Disassembly steps

1. Free wheeling hub cover
2. Gasket
3. Free wheeling hub clutch
4. Compression spring
5. Follower
6. Tension spring

Fig. 111 Manual free-wheeling hub assembly disassembly and assembly components — manual-change free-wheeling 4WD systems

7-78 DRIVE TRAIN

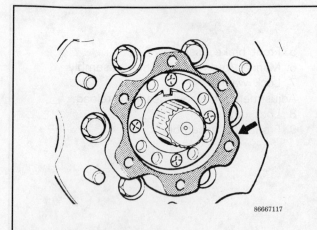

Fig. 112 Apply the sealant to the shaded area in the illustration during the installation procedure — both 4WD systems (manual-change and automatic-change)

Because brake (B) is secured to the knuckle, retainer (A) ceases to rotate, and therefore, the cam, while compressing the return spring, rises out of the cam groove of the retainer (A) and compresses the shift spring. The slide gear is pushed by the shift spring, and then engages with the gear of the housing when the two are in phase and enters the locked state.

The cam turns until the lug of the drive gear contacts the lug of the brake (A). Because of this contact, brake (A) is turned by the drive gear, and therefore, there is also no longer any force of retainer (A) with a tendency to turn brake (A). As a result, there is also no longer any force which presses brake (A) against brake (B) and the drive gear causes brake (A) to turn lightly (there is no friction force).

Because the cam remains meshed, it turns until it contacts the lug of retainer (A), and is locked.

When the transfer is shifted from 4WD to 2WD and the vehicle is driven in reverse, rotation of the gear of the body is transmitted from the slide gear to the cam to the retainer (A) to the brake (A), but retainer (A) ceases to turn, just as when the shift is made from the free state to the locked state. The cam, therefore, turns as far as the cam groove of retainer (A) and is pushed into the cam groove by the return spring.

The slide gear moves with the cam, disengages from the gear of the body, and enters a free state.

DISASSEMBLY & ASSEMBLY

▶ See Figures 112, 114, 115, 116 and 117

1. Place the free-wheeling hub in the free condition. To do this, shift the transfer shift lever to the 2H position, then move the vehicle 4-7 ft. (1-2 m) backwards.
2. Remove the front wheels of the vehicle, and support the raised front end of the vehicle with jackstands.
3. Remove the automatic free-wheeling hub cover. When the cover cannot be loosened by hand, use an oil filter wrench with a protective cloth in between not to damage the cover. Remove the O-ring from the underside of the hub cover. A new O-ring will be needed upon reassembly.
4. Using a pair of snapring pliers, remove the snapring from the driveshaft.
5. Slide the washer (shim) off of the driveshaft.
6. Unscrew the 6 retaining bolts from the automatic freewheeling hub assembly and slide the assembly off of the driveshaft.
7. Remove the rear housing C-clip, which is easily removed by pushing the brake (B) in and using a small prytool to pry it out.
8. Remove brake (B), brake (A) and the brake spring from the housing.
9. Use a small prytool to remove the housing snapring.
10. Using the special tool MB990811-01, lightly push (with a press) the drive gear in and remove the retainer (B) C-ring. Since the return spring relaxes approximately 1.5 in. (40mm) the stroke of the press should be set to more than 1.5 in. (40mm).

✴✴CAUTION

Use a protective cover so as not damage the cover attaching surface of the housing before setting it on the press table. Make sure that the pressing force does not exceed 44.1 lbs. (200 N).

11. Slowly reduce the pressure of the press until the return spring relaxes fully. Remove the drive gear assembly, the slide gear assembly and the return spring.

✴✴CAUTION

When the pressure of the press is removed, make sure that the retainer (A) is not caught by retainer (B).

12. Remove the retainer (B) and the retainer bearing.
13. Remove the drive gear snapring from the drive gear assembly. Separate the retainer (A) and the drive gear.

➡When the drive gear snapring is removed, be sure to replace it with a new one.

14. To remove the slide gear C-ring in the slide gear assembly, push the cam in and remove the slide gear C-ring with the spring compressed. Remove the cam, spring holder, shift spring and the slide gear from the slide gear assembly.

To reassemble:

15. Assemble the slide gear, the shift spring, the spring holder and the cam to make the slide gear assembly. Install a new C-ring to the slide gear assembly.
16. Assemble the drive gear and the retainer (A) to form the drive gear assembly. Install a new drive gear snapring to the drive gear assembly.
17. Grease the following components with multipurpose grease SAE J310, NLGI No. 2:
    a. Retainer (B) — pack the grooves of the retainer (B) with grease.
    b. Brake (B) — pack the grooves of brake (B) with grease.
18. Install the components to the hub housing in the following order: retainer bearing, retainer (B), return spring, slide gear assembly, drive gear assembly, new retainer (B) C-ring, new housing snapring, brake spring, brake (A), brake (B), new

DRIVE TRAIN 7-79

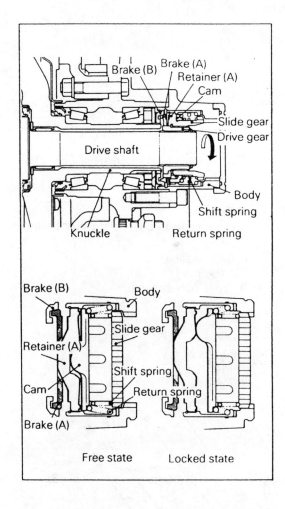
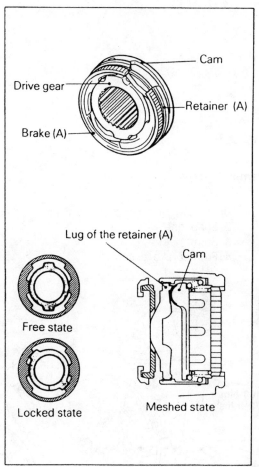

Fig. 113 Internal workings of the automatic-change 4WD free-wheeling hub assembly — all 1983-95 Pick-ups (4WD) and some 1983-95 Monteros

housing C-ring. When installing the return spring, install it with the smaller coil diameter side toward the cam.

19. Using a spring scale, measure the front hub turning resistance without the free-wheeling hub housing installed yet. Note the amount of force needed to turn the hub assembly on the vehicle.

20. Apply a coating of 3M® ART Part No. 8661, or No. 8663, or the equivalent sealant, equally all around and without missing any spots, to the free-wheeling hub body assembly and the front hub contact surfaces.

※※CAUTION

Make sure that there is no excess sealant on the hub outside surface.

21. Align the key of the brake (B) and the keyway of the knuckle spindle, then loosely install the automatic free-wheeling hub assembly to the spindle.

22. Check that the hub proper and the automatic free-wheeling hub assembly are brought into intimate contact when the assembly is forced lightly against the hub proper. If not, turn the hub until intimate contact is achieved.

23. Tighten the free-wheeling hub mounting bolts to 36-43 ft. lbs. (50-60 Nm). Use the spring scale to measure the front hub turning resistance again. Subtract the value measured from before from that now measured to find the turning resistance of the free-wheeling hub. The limit of resistance is 8.7 inch lbs. (1 Nm) or 3.1 lbs. (14 N) on the spring scale. If the free-wheeling hub turning resistance exceeds the limit, disassemble and reassemble the free-wheeling hub again.

24. Install the shim and the snapring to the driveshaft end. Screw the cover onto the hub assembly to 13-25 ft. lbs. (18-35 Nm).

25. Install the wheels to the vehicle. Lower the vehicle to the ground, then finish tightening the wheel lug nuts.

7-80 DRIVE TRAIN

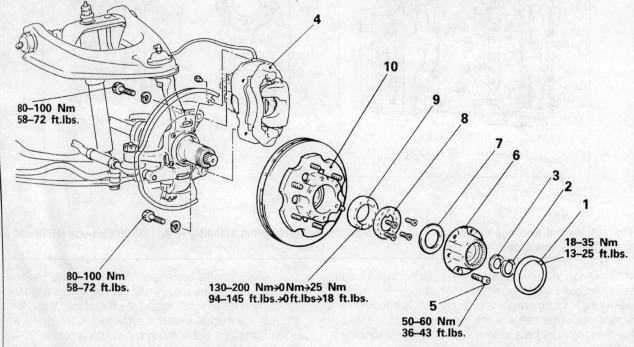

Removal steps
1. Cover
   Adjustment of drive shaft end play
2. Snap ring
3. Shim
4. Front brake assembly
5. Bolts
6. Automatic free-wheeling hub assembly
7. Shim
8. Lock washer
9. Lock nut
10. Front hub assembly

Fig. 114 Axle hub and free-wheeling hub assemblies' removal and installation — automatic-change 4WD systems

DRIVE TRAIN 7-81

Disassembly steps

1. Cover
2. O-ring
3. Housing
4. Housing C ring
5. Brake (B)
6. Brake (A)
7. Brake spring
8. Housing snap ring
9. Retainer (B) C ring
10. Drive gear assembly
11. Slide gear assembly
12. Return spring
13. Retainer (B)
14. Retainer bearing
15. Drive gear snap ring
16. Retainer (A)
17. Drive gear
18. Slide gear C ring
19. Cam
20. Spring holder
21. Shift spring
22. Slide gear

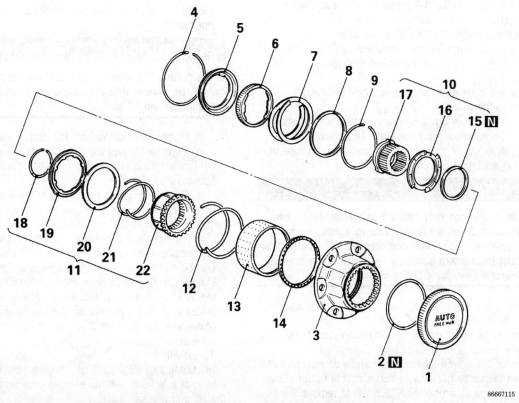

Fig. 115 Automatic free-wheeling hub assembly disassembly and assembly components

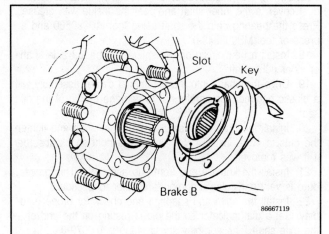

Fig. 116 When assembling the free-wheeling hub assembly onto the spindle, line the key and the key slot up — automatic free-wheeling 4WD systems

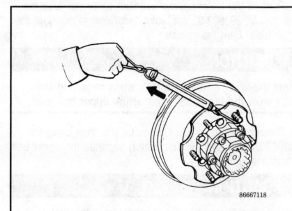

Fig. 117 Use a spring scale to measure the turning resistance both before and after the installation of the free-wheeling hub assembly — automatic free-wheeling 4WD systems

7-82 DRIVE TRAIN

Front Axle Shaft, Bearing And Seal

REMOVAL & INSTALLATION

◆ See Figures 118, 119, 120, 121, 122, 123, 124, 125 and 126

Before beginning, place the locking hubs in the FREE position by placing the transfer lever in the 2H position and moving in reverse for about 3-7 ft. (1-2 m) for the automatic free-wheeling systems. For the manual systems, rotate the handles on the hubs themselves to the FREE position.

1. Slightly loosen the lug nuts while the vehicle is still on the ground (only a little — do not loosen the lug nuts more than ½ of a turn.)
2. Raise the vehicle and support it safely on stands.
3. Remove the wheel.
4. Remove the front brake caliper assembly and position it to the side. Do not disconnect the brake hose; simply move the entire caliper out of the way and suspend it from rope or stiff wire. Do not let it hang by the hose.

※※CAUTION

Brake pads and shoes may contain asbestos, which has been determined to be a cancer causing agent. Never clean the brake surfaces with compressed air! Avoid inhaling any dust from brake surfaces! When cleaning brakes, use commercially available brake cleaning fluids.

5. Remove the free wheeling hub cover assembly and remove the snapring from the axle shaft.
6. Remove the cover gasket (if a manual system) and shim.
7. Remove the tie rod assembly-to-spindle connection cotter pin. Loosen the castle nut (slotted nut) a couple of turns, then use a tie rod separator tool (MB990635-01) to remove the tie rod end from the spindle. Once the tie rod end is loose, remove the castle nut and the tie rod end from the spindle.
8. Remove the lower spindle ball joint cotter pin. Use the special tool MB990809-01 (ball joint separator) to separate the lower ball joint from the spindle.

※※CAUTION

Support the lower arm with a jack when removing the knuckle from the lower ball joint or the upper ball joint.

9. Remove the upper ball joint cotter pin, then using the tool MB990778-01 (ball joint separator) separate the upper ball joint from the knuckle.

※※CAUTION

After the knuckle has been removed, lower the jack slowly.

10. Remove the knuckle and front hub together as a unit.
11. If removing the left side shaft, simply pull the shaft out of the differential carrier assembly. When pulling the left shaft from the differential carrier assembly, be careful that the shaft splines do not damage the oil seal.
12. On 1986 Pick-up only, raise the right lower suspension arm and remove the right shock absorber if removing the right side shaft.
13. Matchmark the right shaft and the inner shaft assembly, then the inner shaft and the housing tube.
14. Disconnect the right shaft from the inner shaft assembly and remove the shaft. Remove the inner shaft from the housing tube and the housing tube from the differential housing.
    a. Remove the inner shaft with an impact puller tool (Mitsubishi tool MB990241-01 and MB990211-01 to remove the inner shaft.

※※CAUTION

When pulling the inner shaft assembly from the differential carrier, be careful that the spline of the inner shaft does not damage the oil seal.

   b. Remove the circlip from the end of the inner shaft.
   c. Remove the self locking nut from the differential mounting bracket, then remove the mounting bracket from the housing tube.
   d. Remove the housing tube from the differential housing.
15. Press the bearing and seal off of the inner shaft and remove the dust seal from the tube.
    a. To remove the bearing, bend the outside periphery of the dust cover inward with a hammer. Install the general service tool MB998348-01 and tighten the nut of tool until the portion A of the tool touches the bearing outer race. Set the inner shaft with the tool attached onto two wooden blocks and drive the inner shaft down to remove the bearing. Make sure the inner shaft is not allowed to fall on a hard surface.

To install:
16. Using tool MB990955-01, install a new dust seal to the tube until it is flush with the housing tube and face, and coat the lip with grease.
17. Using a suitable long steel pipe (outside diameter: 2.95 in. (75 mm), wall thickness: 0.16 in. (4 mm)), install a new dust cover to the inner shaft and coat the inside with grease. Press the bearing onto the shaft using tool MD990560 and a press or tool MB998348-01.
18. Install a new circlip on the splines of the left side shaft or inner right shaft.
19. Drive the shafts into the differential carrier assembly with a plastic hammer. Be careful not to damage the lip of the oil seal.
20. Install the right outer shaft to the inner shaft and tighten the nuts to 40 ft. lbs. (54 Nm). Install the right shock absorber if it was removed.
21. Install the knuckle and front hub assembly. This procedure is very particular and important, refer to Section 8.
22. Install the shim and snapring and check for proper endplay. Set a dial indicator so the pin is resting on the end of the axle shaft. The end-play specification is 0.0079-0.02 in. (0.2-0.5mm). If not within specifications, adjust by adding or removing shims.
23. Install the hub cover. Tighten the retaining bolts to 7-10 ft. lbs. (10-14 Nm).

DRIVE TRAIN 7-83

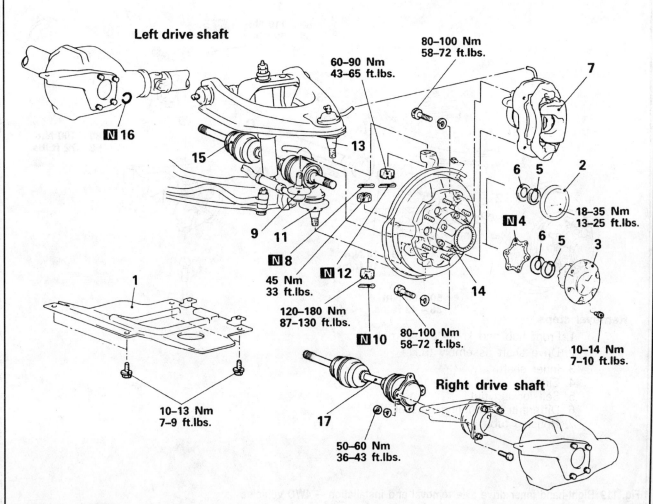

Removal steps

1. Under cover
2. Cover
   (automatic free-wheeling hub)
3. Free-wheeling hub cover
   (manual free-wheeling hub)
4. Gasket (manual free-wheeling hub)

Adjustment of driveshaft end play

5. Snap ring
6. Shim
7. Front brake assembly
8. Cotter pin
9. Connection of tie rod assembly and knuckle
10. Cotter pin
11. Connection of lower ball joint and knuckle
12. Cotter pin
13. Connection of upper ball joint and knuckle
14. Front hub and knuckle assembly
15. Left drive shaft
16. Circlip
17. Right drive shaft

Fig. 118 Front drive axle removal and installation — 4WD vehicles

7-84 DRIVE TRAIN

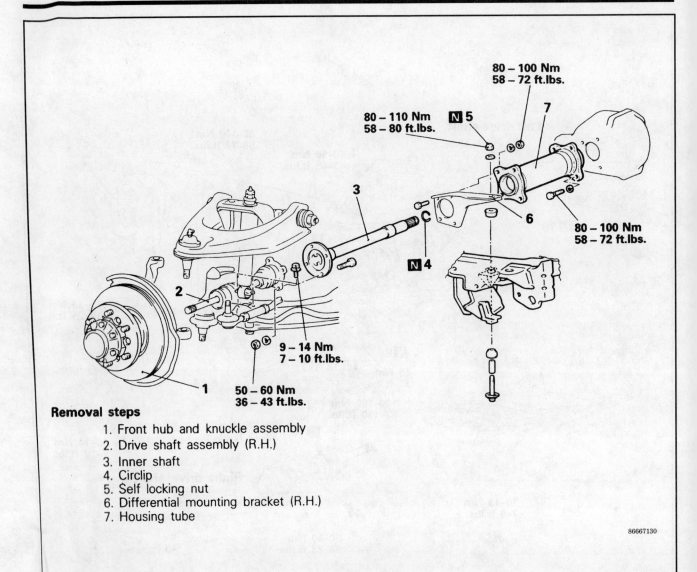

Removal steps
1. Front hub and knuckle assembly
2. Drive shaft assembly (R.H.)
3. Inner shaft
4. Circlip
5. Self locking nut
6. Differential mounting bracket (R.H.)
7. Housing tube

Fig. 119 Right-hand inner drive axle removal and installation — 4WD vehicles

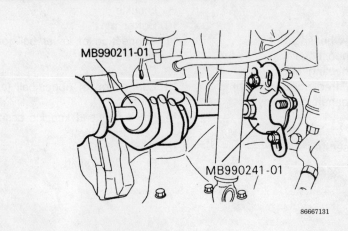

Fig. 120 Remove the inner drive axle from the differential unit with an impact puller — Mitsubishi tools MB990211-01 and MB990241-01

DRIVE TRAIN   7-85

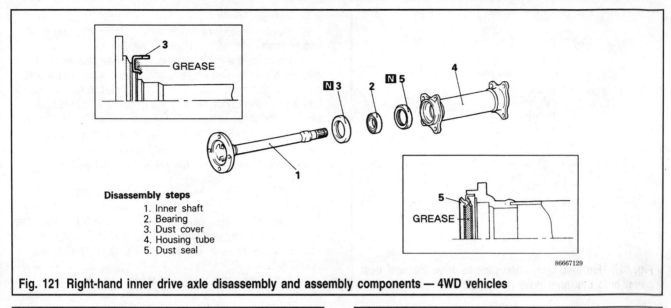

Fig. 121 Right-hand inner drive axle disassembly and assembly components — 4WD vehicles

Disassembly steps
1. Inner shaft
2. Bearing
3. Dust cover
4. Housing tube
5. Dust seal

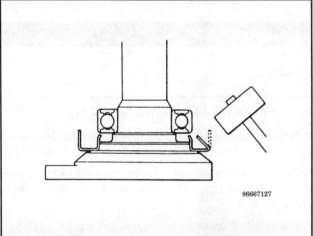

Fig. 122 To remove the drive axle shaft bearing, use a plastic mallet to bend the outside periphery of the dust cover inward

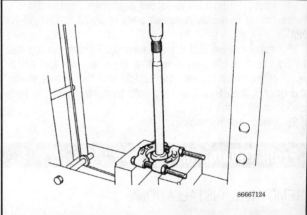

Fig. 124 Set the inner shaft with the tool installed onto two wooden blocks, as shown in the illustration, and use a hammer to drive the shaft downward — make sure to not allow the inner shaft to fall

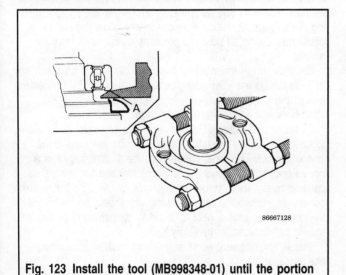

Fig. 123 Install the tool (MB998348-01) until the portion A touches the bearing outer race

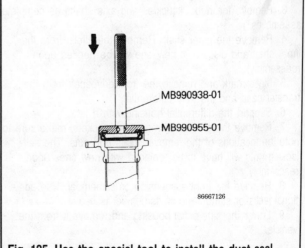

Fig. 125 Use the special tool to install the dust seal into the housing tube

7-86 DRIVE TRAIN

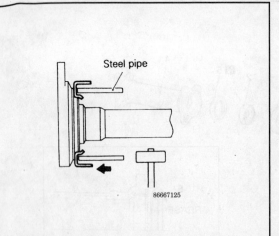

Fig. 126 Use a suitably-sized pipe to drive the new dust cover onto the inner drive axle shaft

24. Install the front brake caliper assembly. Tighten the brake caliper assembly mounting bolts to 58-72 ft. lbs. (80-100 Nm).
25. Install the tire and wheel assembly. Tighten the lug nuts as tight as possible considering that the wheel will turn since it is off of the ground. Lower the vehicle until the wheels touch the ground, then tighten the lug nuts completely. Let the vehicle down all of the way.
26. Road test the vehicle.

Front Differential Carrier

REMOVAL & INSTALLATION

♦ See Figure 127

1. Raise the vehicle and support it safely on stands. Make certain the front stands will not interfere with the housing removal.
2. Remove the under cover and drain the differential.
3. Remove the hubs, knuckles and axle shafts as complete assemblies.
4. Remove the inner shaft. Remove the circlip from the inner shaft and discard (a new one will be needed upon reassembly).
5. Matchmark and remove the front driveshaft from the transfer case and the front differential.
6. Support the differential housing with a jack.
7. Remove the differential mounting brackets, make sure to note the locations of the removed bolts and nuts. The self-locking nuts will have to be replaced with new ones upon reassembly.
8. Remove the front suspension crossmember. Use additional jacks or stands; the crossmember is heavy.
9. Lower the differential housing and remove it from the vehicle.

To install:
10. Mount the housing safely on a suitable jack and raise into position.
11. Lubricate the bushings and install the crossmember and housing brackets. The crossmember mounting nuts and bolts should be tightened to 72-87 ft. lbs. (100-120 Nm). Tighten the housing bracket-to-differential bolts to 58-72 ft. lbs. (80-100 Nm) and the housing brackets-to-frame mounting self locking nuts and bolts to 58-80 ft. lbs. (80-110 Nm). It is important that new self locking nuts are used during installation.
12. Replace the circlips and install the inner axle and axle, knuckle and hub assemblies.
13. Install the front driveshaft — refer to the previous procedures.
14. Tighten the drain plug to 43-51 ft. lbs. (60-70 Nm). Level the vehicle and fill the differential with gear until it just starts to spill out of the fill hole. Tighten the fill plug to 29-43 ft. lbs. (40-60 Nm).
15. Install the under cover and tighten the under cover mounting bolts to 7-9 ft. lbs. (10-13 Nm). Lower the vehicle.
16. Perform a road test and check the differential for leaks.

CV-Joints

OVERHAUL

♦ See Figures 128, 129, 130, 131, 132 and 133

Before an overhaul is possible, remove the front drive axles — refer to the removal procedures described earlier.

➡ The front axle shafts of Montero and 4WD Pick-ups have the serviceable joints (TJ or DOJ) at the outside and the BJ or RJ at the inner end.

The drive axle assembly is a flexible unit consisting of an inner and outer constant velocity (CV) joint joined by the axle shaft. Some cars may have a center bearing or carrier but the inner and outer joint remain. The inner joint, because of its design, is completely flexible and allows the joint to flex left/right, up/down and compress/extend while the joint is turning. This range of motion is necessary to allow the car to accelerate during all possible positions of the car's front wheels.

Because of the inner joints' flexibility, care must be taken not to over-extend the joint during repairs or handling. When either end of the shaft is disconnected from the car, support the shaft with wire to prevent joint damage.

CV-joints are protected by rubber boots or seals, designed to keep the high temperature grease in and the road grime and water out. The most common cause of joint failure is a ripped boot which allows the lubricant to leave the joint, thus causing heavy wear, noise and ultimate failure. The boots are constantly exposed to road hazards and should be inspected frequently. Any time a boot is found to be damaged or slit, it should be replaced immediately.

This is another very good example of maintenance being cheaper than repair.

A replacement boot kit is generally under $35 while a new joint can range above $150.

DRIVE TRAIN 7-87

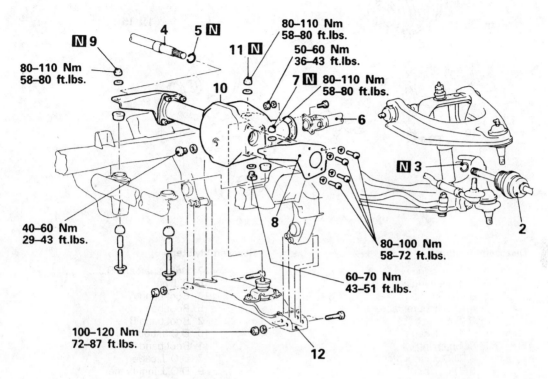

Removal steps
1. Under cover
2. Drive shaft
3. Circlip
4. Inner shaft
5. Circlip
6. Front propeller shaft
7. Self-locking nut
8. Differential mounting bracket (L.H.)
9. Self-locking nut
10. Front suspension crossmember and front differential carrier assembly
11. Self-locking nut
12. Front suspension crossmember
13. Differential mounting bracket (R.H.)
14. Housing tube
15. Front differential carrier assembly

Fig. 127 Front differential carrier removal and installation — 4WD vehicles only

7-88 DRIVE TRAIN

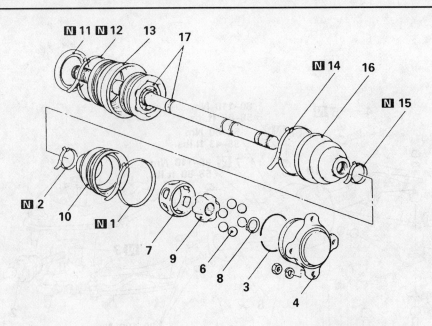

Disassembly steps

1. Boot band A
2. Boot band B
3. Circlip
4. D.O.J. outer race
6. Balls
7. D.O.J. cage
8. Snap ring
9. D.O.J. inner race
10. D.O.J. boot
11. Dust cover
12. Boot protector band
13. Boot protector
14. Boot band A
15. Boot band B
16. B.J. boot
17. Drive shaft and B.J.

Reassembly steps

17. Drive shaft and B.J.
16. B.J. boot
14. Boot band A
15. Boot band B
2. Boot band B
10. D.O.J. boot
1. Boot band A
7. D.O.J. cage
9. D.O.J. inner race
8. Snap ring
6. Balls
4. D.O.J. outer race
3. Circlip
13. Boot protector
12. Boot protector band
11. Dust cover

Fig. 128 Right driveshaft disassembly and reassembly components

Mitsubishi uses no fewer than 4 types of CV-joint on its cars; they appear in various combinations on various driveshafts. The types of joints and their abbreviations are:
- Tripod joint (TJ) — commonly found as an inboard joint, the three sided spider inside can be replaced on the bench.
- Rzeppa joint (RJ) — named for its inventor, it almost always shows up as an outer joint. The joint is not replaceable under most conditions but comes installed on a new axle.
- Double Offset joint (DOJ) — identified by its ball-in-cage arrangement, this highly flexible inner joint can be dismantled and rebuilt.
- Birfield joint (BJ) — outer joint with good reliability. Cannot be disassembled from the axle. A new joint will include a new shaft.

This section will use these abbreviations for each joint. Additionally, any reference to an outer or outboard joint indicates the joint at or near the wheel/hub/knuckle assembly. Not surprisingly, inner or inboard refers to the joint at or closest to the transaxle.

✱✱WARNING

Bands and snaprings are not reusable. Parts are not interchangeable. Each band and boot for each model and type of joint has an identifying code on it. Obtain this code and shop for exact replacements. A few cents saved on the wrong part may cost you a second repair when the boot doesn't seal correctly. The grease that comes in the repair kit should be used throughout the joint. Do not substitute other lubricants; they will not withstand the heat and pressure within the joint.

Tripod Joint

1. Remove the driveshaft from the car, following directions outlined earlier.
2. Remove the large band holding the boot to the joint and the smaller band holding the boot to the shaft. Slide the boot off the joint and down the shaft, out of the way.

DRIVE TRAIN 7-89

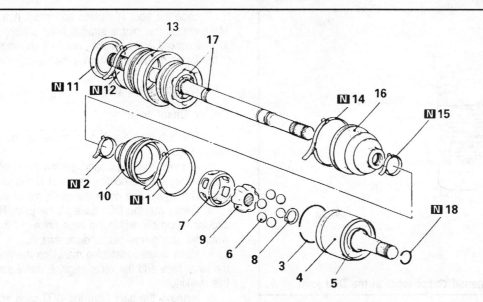

Disassembly steps

1. Boot band A
2. Boot band B
3. Circlip
4. D.O.J. outer race
5. Dust cover
6. Balls
7. D.O.J. cage
8. Snap ring
9. D.O.J. inner race
10. D.O.J. boot
11. Dust cover
12. Boot protector band
13. Boot protector
14. Boot band A
15. Boot band B
16. B.J. boot
17. Drive shaft and B.J.
18. Circlip

Reassembly steps

17. Drive shaft and B.J.
16. B.J. boot
14. Boot band A
15. Boot band B
2. Boot band B
10. D.O.J. Boot
1. Boot band A
7. D.O.J. cage
9. D.O.J. inner race
8. Snap ring
6. Balls
5. Dust cover
4. D.O.J. outer race
3. Circlip
18. Circlip
13. Boot protector
12. Boot protector band
11. Dust cover

Fig. 129 Left driveshaft disassembly and reassembly components

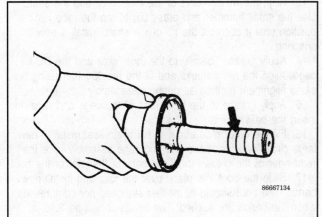

Fig. 130 Wrap vinyl tape around the spline part of the DOJ side of the driveshaft so that the DOJ boots are not damaged when removed

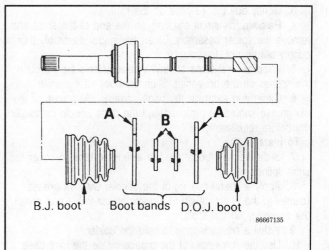

Fig. 131 Install the BJ and DOJ boots and bands in the order as shown in the illustration

7-90 DRIVE TRAIN

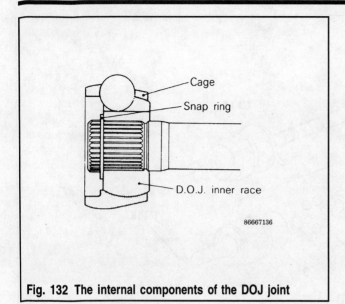

Fig. 132 The internal components of the DOJ joint

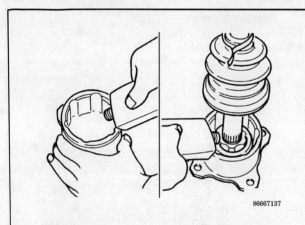

Fig. 133 Apply 1.9 oz. (55 gr) of Repair kit grease to the DOJ outer race, insert the driveshaft into the race, then apply the same amount of grease on top of the race

3. Gently pull the TJ case off the joint.
4. Remove the small snapring on the end of the shaft and remove the spider assembly. Clean the spider assembly thoroughly but do not disassemble it.
5. Wrap several turns of vinyl or electrician's tape around the splines on the driveshaft. Slide the boot off the shaft.
6. Clean and examine the parts carefully. All traces of the old grease should be removed and the parts should be totally dry before reinstallation.

To install:

7. Slide the new boot onto the axle. Use tape to cover the shaft splines.
8. Apply a liberal coating of the special CV-joint grease (comes in the overhaul kit) to the spider assembly and install the assembly on the shaft.
9. Install a new snapring to hold the spider.
10. Use the remainder of the grease inside the joint case and fit the case over the spider.

11. Slide the boot into place and install new boot clamps. Make certain the joint is straight when placing the boot; if the joint is cocked, the boot may not seal correctly.
12. Install a new snapring on the transaxle end of the inner joint.
13. Reinstall the driveshaft.

Double Offset Joint

1. Remove the driveshaft from the car, following directions outlined earlier.
2. Remove the small band holding the boot to the shaft and the larger band holding the boot to the joint. Slide the boot off the joint and down the shaft, out of the way.
3. Gently pull the DOJ case off the joint. There is a large circular spring clip within the case holding it in place. Carefully work this spring free before disassembly.
4. Make reliable matching marks on the end of the shaft, the inner race and the outer cage. A metal scribe is handy for this marking.
5. Remove the balls from the DOJ cage.
6. Remove the cage from the inner race. Turn the cage so that the projections of the inner race align with the recesses of the cage. Slide the cage toward the center of the shaft, not to the outside.
7. Remove the snapring from the shaft. Mount the shaft in a vise with protected jaws.
8. Use a brass bar and small hammer, tap evenly all around the inner race and remove it from the shaft.
9. Apply several turns of tape over the shaft splines. Remove the cage and boot from the shaft.
10. Check the shaft and splines for damage or wear. Inspect the cage, race and balls for any sign of corrosion, wear, cracking or damage. Clean all the parts thoroughly and air dry them completely before installation. Any remaining cleaning solvent can dissolve the lubricating grease.

To install:

11. Wrap the splines in tape; install a new boot and slide it down the shaft.
12. Place the cage onto the driveshaft so that the smaller diameter side is installed first.
13. Align the matchmarks of the inner race and the shaft. Use the small hammer and brass bar to tap the race into position until it contacts the rib of the shaft. Install a new snapring.
14. Apply grease liberally to the inner race and the DOJ cage. Align the matchmarks and fit the two together using the same alignment method as during disassembly.
15. Apply grease to the ball areas of the cage and race and insert the balls.
16. Fill the outer race about 1/3 full of grease. Install a new large circlip and install the ball and cage assembly. Use the remainder of the grease to fill the back of the joint or the boot.
17. Slide the boot into place over the joint and install new bands. The boot should be neither stretched nor compressed when the bands are applied. The bands should be 3.15 in. (80mm) apart. Too much or too little air within the boot may cause premature boot failure.
18. Install the driveshaft into the car.

DRIVE TRAIN 7-91

BOOT REPLACEMENT

Inner Joint

Boot replacement is accomplished by following the appropriate overhaul procedure. The joint components must be removed from the shaft to allow the boot to pass. Even if only the boot is being replaced, the other components should be inspected. Remove as much old grease as possible and reassemble the joint packed with fresh grease.

Outer Joint

1. Remove the driveshaft from the car.
2. The outer boot cannot be removed over the outer end of the shaft. You will need 2 repair kits since the inner joint must also be disassembled.
3. Disassemble the inner joint following overhaul procedures listed earlier.
4. If the shaft has a dynamic damper (usually on left shafts), remove its band and slide the damper over the tape-protected splines.
5. Remove the bands holding the boot at the outer joint and slide the boot all the way down the shaft and off.
6. The exposed joint — either RJ or BJ type — should be inspected for signs of wear if possible. Do NOT disassemble this joint.
7. Fit the new boot onto the shaft and into place near the joint.
8. Use about ½ the grease in the kit to pack the outer joint and the other ½ to fill the boot.
9. Work the boot into place and apply the new bands. Make sure the joint is not at an angle during the securing of the boot.
10. Install the dynamic damper if one was removed. Use a new band to hold it.
11. Install the new boot for the inner joint and reassemble the joint following the overhaul instructions given earlier. Note that the dynamic damper, if used, must be kept free of grease.
12. Install the driveshaft in the car.

Pinion Seal

The front pinion seal cannot be replaced without complete disassembly of the front differential. Because of the number of special tools required and the need for very precise measurement during reassembly, this is a job best left to an experienced repair facility.

REAR AXLE

Identification

The vehicle information code plate, located on the firewall in the engine compartment, lists the axle ratio. On the third line, labeled "Transmission", there is a five digit transmission code, followed by a blank space, followed by a four digit code representing the axle ratio, i.e. 3545 represents 3.545:1.

Understanding Drive Axles

The drive axle is a special type of transmission that reduces the speed of the drive from the engine and transmission and divides the power to the wheels. Power enters the axle from the driveshaft via the companion flange. The flange is mounted on the drive pinion shaft. The drive pinion shaft and gear which carry the power into the differential turn at engine speed. The gear on the end of the pinion shaft drives a large ring gear the axis of rotation of which is 90 degrees away from the of the pinion. The pinion and gear reduce the gear ratio of the axle, and change the direction of rotation to turn the axle shafts which drive both wheels. The axle gear ratio is found by dividing the number of pinion gear teeth into the number of ring gear teeth.

The ring gear drives the differential case. The case provides the two mounting points for the ends of a pinion shaft on which are mounted two pinion gears. The pinion gears drive the two side gears, one of which is located on the inner end of each axle shaft.

By driving the axle shafts through the arrangement, the differential allows the outer drive wheel to turn faster than the inner drive wheel in a turn.

The main drive pinion and the side bearings, which bear the weight of the differential case, are shimmed to provide proper bearing preload, and to position the pinion and ring gears properly.

✱✱WARNING

The proper adjustment of the relationship of the ring and pinion gears is critical. It should be attempted only by those with proper equipment and experience.

Limited-slip differentials include clutches which tend to link each axle shaft to the differential case. Clutches may be engaged either by spring action or by pressure produced by the torque on the axles during a turn. During turning on a dry pavement, the effects of the clutches are overcome, and each wheel turns at the required speed. When slippage occurs at either wheel, however, the clutches will transmit some of the power to the wheel which has the greater amount of traction. Because of the presence of clutches, limited-slip units require a special lubricant.

Determining Axle Ratio

The drive axle is said to have a certain axle ratio. This number (usually a whole number and a decimal fraction) is actually a comparison of the number of gear teeth on the ring gear and the pinion gear. For example, a 4.11 rear means that theoretically, there are 4.11 teeth on the ring gear and one tooth on the pinion gear or, put another way, the driveshaft must turn 4.11 times to turn the wheels once. Actually, on a 4.11 rear, there might be 37 teeth on the ring gear and 9 teeth on the pinion gear. By dividing the number of teeth on the ring gear, the numerical axle ratio (4.11) is obtained. This also provides a good method of ascertaining exactly which axle ratio one is dealing with.

Another method of determining gear ratio is to jack up and support the truck so that both rear wheels are off the ground. Make a chalk mark on the rear wheel and the driveshaft. Put the transmission in neutral. Turn the rear wheel one complete turn and count the number of turns that the driveshaft makes. The number of turns that the driveshaft makes in one complete revolution of the rear wheel is an approximation of the rear axle ratio.

Rear Axle Shafts, Bearings And Seals

REMOVAL & INSTALLATION

1983-91 Pick-ups and Monteros

◆ See Figures 134, 135, 136, 137, 138 and 139

1. Raise the vehicle and support it safely on stands.
2. Remove the rear wheel.
3. Remove the rear brake drum.

✱✱CAUTION

Brake pads and shoes may contain asbestos, which has been determined to be a cancer causing agent. Never clean the brake surfaces with compressed air! Avoid inhaling any dust from brake surfaces! When cleaning brakes, use commercially available brake cleaning fluids.

4. Disconnect and plug the brake line(s) at the wheel cylinder. A small container will be needed to catch the leaking brake fluid before the line is plugged. If a plug for the line is not available, drain the brake fluid into a container.
5. Disconnect the parking brake cable from the shoes and remove the cable from the backing plate.
6. Remove the 4 nuts behind the brake backing plate holding the bearing case to the axle housing assembly.
7. Remove the backing plate, bearing case and the axle shaft as an assembly. If this is not possible by hand, use a slide hammer to remove the assembly.
8. Remove the O-ring and the bearing preload shims. Save the preload shims for reassembly.
9. Remove the oil seal from the axle tube with a hooked slide hammer.
10. To remove the axle shaft bearing, remove the notched locknut with tool MB990785-01 or a brass drift.
11. Remove the lockwasher and flat washer.
12. Screw the locknut back on to the axle shaft about 3 turns.
13. If tool MB990787-01 is not available, it will be necessary to fabricate a metal plate that fits over the axle shaft and abuts the locknut. Drill 4 holes in the plate that align with the bearing case studs and fit the plate onto them. Refit 2 nuts and washers to the bearing case studs diagonally across from each other and tighten them evenly to free the bearing case and the bearing.
14. Use a hammer and drift to remove the bearing outer race from the bearing case.
15. Remove the outer oil seal from the bearing case.

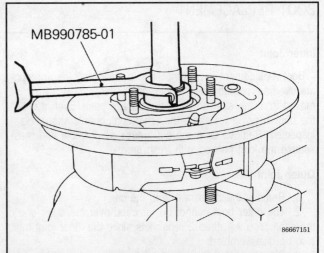

Fig. 134 Use the special tool MB990785-01 to remove the locknut — all vehicles except for 1992-95 Monteros

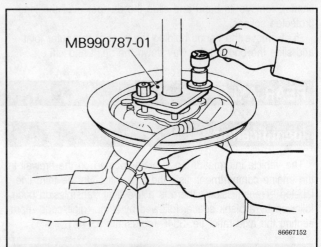

Fig. 135 Use the tool MB990787-01 to remove the axle shaft from the bearing case — all vehicles except for 1992-95 Monteros

To install:

16. Apply grease to the outer surface on the bearing outer race and to the lip of the outer oil seal. Drive them into the bearing case from each side.
17. Slide the bearing case and bearing over the rear axle shaft. Apply grease on the bearing rollers and fit the inner race by pressing it into place. Be careful not to damage or deform the dust cover.
18. Pack the bearing with grease.
19. Install the washer, the crowned lockwasher and the locknut in that order and tighten the locknut to 148 ft. lbs (200 Nm).
20. Bend the tab on the lockwasher into the groove on the locknut. If the tab and the groove do not line up, slightly tighten the locknut until they do.
21. Drive the new inner oil seal into place after greasing it and refit the assembly.
22. Install a new O-ring, install the shims and apply silicone rubber sealant to the face of the bearing case.

DRIVE TRAIN 7-93

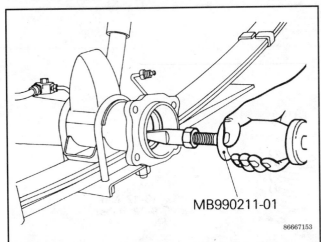

Fig. 136 Use a slide hammer (impact puller) with a hooked end to remove the axle housing tube oil seal — all vehicles except for 1992-95 Monteros

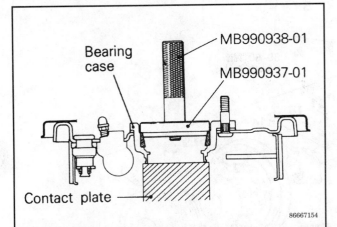

Fig. 137 Install the new bearing case race with the special Mitsubishi tool or another correctly-sized race driving tool — all vehicles except for 1992-95 Monteros

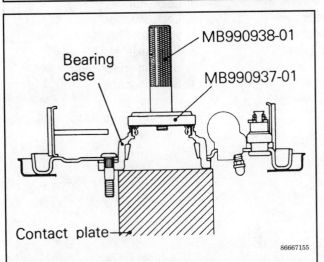

Fig. 138 Install the oil seal with the same tools as with the race bearing — install the seal until it sits flush

23. Install the entire assembly to the axle housing. Torque the retaining nuts to 40 ft. lbs. (54 Nm).
24. Check the axle shaft end-play. If it is not between 0.0020-0.0079 in. (0.05-0.20mm), proceed with the axle shaft end-play adjustment procedure.
25. If the end-play is within specifications, install all removed brake parts and bleed the system.
26. Install the wheel and lower the vehicle.
27. Road test the vehicle. Check for leaks.

1992-95 Montero

♦ See Figures 136, 140, 141, 142, 143, 144, 145, 146, 147 and 148

1. Loosen the wheel lug nuts only $1/2$ a turn.
2. Raise and support the vehicle on jackstands.
3. Remove the wheel(s) from the vehicle.
4. Loosen the bleeder valve on the right rear calliper and drain the brake fluid into a container. Disconnect the rear brake calliper brake hose from the hard line on the frame.
5. Pull the rear brake calliper off of the rear disc and remove from the vehicle.
6. Pull the rear disc off of the rear axle.
7. Remove the parking cable attaching bolt and disconnect the parking brake cable end from the brake assembly. Remove the parking brake assembly from the end of the axle — refer to Section 9 for this procedure.
8. On vehicles with the ABS system, remove the speed sensor.
9. Pull the rear axle shaft out of the axle housing. If the rear axle shaft is difficult to remove, use a slide hammer (impact puller) to remove it.

✲✲WARNING

Do not damage the oil seal during removal.

10. Remove the snapring from the inside end of the axle shaft. Remove one retainer bolt from the backing plate with a plastic mallet. Apply gummed cloth tape around the edge of the bearing case for protection. Fix the axle shaft in a vise or with a similar method. Using a grinder, grind down the retainer flat, on one side, until the thickness of the retainer is only 0.04-0.08 in. (1-2mm). That is that the retainer is ground down toward the axle shaft, not toward the flange. Cut, with a chisel, the place where the retainer ring has been shaven down and remove the retainer.

✲✲CAUTION

Be careful not to damage the bearing case and the axle shaft.

→Only the retainer ring is to be ground down — NOT the axle shaft, the axle flange, the bearing or any other component.

11. Grind the plate of special tool MB990861 with a grinder (see illustration) so that there will be no interference between the plate and the bearing case. While adjusting the height of

7-94 DRIVE TRAIN

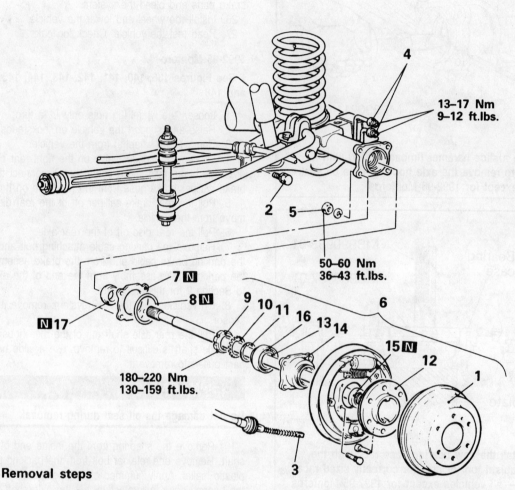

Removal steps

1. Brake drum
2. Parking brake cable attaching bolts
3. Connection of parking brake cable end and brake shoe assembly
4. Connection of brake tubes
5. Nuts
6. Rear axle shaft assembly (with parking brake cable)
7. Shims
8. O-ring
   Adjustment of axle shaft end play
9. Lock nut
10. Lock washer
11. Washer
12. Rear axle shaft
13. Bearing inner race
14. Bearing case
15. Oil seal
16. Bearing outer race
17. Oil seal

Fig. 139 Rear axle shaft, bearings and races removal and installation — all vehicles except for 1992-95 Monteros

DRIVE TRAIN   7-95

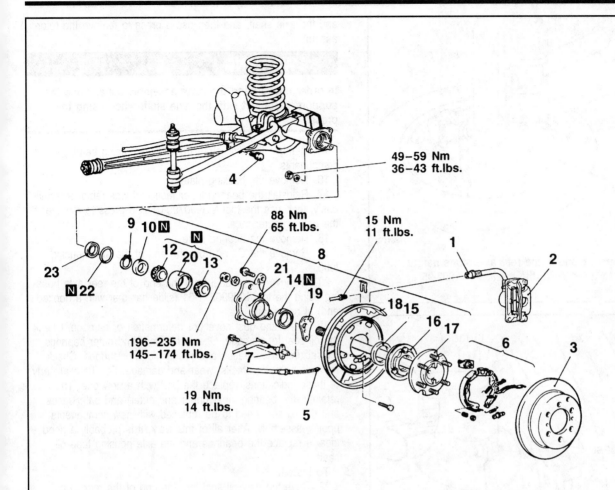

Removal steps

1. Connection for brake pipe
2. Rear brake assembly
3. Brake disc
4. Parking brake cable attaching bolt
5. Parking cable end
6. Parking brake assembly
7. Speed sensor <Vehicles with ABS>
8. Axle shaft assembly
9. Snap ring
10. Retainer
11. Axle shaft sub assembly
    (Parts from step 13 to step 17)
12. Bearing inner race (inner)
13. Bearing inner race (outer)
14. Oil seal
15. Dust cover <Vehicles without ABS>
16. Rotor assembly <Vehicles with ABS>
17. Axle shaft
18. Backing plate
19. Speed sensor bracket
    <Vehicles with ABS>
20. Bearing outer race
21. Bearing case
22. O-ring
23. Oil seal

Installation steps

23. Oil seal
22. O-ring
21. Bearing case
20. Bearing outer race
19. Sped sensor bracket
    <Vehicles with ABS>
18. Backing plate
17. Axle shaft
16. Rotor assembly <Vehicles with ABS>
15. Dust cover <Vehicles with ABS>
13. Bearing inner race (outer)
14. Oil seal
12. Bearing inner race (inner)
10. Retainer
9. Snap ring
8. Axle shaft assembly
7. Speed sensor <Vehicles with ABS>
   (Refer to GROUP 35C – Wheel Speed Sensor.)
6. Parking brake assembly (Refer to GROUP 36 – Parking Brake Drum.)
5. Parking cable end
3. Brake disc
2. Rear brake assembly
1. Connection for brake pipe

Fig. 140 Rear axle shaft, bearings and races removal and installation — 1992-95 Monteros

7-96 DRIVE TRAIN

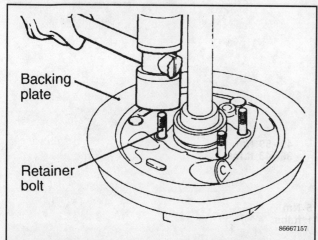

Fig. 141 Remove one of the rear axle studs before attempting to grind down the retainer — 1992-95 Monteros

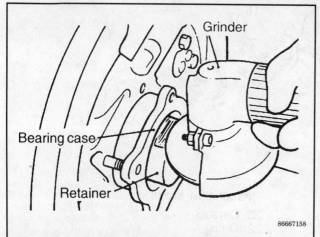

Fig. 142 Using a grinder, grind the retainer, on one side, down to 1-2mm (0.04-0.08 in.) thickness — 1992-95 Monteros

the hanger, secure the washers, plate and nuts in order so that the processed plate is as shown in the illustration.

➡The washers are used to eliminate the difference in height of the bearing case so that the plate and the bearing case are parallel.

Place the end of the bolt against the center of the axle shaft, and then tighten the nuts to remove the axle shaft from the bearing case assembly.

➡The hanger and plate must be placed so that they are parallel.

12. Remove the bearing inner race and the bearing outer race. To remove the races, install the tool MB990560 and use a press to remove the bearing race from the axle shaft.
13. Remove the oil seal and the dust cover (vehicles without ABS).
14. On vehicles without ABS, insert an iron plate of approximately 0.04 in. (1mm) thickness between the rotor assembly

and the axle shaft, and then use a press to remove the rotor assembly.

✱✱WARNING

In order not to bend the rotor assembly plate, place the support in contact with the axle shaft when using the press.

15. Remove the axle shaft from the remaining bearings and components.
16. Remove the backing plate.
17. Reinstall the bearing inner race that was removed previously, then use the tool MB990799-01 and press to remove the bearing outer race.
18. Remove the bearing case.
19. Remove the O-ring from the end of the axle housing tube.
20. Remove the oil seal from the end of the rear axle housing using the tool MB990211-01 (slide hammer with a hooked end), if necessary.
21. Check the dust cover for deformation or damage. Check the oil seal for damage. Check the inner and outer bearings for seizure, discoloration and rough raceway surface. Check the axle shaft for cracks, wear and damage. For there are any of these indications, replace the part with a new one. The retainer, the bearing inner (inner and outer) and outer races and the oil seal need to be replaced with new components upon reassembly. After all of this work, it is probably a good idea to replace the bearings and the axle housing tube oil seals.

To install:
22. Drive the new oil seal into the end of the rear axle housing using the tools MB990932-01 and MB990938-01, if necessary.
23. Install the new O-ring into the axle tube.
24. Apply multipurpose grease to the external surface of the bearing out race. Press-fit the bearing outer race into the bearing case by using the toll MB990890-01.
25. Install the speed sensor bracket to the back of the backing plate.

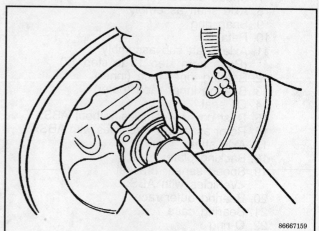

Fig. 143 On the ground down spot on the retainer, use a chisel to split the retainer then remove it — 1992-95 Monteros

DRIVE TRAIN 7-97

26. Install the rotor assembly to the axle shaft by press-fitting (plastic mallet will also work) it on using the special tool MB991388. Slide the backing plate onto the axle shaft.
27. Install the dust cover to the backing plate if the vehicle is equipped with ABS.
28. Install the bearing inner race (outer) to the bearing case. Install the oil seal to the front end of the bearing case. To do this, apply multipurpose grease to the outside of the oil seal. Use the special tools MB990936-01 and MB990938-01 to press-fit the oil seal until it is flush with the end of the bearing case. Apply multipurpose grease to the lip of the oil seal.
29. Pass the axle shaft through the bearing inner race, the bearing case and the second bearing inner race in that order. Use the special tool MB990799 to press-fit the bearing inner race to the axle shaft.

✽✽WARNING

Both bearing inner race sets should be press-fitted together. The left and right lengths of the axle shaft are different in vehicles with rear differential locks. The right side is longer, be careful when installing.

30. Use the tool MB990799-01 to press-fit the retainer onto the axle shaft, while checking that the press-fitting force is at the following values:
    a. Initial press-fitting force is 11,023 lbs. (50,000 N) or more.
    b. Final press-fitting force is 22,046-24,251 lbs. (100,000-110,000 N).
31. If the initial press-fitting force is less than the standard value, replace the axle shaft.
32. After installing the snapring, measure the clearance between the snapring and the retainer with a thickness gauge, and check that it is within the standard values. The standard value is 0.0065 in. (0.166mm) or less. If the clearance exceeds the standard value, change the snapring so that the

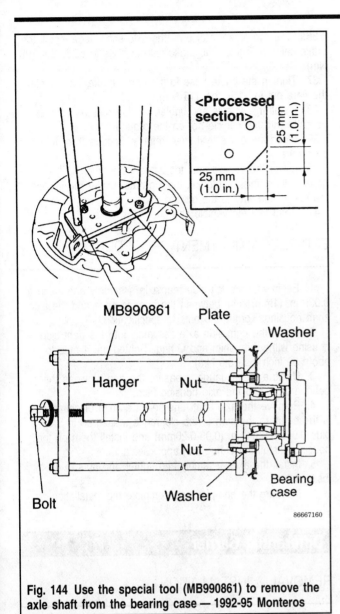

Fig. 144 Use the special tool (MB990861) to remove the axle shaft from the bearing case — 1992-95 Monteros

Fig. 145 Use a iron plate and supports to remove the rotor assembly — 1992-95 Monteros

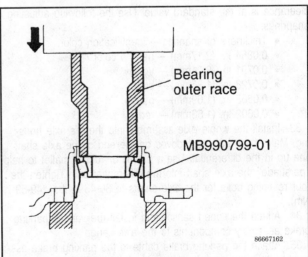

Fig. 146 Use the tool (MB990799-01) to install and remove the bearing races — 1992-95 Monteros

7-98 DRIVE TRAIN

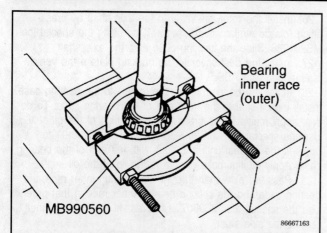

Fig. 147 Use the MB990560 tool to hold the bearing inner race (outer), then use a plastic hammer to drive the axle out of the race — do not let the axle fall onto a hard floor

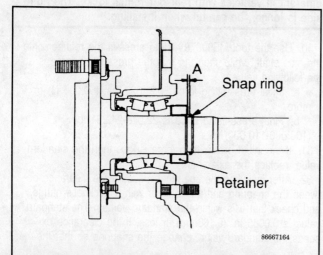

Fig. 148 Measure the clearance (A) between the snapring and the retainer edge — 1992-95 Monteros

clearance is at the standard value. Use the following adjusting snaprings:

- Thickness of snapring — identification color.
- 0.0854 in. (2.17mm) — has no color.
- 0.0791 in. (2.01mm) — yellow.
- 0.0728 in. (1.85mm) — blue.
- 0.0665 in. (1.69mm) — purple.
- 0.0602 in. (1.53mm) — red.

33. Install the whole axle assembly into the the axle housing. Make sure that the grooves on the end of the axle shaft line up in the differential. Use a plastic or rubber mallet to help "persuade" the axle shaft into the differential unit. Tighten the four retaining bolts for the axle shafts to 36-43 ft. lbs. (49-59 Nm).
34. Attach the speed sensor and install the various parking brake assembly components to the axle flange.
35. Attach the parking brake cable to the parking brake assembly, then secure it in place with the cable bracket.
36. Slide the brake rotor onto the axle shaft, then install the brake calliper. Tighten the brake calliper bolts to 65 ft. lbs. (88 Nm).
37. Tighten the brake hose to the frame brake line. Tighten the flare nut to 11 ft. lbs. (15 Nm).
38. Install the wheels and tighten the lug nuts as tight as possible with the vehicle not on the ground.
39. Bleed the brake system — refer to Section 9 for this procedure.
40. Lower the vehicle until the wheels are touching the ground, then finish tightening the lug nuts. Lower the vehicle the rest of the way to the ground.
41. Road test the vehicle and check for leaks.

END-PLAY ADJUSTMENT PROCEDURE

1. Begin with the left side rear axle assembly and insert a 0.039 in. (1mm) shim between the bearing case and the axle shaft housing. Torque the nuts to specification.
2. Install the right side axle assembly into the right side housing without its shim and O-ring. Tighten the 4 nuts to about 50 inch lbs. (5.6 Nm).
3. Using a feeler gauge, measure the gap between the bearing case and the axle housing face.
4. Remove the axle shaft and select a shim or shims that is the equal to the sum of the clearance measured in Step 3 plus 0.002-0.0079 in. (0.05-0.20mm) and install them on the housing. Install the O-ring and apply sealant.
5. Install the axle assembly and tighten the nuts to 40 ft. lbs. (54 Nm).
6. Measure the end-play and complete the installation procedure.

Differential Carrier

REMOVAL & INSTALLATION

1983-95 Pick-Ups and 1983-91 Monteros
▶ See Figures 149, 150 and 151

1. Elevate and safely support the vehicle on stands.
2. Remove the rear wheels.
3. Drain the differential assembly.
4. Remove the brake drum and disconnect the parking brake cable at the brake assembly. It may be necessary to remove or loosen some of the cable retaining brackets.
5. Disconnect the brake lines running to the wheel cylinder.
6. Remove the rear axle retaining nuts.
7. Remove the axles. A slide hammer will be needed for the axles' removal. Note that the axles need only be pulled free of the differential (about 3 in./76mm) not completely removed.
8. Matchmark and disconnect the driveshaft flange at the differential case.
9. Support the differential carrier with a floor jack.
10. Remove the attaching nuts around the circumference of the axle housing. Work the carrier forward and out of the

DRIVE TRAIN 7-99

housing. Lower the jack with the unit when it is clear of the axle.

➡ The carrier may be stuck in place. Leave one nut in place but loosened on the upper stud. Use a piece of square lumber under the carrier case to push or tap upward. This will break the seal around the casing.

11. Make certain that the carrier is balanced on the jack. Elevate the jack to lift the carrier off the insulator studs. As soon as the unit clears the studs, move the jack forward and lower it to remove the carrier.

To install:

12. Install 3M® ART Part No. 8663, 8661 or the equivalent to the differential carrier mounting surface of the axle housing.
13. Reinstall the unit by mounting it in position and securing the retaining bolts and nuts. The ring of nuts should be tightened evenly and alternately to 20 ft. lbs. (27 Nm).
14. Install the front to rear driveshaft and connect it to the differential. The flange joint bolts should be tightened to 41 ft. lbs. (55 Nm).
15. Install the axles following procedures outlined earlier in this section. The axle retaining bolts should be tightened to 41 ft. lbs. (55 Nm).
16. Connect the brake lines to the wheel cylinder, connect the parking brake cable and install the drum. Don't forget to secure the parking brake cable brackets if any were loosened or removed.
17. Bleed the brake system by following the procedures outlined in Section 9.
18. Install the differential drain bolt to 47 ft. lbs. (65 Nm). Fill the differential with the correct amount of lubricant.
19. Install the wheels.
20. Lower the vehicle to the ground.
21. Test drive the vehicle and check for leaks.

1992-95 Montero

▶ See Figures 150, 151 and 152

1. Remove the rear brake calliper mounting bolts, then pull the calliper off of the read disc. Hang the calliper from the vehicle frame using strong wire or cord. Do not disconnect the brake lines if it can be helped.
2. Pull the rear brake disc off of the axle.
3. Disconnect the rear parking brake cable from the brake assembly under where the brake disc was located.
4. Remove the rear axle retaining nuts and pull the rear axles out only 3 in. (70mm). Do not remove the axles entirely.
5. Matchmark and remove the rear driveshaft from the rear differential housing. It may not be necessary to remove the driveshaft from the transmission if it is not in the way.
6. Remove the rear differential lock position harness connector.
7. Disconnect the differential front hose, if so equipped.
8. Remove the attaching nuts around the circumference of the axle housing. Work the carrier forward and out of the housing. Lower the jack with the unit when it is clear of the axle.

➡ The carrier may be stuck in place. Leave one nut in place but loosened on the upper stud. Use a piece of square lumber under the carrier case to push or tap upward. This will break the seal around the casing.

9. Make certain that the carrier is balanced on the jack. Elevate the jack to lift the carrier off the insulator studs. As soon as the unit clears the studs, move the jack forward and lower it to remove the carrier.

To install:

10. Install 3M® ART Part No. 8663, 8661 or the equivalent to the differential carrier mounting surface of the axle housing.
11. Reinstall the unit by mounting it in position and securing the retaining bolts and nuts. The ring of nuts should be tightened evenly and alternately to 35 ft. lbs. (47 Nm).
12. Connect the rear differential lock position harness and the differential front hose.
13. Install the front to rear driveshaft and connect it to the differential. The flange joint bolts should be tightened to 41 ft. lbs. (55 Nm) for the 3.0L engines and to 76 ft. lbs. (103 Nm) for the 3.5L engines.
14. Install the axles following procedures outlined earlier in this section. The axle retaining bolts should be tightened to 41 ft. lbs. (55 Nm).
15. Connect the parking brake cable and install the rear brake disc. Don't forget to secure the parking brake cable brackets if any were loosened or removed.
16. Install the brake calliper to the disc and tighten the mounting bolts.
17. Install the differential drain bolt to 47 ft. lbs. (65 Nm). Fill the differential with the correct amount of lubricant.
18. Install the wheels.
19. Lower the vehicle to the ground.
20. Test drive the vehicle and check for leaks.

Pinion Seal

REMOVAL & INSTALLATION

▶ See Figure 153

1. Raise the vehicle and support it safely on jackstands.
2. Matchmark and remove the driveshaft.
3. Check the turning torque of the pinion before proceeding. It should be 3.5-4.5 inch lbs. (0.4-0.5 Nm). This is the torque that must be reached during installation of the pinion nut.
4. Using a suitable pinion flange holding tool, remove the pinion nut and washer.
5. Remove the companion flange from the drive pinion.
6. Pry the pinion seal out of the differential carrier.

To install:

7. Clean and inspect the sealing surface of the housing.
8. Using a seal driver, drive the new seal into the housing until the flange on the seal is flush with the carrier.
9. With the seal installed, the pinion bearing preload must be set.
10. Tighten the pinion nut (a new self-locking pinion nut must be used) while holding the flange, until the turning torque is the same as before removal. The final pinion nut torque must be between 137-181 ft. lbs. (190-250 Nm).
11. Align the matchmarks and install the driveshaft.
12. Check the level of the differential lubricant when finished.

7-100 DRIVE TRAIN

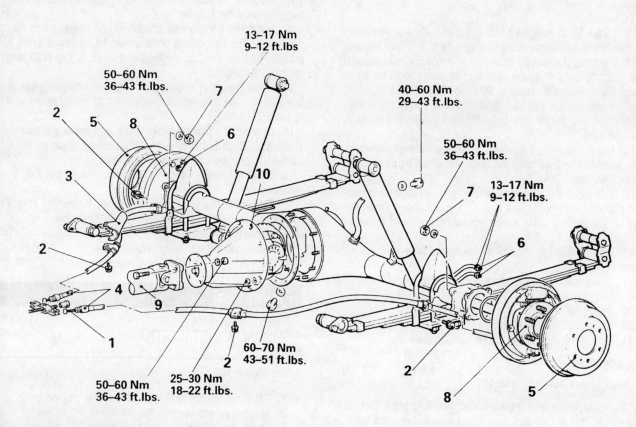

Removal steps

1. Adjuster (Parking brake cable)
2. Parking brake cable attaching nuts
3. Parking brake cable heat protector (right side only)
4. Connection of parking brake cable and equalizer
5. Brake drums
6. Connection of brake tubes
7. Nuts
8. Rear axle shaft assembly
9. Rear propeller shaft
10. Differential carrier

Fig. 149 Rear differential carrier removal and installation — all trucks except for 1992-95 Monteros

DRIVE TRAIN 7-101

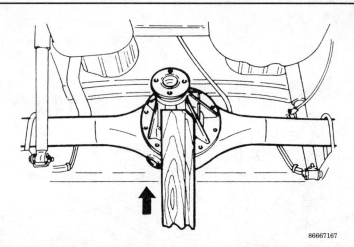

Fig. 150 Use a piece of wood to tap the carrier loose — make sure to leave one nut on the top stud to keep the carrier form falling

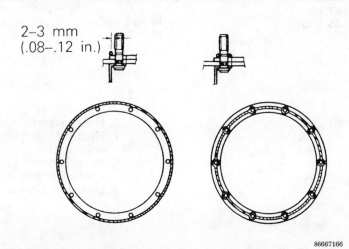

Fig. 151 Apply the sealant to the differential carrier using either one of the two styles shown in the illustration

Rear Axle Housing

REMOVAL & INSTALLATION

Axle With Leaf Springs
▶ See Figures 154, 155 and 156

1. Loosen the lug nuts on the two rear wheels. Only loosen them $1/2$ of a turn at the most.
2. Raise the vehicle and support it safely on jackstands. Make sure the jackstands are under the frame rails, not the rear axle.
3. Drain the differential fluid into a pan or similar container.
4. Remove the wheels. Remove the brake drums.

✱✱CAUTION

Brake pads and shoes may contain asbestos, which has been determined to be a cancer causing agent. Never clean the brake surfaces with compressed air! Avoid inhaling any dust from brake surfaces! When cleaning brakes, use commercially available brake cleaning fluids.

5. Remove the parking brake cable attaching bolts, disconnect the cables from the shoes and unclip them from the backing plates. Disconnect the cables from the parking brake equalizer bracket. Remove the right side parking cable heat protector plate.
6. Drain the brake fluid from the bleeder screw at the right side of the rear brake into a suitable container. Disconnect the brake hose at the T-fitting.
7. Slide the spring clip back on the breather hose, then pull the hose off of the differential case nipple.

7-102 DRIVE TRAIN

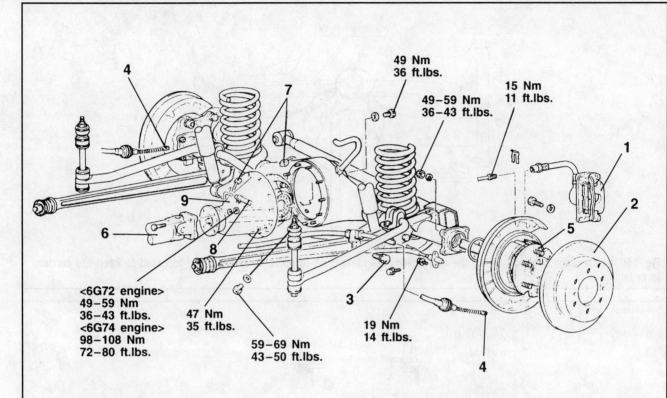

Removal steps
1. Rear brake assembly
2. Brake disc
3. Parking brake cable attaching nut
4. Parking brake cable end
5. Rear axle shaft assembly
6. Rear propeller shaft
7. Rear differential lock position harness connector
8. Hose connection <Vehicles with rear differential lock>
9. Differential carrier

Fig. 152 Rear differential carrier removal and installation — 1992-95 Monteros with rear disc brakes

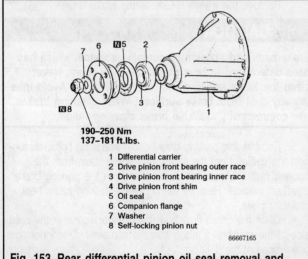

Fig. 153 Rear differential pinion oil seal removal and installation — all Pick-ups and Monteros

8. Remove the load sensing proportioning valve spring support, if equipped.
9. Matchmark and remove the rear driveshaft.
10. Place a suitable jack under the center of the differential housing and support it slightly. If two jacks or two more stands are available, place them supporting the left and right axle tubes. Unbolt the shocks from their lower mounts.
11. Remove the U-bolts and the bump stoppers. Loosen and disconnect the rear shackle assemblies and lower the rear end of the leaf springs.
12. Lower the axle and remove it from the vehicle. If only one jack is used, the unit will be balanced on the differential. Have a helper steady the unit while it is removed.

To install:

13. Raise the axle assembly into position and install the rear end of the leaf springs. Make sure the shackle nuts are on the inside of the shackles and that the bushings are in good condition, otherwise replace them with new ones. Tighten the nuts fully when the vehicle has been lowered to the ground.
14. Position the rear axle over the mounting pads, then install the U-bolts and bump stoppers over the axle. Tighten the

DRIVE TRAIN 7-103

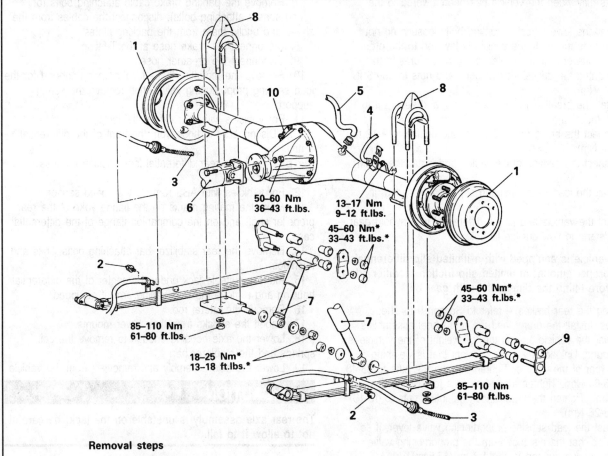

Fig. 154 Rear axle assembly removal and installation — Pick-ups and Monteros with leaf springs

Removal steps

1. Brake drums
2. Parking brake cable attaching bolts
3. Connection of parking brake cable end and brake shoe assembly
4. Connection of brake hose
5. Connection of breather hose
6. Rear propeller shaft
7. Connection of shock absorbers (lower part only)
8. U-bolts and bump stopper
9. Shackle assembly
10. Axle assembly

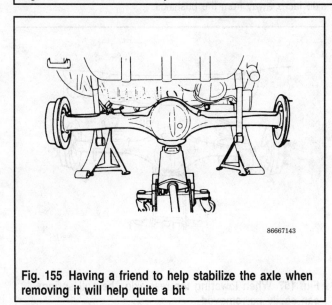

Fig. 155 Having a friend to help stabilize the axle when removing it will help quite a bit

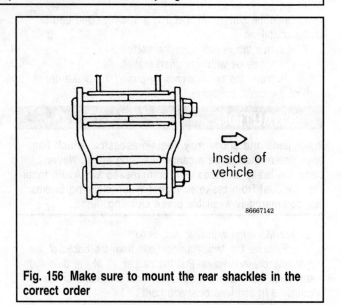

Fig. 156 Make sure to mount the rear shackles in the correct order

7-104 DRIVE TRAIN

U-bolt nuts fully when the vehicle has been lowered to the ground.

15. Install the lower shock mounting bolts loosely; do not tighten the nuts fully until the vehicle is lowered to the ground.

16. Install the driveshaft — refer to the procedures in this section. Tighten the driveshaft rear bolts and nuts to 36-43 ft. lbs. (50-60 Nm).

17. Install the breather hose, if equipped, and secure it in place with the spring clip.

18. Connect the brake hose flare nut and tighten to 9-12 ft. lbs. (13-17 Nm).

19. Connect the parking brake cables and install the retaining bolts.

20. Install the load sensing spring support and spring, if equipped.

21. Level the vehicle and fill the differential with hypoid gear oil until it starts to flow out the fill hole.

➡ If the vehicle is equipped with a limited slip differential, add the proper amount of limited slip friction modifier additive before filling the differential with gear oil.

22. Bleed the rear brakes — refer to Section 9 for these procedures. Install the drums and tire and wheel assemblies.

23. Lower the vehicle so that the full weight of the vehicle is on the ground. Unload any excess weight that is weighing down the rear of the vehicle. Tighten the shackle nuts to 33-43 ft. lbs. (45-60 Nm). Tighten the U-bolt nuts to 61-80 ft. lbs. (85-110 Nm). Tighten the lower shock mounting nuts to 13-18 ft. lbs. (18-25 Nm).

24. Adjust the load sensing proportioning valve lever, if so equipped, so that the distance from the proportioning valve lever to the spring support is about 7 in. (17.8cm).

25. Road test the vehicle and check for leaks.

Axle With Coil Springs

▶ See Figures 157, 158, 159 and 160

1. Loosen the wheel lug nuts. Only loosen the lug nuts at the most $1/2$ of a turn.
2. Raise the vehicle and support it safely with jackstands under the frame rails.
3. Drain the differential fluid into a suitably-sized pan or similar container.
4. Remove the wheels from the vehicle.
5. For Monteros with rear drum brakes:
   a. Remove the brake drums by pulling the brake drums off of the brake assembly.

✸✸CAUTION

Brake pads and shoes may contain asbestos, which has been determined to be a cancer causing agent. Never clean the brake surfaces with compressed air! Avoid inhaling any dust from brake surfaces! When cleaning brakes, use commercially available brake cleaning fluids.

6. For Monteros with rear disc brakes:
   a. Remove the two retaining bolts from the backs of the rear disc brake calipers. Pull the caliper off of the disc and suspend, with the brake attached, from the frame of the vehicle with stiff wire or strong cord.
   b. Pull the rear brake discs off of the assemblies.

7. Remove the parking brake cable attaching bolts (or speed sensor attaching bolts), disconnect the cables from the shoes and unclip them from the backing plates.
8. Disconnect the brake hose at the T-fitting.
9. Disconnect the breather hose.
10. Remove the mounting bolts for the spring support for the load sensing proportioning valve, then remove the spring support.
11. For vehicles with rear differential lock:
    a. Disconnect the hose from the front of the differential unit.
    b. Unplug the rear differential lock position harness connector.
12. For vehicles with ABS remove the speed sensor.
13. Make the mating marks on the flange yoke of the rear propeller shaft and on the companion flange of the differential case.
14. Remove the rear stabilizer bar attaching bolts, links and bushings.
15. Place a suitable jack under the center of the differential housing and remove the rear trailing arm, if equipped.
16. Remove the lateral rod.
17. Unbolt the shocks from their lower mounts.
18. Lower the axle housing enough to remove the coil springs and the stabilizer bar.
19. Lower the axle assembly and remove it from the vehicle.

✸✸CAUTION

The rear axle assembly is unstable on the jack; be careful not to allow it to fall.

To install:

20. With the coil springs and stabilizer bar in place, raise the axle assembly into place and install the lower shock mounting bolts.
21. Install the lateral rod but do not tighten the nuts yet. Install it from the axle housing side.
22. Assemble the trailing arm with its front mounting spacers, bushings and nuts. Make sure the washer's concave side faces away from the bushings.

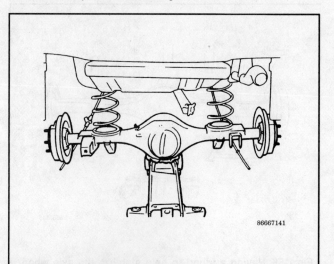

Fig. 157 When lowering the rear axle, the coil springs can easily be removed

DRIVE TRAIN 7-105

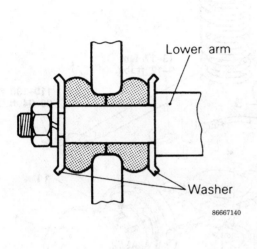

Fig. 158 Install the lower arm washers so that the concave (dished) side faces out

23. Install the trailing arm and tighten the rear mount nuts to 90 ft. lbs. (122 Nm) for rear drum brake equipped Monteros or to 159-181 ft. lbs. (216-245 Nm) for rear disc brake equipped Monteros. Do not tighten the front mounting nuts yet.

24. Install the stabilizer bar and tighten the mounting nuts until the diameter of the bushing is the same as the diameter of the washers (the bushing will squash until the edges are lined up with the edges of the washers).

25. Install the driveshaft (refer to the procedures in this section) and breather hose.

26. Connect the parking brake cables and install the retaining bolts.

27. Attach the speed sensor, the rear differential lock position harness connector and the spring support for the load sensing proportioning valve, if equipped.

28. Level the vehicle and fill the differential with gear oil until it starts to flow out the fill hole.

➡ **If the vehicle is equipped with a limited slip differential, add the proper amount of limited slip friction modifier additive before filling the differential with gear oil.**

29. Bleed the rear brakes (refer to Section 9 for this procedure). Install the drums or discs and rear callipers.

30. Install the wheels. Tighten the lug nuts as tight as possible with the wheels off the ground.

31. Lower the vehicle so that the full weight of the vehicle is on the ground. Torque the lateral rod mounting nuts to 90 ft. lbs. (122 Nm) (170 ft.lbs (230 Nm) for rear disc brake equipped Monteros) and the front trailing arm mounting nuts to 100 ft. lbs. (150 Nm). Tighten the wheel lug nuts fully at this time.

32. Road test the vehicle and check for leaks.

7-106 DRIVE TRAIN

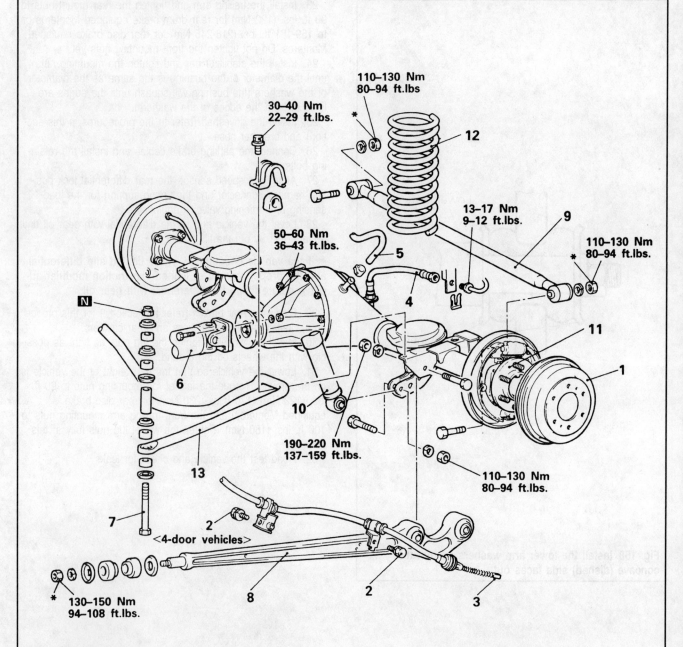

Removal steps

1. Brake drum
2. Parking brake cable attaching bolts
3. Connection of parking brake cable end and brake shoe assembly
4. Connection of brake hose
5. Connection of breather hose
6. Rear propeller shaft
7. Stabilizer bar installation bolt
8. Lower arm
9. Lateral rod
10. Connection of shock absorbers (lower part only)
11. Axle assembly
12. Coil spring
13. Stabilizer bar

Fig. 159 Rear axle assembly removal and installation — Monteros with rear drum brakes and coil springs

DRIVE TRAIN   7-107

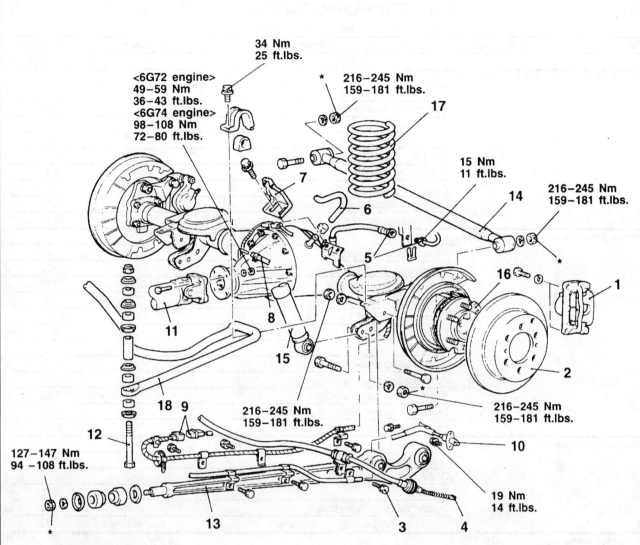

Removal steps

1. Rear brake assembly
2. Brake disc
3. Parking brake cable or speed sensor attaching bolt
4. Connection for parking brake cable end
5. Brake hose connection
6. Breather hose connection
7. Spring support for load sensing proportioning valve
8. Hose connection <Vehicles with rear differential lock>
9. Rear differential lock position harness connector <Vehicles with rear differential lock>
10. Speed sensor <Vehicles with ABS>
11. Rear propeller shaft
12. Stabilizer bar installation bolt
13. Lower arm
14. Lateral rod
15. Shock absorber connection (lower part only)
16. Axle assembly
17. Coil spring
18. Stabilizer bar

NOTE
The part with * must be tightened with the vehicle lowered to the ground.

Fig. 160 Rear axle assembly removal and installation — Monteros with rear disc brakes and coil brakes

TORQUE SPECIFICATIONS
Drive Train Components

Item	Years	Component	Nm	ft.lbs.
Manual Transmission - RWD	1983-95			
		Transmission mounting bolts	43-55	32-39
		Starting motor mounting bolts	22-32	16-23
		Oil drain plug	60	43
		Oil filler plug	30-35	22-25
		Shift knob	4.5-7.5	3.3-5.4
		Stopper plate to control housing	7-11	5-8
		Exhaust pipe mounting bracket mounting bolts	20-27	15-20
		Rear engine support insulator to transmission	20-24	14-17
		No. 2 cross member mounting bolts	40-50	29-36
Manual Transmission - 4WD	1983-95			
		Transmission mounting bolts (A)	65-85	47-61
		Transmission mounting bolts (B)	80-100	58-72
		Starting motor mounting bolts	27-34	20-25
		Transmission to transmission stay	30-42	22-30
		Clutch release cylinder mounting bolts	31-42	22-29
		Oil filler plug and rain plug	55-85	40-61
		Shift knob	4.5-7.5	3.3-5.4
		Control housing cover mounting bolts	15-22	11-15
		Cpntrol housing mounting bolts	15-22	11-15
		Stopper plate to control housing	15-22	11-15
		Rear engine support insulator to transmission	18-25	13-18
		No. 2 cross member mounting bolts	55-75	40-54
Transfer Case Assembly	1983-95			
		Control housing bolt	15-22	11-15
		Oil filler plug	30-35	22-25
		Drain plug	30-35	22-25
		Transfer mounting bracket to pipe	35-55	25-40
		Transfer mounting bracket mounting bolts	18-25	13-18
Automatic Transmission	1983-95			
		Steering wheel lock nut	35-45	25-33
		Self locking flange nut	17-26	12-19
		Cable locking nut	17-26	12-19
		Transmission control arm mounting nut	19-28	14-20
		Shift lock cable mounting nut	4-6	3-4
		Shift lock cable locking nut	9-14	7-10
		Key interlock rod connector locking nut	2-3	1.4-2.2
		Selector lever rod mounting nut	9-14	7-10
		Transmission oil cooler flare nut	40-50	29-36
		Transmission oil cooler eye bolt	30-35	22-25
		Exhaust pipe mounting bracket	20-27	14-20
		Torque converter to flywheel bolts	46-53	33-38
		Bell housing cover	8-10	6-7
		Transmission mounting bolt 10x50 mm (.4x2.0 in.)	43-55	31-40
		Transmission mounting bolt 10x70 mm (.4x2.8 in.)	43-55	31-40
		Transmission mounting bolt 10x16 mm (.4x.6 in.)	30-42	21-30
		Bell crank bracket assembly	27-34	20-25

DRIVE TRAIN 7-109

TORQUE SPECIFICATIONS
Drive Train Components

Item	Years	Component	Nm	ft.lbs.
		Transmission control rod	15-20	11-14
		Automatic transmission cooler tube	40-50	29-36
		No. 2 cross member to transmission	20-25	14-18
		No. 2 cross member to frame	40-55	29-40
Clutch	1983-95			
		Brake pedal to pedal support member	25-35	18-25
		Pedal support member mounting bolt	8-12	6-9
		Clutch master cylinder to firewall	10-15	7-10
		Clutch tube flare nut	13-17	10-12
		Bleeder plug	9-13	7-9
		Eye bolt	20-25	14-18
		Turnover spring bracket mounting nut	9-14	7-10
Front Differential Carrier - 4WD	1983-95			
		Differential mounting bracket to frame	80-100	58-80
		Filler plug	40-60	29-43
		Front suspension cross member to frame	100-120	72-87
		Drain plug	60-70	43-51
		Front drive shaft to differential carrier	50-60	36-43
		Differential mounting bracket to differential carrier	80-110	58-80
		Differential cover	15-22	11-16
		Differentiqal case to drive gear	80-90	58-65
		Bearing cap	55-65	40-47
		Companion flange self locking nut	190-250	137-181
Rear Axle - Leaf Spring	1983-95			
		Shackle assembly attaching nut	45-60	33-43
		Shock absorber attaching nut	18-25	13-18
		U-bolt attaching nut	100-120	72-87
		Brake tube flare nut	13-17	9-12
		Drive shaft attaching nut	50-60	36-43
		Breather hose installation bolt	8-12	6-9
		Bearing case to rear axle housing	50-60	36-43
		Rear axle bearing lock nut	180-220	130-159
		Filler plug	40-60	29-43
		Drain plug	60-70	43-50
		Differential carrier to rear axle housing - RWD	25-30	18-22
		Companion flange	190-250	137-181
		Differential case to drive gear	80-90	58-65
		Bearing cap	55-65	40-47
		Lock plate	15-22	11-16
Rear Axle - Coil Spring	1983-91			
		Lower arm to axle mounting bolts	190-220	137-159
		Shock absorber attaching nut	110-130	80-94
		Lateral rod mounting nuts	110-130	80-94
		Stabilizer bar retaining bracket bolts	30-40	22-29
		Lower arm front mounting nut	130-150	94-108
		Brake tube flare nut	13-17	9-12
		Drive shaft attaching nut	50-60	36-43
		Drive shaft attaching nut - 3.5L engines	98-108	72-80

86667c04

TORQUE SPECIFICATIONS
Drive Train Components

Item	Years	Component	Nm	ft.lbs.
		Filler plug	40-60	29-43
		Drain plug	60-70	43-50
		Differential carrier to rear axle housing - 2.6L engines	25-30	18-22
		Differential carrier to rear axle housing - 3.0L or 3.5L engines	40-55	29-40
	1992-95	Lower arm to axle mounting bolts	216-245	159-181
		Shock absorber attaching nut	216-245	159-181
		Lateral rod mounting nuts	216-245	159-181
		Stabilizer bar retaining bracket bolts	34	25
		Lower arm front mounting nut	127-147	94-108
		Brake tube flare nut	15	11
		Drive shaft attaching nut	49-59	36-43
		Drive shaft attaching nut - 3.5L engines	98-108	72-80
		Filler plug	49	36
		Drain plug	59-69	43-50
		Differential carrier to rear axle housing - 3.0L and 3.5L engines	47	35

86667c05

FRONT SUSPENSION
  COIL SPRING 8-3
  FRONT END ALIGNMENT 8-21
  FRONT WHEEL BEARING 8-19
  KNUCKLE AND SPINDLE 8-16
  LOWER BALL JOINTS 8-8
  LOWER CONTROL ARM AND
    BUSHINGS 8-14
  SHOCK ABSORBER 8-4
  STABILIZER BAR (SWAY BAR) 8-8
  STRUT ROD 8-10
  TORSION BAR 8-3
  UPPER BALL JOINT 8-6
  UPPER CONTROL ARM 8-12
REAR SUSPENSION
  COIL SPRINGS 8-23
  LEAF SPRINGS 8-23
  REAR STABILIZER BAR (SWAY
    BAR) 8-24
  REAR WHEEL ALIGNMENT 8-25
  SHOCK ABSORBERS 8-24
SPECIFICATIONS CHARTS
  TORQUE SPECIFICATIONS 8-44
  WHEEL ALIGNMENT
    SPECIFICATIONS 8-22
STEERING
  IGNITION LOCK 8-29
  IGNITION SWITCH 8-27
  MANUAL STEERING GEAR BOX 8-38
  POWER STEERING GEAR BOX 8-38
  POWER STEERING PUMP 8-40
  STEERING COLUMN 8-30
  STEERING LINKAGE 8-31
  STEERING WHEEL 8-27
  TURN SIGNAL AND WIPER SWITCH
    (COLUMN SWITCH) 8-27
WHEELS
  CLEANING 8-2
  FRONT AND REAR WHEELS 8-2
  INSPECTION 8-2
  WHEEL LUG STUDS 8-2

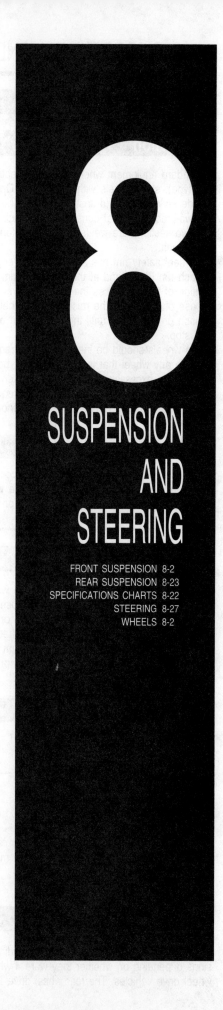

8

SUSPENSION AND STEERING

FRONT SUSPENSION 8-2
REAR SUSPENSION 8-23
SPECIFICATIONS CHARTS 8-22
STEERING 8-27
WHEELS 8-2

8-2 SUSPENSION AND STEERING

WHEELS

Inspection

Standard equipment wheels are drop center (depending on year/type), steel wheels with safety rims. Optional wheels include styled steel and cast aluminum. The steel wheels are the two-piece type that consist of a rim and center section. The two sections are welded together to form a seamless, airtight unit.

A wheel safety rim has a ridge (raised edge) located inboard of each rim flange and at the top of the rim well. Initial inflation of the tire forces the tire bead over the ridge sections. In case of tire failure the raised sections help hold the tire in position on the wheel until the vehicle can be brought to a safe stop.

The wheels should be inspected on a frequent basis. Replace any wheel that is either cracked, bent, severely dented, has excessive runout or has broken welds. The tire inflation valve should also be inspected frequently for wear, leakage, cuts and looseness. The valve should be replaced if defective or its condition is doubtful.

✻✻CAUTION

Some aftermarket wheels may not be compatible with these vehicles. The use of incompatible wheels may result in equipment failure and possible personal injury! Use only approved wheels.

Cleaning

Clean the wheels with a mild soap and water solution only and rinses thoroughly with water. Never use abrasive or caustic materials, especially on aluminum or chrome-plated wheels because the surface could be etched or the plating severely damaged. After cleaning aluminum or chrome-plated wheels, apply a coat of protective wax to preserve the finish and luster.

Front and Rear Wheels

REMOVAL & INSTALLATION

♦ See Figure 1

1. Park the vehicle on a level surface, set the parking brake and then block the opposite wheel.
2. On models with an automatic transmission, move the gear selector lever to **P**. On those with a manual transmission, position the gear shift lever in reverse.
3. If equipped, remove the wheel cover. When removing the wheel cover or center cap, be careful because some center caps must be removed from the rear of the tire assembly (after the wheel has been removed from the vehicle).
4. Break the lug nuts loose by turning them counterclockwise no more than ½ of a turn.
5. Raise the truck until the tire is clear of the ground.
6. Remove the lug nuts and then pull off the wheel.

To install:

7. Clean the wheel lugs and brake drum or hub of all foreign matter.
8. Position the wheel on the brake drum or hub and hand-tighten the lug nuts. Always make sure that the coned ends face inward.
9. Using a lug wrench, tighten all lug nuts in a crisscross pattern until they are snug.
10. Lower the truck. Tighten the lug nuts, in the crisscross pattern, to 87-108 ft. lbs. (120-140 Nm).

Wheel Lug Studs

REMOVAL & INSTALLATION

Wheel lug studs on the front disc rotors usually require the services of a machine shop when replacement is required. Rear axle flange mounted wheel studs can sometimes be hammered out and the new stud pulled in (clearance providing) when replacement is required. Once again, axle flange (axle removed from vehicle) or rear drum mounted studs may require the services of a machine shop for replacement.

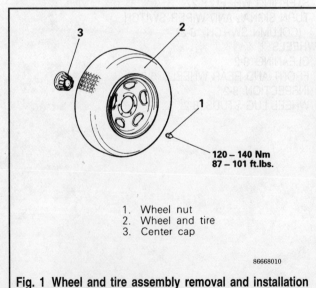

1. Wheel nut
2. Wheel and tire
3. Center cap

120 – 140 Nm
87 – 101 ft.lbs.

Fig. 1 Wheel and tire assembly removal and installation

FRONT SUSPENSION

The Pick-ups came equipped with two different front suspensions depending on whether they were 4-wheel drive or rear wheel drive vehicles. The rear wheel drive Pick-ups came equipped with coil spring front suspensions, and the 4-wheel drive Pick-ups came equipped with torsion bar front suspen-

SUSPENSION AND STEERING 8-3

sions. Generally, most components have similar procedures and where they differ, it will be noted.

The Monteros came equipped only with torsion bar suspensions, which are the same as the 4-wheel drive Pick-up torsion bar suspensions.

Coil Spring

The coil spring front suspension is only equipped on rear wheel drive Pick-up trucks.

REMOVAL & INSTALLATION

▶ See Figures 2 and 3

1. Elevate the vehicle and support safely on jackstands.
2. Remove the wheel.
3. Remove the shock absorber.
4. Remove the bump stop and disconnect the stabilizer bar from the lower control arm. Note that the rubber bushings for the stabilizer are not all the same. They are NOT interchangeable and should be labeled or identified for correct reassembly.
5. Install spring compressor tool (MB-990792 or equivalent) to the coil spring and to compress the spring. Make absolutely sure the spring compressor is correctly installed before compressing the spring.
6. Remove the cotter pin and lower ball joint nut. Place a floor jack under the control arm; elevate the jack to within 1 in. (25mm) of the arm.
7. Release the lower ball joint from the knuckle using a ball joint separator such as MB 990809-01 or equivalent tool. The arm will drop onto the jack as the joint comes loose.
8. Slowly release the compressor tool from the coil spring.
9. Lower the jack as necessary, pull the arm down and remove the spring with the rubber isolation pad and upper spring seat from the vehicle.
10. Install the spring with the rubber isolator. Make sure the spring seats correctly in its upper and lower mounts. Install the compressor tool and compress it enough so the lower ball joint can be inserted through the knuckle. Use the jack to support and adjust the height of the lower control arm during installation.
11. Tighten the lower ball joint nut to 100 ft. lbs. (136 Nm). Install a new cotter pin. Remove the spring compressor.
12. Install the bump stop and install the shock absorber. The shock absorber mounting bolts (upper and lower) are tightened to only 10 ft. lbs. (13 Nm). The bump stopper mounting bolts should be tightened to 51-61 ft. lbs. (70-85 Nm).
13. Install the stabilizer bar, remembering to install the rubber bushings and hardware in the correct locations. Tighten the sway bar bolt until the bolt threads project 0.87-0.94 in. (22-24mm) beyond the nut.
14. Install the wheel and lower the vehicle to the ground.

Torsion Bar

The 4-wheel drive Pick-ups and Monteros are equipped with a torsion bar front suspension.

REMOVAL & INSTALLATION

▶ See Figures 4, 5, 6 and 7

1. Raise the vehicle and support it safely on jackstands. Make sure the stands are placed on the frame rails.
2. Remove the wheel from the side to be repaired.
3. Support the lower control arm with a floor jack. Position the jack at a point away from where the torsion bar attaches to the arm.
4. Fold the dust covers back and slide them away from the ends of the bar.
5. If the bars are to be reused, matchmark the torsion bar at both ends to the anchor and identify left from right.
6. Paint or measure the distance of the exposed threads of the rear mounting bolt down to the nut to aid in adjustment when installing. Remove the rear anchor arm mounting nut and bolt.
7. Loosen the adjusting nut and pull the torsion bar from the lower arm assembly.
8. Check the torsion bar for bends or damage. Inspect the dust covers for cracks. Check the anchor bolt for bending or distortion. If the torsion bar has received a heavy impact — such as during off-road use — it is recommended that the bar be professionally crack-checked using electric methods. The peace of mind is worth the cost.
9. If the bar is being reused, lubricate the ends and install the torsion bar aligning the matchmarks. If a new bar is being used, align the white stripe on the front splines with the mark on the anchor. There is a mark on the front of the torsion bar to differentiate between left and right. Do not install the bar with the mark facing the rear.
10. Install the torsion bar to the rear anchor so that the length of the mounting bolt from the nut to the head of the bolt is the specified length with the rebound bumper in contact with the crossmember (distance A). Reposition the bar as required to meet the specifications. The specifications are:
  - 1983-91 Montero left side — 5.315-5.590 in. (13.5-14.2cm)
  - 1983-91 Montero right side — 4.882-5.196 in. (12.4-13.2cm)
  - 1992-95 Montero left side — 5.866 in. (14.9cm)
  - 1992-95 Montero right side — 5.846 in. (14.85cm)
  - 1983-91 Pick-up left side — 5.512-5.827 in. (14.0-14.8cm)
  - 1983-91 Pick-up right side — 5.315-5.630 in. (13.5-14.3cm)
  - 1992-95 Pick-up left side — 5.866 in. (14.9cm)
  - 1992-95 Pick-up right side — 5.846 in. (14.85cm)

11. To initially set the ride height, tighten the rear anchor mounting nut to the same point at which it was removed, if the old bar is being reused. If a new bar has been installed, tighten the nut so that the exposed length of the bolt threads is to specification (distance B):
  - 1983-91 Montero left side — 2.40 in. (6.1cm) or less
  - 1983-91 Montero right side — 2.79 in. (7.1cm) or less
  - 1992-95 Montero left and right sides — 3.15 in. (8.0cm) or less

8-4 SUSPENSION AND STEERING

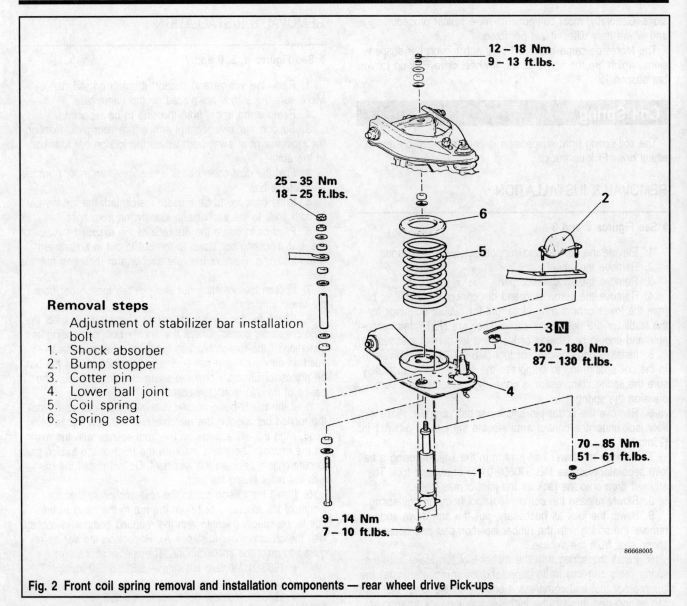

Removal steps
Adjustment of stabilizer bar installation bolt
1. Shock absorber
2. Bump stopper
3. Cotter pin
4. Lower ball joint
5. Coil spring
6. Spring seat

Fig. 2 Front coil spring removal and installation components — rear wheel drive Pick-ups

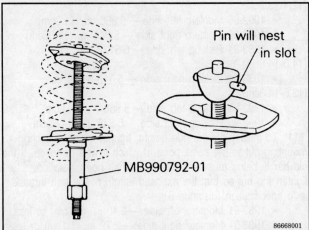

Fig. 3 Use the special tool (MB990792-01) to compress the coil spring for removal and installation — rear wheel drive Pick-ups

- 1986 Pick-up left side — 3 in. (7.6cm) or less
- 1986 Pick-up right side — 2.67 in. (6.85cm) or less
- 1987-95 Pick-up left side — 3.94 in. (10.0cm) or less
- 1987-95 Pick-up right side — 3.39 in. (8.6cm) or less

12. Fill the dust covers with grease and fold them back into position.
13. Adjust the torsion bar to the correct riding height.
14. Install the wheel and lower the vehicle to the ground. With the vehicle on the ground but unladen, measure the distance from the bottom of the bump stopper to the contact face of the bump stop bracket. Correct distance is 2.79-3.11 in. (71-79mm). On the 1992-95 Monteros, measure the space from the top tip of the bump stopper to the bump stopper bracket, which should be 0.83-0.91 in. (21-23mm). Adjust this distance by turning the adjusting nut on the anchor bolt.

Shock Absorber

➡ This procedure is for shock absorbers mounted separately from the spring. It does not apply to MacPherson or other strut systems.

SUSPENSION AND STEERING 8-5

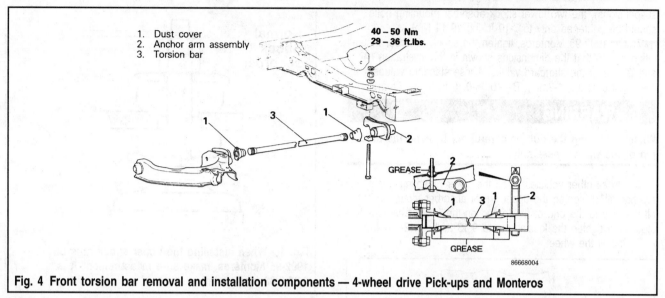

Fig. 4 Front torsion bar removal and installation components — 4-wheel drive Pick-ups and Monteros

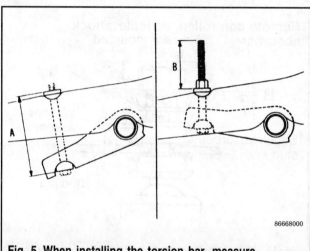

Fig. 5 When installing the torsion bar, measure distances A and B — 4-wheel drive Pick-ups and Monteros

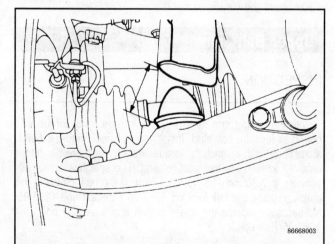

Fig. 6 Measure the bump stopper distance from the bottom of the stopper to the bracket — 1983-91 Monteros and 1983-95 4-wheel drive Pick-ups

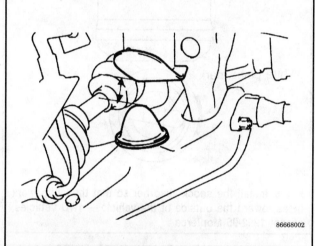

Fig. 7 On 1992-95 Monteros, measure from the top tip of the stopper to the bracket

REMOVAL & INSTALLATION

▶ See Figures 8, 9, 10 and 11

1. Support the vehicle safely on jackstands. Support it at a height which will allow sufficient room to reach the upper shock absorber mount.
2. Remove the wheel.
3. Remove the upper shock nut, washer and bushing.
4. Remove the lower mounting bolt(s) and remove the shock from the vehicle.
5. Install the shock absorber in the vehicle and loosely tighten the upper shock absorber nut. Make sure the install the bushings and washers before installing the upper nut.
6. For 4WD vehicles, install the shock absorber so that the white paint mark at the lower side of the shock absorber faces the outer side of the vehicle. Install the lower shock absorber bolt(s). On 1992-95 Montero torsion bar suspensions, tighten the single lower mounting bolt and nut to 65-76 ft. lbs. (88-103 Nm). On all other vehicles (whether torsion bar or coil spring

8-6 SUSPENSION AND STEERING

suspensions), the two lower shock absorber mounting bolts should be tightened only to 7-10 ft. lbs. (9-14 Nm).

7. On 1992-95 Monteros, tighten the shock absorber installation nut so that the dimensions shown in the illustration (**A** and **B**) are at the standard values. These standard values are: A — 0.04-0.08 in. (1-2mm), B — 0.06-0.10 in. (1.5-2.5mm).

❊❊WARNING

When tightening the nut, be careful not to bend the stud pin of the washer assembly.

8. On all other vehicles, once the lower mounting bolt(s) are tightened, tighten the upper shock absorber mounting nut all the way to the end of the thread on the shock absorber stud, then tighten the locking nut to 9-13 ft. lbs. (12-18 Nm).

9. Install the wheel.

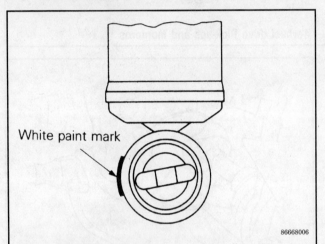

Fig. 8 Install the shock absorber so that the white mark faces toward the outside of the vehicle — 4WD vehicles, except 1992-95 Monteros

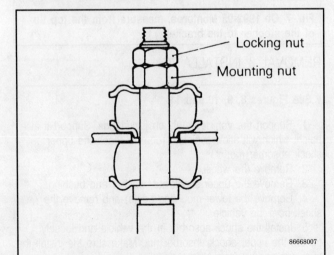

Fig. 9 Install the bushings, washers and nuts in the order shown — except 1992-95 Monteros

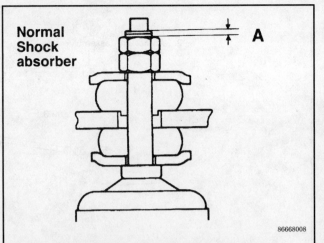

Fig. 10 When installing the upper shock nuts on 1992-95 Monteros, make sure measurement A is 0.04-0.08 in. (1-2mm)

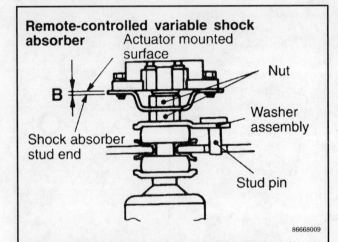

Fig. 11 When installing the upper shock assembly, make sure its components are positioned as shown — 1992-95 Monteros

10. Lower the vehicle to the ground.

Upper Ball Joint

INSPECTION

With the control arm removed, check the starting torque required to turn the ball stud. Install the nut back onto the ball stud. Then, with an inch lb. torque wrench, measure the torque required to start the ball joint rotating. The specification for all vehicles is 7-30 inch lbs. (0.8-3.5 Nm). If the ball joint fails this check, replace the ball joint on 1983-86 Pick-ups and 1983-95 Monteros, or replace the upper control arm assembly on 1987-95 Pick-ups.

SUSPENSION AND STEERING 8-7

REMOVAL & INSTALLATION

◆ See Figures 12, 13, 14 and 15

➡The upper ball joint and upper control arm must be replaced as an assembly on 1987-95 Pick-ups.

1. Elevate the vehicle and support it safely on jackstands.
2. Remove the upper control arm — refer to the procedures in this section.
3. Remove the ring and boot from the ball joint.
4. Remove the snapring from the upper ball joint by using a pair of snapring pliers.
5. Press the upper ball joint out of the upper control arm by using tools MB 990800-2-01, MB 990800-3-01, MB 990800-4-01 and MB 990799-01, or equivalent installing tools.
6. Align the mating mark on the upper ball joint with the upper arm center line.

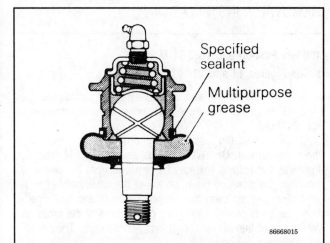

Fig. 12 The internal construction of the upper ball joint — all Pick-ups and Monteros except for 1987-95 Pick-ups

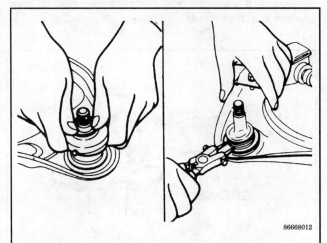

Fig. 13 Remove the dust cover, then use a pair of snapring pliers to remove the snapring from the upper ball joint

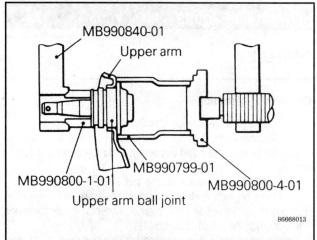

Fig. 14 Use special Mitsubishi tools, or their equivalent, to remove the upper ball joint from the control arm — all Pick-ups and Monteros except for 1987-95 Pick-ups

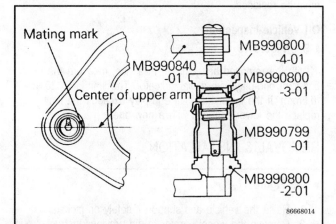

Fig. 15 Install the upper ball joint so that the mark on its side lines up with the center line of the control arm — all Pick-ups and Monteros except for 1987-95 Pick-ups

7. Press the ball joint in using the same tools that were used to remove the ball joint.
8. Using a pair of snapring pliers, fit the snapring securely in the groove of the joint case. If the snapring is loose, install a new one.

➡Limit the opening of the snapring to a minimum.

9. Apply the multipurpose grease to both the interior of the dust cover and the upper ball joint. Apply 3M ART Part No. 8663 or equivalent sealant to the grooves in the upper ball joint. Secure the dust cover to the upper ball joint with a ring.
10. Install the upper control arm — refer to the procedures in this section.
11. Lubricate the ball joint with a grease gun.
12. Adjust the riding height if equipped with a torsion bar and align the front end.

8-8 SUSPENSION AND STEERING

Lower Ball Joints

INSPECTION

On Vehicle Inspection

1. Support the vehicle on stands.
2. Disconnect the ball joint at the lower end of the strut as described below; you need not remove the lower control arm entirely.
3. Install the nut back onto the ball stud.
4. Using an inch lb. torque wrench, measure the torque required to start the ball joint rotating. Acceptable specifications are 9-30 inch lbs. (1.01-3.39 Nm).

If the figures are within specification, the ball joint is satisfactory. If the figure is too high, the joint should be replaced. If the figure is too low, you can still reuse the joint provided its rotation is smooth and even. If there is roughness or play, it must be replaced.

Off Vehicle Inspection
▶ See Figure 16

With the removed ball joint set in a vise, measure the lower ball joint end-play with a dial indicator. The limit is 0.020 in. (0.5mm). If the lower ball joint end-play exceeds the limit, replace the lower ball joint with a new one.

REMOVAL & INSTALLATION

▶ See Figure 17

1. Raise the vehicle and support it safely on jackstands.
2. Remove the wheel — refer to the previous instructions in this section.
3. Remove the lower control arm — refer to the procedures later in this section.
4. Remove the ball joint retaining nuts and bolts and remove the ball joint from the arm.
5. Install the new ball joint. Tighten the ball joint retaining nuts and bolts to the following amounts:
   - 1983-95 4WD Pick-ups — 39-54 ft. lbs. (54-74 Nm)
   - 1983-95 2WD Pick-ups — 22-30 ft. lbs. (30-42 Nm)
   - 1983-91 Monteros — 39-54 ft. lbs. (54-74 Nm)
   - 1992-95 Monteros — 60 ft. lbs. (81 Nm)
6. Install the lower control arm and connect the ball joint to the knuckle. Tighten the ball stud nut to 87-130 ft. lbs. (120-180 Nm) and install a new cotter pin.
7. Lubricate the ball joint with a grease gun.
8. Install the wheel and lower the vehicle back to the ground.
9. Adjust the riding height, if equipped with a torsion bar, and align the front end.

Stabilizer Bar (Sway Bar)

REMOVAL & INSTALLATION

1983-95 Pick-up and 1983-91 Montero
▶ See Figures 18 and 19

1. Raise the vehicle and support safely on jackstands.
2. Remove the front wheels. Remove the under cover or splash shield.
3. Disconnect the bracket or clamp holding the sway bar to the crossmember. On 4WD Pick-ups and 1983-91 Monteros, the sway bar holding link may be removed from the crossmember; the retaining nut is on top of the crossmember.
4. Remove the sway bar support brackets and bushings from the lower control arm. Take careful note of the order and placement of the washers, bushings and spacers. They must be reinstalled correctly.
5. Remove the sway bar from the vehicle.
6. Inspect the bar for cracking or bending. Check the bushings for cracking or deformation. If any fault is found, replace components. It is recommended to replace the rubber bushings any time they are removed.

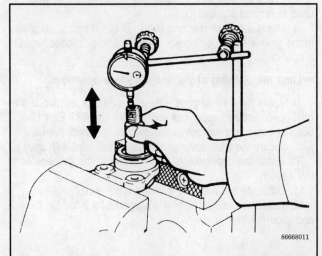

Fig. 16 Measure the end-play of the lower ball joint with a dial indicator — all Pick-ups and Monteros

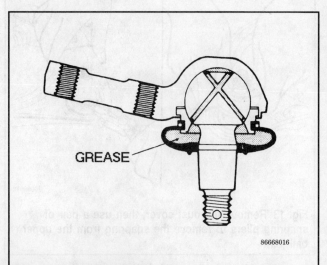

Fig. 17 The internal construction of a lower ball joint found on some Pick-ups and Monteros

SUSPENSION AND STEERING   8-9

7. Install the bar into place and attach all the hardware and brackets loosely. Make sure the washers and bushings are correctly installed.

8. For 1983-91 Monteros and all 4WD Pick-ups, use a new nut on the upper link mount (above the crossmember) and tighten it until 0.24-0.31 in. (6-8mm) of threads projects through the nut.

9. Tighten the bracket or clamp bolts to 8 ft. lbs. (10 Nm).

10. Install new nuts at each lower control arm. Refer to the following directions for each vehicle:

• 1983-91 Monteros and 4WD Pick-ups: tighten the nuts until 0.24-0.35 in. (6-9mm) of threads project through the nut.

• 1992-95 4WD Pick-ups: Measure the length of the lower control arm sway bar mounting bolt. If the bolt measures 2.1 in. (55mm) long, then 0.24-0.31 in. (6-8mm) of threads should protrude past the nut. If the bolt measured 2.3 in. (60mm) in length, there should be 0.31-0.35 in. (8-9mm) of threads projecting past the nut.

• 2wd Pick-ups, the bolt must project 0.90 in. (23mm) through the nut.

11. Install the under cover or splash shield.

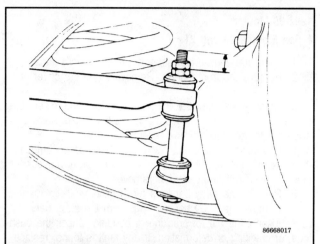

Fig. 19 When installing the stabilizer end links, make sure that the measurement is correct — 1983-95 Pick-ups and 1983-91 Monteros

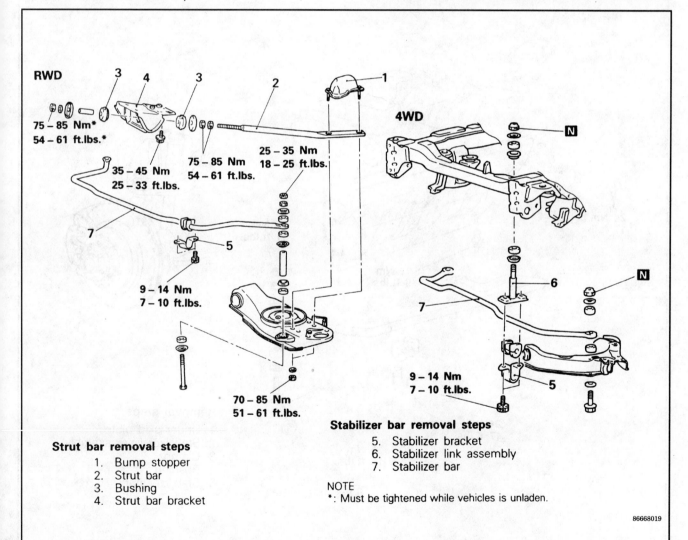

Strut bar removal steps
1. Bump stopper
2. Strut bar
3. Bushing
4. Strut bar bracket

Stabilizer bar removal steps
5. Stabilizer bracket
6. Stabilizer link assembly
7. Stabilizer bar

NOTE
*: Must be tightened while vehicles is unladen.

Fig. 18 Stabilizer bar and strut bar removal and installation components — 1983-95 Pick-ups and 1983-91 Monteros

8-10 SUSPENSION AND STEERING

12. Install the wheels and lower the vehicle to the ground.

1992-95 Montero

▶ See Figures 20 and 21

1. Raise the vehicle and support safely on jackstands.
2. Remove the front wheels. Remove the under cover or splash shield.
3. Remove the stabilizer link assembly from the lower control arms. Loosen and remove the nut and washers holding the sway bar to the stabilizer link assembly.
4. loosen and remove the sway bar clamp bolts, remove the clamps from the vehicle. Removal of the sway bar bushings is easier with the sway bar removed.
5. Remove the sway bar from the vehicle.
6. Remove the sway bar bushings from the sway bar.
7. Inspect the bar for cracking or bending. Check the bushings for cracking or deformation. If any fault is found, replace components. It is recommended to replace the rubber bushings any time they are removed.
8. Install the bar into place and attach all the hardware and brackets loosely. Make sure the washers and bushings are correctly installed.
9. Tighten the bracket or clamp bolts to 17 ft. lbs. (24 Nm).
10. Attach the sway bar ends to the stabilizer link assemblies and the stabilizer link assemblies to the lower control arms. Tighten the stabilizer link assemblies' lower-control-arm-mounting nuts to 25 ft. lbs. (33 Nm). Tighten the sway bar-to-stabilizer link assembly mounting nuts to 11 ft. lbs. (15 Nm) on 1992-93 Monteros, and to 69 ft. lbs. (93 Nm) on 1994-95 Monteros.
11. Install the under cover or splash shield.
12. Install the wheels and lower the vehicle to the ground.

Strut Rod

➡ Only rear wheel drive Pick-ups came equipped with strut bars.

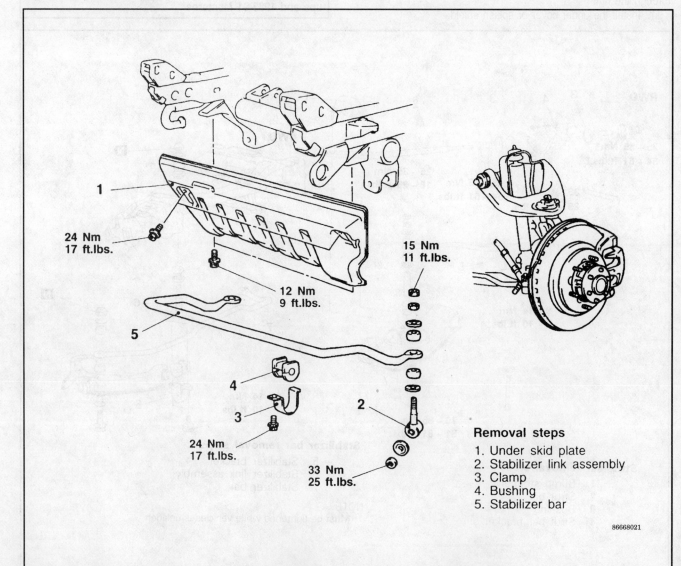

Fig. 20 Stabilizer bar removal and installation components — 1992-93 Monteros

SUSPENSION AND STEERING 8-11

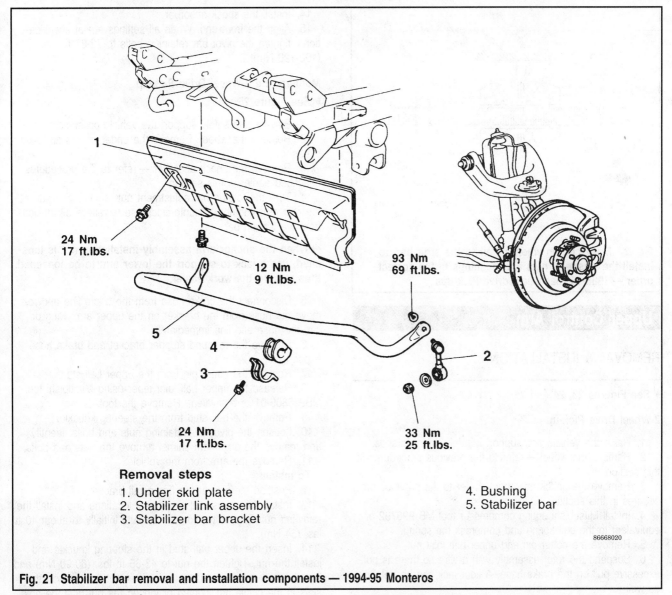

Removal steps
1. Under skid plate
2. Stabilizer link assembly
3. Stabilizer bar bracket
4. Bushing
5. Stabilizer bar

Fig. 21 Stabilizer bar removal and installation components — 1994-95 Monteros

REMOVAL & INSTALLATION

▶ See Figures 18 and 22

1. Elevate and safely support the truck on stands.
2. Remove the front wheels. Remove the under covers.
3. Remove the front locknut from the strut rod.
4. Remove the two bolts holding the strut rod to the control arm. Remove the bump stopper from the lower control arm, then pull the rod rearward to remove it.

✸✸WARNING

Take careful note of the bushings and washers at the front of the rod. They must be reinstalled correctly. The pieces are not interchangeable.

5. If both strut rods are removed or one is being replaced, take note that the bars are different for left and right sides. The left side rod will have either an **L** or a white mark on the bar. Right side bars are identified by an **R** or no marking.
6. When reinstalling, set the rearmost nut on the strut rod to the correct distance of 3.28 in. (8.35cm) from the tip of the rod. This creates the approximately correct effective length for the strut rod.
7. Install the front of the rod with the bushings and spacers correctly placed. Install the end nut snug but do not final tighten it.
8. Attach the strut rod and bump stopper to the control arm, tightening the nuts to 51-61 ft. lbs. (70-85 Nm).
9. Install the wheels and lower the vehicle to the ground.
10. Bounce the vehicle two or three times to stabilize the suspension. Make sure the nuts on the front of the strut rod are snug. With the truck free of passengers and cargo, tighten the front and rear nuts on the strut rod to 54-61 ft. lbs. (75-85 Nm). Tighten the rear nut's lock nut also to 54-61 ft. lbs. (75-85 Nm).
11. Reinstall the under covers.
12. Have the alignment checked at a reliable repair facility.

8-12 SUSPENSION AND STEERING

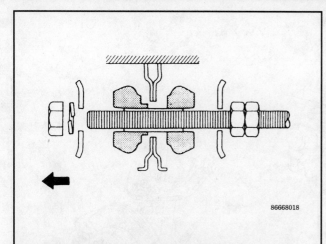

Fig. 22 Make certain when installing the strut bar, to install the washers, nuts and bushings in the correct order — 1983-95 rear wheel drive Pick-ups

Upper Control Arm

REMOVAL & INSTALLATION

▸ See Figures 23, 24 and 25

2-Wheel Drive Pick-up

1. Raise the vehicle and support it safely on jackstands.
2. Remove the wheel — refer to the previous procedures in this section.
3. Remove the shock absorber — refer to the previous procedures in this section.
4. Install Mitsubishi spring compressor tool MB 990792 or equivalent to the coil spring and compress the spring.
5. Remove the cotter pin and upper ball joint nut.
6. Suspend the rotor assembly with a wire so there is not excessive pull on the brake hose. A floor jack can be helpful as well.
7. Release the upper ball joint taper using Mitsubishi tool MB990809-01 or the equivalent.
8. Remove the tool and remove the ball stud from the knuckle.
9. Loosen the pivot bar retaining nuts and bolts, identify and remove the alignment shims.
10. Remove the nuts and bolts and remove the arm from the vehicle.
11. Install the arm to the frame rail bracket, install the shims in their original locations and install the retaining nuts and bolts. Tighten the nuts initially to about 40 ft. lbs. (54 Nm).
12. Install the ball joint to the knuckle and tighten the ball joint nut to 43-65 ft. lbs. (60-90 Nm). Install a new cotter pin through the castle nut. If the hole does not line up with the slots in the castle nut, tighten or loosen the nut until the hole lines up, but make sure that the nut is still within the torque range.
13. Seat the spring correctly and slowly loosen the compressor. Make certain both ends of the spring are correctly placed in the spring indentations in the arm and frame. Remove the spring compressor.

14. Install the shock absorber.
15. Align the front end. When all settings are at specifications, tighten the pivot bar retaining bolts to 72-87 ft. lbs. (100-120 Nm).

Montero and 4-Wheel Drive Pick-up

▸ See Figure 26

1. Elevate and safely support the vehicle on stands.
2. Remove the wheel. Remove the under covers and skid plates.
3. Remove the shock absorber — refer to the procedures described earlier in this section.
4. Turn the torsion bar adjustment nut (anchor bolt nut, at the rear crossmember) counterclockwise to relieve all tension from the torsion bar.

➡When the anchor arm assembly installation nut is loosened, use a jack to support the lower arm to be loosened, thus making the work easier.

5. Disconnect the brake hose from the brake line and remove the hose from the bracket on the upper arm. Plug or cap the hose and line immediately.
6. Remove the rebound stopper bracket and brake hose support.
7. Remove the cotter pin from the upper ball stud.
8. Release the upper ball joint taper using Mitsubishi tool MB990809-01 or equivalent. Remove the tool.
9. Remove the ball stud from the steering knuckle.
10. Loosen the pivot bar retaining nuts and bolts, identify and remove the alignment shims, remove the nuts and bolts.
11. Remove the arm from the vehicle.

To install:

12. Position the arm at the frame rail bracket.
13. Install the shims in their original locations and install the retaining nuts and bolts. Torque the nuts initially to about 40 ft. lbs. (54 Nm).
14. Insert the upper ball stud in the steering knuckle and install the nut. Tighten the nut to 43-65 ft. lbs. (60-90 Nm) and install a new cotter pin. If the hole does not line up with the slots in the castle nut, tighten or loosen the nut until the hole lines up, but make sure that the nut is still within the torque range.
15. Install the shock absorber — refer to the same shock absorber procedures that were referred to during disassembly.
16. Install the rebound stopper bracket and brake hose support. Tighten the mounting bolts to 7-10 ft. lbs. (9-14 Nm).
17. Remove the plugs and carefully attach the brake hose to the line. Don't crossthread the fitting. Make certain the retaining clip is seated.
18. Bleed the brake system.
19. Turn the torsion bar adjustment nut clockwise to its approximate previous position. This will apply a load on the bar.
20. Install the wheel and lower the vehicle.
21. Set the ride height. Refer to the torsion bar procedure described earlier in this section. Have the front end alignment checked.
22. Apply final torque to the upper arm retaining bolts, setting them to 72-87 ft. lbs. (100-120 Nm).
23. Reinstall the skid plates and under covers.

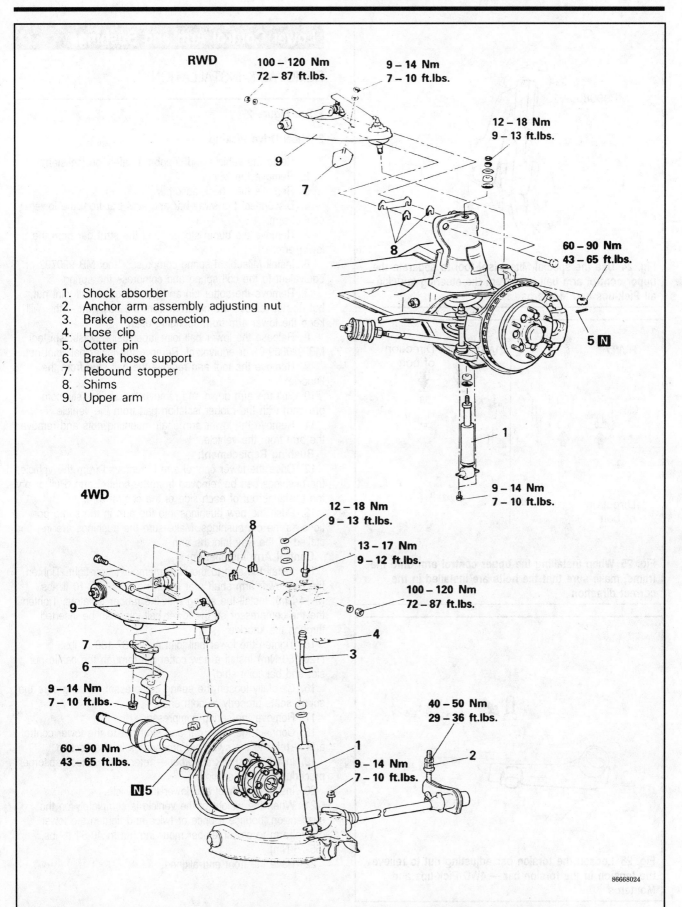

Fig. 23 Upper control arm removal and installation components — 2WD and 4WD Pick-ups and Monteros

8-14 SUSPENSION AND STEERING

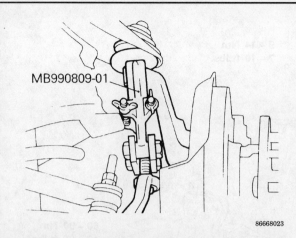

Fig. 24 Use the special Mitsubishi tool to separate the upper control arm ball joint from the steering knuckle — all Pick-ups and Monteros

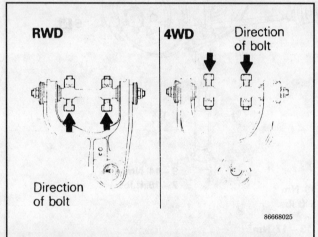

Fig. 25 When installing the upper control arm onto the frame, make sure that the bolts are installed in the correct direction

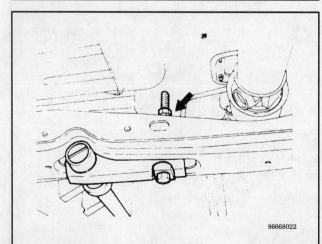

Fig. 26 Loosen the torsion bar adjusting nut to relieve the tension in the torsion bar — 4WD Pick-ups and Monteros

Lower Control Arm and Bushings

REMOVAL & INSTALLATION

▶ See Figure 27

2-Wheel Drive Pick-up

1. Raise the vehicle and support it safely on jackstands.
2. Remove the wheel.
3. Remove the shock absorber.
4. Disconnect the sway bar and strut bar from the lower control arm.
5. Remove the bump stopper and the strut bar from the lower arm.
6. Install Mitsubishi spring compressor tool MB 990792 or equivalent to the coil spring and compress the spring.
7. Remove the cotter pin and loosen lower ball joint nut, but do not remove the nut from from the ball joint — this will keep the lower arm from springing open.
8. Release the lower ball joint taper using Mitsubishi tool MB 990809-01 or equivalent. Remove the lower ball joint nut.
9. Remove the tool and remove the ball stud from the knuckle.
10. Pull the arm down and remove the spring (still compressed) with the rubber isolation pad from the vehicle.
11. Remove the lower arm shaft mounting nuts and remove the arm from the vehicle.

Bushing Replacement:

12. Once the lower control arm is removed from the vehicle, the bushings can be removed from the control arm. Pull or pry the bushings out of each side of the control arm.
13. Push the new bushings into the arm in the same positions as the old bushings. Make sure the bushings are installed all the way into the arm.

Control Arm Installation:

14. Install the arm to the crossmember finger-tight. Tighten the two lower arm shaft rear mounting nuts to 7-10 ft. lbs. (9-14 Nm). Install the spring with the rubber isolators. Tighten the the compressor so the lower ball joint can be inserted through the knuckle.
15. Tighten the lower ball joint nut to 87-130 ft. lbs. (120-180 Nm) Install a new cotter pin through the castle nut slot and ball joint stud.
16. Carefully loosen the spring compressor, making sure the spring seats properly at both ends.
17. Remove the spring compressor.
18. Connect the sway bar and strut bar to the lower control arm — refer to the sway bar procedures in this section.
19. Install the shock absorber — refer to the shock absorber procedures in this section.
20. Install the wheel and lower the vehicle.
21. When the weight of the vehicle is completely on the suspension, bounce it once or twice and tighten the lower control arm to crossmember mounting nut to 40-54 ft. lbs. (55-75 Nm).
22. Have the front end aligned.

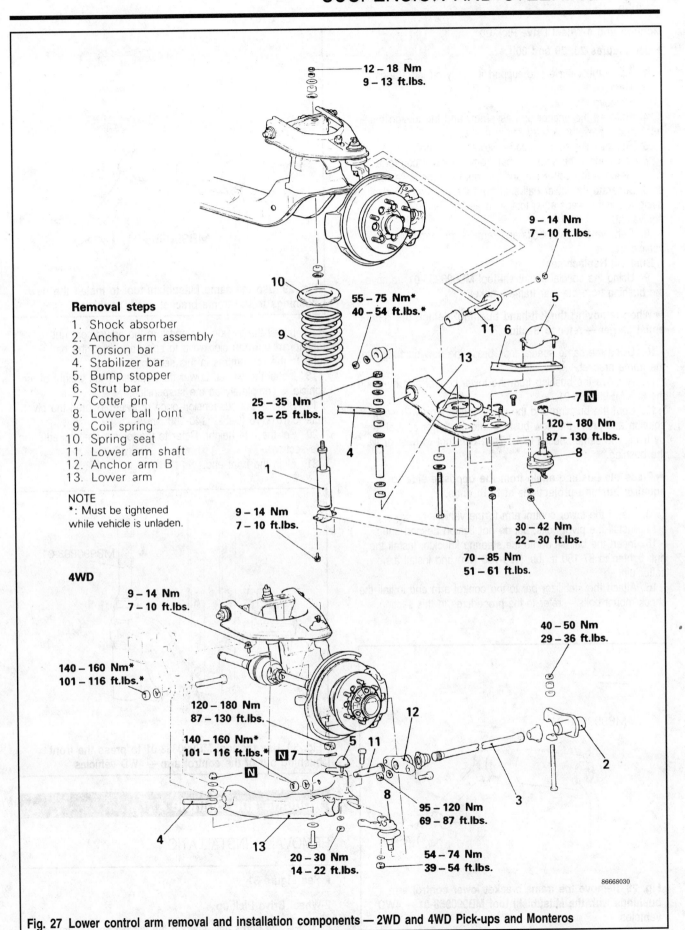

Fig. 27 Lower control arm removal and installation components — 2WD and 4WD Pick-ups and Monteros

8-16 SUSPENSION AND STEERING

Montero and 4-Wheel Drive Pick-up

♦ See Figures 28, 29 and 30

1. Raise the vehicle and support it safely on jackstands.
2. Remove the wheels.
3. Remove the skid plate.
4. Remove the anchor arm assembly and the torsion bar — refer to the previous procedures in this section.
5. Remove the shock absorber lower attaching bolt.
6. Disconnect the stabilizer bar from the lower control arm.
7. Remove the cotter pin and the nut from the lower ball stud. Separate the lower ball stud from the steering knuckle using a suitable separating tool such as MB 990809-01 or equivalent.
8. Remove the pivot bolts and remove the arm from the vehicle.

Bushing Replacement:

9. Using the special Mitsubishi tool MB990958-01, remove the bushing from the rear frame bracket.

➡ **When removing the left-hand bushing, detach the differential carrier — refer to Section 7.**

10. Using the same special tool, press the new bushing into the frame bracket.
11. Remove the bushing from the lower control arm by using the tool MB990883-01.
12. Coat the bushing and the lower control arm with soap solution and press the new bushing into the lower control arm by using the same special tools. Take care not to twist or tilt the bushing.

➡ **Press the bushing again from the opposite side to equalize bushing projections at both ends.**

13. Install the lower control arm to the vehicle.
14. Install the pivot bolts, but do not tighten beyond snug.
15. Insert the ball stud into the steering knuckle. Install the nut, tighten to 87-130 ft. lbs. (120-180 Nm) and install a new cotter pin.
16. Attach the stabilizer bar to the control arm and install the shock mount bolts — refer to the procedures in this section.

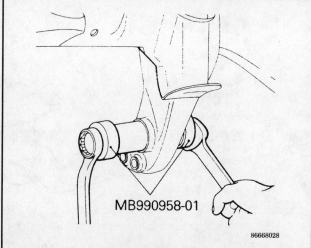

Fig. 29 Use the same Mitsubishi tool to install the new bushings to the frame bracket — 4WD vehicles

17. Install the torsion bar and turn the adjustment nut (anchor mount nut) clockwise to apply a load to the bar — refer to the procedures in this section.
18. Install the wheel. Lower the vehicle so the weight of the vehicle is completely on the suspension.
19. Bounce the suspension once or twice. Tighten the pivot nuts to 101-116 ft. lbs. (140-160 Nm).
20. Set the ride height. Refer to the torsion bar procedure in this section.
21. Align the front end.

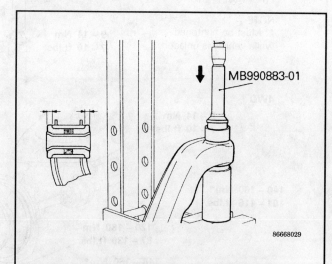

Fig. 30 Use the tool MB990883-01 to press the front bushing out of the control arm — 4WD vehicles

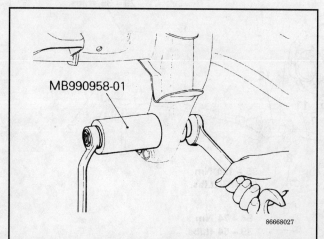

Fig. 28 Remove the frame bracket lower control arm bushings with the Mitsubishi tool MB990958-01 — 4WD vehicles

Knuckle and Spindle

REMOVAL & INSTALLATION

♦ See Figure 31

2-Wheel Drive Pick-up

1. Elevate and safely support the vehicle on jackstands.

SUSPENSION AND STEERING 8-17

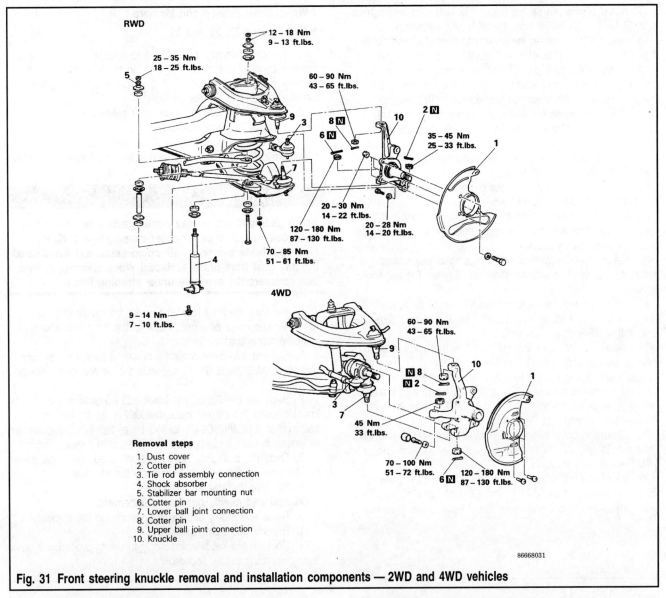

Fig. 31 Front steering knuckle removal and installation components — 2WD and 4WD vehicles

Removal steps
1. Dust cover
2. Cotter pin
3. Tie rod assembly connection
4. Shock absorber
5. Stabilizer bar mounting nut
6. Cotter pin
7. Lower ball joint connection
8. Cotter pin
9. Upper ball joint connection
10. Knuckle

2. Remove the wheel.
3. Remove the brake caliper and suspend it from wire out of the way. Do not disconnect or loosen the hose; simply move the whole assembly out of the way. Do not allow the caliper to hang by the hose.

✱✱CAUTION

Brake pads and shoes may contain asbestos, which has been determined to be a cancer causing agent. Never clean the brake surfaces with compressed air! Avoid inhaling any dust from brake surfaces! When cleaning brakes, use commercially available brake cleaning fluids.

4. Remove the small center cap.
5. Remove the cotter pin, lock cap and spindle nut. Remove the large flat washer and the outer wheel bearing. The flat washer and bearing may also be held in place with a thumb during removal and then separated later.
6. Pull the brake rotor and hub off the spindle.
7. Unbolt and remove the round metal splash shield.

8. Remove the cotter pin and loosen but do not remove the nut on the tie rod end. Use the correct tool to separate the tie rod joint. Remove the tie rod end nut from the tie rod.
9. Disconnect the shock absorber at its upper mount. Disconnect it at its lower mount and remove the shock absorber.
10. Install a spring compressor such as MB 990792-01 or its equivalent and compress the spring.
11. Remove the stabilizer bar retaining bolt (link) from the lower arm.
12. Remove the cotter pins and loosen the nuts on the upper and lower ball joints. Use the correct puller — MB 990809-01 or its equivalent — to separate the upper and lower joints from the knuckle.
13. With a floor jack supporting the lower control arm, the spring can be decompressed at this time.
14. Remove the knuckle and spindle.

To install:
15. Install the knuckle in position.
16. Make certain the spring compressor is correctly installed, then compress the spring again.

8-18 SUSPENSION AND STEERING

17. Install a new nut on the upper ball joint stud and tighten it to 43-65 ft. lbs. (60-90 Nm). Install a new cotter pin.
18. Install a new nut on the lower ball joint stud and tighten it to 87-130 ft. lbs. (120-180 Nm). Install a new cotter pin.
19. Connect the stabilizer bar to the control arm, making certain the spacers and washers are correctly assembled on the bolt. Refer to the sway bar procedures in this section.
20. Carefully loosen the spring compressor, allowing the spring into place. Make certain that it is correctly seated. When the spring is in place, remove the compressor. Lower the floor jack and remove from under the vehicle.
21. Install the shock absorber.
22. Assemble the tie rod to knuckle joint. Use a new nut and tighten it to 30 ft. lbs. (40 Nm).
23. Install the splash shield to the steering knuckle.
24. Fit the rotor and hub assembly onto the spindle with the bearings and seal in place.
25. Making certain that the outer bearing is well seated, install the flat washer and the shaft nut. Tighten the nut finger-tight.
26. Using a torque wrench, tighten the nut to 22 ft. lbs. (30 Nm), back the nut off to 0 ft. lbs. (0 Nm), then final tighten to 6 ft. lbs. or 72 inch lbs. (8 Nm).
27. Install the lock cap and cotter pin. If the holes in the lockcap do not align with the castle nut on the shaft, turn the lock cap and try again; the lock cap has offset holes to compensate for this occurrence. In the unlikely event that it still will not align, back off the shaft nut by NO MORE than 30° of rotation (30° is equal to ½ of a flat side on the nut.).
28. Clean the outer edge of the small hub cover thoroughly. Coat the flange with 3M® 8663 sealant or equivalent and install the small hub cover.
29. Install the brake caliper and pads. Tighten the front calliper mounting bolts to 58-72 ft. lbs. (80-100 Nm).
30. Install the wheel and lower the vehicle to the ground.

4-Wheel Drive Pick-up and Montero
♦ See Figures 32, 33 and 34

1. Before beginning, place the free-wheeling hub in the free position by placing the transfer lever in the 2H position and moving in reverse for about 6-7 ft. (2 m) or by shifting the transfer shifter into the 2H position.
2. Raise the vehicle and support it safely on jackstands.
3. Remove the front wheel(s).
4. Remove the front brake caliper, pads and adaptor. Position them out of the way, suspended from stiff wire; do not let the caliper hang by the hose.

※※CAUTION

Brake pads and shoes may contain asbestos, which has been determined to be a cancer causing agent. Never clean the brake surfaces with compressed air! Avoid inhaling any dust from brake surfaces! When cleaning brakes, use commercially available brake cleaning fluids.

5. Remove the front hub and brake rotor assembly, with the inner and outer bearings. Refer to the front hub assembly removal procedures in Section 7.
6. Support the lower control arm with a jack. Use the correct tools (MB990635-01) to separate the tie rod joint from the knuckle.
7. Separate the upper and lower ball joints from the knuckle using the correct tools (MB990778-01 for the upper ball joint and MB990809-01 for the lower ball joint). Remember to loosen the nut and separate the joint, then remove the nut.
8. Once the ball joints have been released, lower the control arm slowly just enough to remove the knuckle.
9. Remove the steering knuckle.

Oil seal and needle bearing replacement:
10. Raise and safely support the vehicle on jackstands.
11. Remove the front wheel(s).
12. Remove the knuckle assembly following procedures outlined previously in this section.

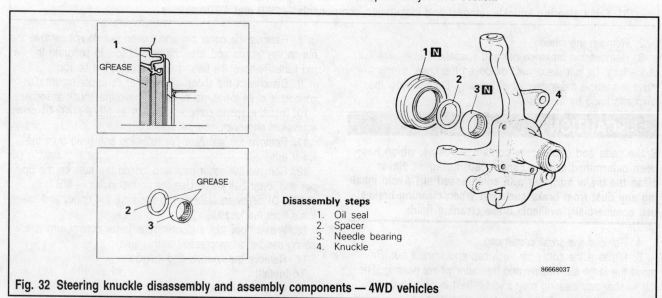

Disassembly steps
1. Oil seal
2. Spacer
3. Needle bearing
4. Knuckle

Fig. 32 Steering knuckle disassembly and assembly components — 4WD vehicles

13. Remove the oil seal from the back of the knuckle and remove the spacer.

SUSPENSION AND STEERING

14. Drive out the needle bearing with a brass punch by tapping uniformly around the housing.

> **WARNING**
> Once the needle bearing has been removed, it cannot be reused.

15. Apply multipurpose grease to the roller surface of the new needle bearing.
16. Use the correct bearing installation tools (MB 990938-01 and 990956-01 or equivalents) to carefully install the bearing until it is flush with the knuckle face. Use care not to insert the bearing too far.
17. Apply multipurpose grease to the knuckle face of the spacer. Install the spacer to the knuckle so that the chamfered (tapered) side is towards the center of the vehicle.
18. Use the correct seal driver and install a new oil seal. Tap in the seal until it is flush with the knuckle end face.
19. Pack multipurpose grease inside the oil seal and lip

Knuckle installation:

20. Install the knuckle to the ball joint studs. Tighten the lower ball joint nuts to 87-130 ft. lbs. (120-180 Nm). Tighten the upper ball joint nut to 43-65 ft. lbs (60-90 Nm). Install new cotter pins.
21. Connect the tie rod end to the knuckle. Tighten the nut to 33 ft. lbs. (45 Nm) and install a new cotter pin.
22. Refer to Section 7 for the hub and rotor assembly installation procedures.
23. Install the brake caliper and pads. Tighten the brake calliper mounting bolts to 58-72 ft. lbs. (80-100 Nm).
24. Install the wheel and lower the vehicle to the ground.

Front Wheel Bearing

REMOVAL & INSTALLATION

♦ See Figures 35, 36, 37, 38 and 39

1. Remove the front hub/rotor assembly from the steering knuckle as previously described.
2. Remove the outer bearing from the hub. Remove the splash shield.
3. Remove the nuts and bolts that attach the hub to the rotor and separate them. Clean out all of the old grease from the inside of the hub.
4. The race for the outer bearing may be removed with a brass rod and mallet. Drive the race out from the inside but do not mar the inner surface of the hub.

> **WARNING**
> Do not remove the bearing races if the same bearing is to be reused after cleaning or repacking. If a bearing is replaced, the race MUST be replaced. Don't use a new bearing on an old race.

5. Use a seal puller on the back of the hub to remove the inner bearing seal.

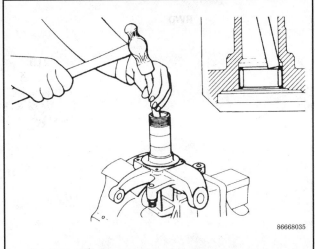

Fig. 33 Tap the knuckle bearing out of the steering knuckle with a brass punch — 4WD vehicles

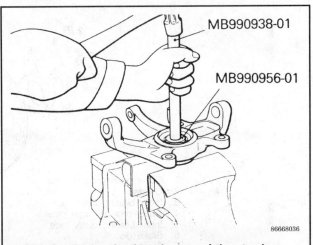

Fig. 34 Set the bearing into the rear of the steering knuckle with the Mitsubishi tools MB990938-01 and MB990956-01 — 4WD vehicles

6. Once the outer bearing race is removed, flip the hub assembly over and tap the inner race out using the same procedure as with the outer race.

To install:

7. Use bearing drivers of the proper size and drive the new races into the hub. Don't try to install the race with the brass rod; chances are high that the race will be crooked or damaged.
8. Assemble the rotor and front hub. Tighten the nuts to 40 ft. lbs. (54 Nm).
9. Apply wheel bearing grease to the inside of the front hub.
10. Pack the inner bearing and install to the hub.
11. Install a new oil seal into the hub so it is flush with the hub end face.
12. Install the assembly onto the steering knuckle.
13. Pack and install the outer bearing.
14. Continue with reinstallation of the hub as outlined previously.

8-20 SUSPENSION AND STEERING

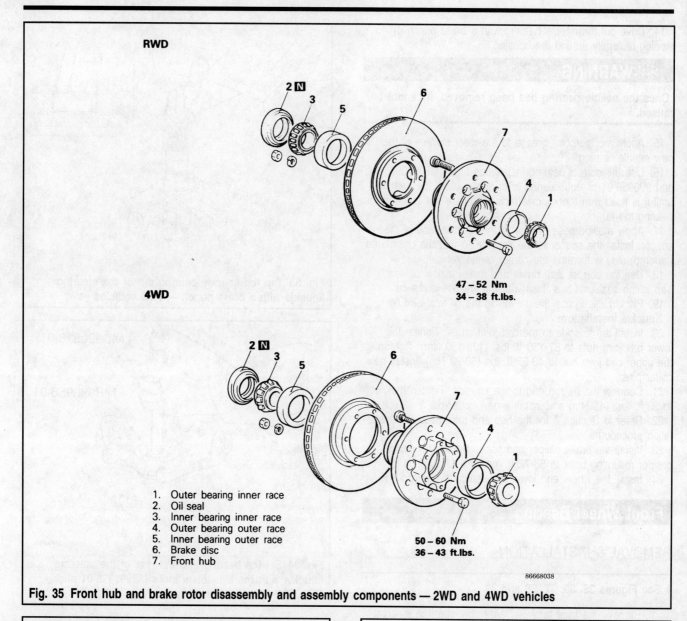

1. Outer bearing inner race
2. Oil seal
3. Inner bearing inner race
4. Outer bearing outer race
5. Inner bearing outer race
6. Brake disc
7. Front hub

Fig. 35 Front hub and brake rotor disassembly and assembly components — 2WD and 4WD vehicles

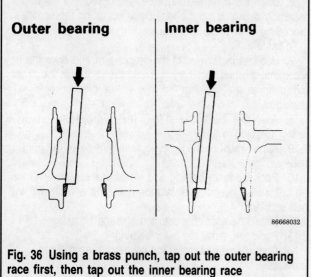

Fig. 36 Using a brass punch, tap out the outer bearing race first, then tap out the inner bearing race

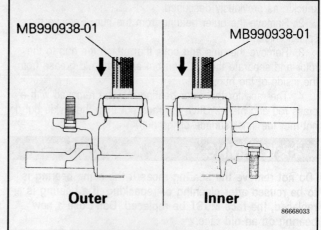

Fig. 37 Use the tool MB990938-01 or another suitably-sized bearing race installing tool to tap the races in place in the hub

SUSPENSION AND STEERING

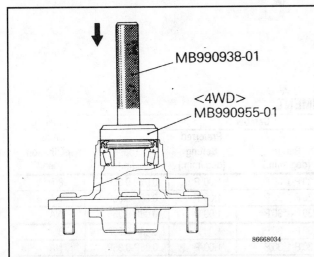

Fig. 38 Use the same tool to set the inner oil seal into the hub until it is flush with the hub edge

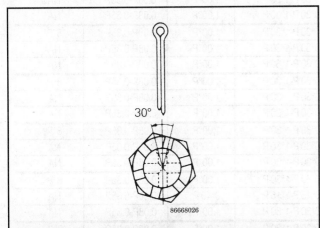

Fig. 39 When installing the cotter pin to the knuckle spindle nut, if the slot on the castle nut and the hole in the spindle do not match up, back the nut up a maximum of 30° until a slot does line up

Front End Alignment

Alignment of the front wheels is essential if your truck is to go, stop and turn as designed. Alignment can be altered by collision, overloading, poor repair or bent components.

If your vehicle is exhibiting bizarre handling and/or poor road manners, the first place to look is the tires. Although the tires may wear as a result of an alignment problem, worn or poorly inflated tires can make you chase alignment problems which don't exist.

Once you have eliminated all other causes, unload everything from the rear of the Pick-up or Montero except the spare tire, set the tire pressures to the correct level and take the vehicle to a reputable alignment facility. Since the alignment settings are measured in very small increments, it is almost impossible for the home mechanic to accurately determine the settings. The explanations that follow will help you understand the three dimensions of alignment: caster, camber and toe.

CASTER

Caster is the tilting of the steering axis either forward or backward from the vertical, when viewed from the side of the vehicle. A backward tilt is said to be positive and a forward tilt is said to be negative. Changes in caster affect the straight line tendency of the vehicle and the "return to center" of the steering after a turn. If the caster is radically different between the left and right wheels (such as after hitting a sizeable pothole), the vehicle will exhibit a nasty pull to one side.

On the Pick-ups and Monteros, caster is fully adjustable, either by adjusting the upper arm cross-shaft (Montero) or by changing shim thickness (Pick-up).

CAMBER

Camber is the tilting of the wheels from the vertical (leaning in or out) when viewed from the front of the vehicle. When the wheels tilt outward at the top, the camber is said to be positive. When the wheels tilt inward at the top the camber is said to be negative. The amount of tilt is measured in degrees from the vertical. This measurement is called camber angle.

Camber affects the position of the tire on the road surface during vertical suspension movement and cornering. Changes in camber affect the handling and ride qualities of the vehicle, as well as tire wear. Many tire wear patterns indicate camber related problems from misalignment, overloading or poor driving habits.

Camber is adjustable on the Pick-ups and Monteros; both require the replacement of shims in the front suspension.

TOE

Toe is the turning in or out (parallelism) of the wheels. The actual amount of toe setting is normally only a fraction of a centimeter. The purpose of toe-in (or out) specification is to ensure parallel rolling of the wheels. Toe-in also serves to offset the small deflections of the steering support system which occur when the vehicle is rolling forward.

Changing the toe setting will radically affect the overall "feel" of the steering, the behavior of the truck under braking, tire wear and even fuel economy. Excessive toe (in or out) causes excessive drag or scrubbing on the tires.

Toe is adjustable on all Mitsubishi vehicles and is the most common value adjusted during an alignment. It is generally measured in decimal millimeters or degrees. It is adjusted by loosening the locknut on each tie rod end and turning the rod until the correct reading is achieved. The rods on the left and right must remain equal in length during all adjustments.

SUSPENSION AND STEERING

WHEEL ALIGNMENT

Year	Model	Caster Range (deg./min.)	Caster Preferred Setting (deg./min.)	Camber Range (deg./min.)	Camber Preferred Setting (deg./min.)	Toe-in (in.)	Steering Axis Inclination (deg.)
1983	Montero	2°00'P-3°00'P	2°30'P	0°30'P-1°30'P	1.00°P	0.08P-0.35P	8°P
	2WD Pick-up	1°30'P-3°30'P	2°30'P	0°30'P-1°30'P	1.00°P	0.08P-0.35P	NA
	4WD Pick-up	1°00'P-3°00'P	2°00'P	0°30'P-1°30'P	1.00°P	0.08P-0.35P	NA
1984	Montero	2°00'P-3°00'P	2°30'P	0°30'P-1°30'P	1.00°P	0.08P-0.35P	8°P
	2WD Pick-up	1°30'P-3°30'P	2°30'P	0°30'P-1°30'P	1.00°P	0.08P-0.35P	NA
	4WD Pick-up	1°00'P-3°00'P	2°00'P	0°30'P-1°30'P	1.00°P	0.08P-0.35P	NA
1985	Montero	2°00'P-3°00'P	2°30'P	0°30'P-1°30'P	1.00°P	0.08P-0.35P	8°P
	2WD Pick-up	1°30'P-3°30'P	2°30'P	0°30'P-1°30'P	1.00°P	0.08P-0.35P	NA
	4WD Pick-up	1°00'P-3°00'P	2°00'P	0°30'P-1°30'P	1.00°P	0.08P-0.35P	NA
1986	Montero	2°00'P-3°00'P	2°30'P	0°30'P-1°30'P	1.00°P	0.08P-0.35P	8°P
	2WD Pick-up	1°30'P-3°30'P	2°30'P	0°30'P-1°30'P	1.00°P	0.08P-0.35P	NA
	4WD Pick-up	1°00'P-3°00'P	2°00'P	0°30'P-1°30'P	1.00°P	0.08P-0.35P	NA
1987	Montero	2°00'P-3°00'P	2°30'P	0°30'P-1°30'P	1.00°P	0.08P-0.35P	8°P
	2WD Pick-up	1°30'P-3°30'P	2°30'P	0°30'P-1°30'P	1.00°P	0.08P-0.35P	NA
	4WD Pick-up	1°00'P-3°00'P	2°00'P	0°30'P-1°30'P	1.00°P	0.08P-0.35P	NA
1988	Montero	2°00'P-3°00'P	2°30'P	0°30'P-1°30'P	1.00°P	0.08P-0.35P	8°P
	2WD Pick-up	1°30'P-3°30'P	2°30'P	0°30'P-1°30'P	1.00°P	0.08P-0.35P	NA
	4WD Pick-up	1°00'P-3°00'P	2°00'P	0°30'P-1°30'P	1.00°P	0.08P-0.35P	NA
1989	Montero	2°00'P-4°00'P	3°00'P	0°30'P-1°30'P	1.00°P	0.08P-0.35P	8°P
	2WD Pick-up	1°30'P-3°30'P	2°30'P	0°15'P-1°15'P	0°40'P	0.08P-0.35P	NA
	4WD Pick-up	1°00'P-3°00'P	2°00'P	0°30'P-1°30'P	1.00°P	0.08P-0.35P	NA
1990	Montero	1°55'P-3°55'P	2°55'P	0°30'P-1°30'P	1.00°P	0.08P-0.35P	8°P
	2WD Pick-up	1°30'P-3°30'P	2°30'P	0°10'P-1°10'P	0°40'P	0.08P-0.35P	NA
	4WD Pick-up	1°00'P-3°00'P	2°00'P	0°30'P-1°30'P	1.00°P	0.08P-0.35P	NA
1991	Montero	1°55'P-3°55'P	2°55'P	0°30'P-1°30'P	1.00°P	0.08P-0.35P	8°P
	2WD Pick-up	1°30'P-3°30'P	2°30'P	0°10'P-1°10'P	0°40'P	0.08P-0.35P	NA
	4WD Pick-up	1°00'P-3°00'P	2°00'P	0°30'P-1°30'P	1.00°P	0.08P-0.35P	NA
1992	Montero	2°00'P-4°00'P	3°00'P	0°10'P-1°10'P	0°40'P	0.00P-0.28P	NA
	2WD Pick-up	1°30'P-3°30'P	2°30'P	0°10'P-1°10'P	0°40'P	0.08P-0.35P	NA
	4WD Pick-up	1°00'P-3°00'P	2°00'P	0°30'P-1°30'P	1.00°P	0.08P-0.35P	NA
1993	Montero	2°00'P-4°00'P	3°00'P	0°10'P-1°10'P	0°40'P	0.00P-0.28P	NA
	2WD Pick-up	1°30'P-3°30'P	2°30'P	0°10'P-1°10'P	0°40'P	0.08P-0.35P	NA
	4WD Pick-up	1°00'P-3°00'P	2°00'P	0°30'P-1°30'P	1.00°P	0.08P-0.35P	NA
1994	Montero	2°00'P-4°00'P	3°00'P	0°10'P-1°10'P	0°40'P	0.00P-0.28P	NA
	2WD Pick-up	1°30'P-3°30'P	2°30'P	0°10'P-1°10'P	0°40'P	0.08P-0.35P	NA
	4WD Pick-up	1°00'P-3°00'P	2°00'P	0°30'P-1°30'P	1.00°P	0.08P-0.35P	NA
1995	2WD Pick-up	1°50'P-3°50'P	2°50'P	0°17'P-1°17'P	0°67'P	0.08P-0.35P	NA
	Montero	2°00'P-4°00'P	3°00'P	0°17'P-1°17'P	0°67'P	0.00P-0.28P	NA

NA - Not Available

SUSPENSION AND STEERING

REAR SUSPENSION

Coil Springs

REMOVAL & INSTALLATION

1989-95 Montero

1. Raise the vehicle and support it safely on jackstands.
2. Remove the rear wheels.
3. Remove the parking brake cable attaching bolts at each side.
4. Use 2 floor jacks or an axle cradle to support the weight of the axle. Do not try to balance the axle on 1 jack.
5. Remove the bolt that attaches the lateral rod (Panhard rod) to the body.
6. Remove the lower shock mounting bolts.
7. Slowly lower the axle and remove the coil springs with their seats.

※※WARNING

When lowering the jack, take great care not to damage the brake lines running between the main line and the axle housing.

8. Reinstall the springs and elevate the axle into position. Make certain that both the spring and seat are correctly positioned.
9. Install the lower shock mounting bolt, tightening it to 80-94 ft. lbs. (110-130 Nm) for 1989-91 Monteros. Tighten the lower shock mounting bolts to 159-181 ft. lbs. (216-245 Nm) for 1992-95 Monteros.
10. Connect the lateral rod to the body and tighten the bolt to 80-94 ft. lbs. (110-130 Nm) for 1989-91 Monteros. Tighten the lateral rod mounting bolt to 159-181 ft. lbs. (216-245 Nm) for 1992-95 Monteros.
11. Remove the apparatus supporting the axle.
12. Reinstall the brackets and bolts holding the parking brake cables.
13. Install the rear wheels and lower the vehicle to the ground.

Leaf Springs

REMOVAL & INSTALLATION

Pick-up and 1983-89 Montero
▶ See Figure 40

1. Raise the vehicle and support it safely on jackstands.
2. Remove the rear wheels.
3. Remove the parking brake cable bracket bolts.
4. Use 2 floor jacks or an axle cradle to support the weight of the axle. Do not try to balance the axle on 1 jack.
5. Disconnect and remove the shock absorber lower mounting nuts, washers and bushings from the lower leaf spring brackets.
6. Remove the nuts, washers and U-bolts attaching the springs to the axle housing. Remove the seat and spacer.
7. Remove the spring shackle bolts and shackle from the rear of the spring, then remove the spring front bolt.
8. Remove the springs from the vehicle.
9. When reinstalling, the retaining bolts and shackles must be installed from the outside of the vehicle to the inside. (Threaded end towards the center of the vehicle.) Make sure the rubber bushings are in place and in good condition, replacing any that are compressed or damaged. Tighten the nuts on the rear shackle and front bolt just snug. They will be final tightened later.
10. Install the U-bolts and bottom plate, making sure the hole in the bottom plate aligns with the bolt through the spring leaves. Tighten the U-bolts nuts evenly to 72-87 ft. lbs. (100-120 Nm).
11. Install the shock absorber, tightening the mounting bolts just snug.
12. Remove the axle support apparatus and install the parking brake cable retaining brackets.
13. Install the wheel and lower the vehicle to the ground.
14. Bounce the rear of the vehicle once or twice to stabilize the suspension. Tighten the shock absorber mounting nuts to 13-18 ft. lbs. (18-25 Nm).
15. Tighten the front spring retaining nut and bolt to 104 ft. lbs. (140 Nm).
16. Tighten the rear spring shackle nuts to 33-43 ft. lbs. (45-60 Nm).

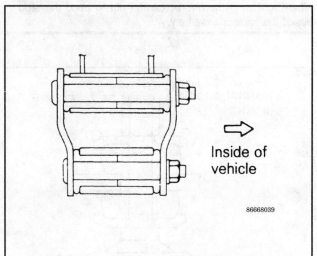

Fig. 40 Install the rear leaf spring shackles so that the mounting nuts are toward the inside of the vehicle

SUSPENSION AND STEERING

Shock Absorbers

REMOVAL & INSTALLATION

♦ See Figure 41

1. Loosen the lug nuts on the wheels where the shock absorbers are going to be removed. Only loosen the lug nuts a maximum of $\frac{1}{2}$ of a turn.
2. Raise and support the vehicle with jackstands under the vehicle's frame.
3. Remove the rear wheel(s).
4. Jack the rear axle up slightly, then remove the upper mounting nuts from the shock absorber. On the leaf spring rear-ends, pull the shock absorber upper mount off of the mounting stud. Make sure to note the locations of the various washers, bushings and nuts.
5. Lower the rear axle. On coil spring rear-ends, this will pull the upper shock absorber mounting stud out of the mounting bracket. Remove the lower mounting bolt from the shock absorber.
6. Remove the shock absorber from the vehicle.
7. To install, mount the shock absorber with the lower mounting bolt. Only tighten the lower bolt until snug.
8. Raise the rear axle with the floor jack until the shock absorber upper mounting stud or boss is in line with the mounting component. On leaf spring rear-ends, only tighten the upper mounting bolt until snug. On coil spring rear-ends, tighten the upper shock absorber mounting nut so that the amount of threads showing above the second nut is 0.04-0.08 in. (1-2mm). Make sure the two uppermost of the two upper mounting nuts is tightened down onto the lower nut to 11 ft. lbs. (15 Nm).

✳✳WARNING

When tightening the nut, be careful not to bend the stud pin of the washer assembly.

9. Lower the rear axle, install the wheel and lower the vehicle to the ground.
10. Push the vehicle down a couple of times to settle the suspension.
11. On coil spring rear-ends, tighten the lower mounting nut to 80-94 ft. lbs. (110-130 Nm) for 1989-91 Monteros and to 159-181 ft. lbs. (216-245 Nm) for 1992-95 Monteros. On leaf spring rear-ends, tighten the upper and lower mounting nuts to 13-18 ft. lbs. (18-25 Nm).

Rear Stabilizer Bar (Sway Bar)

REMOVAL & INSTALLATION

Coil Spring Rear Suspension
♦ See Figures 42, 43 and 44

1. Loosen the rear wheel lug nuts $\frac{1}{2}$ of a turn.
2. Raise and support the rear of the vehicle on jackstands.
3. Remove the rear wheels.
4. Disconnect the parking brake cable attaching bolts.
5. Remove the rear differential lock position harness attaching bolt and the rear sensor attaching bolt (vehicles with ABS), if so equipped.
6. Support the rear axle with a floor jack. Disconnect the lower shock absorber mounting bolts.
7. Remove the sway bar mounting brackets and bushings from the rear axle assembly.
8. Loosen, then remove the nuts, washers and bushings from the two front ends of the stabilizer bar. Make certain that the order of the various parts is known for reassembly.
9. Slowly lower the rear axle being aware that the rear coil springs may fall out of their mounting positions. If this happens it is no big problem, simply refer to the rear coil spring installation procedures after the stabilizer bar has been installed back into the vehicle. Remove the stabilizer bar from the right side of the vehicle.

✳✳WARNING

When lowering the jack, take care not to damage the rear brake pipe between the main brake pipe and the rear axle housing.

10. To install, place the sway bar in the vehicle and assemble the joint cups and rubber bushings in the order they were removed. Install the end link nut on the stabilizer bar mounting bolt and tighten until 0.59-0.67 in. (15-17mm) of threads protrudes past the top nut.
11. Install the mounting bracket bushings to the sway bar, then tighten the brackets to the rear axle. Tighten the bracket bolts to 25 ft. lbs. (34 Nm).
12. If the coil springs fell out during removal of sway bar, install them now — refer to the coil spring procedures in this section.
13. Install the shock absorber lower mounting bolts only until snug.
14. Install and tighten the rear sensor attaching bolt, the rear differential lock position harness attaching bolt and the parking brake cable attaching bolts.

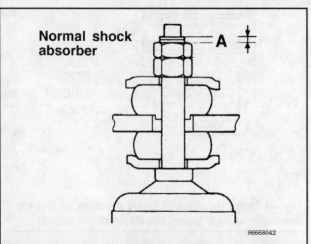

Fig. 41 On the coil spring rear-end Monteros, install the upper shock absorber washers, bushings and nuts in the correct order

SUSPENSION AND STEERING   8-25

15. Install the wheels and lower the vehicle to the ground.
16. Push the rear of the vehicle up and down a few times to help the suspension settle. Tighten the lower shock absorber mounting bolts to the following amounts:
    - 1989-91 Monteros — 80-94 ft. lbs. (110-130 Nm)
    - 1992-95 Monteros — 159-181 ft. lbs. (215-245 Nm)

Rear Wheel Alignment

All Mitsubishi models covered by this manual have rear suspension systems on which bent or damaged components can cause unusual tire wear and/or handling problems. However, no adjustments can be made to the rear alignment; the critical dimensions are determined by correct installation and location of components.

Even though the rear alignment cannot be adjusted on these vehicles, this does not mean it should not be checked occasionally. The dimensions of toe and camber can be easily read on an alignment rack and compared to the charts available in a reputable alignment shop. This is a handy, low-cost diagnostic tool when investigating tire wear at the rear. Additionally, the rear alignment should be checked after every heavy impact involving the rear wheels or suspension. Finding and replacing a bent component quickly can save the cost of two tires and eliminate the risk of bad handling and braking.

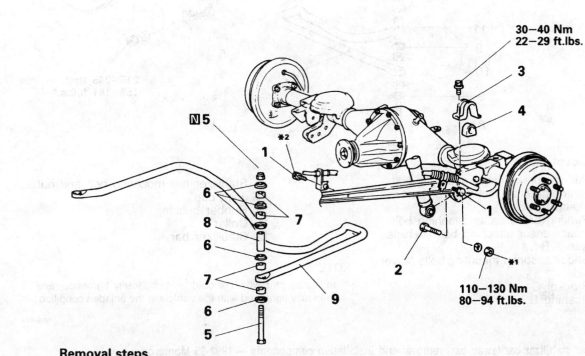

Removal steps
1. Parking brake cable attaching bolt
2. Shock absorber mounting bolts (Lower side)
3. Bracket C
4. Bushing B
5. Stabilizer bar mounting bolt and nut
6. Joint cup
7. Rubber bushing
8. Collar
9. Stabilizer bar

NOTE
(1) *¹ : Tighten when the vehicle is unloaded.
(2) The part with *² is available only for the 4-door models.

Fig. 42 Rear stabilizer bar (sway bar) removal and installation components — 1989-91 Monteros

8-26 SUSPENSION AND STEERING

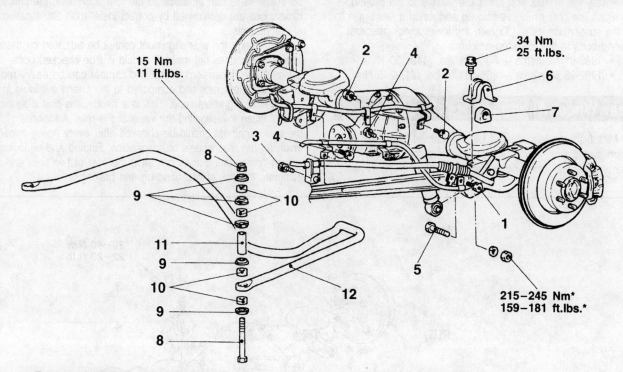

Removal steps

1. Parking brake cable attaching bolt
2. Rear differential lock position harness attaching bolt
3. Parking brake cable attaching bolt
4. Rear sensor attaching bolt (Vehicles with A.B.S.)
5. Shock absorber mounting bolts (lower side)
6. Bracket C
7. Bushing B
8. Stabilizer bar mounting bolt and nut
9. Joint cup
10. Rubber bushing
11. Collar
12. Stabilizer bar

NOTE
*: Indicates part which should be temporarily tightened, and then fully tightened with the vehicle in the unladen condition.

Fig. 43 Rear stabilizer bar (sway bar) removal and installation components — 1992-95 Monteros

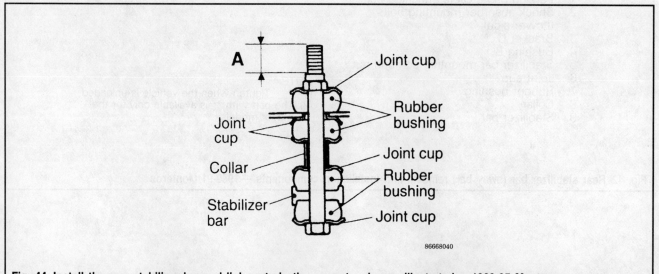

Fig. 44 Install the rear stabilizer bar end link parts in the correct order, as illustrated — 1989-95 Monteros

SUSPENSION AND STEERING 8-27

STEERING

Steering Wheel

REMOVAL & INSTALLATION

Vehicles Without Air Bags
▶ See Figures 45 and 46

1. Disconnect the negative battery cable from the battery.
2. Pull off the horn cover at the center of the wheel by grasping the upper edge with your fingers to release it. On some models the pad is screwed on from behind the wheel. Disconnect the horn wire connector.
3. Remove the steering wheel retaining nut. Matchmark the relationship between the wheel and shaft.
4. Screw the two bolts of a steering wheel puller into the wheel. Turn the bolt at the center of the puller to force the wheel off the steering shaft. Do not pound on the wheel to remove it, or the collapsible steering shaft may be damaged.
5. Install the wheel and push it onto the shaft splines by hand far enough to start the retaining nut.
6. Install the retaining nut and torque it to 30 ft. lbs. (40 Nm).
7. Connect the horn wire and reinstall the horn pad. If it is the snap-on type, make sure all the clips are engaged.
8. Connect the negative cable to the battery.

Vehicles With Air Bags

For steering wheel removal and installation procedures covering vehicles with a Supplemental Restraint System (SRS), refer to Section 6. The SRS is a potentiallly dangerous system

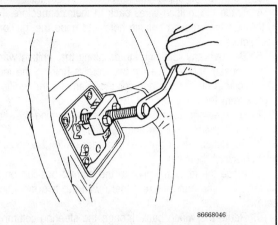

Fig. 46 Use a puller to remove the steering wheel from the steering column — do not hammer on the steering column, it will damage the collapsible steering assembly

to work on; read the warnings and precautions in Section 6 before proceeding.

✶✶CAUTION

Before removing the steering wheel of a vehicle equipped with an air bag, read the Supplemental Restraint System precautions and disarming/arming procedures in Section 6.

Turn Signal and Wiper Switch (Column Switch)

REMOVAL & INSTALLATION

Refer to the instrument and switch part of Section 6 for the removal and installation procedures for this component.

Ignition Switch

The removal and installation procedures are essentially the same for all models of the Pick-up and Montero. The 1992-95 Montero also has a key hole illumination ring, a key hole illumination light timer, and a buzzer assembly.

The key hole illumination ring is located on the tip of the steering lock cylinder. The key hole illumination light timer is mounted to the right-hand side center reinforcement, behind the center console. The buzzer assembly is mounted to the left-hand side center reinforcement.

REMOVAL & INSTALLATION

▶ See Figures 47, 48 and 49

1. Disconnect the negative (-) battery cable from the battery.

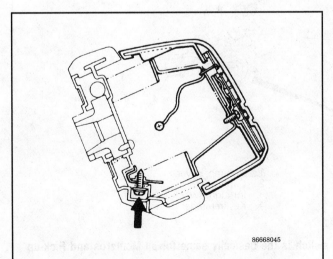

Fig. 45 Some models have their horn pads retained with a screw located in the rear of the steering wheel

8-28 SUSPENSION AND STEERING

2. On 1992-95 Monteros, remove the instrument under cover.

3. Remove the upper and lower column covers.

4. Trace the ignition wires back to their connectors. Reach in through the steering column hole and undo the ignition wire connectors.

5. Remove any retaining straps along the ignition wires.

6. Loosen and remove the two screws holding the ignition switch clamp on the steering column. Remove the ignition switch from the column.

7. Unscrew the screws holding the key reminder switch to the ignition switch housing and remove it.

8. To install, screw the key reminder switch back onto the ignition switch housing.

9. Place the ignition switch in the correct position on the steering column and secure it there with the two mounting screws.

10. Route the wiring back through the steering column hole and plug the connectors back together.

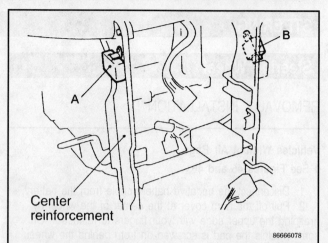

Fig. 49 The 1992-95 Monteros have a buzzer (A) and a key hole illumination timer (B) mounted to the center reinforcements

11. Secure the ignition wires to the steering column with new wire straps.

12. Install the upper and lower column covers.

13. On 1992-95 Monteros, install the instrument under cover.

14. Connect the battery cable back to the battery.

TESTING

▶ See Figures 50, 51, 52, 53, 54 and 55

1. Disconnect the wiring plug from the ignition switch, and connect an ohmmeter to the switch side plug.

2. Operate the switch, and check the continuity between the terminals.

3. It may be necessary to remove and reinsert the ignition key, and check the operation of the key reminder switch.

4. If the ignition switch fails any of the tests, replace it with a new switch.

1 Ignition switch
2 Key reminder switch

Fig. 47 The ignition switch assembly is, except for a few minor additions, essentially the same for all Monteros and Pick-ups

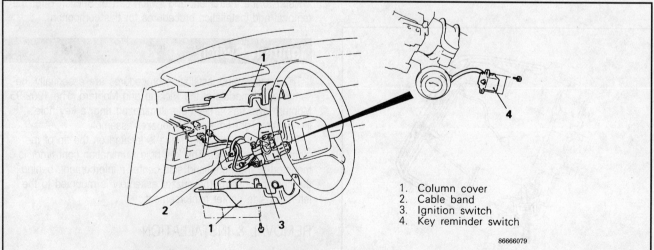

1. Column cover
2. Cable band
3. Ignition switch
4. Key reminder switch

Fig. 48 The procedure for removing and installing the ignition switch is the basically same for all Monteros and Pick-up trucks

SUSPENSION AND STEERING 8-29

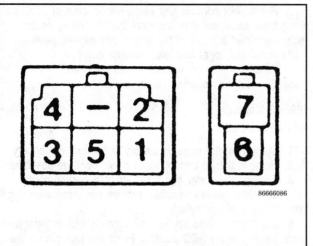

Fig. 50 The ignition switchside connector's terminal configuration — 1983-86 Monteros and Pick-ups

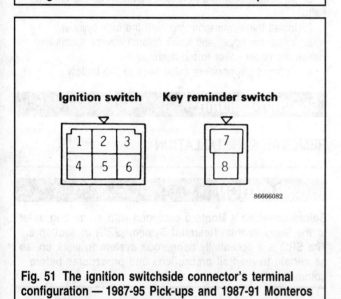

Fig. 51 The ignition switchside connector's terminal configuration — 1987-95 Pick-ups and 1987-91 Monteros

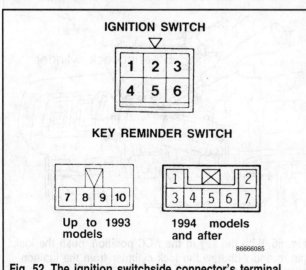

Fig. 52 The ignition switchside connector's terminal configuration — 1992-95 Monteros

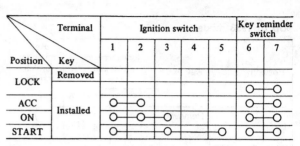

Fig. 53 Check the continuity between the terminals when the ignition switch is in the various positions — 1983-86 Monteros and Pick-ups

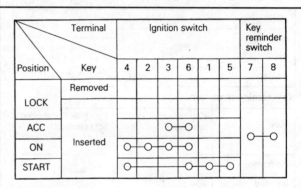

Fig. 54 The continuity table shows which terminals should be checked and which position the ignition switch should be in — 1987-95 Pick-ups and 1987-91 Monteros

Ignition Lock

REMOVAL & INSTALLATION

1983-95 Pick-up and 1983-91 Montero

1. Disconnect the negative battery cable.
2. If the vehicle is equipped with a tilt steering column, put the column in its lowest position.
3. Remove the upper and lower column covers.

➡While removal of the steering wheel is not usually required, it makes the job much easier.

4. Remove the wiring harness bands or clips. Disconnect the ignition switch harness.
5. If the ignition switch (electrical part) can be removed from the lock, remove it. Older vehicles use one-piece units.

8-30 SUSPENSION AND STEERING

Up to 1993 models

Position	Key	Ignition switch terminal						Key reminder switch terminal	
		1	2	3	4	5	6	1	4
LOCK	Removed								
ACC	Inserted			O—	—	—O			O—	—O
ON			O—O—O						
START			O—	O—O—O					

1994 models and after
Ignition switch

Position	Terminal					
	1	2	3	4	5	6
LOCK						
ACC		O—				—O
ON		O—O			—O	
START		O—O—O—			—O	

Key reminder switch

Key	Key reminder switch terminal		Key hole illumination light terminal	
	4	6	1	2
Removed	O—|—O		O—⊗—O	
Inserted				

Fig. 55 The continuity inspection chart for 1992-95 Monteros — if the ignition switch fails any of the tests, replace it with a new switch

6. The installation bolts have no head (they were broken off during installation as a theft deterrent); use a hacksaw blade to cut a groove into the head of the special mounting bolts. Use a screwdriver placed in the new groove to remove the bolts.

7. Remove the lock assembly from the steering column.

To install:

8. With the key inserted in the new lock, install the lock assembly to the steering column and hand-tighten new bolts. Turn the key from time to time as the bolts are tightened, making sure the key does not bind.

9. Once the lock is in place and confirmed in working order, use a socket wrench to evenly tighten the bolts. Again check the key operation during the tightening. Continue tightening the bolts until the heads break off.

10. Connect the switch if it was removed from the lock assembly.

11. Connect the wiring harness and check the assembly for proper operation.

12. Reinstall or connect any wiring bands, clips or retainers which were loosened. It is important that the wiring be correctly contained and out of the way of any moving parts.

13. Install the upper and lower column covers.

14. Install the steering wheel if it was removed.

15. Connect the negative battery cable.

1992-95 Montero

▶ See Figure 56

1. Disconnect the negative battery cable from the battery.
2. Remove the instrument under cover from the dashboard.
3. Remove the upper and lower steering column covers.
4. On 1994 and later models, remove the illumination ring from the ignition lock cylinder.
5. Insert the ignition key into the steering lock cylinder and place the key in the ACC position. Press the lock pin on the side of the lock cylinder down with a small punch or prytool. Pull the cylinder out of the ignition switch.
6. To install, push the new lock cylinder into the ignition switch.
7. Install the illumination ring onto the lock cylinder.
8. Install the upper and lower column covers. Install the instrument under cover to the dashboard.
9. Connect the negative cable back to the battery.

Steering Column

REMOVAL & INSTALLATION

✱✱CAUTION

Before servicing a Montero equipped with an air bag, refer to the Supplemental Restraint System (SRS) in Section 6. The SRS is a potentially dangerous system to work on, so be certain to read all precautions and procedures before commencing.

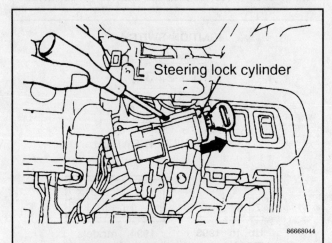

Fig. 56 With the key in the ACC position, push the lock pin in and withdraw the lock cylinder from the ignition switch — 1992-95 Monteros

SUSPENSION AND STEERING 8-31

Montero

◆ See Figures 57, 58 and 59

1. Disconnect the negative battery cable from the battery.
2. Remove the pinchbolt holding the steering joint to the steering gear box.
3. Remove the horn pad, air bag and steering wheel.
4. Remove the lower and upper column covers.
5. Remove the lap heater duct.
6. Disconnect the band or clip holding the wiring harnesses to the column.
7. Remove the column switch and disconnect its wiring harness.
8. Remove the small screws holding the ignition switch (electrical switch) to the key assembly. Disconnect the harness and remove the switch.
9. Disconnect and remove the key reminder switch.
10. Unhook the brake pedal return spring from the column bracket.
11. Remove the bolts holding the dust cover boot to the floorboard.
12. Remove the bolts holding the column bracket to the dash. Support the column while removing the last bolt and remove the column from the car.

To install:

13. When reassembling, fit the column into place and make certain the steering joint is seated on the steering gearbox. Install the bolts holding the column bracket to the dash and finger-tighten them.
14. Double check positioning and alignment of the column from end to end. When the column is correctly located, tighten the bracket bolts to 16 ft. lbs. (22 Nm).
15. Install the dust cover at the floor and apply a coating of sealant to the bolt holes from inside the vehicle. Tighten the bolts.

✱✱WARNING

Do not loosen the column tube clamp bolts. if the clamp bolts should be loosened, retighten them securely while pulling the steering shaft out fully toward the interior of the car.

16. Connect the brake pedal return spring.
17. Install the key reminder switch, the ignition electrical switch and the column switch. Route the harnesses carefully and connect each to its mate.
18. Install new bands or clips to hold the cables in place.
19. Install the lap heater duct.
20. Replace the upper and lower steering column covers.
21. Install the steering wheel and horn pad — refer to procedures in this section.
22. Install the pinchbolt at the steering gearbox and tighten the bolt to 22-25 ft. lbs. (29-34 Nm).
23. Connect the negative cable to the battery.

Pick-up

◆ See Figures 58, 59 and 60

1. Disconnect the negative battery cable from the battery.
2. Disconnect the pinchbolt holding the steering shaft to the steering gear box.
3. Remove the small screw holding the horn pad to the wheel and remove the horn pad by pulling the lower part toward you.
4. Remove the steering wheel assembly — refer to procedures in this section.
5. Remove the lower and upper steering column covers.
6. Remove the column switch and disconnect the harness connector.
7. Unplug the wiring harness to the ignition switch.
8. On vehicles with automatic transmissions, disconnect the wiring to the overdrive **OFF** switch.
9. If equipped with automatic transmission, disconnect the gear shift control cable at the column.
10. Unhook the brake pedal return spring from the steering column bracket.
11. Remove the small bolts holding the dust cover boot to the floorboard.
12. Remove the 4 bolts holding the column bracket to the dashboard frame. Support the column and remove it from the vehicle.

To install:

13. When reassembling, fit the column into place and make certain the steering joint is seated on the steering gearbox. Install the bolts holding the column bracket to the dash and finger-tighten them.
14. Double check positioning and alignment of the column from end-to-end. When the column is correctly located, tighten the bracket bolts to 6-9 ft. lbs. (8-12 Nm).
15. Install the dust cover at the floor and apply a coating of sealant to the bolt holes from inside the vehicle, then insert and tighten the bolts. Do not overtighten these bolts.
16. Connect the brake pedal return spring.
17. For vehicles with automatic transmission, connect the gear shift selector cable. Install the overdrive switch and connect its harness.
18. Connect the ignition switch harness.
19. Install the column switch and connect the wiring connector.
20. Install the upper and lower steering column cover.
21. Install the steering wheel and horn pad assemblies.
22. Install the pinchbolt, connecting the steering column to the gear box, and tighten to 11-14 ft. lbs. (15-20 Nm) for rear wheel drive Pick-ups, and to 22-25 ft. lbs. (30-35 Nm) for 4-wheel drive Pick-ups.
23. Connect the negative cable to the battery.

Steering Linkage

REMOVAL & INSTALLATION

◆ See Figures 61, 62, 63 and 64

Pitman Arm

➡ It is possible to remove the Pitman arm with the steering box on the car. The job is much easier if the steering gearbox is removed.

1. Elevate and safely support the vehicle on jackstands.
2. Remove the left front wheel.

8-32 SUSPENSION AND STEERING

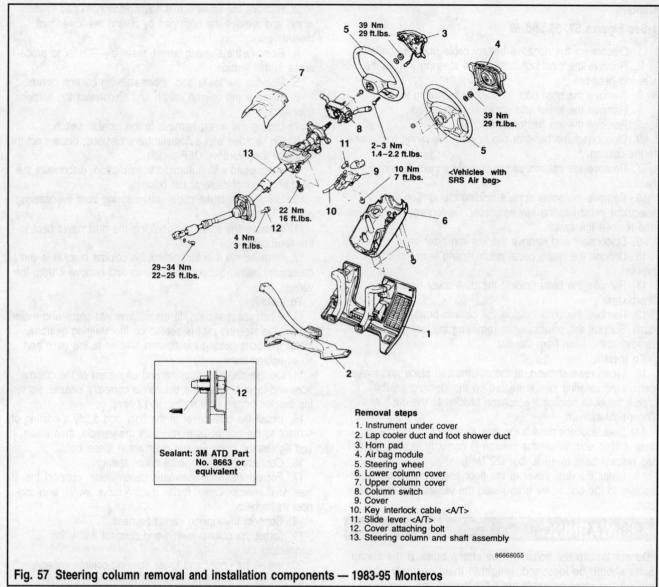

Removal steps
1. Instrument under cover
2. Lap cooler duct and foot shower duct
3. Horn pad
4. Air bag module
5. Steering wheel
6. Lower column cover
7. Upper column cover
8. Column switch
9. Cover
10. Key interlock cable <A/T>
11. Slide lever <A/T>
12. Cover attaching bolt
13. Steering column and shaft assembly

Fig. 57 Steering column removal and installation components — 1983-95 Monteros

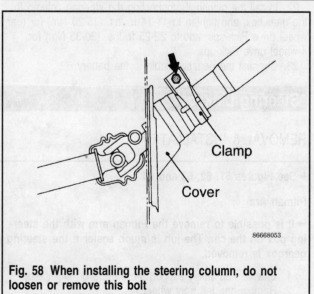

Fig. 58 When installing the steering column, do not loosen or remove this bolt

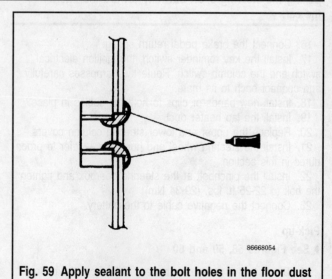

Fig. 59 Apply sealant to the bolt holes in the floor dust cover before inserting and tightening the bolts

SUSPENSION AND STEERING 8-33

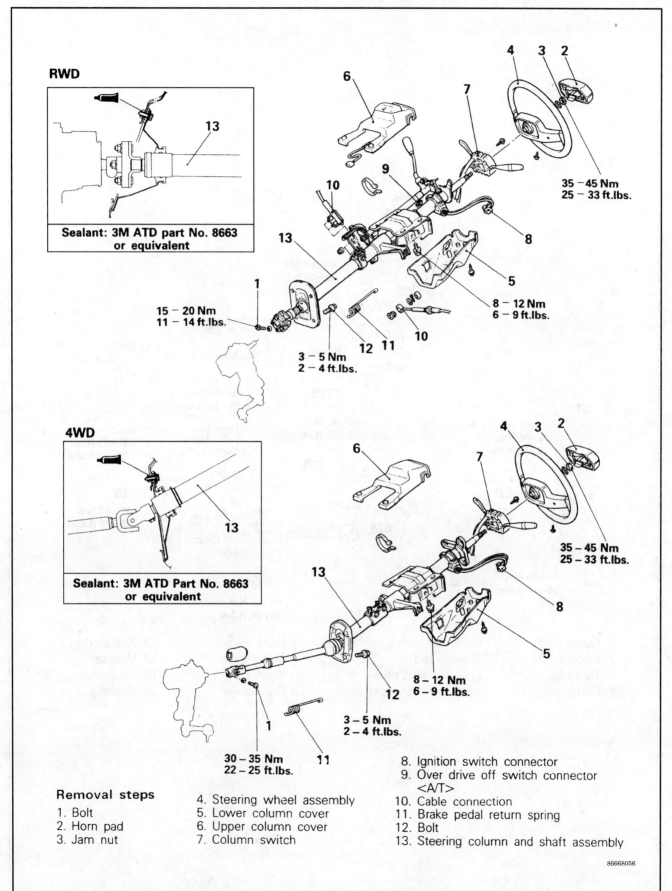

Fig. 60 Steering column removal and installation components — 1983-95 Pick-ups

8-34 SUSPENSION AND STEERING

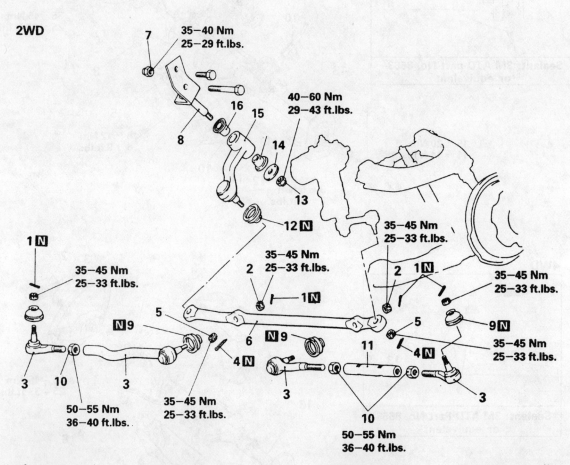

1. Cotter pin
2. Slotted nuts
3. Tie rod ends
4. Cotter pin
5. Slotted nuts
6. Relay rod
7. Self-locking nuts
8. Idler arm support
9. Dust covers
10. Nuts
11. Pipes
12. Dust cover
13. Self-locking nut
14. Washer
15. Idler arm
16. Bushing

Fig. 61 Steering linkage components and torque values for 1983-91 rear wheel drive vehicles

SUSPENSION AND STEERING 8-35

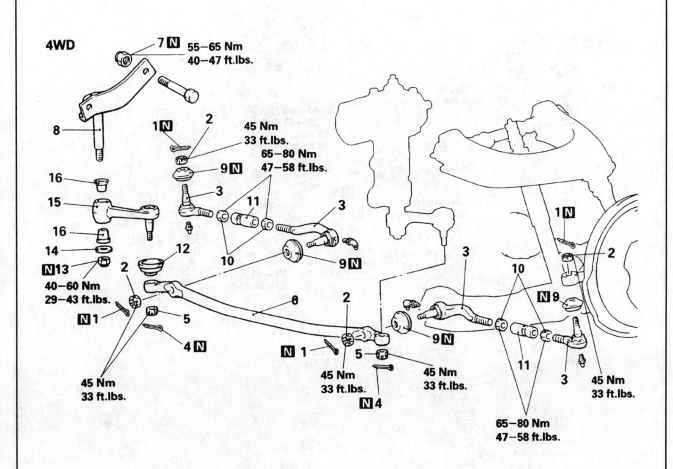

Fig. 62 Steering linkage components and torque values for 1983-91 4-wheel drive vehicles

8-36 SUSPENSION AND STEERING

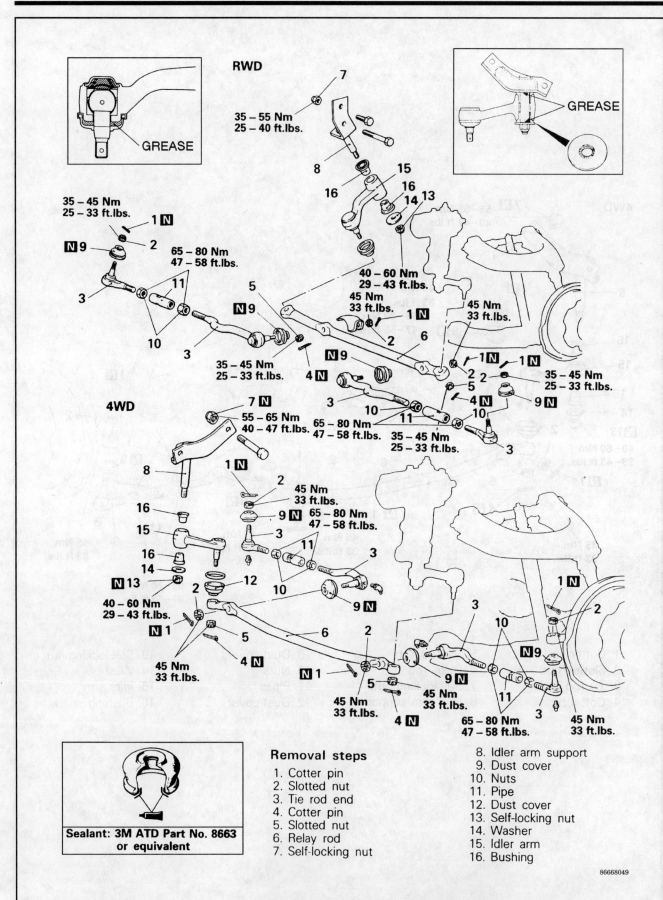

Fig. 63 Steering linkage components and torque values for 1992-95 Pick-ups and Monteros

SUSPENSION AND STEERING 8-37

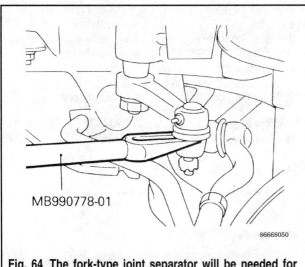

Fig. 64 The fork-type joint separator will be needed for removal of some of the front steering components

3. Remove the cotter pin and castle nut holding the Pitman arm to the relay rod.
4. Use a screw-type joint separator to pull the relay rod off the Pitman arm joint. Do not hammer on the joint and do not use a fork-type separator.
5. Remove the locknut and washer from the bottom of the steering gearbox.
6. Using a Pitman arm puller (MB 990809-01 or equivalent), remove the arm from the steering box.
7. When installing the arm, note that there are matchmarks on both the arm and the gearbox output shaft. These marks MUST be aligned at reinstallation.
8. Install the locknut and washer to the steering pump shaft. Tighten the nut to 190 ft. lbs. (140 Nm).
9. Fit the Pitman arm to the relay rod joint. Use a new boot. Install a new castle nut, tightening it to 30 ft. lbs. (40 Nm). Install a new cotter pin.
10. Install the left wheel. Lower the vehicle to the ground.

Idler Arm

1. Elevate and safely support the vehicle on stands.
2. Remove the front wheels.
3. Remove the cotter pin and the castle nut holding the relay rod to the idler arm.
4. Use a screw-type joint separator to split the ball joint at the relay rod.
5. Remove the nut holding the idler arm to its pivot. Remove the arm, noting the placement of the spacers, washers and the plastic bushings. Alternatively, the idler arm pivot may be unbolted from the vehicle and the arm disassembled on the workbench.
6. Remove the dust cover and O-ring from the end of the tie rod. Apply grease (Multipurpose grease SAE J310, NLGI No. 2) to the lip portion of the dust cover and inside the cover. Apply 3M® ART Part No. 8663 or the equivalent to the dust cover installation surface, then press it in.
7. Apply grease to the inside surface of the bushing and the idler arm support shaft. Insert the bushing into the idler arm, then insert the idler arm support into the idler arm. Insert so that the knurled surface of the washer is facing the bushing side. Tighten the self locking nut to 36 ft. lbs. (50 Nm).

8. If the entire pivot and bracket was removed, tighten the mounting bolts to 40 ft. lbs. (55 Nm).
9. Connect the idler arm to the relay rod and install the nut on the ball stud. Tighten the nut to 33 ft. lbs. (45 Nm); install a new cotter pin.
10. Lubricate the front end components.
11. Install the wheels and lower the vehicle to the ground.

Center Link

Mitsubishi refers to this component as the Relay Rod.
1. Disconnect the tie-rod ends from the relay rod on each side.
2. Disconnect the relay rod at the Pitman arm, using a screw-type puller to separate the joint.
3. Position a jack or jackstand under the relay rod at the Pitman arm to support it; allowing the linkage to hang free may damage one of the joints.
4. Disconnect the relay rod from the idler arm.
5. Remove the linkage assembly from the car.
6. Individual pieces of the linkage may be separated on the workbench. Do not reuse any dust cover, O-ring or cotter pin.

To install:
7. Reassemble the linkage and position it under the car, using the jack to assist you.
8. Install the joint to the idler arm first, then connect the Pitman arm joint and install the tie rod ends to the knuckle. Each castellated nut should be tightened to 33 ft. lbs. (45 Nm) and be secured with a new cotter pin.
9. Lubricate the front end steering components.

Tie Rod Ends

▶ See Figures 65 and 66

1. Raise the vehicle and support it safely on jackstands.
2. Remove the wheel on the same side as the tie rod end which needs to be replaced.
3. Remove the cotter pin from the tie rod end. Loosen, but do not remove, the nut on the tie rod.
4. Use a fork-type tie rod separator, remove the tie rod from the steering knuckle. Remove the nut and pull the outer tie rod out of the knuckle. Use a screw-type joint separator to remove the inner tie rod end from the relay rod (center link).
5. Loosen the sleeve clamp nut, if equipped, and unscrew the tie rod end from the sleeve or inner tie rod. Count the number of turns required to remove the end, or measure, before removal, the amount of threads protruding from the sleeve. Do the same for the inner tie rod end.
6. If both ends of the tie rod were removed, there is a third way to measure the correct location of the tie rods. Install the tie rods into the sleeve and continue to tighten until the distance between the centers of each tie rod end is equal to the following dimensions:
   • 1983-91 Rear wheel drive vehicles — 14.8-14.9 in. (37.6-37.8cm)
   • 1992-95 Rear wheel drive vehicles — 11.988-12.067 in. (30.45-30.65cm)
   • 1983-95 4-wheel drive vehicles — 10.9-11.0 in. (27.7-27.9cm)
7. Install the new end and turn it in the same number of turns or until the same amount of threads is left protruding from the sleeve. This will approximate the original setting.

8-38 SUSPENSION AND STEERING

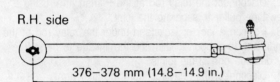

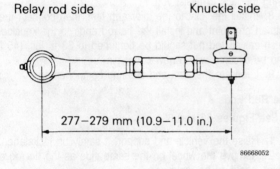

Fig. 65 Tie rod end measurements for reassembly — 1983-91 vehicles

Fig. 66 Tie rod end measurements for reassembly — 1992-95 vehicles

8. Install the tie rod joint to the knuckle. Tighten the nut to 25-33 ft. lbs. (35-45 Nm). Install a new cotter pin.
9. Install the inner tie rod end to the relay rod, then tighten to the castle nut to 33 ft. lbs. (45 Nm). Install a new cotter pin.
10. Lubricate the front end.
11. Install the wheels and lower the vehicle to the ground.
12. Have the front end aligned at an automotive shop.

Manual Steering Gear Box

REMOVAL & INSTALLATION

Pick-up
♦ See Figures 67 and 68

1. Disconnect the negative battery cable.
2. Raise the vehicle and support it safely on jackstands.
3. Remove the wheels if additional access is desired. This is, however, not necessary.
4. Remove the pinch bolt that holds the steering column shaft to the steering gear main shaft.
5. Remove the cotter pin, castle nut and remove the steering linkage from the Pitman arm. Use a screw type puller to separate the joint — refer to the procedures in this section.
6. Remove the steering gear box mounting nuts and remove the gear from the vehicle. New mounting nuts will be needed upon installation.
7. Install the steering gear box to the frame and tighten the new mounting nuts to 25-40 ft. lbs. (35-55 Nm).
8. Connect the Pitman arm to the relay rod (center link) and tighten the nut to 33 ft. lbs. (45 Nm). Install a new cotter pin.
9. Connect the steering universal joint to the gearbox and tighten the pinchbolt to 11-14 ft. lbs. (15-20 Nm).
10. Install the wheels, if removed, and lower the vehicle to the ground.
11. Connect the negative battery cable. Have the front alignment inspected and adjusted as necessary.

Power Steering Gear Box

REMOVAL & INSTALLATION

Pick-up and Montero
♦ See Figures 68 and 69

1. Disconnect the negative battery cable.
2. Raise the vehicle and support it safely on jackstands. Remove the front wheels.
3. Fold back the dust boot covering the steering shaft joint; remove the pinch bolt that holds the steering column shaft to the steering gear box.
4. Remove the cotter pin, castle nut and disconnect the steering linkage from the Pitman arm. Use a screw-type puller to separate the joint.
5. Disconnect and plug the fluid lines from the steering gear box.

SUSPENSION AND STEERING

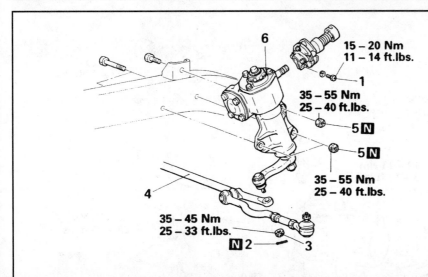

Removal steps
1. Bolt
2. Split pin
3. Slotted nut
4. Relay rod
5. Self locking nut
6. Steering gear box

Fig. 67 Manual steering gear box removal and installation — only available on Pick-up trucks

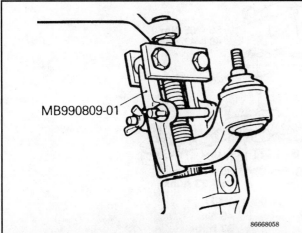

Fig. 68 Use a screw-type joint separator/puller to remove the relay rod (center link) end from the Pitman arm

6. Remove the steering gear mounting nuts and remove the gear from the vehicle. New mounting nuts will be needed for installation.
7. To install, fit the steering gearbox into position on the vehicle frame and tighten the mounting bolts to 40-47 ft. lbs. (55-65 Nm).
8. Connect the pressure and return lines to the steering gearbox. Tighten the pressure hose the 9-13 ft. lbs. (12-18 Nm).
9. Connect the Pitman arm to the relay rod (center link) and tighten the nut to 33 ft. lbs. (44 Nm). Install a new cotter pin.
10. Connect the steering column shaft to the steering gearbox and tighten the pinchbolt to 22-25 ft. lbs. (30-35 Nm) for 4-wheel drive vehicles, and to 11-14 ft. lbs. (15-20 Nm) for rear wheel drive vehicles.

11. Install the front wheels and lower the vehicle to the ground.
12. Pour power steering fluid into the reservoir to the lower mark.
13. With the coil high tension line disconnected at the coil, operate the starter intermittently (15 or 20 seconds each time) and turn the steering wheel all the way from left to right and back several times. Do this 5 or 6 times while an assistant keeps watch on the fluid level in the reservoir.

✴✴WARNING

Do not be tempted to bleed the system with the engine idling. The air in the system will be broken up and absorbed into the fluid, resulting in a noisy, foamy mess. During the bleeding procedure, replenish the fluid so that the level never falls below the lower level on the reservoir.

14. Once the preliminary bleeding is done, the fluid level should stay fairly constant as the wheel is turned while cranking the motor. When this is true, the coil wire may be reconnected and the engine started. Only run the engine at idle; high rpm will not help.
15. With the engine running, turn the wheel from lock to lock several times and observe the fluid level in the reservoir. The level should not change more than about 0.20 in. (5mm) throughout the full travel of the wheel. The fluid should not look cloudy or milky.

➡**If the system still contains air, the pump will be noisy during operation. This noisy operation can damage the pump. If the fluid level changes markedly during steering wheel movement, the system still contains air. If the fluid level rises suddenly after the engine has been shut off, the system still contains air.**

16. Inspect all the fluid line connections for any sign of leakage.
17. Have the front alignment checked and adjusted by a reputable repair facility.

8-40 SUSPENSION AND STEERING

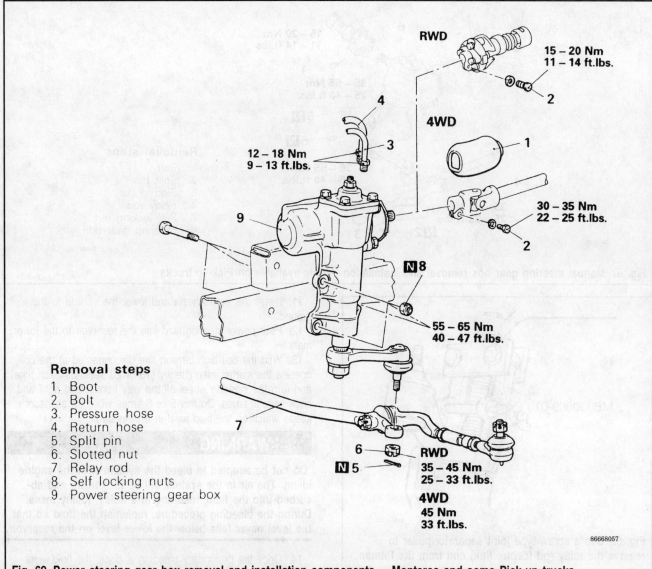

Removal steps
1. Boot
2. Bolt
3. Pressure hose
4. Return hose
5. Split pin
6. Slotted nut
7. Relay rod
8. Self locking nuts
9. Power steering gear box

Fig. 69 Power steering gear box removal and installation components — Monteros and some Pick-up trucks

Power Steering Pump

REMOVAL & INSTALLATION

▶ See Figures 70, 71 and 72

1. Disconnect the negative battery cable.
2. Remove the reservoir cap and disconnect the return (lower) hose from the fluid reservoir. Drain the fluid into a clean pan.
3. Depending on the access for your particular vehicle, raise the vehicle and support it safely on jackstands, if necessary.
4. If the pulley is to be removed from the pump, loosen the bolts now. Leave them finger-tight in place.
5. Loosen the pump mounting bolts, move the pump towards the engine and remove the belt. On 3.0L (24 valve) and 3.5L engine-equipped Monteros, loosen the drive belt tensioner adjusting bolt, then remove the belt. Turn the pump by hand to eliminate the fluid from the system.
6. Disconnect the pressure hose at the top of the pump and the suction hose at the side of the pump. Allow the fluid to drain into the container.
7. Remove the pump attaching bolts and lift the pump from the brackets.
8. When reinstalling, make certain the bracket-to-engine bolts are tight, then install the pump to the brackets.
9. If the pulley was removed, reinstall it and tighten its nuts securely.
10. Install the drive belt and adjust its tension correctly — refer to Section 1.
11. Connect the pressure and return hoses. Make sure the hoses are not twisted and do not contact any other parts so as to avoid chafing. When installing the pressure hose, make certain that the notch of the fitting contacts the suction connector of the oil pump. The pressure hose may use O-rings to help it seal, remove the old ones and install new ones before installation.
12. Pour power steering fluid into the reservoir to the lower mark, then bleed the system.

SUSPENSION AND STEERING 8-41

13. Inspect all the fluid line connections for any sign of leakage.

SYSTEM BLEEDING

1. Remove the reservoir cap and pour fluid in until it reaches the lower mark, if needed.

2. With the coil high tension line disconnected at the coil, operate the starter intermittently (15 or 20 seconds each time) and turn the steering wheel all the way from left to right and back several times. Do this 5 or 6 times while an assistant keeps watch on the fluid level in the reservoir.

⁕⁕WARNING

Do not be tempted to bleed the system with the engine idling. The air in the system will be broken up and absorbed into the fluid, resulting in a noisy, foamy mess. During the bleeding procedure, replenish the fluid so that the level never falls below the lower level on the reservoir.

3. Once the preliminary bleeding is done, the fluid level should stay fairly constant as the wheel is turned while cranking the motor. When this is true, the coil wire may be reconnected and the engine started. Only run the engine at idle; high rpm will not help.

4. With the engine running, turn the wheel from lock to lock several times and observe the fluid level in the reservoir. The level should not change more than about 0.20 in. (5mm) throughout the full travel of the wheel. The fluid should not look cloudy or milky.

➡ If the system still contains air, the pump will be noisy during operation. This noisy operation can damage the pump. If the fluid level changes markedly during steering wheel movement, the system still contains air. If the fluid level rises suddenly after the engine has been shut off, the system still contains air.

5. Shut the engine off, fill the reservoir up to the full line, if needed. Install the reservoir cap.

8-42 SUSPENSION AND STEERING

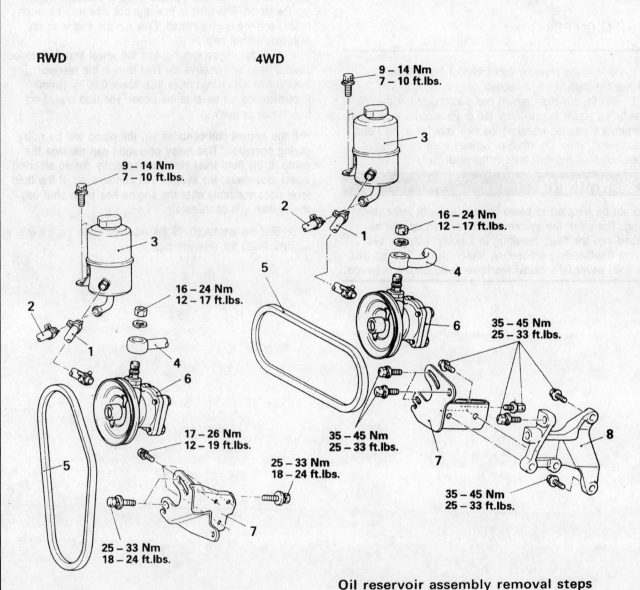

Oil reservoir assembly removal steps
1. Return hose
2. Suction hose
3. Reservoir assembly

Oil pump removal steps
1. Return hose
2. Suction hose
4. Pressure hose
5. Drive belt
6. Oil pump
7. Oil pump bracket
8. Oil pump mount bracket

Fig. 70 Power steering oil pump removal and installation components — 1983-95 Pick-up trucks and 1983-89 Monteros

SUSPENSION AND STEERING 8-43

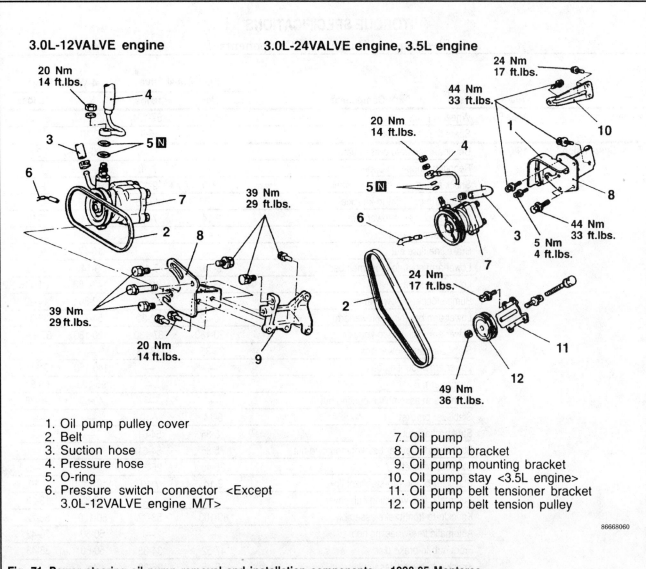

Fig. 71 Power steering oil pump removal and installation components — 1990-95 Monteros

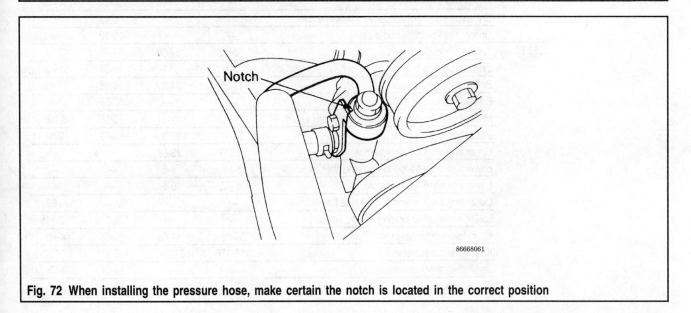

Fig. 72 When installing the pressure hose, make certain the notch is located in the correct position

8-44 SUSPENSION AND STEERING

TORQUE SPECIFICATIONS
Steering And Suspension Components ①

Item	Years	Component	Rear Wheel Drive Nm	Rear Wheel Drive ft. lbs.	4-Wheel Drive Nm	4-Wheel Drive ft. lbs.
Wheels	1983-95	Wheel lug nuts	120-140	87-101	120-140	87-101
		Spare tire carrier-to-body bolts	8-12	6-9	8-12	6-9
Front Suspension	1987-95	Upper arm shaft-to-crossmember	100-120	72-87	100-120	72-87
		Rebound stopper-to-upper arm	9-14	7-10	9-14	7-10
		Shock absorber-to-crossmember	12-18	9-13	12-18	9-13
		Upper arm ball joint-to-knuckle	60-90	43-65	60-90	43-65
		Shock absorber-to-lower arm	9-14	7-10	9-14	7-10
		Anchor arm locknut	—	—	40-50	29-36
		Brake line flare nut	—	—	13-17	9-12
		Lower arm shaft-to-crossmember	9-14	7-10	9-14	7-10
		Lower arm shaft ②	55-75	40-54	140-160	101-116
		Bump stopper-to-lower arm	70-85	51-61	20-30	14-22
		Lower arm ball joint-to-lower arm	30-42	22-30	54-75	39-54
		Lower arm ball joint-to-knuckle	120-180	87-130	120-180	87-130
		Stabilizer bar-to-lower arm	25-35	18-25	—	—
		Lower arm mounting bolt ②	—	—	140-160	101-116
		Anchor arm B	—	—	95-120	69-87
		Anchor arm assembly mounting nut	—	—	40-50	29-36
		Stabilizer bracket	9-14	7-10	—	—
		Strut bar locknut ②	75-85	54-61	—	—
		Strut bar-to-strut bar bracket mounting nut	75-85	54-61	—	—
		Strut bar bracket	35-45	25-33	—	—
		Stabilizer bracket-to-stabilizer link	9-14	7-10	9-14	7-10
		Automatic free-wheeling hub cover	—	—	18-35	13-25
		Knuckle-to-front brake assembly	80-100	58-72	80-100	58-72
		Automatic free-wheeling hub	—	—	50-60	36-43
		Front hub-to-brake disc	47-52	34-38	50-60	36-43
		Knuckle-to-tie rod assembly	35-45	25-33	45	33
		Right drive shaft-to-inner shaft	—	—	50-60	36-43
		Undercover-to-front suspension crossmember	—	—	10-13	7-9
		Undercover-to-under skid plate	—	—	10-13	7-9
	1983-86	Tie rod turnbuckle locknut	50-53	37-39	64-78	48-57
		Jam nut on stop bolt	20	14	69-98	51-72
		Brake disc-to-front hub	47-50	34-37	50-58	37-43
		Free-wheeling hub body assembly	—	—	50-58	37-43
		Manual free-wheeling hub cover assembly	—	—	10-13	8-10
		Lower ball joint-to-lower arm	30-41	22-30	53-73	40-54
		Upper arm shaft-to-crossmember	99-117	73-86	—	—
		Lower arm shaft flange-to-crossmember	8-11	6-8	—	—
		Lower arm shaft-to-crossmember	54-73	40-54	—	—
		Upper arm shaft-to-arm post of side frame	—	—	99-117	73-86
		Lower arm shaft-to-bracket of side frame	—	—	138-156	102-115
		Knuckle-to-upper ball joint	59-88	44-65	59-88	44-65
		Knuckle-to-lower ball joint	118-176	87-130	118-176	87-130
		Jam nut of shock absorber	—	—	12-17	9-13
		Shock absorber-to-lower arm	8-11	6-8	8-11	6-8

SUSPENSION AND STEERING 8-45

TORQUE SPECIFICATIONS
Steering And Suspension Components ①

Item	Years	Component	Rear Wheel Drive Nm	Rear Wheel Drive ft. lbs.	4-Wheel Drive Nm	4-Wheel Drive ft. lbs.
Front Suspension – (continued)	1983-86	Shock absorber-to-crossmember	12-17	9-13	—	—
		Strut bar-to-lower arm	69-83	51-61	—	—
		Strut bar-to-strut bar bracket	74-83	55-61	—	—
		Strut bar bracket	35-44	26-32	—	—
		Stabilizer bracket	8-11	6-8	8-11	6-8
		Jam nut of anchor bolt	—	—	40-49	29-36
		Front suspension crossmember	—	—	99-117	73-86
		Front suspension crossmember-to-bracket	—	—	30-41	22-30
Rear Suspension – with Leaf Springs	1987-95	Shackle assembly attaching nut ②	45-60	33-43	45-60	33-43
		Leaf spring front mounting bolt ②	120-160	87-116	120-160	87-116
		Front pin assembly attaching nut (Montero)	—	—	45-60	33-43
		Front pin assembly attaching bolt (Montero)	—	—	14-20	10-14
		Shock absorber attaching nut ②	18-25	13-18	18-25	13-18
		U-bolt attaching nut	100-120	72-87	100-120	72-87
	1983-86	Spring pin to hanger bracket	14-19	11-14	14-19	11-14
		Spring U-bolt	84-107	62-79	84-107	62-79
		Shock absorber	18-24	14-18	18-24	14-18
		Spring pin and shackle pin	45-58	33-43	45-58	33-43
Rear Suspension – with Coil Springs	1992-95	Lower arm-to-body	—	—	130-150	94-108
		Lower arm-to-rear axle housing	—	—	216-245	159-181
		Shock absorber-to-body	—	—	40-50	29-36
		Shock absorber-to-rear axle housing	—	—	216-245	158-181
		Lateral rod mounting nut	—	—	216-245	159-181
		Axle bumper-to-body	—	—	8-12	6-9
		Stabilizer bar-to-rear axle housing	—	—	30-40	22-29
	1989-91	Lower arm-to-body	—	—	130-150	94-108
		Lower arm-to-rear axle housing	—	—	190-220	137-159
		Shock absorber-to-body	—	—	40-50	29-36
		Shock absorber-to-rear axle housing	—	—	110-130	80-94
		Lateral rod mounting nut	—	—	110-130	80-94
		Axle bumper-to-body	—	—	8-12	6-9
		Stabilizer bar-to-rear axle housing	—	—	30-40	22-29
Steering Components	1983-95	Steering wheel locknut	35-45	25-33	35-45	25-33
		Column tube clamp bolt	4-6	3-4	4-6	3-4
		Socket assembly-to-yoke	20-25	14-18	—	—
		Dash panel cover	3-5	2-4	3-5	2-4
		Column shaft assembly-to-steering gear box	15-20	11-14	—	—
		Joint assembly-to-steering gear box	—	—	30-35	22-25
		Tilt bracket bolt	8-12	6-9	8-12	6-9
		Manual gear box installation	35-55	25-40	—	—
		Manual gear box-to-column shaft assembly	15-20	11-14	—	—
		Pitman arm-to-relay rod	35-45	35-45	35-45	35-45
		Manual side cover	25-35	18-25	—	—
		Manual adjusting cover locknut	110-150	80-108	—	—
		Manual locknut	25-35	18-25	—	—

8-46 SUSPENSION AND STEERING

TORQUE SPECIFICATIONS
Steering And Suspension Components ①

Item	Years	Component	Rear Wheel Drive Nm	Rear Wheel Drive ft. lbs.	4-Wheel Drive Nm	4-Wheel Drive ft. lbs.
Steering Components – (continued)	1983-95	Manual end cover	35-45	25-33	—	—
		Manual Pitman arm installation	130-150	94-108	—	—
		Power gear box installation	55-65	40-47	55-65	40-47
		Power gear box-to-column shaft assembly	15-20	11-14	—	—
		Power gear box-to-joint assembly	—	—	30-35	22-25
		Power gear box-to-pressure hose	12-18	9-13	12-18	9-13
		Power gear box-to-return hose	12-18	9-13	12-18	9-13
		Power side cover	45-55	33-40	45-55	33-40
		Power adjusting bolt locknut	30-45	22-33	30-45	22-33
		Power breather plug	3-4	2-3	3-4	2-3
		Power Pitman arm installation	130-150	94-108	130-150	94-108
		Power valve housing	45-55	33-40	45-55	33-40
		Power locknut	180-230	130-166	180-230	130-166
		Oil pump bracket-to-engine – front	25-33	18-24	35-45	25-33
		Oil pump bracket-to-engine – right side	17-26	12-19	35-45	25-33
		Oil pump bracket-to-engine – rear	25-33	18-24	35-45	25-33
		Oil pump mounting bracket	—	—	33-45	25-33
		Oil pump-to-pressure hose	16-24	12-17	16-24	12-17
		Oil reservoir assembly	9-14	7-10	9-14	7-10
		Oil pump body-to-oil pump bracket	25-33	18-24	35-45	25-33
		Connector assembly	70-80	51-58	70-80	51-58
		Pump cover	33-43	24-31	33-43	24-31
		Suction connector	14-18	10-13	14-18	10-13
		Tie rod end-to-knuckle	35-45	25-33	45	33
		Tie rod end-to-relay rod	45	33	45	33
		Tie rod end-to-pipe	65-80	47-58	65-80	47-58
		Relay rod-to-Pitman arm	35-45	25-33	45	33
		Relay rod-to-idler arm	35-45	25-33	45	33
		Idler arm-to-idler arm support	40-60	29-43	40-60	29-43
		Idler arm support-to-frame	35-40	25-29	35-40	25-29

① If the torque values of the chart differ from either the illustrations or the text procedure torque values, use the values in the illustrations or the text.
② Must be tightened while vehicle is resting on the ground.

ANTI-LOCK BRAKE SYSTEM (ABS) — FOUR WHEEL
  ELECTRONIC CONTROL UNIT (ECU) 9-49
  G-SENSOR 9-48
  HYDRAULIC UNIT 9-47
  WHEEL SPEED SENSOR 9-49
ANTI-LOCK BRAKE SYSTEM (ABS) — REAR WHEEL
  DESCRIPTION OF SYSTEM 9-52
  G-SENSOR AND SYSTEM CONTROL UNIT 9-54
  LOAD SENSING PROPORTIONING VALVE 9-55
  MODULATOR UNIT 9-54
  SPEED SENSOR 9-55
BRAKE SYSTEM
  ADJUSTMENT 9-4
  BLEND PROPORTIONING VALVE 9-13
  BRAKE HOSES AND LINES 9-15
  BRAKE LIGHT SWITCH 9-5
  BRAKE PEDAL 9-5
  BRAKE SYSTEM BLEEDING 9-15
  LOAD SENSING PROPORTIONING VALVE 9-14
  MASTER CYLINDER 9-6
  OPERATION AND SYSTEM DESCRIPTION 9-2
  POWER BRAKE BOOSTER 9-12
  VACUUM CHECK VALVE 9-15
FRONT DISC BRAKES
  BRAKE CALIPER 9-20
  BRAKE DISC (ROTOR) 9-24
  BRAKE PADS 9-18
PARKING BRAKE
  BRAKE LEVER 9-43
  BRAKE SHOES 9-43
  CABLES 9-38
REAR DISC BRAKES
  BRAKE CALIPER 9-36
  BRAKE DISC (ROTOR) 9-37
  BRAKE PADS 9-33
REAR DRUM BRAKES
  BRAKE BACKING PLATE 9-32
  BRAKE DRUMS 9-26
  BRAKE SHOES 9-26
  WHEEL CYLINDERS 9-30
SPECIFICATIONS CHARTS
  BRAKE SPECIFICATIONS 9-57
  TORQUE SPECIFICATIONS 9-58

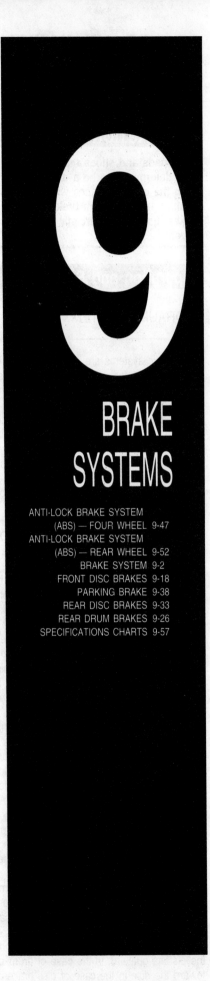

9
BRAKE SYSTEMS

ANTI-LOCK BRAKE SYSTEM (ABS) — FOUR WHEEL 9-47
ANTI-LOCK BRAKE SYSTEM (ABS) — REAR WHEEL 9-52
BRAKE SYSTEM 9-2
FRONT DISC BRAKES 9-18
PARKING BRAKE 9-38
REAR DISC BRAKES 9-33
REAR DRUM BRAKES 9-26
SPECIFICATIONS CHARTS 9-57

BRAKE SYSTEM

> **⁂⁂CAUTION**
>
> Brake pads and shoes may contain asbestos, which has been determined to be a cancer causing agent. Never clean the brake surfaces with compressed air! Avoid inhaling any dust from brake surfaces! When cleaning brakes, use a commercially available brake cleaning solvent.

Operation and System Description

HYDRAULIC SYSTEM

Hydraulic systems are used to actuate the brakes of all modern automotive vehicles. A hydraulic system rather than a mechanical system is used for two reasons. First, fluid under pressure can be carried to all parts of an automobile by small hoses — some of which are flexible — without taking up a significant amount of room or posing routing problems. Second, a great mechanical advantage can be given to the brake pedal, and the foot pressure required to actuate the brakes can be reduced by making the surface area of the master cylinder pistons smaller than that of any of the pistons in the wheel cylinders or calipers.

The master cylinder consists of a fluid reservoir and a single or double cylinder and piston assembly. Double (or dual) master cylinders are designed to separate the front and rear braking systems hydraulically in case of a leak. The master cylinder coverts mechanical motion from the pedal into hydraulic pressure within the lines. This pressure is translated back into mechanical motion at the wheels by either the wheel cylinder (drum brakes) or the caliper (disc brakes). Since these components receive the pressure from the master cylinder, they are generically classed as slave cylinders in the system.

Steel lines carry the brake fluid to a point on the vehicle's frame near each of the vehicle's wheels. The fluid is then carried to the slave cylinders by flexible tubes in order to allow for suspension and steering movements.

Each wheel cylinder contains two pistons, one at either end, which push outward in opposite directions and force the brake shoe into contact with the drum. In disc brake systems, the slave cylinders are part of the calipers. One, two or four cylinders are used to force the brake pads against the disc, but all cylinders contain one piston only. All slave cylinder pistons employ some type of seal, usually made of rubber, to minimize the leakage of fluid around the piston. A rubber dust boot seals the outer end of the cylinder against dust and dirt. The boot fits around the outer end of either the piston or the brake actuating rod.

When at rest the entire hydraulic system, from the piston(s) in the master cylinder to those in the wheel cylinders or calipers, is full of brake fluid. Upon application of the brake pedal, fluid trapped in front of the master cylinder piston(s) is forced through the lines to the slave cylinders. Here it forces the pistons outward, in the case of drum brakes, and inward toward the disc in the case of disc brakes. The motion of the pistons is opposed by return springs mounted outside the cylinders in drum brakes, and by internal springs or spring seals, in disc brakes.

Upon release of the brake pedal, a spring located inside the master cylinder immediately returns the master cylinder piston(s) to the normal position. The pistons contain check valves and the master cylinder has compensating ports drilled within it. These are uncovered as the pistons reach their normal position. The piston check valves allow fluid to flow toward the wheel cylinders or calipers as the pistons withdraw. As the return springs force the brake pads or shoes into the released position, the excess fluid in the lines is allowed to re-enter the reservoir through the compensating ports.

Dual circuit master cylinders employ two pistons, located one behind the other, in the same cylinder. The primary piston is actuated directly by mechanical linkage from the brake pedal. The secondary piston is actuated by fluid trapped between the two pistons. If a leak develops in front of the secondary pistons, it moves forward until it bottoms against the front of the master cylinder, and the fluid trapped between the pistons will operate the rear brakes. If the rear brakes develop a leak, the primary piston will move forward until direct contact with the secondary piston takes place, and it will force the secondary piston to actuate the front brakes. In either case, the brake pedal moves farther when the brakes are applied and less braking power is available.

All dual-circuit systems incorporate a switch which senses either line pressure or fluid level. This system will warn the driver when only half of the brake system is operational.

In some disc brake systems, this valve body also contains a metering valve and, in some cases, a proportioning valve. The metering valve keeps pressure from traveling to the disc brakes on the front wheels until the brake shoes on the rear wheels have contacted the drum, ensuring that the front brakes will never be used alone. The proportioning valve controls the pressure to the rear brakes avoiding rear wheel lock-up during very hard braking.

DISC BRAKES

> **⁂⁂CAUTION**
>
> Brake pads may contain asbestos, which has been determined to be a cancer causing agent. never clean the brake surfaces with compressed air! Avoid inhaling any dust from any brake surface! When cleaning brake surfaces, use a commercially available brake cleaning fluid.

Instead of the traditional expanding brakes that press outward against a circular drum, disc brake systems employ a cast iron disc with brake pads positioned on either side of it. An easily seen analogy is the hand brake arrangement on a bicycle. The pads squeeze onto the rim of the bike wheel, slowing its motion. Automobile disc brakes use the identical principal but apply the braking effort to a separate disc instead of the wheel.

BRAKE SYSTEMS

The disc or rotor is a one-piece casting mounted just inside the wheel. Some discs are one solid piece while others have cooling fins between the two braking surfaces. These vented rotors enable air to circulate between the braking surfaces cooling them quicker and making them less sensitive to heat buildup and fade. Disc brakes are only slightly affected by dirt and water since contaminants are thrown off by the centrifugal action of the rotor or scraped off by the pads. Also, the equal clamping action of the two brake pads tend to ensure uniform, straight-line stops, although unequal application of the pads between the left and right wheels can cause a vicious pull under braking. All disc brakes are inherently self-adjusting.

There are three general types of disc brakes:
1. Fixed caliper
2. Sliding caliper
3. Floating caliper

The fixed caliper design uses two pistons mounted on either side of the rotor (in each side of the caliper). The caliper is mounted rigidly and does not move. This is a very efficient brake system but the size of the caliper and its mounts adds weight and bulk to the car.

The sliding and floating designs are quite similar. In fact, these two types are often lumped together. In both designs, one pad is moved into contact with the rotor by hydraulic force. The caliper, which is not held in a fixed position, moves slightly on its mount, bringing the other pad into contact with the rotor. There are various methods of attaching floating calipers. Some pivot at the bottom or top, and some slide on mounting bolts. Many uneven brake wear problems can be caused by dirty or seized slides and pivots.

DRUM BRAKES

> **✼✼CAUTION**
>
> Brake shoes may contain asbestos, which has been determined to be a cancer causing agent. never clean the brake surfaces with Compressed air! Avoid inhaling any dust from any brake surface! When cleaning brake surfaces, use a commercially available brake cleaning fluid.

Drum brakes employ two brake shoes mounted on a stationary backing plate. These shoes are positioned inside a circular cast iron drum which rotates with the wheel. The shoes are held in place by springs; this allows them to slide toward the drum (when they are applied) while keeping the linings and drums in alignment.

The shoes are actuated by a wheel cylinder which is mounted at the top of the backing plate. When the brakes are applied, hydraulic pressure forces the wheel cylinder's two actuating links outward. Since these links bear directly against the top of the brake shoes, the tops of the shoes are then forced outward against the inside of the drum. This action forces the bottoms of the two shoes to contact the brake drum by rotating the entire assembly slightly (known as servo action). When the pressure within the wheel cylinder is relaxed, return springs pull the shoes away from the drum.

Most modern drum brakes are designed to self-adjust during application when the vehicle is moving in reverse. This motion causes both shoes to rotate very slightly with the drum, rocking an adjusting lever and thereby causing rotation of the adjusting screw via a star wheel. This on-board adjustment system reduces the need for maintenance adjustments but most drivers don't back up enough to keep the brakes properly set.

POWER BOOSTER

Power brakes operate just as non-power brake systems except in the actuation of the master cylinder pistons. A vacuum diaphragm is located on the front of the master cylinder and assists the driver in applying the brakes, reducing both the effort and travel he must put into moving the brake pedal.

The vacuum diaphragm housing is connected to the intake manifold by a vacuum hose. A check valve is placed at the point where the hose enters the diaphragm housing, so that during periods of low manifold vacuum brake assist vacuum will not be lost.

Depressing the brake pedal closes off the vacuum source and allows atmospheric pressure to enter on one side of the diaphragm. This causes the master cylinder pistons to move and apply the brakes. When the brake pedal is released, vacuum is applied to both sides of the diaphragm, and return springs return the diaphragm and master cylinder pistons to the released position. If the vacuum fails, the brake pedal rod will butt against the end of the master cylinder actuating rod, and direct mechanical application will occur as the pedal is depressed.

The hydraulic and mechanical problems that apply to conventional brake systems also apply to power brakes, and should be checked for if the tests below do not reveal the problem. **Test for a system vacuum leak as described below:**

1. Operate the engine at idle without touching the brake pedal for at least one minute.
2. Turn off the engine, and wait one minute.
3. Test for the presence of assist vacuum by depressing the brake pedal and releasing it several times. Light application will produce less and less pedal travel, if vacuum was present. If there is no vacuum, air is leaking into the system somewhere.

Test for system operation as follows:

4. Pump the brake pedal (with engine off) until the supply vacuum is entirely gone.
5. Put a light, steady pressure on the pedal.
6. Start the engine, and operate it at idle. If the system is operating, the brake pedal should fall toward the floor if constant pressure is maintained on the pedal.

Power brake systems may be tested for hydraulic leaks just as ordinary systems are tested.

9-4 BRAKE SYSTEMS

Adjustment

DISC BRAKES

Disc brakes are inherently self-adjusting, and set themselves automatically after linings are replaced when the brake pedal is applied.

DRUM BRAKES

Drum brakes do not require adjustment unless self adjusters give problems or linings are replaced. Adjustment for those models with rear drum brakes is given below.

With the vehicle safely supported and the rear wheels removed, turn the star wheel on the adjuster to adjust the outer diameter of the brake shoes. Measure accurately across the diameter of the outer brake surfaces. Correct distance is 9.96-9.98 in. (25.30-25.35cm) for both Montero and Pick-up. Notice that it is necessary to measure accurately to the tenth of a millimeter; use the correct tool.

Make certain the parking brake lever is fully released and the cable is not applying tension. Install the brake drum properly. Apply the parking brake via the lever and the service brakes via the pedal alternately until the shoes are adjusted. The "feel" of the pedal and the lever should remain constant.

BRAKE PEDAL

▶ See Figures 1, 2, 3, 4 and 5

1. Measure the distance (height) from the top surface of the floor, not including the carpet, to the top surface of the brake pedal. Correct distance is:
   - 1983-91 Montero: 7.51-7.71 in. (19.1-19.6cm)
   - 1992-95 Montero: 7.3-7.51 in. (18.6-19.1cm)
   - Pick-up: 6.54-6.73 in. (16.6-17.1cm)
2. If the pedal height is not correct, move the brake light switch out of contact with the pedal arm.
3. Loosen the lock nut on the pedal rod. Use pliers to turn the pedal rod, either lengthening it or shortening it, until the correct pedal height is obtained.
4. Tighten the locknut. Reposition the brake light switch so that the distance between the outer case of the switch and the pedal is 0.019-0.039 in. (0.5-1.0mm). Note that the switch tab or prong must press against the pedal to keep the brake lights off.
5. Tighten the locknut on the brake light switch.
6. Check the free-play in the pedal. With the engine off, depress the brake pedal fully several times. Then check the movement in the pedal before any resistance is encountered. Correct free-play is:
   - Monteros and Pick-ups: 0.12-0.31 in. (3-8mm)
7. If there is no free-play in the pedal, check the adjustment of the brake light switch. If there is excessive free-play, look for wear or play in the clevis pin and brake pedal arm.

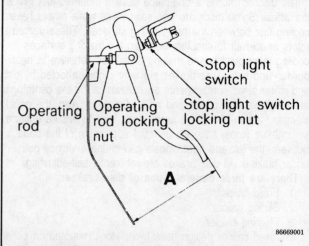

Fig. 1 The brake pedal operating components — measure the distance from the pedal to the floor (A)

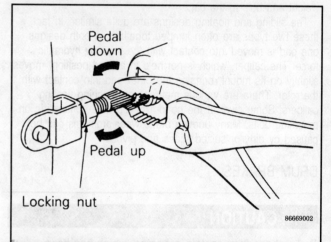

Fig. 2 Use a pair of pliers to adjust the brake pedal operating rod to adjust measurement (A) to the standard specifications

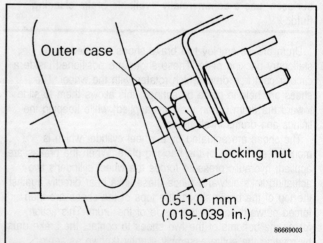

Fig. 3 Adjust the brake light switch so that there is 0.019-0.039 in. (0.5-1.0mm) between the outer case and the brake pedal

BRAKE SYSTEMS 9-5

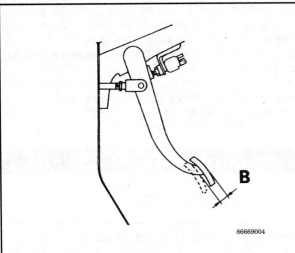

Fig. 4 Measure the distance (B) between the pedal resting position and the point when resistance is felt

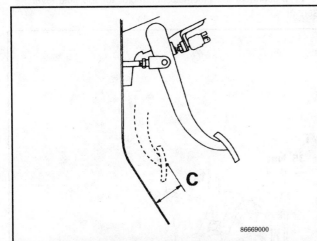

Fig. 5 Measure distance (C) after depressing the brake pedal with approximately 110 lbs. (500 N) of force while the engine is running

8. Start the engine, depress the brake pedal with approximately 110 lbs. (500 N) of force, and measure the clearance between the brake pedal and the firewall.
- Pick-ups: 3.15 in. (80mm) or more.
- 1983-91 Montero: 3.74 in. (95mm) or more.
- 1992-95 Montero: 3.94 in. (100mm) or more.

9. If the clearance is less than the standard value, check for air in the brake line or brake fluid leakage, and check the brakes themselves (for excessive shoe clearance caused by a malfunction of the automatic adjuster mechanism), and repair where necessary.

Brake Light Switch

REMOVAL & INSTALLATION

▶ See Figure 3

1. Disconnect the wiring at the back of the switch.
2. Loosen the locknut holding the switch to the bracket. Remove the locknut and remove the switch.
3. Install the new switch and install the locknut, tightening it just snug.
4. Reposition the brake light switch so that the distance between the outer case of the switch and the pedal is 0.020-0.040 in. (0.5-1.0mm). Note that the switch tab or prong must press against the pedal to keep the brake lights off. As the pedal moves away from the switch, the tab is released and closes the switch which turns on the lights.
5. Hold the switch in the correct position and tighten the locknut.
6. Connect the wiring to the switch.
7. Check the operation of the switch. Turn the ignition key to **ON** but do not start the engine. Have an assistant observe the brake lights at the rear while you push lightly on the brake pedal. The lights should come on just as the brake pedal passes the point of free-play.
8. Adjust the switch as necessary to get the correct response from the lights. The small amount of free-play in the pedal should not trigger the brake lights; if the switch is set incorrectly, the brake lights will flicker due to pedal vibration on road bumps.

Brake Pedal

REMOVAL & INSTALLATION

▶ See Figures 6 and 7

Brake pedal removal procedures from 1983-95 have changed very little and, although there may be slight differences, this one procedure covers all Pick-ups and Monteros.

1. Disconnect the negative battery cable from the battery.
2. Remove the stop light (brake) switch from the pedal support member.
3. Disconnect the brake pedal return spring from the brake pedal and the dash-mounted bracket.
4. Remove the cotter pin and flat washer, which hold the clevis pin in place on the brake pedal. The clevis pin holds the brake actuating rod onto the brake pedal.
5. Disconnect the actuating rod from the brake pedal once the clevis pin has been removed.
6. Remove the cotter pin and flat washer from the stud holding the shift lock cable (if so equipped) to the brake pedal. Disconnect the end of the shift lock cable from the brake pedal.
7. Remove the brake pedal shaft, nut and washer. Pull the brake pedal down and out of the pedal support member.
8. If need be, remove the two brake pedal bushings and the spacer from the upper end of the brake pedal.
9. To install, mount the two bushings and the spacer back onto the upper end of the brake pedal, if removed.
10. Slide the brake pedal up into its position in the pedal support member, then install the brake pedal shaft, washer and nut. Tighten the nut to 18-25 ft. lbs. (25-35 Nm).
11. Hook the end of the shift lock cable onto its mounting stud on the brake pedal, then instal the washer and a new cotter pin.

9-6 BRAKE SYSTEMS

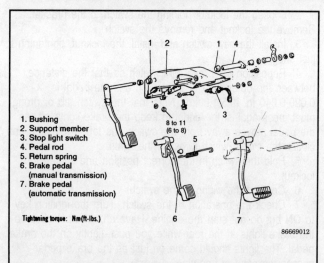

Fig. 6 Brake pedal assembly removal and installation components — 1983-86 Pick-ups and Monteros

12. Install the clevis pin into the brake pedal, after positioning the brake actuating rod around the brake pedal arm. Install the washer and a new cotter pin to the clevis pin.
13. Hook the return spring to the dash-mounted bracket, then to the brake pedal arm.
14. Install the stop light (brake) switch to the pedal support member. Adjust the distance of the stop light switch — refer to the procedures in this section.
15. Connect the negative battery cable.

Master Cylinder

REMOVAL & INSTALLATION

▶ See Figures 8, 9, 10, 11, 12, 13, 14 and 15

1. Use a clean suction tube (a turkey baster works well) to remove as much fluid as possible from the reservoir.
2. Disconnect the electrical connector for the fluid level sensor.

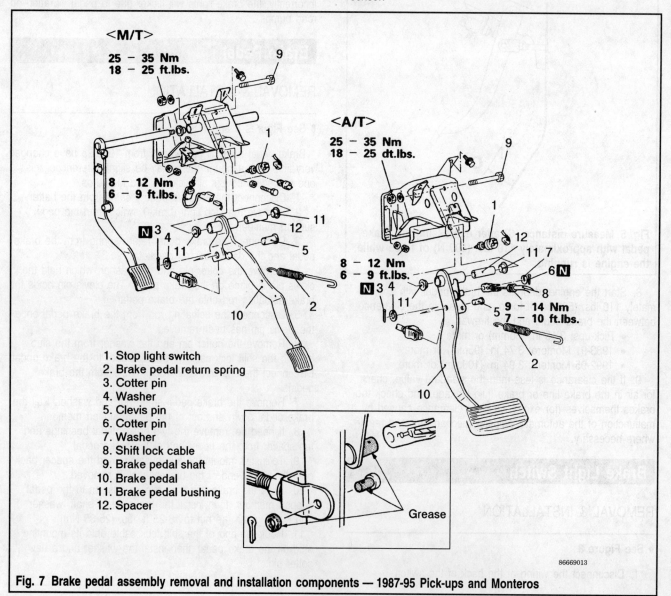

Fig. 7 Brake pedal assembly removal and installation components — 1987-95 Pick-ups and Monteros

BRAKE SYSTEMS 9-7

3. Using only a wrench of the correct size (no pliers — ever!), carefully disconnect each brake line from the master cylinder. Label each line with a piece of tape. Plug or tape the end of the line to keep dirt and moisture out.

✱✱WARNING

Do not bend or crimp the steel brake lines. Handle them with extreme care! If damaged, they must be replaced.

4. Remove the nuts holding the master cylinder to the brake booster and remove the cylinder. Some fluid will remain within the cylinder; take care not to spill it on painted surfaces.

5. Install the master cylinder to the brake booster and tighten the nuts to 9 ft. lbs. (12 Nm).

6. Carefully connect each brake line to its port on the master cylinder. Start each by hand, making sure that the fitting is at 90° to the port before starting to turn it. Once threaded one or two turns by hand, the wrench may be used to tighten each fitting to 15 ft. lbs. (20 Nm).

✱✱WARNING

Do not overtighten the flare nuts. Overtightening will cause damage to the nut and/or the master cylinder!

7. Connect the wiring connectors to the master cylinder.
8. For cars with remote reservoirs, connect the fluid tubes to the master cylinder. Make sure the lines are firmly seated and the clamps are firmly set.
9. Fill the system with clean, fresh brake fluid.
10. Bleed the brake system at all four wheels.

OVERHAUL

▶ See Figures 16, 17, 18, 19, 20, 21, 22, 23, 24, 25, 26, 27 and 28

➡ Special measuring tools are required for this procedure.

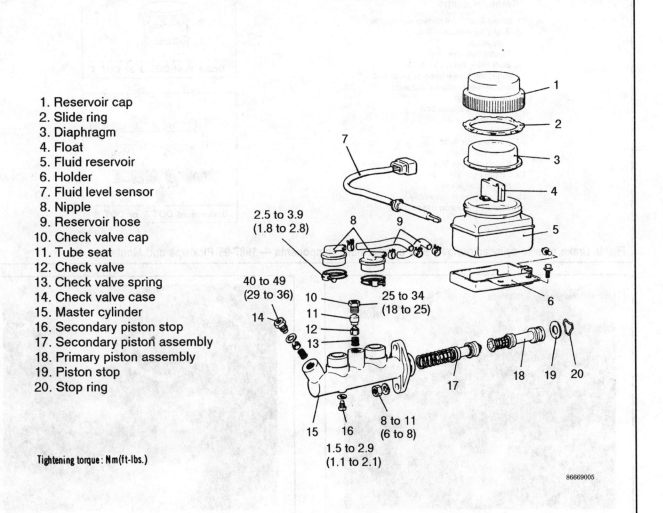

1. Reservoir cap
2. Slide ring
3. Diaphragm
4. Float
5. Fluid reservoir
6. Holder
7. Fluid level sensor
8. Nipple
9. Reservoir hose
10. Check valve cap
11. Tube seat
12. Check valve
13. Check valve spring
14. Check valve case
15. Master cylinder
16. Secondary piston stop
17. Secondary piston assembly
18. Primary piston assembly
19. Piston stop
20. Stop ring

Tightening torque: Nm(ft-lbs.)

Fig. 8 Brake master cylinder removal and disassembly components — 1983-86 Pick-ups and Monteros

9-8 BRAKE SYSTEMS

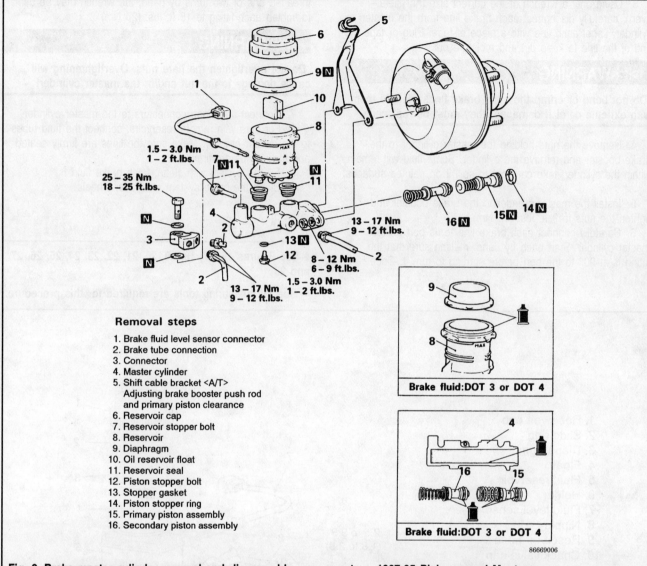

Fig. 9 Brake master cylinder removal and disassembly components — 1987-95 Pick-ups and Monteros

Fig. 10 Using a turkey baster to remove the brake fluid from the reservoir works well

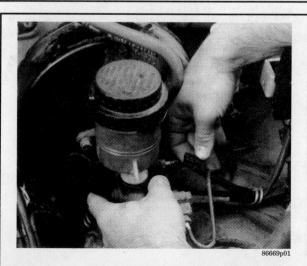

Fig. 11 Disconnect the electrical wiring from the master cylinder

BRAKE SYSTEMS 9-9

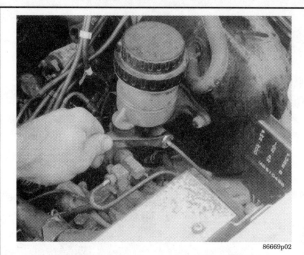

Fig. 12 Loosen and remove the hard lines from the master cylinder — be careful not to bend the lines

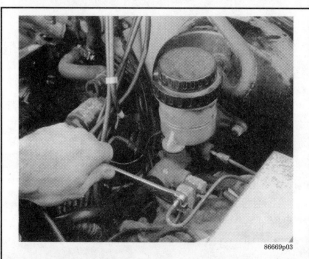

Fig. 13 Use the hard lines which have the flares in the line are machine made

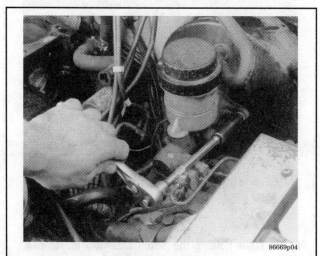

Fig. 14 Remove the mounting nuts holding the brake master cylinder to the power booster

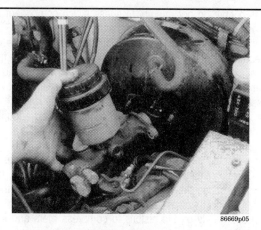

Fig. 15 Remove the master cylinder from the power brake booster — be careful not to spill any brake fluid on painted components, since the fluid will damage paint

1. Support the master cylinder in a vise. Clamp the vise jaws to the bolt flange, not to the body of the cylinder.
2. If the fluid reservoir is attached to the cylinder, remove the reservoir retaining screw and gently lift the reservoir clear. If the reservoir is a remotely mounted one (1983-86 vehicles), remove the two screw clamps holding the nipples in place. Then disconnect the reservoir hoses from the nipples and from the reservoir itself.
3. Remove the small seals under the reservoir.
4. Use a long thin tool to press inward on the piston and remove the piston stopper bolt and gasket. The piston need only be pushed enough to take tension off the bolt.
5. Again pushing gently on the piston, use snapring pliers to remove the piston retaining clip at the end of the cylinder. Be ready to catch the piston as you release pressure on it.
6. Remove the primary piston assembly and set is aside. Remove the secondary piston assembly. If this is difficult, gradually apply a stream of compressed air from the outer port on the secondary end of the master cylinder.

✶✶WARNING

Do NOT disassemble the primary or secondary pistons!

7. Clean the components with either clean brake fluid or a special brake cleaning solvent. Do NOT use petroleum based solvents and do NOT use water.
8. Check the inner surface of the master cylinder for rust or pitting. Check the pistons for any signs of rust, scoring or wear.
9. Using a cylinder micrometer, measure the cylinder bore in six locations. The locations should be the approximate bottom, center and top of the bore and measurements should be taken of the height and width of the bore at each position. All six measurements should be virtually identical. Variations of more than 0.00098 in. (0.025mm) disqualify the unit from use. Record the bore diameter.
10. Measure each piston in two dimensions. Again, measurements must be virtually identical. Subtract the piston diameter from the bore diameter. If the difference is greater than 0.0059 in. (0.15mm), there is enough wear to warrant replacing both the cylinder and pistons. Always replace the cylinder and pis-

BRAKE SYSTEMS

tons as a unit; do not attempt to guess which component is worn and do not mix old and new parts.

11. If the cylinder measurements are proper, the inside may be cleaned with a piece of crocus cloth soaked in brake fluid. This will remove any glaze or very light scratches. Honing the master cylinder is specifically not recommended as it will affect the piston to wall clearance measured in Steps 9 and 10 above. If the cylinder bore is scratched or corroded enough to warrant honing, the unit should be replaced.

12. Before reassembly, the piston assemblies and the cylinder bore should be liberally coated with clean fresh brake fluid.

13. Carefully insert the secondary piston into the bore, taking care not to gouge the bore or damage the seals. Insert the primary piston in similar fashion.

14. Push gently on the pistons to compress them and install the snapring.

15. Continue pushing on the pistons and install the piston stop bolt with a new O-ring (or gasket). Tighten the piston stopper bolt only to 1-2 ft. lbs. (1.5-3.0 Nm).

16. Install new reservoir seals on the cylinder and install either the fluid inlet ports or the reservoir. Tighten the reservoir retaining bolt to 2.2 ft. lbs. (3 Nm).

17. Before reinstallation, adjust the clearance between the brake booster pushrod and the the master cylinder primary piston. This dimension **A** is critical to the correct release of the master cylinder. Improper clearance will result in the brakes being partially applied under all conditions.

   a. Measure the distance between the master cylinder end face and the piston. This is most easily done by taking the measurement with a Vernier caliper and a straight scale (a piece of metallic bar stock whose thickness is uniform, such as a square) placed on the master cylinder end face. After measuring, subtract the thickness of the straight scale and record the measurement. Call this measurement **B**.

   b. Find the distance between the contact face of the master cylinder mounting flange and the end face of the master cylinder. Record this distance and call it dimension **C**.

   c. Measure the distance between the master cylinder mounting surface on the brake booster and the end of the

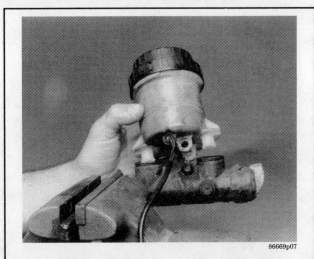

Fig. 17 Unfasten the reservoir mounting bolt, then remove the reservoir from the brake master cylinder

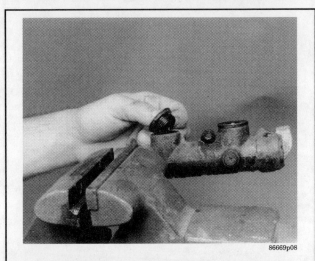

Fig. 18 Remove the reservoir mounting seals — new seals will be needed for reassembly

Fig. 16 Mount the brake master cylinder in a vise so that only the mounting flange (and not the cylinder body) is in the jaws

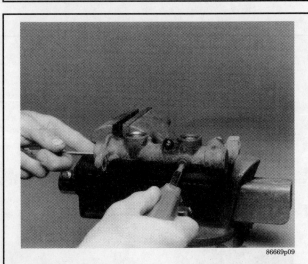

Fig. 19 While holding the piston assembly in, remove the piston stopper bolt

BRAKE SYSTEMS 9-11

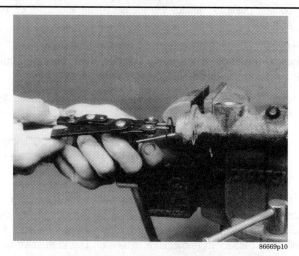

Fig. 20 Using snapring pliers, remove the snapring from the piston hole in the master cylinder

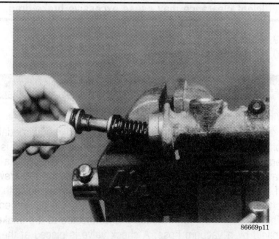

Fig. 21 Remove the primary piston assembly from the master cylinder — do not disassemble the piston assemblies themselves

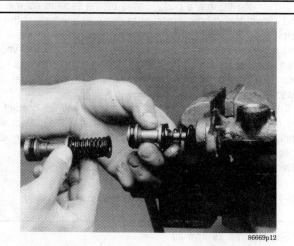

Fig. 22 Remove the secondary piston assembly — if the secondary piston is difficult to remove, apply compressed air to the secondary side outlet port

Fig. 23 Make certain to note which components are mounted in which position

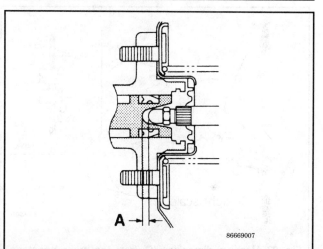

Fig. 24 Check the clearance (A) between the brake booster pushrod and the primary piston — follow the special procedure

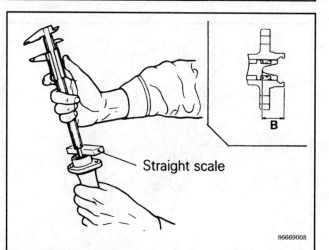

Fig. 25 Using a Vernier caliper and straight scale (along the edge of the master cylinder), obtain dimension (B). Be sure to subtract the thickness of the scale

9-12 BRAKE SYSTEMS

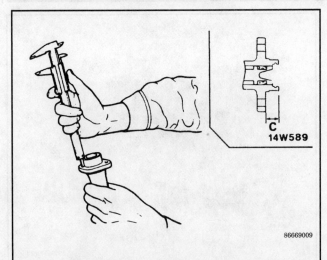

Fig. 26 Measure dimension (C) from the master cylinder's brake booster installation surface to the edge

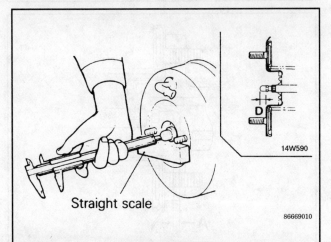

Fig. 27 Measure dimension (D) from the brake booster's master cylinder installation surface to the end of the pushrod

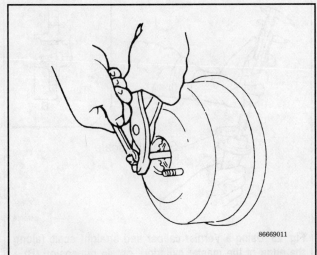

Fig. 28 If clearance is not within the standard value range, turn the pushrod screw to achieve desired length

booster pushrod. Record this measurement and call it dimension **D**.

→Obtain dimension (D) by first placing a straight scale against the edge of the brake booster, then measuring and subtracting the thickness of the straight scale.

d. Critical dimension **A** is found by using the equation A= B-C-D. The correct value for **A** is:
- Montero: 0.004-0.020 in. (0.1-0.5mm)
- 1983-88 2WD Pick-up: 0.015-0.031 in. (0.4-0.8mm)
- 1987-88 4WD Pick-up: 0.028-0.043 in. (0.7-1.1mm)
- 1989-91 Pick-up: 0.028-0.043 in. (0.7-1.1mm)
- 1992-95 Pick-up: 0.004-0.020 in. (0.1-0.5mm)

18. If the free-play dimension is not within the acceptable range, adjust the length of the pushrod by carefully turning the adjustable (threaded) tip with pliers. After adjusting the rod, it will be necessary to re-measure dimension **D** and re-solve the equation.

※※WARNING

Insufficient clearance may cause excessive brake drag.

19. Reinstall the master cylinder.
20. Bleed the brake system.

Power Brake Booster

Virtually all vehicles today use a vacuum assisted power brake system to multiply the braking force and reduce pedal effort. Since vacuum is always available when the engine is operating, the system is simple and efficient. A vacuum diaphragm is located between the master cylinder and the firewall and assists the driver in applying the brakes, reducing both the effort and travel that must be put into moving the brake pedal.

The vacuum diaphragm housing is connected to the intake manifold by a vacuum hose. A check valve is placed at the point where the hose enters the diaphragm housing, so that during periods of low manifold vacuum brake assist will not be lost.

Depressing the brake pedal closes off the vacuum source and allows atmospheric pressure to enter on one side of the diaphragm. This causes the master cylinder pistons to move and apply the brakes. When the brake pedal is released, vacuum is applied to both sides of the diaphragm and springs return the diaphragm and master cylinder pistons to the released position.

If the vacuum supply fails, the brake pedal rod will contact the end of the master cylinder actuator rod and the system will apply the brakes without any power assistance. The driver will notice that much higher pedal effort is needed to stop the vehicle and that the pedal feels "harder" than usual.

REMOVAL & INSTALLATION

▸ See Figures 29 and 30

1. Slide back the clip and disconnect the vacuum supply line at the brake booster. Pull the hose off gently in order to avoid damaging the check valve.
2. Remove the master cylinder as described previously.

BRAKE SYSTEMS 9-13

3. Disconnect the pushrod at the brake pedal. Pull the cotter pin or lockpin out of the pedal clevis. Separate the rod from the pedal.

4. Remove the mounting bolts and nuts from the firewall and remove the booster.

5. On the 1983-85 diesel powered Pick-ups, the 1983-86 gasoline powered Pick-ups and the 1983-95 Monteros, remove the two gaskets and the booster spacer from the firewall.

6. Install the booster and tighten the nuts and bolts to 9 ft. lbs. (12 Nm). Do not overtighten these bolts.

➡ **Some boosters use a spacer block and gaskets between the firewall and the booster. If present, they must be reinstalled in the correct order and position.**

7. Connect the brake pedal to the pushrod and install a new cotter pin. A light coat of multipurpose grease on the clevis pin and washer is recommended.

8. Before reinstalling the master cylinder, the booster pushrod clearance must be calculated and adjusted. This is particularly important if the booster has been replaced. Please refer to the master cylinder procedure given previously and perform the necessary steps.

9. Install the master cylinder and bleed the brake system completely.

Blend Proportioning Valve

The Mitsubishi 1983-86 Pick-ups and the 1983-91 Monteros are equipped with a Blend Proportioning Valve (BPV). The BPV is located on the left rear frame near the rear axle for the 1983-86 Pick-ups. Not all 1983-86 Pick-ups are equipped with BPV's; some Pick-ups are equipped with a load sensing proportioning valve. The 1983-90 Monteros with the 2.6L engines also have the BPV mounted back by the rear axle. The 1989-91 Monteros with the 3.0L engines have it mounted just below the brake booster.

The BPV cannot be adjusted, however, if the vehicle is exhibiting strange or dangerous braking tendencies, have the BPV tested at a reputable automotive service center. If the BPV is faulty it will have to be replaced with a new unit.

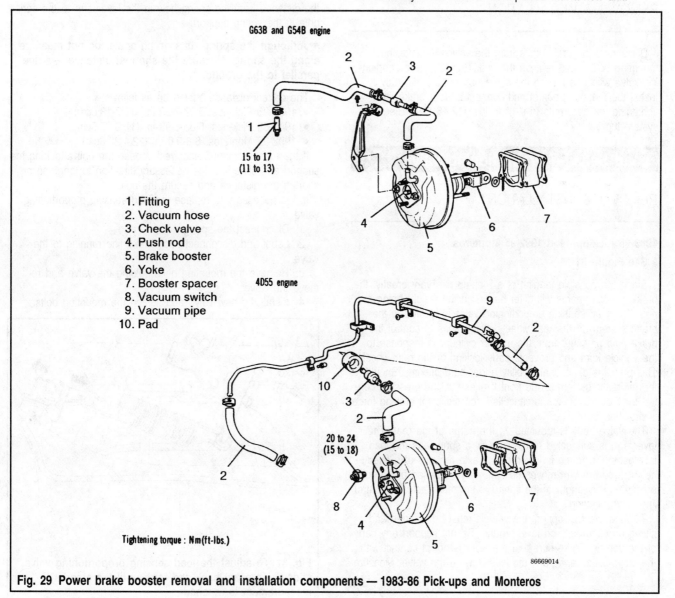

Fig. 29 Power brake booster removal and installation components — 1983-86 Pick-ups and Monteros

9-14 BRAKE SYSTEMS

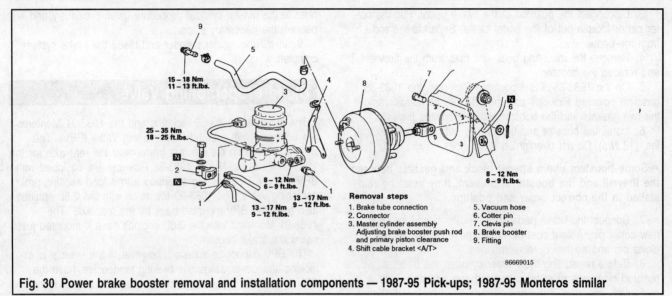

Fig. 30 Power brake booster removal and installation components — 1987-95 Pick-ups; 1987-95 Monteros similar

REMOVAL & INSTALLATION

Disconnect the brake lines from the valve. Remove the mounting bolt(s) and remove the unit. Replace the proportioning valve with one having the same numerical code, only. Install the mounting bolt(s) and connect all four brake lines, tightening the flare nuts to 9-12 ft. lbs. (12-16 Nm). Bleed the system thoroughly.

Load Sensing Proportioning Valve

REMOVAL & INSTALLATION

1983-95 Pick-ups and 1992-95 Monteros
▶ See Figure 31

Since the possible loading of a Pick-up can vary greatly, the braking requirements will differ based on the weight being carried. The load sensing proportioning valve is located in the brake system for the rear wheels. It functions to control the brake fluid pressure from the master cylinder in response to the vehicle load and prevents early locking of the rear wheels. This provides directional stability under hard braking. Should the hydraulic system to the front brakes fail, the pressure control for the rear brakes is cancelled and sufficient braking force is provided.

The valve body is mounted on the frame at the rear. The lever arm is connected to the axle by a spring. As the load increases, the frame is lowered (the axle remains at a fixed height) and the arm moves out, allowing increased brake fluid pressure. The length of the spring is the main factor in controlling valve function.

To measure this spring, the vehicle must be unladen and sitting on its wheels on level ground. Do not support any part of the vehicle. Make sure that the lever is not in contact with the stopper bolt and push the lever toward the valve. Measure the distance from the spring hole in the lever arm to the spring hole in the spring support.

➡ **Although the spring runs on an angle, do not measure along the spring. Measure the shortest distance — a line parallel to the ground.**

The correct distance should be as follows:
- 1989-95 Pick-ups: 6.97-7.09 in. (17.7-18.0cm)
- 1983-88 Pick-ups: 6.85-6.93 in. (17.2-17.6cm)
- 1992-95 Monteros: 8.8-9.0 in. (22.4-22.8cm)

If the measurement is incorrect, loosen the bolt attaching the support and slide it in the necessary direction to lengthen or shorten the distance and secure the bolt.

If it is necessary to replace the load sensing proportioning valve:

1. Disconnect the spring.
2. Label and disconnect each brake line running to the valve.
3. Remove the mounting bolts holding the valve and remove the valve.
4. Install the new valve and tighten the mounting bolts.

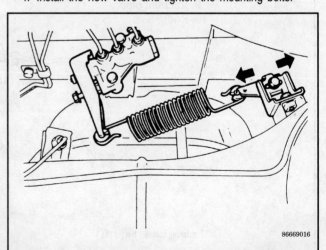

Fig. 31 To adjust the load sensing proportioning valve, slide the bracket back and forth until the standard measurement is reached

BRAKE SYSTEMS 9-15

5. Connect each brake line to the correct port and tighten the fitting to 12 ft. lbs. (17 Nm).
6. Install the spring.
7. Measure the effective distance as outlined above.
8. Bleed the brake system.

Vacuum Check Valve

REMOVAL & INSTALLATION

The vacuum check valve found in the brake booster line is a simple one-way valve. It is designed and placed to hold vacuum within the booster for use during periods of low engine vacuum.

To test the valve, blow into each end. Air should NOT pass when you blow from the engine side but should pass when you blow from the booster side. Make sure the valve is installed correctly in the vacuum hose when reinstalling.

Brake Hoses and Lines

REMOVAL & INSTALLATION

◆ See Figure 32

Metal lines and rubber brake hoses should be checked frequently for leaks and external damage. Metal lines and particularly prone to crushing and kinking under the car. Any such deformation can restrict the proper flow of fluid and therefore impair braking at the wheels. Rubber hoses should be checked for cracking or scraping; such damage can create a weak spot in the hose and it could fail under pressure.

Any time the lines are removed or disconnected, extreme cleanliness must be observed. The slightest bit of dirt in the system can plug a fluid port and render the brakes defective. Clean all joints and connections before disassembly (use a stiff brush and clean brake fluid) and plug the lines and ports as soon as they are opened. New lines and hoses should be blown or flushed clean before installation to remove any contamination with clean brake fluid. To replace a line or hose:

1. Raise and safely support the vehicle on jackstands.
2. Remove wheel(s) as necessary for access.
3. Clean the surrounding area on the joints to be disconnected.
4. Place a catch pan under the joint to be disconnected. To reduce fluid spillage during the repair, either plug the vent hole in the reservoir cap or substitute a cap with no vent hole.
5. Using two wrenches — one to hold the joint and one to turn the fitting — disconnect the hose or line to be replaced.
6. Disconnect the other end of the line or hose, moving the drain pan if necessary. Always use two wrenches if possible.
7. Disconnect any retaining clips or brackets holding the line and remove the line.
8. If the system is to remain open for more time than it takes to swap hoses, tape or plug each remaining line and port to keep dirt out and fluid in.
9. Install the new line or hose, starting the end farthest from the master cylinder first. Connect the other end and confirm that both fittings are correctly threaded and turn smoothly with the fingers. Make sure the new line will not rub against any other part. Brake lines must be at least ½ in. (13mm) from the steering column and other moving parts. Any protective shielding or insulators must be reinstalled in the original location.

➡If the new metal line requires bending, do so gently using a pipe bending tool. Do not attempt to bend the tubing by hand; it will kink the pipe and render it useless.

10. Using two wrenches as before, tighten each fitting to 9-12 ft. lbs. (13-17 Nm).
11. Install any retaining clips or brackets on the lines.
12. Refill the brake reservoir and replace the unvented cap if one was substituted.
13. Bleed the brake system beginning with the wheel closest to the replaced hose.
14. Install the wheels and lower the vehicle to the ground.

Brake System Bleeding

STANDARD SYSTEM

◆ See Figures 33, 34 and 35

The brake system is bled through the bleeder screws on the calipers and/or rear wheel cylinders. The system must be bled whenever air enters, whether due to disconnection of a hydraulic line or a leak. A spongy brake pedal indicates bleeding is needed even when no repairs have been made. Also check carefully for leaks and repair them as necessary.

Bleeding pumps fluid into the system via the master cylinder and bleeds air out through the bleed valves. You must keep the master cylinder reservoir full of fluid at all times to provide a continuous flow of air-free fluid into the system. You'll also need an assistant who will keep constant pressure on the brake pedal at all times to keep air from being drawn back into the system.

✱✱WARNING

Do not allow brake fluid to splash or spill onto painted surfaces; the paint will be damaged. If spillage occurs, flush the area immediately with clean water.

1. Fill the master cylinder reservoir to the "MAX" line with brake fluid and keep it at least half full throughout the bleeding procedure.
2. If the master cylinder has been removed or disconnected, it must be bled before any brake unit is bled. To bleed the master cylinder:
   a. Disconnect the front brake line from the master cylinder and allow fluid to flow from the front connector port.
   b. Reconnect the line to the master cylinder and tighten it until it is fluid tight.
   c. Have a helper press the brake pedal down one time and hold it down.
   d. Loosen the front brake line connection at the master cylinder. This will allow trapped air to escape, along with some fluid. (Have a rag or small container handy to catch the fluid.)

9-16 BRAKE SYSTEMS

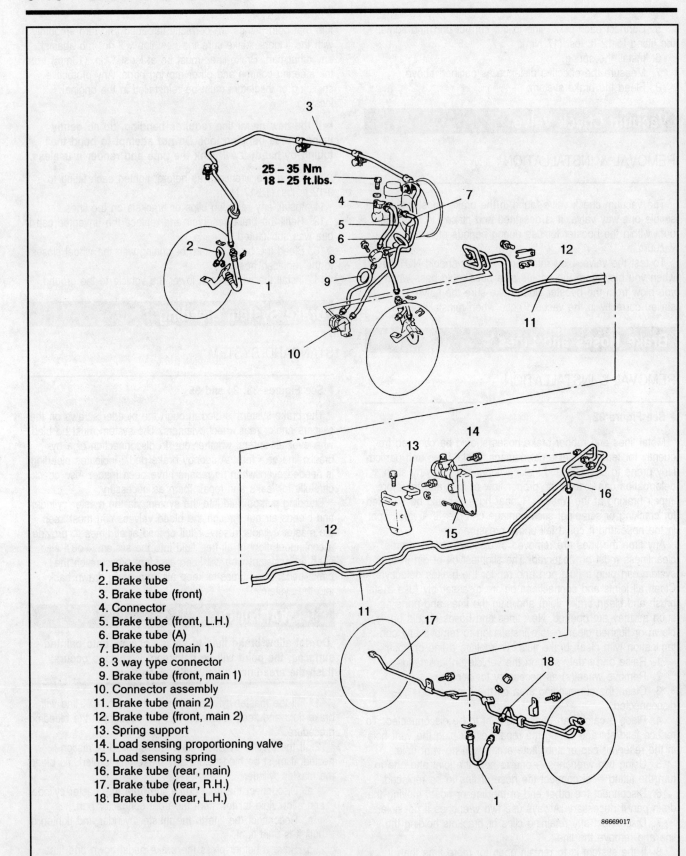

1. Brake hose
2. Brake tube
3. Brake tube (front)
4. Connector
5. Brake tube (front, L.H.)
6. Brake tube (A)
7. Brake tube (main 1)
8. 3 way type connector
9. Brake tube (front, main 1)
10. Connector assembly
11. Brake tube (main 2)
12. Brake tube (front, main 2)
13. Spring support
14. Load sensing proportioning valve
15. Load sensing spring
16. Brake tube (rear, main)
17. Brake tube (rear, R.H.)
18. Brake tube (rear, L.H.)

Fig. 32 Brake line routing for 1990-95 Pick-ups — although slight differences are present, other models and years are similar

BRAKE SYSTEMS 9-17

e. Again tighten the line, release the pedal slowly and repeat the sequence (Steps 2c, d, and e) until only fluid runs from the port. No air bubbles should be present in the fluid.

f. Final tighten the line fitting at the master cylinder to 11 ft. lbs. (15 Nm).

g. After all the air has been bled from the front connection, bleed the master cylinder at the rear connection by repeating Steps 2a through e.

3. Start with the right rear brake caliper or wheel cylinder. Place the correct size box-end or line wrench over the bleeder valve and attach a tight-fitting transparent hose over the bleeder. Allow the tube to hang submerged in a transparent container of clean brake fluid. The fluid must remain above the end of the hose at all times, otherwise the system will ingest air instead of fluid.

4. Have an assistant pump the brake pedal several times slowly and hold it down.

5. Open the bleed point (about ¼-½ turn is usually enough) and watch for bubbles in the brake fluid flow into the

Fig. 35 Do the same for the front calipers — have a helper work the pedal while loosening and tightening the bleeder valve

container. You must close the bleed valve before the brake pedal bottoms in its travel. Once the assistant has the feel of the pedal, he can warn you in advance. The more the bleed valve is opened, the quicker the pedal travels downward. With the bleed point closed, the assistant should release the pedal so the master cylinder will admit fresh, air-free fluid into the system. Once he has put pressure back on the pedal, you may re-open the bleed point, again closing it before the pedal bottoms out. Repeat this process until the fluid flowing from the bleed point is entirely air-free.

6. After two or three bleedings, check and replenish the fluid in the reservoir.

7. Repeat the process, in the correct order, at each remaining wheel. The general theory is to bleed the longest lines first and work towards the master cylinder — right rear, left rear, right front, left front.

8. When all four wheels have been bled and no air is evident in the fluid, refill the master cylinder reservoir and attach the cap tightly.

VEHICES WITH LOAD SENSING PROPORTIONING VALVE

▶ See Figure 36

After bleeding the right rear wheel, proceed to bleed the load sensing valve located on the frame at the rear of the vehicle. Use the bleeder fitting on the valve; do not attempt to loosen the lines entering the unit. Follow the bleeding procedure for normal systems and bleed the proportioning valve as you would any other cylinder or caliper.

VEHICLES WITH ABS

▶ See Figure 37

After bleeding the master cylinder and the four wheels, bleed the brake lines going to the hydraulic unit or modulator unit. Start the engine before bleeding the brake system on vehicles

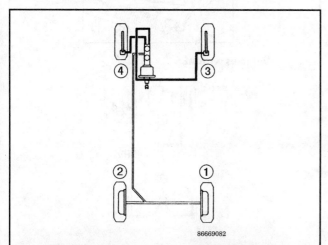

Fig. 33 Bleed the brake system in the order shown above — Pick-ups and Monteros with no load sensing valve or ABS systems

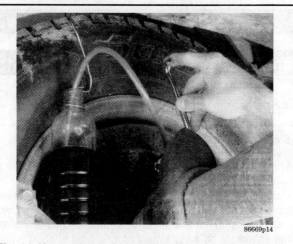

Fig. 34 Use a piece of hose with one end attached over the bleeder valve and the other end submerged in a clear, unbreakable container of brake fluid

9-18 BRAKE SYSTEMS

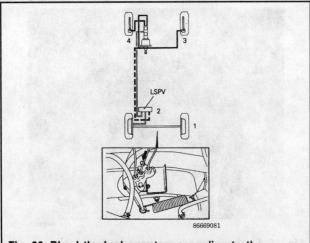

Fig. 36 Bleed the brake system according to the sequence shown in the illustration — Pick-ups with load sensing proportioning valves

with ABS. Otherwise the system for bleeding the brakes is identical to the system for vehicles without ABS.

✱✱CAUTION

When supplying brake fluid for vehicles with ABS, the filter should be installed to the master cylinder reserve tank.

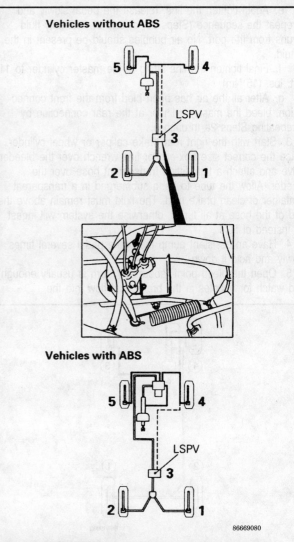

Fig. 37 Bleed the brake system according to the numbered steps in the illustration — Monteros with and without ABS systems

FRONT DISC BRAKES

✱✱CAUTION

Brake pads and shoes may contain asbestos, which has been determined to be a cancer causing agent. Never clean the brake surfaces with compressed air! Avoid inhaling any dust from brake surfaces! When cleaning brakes, use commercially available brake cleaning fluids.

✱✱WARNING

If the vehicle has been recently driven before repairs, all the brake surfaces and components may be very hot. Work carefully and wear gloves.

Brake Pads

WEAR INDICATORS

▶ See Figure 38

The front disc brake pads are equipped with a metal tab which will come into contact with the disc after the friction surface material has worn near its usable minimum. The wear indicators make a constant, distinct metallic sound that should be easily heard. (The sound has been described as similar to either fingernails on a blackboard or a field full of crickets.) The key to recognizing that it is the wear indicators and not some other brake noise is that the sound is heard when the vehicle is being driven WITHOUT the brakes applied. It may or may not be present under braking during normal driving.

BRAKE SYSTEMS 9-19

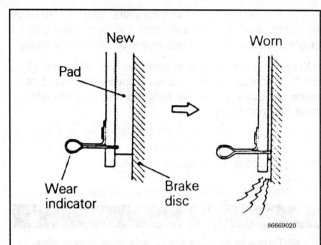

Fig. 38 The wear indicators on the brake pads indicates when the brake pads have worn down to the point when pad replacement is necessary

It should also be noted that any disc brake system, by its design, cannot be made to work silently under all conditions. Each system includes various shims, plates, cushions and brackets to suppress brake noise but no system can completely silence all noises. Some brake noise — either high or low frequency — can be considered normal under some conditions. Such noises can be controlled and perhaps lessened, but cannot be totally eliminated.

INSPECTION

▶ See Figure 39

The front brake pads may be inspected without removal. With the front end elevated and supported, remove the wheel(s). Unlock the steering column lock and turn the wheel so that the brake caliper is out from under the fender.

View the pads — inner and outer — through the cut-out in the center of the caliper. Remember to look at the thickness of the pad friction material (the part that actually presses on the disc) rather than the thickness of the backing plate which does not change with wear.

Remember that you are looking at the profile of the pad, not the whole thing. Brake pads can wear on a taper which may not be visible through the window. It is also not possible to check the contact surface for cracking or scoring from this position. This quick check can be helpful only as a reference; detailed inspection requires pad removal.

After the pads are removed, measure the thickness of the LINING portion (NOT the backing) of the pads. It must be at least 0.04 in. (1mm) for all 1983-86 vehicles and 0.08 in. (2mm) for all vehicles built after 1987. This is a Mitsubishi factory minimum; please note that local inspection standards enforced by your state must be given precedence if they require a greater thickness.

REMOVAL & INSTALLATION

▶ See Figures 40, 41 and 42

➡Whenever brake pads are replaced, replace them in complete sets; that is, replace the pads on both front wheels even if only one side is worn.

There are several combinations of shims, spacers and clips in use on Mitsubishi vehicles. When disassembling, work on one side at a time and pay attention to placement of these components. If you become confused during reassembly, refer to the other side for correct placement. To remove the brake pads:

1. Raise the vehicle and support it safely on stands.
2. Remove the front wheels.
3. Unscrew and remove the lockpin without disturbing the grease coating. Place the pin in a clean spot where grease will not pick up dust. Lift up the caliper body by using the guide pin bolt as a fulcrum. Fasten the caliper in the raised position with wire.

✱✱WARNING

There is a coating of special grease on the lockpin. Be careful that this grease is not removed, and that dirt does not contaminate the grease on the pin.

4. Remove the shims, brake pads and spring clips in order. Depending on the number of such components, either diagram their placement or lay them aside in order.
5. Clean the surface of the piston with a clean, damp rag. Make sure there is room in the master cylinder reservoir for more fluid (fluid will be forced back into the reservoir in the next step). If necessary, remove some fluid with a clean squeeze ball syringe.
6. Use a large C-clamp or similar tool to depress the caliper pistons back into the calipers. The pistons will need to be almost flush with the case to fit over the new, thicker pads.

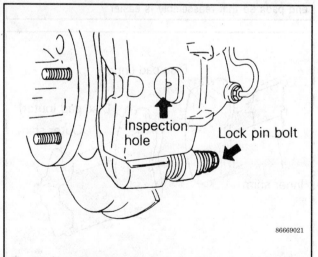

Fig. 39 Use the inspection hole to check the thickness of the front caliper brake pads

BRAKE SYSTEMS

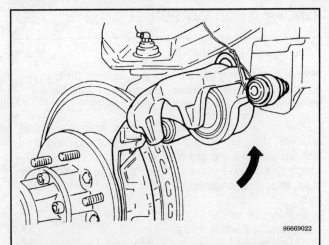

Fig. 40 Remove the lockpin bolt from the caliper, lift the caliper up and secure it in place with a wire or strong cord

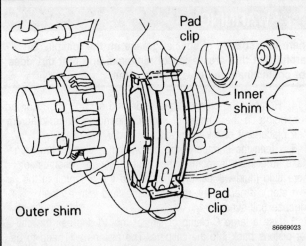

Fig. 41 Note the positions of the various shims, clips and pads so that reassembly is easier

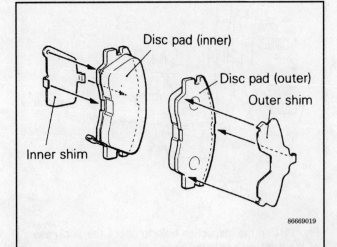

Fig. 42 Before installing the new brake pads to the caliper, install the outer and inner shims to the pads

7. Install the brake pads with the shims, clips and fittings in the correct order. Vehicles with upper and lower spring clips for each pad should have these clips replaced with the pads.

➡ Keep the brake surface of each pad free of grease, oil and fluids during the installation. A greasy fingerprint or similar light contact may be removed with a commercial brake cleaning spray.

8. Remove the wire, lower the caliper into position, and screw in the lower lockpin. Make certain the rubber boot is correctly placed on the lockpin and is not pinched or deformed.
9. Replace the wheels and lower the vehicle.

✲✲WARNING

Do NOT attempt to drive the vehicle immediately after lowering it to the ground. The first two or three brake pedal applications may not provide any brake response.

10. Pump the brake pedal several times with the engine off. The first two or three pedal strokes may be longer than usual as the pistons return from their compressed position and drive the pads inward. After a reasonable feel has been achieved, check the master cylinder reservoir and top up the fluid as needed. Start the engine and pump the brake pedal again, checking for the proper feel and engagement point.
11. Since the brake hoses and hydraulic system was not opened during the repair, it is usually not necessary to bleed the brake system after pad replacement. Use good judgement in determining pedal feel; bleeding may be necessary for other reasons.

➡ Braking should be moderate for the first 10 miles (16 km) or so until the new pads seat correctly. The new pads will bed best if put through several moderate heating and cooling cycles. Avoid hard braking until the brakes have experienced several long, slow stops with time to cool in between. Taking the time to properly bed the brakes will yield quieter operation, more efficient stopping and contribute to extended brake life.

Brake Caliper

➡ For the dual piston calipers found on 1992-95 Monteros, follow the same procedures for each piston in the caliper.

REMOVAL & INSTALLATION

▶ See Figures 43, 44, 45, 46, 47, 48, 49 and 50

1. Raise and safely support the front of the vehicle on jackstands. Set the parking brake and block the rear wheels.
2. Siphon a sufficient amount of brake fluid from the master cylinder reservoir to prevent the brake fluid from overflowing when removing or installing the caliper. This is necessary as the piston must be forced into the cylinder bore to provide clearance to install the caliper.

BRAKE SYSTEMS 9-21

3. Remove the wheel.

➡ Disassemble brakes one wheel at a time. This will prevent parts confusion and also prevent the opposite piston from popping out during installation.

4. Disconnect the brake line at the clip on the strut and disconnect the hose union at the caliper. Use a pan to catch any spilled fluid and immediately plug the hose end.

5. Remove the two caliper mounting bolts and then remove the caliper from the mounting bracket. Alternatively, remove the two bolts (screwed directly into the backing plate) which fasten the caliper support to the backing plate, and remove the entire assembly. Remove the brake pads and clips from the support as described above. Remove the lockpin (lower) and guide pin (upper) from the caliper support, and remove the caliper from the support.

To install:

6. Use a caliper compressor, a C-clamp or a large pair of pliers to slowly press the caliper piston back into the caliper.

Fig. 45 Place the pivot (mounting) bolts in a clean spot to keep them from getting contaminated by dirt

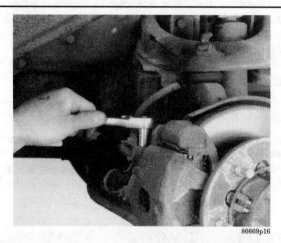

Fig. 43 To remove the caliper assemblies, start by removing the brake line retaining bracket from the caliper

Fig. 46 After disconnecting and plugging the brake line to the caliper, pull the caliper off of the rotor and brake bracket assembly

Fig. 44 Remove the two mounting bolts holding the caliper to the caliper bracket

Fig. 47 Remove the outer brake pad with the outer shim from the brake bracket

9-22 BRAKE SYSTEMS

Fig. 48 Once the pad is removed, the shim can be separated from the pad — note the positions of the shims

Fig. 49 Do the same with the rear brake pad

Fig. 50 Remove the brake pad upper and lower spring clips from the brake bracket

7. If the mounting plate was removed from the backing plate, reinstall it and coat the bolts with an anti-seize compound. Tighten the bolts to 66 ft. lbs. (90 Nm).

8. Install the pads onto the mount and install the caliper. Make sure the slide bushings and bolts are clean and properly lubricated. Do not use regular grease or spray lubricants; they cannot withstand the extreme heat generated by the brakes. Tighten the lower bolt to 23-30 ft. lbs. (32-42 Nm) and the upper bolt to 29-36 ft. lbs. (40-50 Nm).

9. Connect the fluid hose to the caliper and then to the hard line at the body. Tighten each fitting to 12 ft. lbs. (17 Nm) and make sure the hose is correctly routed and not kinked. Remember to install any clips or retainers holding the line.

10. Refill the master cylinder reservoir to the MAX line. Bleed the brake system. Keep a close eye on the reservoir, maintaining its level at at least half during the bleeding. Since the caliper was emptied, a fair amount of brake fluid may need to be added.

11. Reinstall the wheel and lower the vehicle to the ground.

12. Check the level in the fluid reservoir and top it off as needed. Before driving the vehicle, start the engine and operate the brake pedal several times to check the feel and engagement point of the brakes.

OVERHAUL

▶ See Figures 51, 52, 53, 54, 55 and 56

➡ To overhaul the brake calipers, you must have a controlled source of compressed air to force the pistons out. You should also have a generous supply of clean brake fluid or commercial brake cleaning spray to clean parts.

1. Drain the remaining fluid from the caliper.

2. Carefully remove the dust boot from around the piston by prying the boot ring out of its groove. Remove the piston and piston boot by applying compressed air through the brake hose fitting hole.

✲✲CAUTION

Do not hold the caliper during this procedure, and do not place fingers in front of the piston in an attempt to catch it or protect it when applying compressed air. Place a piece of cloth in front of the piston, and slowly increase the force of the compressed air to prevent the piston from springing out abruptly; use just enough air pressure to ease the piston out.

3. Remove the seal from the inside of the caliper bore. Clean the caliper bore with brake cleaner, alcohol or brake fluid.

✲✲WARNING

Never use a screwdriver or similar tools, because doing so could damage the cylinder surface.

4. Check all the parts for wear, deterioration, cracking or other abnormal conditions. Corrosion, generally caused by water in the system, will appear as white deposits on the metal, similar to what may be found on an old aluminum storm

door around the house. Pay close attention to the condition of the inside of the caliper bore and the outside of the piston. Any sign of corrosion or scoring requires new parts; do not attempt to clean or resurface either face if there is damage present.

5. The caliper overhaul kit will, at minimum, contain new seals and dust boots. A good kit will contain a new piston as well, but you may have to buy the piston separately. Any time the caliper is disassembled, a new piston is highly recommended in addition to the seals.

6. Clean all the components to be reused with a brake cleaner, alcohol or clean brake fluid and dry them thoroughly. Take any steps necessary to eliminate moisture or water vapor from the parts.

7. Coat all the caliper components with fresh brake fluid from an unopened can.

➡Some kits come with special assembly lubricants for the piston seals and slides or guidepins. Use these lubricants according to the directions within the kit.

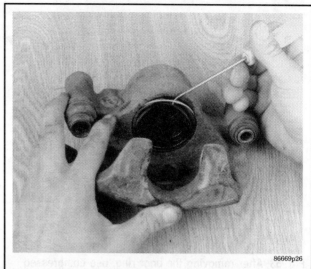
Fig. 52 Remove the boot ring with a pointed prytool

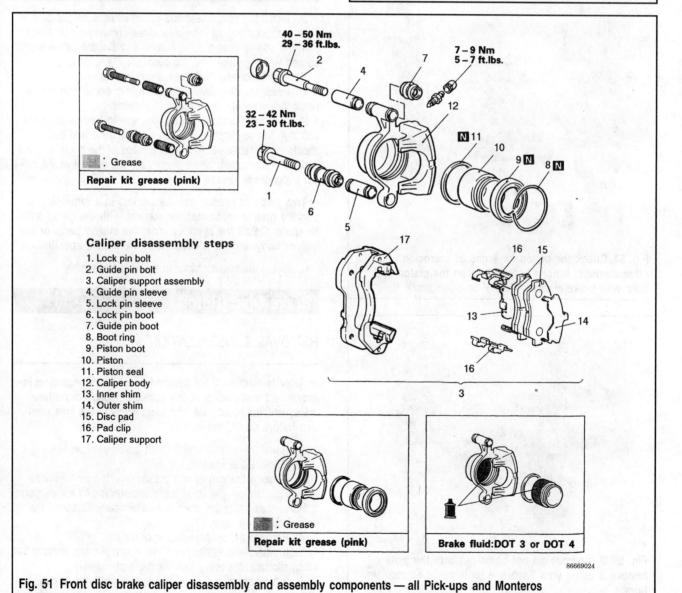

Caliper disassembly steps
1. Lock pin bolt
2. Guide pin bolt
3. Caliper support assembly
4. Guide pin sleeve
5. Lock pin sleeve
6. Lock pin boot
7. Guide pin boot
8. Boot ring
9. Piston boot
10. Piston
11. Piston seal
12. Caliper body
13. Inner shim
14. Outer shim
15. Disc pad
16. Pad clip
17. Caliper support

Fig. 51 Front disc brake caliper disassembly and assembly components — all Pick-ups and Monteros

9-24 BRAKE SYSTEMS

Fig. 53 After removing the boot ring, use compressed air to push the boot and piston out of the caliper

Fig. 54 Check the piston for signs of corrosion and other damage; if none is found, clean the piston and bore with brake cleaner, alcohol or clean brake fluid

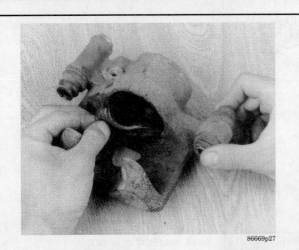

Fig. 55 If the boot did not come out with the piston, remove it using your fingers — tools could scratch the bore

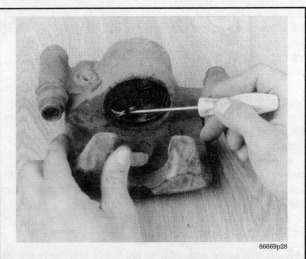

Fig. 56 If using a prytool to remove the piston seal, be extremely careful not to scratch the piston bore

8. Install the piston seal and piston into the caliper bore. This is an exacting job; the clearances are very small. Make sure the seal is seated in its groove and that the piston is not cocked when inserted into the bore.
9. Install the new dust boot and new boot ring.
10. Install the slide bushings and rubber boots onto the caliper if they were removed during disassembly.
11. The guide and lockpin sleeves should be coated inside with the grease supplied with the rebuilding kit, and boots should all be replaced. Also coat the lips of the boots and the surface of the caliper which bears (and turns) against the caliper support with grease.

➡Two kinds of grease may be packed in a rebuilding kit; use the grease recommended for use with the guide and lockpins. On all the systems, coat the sliding parts of the caliper body and the sleeves with the grease specified.

12. Install the brake caliper to the wheel assembly.

Brake Disc (Rotor)

REMOVAL & INSTALLATION

➡Refer to Section 8 for detailed procedures regarding removal and installation of the hub assembly. Four wheel drive vehicles must have the front hubs in the free position before disassembly.

1. Raise the vehicle and support it safely on stands.
2. Remove the wheel.
3. Remove the caliper and brake pads. It may be necessary to disconnect the small brake line running to the caliper. If this is done, plug the open line immediately. Suspend the caliper out of the way.
4. Remove the caliper adaptor or mount.
5. If the vehicle is equipped with 4 wheel drive, remove the automatic hub, the shim, lock washer and locknut.
6. If the vehicle is 2-wheel drive, remove the dust cap, cotter pin, nut lock, nut, washer and outer bearing.

BRAKE SYSTEMS 9-25

7. Remove the front hub and rotor assembly. Provide matchmarks on the rotor and hub; unbolt the rotor from the hub.

8. Install the rotor to the hub and tighten the attaching nuts and bolts to 38 ft. lbs. (52 Nm).

9. Install the assembly and retaining parts to the spindle — refer to Sections 7 and 8.

10. Install the caliper and brake assembly — refer to the preceding procedures in this section.

11. Reconnect the brake line if it was disconnected and bleed the brakes.

12. Install the wheel and lower the vehicle to the ground.

INSPECTION

➡ **Mitsubishi does not provide resurfacing specifications. Any disc which is gouged, scored, worn or warped should be replaced.**

Run-out

Before measuring the run-out on the front discs, confirm that the front wheel bearing play is within specification.

1. Raise and safely support the vehicle on jackstands. If only the front is supported, set the parking brake and block the rear wheels.

2. Remove the wheel.

3. Remove the brake caliper from its mount and suspend it out of the way by a piece of wire. Don't disconnect the fluid hose and don't allow the caliper to hang by the hose. Remove the brake pads with all the clips, springs, shims, etc.

4. Inspect the disc surface for grooves, cracks and rust. Clean the disc thoroughly and remove all rust.

5. Mount a dial indicator with a magnetic or universal base somewhere on the vehicle so that the tip of the indicator contacts the rotor about 0.20 in. (5mm) from the outer edge.

6. Zero the dial indicator. Turn the rotor one full revolution and observe the total indicated run-out. Refer to the Brake Specifications chart in this section for the specifications for your vehicle.

7. If the run-out exceeds the maximum allowed, the rotor needs replacing.

Thickness

The thickness of the rotor or disc partially determines its ability to withstand heat and provide adequate stopping force. Every rotor has a minimum thickness established by the manufacturer. This minimum measurement must NOT be exceeded under any condition. A rotor which is too thin may crack under braking; if this occurs the wheel can lock instantly, resulting in loss of control and a possible collision.

If any part of the rotor measures below minimum, the disc must be replaced. Since the allowable wear from brand new to minimum is only 0.04-0.06 in. (1.0-1.5mm), resurfacing is not recommended.

Thickness and thickness variation can be measured with a micrometer capable of reading to 0.00010 in. (0.0025mm). All measurements must be made at the same distance in from the edge of the rotor. Measure at four equally spaced points around the disc and record each measurement. Compare the measurements to the minimum specifications in the chart. A rotor varying by more than 0.0010 in. (0.025mm) can cause pedal vibration and/or front end vibration during stops. A rotor not meeting thickness requirements, or with varying thicknesses, must be replaced.

Condition

A new rotor will have a smooth, even surface which rapidly changes with use. It is not uncommon for a rotor to develop very fine concentric scores (like the grooves on a record) due to dust and grit being trapped by the brake pads. This slight irregularity is normal, but as the grooves deepen, wear and noise increase and stopping may be affected. As a general rule, any groove deep enough to snag a fingernail during inspection is cause for rotor replacement.

Any sign of blue spots, discoloration, heavy rusting or outright gouging require replacement. If you are checking the disc on the truck (such as during pad replacement or tire rotation) remember to turn the disc and check both the inner and outer faces completely. A small mirror and a bright light can be very helpful in seeing the inner face of the rotor. If anything looks questionable or requires consideration, choose the safer alternative and replace the rotor. The front brakes are a critical system and must be maintained at 100% efficiency.

Any time a rotor is replaced, the pads should also be replaced so that the surfaces mate properly. Since brake pads should be replaced in sets (both front or rear wheels), consider replacing both rotors instead of just one. The restored feel and accurate stopping make the extra investment well worthwhile.

BRAKE SYSTEMS

REAR DRUM BRAKES

Brake Drums

REMOVAL & INSTALLATION

♦ See Figure 57

> **✻✻CAUTION**
>
> Brake pads and shoes may contain asbestos, which has been determined to be a cancer causing agent. Never clean the brake surfaces with compressed air! Avoid inhaling any dust from brake surfaces! When cleaning brakes, use commercially available brake cleaning fluids.

1. Block the front wheels, raise the vehicle at the rear and support it safely on jackstands. Make sure the parking brake is released.
2. Remove the rear wheels.
3. Remove the brake drum by pulling it towards you. If the drum is difficult to remove, first check that the parking brake is fully released and that the cables to the wheels are not binding. Then follow either of these procedures:

    a. Remove the cover from the adjustment hole at the rear or back face of the backing plate.

    b. Insert a small prytool into the adjustment hole and use it to separate the adjustment lever from the adjuster.

    c. Use a brake adjusting tool or prytool to turn the star wheel and loosen the brake adjustment. Turning the wheel in the correct direction will contract the shoes away from the drum.

    d. If the drum has additional holes drilled in it (not all do), insert two M8x1.25 bolts.

    e. Turn the bolts tighter; they will press on the hub and force the drum off the brake shoes. Remember to turn the bolts alternately and evenly; if the drum is cocked, it can cause damage as it comes off (if it comes off at all).

4. Before reinstalling the drum, the brake shoes must be in the correct position. If the shoes are expanded too far the drum simply won't go over them. Refer to the brake shoe removal and installation procedures in this section for detailed instructions on setting the shoe clearance.
5. Install the drum onto the rear axle lug studs. Push it until it is set flat and completely in as far as it will go.

INSPECTION

♦ See Figure 58

Measure the inside diameter of the drum with an inside micrometer at a number of different positions. Compare it with the specifications listed in the Brake Specifications chart in this section. If the drum is worn past the limit at any point, it must be replaced. Additionally, the diameter should be constant at all points measured. An out-of-round condition may cause vibration under braking.

Mitsubishi gives no limits for machining drums to remove uneven wear other than the absolute maximum diameter (wear limit). If the drum is worn unevenly around its diameter or has wear grooves, you may be able to salvage it by having it machined BUT machining must not increase the diameter past the specified limits. If it does, the drum must be replaced.

Drums, particularly the larger ones, can crack. Check the inner and outer surfaces and the area around the wheel lug holes for any sign of hairline cracks. Obviously, if any are found, the drum is useless.

Brake Shoes

INSPECTION

The rear brake drums must be removed in order for the shoes may be inspected and measured for remaining thickness. Use a cloth dampened with water to wipe down the components or use an aerosol brake cleaner. Do not use a dry brush or air to clean the shoes. Measure the friction material (not the backing plate) at the thinnest point. Refer to the Brake Specifications chart for minimum thicknesses.

➡It is a characteristic of brake shoes to show slightly uneven wear or for each shoe to wear differently. A shoe may have very different thicknesses at opposite ends; if the rest of the system is in good condition, this is not a sign of a problem.

REMOVAL & INSTALLATION

♦ See Figures 59, 60, 61, 62, 63, 64, 65, 66, 67, 68, 69, 70, 71 and 72

Disassemble the brakes one wheel at a time. This will prevent parts confusion and also prevent the opposite wheel cylinder pistons from popping out during repair. While this work can be accomplished with common hand tools, there are several specialty brake tools available in automotive parts and retail stores which can make the job easier. The brake spring tool, in particular, is strongly recommended.

1. Raise and safely support the vehicle on jackstands.
2. Remove the rear brake drums as described above.
3. Using brake spring pliers, disconnect the shoe retainer spring (the bottom spring connected to both brake shoes) and disconnect the shoe return spring (the spring which is wrapped around the adjuster) and brake shoe adjuster.
4. Remove the shoe hold-down spring on either side by first depressing slightly and then twisting the retaining cap (there is a common brake tool that is designed to make this almost effortless). The pins can rotate in the backing plate, so reach around with one hand to keep the pin from spinning while twisting the retaining caps. The tangs on the retaining post must line up with the notches in the cap; release the cap in this way and then remove the cap, spring, and spring seat. Remove shoe retaining spring with the pliers.
5. Remove the shoe, to which the parking brake cable is not attached.

BRAKE SYSTEMS 9-27

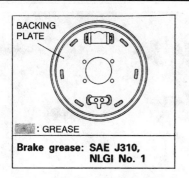
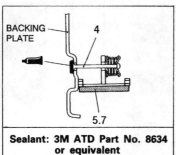
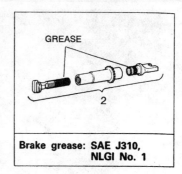
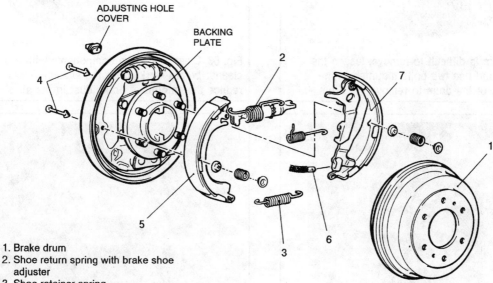

1. Brake drum
2. Shoe return spring with brake shoe adjuster
3. Shoe retainer spring
4. Shoe hold-down pin
5. Shoe and lining assembly
6. Parking brake rear cable end connection
7. Shoe and lever assembly

Fig. 57 Rear brake drum and shoe removal and installation components — rear drum brake equipped Pick-ups and Monteros

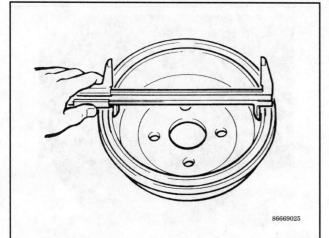

Fig. 58 Use a micrometer or ruler to measure the inside diameter of the brake drum at several locations around the circumference

Fig. 59 Remove the wheel and pull the drum off of the lug studs

9-28 BRAKE SYSTEMS

Fig. 60 If the drum is difficult to remove, lessen the adjuster tension and use two bolts mounted in the holes on the face of the drum to remove the drum

Fig. 61 Tighten the bolts to push the drum off of the lug studs — always tighten the bolts equally to prevent the drum from cocking to one side

Fig. 62 Once the drum is removed, the brake components can be seen and identified

Fig. 63 Clean the brake components with a brake cleaner before working on the system — this will help reduce the amount of brake dust in the air

Fig. 64 Using the brake tool, remove the shoe return spring from both brake shoes

Fig. 65 Using another special brake tool (available at most automotive stores), twist the spring retainer to unlock it from the brake shoe's hold-down pin

BRAKE SYSTEMS 9-29

Fig. 66 After the pin has been loosened, remove the hold-down spring and the two metal retaining caps

Fig. 69 Remove the other brake shoe using the same method as the first — after removal, only the wheel cylinder and the parking brake cable will remain

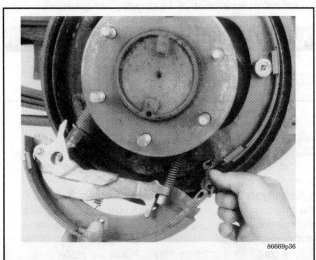

Fig. 67 Remove one brake shoe and disconnect the shoe retainer spring from the bottom of both shoes

Fig. 70 Disconnect the parking brake cable from the backing plate and pull out through the back

Fig. 68 Disconnect the parking brake cable from the shoe — this is easier when the shoe has been removed from the backing plate

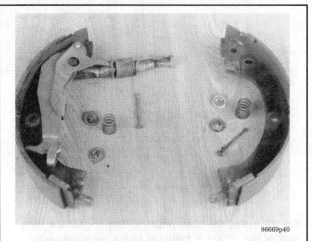

Fig. 71 Since the brake parts can easily be mixed up, lay them out on a flat surface in the positions from which they were removed

9-30 BRAKE SYSTEMS

6. Remove the other shoe then disconnect the parking brake cable at the shoe. It's easier to disconnect the cable with the shoe removed from the backing plate, but can be done otherwise if you wish.

7. If the parking brake cable is to be removed entirely, you can now remove the snapring at the backing plate and pull the end of the cable through.

8. To install, first grease the surfaces of the shoe that will contact the backing plate (this means the side edge of the shoe, NOT the friction surface), the backing plate contact points, and the working surfaces of the anchor plate and wheel cylinder pistons. Use a grease recommended for precisely this application SAE J310, NLGI No. 1); not all lubricants are appropriate.

9. Assemble hold down pins and springs and install the parking brake cable to the shoe. Install the shoe and secure the spring pin with the push-and-twist motion. (This can become a three-handed job; an assistant may be required.).

10. Install the other shoe and secure it.

11. Install the adjusting mechanism and springs, making sure the springs are in the correct position. It is quite possible to install the springs backwards; this will cause impaired function and noise. When installing the shoe retainer spring and the shoe return spring, screw the adjuster in to its narrowest setting. To correctly install the shoe return spring and adjuster, follow this procedure:

   a. Install the right-hand thread brake adjuster to the left-hand side brake, and the left-hand thread brake adjuster to the right-hand side brake.

   b. Install the brake shoe adjuster, so that the identification grooves face outward.

   c. Attach the longer end of the shoe return spring to the shoe and lever assembly.

   d. Turn the brake shoe adjuster to adjust the outside diameter of the brake shoe assembly.

➡ **The adjuster on Montero and Pick-up has an identification groove on the non-forked end. This groove must face outward at installation.**

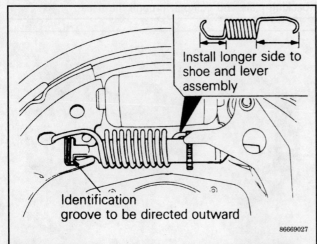

Fig. 72 Install the shoe return spring so that the long side of the spring is mounted to the lever assembly shoe

12. Double check the placement and installation of all components. Everything must be accurately placed and securely fastened.

13. Use the adjuster to set the brakes to the correct diameter. Use an accurate measuring device to measure the outer diameter of the brake shoes. This measurement must be set correctly; otherwise either the drum will not go over the shoes or the adjustment will be too loose. Correct measurements are:
   - Montero: 9.96-25.30-25.35cm
   - Pick-up, 2WD 1983-86: 24.00-24.05cm
   - Pick-up, all others: 25.30-25.35cm

14. Install the brake drum as described above.

15. Install the wheels. Make sure you depress the brake pedal repeatedly to fully adjust the self-adjusters before operating the vehicle. Operate the vehicle at very low speed in both forward and reverse gears, pumping the brake pedal repeatedly. Bring the vehicle to a full stop and operate the parking brake several times. Pedal feel should be normal, but the vehicle should roll freely (with no drag) when the pedal is released. If any condition is abnormal, the brake adjustment and/or components must be investigated and corrected.

Wheel Cylinders

ON-VEHICLE INSPECTION

The wheel cylinder seals are prone to leakage from deterioration. This condition may be checked on the truck without disassembly. After the truck is elevated and supported safely, remove the rear wheels. Remove the brake drum on each side.

Carefully lift the edge of each boot away from the body of the cylinder. Inspect the inside of the boot and the edge/end of the cylinder. You can expect a very slight (normal) moistness in the area — usually covered in dust — but any sign of wet fluid is cause for immediate repair. A leak, no matter how slight, can significantly reduce the braking effort on that wheel. It can also admit air into the brake system, causing poor pedal feel.

REMOVAL & INSTALLATION

♦ See Figures 73, 74, 75 and 76

1. Remove the brake drums and shoes as described previously. Although some vehicles will allow the cylinder to be removed with the brake shoes in place, this is not recommended. Chances are good that the brake shoes will become soaked with brake fluid. Contaminated brake shoes will need replacement so you'll have to take them off anyway.

2. Place a drain pan under the work area and disconnect the brake line at its connection to the wheel cylinder (behind the backing plate). Plug the open end of the brake line.

➡ **Work carefully and use a line wrench on the fitting. This fitting is usually rusted or otherwise difficult to turn. Don't damage the brake line.**

BRAKE SYSTEMS 9-31

3. Remove the two bolts that fasten the wheel cylinder to the backing plate and remove the wheel cylinder. Again, these bolts may be harder than usual to turn.

4. Install the cylinder and tighten the mounting bolts to 13 ft. lbs. (18 Nm). This is just enough to hold the cylinder in place; overtightening will strip the bolt holes.

5. Connect the brake line to the cylinder.

6. Install the shoes and drums correctly. Make sure the self-adjusters have taken up play so the brakes actuate normally.

7. Bleed the system thoroughly before operating the vehicle.

OVERHAUL

➡This procedure requires a special piston cup installing tool MB990623-01 for 2-wheel drive Pick-ups and all Monteros, 4-wheel drive Pick-ups need MB 990621-01. A generous supply of DOT 3 brake fluid, alcohol or aerosol brake cleaner should be available for cleaning purposes.

Fig. 74 Disconnect the brake line and the retaining bolts from the rear of the wheel cylinder

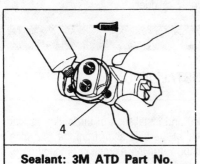

Sealant: 3M ATD Part No. 8634 or equivalent

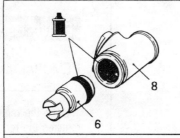

Brake fluid: DOT 3 or DOT 4

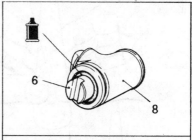

Repair kit grease (pink)

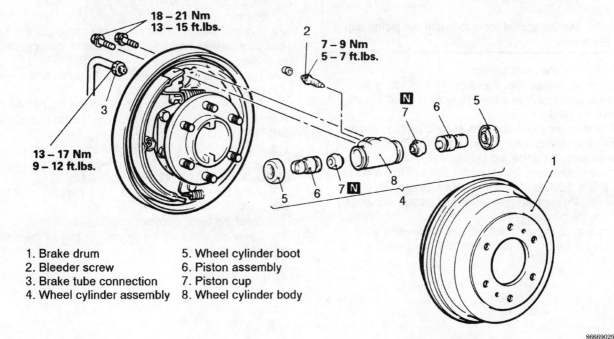

1. Brake drum
2. Bleeder screw
3. Brake tube connection
4. Wheel cylinder assembly
5. Wheel cylinder boot
6. Piston assembly
7. Piston cup
8. Wheel cylinder body

Fig. 73 Rear brake wheel cylinder removal and installation components — all rear drum brake Pick-ups and Monteros

9-32 BRAKE SYSTEMS

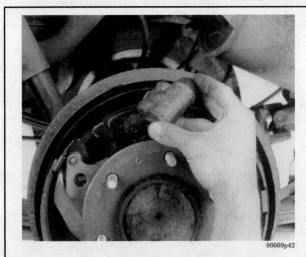

Fig. 75 Remove the wheel cylinder from the backing plate to rebuild or replace with a new unit

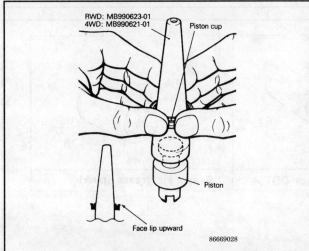

Fig. 76 Use the special tools to install the piston cup to the piston

1. Remove the wheel cylinder.
2. On either side, work the rubber boot off with a soft, blunt instrument such as a rounded piece of wood and then remove the piston.
3. Inspect the piston and cylinder walls for corrosion or scoring and replace parts that are defective. Check the clearance between the piston and cylinder wall. (Subtract the outer diameter of the piston from the inner diameter of the bore.) The limit is 0.006 in. (0.15mm). If clearance exceeds that value, replace the parts. Use inside and outside micrometers to make the measurements.
4. Clean all the parts with either brake cleaning spray or clean brake fluid. Clean the pistons and cylinder wall with DOT 3 brake fluid. If the piston/cylinder can be reused, replace the piston cup as follows:

   a. Work the cup off the end of the piston without causing damage.
   b. Apply the special grease included in the rebuild kit to the new piston cup and the special tool. If no grease is provided, use clean brake fluid.
   c. Set the piston on the workbench with one end up. Mount the special tool on top of the piston. Put the piston cup over the special tool with its lip facing upward.
   d. Slide the cup slowly and steadily down the special tool and into the groove of the piston. Be careful not to stop partway down! The idea is one smooth motion without stretching or ripping the cup.
   e. Repeat the operation for the other piston.

5. Coat the wheel cylinder bore and and piston cups either with the corrosion preventive agent contained in the repair kit or with clean brake fluid.
6. Carefully install the pistons into the wheel cylinder.
7. Apply the grease included in the repair kit to both of the piston ends and then install new rubber boots on both ends. Do not use other greases; the seals may be damaged.
8. Install the wheel cylinder.

Brake Backing Plate

REMOVAL & INSTALLATION

Since the removal and installation of the brake backing plate requires the removal of the axle shaft from the rear axle housing, referral to Section 7 for the extraction procedures of the axle shaft.

1. Raise the rear of the vehicle and support with jackstands. Make sure to chock the front wheels with wheel blocks and that the parking brake is not engaged.
2. Remove the wheels.
3. Remove the rear drum, the brake components and the wheel cylinder from the backing plate — refer to procedures in this section.
4. Remove the rear axle shaft from the axle housing — refer to Section 7 for this procedure.
5. Remove the four retaining bolts holding the backing plate on the axle shaft, then remove the backing plate from the axle shaft.
6. Install the backing plate onto the axle shaft, then install the axle shaft into the axle housing. Tighten the axle flange-to-axle housing bolts to 36-43 ft. lbs. (50-60 Nm).
7. Install the wheel cylinder, the brake components and the drums onto the rear axles.
8. Bleed the brake system and reinstall the wheels.
9. Lower the vehicle to the ground and remove the front wheel chocks.

BRAKE SYSTEMS 9-33

REAR DISC BRAKES

✳✳CAUTION

Brake pads and shoes may contain asbestos, which has been determined to be a cancer causing agent. Never clean the brake surfaces with compressed air! Avoid inhaling any dust from brake surfaces! When cleaning brakes, use commercially available brake cleaning fluids.

✳✳WARNING

If the vehicle has been recently driven before repairs, all the brake surfaces and components may be very hot. Work carefully and wear gloves.

Brake Pads

WEAR INDICATORS

Some rear disc brake pads are equipped with a metal tab which will come into contact with the disc after the friction surface material has worn near its usable minimum. The wear indicators make a constant, distinct metallic sound that should be easily heard. (The sound has been described as similar to either fingernails on a blackboard or a field full of crickets.) The key to recognizing that it is the wear indicators and not some other brake noise is that the sound is heard when the truck is being driven WITHOUT the brakes applied. It may or may not be present under braking during normal driving.

INSPECTION

The rear brake pads may be inspected without removal. With the rear end elevated and supported, remove the wheel(s). View the pads, inner and outer, through the cut-out in the center of the caliper. Remember to look at the thickness of the pad friction material (the part that actually presses on the disc) rather than the thickness of the backing plate which does not change with wear.

Remember that you are looking at the profile of the pad, not the whole thing. Brake pads can wear on a taper which may not be visible through the window. It is also not possible to check the contact surface for cracking or scoring from this position. This quick check can be helpful only as a reference; detailed inspection requires pad removal.

After the pads are removed, measure the thickness of the LINING portion (NOT the backing) of the pads. It must be at least 0.08 in. (2mm) for all 1992-95 Monteros. This is a Mitsubishi factory minimum; please note that local inspection standards enforced by your state must be given precedence if they require a greater thickness.

REMOVAL & INSTALLATION

▶ See Figures 77, 78, 79, 80, 81 and 82

➡Whenever brake pads are replaced, replace them in complete sets; that is, replace the pads on both front wheels even if only one side is worn.

There are several combinations of shims, spacers and clips in use on Mitsubishi vehicles. When disassembling, work on one side at a time and pay attention to placement of these components. If you become confused during reassembly, refer to the other side for correct placement. To remove the brake pads:

1. Raise the vehicle and support it safely on jackstands.
2. Remove the front wheels.
3. Siphon a sufficient amount of brake fluid from the master cylinder reservoir to prevent the brake fluid from overflowing when installing new brake pads. This is necessary as the piston must be forced into the cylinder bore to provide clearance to install the caliper with the new pads.
4. Unscrew and remove the lockpin without disturbing the grease coating. Place the pin in a clean spot where grease will not pick up dust. Lift up the caliper body by using the guide pin bolt as a fulcrum. Fasten the caliper in the raised position with wire.

✳✳WARNING

There is a coating of special grease on the lockpin. Be careful that this grease is not removed, and that dirt does not contaminate the grease on the pin.

5. Remove the shims, brake pads and spring clips in order. Depending on the number of such components, either diagram their placement or lay them aside in order.
6. Clean the surface of the piston with a clean, damp rag. Make sure there is room in the master cylinder reservoir for more fluid (fluid will be forced back into the reservoir in the next step). If necessary, remove some fluid with a clean squeeze ball syringe.
7. Use a large C-clamp or similar tool to depress the caliper pistons back into the calipers. The pistons will need to be almost flush with the case to fit over the new, thicker pads.
8. Install the brake pads with the shims, clips and fittings in the correct order. Vehicles with upper and lower spring clips for each pad should have these clips replaced with the pads.

➡Keep the brake surface of each pad free of grease, oil and fluids during the installation. A greasy fingerprint or similar light contact may be removed with a commercial brake cleaning spray.

9. Remove the wire, lower the caliper into position, and screw in the lower lockpin. Make certain the rubber boot is correctly placed on the lockpin and is not pinched or deformed.

9-34 BRAKE SYSTEMS

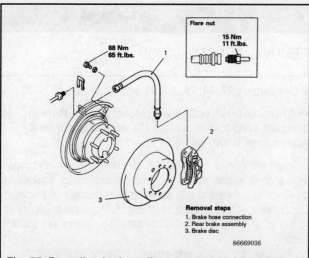

Fig. 77 Rear disc brake caliper removal and installation components — 1992-95 Monteros

10. Replace the wheels and lower the vehicle.

✲✲WARNING

Do NOT attempt to drive the vehicle immediately after lowering it to the ground. The first two or three brake pedal applications may not provide any brake response.

11. Pump the brake pedal several times with the engine off. The first two or three pedal strokes may be longer than usual as the pistons return from their compressed position and drive the pads inward. After a reasonable feel has been achieved, check the master cylinder reservoir and top up the fluid as needed. Start the engine and pump the brake pedal again, checking for the proper feel and engagement point.

12. Since the brake hoses and hydraulic system was not opened during the repair, it is usually not necessary to bleed the brake system after pad replacement. Use good judgement

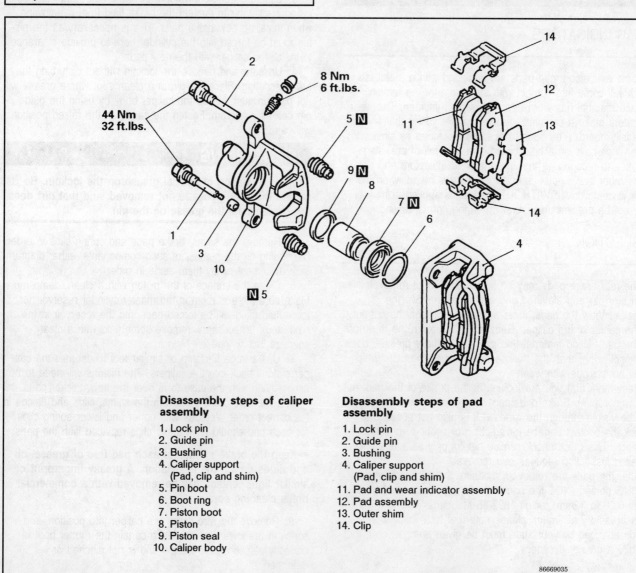

Fig. 78 Rear brake caliper disassembly and reassembly components — 1992-95 Monteros

BRAKE SYSTEMS 9-35

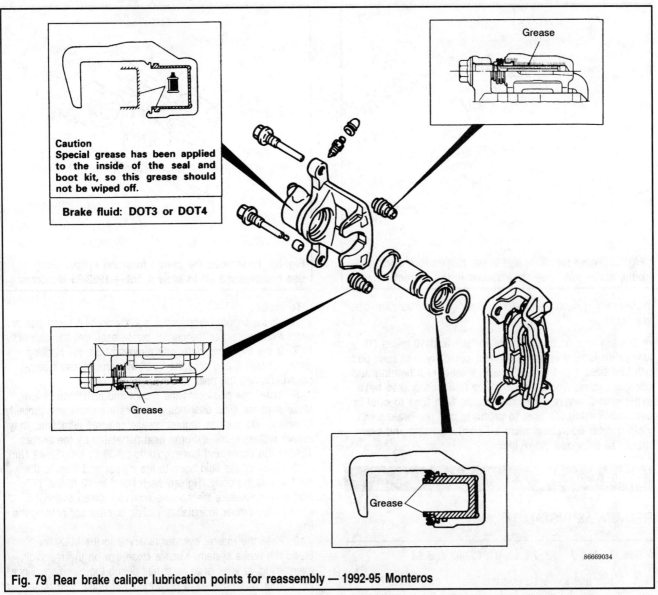

Fig. 79 Rear brake caliper lubrication points for reassembly — 1992-95 Monteros

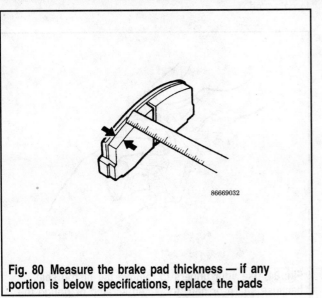

Fig. 80 Measure the brake pad thickness — if any portion is below specifications, replace the pads

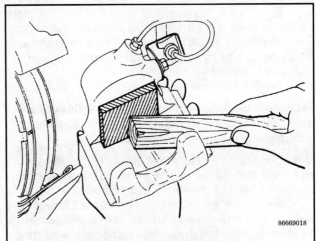

Fig. 81 Use a C-clamp to push the cylinder back into the caliper, but two pieces of wood can be used if no C-clamp is available

9-36 BRAKE SYSTEMS

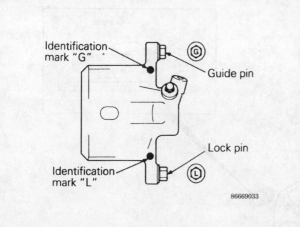

Fig. 82 When installing the brake caliper lock and guide pins, make sure they are installed in the correct holes

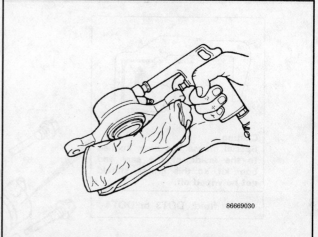

Fig. 83 To remove the piston from the caliper body, use compressed air to blow it out — 1992-95 Monteros

in determining pedal feel; bleeding may be necessary for other reasons.

➡Braking should be moderate for the first 10 miles (16 km) or so until the new pads seat correctly. The new pads will bed best if put through several moderate heating and cooling cycles. Avoid hard braking until the brakes have experienced several long, slow stops with time to cool in between. Taking the time to properly bed the brakes will yield quieter operation, more efficient stopping and contribute to extended brake life.

Brake Caliper

REMOVAL & INSTALLATION

◆ See Figures 77, 78, 79, 80, 81, 82, 83 and 84

1. Raise and safely support the front of the vehicle on jackstands. Set the parking brake and block the rear wheels.
2. Siphon a sufficient amount of brake fluid from the master cylinder reservoir to prevent the brake fluid from overflowing when removing or installing the caliper. This is necessary as the piston must be forced into the cylinder bore to provide clearance to install the caliper.
3. Remove the wheel.

➡Disassemble brakes one wheel at a time. This will prevent parts confusion and also prevent the opposite piston from popping out during installation.

4. Disconnect the brake line at the clip on the strut and disconnect the hose union at the caliper. Use a pan to catch any spilled fluid and immediately plug the hose end.
5. Remove the two caliper mounting bolts and then remove the caliper from the mounting bracket. Alternatively, remove the two bolts (screwed directly into the backing plate) which fasten the caliper support to the backing plate, and remove the entire assembly. Remove the brake pads and clips from the support as described above. Remove the lockpin (lower) and guide pin (upper) from the caliper support, and remove the caliper from the support.

To install:

6. Use a caliper compressor, a C-clamp or a large pair of pliers to slowly press the caliper piston back into the caliper.
7. If the mounting plate was removed from the backing plate, reinstall it and coat the bolts with an anti-seize compound. Tighten the bolts to 66 ft. lbs. (90 Nm).
8. Install the pads onto the mount and install the caliper. Make sure the slide bushings and bolts are clean and properly lubricated. Do not use regular grease or spray lubricants; they cannot withstand the extreme heat generated by the brakes. Tighten the upper and lower bolts to 23-30 ft. lbs. (32-42 Nm).
9. Connect the fluid hose to the caliper and then to the hard line at the body. Tighten each fitting to 12 ft. lbs. (17 Nm) and make sure the hose is correctly routed and not kinked. Remember to install any clips or retainers holding the line.
10. Refill the master cylinder reservoir to the MAX line. Bleed the brake system. Keep a close eye on the reservoir, maintaining its level at at least half during the bleeding. Since the caliper was emptied, a fair amount of brake fluid may need to be added.

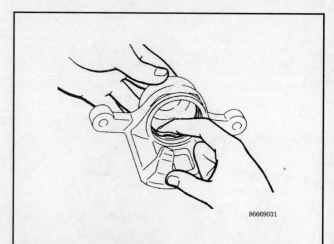

Fig. 84 Use your fingers to remove the piston seal, because sharp or hard tools can scratch or gouge the piston bore

BRAKE SYSTEMS 9-37

11. Reinstall the wheel and lower the vehicle to the ground.
12. Check the level in the fluid reservoir and top it off as needed. Before driving the car, start the engine and operate the brake pedal several times to check the feel and engagement point of the brakes.

OVERHAUL

♦ See Figures 77, 78, 79, 80, 81, 82, 83 and 84

➡ To overhaul the brake calipers, you must have a controlled source of compressed air to force the pistons out. You should also have a generous supply of clean brake fluid or commercial brake cleaning spray to clean parts.

1. Drain the remaining fluid from the caliper.
2. Carefully remove the dust boot from around the piston by prying the boot ring out of its groove. Remove the piston and piston boot by applying compressed air through the brake hose fitting hole.

✲✲CAUTION

Do not hold the caliper during this procedure, and do not place fingers in front of the piston in an attempt to catch it or protect it when applying compressed air. Place a piece of cloth in front of the piston, and slowly increase the force of the compressed air to prevent the piston from springing out abruptly; use just enough air pressure to ease the piston out.

3. Remove the seal from the inside of the caliper bore. Clean the caliper bore with brake cleaner, alcohol or brake fluid.

✲✲WARNING

Never use a screwdriver or similar tools, because doing so could damage the cylinder surface.

4. Check all the parts for wear, deterioration, cracking or other abnormal conditions. Corrosion, generally caused by water in the system, will appear as white deposits on the metal, similar to what may be found on an old aluminum storm door around the house. Pay close attention to the condition of the inside of the caliper bore and the outside of the piston. Any sign of corrosion or scoring requires new parts; do not attempt to clean or resurface either face if there is damage present.
5. The caliper overhaul kit will, at the minimum, contain new seals and dust boots. A good kit will contain a new piston as well, but you may have to buy the piston separately. Any time the caliper is disassembled, a new piston is highly recommended in addition to the seals.
6. Clean all the components to be reused with a brake cleaner, alcohol or clean brake fluid and dry them thoroughly. Take any steps necessary to eliminate moisture or water vapor from the parts.
7. Coat all the caliper components with fresh brake fluid from an unopened can.

➡ Some kits come with special assembly lubricants for the piston seals and slides or guidepins. Use these lubricants according to the directions within the kit.

8. Install the piston seal and piston into the caliper bore. This is an exacting job; the clearances are very small. Make sure the seal is seated in its groove and that the piston is not cocked when inserted into the bore.
9. Install the new dust boot and new boot ring.
10. Install the slide bushings and rubber boots onto the caliper if they were removed during disassembly.
11. The guide and lockpin sleeves should be coated inside with the grease supplied with the rebuilding kit, and boots should all be replaced. Also coat the lips of the boots and the surface of the caliper which bears (and turns) against the caliper support with grease.

➡ Two kinds of grease may be packed in a rebuilding kit; use the grease recommended for use with the guide and lockpins. On all the truck systems, coat the sliding parts of the caliper body and the sleeves with the grease specified.

12. Install the brake caliper to the wheel assembly.

Brake Disc (Rotor)

REMOVAL & INSTALLATION

♦ See Figure 77

1. Raise and safely support the vehicle on jackstands.
2. Remove the wheel(s).
3. Remove the rear brake caliper and mount (brake assembly) following procedures given previously in this section, but without disconnecting the hydraulic line. Suspend the assembly with wire; do not allow the unit to hang by the brake line.
4. Make mating marks to show the relationship between the brake disc and hub; pull the disc off the hub.
5. Slide the rear brake disc onto the rear axle lug studs so that the matchmarks are aligned.
6. Install the rear brake caliper to the disc and brake mount — refer to the procedures found in this section.

INSPECTION

Brake discs should be inspected for thickness at a number of spots around the braking surface with a micrometer. If the thickness AT ANY POINT is less than the minimum given in the Brake Specifications Chart, the disc must be replaced. Roughness or significant grooving is also reason to replace the disc.

The disc should also be checked for run-out. With the disc installed (hold it with two lug nuts if necessary), mount a dial indicator to the shock and zero it with the tip in contact with the braking surface on the disc about 0.60 in. (15mm) from the edge. Rotate the disc slowly. Read the indicator; the limit for

9-38 BRAKE SYSTEMS

run-out is 0.006 in. (0.15mm). The disc must be replaced if run-out exceeds this amount.

PARKING BRAKE

Cables

ADJUSTMENT

1983-91 Montero

❉❉WARNING

On vehicles with rear drum brakes, the shoe-to-drum clearance must be adjusted before adjusting the cable. If the brake clearance is out of adjustment, the cable adjustment will not cure the problem.

1. Apply the brake with normal pressure (about 40 lbs. or 182 N) and count the number of clicks required to bring the lever tight enough to hold the vehicle. Don't try to pull the lever into the back seat, 4-5 clicks are sufficient. If the number of clicks is incorrect, proceed with the remaining steps.
2. Release the brake lever and tighten the cable adjuster (located under the vehicle, at the center of the cable equalizer) until all cable slack is just removed.
3. Start the engine, allowing it to idle and apply the footbrake and release it. Then apply the handbrake and release it, apply the footbrake and release it, etc. in a continuous cycle until the automatic adjusters at the rear stop clicking.
4. Recheck the number of clicks required to apply the brake. Adjust the cable adjuster and repeat the check until the number of clicks required is correct.
5. Raise the rear of the vehicle and support it safely. Release the handbrake and rotate each rear wheel to make sure the brakes are not dragging.

1992-95 Montero

▶ See Figures 85 and 86

1. Pull the parking brake lever and pull up until the parking brakes will hold the vehicle from moving. The handle should only have to be pulled up 4-6 clicks.
2. If the handle stroke is not within the standard value range, make the following adjustments.
3. Loosen the adjuster to slacken the parking brake cable.
4. Remove the adjustment hole plug on the rear brake assemblies, and then use flat-tipped prytool to turn the adjuster in the direction of the arrow (the direction which expands the shoe) so that the disc will not rotate.
5. Return the adjuster 3-4 notches in the direction opposite to the direction of the arrow.
6. Turn the adjusting nut to adjust the parking brake lever stroke to within the standard value range.

➡If the number of brake lever notches engaged is less than the standard value, the cable has been pulled excessively. Be sure to adjust it to within the standard value.

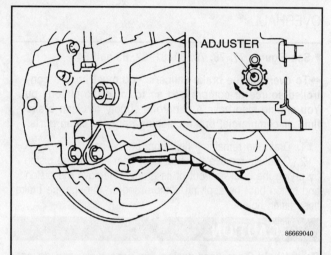

Fig. 85 Turn the adjuster in the direction of the arrow so that the disc will not rotate — 1992-95 Montero

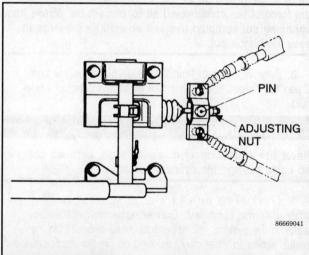

Fig. 86 The adjustment components on 1992-95 Montero parking brakes

7. After making the adjustment, check to be sure that there is no play between the adjusting nut and the pin. Also check that the adjusting nut is securely held at the nut holder.
8. After adjusting the lever stroke, jack up the rear of the vehicle and support the vehicle with jackstands.
9. With the parking brake lever in the released position, turn the rear wheel to confirm that the rear brakes are not dragging.
10. Lower the vehicle back to the ground.

BRAKE SYSTEMS 9-39

Pick-up

♦ See Figures 87, 88 and 89

1. Pull the lever up with a force of about 66 lbs. (300 N) and count the number of notches or clicks. Correct number is 16 or 17.
2. If the number of clicks is incorrect, elevate and safely support the vehicle on stands; block the front wheels.
3. Under the vehicle, loosen the adjusting nuts to slacken each brake cable (one runs to each rear wheel from the equalizer).
4. Tighten the adjusting nuts just to the point of taking the slack out of the cable. Repeatedly pull and release the brake lever to adjust the rear brake shoes.
5. Tighten the adjusting nuts until the lever has the correct number of clicks.

➡ Make certain that the joint and equalizer are at right angles when the adjustment is finished. An angle other than 90° will cause uneven brake application.

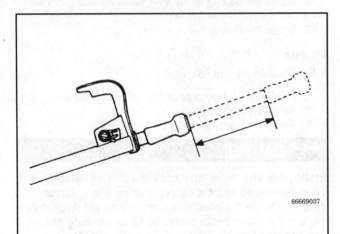

Fig. 87 Pull the parking brake handle out until the parking brakes will hold the vehicle from rolling — Pick-ups

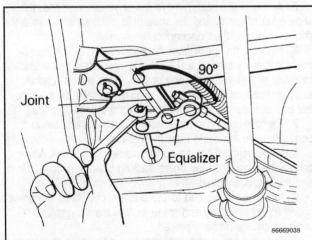

Fig. 88 Under the truck, tighten the nuts to the cables so that the equalizer forms a 90° angle with the bracket — Pick-up trucks

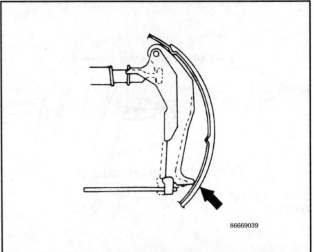

Fig. 89 Remove the rear drums and check to make sure that the brake lever is just touching the brake shoe

6. Fully release the parking brake lever and remove each rear wheel and brake drum. Make certain the brake lever is just touching the shoe.

✵✵WARNING

If the parking brake cable is pulled too far, the adjuster lever will not fit the adjuster, resulting in faulty operation.

7. Reinstall the drum and install the wheel.
8. Apply and release the parking brake once or twice. With the brake fully released, spin each rear wheel by hand and check for dragging brakes.

REMOVAL & INSTALLATION

Pick-up

FRONT CABLE

♦ See Figure 90

1. Elevate and safely support the vehicle on stands.
2. Inside the vehicle, under the dash, identify and unplug the parking brake indicator switch. Remove the switch and bracket.
3. The end of the cable sheath is held in place by a snapring. Remove it. You are reminded that these snaprings tend to vanish if not removed carefully.
4. Lift the pawl from the ratchet and push the parking brake lever all the way in. In this position, the cable may be disconnected from the lever.
5. Under the vehicle, trace the cable and remove the clips and brackets holding it to the body.
6. Disconnect the cable from the lever assembly. Remove the cable from the vehicle. Don't damage the body grommet while removing it.
7. Route the new cable into position. It's usually easier to attach the inner end to the lever first, then deal with the equalizer end. Loosen the adjusting nuts on the rear cables if necessary.

9-40 BRAKE SYSTEMS

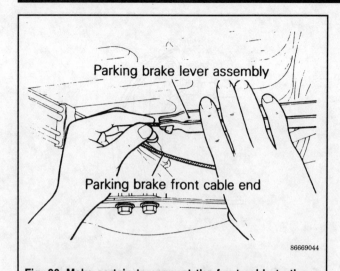

Fig. 90 Make certain to connect the front cable to the lever assembly correctly — Pick-up trucks

8. Once both ends are secured, install the clips and brackets holding the cable to the body.
9. Fit the rubber body grommet into place and make certain it is secure.
10. Install the snapring on the end of the cable.
11. Install the brake indicator switch in place and connect the wiring but do not final tighten the switch mount.
12. Adjust the parking brake lever stroke according to procedures given earlier in this section.
13. With the cable correctly adjusted, adjust the switch so that the warning light illuminates when the lever is pulled one notch. Tighten the switch mount.

REAR CABLE
▶ See Figures 91, 92 and 93

1. Elevate and safely support the vehicle on stands.
2. Remove the rear wheels.
3. Remove the brake drum.

✱✱CAUTION

Brake pads and shoes may contain asbestos, which has been determined to be a cancer causing agent. Never clean the brake surfaces with compressed air! Avoid inhaling any dust from brake surfaces! When cleaning brakes, use commercially available brake cleaning fluids.

4. At the equalizer, disconnect the return spring.
5. With the parking brake released, remove the adjusting nut from the cable to be removed. If only one cable is being removed, loosen the other adjuster to make the job easier. If both cables are being removed, loosen the nuts alternately and evenly.
6. Disconnect the brake cable from the brake shoe. You may find it easier to remove the shoes as a unit, then disconnect the cable.
7. Remove the cable from the backing plate. This can be made easier by feeding an offset box wrench (12mm) over the end of the cable and pushing it up until it reaches the stopper.

Push on the wrench to release the tabs on the stopper and pull the cable free from behind the backing plate.
8. Remove the clips and brackets holding the cable to the body. Remove the cable from the vehicle.
9. When reinstalling, thread the cable into position. Make certain the stopper at the brake backing plate is securely fitted and the tab is engaged.
10. Install the cable end to the brake shoe. If the brakes were removed, reinstall them.
11. Turn the brake shoe adjuster to set the outside diameter of the shoes to 9.45-9.47 in. (240.0-240.5mm) for 1983-86 rear-wheel drive and 9.96-9.98 in. (253.0-253.5mm) for all other Pick-ups. This is an important dimension and must be observed.
12. Install the brake drum.
13. Connect the brake cable to the equalizer and tighten the adjuster nut just enough to remove slack from the cable.
14. Install the return spring correctly.
15. Install the rear wheels.
16. Adjust the parking brake lever travel according to instructions given previously in this section.
17. Lower the vehicle to the ground.

Montero
▶ See Figures 94 and 95

1. Elevate and safely support the vehicle on stands.
2. Remove the rear wheels.
3. Remove the rear brake drum or brake disc.

✱✱CAUTION

Brake pads and shoes may contain asbestos, which has been determined to be a cancer causing agent. Never clean the brake surfaces with compressed air! Avoid inhaling any dust from brake surfaces! When cleaning brakes, use commercially available brake cleaning fluids.

4. At the equalizer, remove the adjusting nut. Disconnect the cable to be replaced from the equalizer yoke. Remove the small heat shield from the cable.
5. At the rear wheel, remove the shoe return spring, the shoe retainer spring and the shoe hold-down pins. Remove the shoes as a unit, then disconnect the cable.
6. Remove the cable from the backing plate. This can be made easier by feeding a 0.47 in. (12mm) offset box wrench over the end of the cable and pushing it up until it reaches the stopper. Push on the wrench to release the tabs on the stopper and pull the cable free from behind the backing plate.
7. Remove the clips and brackets holding the cable to the body. Remove the cable from the vehicle.
8. When reinstalling, thread the cable into position. Make certain the stopper at the brake backing plate is securely fitted and the tab is engaged.
9. Install the cable end to the brake shoe. Install the brake shoes as an assembly and make certain the springs and retaining clips are properly placed.
10. Turn the brake shoe adjuster to set the outside diameter of the shoes to 9.96-9.98 in. (253.0-253.5mm) on rear drum brake systems. This is an important dimension and must be observed.
11. Install the brake drum or brake disc.

BRAKE SYSTEMS 9-41

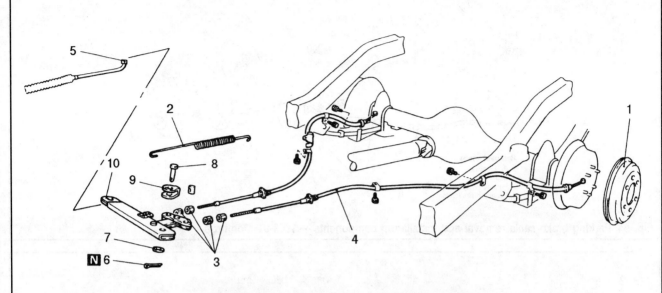

Parking brake rear cable removal steps

1. Brake drum
2. Return spring
3. Adjusting nut
4. Parking brake rear cable

Lever assembly removal steps

2. Return spring
5. Parking brake front cable end
3. Adjusting nut
6. Cotter pin
7. Plain washer
8. Clevis pin
9. Spacer
10. Lever assembly

Fig. 91 Rear parking brake cable removal and installation components — 1983-95 Pick-ups

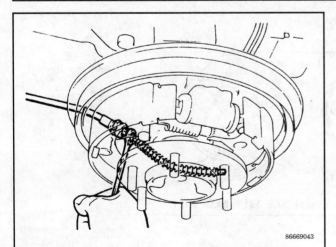

Fig. 92 Use a box end wrench to remove the parking brake cable from the rear brake assembly — 1983-95 Pick-up trucks

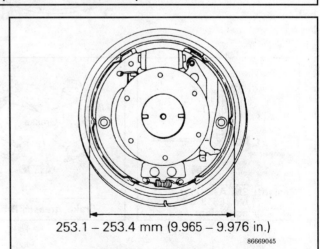

253.1 – 253.4 mm (9.965 – 9.976 in.)

Fig. 93 Adjusting the rear brake shoes to specification will allow the brakes to properly adjust themselves after installation — 1987-95 Pick-ups

9-42 BRAKE SYSTEMS

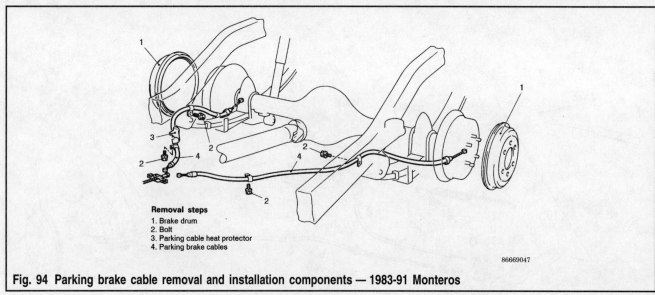

Fig. 94 Parking brake cable removal and installation components — 1983-91 Monteros

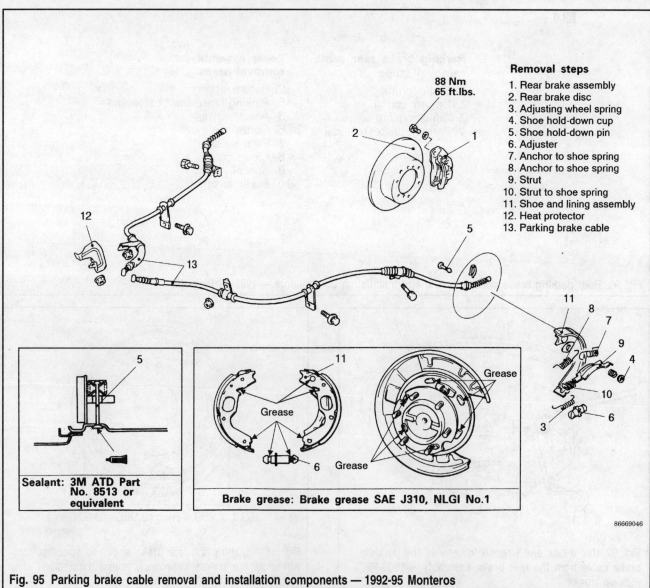

Fig. 95 Parking brake cable removal and installation components — 1992-95 Monteros

BRAKE SYSTEMS 9-43

12. Connect the brake cable to the equalizer yoke. Install and tighten the adjuster nut just enough to remove slack from the cable.
13. Install the rear wheels.
14. Install the cable retaining clips and install the small cable heat shields.
15. Adjust the parking brake lever travel according to instructions given previously in this section.
16. Lower the vehicle to the ground.

Brake Lever

REMOVAL & INSTALLATION

Pick-up
♦ See Figures 96, 97 and 98

The parking brake lever assembly is located under the center of the dashboard.

1. Remove the mounting bolt holding the parking brake switch in place on the lever assembly. Remove the switch away from the lever assembly; disconnection of the electrical wires to the switch is not necessary.
2. Disconnect the accelerator cable from the accelerator pedal assembly, then remove the accelerator pedal assembly from the parking lever bracket by removing the mounting nut and bolt from the pedal assembly.
3. Remove the snapring holding the parking brake cable to the pedal bracket. Be careful not to lose the snapring.
4. Remove the pawl from the ratchet mechanism in the brake lever. With the parking brake pull rod pushed in, disconnect the parking brake front cable end from the parking brake pull rod.
5. Remove the parking brake pull rod and bracket from the firewall.
6. To install, mount the parking brake pull rod bracket to the firewall with the one bolt.
7. Attach the front parking brake cable to the parking brake pull rod.
8. Install the front cable to the pull rod bracket, and retain with the snapring.
9. Mount the accelerator pedal bracket to the pull rod bracket with the mounting nut and bolt.
10. Attach the accelerator cable to the accelerator pedal assembly.
11. Install the parking brake switch. Adjust the parking brake switch position so that the warning lamp goes out when the parking brake pull rod is fully returned to the in position and will light when the parking brake pull rod is pulled one notch.
12. After adjusting the parking brake pull rod stroke, jack up the rear of the vehicle.
13. With the parking brake pull rod released, turn the rear wheels to verify that the rear brakes are not dragging.

Montero
♦ See Figure 99

1. Raise and safely support the rear of the Montero and block the front wheels. Make sure that the parking brake is not applied.

2. From under the Montero, disconnect the two rear parking brake cables from the equalizer assembly.
3. Inside the vehicle, remove the rear console panel, the inner box and the floor console assembly. Remove the floor console bracket.
4. Remove the retaining bolts from the parking brake stay, then remove the stay itself. Remove the bushing from inside of the parking brake stay.
5. Remove the parking brake shaft cover.
6. Disconnect the parking brake switch harness, then remove the parking brake switch from the parking brake lever assembly.
7. Remove the parking brake lever from the floor of the Montero.
8. To install, mount the parking brake lever to the floor with the mounting bolts.
9. Install the parking brake switch and plug the harness connectors back together.
10. Install the parking brake shaft cover with the retaining bolts.
11. Install the parking brake stay bushing to the stay, then mount the stay to the parking brake lever assembly and the floor.
12. From under the vehicle, connect the two rear parking brake cables to the equalizer assembly. Adjust the cables using the procedures described previously in this section.
13. Rotate the rear wheels to make certain that the rear brakes are not rubbing.
14. Lower the vehicle back down to the ground and remove the wheel blocks.

Brake Shoes

Only the 1992-95 Monteros are equipped with rear disc brakes and, therefore, have separate brake pads for the parking brakes. Other vehicles use the brake shoes in their drum brake assemblies; for those vehicles, refer to the procedures for drum brake shoe removal and installation.

✱✱CAUTION

Brake pads and shoes may contain asbestos, which has been determined to be a cancer causing agent. Never clean the brake surfaces with compressed air! Avoid inhaling any dust from brake surfaces! When cleaning brakes, use commercially available brake cleaning fluids.

REMOVAL & INSTALLATION

1992-95 Montero
♦ See Figures 100, 101, 102 and 103

➡This procedure requires the disassembly of easily confusing springs and components. Make certain that the direction as well as the location of the various parts is known before disassembly. Work on each side one at a time so that if confusion is created as to the locations of components during reassembly the other wheel can be used as an example.

9-44 BRAKE SYSTEMS

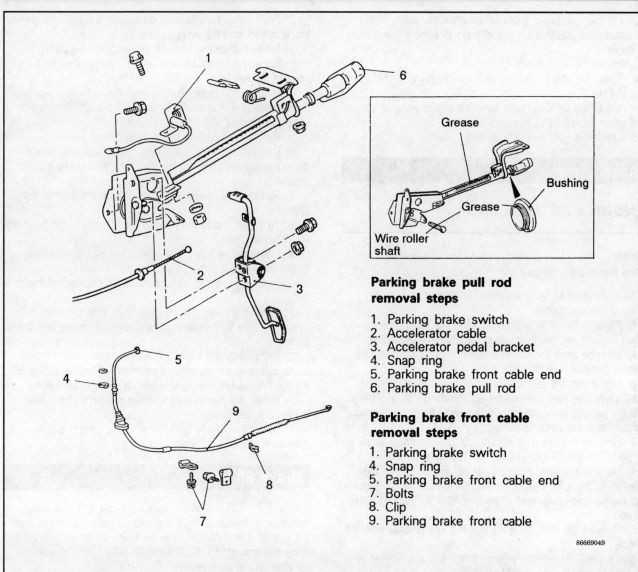

Parking brake pull rod removal steps

1. Parking brake switch
2. Accelerator cable
3. Accelerator pedal bracket
4. Snap ring
5. Parking brake front cable end
6. Parking brake pull rod

Parking brake front cable removal steps

1. Parking brake switch
4. Snap ring
5. Parking brake front cable end
7. Bolts
8. Clip
9. Parking brake front cable

Fig. 96 Parking brake pull rod and front brake cable removal and installation — 1983-95 Pick-ups

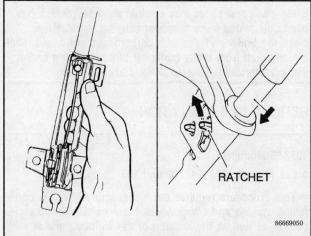

Fig. 97 To remove the front end of the front parking brake cable, remove the pawl from the ratchet, then push the rod all the way in — Pick-ups

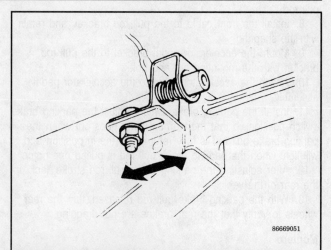

Fig. 98 Adjust the parking switch by moving it back and forth, then tighten the mounting nut to secure it in place — 1983-95 Pick-ups

BRAKE SYSTEMS 9-45

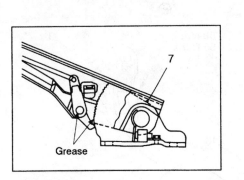
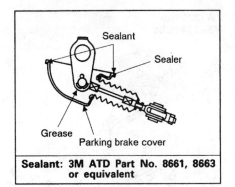
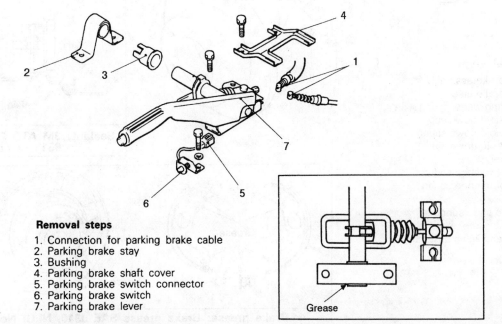

Removal steps
1. Connection for parking brake cable
2. Parking brake stay
3. Bushing
4. Parking brake shaft cover
5. Parking brake switch connector
6. Parking brake switch
7. Parking brake lever

Fig. 99 Parking brake lever assembly removal and installation components — Monteros

1. Raise and safely support the Montero on jackstands and remove the rear wheel(s). Block the front wheels and make sure that the parking brake is not applied.
2. Remove the rear brake caliper from the disc. Do not disconnect the brake hose from the caliper; suspend the caliper from the frame of the vehicle with stiff wire or cord.
3. Matchmark the disc and one of the rear lug studs, then pull the rotor off of the axle.
4. Check the brake shoes and rotor:
   a. Measure the thickness of the brake lining at several places. The standard value is 0.256 in. (6.5mm) and the minimum is 0.177 in. (4.5mm). Replace the brake shoes if any part of the thickness of the brake lining is the minimum value or less.
   b. Measure the brake disc drum inner diameter at two or more places. The standard value is 7.756 in. (197mm) and the minimum value is 7.795 in. (198mm). Replace the brake disc if the drum inner diameter is the limit value or more.
5. Disconnect the adjusting wheel spring from the bottom of both of the brake shoes.
6. Push the shoe hold-down spring in and twist it so that the slots in the shoe hold-down cup align with the tangs on the shoe hold-down pin. Release the spring and remove the spring, the cup and the pin from each shoe.
7. Remove the adjuster from in between the bottom of the two shoes.
8. Remove the two anchor to shoe springs from the top of the assembly.
9. Remove the centrally located strut and strut spring from in between the shoes.
10. Remove the retainer clip holding the parking brake cable into the back of the brake backing plate.
11. Remove each of the shoes, making sure to disconnect the rear parking cable from the one shoe.
12. To install, attach the parking brake cable to the one shoe, then install the retainer clip to hold the parking brake cable in the backing plate.
13. Install the two shoes to the backing plate, after greasing the backing plate and shoes in the specified areas (illustration).

9-46 BRAKE SYSTEMS

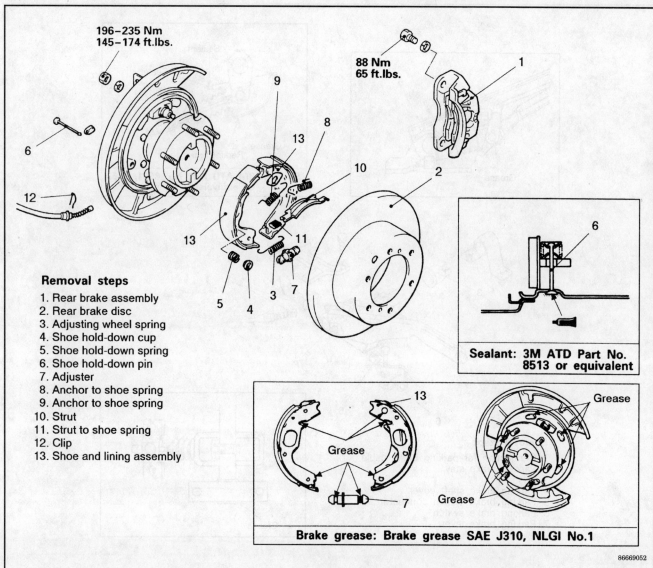

Fig. 100 Parking brake drum assembly removal and installation components — 1992-95 Monteros

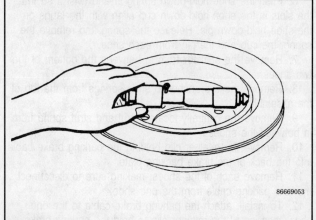

Fig. 101 Use a micrometer to measure at least two positions around the inside of the drum — replace the rear brake disc if any distance is greater than 7.795 in. (198mm)

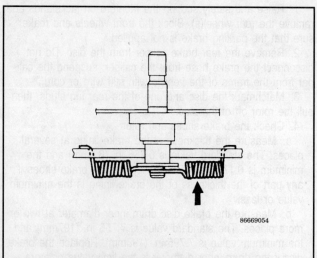

Fig. 102 Mount the painted anchor to shoe spring where the arrow indicates — 1992-95 Monteros

BRAKE SYSTEMS

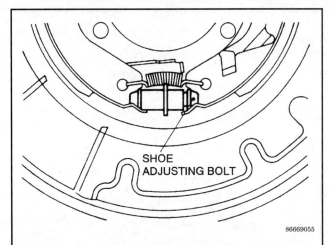

Fig. 103 Install the shoe adjusting bolt so that on the left-hand wheel it is towards the rear and on the right-hand wheel it is toward the front — 1992-95 Monteros

14. Install the strut spring to the strut, then install the strut to the two shoes. Make sure the strut is installed in the correct direction.
15. Since the load is different for each anchor to shoe spring, the spring which is to be installed on the outside (right side spring) of the other (left side spring) is painted for identification purposes.
16. Install the adjuster so that the shoe adjusting bolt for the left-hand wheel is towards the rear of the vehicle, and the shoe adjusting bolt for the right-hand wheel is towards the front of the vehicle.
17. Install the two shoe hold-down springs (along with the cups and pins) to both shoes.
18. Install the adjusting wheel spring to both of the shoes.
19. Slide the rear brake disc onto the lug studs so that the matchmarks line up.
20. Install the caliper to the disc and tighten the mounting bolts to 65 ft. lbs. (88 Nm).
21. Adjust the parking brake — refer to the procedures in this section.
22. Install the wheels and make sure that the rear brakes are not rubbing too much.
23. Lower the vehicle to the ground and remove the wheel blocks.

ADJUSTMENT

For the rear wheel parking brake adjustment, refer to the procedures for adjusting the parking brake lever.

ANTI-LOCK BRAKE SYSTEM (ABS) — FOUR WHEEL

The anti-lock system found on 1992-95 is a sophisticated electronic and hydraulic system. Sensors at each wheel generate a signal relative to the speed of that wheel; the signal is transmitted to a computer which compares the rolling speed of each wheel. If one (or more) wheels begin to slow out of proportion to the others (as would happen if one began to lock under braking), the computer signals the ABS hydraulic unit to reduce brake pressure to that wheel.

Braking effort is reduced and the wheel is allowed to roll until it approaches the same speed as the others. Stability is thus improved by eliminating wheel lock. Because of the complexity of the system, pressure changes can occur several time per second in a panic stop as the system reacts to varying signals from all four wheels.

Should any part of the ABS system fail, an indicator light on the dash will illuminate; this DOES NOT indicate total brake failure, but only a loss of anti-lock function. The brake system will still function as a normal, non-ABS system.

Models equipped with this system may exhibit various behaviors which are not seen on normal brake systems. A pulsing, felt in the brake pedal or steering, may be noticed when the system is activated by hard braking, particularly on slippery surfaces. Although surprising, this pulsing is normal and a sign that the system is reacting to conditions. Additionally, as the vehicle reaches about 4 miles per hour after starting the engine, a whining or motor whirr may be heard in the engine compartment. This is from the system performing a self-check and building pressure; again, quite normal.

As complicated as an ABS system is, it rarely fails. Extensive development has made the system reliable and long lived. Troubleshooting and diagnosing an ABS system requires test equipment for both the electrical and computer portion as well as sophisticated hydraulic test equipment. It is a job best left to those well-trained in the field.

Hydraulic Unit

TESTING

To test the hydraulic unit a special Mitsubishi scan tool and an oscilloscope would be needed. Therefore, if the ABS warning light comes on or a problem with the ABS is evident, have the system diagnosed by an automotive mechanic familiar with the Mitsubishi ABS systems. After the vehicle has been diagnosed and the cause is known, the defective part can be replaced at home.

REMOVAL & INSTALLATION

▶ See Figure 104

✱✱WARNING

Removal of this unit requires extreme care in handling and operation under the cleanest possible conditions. The slightest bit of dirt can foul the system. Do not disassemble the lines unless suitable plugs for both the lines and ports are immediately available. The need for extreme cleanliness cannot be overstated!

9-48 BRAKE SYSTEMS

The hydraulic unit is located in the engine compartment, attached to the right-hand fender by the firewall.
1. Disconnect the negative battery cable.
2. Drain the brake fluid from the system.
3. Remove the retaining screws and remove the connector bracket from the top of the relay and hydraulic unit.
4. Loosen and detach the five brake lines from the hydraulic unit. Make sure to not bend or reform any of the hard lines. Have rags handy since some brake fluid will have been retained in the unit. Make sure to keep the fluid off of any painted surfaces.
5. Unplug the engine compartment wire harness from the hydraulic unit.
6. Remove the hydraulic unit bracket from the fender and disconnect the ground cable from the bracket.
7. Remove the hydraulic unit and bracket from the vehicle.

✳✳CAUTION

The hydraulic unit is heavy, so care should be taken when removing it. The hydraulic unit is not to be disassembled; its nuts and bolts should absolutely not be loosened. The unit must not be dropped or otherwise subjected to impact shocks. It must not be turned upside down or laid on its side.

8. At this time the relay box cover, valve relay and motor relay can be removed from the top of the hydraulic unit.

To install:
9. Affix the motor relay, the valve relay and the relay box cover to the top of the hydraulic unit.
10. Install the hydraulic unit into the hydraulic unit bracket and secure in position with the retaining bolts and nuts.
11. Mount the hydraulic unit bracket into the engine compartment and tighten the mounting bolts until snug. Make sure to reattach the ground cable.
12. Plug the engine compartment wire harness connectors back together.
13. Install the brake pipes to the hydraulic unit. Tighten the brake pipe flare nuts to 11 ft. lbs. (15 Nm).
14. Set the connector bracket in place and secure it there with the retaining screws.
15. Connect the negative cable back to the battery.
16. Fill the master cylinder with fluid and bleed the system. Make sure that the master cylinder never falls below half full.
17. Have hydraulic unit tested by a mechanic familiar with Mitsubishi vehicles.

G-Sensor

TESTING

To test the G-sensor a special Mitsubishi scan tool would be needed. Therefore, if the ABS warning light comes on or a problem with the ABS is evident, have the system diagnosed by an automotive mechanic familiar with the Mitsubishi ABS systems. After the vehicle has been diagnosed and the cause is known, the defective part can be replaced at home.

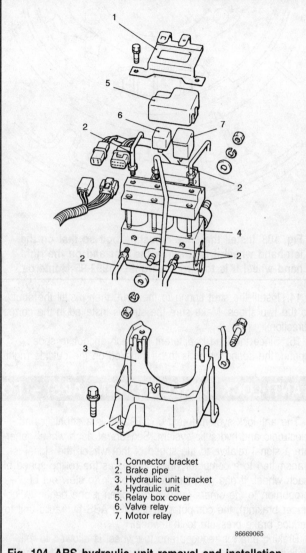

1. Connector bracket
2. Brake pipe
3. Hydraulic unit bracket
4. Hydraulic unit
5. Relay box cover
6. Valve relay
7. Motor relay

Fig. 104 ABS hydraulic unit removal and installation components — 1992-95 Monteros

REMOVAL & INSTALLATION

▶ See Figure 105

On all 1992-95 Monteros, the G-sensor is located under the rear center console next to the parking brake hand lever.
1. Disconnect the negative battery cable from the battery.
2. Remove the rear center console to gain access to the G-sensor.
3. Unplug the wiring harness connector to the G-sensor.
4. Remove the two retaining bolts and extract the G-sensor from the vehicle.

✳✳WARNING

When removing the G-sensor, take care not to drop it or subject it to a severe impact.

5. To install, set the G-sensor in place and secure it in place with the retaining bolts.
6. Plug the wiring harness connectors back together.

BRAKE SYSTEMS 9-49

7. Install the rear center console.
8. Connect the negative cable to the battery.
9. Have the ABS system re-checked to make certain that the new G-sensor functions correctly.

Electronic Control Unit (ECU)

TESTING

To test the ECU, a special Mitsubishi scan tool and oscilloscope would be needed. Therefore, if the ABS warning light comes on or a problem with the ABS is evident, have the system diagnosed by an automotive mechanic familiar with the Mitsubishi ABS systems. After the vehicle has been diagnosed and the cause is known, the defective part can be replaced at home.

REMOVAL & INSTALLATION

▶ See Figure 106

1. Disconnect the negative battery cable from the battery.
2. For Monteros with an optional third seat, remove the third seat.
3. Remove the right-hand, rear, lower quarter trim to gain access to the ECU.
4. Remove the large bracket by removing the four mounting screws.
5. Remove the two mounting screws for the smaller bracket. Remove the ECU and small bracket from the large bracket.
6. Unplug the wiring harness connector from the ECU. Remove the two ECU mounting bolts from the small bracket.
7. To install, mount the ECU to the small bracket with the two mounting screws.
8. Plug the wiring harness connector back into the ECU.
9. Affix the small bracket to the large bracket with the two mounting bolts. Mount the entire bracket/ECU assembly to the inner fender with the four mounting bolts.
10. Install the lower quarter trim. Install the third seat, if removed.
11. Connect the negative cable to the battery.

Wheel Speed Sensor

TESTING

▶ See Figures 107, 108 and 109

1. Remove the speed sensor and toothed rotor.
2. Check whether any metallic foreign material has adhered to the pole piece at the speed sensor tip, and if so, remove it.

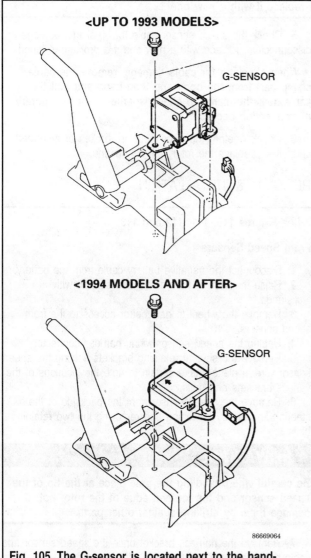

Fig. 105 The G-sensor is located next to the hand-operated parking brake lever — 1992-95 Monteros

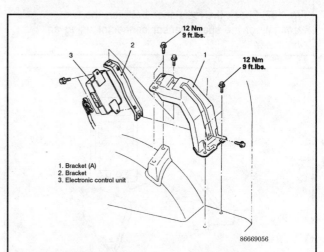

Fig. 106 Electronic control unit is located under the right rear quarter panel, mounted to the fender — 1992-95 Monteros

9-50 BRAKE SYSTEMS

Also check whether the pole piece is damaged, and, if so, replace it with a new one.

➡ The pole piece can become magnetized because of the magnet built into the speed sensor, with the result that metallic foreign material easily adheres to it. Moreover, the pole piece may not be able to function to correctly sense the wheel rotation speed if it is damaged.

3. Measure the resistance between the speed sensor terminals. The resistance for the front should be 0.9-1.1 kilo-ohms and the rear should be 1.3-2.1 kilo-ohms. If the internal resistance of the speed sensor is not within the standard value, replace it with a new speed sensor.

4. Remove all connections from the speed sensor, and then measure the resistance between terminals 1 and 2 and the body of the speed sensor. The resistance should be 100 kilo-ohms or more. If the speed sensor insulation resistance is outside the standard value range, replace with a new speed sensor.

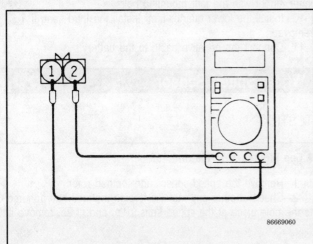

Fig. 107 Check the resistance between the two terminals of the speed sensor connector using an ohmmeter

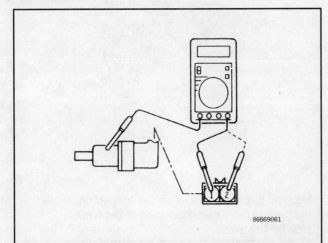

Fig. 108 Check the resistance from terminal 1 to the sensor body, then test the resistance between terminal 2 and the sensor body

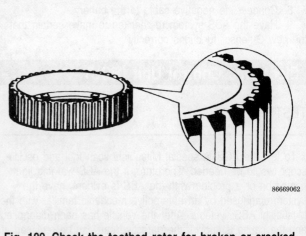

Fig. 109 Check the toothed rotor for broken or cracked teeth — if the rotor has any cracks or broken teeth, replace it with a new one

5. Check the speed sensor cable for breakage, damage or disconnection; replace with a new one if a problem is found.

➡ When checking for cable damage, remove the cable clamp part from the body and then bend and pull the cable near the clamp to check whether or not temporary disconnection occurs.

6. Check whether the rotor teeth are broken or deformed, and, if so, replace the rotor with a new one.

REMOVAL & INSTALLATION

▶ See Figures 110, 111, 112 and 113

Front Speed Sensors

1. Disconnect the negative battery cable from the battery.
2. Raise the front of the vehicle and support with jackstands.
3. Remove the wheel to gain better access to the front speed sensors.
4. Unplug the speed sensor wiring harness connector.
5. Remove the several retaining brackets holding the speed sensor wire in place. Make certain to note the locations of the various brackets and bolts.
6. Remove the speed sensor from the backside of the steering knuckle by loosening and removing the two retaining bolts.

✱✱WARNING

Be careful when handling the pole piece at the tip of the speed sensor and the toothed edge of the rotor not to damage them by striking against other parts.

7. Remove the harness bracket once the speed sensor and cable has been removed from the vehicle.
8. In order to remove the front speed sensor rotor, the front hubs and brake rotors will need to be removed from the steering knuckle — refer to Section 7 for that procedure.
9. Once the front brake rotor has been removed from the hub assembly, the speed sensor rotor can be removed from

BRAKE SYSTEMS 9-51

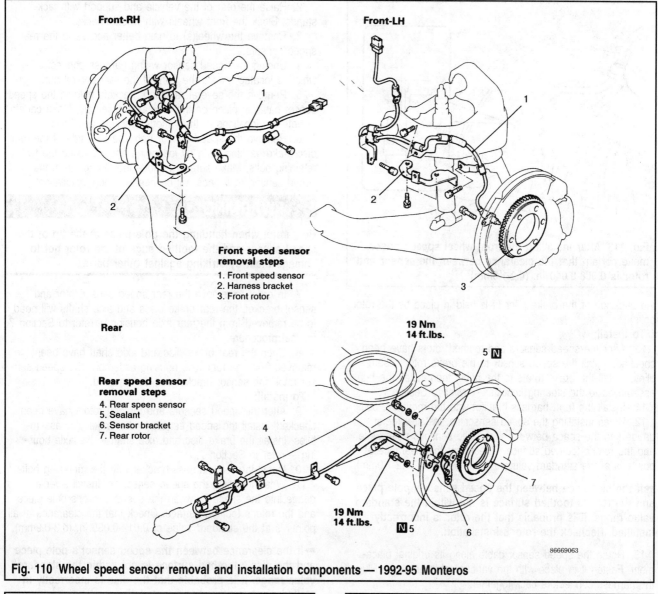

Fig. 110 Wheel speed sensor removal and installation components — 1992-95 Monteros

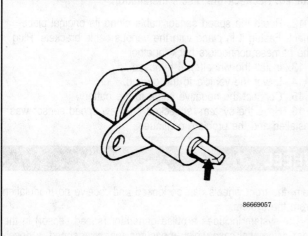

Fig. 111 Be careful when handling the speed sensor not to damage it by accidentally striking it against other parts

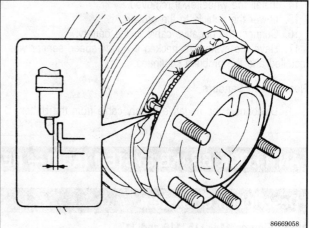

Fig. 112 When installing the rear wheel speed sensor, make sure that the clearance is 0.012-0.036 in. (0.3-0.9mm)

9-52 BRAKE SYSTEMS

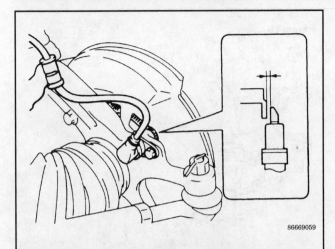

Fig. 113 After installing the front wheel speed sensor, make certain that the clearance between the sensor and rotor is 0.008-0.040 in. (0.2-1.0mm)

the backside of the brake rotor (it is held in place by the rotor bolts).

To install:

10. After the speed sensors and toothed rotors have been checked, install the speed sensor to the brake rotor assemblies. Install the brake rotors to the hubs, then install the hub assemblies to the steering knuckles — refer to Section 7.
11. Install the front harness bracket with the mounting bolts.
12. When installing the speed sensor tip, install a feeler gauge into the space between the speed sensor's pole piece and the rotor's toothed surface. Check that the clearance at all points is at the standard value of 0.008-0.039 in. (0.2-1.0mm).

➡ If the clearance between the speed sensor's pole piece and the rotor's toothed surface is not within the standard value range, it is probable that the rotor is incorrectly installed. Recheck the rotor's installation.

13. Route the speed sensor cable along its original placement. Fasten it in place with the various cable brackets. Plug the harness connectors back together.
14. Install the wheel(s), if removed.
15. Lower the vehicle to the ground.
16. Connect the negative cable to the battery.
17. Have the system rechecked if a new speed sensor was installed and the problem continues.

Rear Speed Sensors

1. Disconnect the negative battery cable from the battery.
2. Raise the rear of the vehicle and support with jackstands. Block the front wheels with wheel chocks.
3. Remove the wheel(s) to gain better access to the rear speed sensors.
4. Unplug the speed sensor wiring harness connector, which is located toward the front of the rear control arm.
5. Remove the several retaining brackets holding the speed sensor cable in place on the rear axle assembly. Make certain to note the locations of the various brackets and bolts.
6. Remove the speed sensor from the backside of the rear disc brake backing plate by loosening and removing the two retaining bolts. Make sure to remove the O-rings from the speed sensor; new ones will be needed upon installation.

✱✱WARNING

Be careful when handling the pole piece at the tip of the speed sensor and the toothed edge of the rotor not to damage them by striking against other parts.

7. In order to remove the rear speed sensor rotor and sensor bracket, the rear brake discs and axle shafts will need to be removed from the rear axle housing — refer to Section 7 for that procedure.
8. Once the rear brake disc and axle shaft have been removed from the rear axle housing assembly, the speed sensor rotor and sensor bracket can be removed.

To install:

9. After the speed sensors and toothed rotors have been checked, install the speed sensor to the brake disc assemblies. Install the brake disc and axle shaft to the axle housing — refer to Section 7.
10. Install the rear harness bracket with the mounting bolts.
11. When installing the speed sensor tip, install a feeler gauge into the space between the speed sensor's pole piece and the rotor's toothed surface. Check that the clearance at all points is at the standard value of 0.012-0.035 in. (0.3-0.9mm).

➡ If the clearance between the speed sensor's pole piece and the disc's toothed surface is not within the standard value range, it is probable that the disc is incorrectly installed. Recheck the disc's installation.

12. Route the speed sensor cable along its original placement. Fasten it in place with the various cable brackets. Plug the harness connectors back together.
13. Install the wheel(s), if removed.
14. Lower the vehicle to the ground.
15. Connect the negative cable to the battery.
16. Have the system rechecked if a new speed sensor was installed and the problem continues.

ANTI-LOCK BRAKE SYSTEM (ABS) — REAR WHEEL

Description of System

♦ See Figures 114, 115, 116 and 117

Found on 1992-95 Pick-ups, this system is a variant of ABS which acts only on the rear wheels. It is designed to allow maximum rear wheel braking on slick surfaces, thus preventing locked wheel skids and possible loss of control. On this system, the front wheels can be locked and receive no modulation from the system.

The system includes a pulse generator (speed sensor) in the rear differential case which measures rear axle speed, a deceleration or G-sensor generating a signal according to vehicle speed, a load sensing proportioning valve, a control unit or computer which interprets the signals and a modulator which controls the pressure of the brake fluid to the rear wheels.

BRAKE SYSTEMS 9-53

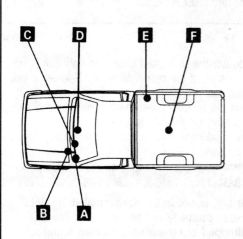

Name	Symbol
G sensor	D
Modulator	E
Rear anti-lock brake system control unit	D
Rear anti-lock brake system warning light	C
Data link connector	A
Speed sensor	F
Stop light switch	B

Fig. 114 Component locations of the rear-wheel ABS system — 1992-95 Pick-up

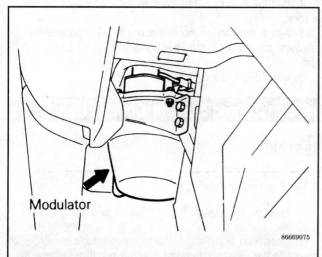

Fig. 115 The modulator unit is located under the right-hand side of the truck bed, mounted onto the frame

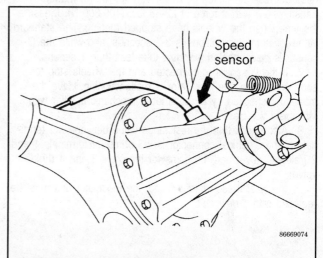

Fig. 117 The speed sensor is mounted in the top of the rear differential — 1992-95 Pick-ups

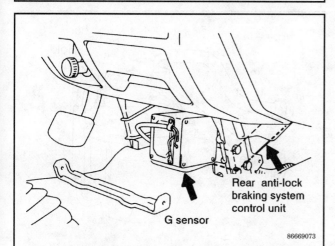

Fig. 116 The G-sensor and rear anti-lock baking system control unit locations inside the vehicle and under the front center console

Should any of the components fail, the control unit enters the fail-safe mode, illuminating the dash warning light and disabling the lock-up control. Normal braking is available.

A simple test of this system can be performed from the driver's seat. With the truck stationary, run the engine for at least 5 seconds. Turn the ignition key to **LOCK**, depress the brake pedal and while your foot is still on the brake, turn the key back to the **ON** (not **START**) position. If you hear a muffled clicking or tapping, the modulator solenoid valve is working and the system is operating fully.

9-54 BRAKE SYSTEMS

Modulator Unit

TESTING

▶ See Figure 118

1. Raise and safely support the vehicle so that all four wheels are off of the ground on jackstands. Keep the vehicle as close as possible to a horizontal state.
2. Confirm that there is no dragging in the rear brake.
3. Warm up the engine.
4. Keep the 4WD vehicle in the 2H condition.
5. Keep the engine in an idling condition. In case of a manual transmission, put the shift lever in the 4th gear. In the case of an automatic transmission, put the shift lever in the D range and after depressing the accelerator pedal once, allow the engine to idle again.
6. While running the rear wheels, depress the brake pedal (also depress the clutch pedal for vehicles with a manual transmission) with a force of 22-66 lbs. (100-300 N). Measure the time when the wheels have stopped running. The standard value for normal conditions is 3-5 seconds, and while the G-sensor is disconnected it should take less than 1 second.
7. If the G-sensor is connected and the wheels stop too soon or too late, the modulator is malfunctioning. Have the modulator checked by an automotive mechanic familiar with Mitsubishi vehicles.
8. Also check that the resistance value between the terminals of the electrical connector is 4.3 ohms for terminals 1 and 2 (release valve) and 5.0 ohms for terminals 3 and 4 (hold valve).
9. If any of these tests is failed, the modulator unit must be replaced with a new unit.

REMOVAL & INSTALLATION

▶ See Figure 119

1. The modulator unit is located under the right-hand bed of the Pick-up, in front of the rear right wheel. Remove a heat shield, if so equipped.
2. Disconnect the vacuum hose, brake fluid lines and electrical connectors from the modulator unit.
3. Unbolt the modulator bracket from the frame and remove the unit with the bracket.

✴✴CAUTION

The modulator unit is not to be disassembled; its nuts and bolts should absolutely not be loosened. The unit must not be dropped or otherwise subjected to impact shocks. It must also not be turned upside down or laid on its side.

4. Bolt the modulator unit back to the fender with the bracket.
5. Attach the vacuum hose, the two brake lines and the electrical connector. Tighten the brake lines to 12 ft. lbs. (17 Nm).
6. Bleed the brake system.

G-Sensor and System Control Unit

TESTING

▶ See Figures 120 and 121

1. Remove the front center console to gain access to the G-sensor.
2. Connect a positive terminal of a controllable power source to the G-sensor terminal 1 and a negative terminal to G-sensor terminal 3, then apply 7.3V (the limits of voltage guaranteed for operation is 7.0-7.5V).

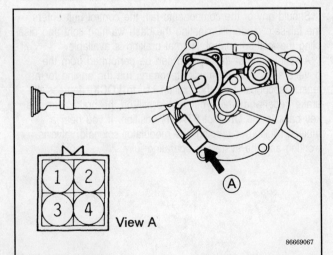

Fig. 118 Test the resistance between terminals 1 and 2, and between terminals 3 and 4 — 1992-95 Pick-ups

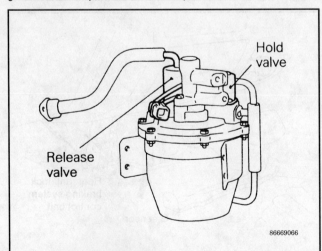

Fig. 119 The modulator unit also contains the release and hold valves — 1992-95 Pick-ups

BRAKE SYSTEMS

3. Connect a voltmeter to terminal 2 and measure the variations of voltage when the G-sensor is tipped up and down (tip the G-sensor from the normal mounting position to the position where the front face is facing straight down).

4. When the G-sensor is sitting in the position in which it is normally mounted the voltage should be 1.1-1.5V and when the G-sensor is tipped down, the voltage should be 4.6-5.0V.

5. If the G-sensor does not register these voltages, replace the G-sensor with a new unit.

REMOVAL & INSTALLATION

1. Disconnect the negative battery cable from the battery.
2. Remove the front center console from the Pick-up truck to gain access to the G-sensor and the System Control Unit (SCU).
3. Disconnect the electrical wiring to the G-sensor, then loosen and remove any retaining screws from the G-sensor. Remove the G-sensor from the vehicle.

4. The SCU is located behind the G-sensor. Disconnect the electrical wiring to the SCU, then remove the mounting screws. Extract the SCU from the vehicle.

5. To install, connect the wiring to the SCU and mount in place with the retaining screws.

6. Install the G-sensor with the mounting screws and connect the wiring.

7. Install the front center console to the vehicle, then connect the negative cable to the battery.

Load Sensing Proportioning Valve

For the removal and installation procedures for this component refer to the portion of this section which deals with proportioning valves.

Speed Sensor

The speed sensor on these vehicles is located on the top of the rear differential unit.

TESTING

▶ See Figure 122

1. Remove the speed sensor from the rear differential.
2. Confirm that the resistance between the connector terminals is 420-520 ohms.
3. Measure the insulation resistance between the terminals 1, 2 and the speed sensor body. The standard value is 100 kilo-ohms or more.
4. If the speed sensor fails either of these two tests, replace the speed sensor with a new unit.

REMOVAL & INSTALLATION

1. Disconnect the negative battery cable from the battery.
2. Raise the rear of the vehicle and support with jackstands. Block the front wheels with wheel chocks.

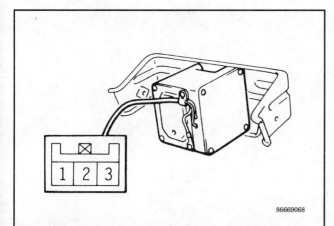

Fig. 120 Measure the voltage from terminal 2 while supplying 7.3V to terminals 1 (positive) and terminal 3 (negative)mdash;1992-95 Pick-ups

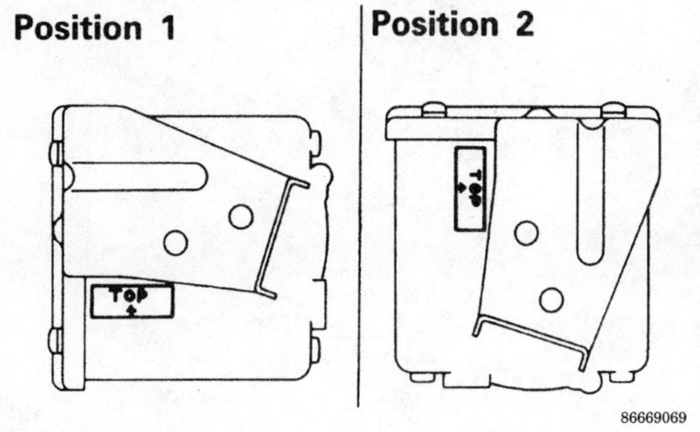

Fig. 121 With the power supply and voltmeter hooked to the connector, tilt the G-sensor as shown and measure the voltage change

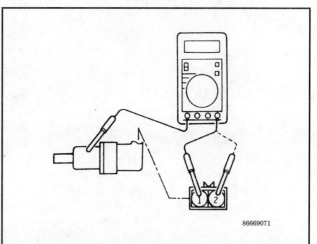

Fig. 122 Test the resistance between the two terminals, then test the resistance between each terminal and the sensor body — 1992-95 Pick-ups

BRAKE SYSTEMS

3. Locate the speed sensor mounting position on the differential and trace the cable back along its route until the connector is reached. Unplug the speed sensor wiring harness connector.

4. Remove the several retaining brackets holding the speed sensor cable in place on the rear axle assembly. Make certain to note the locations of the various brackets and bolts.

5. Loosen the retaining ring of the speed sensor on the differential by unscrewing it. Once the retaining ring has been loosened, pull the speed sensor out of the rear differential.

✹✹WARNING

Be careful when handling the pole piece at the tip of the speed sensor not to damage it by striking it against other parts.

To install:

6. Insert the speed sensor into the rear differential unit and tighten the retaining ring until snug.

7. Route the speed sensor cable along its original placement. Fasten it in place with the various cable brackets. Plug the harness connectors back together.

8. Lower the vehicle to the ground.

9. Connect the negative cable to the battery.

10. Have the system rechecked if a new speed sensor was installed and the problem continues.

BRAKE SPECIFICATIONS

All measurements in inches unless noted

Year	Model		Master Cylinder Bore	Brake Disc Original Thickness	Brake Disc Minimum Thickness	Brake Disc Maximum Run-out	Brake Drum Diameter Original Inside Diameter	Brake Drum Diameter Max. Wear Limit	Brake Drum Diameter Maximum Machine Diameter	Minimum Lining Thickness Front	Minimum Lining Thickness Rear
1983	Montero		0.870	0.790	0.720	0.006	10.00	10.08	—	0.039	0.039
	Pick-up		0.870	0.790	0.720	0.006	9.49	9.57	—	0.039	0.039
1984	Montero		0.870	0.790	0.720	0.006	10.00	10.08	—	0.039	0.039
	Pick-up		0.870	0.790	0.720	0.006	9.49	9.57	—	0.039	0.039
1985	Montero		0.875	0.790	0.724	0.006	10.00	10.08	—	0.039	0.039
	Pick-up	1	0.875	0.790	0.720	0.006	9.49	9.57	—	0.039	0.039
	Pick-up	2	0.875	0.790	0.720	0.006	10.00	10.08	—	0.039	0.039
1986	Montero		0.875	0.790	0.724	0.006	10.00	10.08	—	0.039	0.039
	Pick-up	1	0.875	0.790	0.720	0.006	9.49	9.57	—	0.039	0.039
	Pick-up	2	0.875	0.790	0.720	0.006	10.00	10.08	—	0.039	0.039
1987	Montero		0.875	0.866	0.720	0.006	10.00	10.08	—	0.039	0.039
	Pick-up	1	0.875	0.866	0.803	0.006	10.00	10.08	—	0.079	0.039
	Pick-up	2	0.938	0.866	0.803	0.006	10.00	10.08	—	0.079	0.039
1988	Montero		0.938	0.866	0.803	0.006	10.00	10.08	—	0.039	0.039
	Pick-up	1	0.875	0.866	0.803	0.006	10.00	10.08	—	0.079	0.039
	Pick-up	2	0.938	0.866	0.803	0.006	10.00	10.08	—	0.079	0.039
1989	Montero		0.938	0.866	0.803	0.006	10.00	10.08	—	0.079	0.039
	Pick-up		0.938	0.866	0.803	0.006	10.00	10.08	—	0.079	0.039
1990	Montero		0.938	0.866	0.803	0.006	10.00	10.08	—	0.079	0.039
	Pick-up		0.938	0.866	0.803	0.006	10.00	10.08	—	0.079	0.039
1991	Montero		0.938	0.866	0.803	0.006	10.00	10.08	—	0.079	0.039
	Pick-up		0.938	0.866	0.803	0.006	10.00	10.08	—	0.079	0.039
1992	Montero		0.938	3	4	0.003	—	7.80	—	0.079	0.079 5
	Pick-up		0.938	0.866	0.803	0.006	10.00	10.08	—	0.079	0.039
1993	Montero		0.938	3	4	0.003	—	7.80	—	0.079	0.079 5
	Pick-up		0.938	0.866	0.803	0.006	10.00	10.08	—	0.079	0.039
1994	Montero		0.938	3	4	0.003	—	7.80	—	0.079	0.079 5
	Pick-up		0.938	0.866	0.803	0.006	10.00	10.08	—	0.079	0.039
1995	Pick-up		0.938	0.866	0.803	0.006	10.00	10.08	—	0.079	0.039
	Montero		0.938	3	4	0.003	—	7.80	—	0.079	0.079 6

1 2WD
2 4WD
3 Front: 0.940; Rear: 0.710
4 Front: 0.880; Rear: 0.330
5 Drum shoe: 0.177
6 Brake shoe: 0.040

BRAKE SYSTEMS

TORQUE SPECIFICATIONS
Basic Brake System Components

Component	ft. lbs.	Nm
1983-86 Pick-ups and Monteros		
Support member-to-brake booster	6-8	8-11
Check valve cap	18-25	25-35
Check valve case	29-36	40-50
Secondary piston stop	1.1-2.1	1.5-3.0
Reservoir band	1.8-2.8	2.5-3.9
Master cylinder-to-brake booster	6-8	8-11
Master cylinder connector bolt tightening	18-25	25-35
Fitting	11-13	15-17
Vacuum switch	15-18	20-24
Bleeder screw	5-6	7-8
Brake tube flare nut	10-12	13-16
Caliper support-to-knuckle - Rear-wheel drive models	51-65	69-88
Caliper support-to-knuckle - 4-wheel drive models	58-72	79-98
Brake disc-to-front hub - Rear-wheel drive models	34-38	46-51
Brake disc-to-front hub - 4-wheel drive models	36-44	49-59
Dust cover-to-knuckle - Rear-wheel drive models	15-21	20-29
Wheel cylinder-to-backing plate bolts	13-15	18-20
1987-95 Pick-ups and Monteros		
Brake booster-to-pedal support member	6-9	8-12
Brake pedal shaft	18-25	25-35
Shift lock cable fixing nut	7-10	9-14
Reservoir stopper bolt	1-2	1.5-3.0
Piston stopper	1-2	1.5-3.0
Master cylinder-to-brake booster	6-9	8-12
Fitting	11-13	15-18
Master cylinder-to-brake line connector	18-25	25-35
Brake line flare nut	9-12	13-17
Bleeder screw	5-7	7-9
Caliper support-to-knuckle mounting bolt	58-72	80-100
Guide pin bolt (Upper)	29-36	40-50
Lock pin bolt (Lower)	23-30	32-42
Wheel cylinder-to-backing plate	13-15	18-21
Pedal support member installation bolt	13-18	18-25
Steering column assembly installation bolt	13-18	18-25

EXTERIOR
  ANTENNA 10-13
  BUMPERS 10-6
  CHASSIS AND CAB MOUNTING
    BUSHINGS 10-21
  DOORS 10-2
  FENDERS 10-16
  GRILLE 10-13
  HOOD 10-4
  OUTSIDE MIRRORS 10-13
  PICK-UP TRUCK BED 10-19
  POWER SUNROOF 10-27
  REAR DOOR AND TAILGATE 10-6
  SPARE TIRE CARRIER 10-28
INTERIOR
  BACK DOOR LOCK 10-48
  CONSOLES 10-37
  DOOR GLASS, REGULATOR, AND
    POWER WINDOW MOTOR 10-48
  DOOR LOCKS 10-40
  DOOR PANELS (DOOR PADS OR
    LINERS) 10-37
  HEADLINER 10-40
  INSIDE MIRROR 10-68
  INSTRUMENT
    PANEL/DASHBOARD 10-29
  INTERIOR TRIM PANELS 10-38
  REAR WINDOWS 10-64
  SEAT BELT SYSTEMS 10-71
  SEATS 10-68
  VENT WINDOWS 10-64
  WINDSHIELD GLASS 10-55
SPECIFICATIONS CHARTS
  TORQUE SPECIFICATIONS 10-71

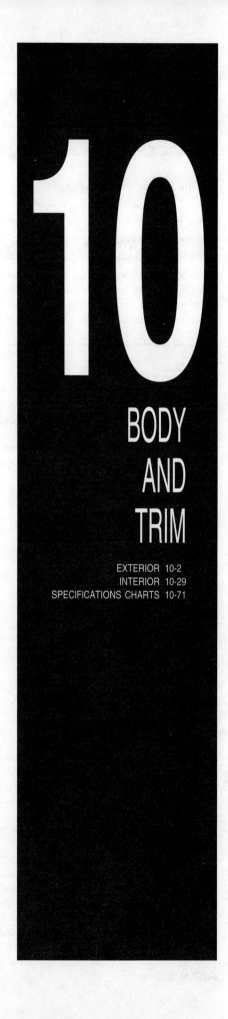

10

BODY AND TRIM

EXTERIOR 10-2
INTERIOR 10-29
SPECIFICATIONS CHARTS 10-71

10-2 BODY AND TRIM

EXTERIOR

Doors

REMOVAL & INSTALLATION

♦ See Figure 1

⁕⁕WARNING

The doors are heavier than they appear. Support the door from the bottom and use a helper during removal and installation. Do not allow the door to sag while partially attached and do not subject the door to impact or twisting motions.

Front Door

1. Remove the door opening trim on Monteros.
2. On all Monteros, removal of the cowl side trim (kick panel) just in front of the door may be necessary for access to the wiring connectors.
3. Disconnect the wiring harness (if any) running between the door and the body. Carefully pull the wiring free of the body.
4. Use a floor jack padded with rags or soft lumber to support the door at its lower midpoint. Use a felt tip marker to outline the hinge position on the door.
5. Remove the upper and lower hinge covers if any are present. Remove the small pin holding the door check arm in place.
6. Have a helper support the door, keeping it upright at all times. Remove the bolts holding the upper and lower hinge to the door. There will probably be alignment shims between the hinge and the door panel; take note of their location and placement for reassembly.
7. Using two people, lift the door clear of the car.
8. The door hinge may be removed from the body if necessary. On Pick-ups, the bolts are concealed. Use special tool MB990900-01 or equivalent. This is a specially shaped wrench built to do the job; removing the bolts will be difficult without this tool.

To install:

9. If the door hinge is removed from the body, reinstall it and tighten the bolts to 25-40 ft. lbs. (35-55 Nm).
10. Place the door in position and support it. Use the jack to fine tune the position until the bolt holes and matchmark (hinge outline) align.
11. Install the hinge bolts and nuts, and the alignment shims. Tighten the bolts to 12-19 ft. lbs. (17-26 Nm).
12. If all has gone well, the door should almost be in the original position. Refer to the door adjustment procedures to align the door and body. It may be necessary to loosen the hinge bolts and reposition the door; remember to retighten them each time or the door will shift out of place.
13. Once adjusted, connect the door check lever and install the pin. Install the hinge covers if any were removed.
14. Route the wiring harness(es) into the body and connect them to their leads.
15. Test the operation of any electrical components in the door (locks, mirrors, speakers, etc.) and test drive the car, checking the door for air leaks and rattles.

Rear Door

1. Remove the rear door scuff plate (sill plate).
2. Remove the lower trim from the center pillar. Disconnect the wiring harness running from the door to the pillar. Carefully pull the wire clear of the pillar.
3. Use a floor jack padded with rags or soft lumber to support the door at its lower midpoint. Use a felt tip marker to outline the hinge position on the door.
4. Remove the small pin holding the door check arm in place.
5. Have a helper support the door, keeping it upright at all times. Remove the bolts holding the upper and lower hinge to the door. Using two people, lift the door clear of the car.
6. The door hinge may be removed from the body if necessary.

To install:

7. If the door hinge is removed from the body, reinstall it and tighten the bolts to 33 ft. lbs. (45 Nm).
8. Place the door in position and support it. Use the jack to fine tune the position until the bolt holes and matchmark (hinge outline) align.
9. Install the hinge bolts and the alignment shims. Tighten the bolts to 16 ft. lbs. (22 Nm).
10. The door should be in almost the original position. Refer to the door adjustment procedures to align the door and body. It may be necessary to loosen the hinge bolts and reposition the door; remember to retighten them each time or the door will shift out of place.
11. Once adjusted, connect the door check lever and install the pin. Install the hinge covers if any were removed.
12. Route the wiring harness(es) into the body and connect them to their leads.
13. Install the lower pillar trim and the sill plate.
14. Test the operation of any electrical components in the door (locks, speakers, etc.) and test drive the car, checking the door for air leaks and rattles.

ADJUSTMENT

♦ See Figures 2 and 3

When checking door alignment, look carefully at each seam between the door and body. The gap should be constant and even all the way around the door. Pay particular attention to the door seams at the corners farthest from the hinges; this is the area where errors will be most evident. Additionally, the door should pull against the weatherstrip when latched to seal out wind and water. The contact should be even all the way around and the stripping should be about half compressed.

The position of the door can be adjusted in three dimensions: fore and aft, up and down, in and out. The primary adjusting points are the hinge-to-body bolts.

Apply tape to the fender and door edges to protect the paint. Two layers of common masking tape works well. Loosen

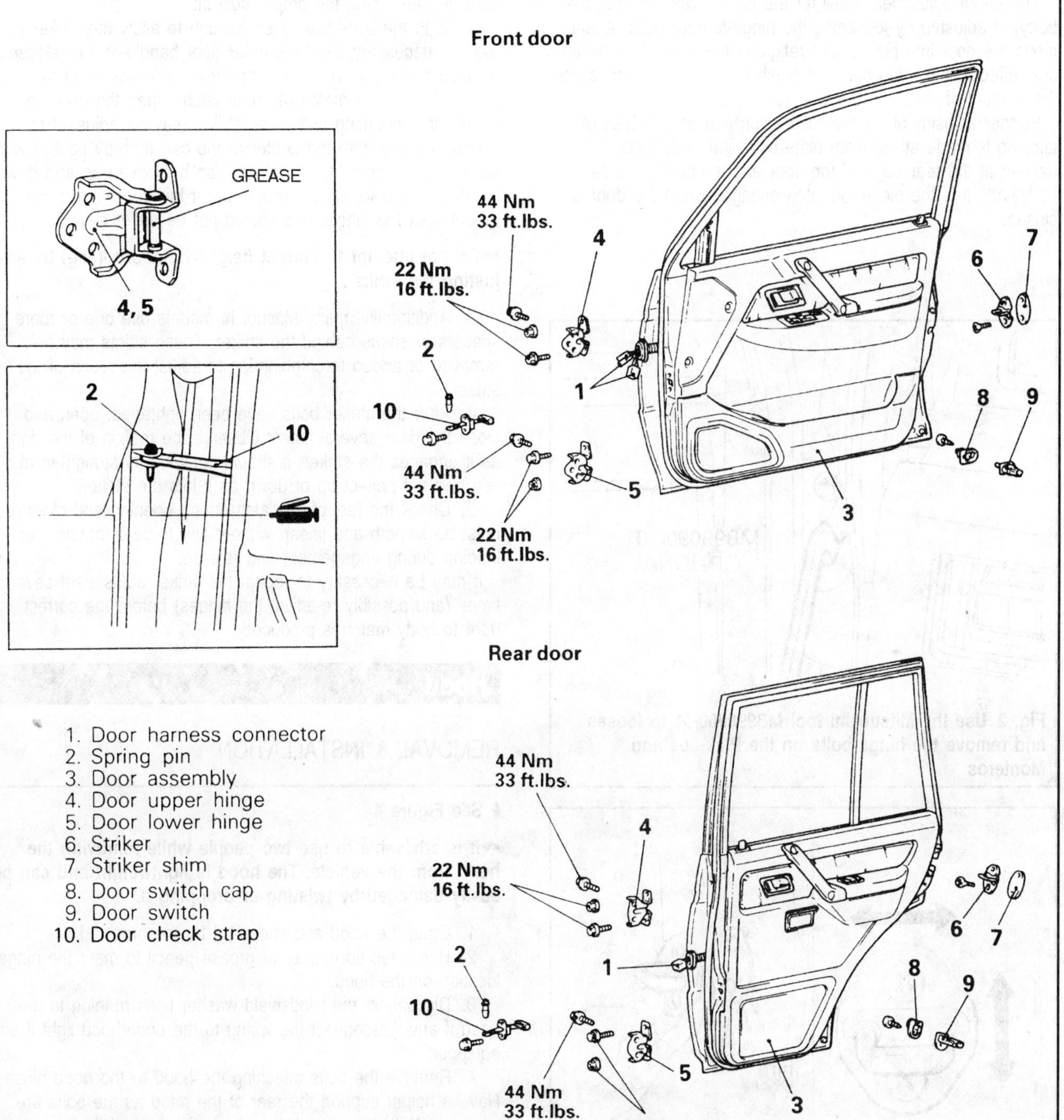

BODY AND TRIM 10-3

Front door

- 44 Nm / 33 ft.lbs.
- 22 Nm / 16 ft.lbs.
- 44 Nm / 33 ft.lbs.
- 22 Nm / 16 ft.lbs.

GREASE

Rear door

- 44 Nm / 33 ft.lbs.
- 22 Nm / 16 ft.lbs.
- 44 Nm / 33 ft.lbs.
- 22 Nm / 16 ft.lbs.

1. Door harness connector
2. Spring pin
3. Door assembly
4. Door upper hinge
5. Door lower hinge
6. Striker
7. Striker shim
8. Door switch cap
9. Door switch
10. Door check strap

Fig. 1 Front and rear door assembly on the Monteros. The front door assemblies on the Pick-ups are similar

10-4 BODY AND TRIM

the bolts (using special tool MB 990900-01 if necessary) just enough to allow the hinge to move. With the help of an assistant, position the door and retighten the bolts. Inspect the door seams carefully and repeat the adjustment until correctly aligned.

The in-out adjustment (how far the door sticks out from the body) is adjusted by loosening the hinge-to-door bolts. Again, move the door into place, then retighten the bolts. This dimension affects both the amount of crush on the weatherstrips and the amount of "bite" on the striker.

Further adjustment for closed position and smoothness of latching is made at the latch plate or striker. This piece is located at the rear edge of the door and is attached to the bodywork; it is the piece the latch engages when the door is closed.

Although the striker size and style may vary between models or from front to rear, the method of adjusting it is the same:

1. Loosen the large Phillips head screw(s) holding the striker. Know in advance that these bolts will be very tight; an impact screwdriver is a handy tool to have for this job. Make sure you are using the proper size bit.
2. With the bolts just loose enough to allow the striker to move if necessary, hold the outer door handle in the released position and close the door. The striker will move into the correct location to match the door latch. Open the door and tighten the mounting bolts. The striker may be adjusted towards or away from the center of the car, thereby tightening or loosening the door fit. The striker can be moved up and down to compensate for door position, but if the door is correctly mounted at the hinges this should not be necessary.

➡ **Do not attempt to correct height variations (sag) by adjusting the striker.**

3. Additionally, many Mitsubishi models use one or more spacers or shims behind the striker. These shims may be removed or added in combination to adjust the reach of the striker.
4. After the striker bolts have been tightened, open and close the door several times. Observe the motion of the door as it engages the striker; it should continue its straight-in motion and not deflect up or down as it hits the striker.
5. Check the feel of the latch during opening and closing. It must be smooth and linear, without any trace of grinding or binding during engagement and release.

It may be necessary to repeat the striker adjustment several times (and possibly re-adjust the hinges) before the correct door to body match is produced.

Hood

REMOVAL & INSTALLATION

▶ See Figure 4

➡ **It is advisable to use two people while removing the hood from the vehicle. The hood is lightweight and can be easily damaged by twisting or dropping it.**

1. Open the hood and support it by the prop rod.
2. Use a felt tip marker or grease pencil to mark the hinge location on the hood.
3. Disconnect the windshield washer lines running to the hood if any. Disconnect the wiring to the underhood light if so equipped.
4. Remove the bolts attaching the hood to the hood hinges. Have a helper support the rear of the hood as the bolts are removed. Without support, the hood will slide off the hinges and hit the upper bodywork.
5. Remove the hood assembly, lifting it off the hood prop. Place the hood out of the work area, resting on its side on a protected surface.
6. Reinstall the hood by carefully placing it in position and installing the hinge bolts finger-tight. Install the hood prop im-

Fig. 2 Use the Mitsubishi tool MB990900-01 to loosen and remove the hinge bolts on the Pick-ups and Monteros

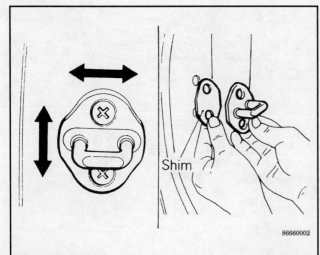

Fig. 3 The striker plate can be moved up and down or in and out to adjust the latching alignment

BODY AND TRIM 10-5

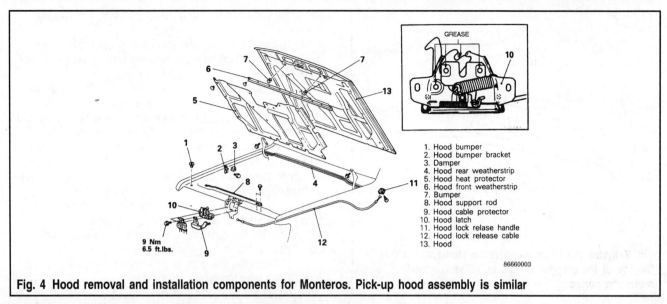

Fig. 4 Hood removal and installation components for Monteros. Pick-up hood assembly is similar

1. Hood bumper
2. Hood bumper bracket
3. Damper
4. Hood rear weatherstrip
5. Hood heat protector
6. Hood front weatherstrip
7. Bumper
8. Hood support rod
9. Hood cable protector
10. Hood latch
11. Hood lock release handle
12. Hood lock release cable
13. Hood

mediately. Move the hood around on the hinges until the matchmarks (felt tip marker) align.

❊❊WARNING

Take great care to prevent the hood from bumping the windshield. Not only will the hood be damaged, the windshield can be broken by careless installation.

7. Adjust the hood. When the hood is in final alignment, tighten the hinge bolts to 10 ft. lbs. (14 Nm).
8. Reconnect the washer lines and any electrical wires which were disconnected.

ADJUSTMENT

◆ See Figures 5, 6 and 7

1. Open the hood and support with the hood prop.
2. Loosen the hinge bolts until they are just finger-tight.

❊❊WARNING

Take great care to prevent the hood from bumping the windshield. Not only will the hood be damaged, the windshield can be broken by careless installation.

3. Close the hood — at least engaging the first latch — and check the hood-to-fender and hood-to-cowl seams and alignment. Re-open the hood and make adjustments as needed to give equal seams all around. When the hood is in final alignment, tighten the hinge bolts to 10 ft. lbs. (14 Nm).
4. Check the hood latch operation; it should engage easily and smoothly. The latch should release without the use of excessive force. The hood latch mounting bolts may be loosened; the latch can be adjusted left/right or up/down to compensate for the hood striker position. On Montero, the striker (on the hood) can also be adjusted by loosening the bolts.
5. With the hood closed and latched, check the height of the hood. It should sit even with the fender tops; this dimension may be adjusted by turning the hood stops to a higher or

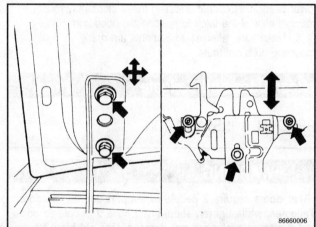

Fig. 5 The hood latch and hinges can be moved to adjust the position of the hood and closing characteristics

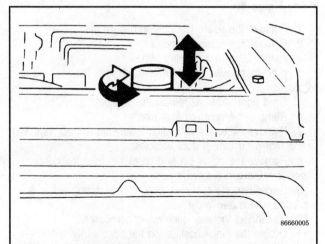

Fig. 6 To adjust the height of the closed hood in relation to the fenders, the hood stoppers can be raised or lowered

10-6 BODY AND TRIM

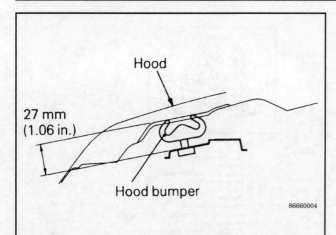

Fig. 7 Install the hood stoppers on Monteros so that the top of the stopper is 1.06 in. (27mm) from its mounting surface

lower position. Note that adjusting these stops may require readjustment of the latch to position the hood correctly.

6. Make sure, after all adjustments are made, to tighten all mounting bolts and nuts.

Rear Door and Tailgate

REMOVAL & INSTALLATION

※※WARNING

Rear doors require 2 people to support and lift the door. Tailgates, while lighter, should still have 2 people to control the tailgate and prevent damage. Use old blankets, cloths or other padding to protect the painted and glass surfaces.

Montero
▶ See Figure 8

1. Remove the spare tire lock and spare tire.
2. Remove the spare tire carrier.
3. Remove the rear opening trim.
4. If equipped with rear seat belts, remove the rear retractor cover.
5. On 4 door vehicles, remove the speaker.
6. Remove the quarter panel trim.
7. Remove either the rear pillar lower trim (2 door vehicles) or the rear pillar trim (4 door vehicles).
8. Remove the cotter pin and clevis pin from the door check rod. Discard the cotter pin.
9. Disconnect the rear washer fluid tube if the vehicle is equipped with a rear wiper.
10. Disconnect the rear door wiring harnesses.
11. Outline the hinge position on the door with a felt-tip marker. Have at least one other person support the door. Remove the bolts holding the hinge to the door. Remove the door.
12. Reinstall the door, observing the marks made earlier, and attach the hinge bolts. Tighten the bolts to 36 ft. lbs. (49 Nm).
13. If all went well, the door alignment should be very close to the original. Check the seams all around for equal spacing. If adjustment is necessary, loosen the hinge-to-body bolts and reposition the door. Tighten the bolts and recheck.
14. Further adjustments may be made at the latch and/or striker. The angle of the door (in at the top/out at the bottom or similar dimension) must be adjusted by loosening the hinge-to-door bolts and moving the door accordingly. This should only be done as a last resort; changes in this dimension may cause changes in seam width or body alignment.
15. Connect the door check clevis pin and install a new cotter pin.
16. Connect the electrical leads; connect the washer hose tube, if one was removed.
17. Install the pillar trim and install the quarter panel trim pieces.
18. Install the speaker and the seat belt retractor covers, if they were removed.
19. Fit the rear opening trim without distorting it.
20. Install the spare tire holder 17 ft. lbs. (23 Nm) and install the spare tire with its lock.
21. Using a grease gun with a fine tip, lubricate the rear door hinges by introducing multi-purpose grease into the small hole in each hinge. Move the door back and forth several times to circulate the grease.

Pick-Up
▶ See Figures 9 and 10

1. Remove the special bolts holding the safety wire on each side of the tailgate.
2. Have a helper support the gate; remove the 4 bolts holding the hinge to the truck bed.
3. Remove the tailgate.
4. After reinstalling the tailgate and securing the bolts, check the tailgate for correct alignment in the "normally open" and fully closed positions. Adjust the tailgate by loosening the hinge bolts if necessary. Retighten the bolts.
5. Install the special bolts with the safety wires. Make certain the spacers and washers are correctly installed. Tighten the bolts.

Bumpers

REMOVAL & INSTALLATION

The bumpers, front or rear, are actually assemblies. The outer surface is the cover or fascia, behind which may be found absorbent materials (foam block or honeycomb lattice) and the reinforcing steel bar. The components cannot be removed separately; the entire assembly must be removed from the car and then broken down into individual pieces.

BODY AND TRIM 10-7

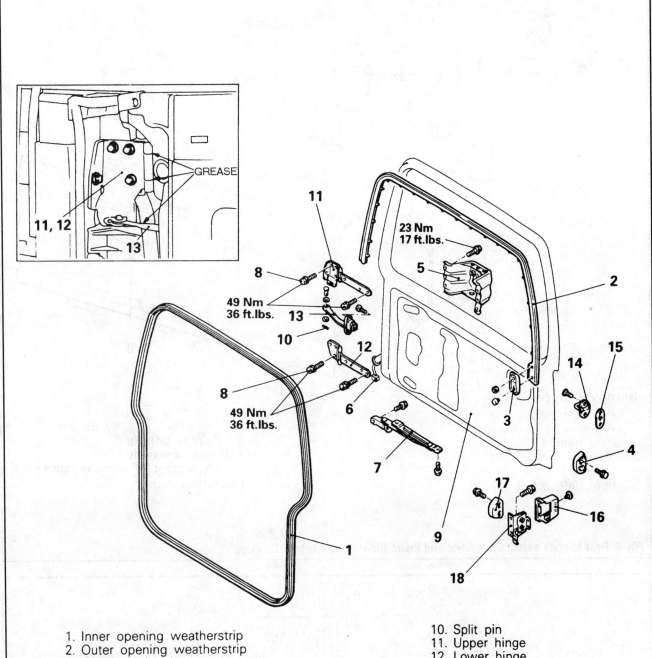

1. Inner opening weatherstrip
2. Outer opening weatherstrip
3. Weatherstrip plate
4. Bumper rubber
5. Spare tyre carrier
6. Harness connector
7. Back door stopper
8. Hinge attaching bolt
9. Back door
10. Split pin
11. Upper hinge
12. Lower hinge
13. Door check strap
14. Striker
15. Shim
16. Back door bumper cover
17. Back door bumper female
18. Back door bumper bracket

Fig. 8 Back door assembly removal and installation components — Monteros

10-8 BODY AND TRIM

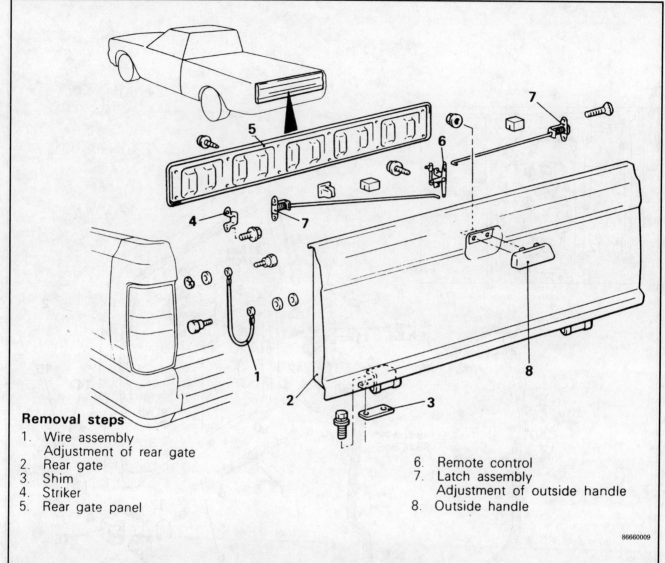

Removal steps
1. Wire assembly
   Adjustment of rear gate
2. Rear gate
3. Shim
4. Striker
5. Rear gate panel
6. Remote control
7. Latch assembly
   Adjustment of outside handle
8. Outside handle

Fig. 9 Rear tailgate assembly removal and installation components — Pick-ups

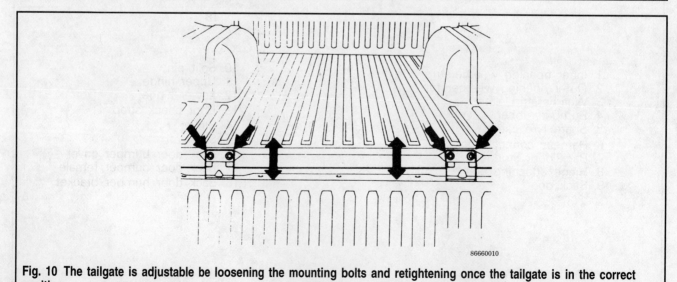

Fig. 10 The tailgate is adjustable be loosening the mounting bolts and retightening once the tailgate is in the correct position

BODY AND TRIM 10-9

> ✱✱**CAUTION**
>
> The bumper assemblies are heavy! Always support the bumper from below in at least two places before removing the last bolts. Never lie under the bumper while removing it.

The bumper covers are generally made of urethane. If this heavy, flexible plastic receives minor damage, generally under 1 in. (25mm), it can be repaired by a reputable body shop using established industry methods. Like any body repair, the bumper will require painting and should be removed from the car. In the event of more extensive damage, the cover can be replaced individually after the bumper is removed. This will save the cost of a complete bumper assembly.

Montero

▶ See Figures 11 and 12

Straightforward and simple, the Montero bumpers are removed from either end by removing the 4 bolts holding the unit to the mounts. The rear bumper has 4 additional bolts — 2 at each side — at the leading edge of the bumper corner trim. Although the bumpers are simple to remove, they are still heavy; be careful!

1. On the front bumper, remove the lower skid plate from the vehicle. The skid plate is held in place by four or more retaining bolts. Remove the five bumper garnish retaining nuts from the underside of the top of the bumper, then remove the bumper garnish.
2. Some vehicles are equipped with a bumper mounted outside temperature sensor. Remove this before complete removal of the bumper.
3. Loosen and remove the bumper retaining bolts (four total for the front, ten or eight total for the rear). Have an assistant or a floor jack support the bumper while removing the mounting bolts.

> ✱✱**CAUTION**
>
> Be careful when removing the mounting bolts from the bumper, the bumper assemblies are heavy and can cause personal harm.

To install:

4. Using a floor jack or an assistant, position the bumper in place. Install the mounting bolts and tighten front bumper bolts to 60 ft. lbs. (81 Nm), rear bumper bolts to 33 ft. lbs. (44 Nm).
5. Install the outside temperature sensor, if the vehicle is equipped with one.
6. Install the bumper garnish and the skid plate on the front bumper. Tighten the skid plate front bolts to 17 ft. lbs. (24 Nm) and the rear bolts to 8 ft. lbs. (12 Nm).

Pick-Up

FRONT

▶ See Figures 13 and 14

1. Remove the 5 bolts holding the front skirt panel to the bumper.
2. Disconnect the wiring to the front turn signal lights. The lights may stay in the bumper if you wish.
3. Disconnect the front bumper from the side stays on each side.
4. Remove the bolts attaching the front bumper support to the frame and remove the front bumper.

To install:

5. When reinstalling, place the bumper on the truck and tighten the retaining bolts finger-tight. Align the holes and install the side bolts to the side stays.
6. Check the bumper for correct alignment; it should not sag or hang on an angle. Adjust its position as needed and tighten the bolts.
7. Connect the wiring for the turn signals.
8. Install the front skirt panel.

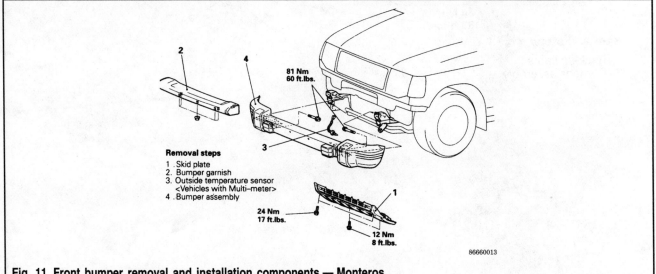

Fig. 11 Front bumper removal and installation components — Monteros

10-10 BODY AND TRIM

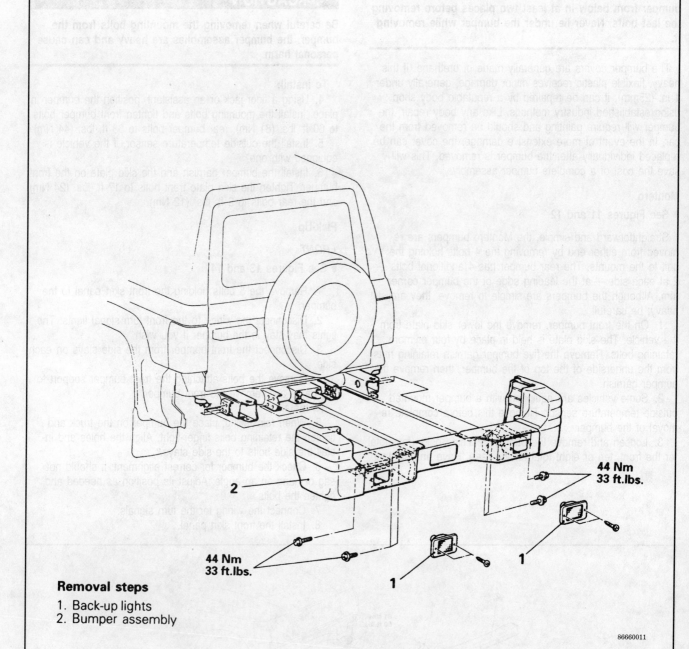

Removal steps
1. Back-up lights
2. Bumper assembly

Fig. 12 Rear bumper removal and installation components — Monteros

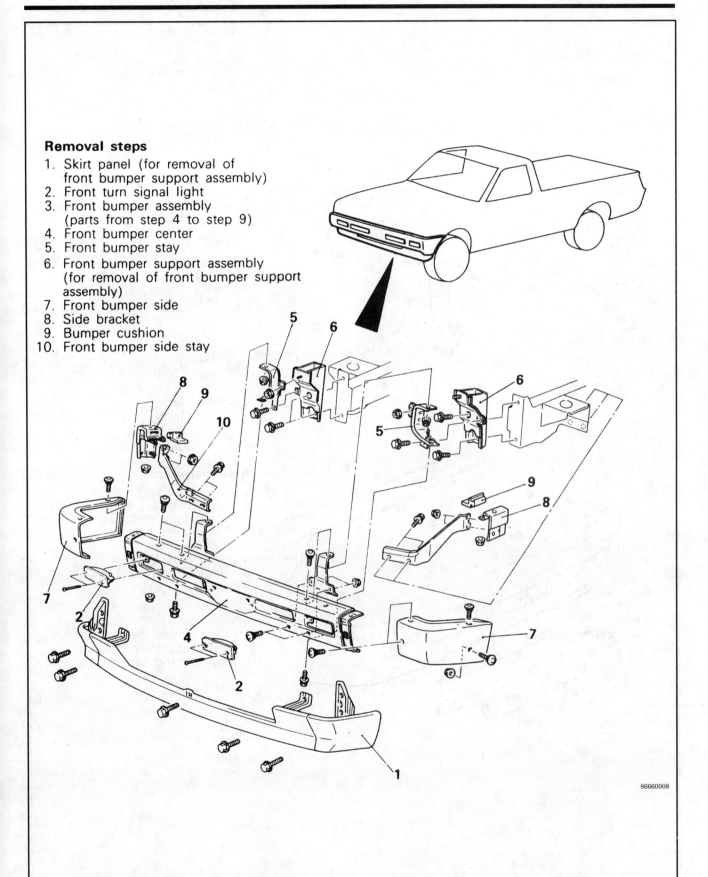

Fig. 13 Front bumper removal and installation components — 1983-92 Pick-ups

10-12 BODY AND TRIM

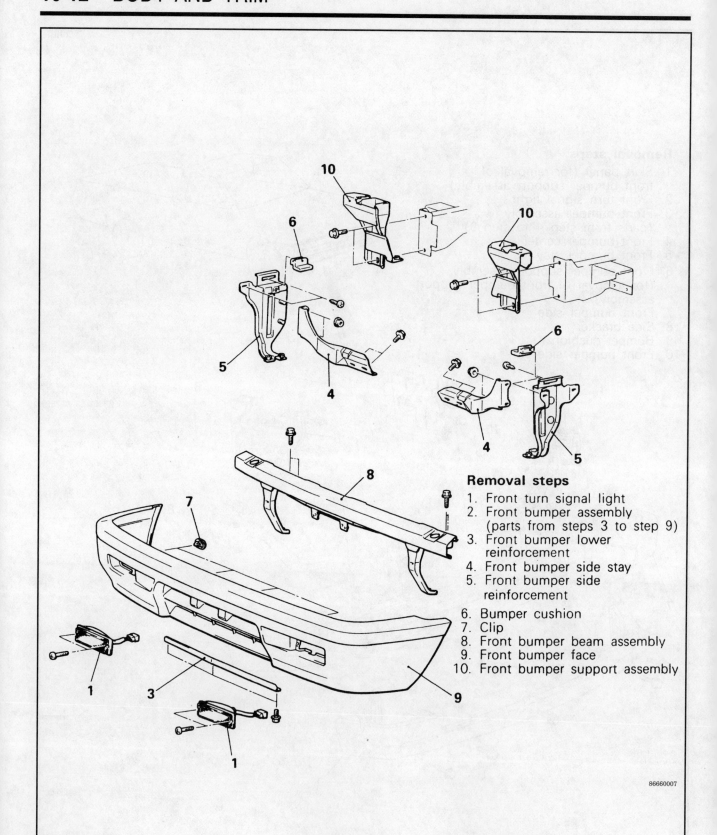

Fig. 14 Front bumper removal and installation components — 1993-95 Pick-ups

BODY AND TRIM 10-13

REAR

Not all Pick-ups are delivered with rear bumpers. Those that do have bumpers at the rear may have either the optional Mitsubishi bumper or one of a number of aftermarket products. In general, the rear bumper simply bolts to the frame, but the style and manufacture of the bumper may affect its installation.

Removal of these bumpers is generally straightforward, but be warned that some bumpers, notably the plate steel ones, are extremely heavy! Use a floor jack to help support the unit and have an assistant handy throughout the removal or installation. If the bumper contains any lighting or electrical equipment, remember to disconnect or remove it before separating the bumper from the truck.

Grille

REMOVAL & INSTALLATION

♦ See Figure 15

In all cases, the grille may be removed without removing other parts. Depending on the model, the grille is held by a variety of clips and screws. The problem is usually not removing the fittings but finding all of them. Raise the hood and look for screws placed vertically in the front metalwork. Small screws may be used to hold part of the grille.

Some grilles are retained by screws placed from the front. These will be black and hard to see in the recesses of the plastic. Most grilles incorporate plastic spring clips; these must be released correctly or they will break, becoming unusable. A narrow probe or small flat-tip prytool may be used to release the lock tabs.

When you think all the mounts and clips are removed, pull gently outward on the grille. If it is still retained anywhere, it will be immediately noticeable. Remove the attachment and remove the grille. Once fully released, the grill should almost come off unassisted; if anything but the lightest pull is required, it is most likely still attached.

When reinstalling, remember that the condenser or radiator is right behind the grille; be very careful not to damage anything with either the grille or any tools. Make sure any spring clips are properly engaged and install the mounting screws.

Outside Mirrors

REMOVAL & INSTALLATION

♦ See Figure 16

The mirrors on Mitsubishi automobiles can be removed from the door without disassembling the door liner or other components. Both left and right outside mirrors may be either manual, manual remote (small lever on the inside to adjust the mirror) or electric remote. If the mirror glass is damaged, replacements may be available through your dealer or a reputable glass shop in your area. If the plastic housing is damaged or cracked, the entire unit will need to be replaced.

1. If the mirror is manual remote, check to see if the adjusting handle is retained by a hidden screw. If so, remove the screw and remove the handle.
2. Remove the delta cover; that's the triangular black inner cover. It can be removed with a blunt plastic or wooden tool. Don't use a metal prytool; the plastic will be marred. Some models' delta covers are held in place with a retaining screws.
3. Depending on the model of car and style of mirror, there may be concealment plugs or other minor parts under the delta cover — remove them. If the electric connectors are present, unplug them.
4. Support the mirror housing from the outside and remove the three bolts or nuts holding the mirror to the door.
5. If the wiring to the electric mirror was not disconnected previously, disconnect it now. Some connectors can only be reached after the mirror is free of the door. Remove the mirror assembly.

To install:

6. Fit the mirror to the door and install the nuts and bolts to hold it. Connect the wiring harnesses if they are on the outside of the door. Pay particular attention to the placement and alignment of any gaskets or weatherstrips around the mirror; serious wind noises may result from careless work.
7. If the wiring connectors are on the inside of the door, plug them back together and install any concealment plugs, dust boots or seals which were removed.
8. Install the delta cover and install the control lever, if it was removed.

Antenna

REMOVAL & INSTALLATION

Pick-Up

♦ See Figures 17 and 18

1. Inside the vehicle, release the glove box stopper by pressing it towards the center of the cab and swinging the door downwards.
2. Reach behind the radio and disconnect the antenna cable.

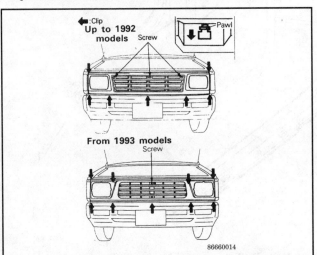

Fig. 15 Make certain to detach all retaining screws and clips before removing the front grille from the vehicle

10-14 BODY AND TRIM

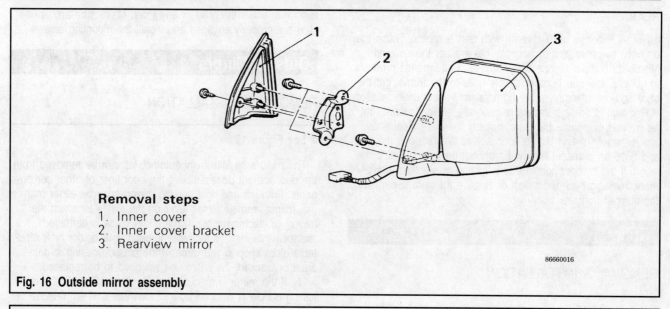

Removal steps
1. Inner cover
2. Inner cover bracket
3. Rearview mirror

Fig. 16 Outside mirror assembly

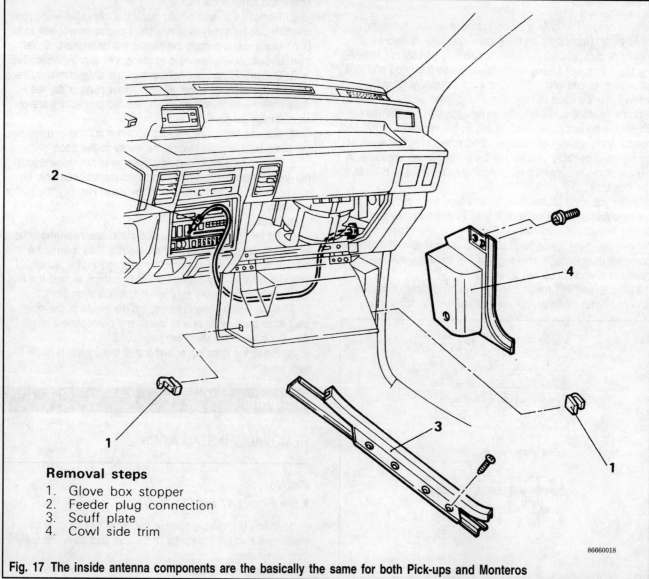

Removal steps
1. Glove box stopper
2. Feeder plug connection
3. Scuff plate
4. Cowl side trim

Fig. 17 The inside antenna components are the basically the same for both Pick-ups and Monteros

BODY AND TRIM

3. Remove the right sill plate and remove the cowl side trim (kick panel).

4. Outside the vehicle, remove the inner liner from the right front wheel arch. On 4 — wheel drive vehicles, remove the mud flap first. This job is easier with the front end elevated and supported with the wheel removed.

5. Remove the antenna mast by unscrewing it from the base.

6. Remove the upper mounting nut and the lower retaining bolt.

7. Remove the antenna base from the fender area and carefully extract the cable.

To install:

8. Install the new antenna base and secure the top nut and bottom bolt.

9. Route the cable into the cab and route it correctly to the radio.

10. Install the splash shield using two-sided adhesive tape along with the mounting screws. Install the mud flap if it was removed. Replace the wheel and lower the vehicle to the ground.

11. Check the cable routing and position it so that it is not in the way of anything including the passenger's feet. Connect the cable to the radio.

12. Install the kick panel and the sill plate.

13. Swing the glove box into position and engage the stopper.

Montero

▶ See Figure 19

This procedure is for the removal of the front fender mounted antenna. To remove the whip antenna feeder cable, the front and rear console, the front passenger's seat, the rear seat, the cowl side trim, center pillar trim, the quarter trim lower and the carpet must be removed from the vehicle to gain access to the feeder cable. The cable itself plugs in at the radio and back by the rear fender. Replacing the antenna base on 1983-91 Monteros requires removal of the right front fender. While not extremely difficult, it may be more work than you wish to undertake. The mast may be replaced separately by simply unscrewing it from the base. The following procedure is for replacing the base and cable as a unit.

1. Loosen or remove the radio or center console and disconnect the antenna cable.

2. Release the glove box stopper by pressing it towards the center of the cab and swinging the door downwards.

3. Reach up through the glovebox opening and detach the antenna retaining clips.

4. Remove the upper mounting nut from the fender top. Unscrew the antenna mast from the base.

5. For 1983-91 Monteros, remove the front fender.

6. On 1992-95 Monteros, remove the splash shield under the right front fender wheel well.

7. Remove the upper part of the antenna base and carefully pull the cable through the hole. Disconnect the electrical harness if the Montero is equipped with an electrical antenna.

8. Disconnect and remove the antenna base and electric motor, if so equipped.

9. Install the antenna base (and motor) and the new ground base (upper part). Route the cable into the vehicle and to the radio. Make sure it will not interfere with any moving parts or the passenger's feet.

10. Plug the electrical harness connectors together, if so equipped.

11. Reinstall the fender, if removed.

12. Install the wheel well splash shield. Apply butyl rubber tape to the flange part of the fender before installing the splash shield. Use 3M® No. 8626 or No. EC-5310 or equivalent.

13. Install the antenna mast and tighten the upper retaining nut.

14. Route the antenna cable into the retaining clips.

15. Install the glovebox and stopper.

16. Reinstall the radio and/or center console.

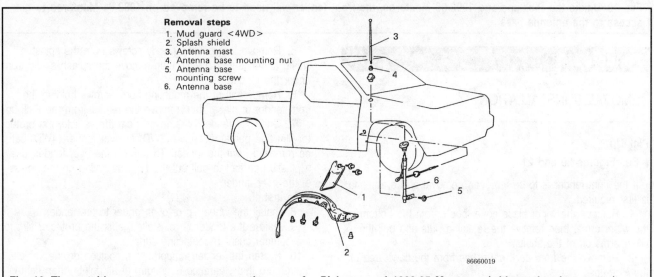

Fig. 18 The outside antenna components are the same for Pick-ups and 1992-95 Monteros (without electric antenna)

Removal steps
1. Mud guard <4WD>
2. Splash shield
3. Antenna mast
4. Antenna base mounting nut
5. Antenna base mounting screw
6. Antenna base

10-16 BODY AND TRIM

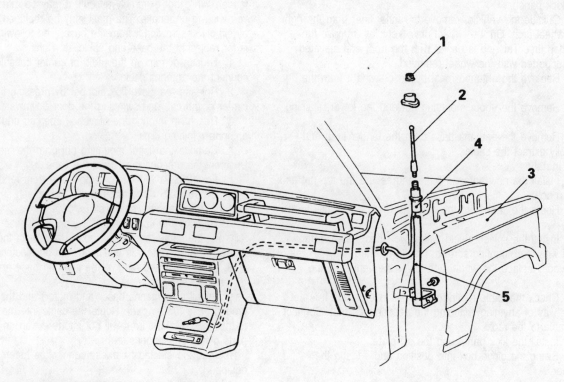

Removal steps
1. Mounting nut
2. Antenna mast
3. Front fender panel
4. Ground base
5. Antenna base

Fig. 19 Unlike the Pick-ups and 1992-95 Monteros, the front fender will need to be removed on 1983-91 Monteros to gain access to the antenna base

Fenders

REMOVAL & INSTALLATION

Pick-Up

♦ See Figures 20 and 21

If the right fender is to be removed, removal of the antenna is first required.

1. Remove the wiper blade cover boots from the bottom of the wiper arms, then remove the retaining nuts and pull the wiper arms off of the shafts.
2. Remove the front deck garnishes from the cowl area of the Pick-up.
3. Remove the front combination lights — refer to Section 6 for this procedure.
4. On 4WD Pick-ups, remove the front mud flap from the lower fender.
5. Remove the mounting body screws from the splash shield (inner fender cover), then remove the splash shield from under the fender.
6. Remove all of the mounting body screws holding the front fender in place, then remove the fender from the Pick-up.
7. Once the fender is removed from the vehicle, the protector film and partial wheel lip (2WD) or over fender (4WD) can be removed from the fender. To remove the over fender, use a plastic trim tool to pull out the clips by prying on the edges of the over fender.

To install:

8. Install the wheel lip or over fender to the fender.
9. Before the fender is reinstalled, affix the protector film to the bumper cushion contacting surface.
10. Fasten the fender temporarily in position on the vehicle, make sure that clearance is uniform at all points, then tighten the mounting bolts. If the Pick-up was equipped with fender seal, make certain to reinstall the seal.
11. Install the splash shield to the under side of the fender, then secure it in place with the mounting screws. Apply 3M®

BODY AND TRIM 10-17

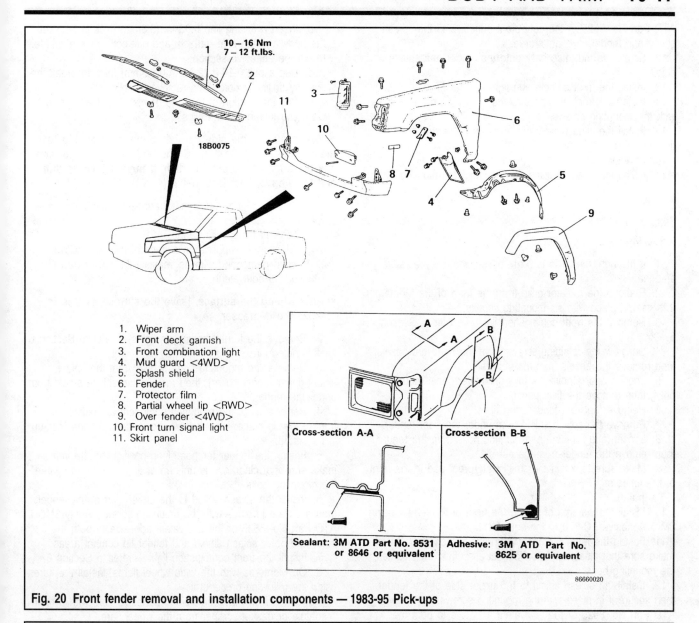

Fig. 20 Front fender removal and installation components — 1983-95 Pick-ups

1. Wiper arm
2. Front deck garnish
3. Front combination light
4. Mud guard <4WD>
5. Splash shield
6. Fender
7. Protector film
8. Partial wheel lip <RWD>
9. Over fender <4WD>
10. Front turn signal light
11. Skirt panel

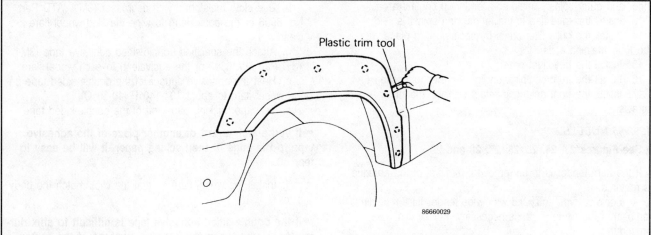

Fig. 21 Use a plastic trim tool to pry the clips lose — do not use a metal tool, the paint or the moulding could become damaged

10-18 BODY AND TRIM

ATD Part No. 8625 or the equivalent adhesive to the splash shield and fender lip contact area.

12. Screw the mud flap back onto the fender and splash shield.
13. Install the front combination light — refer to Section 6.
14. Install the front garnishes to the cowl area of the Pick-up with the mounting screws.
15. Install the windshield wiper arms onto the shafts and tighten the mounting nuts to 7-12 ft. lbs. (10-16 Nm).
16. Affix the wiper arm nut covers back onto the arms.
17. Install the antenna if the right fender was removed.

Montero

1983-91 MODELS

▶ See Figure 22

1. If the right fender is to be removed, remove the radio antenna.
2. Remove the radiator grille from the front of the Montero. It is held in place with six mounting screws.
3. Remove the front combination lights — refer to Section 6.
4. Unscrew the retaining screws from the headlight bezels, then remove the bezels from the vehicle.
5. Loosen and remove the four grille filler panel mounting bolts, then remove the filler panel.
6. Remove the front fender mud flaps.
7. Remove the splash shield from under the front fender.
8. Loosen and remove the mounting bolts from the fender, then remove the fender from the vehicle.
9. Make sure to retain the small waterproof pad in the front of the fenders.

To install:

10. Install the waterproof pad to the front of the fender using 3M® Adhesive EC-870 or the equivalent.
11. Fasten the fender temporarily in position on the vehicle, make sure that clearance is uniform at all points, then tighten the mounting bolts.
12. Install the splash shield to the under side of the fender, then secure it in place with the mounting screws. Apply 3M® ATD Part No. 8625 or the equivalent adhesive to both the inner and outer splash shield and fender lip contact areas.
13. Install the mud flap to the fender and splash shield.
14. Install the grille filler panel by securing it in place with the four retaining bolts.
15. Install the headlight bezels.
16. Install the front combination light — refer to Section 6.
17. Install the front grille and retain with the mounting screws.

1992-95 MODELS

▶ See Figures 23, 24, 25, 26, 27, 28 and 29

If the right fender is to be removed, removal of the antenna is required.

1. For Monteros equipped with wide fender flare mouldings, removal of the wheel flares is necessary. Follow this procedure:

   a. Apply masking tape to the outside circumference of each side garnish.
   b. Insert a fishing line 0.03 in. (0.8mm) thick in between the body and the side garnish, and pull both ends alternately to cut the adhesive section.
   c. Use a drill 0.16-0.22 in. (4.0-5.5mm) thick to break the rivet by drilling a hole, then remove the blind rivet.
   d. Pull the section of the side garnish with the clips toward you to remove the clips.

➡ When reusing the side garnish, remove by pulling the fishing line along the edge of the body so as not to damage the edge of the side garnish. If the adhesive is difficult to remove, heat it to 104°F (40°C).

   e. Scrape off the double-sided adhesive tape with a resin spatula.
   f. Tear off the masking tape.
   g. Wipe off the application surface of the body with a clean cloth dampened with degreaser (3M® ATD Part No. 8906 or the equivalent).

➡ After wiping the surface, leave the surface as it is to volatize the degreaser.

2. Remove the front combination lights — refer to Section 6 for this procedure.
3. Remove the mounting body screws from the splash shield (inner fender cover), then remove the splash shield from under the fender.
4. Remove all of the mounting body screws holding the front fender in place, then remove the fender from the Pick-up.

To install:

5. Fasten the fender temporarily in position on the vehicle, make sure that clearance is uniform at all points, then tighten the mounting bolts.
6. Install the splash shield to the under side of the fender, then secure it in place with the mounting screws. Apply 3M® ATD Part No. 8625 or the equivalent adhesive to both the inner and outer splash shield and fender lip contact areas.
7. Install the front combination light — refer to Section 6.
8. On Monteros with the wide wheel flares, install the flares using the following procedure:

   a. Scrape off the double-sided adhesive tape with a resin spatula or gasket scraper from the wheel flare.
   b. Use cloth moistened with degreaser (3M® ATD Part No. 8906 or the equivalent) to wipe the side wheel flare clean.
   c. Attach the specified double-sided adhesive tape (3M® ATD Part No. 6382 or the equivalent) to each wheel flare.
   d. Heat the adhesive surface of the double-sided tape on the wheel flare to about 104-140°F (40-60°C).
   e. Tear the backing paper off of the double-sided tape.

➡ If you attach a part of another piece of the adhesive tape to the edge of the backing paper, it will be easy to tear off.

   f. Install the wheel flare so that the clips match the body holes.

➡ If the double-sided adhesive tape is difficult to affix during the winter, warm the bonding surfaces of the body and the wheel flare to 104-140°F (40-60°C) before affixing the tape.

   g. Firmly press in the wheel flare.

BODY AND TRIM 10-19

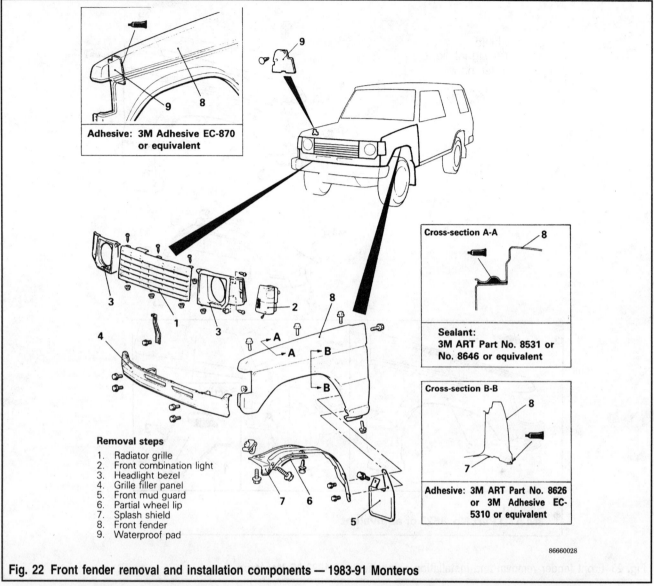

Fig. 22 Front fender removal and installation components — 1983-91 Monteros

h. Use a riveter to install new rivets into the rivet holes.
9. Install the antenna if the right fender was removed.

Pick-Up Truck Bed

REMOVAL & INSTALLATION

▶ See Figure 30

1. Elevate and safely support the vehicle on stands. Make certain all four stands are securely placed on both the ground and the truck. The support must be extremely solid.
2. Remove the filler neck cover or protective panel.
3. Disconnect the filler neck at the body end.
4. Disconnect the rear wiring harness.
5. Loosen and remove the rear body mounting nuts; discard the nuts and use new ones at reassembly. Disassemble one mount at a time and pay great attention to the order and placement of washers, pads, spacers etc. They must be reassembled correctly.
6. Either with the use of an overhead lift or several helpers, remove the cargo bed from the truck frame.
7. When reinstalling, lift the bed into place and align the bolt holes. Assemble each mounting bolt with the proper hardware and install it properly. Leave each finger-tight so that the bed may be moved slightly if hole alignment is needed.
8. Tighten each nut to 20-23 ft. lbs. (28-32 Nm). Resist the temptation to tighten these nuts to extremes; overtightening crushes the pads causing both excessive noise and reducing the natural motion of the bed.
9. Plug the rear wiring harness connectors back together.
10. Install the filler neck and neck cover.
11. Lower the Pick-up back to the ground.

10-20 BODY AND TRIM

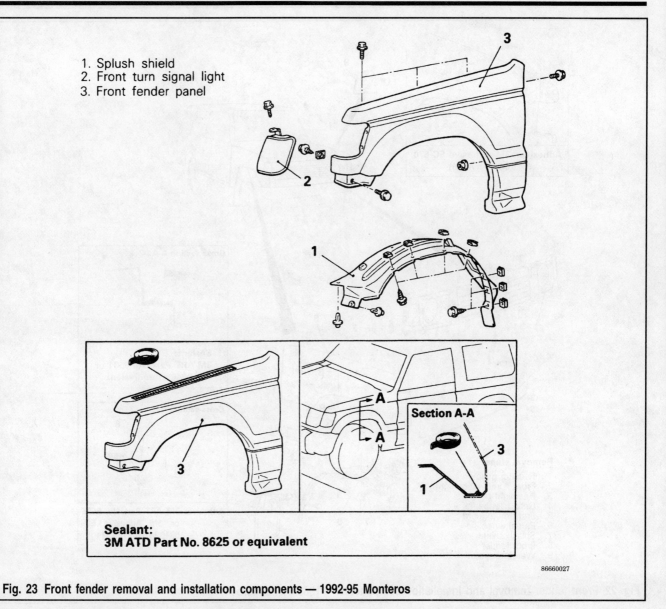

Fig. 23 Front fender removal and installation components — 1992-95 Monteros

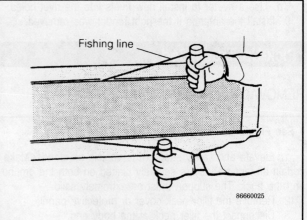

Fig. 24 Apply masking tape to either edge of the piece of molding which will be removed (wheel flare) — the tape is to protect the paint during removal

Fig. 25 Use a piece of fishing line attached to two pieces of wood to pull under the piece of molding, which will cut the adhesive tape

BODY AND TRIM 10-21

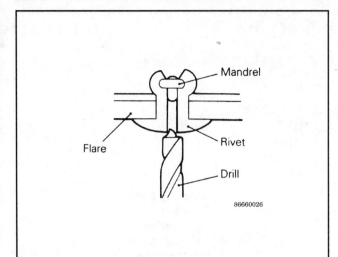

Fig. 26 Drill a hole into the rivet to hollow it out, then remove what is left of the rivet

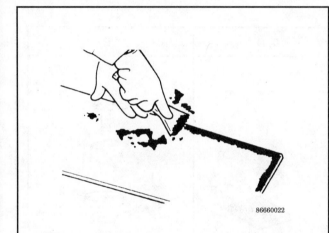

Fig. 27 Use a plastic resin spatula to scrape the old adhesive tape off of the vehicle body and wheel flare

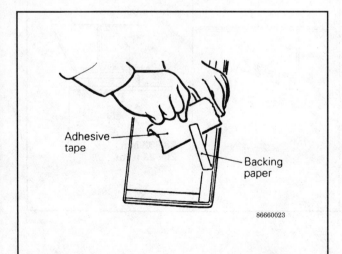

Fig. 28 To pull the backing paper off of the double-sided tape, use another piece of tape to grip the paper

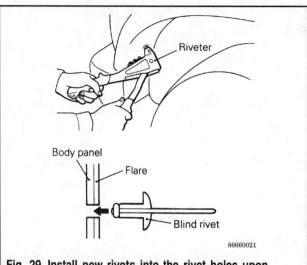

Fig. 29 Install new rivets into the rivet holes upon installation

Chassis and Cab Mounting Bushings

REMOVAL & INSTALLATION

♦ See Figures 31, 32, 33 and 34

If certain procedures covered by this manual require the cab/body mount bolts to be removed so that the body may be raised, it is critical that the body NOT be raised any more than necessary for access. If the cab/body is lifted too high, twisting and damage could occur. The body must be properly supported at all times.

If a cab/body mount must be removed completely for replacement, ALL mounts on that side of the vehicle MUST be unbolted in order to prevent damage. Again, the body must be suitably supported at multiple points (near each of the mounts) in order to prevent damage.

1. Raise and support the vehicle safely by placing jackstands under the body at each of the mounts on that side. Protect the body from the jackstands using wooden blocks. The blocks will also help distribute the vehicle's weight on a larger area than the lips of the jackstand. Place at least one jackstand under the frame temporarily for safety.

2. Loosen and remove the mount bolts on that side of the vehicle. If possible, remove the retainer and lower cushion as well.

3. Remove the jackstand from the FRAME only, then slowly lower the frame away from the body on that side. Make sure you remain clear of the frame during this step. Also keep small children and pets away. (If you think ethylene glycol antifreeze is bad for them, a descending frame is even worse).

➡Do not lower the frame away from the body any more than absolutely necessary.

4. Once the frame is lowered sufficiently, support it at this new height using a jackstand for safety.

5. Remove the upper cushion, spacer and shims, as applicable.

6. During installation remember that at NO point should you work under the vehicle unless both the body and frame are

10-22 BODY AND TRIM

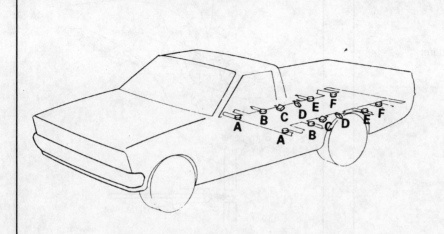

1. Mounting bolt
2. Body shim (A)
3. Plain washer
4. Self locking nut
5. Mounting shim
6. Body shim (B)

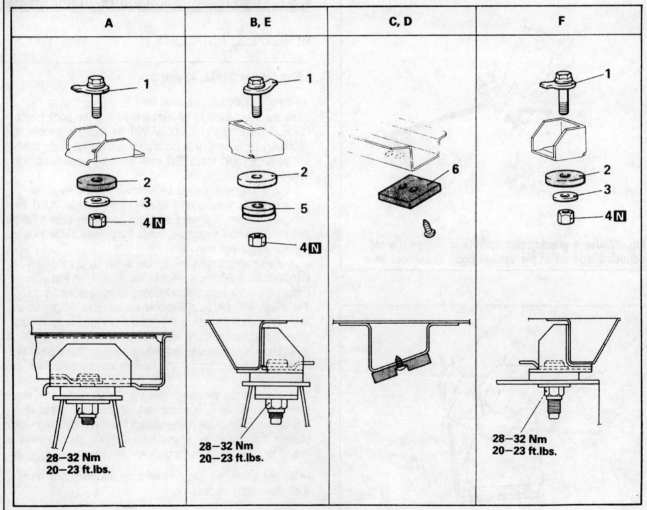

Fig. 30 Make certain to install the correct nuts, washers and bushings on each mounting pad of the truck bed

BODY AND TRIM 10-23

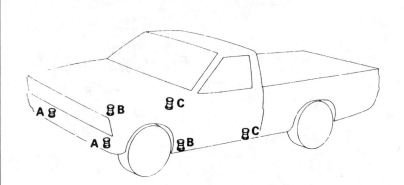

1. Special bolt
2. Plain washer
3. Body mounting rubber (A)
4. Spacer
5. Body mounting rubber (B)
6. Washer
7. Self locking flange nut
8. Plain washer
9. Self locking nut

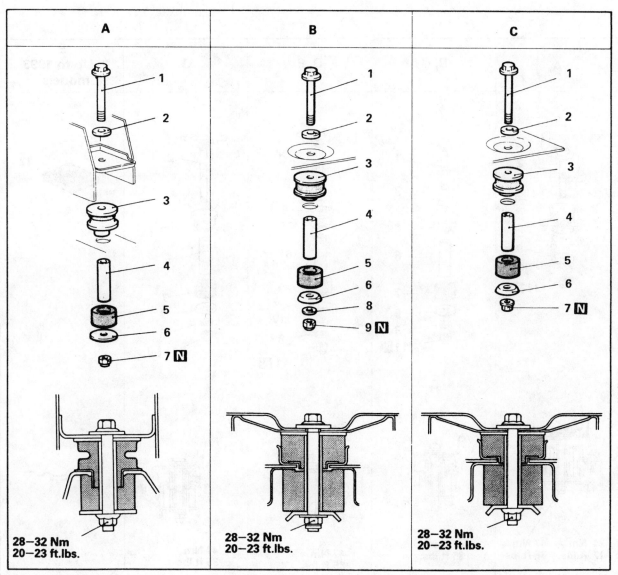

Fig. 31 Body mounting positions and components — 1983-95 Pick-ups

10-24 BODY AND TRIM

1. Special bolt
2. Mounting bolt
3. Plain washer
4. Body mounting rubber
5. Body mounting rubber A
6. Spacer
7. Body mounting rubber B
8. Plate
9. Washer
10. Body mount stopper
11. Self locking nut
12. Body shim

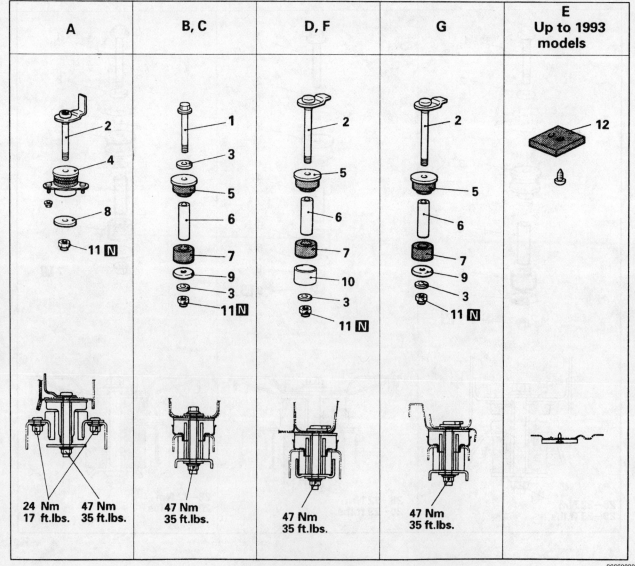

Fig. 32 Body mounting positions and components — 1992-95 Monteros

BODY AND TRIM 10-25

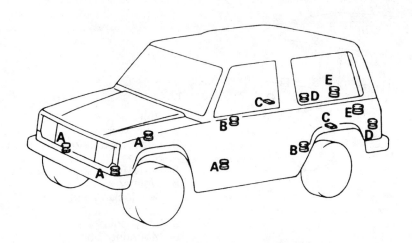

1. Special bolt
2. Plain washer
3. Body mounting rubber (A)
4. Spacer
5. Body mounting rubber (B)
6. Washer
7. Self locking nut
8. Body shim
9. Mounting bolt

NOTE
N : Non-reusable parts

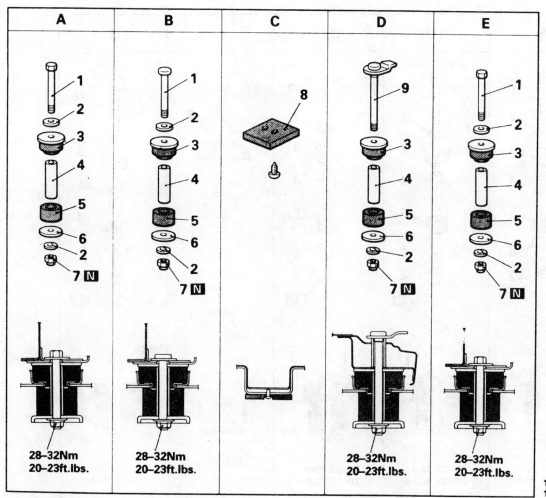

Fig. 33 Body mounting positions and components — 1983-91 2-door Monteros

10-26 BODY AND TRIM

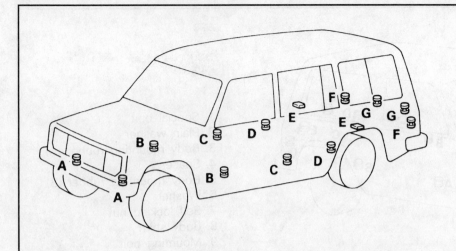

1. Bolt
2. Plain washer
3. Body mounting rubber (A)
4. Spacer
5. Body mounting rubber (B)
6. Washer
7. Self locking nut
8. Special bolt
9. Body shim
10. Mounting bolt

NOTE
N : Non-reusable parts

A	B,C	D,F	E	G
1, 2, 3, 4, 5, 6, 2, 7N	8, 2, 3, 4, 5, 6, 2, 7N	10, 2, 3, 4, 5, 6, 2, 7N	9	10, 3, 4, 5, 6, 2, 7N
28–32 Nm 20–23 ft.lbs.	28–32 Nm 20–23 ft.lbs.	28–32 Nm 20–23 ft.lbs.		28–32 Nm 20–23 ft.lbs.

Fig. 34 Body mounting positions and components — 1983-91 4-door Monteros

BODY AND TRIM 10-27

supported using jackstands. The floor jack is NOT sufficient to assure safety, even for short periods of time.

7. Install the upper body bushing, spacers and shims that were removed.

8. Raise the frame with the floor jack until it contacts the body. Support the frame with a jackstand at this height. Install the lower bushings, spacers, washers and new nuts.

9. Tighten the nuts to 20-23 ft. lbs. (28-32 Nm) for 1983-95 Pick-ups and 1983-91 Monteros, and tighten to 35 ft. lbs. (47 Nm) for 1992-95 Monteros.

10. Once all of the removed body mounts has been tightened, lower the vehicle to the ground. Repeat for the other side, if needed.

Power Sunroof

REMOVAL & INSTALLATION

♦ See Figures 35, 36, 37 and 38

Although there are slight differences between the power sunroofs in the 1989 Monteros and 1990-95 Monteros, this procedure is for all Monteros with power sunroofs.

1. Disconnect the negative battery cable from the battery.
2. Remove the sunroof switch off of the switch, then disconnect the electrical harness plug.
3. Remove the headliner from the Montero — refer to the headliner procedures found later in this section.
4. Close the sunroof fully, then remove the motor.

➡ If the sunroof does not move or can not be closed, matchmark the roof lid and guide rail.

5. Place matchmarks on the motor intermittent gear and bracket.
6. Remove the control unit.
7. Remove the rear set bracket, then extract the sunroof assembly. After removal of the sunroof assembly, remove the front set bracket.
8. Unless necessary, do not remove the drain hoses. To remove the hoses, however, tie a rope to the end of the drain

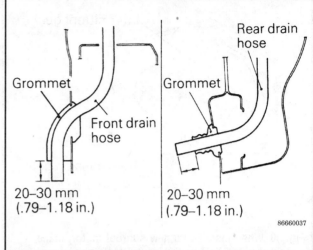

Fig. 37 Install the new drain hoses to the dimensions shown in the illustration

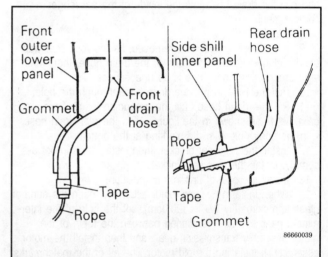

Fig. 36 Use a piece of rope and tape to pull the old drain hoses out of their holes and grommets

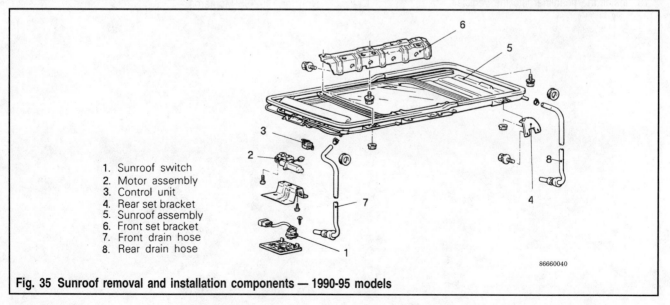

1. Sunroof switch
2. Motor assembly
3. Control unit
4. Rear set bracket
5. Sunroof assembly
6. Front set bracket
7. Front drain hose
8. Rear drain hose

Fig. 35 Sunroof removal and installation components — 1990-95 models

10-28 BODY AND TRIM

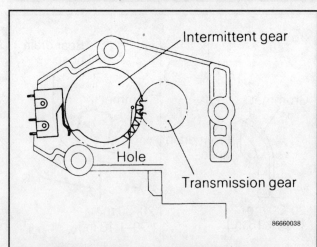

Fig. 38 When installing a new sunroof motor, make certain that the hole of the intermittent gear is in between two teeth of the transmission gear

hose. Wind a tape around it so that there is no unevenness, and pull the drain hose into the inside of the passenger compartment.

To install:

9. If the drain hoses were removed, tie the rope that was used during removal to the end of the drain hose, then wind tape around it so that there is no unevenness.
10. Pull the rope to pull the drain hose through the hole.
11. Pull the drain hose until the protruding length from the grommet is as shown in the figure. Align the rear drain hose with the body hole with the marking at the bottom.
12. Install the front set bracket, then install the sunroof assembly. Install the rear set bracket.
13. Install the control unit into its position.
14. When installing a new motor assembly, open the sunroof glass approximately 7.9 in. (20.0cm), set the hole of the intermittent gear so that it is aligned between the teeth of the motor assembly transmission gear, and then install the motor assembly. If installing the old motor, line all of the matchmarks up and install the motor. Apply a small amount of multipurpose grease to the smaller motor gear.
15. Install the headliner into the vehicle.
16. Plug the electrical harness connectors back together and install the sunroof switch. Install the sunroof switch cover.
17. Connect the negative cable to the battery.

Spare Tire Carrier

REMOVAL & INSTALLATION

Pick-Up

▶ See Figures 39 and 40

1. Raise and support the vehicle safely using jackstands for access.
2. Lower the spare tire and remove it from the carrier.
3. Remove the cotter pin and bolts attaching the spare tire hoist to the frame.
4. Remove the spare tire hoist from the frame.

To install:

5. Align the tire carrier with the crossmember and shaft, then connect it with the pin, retainer and bolts. Tighten the bolts to 6-9 ft. lbs. (8-12 Nm).
6. Mount the spare tire onto the carrier so that its outer sidewall is upward, then tighten to 14 ft. lbs. (20 Nm) of torque using the jack handle and wheel nut wrench.
7. Lower the vehicle back to the ground.

Montero

▶ See Figure 41

1. Loosen and remove the spare tire mounting lug nuts from the tire carrier.
2. Pull the spare tire off of the carrier and set aside.
3. Unscrew the six spare tire carrier mounting bolts, then remove the carrier from the back door.
4. To install, set the carrier on the back door and align the mounting holes.
5. Insert the mounting bolts and tighten them until finger-tight.
6. Tighten the mounting bolts in a crisscross fashion to 6-7 ft. lbs. (8-10 Nm) for 1983-91 Monteros or to 17 ft. lbs. (23 Nm) for 1992-95 Monteros.

BODY AND TRIM 10-29

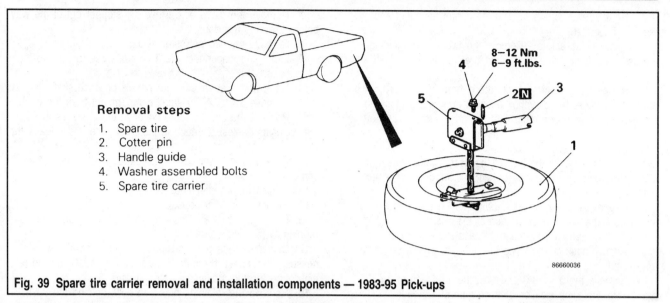

Fig. 39 Spare tire carrier removal and installation components — 1983-95 Pick-ups

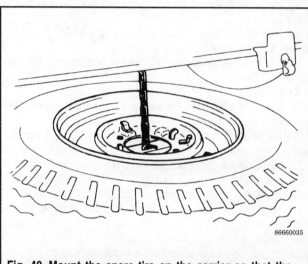

Fig. 40 Mount the spare tire on the carrier so that the outer sidewall of the tire is facing upward

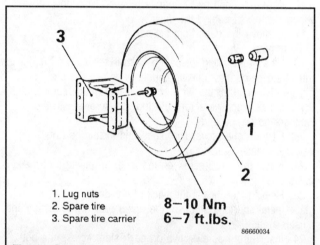

Fig. 41 Spare tire carrier removal and installation components — 1983-91 Monteros, 1992-95 Monteros are similar

INTERIOR

Instrument Panel/Dashboard

REMOVAL & INSTALLATION

Each dashboard uses a variety of nuts and bolts of different sizes and lengths. Correct cataloging and reinstallation is required if the dash is to be free of annoying squeaks and rattles. Additionally, wires and hoses should be labeled at the time of removal. The amount of time saved during reassembly makes the extra effort well worthwhile. Removal of the dash is a complicated, time consuming task. Do not attempt this procedure if you are not comfortable with extended projects. Additional helpful information may be found in other parts of this book dealing with specific assemblies. For example, if not certain how to remove the radio, refer to "Radio Removal and Installation" in this section.

Pick-Up

1983-86 MODELS

1. Disconnect the negative battery cable.
2. Loosen the tilt wheel lock and lower the wheel to the lowest position.
3. Remove the steering wheel.
4. Remove the upper and lower steering column covers.
5. Remove the instrument cluster. Disconnect the gauge wiring and the speedometer cable. Once the cluster is out, remove the single screw in the upper center of the mounting space in the dashboard.
6. Remove the radio; disconnect the diesel indicator if so equipped.
7. Remove the glove box.
8. Remove the floor console and/or center console, depending on equipment

9. Remove the screws holding the center air outlet and remove the outlet.
10. Remove the ashtray. Remove the two screws holding the ashtray holder. Slide the holder (bracket) into the dash and remove it.
11. Remove the 4 screws from the bottom of the dash. Two are located under the glovebox opening and two are located approximately under the dash light dimmer switch on the left side.
12. Reach in through the opening in the center of the dash and remove the nut at the rear of the dash. Make certain you identify the correct nut; it may be either a wing nut or a hex nut.
13. Reach just inside the glovebox opening and remove the screw from the lower inner surface of the dash.
14. From underneath the dash, release the 4 hex nuts holding the upper dash cover. Once the nuts are off, lift the cover up and off.
15. Remove the 5 screws and bolts from the top of the dashboard. Note the location of each type of fastener.
16. Work the dash away from its mounts. Disconnect the wiring harnesses behind the dash.
17. Remove the dash from the car.

To install:
18. When reinstalling, place the dash in the car and connect the wiring harnesses, making certain none are stretched or pinched. Check that the air ducts align and match correctly.
19. Install 5 screws across the top; the screws with the washers go on the ends.
20. Fit the upper cover into place and install the nuts from below the dash.
21. Install the small screw on the inner lower edge of the glove box.
22. Reinstall the nut at the back of the dash.
23. Install the lone screw at the upper rear of the instrument cluster area.
24. Install the 4 screws, two on each side, across the bottom of the dash.
25. Slide the ashtray holder into place and install the screws. Install the ashtray.
26. Install the center air outlet.
27. Replace the glove box and tighten the screws.
28. Install the radio; connect the diesel indicator wiring if so equipped.
29. Install the instrument cluster; connect the wiring and the speedometer cable.
30. Install the steering column covers.
31. Elevate the tilt column and install the steering wheel.
32. Connect the negative battery cable.

1987-95 MODELS

▶ See Figures 42, 43, 44 and 45

1. Disconnect the negative battery cable.
2. Remove the hazard flasher switch.
3. Pop out the hole cover on the right side of the instrument hood.
4. Remove the instrument hood.
5. Remove the instrument cluster, remembering to disconnect the wiring and speedometer cable at the rear.
6. Remove the fusebox cover. Remove the fusebox from the dashboard and let it hang.
7. Remove the glove box. Move the stopper out of the way before lowering the door.
8. Remove the two defroster ducts.
9. Label and disconnect the 3 control wires running to the heater unit.
10. Remove the speaker cover panels on each side.
11. Remove the clock or small compartment from the dash.
12. Pop out the small square cover at the upper center of the dash. Its easiest to reach in the clock hole and release the clips with your fingers.
13. Remove the lower center faceplate or cover from the center of the dash.
14. Remove the shifter knob.
15. Remove the floor console assembly.
16. Disconnect the left side and front harness connectors running to the dashboard.
17. Disconnect the air conditioner and heater wiring harnesses. Disconnect the instrument cluster harness and the radio connectors, including the antenna wire.
18. Remove the retaining nuts and bolts holding the dash. They are located across the top, along the sides and at the bottom.
19. Use the tilt release lever and move the steering column all the way downward.
20. Remove the dash carefully; several components are still connected to it.

To install:
21. Place the dash into the vehicle carefully and connect the main electrical harnesses.
22. Install all the retaining nuts and bolts but do not tighten any of them until the alignment of the dash is correct.
23. Reinstall the floor console, the shifter knob and the lower center faceplate.
24. Install the upper bolt cover in the center of the dash.
25. Replace the clock or small pocket. Don't forget to plug wiring together.
26. Install the speaker covers.
27. Route the heater control cables properly and connect them. Refer to "Heater Core Removal and Installation" for the correct alignment procedure.
28. Install the defroster ducts.
29. Install the glove box.
30. Remount the fusebox on the dash and install the lid.
31. Install the instrument cluster. Connect the speedometer cable and wiring harnesses before tightening the screws.
32. Install the instrument cluster hood and install the bolt cover.
33. Install the hazard warning switch.
34. Connect the negative battery cable.

Montero

1983-86 MODELS

1. Disconnect the negative battery cable.
2. Remove the steering wheel.
3. Remove the center console.
4. Remove the meter assembly. Carefully disconnect the wiring and speedometer cable.
5. Remove the gauge assembly and disconnect its wiring.
6. Remove the lap heater ducts below the dash and disconnect the release cable bracket from the dash.

BODY AND TRIM 10-31

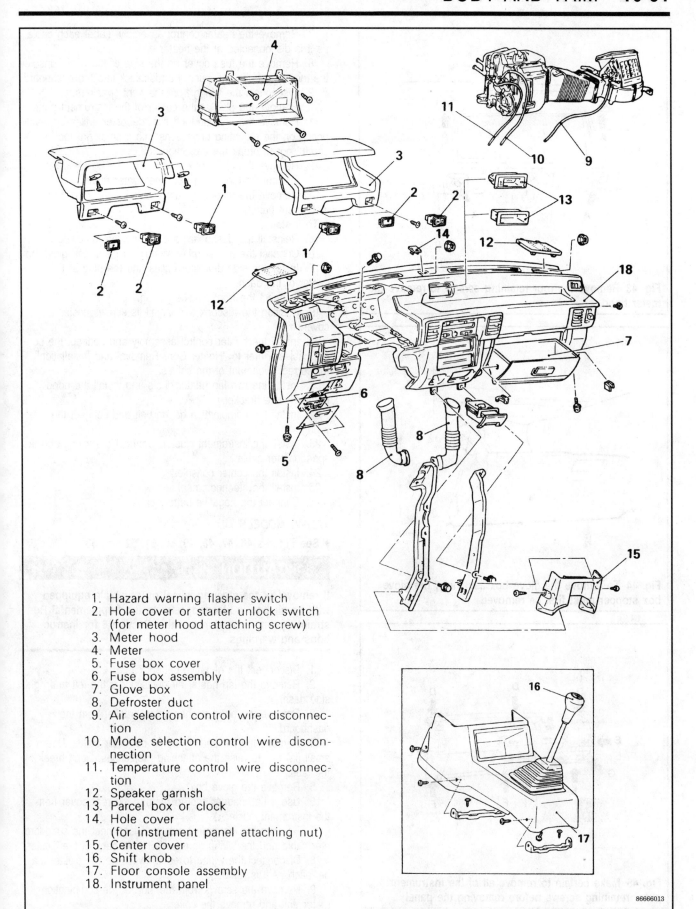

1. Hazard warning flasher switch
2. Hole cover or starter unlock switch (for meter hood attaching screw)
3. Meter hood
4. Meter
5. Fuse box cover
6. Fuse box assembly
7. Glove box
8. Defroster duct
9. Air selection control wire disconnection
10. Mode selection control wire disconnection
11. Temperature control wire disconnection
12. Speaker garnish
13. Parcel box or clock
14. Hole cover (for instrument panel attaching nut)
15. Center cover
16. Shift knob
17. Floor console assembly
18. Instrument panel

Fig. 42 Instrument panel (dashboard) removal and installation components

10-32 BODY AND TRIM

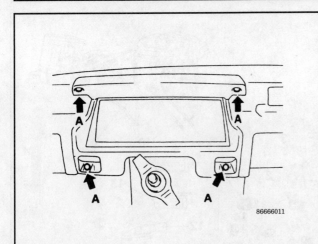

Fig. 43 Remove the four retaining screws to remove the meter hood

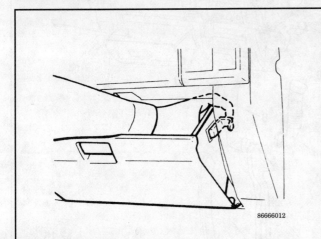

Fig. 44 To remove the glove box assembly, the glove box stopper must first be removed

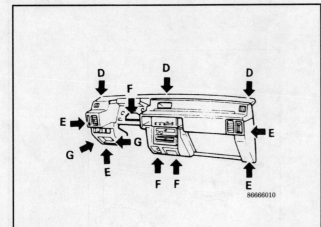

Fig. 45 Make certain to remove all of the instrument panel retaining screws before removing the panel

7. Remove the heater control assembly. Label each cable as it is disconnected at the heater end.
8. Remove the fuse cover on the side of the dash. unscrew the mounting bolts and push the fuseblock into the dashboard.
9. Disconnect the wiring from the front speakers.
10. Remove the plug at the center of the instrument panel.
11. Remove the right and left side defroster grilles by gently prying on the mounting projections with a small flat tool. Be careful not to break the projections off.
12. Remove the glove box.
13. Disconnect the wiring from the heater relay.
14. Remove the mounting nuts and bolts and remove the instrument panel.

To install:

15. Reinstall the dashboard and secure the mounting bolts.
16. Connect the heater relay wiring and install the glove box.
17. Install the side defroster grilles and the plug at the center of the dash.
18. Connect the front speaker wiring.
19. Position the fusebox and install its screws; install the cover.
20. Install the heater control assembly and connect the control cables. Refer to "Heater Core Removal and Installation" for correct alignment of the cables.
21. Install the two lap heater ducts and install the hood release cable bracket.
22. Install the combination gauge unit and connect the wiring.
23. Install the instrument cluster. Connect the wiring and the speedometer cable.
24. Install the center console.
25. Install the steering wheel.
26. Connect the negative battery cable.

1987-91 MODELS

▶ See Figures 46, 47, 48, 49, 50, 51, 52 and 53

✱✱CAUTION

If removal of the steering wheel on a vehicle equipped with an air bag is planned, refer to the Supplemental Restraint System procedures in this Section 6 for instructions and warnings

1. Disconnect the negative battery cable.
2. Remove the lap heater air ducts under the left and right side dash.
3. Remove the hood release cable bracket from the dashboard.
4. Remove the left and right side defroster grilles. Use a small flat tool to raise the attaching projections. Don't break the tabs.
5. Remove the glove box.
6. Use a flat, padded tool to remove the rear cover from the instrument cluster.
7. Remove the instrument cluster. Disconnect the speedometer cable and the wiring harnesses to the back of the cluster.
8. Disconnect the wiring to each switch on the cluster. Label each as it is removed.
9. Remove the screws from the side of the combination meter unit and remove the cover.

BODY AND TRIM 10-33

Fig. 46 Use a prytool and a piece of rubber or a rag to remove the cluster rear cover

Fig. 47 Remove the fuse box cover and the fuse box itself

Fig. 48 Remove the retaining screws securing the instrument cluster

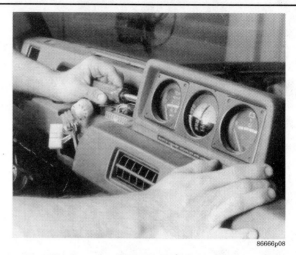

Fig. 49 Removing the combination meter case from the instrument panel — 1987-91 Monteros

10. Remove the 4 screws at the base of the combination meter and gently remove the meter. Disconnect the wire harnesses.

❋❋WARNING

The inclinometer can be damaged by dropping or bumping it. Do not tilt the unit so far as to exceed the maximum indication on the scale. Resist the temptation to play with the unit.

11. Remove the center panel or lower console.
12. Disconnect the 3 heater control cables running to the heater unit. Label each one.
13. Remove the center reinforcement and bring it out as a unit with the radio or stereo. Disconnect the wiring when the unit is free.
14. Remove the horn pad and remove the steering wheel.
15. Remove the fuse box cover and release the fusebox from the dashboard.

Fig. 50 The center panel and all of its components will have to also be removed

10-34 BODY AND TRIM

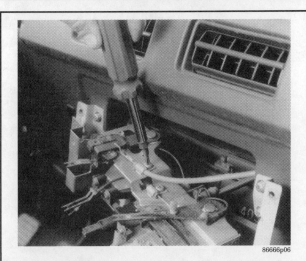

Fig. 51 Unhook the heater control assembly cables from the heater control levers

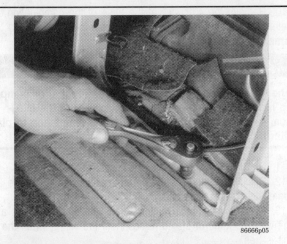

Fig. 52 Remove all of the mounting bolts from the center reinforcement, then remove the reinforcement itself

Fig. 53 After removing all of its mounting screws, remove the instrument panel from the vehicle

16. Remove the dashboard mounting nuts and bolts and remove the dashboard.

To install:

17. When reinstalling, position the dash and start each nut and bolt. Make certain the dash is aligned and then tighten the fittings.
18. Install the fusebox assembly and its cover.
19. Install the steering wheel and horn pad.
20. Install the radio and center reinforcement. Connect the wiring before tightening the bolts.
21. Carefully route and connect the heater control cables. Refer to "Heater Core Removal and Installation" for correct alignment of the cables.
22. Install center panel and secure the heater controls.
23. Install the combination gauges and connect the wiring.
24. Install the cover for the combination gauges.
25. Install the instrument cluster, connecting the speedometer cable and the many wiring connectors.
26. Install the cluster cover; make sure it is firmly seated in place.
27. Replace the glove box.
28. Reinstall the side defroster grilles.
29. Connect the hood release cable bracket.
30. Install the two lap heater ducts.
31. Connect the negative battery cable.

1992-95 MODELS

▶ See Figures 54 and 55

❈❈CAUTION

When installing or removing the instrument panel, do not allow any impact or shock to the SRS diagnosis unit. Always disable the SRS system when working on or near the SRS components. Refer to Section 6 for SRS precautions.

1. Disconnect the negative (-) battery cable from the battery.
2. Wait at least 60 seconds after disconnecting the battery cable to make certain that the SRS is completely disarmed before servicing the vehicle.
3. Remove the floor console.
4. Remove the hood lock release and the fuel filler door lock release handles.
5. Loosen and remove the retaining screws, then the instrument under and corner covers.
6. Remove the glove box stopper and then the glove box.
7. Unscrew the mounting screws and remove the center panel A.
8. Extract the heater control assembly.
9. Take the radio and tape player out.
10. Pop the meter hood plug out, then remove the meter bezel assembly by loosening and removing the mounting screws.
11. Remove the screws holding the combination meter (instrument cluster) in place and extract the meter.
12. Disconnect and remove the speedometer cable adapter. First disconnect the speedometer cable at the transmission end of the cable, then remove the lock of the speedometer cable adapter from the instrument panel. Pull the speedometer cable

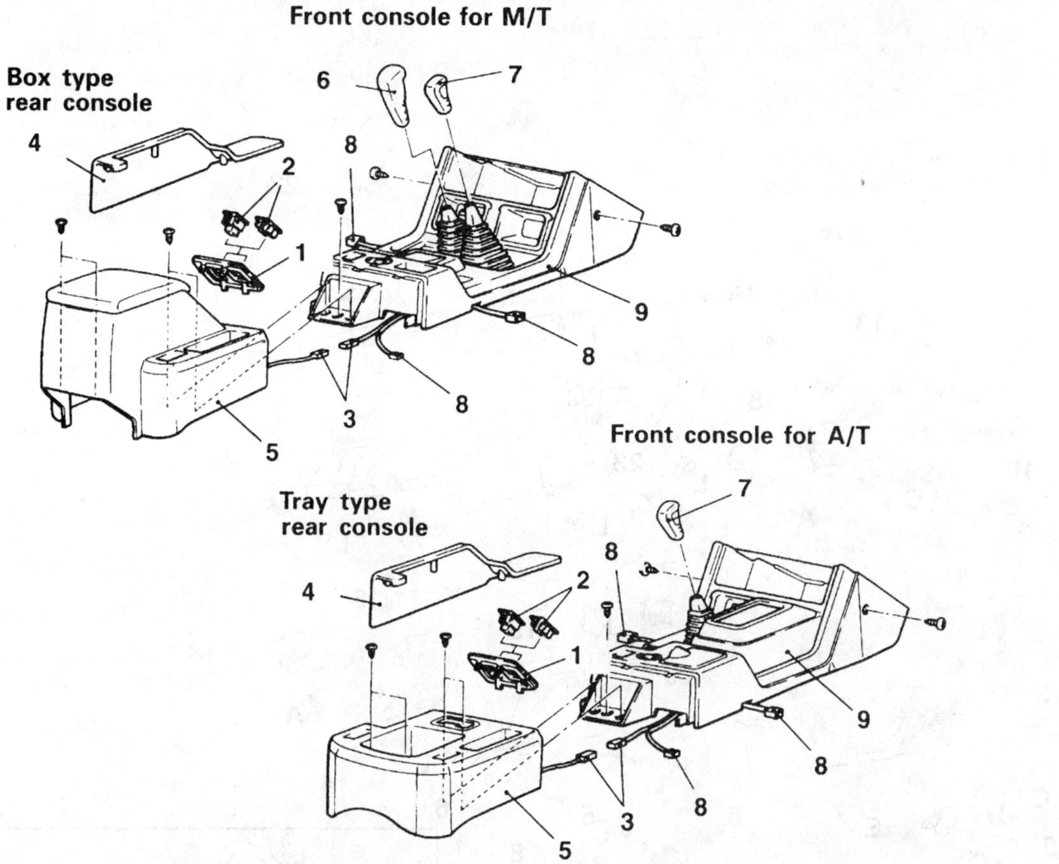

BODY AND TRIM 10-35

Fig. 54 Floor console removal and installation component locations — 1992-95 Monteros

1. Switch panel
2. Suspension control switch or hole cover
3. Rear console harness connector
4. Side panel A
5. Rear console assembly
6. M/T shift lever knob
7. Transfer shift lever knob
8. Floor console harness connector
9. Front console assembly

slightly toward the vehicle interior, then remove the speedometer cable adapter.

13. Remove the four screws holding the column cover in place, then remove the cover itself.
14. Remove the clock or clock plug.
15. Remove the side defroster garnish, the door mirror control switch, the front speaker, the rheostat, the rear wiper and washer switch and the door lock switch.
16. Disconnect the ventilation control wire and the wiring harness.
17. Unscrew the steering column installation bolts.
18. Unscrew all the retaining screws holding the dashboard (instrument panel assembly) in place and remove the assembly.

To install:
19. Hold the instrument panel the assembly in position and secure it in place with its mounting screws.
20. Tighten the steering column installation bolts and plug the harness connector back together.
21. To install the ventilation control wire, set the cool air bypass dial to the closed position. Close the cool air bypass lever at the heater unit side (lever is lightly hit against the stopper). Install the ventilation control wire and secure it with the clip.
22. Install the door lock switch, the rear wiper and washer switch, the rheostat, the front speaker, the door mirror control switch, the side defroster garnish, the clock or clock plug and the column cover.
23. Hook the speedometer cable adapter back up the back of the combination meter (instrument cluster).
24. Install the combination meter, the meter bezel assembly and the meter hood plug.
25. hook the radio and tape player back up and install them in the dashboard.
26. Install the heater control assembly, center panel A, the glove box assembly and the glove box stopper.
27. Install the instrument under and corner covers, the fuel filler door lock release handle and the hood lock release handle.
28. Install the floor console.
29. Connect the negative (-) battery cable to the battery and enable the SRS.

10-36 BODY AND TRIM

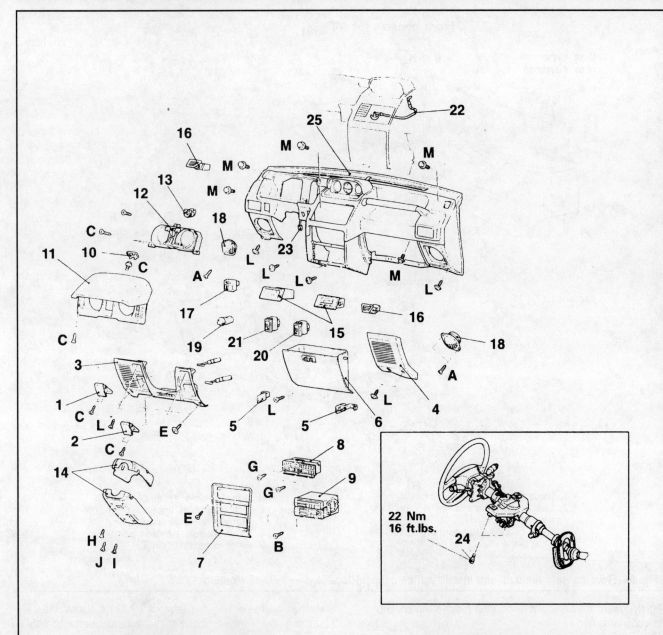

1. Hood lock release handle
2. Fuel filler door lock release handle
3. Instrument under cover
4. Instrument corner cover
5. Glove box stopper
6. Grove box assembly
7. Center panel A
8. Heater control assembly
9. Radio and tape player
10. Meter hood plug
11. Meter bezel assembly
12. Combination meter
13. Speedometer cable adapter <Up to 1993 models>
14. Column cover
15. Clock or clock plug
16. Side defroster garnish
17. Door mirror control switch
18. Front speaker
19. Rheostat
20. Rear wiper and washer switch
21. Door lock switch
22. Ventilation control wire
23. Harness connector
24. Steering column installation bolts
25. Instrument panel assembly

Fig. 55 Instrument panel removal and installation component locations — 1992-95 Monteros

BODY AND TRIM 10-37

Consoles

♦ See Figure 54

The consoles in Mitsubishi vehicles are generally easily removed. Components mounted in the console, gauges, clock, radio, etc., should be removed before the console is loosened. Removing the floor console almost always requires the removal of the manual shifter knob. Do not remove the automatic shifter handle; the console can be turned and removed over the handle.

The biggest problem in removing the console is finding all the concealed screws holding it in place. Quite often these screws are located behind an ash tray or below a cargo pocket. Proceed carefully and don't force anything loose. Quite often the screws are behind a small cap which must be popped out.

Access to the screws along the sides of the floor console will be easier if the seats are removed. The empty interior is particularly handy when removing the consoles for access to other components such as the dashboard or heater core.

After the console has been loosened all around, carefully lift it away from its mounts and check for wiring which is still connected. Disconnect it and place the connector out of harm's way while the other work progresses. Some floor or rear consoles have the seat belts routed through slots. If it is necessary to remove the console completely, the belts must be threaded back through the holes and remain in the car. Do NOT unbolt the seat belts to get the console out.

With the seats and the consoles removed, this is a perfect time to vacuum the interior carpets. You'll be able to reach places not usually available and you'll probably find enough loose change to make the project worthwhile. Give the console a thorough cleaning and a light coat of vinyl protectorant; again, you can now reach those impossible locations.

Reinstalling the consoles requires accurate positioning. All the bolt holes must align and all the retaining screws should be started one or two turns until the console is in its final position. The screws may be tightened after the console is correctly located. Remember to recover the wires that may have been placed under the carpet or behind the dash. Connect them to their components. Reinstall the various devices which mount in the console. Check the operation of each item and test drive the car after the seats are reinstalled to check for squeaks and rattles.

Door Panels (Door Pads or Liners)

REMOVAL & INSTALLATION

♦ See Figures 56, 57 and 58

➡This is a general procedure. Depending on vehicle and model, the order of steps may need to be changed slightly.

1. Remove the inner mirror control knob (if manual remote) and remove the inner delta cover from the mirror mount.
2. Remove the screws holding the armrest and remove the armrest. The armrest screws may be concealed behind plastic caps which may be popped out with a non-marring tool. If the door has a pull handle or grab handle, remove it also.
3. Remove the surround or cover for the inside door handle. Again, seek the hidden screw; remove it and slide the cover off over the handle.
4. If not equipped with electric windows, remove the window winder handle. This can be tricky, but not difficult. Install a piece of tape on the door pad to show the position of the handle before removal. The handle is held onto the winder axle by a spring clip shaped like the Greek letter omega: Ω. The clip is located between the back of the winder handle and the door pad. It is correctly installed with the legs pointing along the length of the winder handle. There are three common ways of removing the clip:

    a. Use a door handle removal tool. This inexpensive slotted and tooted tool can be fitted between the winder and the panel and used to push the spring clip free.

    b. Use a rag or piece of cloth and work it back and forth between the winder and door panel. If constant upward tension is kept, the clip will be forced free. Keep watch on the clip as it pops out; it may get lost.

    c. Straighten a common paper clip and bend a very small J-hook at the end of it. Work the hook down from the top of the winder and engage the loop of the spring clip. As you pull the clip free, keep your other hand over the area. If this is not done the clip will vanish to an undisclosed location, never to be seen again.

5. In general, power door lock and window switches may remain in place until the pad is removed. Some cannot be removed until the doorpad is off the door.
6. If the car has manual vertical door locks, remove the lock knob by unscrewing it. If this is impossible (because they're in square housings) wait until the pad is lifted free. Check the door liner for any remaining screws or attaching bolts. Some will have a screw or two at the front edge of the pad at the top and bottom; remove them.
7. Using a broad, flat-bladed tool — not a screwdriver — begin gently prying the door pad away from the door. You are releasing plastic inserts from plastic seats. There will be 6 to 12 of them around the door. With care, the plastic inserts can be reused several times.
8. When all the clips are loose, lift up on the panel to release the lip at the top of the door. This may require a bit of jiggling to loosen the panel; do so gently and don't damage the panel. The upper edge (at the window sill) is attached by a series of retaining clips.
9. Once the panel is free, keep it close to the door and check behind it. Disconnect any wiring for switches, lights or speakers which may be attached.

➡**Behind the panel is a plastic or paper sheet taped or glued to the door. This is a water shield and must be intact to prevent water entry into the car. It must be securely attached at its edges and not be ripped or damaged. Small holes or tears can be patched with waterproof tape applied to both sides of the liner.**

10. When reinstalling, connect any wiring harnesses and align the upper edge of the panel along the top of the door first. Make sure the left-right alignment is correct; tap the top of the panel into place with the heel of your hand.
11. Make sure the plastic clips align with their holes; pop each retainer into place with gentle pressure.

10-38 BODY AND TRIM

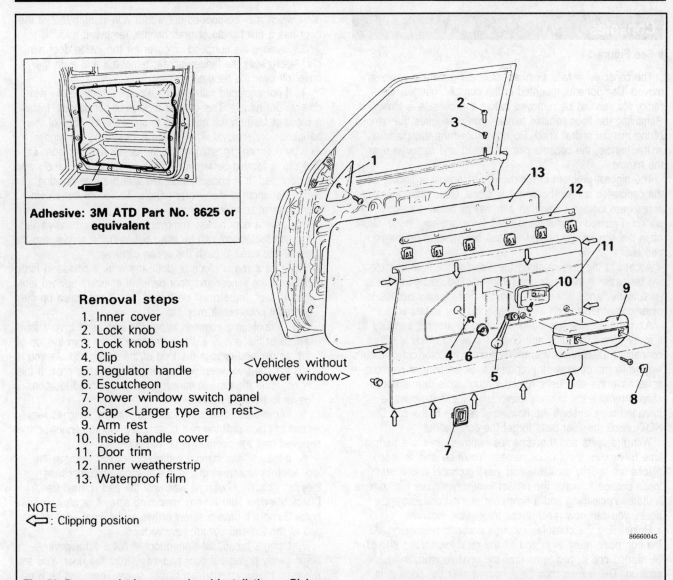

Adhesive: 3M ATD Part No. 8625 or equivalent

Removal steps
1. Inner cover
2. Lock knob
3. Lock knob bush
4. Clip
5. Regulator handle
6. Escutcheon   <Vehicles without power window>
7. Power window switch panel
8. Cap <Larger type arm rest>
9. Arm rest
10. Inside handle cover
11. Door trim
12. Inner weatherstrip
13. Waterproof film

NOTE
⇐ : Clipping position

Fig. 56 Door panel trim removal and installation — Pick-ups

12. Install the armrest and door handle bezel, remembering to install any caps or covers over the screws.
13. Install the window winder handle on cars with manual windows. Place the spring clip into the slot on the handle, remembering that the legs should point along the long dimension of the handle.
14. Align the handle with the tape mark made earlier and put the winder over the end of the axle. Use the heel of your hand to give the center of the winder a short, sharp blow. This will cause the winder to move inward and the spring will engage its locking groove. The secret to this trick is to push the winder straight on; if it's crooked, it won't engage and you may end up looking for the spring clip.
15. Install any remaining parts or trim pieces which may have been removed earlier — Map pockets, speaker grilles, etc.
16. Install the delta cover and the remote mirror handle if they were removed.

Interior Trim Panels

REMOVAL & INSTALLATION

♦ See Figures 59, 60, 61, 62, 63, 64 and 65

The interior trim panels in Mitsubishi Pick-ups and Monteros are held in place with retaining screws and retaining clips. To remove the retaining clips follow this procedure:
1. Use a Phillips screwdriver to push the pin (at the center of the trim clip) inward to a depth of about 0.08 in. (2mm).
2. Pull the trim clip outward to remove it.

✻✻WARNING

Do not push the pin inward more than necessary because it may damage the grommet, or the pin may fall in, if pushed too far.

BODY AND TRIM 10-39

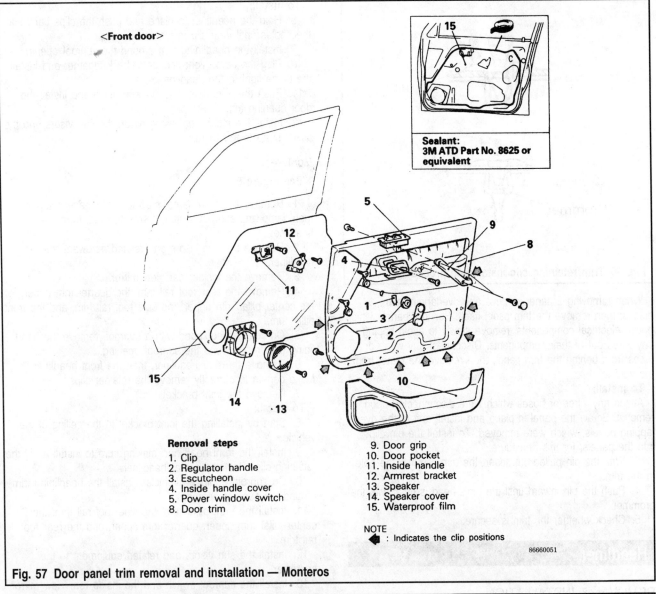

Fig. 57 Door panel trim removal and installation — Monteros

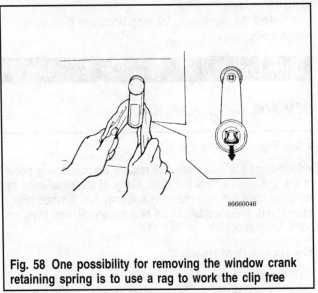

Fig. 58 One possibility for removing the window crank retaining spring is to use a rag to work the clip free

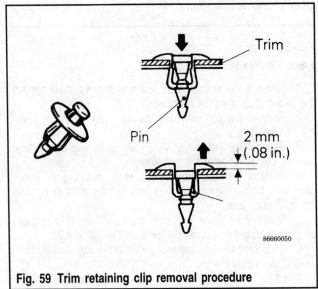

Fig. 59 Trim retaining clip removal procedure

10-40 BODY AND TRIM

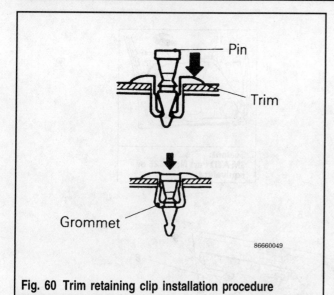

Fig. 60 Trim retaining clip installation procedure

When removing a panel, remove any overlapping trim pieces, then remove the trim panel itself. For trim panels which house electrical components, remove slowly to avoid ripping any wires out of their components. Disconnect any wires or tubes from behind the trim panel, then remove the panel from the vehicle.

To install:

Attach any wires or hoses which were disconnected upon removal. Screw the panel in place and install any of the overlapping panels, which were removed. To install the pin clips into the panels, for this procedure:

3. With the pin pulled out, insert the trim clip into the hole in the trim.
4. Push the pin inward until the pin's head is flush with the grommet.
5. Check whether the trim is secure.

Headliner

REMOVAL & INSTALLATION

Pick-Up

♦ See Figures 66, 67 and 68

1. Remove the assist strap's retaining screws, then remove the assist strap from the headliner.
2. Remove the sun visor and the sun visor hardware from the headliner.
3. Remove the inside rear view mirror retaining screws and remove the mirror from the headliner.
4. Remove the door opening trim which is facing the front pillar trim. With a plastic trim stick, pry out the clips to remove the front pillar trim.
5. Remove the dome light from the headliner. Unplug the electrical wiring harness connector from the dome light.
6. On vehicles with a sunroof, remove the headlining trim around the sunroof.
7. Insert the trim stick into the headlining and pry out the headlining attaching clips to remove it.

To install:

8. Hold the headliner in place and push the clips back into their holes until they clip in place.
9. Install the headlining trim around the sunroof opening.
10. Plug the dome light connector back together and install the dome light to the headliner.
11. Push the front pillar trim back in place and install the door opening trim.
12. Install the inside rear view mirror, the sun visors and the assist strap to the headliner.

Montero

♦ See Figure 69

1. Remove the room lamp, map lamp, luggage compartment lamp and sunroof switch, if so equipped, from the headliner.
2. Remove the sun visors and related hardware from the headliner.
3. Remove the inside rear view mirror.
4. Remove the rear roof rail trim, the quarter trim upper, the center pillar trim upper, the side roof rail trim, and the front pillar trim.
5. On vehicles equipped with a sunroof, remove the headlining trim from around the sunroof opening.
6. Remove the rear headliner, then the front headliner. Make certain to correctly remove the retainer clips.
7. Remove the joint bracket.

To install:

8. Start by installing the joint bracket to the ceiling of the vehicle.
9. Install the front headliner, making sure to install all of the retaining clips. Install the rear headliner.
10. On sunroof equipped vehicles, install the headlining trim around the sunroof opening.
11. Install the front pillar trim, the side roof rail trim, the center pillar trim upper, quarter trim upper, and the rear roof rail trim.
12. Install the sun visors and related equipment to the headliner.
13. Install the sunroof switch and the inside rear view mirror to the headliner.
14. Install the room lamp, the map lamp, and the luggage compartment lamps.

Door Locks

REMOVAL & INSTALLATION

♦ See Figures 70 and 71

➡Removing the door lock will require disconnecting some of the link rods within the door. Many of the clips used to hold the rods are non-reusable and may break when disassembled. Make certain there is a supply of new clips on hand for reassembly.

Key Lock in Door Handle

1. Remove the inner door liner. Make sure the window glass is in the down position before removing the window controls.

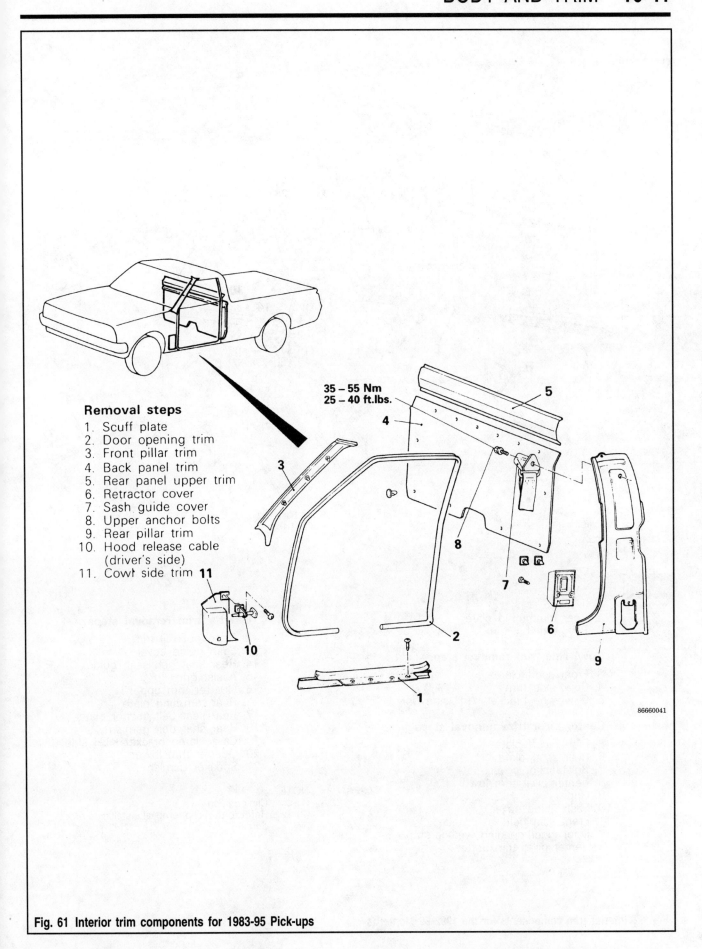

Fig. 61 Interior trim components for 1983-95 Pick-ups

10-42 BODY AND TRIM

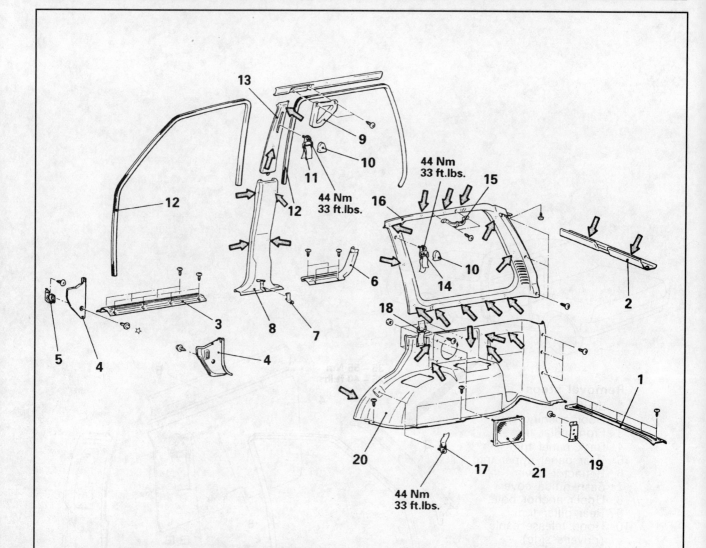

1. Rear trimmming plate
2. Rear roof rail trim

Cowl side trim removal steps

3. Front scuff plate
4. Cowl side trim
5. Cowl side bracket <RH side only>

Center pillar trim removal steps

3. Front scuff plate
6. Rear scuff plate
7. Belt anchor cover
8. Center pillar trim lower
9. Grip
10. Sash guide cover
11. Front seat belt sash guide
12. Door inner opening weatherstrip
13. Center pillar trim upper

Quarter trim removal steps

2. Rear roof rail trim
10. Sash guide cover
14. Rear seat belt sash guide
15. Assist grip
16. Quarter trim upper
1. Rear trimming plate
17. Rear seat belt anchor plate
18. Rear seat belt garnish
19. Cargo lamp bracket <LH side>
20. Quarter trim lower
21. Speaker garnish

NOTE
(1) ⇦ : Clip position
(2) ☆ : Refer to trim clip removal/installation procedures

Fig. 62 Interior trim components for the 1992-95 Monteros

BODY AND TRIM   10-43

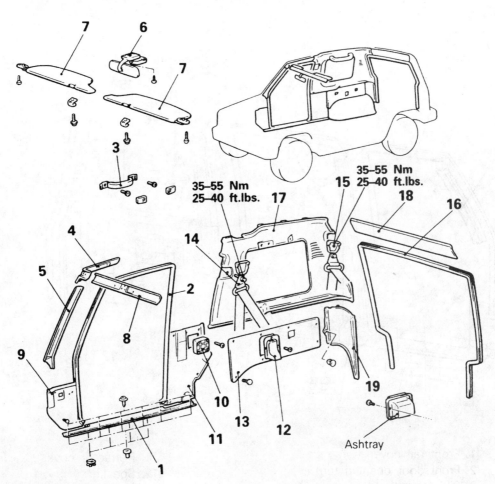

1. Front scuff plate
2. Door opening trim
3. Assist grip
4. Front side roof rail trim
5. Front pillar trim
6. Inside rear view mirror
7. Sun visor
8. Front roof rail trim
9. Cowl side trim
10. Speaker
11. Center pillar trim
12. Retractor cover
13. Quarter trim
14. Front seat belt anchor plate
15. Rear seat belt anchor plate <Vehicles with rear seat belts>
16. Back door opening trim
17. Upper quarter trim
18. Rear roof rail trim
19. Rear pillar lower trim

Fig. 63 Interior trim pieces for 2-door 1983-91 Monteros

10-44 BODY AND TRIM

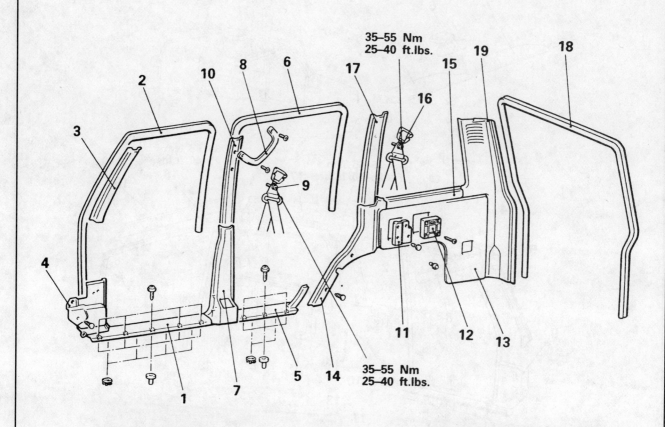

1. Front rail cover
2. Front door opening trim
3. Pillar trim (front)
4. Cowl side trim
5. Rear rail cover
6. Rear door opening trim
7. Center pillar trim
8. Get on and off grip
9. Front seat belt installation bolt
10. Pillar trim (center)
11. Retractor cover
12. Speaker
13. Quarter trim
14. Rear side trim
15. Quarter trim, upper
16. Rear seat belt installation bolt
17. Pillar trim (quarter)
18. Back door opening trim
19. Rear pillar trim

Fig. 64 Interior trim pieces for 4-door 1983-91 Monteros

BODY AND TRIM 10-45

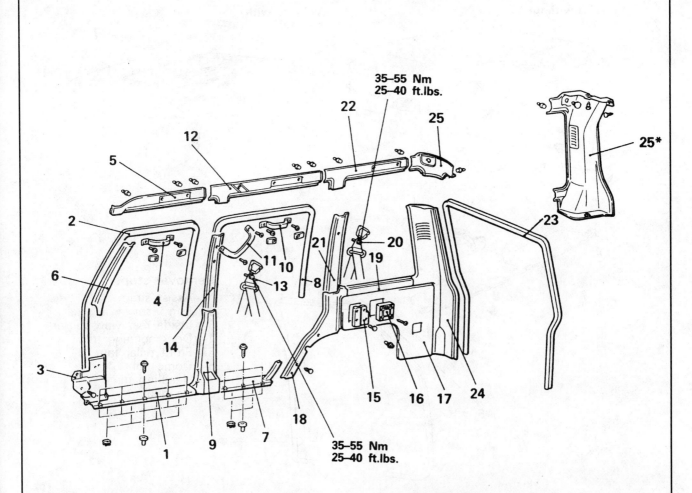

1. Front rail cover
2. Front door opening trim
3. Cowl side trim
4. Assist grip
5. Front roof side trim
6. Front pillar trim
7. Rear rail cover
8. Rear door opening trim
9. Center pillar trim
10. Assist grip
11. Get on and off grip
12. Center roof side trim
13. Front seat belt installation bolt
14. Pillar trim (center)
15. Retractor cover
16. Speaker
17. Quarter trim
18. Rear side trim
19. Quarter trim, upper
20. Rear seat belt installation bolt
21. Pillar trim (quarter)
22. Rear roof side trim
23. Back door opening trim
24. Rear pillar trim
25. Rear pillar trim (upper)

NOTE
The * symbol indicates the rear pillar trim (right) when the dual air conditioner is installed.

Fig. 65 Interior trim pieces for 4-door 1983-91 Monteros equipped with sunroofs

10-46 BODY AND TRIM

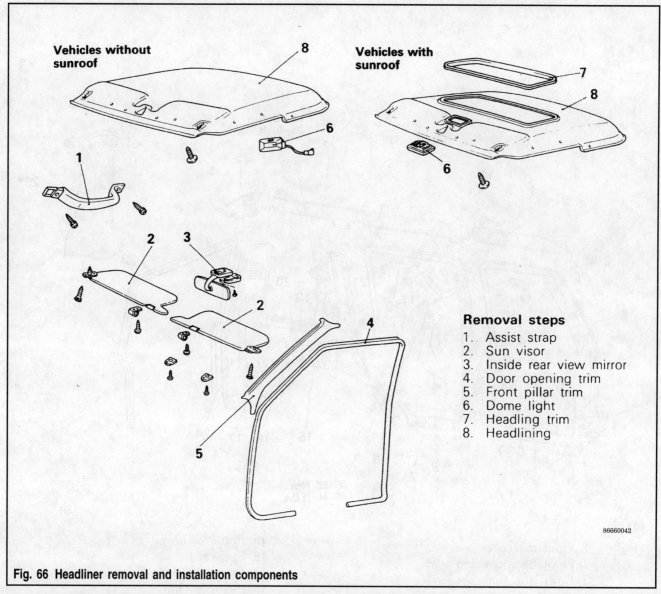

Fig. 66 Headliner removal and installation components

Removal steps
1. Assist strap
2. Sun visor
3. Inside rear view mirror
4. Door opening trim
5. Front pillar trim
6. Dome light
7. Headling trim
8. Headlining

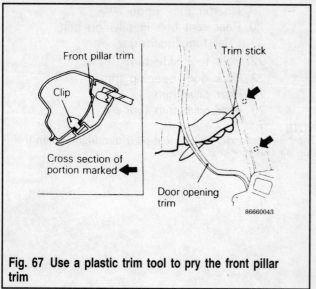

Fig. 67 Use a plastic trim tool to pry the front pillar trim

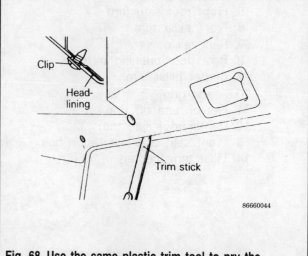

Fig. 68 Use the same plastic trim tool to pry the headliner clips loose to remove the headliner

BODY AND TRIM 10-47

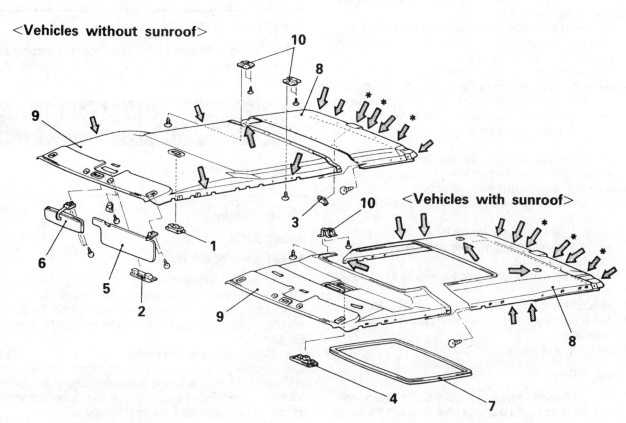

Removal steps
<Vehicles without sunroof>

1. Room lamp
2. Map lamp
3. Luggage compartment lamp
5. Sunvisors
6. Inside rear view mirror
• Rear roof rail trim
• Quarter trim upper
• Center pillar trim upper
• Side roof rail trim
• Front pillar trim
8. Rear headlining
9. Front headlining
10. Joint bracket

Removal steps
<Vehicles with sunroof>

1. Room lamp
2. Map lamp
3. Luggage compartment lamp
4. Sunroof switch
5. Sunvisors
6. Inside rear view mirror
• Rear roof rail trim
• Quarter trim upper
• Center pillar trim upper
• Side roof rail trim
• Front pillar trim
7. Headlining trim
8. Rear headlining
9. Front headlining
10. Joint bracket

NOTE
(1) ⇐ : Clip position
(2) * : Vehicles with partial trim

Fig. 69 Headliner removal and installation components — Monteros with and without sunroofs

BODY AND TRIM

2. Carefully remove the inner moisture liner. Take your time and do not rip or damage the liner.

3. Disconnect or release the clips holding the link rods to both the lock cylinder and the door handle. Depending on the model, it may be easier to disconnect the other end of the rod (at the latch assembly) first.

4. Disconnect any wiring harnesses running to the lock or handle. Generally these cables have connectors in the line; do not try to disconnect the wiring right at the lock.

5. Remove the retaining nuts or bolts holding the handle assembly to the door and remove the handle. The lock portion can be removed by a competent locksmith. Disassembly by the owner/mechanic is not recommended due to the number of small parts and springs within the lock.

To install:

6. The lock assembly must be installed in the door handle and the small lever (arm) attached. Place the handle in the door and secure the mounting nuts and bolts.

7. Connect the wiring to the handle or lock if any was removed.

8. Carefully connect the link rods to the handle and lock, using new clips where necessary. Reconnect the rods to the latch, if any were removed.

9. Reinstall the window glass (and the window track) if they were removed.

10. Reinstall the moisture liner. Apply a bead of waterproof sealer to the outer edge all the way around and align the plastic carefully.

11. Install the door liner and trim pieces.

Key Lock in Door

1. Remove the inner door liner. Make sure the window glass is in the up or closed position before removing the window controls.

2. Carefully remove the inner moisture liner. Take your time and do not rip or damage the liner.

3. Disconnect or release the clips holding the link rod to the lock cylinder. Depending on the model, it may be easier to disconnect the other end of the rod (at the latch assembly) first.

4. Disconnect any wiring harnesses running to the lock. Generally these cables have connectors in the line; do not try to disconnect the wiring right at the lock.

5. The lock cylinder is held to the door by a horseshoe-shaped spring clip. Remove the clip and remove the lock cylinder. The cylinder may be repaired by a competent locksmith. Disassembly by the owner/mechanic is not recommended due to the number of small parts and springs within the lock.

To install:

6. Install the cylinder into the door and fit the horseshoe clip. Make sure the cylinder is held firmly in place.

7. Connect any other wiring which was removed.

8. Connect the link rod, using new clips as necessary.

9. Reinstall the moisture liner. Apply a bead of waterproof sealer to the outer edge all the way around and align the plastic carefully.

10. Install the door liner and trim pieces.

Back Door Lock

REMOVAL & INSTALLATION

▶ See Figures 72 and 73

Removal and installation for the back (luggage compartment) door locks is the same as that for removal and installation of the other door locks — refer to the preceding procedures.

Door Glass, Regulator, and Power Window Motor

REMOVAL & INSTALLATION

1983-91 Montero

FRONT DOOR

▶ See Figures 74 and 75

1. Lower the window.
2. Remove the door trim pad and the moisture barrier.
3. Using a blunt, non-marring instrument, protect the surrounding surface and gently pry the outer weatherstrip free of the door.
4. Remove the inner weatherstrip.
5. Remove the center sash protector. Remove the two bolts in the door and the one bolt at the leading edge of the door holding the vent window assembly. Carefully remove the vent window and its frame as a unit from the door.
6. Disconnect the glass holder bolts from the regulator arms and lift the glass out of the door.
7. Remove the regulator retaining bolts. Unplug the connector to the electric window motor, if there is one.
8. Remove the window regulator. The motor for power windows may be removed once the regulator is out of the door.

✱✱ CAUTION

The tension of the spring may cause the regulator arm to jump up when the motor is removed. Remove the regulator spring BEFORE removing the motor. Failure to do so can result in injuries.

To install:

9. When reassembling, install the spring and then the motor.

10. The regulator and motor must be placed in the door as a unit. Tighten the retaining bolts to 51 inch lbs. (6 Nm). Connect the wiring for the electric motor.

11. The metal regulator tracks and guides should receive a light coat of multipurpose grease during reassembly. Do not get any grease on the glass or the felt/rubber channels and weatherstrips.

12. Install the window glass and holders as a unit. If the glass was removed from the holders, it must be reassembled so that the distance (Y) from the leading edge to the center of the leading holder bolt hole is 3.012-3.051 in. (76.5-77.5mm). Once that distance is established, the second holder is in-

BODY AND TRIM

1. Inside lock knob
2. Inside lock knob bush
3. Door trim and waterproof film
4. Rod snap
5. Outside handle
6. Door latch
7. Inside handle
8. Door lock actuator

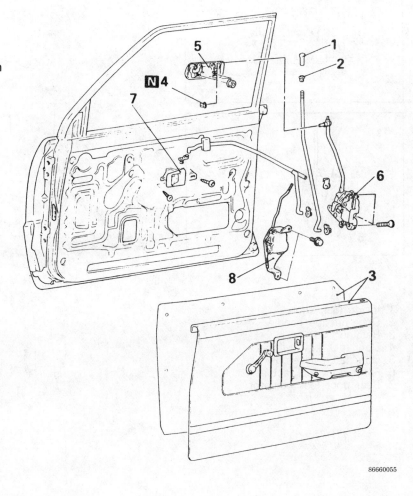

Fig. 70 Door handle and lock assembly removal and installation components — Pick-ups

10-50 BODY AND TRIM

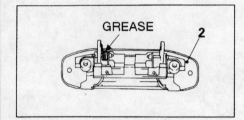

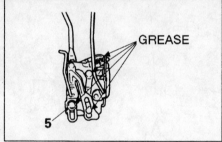

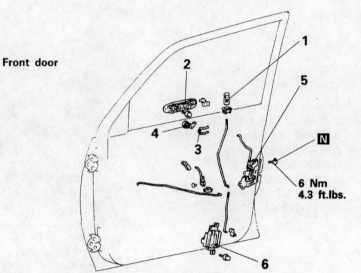

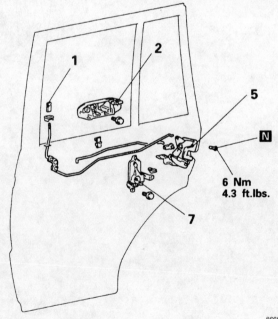

1. Inside lock knob
2. Door outside handle
3. Retainer
4. Door lock key cylinder
5. Door latch assembly
6. Front door lock actuator
7. Rear door lock actuator

Fig. 71 Door handle and lock assembly removal and installation components — Monteros

BODY AND TRIM 10-51

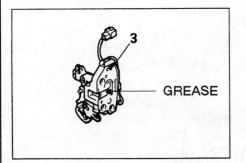

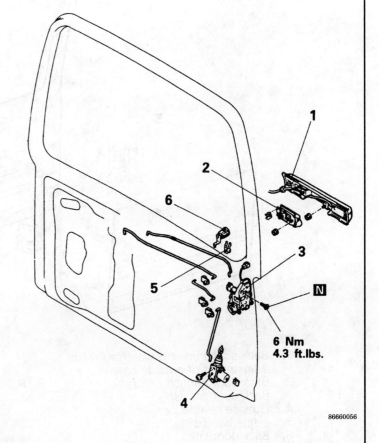

1. License plate light garnish
2. Door outside handle
3. Back door latch assembly
4. Back door lock actuator
5. Retainer
6. Back door key cylinder

Fig. 72 Back door handle and lock assembly removal and installation components — 1992-95 Monteros

10-52 BODY AND TRIM

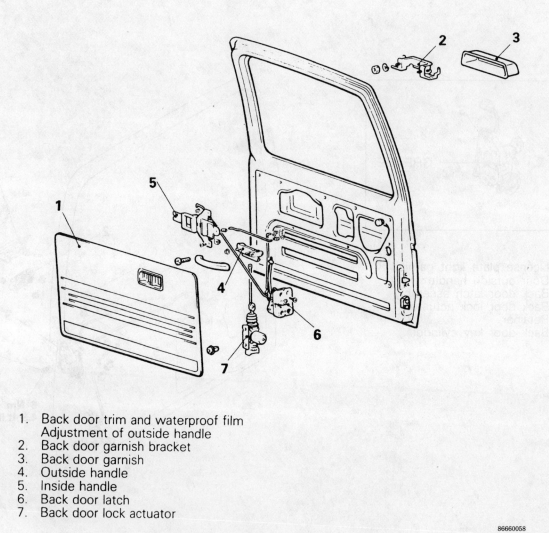

1. Back door trim and waterproof film
   Adjustment of outside handle
2. Back door garnish bracket
3. Back door garnish
4. Outside handle
5. Inside handle
6. Back door latch
7. Back door lock actuator

Fig. 73 Back door handle and lock assembly removal and installation components — 1983-91 Monteros

BODY AND TRIM 10-53

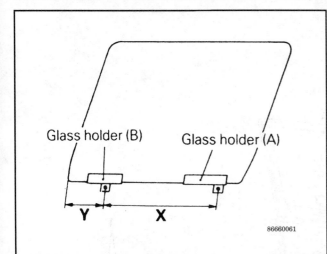

Fig. 74 Be sure the Y and X measurements are correct when installing the glass

stalled (distance X) with the centerline of the bolt hole 18.366-18.406 in. (466.5-467.5mm) from the leading holder.
  13. Install the vent window assembly and tighten the two bolts to 51 inch lbs. (6 Nm). Tighten the screw in the upper frame snug.
  14. Install the center sash protector.
  15. Install the inner and outer weather strips.
  16. Raise the window fully. Loosen the screws and bolts holding the window and the rear lower track (sash). Move the window, rear sash, and sub-roller assembly (if no vent window) to adjust the glass position. The glass must sit squarely and fit evenly at all the edges. When positioned correctly, retighten the bolts.
  17. Install the moisture barrier and the door trim pad.

REAR DOOR

♦ See Figures 76 and 77

  1. Lower the window.
  2. Remove the door panel and the moisture barrier.
  3. Using a blunt, non-marring instrument, protect the surrounding surface and gently pry the outer weatherstrip free of the door.
  4. Remove the inner weatherstrip.
  5. Remove the felt and rubber runchannel. Don't abuse it; the channel will be reinstalled.
  6. Remove the small retaining screw on top of the door and the two bolts holding the center sash to the door. Remove the center sash (track) from the door by lifting it out of the door.
  7. Remove the fixed quarter glass with its weatherstrip.
  8. Unbolt the glass holder from the regulator and lift the glass out of the door.
  9. If equipped with power windows, disconnect the wiring harness to the window motor. Remove the regulator mounting bolts and remove the regulator. The electric motor may be removed if desired.

✻✻CAUTION

The tension of the spring may cause the regulator arm to jump up when the motor is removed. Remove the regulator spring BEFORE removing the motor. Failure to do so can result in injuries.

  10. When reassembling, install the spring and then the motor. The regulator and motor must be placed in the door as a unit. Tighten the retaining bolts to 51 inch lbs. (6 Nm). Connect the wiring for the electric motor.
  11. The metal regulator tracks and guides should receive a light coat of multipurpose grease during reassembly. Do not get any grease on the glass or the felt/rubber channels and weatherstrips.
  12. Install the window glass and holder as a unit. If the glass was removed from the holder, it must be reassembled so that the distance from the leading edge to the leading edge of the slotted track (not the glass holder frame) is 9.1-9.4 in. (23.1-23.9cm).
  13. Install the fixed glass, making sure the weatherstrip is not twisted or out of place.
  14. Install the center track or sash and tighten the screws and bolts.
  15. Raise the glass fully and inspect the fit at the top and sides. If adjustment is needed, loosen the center track bolts and move the sash to adjust the glass. Tighten the bolts when the glass is correctly located.
  16. Reinstall the moisture barrier and the door trim pad.

1992-95 Montero

♦ See Figure 78

FRONT DOOR

  1. Remove the inner door panel and the plastic waterproofing film.
  2. Lower the window until the mounting bolts for the window can be seen, then remove the mounting bolts on the bottom of the piece of glass.
  3. Remove the door glass holders from the door.
  4. Pull the glass up and turn it so that the rear edge comes out first. Remove the glass from the door.
  5. Remove the window regulator assembly form the door. If the vehicle is equipped with power windows, unplug the electrical wires to the motor before removal.

To install:

  6. Mount the window regulator in the door and secure in place with the mounting bolts. Plug the electrical wires back together, if so equipped.
  7. Install the glass holders.
  8. Slide the piece of glass into the door with the front edge going in first, then rotate the glass to sit proper once it is installed in the door.
  9. Let the glass sit in the pocket on the regulator assembly, then install the mounting bolts to hold the window in place.
  10. Install the waterproofing film to the door and install the inner door panel.

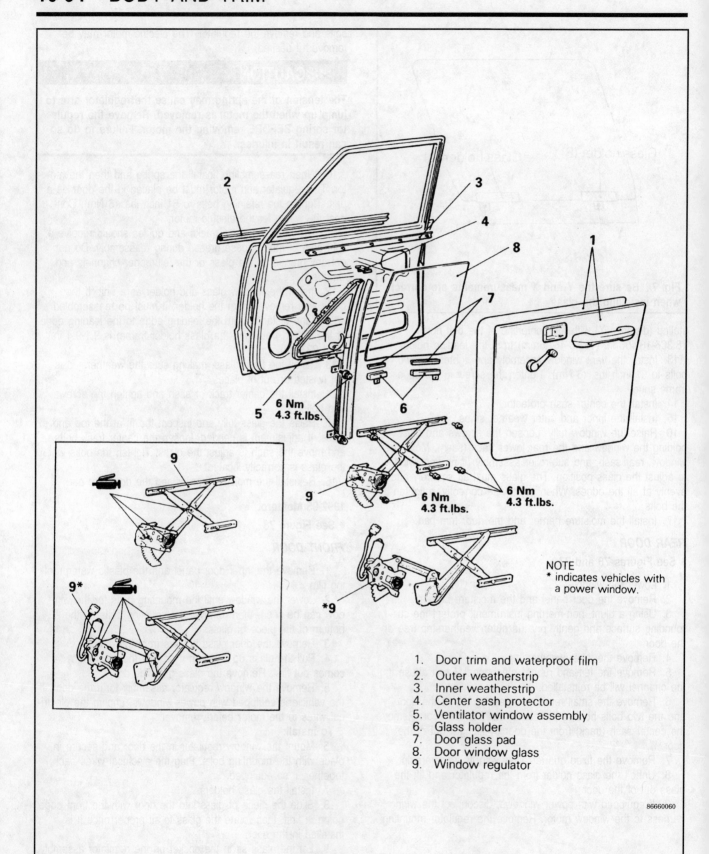

Fig. 75 Front door glass assembly removal and installation components — 1983-91 Monteros

BODY AND TRIM   10-55

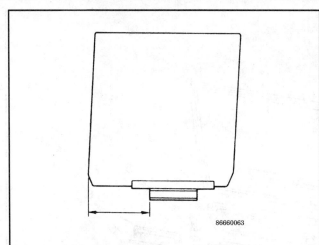

Fig. 76 When installing a new rear window, make certain that the slotted track is the correct distance from the front edge

REAR DOOR

1. Remove the inner door panel and the plastic waterproofing film.
2. Remove the rear door center sash by removing the upper and lower mounting bolts.
3. Remove the door glass holders from the door.
4. Remove the glass from the door.
5. Remove the stationary window glass and weatherstrip together as one piece.
6. Remove the window regulator assembly form the door. If the vehicle is equipped with power windows, unplug the electrical wires to the motor before removal.

To install:
7. Mount the window regulator in the door and secure in place with the mounting bolts. Plug the electrical wires back together, if so equipped.
8. Install the stationary window glass and weatherstrip into its position.
9. Install the glass holders.
10. Install the window glass into the door.
11. Let the glass sit in the pocket on the regulator assembly, then install the mounting bolts to hold the window in place.
12. Install the rear door center sash by inserting and tightening the mounting bolts.
13. Install the waterproofing film to the door and install the inner door panel.

Pick-Up

▶ See Figures 79 and 80

1. Raise the window.
2. Remove the door pad and the moisture barrier. Remove the outside rear view mirror.
3. Using a blunt, non-marring instrument, protect the surrounding surface and gently pry the outer weatherstrip free of the door, then remove it rearward.
4. Remove the lower rear window sash or track. Remove the bolts and then turn the track so that it will come out the access hole.

5. Lower the window so that the glass mount bolts are visible at the access holes. Support the glass and remove the glass mounting bolts.
6. Pull the door glass upward and carefully turn it so that the rear edge of the glass comes out the top of the door. Remove the glass.
7. Support the regulator and remove the regulator mounting bolts. An assistant can be useful here. Unplug any electrical wiring going to the regulator.
8. Remove the regulator through the access hole.

To install:
9. Reinstall the regulator and secure the bolts. Plug the electrical wire connectors back together, if removed.
10. Install glass assembly with the holders attached. Use the same half-turn method to install the glass as was used to remove it. If the glass was changed or removed from the holders, it must be reassembled so that the distance (Y) from the rear edge of the glass to the center of the nearest bolt hole (in the holder) is 3.9-4.0 inches. (99-100mm). Once that distance is established, the second holder is installed so that the distance (X) between the center of the bolt holes is 18.3-18.4 in. (465-467mm).
11. Install the rear sash and tighten its bolts.
12. Install the outer weatherstrip.
13. Install the outer mirror.
14. Roll the window all the way up. Check the fit of the glass at the top and sides. Loosen the roller guide mounting volts and then adjust the slant of the glass front-to-back. lower the door glass until the holder reaches the access holes and loosen the glass holder mounting bolts. Adjust the front to rear position of the glass. Loosen the bolts holding the lower vertical window track and adjust the glass position.

➡ Any one adjustment may affect the others and more than one pass may be necessary. Operate the window bit by bit, checking the fit at every location in every position. Once the window is correctly placed, tighten the retaining bolts.

15. Install the moisture barrier and the door trim.

Windshield Glass

REMOVAL & INSTALLATION

▶ See Figures 81, 82, 83, 84, 85, 86, 87, 88, 89, 90, 91, 92 and 93

1. For all Pick-ups, remove the sun visors, the inside rear view mirror, the door opening trim, and the front pillar trim. Remove the headlining from the roof panel to provide space for the wire which will be used to cut the windshield free for the adhesive. Also remove the windshield wiper arms and front deck garnishes.
2. Using the special window moulding tool to remove the front side and upper mouldings.
3. Remove the moulding clips.
4. Remove the lower spacer, if so equipped, and dispose of it. A new spacer will be needed for installation.
5. Cover the painted body surfaces near the windshield with cloth tape. Then, using a sharp-pointed drill, make a hole

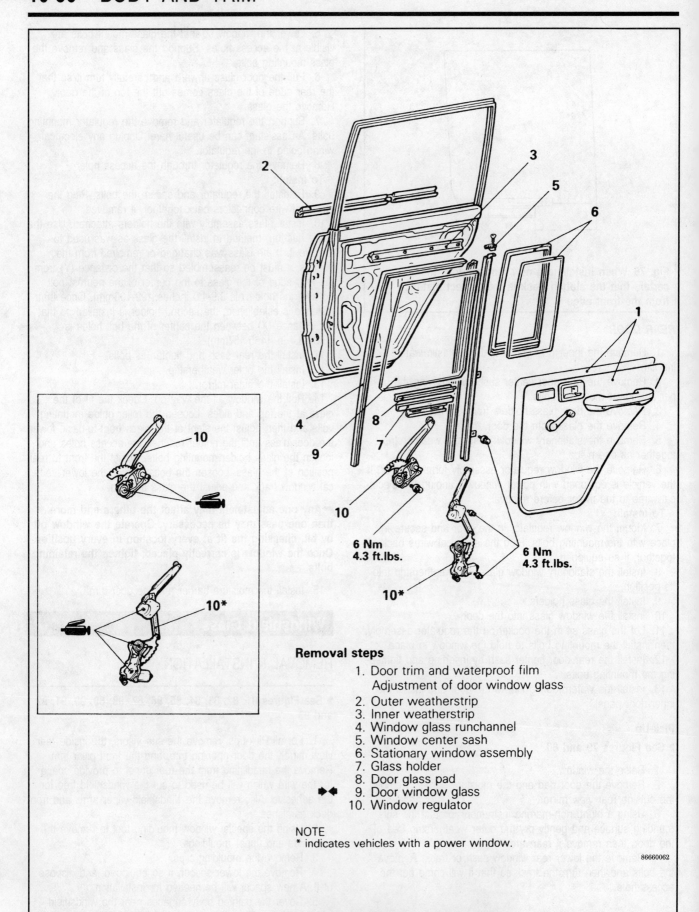

Fig. 77 Rear door glass assembly removal and installation components — 1983-91 Monteros

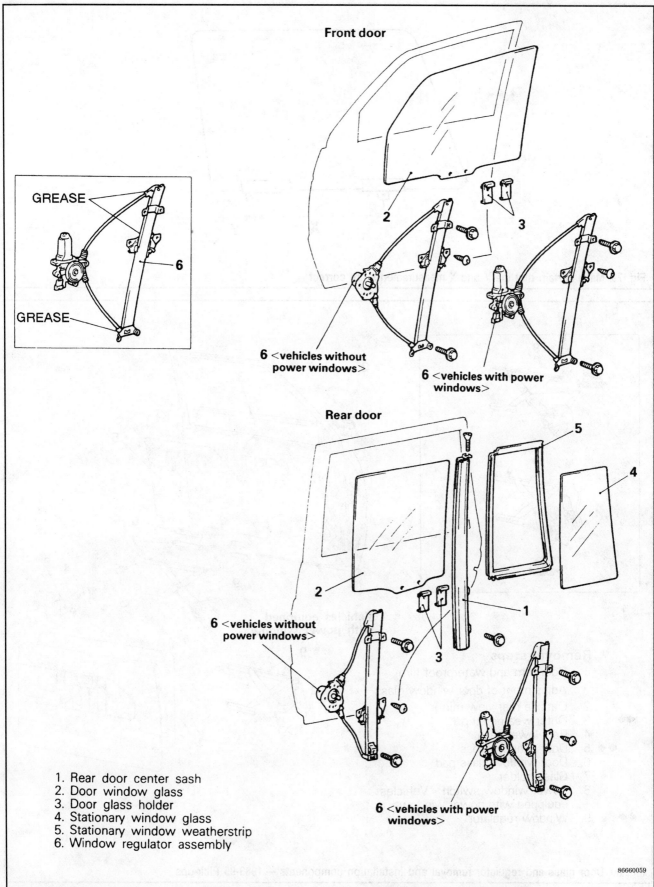

Fig. 78 Front and rear door glass assembly removal and installation components — 1992-95 Monteros

10-58 BODY AND TRIM

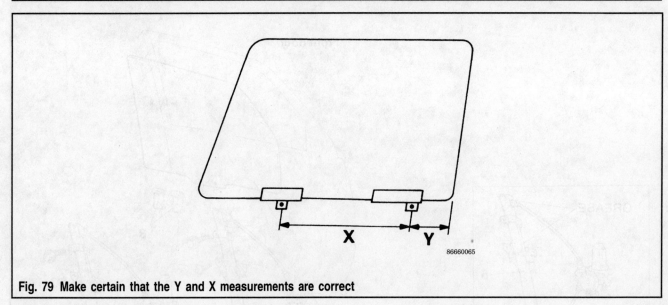

Fig. 79 Make certain that the Y and X measurements are correct

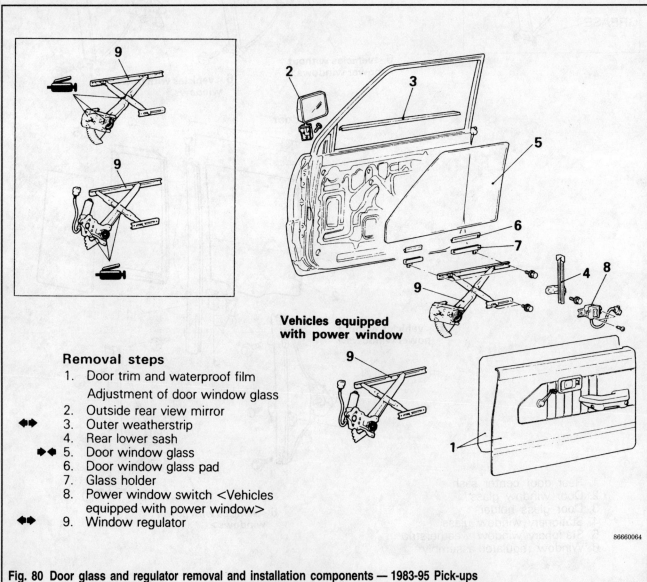

Removal steps
1. Door trim and waterproof film
   Adjustment of door window glass
2. Outside rear view mirror
◆◆ 3. Outer weatherstrip
4. Rear lower sash
▶◆ 5. Door window glass
6. Door window glass pad
7. Glass holder
8. Power window switch <Vehicles equipped with power window>
◆◆ 9. Window regulator

Fig. 80 Door glass and regulator removal and installation components — 1983-95 Pick-ups

BODY AND TRIM   10-59

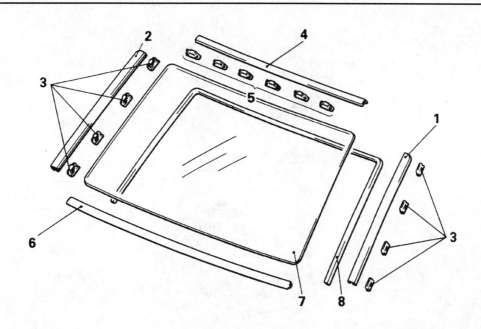

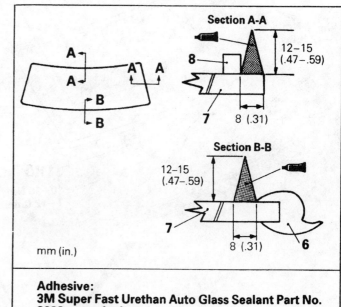

Pre-removal and Post-installation Operation
Removal and Installation
- Instrument Panel
- Front Pillar Trim
- Front Deck Garnish

Adhesive:
3M Super Fast Urethan Auto Glass Sealant Part No. 8609 or equivalent

Removal steps
1. Windshield side moulding (LH)
2. Windshield side moulding (RH)
3. Windshield clip
4. Windshield upper moulding
5. Windshield clip
6. Windshield lower moulding
7. Windshield glass
8. Window spacer

Fig. 81 Windshield assembly removal and installation components — Monteros

10-60 BODY AND TRIM

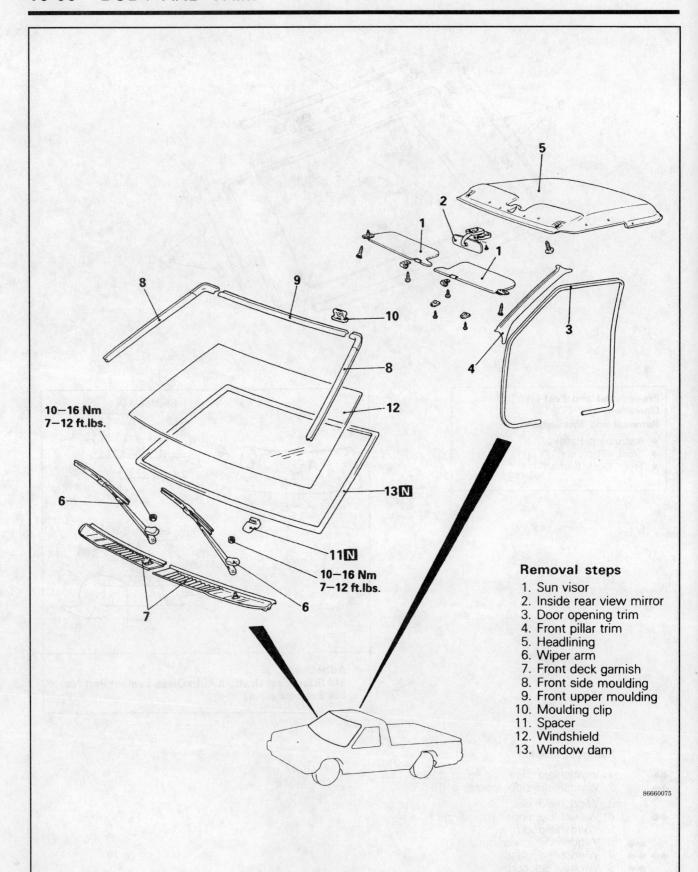

Fig. 82 Windshield removal and installation components — Pick-ups

BODY AND TRIM 10-61

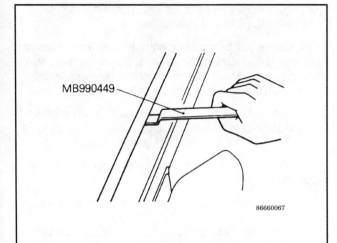

Fig. 83 Use the special Mitsubishi clip removing tool (MB990449) to remove the trim moulding

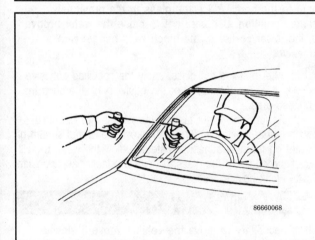

Fig. 84 Have an assistant help with cutting the adhesive to remove the windshield

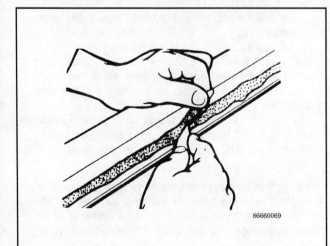

Fig. 85 Use a sharp knife to cut the old adhesive off of the body mounting flange

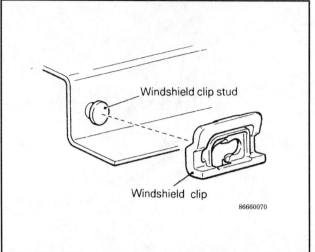

Fig. 86 Replace any broken moulding retaining clips with new ones

in the adhesive area of the windshield. Insert a wire through the hole from the inside.

6. From inside and outside (an assistant is necessary) of the vehicle, pull the wire along the edge of the windshield, thus cutting the adhesive.

7. Make mating marks on the windshield and body. Take out the windshield by using the glass holders.

➡ If the windshield being removed is to be reinstalled, place it on a protected bench or holding fixture.

8. Remove the window dam or spacer from the window opening.

To install:

9. Using isopropyl alcohol, remove the grease from the glass and body surfaces to which the adhesive will be applied. After removing the grease, allow the cleaned parts to dry for more than three minutes.

10. Remove the back paper from the window dam, and attach the window dam to the glass. The window dam should be

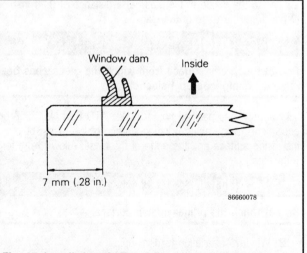

Fig. 87 Install the window dam at the position shown in the illustration — Pick-ups

10-62 BODY AND TRIM

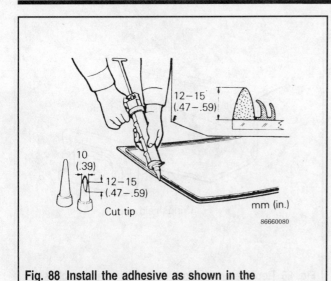

Fig. 88 Install the adhesive as shown in the illustration — Pick-ups

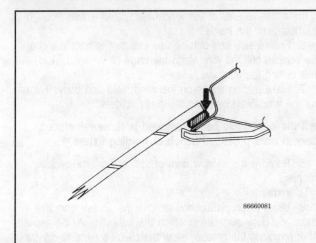

Fig. 89 Make certain that the adhesive and window dam are in the correct position — Pick-ups

applied to the inside of the glass at a distance of 0.28 in. (7mm) from the edge of the glass.

✴✴CAUTION

Do not touch any surface from which the grease has been removed using isopropyl alcohol.

11. Apply the specified primer (3M® ATD Part No. 8608 or equivalent) to the entire bonding surface of the glass, both to the inside surface and the edge of the glass. Allow to dry for 5 minutes.

✴✴CAUTION

Do not touch the primer coated surface.

12. Window installation for Pick-ups:
    a. Using a sharp knife, CAREFULLY remove the old adhesive on the body opening pinch-weld flange evenly to a thickness of within 0.08 in. (2mm) all around. Finish the flange surfaces so that they are smooth.

➡ **Be careful not to remove more adhesive than necessary, and also not to damage the paint work on the body surface with the knife. If the paint work is damaged, repair the damaged area with touch-up paint or Tectyl.**

   b. Using an adhesive gun, apply the specified adhesive to the window glass mounting surface. The specified adhesive is 3M® ATD Part No. 8609 or the equivalent.

➡ **Cut the nozzle tip into a V shape to facilitate adhesive application.**

   c. Using a glass holder, place the glass on the body opening.
   d. Press the glass gently so that no adhesive appears. After bonding the glass to the body, install the mouldings securely before the adhesive hardens.

➡ **Do not move the glass and mouldings after installing them to the body. Place the glass in the previously marked position. Use care not to close the water groove (in the lower corner of the pinch weld flange) with adhesive.**

   e. After bonding the glass, apply the specified adhesive (3M® ATD Part No. 8609 or the equivalent) all around the bonded area.
   f. Perform the water test for the windshield. Use cold water spray, being careful not to direct a powerful stream of water on the new adhesive material. Allow the water to spill over the edges of the glass. If there are leaks, apply sealant at the leak point(s).

✴✴WARNING

If it is necessary to move the vehicle, move it slowly.

   g. Using a clean, lint-free cloth liberally dampened with Glow Chemical Solvent GS — 35, 3M® ATD Part No. 8984 or the equivalent, wipe away the dirt from the glass perimeter and the body.
   h. Install the spacers and moulding clips, if any were removed.
13. Window installation for Monteros:
    a. When replacing the glass with a new windshield, temporarily set the glass against the body where they match.
    b. Install the windshield lower moulding onto the windshield glass.
    c. Soak a sponge in the primer (3M® Super Fast Urethane Primer Part No. 8608 or the equivalent), and apply evenly to the glass and the body in the places shown in the figure.

➡ **The primer increases the adhesive strength, so be sure to apply it evenly around the entire circumference. Also, a too thick of an application will cause a decrease of the adhesive strength. Do not touch the coated surface.**

   d. After applying the primer allow it to dry for 3-30 minutes.
   e. Within 30 minutes of applying the primer, fill the sealant gun with adhesive (3m® Super Fast Urethane Auto Glass Sealant Part No. 8609 or the equivalent) and apply

BODY AND TRIM 10-63

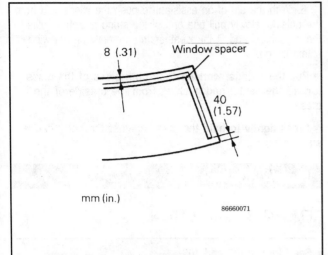

Fig. 90 Install the new windshield spacer onto the window with the dimensions shown — Monteros

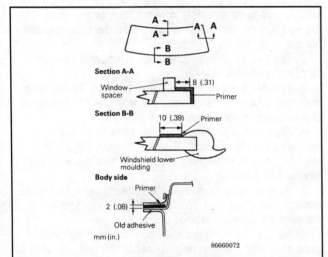

Fig. 91 Install the primer to the windshield in the areas shown in the illustration — Monteros

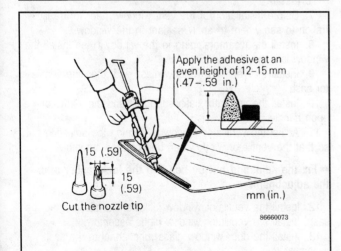

Fig. 92 Apply the adhesive at an even height and 0.47-.59 in. (12-15mm) high — Monteros

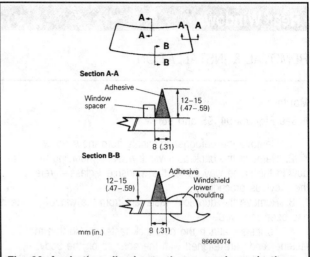

Fig. 93 Apply the adhesive to the areas shown in the illustration — Monteros

the adhesive evenly around the entire circumference of the windshield.

➡ Cut the nozzle tip of the sealant gun into a V shape to facilitate the adhesive application.

   f. After applying the adhesive, match up the mating marks on the glass and the body, lightly press the windshield glass evenly so that it adheres completely.

   g. After removing any adhesive that is sticking out or adhering to the body or glass with a spatula, clean off the spot with 3M® ATD Part No. 8906 or the equivalent. After completion of this operation (after installing the glass), place it somewhere where it will not be disturbed, until the adhesive sets.

✳✳WARNING

If heat is applied with an infra-red lamp to shorten the setting time, keep the surface temperature of the adhesive below 212°F (100°C).

   h. After attaching the windshield glass to the body, let it stand for 30 minutes or more, and then test for water leakage.

➡ If moving the vehicle, it should be done gently. When testing for water leakage, do not pinch the end of the hose to spray the water.

14. Install the window mouldings, if not already done.
15. For Pick-up trucks, reinstall the front deck garnishes, the wiper arms, and the interior pieces which were removed during disassembly.

10-64 BODY AND TRIM

Rear Windows

REMOVAL & INSTALLATION

Montero

♦ See Figures 94, 95 and 96

1. Remove the defogger terminals from the window.
2. Remove the back door window glass moulding and window in the same way as for the windshield glass — refer to the previous procedures.
3. Remove the dual lock fasteners from the window, once it has been removed.
4. To install, attach the dual lock fasteners so that the fastener ends are aligned with the notches on the body.
5. Install the rear window in the same manner as the windshield — refer to the preceding procedures.
6. Install the defogger terminals to the window.

Pick-Up

♦ See Figures 97, 98, 99, 100 and 101

1. Use a prytool to pry up the lip of the weatherstrip and remove the rear window glass.
2. Once the window is removed from the vehicle, the window assembly can disassembled.

To install:

3. Install the new weatherstrip to the window. Set two strings around the window in the weatherstrip groove.

➡ Make sure that the strings overlap each other at both ends.

4. Apply a soap solution to the entire surface of the body flange.
5. Place the windshield in position from the outside with the strings placed inside the cabin.

6. With the aid of an assistant to push the windshield from the outside, slowly pull one end of the string at right angles to the windshield and fit the weatherstrip correctly on the windshield flange.

➡ Pull the strings, working from both sides of the glass toward the center and pushing from the outside of the glass.

➡ Press lightly to hold the glass against the body flange surface.

Vent Windows

REMOVAL & INSTALLATION

♦ See Figures 102 and 103

1. Refer to the procedures for the removal and installation of the door window glass replacement, and remove the vent window and frame assembly from the door as described. With the vent window and frame removed from the door, disassembly can easily be accomplished.
2. Remove the door window glass runchannel from the ventilator sash assembly.
3. Remove the vent window hinge assembly by removing the screws. Make certain to note the locations and placement of the screws, hinges, packing and washers.
4. Remove the pin holding the ventilator window lock handle onto the ventilator sash. Remove the push button, the push button spring and the wave washer from the ventilator window lock handle.
5. Remove the screws from the ventilator open handle. Remove the ventilator sash spring kit and note the positions of the various washers, bushings and nuts.
6. Remove the ventilator sash from the ventilator window glass pad. Slide the window glass out of the ventilator sash assembly frame, then remove the weatherstrip from the window.

To install:

7. Before installation of the vent window, refer to the illustration to see where to apply sealant to the window.
8. Install the weatherstriping to the window, then install the window into the frame.
9. Install the ventilator window glass pad, then the ventilator sash.
10. Install the ventilator sash spring kit and the ventilator open handle.
11. Adjust the ventilator window by turning the adjusting nut so that the ventilator window can be operated smoothly.

➡ Fix the adjusting nut by bending the lockwasher after the adjustment.

12. Install the ventilator window lock handle assembly to the sash. Install the ventilator window hinge assembly.
13. Install the door window glass runchannel to the sash assembly.
14. Install the vent window assembly into the door — refer to the door window glass removal and installation procedures.

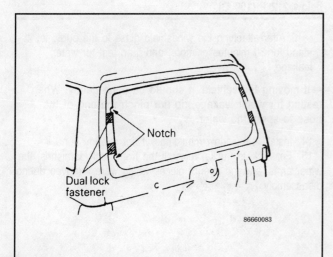

Fig. 94 Make certain that the dual lock fasteners are aligned with the notches on the body — Monteros

BODY AND TRIM 10-65

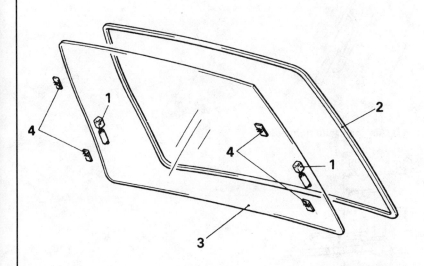

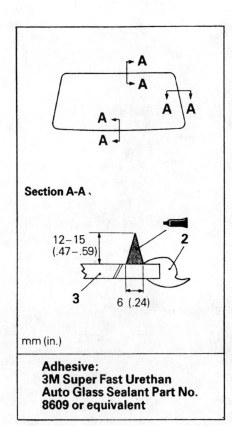

Section A-A

mm (in.)

**Adhesive:
3M Super Fast Urethan
Auto Glass Sealant Part No.
8609 or equivalent**

1. Defogger terminal
2. Back door window glass moulding
3. Back door window glass
4. Dual lock fastener

Fig. 95 Back door window removal and installation components — Monteros

10-66 BODY AND TRIM

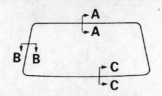

<Application of primer>
Section A-A, B-B, C-C

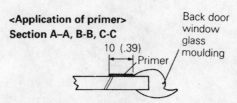

Body side
Section A-A

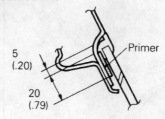

Section B-B (Also paint the right side)

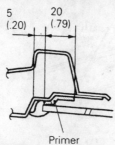

Section C-C

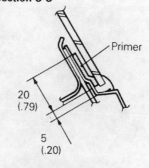

<Application of adhesive>
Section A-A, B-B, C-C

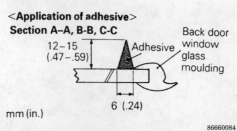

mm (in.)

Fig. 96 Apply the primer in the positions at the areas shown in the illustration — Monteros

BODY AND TRIM 10-67

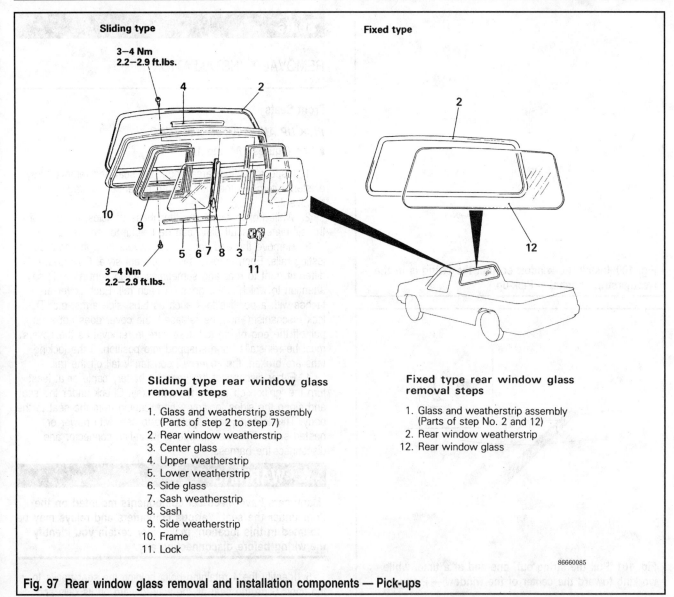

Fig. 97 Rear window glass removal and installation components — Pick-ups

Sliding type rear window glass removal steps

1. Glass and weatherstrip assembly (Parts of step 2 to step 7)
2. Rear window weatherstrip
3. Center glass
4. Upper weatherstrip
5. Lower weatherstrip
6. Side glass
7. Sash weatherstrip
8. Sash
9. Side weatherstrip
10. Frame
11. Lock

Fixed type rear window glass removal steps

1. Glass and weatherstrip assembly (Parts of step No. 2 and 12)
2. Rear window weatherstrip
12. Rear window glass

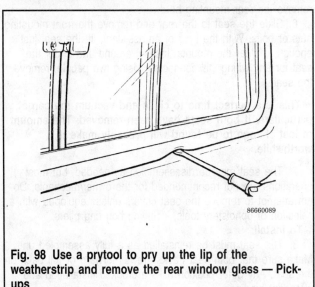

Fig. 98 Use a prytool to pry up the lip of the weatherstrip and remove the rear window glass — Pick-ups

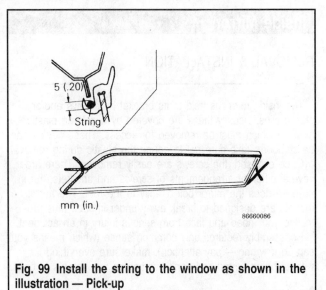

Fig. 99 Install the string to the window as shown in the illustration — Pick-up

10-68 BODY AND TRIM

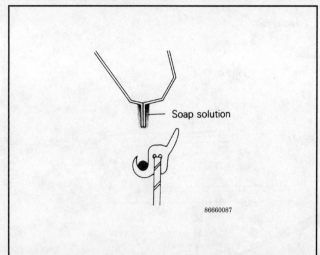

Fig. 100 Install the window so that the string is in the weatherstrip channel — Pick-up

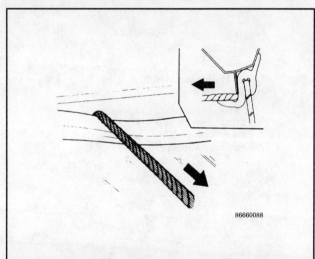

Fig. 101 Pull the string out, one end at a time, while working toward the center of the window — Pick-ups

Inside Mirror

REMOVAL & INSTALLATION

The inside mirror is held to its bracket by screws and/or plastic clips. Usually these are covered by a colored plastic housing which must be removed for access. These covers can be stubborn; take care not to gouge the plastic during removal.

Once exposed, the screws are easily removed. There are several different arrangements of screws and brackets, but in no case does the mirror bolt directly to the roof. The mirror mounts are designed to break away under impact, thus protecting your head and face from serious injury in an accident.

Reassembly requires only common sense (which means you can do it wrong — pay attention); make sure everything fits without being forced and don't overtighten any screws or bolts.

Seats

REMOVAL & INSTALLATION

Front Seats

PICK-UP AND 1992-95 MONTERO

▶ See Figures 104 and 105

1. Inspect the floor area under the seat and remove stray or stored items. If the seat has an under-seat storage tray, remove it.
2. While not required on other these vehicles, removal of the sill plate or scuff plate can be helpful to remove it.
3. Remove the end caps or rail covers from the seat adjusting rails. Each model uses a different style. Covers are different front to rear and sometimes left to right as well; pay attention to which cover goes on each rail. Each cover attaches with a positive lock such as tabs, side arms, etc. This lock mechanism must be released; the cover does not simply pull off the end of the rail. Use care in removal as the covers must be reinstalled and snapped into position. If the locking tabs are broken, the cover will constantly fall off the rail.
4. Either disconnect the negative battery cable or at least turn the ignition off and remove the key. Check under the seat and along the sides for any wiring running from the seat to the body. This is particularly applicable to cars with power or heated seats. Trace each wire to its in-line connector and disconnect the harness.

※※WARNING

Many cars have electrical components mounted on the floor under the seat. Various computers and relays may be mounted in this location; make very certain you identify the wiring before disconnecting any.

5. Identify the 4 retaining bolts holding the seat tracks to the floor. Slide the seat all the way forward on its rails and remove the rear mounting hardware.
6. Slide the seat to the rear and remove the front mounting nuts or bolts. With the help of an assistant, lift the seat just enough to clear the mounts. Inspect around and under the seat for any wiring still connected. Using two people, remove the seat.

➡**This is a perfect time to clean and vacuum the carpet, particularly if both seats have been removed. The amount of lost change to be found will probably make it worthwhile.**

7. The seat may be disassembled as needed, but most operations are not recommended for the owner/mechanic. Do not attempt to remove the seat covers unless equipped with a selection of upholstery tools, including hog ring pliers.

To install:

8. The seat must be reinstalled as a fully assembled unit. Make sure that both seat tracks are full forward or full back on the seat. Any other position risks a crooked installation and attendant binding.

BODY AND TRIM 10-69

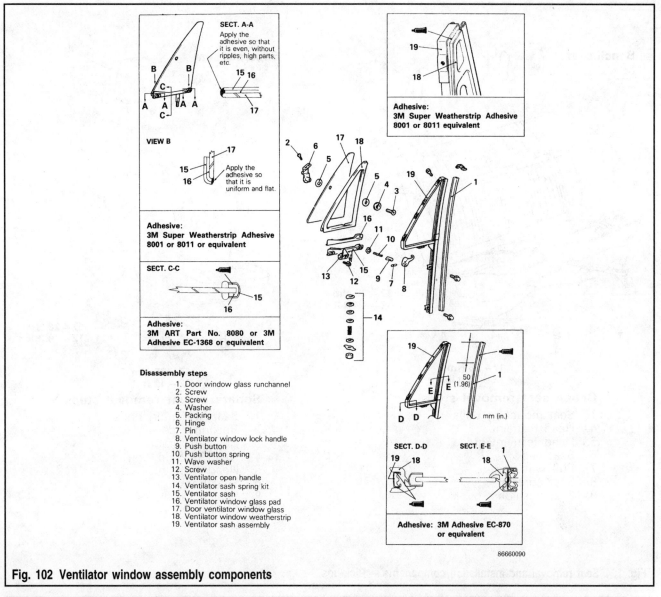

Fig. 102 Ventilator window assembly components

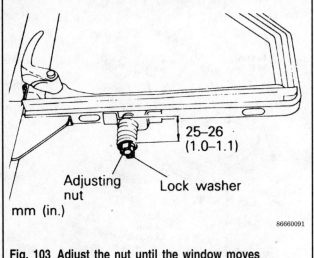

Fig. 103 Adjust the nut until the window moves smoothly, then secure in place with the lockwasher

9. Lift the seat into the car and position it so that the mounts align. Install the nuts and bolts finger-tight only.
10. Move the seat through its full range of travel, checking that it will lock securely in each position.
11. Tighten the mounting hardware to the proper torque value. Correct tightening order must be observed:
 • Inner rear, outer rear, inner front and outer front.
The bolts should be tightened to 9 ft. lbs (12 Nm) for 1983-91 Pick-ups, to 12-19 ft. lbs. (17-26 Nm) for 1992-95 Pick-ups, and to 33 ft. lbs. (44 Nm) for 1992-95 Monteros. Check the seat motion and locking function after tightening the hardware.
12. Connect any wiring which was disconnected.
13. Install the end caps on each seat track. Make certain they are securely in place.
14. Install any trays or components removed from below the seat.
15. Install the scuff plate if it was removed.

1983-91 MONTERO

The Montero follows the steps above with one notable exception. Because of the suspension unit between the seat and

10-70 BODY AND TRIM

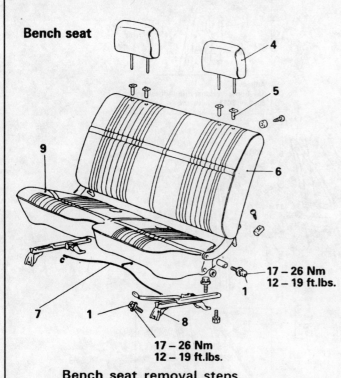
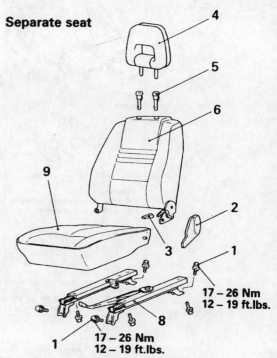

Fig. 104 Seat removal and installation components — Pick-ups

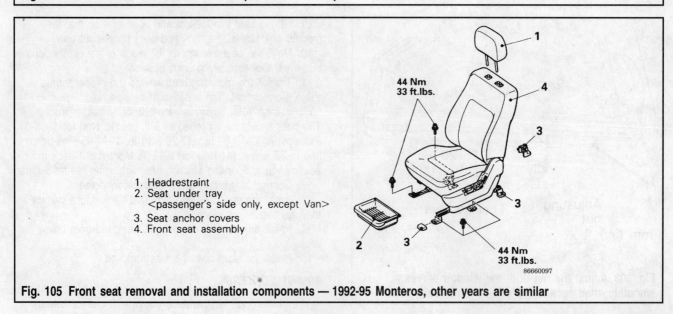

Fig. 105 Front seat removal and installation components — 1992-95 Monteros, other years are similar

BODY AND TRIM 10-71

the floor, the seat should be unbolted from the suspension frame. Attempting to remove both at once increases weight and bulk which could lead to damage. In most cases, removing only the seat will suffice to provide access to other areas or allow the seat to be replaced or repaired.

If it is necessary to remove the seat suspension frame, it must be removed from the floor as a unit (4 bolts). Do not attempt to disassemble or adjust the suspension unit. Reinstall the frame first and then install the seat. All mounting bolts for the seat and the suspension unit should be tightened to 34 ft. lbs. (45 Nm).

After reinstallation, check both the fore and aft locking of the seat and the function of the suspension unit.

Rear Seat and Seatback

MONTERO

▶ See Figures 106 and 107

1. Disconnect either the rear seat leg bracket (2 door) or the seat track (4 door) from the floor. This may require repositioning the seat for access.
2. Remove the plastic cover from the recliner mechanism. Remove the two bolts holding the seatback to the seat bottom. Make certain you are not disconnecting the adjuster; the spring is under tension.
3. Remove the rear seat back from the cushion.
4. Remove the seat cushion.
5. Reassemble in reverse order. Install all the retaining bolts just snug. Check the operation of the seat and seatback through all functions and positions. Tighten the seatback to seat retaining bolts to 39 ft. lbs. (52 Nm). On 2-door vehicles, tighten the seat bracket floor bolts to 9 ft. lbs. (12 Nm); on 4 door vehicles, tighten the rear seat track retaining bolt to 39 ft. lbs (52 Nm) and the front retaining bolt to 9 ft. lbs. (12 Nm).

Seat Belt Systems

REMOVAL & INSTALLATION

▶ See Figures 108 and 109

All the seat belt systems found on the Mitsubishi trucks (Pick-ups and Monteros) are removed and installed in generally the same manner. The system, whether the front seat belt system or the rear seat system, starts at the seat belt retractor, is routed up and through a seat belt bracket. The belt then travels to the seat, where it is attached to the seat belt buckle. The belt is routed through the buckle and then is attached to the vehicle floor. The other belt is fastened to the rear of the seat frame or to the floor, and is routed up through or around (depending on whether the seat is a bench type seat or a bucket type seat) the seat.

The seat belt mounting components (i.e. the retractor, the brackets, the floor attachment) are held in place with large bolts. To remove the system, remove any piece of trim which may be concealing the retractor. Remove the bolts mounting the retractor to the vehicle body. Remove the bolts holding the upper seat belt bracket to the upper body and the bolts holding the seat belt's other end to the floor. Remove the other belt's floor attachment bolt, and either pull the belt our through the seat, or simply remove the belt from the vehicle. If any wires are attached to the seat belt buckle, disconnect and label them.

To install:

To install, attach any wiring to the buckle, and route the belt back through the seat. Attach the mounting components with the bolts and refer to the torque chart for the correct amount to tighten them. Install any trim pieces removed during disassembly.

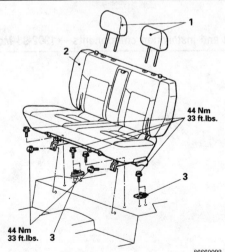

1. Head restraint
2. Rear seat assembly
3. Striker

Fig. 106 Second seat removal and installation components — 1992-95 Monteros, other years similar

10-72 BODY AND TRIM

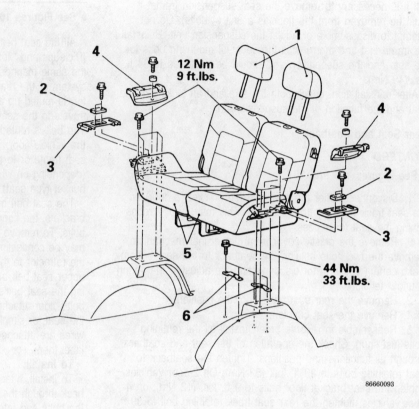

1. Head restraint
2. Damper
3. Seat anchor cover (A)
4. Seat anchor cover (B)
5. Third seat assembly
6. Striker

Fig. 107 Third seat removal and installation components — 1992-95 Monteros

BODY AND TRIM　10-73

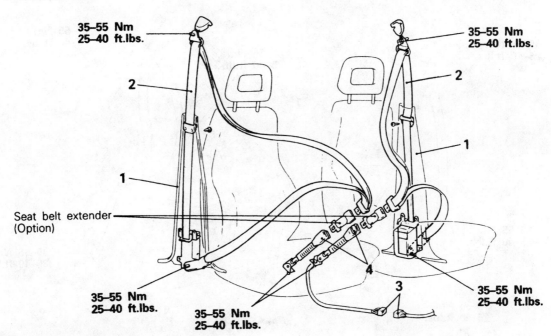

Fig. 108 Front seat belt routing on 1990-92 Monteros — other years and models are similar

10-74 BODY AND TRIM

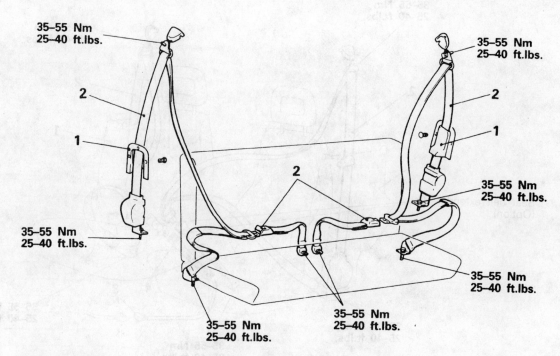

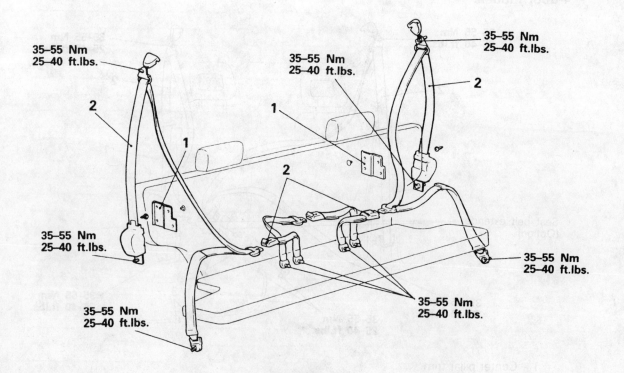

1. Retractor cover
2. Rear seat belt assembly

Fig. 109 Rear seat belt routing on 1990-91 Monteros — other years are similar

BODY AND TRIM 10-75

TORQUE SPECIFICATIONS
Body And Trim Components

Component	ft. lbs.	Nm
1983-86 Pick-ups and Monteros		
Cabin mounting bolts	21-23	28-31
Rear body mounting bolts (pick-up)	21-23	28-31
Hood lock	6.6-7.9	9-10
Release cable lock nut	2.6-2.8	3.5-3.9
Hood hook bolt adjusting nut	10-13	13-18
Spare tire carrier mounting bolts	6-8	8-11
Door hinge attaching bolts	10-18	13-25
Seat belt anchor bolts	17	23
1987-95 Pick-ups		
Cab body	20-23	28-32
Rear body	20-23	28-32
Hood hinge	6.5-10	9-14
Hood latch	6.5-8	9-11
Weatherstrip (sash) to sliding glass frame	2.2-2.9	3-4
Windshield wiper arm locking nut	7-12	10-16
Door hinge-to-door	12-19	17-26
Door hinge-to-body	25-40	35-55
Under cover-to-under skid plate (4WD)	13-18	18-25
Under cover (4WD with automatic transmission)	7-9	10-13
Transfer case protector (4WD)	25-40	35-55
Sub pivot mounting nut	2.8-4.2	4-6
Seat mounting nut	12-19	17-26
Seat belt mounting bolt	25-40	35-55
Retractor mounting bolt	25-40	35-55
1987-95 Monteros		
Body-to-frame	20-23	28-32
Hood hook-to-hood	2.9-4.3	4-6
Hood latch-to-body	2.9-4.3	4-6
Hood latch release cable	2.5-2.9	3.5-4.0
Front door hinge-to-door panel	12-19	17-26
Front door hinge-to-body	25-40	35-55
Rear door hinge-to-door panel	12-19	17-26
Rear door hinge-to-body	25-40	35-55
Ventilator window assembly-to-door panel	4.3	6
Window regulator-to-door panel	4.3	6
Door lower sash-to-door panel	4.3	6
Back door hinge-to-body	25-40	35-55
Back door-to-back door hinge	25-40	35-55
Spare tire bracket-to-back door	6-7	8-10
Seat anchor bolts-head marked 8	6.5-10	9-14
Seat anchor bolts-head marked 10	25-40	35-55
Front seat cushion-to-seat adjuster	12-19	17-26
Front seat back-to-seat cushion (reclining adjuster side)	33-43	45-60
Front seat back-to-seat cushion	7-11	10-15
Rear seat leg bracket-to-body	6.5-10	9-14
Rear seat back-to-seat cushion	33-43	45-60
All seat belt tightening (anchor) bolts	25-40	35-55

GLOSSARY

AIR/FUEL RATIO: The ratio of air-to-gasoline by weight in the fuel mixture drawn into the engine.

AIR INJECTION: One method of reducing harmful exhaust emissions by injecting air into each of the exhaust ports of an engine. The fresh air entering the hot exhaust manifold causes any remaining fuel to be burned before it can exit the tailpipe.

ALTERNATOR: A device used for converting mechanical energy into electrical energy.

AMMETER: An instrument, calibrated in amperes, used to measure the flow of an electrical current in a circuit. Ammeters are always connected in series with the circuit being tested.

AMPERE: The rate of flow of electrical current present when one volt of electrical pressure is applied against one ohm of electrical resistance.

ANALOG COMPUTER: Any microprocessor that uses similar (analogous) electrical signals to make its calculations.

ARMATURE: A laminated, soft iron core wrapped by a wire that converts electrical energy to mechanical energy as in a motor or relay. When rotated in a magnetic field, it changes mechanical energy into electrical energy as in a generator.

ATMOSPHERIC PRESSURE: The pressure on the Earth's surface caused by the weight of the air in the atmosphere. At sea level, this pressure is 14.7 psi at 32°F (101 kPa at 0°C).

ATOMIZATION: The breaking down of a liquid into a fine mist that can be suspended in air.

AXIAL PLAY: Movement parallel to a shaft or bearing bore.

BACKFIRE: The sudden combustion of gases in the intake or exhaust system that results in a loud explosion.

BACKLASH: The clearance or play between two parts, such as meshed gears.

BACKPRESSURE: Restrictions in the exhaust system that slow the exit of exhaust gases from the combustion chamber.

BAKELITE: A heat resistant, plastic insulator material commonly used in printed circuit boards and transistorized components.

BALL BEARING: A bearing made up of hardened inner and outer races between which hardened steel balls roll.

BALLAST RESISTOR: A resistor in the primary ignition circuit that lowers voltage after the engine is started to reduce wear on ignition components.

BEARING: A friction reducing, supportive device usually located between a stationary part and a moving part.

BIMETAL TEMPERATURE SENSOR: Any sensor or switch made of two dissimilar types of metal that bend when heated or cooled due to the different expansion rates of the alloys. These types of sensors usually function as an on/off switch.

BLOWBY: Combustion gases, composed of water vapor and unburned fuel, that leak past the piston rings into the crankcase during normal engine operation. These gases are removed by the PCV system to prevent the buildup of harmful acids in the crankcase.

BRAKE PAD: A brake shoe and lining assembly used with disc brakes.

BRAKE SHOE: The backing for the brake lining. The term is, however, usually applied to the assembly of the brake backing and lining.

BUSHING: A liner, usually removable, for a bearing; an anti-friction liner used in place of a bearing.

CALIPER: A hydraulically activated device in a disc brake system, which is mounted straddling the brake rotor (disc). The caliper contains at least one piston and two brake pads. Hydraulic pressure on the piston(s) forces the pads against the rotor.

CAMSHAFT: A shaft in the engine on which are the lobes (cams) which operate the valves. The camshaft is driven by the crankshaft, via a belt, chain or gears, at one half the crankshaft speed.

CAPACITOR: A device which stores an electrical charge.

CARBON MONOXIDE (CO): A colorless, odorless gas given off as a normal byproduct of combustion. It is poisonous and extremely dangerous in confined areas, building up slowly to toxic levels without warning if adequate ventilation is not available.

CARBURETOR: A device, usually mounted on the intake manifold of an engine, which mixes the air and fuel in the proper proportion to allow even combustion.

CATALYTIC CONVERTER: A device installed in the exhaust system, like a muffler, that converts harmful byproducts of combustion into carbon dioxide and water vapor by means of a heat-producing chemical reaction.

CENTRIFUGAL ADVANCE: A mechanical method of advancing the spark timing by using flyweights in the distributor that react to centrifugal force generated by the distributor shaft rotation.

GLOSSARY 10-77

CHECK VALVE: Any one-way valve installed to permit the flow of air, fuel or vacuum in one direction only.

CHOKE: A device, usually a moveable valve, placed in the intake path of a carburetor to restrict the flow of air.

CIRCUIT: Any unbroken path through which an electrical current can flow. Also used to describe fuel flow in some instances.

CIRCUIT BREAKER: A switch which protects an electrical circuit from overload by opening the circuit when the current flow exceeds a predetermined level. Some circuit breakers must be reset manually, while most reset automatically.

COIL (IGNITION): A transformer in the ignition circuit which steps up the voltage provided to the spark plugs.

COMBINATION MANIFOLD: An assembly which includes both the intake and exhaust manifolds in one casting.

COMBINATION VALVE: A device used in some fuel systems that routes fuel vapors to a charcoal storage canister instead of venting them into the atmosphere. The valve relieves fuel tank pressure and allows fresh air into the tank as the fuel level drops to prevent a vapor lock situation.

COMPRESSION RATIO: The comparison of the total volume of the cylinder and combustion chamber with the piston at BDC and the piston at TDC.

CONDENSER: 1. An electrical device which acts to store an electrical charge, preventing voltage surges. 2. A radiator-like device in the air conditioning system in which refrigerant gas condenses into a liquid, giving off heat.

CONDUCTOR: Any material through which an electrical current can be transmitted easily.

CONTINUITY: Continuous or complete circuit. Can be checked with an ohmmeter.

COUNTERSHAFT: An intermediate shaft which is rotated by a mainshaft and transmits, in turn, that rotation to a working part.

CRANKCASE: The lower part of an engine in which the crankshaft and related parts operate.

CRANKSHAFT: The main driving shaft of an engine which receives reciprocating motion from the pistons and converts it to rotary motion.

CYLINDER: In an engine, the round hole in the engine block in which the piston(s) ride.

CYLINDER BLOCK: The main structural member of an engine in which is found the cylinders, crankshaft and other principal parts.

CYLINDER HEAD: The detachable portion of the engine, usually fastened to the top of the cylinder block and containing all or most of the combustion chambers. On overhead valve engines, it contains the valves and their operating parts. On overhead cam engines, it contains the camshaft as well.

DEAD CENTER: The extreme top or bottom of the piston stroke.

DETONATION: An unwanted explosion of the air/fuel mixture in the combustion chamber caused by excess heat and compression, advanced timing, or an overly lean mixture. Also referred to as "ping".

DIAPHRAGM: A thin, flexible wall separating two cavities, such as in a vacuum advance unit.

DIESELING: A condition in which hot spots in the combustion chamber cause the engine to run on after the key is turned off.

DIFFERENTIAL: A geared assembly which allows the transmission of motion between drive axles, giving one axle the ability to turn faster than the other.

DIODE: An electrical device that will allow current to flow in one direction only.

DISC BRAKE: A hydraulic braking assembly consisting of a brake disc, or rotor, mounted on an axle, and a caliper assembly containing, usually two brake pads which are activated by hydraulic pressure. The pads are forced against the sides of the disc, creating friction which slows the vehicle.

DISTRIBUTOR: A mechanically driven device on an engine which is responsible for electrically firing the spark plug at a predetermined point of the piston stroke.

DOWEL PIN: A pin, inserted in mating holes in two different parts allowing those parts to maintain a fixed relationship.

DRUM BRAKE: A braking system which consists of two brake shoes and one or two wheel cylinders, mounted on a fixed backing plate, and a brake drum, mounted on an axle, which revolves around the assembly.

DWELL: The rate, measured in degrees of shaft rotation, at which an electrical circuit cycles on and off.

ELECTRONIC CONTROL UNIT (ECU): Ignition module, module, amplifier or igniter. See Module for definition.

ELECTRONIC IGNITION: A system in which the timing and firing of the spark plugs is controlled by an electronic control unit, usually called a module. These systems have no points or condenser.

END-PLAY: The measured amount of axial movement in a shaft.

Glossary

ENGINE: A device that converts heat into mechanical energy.

EXHAUST MANIFOLD: A set of cast passages or pipes which conduct exhaust gases from the engine.

FEELER GAUGE: A blade, usually metal, of precisely predetermined thickness, used to measure the clearance between two parts.

FIRING ORDER: The order in which combustion occurs in the cylinders of an engine. Also the order in which spark is distributed to the plugs by the distributor.

FLOODING: The presence of too much fuel in the intake manifold and combustion chamber which prevents the air/fuel mixture from firing, thereby causing a no-start situation.

FLYWHEEL: A disc shaped part bolted to the rear end of the crankshaft. Around the outer perimeter is affixed the ring gear. The starter drive engages the ring gear, turning the flywheel, which rotates the crankshaft, imparting the initial starting motion to the engine.

FOOT POUND (ft. lbs. or sometimes, ft.lb.): The amount of energy or work needed to raise an item weighing one pound, a distance of one foot.

FUSE: A protective device in a circuit which prevents circuit overload by breaking the circuit when a specific amperage is present. The device is constructed around a strip or wire of a lower amperage rating than the circuit it is designed to protect. When an amperage higher than that stamped on the fuse is present in the circuit, the strip or wire melts, opening the circuit.

GEAR RATIO: The ratio between the number of teeth on meshing gears.

GENERATOR: A device which converts mechanical energy into electrical energy.

HEAT RANGE: The measure of a spark plug's ability to dissipate heat from its firing end. The higher the heat range, the hotter the plug fires.

HUB: The center part of a wheel or gear.

HYDROCARBON (HC): Any chemical compound made up of hydrogen and carbon. A major pollutant formed by the engine as a byproduct of combustion.

HYDROMETER: An instrument used to measure the specific gravity of a solution.

INCH POUND (inch lbs.; sometimes in.lb. or in. lbs.): One twelfth of a foot pound.

INDUCTION: A means of transferring electrical energy in the form of a magnetic field. Principle used in the ignition coil to increase voltage.

INJECTOR: A device which receives metered fuel under relatively low pressure and is activated to inject the fuel into the engine under relatively high pressure at a predetermined time.

INPUT SHAFT: The shaft to which torque is applied, usually carrying the driving gear or gears.

INTAKE MANIFOLD: A casting of passages or pipes used to conduct air or a fuel/air mixture to the cylinders.

JOURNAL: The bearing surface within which a shaft operates.

KEY: A small block usually fitted in a notch between a shaft and a hub to prevent slippage of the two parts.

MANIFOLD: A casting of passages or set of pipes which connect the cylinders to an inlet or outlet source.

MANIFOLD VACUUM: Low pressure in an engine intake manifold formed just below the throttle plates. Manifold vacuum is highest at idle and drops under acceleration.

MASTER CYLINDER: The primary fluid pressurizing device in a hydraulic system. In automotive use, it is found in brake and hydraulic clutch systems and is pedal activated, either directly or, in a power brake system, through the power booster.

MODULE: Electronic control unit, amplifier or igniter of solid state or integrated design which controls the current flow in the ignition primary circuit based on input from the pick-up coil. When the module opens the primary circuit, high secondary voltage is induced in the coil.

NEEDLE BEARING: A bearing which consists of a number (usually a large number) of long, thin rollers.

OHM:(Ω) The unit used to measure the resistance of conductor-to-electrical flow. One ohm is the amount of resistance that limits current flow to one ampere in a circuit with one volt of pressure.

OHMMETER: An instrument used for measuring the resistance, in ohms, in an electrical circuit.

OUTPUT SHAFT: The shaft which transmits torque from a device, such as a transmission.

OVERDRIVE: A gear assembly which produces more shaft revolutions than that transmitted to it.

OVERHEAD CAMSHAFT (OHC): An engine configuration in which the camshaft is mounted on top of the cylinder head and operates the valve either directly or by means of rocker arms.

OVERHEAD VALVE (OHV): An engine configuration in which all of the valves are located in the cylinder head and the camshaft is located in the cylinder block. The camshaft operates the valves via lifters and pushrods.

GLOSSARY 10-79

OXIDES OF NITROGEN (NOx): Chemical compounds of nitrogen produced as a byproduct of combustion. They combine with hydrocarbons to produce smog.

OXYGEN SENSOR: Used with the feedback system to sense the presence of oxygen in the exhaust gas and signal the computer which can reference the voltage signal to an air/fuel ratio.

PINION: The smaller of two meshing gears.

PISTON RING: An open-ended ring which fits into a groove on the outer diameter of the piston. Its chief function is to form a seal between the piston and cylinder wall. Most automotive pistons have three rings: two for compression sealing; one for oil sealing.

PRELOAD: A predetermined load placed on a bearing during assembly or by adjustment.

PRIMARY CIRCUIT: The low voltage side of the ignition system which consists of the ignition switch, ballast resistor or resistance wire, bypass, coil, electronic control unit and pick-up coil as well as the connecting wires and harnesses.

PRESS FIT: The mating of two parts under pressure, due to the inner diameter of one being smaller than the outer diameter of the other, or vice versa; an interference fit.

RACE: The surface on the inner or outer ring of a bearing on which the balls, needles or rollers move.

REGULATOR: A device which maintains the amperage and/or voltage levels of a circuit at predetermined values.

RELAY: A switch which automatically opens and/or closes a circuit.

RESISTANCE: The opposition to the flow of current through a circuit or electrical device, and is measured in ohms. Resistance is equal to the voltage divided by the amperage.

RESISTOR: A device, usually made of wire, which offers a preset amount of resistance in an electrical circuit.

RING GEAR: The name given to a ring-shaped gear attached to a differential case, or affixed to a flywheel or as part of a planetary gear set.

ROLLER BEARING: A bearing made up of hardened inner and outer races between which hardened steel rollers move.

ROTOR: 1. The disc-shaped part of a disc brake assembly, upon which the brake pads bear; also called, brake disc. 2. The device mounted atop the distributor shaft, which passes current to the distributor cap tower contacts.

SECONDARY CIRCUIT: The high voltage side of the ignition system, usually above 20,000 volts. The secondary includes the ignition coil, coil wire, distributor cap and rotor, spark plug wires and spark plugs.

SENDING UNIT: A mechanical, electrical, hydraulic or electromagnetic device which transmits information to a gauge.

SENSOR: Any device designed to measure engine operating conditions or ambient pressures and temperatures. Usually electronic in nature and designed to send a voltage signal to an on-board computer, some sensors may operate as a simple on/off switch or they may provide a variable voltage signal (like a potentiometer) as conditions or measured parameters change.

SHIM: Spacers of precise, predetermined thickness used between parts to establish a proper working relationship.

SLAVE CYLINDER: In automotive use, a device in the hydraulic clutch system which is activated by hydraulic force, disengaging the clutch.

SOLENOID: A coil used to produce a magnetic field, the effect of which is to produce work.

SPARK PLUG: A device screwed into the combustion chamber of a spark ignition engine. The basic construction is a conductive core inside of a ceramic insulator, mounted in an outer conductive base. An electrical charge from the spark plug wire travels along the conductive core and jumps a preset air gap to a grounding point or points at the end of the conductive base. The resultant spark ignites the fuel/air mixture in the combustion chamber.

SPLINES: Ridges machined or cast onto the outer diameter of a shaft or inner diameter of a bore to enable parts to mate without rotation.

TACHOMETER: A device used to measure the rotary speed of an engine, shaft, gear, etc., usually in rotations per minute.

THERMOSTAT: A valve, located in the cooling system of an engine, which is closed when cold and opens gradually in response to engine heating, controlling the temperature of the coolant and rate of coolant flow.

TOP DEAD CENTER (TDC): The point at which the piston reaches the top of its travel on the compression stroke.

TORQUE: The twisting force applied to an object.

TORQUE CONVERTER: A turbine used to transmit power from a driving member to a driven member via hydraulic action, providing changes in drive ratio and torque. In automotive use, it links the driveplate at the rear of the engine to the automatic transmission.

TRANSDUCER: A device used to change a force into an electrical signal.

GLOSSARY

TRANSISTOR: A semi-conductor component which can be actuated by a small voltage to perform an electrical switching function.

TUNE-UP: A regular maintenance function, usually associated with the replacement and adjustment of parts and components in the electrical and fuel systems of a vehicle for the purpose of attaining optimum performance.

TURBOCHARGER: An exhaust driven pump which compresses intake air and forces it into the combustion chambers at higher than atmospheric pressures. The increased air pressure allows more fuel to be burned and results in increased horsepower being produced.

VACUUM ADVANCE: A device which advances the ignition timing in response to increased engine vacuum.

VACUUM GAUGE: An instrument used to measure the presence of vacuum in a chamber.

VALVE: A device which control the pressure, direction of flow or rate of flow of a liquid or gas.

VALVE CLEARANCE: The measured gap between the end of the valve stem and the rocker arm, cam lobe or follower that activates the valve.

VISCOSITY: The rating of a liquid's internal resistance to flow.

VOLTMETER: An instrument used for measuring electrical force in units called volts. Voltmeters are always connected parallel with the circuit being tested.

WHEEL CYLINDER: Found in the automotive drum brake assembly, it is a device, actuated by hydraulic pressure, which, through internal pistons, pushes the brake shoes outward against the drums.

AIR CONDITIONER
  AIR CONDITIONING CONTROL UNIT
    REMOVAL & INSTALLATION 6-50
  AIR CONDITIONING SWITCH
    REMOVAL & INSTALLATION 6-46
  COMPRESSOR
    REMOVAL & INSTALLATION 6-34
  CONDENSER FAN MOTOR
    REMOVAL & INSTALLATION 6-40
  CONDENSER
    GENERAL INFORMATION 6-36
    REMOVAL & INSTALLATION 6-37
  EVAPORATOR CORE
    PRECAUTIONS 6-41
    REMOVAL & INSTALLATION 6-42
  EXPANSION VALVE
    REMOVAL & INSTALLATION 6-50
  REAR EVAPORATOR ASSEMBLY
    REMOVAL & INSTALLATION 6-45
  RECEIVER/DRIER
    REMOVAL & INSTALLATION 6-50
  REFRIGERANT LINES
    GENERAL INFORMATION 6-50
    O-RING & FITTING INSTALLATION 6-52
  SUB-CONDENSER AND FAN MOTOR
    REMOVAL & INSTALLATION 6-40
ANTI-LOCK BRAKE SYSTEM (ABS) — FOUR WHEEL
  ELECTRONIC CONTROL UNIT (ECU)
    REMOVAL & INSTALLATION 9-49
    TESTING 9-49
  G-SENSOR
    REMOVAL & INSTALLATION 9-48
    TESTING 9-48
  HYDRAULIC UNIT
    REMOVAL & INSTALLATION 9-47
    TESTING 9-47
  WHEEL SPEED SENSOR
    REMOVAL & INSTALLATION 9-50
    TESTING 9-49
ANTI-LOCK BRAKE SYSTEM (ABS) — REAR WHEEL
  DESCRIPTION OF SYSTEM 9-52
  G-SENSOR AND SYSTEM CONTROL UNIT
    REMOVAL & INSTALLATION 9-55
    TESTING 9-54
  LOAD SENSING PROPORTIONING VALVE 9-55
  MODULATOR UNIT
    REMOVAL & INSTALLATION 9-54
    TESTING 9-54
  SPEED SENSOR
    REMOVAL & INSTALLATION 9-55
    TESTING 9-55
AUDIO SYSTEM
  RADIO, TAPE PLAYER AND CD PLAYER
    REMOVAL & INSTALLATION 6-63
  SPEAKERS
    REMOVAL & INSTALLATION 6-64
AUTOMATIC TRANSMISSION
  ADJUSTMENTS
    BAND ADJUSTMENT 7-42
    KICKDOWN LINKAGE ADJUSTMENT 7-43
    SHIFTER CONTROL LINKAGE 7-42
    THROTTLE PRESSURE ADJUSTMENTS 7-45
  EXTENSION HOUSING SEAL 7-55
  FLUID, PAN AND FILTER SERVICE 7-36
  IDENTIFICATION 7-35
  PARK/NEUTRAL POSITION (INHIBITOR) SWITCH
    ADJUSTMENT 7-47
    REMOVAL & INSTALLATION 7-46
  SHIFTER LINKAGE
    REMOVAL & INSTALLATION 7-48
  TRANSMISSION
    REMOVAL & INSTALLATION 7-57

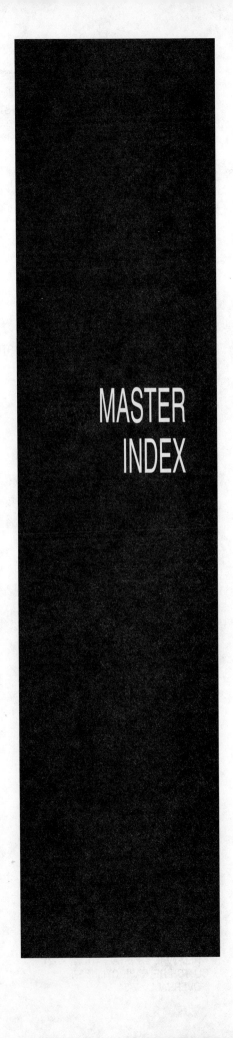

MASTER INDEX

INDEX

BASIC ELECTRICITY
BATTERY, STARTING AND CHARGING SYSTEMS
BASIC OPERATING PRINCIPLES 3-3
UNDERSTANDING BASIC ELECTRICITY 3-2
BRAKE SYSTEM
ADJUSTMENT
BRAKE PEDAL 9-4
DISC BRAKES 9-4
DRUM BRAKES 9-4
BLEND PROPORTIONING VALVE
REMOVAL & INSTALLATION 9-14
BRAKE HOSES AND LINES
REMOVAL & INSTALLATION 9-15
BRAKE LIGHT SWITCH
REMOVAL & INSTALLATION 9-5
BRAKE PEDAL
REMOVAL & INSTALLATION 9-5
BRAKE SYSTEM BLEEDING
STANDARD SYSTEM 9-15
VEHICES WITH LOAD SENSING PROPORTIONING VALVE 9-17
VEHICLES WITH ABS 9-17
LOAD SENSING PROPORTIONING VALVE
REMOVAL & INSTALLATION 9-14
MASTER CYLINDER
OVERHAUL 9-7
REMOVAL & INSTALLATION 9-6
OPERATION AND SYSTEM DESCRIPTION
DISC BRAKES 9-2
DRUM BRAKES 9-3
HYDRAULIC SYSTEM 9-2
POWER BOOSTER 9-3
POWER BRAKE BOOSTER
REMOVAL & INSTALLATION 9-12
VACUUM CHECK VALVE
REMOVAL & INSTALLATION 9-15
CARBURETED ENGINE CONTROLS
FEEDBACK CARBURETOR SYSTEM (FBC) 4-21
CARBURETED FUEL SYSTEM
CARBURETORS
ADJUSTMENTS 5-3
OVERHAUL 5-6
REMOVAL & INSTALLATION 5-5
MECHANICAL FUEL PUMP
REMOVAL & INSTALLATION 5-2
TESTING 5-3
CIRCUIT PROTECTION
COMPONENT LOCATIONS
1983-89 MONTERO 6-97
1990-95 MONTERO 6-97
PICK-UP 6-97
FUSES
REMOVAL & INSTALLATION 6-97
FUSIBLE LINKS
REMOVAL & INSTALLATION 6-96
CLUTCH
ADJUSTMENTS
PEDAL HEIGHT 7-21
PEDAL PLAY (FREE-PLAY) 7-21
CLUTCH LINKAGE
REMOVAL & INSTALLATION 7-26
CLUTCH MASTER CYLINDER
OVERHAUL 7-33
REMOVAL & INSTALLATION 7-33
CLUTCH PEDAL
REMOVAL & INSTALLATION 7-21
CLUTCH RELEASE CYLINDER
OVERHAUL 7-34
REMOVAL & INSTALLATION 7-34
SYSTEM BLEEDING 7-34
DISC AND PRESSURE PLATE
REMOVAL & INSTALLATION 7-32
COMPONENT LOCATION AND VACUUM DIAGRAMS 4-80
CRUISE CONTROL
CLUTCH PEDAL POSITION SWITCH
INSPECTION 6-56
COLUMN SWITCH
TESTING 6-54
MAIN SWITCH
INSPECTION 6-54
PARK/NEUTRAL POSITION (INHIBITOR) SWITCH
INSPECTION 6-56
SPEED SENSOR
REMOVAL & INSTALLATION 6-55
TESTING 6-55
STOP LIGHT/BRAKE SWITCH
INSPECTION 6-55
VACUUM SYSTEM COMPONENTS
REMOVAL & INSTALLATION 6-58
TESTING 6-56
DIAGNOSTIC TROUBLE CODES
CLEARING DIAGNOSTIC TROUBLE CODES
ALL MODELS 4-69
GENERAL INFORMATION 4-68
READING DIAGNOSTIC TROUBLE CODES
MONTEROS 4-68
PICK-UP TRUCKS 4-68
DIESEL ENGINES EMISSION CONTROLS
AIR CONDITIONING IDLE SPEED COMPENSATOR 4-21
CRANKCASE VENTILATION SYSTEM
GENERAL INFORMATION 4-20
EVAPORATIVE EMISSION CONTROLS
GENERAL INFORMATION 4-20
EXHAUST GAS RECIRCULATION (EGR)
OPERATION 4-20
REMOVAL & INSTALLATION 4-21
TESTING 4-20
FUEL INJECTION 4-21
HIGH ALTITUDE COMPENSATION SYSTEM 4-21
DIESEL FUEL SYSTEM
GLOW PLUGS
REMOVAL & INSTALLATION 5-45
IDLE SPEED ADJUSTMENT 5-45
IDLE-UP ADJUSTMENT 5-45
INJECTION LINES
REMOVAL & INSTALLATION 5-40
INJECTION PUMP
INJECTION TIMING ADJUSTMENT 5-44
REMOVAL & INSTALLATION 5-40
INJECTORS
REMOVAL & INSTALLATION 5-40
TESTING 5-40
DRIVELINE
CENTER BEARING
REMOVAL & INSTALLATION 7-73
FRONT DRIVESHAFT AND U-JOINTS
REMOVAL & INSTALLATION 7-70
UNIVERSAL JOINT REPLACEMENT 7-70
REAR DRIVESHAFT AND U-JOINTS
REMOVAL & INSTALLATION 7-70
UNIVERSAL JOINT REPLACEMENT 7-73
ELECTRONIC IGNITION
ADJUSTMENTS
RELUCTOR GAP 2-11
COMPONENT REPLACEMENT
REMOVAL & INSTALLATION 2-15

INDEX 10-83

DESCRIPTION AND OPERATION 2-11
DIAGNOSIS AND TESTING
    CRANK ANGLE SENSOR 2-12
    IGNITION COIL AND BALLAST RESISTOR 2-12

ENGINE ELECTRICAL
ALTERNATOR
    ALTERNATOR PRECAUTIONS 3-7
    REMOVAL & INSTALLATION 3-7
BATTERY
    REMOVAL & INSTALLATION 3-12
DISTRIBUTOR
    INSTALLATION 3-4
    REMOVAL 3-4
SENDING UNITS AND SENSORS
    REMOVAL & INSTALLATION 3-23
STARTER
    DIAGNOSIS 3-14
    REMOVAL & INSTALLATION 3-14
    SOLENOID REPLACEMENT 3-14
    STARTER OVERHAUL 3-16
VOLTAGE REGULATOR
    REMOVAL & INSTALLATION 3-12

ENGINE MECHANICAL
CAMSHAFTS AND BEARINGS
    INSPECTION AND MEASUREMENT 3-156
    REMOVAL & INSTALLATION 3-146
CHECKING ENGINE COMPRESSION
    DIESEL ENGINES 3-28
    GASOLINE ENGINES 3-28
COMBINATION MANIFOLD
    REMOVAL & INSTALLATION 3-81
CORE (FREEZE) PLUGS
    REPLACEMENT 3-166
CRANKSHAFT AND MAIN BEARINGS
    INSPECTION AND CLEANING 3-169
    INSTALLATION 3-169
    REMOVAL 3-168
CYLINDER HEAD AND GASKET
    CYLINDER HEAD CLEANING AND INSPECTION 3-107
    REMOVAL & INSTALLATION 3-98
    RESURFACING 3-108
ENGINE FAN
    REMOVAL & INSTALLATION 3-89
ENGINE OIL COOLER
    REMOVAL & INSTALLATION 3-88
ENGINE OVERHAUL TIPS
    INSPECTION TECHNIQUES 3-26
    OVERHAUL TIPS 3-26
    REPAIRING DAMAGED THREADS 3-26
    TOOLS 3-25
ENGINE
    REMOVAL & INSTALLATION 3-39
EXHAUST MANIFOLD
    REMOVAL & INSTALLATION 3-76
FLYWHEEL AND RING GEAR
    REMOVAL & INSTALLATION 3-171
    RING GEAR REPLACEMENT 3-172
HYDRAULIC LASH ADJUSTERS
    INSPECTION 3-115
    REMOVAL & INSTALLATION 3-114
INTAKE MANIFOLD
    REMOVAL & INSTALLATION 3-68
OIL PAN
    REMOVAL & INSTALLATION 3-137
OIL PUMP AND SILENT SHAFTS
    REMOVAL & INSTALLATION 3-139
PISTONS AND CONNECTING RODS
    INSPECTION 3-159
    INSTALLATION 3-161
    PISTON PIN (WRIST PIN) REPLACEMENT 3-163
    REMOVAL 3-157
    ROD BEARING REPLACEMENT 3-164
RADIATOR
    REMOVAL & INSTALLATION 3-85
REAR MAIN SEAL
    REMOVAL & INSTALLATION 3-166
ROCKER ARMS AND SHAFTS
    REMOVAL & INSTALLATION 3-62
THERMOSTAT
    REMOVAL & INSTALLATION 3-66
TIMING BELT AND COVERS
    REMOVAL & INSTALLATION 3-117
TIMING CHAIN AND COVER
    REMOVAL & INSTALLATION 3-134
TURBOCHARGER
    REMOVAL & INSTALLATION 3-84
VALVE COVER (ROCKER ARM COVER)
    REMOVAL & INSTALLATION 3-58
VALVE GUIDES
    REMOVAL & INSTALLATION 3-114
VALVE SEATS
    REMOVAL & INSTALLATION 3-113
VALVES AND VALVE SPRINGS
    INSPECTION 3-110
    INSTALLATION 3-113
    REMOVAL 3-109
WATER PUMP
    REMOVAL & INSTALLATION 3-90

EXHAUST SYSTEM
CATALYTIC CONVERTER
    REMOVAL & INSTALLATION 3-174
CENTER PIPE
    REMOVAL & INSTALLATION 3-174
FRONT PIPE
    REMOVAL & INSTALLATION 3-174
INSPECTION 3-173
SAFETY PRECAUTIONS 3-173
SPECIAL TOOLS 3-173
TAILPIPE AND MUFFLER
    REMOVAL & INSTALLATION 3-174

EXTERIOR
ANTENNA
    REMOVAL & INSTALLATION 10-13
BUMPERS
    REMOVAL & INSTALLATION 10-6
CHASSIS AND CAB MOUNTING BUSHINGS
    REMOVAL & INSTALLATION 10-21
DOORS
    ADJUSTMENT 10-2
    REMOVAL & INSTALLATION 10-2
FENDERS
    REMOVAL & INSTALLATION 10-16
GRILLE
    REMOVAL & INSTALLATION 10-13
HOOD
    ADJUSTMENT 10-5
    REMOVAL & INSTALLATION 10-4
OUTSIDE MIRRORS
    REMOVAL & INSTALLATION 10-13
PICK-UP TRUCK BED
    REMOVAL & INSTALLATION 10-19
POWER SUNROOF
    REMOVAL & INSTALLATION 10-27
REAR DOOR AND TAILGATE
    REMOVAL & INSTALLATION 10-6

10-84 INDEX

SPARE TIRE CARRIER
  REMOVAL & INSTALLATION 10-28
FIRING ORDERS 2-7
FLUIDS AND LUBRICANTS
  AUTOMATIC TRANSMISSION
    DRAIN, REFILL AND FILTER REPLACEMENT 1-39
    LEVEL CHECK AND FLUID RECOMMENDATIONS 1-39
    PAN AND FILTER SERVICE 1-39
  BODY DRAIN HOLES 1-46
  BRAKE AND CLUTCH MASTER CYLINDERS
    LEVEL CHECK AND FLUID RECOMMENDATIONS 1-44
  CHASSIS GREASING 1-45
  COOLING SYSTEM
    DRAINING, FLUSHING AND REFILLING 1-43
    FLUID RECOMMENDATIONS 1-42
    LEVEL CHECK 1-42
  DOOR HINGES 1-46
  DRIVE AXLES
    DRAIN AND REFILL 1-41
    LEVEL CHECK AND FLUID RECOMMENDATIONS 1-40
  ENGINE OIL RECOMMENDATIONS
    DIESEL ENGINES 1-33
    NORMALLY ASPIRATED ENGINES 1-33
    SYNTHETIC OIL 1-34
  ENGINE
    OIL AND FILTER CHANGE 1-35
    OIL LEVEL CHECK 1-34
  FLUID DISPOSAL 1-32
  FUEL RECOMMENDATIONS
    DIESEL ENGINES 1-32
    GASOLINE ENGINES 1-32
  HOOD LATCH AND HINGES 1-46
  LOCK CYLINDERS 1-46
  MANUAL STEERING GEAR 1-45
  MANUAL TRANSMISSION
    DRAIN AND REFILL 1-38
    LEVEL CHECK AND FLUID RECOMMENDATIONS 1-38
  POWER STEERING PUMP
    LEVEL CHECK AND FLUID RECOMMENDATIONS 1-44
  TRANSFER CASE 1-39
  WHEEL BEARINGS
    REMOVAL, PACKING & INSTALLATION 1-47
FRONT DISC BRAKES
  BRAKE CALIPER
    OVERHAUL 9-22
    REMOVAL & INSTALLATION 9-20
  BRAKE DISC (ROTOR)
    INSPECTION 9-25
    REMOVAL & INSTALLATION 9-24
  BRAKE PADS
    INSPECTION 9-19
    REMOVAL & INSTALLATION 9-19
    WEAR INDICATORS 9-18
FRONT DRIVE AXLE — 4-WHEEL DRIVE VEHICLES
  AUTOMATIC LOCKING HUBS
    DISASSEMBLY & ASSEMBLY 7-78
  CV-JOINTS
    BOOT REPLACEMENT 7-91
    OVERHAUL 7-86
  FRONT AXLE SHAFT, BEARING AND SEAL
    REMOVAL & INSTALLATION 7-82
  FRONT DIFFERENTIAL CARRIER
    REMOVAL & INSTALLATION 7-86
  IDENTIFICATION 7-75
  MANUAL LOCKING HUBS
    DISASSEMBLY & ASSEMBLY 7-75
  PINION SEAL 7-91

FRONT SUSPENSION
  COIL SPRING
    REMOVAL & INSTALLATION 8-3
  FRONT END ALIGNMENT
    CAMBER 8-21
    CASTER 8-21
    TOE 8-21
  FRONT WHEEL BEARING
    REMOVAL & INSTALLATION 8-19
  KNUCKLE AND SPINDLE
    REMOVAL & INSTALLATION 8-16
  LOWER BALL JOINTS
    INSPECTION 8-8
    REMOVAL & INSTALLATION 8-8
  LOWER CONTROL ARM AND BUSHINGS
    REMOVAL & INSTALLATION 8-14
  SHOCK ABSORBER
    REMOVAL & INSTALLATION 8-5
  STABILIZER BAR (SWAY BAR)
    REMOVAL & INSTALLATION 8-8
  STRUT ROD
    REMOVAL & INSTALLATION 8-11
  TORSION BAR
    REMOVAL & INSTALLATION 8-3
  UPPER BALL JOINT
    INSPECTION 8-6
    REMOVAL & INSTALLATION 8-7
  UPPER CONTROL ARM
    REMOVAL & INSTALLATION 8-12
FUEL INJECTION ENGINE CONTROLS
  BAROMETRIC PRESSURE SENSOR
    COMPONENT TESTING 4-32
    HARNESS TESTING 4-31
    OPERATION 4-31
    REMOVAL & INSTALLATION 4-32
  CAMSHAFT POSITION SENSOR
    COMPONENT TESTING 4-50
    HARNESS TESTING 4-48
    OPERATION 4-48
    REMOVAL & INSTALLATION 4-50
  CHECK ENGINE/MALFUNCTION INDICATOR LAMP 4-22
  CLOSED THROTTLE POSITION SWITCH
    COMPONENT ADJUSTING 4-40
    COMPONENT TESTING 4-40
    HARNESS TESTING 4-39
    OPERATION 4-39
    REMOVAL & INSTALLATION 4-40
  COMPONENT INSPECTION PROCEDURES 4-22
  CRANKSHAFT POSITION SENSOR
    COMPONENT TESTING 4-52
    HARNESS TESTING 4-51
    OPERATION 4-50
    REMOVAL & INSTALLATION 4-52
  EGR TEMPERATURE SENSOR
    COMPONENT TESTING 4-53
    HARNESS TESTING 4-52
    OPERATION 4-52
    REMOVAL & INSTALLATION 4-54
  ENGINE CONTROL MODULE (ECM) POWER GROUND
    OPERATION 4-27
    REMOVAL & INSTALLATION 4-27
    TESTING 4-27
  ENGINE CONTROL MODULE (ECM)
    TERMINAL VOLTAGE TESTING 4-57
  ENGINE COOLANT TEMPERATURE SENSOR
    COMPONENT TESTING 4-33
    HARNESS TESTING 4-32
    OPERATION 4-32

INDEX 10-85

REMOVAL & INSTALLATION 4-34
EVAPORATIVE EMISSION PURGE SOLENOID
  COMPONENT TESTING 4-55
  HARNESS TESTING 4-54
  OPERATION 4-54
  REMOVAL & INSTALLATION 4-55
HARNESS CHECKING PRECAUTIONS 4-22
IDLE AIR CONTROL MOTOR (DC MOTOR)
  COMPONENT TESTING 4-43
  HARNESS TESTING 4-43
  OPERATION 4-42
  REMOVAL & INSTALLATION 4-43
IDLE AIR CONTROL MOTOR (STEPPER MOTOR)
  COMPONENT TESTING 4-44
  HARNESS TESTING 4-43
  OPERATION 4-43
  REMOVAL & INSTALLATION 4-45
IDLE SPEED CONTROL MOTOR POSITION SENSOR
  COMPONENT TESTING 4-42
  HARNESS TESTING 4-41
  OPERATION 4-41
  REMOVAL & INSTALLATION 4-42
IDLE SPEED CONTROL MOTOR
  COMPONENT TESTING 4-41
  HARNESS TESTING 4-41
  OPERATION 4-40
  REMOVAL & INSTALLATION 4-41
INTAKE AIR TEMPERATURE SENSOR
  COMPONENT TESTING 4-30
  HARNESS TESTING 4-30
  OPERATION 4-30
  REMOVAL & INSTALLATION 4-31
KNOCK SENSOR
  COMPONENT TESTING 4-55
  HARNESS TESTING 4-55
  OPERATION 4-55
  REMOVAL & INSTALLATION 4-55
MULTI-PORT FUEL INJECTION RELAY
  COMPONENT INSPECTION 4-24
  HARNESS TESTING 4-23
  OPERATION 4-23
  REMOVAL & INSTALLATION 4-26
OXYGEN SENSOR
  COMPONENT TESTING 4-47
  HARNESS TESTING 4-46
  OPERATION 4-45
  REMOVAL & INSTALLATION 4-47
THROTTLE POSITION SENSOR
  ADJUSTMENT 4-35
  COMPONENT TESTING 4-35
  HARNESS TESTING 4-34
  OPERATION 4-34
  REMOVAL & INSTALLATION 4-37
TROUBLESHOOTING PROCEDURES 4-22
VARIABLE INDUCTION CONTROL SOLENOID
  COMPONENT TESTING 4-56
  HARNESS TESTING 4-56
  OPERATION 4-56
  REMOVAL & INSTALLATION 4-56
VOLUME AIR FLOW SENSOR
  COMPONENT TESTING 4-29
  HARNESS TESTING 4-28
  OPERATION 4-28
  REMOVAL & INSTALLATION 4-29
  TROUBLESHOOTING HINTS 4-28
FUEL TANK
  TANK ASSEMBLY
    REMOVAL & INSTALLATION 5-46

GASOLINE ENGINE EMISSION CONTROLS
  EVAPORATIVE EMISSION CONTROL SYSTEM
    OPERATION 4-3
    TESTING 4-3
  EXHAUST GAS RECIRCULATION (EGR) SYSTEM
    OPERATION 4-7
    TESTING 4-7
  HIGH ALTITUDE COMPENSATION SYSTEM
    INSPECTION — CALIFORNIA MODELS 4-16
    INSPECTION — NON-CALIFORNIA MODELS 4-13
    OPERATION 4-13
  INTAKE AIR TEMPERATURE CONTROL SYSTEM
    INSPECTION 4-17
  JET VALVE
    OPERATION 4-12
  MAINTENANCE REMINDER LIGHTS
    RESETTING 4-12
  MIXTURE CONTROL VALVE (MCV)
    INSPECTION 4-18
  POSITIVE CRANKCASE VENTILATION (PCV) SYSTEM
    REMOVAL & INSTALLATION 4-2
    SYSTEM OPERATION 4-2
    TESTING 4-2
  SECONDARY AIR SUPPLY SYSTEM
    OPERATION 4-10
    TESTING 4-10
  VACUUM REGULATOR VALVE
    INSPECTION 4-13
  VARIABLE INDUCTION SYSTEM (VIS)
    INSPECTION 4-18
    OPERATION 4-18
HEATER
  BLOWER MOTOR
    REMOVAL & INSTALLATION 6-13
  CONTROL CABLES
    REMOVAL & INSTALLATION 6-34
  CONTROL PANEL AND BLOWER SWITCH
    REMOVAL & INSTALLATION 6-30
  HEATER CORE
    REMOVAL & INSTALLATION 6-15
HOW TO USE THIS BOOK 1-2
IDLE SPEED AND MIXTURE ADJUSTMENT
  IDLE SPEED ADJUSTMENT
    CARBURETED ENGINES 2-21
    DIESEL ENGINES 2-22
    FUEL INJECTED ENGINES 2-22
  MIXTURE ADJUSTMENT
    CARBURETED ENGINES 2-22
    FUEL INJECTED ENGINES 2-23
IGNITION TIMING
  GENERAL INFORMATION
    INSPECTION AND ADJUSTMENT 2-17
INSTRUMENTS AND SWITCHES
  CLOCK
    REMOVAL & INSTALLATION 6-85
  COMBINATION SWITCH (HEADLIGHT/TURN SIGNAL/WINDSHIELD WIPER SWITCH)
    REMOVAL & INSTALLATION 6-85
  FUEL GAUGE
    REMOVAL & INSTALLATION 6-82
  INSTRUMENT CLUSTER/COMBINATION METER
    REMOVAL & INSTALLATION 6-78
  MULTI-METER
    REMOVAL & INSTALLATION 6-83
  OIL PRESSURE GAUGE
    REMOVAL & INSTALLATION 6-82
  PRINTED CIRCUIT BOARD
    REMOVAL & INSTALLATION 6-82

10-86 INDEX

SPEEDOMETER CABLE
  REMOVAL & INSTALLATION 6-80
TEMPERATURE GAUGE
  REMOVAL & INSTALLATION 6-82
VOLTMETER
  REMOVAL & INSTALLATION 6-82
WINDSHIELD AND REAR WINDOW WIPER SWITCHES
  REMOVAL & INSTALLATION 6-83
INTERIOR
  BACK DOOR LOCK
    REMOVAL & INSTALLATION 10-48
  CONSOLES 10-37
  DOOR GLASS, REGULATOR, AND POWER WINDOW MOTOR
    REMOVAL & INSTALLATION 10-48
  DOOR LOCKS
    REMOVAL & INSTALLATION 10-40
  DOOR PANELS (DOOR PADS OR LINERS)
    REMOVAL & INSTALLATION 10-37
  HEADLINER
    REMOVAL & INSTALLATION 10-40
  INSIDE MIRROR
    REMOVAL & INSTALLATION 10-68
  INSTRUMENT PANEL/DASHBOARD
    REMOVAL & INSTALLATION 10-29
  INTERIOR TRIM PANELS
    REMOVAL & INSTALLATION 10-38
  REAR WINDOWS
    REMOVAL & INSTALLATION 10-64
  SEAT BELT SYSTEMS
    REMOVAL & INSTALLATION 10-71
  SEATS
    REMOVAL & INSTALLATION 10-68
  VENT WINDOWS
    REMOVAL & INSTALLATION 10-64
  WINDSHIELD GLASS
    REMOVAL & INSTALLATION 10-55
JACKING AND HOISTING
  EMERGENCY JACKING 1-50
  SERVICE JACKING 1-51
JUMP STARTING
  JUMP STARTING PRECAUTIONS 1-52
  JUMP STARTING PROCEDURE 1-52
LIGHTING
  HEADLIGHTS
    AIMING 6-88
    REMOVAL & INSTALLATION 6-87
  SIGNAL AND MARKER LIGHTS
    REMOVAL & INSTALLATION 6-92
MANUAL TRANSMISSION
  ADJUSTMENTS
    CLUTCH SWITCH 7-10
    LINKAGE 7-10
    SHIFTER 7-10
  BACK-UP LIGHT SWITCH
    REMOVAL & INSTALLATION 7-14
  EXTENSION HOUSING SEAL
    REMOVAL & INSTALLATION 7-14
  GEARSHIFT LEVER ASSEMBLY
    REMOVAL & INSTALLATION 7-12
  IDENTIFICATION 7-2
  TRANSMISSION
    REMOVAL & INSTALLATION 7-15
MULTI-PORT FUEL INJECTION SYSTEM (MFI)
  ELECTRIC FUEL PUMP
    REMOVAL & INSTALLATION 5-20
    TESTING 5-20
  FUEL FILTER
    REMOVAL & INSTALLATION 5-34

FUEL INJECTORS
  REMOVAL & INSTALLATION 5-26
  TESTING 5-29
FUEL PRESSURE REGULATOR
  REMOVAL & INSTALLATION 5-30
PRESSURE RELIEF VALVE
  REMOVAL & INSTALLATION 5-33
  TESTING 5-33
RELIEVING FUEL SYSTEM PRESSURE 5-19
THROTTLE BODY
  ADJUSTMENTS 5-23
  REMOVAL & INSTALLATION 5-21
PARKING BRAKE
  BRAKE LEVER
    REMOVAL & INSTALLATION 9-43
  BRAKE SHOES
    ADJUSTMENT 9-47
    REMOVAL & INSTALLATION 9-43
  CABLES
    ADJUSTMENT 9-38
    REMOVAL & INSTALLATION 9-39
REAR AXLE
  DETERMINING AXLE RATIO 7-91
  DIFFERENTIAL CARRIER
    REMOVAL & INSTALLATION 7-98
  IDENTIFICATION 7-91
  PINION SEAL
    REMOVAL & INSTALLATION 7-99
  REAR AXLE HOUSING
    REMOVAL & INSTALLATION 7-101
  REAR AXLE SHAFTS, BEARINGS AND SEALS
    END-PLAY ADJUSTMENT PROCEDURE 7-98
    REMOVAL & INSTALLATION 7-92
  UNDERSTANDING DRIVE AXLES 7-91
REAR DISC BRAKES
  BRAKE CALIPER
    OVERHAUL 9-37
    REMOVAL & INSTALLATION 9-36
  BRAKE DISC (ROTOR)
    INSPECTION 9-37
    REMOVAL & INSTALLATION 9-37
  BRAKE PADS
    INSPECTION 9-33
    REMOVAL & INSTALLATION 9-33
    WEAR INDICATORS 9-33
REAR DRUM BRAKES
  BRAKE BACKING PLATE
    REMOVAL & INSTALLATION 9-32
  BRAKE DRUMS
    INSPECTION 9-26
    REMOVAL & INSTALLATION 9-26
  BRAKE SHOES
    INSPECTION 9-26
    REMOVAL & INSTALLATION 9-26
  WHEEL CYLINDERS
    ON-VEHICLE INSPECTION 9-30
    OVERHAUL 9-31
    REMOVAL & INSTALLATION 9-30
REAR SUSPENSION
  COIL SPRINGS
    REMOVAL & INSTALLATION 8-23
  LEAF SPRINGS
    REMOVAL & INSTALLATION 8-23
  REAR STABILIZER BAR (SWAY BAR)
    REMOVAL & INSTALLATION 8-24
  REAR WHEEL ALIGNMENT 8-25
  SHOCK ABSORBERS
    REMOVAL & INSTALLATION 8-24

INDEX 10-87

ROUTINE MAINTENANCE
AIR CLEANER
  FILTER REPLACEMENT 1-13
AIR CONDITIONING SYSTEM
  DISCHARGING, EVACUATING AND CHARGING THE SYSTEM 1-29
  GENERAL SERVICING PROCEDURES 1-26
  INSPECTION 1-28
  LEAK TESTING 1-29
  PREVENTIVE MAINTENANCE CHECKS 1-27
  SAFETY PRECAUTIONS 1-28
  SERVICE PRECAUTIONS 1-27
  TEST GAUGES 1-28
BATTERY
  CABLES AND CLAMPS 1-23
  CHARGING AND REPLACEMENT 1-23
  GENERAL MAINTENANCE 1-22
  MAINTENANCE-FREE BATTERIES 1-22
  REPLACEMENT BATTERIES 1-22
BELTS
  CHECKING AND ADJUSTING TENSION 1-24
  REMOVAL & INSTALLATION 1-26
CV-BOOTS
  INSPECTION 1-26
EVAPORATIVE CANISTER
  SERVICE 1-21
FUEL FILTER
  REMOVAL & INSTALLATION 1-16
HOSES
  INSPECTION 1-26
  REMOVAL & INSTALLATION 1-26
PCV VALVE
  REMOVAL & INSTALLATION 1-19
TIMING BELT
  INSPECTION 1-26
TIRES AND WHEELS
  CARE OF SPECIAL WHEELS 1-32
  STORAGE 1-32
  TIRE DESIGN 1-31
  TIRE ROTATION 1-31
WINDSHIELD WIPERS
  WIPER REFILL REPLACEMENT 1-30
SERIAL NUMBER IDENTIFICATION
CHASSIS NUMBER 1-8
ENGINE 1-9
TRANSMISSION 1-13
UNDERHOOD LABELS 1-13
VEHICLE IDENTIFICATION NUMBER 1-6
VEHICLE INFORMATION CODE PLATE 1-7
VEHICLE SAFETY CERTIFICATION LABEL 1-8
SERVICING YOUR VEHICLE SAFELY
DO'S 1-6
DON'TS 1-6
SPECIFICATIONS CHARTS
ALTERNATOR SPECIFICATIONS 3-13
AUTOMATIC TRANSMISSION IDENTIFICATION 7-35
BRAKE SPECIFICATIONS 9-57
CAMSHAFT SPECIFICATIONS 3-34
CAPACITIES 1-53
CARBURETOR SPECIFICATIONS 5-18
CRANKSHAFT AND CONNECTING ROD SPECIFICATIONS 3-36
DIESEL ENGINE TUNE-UP CHART 2-23
ENGINE IDENTIFICATION 1-12
GASOLINE ENGINE TUNE-UP CHART 2-24
GENERAL ENGINE SPECIFICATIONS 3-29
MAINTENANCE INTERVALS 1-53
MANUAL TRANSMISSION IDENTIFICATION 7-2
PISTON AND RING SPECIFICATIONS 3-32

RECOMMENDED LUBRICANTS 1-53
STARTER SPECIFICATIONS 3-22
TORQUE SPECIFICATIONS 3-38, 8-44, 9-58, 10-71, 7-105
VALVE SPECIFICATIONS 3-30
VEHICLE IDENTIFICATION 1-7
WHEEL ALIGNMENT SPECIFICATIONS 8-22
STEERING
IGNITION LOCK
  REMOVAL & INSTALLATION 8-29
IGNITION SWITCH
  REMOVAL & INSTALLATION 8-27
  TESTING 8-28
MANUAL STEERING GEAR BOX
  REMOVAL & INSTALLATION 8-38
POWER STEERING GEAR BOX
  REMOVAL & INSTALLATION 8-38
POWER STEERING PUMP
  REMOVAL & INSTALLATION 8-40
  SYSTEM BLEEDING 8-41
STEERING COLUMN
  REMOVAL & INSTALLATION 8-30
STEERING LINKAGE
  REMOVAL & INSTALLATION 8-31
STEERING WHEEL
  REMOVAL & INSTALLATION 8-27
TURN SIGNAL AND WIPER SWITCH (COLUMN SWITCH)
  REMOVAL & INSTALLATION 8-27
SUPPLEMENTAL RESTRAINT SYSTEM (AIR BAG)
AIR BAG MODULE AND CLOCK SPRING
  REMOVAL & INSTALLATION 6-12
  SYSTEM PRECAUTIONS 6-12
GENERAL INFORMATION
  DISARMING AND ENABLING THE SYSTEM 6-12
  SERVICE PRECAUTIONS 6-11
  SYSTEM OPERATION AND COMPONENTS 6-11
TOOLS AND EQUIPMENT
SPECIAL TOOLS 1-5
TOWING THE VEHICLE 1-49
TRAILER TOWING
COOLING
  ENGINE 1-49
  TRANSMISSION 1-49
HITCH WEIGHT 1-48
TRAILER WEIGHT 1-48
WIRING 1-49
TRAILER WIRING 6-96
TRANSFER CASE
ADJUSTMENTS 7-68
FRONT AND REAR OUTPUT SHAFT SEALS 7-68
IDENTIFICATION 7-68
TRANSFER CASE
  REMOVAL & INSTALLATION 7-68
TUNE-UP PROCEDURES
SPARK PLUG WIRES
  CHECKING AND REPLACEMENT 2-6
SPARK PLUGS
  INSPECTION 2-6
  INSTALLATION 2-6
  REMOVAL 2-5
UNDERSTANDING AND TROUBLESHOOTING ELECTRICAL SYSTEMS
MECHANICAL TEST EQUIPMENT
  HAND VACUUM PUMP 6-11
  VACUUM GAUGE 6-10
SAFETY PRECAUTIONS 6-3
TROUBLESHOOTING AND DIAGNOSIS
  ORGANIZED TROUBLESHOOTING 6-3
  SPECIFIC TROUBLESHOOTING 6-4

10-88 INDEX

TEST EQUIPMENT 6-4
WIRING HARNESSES
  WIRE GAUGE 6-8
  WIRING REPAIR 6-9
UNDERSTANDING BASIC ELECTRICITY 6-2
VALVE LASH
  GENERAL INFORMATION
    ADJUSTMENT 2-20
WHEELS
  CLEANING 8-2
  FRONT AND REAR WHEELS
    REMOVAL & INSTALLATION 8-2
  INSPECTION 8-2

WHEEL LUG STUDS
  REMOVAL & INSTALLATION 8-2
WINDSHIELD WIPERS
  INTERMITTENT WIPER RELAY
    REMOVAL & INSTALLATION 6-76
  REAR WIPER AND WASHER MOTOR
    REMOVAL & INSTALLATION 6-74
  WINDSHIELD WASHER MOTOR AND FLUID RESERVOIR
    REMOVAL & INSTALLATION 6-71
  WINDSHIELD WIPER MOTOR AND LINKAGE
    REMOVAL & INSTALLATION 6-70
  WIPER BLADE AND ARM
    REPLACEMENT 6-70
WIRING DIAGRAMS 6-103

Don't Miss These Other Important Titles From NP/CHILTON'S

TOTAL CAR CARE MANUALS
The ULTIMATE in automotive repair manuals

ACURA
Coupes and Sedans 1986-93
PART NO. 8426/10300

AMC
Coupes/Sedans/Wagons 1975-88
PART NO. 14300

BMW
Coupes and Sedans 1970-88
PART NO. 8789/18300
318/325/M3/525/535/M5 1989-93
PART NO. 8427/18400

CHRYSLER
Aspen/Volare 1976-80
PART NO. 20100
Caravan/Voyager/Town & Country 1984-95
PART NO. 8155/20300
Caravan/Voyager/Town & Country 1996-99
PART NO. 20302
Cirrus/Stratus/Sebring/Avenger 1995-98
PART NO. 20320
Colt/Challenger/Conquest/Vista 1971-89
PART NO. 20340
Colt/Vista 1990-93
PART NO. 8418/20342
Concorde/Intrepid/New Yorker/LHS/Vision 1993-97
PART NO. 8817/20360
Front Wheel Drive Cars-4 Cyl 1981-95
PART NO. 8673/20382
Front Wheel Drive Cars-6 Cyl 1988-95
PART NO. 8672/20384
Full-Size Trucks 1967-88
PART NO. 8662/20400
Full-Size Trucks 1989-96
PART NO. 8166/20402
Full-Size Vans 1967-88
PART NO. 20420
Full-Size Vans 1989-98
PART NO. 8169/20422
Neon 1995-99
PART NO. 20600
Omni/Horizon/Rampage 1978-89
PART NO. 8787/20700
Ram 50/D50/Arrow 1979-93
PART NO. 20800

FORD
Aerostar 1986-96
PART NO. 8057/26100
Aspire 1994-97
PART NO. 26120
Contour/Mystique/Cougar 1995-99
PART NO. 26170
Crown Victoria/Grand Marquis 1989-94
PART NO. 8417/26180
Escort/Lynx 1981-90
PART NO. 8270/26240
Escort/Tracer 1991-99
PART NO. 26242
Fairmont/Zephyr 1978-83
PART NO. 26320
Ford/Mercury Full-Size Cars 1968-88
PART NO. 8665/26360
Full-Size Vans 1961-88
PART NO. 26400

Features:
- Based on actual teardowns
- Each manual covers all makes and models (unless otherwise indicated)
- Expanded photography from vehicle teardowns
- Actual vacuum and wiring diagrams—not general representations
- Comprehensive coverage
- Maintenance interval schedules
- Electronic engine and emission controls

Full-Size Vans 1989-96
PART NO. 8157/26402
Ford/Mercury Mid-Size Cars 1971-85
PART NO. 8667/26580
Mustang/Cougar 1964-73
PART NO. 26600
Mustang/Capri 1979-88
PART NO. 8580/26604
Mustang 1989-93
PART NO. 8253/26606
Mustang 1994-98
PART NO. 26608
Pick-Ups and Bronco 1976-86
PART NO. 8576/26662
Pick-Ups and Bronco 1987-96
PART NO. 8136/26664
Pick-Ups/Expedition/Navigator 1997-00
PART NO. 26666
Probe 1989-92
PART NO. 8266/26680
Probe 1993-97
PART NO. 8411/46802
Ranger/Bronco II 1983-90
PART NO. 8159/26686
Ranger/Explorer/Mountaineer 1991-97
PART NO. 26688
Taurus/Sable 1986-95
PART NO. 8251/26700
Taurus/Sable 1996-99
PART NO. 26702
Tempo/Topaz 1984-94
PART NO. 8271/26720
Thunderbird/Cougar 1983-96
PART NO. 8268/26760
Windstar 1995-98
PART NO. 26840

GENERAL MOTORS
Astro/Safari 1985-96
PART NO. 8056/28100
Blazer/Jimmy 1969-82
PART NO. 28140
Blazer/Jimmy/Typhoon/Bravada 1983-93
PART NO. 8139/28160
Blazer/Jimmy/Bravada 1994-99
PART NO. 8845/28862
Bonneville/Eighty Eight/LeSabre 1986-99
PART NO. 8423/28200
Buick/Oldsmobile/Pontiac Full-Size 1975-90
PART NO. 8584/28240
Cadillac 1967-89
PART NO. 8587/28260
Camaro 1967-81
PART NO. 28280
Camaro 1982-92
PART NO. 8260/28282

Camaro/Firebird 1993-98
PART NO. 28284
Caprice 1990-93
PART NO. 8421/28300
Cavalier/Sunbird/Skyhawk/Firenza 1982-94
PART NO. 8269/28320
Cavalier/Sunfire 1995-00
PART NO. 28322
Celebrity/Century/Ciera/6000 1982-96
PART NO. 8252/28360
Chevette/1000 1976-88
PART NO. 28400
Chevy Full-Size Cars 1968-78
PART NO. 28420
Chevy Full-Size Cars 1979-89
PART NO. 8531/28422
Chevy Mid-Size Cars 1964-88
PART NO. 8594/28440
Citation/Omega/Phoenix/Skylark/XII 1980-85
PART NO. 28460
Corsica/Beretta 1988-96
PART NO. 8254/28480
Corvette 1963-82
PART NO. 28500
Corvette 1984-96
PART NO. 28502
Cutlass RWD 1970-87
PART NO. 8668/28520
DeVille/Fleetwood/Eldorado/Seville 1990-93
PART NO. 8420/28540
Electra/Park Avenue/Ninety-Eight 1990-93
PART NO. 8430/28560
Fiero 1984-88
PART NO. 28580
Firebird 1967-81
PART NO. 28600
Firebird 1982-92
PART NO. 8534/28602
Full-Size Trucks 1970-79
PART NO. 28620
Full-Size Trucks 1980-87
PART NO. 8577/28622
Full-Size Trucks 1988-98
PART NO. 8055/28624
Full-Size Vans 1967-86
PART NO. 28640
Full-Size Vans 1987-97
PART NO. 8040/28642
Grand Am/Achieva 1985-98
PART NO. 8257/28660
Lumina/Silhouette/Trans Sport/Venture 1990-99
PART NO. 8134/28680
Lumina/Monte Carlo/Grand Prix/Cutlass Supreme/Regal 1988-96
PART NO. 8258/28682

Metro/Sprint 1985-99
PART NO. 8424/28700
Nova/Chevy II 1962-79
PART NO. 28720
Pontiac Mid-Size 1974-83
PART NO. 28740
Chevrolet Nova/GEO Prizm 1985-93
PART NO. 8422/28760
Regal/Century 1975-87
PART NO. 28780
Chevrolet Spectrum/GEO Storm 1985-93
PART NO. 8425/28800
S10/S15/Sonoma Pick-Ups 1982-93
PART NO. 8141/28860
S10/Sonoma/Blazer/Jimmy/Bravada Hombre 1994-99
PART NO. 8845/28862

HONDA
Accord/Civic/Prelude 1973-83
PART NO. 8591/30100
Accord/Prelude 1984-95
PART NO. 8255/30150
Civic, CRX and del SOL 1984-95
PART NO. 8256/30200

HYUNDAI
Coupes/Sedans 1986-93
PART NO. 8412/32100
Coupes/Sedans 1994-98
PART NO. 32102

ISUZU
Amigo/Pick-Ups/Rodeo/Trooper 1981-96
PART NO. 8686/36100
Cars and Trucks 1981-91
PART NO. 8069/36150

JEEP
CJ 1945-70
PART NO. 40200
CJ/Scrambler 1971-86
PART NO. 8536/40202
Wagoneer/Commando/Cherokee 1957-83
PART NO. 40600
Wagoneer/Comanche/Cherokee 1984-98
PART NO. 8143/40602
Wrangler/YJ 1987-95
PART NO. 8535/40650

MAZDA
Trucks 1972-86
PART NO. 46600
Trucks 1987-93
PART NO. 8264/46602
Trucks 1994-98
PART NO. 46604
323/626/929/GLC/MX-6/RX-7 1978-89
PART NO. 8581/46800
323/Protege/MX-3/MX-6/626 Millenia/Ford Probe 1990-98
PART NO. 8411/46802

MERCEDES
Coupes/Sedans/Wagons 1974-84
PART NO. 48300

MITSUBISHI
Cars and Trucks 1983-89
PART NO. 7947/50200

3P1VerB

Total Car Care, continued

Eclipse 1990-98
PART NO. 8415/50400

Pick-Ups and Montero 1983-95
PART NO. 8666/50500

NISSAN
Datsun 210/1200 1973-81
PART NO. 52300

Datsun 200SX/510/610/710/810/Maxima 1973-84
PART NO. 52302

Nissan Maxima 1985-92
PART NO. 8261/52450

Maxima 1993-98
PART NO. 52452

Pick-Ups and Pathfinder 1970-88
PART NO. 8585/52500

Pick-Ups and Pathfinder 1989-95
PART NO. 8145/52502

Sentra/Pulsar/NX 1982-96
PART NO. 8263/52700

Stanza/200SX/240SX 1982-92
PART NO. 8262/52750

240SX/Altima 1993-98
PART NO. 52752

Datsun/Nissan Z and ZX 1970-88
PART NO. 8846/52800

RENAULT
Coupes/Sedans/Wagons 1975-85
PART NO. 58300

SATURN
Coupes/Sedans/Wagons 1991-98
PART NO. 8419/62300

SUBARU
Coupes/Sedan/Wagons 1970-84
PART NO. 8790/64300

Coupes/Sedans/Wagons 1985-96
PART NO. 8259/64302

SUZUKI
Samurai/Sidekick/Tracker 1986-98
PART NO. 66500

TOYOTA
Camry 1983-96
PART NO. 8265/68200

Celica/Supra 1971-85
PART NO. 68250

Celica 1986-93
PART NO. 8413/68252

Celica 1994-98
PART NO. 68254

Corolla 1970-87
PART NO. 8586/68300

Corolla 1988-97
PART NO. 8414/68302

Cressida/Corona/Crown/MkII 1970-82
PART NO. 68350

Cressida/Van 1983-90
PART NO. 68352

Pick-ups/Land Cruiser/4Runner 1970-88
PART NO. 8578/68600

Pick-ups/Land Cruiser/4Runner 1989-98
PART NO. 8163/68602

Previa 1991-97
PART NO. 68640

Tercel 1984-94
PART NO. 8595/68700

VOLKSWAGEN
Air-Cooled 1949-69
PART NO. 70200

Air-Cooled 1970-81
PART NO. 70202

Front Wheel Drive 1974-89
PART NO. 8663/70400

Golf/Jetta/Cabriolet 1990-93
PART NO. 8429/70402

VOLVO
Coupes/Sedans/Wagons 1970-89
PART NO. 8786/72300

Coupes/Sedans/Wagons 1990-98
PART NO. 8428/72302

General Interest / Recreational Books

We offer specialty books on a variety of topics including Motorcycles, ATVs, Snowmobiles and automotive subjects like Detailing or Body Repair. Each book from our General Interest line offers a blend of our famous Do-It-Yourself procedures and photography with additional information on enjoying automotive, marine and recreational products. Learn more about the vehicles you use and enjoy while keeping them in top running shape.

ATV Handbook
PART NO. 9123

Auto Detailing
PART NO. 8394

Auto Body Repair
PART NO. 7898

Briggs & Stratton Vertical Crankshaft Engine
PART NO. 61-1-2

Briggs & Stratton Horizontal Crankshaft Engine
PART NO. 61-0-4

Briggs & Stratton Overhead Valve (OHV) Engine
PART NO. 61-2-0

Easy Car Care
PART NO. 8042

Motorcycle Handbook
PART NO. 9099

Snowmobile Handbook
PART NO. 9124

Small Engine Repair (Up to 20 Hp)
PART NO. 8325

Total Service Series

These innovative books offer repair, maintenance and service procedures for automotive related systems. They cover today's complex vehicles in a user-friendly format, which places even the most difficult automotive topic well within the reach of every Do-It-Yourselfer. Each title covers a specific subject from Brakes and Engine Rebuilding to Fuel Injection Systems, Automatic Transmissions and even Engine Trouble Codes.

Automatic Transmissions/Transaxles Diagnosis and Repair
PART NO. 8944

Brake System Diagnosis and Repair
PART NO. 8945

Chevrolet Engine Overhaul Manual
PART NO. 8794

Engine Code Manual
PART NO. 8851

Ford Engine Overhaul Manual
PART NO. 8793

Fuel Injection Diagnosis and Repair
PART NO. 8946

Collector's Hard-Cover Manuals

Chilton's Collector's Editions are perfect for enthusiasts of vintage or rare cars. These hard-cover manuals contain repair and maintenance information for all major systems that might not be available elsewhere. Included are repair and overhaul procedures using thousands of illustrations. These manuals offer a range of coverage from as far back as 1940 and as recent as 1997, so you don't need an antique car or truck to be a collector.

Auto Repair Manual 1993-97
PART NO. 7919

Auto Repair Manual 1988-92
PART NO. 7906

Auto Repair Manual 1980-87
PART NO. 7670

Auto Repair Manual 1972-79
PART NO. 6914

Auto Repair Manual 1964-71
PART NO. 5974

Auto Repair Manual 1954-63
PART NO. 5652

Auto Repair Manual 1940-53
PART NO. 5631

Import Car Repair Manual 1993-97
PART NO. 7920

Import Car Repair Manual 1988-92
PART NO. 7907

Import Car Repair Manual 1980-87
PART NO. 7672

Truck and Van Repair Manual 1993-97
PART NO. 7921

Truck and Van Repair Manual 1991-95
PART NO. 7911

Truck and Van Repair Manual 1986-90
PART NO. 7902

Truck and Van Repair Manual 1979-86
PART NO. 7655

Truck and Van Repair Manual 1971-78
PART NO. 7012

"...and even more from CHILTON"

System-Specific Manuals

Guide to Air Conditioning Repair and Service 1982-85
PART NO. 7580

Guide to Automatic Transmission Repair 1984-89
PART NO. 8054

Guide to Automatic Transmission Repair 1984-89
Domestic cars and trucks
PART NO. 8053

Guide to Automatic Transmission Repair 1980-84
Domestic cars and trucks
PART NO. 7891

Guide to Automatic Transmission Repair 1974-80
Import cars and trucks
PART NO. 7645

Guide to Brakes, Steering, and Suspension 1980-87
PART NO. 7819

Guide to Fuel Injection and Electronic Engine Controls 1984-88
Domestic cars and trucks
PART NO. 7766

Guide to Electronic Engine Controls 1978-85
PART NO. 7535

Guide to Engine Repair and Rebuilding
PART NO. 7643

Guide to Vacuum Diagrams 1980-86
Domestic cars and trucks
PART NO. 7821

Multi-Vehicle Spanish Repair Manuals

Auto Repair Manual 1992-96
PART NO. 8947

Import Repair Manual 1992-96
PART NO. 8948

Truck and Van Repair Manual 1992-96
PART NO. 8949

Auto Repair Manual 1987-91
PART NO. 8138

Auto Repair Manual 1980-87
PART NO. 7795

Auto Repair Manual 1976-83
PART NO. 7476

SELOC MARINE MANUALS

OUTBOARDS

Chrysler Outboards, All Engines 1962-84
PART NO. 018-7(1000)

Force Outboards, All Engines 1984-96
PART NO. 024-1(1100)

Honda Outboards, All Engines 1988-98
PART NO. 1200

Johnson/Evinrude Outboards, 1.5-40HP, 2-Stroke 1956-70
PART NO. 007-1(1300)

Johnson/Evinrude Outboards, 1.25-60HP, 2-Stroke 1971-89
PART NO. 008-X(1302)

Johnson/Evinrude Outboards, 1-50 HP, 2-Stroke 1990-95
PART NO. 026-8(1304)

Johnson/Evinrude Outboards, 50-125 HP, 2-Stroke 1958-72
PART NO. 009-8(1306)

Johnson/Evinrude Outboards, 60-235 HP, 2-Stroke 1973-91
PART NO. 010-1(1308)

Johnson/Evinrude Outboards, 80-300 HP, 2-Stroke 1992-96
PART NO. 040-3(1310)

Mariner Outboards, 2-60 HP, 2-Stroke 1977-89
PART NO. 015-2(1400)

Mariner Outboards, 45-220 HP, 2 Stroke 1977-89
PART NO. 016-0(1402)

Mercury Outboards, 2-40 HP, 2-Stroke 1965-91
PART NO. 012-8(1404)

Mercury Outboards, 40-115 HP, 2-Stroke 1965-92
PART NO. 013-6(1406)

Mercury Outboards, 90-300 HP, 2-Stroke 1965-91
PART NO. 014-4(1408)

Mercury/Mariner Outboards, 2.5-25 HP, 2-Stroke 1990-94
PART NO. 035-7(1410)

Mercury/Mariner Outboards, 40-125 HP, 2-Stroke 1990-94
PART NO. 036-5(1412)

Mercury/Mariner Outboards, 135-275 HP, 2-Stroke 1990-94
PART NO. 037-3(1414)

Mercury/Mariner Outboards, All Engines 1995-99
PART NO. 1416

Suzuki Outboards, All Engines 1985-99
PART NO. 1600

Yamaha Outboards, 2-25 HP, 2-Stroke and 9.9 HP, 4-Stroke 1984-91
PART NO. 021-7(1700)

Yamaha Outboards, 30-90 HP, 2-Stroke 1984-91
PART NO. 022-5(1702)

Yamaha Outboards, 115-225 HP, 2-Stroke 1984-91
PART NO. 023-3(1704)

Yamaha Outboards, All Engines 1992-98
PART NO. 1706

STERN DRIVES

Marine Jet Drive 1961-96
PART NO. 029-2(3000)

Mercruiser Stern Drive Type 1, Alpha, Bravo I, II, 1964-92
PART NO. 005-5(3200)

Mercruiser Stern Drive Alpha 1 Generation II 1992-96
PART NO. 039-X(3202)

Mercruiser Stern Drive Bravo I, II, III 1992-96
PART NO. 046-2(3204)

OMC Stern Drive 1964-86
PART NO. 004-7(3400)

OMC Cobra Stern Drive 1985-95
PART NO. 025-X(3402)

Volvo/Penta Stern Drives 1968-91
PART NO. 011-X(3600)

Volvo/Penta Stern Drives 1992-93
PART NO. 038-1(3602)

Volvo/Penta Stern Drives 1992-95
PART NO. 041-1(3604)

INBOARDS

Yanmar Inboard Diesels 1988-91
PART NO. 7400

PERSONAL WATERCRAFT

Kawasaki 1973-91
PART NO. 032-2(9200)

Kawasaki 1992-97
PART NO. 042-X(9202)

Polaris 1992-97
PART NO. 045-4(9400)

Sea Doo/Bombardier 1988-91
PART NO. 033-0(9000)

Sea Doo/Bombardier 1992-97
PART NO. 043-8(9002)

Yamaha 1987-91
PART NO. 034-9(9600)

Yamaha 1992-97
PART NO. 044-6(9602)

FOR THE TITLES LISTED, PLEASE VISIT YOUR LOCAL CHILTON RETAILER

For a FREE catalog of Chilton's complete line of Automotive Repair Manuals, or to order direct Call toll-free 877-4CHILTON

NP/CHILTON'S 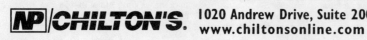 1020 Andrew Drive, Suite 200 • West Chester, PA 19380-4291
www.chiltononline.com